CALCULUS
WITH EARLY VECTORS

EQUATIONS OF LINES IN THE PLANE

Slope of the line
$$m = \frac{y_1 - y_0}{x_1 - x_0}$$

Slope intercept form
$$y = mx + b$$

Point slope form
$$y - y_0 = m(x - x_0)$$

Two point form
$$y - y_0 = \frac{y_1 - y_0}{x_1 - x_0}(x - x_0)$$

LINES AND PLANES

$ax + by = c$ is an equation of a line in the plane.

$ax + by + cz = d$ is an equation of a plane in 3–space.

$\vec{r} \cdot \vec{n} = \vec{r}_0 \cdot \vec{n}$ is an equation of a line in the plane and an equation of a plane in 3–space.

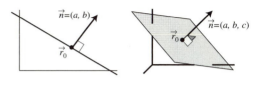

TRIGONOMETRY

TRIGONOMETRIC FUNCTIONS

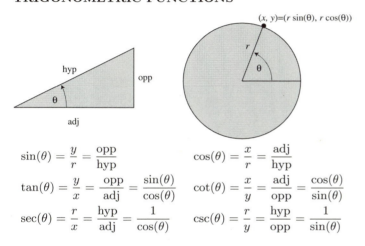

$$\sin(\theta) = \frac{y}{r} = \frac{\text{opp}}{\text{hyp}} \qquad \cos(\theta) = \frac{x}{r} = \frac{\text{adj}}{\text{hyp}}$$

$$\tan(\theta) = \frac{y}{x} = \frac{\text{opp}}{\text{adj}} = \frac{\sin(\theta)}{\cos(\theta)} \qquad \cot(\theta) = \frac{x}{y} = \frac{\text{adj}}{\text{opp}} = \frac{\cos(\theta)}{\sin(\theta)}$$

$$\sec(\theta) = \frac{r}{x} = \frac{\text{hyp}}{\text{adj}} = \frac{1}{\cos(\theta)} \qquad \csc(\theta) = \frac{r}{y} = \frac{\text{hyp}}{\text{opp}} = \frac{1}{\sin(\theta)}$$

BASIC IDENTITIES

$$\sin(-\theta) = -\sin(\theta) \qquad \cos(-\theta) = -\cos(\theta)$$
$$\sin^2(\theta) + \cos^2(\theta) = 1 \qquad 1 + \tan^2(\theta) = \sec^2(\theta)$$

ADDITION FORMULAS

$$\sin(\theta + \phi) = \sin(\theta)\cos(\phi) + \cos(\theta)\sin(\phi)$$
$$\cos(\theta + \phi) = \cos(\theta)\cos(\phi) - \sin(\theta)\sin(\phi)$$
$$\tan(\theta + \phi) = \frac{\tan(\theta) + \tan(\phi)}{1 - \tan(\theta)\tan(\phi)}$$

HALF-ANGLE FORMULAS

$$\sin^2(\theta) = \frac{1 - \cos(2\theta)}{2} \qquad \cos^2(\theta) = \frac{1 + \cos(2\theta)}{2}$$

LAW OF COSINES

$$a^2 = b^2 + c^2 - 2bc\cos(A)$$

LAW OF SINES

$$\frac{\sin(A)}{a} = \frac{\sin(B)}{b} = \frac{\sin(C)}{c}$$

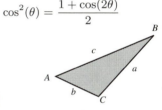

COORDINATE TRANSFORMATIONS

THE POLAR TRANSFORMATION

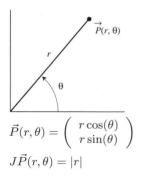

$$\vec{P}(r, \theta) = \begin{pmatrix} r\cos(\theta) \\ r\sin(\theta) \end{pmatrix}$$

$$J\vec{P}(r, \theta) = |r|$$

THE CYLINDRICAL TRANSFORMATION

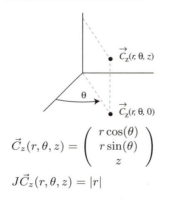

$$\vec{C}_z(r, \theta, z) = \begin{pmatrix} r\cos(\theta) \\ r\sin(\theta) \\ z \end{pmatrix}$$

$$J\vec{C}_z(r, \theta, z) = |r|$$

THE SPHERICAL TRANSFORMATION

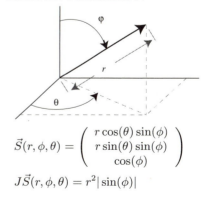

$$\vec{S}(r, \phi, \theta) = \begin{pmatrix} r\cos(\theta)\sin(\phi) \\ r\sin(\theta)\sin(\phi) \\ \cos(\phi) \end{pmatrix}$$

$$J\vec{S}(r, \phi, \theta) = r^2|\sin(\phi)|$$

CALCULUS WITH EARLY VECTORS

Phillip Zenor
Edward E. Slaminka
Donald Thaxton

Auburn University

PRENTICE HALL, Upper Saddle River, New Jersey 07458

of Congress Cataloging-in-Publication Data

hillip.
lculus with early vectors / Phillip Zenor, Edward E. Slaminka,
ld Thaxton.
p.   cm.
cludes index.
SBN: 0-13-791203-X
1. Calculus. 2. Vector analysis. 3. Mathematical physics.
4. Engineering mathematics. I. Slaminka, Edward E. II. Thaxton,
Donald. III. Title.
QA303.Z44   1998
515—dc21   94-25848
                                                    CIP

Acquisition Editor: George Lobell
Editorial Assistant: Gale Epps
Editorial Director: Tim Bozik
Editor-in-Chief: Jerome Grant
Assistant Vice President of Production and Manufacturing: David W. Riccardi
Editorial/Production Supervision: Richard DeLorenzo
Managing Editor: Linda Mihatov Behrens
Executive Managing Editor: Kathleen Schiaparelli
Manufacturing Buyer: Alan Fischer
Manufacturing Manager: Trudy Pisciotti
Marketing Manager: Melody Marcus
Marketing Assistant: Amy Lysik
Creative Director: Paula Maylahn
Art Director: Jayne Conte
Cover Designer: Bruce Kenselaar
Cover Photo Credit: National Aeronautics and Space Administration

©1999 by Prentice-Hall, Inc.
Simon & Schuster / A Viacom Company
Upper Saddle River, New Jersey 07458

Printed in the United States of America

10 9 8 7 6 5 4 3 2 1

ISBN 0-13-791203-X

Prentice-Hall International (UK) Limited, London
Prentice-Hall of Australia Pty. Limited, Sydney
Prentice-Hall Canada Inc., Toronto
Prentice-Hall Hispanoamericana, S.A., Mexico
Prentice-Hall of India Private Limited, New Delhi
Prentice-Hall of Japan, Inc., Tokyo
Simon & Schuster Asia Pte. Ltd., Singapore
Editora Prentice-Hall do Brasil, Ltda., Rio de Janeiro

Contents

Preface

Most calculus texts currently available try to cover a standard set of topics for for an increasingly diverse group of users. The result is an encyclopedic approach where much of the beauty and power of the subject is lost. There is, of course, a common base of knowledge that is important to any student of calculus, regardless of the student's plan of study. Most of the popular calculus texts do an effective job of presenting this common material: techniques of integration and differentiation of real valued functions and some of the standard applications such as graphing real valued functions and finding areas (or volumes) of very simple regions, sequences and series, etc. However, differentiating and integrating real valued functions seem to be the main objective of these texts. Other topics seem, at best, to be inserted randomly, ignoring their relationships with interconnected topics. Important topics, such as vectors and multivariate calculus, are treated as one would treat optional topics. This is not to say that these topics are not well exposited; but rather, the student's intuition and sense of geometry in higher dimensions is not given the chance to develop and mature before the chapters on multivariate calculus are encountered.

Long ago and far away – the 1960's – most high school precalculus students also took trigonometry based physics, and in their freshmen year at college took both calculus and calculus-based physics. Calculus is a much easier and more lucid subject when it focuses on the foundations for physics and engineering. We hope that this text sensitizes mathematics departments to reexamine their relationship with physics and engineering departments.

Calculus with Early Vectors is directed towards students who are required to take a calculus based physics course, primarily engineering, science, and mathematics majors. By focusing on the requirements of a specific group of students, we have been able to structure the text so that calculus is presented as a single subject rather than a collection of topics. This text is a result of over 20 years of continuous classroom experimentation, development, and testing. While vector calculus is covered early (indeed vectors are introduced in Chapter 1 in the precalculus review), the result is not a cut-and-paste job of moving forward a few topics tradition-ally covered later. The material is organized so that vector calculus is thoroughly covered over a four-quarter or three-semester sequence without neglecting topics usually covered in a calculus course. We present vector calculus and the calculus of real variables as a single subject; vector calculus is an integral part of the course from the beginning. We lay a solid mathematical foundation for understand-ing Newtonian physics and electricity and magnetism. We do not attempt to teach physics, but to introduce those mathematical and physical concepts which are essential in understanding physics.

We use innovative approaches only when they are pedagogically useful for reaching the goals of the text. We employ standard and popular approaches for topics whenever it is consistent with our strat-egy to do so; for example, our presentations of topics such as the rules for differentiation, techniques of integration, exponentials, log-arithms, etc. are quite standard. However, **Calculus with Early Vectors** departs from the standard and new calculus texts in both its organization and presentation. Points of departure include:

> We introduce vector valued functions in the beginning and we treat vector calculus and real valued calculus as a single subject throughout the text.

> We introduce enough physics so that, throughout the text, we are able to use ideas such as velocity, acceleration, force, en-ergy, and work for illustrations and motivation. Thus there is an underlying theme. We are building mathematical tools to model and solve physical and geometric problems.

> We present line, surface, and volume integrals, and the gradi-ent, curl, and divergence operators in the flow of the text rather than treat them as optional isolated topics. The material is or-ganized so that these topics are covered with a thoroughness and depth not matched even in many upper level vector cal-culus texts. This is accomplished without neglecting the usual topics involving real variable calculus.

> We introduce the Jacobian as the rate of change of length, area, or volume. This permits us to do line integrals, surface integrals, and volume integrals by first parametrizing the region and then using change of variables theorems to translate the integral to one over an interval, rectangle, or a rectangular box.

We emphasize the power of calculus as a tool for modeling complex physical problems. As a result, the methods of differentiation and integration, while not neglected, are presented primarily as necessary skills needed to solve problems that arise from mathematical models.

Sequences and series are integrated into related topics and are treated together with improper integrals. Taylor polynomials are early applications of derivatives and the use of the Remainder Theorem is an application of max/min.

We approach the theoretical aspects of calculus with the belief that, at the introductory level, it is important to understand the geometric basis for theorems and to develop an intuitive understanding for the statements of the theorems and their implications. We, at times, give geometric and/or intuitive arguments for theorems in a way that permits the students to "see why the theorem is true" rather than a formal proof. However, only in rare cases do we ask the student to take the truth of a theorem solely on faith.

Expository Features

We have written this text with the assumption that it will be read by the student. It is our intent that the book be "user friendly." The students in our experimental classes have always had diverse backgrounds. Some of them have only had algebra and trigonometry in high school while others have had a full year of calculus. Some of them know a lot of geometry while others know very little. As a result, we have found it worthwhile to start off with a rather gentle approach and let the presentation grow leaner as the class matures. Other expository features included are:

- The exposition is concise, correct, and to-the-point.

- There are ample examples and illustrations.

- Definitions, theorems, and corollaries are set off in boxes to aid in finding important information.

- There is coherence in the order of the topics and their exposition. This was absolutely necessary to accomplish the goals of the text.

Technology

Many of the current and new calculus texts focus on the use of computers and calculators to facilitate the teaching of calculus.

This trend toward the integrated use of calculators and computers in calculus courses will, and should, continue. While this book does not require the use of technology, we have freely included calculator/computer oriented exercises when they arise naturally in context and when they are pedagogically useful.

Supplementary Materials

In addition to student and instructor solutions manuals, and test banks, the authors are committed to the development and maintenance of a WWW site as a dynamic information source for the students and instructors using this text. The web site will include:

- Supplementary problems and projects applying the text material to suitable engineering and physics topics,

- Mathematica/copyright and Maple/copyright worksheets designed to complement and amplify the text, and

- Links to web pages of current interest.

The authors encourage contributions to this site from users of our text. Please try the web page at

```
http://www.auburn.edu/~zenorpl/EarlyVectors/
```

and send the authors your comments and suggestions. Their addresses are

Phillip Zenor and Edward E. Slaminka
Department of Mathematics
218 Parker Hall
Auburn University, AL 36849

e-mail addresses:
Phillip Zenor: zenorpl@mail.auburn.edu
Edward E. Slaminka: slam@mail.auburn.edu

Description of Contents

Chapters 1 and 2 are devoted to geometric topics in one, two, and three dimensions. These chapters introduce some material that is new to the students and also serve as a review of algebra and trigonometry. In Chapter 1 we begin with an introduction to vector algebra and geometry in two and three dimensions. In Chapter 2 we review the notion of function and then introduce the concept of vector valued functions. Sequences are also introduced as an example.

In Chapters 3, 4, and 5 we develop the concept of velocity (as a vector valued function of time) to motivate the introduction of the derivative. Chapter 3 focuses on limits and the definition of the derivative. First order Taylor polynomials and Newton's methods are early applications. As we develop the standard rules for differentiation, we incorporate related rates problems as immediate applications of the differential calculus. Taylor polynomials are constructed in a very natural way expanding upon the idea of linear approximations. In Chapter 5 we consider the geometry of functions and curves starting with vertical and horizontal asymptotes and limits of sequences. We treat such topics as increasing and decreasing functions and curves in the plane, concavity, max/min problems (where we also prove the remainder theorem for Taylor polynomials), curvature, the normal, tangential, and binormal components of acceleration.

In Chapter 6 we introduce the definite integral. Our approach is very standard. However, we emphasize using Riemann sums to approximate geometric properties such as volume and area, and then determine the appropriate integral. Immediately after we introduce the Fundamental Theorem of Calculus, we take up logarithms, exponentials, and inverse trigonometric functions. We present methods of integration in Chapter 10.

Chapter 8 is an optional section that treats first order linear differential equations with constant coefficients. In Chapter 9 we present L'Hôpital's Rule, improper integrals, and infinite series. We take advantage of the intimate relationship between the improper integral and infinite series. We view an infinite series as an improper integral of an associated function defined on a terminal segment of the reals to the reals. We then develop the standard tests for convergence and expound upon power series and radii of convergence.

In Chapter 11 we introduce and thoroughly develop the concept of work, beginning with a discussion of a constant force acting on a mass traveling through a straight line displacement. This idea is extended, using Riemann sums and parametrizations of paths, to allow variable forces and arbitrary paths. We carefully develop the Work-Energy Theorem, which explains the relationship between the change of the kinetic energy and the work done by the sum of all the forces acting on the mass as it moves along a trajectory in $\mathbb{R}^3$. This sets the stage for the development of line integrals and, in turn, leads naturally to the introduction of potential functions and conservative forces. We thoroughly explore the relationships between potential functions and conservative forces, the gradient, and their related force fields. We develop line integrals of functions from $\mathbb{R}^n$ to $\mathbb{R}$ over fundamental curves in $\mathbb{R}^n$ as well as line integrals in the context of a vector field.

In Chapter 12 we take advantage of the student's knowledge of potential functions (real valued functions defined on subsets of $\mathbb{R}^n$) to consider optimization problems.

In Chapter 13 we introduce linear functions from $\mathbb{R}^n$ to $\mathbb{R}^m$ and their associated matrices and polar, spherical, and cylindrical transformations. The Jacobian is introduced as the rate of change of area (via cross products) or the rate of change of volume (via triple products), setting the stage for multiple integrals and their change of variables theorems. The derivative of functions from $\mathbb{R}^n$ to $\mathbb{R}^m$ is introduced as a matrix. This allows us to define parametrizations for surfaces and solids and to derive their Jacobians.

In Chapter 14 we evaluate surface integrals by using the change of variables theorems to translate the surface integrals to integrals over rectangles. Similarly, we evaluate integrals over volumes by translating volume integrals to integrals over a rectangular box. It is important to emphasize that this approach unifies line integrals, surface integrals, and integrals over solids.

Applications include computing the mass and center of mass of surfaces and solids as well as computing the kinetic energy of rotating surfaces and solids. We also compute the flux through oriented surfaces. Standard problems such as finding the area bounded by curves given by polar equations, surface areas of graphs of functions and of surfaces of revolution, volumes of solids of revolution, etc. are simple applications of the techniques we employ.

We introduce Gaussian surfaces to pave the way for the Divergence Theorem.

In Chapter 15 we are able to take full advantage of the ground work we have laid. We develop the divergence and curl operators, present Gauss' Theorem, Stokes' Theorem, and Green's Theorem, and complete the picture of conservative forces. This accomplishes one of the major objectives of the text.

Notation

We have tried to keep notation to a minimum. We use $\mathbb{R}^n$ to denote n–dimensional space and we use calligraphic letters, such as $\mathcal{C}$, to denote graphs, partitions, refinements, etc. We have refrained from using boldfaced letters for vectors, and, instead, use $\vec{v}$ for vectors (except for the case where we are referring to the standard unit vectors, which we denote by $\hat{\imath}, \hat{\jmath}$, and $\hat{k}$). We only use symbols from mathematical logic in a few cases for expedience and presentation. For example, we do use the implication symbol $\Rightarrow$. However, we explain its meaning when it is first used. In the exercises we employ the following symbols to denote "technology" problems.

⊟   Denotes an exercise that requires a standard calculator for its solution.

⊞   Denotes an exercise that would be aided by a graphics calculator.

▣   Denotes an exercise that could be solved with a calculator but is greatly simplified with the use of a computer and either a high level programming language or a mathematical package such as Mathmatica.

We also use the notation Section 5.4 to indicate Chapter 5, Section 4. Similarly, Theorem 3.4.5 refers to Theorem 5 of Chapter 3, Section 4.

Flexibility

It is our hope that this text might be used in a variety of environments. At Auburn University, it was originaly used as part of an integrated calculus/physics course. It is currently being used as the text for a "stand alone" calculus course for science and engineering majors, and it is also being used as the foundation for a two year sequence of courses that integrate topics usually included in a pre-engineering curriculum. Following is a chapter dependency chart. Numerous paths through this chart have been succesfully implemented by the authors and their colleagues.

Acknowledgments

We would like to thank the following reviewers for their insightful comments:

Mark W. Coffee, University of Colorado–Denver;
Charles Schwartz, New Mexico State University;
Andre Adler, Illinois Institute of Technology;
R. Dante DeBlassie, Texas A & M University;
Donald Johason, North Central College;
William Mahavier, Nicholls State University; and
Cecilia Knoll, Florida Institute of Technology.

We would also like to thank the editorial staff at Prentice Hall, especially George Lobell, Richard DeLorenzo, and Joan Eurell.

And finally we wish to thank our colleagues and graduate students at Auburn University who taught calculus classes using preliminary versions of this text, and the many students who took our course. Their comments and suggestions were invaluable.

Chapter Dependency Flow Chart

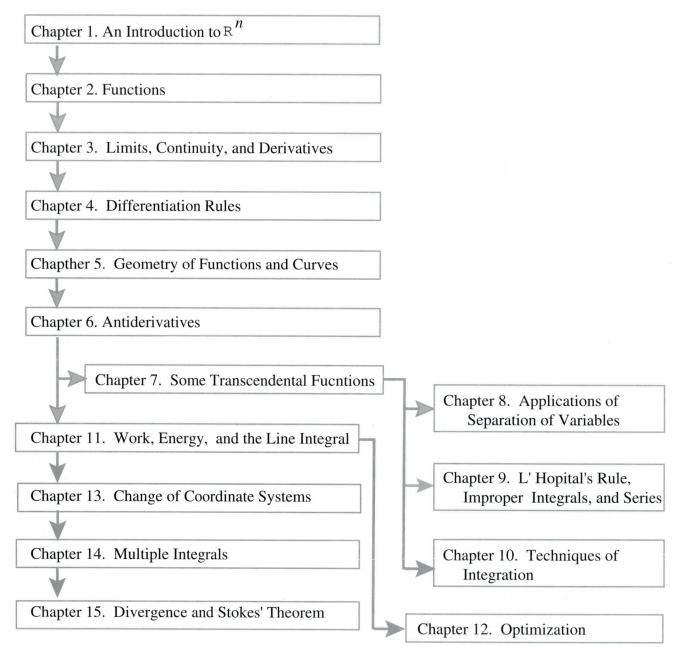

Chapter 1. An Introduction to $\mathbb{R}^n$

Chapter 2. Functions

Chapter 3. Limits, Continuity, and Derivatives

Chapter 4. Differentiation Rules

Chapther 5. Geometry of Functions and Curves

Chapter 6. Antiderivatives

Chapter 7. Some Transcendental Fucntions

Chapter 8. Applications of Separation of Variables

Chapter 11. Work, Energy, and the Line Integral

Chapter 9. L' Hopital's Rule, Improper Integrals, and Series

Chapter 13. Change of Coordinate Systems

Chapter 10. Techniques of Integration

Chapter 14. Multiple Integrals

Chapter 15. Divergence and Stokes' Theorem

Chapter 12. Optimization

Chapter 1

N–Dimensional Space

1.1 An Introduction to $\mathbb{R}^n$

Up to now, your mathematical experience has probably focused on
the set of real numbers, which we realize geometrically as a straight
line, "*the real line*." However, most mathematical descriptions require more "dimensions." To locate a position on the earth's surface
we need two dimensions, an ordered pair of real numbers representing the latitude and the longitude. It will take three dimensions, or
an ordered triple of real numbers (latitude, longitude, altitude), to
locate a position above or below the earth's surface. Similarly, we
can describe the shape of a box with an ordered triple of numbers:
width, height, depth. The purpose of Chapter 1 is to begin to build
an intuition and sense of geometry in two and three dimensions, and
to begin to develop the tools to build a mathematical model of the
physical world.

Points on the Real Line

When we speak about the real line, we actually have a model in
mind that is obtained by starting with a straight line. We pick a
position to represent the "origin" and label it '0' and we pick a scale
to give us a notion of distance. Then we identify the number x with
the point or position on the line that is $|x|$ units[1] from '0' in the
positive direction if x is positive and in the negative direction if x is
negative. (See Figure 1.)

---

[1]$|x|$ is used to denote the absolute value or the magnitude of the number x.
Thus, $|x| = x$ if $x \geq 0$ and $|x| = -x$ if $x < 0$.

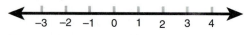

Figure 1. *The real line.*

Points on The Plane

The idea of representing the set of ordered pairs of real numbers as a plane should not be new. We draw two mutually perpendicular lines and call the position where these lines meet the *origin*. Usually, we draw one of these lines horizontally and the other vertically. Scales are chosen for the lines. Then we choose one direction to be the positive direction on each of the lines. Typically, the positive direction for the horizontal line is to the right and the positive direction for the vertical line is up. One of these lines (usually the horizontal one) is identified as the "first axis" and the other is the "second axis."

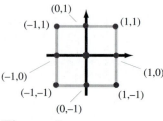

Figure 2. *The points* $(0,0)$, $(1,1)$, $(-1,1)$, $(1,-1)$, $(-1,-1)$, $(1,0)$, $(0,1)$, $(-1,0)$, *and* $(0,-1)$ *located in the plane.*

The ordered pair (x,y) represents the point in the plane that is $|x|$ units from the "second axis" (in the positive direction if x is positive and in the negative direction if x is negative) and which is $|y|$ units from the "first axis" (in the positive direction if y is positive and in the negative direction if y is negative). Clearly then, $(0,0)$ denotes the origin. (See Figure 2.)

Often, we refer to the plane as the xy–plane, where x represents the first coordinate of a position in the plane and y represents the second coordinate. Similarly, if we refer to the uv–plane, u represents the first coordinate and v represents the second coordinate of the point in the plane.

As illustrated in Figure 3, the axes divide the plane into 4 parts or *quadrants*.

Figure 3.
$Q_1 = $ *Quadrant I: The set of all* (x,y) *with* $x > 0$ *and* $y > 0$.
$Q_2 = $ *Quadrant II: The set of all* (x,y) *with* $x < 0$ *and* $y > 0$.
$Q_3 = $ *Quadrant III: The set of all* (x,y) *with* $x < 0$ *and* $y < 0$.
$Q_4 = $ *Quadrant IV: The set of all* (x,y) *with* $x > 0$ *and* $y < 0$.

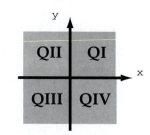

As the next example shows, sometimes we have to work a little to get the coordinates of a point from the information that is given.

EXAMPLE 1: The minute hand on a clock is six inches long. Find the coordinates of the tip of the minute hand at four minutes past four. Assume that the axes are as drawn in Figure 4.b.

Figure 4.a *The clock.*

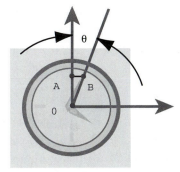

Figure 4.b *The placement of the coordinate axes.*

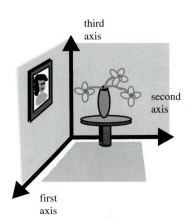

Figure 5.a *Realizing the coordinate axes as the corner of a room.*

SOLUTION: The minute hand is 4/60 of the way around the clock and there are 2π radians in a circle. It follows that $\theta = (4/60)(2\pi) = 2\pi/15$ radians. Referring to the triangle AOB in Figure 4.b, we see that the y–coordinate is

$$(\text{hypotenuse}) \times \cos\theta = 6\cos\theta = 6\cos\left(\frac{2\pi}{15}\right) \approx 5.48.$$

Similarly, the x–coordinate is

$$(\text{hypotenuse}) \times \sin\theta = 6\sin\theta = 6\sin(2\pi/15) \approx 2.44. \qquad \blacksquare$$

A little more difficult to draw, but still geometrically realizable, is the collection of ordered triples of real numbers. We draw three mutually perpendicular lines with a point in common. (See Figure 5.a.) A good model for this might be a corner of a room. For future purposes, we will be careful to pick a particular *orientation* when we choose "positive" directions for each of the lines. Our choice will be based on the *right hand rule*. We pick the positive directions on the lines so that if we hold our right hand with the "little finger side" flat on the plane containing the first and second coordinate axes, with our fingers curling from the positive half of the first axis toward the positive half of the second axis, then our thumb points in the positive direction of the third axis. (See Figure 5.b.) Thus, once we have chosen a positive direction on two of the axes, the positive direction on the third axis can be determined.

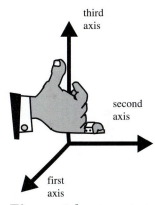

Figure 5.b *The right hand rule.*

The ordered triple (a, b, c) represents the position in 3–space that is $|a|$ units from the origin along the first axis (in the positive direction if a is positive and in the negative direction if a is negative), $|b|$ units from the origin along the second axis (in the appropriate direction), and $|c|$ units along the third axis. We can represent the point (a, b, c) graphically as the corner of a box. (See Figure 6.)

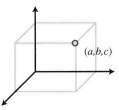

Figure 6. *The point (a, b, c).*

Following the convention established in the plane, if we talk about a point in xyz–space, then x represents the first coordinate, y, the second coordinate, and z, the third coordinate. The first axis is referred to as the x–axis, etc.

The symbol $\mathbb{R}^n$ denotes the set of all ordered n–tuples. Thus, $\mathbb{R}^1$ is the set of all 1–tuples, which is just the set of all real numbers, $\mathbb{R}^2$ is the set of all ordered pairs of real numbers, and $\mathbb{R}^3$ is the set of all ordered triples of real numbers. We call members of $\mathbb{R}^n$ *n–dimensional vectors*. When the context permits, we will be a little sloppy and refer to an n–dimensional vector as simply a vector. Thus, if in context it is clear that we are working in $\mathbb{R}^2$, then a vector will be an ordered pair (x, y) of real numbers. In general, when a symbol represents a vector, we so indicate by putting an arrow above the symbol; e.g., $\vec{a}$, $\vec{x}$, $\vec{v}$, $\vec{\alpha}$, $\vec{\chi}$, $\vec{\tau}$, etc. The vector $\vec{0}$ will always mean the vector $(0, 0, ..., 0)$. We use symbols with no arrow to denote real numbers. In $\mathbb{R}^2$ we often use $\hat{\imath}$ to denote $(1, 0)$ and $\hat{\jmath}$ to denote $(0, 1)$. Similarly, in $\mathbb{R}^3$ we use $\hat{\imath}$ to denote $(1, 0, 0)$, $\hat{\jmath}$ to denote $(0, 1, 0)$, and $\hat{k}$ to denote $(0, 0, 1)$.

NOTE: It is often more convenient to write a vector vertically rather than horizontally. Thus, (a, b, c) and $\begin{pmatrix} a \\ b \\ c \end{pmatrix}$ denote the same point. This notation is first used in Exercise 22 at the end of this section. We begin using the vertical representation of vectors extensively when we start doing algebra on vectors in Section 1.3.

EXAMPLE 2: A door 3 ft wide and 7 ft high is opened 35°. If the coordinate axes are placed as in Figure 7.a, what are the coordinates of the top right corner of the door?

SOLUTION: This is a 3–dimensional variation of the clock problem in Example 1. Notice that the third coordinate does not change. The second coordinate is the length of side AB of triangle ABC (Figure 7.b) while the first coordinate is the length of side BC. Thus, the coordinates of the corner are $(3\sin 35°, 3\cos 35°, 7) \approx (1.721, 2.227, 7)$ ft. ∎

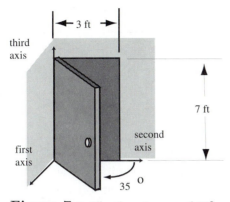
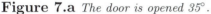

Figure 7.a *The door is opened 35°.*

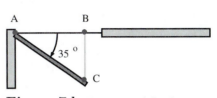

Figure 7.b *Top view of the door.*

EXAMPLE 3: The location of a point on the earth's surface is often given in terms of its longitude and latitude. The longitudes are semicircles that pass through the poles, and the latitudes are circles on the surface that are parallel to the equator. The latitudes and longitudes are given in terms of angles measured in degrees. The 0° longitude passes through Greenwich, a district in London. The longitudes and latitudes are measured as in Figure 8.c. We use θ to denote the longitude and ϕ to denote the latitude, where $-180° \leq \theta \leq 180°$ and $-90° \leq \phi \leq 90°$. The 180°E and 180°W longitudes are the same, and the international date line approximately follows this longitude. The position 25°E, 40°N is denoted by $(25, 40)$ while the position 25°W, 40°S is denoted by $(-25, -40)$.

EXAMPLE 4: Place the coordinate axes as illustrated in Figure 8.a. Find the xyz–coordinates of the position 25°E, 40°N.

SOLUTION: We assume that the earth is a sphere with radius $R = 6400$ km. The magnitude of the z–coordinate is the length of the side opposite the angle ϕ of the triangle $\vec{OPD}$ in Figure 8.c. So, $z = 6400 \sin 40° \approx 4114$ km. The magnitude of the x–coordinate is the length of the side adjacent the angle $\theta = 25°$ in the triangle $\vec{OAB}$ in Figure 8.c and the magnitude of the y–coordinate is the length of the opposite side of the same triangle. The length of the hypotenuse of the triangle $\vec{OAB}$ is $R\cos\phi = R\cos 40°$. It follows that $x = R\cos\phi\cos\theta = R\cos 40° \cos 25° \approx 4443$ km and $y = R\cos 40° \sin 25° \approx 2072$ km. ∎

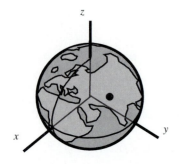

Figure 8.a *The origin is at the earth's center. The x–axis radiates through the intersection of the equator with the $0°$ longitude, the z–axis is drawn through the north pole, and the y–axis is drawn so that the right-hand rule applies.*

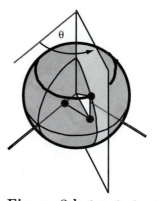

Figure 8.b *θ is the longitudinal coordinate of any point on the intersection of the half plane with the earth's surface.*

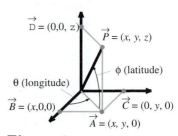

Figure 8.c *The point on the surface of the earth with longitude θ and latitude ϕ.*

The Distance Formula

We need the notion of distance between points in $\mathbb{R}^n$. Our definition of distance is motivated by the Pythagorean Theorem, which states that if a and b are the lengths of the legs of a right triangle and c is the length of the hypotenuse, then $a^2 + b^2 = c^2$.

Definition: The Distance Between Two Points

If $\vec{x} = (x_1, x_2, \ldots, x_n)$ and $\vec{y} = (y_1, y_2, \ldots, y_n)$, then the distance from $\vec{x}$ to $\vec{y}$ is given by

$$\text{dist}(\vec{x}, \vec{y}) = \sqrt{(x_1 - y_1)^2 + (x_2 - y_2)^2 + \ldots + (x_n - y_n)^2}.$$

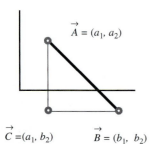

Figure 9.a *The right triangle ABC.*

To see that this definition of distance agrees with the Pythagorean Theorem in the plane, let $\vec{A} = (a_1, a_2)$ and $\vec{B} = (b_1, b_2)$ be two points in the plane. If we let $\vec{C} = (a_1, b_2)$, then the triangle with endpoints $\vec{A}, \vec{B}$, and $\vec{C}$ is a right triangle with legs the segments from $\vec{A}$ to $\vec{C}$ and $\vec{B}$ to $\vec{C}$ and hypotenuse the segment from $\vec{A}$ to $\vec{B}$. Denote the lengths of these segments by AC, BC, and AB, respectively. See Figure 9.a.

Since AC equals $|a_2 - b_2|$ and BC equals $|a_1 - b_1|$ we have that

$$AB = \sqrt{(a_1 - b_1)^2 + (a_2 - b_2)^2}.$$

Similarly, in $\mathbb{R}^3$, if AB is the length of the line segment with end points $\vec{A} = (a_1, a_2, a_3)$ and $\vec{B} = (b_1, b_2, b_3)$, then AB is the length of the hypotenuse of the right triangle with legs of length AD and DB, where $\vec{D} = (a_1, a_2, b_3)$. (See Figure 9.b.) AD is $|a_3 - b_3|$ and DB is the length of the hypotenuse of the right triangle with legs of length DC and CB, where $\vec{C} = (a_1, b_2, b_3)$. We see that

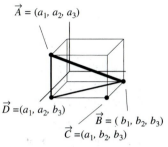

$$DB = \sqrt{(a_1 - b_1)^2 + (a_2 - b_2)^2}.$$

Thus,

$$
\begin{aligned}
AB &= \sqrt{(\text{length of } DB)^2 + (\text{length of } AD)^2} \\
&= \sqrt{(a_1 - b_1)^2 + (a_2 - b_2)^2 + (a_3 - b_3)^2},
\end{aligned}
$$

Figure 9.b *The right triangle ADB.*

which fits our definition of the distance between $\vec{A}$ and $\vec{B}$. ∎

EXAMPLE 5: Find the distance between $\vec{P_1} = (1, 2)$ and $\vec{P_2} = (-3, 1)$.

SOLUTION:

$$\text{dist}(\vec{P_1}, \vec{P_2}) = \sqrt{(1 - (-3))^2 + (2 - 1)^2} = \sqrt{17}.$$ ∎

EXERCISES 1.1

In Exercises 1–3, the points $\vec{p}_1, \ldots, \vec{p}_7$ are vertices of a rectangular box as illustrated in Figure 10.

1. Given that $\vec{p}_1 = (1, 2, 3)$, find $\vec{p}_2$, $\vec{p}_3$, $\vec{p}_4$, $\vec{p}_5$, $\vec{p}_6$, and $\vec{p}_7$.

2. Given that $\vec{p}_2 = (1, 0, 3)$ and $\vec{p}_5 = (0, 2, 0)$, find $\vec{p}_1$, $\vec{p}_3$, $\vec{p}_4$, $\vec{p}_6$, and $\vec{p}_7$.

3. Given that $\vec{p}_4 = (1, 3, 0)$ and $\vec{p}_7 = (0, 0, 3)$, find $\vec{p}_1$, $\vec{p}_2$, $\vec{p}_3$, $\vec{p}_5$, and $\vec{p}_6$.

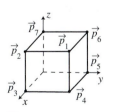

Figure 10. *Figure for Exercises 1–3.*

Find the distance between the pairs of points given in Exercises 4–9.

4. $\vec{P_1} = (1, 3)$, $\vec{P_2} = (0, 1)$.

5. $\vec{P_1} = (-1, 6)$, $\vec{P_2} = (1, -3)$.

6. $\vec{P_1} = (2\sqrt{2}, 3\sqrt{3})$, $\vec{P_2} = (3\sqrt{2}, 5\sqrt{3})$.

7. $\vec{P_1} = (1/2, \sqrt{2})$, $\vec{P_2} = (2, 1/\sqrt{2})$.

8. $\vec{P_1} = (-1, 2, 0)$, $\vec{P_2} = (0, 1, 0)$.

9. $\vec{P_1} = (1/\sqrt{2}, 0, \sqrt{2})$, $\vec{P_2} = (\sqrt{2}, 0, 1/\sqrt{2})$.

10. If $\vec{A}, \vec{B}$ and $\vec{C}$ are three points in the plane, $d(\vec{A}, \vec{B}) = 3$, and $d(\vec{B}, \vec{C}) = 2$, what can you say about $d(\vec{A}, \vec{C})$?

11. Let $\vec{P_1} = (-1, 0)$ and $\vec{P_2} = (1, 0)$. Find two points $\vec{P_3}$ and $\vec{P_4}$ such that $\{\vec{P_1}, \vec{P_2}, \vec{P_3}\}$ and $\{\vec{P_1}, \vec{P_2}, \vec{P_4}\}$ are the vertices of an equilateral triangle.

12. Let $\vec{P}_1 = (-1, 2)$ and $\vec{P}_2 = (0, 0)$. Find the point, $\vec{P}_3$, in Quadrant 1 such that $\{\vec{P}_1, \vec{P}_2, \vec{P}_3\}$ are vertices of an equilateral triangle.

13. Let $\vec{P}_1 = (1, 3)$ and $\vec{P}_2 = (-2, -1)$. Find numbers a_1 and a_2 so that $\{\vec{P}_1, \vec{P}_2, (a_1, 0)\}$ and $\{\vec{P}_1, \vec{P}_2, (0, a_2)\}$ are vertices of right triangles with hypotenuse the line segment joining $\vec{P}_1$ and $\vec{P}_2$.

14. Let $\vec{P}_1 = (-1, 0, 0)$ and $\vec{P}_2 = (0, 1, 0)$. Find a positive number a so that $\{\vec{P}_1, \vec{P}_2, (-a, a, a)\}$ are vertices of an equilateral triangle.

A ring is attached to a 3ft rope a distance A from one end. The rope is suspended at its ends as in Figure 11. In Exercises 15 and 16, find the coordinates of the ring using the given information.

15. $A = 1$ ft, $B = 2$ ft.   16. $A = 1.5$ ft, $B = 2$ ft.

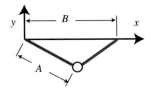

Figure 11. *Illustration for Exercises 15 and 16.*

In Exercises 17–19, The length of the minute hand of a clock is 3 in and the hour hand is 2 in long. Place the coordinate axes as in Figure 4.b. The time is 4:35.

17. Find the xy–coordinates of the tip of the minute hand.

18. Find the xy–coordinates of the tip of the hour hand.

19. Find the distance between these points.

Three ropes of lengths R1, R2, and R3 are suspended from the ceiling at positions $\vec{A}$, $\vec{B}$, and $\vec{C}$ as in Figure 12. The other ends are tied to a weight at $\vec{P}$ that pulls the ropes tight. In Exercises 20–21, find the coordinates of $\vec{P}$.

20. $R1 = 1$, $R2 = 1$, $R3 = 1$, $\vec{A} = (1, .5, 0)$, $\vec{B} = (0, 1, 0)$, and $\vec{C} = (1, 1, 0)$, all in feet.

21. $R1 = 2$, $R2 = 1.5$, $R3 = 1.5$, $\vec{A} = (1, 0, 0)$, $\vec{B} = (0, 1, 0)$, and $\vec{C} = (1, 1, 0)$, all measured in feet.

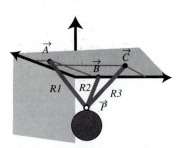

Figure 12. *Illustration for Exercises 20 and 21.*

22. Use the following definition of the center of mass of a system of point masses:
If $P_1, P_2, \ldots, P_n$ is a system of "point" masses such that
P_1 has mass m_1 and is located at (x_1, y_1, z_1)
P_2 has mass m_2 and is located at (x_2, y_2, z_2)
$\qquad \vdots \qquad\qquad \vdots \qquad\qquad\qquad \vdots$
P_n has mass m_n and is located at (x_n, y_n, z_n),
then

$$\begin{pmatrix} x \\ y \\ z \end{pmatrix} = \frac{1}{m_1 + \cdots + m_n} \begin{pmatrix} m_1 x_1 + \cdots + m_n x_n \\ m_1 y_1 + \cdots + m_n y_n \\ m_1 z_1 + \cdots + m_n z_n \end{pmatrix}$$

is called the *center of mass* of the system.
Given the system:
P_1 is a 2kg mass located at $(1, 2, 3)$ m,
P_2 is a 3kg mass located at $(1, -2, 0)$ m, and
P_3 is a 1kg mass located at $(2, 0, 0)$ m,
find the center of mass of the system.

In Exercises 23–25, find the xyz–coordinates of $\vec{p}$, the corner point of the door.

23. See Figure 13.a.

   a. $W = 3$ ft, $H = 4$ ft, and $\theta = \pi/6$ radians.

   b. $W = 4$ ft, $H = 6$ ft, and $\theta = \pi/3$ radians.

24. See Figure 13.b.

   a. $W = 3$ ft, $H = 7$ ft, and $\theta = 1.45$ radians.

   b. $W = 2$ ft, $H = 3$ ft, and $\theta = \pi/8$ radians.

25. See Figure 13.c.

   a. $W = 3$ ft, $H = 7$ ft, and $\theta = 1.45$ radians.

   b. $W = 2$ ft, $H = 3$ ft, and $\theta = \pi/8$ radians.

Prague is at $14.5° E, 50° N$. Atlanta, Georgia is at $84.3° W\ 33.8° N$. In Exercises 25–27, place the coordinate axes as in Figure 8.a.

26. Find the xyz–coordinates for Prague.

27. Find the xyz–coordinates for Atlanta.

28. Find the straight-line distance between the cities.

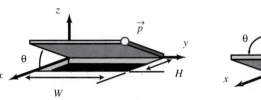

Figure 13.a *Exercise 23.*

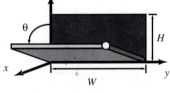

Figure 13.b *Exercise 24.*

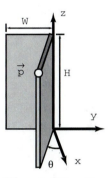

Figure 13.c *Exercise 25.*

1.2   Graphs in $\mathbb{R}^2$ and $\mathbb{R}^3$ and Their Equations

We start this section with some examples that are hopefully familiar to you from high school.

EXAMPLE 1:

• The set of points in the plane satisfying the equation $y = 3x + 2$ is the line with slope 3 that contains the point $(0, 2)$. See Figure 1.a.

• The set of points in the plane satisfying the equation $x^2 + y^2 = 4$ is the circle with radius 2 centered at the origin. See Figure 1.b.

• The set of points in $\mathbb{R}^3$ satisfying $x^2 + y^2 + z^2 = 4$ is the sphere with radius 2 centered at the origin. See Figure 1.c.   ∎

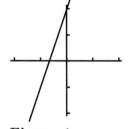

Figure 1.a *The graph of* $y = 3x + 2$.

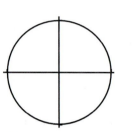

Figure 1.b *The graph of* $x^2 + y^2 = 4$.

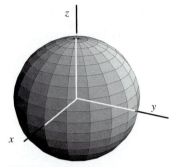

Figure 1.c *The graph of* $x^2 + y^2 + z^2 = 4$.

In general, a *graph* in $\mathbb{R}^n$ is simply a subset of $\mathbb{R}^n$. If $\mathcal{E}$ is an algebraic expression involving two unknowns, say x and y, then the graph $\mathcal{G}$ of $\mathcal{E}$ is the set of all points (x, y) in the plane that satisfy $\mathcal{E}$, in this case, $\mathcal{E}$ is called an equation for $\mathcal{G}$. Similarly, if $\mathcal{E}$ is an algebraic expression involving three unknowns, say x, y and z, the graph $\mathcal{G}$ of $\mathcal{E}$ is the set of all points (x, y, z) in $\mathbb{R}^3$ that satisfy $\mathcal{E}$.

The Line in the Plane

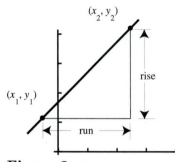

Figure 2. *The slope of a line is the rise over the run.*

Probably the most familiar examples of graphs and their equations are lines in the plane. Recall that any nonvertical line in the plane can be written as

$$y = mx + b \quad \text{(Slope–intercept form)}$$

where m is the slope of the line and $(0, b)$ is a point on the line. (b is called the *y–intercept* of the line.) If $\vec{P}_1 = (x_1, y_1)$ and $\vec{P}_2 = (x_2, y_2)$ are distinct points on a nonvertical line, then the slope of the line containing $\vec{P}_1$ and $\vec{P}_2$ is given by

$$m = \frac{y_2 - y_1}{x_2 - x_1} = \frac{\text{rise}}{\text{run}} .$$

Of course, if $\vec{P}_1 = (x_1, y_1)$ and $\vec{P}_2 = (x_2, y_2)$ are points on a vertical line, then $x_1 = x_2$ and the slope of the vertical line is undefined. An equation for the line containing $\vec{P}_1$ with slope m is

$$y - y_1 = m(x - x_1) \quad \text{(Point–slope form)}.$$

Finally, recall that

(a) Two lines are *parallel* if and only if they have the same slope or they are both vertical.

(b) A vertical line has no slope, and the equation for a vertical line is of the form

$$x = x_0.$$

(c) Two lines are *perpendicular* if and only if either:

  (i) one is horizontal and the other is vertical; or

  (ii) the slope of one is the negative reciprocal of the slope of the other (*i.e.*, if m_1 and m_2 are the slopes of the two lines, then $m_1 = -\frac{1}{m_2}$).

Standard Planes in $\mathbb{R}^3$

We will do a more extensive study of planes in $\mathbb{R}^3$ later; however, there are some important planes that we can describe now.

Recall from geometry that if L_1 and L_2 are two distinct lines in $\mathbb{R}^3$ that intersect, then there is exactly one plane in $\mathbb{R}^3$ that contains both L_1 and L_2. The plane in xyz–space that contains the x–coordinate axis and the y–coordinate axis is called the *xy–coordinate plane*, the plane that contains the x– and z–coordinate axes is called the *xz–coordinate plane*, and the plane that contains the y– and z–coordinate axes is called the *yz–coordinate plane*.

The xy–coordinate plane is the set of all points of the form $(x, y, 0)$; so $z = 0$ is the equation for the xy–coordinate plane. Indeed, if c is any number, then the set of all points in xyz–space of the form (x, y, c) is a plane parallel to the xy–plane. The set of all points in xyz–space of the form (x, c, z) is a plane parallel to the xz–plane. The set of all points of the form (c, y, z) is a plane parallel to the yz–plane. Thus, $x = c$ is an equation for the plane parallel to the yz–plane containing the point $(c, 0, 0)$ (Figure 3.a), $y = c$ is an equation for the plane parallel to the xz–plane containing the point $(0, c, 0)$ (Figure 3.b), and $z = c$ is an equation for the plane parallel to the xy–plane (Figure 3.c).

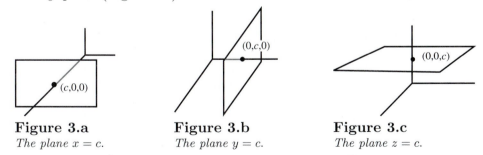

Figure 3.a   **Figure 3.b**   **Figure 3.c**
The plane $x = c$.   *The plane $y = c$.*   *The plane $z = c$.*

Now, in $\mathbb{R}^2$, $x = c$ is an equation for a vertical line, while in xyz–space the same equation $x = c$ is an equation for a plane. We can carry this idea further: consider the equation $y = -2x + 3$. In the xy–plane, this is an equation of a line. However, in xyz–space, the equation $y = -2x + 3$ puts no restriction on the z–coordinate. The point (x, y, z) satisfies $y = -2x + 3$ if and only if (x, y) satisfies

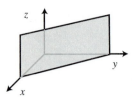

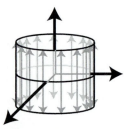

Figure 4. *The graph of* $y = -2x + 3$ *in* $\mathbb{R}^3$.

Figure 5. *The graph of* $\sqrt{(x-a)^2 + (y-b)^2} = r$ *in* $\mathbb{R}^3$.

$y = -2x + 3$. As we illustrate in Figure 4, the equation $y = -2x + 3$ describes a plane in xyz–space parallel to the z–axis, which cuts the xy–plane in the line described by $y = -2x + 3$.

Other Examples of Graphs and Expressions That Describe Them

EXAMPLE 2: An equation for the circle in the xy–plane with center (a, b) and radius r is $\sqrt{(x-a)^2 + (y-b)^2}$. However, the graph of the same equation in xyz–space will be a cylinder with axis the z–axis as shown in Figure 5. ∎

EXAMPLE 3: The graph of $y > 2x$ in the xy–plane is the set of points in the plane that lie above the line $y = 2x$, while in xyz–space, the graph of the same equation is the set of all points in 3–space to the "right" of the plane $y = 2x$. See Figures 6.a and 6.b. ∎

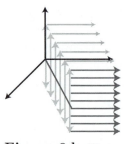

Figure 6.a *The graph of* $y > 2x$ *in the plane.*

Figure 6.b *The graph of* $y > 2x$ *in* $\mathbb{R}^3$.

EXAMPLE 4: The graph of $x^2 = 1$ in the xy–plane consists of two parallel lines $x = 1$ and $x = -1$, while in xyz–space, the graph of the same equation consists of two planes. See Figures 7.a and b. ∎

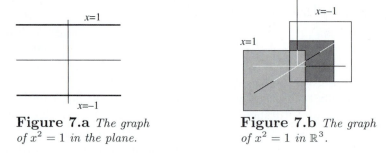

Figure 7.a *The graph of* $x^2 = 1$ *in the plane.*

Figure 7.b *The graph of* $x^2 = 1$ *in* $\mathbb{R}^3$.

EXAMPLE 5: The graph of $(x - a)^2 - (y - b)^2 = 0$ in the xy–

plane consists of two lines $x - a = y - b$ and $x - a = -(y - b)$, while in xyz–space, the graph of the same equation consists of two planes intersecting in a line parallel to the z–axis. See Figure 8.   ∎

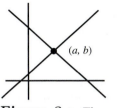

Figure 8.a *The graph of $(x - a)^2 - (y - b)^2 = 0$ in the plane.*

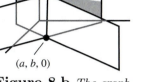

Figure 8.b *The graph of $(x - a)^2 - (y - b)^2 = 0$ in $\mathbb{R}^3$.*

EXERCISES 1.2

1. Can a line in $\mathbb{R}^2$ contain the points $(1, 1), (1, 2)$, and $(2, 3)$?

2. Can a plane in $\mathbb{R}^3$ contain the points $(1, 1, 0), (1, 2, 0)$, and $(2, 3, 0)$?

3. How many planes in $\mathbb{R}^3$ contain the points $(1, 0, 0)$ and $(0, 1, 0)$? and also $(0, 0, 0)$?

4. Find an equation for the circle in the plane with center $(-1, 2)$ and radius 6.

5. Find an equation for the sphere (the surface of a ball) in xyz–space with center $(-1, 2, 3)$ and radius 6.

In Exercises 6–11, find an equation or an inequality that describes the geometric object.

6. The disk of radius 2 and center $(1, 1)$ in the uv–plane. (A disk is a circle together with its interior.)

7. The exterior of the circle with radius 2 and center $(1, 1)$ in the xy–plane.

8. The interior of a sphere in uvw–space with center $(1, 1, 1)$ and radius 2.

9. The plane in uvw–space tangent to the sphere in Exercise 8 at the point $(1, 1, 3)$.

10. The circle in the plane of radius r and center (h, k).

11. The sphere in 3-space of radius r and center (h, k, l).

In Exercises 12–14, find the center and radius of the circle in the xy–plane.

12. $x^2 + y^2 = 4$.

13. $x^2 + 4x + y^2 + 2y = 2$.

14. $x^2 - 5x + y^2 + 3y = 0$.

In Exercises 15–17, find the center and radius of the sphere in $\mathbb{R}^3$.

15. $x^2 + y^2 + z^2 = 4$.

16. $x^2 - 4x + y^2 + z^2 = 0$.

17. $x^2 - 5x + 6 + y^2 + 3y + z^2 + 3z = 0$.

In Exercises 18–23, find an equation for the line in the xy-plane satisfying the given conditions.

18. Slope 2 and contains $(1, 3)$.

19. Slope 3 and contains $(5, 4)$.

20. Parallel to $y = 3x + 2$ and contains $(0, 1)$.

21. Perpendicular to $y = -x + 1$ and contains $(3, 1)$.

22. Contains the points $(3, 1)$ and $(4, 2)$.

23. Contains the points $(-1, 6)$ and $(0, 2)$.

Find an equation for the plane in xyz–space that satisfies the conditions given in Exercises 24–30

24. Parallel to the xy–plane and contains $(3, 1, -1)$.

25. Parallel to the yz–plane and contains $(1, 2, 0)$.

26. Parallel to the z–axis and contains the points $(0, 1, 0)$ and $(1, 0, 0)$.

27. Parallel to the z–axis and contains the points $(0, 1, 3)$ and $(1, 0, 25)$.

28. Parallel to the plane $y = 2x + 3$ and contains $(0, 0, 0)$.

29. Parallel to the plane $y = 2z - 1$ and contains $(1, 2, 1)$.

30. Parallel to the plane $z = 3$ and contains the point $(3, 2, 1)$.

In Exercises 31–37, describe the graphs of the given equation in xyz–space.

31. $x^2 + y^2 + z^2 = 0$.

32. $x^2 + y^2 = 10$.

33. $x^2 + 4x + y^2 + 2y = 6$.

34. $z^2 = 1$.

35. $z^2 + x^2 = 5$.

36. $y^2 - z^2 = 0$.

37. $y^2 + 2y + z^2 + 4z = 0$.

38. Find an equation for the set of points in $\mathbb{R}^2$ that are equidistant from $(1, 3)$ and $(-1, 4)$.

39. Find an equation for the set of points in $\mathbb{R}^3$ that are equidistant from $(3, 2, 0)$ and $(-1, 0, 2)$.

40. Show that the set of all points in $\mathbb{R}^2$ that are equidistant from (x_1, y_1) and (x_2, y_2) is a straight line.

41. Show that the set of all points in $\mathbb{R}^3$ that are equidistant from (x_1, y_1, z_1) and (x_2, y_2, z_2) has an equation of the form $ax + by + cz + d = 0$.

1.3   Algebra in $\mathbb{R}^n$

In this section, we introduce some basic algebraic tools for n–dimensional space: addition and subtraction of vectors and the multiplication of a vector by a scalar. These operations have very useful geometric and physical interpretations.

Recall that we have seen how to indicate a vector vertically. Thus, if $\vec{p}$ is in $\mathbb{R}^n$, we can represent $\vec{p}$ by the equivalent notations:

$$\vec{p} = (p_1, p_2, \ldots, p_n) = \begin{pmatrix} p_1 \\ p_2 \\ \vdots \\ p_n \end{pmatrix}.$$

The vertical representation of a vector can be quite useful when we are writing vectors involving complicated algebraic expressions.

Vector Addition, Multiplication by Scalars, and the Norm

Let $\vec{p} = (p_1, p_2, \ldots, p_n)$ and $\vec{q} = (q_1, q_2, \ldots, q_n)$ be vectors in $\mathbb{R}^n$, and let c be a real number. Then define

$$\vec{p} + \vec{q} = \begin{pmatrix} p_1 + q_1 \\ p_2 + q_2 \\ \vdots \\ p_n + q_n \end{pmatrix}, \quad \vec{p} - \vec{q} = \begin{pmatrix} p_1 - q_1 \\ p_2 - q_2 \\ \vdots \\ p_n - q_n \end{pmatrix}, \quad c\vec{p} = \begin{pmatrix} cp_1 \\ cp_2 \\ \vdots \\ cp_n \end{pmatrix}.$$

The *norm* of $\vec{p}$ or the *length* of $\vec{p}$ is the length of the line segment joining $\vec{p}$ with the origin. The norm of $\vec{p}$ will be denoted by $\|\vec{p}\|$. We have that

$$\|\vec{p}\| = \sqrt{p_1^2 + p_2^2 + \ldots + p_n^2}.$$

If $\|\vec{p}\| = 1$, then $\vec{p}$ is called a *unit vector*.

EXAMPLE 1: Let $\vec{p} = (1, 2, -1, 5)$, $\vec{q} = (2, \frac{1}{2}, -2, -3)$, and $c = 3$. Then

$$\vec{p} + \vec{q} = \begin{pmatrix} 1 + 2 \\ 2 + \frac{1}{2} \\ (-1) + (-2) \\ 5 + (-3) \end{pmatrix} = \begin{pmatrix} 3 \\ 2.5 \\ -3 \\ 2 \end{pmatrix},$$

$$\vec{p} - \vec{q} = \begin{pmatrix} -1 \\ 1.5 \\ 1 \\ 8 \end{pmatrix}, c\vec{p} = \begin{pmatrix} 3 \\ 6 \\ -3 \\ 15 \end{pmatrix}$$

$$\|\vec{p}\| = \sqrt{1 + 4 + 1 + 25} = \sqrt{31}. \qquad \blacksquare$$

The following theorem tells us that vector addition and multiplication of a vector by a real number behave as we would expect.

Theorem 1 *Let $\vec{p}$, $\vec{q}$ and $\vec{r}$ be vectors, and let a and b be numbers. Then*

   (a) $(\vec{p} + \vec{q}) + \vec{r} = \vec{p} + (\vec{q} + \vec{r})$;

   (b) $\vec{p} + \vec{q} = \vec{q} + \vec{p}$;

(c) $a(\vec{p} + \vec{q}) = a\vec{p} + a\vec{q}$;

(d) $(a + b)\vec{p} = a\vec{p} + b\vec{p}$.

Proof: Let $\vec{p} = (p_1, p_2, \ldots, p_n)$, $\vec{q} = (q_1, q_2, \ldots, q_n)$, and $\vec{r} = (r_1, r_2, \ldots, r_n)$ be vectors in $\mathbb{R}^n$. To prove Part 1, we simply perform the operations as defined.

$$(\vec{p} + \vec{q}) + \vec{r}$$

$$= ((p_1, p_2, \ldots, p_n) + (q_1, q_2, \ldots, q_n)) + (r_1, r_2, \ldots, r_n)$$

$$= (p_1 + q_1, p_2 + q_2, \ldots, p_n + q_n)) + (r_1, r_2, \ldots, r_n)$$

$$= ((p_1 + q_1) + r_1, (p_2 + q_2) + r_2, \ldots, (p_n + q_n) + r_n)$$

$$= (p_1 + (q_1 + r_1), p_2 + (q_2 + r_2), \ldots, p_n + (q_n + r_n))$$

$$= \vec{p} + (\vec{q} + \vec{r}).$$ ■

The proofs to Parts 2, 3, and 4 are equally straightforward applications of the definitions, and they are left to the reader.

Some Helpful Geometry

Up to this point, we have been considering a vector to be a position in $\mathbb{R}^n$. There is another interpretation that proves to be very helpful.

Consider a vector $\vec{A}$ in $\mathbb{R}^2$ starting at the position $\vec{P}$ and terminating at the position $\vec{Q}$. Let θ denote the angle that $\vec{A}$ makes with the horizontal, measured in a "counterclockwise" fashion from the horizontal line drawn from $\vec{P}$. Then the ordered pair ((length of $\vec{A}$) $\cos(\theta)$, (length of $\vec{A}$) $\sin(\theta)$) tells us both the length of $\vec{A}$ and the direction of $\vec{A}$. If $\vec{p} = (p_1, p_2)$ is a vector in the plane, then it can be represented by *any* directed line segment that extends p_1 units in the x–direction and p_2 units in the y–direction. In the same way, to get a geometric representation of an element of $\mathbb{R}^n$, $\vec{p} = (p_1, p_2, \ldots, p_n)$, we can choose any line segment that extends p_1 units in the first direction, p_2 units in the second direction, etc.

EXAMPLE 2: The vector $(1, 1)$ can be represented by any directed line segment that runs 1 unit to the right and 1 unit up. (See Figure 2.)   ■

EXAMPLE 3: All of the directed line segments in Figure 3 represent the vector $(1, -1, 1/2)$.   ■

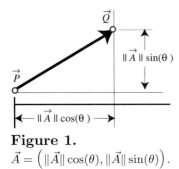

Figure 1.
$\vec{A} = \left(\|\vec{A}\| \cos(\theta), \|\vec{A}\| \sin(\theta) \right).$

Figure 2. *Directed line segments representing* $(1, 1)$ *in the plane.*

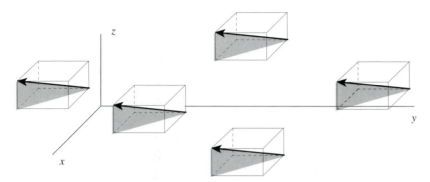

Figure 3. *Directed line segments representing $(1, -1, 1/2)$ in $\mathbb{R}^3$.*

The directed line segment realization of a vector is extremely helpful in giving some geometric meaning to the algebraic operations we have introduced in this section.

Multiplication by a Scalar (See Figure 4.)

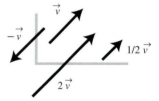

Figure 4. *Scaled versions of the vector $\vec{v}$.*

Since $c(p_1, p_2, \ldots, p_n) = (cp_1, cp_2, \ldots, cp_n)$, multiplication of a vector by a real number has the effect of magnifying or shrinking the vector, and/or reversing the direction of the vector. In the context of vector algebra, we often refer to real numbers as *scalars*.

- If $0 < c < 1$ then multiplication of $\vec{v}$ by c will shrink $\vec{v}$. For example, $\frac{1}{2}\vec{v}$ is a vector that points in the same direction as $\vec{v}$ but it is only half the length of $\vec{v}$.

- If $1 < c$, then $c\vec{v}$ points in the same direction as $\vec{v}$ but is longer than $\vec{v}$. For example, $2\vec{v}$ points in the same direction but is twice as long as $\vec{v}$.

- If $c < 0$, then $c\vec{v}$ has the same length as $|c|\vec{v}$ but points in the opposite direction of $\vec{v}$.

Of particular interest is scaling a nonzero vector $\vec{v}$ to obtain a unit vector pointing in the same direction as $\vec{v}$. The next theorem tells us how to do this.

Theorem 2 *Suppose that $\vec{v}$ is a vector and c is a number. Then*

(a) $\|c\vec{v}\| = |c|\,\|\vec{v}\|$, *and*

(b) *If $\|\vec{v}\| \neq \vec{0}$, then $\vec{v}/\|\vec{v}\|$ is a unit vector that points in the same direction as $\vec{v}$. (Recall that a unit vector is one that has length exactly one unit.)*

Proof:

(a) Let $\vec{v} = (v_1, \ldots v_n)$. Then

$$\|c\vec{v}\| = \sqrt{(cv_1)^2 + \cdots + (cv_n)^2} = |c|\sqrt{(v_1)^2 + \cdots + (v_n)^2} = |c|\,\|\vec{v}\|.$$

(b) Let $c = 1/\|\vec{v}\|$. Then by Part (a),

$$\left\|\frac{\vec{v}}{\|\vec{v}\|}\right\| = \frac{1}{\|\vec{v}\|}\|\vec{v}\| = 1.$$

■

EXAMPLE 4: The vector $\left(\frac{1}{\sqrt{3}}, \frac{1}{\sqrt{3}}, \frac{1}{\sqrt{3}}\right)$ is a unit vector with the same direction as $(1, 1, 1)$.   ■

Addition and Subtraction of Vectors

Suppose that $\vec{p} = (p_1, p_2, \ldots, p_n)$ and $\vec{q} = (q_1, q_2, \ldots, q_n)$. Then $\vec{p} + \vec{q} = (p_1 + q_1, p_2 + q_2, \ldots, p_n + q_n)$ is represented by a line segment that runs $p_1 + q_1$ units in the first direction, $p_2 + q_2$ units in the second direction, and so on. Let **P** be any directed line segment representing $\vec{p}$ and then let **Q** be the directed line segment representing $\vec{q}$ that is drawn emanating from the terminal end of **P**. Then the line segment drawn emanating from the initial end of **P** and terminating at the terminal end of **Q** is a representative of $\vec{p} + \vec{q}$. (See Figures 5.a and b.)

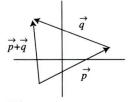

Figure 5.a *A graphical representation of the sum of two vectors.*

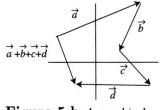

Figure 5.b *A graphical representation of the sum of several vectors.*

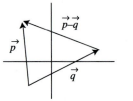

Figure 5.c *A graphical representation of $\vec{p} - \vec{q}$ by the segment drawn from the end of $\vec{q}$ to the end of $\vec{p}$.*

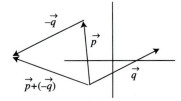

Figure 5.d *$\vec{p} - \vec{q}$ represented by adding $-\vec{q}$ to $\vec{p}$.*

To get a geometric picture of $\vec{p} - \vec{q}$, observe that $(\vec{p} - \vec{q}) + \vec{q} = \vec{p}$. And so, if we draw

(a) a representative of $\vec{q}$

(b) the representative of $\vec{p} - \vec{q}$ emanating from the terminal end of $\vec{q}$, and

(c) the representation of $\vec{p}$ emanating from initial point of $\vec{q}$,

then the terminal end of $\vec{p} - \vec{q}$ must also be the terminal end of $\vec{p}$. (See Figures 5.c and d.)

Physical Realizations of Vectors

Any quantity that has direction and magnitude can be described using vectors. An important example is the notion of force. Intuitively, we can see that force involves both direction and magnitude. When we push or pull on something, the direction of our exertion is as important as how hard we are pushing or pulling.

In the metric system, forces are measured in Newtons and in the English system, forces are measured in pounds. The magnitude of the gravitational force acting on a mass is its weight. In the metric system, mass is measured in kilograms (kg) and the magnitude of the force due to the earth's gravity (at sea level) of an x–kilogram mass is approximately $9.8x$ Newtons (N). Thus, the gravitational force acting on a 10 kg mass is 98 N.

Figure 6. *The effect of pushing on the block depends not only on how hard it is pushed, but also on its direction and where the force is applied.*

Force

An important law given to us by Newton states that the motion of the object does not change if the sum of all forces acting on the object is $\vec{0}$.

EXAMPLE 5: A block is resting on level ground. The magnitude of the gravitational force acting on the block is 20 N.

(a) What is the gravitational force acting on the block?

(b) What force is the ground exerting on the block?

(c) A rope is pulling straight up on the block with a force of magnitude 10 N. What force is the ground exerting on the block?

SOLUTION:

(a) Since it acts straight down and its magnitude is 20 N, the gravitational force is $(0, -20)$ N.

(b) Let $\vec{F}$ denote the force exerted by the ground. Since the sum of the forces is $\vec{0}$, we have $\vec{F} + (0, -20) = 0$, or $\vec{F} = (0, 20)$ N. See Figure 7.a.

(c) It is often helpful to make a vector diagram of the forces acting on the block as we have done in Figures 7.a and c. Such a diagram is called a *free body diagram*. $\vec{F_1} = (0, 10)$ and $\vec{F_3} = (0, -20)$. The equation $\vec{F_1} + \vec{F_2} + \vec{F_3} = 0$ gives us that $\vec{F_2} = (0, 10)$ N. ∎

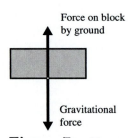

Figure 7.a *Forces acting on a block resting on the ground.*

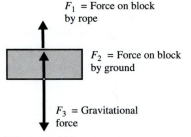

Figure 7.b *The weight is partially supported with a rope.*

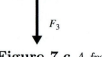

Figure 7.c *A free body diagram.*

EXAMPLE 6: The sum of the forces in Figure 8 is $\vec{0}$. Given that $\vec{F_3} = (0, -20)$ N, find $\vec{F_1}$ and $\vec{F_2}$.

SOLUTION: $(-\cos 30°, \sin 30°)$ and $(\cos 45°, \sin 45°)$ are unit vectors that point in the direction of $\vec{F_1}$ and $\vec{F_2}$, respectively. Thus, $\vec{F_1} = \|\vec{F_1}\|(-\cos 30°, \sin 30°)$ and $\vec{F_2} = \|\vec{F_2}\|(\cos 45°, \sin 45°)$. Expanding the equation

$$\vec{F_1} + \vec{F_2} + \vec{F_3} = 0$$

gives us

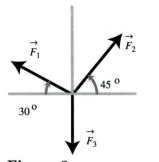

Figure 8.
The sum $\vec{F_1} + \vec{F_2} + \vec{F_3} = \vec{0}$.

$$\|\vec{F_1}\| \begin{pmatrix} -\cos 30° \\ \sin 30° \end{pmatrix} + \|\vec{F_2}\| \begin{pmatrix} \cos 45° \\ \sin 45° \end{pmatrix} + \begin{pmatrix} 0 \\ -20 \end{pmatrix}$$

$$= \|\vec{F_1}\| \begin{pmatrix} -\frac{\sqrt{3}}{2} \\ \frac{1}{2} \end{pmatrix} + \|\vec{F_2}\| \begin{pmatrix} \frac{\sqrt{2}}{2} \\ \frac{\sqrt{2}}{2} \end{pmatrix} + \begin{pmatrix} 0 \\ -20 \end{pmatrix}$$

$$= \vec{0}.$$

Setting the x–coordinates equal and the y–coordinates equal we get two equations:

$$-\sqrt{3}\|\vec{F_1}\| + \sqrt{2}\|\vec{F_2}\| = 0 \tag{1}$$

and

$$\|\vec{F_1}\| + \sqrt{2}\|\vec{F_2}\| = 40. \tag{2}$$

Subtracting Equation (1) from Equation (2), we get

$$(1 + \sqrt{3})\|\vec{F_1}\| = 40 \quad \text{or} \quad \|\vec{F_1}\| = \frac{40}{1 + \sqrt{3}}.$$

Substituting this value for $\|\vec{F_1}\|$ into Equation (1) we have $\vec{F_2} = \frac{40}{\sqrt{2}}\left(\frac{\sqrt{3}}{1+\sqrt{3}}\right)$. Thus, we have

$$\vec{F_1} = \left(\frac{40}{1 + \sqrt{3}}\right)\left(-\frac{\sqrt{3}}{2}, \frac{1}{2}\right) \quad \text{and}$$

$$\vec{F_2} = \left(\frac{40}{\sqrt{2}}\right)\left(\frac{\sqrt{3}}{1 + \sqrt{3}}\right)\left(\frac{\sqrt{2}}{2}, \frac{\sqrt{2}}{2}\right). \qquad \blacksquare$$

EXERCISES 1.3

1. If 3 vectors $\vec{A}, \vec{B}$ and $\vec{C}$ in $\mathbb{R}^3$ satisfy the equation $3\vec{A} + 2\vec{B} = \vec{C}$, express $\vec{A}$ and $\vec{B}$ in terms of the other two vectors.

In Exercises 2 and 3, compute $\|\vec{r}\|$, $c\vec{r}$, $\vec{r} + \vec{v}$, and $\vec{r} - \vec{v}$.

2. $c = 2$, $\vec{r} = (3, -1, 2)$, $\vec{v} = (1, 1, -2)$.

3. $c = \sqrt{2}$, $\vec{r} = (-1, 1, \sqrt{2})$, $\vec{v} = (1, \sqrt{2}, 3)$.

4. Let $\vec{v} = (1, 3)$. Draw the vectors $\vec{v}$, $2\vec{v}$, $\frac{1}{2}\vec{v}$, and $-2\vec{v}$ emanating from the origin.

5. Let $\vec{v} = (-1, 2)$. Draw the vectors $\vec{v}$, $3\vec{v}$, $(1/3)\vec{v}$ and $-3\vec{v}$ emanating from the origin.

Use your calculator in Exercises 6–8 to find the norm of $\vec{r}$. (Many calculators find the norm of a vector as a built in function.)

6. $\vec{r} = (2.3, 7.123, -6.53)$

7. $\vec{r} = (-0.002, 0.00262, 0.0035)$

8. $\vec{r} = (1232.3, 5677.153, -3206.63)$

In Exercises 9 and 10, let $\vec{u} = (1, 2, -6)$ and $\vec{t} = (-2, 1, 4)$. Solve the equations for $\vec{s}$.

9. $6\vec{u} + 2\vec{t} = 3\vec{s}$.   10. $2\vec{s} + 2\vec{t} = 3\vec{s} + 4\vec{t}$.

11. Let $\vec{A}$ be the directed line segment from $(1, 4, 2)$ to $(-1, 3, -2)$. What ordered triple does $\vec{A}$ represent?

Let $\vec{a} = (2, 3)$, $\vec{b} = (-1, 1)$, $\vec{c} = (3, -1)$, and $\vec{d} = (2, -1)$. Graphically compute the results of the operations in Exercises 12–19.

12. $\vec{a} + \vec{b} + \vec{c}$.   13. $\vec{b} + \vec{c} + \vec{d}$.   14. $\vec{a} - \vec{b}$.

15. $\vec{b} - \vec{c}$.   16. $2\vec{a}$.   17. $(1/2)\vec{a}$.

18. $-\vec{a}$.   19. $\vec{b} + 2\vec{c} + 3\vec{d}$.

In Exercises 20–22, find numbers a, b, and c so that
$\vec{v} = a\hat{\imath} + b\hat{\jmath} + c\hat{k}$.

20. $\vec{v} = (1, 3, 5)$.   21. $\vec{v} = (-3, 2, 1)$.

22. $\vec{v} = (\pi, 2, -3)$.

In Exercises 23 and 24, express the following vectors as ordered triples.

23. $3\hat{\imath} + 4\hat{\jmath} - 4\hat{k}$.

24. $-2\hat{\imath} + 16\hat{\jmath}$.

In Exercises 25–27, find the unit vector pointing in the same direction as $\vec{v}$.

25. $\vec{v} = (2, -1, 5)$.   26. $\vec{v} = (-3, 2, 1, 6)$.

27. $\vec{v} = (-1, 2, \pi, 2, -3)$.

In Exercises 28 and 29, use your calculator to find a unit vector with the same direction os $\vec{r}$.

28. $\vec{r} = (256.32, 521.61, 625)$.

29. $\vec{r} = -(256.32, 521.61, 625)$.

30. Let C be a clock with the coordinate axes placed as in Figure 4.b, Section 1. The minute hand is 3 in long and the length of the hour hand is 2 in. Find a unit vector that points from the tip of the minute hand toward the tip of the hour hand at 11:40 o'clock.

The vectors $\vec{F_1}$, $\vec{F_2}$, and $\vec{F_3}$ are as illustrated below. In Exercises 31 and 32, find $\vec{F_1}$ and $\vec{F_2}$ given that $\vec{F_3} = (0, -20)$ and that $\vec{F_1} + \vec{F_2} + \vec{F_3} = 0$.

31. $\alpha = \frac{\pi}{4}$ and $\beta = \frac{\pi}{4}$.

32. $\alpha = \frac{\pi}{6}$ and $\beta = \frac{\pi}{3}$.

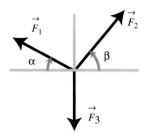

Figure 9. *Figure for Exercises 31 and 32.*

33. The coordinate axes are situated as in Figure 8 of Section 1. Find a unit vector that points from Prague toward Atlanta. See Exercises 27 and 28 of Section 1.1 for the coordinates of Prague and Atlanta.

As in Exercises 15 and 16 of Section 1.1, a 3lb weight is hung from a ring that is attached to a 3ft rope which is suspended from the ceiling. See Figure 10. In Exercises 34–37, A = 1 ft and B = 2 ft.

34. Find a unit vector that points from $\vec{R}$ to $\vec{P}$.

35. Find a unit vector that points from $\vec{R}$ to $\vec{Q}$.

36. Find $\vec{F_1}$.

37. Find $\vec{F_3}$.

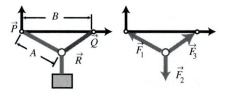

Figure 10. *Illustration for Exercises 34–37.*

1.4   The Dot Product

Definition: The Dot or Inner Product

Let $\vec{u} = (u_1, u_2, \ldots, u_n)$ and $\vec{v} = (v_1, v_2, \ldots, v_n)$ be vectors in $\mathbb{R}^n$. We define the *dot product* or *inner product* of $\vec{u}$ and $\vec{v}$ to be

$$\vec{u} \cdot \vec{v} = u_1 v_1 + u_2 v_2 + \cdots + u_n v_n.$$

It is important to remember that the dot product of two vectors is a real number. As the following example demonstrates, the computation of the dot product of two vectors is quite straightforward.

EXAMPLE 1:

(a) If $\vec{u} = (3,2)$ and $\vec{v} = (1,4)$, then $\vec{u} \cdot \vec{v} = (3)(1) + (2)(4) = 11$.

(b) If $\vec{s} = (3,4,-1,2)$ and $\vec{t} = (1,-1,0,3)$, then $\vec{s} \cdot \vec{t} = (3)(1) + (4)(-1) + (-1)(0) + (2)(3) = 5$. ∎

The following theorem is fundamental in understanding the geometric meaning of the dot product.

Theorem 1 *If $\vec{u}$ and $\vec{v}$ are nonzero vectors, and θ is the angle between them when drawn from a common point, then*

$$\vec{u} \cdot \vec{v} = \|\vec{u}\| \, \|\vec{v}\| \cos(\theta).$$

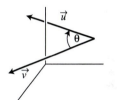

Figure 1. *The angle between $\vec{u}$ and $\vec{v}$ is θ.*

Before proving Theorem 1, it may be helpful to list three of its consequences. Let $\vec{u}$ and $\vec{v}$ be nonzero vectors, drawn emanating from a common point, as in Figure 1, and let θ be the angle between them. Then

1. $\cos(\theta) = \dfrac{\vec{u} \cdot \vec{v}}{\|\vec{u}\| \, \|\vec{v}\|}$.

2. The vectors $\vec{u}$ and $\vec{v}$ are perpendicular if and only if $\vec{u} \cdot \vec{v} = 0$ (the angle between them is $\pm \frac{\pi}{2}$).

3. $\|\vec{v}\|^2 = \vec{v} \cdot \vec{v}$.

Definition: Orthogonal Vectors

If two vectors are perpendicular, we say that they are *orthogonal*.

EXAMPLE 2: The vectors $\vec{u} = (1,0,3)$ and $\vec{v} = (-3,1,1)$ are orthogonal since $\vec{u} \cdot \vec{v} = 0$. ∎

EXAMPLE 3: Let $\vec{u} = (1,2,-1)$ and $\vec{v} = (0,-1,1)$. Then

$$\|\vec{u}\| = \sqrt{1+4+1} = \sqrt{6},$$
$$\|\vec{v}\| = \sqrt{0+1+1} = \sqrt{2}, \text{ and}$$
$$\vec{u} \cdot \vec{v} = -3.$$

Thus, the cosine of the angle between $\vec{u}$ and $\vec{v}$ is given by

$$\cos\theta = \frac{\vec{u}\cdot\vec{v}}{\|\vec{u}\|\,\|\vec{v}\|} = \frac{-3}{\sqrt{12}} = -\frac{\sqrt{3}}{2},$$

or

$$\theta = \frac{5\pi}{6} \text{ radians.} \qquad\blacksquare$$

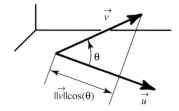

Figure 2.a *The length of the adjacent side of the triangle is* $\|\vec{v}\|\cos(\theta)$.

Proof of Theorem 1 (for $n=3$): Consider vectors $\vec{u} = (u_1, u_2, u_3)$ and $\vec{v} = (v_1, v_2, v_3)$ drawn emanating from a common point. If we can find a number t such that $t\vec{u}$ and $\vec{v} - t\vec{u}$ are perpendicular, then $\|\vec{v} - t\vec{u}\|$ and $\|t\vec{u}\|$ are the lengths of the legs of a right triangle with hypotenuse of length $\|\vec{v}\|$. Let θ be the angle between $\vec{u}$ and $\vec{v}$. See Figure 2.

$$\|t\vec{u}\| = \|\vec{v}\|\cos\theta \quad\text{and} \tag{3}$$

$$\|t\vec{u}\|^2 + \|\vec{v} - t\vec{u}\|^2 = \|\vec{v}\|^2. \tag{4}$$

Expanding Equation (4), we obtain

$$t^2(u_1^2 + u_2^2 + u_3^2) + (v_1^2 + v_2^2 + v_3^2) - 2t(u_1v_1 + u_2v_2 + u_3v_3) + t^2(u_1^2 + u_2^2 + u_3^2)$$

$$= (v_1^2 + v_2^2 + v_3^2),$$

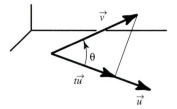

Figure 2.b *Choose t so that* $t\|v\| = \|\vec{v}\|\cos(\theta)$.

or

$$t(u_1^2 + u_2^2 + u_3^2) = (u_1v_1 + u_2v_2 + u_3v_3),$$

which is the same as

$$t\|\vec{u}\|^2 = \vec{u}\cdot\vec{v}. \tag{5}$$

Now, substituting Equation (3) into Equation (5), we have

$$\vec{u}\cdot\vec{v} = t\|\vec{u}\|^2 = (t\|\vec{u}\|)(\|\vec{u}\|) = \|\vec{u}\|\,\|\vec{v}\|\cos\theta. \qquad\blacksquare$$

Projection of a Vector

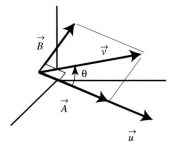

Figure 3. $\vec{v} = \vec{A} + \vec{B}$.

If $\vec{v}$ and $\vec{u}$ are vectors, then the dot product gives us a very useful method for decomposing $\vec{v}$ into the sum of two vectors, $\vec{A}$ and $\vec{B}$, so that $\vec{A}$ points in the direction of $\vec{u}$ and $\vec{B}$ is orthogonal to $\vec{u}$. As in Figure 3, draw $\vec{u}$ and $\vec{v}$ emanating from a common point. Then

- $\vec{v}$ forms the hypotenuse of a right triangle that has one leg lying on the vector $\vec{u}$. The length of that leg is

$$\|\vec{A}\| = \begin{cases} \|\vec{v}\|\cos\theta, & \text{if } 0 \le \theta \le \pi/2 \\ -\|\vec{v}\|\cos\theta, & \text{if } \pi/2 < \theta \le \pi. \end{cases}$$

- $\dfrac{\vec{u}\cdot\vec{v}}{\|\vec{u}\|} = \|\vec{v}\|\cos\theta = \begin{cases} \|\vec{A}\|, & \text{if } 0 \le \theta \le \pi/2 \\ -\|\vec{A}\|, & \text{if } \pi/2 < \theta \le \pi. \end{cases}$

- $\dfrac{\vec{u}}{\|\vec{u}\|}$ is a unit vector that points in the direction of $\vec{u}$. Therefore, it is also a unit vector that points in the direction of $\vec{A}$.

Thus, $\vec{A} = \left(\dfrac{\vec{u} \cdot \vec{v}}{\|\vec{u}\|}\right)\left(\dfrac{\vec{u}}{\|\vec{u}\|}\right) = \dfrac{\vec{u} \cdot \vec{v}}{\|\vec{u}\|^2}\vec{u} = \left(\dfrac{\vec{u} \cdot \vec{v}}{\vec{u} \cdot \vec{u}}\right)\vec{u}.$

By construction, $\vec{A}$ points in the direction of $\vec{u}$. Since we want $\vec{A} + \vec{B} = \vec{v}$, it must be the case that $\vec{B} = \vec{v} - \vec{A}$. We leave it as an exercise (Exercise 40) to show that $\vec{B}$ is orthogonal to $\vec{v}$. This development leads to the following definition.

Definition: $\mathrm{proj}_{\vec{u}}(\vec{v})$ and $\mathrm{orth}_{\vec{u}}(\vec{v})$

If $\vec{u}$ and $\vec{v}$ are nonzero vectors, then

$$\mathrm{proj}_{\vec{u}}(\vec{v}) = \left(\dfrac{\vec{u} \cdot \vec{v}}{\vec{u} \cdot \vec{u}}\right)\vec{u}$$

is called the *projection of $\vec{v}$ in the direction of $\vec{u}$.*

$$\mathrm{orth}_{\vec{u}}(\vec{v}) = \vec{v} - \mathrm{proj}_{\vec{u}}(\vec{v}) = \vec{v} - \left(\dfrac{\vec{u} \cdot \vec{v}}{\vec{u} \cdot \vec{u}}\right)\vec{u}$$

is called the *component of $\vec{v}$ orthogonal to $\vec{u}$.*

EXAMPLE 4: Let $\vec{p} = (1, 3, 2)$ and $\vec{q} = (1, 0, 1)$. The projection of $\vec{p}$ in the direction of $\vec{q}$ is

$$\left(\dfrac{\vec{p} \cdot \vec{q}}{\vec{q} \cdot \vec{q}}\right)\vec{q} = \dfrac{3}{2}(1, 0, 1) = \left(\dfrac{3}{2}, 0, \dfrac{3}{2}\right).$$

The component of $\vec{p}$ which is orthogonal to $\vec{q}$ is $(1, 3, 2) - \left(\frac{3}{2}, 0, \frac{3}{2}\right)$ $= \left(-\frac{1}{2}, -3, \frac{1}{2}\right).$ ∎

EXERCISES 1.4

1. If $\vec{u} = (1, 0)$, $\vec{v} = (1, 1)$, and $\vec{w} = (2, 1)$ are three vectors in $\mathbb{R}^2$, what is the value of $(\vec{u} \cdot \vec{v}) \cdot \vec{w}$?

2. If $\vec{u}$ and $\vec{v}$ are two vectors in $\mathbb{R}^2$, that point in the same direction, is $\vec{u} \cdot \vec{v}$ positive or negative? If $\vec{u}$ and $\vec{v}$ point in opposite directions, is $\vec{u} \cdot \vec{v}$ positive or negative?

3. In each part of Figure 4 (see next page), determine if $\vec{A} \cdot \vec{B}$ is positive, negative, or zero.

In Exercises 4–13, compute $\vec{p} \cdot \vec{q}$.

4. $\vec{p} = (1, 3, -6)$, $\vec{q} = (5, 3, 2)$.

5. $\vec{p} = (5, \pi, 2)$, $\vec{q} = (\frac{1}{5}, 0, \pi)$.

6. $\vec{p} = (5, 2, -1)$, $\vec{q} = (1, -1, 3)$.

7. $\vec{p} = (-1, 5, 2, 0)$, $\vec{q} = (1, 2, 0, -3)$.

8. $\vec{p} = \hat{\imath} - 3\hat{\jmath} + \hat{k}$, $\vec{q} = -\hat{\imath} + 2\hat{\jmath} + 5\hat{k}$.

9. $\vec{p} = 2\hat{\imath} + \hat{k}$, $\vec{q} = -\hat{\imath} - 3\hat{\jmath} + \hat{k}$.

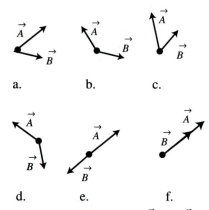

a.   b.   c.

d.   e.   f.

Figure 4. *The vectors $\vec{A}$ and $\vec{B}$ drawn enamating from a common point.*

10. $\vec{p} = 2\hat{\imath} + \hat{k}$, $\vec{q} = -3\hat{\jmath}$.

11. $\vec{p} = -2\hat{k}$, $\vec{q} = -3\hat{\jmath}$.

12. $\|\vec{p}\| = 5$, $\|\vec{q}\| = 2$, and the angle between $\vec{p}$ and $\vec{q}$ is $45°$.

13. $\|\vec{p}\| = 2$, $\|\vec{q}\| = 2\sqrt{2}$, and the angle between $\vec{p}$ and $\vec{q}$ is $\frac{\pi}{6}$.

14. Find the cosine of the angle between $\vec{p} = (1, 3, 0)$, $\vec{q} = (1, 3, 2)$.

15. Find the cosine of the angle between $\vec{p} = (1, -1, 1)$, $\vec{q} = (0, -1, -1)$.

16. Determine if the angle between $(1, -2, 4)$ and $(-1, 0, 3)$ is acute or obtuse.

17. Determine if the angle between $(-1, -2, 0)$ and $(-1, 3, 5)$ is acute or obtuse.

Most calculators have a built-in function to calculate the dot product. Use this function to calculate $\vec{u} \cdot \vec{v}$ in Exercises 18 and 19.

18. $\vec{u} = (1.24, 2.561, 0.124)$ and $\vec{v} = (-9.12, 3.6445, 2.987)$.

19. $\vec{u} = (125.32, 92.487, 55.893)$ and $\vec{v} = (950.3, 9987.2, 22.534)$.

20. Find the angle between $\vec{p} = (2.24, 1.24, -3.2)$ and $\vec{q} = (1.12, 3.23, 2.1)$.

21. Find the angle between $\vec{p} = (-0.014, 1.124, 0.12)$ and $\vec{q} = (0.12, 0.23, 0.11)$.

In Exercises 22–25, find the cosine of the angle between the vector $\vec{p}$ and each of the coordinate axis.

22. $\vec{p} = (1, 2, -1)$.   23. $\vec{p} = (-1, 3, 2)$.

24. $\vec{p} = \hat{\imath} - 2\hat{\jmath}$.   25. $\vec{p} = -2\hat{\jmath} + 3\hat{k}$.

26. In each of the figures for Exercise 3, sketch the projection of $\vec{A}$ in the direction of $\vec{B}$.

27. In each of the figures for Exercise 3, sketch the component of $\vec{A}$ that is orthogonal to $\vec{B}$.

In Exercises 28–33, find the projection of $\vec{p}$ in the direction of $\vec{q}$ and the component of $\vec{p}$ orthogonal to $\vec{q}$.

28. $\vec{p} = (1, 3, -6)$, $\vec{q} = (5, 3, 2)$.

29. $\vec{p} = (5, 2, -1)$, $\vec{q} = (1, -1, 3)$.

30. $\vec{p} = (-1, 5, 2, 0)$, $\vec{q} = (1, 2, 0, -3)$.

31. $\vec{p} = \hat{\imath} - 3\hat{\jmath} + \hat{k}$, $\vec{q} = -\hat{\imath} + 2\hat{\jmath} + 5\hat{k}$.

32. $\vec{p} = (325.2, 950.1, -620.9)$, $\vec{q} = (829.3, 230.5, 521.2)$.

33. $\vec{p} = (0.232, -0.910, 0.350)$, $\vec{q} = (-0.101, 0.222, 0.350)$.

In Exercises 34–36, find two unit vectors in $\mathbb{R}^3$ that are orthogonal to both $\vec{p}$ and $\vec{q}$.

34. $\vec{p} = \hat{\imath}$, $\vec{q} = \hat{\jmath}$ 35. $\vec{p} = (1, 3, 5)$, $\vec{q} = (1, 3, 6)$.

36. $\vec{p} = (1, 1, 1)$, $\vec{q} = (5, 3, 2)$.

37. Suppose that $\vec{A} = (a, b)$, $\vec{B}$ is orthogonal to $\vec{A}$, and $\|\vec{A}\| = \|\vec{B}\|$. Show that either $\vec{B} = (-b, a)$ or $\vec{B} = (b, -a)$.

38. Show that if $\vec{A}$, $\vec{B}$, and $\vec{C}$ are vectors, then:
    a. $\vec{A} \cdot (\vec{B} + \vec{C}) = \vec{A} \cdot \vec{B} + \vec{A} \cdot \vec{C}$.
    b. $\vec{A} \cdot \vec{B} = \vec{B} \cdot \vec{A}$.
    c. $|\vec{A} \cdot \vec{B}| \leq \|\vec{B}\|\,\|\vec{A}\|$.

39. Use Exercise 38 to show that if a, b, c and d are numbers, then $(ac + bd)^2 \leq (a^2 + b^2)(c^2 + d^2)$.

40. Show that $\operatorname{orth}_{\vec{v}}(\vec{u}) \cdot \vec{v} = 0$. This shows that $\operatorname{orth}_{\vec{v}}\vec{u}$ and $\vec{v}$ are orthogonal.

41. Show that if $\vec{u}$, $\vec{v}$ and $\vec{w}$ are vectors, then

$$\operatorname{proj}_{\vec{u}}(\vec{v} + \vec{w}) = \operatorname{proj}_{\vec{u}}\vec{v} + \operatorname{proj}_{\vec{u}}\vec{w}.$$

42. Recall that the length of the arc on a circle of radius r has length $r\theta$ where θ is the inscribed angle measured in radians. Assuming a spherical earth, find the surface distance from Prague to Atlanta. See Exercise 26 of Section 1.1 and Exercise 33 of Section 1.3.

43. Let θ_x denote the angle that the vector (x, y, z) makes with the x–axis, θ_y denote the angle that the vector (x, y, z) makes with the y–axis, and θ_z denote the angle that the vector (x, y, z) makes with the z–axis. The numbers $\alpha_x = \cos(\theta_x)$, $\alpha_y = \cos(\theta_y)$, and $\alpha_z = \cos(\theta_z)$ are called the *direction cosines* for the vector (x, y, z). Show that $\alpha_x^2 + \alpha_y^2 + \alpha_z^2 = 1$.

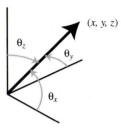

Figure 5.
The direction cosines.

1.5   Determinants, Areas and Volumes

Let $\vec{A} = (a_x, a_y)$ and $\vec{B} = (b_x, b_y)$ be two vectors in the plane drawn emanating from a common point $\vec{C}$. They form adjacent edges of a parallelogram $\mathcal{P}$ as in Figure 1.a. In order to find the area of $\mathcal{P}$, recall from geometry that the area of a parallelogram is the product of its height with the length of its base. We will take $\|\vec{A}\|$ to be the length of the base of $\mathcal{P}$. Now, $\vec{A}_{perp} = (-a_y, a_x)$ is a vector that is orthogonal (a_x, a_y) and $\|\vec{A}_{perp}\| = \|\vec{A}\|$. As illustrated in Figure 1.c, the height of $\mathcal{P}$ is the length of the projection of $\vec{B}$ in the direction of $\vec{A}_{perp}$, which is given by

Figure 1.a *The parallelogram with adjacent edges $\vec{A}$ and $\vec{B}$.*

$$\|\mathrm{proj}_{\vec{A}_{perp}} \vec{B}\| = \frac{|\vec{B} \cdot \vec{A}_{perp}|}{\|\vec{A}_{perp}\|^2} \|\vec{A}_{perp}\| = \frac{|-b_x a_y + b_y a_x|}{\|\vec{A}\|}.$$

Thus, the area of the parallelogram is

$$\text{height} \times \text{base} = \|\vec{A}\| \|\mathrm{proj}_{\vec{A}_{perp}} \vec{B}\| = |-b_x a_y + b_y a_x|.$$

This leads to the definition of the determinant of a 2×2 matrix.

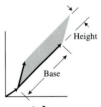

Figure 1.b *The base and the height of $\mathcal{P}$.*

Definition: Determinant of a 2×2 matrix

If $A = \begin{pmatrix} a_1 & a_2 \\ b_1 & b_2 \end{pmatrix}$ is a 2×2 square matrix of real numbers, then the determinant of A, denoted by $\det(A) = \begin{vmatrix} a_1 & a_2 \\ b_1 & b_2 \end{vmatrix}$, is defined by $\det(A) = a_1 b_2 - a_2 b_1$.

We have the following theorem.

Theorem 1 *The absolute value of the determinant of a 2×2 matrix $A = \begin{pmatrix} a_1 & a_2 \\ b_1 & b_2 \end{pmatrix}$ is the area of a parallelogram with adjacent edges the row vectors (a_1, a_2) and (b_1, b_2).*

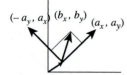

Figure 1.c *The vectors $\vec{A}$, $\vec{B}$ and $\vec{A}_{perp}$.*

EXAMPLE 1: Let $A = \begin{pmatrix} 1 & 2 \\ 3 & 4 \end{pmatrix}$. Then $\det(A) = (1)(4) - (3)(2) = -2$. The area of the parallelogram with adjacent edges $(1, 2)$ and $(3, 4)$ is $|\det(A)| = 2$. ∎

Now, consider two 3–dimensional vectors $\vec{A} = (a_x, a_y, a_z)$ and $\vec{B} = (b_x, b_y, b_z)$ drawn emanating from a common point as illustrated in Figure 2. These vectors are adjacent edges of a parallelogram $\mathcal{P}$ with base of length $\|\vec{A}\|$ and height $\|\vec{B}\| \sin \theta$. Thus, the area of $\mathcal{P}$ is $\|\vec{A}\| \|\vec{B}\| \sin \theta$. In the pursuit of $\|\vec{A}\| \|\vec{B}\| \sin \theta$, we will introduce the *cross product* of two vectors.

Let $\vec{A} = (a_x, a_y, a_z)$ and $\vec{B} = (b_x, b_y, b_z)$, and let M be the matrix with two rows, the first being $\vec{A}$ and the second $\vec{B}$. That is,

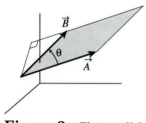

Figure 2. *The parallelogram with adjacent sides $\vec{A}$ and $\vec{B}$.*

$$M = \begin{pmatrix} a_x & a_y & a_z \\ b_x & b_y & b_z \end{pmatrix}.$$

We will let M_1 be the 2×2 matrix obtained by eliminating the first column from M, M_2 be the 2×2 matrix obtained by eliminating the second column from M, and M_3 be the matrix obtained by eliminating the third column from M.

$$M_1 = \begin{pmatrix} a_y & a_z \\ b_y & b_z \end{pmatrix}, \quad M_2 = \begin{pmatrix} a_x & a_z \\ b_x & b_z \end{pmatrix}, \quad \text{and } M_3 = \begin{pmatrix} a_x & a_y \\ b_x & b_y \end{pmatrix}.$$

We are now in a position to define the cross product of two vectors. Theorem 2 will explain our current interest in this operation.

Definition: The Cross Product

Let $\vec{A} = (a_x, a_y, a_z)$ and $\vec{B} = (b_x, b_y, b_z)$ and let

$$M = \begin{pmatrix} a_x & a_y & a_z \\ b_x & b_y & b_z \end{pmatrix}.$$

Then we define

$$\vec{A} \times \vec{B} = (\det(M_1), -\det(M_2), \det(M_3))$$

$$= \left(\begin{vmatrix} a_y & a_z \\ b_y & b_z \end{vmatrix}, -\begin{vmatrix} a_x & a_z \\ b_x & b_z \end{vmatrix}, \begin{vmatrix} a_x & a_y \\ b_x & b_y \end{vmatrix} \right).$$

$\vec{A} \times \vec{B}$ is read 'A cross B.'

EXAMPLE 2: Compute $\hat{\imath} \times \hat{k}$.

SOLUTION: Let $M = \begin{pmatrix} 1 & 0 & 0 \\ 0 & 0 & 1 \end{pmatrix}$. Then

$$\det(M_1) = \begin{vmatrix} 0 & 0 \\ 0 & 1 \end{vmatrix} = 0, \ \det(M_2) = \begin{vmatrix} 1 & 0 \\ 0 & 1 \end{vmatrix} = 1, \text{ and } \det(M_3) = \begin{vmatrix} 1 & 0 \\ 0 & 0 \end{vmatrix} = 0.$$

And so, $\hat{\imath} \times \hat{k} = (0, -1, 0) = -\hat{\jmath}$. ∎

EXAMPLE 3: Compute $\vec{A} \times \vec{B}$, where $\vec{A} = (3, 1, 2)$ and $\vec{B} = (1, -3, 1)$.

SOLUTION: Using the expansion given in the definition, we have

$$\vec{A} \times \vec{B} = \begin{pmatrix} (1)(1) - (2)(-3) \\ -((3)(1) - (2)(1)) \\ (3)(-3) - (1)(1) \end{pmatrix} = \begin{pmatrix} 7 \\ -1 \\ -10 \end{pmatrix}. \quad ∎$$

Theorems 2 and 3 provide important geometric information about the cross product of two vectors.

Theorem 2 *If $\vec{A}$ and $\vec{B}$ are vectors in $\mathbb{R}^3$, emanating from the same point, and θ is the angle between them such that $0 \leq \theta \leq 180°$, then*
$$\|\vec{A} \times \vec{B}\| = \|\vec{A}\|\|\vec{B}\| \sin\theta.$$
Thus, if $\mathcal{P}$ is the parallelogram with adjacent edges $\vec{A}$ and $\vec{B}$, then the area of $\mathcal{P}$ is given by $\|\vec{A} \times \vec{B}\|$.

You are asked to prove Theorem 2 in Exercise 48.

EXAMPLE 4: Notice that $\hat{k}$ and $\hat{\imath}$ are perpendicular unit vectors. Thus, they form adjacent edges of a *unit* square. This illustrates Theorem 2 since $\|\hat{\imath} \times \hat{k}\| = \| -\hat{\jmath}\| = 1$. ∎

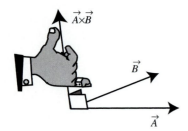

Figure 3.
The right hand rule.

EXAMPLE 5: From Example 3, the area of a parallelogram with adjacent edges $\vec{A} = (3, 1, 2)$ and $\vec{B} = (1, -3, 1)$ drawn emanating from a common point is $|(7, -1, -10)| = \sqrt{150}$.   ∎

Theorem 3 *If $\vec{A}$ and $\vec{B}$ are vectors in $\mathbb{R}^3$, then $\vec{A} \times \vec{B}$ is perpendicular to the plane that contains the vectors $\vec{A}$ and $\vec{B}$ when drawn emanating from a common point. The direction of $\vec{A} \times \vec{B}$ can be determined by the "right hand rule;" i.e., if the little finger side of the right hand is placed flat on the plane determined by $\vec{A}$ and $\vec{B}$, with the fingers curling from $\vec{A}$ to $\vec{B}$, then the extended thumb points in the direction $\vec{A} \times \vec{B}$. (See Figure 3.)*

EXAMPLE 6: Using Theorems 2 and 3, it is quite easy to see that $\hat{\imath} \times \hat{\jmath} = \hat{k}$, $\hat{\jmath} \times \hat{k} = \hat{\imath}$.   ∎

The following are important computational properties of cross products, which you will be asked to prove in the exercises.

Properties of Cross Products

Let $\vec{A}$, $\vec{B}$ and $\vec{C}$ be vectors, and let a and b be real numbers. Then

1. $\vec{A} \times \vec{B} = -(\vec{B} \times \vec{A})$.

2. $(a\vec{A}) \times \vec{B} = \vec{A} \times (a\vec{B}) = a(\vec{A} \times \vec{B})$.

3. $(a + b)\vec{A} = a\vec{A} + b\vec{A}$.

4. $\vec{A} \times (\vec{B} + \vec{C}) = (\vec{A} \times \vec{B}) + (\vec{A} \times \vec{C})$ and $(\vec{A} + \vec{B}) \times \vec{C} = (\vec{A} \times \vec{C}) + (\vec{B} \times \vec{C})$.

Property 1 shows us that the order of multiplication for the cross product does make a difference! Properties 2, 3 and 4 are just like the usual laws of multiplication of real numbers. These rules are particularly helpful if you are using $\{\hat{\imath}, \hat{\jmath}, \hat{k}\}$ notation.

EXAMPLE 7: Using Theorems 2 and 3, it is easy to evaluate the cross products of the various principal unit vectors. For example, knowing that $\hat{\imath} \times \hat{\imath} = \vec{0}$, and that $\hat{\imath} \times \hat{k} = -\hat{\jmath}$, we have   ∎

$$3\hat{\imath} \times (4\hat{\imath} + 2\hat{k}) = 12(\hat{\imath} \times \hat{\imath}) + 6(\hat{\imath} \times \hat{k}) = -6\hat{\jmath}.$$

The Triple Product

We now develop a three dimensional version of Theorem 1. Suppose that $\vec{a}, \vec{b}$ and $\vec{c}$ are vectors in $\mathbb{R}^3$ drawn emanating from a common point. Then $\vec{a}, \vec{b}$ and $\vec{c}$ are adjacent edges of a parallelepiped P as shown in Figure 4. If we take P_1 to be the parallelogram with adjacent sides $\vec{a}$ and $\vec{b}$, then the area of P_1 is, by Theorem 2, simply $\|\vec{a} \times \vec{b}\|$. The length of the projection of $\vec{c}$ in the direction of $\vec{a} \times \vec{b}$ is $|\vec{c} \cdot (\vec{a} \times \vec{b})|/\|\vec{a} \times \vec{b}\|$. This length is the height, relative to the base P_1, of P. Since the volume of a parallelepiped is the area of its base times its height, the volume V of the parallelepiped P is given by $V = |\vec{c} \cdot (\vec{a} \times \vec{b})|$.

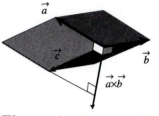

Figure 4. *The volume V of the parallelepiped P is given by $V = |\vec{c} \cdot (\vec{a} \times \vec{b})|$.*

Definition: The Triple Product

If $\vec{a}, \vec{b}$, and $\vec{c}$ are vectors in $\mathbb{R}^3$ drawn emanating from a common point, then the quantity

$$\vec{c} \cdot (\vec{a} \times \vec{b})$$

is referred to as the *triple product of $\vec{c}, \vec{a}$, and $\vec{b}$.*

EXAMPLE 8: Find the volume of $\mathcal{P}$, the parallelepiped with adjacent edges the line segments $[\vec{p}, \vec{q}]$, $[\vec{p}, \vec{r}]$, and $[\vec{p}, \vec{s}]$, where $\vec{p} = (1, 2, 1), \vec{q} = (1, 0, 2), \vec{r} = (0, 2, 0)$, and $\vec{s} = (1, 1, 1)$.

SOLUTION: Let $\vec{a} = \vec{p} - \vec{q}$, $\vec{b} = \vec{p} - \vec{r}$, and $\vec{c} = \vec{p} - \vec{s}$. Then the volume of P is simply $|\vec{c} \cdot (\vec{a} \times \vec{b})|$. It is straightforward to see that $\vec{a} \times \vec{b}$ is $(2, -1, -2)$. Since $\vec{c} = (0, 1, 0)$ we have that the volume of $\mathcal{P}$ is $|\vec{c} \cdot (\vec{a} \times \vec{b})| = 1$. ∎

We are now in a position to define the determinant of a 3×3 matrix.

Definition: The Determinant of a 3×3 Matrix

Let $\vec{C} = (c_x, c_y, c_z)$, $\vec{A} = (a_x, a_y, a_z)$, and $\vec{B} = (b_x, b_y, b_z)$. Let M be the matrix with rows $\vec{C}$, $\vec{A}$, and $\vec{B}$. That is,

$$M = \begin{pmatrix} c_x & c_y & c_z \\ a_x & a_y & a_z \\ b_x & b_y & b_z \end{pmatrix}.$$

Then $\det(M)$ is defined by

$$\det(M) = \begin{vmatrix} c_x & c_y & c_z \\ a_x & a_y & a_z \\ b_x & b_y & b_z \end{vmatrix} = \vec{C} \cdot (\vec{A} \times \vec{B}).$$

Computationally,

$$\det(M) = (c_x, c_y, c_z) \cdot \left(\begin{vmatrix} a_y & a_z \\ b_y & b_z \end{vmatrix}, -\begin{vmatrix} a_x & a_z \\ b_x & b_z \end{vmatrix}, \begin{vmatrix} a_x & a_y \\ b_x & b_y \end{vmatrix} \right)$$

$$= c_x \begin{vmatrix} a_y & a_z \\ b_y & b_z \end{vmatrix} - c_y \begin{vmatrix} a_x & a_z \\ b_x & b_z \end{vmatrix} + c_z \begin{vmatrix} a_x & a_y \\ b_x & b_y \end{vmatrix}.$$

Geometrically, if M is a 3×3 matrix, then $|\det(M)|$ is the volume of a parallelepiped with adjacent edges the *row vectors* of the matrix drawn emanating from a common point.

Observation: The formula for the cross product is easier to remember if we use the following. To compute the cross product of two vectors $\vec{A} = (a_x, a_y, a_z)$ and $\vec{B} = (b_x, b_y, b_z)$, we form the matrix that has as first row the symbols $\hat{\imath}, \hat{\jmath}$, and $\hat{k}$, and we let the second and third rows be $\vec{A}$ and $\vec{B}$. Now, formally find the determinant; that is, the first term will be $\hat{\imath}$ multiplied by the determinant of the 2×2 matrix formed by cancelling the first row and first column of our 3×3 matrix. The second term will be $\hat{\jmath}$ multiplied by the **negative** of the determinant of the 2×2 matrix formed by cancelling the first rwo and second column of our 3×3 matrix. And the third term will be $\hat{k}$ multiplied by the determinant of the 2×2 matrix formed by cancelling the first row and third column of our 3×3 matrix.

$$\begin{bmatrix} \hat{\imath} & \hat{\jmath} & \hat{k} \\ a_x & a_y & a_z \\ b_x & b_y & b_z \end{bmatrix} \qquad \begin{bmatrix} \hat{\imath} & \hat{\jmath} & \hat{k} \\ a_x & a_y & a_z \\ b_x & b_y & b_z \end{bmatrix} \qquad \begin{bmatrix} \hat{\imath} & \hat{\jmath} & \hat{k} \\ a_x & a_y & a_z \\ b_x & b_y & b_z \end{bmatrix}$$

So

$$\vec{A} \times \vec{B} = \hat{\imath} \begin{vmatrix} a_y & a_z \\ b_y & b_z \end{vmatrix} - \hat{\jmath} \begin{vmatrix} a_x & a_z \\ b_x & b_z \end{vmatrix} + \hat{k} \begin{vmatrix} a_x & a_y \\ b_x & b_y \end{vmatrix}$$

$$= \hat{\imath}(a_y b_z - a_z b_y) + \hat{\jmath}(a_z b_x - a_x b_z) + \hat{k}(a_x b_y - a_y b_x).$$

The following are important computational properties of triple products which you will be asked to prove in the exercises.

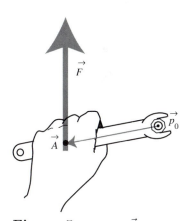

Figure 5. *A force $\vec{F}$ is applied at position $\vec{A}$ on the wrench, which induces a moment on the nut about position $\vec{p}_0$.*

Properties of Triple Products

Let $\vec{A}$, $\vec{B}$ and $\vec{C}$ be vectors and let a be real a number. Then

1. $\vec{C} \cdot (\vec{A} \times \vec{B}) = \vec{B} \cdot (\vec{C} \times \vec{A}) = -\vec{A} \cdot (\vec{B} \times \vec{C})$.

2. $(a\vec{C}) \cdot (\vec{A} \times \vec{B}) = \vec{C} \cdot ((a\vec{A}) \times \vec{B}) = \vec{C} \cdot (\vec{A} \times (a\vec{B})) = a(\vec{C} \cdot (\vec{A} \times \vec{B}))$.

Property 1 states that you change at most the sign of the triple product of three vectors when you change the order of multiplication.

If $\vec{A} = (a_x, a_y)$ and $\vec{B} = (b_x, b_y)$ are vectors in $\mathbb{R}^2$ we can still form their cross product by embedding $\vec{A}$ and $\vec{B}$ into $\mathbb{R}^3$. We do this by expressing $\vec{A}$ as $(a_x, a_y, 0)$ and $\vec{B}$ as $(b_x, b_y, 0)$. Then by using the above method to find $\vec{A} \times \vec{B}$ we obtain $\hat{k}(a_x b_y - a_y b_x)$, a vector perpedicular to the xy–plane.

Torque

If we use a wrench to turn a nut on a bolt, then we are applying a *moment* or *torque* to the nut. In general, if M is a mass, $\vec{p}_0$ and $\vec{A}$ are points, and $\vec{F}$ is a force applied at position $\vec{A}$, then the *torque* or *moment* about the position $\vec{p}_0$ produced by $\vec{F}$ acting at position $\vec{A}$ is defined to be $(\vec{A} - \vec{p}_0) \times \vec{F}$. (See Figure 5.) In your engineering and physics classes, this formula is written as $\vec{R} \times \vec{F}$, where $\vec{R} = (\vec{A} - \vec{p}_0)$. Torque (or moment) is measured in units of (unit of length) $\times$ (unit of force.) In the mks system, it is Newton meter, which is denoted by Nm. In the British system, torque is measured in foot pounds. It is customary to use the Greek letter $\vec{\tau}$ (tau) to denote torque (with an arrow since torque is a vector quantity.)

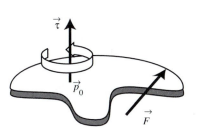

Figure 6. *The torque is a vector that points in the direction of the axis of the induced tendency to rotate.*

According to our definition, neither the point $\vec{A}$ where the force is applied nor the "pivot point" need be on the mass. In our example of a force applied to a wrench to obtain a torque on a nut, the force is applied to the wrench, the mass is the nut, and the pivot point is the center of the bolt (which is not on the nut.)

A torque applied to a mass about a point tends to cause the mass to rotate and the torque is a vector that points in the direction of the axis of the induced tendency to rotate. See Figure 6.

EXAMPLE 9: If the force $\vec{F} = (1, 2)$N is applied to a metal plate at the position $\vec{A} = (1, -1)$m then the torque about the point

$\vec{p}_0 = (0, 1)$m is given by $(\vec{A} - \vec{p}_0) \times \vec{F}$. We embed $\vec{A}, \vec{p}_0$ and $\vec{F}$ in $\mathbb{R}^3$ to obtain

$$
\begin{aligned}
(\vec{A} - \vec{p}_0) \times \vec{F} &= ((1, -1, 0) - (0, 1, 0))\text{m} \times (1, 2, 0)\text{N} \\
&= (1, -2, 0)\text{m} \times (1, 2, 0)\text{N} \\
&= (0, 0, 4)\text{Nm}. \qquad\blacksquare
\end{aligned}
$$

EXAMPLE 10: A 2 N force is applied at the position $(2, 3)$m. What are the components of $\vec{F}$ so that the magnitude of the resulting torque about the point $(1, 0)$m is the maximum possible?

SOLUTION: Let $\vec{R} = (2, 3) - (1, 0) = (1, 3)$m, and let $\vec{F} = (a, b)$. We are given that $\|\vec{F}\| = 2$ N, which gives us the equation

$$a^2 + b^2 = 4 \text{ N}^2. \tag{6}$$

The torque $\vec{\tau}$ that results from applying $\vec{F}$ at $(2, 3)$m is $\vec{\tau} = \vec{R} \times \vec{F}$. Thus,

$$\|\vec{\tau}\| = \|\vec{R}\| \|\vec{F}\| \sin(\theta), \tag{7}$$

where θ is the acute angle between $\vec{R}$ and $\vec{F}$. The maximum value will be obtained when $\sin(\theta) = 1$, or when $\theta = \frac{\pi}{2}$. We have that $\vec{R}$ and $\vec{F}$ are orthogonal and their dot product is zero. This gives us a second equation for a and b.

$$\vec{R} \cdot \vec{F} = (1, 3) \cdot (a, b) = 0 \text{ or } a + 3b = 0. \tag{8}$$

From Equation (8) we have that $a = -3b$. Substituting into Equation (6), we have $9b^2 + b^2 = 4$, or $b = \pm 2/\sqrt{10}$. This gives two solutions to our problem: $\vec{F} = \left(\frac{6}{\sqrt{10}}, \frac{2}{\sqrt{10}} \right) N$ and $\vec{F} = \left(-\frac{6}{\sqrt{10}}, -\frac{2}{\sqrt{10}} \right) N.$ $\blacksquare$

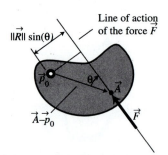

Figure 7. *The line of action of the force $\vec{F}$ acting on a mass at position $\vec{A}$.*

Let $\mathcal{L}$ be the line containing $\vec{F}$ when $\vec{F}$ is drawn emanating from position $\vec{p}_0$, as illustrated in Figure 7. $\mathcal{L}$ is called the *line of action* of the force $\vec{F}$. Equation (7) can be rewritten as $\|\vec{\tau}\| = \|\vec{F}\| \|\vec{R}\| \sin(\theta)$. But $\|\vec{R}\| \sin(\theta)$ is the perpendicular distance from $\vec{p}_0$ to $\mathcal{L}$. Thus, we have the following theorem.

Theorem 4 *The magnitude of the moment on a mass about a point $\vec{p}_0$ induced by a force $\vec{F}$ is $\|\vec{F}\|$ times the perpendicular distance from $\vec{p}_0$ to the line of action of the force.*

EXERCISES 1.5

1. Find $\begin{vmatrix} a & b \\ 0 & 0 \end{vmatrix}$.

2. Find $\begin{vmatrix} a & b \\ a & b \end{vmatrix}$.

3. Find $\begin{vmatrix} 1 & 3 \\ 3 & -1 \end{vmatrix}$.

4. Find $\begin{vmatrix} 0 & 1 \\ 2 & 2 \end{vmatrix}$.

In Exercises 5 and 6, the vectors $\vec{A}$ and $\vec{B}$ are drawn emanating from a common point to form adjacent edges of a parallelogram. Find the area of the parallelogram.

5. $\vec{A} = (1,3)$, $\vec{B} = (2,-1)$.

6. $\vec{A} = (-1,2)$, $\vec{B} = (2,1)$.

In Exercises 7–10, find $\vec{v} \times \vec{r}$ and $\|\vec{v} \times \vec{r}\|$.

7. $\vec{v} = (1,2,0)$, $\vec{r} = (-1,3,1)$.

8. $\vec{v} = (-1,2,-1)$, $\vec{r} = (1,-1,1)$.

9. $\vec{v} = (0,1,2)$, $\vec{r} = (1,0,2)$.

10. $\vec{v} = \hat{\imath} + \hat{\jmath} - 2\hat{k}, \vec{r} = 2\hat{\jmath} + \hat{k}$.

11. Most calculators have the cross product as a built-in function. Use your calculator to calculate $(132, -321, 93) \times (90, 83, -52)$.

12. Most calculators have the cross product as a built-in function. Use your calculator to calculate $(1.32, -1.321, 3.93) \times (-2.90, 1.83, -5.2)$.

13. The vectors $\vec{v}$ and $\vec{r}$ in Exercise 7 are drawn emanating from a common point to form adjacent edges of a parallelogram. Find the area of the parallelogram.

14. The vectors $\vec{v}$ and $\vec{r}$ in Exercise 8 are drawn emanating from a common point to form adjacent edges of a parallelogram. Find the area of the parallelogram.

15. The vectors $\vec{v}$ and $\vec{r}$ in Exercise 9 are drawn emanating from a common point to form adjacent edges of a parallelogram. Find the area of the parallelogram.

16. Find two vectors that are perpendicular to both $\vec{v}$ and $\vec{r}$ in Exercise 7.

17. Find two vectors that are perpendicular to both $\vec{v}$ and $\vec{r}$ in Exercise 8.

18. Find two *unit* vectors orthogonal to $\vec{v}$ and $\vec{r}$ in Exercise 7.

19. Find two *unit* vectors orthogonal to $\vec{v}$ and $\vec{r}$ in Exercise 8.

20. Find the sine of the angle between the vectors in Exercise 7.

21. Find the sine of the angle between the vectors in Exercise 8.

In Exercises 22–24, three vectors are drawn emanating from a common point to form adjacent edges of a parallelepiped. Find the volume of the parallelepiped.

22. $(1,2,1)$, $(1,0,-1)$, and $(1,1,0)$.

23. $(-1,2,1)$, $(0,0,1)$, and $(1,1,1)$.

24. $(0,1,-1)$, $(2,1,4)$, and $(1,-1,3)$.

25. Calculate $\det \begin{pmatrix} 1 & 2 & 1 \\ 1 & 0 & -1 \\ 1 & 1 & 0 \end{pmatrix}$. Compare your result with your solution to Exercise 22.

26. Calculate $\det \begin{pmatrix} -1 & 2 & 1 \\ 0 & 0 & 1 \\ 1 & 1 & 1 \end{pmatrix}$. Compare your result with your solution to Exercise 23.

27. Calculate $\det \begin{pmatrix} -6.11 & 2.21 & 1.03 \\ 1.25 & 2.431 & 2.98 \\ 2.33 & 1.98 & 5.44 \end{pmatrix}$. (Use the built-in determinant function.)

28. Compute $\hat{\imath} \times \hat{\jmath}$, $\hat{\jmath} \times \hat{k}$, and $\hat{k} \times \hat{\imath}$.

29. Let $\vec{u} = (1, 0, 0), \vec{v} = (0, 1, 0)$ and $\vec{w} = (0, 1, 1)$. Compute the following, if possible. Otherwise, explain why the expression is not computable.

    a. $(\vec{u} \cdot \vec{v}) \times \vec{w}$ b. $\vec{u} \cdot (\vec{v} \times \vec{w})$

    c. $(\vec{u} \times \vec{v}) \times \vec{w}$ d. $\vec{u} \times (\vec{v} \times \vec{w})$

30. Let $\vec{u} = (1, 0, 0), \vec{v} = (0, 2, 0)$, and $\vec{w} = (0, 0, 3)$ be vectors in $\mathbb{R}^3$. Find the triple product $\vec{u} \cdot (\vec{v} \times \vec{w})$ by geometric analysis.

31. What must be true about the vectors $\vec{A}$ and $\vec{B}$ if $\vec{A} \times \vec{B} = \vec{0}$?

32. What must be true about the vectors $\vec{A}$ and $\vec{B}$ if $\|\vec{A} \times \vec{B}\| = \|\vec{A}\|\|\vec{B}\|$?

In Exercises 33–36, find the torque about $\vec{p}_0$ induced on a mass by the force $\vec{F}$ applied at the position $\vec{A}$.

33. $\vec{p}_0 = (2, 0)$m, $\vec{A} = (1, 1)$m, and $\vec{F} = (-1, 2)$ N.

34. $\vec{p}_0 = (-1, 1)$m, $\vec{A} = (0, 0)$m, and $\vec{F} = (3, 2)$ N.

35. $\vec{p}_0 = (2, 0, 1)$m, $\vec{A} = (1, 1, 1)$m, and $\vec{F} = (1, -1, 2)$ N.

36. $\vec{p}_0 = (\hat{\imath} + \hat{k})$m, $\vec{A} = (-2\hat{\imath} - \hat{\jmath} + \hat{k})$m, and $\vec{F} = -\hat{k}$ N.

37. Find a unit vector that points in the direction of the axis of the tendency to rotate induced by the torque obtained in Exercise 35.

38. Find a unit vector that points in the direction of the axis of the tendency to rotate induced by the torque obtained in Exercise 36.

39. A force with magnitude 3 N produces a torque about position $\vec{p}$ with magnitude 6 Nm. Find the perpendicular distance from $\vec{p}$ to the line of action of the force.

40. A force produces a torque about position $\vec{p}$ with magnitude 6 Nm. The perpendicular distance from $\vec{p}$ to the line of action of the force is 2 m. What is the magnitude of the force?

In Exercises 41–43, $\mathcal{L}$ denotes the line that contains the vector $\vec{v}$ when $\vec{v}$ is drawn emanating from the point $\vec{A}$. Find the perpendicular distance from the point $\vec{p}_0$ to the line $\mathcal{L}$.

41. $\vec{p}_0 = (1, 1, 1)$, $\vec{v} = (2, -1, 1)$, and $\vec{A} = (0, 0, 0)$.

42. $\vec{p}_0 = (1, 0, 1)$, $\vec{v} = (0, -1, 1)$, and $\vec{A} = (1, 3, -1)$.

43. $\vec{p}_0 = (0, 1, 3)$, $\vec{v} = (2, 0, 1)$, and $\vec{A} = (-1, 1, 0)$.

44. What must be true about the vectors $\vec{A} = (a_x, a_y)$ and $\vec{B} = (b_x, b_y)$ if $\begin{vmatrix} a_x & a_y \\ b_x & b_y \end{vmatrix} = 0$?

45. What must be true about the vectors $\vec{A} = (a_x, a_y)$ and $\vec{B} = (b_x, b_y)$ if $\begin{vmatrix} a_x & a_y \\ b_x & b_y \end{vmatrix} = \|\vec{A}\|\|\vec{B}\|$?

46. What must be true about the vectors $\vec{A} = (a_x, a_y, a_z)$, $\vec{B} = (b_x, b_y, b_z)$ and $\vec{C} = (c_x, c_y, c_z)$ if $\begin{vmatrix} a_x & a_y & a_z \\ b_x & b_y & b_z \\ c_x & c_y & c_z \end{vmatrix} = \|\vec{A}\|\|\vec{B}\|\|\vec{C}\|$?

47. What must be true about the vectors $\vec{A} = (a_x, a_y, a_z)$, $\vec{B} = (b_x, b_y, b_z)$ and $\vec{C} = (c_x, c_y, c_z)$ if $\begin{vmatrix} a_x & a_y & a_z \\ b_x & b_y & b_z \\ c_x & c_y & c_z \end{vmatrix} = 0$?

48. Expand $\|\vec{A} \times \vec{B}\|$, $\|\vec{A}\|^2 \|\vec{B}\|^2$, and $(\vec{A} \cdot \vec{B})^2$, and show that

$$\begin{aligned} \|\vec{A} \times \vec{B}\|^2 &= \|\vec{A}\|^2 \|\vec{B}\|^2 - (\vec{A} \cdot \vec{B})^2 \\ &= \|\vec{A}\|^2 \|\vec{B}\|^2 \sin^2 \theta. \end{aligned}$$

49. Let $\vec{A}$, and $\vec{B}$ be vectors, and let a and b be real numbers. Show that

    a. $\vec{A} \times \vec{B} = -\vec{B} \times \vec{A}$.

    b. $(a\vec{A}) \times \vec{B} = \vec{A} \times (a\vec{B}) = a(\vec{A} \times \vec{B})$.

    c. $(a + b)\vec{A} = a\vec{A} + b\vec{A}$.

    d. $\vec{A} \times \vec{B} = \vec{A} \times (\vec{B} + a\vec{A})$

50. Let $\vec{A}$, $\vec{B}$ and $\vec{C}$ be vectors and a and b be real numbers. Show that

   a. $\vec{C} \cdot (\vec{A} \times \vec{B}) = (\vec{C} \times \vec{A}) \cdot \vec{B}$.

   b. $\vec{C} \cdot (\vec{A} \times \vec{B}) = (\vec{C} + a\vec{A}) \cdot (\vec{A} \times \vec{B}) = (\vec{C} + a\vec{B}) \cdot (\vec{A} \times \vec{B})$.

   c. $(a\vec{C}) \cdot (\vec{A} \times \vec{B}) = \vec{C} \cdot ((a\vec{A}) \times \vec{B}) = \vec{C} \cdot (\vec{A} \times (a\vec{B})) = a(\vec{C} \cdot (\vec{A} \times \vec{B}))$.

51. Prove that the following properties are true.

   a.
$$\begin{vmatrix} a_1 & a_2 & a_3 \\ b_1 & b_2 & b_3 \\ c_1 & c_2 & c_3 \end{vmatrix} = - \begin{vmatrix} b_1 & b_2 & b_3 \\ a_1 & a_2 & a_3 \\ c_1 & c_2 & c_3 \end{vmatrix}$$

and

$$\begin{vmatrix} a_1 & a_2 & a_3 \\ b_1 & b_2 & b_3 \\ c_1 & c_2 & c_3 \end{vmatrix} = - \begin{vmatrix} a_1 & a_2 & a_3 \\ c_1 & c_2 & c_3 \\ b_1 & b_2 & b_3 \end{vmatrix} .$$

51.   b. If d is a number,
$$\begin{vmatrix} a_1 & a_2 & a_3 \\ b_1 & b_2 & b_3 \\ c_1 & c_2 & c_3 \end{vmatrix} = \begin{vmatrix} a_1 & a_2 & a_3 \\ b_1 + da_1 & b_2 + da_2 & b_3 + da_3 \\ c_1 & c_2 & c_3 \end{vmatrix} .$$

  c. If d is a number,
$$d\begin{vmatrix} a_1 & a_2 & a_3 \\ b_1 & b_2 & b_3 \\ c_1 & c_2 & c_3 \end{vmatrix} = \begin{vmatrix} da_1 & da_2 & da_3 \\ b_1 & b_2 & b_3 \\ c_1 & c_2 & c_3 \end{vmatrix}$$
$$= \begin{vmatrix} a_1 & a_2 & a_3 \\ b_1 & b_2 & b_3 \\ dc_1 & dc_2 & dc_3 \end{vmatrix} .$$

  d.
$$\begin{vmatrix} a_1 & 0 & 0 \\ b_1 & b_2 & 0 \\ c_1 & c_2 & c_3 \end{vmatrix} = \begin{vmatrix} a_1 & a_2 & a_3 \\ 0 & b_2 & b_3 \\ 0 & 0 & c_3 \end{vmatrix}$$
$$= a_1 b_2 c_3 .$$

1.6 Equations of Lines and Planes

In this section, we use the inner product to derive equations for lines and planes. We need to add to our vocabulary before getting started.

Definition: Normal Vectors to Lines and Planes

We will say that a vector $\vec{n}$ is *normal* to a line L if it is perpendicular to that line. It is *normal* to a plane P if it is normal to every line lying in P.

Equations for Lines in the Plane

Let $\vec{n}$ be a vector normal to the line L in $\mathbb{R}^2$. Clearly, if you know a point on a line and you also know a vector that is perpendicular to that line, you can describe that line. Let $\vec{r}_0 = (x_0, y_0)$ be a point on the line. As in Figure 1, draw $\vec{n}$ emanating from $\vec{r}_0$. We see that $\vec{r}$ is in L if and only if $\vec{n}$ is orthogonal to $\vec{r} - \vec{r}_0$. Thus, we have the following theorem giving us a vector equation for a line.

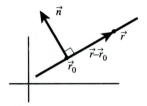

Figure 1. *The vector $\vec{n}$ is normal to the line L and the vector $\vec{r} - \vec{r}_0$ points in the direction of L.*

> **Theorem 1** *Let $\vec{n}$ be a vector in $\mathbb{R}^2$ that is normal to the line L (in $\mathbb{R}^2$), and let r_0 be a point on the line. Then*
>
> $$\vec{n} \cdot (\vec{r} - \vec{r}_0) = 0, \text{ or } \vec{n} \cdot \vec{r} = \vec{n} \cdot \vec{r}_0$$
>
> *is an equation for the line L.*

Let $\vec{n} = (a, b)$ be a vector normal to a line L in $\mathbb{R}^2$, and let $\vec{r}_0 = (x_0, y_0)$ be a point in the line. Then according to the vector equation in Theorem 1, (x, y) is on the line if and only if

$$
\begin{aligned}
(a, b) \cdot ((x, y) - (x_0, y_0)) &= 0, \\
a(x - x_0) + b(y - y_0) &= 0, \text{ or} \\
ax + by &= ax_0 + by_0, \\
ax + by &= c, \text{ where } c = ax_0 + by_0,
\end{aligned}
$$

which is a familiar equation of a line.   ■

Equations for Planes in $\mathbb{R}^3$

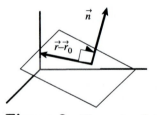

Figure 2. *The vector $\vec{n}$ is normal to the plane P and the vector $\vec{r} - \vec{r}_0$ is parallel to the plane.*

By similar reasoning, we can obtain an equation for a plane in 3–space. Let $\vec{n} = (a, b, c)$ be a vector normal to the plane P, and let $\vec{r}_0 = (x_0, y_0, z_0)$ be a point in the plane. As in Figure 2, draw $\vec{n}$ emanating from $\vec{r}_0$. Then $\vec{r}$ is in the plane if and only if the vectors $\vec{n}$ and $\vec{r} - \vec{r}_0$ are orthogonal.

We have a vector equation for a plane in $\mathbb{R}^3$ that is analogous to the equation for a line in the plane.

> **Theorem 2** *Let $\vec{n}$ be a vector in $\mathbb{R}^3$ that is normal to the plane P, and let r_0 be a point on the plane. Then*
>
> $$\vec{n} \cdot (\vec{r} - \vec{r}_0) = 0, \text{ or } \vec{n} \cdot \vec{r} = \vec{n} \cdot \vec{r}_0$$
>
> *is an equation for the plane P.*

As in the case of a line, the equation for the plane in Theorem 2 can be expanded into a more familiar form. Let $\vec{n} = (a, b, c)$:

$$(a, b, c) \cdot ((x, y, z) - (x_0, y_0, z_0)) = 0,$$

$$a(x - x_0) + b(y - y_0) + (z - z_0) = 0, \text{ or}$$

$$ax + by + cz = ax_0 + by_0 + cz_0, \text{ or}$$

$$ax + by + cz = d, \text{ where } d = ax_0 + by_0 + cz_0. \blacksquare$$

EXAMPLE 1: Find an equation for the plane that is normal to $\vec{n} = (3, -1, 1)$ and contains the point $\vec{r}_0 = (-1, 0, 1)$.

SOLUTION:

$$\vec{n} \cdot \vec{r} = \vec{n} \cdot \vec{r}_0$$

$$(3, -1, 1) \cdot (x, y, z) = (3, -1, 1) \cdot (-1, 0, 1)$$

$$3x - y + z = -2. \qquad\qquad \blacksquare$$

We reverse the process. If $a, b,$ and c are numbers, at least one of which is not zero, and d is a number, then we can show that

$$ax + by + cz = d \qquad\qquad (9)$$

is an equation of a plane. Note that Equation (9) can be written as

$$(a, b, c) \cdot (x, y, z) = d.$$

For the sake of argument, suppose that $a \neq 0$. Let $\vec{r}_0 = (\frac{d}{a}, 0, 0)$. Then the number d from Equation (9) is $(a, b, c) \cdot \vec{r}_0$, so that Equation (9) can be written in the form

$$\vec{n} \cdot \vec{r} = \vec{n} \cdot \vec{r}_0, \text{ where } \vec{n} = (a, b, c).$$

We have the following theorem.

> **Theorem 3** *The set P in xyz–space is a plane if and only if there are numbers $a, b, c,$ and d, with at least one of $a, b,$ or c not zero, such that*
>
> $$ax + by + cz = d$$
>
> *is an equation for P. Furthermore, the vector (a, b, c) is normal to the plane P.*

EXAMPLE 2: Find an equation for the plane P that contains the origin and is parallel to the plane P' with equation $2x + 3y + z = 5$.

SOLUTION: The vector $\vec{n} = (2, 3, 1)$ is normal to the plane P'. Since P is parallel to P', the vector $\vec{n}$ must also be normal to P. Thus, a vector equation for P is $\vec{n} \cdot \vec{r} = \vec{n} \cdot \vec{0}$. Writing this equation in terms of coordinates, we have

$$(2,3,1) \cdot (x,y,z) = 0, \text{ or}$$

$$2x + 3y + z = 0. \qquad \blacksquare$$

EXAMPLE 3: Find an equation for the plane in xyz–space that contains $(1,0,0)$, $(1,1,1)$ and $(-1,1,0)$. We provide two approaches.

SOLUTION: The equation for the plane must be of the form $ax + by + cz = d$. Since each of the points $(1,0,0)$, $(1,1,1)$, and $(-1,1,0)$ must satisfy this equation, we have the following system of linear equations:

$$a = d \qquad (10)$$

$$a + b + c = d \qquad (11)$$

$$-a + b = d. \qquad (12)$$

Adding Equations (10) and (12), we obtain

$$b = 2d. \qquad (13)$$

Substituting Equations (10) and (13) into Equation (11) we get $d + 2d + c = d$, or $c = -2d$. Putting these values for (a,b,c) in the equation $ax + by + cz = d$, we have

$$dx + 2dy - 2dz = d, \text{ or, } x + 2y - 2z = 1. \qquad \blacksquare$$

We can also solve this problem using the cross product.

SOLUTION: Let $\vec{A}$ be the vector from $(1,0,0)$ to $(1,1,1)$, and let $\vec{B}$ be the vector from $(1,0,0)$ to $(-1,1,0)$. Thus $\vec{A} = (0,1,1)$ and $\vec{B} = (-2,1,0)$ are two vectors in the plane emanating from a common point. The cross product $\vec{A} \times \vec{B}$ will serve as our normal vector $\vec{n}$.

$$\vec{n} = \vec{A} \times \vec{B} = (-1,-2,2).$$

Letting $\vec{r}_0 = (1,0,0)$ in the vector equation for a plane from Theorem 2, we have

$$(-1,-2,2) \cdot (x,y,z) = (-1,-2,2) \cdot (1,0,0),$$

which yields
$$x + 2y - 2z = 1.$$

$\blacksquare$

The Distance From a Point to a Line in $\mathbb{R}^2$ and From a Point to a Plane in $\mathbb{R}^3$

We now have the tools to derive the formula for the distance from a point to a line in $\mathbb{R}^2$ as well as the formula for the distance from a point to a plane in $\mathbb{R}^3$.

Let $\vec{r_1} = (x_1, y_1)$ be a point in $\mathbb{R}^2$, and let $ax + by = c$ be a line L. The vector $\vec{n} = (a, b)$ is perpendicular to L. Let $\vec{r_0} = (x_0, y_0)$ be any point on L. We will project the vector $(x_1 - x_0, y_1 - y_0)$ onto the vector (a, b). The length of this projection is the distance from (x_1, y_1) to L. (See Figure 3.)

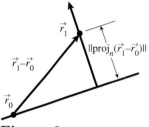

Figure 3. *The normal vector $\vec{n}$ is drawn so that it contains $\vec{r_1}$.*

Thus,

$$\text{distance} = \left\| \frac{(a, b) \cdot (x_1 - x_0, y_1 - y_0)}{(a, b) \cdot (a, b)} (a, b) \right\|$$

$$= \frac{|ax_1 + by_1 - ax_0 - by_0|}{a^2 + b^2} \sqrt{a^2 + b^2}$$

$$= \frac{|ax_1 + by_1 - c|}{\sqrt{a^2 + b^2}}, \text{ since } ax_0 + by_0 = c.$$

We have the following theorem.

Theorem 4 *If (x_1, y_1) is a point, and L is a line in $\mathbb{R}^2$ with equation $ax + by = c$, then the distance from (x_1, y_1) to L is given by*
$$\frac{|ax_1 + by_1 - c|}{\sqrt{a^2 + b^2}}.$$

In Exercise 18, the student is asked to employ the same methods to obtain the formula for the distance from a point to a plane in $\mathbb{R}^3$.

Theorem 5 *If (x_1, y_1, z_1) is a point, and P is a line in $\mathbb{R}^3$ with equation $ax + by + cz = d$, then the distance from (x_1, y_1, z_1) to P is given by* $\dfrac{|ax_1 + by_1 + cz_1 - d|}{\sqrt{a^2 + b^2 + c^2}}.$

Projections of Vectors onto Lines and Planes

Let $\vec{F} = (F_x, F_y)$ be a vector in the plane, and let $\mathcal{L}$ be the line

$ax + by = d$. Then $\vec{n} = (a, b)$ is normal to $\mathcal{L}$, and

$$\text{proj}_{\vec{n}}\vec{F} = \frac{\vec{F} \cdot \vec{n}}{\|\vec{n}\|^2}\vec{n}.$$

The projection of $\vec{F}$ in the direction of $\vec{n}$ is the part of $\vec{F}$ that is normal, or perpendicular, to $\mathcal{L}$. It follows that

$$\text{orth}_{\vec{n}}\vec{F} = \vec{F} - \text{proj}_{\vec{n}}\vec{F}$$

is the part of $\vec{F}$ that points in the direction of $\mathcal{L}$. (See Figure 4.a.) Similarly, let $\vec{F} = (F_x, F_y, F_z)$ be a three–dimensional vector, and let $\mathcal{P}$ be the plane $ax + by + cz = d$. Then $\vec{n} = (a, b, c)$ is normal to $\mathcal{P}$, and the part of $\vec{F}$ normal to the plane is

$$\text{proj}_{\vec{n}}\vec{F} = \frac{\vec{F} \cdot \vec{n}}{\|\vec{n}\|^2}\vec{n},$$

while

$$\text{orth}_{\vec{n}}\vec{F} = \vec{F} - \text{proj}_{\vec{n}}\vec{F}$$

is that part of $\vec{F}$ that points in the direction parallel to $\mathcal{P}$. (See Figure 4.b.) This leads us to the following definition.

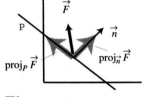

Figure 4.a *The projection of a vector onto a line.*

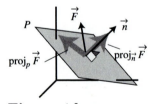

Figure 4.b *The projection of a vector onto a plane.*

Definition: The Projection of a Vector onto a Line or a Plane

Let $\vec{n}$ be a vector normal to $\mathcal{P}$, a line in $\mathbb{R}^2$, or a plane in $\mathbb{R}^3$. Then the *projection of* $\vec{F}$ *onto* $\mathcal{P}$ is defined by

$$\text{proj}_{\mathcal{P}}\vec{F} = \text{orth}_{\vec{n}}\vec{F} = \vec{F} - \text{proj}_{\vec{n}}\vec{F}.$$

We refer to $\text{proj}_{\vec{n}}\vec{F}$ as the *component of* $\vec{F}$ *normal to* $\mathcal{P}$.

EXAMPLE 4: Let $\vec{F} = (1, 2, 3)$, and let $\mathcal{P}$ be the plane $x + y + z = 3$. Find the component of $\vec{F}$ normal to $\mathcal{P}$ and the projection of $\vec{F}$ onto $\mathcal{P}$.

SOLUTION: The vector $\vec{n} = (1, 1, 1)$ is normal to $\mathcal{P}$. Thus, the component of $\vec{F}$ normal to $\mathcal{P}$ is

$$\text{proj}_{\vec{n}}\vec{F} = \frac{\vec{F} \cdot \vec{n}}{\|\vec{n}\|^2}\vec{n} = \frac{6}{3}(1, 1, 1) = (2, 2, 2).$$

The projection of $\vec{F}$ onto $\mathcal{P}$ is

$$\text{proj}_{\mathcal{P}}\vec{F} = (1, 2, 3) - (2, 2, 2) = (-1, 0, 1). \qquad \blacksquare$$

EXAMPLE 5:

(a)

$$\text{proj}_{(1,3)}3(2,4) \;=\; \text{proj}_{(1,3)}(6,12) = \frac{(1,3)\cdot(6,12)}{(1,3)\cdot(1,3)}(1,3) = \frac{42}{10}(1,3).$$

$$3\text{proj}_{(1,3)}(2,4) \;=\; 3\frac{(1,3)\cdot(2,4)}{(1,3)\cdot)(1,3)}(1,3) = 3\frac{14}{10}(1,3) = \frac{42}{10}(1,3).$$

(b)

$$\text{proj}_{(1,3)}\left((2,4)+(5,6)\right) \;=\; \frac{(7,10)\cdot(1,3)}{(1,3)\cdot(1,3)}(1,3) = \frac{37}{10}(1,3).$$

$$\text{proj}_{(1,3)}(2,4) + \text{proj}_{(1,3)}(5,6) =$$

$$\frac{(1,3)\cdot(2,4)}{(1,3)\cdot(1,3)}(1,3) + \frac{(1,3)\cdot(5,6)}{(1,3)\cdot(1,3)}(1,3)$$

$$= \frac{14}{10}(1,3) + \frac{23}{10}(1,3) = \frac{37}{10}(1,3).$$

EXAMPLE 6: A 30–lb block is resting on a frictionless plane that makes an angle of 20° with the horizontal. A rope is used to keep the block at rest. Assume that the rope is parallel to the plane, as in Figure 5.a.

(a) Find force that the plane is exerting on the block.

(b) Find the force exerted on the block by the rope.

SOLUTION: Let $\vec{F}_T$ be the force exerted on the block by the rope, let $\vec{F}_N$ be the force exerted on the block by the plane, and let $\vec{F}_G$ be the gravitational force as illustrated in Figure 5.b. An equation for the plane is $y - \tan(20°)x = 0$, or approximately $y - 0.364x = 0$. The vector normal to the plane is given by $\vec{n} = (-0.364, 1)$. The gravitational force is $\vec{F}_G = (0, -30)$ N. Since $\vec{F}_N$ is orthogonal to the plane and $\vec{F}_T$ is parallel to the plane, $\text{proj}_{\vec{n}}\vec{F}_N = \vec{F}_N$ and $\text{proj}_{\vec{n}}\vec{F}_T = \vec{0}$.

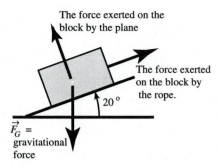

Figure 5.a *The block on the plane and the forces involved.*

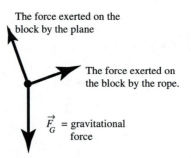

Figure 5.b *A "free body" diagram of the forces acting on the block.*

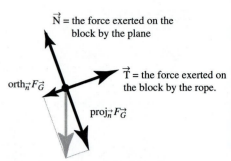

Figure 5.c *A "free body" diagram of the forces acting on the block displaying the components of $\vec{F}_G$ that are parallel and normal to the plane.*

(a) The sum of the forces is zero, or

$$\vec{F}_T + \vec{F}_N + \vec{F}_G = \vec{0}. \tag{14}$$

Projecting both sides of Equation (14) onto $\vec{n}$ and applying Exercise 41 of Section 1.4, we obtain

$$
\begin{aligned}
\vec{0} &= \operatorname{proj}_{\vec{n}}(\vec{F}_T + \vec{F}_N + \vec{F}_G)\\
&= \vec{F}_N + \frac{(0,-30)\cdot(-0.364,1)}{(-0.364,1)\cdot(-0.364,1)}(-0.364,1)\\
&= \vec{N} + (9.64,-26.48)\\
\vec{F}_N &= (-9.64,26.48)\text{N}.
\end{aligned}
$$

(b) The sum of the forces parallel to the plane is zero. Returning to Equation (14),

$$
\begin{aligned}
\vec{0} &= \vec{F}_T + \vec{F}_N + \vec{F}_G\\
&= \vec{F}_T + (-9.64,26.48)\text{N} + (0,-30)\text{N}\\
\vec{F}_T &= (9.64,3.52)\ \text{N}.
\end{aligned}
$$

EXERCISES 1.6

1. Sketch the lines l_1, l_2 and l_3 containing the point $\vec{p}$ such that $\vec{v}_i$ is normal to l_i.

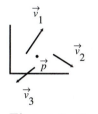

Figure 6. *Exercise 1.*

In Exercises 2–4, find an equation for the plane that is normal to the vector $\vec{n}$ and contains the vector $\vec{p}$.

2. $\vec{n} = (3,1,2)$, $\vec{p} = (1,2,0)$.

3. $\vec{n} = (1,-1,6)$, $\vec{p} = (2,0,-3)$.

4. $\vec{n} = 3\hat{\imath} + 2\hat{\jmath} + \hat{k}$, $\vec{p} = -\hat{\imath} + 2\hat{\jmath}$.

5. Find an equation for the plane parallel to $2x - 3y + z = 2$ that contains the point $(2,1,-6)$.

In Exercises 6–11, find an equation for the plane that contains the three points.

6. $\vec{r} = (1,1,0)$, $\vec{s} = (2,-1,3)$, and $\vec{t} = (3,0,1)$.

7. $\vec{r} = (2,-1,3)$, $\vec{s} = (-1,0,0)$, and $\vec{t} = (1,3,-5)$.

8. $\vec{r} = 2\hat{\imath} + 3\hat{\jmath} + \hat{k}$, $\vec{s} = 2\hat{\imath} + \hat{\jmath}$, and $\vec{t} = \hat{\imath}$.

9. $\vec{r} = (1,2,1)$, $\vec{s} = (1,0,-1)$, and $\vec{t} = (0,0,0)$

10. $\vec{r} = (-1,2,1)$, $\vec{s} = (0,0,1)$, and $\vec{t} = (1,1,1)$

11. $\vec{r} = (0,1,-1)$, $\vec{s} = (2,1,4)$, and $\vec{t} = (1,-1,3)$

12. Let $\vec{P} = (a,b,c)$ be a point in re^3. Find the distance from $\vec{P}$ to each of the coordinate axes.

13. Let $\vec{P} = (a,b,c)$ be a point in $\mathbb{R}^3$. Find the distance from $\vec{P}$ to the xy–plane, the xz–plane, and to the yz–plane.

In Exercises 14–15, find the distance from the point r_1 to the line L.

14. $r_1 = (1,2)$ and L has equation $x + y = 1$.

15. $r_1 = (2,-2)$ and L has equation $x - y = 6$.

In Exercises 16–17, find the distance from the point r_1 to the plane P.

16. $r_1 = (1,2,-1)$ and P has equation $x+y+z = 1$.

17. $r_1 = (-2,-2,3)$ and P has equation $x-y+2z = 10$.

18. If $Ax + By + Cz = D$ is a plane and (x_1,y_1,z_1) is a point in $\mathbb{R}^3$, show that the distance from the point to the plane is given by

$$\frac{|Ax_1 + By_1 + Cz_1 - D|}{\sqrt{A^2 + B^2 + C^2}}.$$

19. In each of the figures below, sketch the projection of the vector $\vec{v}$ onto the line l and the component of $\vec{v}$ that is normal to l.

Figure 7. *Exercise 19.*

In Exercises 20–22, find the component of vector $\vec{F}$ that is normal to the line or plane $\mathcal{P}$, and the projection of $\vec{F}$ onto $\mathcal{P}$.

20. $\vec{F} = (2,3)$ and $\mathcal{P}$ is given by $x - 2y = 64$.

21. $\vec{F} = (-2,3)$ and $\mathcal{P}$ is a line with inclination $-20°$.

22. $\vec{F} = (2,3,-2)$ and $\mathcal{P}$ is given by $x-y+2z = -3$.

A 45 lb block is on a frictionless plane that makes an angle α with the horizontal. It is held at rest by a rope as in the figure. In Exercises 23 and 24, indicate the force exerted on the block by the plane and by the rope.

23. $\alpha = 10°$.   24. $\alpha = 35°$.

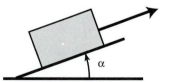

Figure 8. *Exercises 23 and 24*

Chapter 2

Functions

2.1 Functions

In the first chapter, we saw numerous examples of graphs. Of particular importance is a special class of graphs, called functions.

If **A** and **B** are sets, then a function from **A** to **B** assigns to each element of **A** exactly one element of **B**. We have the following formal definition.

Definition: Function

If **A** and **B** are sets, then f is a *function* from **A** into **B** if f is a set of ordered pairs such that:

(a) If (a, b) is in f, then a is in **A** and b is in **B**.

(b) If a is in **A**, then there is one and only one b in **B** such that (a, b) is in f.

We will use the shorthand notation $f : \mathbf{A} \to \mathbf{B}$ to mean that f is a function from **A** into **B**.

There are some standard words related to functions that give us a working vocabulary. The set **A** is called the *domain* of f and **B** is called the *range* of f. If f is a function and (a, b) is in f, then we often write $b = f(a)$, and we say that b is in the *image* of a. The *image* of f is the set of all second terms of f. That is, y is in the image of f if and only if there is an x in the domain of f such that $y = f(x)$. Following conventions, the *graph of a function* refers to the function itself.

The concepts of domain and image of a function are illustrated in Figure 1.

Figure 1. *The Domain and Image.*

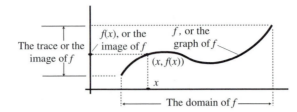

Notice that f is a function from a subset of the real line into the real line if and only if any vertical line L passes through f at no more than one point. (See Figure 2.a and 2.b.)

Intuitively, a function is a rule that assigns a point of its range to each member of the domain. Thus, one might think of a machine so that if you put a point x of the domain of the function into the machine (see Figure 3) then the function machine gives you back exactly one point $f(x)$ of the range of the function.

Figure 2.a *A subset of the plane that is* not *a function.*

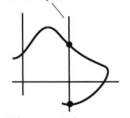

A vertical line that passes through the graph at more than one place

Figure 2.b *A subset of the plane that is* not *a function.*

Figure 3. *The function machine.*

EXAMPLE 1: $f = \{(1,2),(2,3),(3,4)\}$ is a function with domain $\{1,2,3\}$ and range $\{2,3,4\}$. $f(1) = 2$, $f(2) = 3$, $f(3) = 4$. The image of 1 is 2, the image of 2 is 3, and the image of 3 is 4. ∎

EXAMPLE 2: $f = \{(1,2),(2,2),(3,2),(4,2)\}$ is a function with domain $\{1,2,3,4\}$ and range the single value $\{2\}$. $f(1) = f(2) = f(3) = f(4) = 2$. Thus f is a constant function.[1] ∎

EXAMPLE 3: The graph of the equation $y = \sqrt{x^2 + 2x + 1}$ is a function with domain $\mathbb{R}$ and range $\mathbb{R}$. However, the graph of $y^2 = x^2 + 2x + 1$ is not a function. Why? ∎

EXAMPLE 4: Let $\mathbf{G}$ be the graph of $y = \frac{x^2+2x+1}{x^2-2x+1}$. Then $\mathbf{G}$ is a function and $\mathbf{G}(x) = \frac{x^2+2x+1}{x^2-2x+1}$. The domain of $\mathbf{G}$ is the set of all real

---

[1]A function f is said to be a *constant function* if there is a number c such that $f(x) = c$ for all x in the domain of f.

numbers except $x = 1$ (where the denominator $x^2 - 2x + 1 = 0$). ∎

The previous two examples illustrate a feature convention in mathematics. In the definition of function, the domain and range must be stated. Thus the function $f: \mathbb{R} \to \mathbb{R}$ given by $f(x) = 2x$ is not the same as the function $g:$ integers $\to$ integers given by $g(x) = 2x$ since the domain (and the range) are different. However, when we speak about the graph of an algebraic relation, unless specifically stated otherwise, we assume that the domain is the largest subset of $\mathbb{R}$ (or $\mathbb{R}^2$ or $\mathbb{R}^3$) on which that graph can be defined.

EXAMPLE 5: Clearly $f: \mathbb{R} \to \mathbb{R}$ defined by $f(x) = mx + b$ is a function that is a line. Thus, any nonvertical line in the plane is a function from $\mathbb{R}$ into $\mathbb{R}$. A horizontal line $y = c$ is a constant function defined by $f(x) = c$. ∎

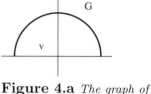

Figure 4.a *The graph of* $G(x) = \sqrt{1 - x^2}$.

EXAMPLE 6: The circle $x^2 + y^2 = 1$ is not a function. However, let G and H be defined by $G(x) = \sqrt{1 - x^2}$ and $H(x) = -\sqrt{1 - x^2}$. Then G and H are functions whose union is the unit circle. (See Figure 4.) The domain of G is the set $\{x \mid -1 \leq x \leq 1\}$ and the image of G is the set $\{y \mid 0 \leq y \leq 1\}$. The image of H is the set $\{y \mid -1 \leq y \leq 0\}$. ∎

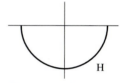

Figure 4.b *The graph of* $H(x) = -\sqrt{1 - x^2}$.

If you are familiar with computer programming languages, you have probably had the experience of inadvertently declaring a real number to be an integer. If the programmer declares a real number to be an integer, then the computer picks out the largest integer less than or equal to the real number. For example, if $\sqrt{2}$ is declared to be an integer, then the computer will give back the number 1. It is common in mathematics texts to use the following function, which has domain all real numbers, and range the integers.

Definition: $[[x]]$

If x is a real number, then we denote by $[[x]]$ the largest integer less than or equal to x.

NOTE: In some programming languages and wordprocessors $[[x]]$ is indicated by $\lfloor \lfloor x \rfloor \rfloor$.

EXAMPLE 7: Let L be the function defined by $L(x) = [[x]]$. The graph of L looks like a long stairway. The domain of L is all of $\mathbb{R}$ but the image of L is the set of all integers. (See Figure 5.) ∎

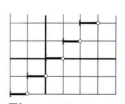

Figure 5. *The function* $L(x) = [[x]]$.

Sequences

Suppose that we have a savings account, and we put some money into and take some money out of the account from time to time. Each month we earn some interest on the money that is in the account. At the end of each month, we receive a statement that tells us the balance in our account. If we let $a(n)$ denote the balance at the end of the nth month, then a is a function from the set of positive integers into the real numbers. This function is just one example of a *sequence*. The following definition formalizes the idea for us.

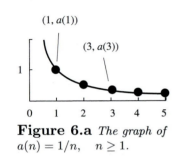

Figure 6.a *The graph of* $a(n) = 1/n$, $\quad n \geq 1$.

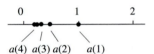

Figure 6.b *The image of* $a(n) = 1/n$, $\quad n \geq 1$.

Definition: Sequences

The statement that a is a sequence means that

(a) a is a function; and

(b) there is a number N such that the domain of a is the set of all integers n with $n \geq N$.

The nth term $a(n)$ of the sequence a is often denoted by a_n.
The number N is called the *initial integer* in the domain of a.

In less formal terms, a sequence is a function whose domain is a consecutive list of integers that starts from some number (e.g., $1, 2, 3, \ldots$ or $13, 14, 15, \ldots$). Unless specifically stated otherwise, our sequences are real-valued functions. Often we specify a sequence by writing the first few terms.

EXAMPLE 8: Let a be the sequence defined by $a(n) = 1/n!, n \geq 0$. Then $a = 1, \dfrac{1}{2}, \dfrac{1}{(3)(2)}, \dfrac{1}{(4)(3)(2)}, \dfrac{1}{(5)(4)(3)(2)}, \cdots = 1, \dfrac{1}{2}, \dfrac{1}{6}, \dfrac{1}{24}, \dfrac{1}{120}, \cdots .$ ∎

While we can draw a portion of the graph of a sequence (see Figure 6.a), we commonly visualize a sequence by its image as in Figure 6.b.

EXERCISES 2.1

In Exercises 1–11, determine whether or not the given set is a function, and if so, give the domain and image.

1. $\{(1, 2), (2, 3), (1, 4)\}$.

2. $\{(2, 2), (3, 3), (4, 4), (5, 5)\}$.

3. $\{(1, 6), (2, 6), (3, 6)\}$.

4. The graph of $x = \sqrt{1 - y^2}$.

5. The graph of $y = \sqrt{x^2 - 1}$.

6. The graph of $x + y^3 = 1$.

7. The graph of $y = x^2$.

8. The graph of $x^2 = -y$.

9. The graph of $y = \dfrac{x}{x^2 + 3x + 2}$. (Find the domain only.)

10. The graph of $y = \dfrac{x}{8x^2 + 3x + 2}$. (Find the domain only.)

11. The graph of $y = \sqrt{x^2 + 3x + 2}$.

12. For each of the following graphs, determine whether it is a function of x and/or a function of y.

  a. If it is a function of x, what is its domain and image?

  b. If it is a function of y, what is its domain and image?

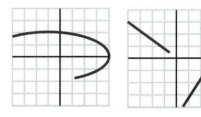

Graph 1   Graph 2

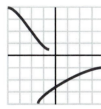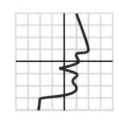

Graph 3   Graph 4

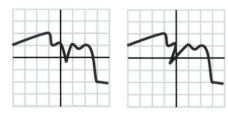

Graph 5   Graph 6

In Exercises 13–15, compute $f(x_0)$, $f(x_1)$, and $f(x_2)$.

13. $f(x) = x^2 + 2x + 1$, $x_0 = 1$, $x_1 = 0, x_2 = \pi$.

14. $f(x) = [[4x]]$, $x_0 = \frac{1}{6}$, $x_1 = \frac{\pi}{4}$, $x_2 = -\frac{1}{2}$.

15. $f(x) = [[x^2]]$, $x_0 = 1$, $x_1 = 2$, $x_2 = -3$.

In Exercises 16–24, sketch the given function and give its domain and image.

16. $f(x) = |x|$, $\left(\text{recall } |x| = \left\{ \begin{array}{ll} x & \text{if } x \geq 0 \\ -x & \text{if } x < 0 \end{array} \right\}\right)$.

17. $f(x) = 4x^2$.

18. $f(x) = \sqrt{1 - \frac{x^2}{4}}$.

19. $f(x) = \sqrt{2x}$.

20. $f(x) = \sqrt{\frac{x^2}{4} - 1}$.

21. $f(x) = |x^3|$.

22. $f(x) = [[x + 3]]$.

23. $f(x) = [[x]] + x$.

24. $f(x) = |[[x]]|$.

25. A farmer wishes to enclose a rectangular garden with 100 yards of fencing. One side of the garden is bounded by a river so that only three sides must be fenced. Find a function that gives the area in terms of the side bounded by the river.

26. A cylindrical tin can is made with volume V. Find an expression that gives the surface area of the can (including the top and bottom) as a function of the radius.

27. A cube is made with volume V. Find an expression that gives the surface area of the cube as a function of the volume of the cube.

28. A rectangle is inscribed in a circle of radius R. Find a function that gives the area of the rectangle in terms of the length of one of its sides.

In Exercises 29 and 30, a 6 foot man is walking away (on level ground) from a light located 15 ft above the ground.

29. Find a function that gives the length of his shadow in terms of his distance from the point on the ground directly below the light.

30. Find the function that gives the length of his shadow in terms of the distance from the top of his head to the light.

In Exercises 31–37, write out the first five terms of the given sequence.

31. $a(n) = \dfrac{(-1)^n}{n}, \quad n \geq 1.$

32. $a(n) = \dfrac{n!}{2^n}, \quad n \geq 0.$[2]

33. $a(n) = \dfrac{n!}{3!(n-3)!}, \quad n \geq 3.$

34. $a(n) = n + (-1)^n n, \quad n \geq 0.$

35. $a(n) = br^n, \quad n \geq 0,$ where $b = 2$ and $r = 0.5.$

36. $a(0) = 1$ and $a(n+1) = (n+1)a(n).$

37. $a(0) = 1,$ $a(1) = 1,$ and $a(n+2) = a(n)+a(n+1)$

If $n \geq m$ are integers, then the symbol $\binom{n}{m}$ is used to denote $\dfrac{n!}{m!(n-m)!}$ *and $\binom{n}{m}$ is called a* binomial coefficient .[3]

38. Evaluate $\binom{5}{3}$ and $\binom{6}{2}$.

39. Show that $(a+b)^2 = \binom{2}{0}a^2 + \binom{2}{1}ab + \binom{2}{2}b^2$.

40. Show that $(a+b)^3 = \binom{3}{0}a^3 + \binom{3}{1}a^2b + \binom{3}{2}ab^2 + \binom{3}{3}b^3$.

---

[2]Recall that $0! = 1.$

[3]$\binom{n}{m}$ is read n *choose* m because it represents the number of ways m objects can be chosen or selected from a collection of n objects.

41. Using the assumption that

$$(a+b)^k$$

$$= \binom{k}{0}a^k + \binom{k}{1}a^{k-1}b + \binom{k}{2}a^{k-2}b^2$$

$$+ \cdots + \binom{k}{k-1}ab^{k-1} + \binom{k}{k}^k,$$

show that

$$(a+b)^{k+1}$$

$$= \binom{k+1}{0}a^{k+1} + \binom{k+1}{1}a^k b$$

$$+ \binom{k+1}{2}a^{k-1}b^2 + \cdots + \binom{k+1}{k}ab^k$$

$$+ \binom{k+1}{k+1}b^{k+1}.$$

You have shown by induction that $\binom{n}{n-m}$ is the coefficient for the term $a^{n-m}b^m$ in the expansion of the binomial $(a+b)^n$.

If a is a sequence and N is the initial integer in the domain of a, then we build a new sequence s called the sequence of partial sums of a by letting $s(1) = a(N),$ $s(2) = a(N)+a(N+1)$ and in general, $s(n) = a(N)+a(N+1)+a(N+2)+\cdots+a(N+n).$ In Exercises 42–44, find the first five terms of the sequence of partial sums of the sequence a.

42. $a(n) = \dfrac{1}{2^n}, \quad n \geq 0.$

43. $a(n) = (-1)^n, \quad n \geq 3.$

44. $a(n) = n!, \quad n \geq 0.$

2.2   Functions and Graphing Technology

Graphing technologies provide powerful visualization tools. Much of the graphics in this text is produced using Mathematica©. As you develop skill with your graphing calculator and/or computer software, you will be able to generate unlimited graphical examples. It takes some practice to learn how to use these tools effectively.

EXAMPLE 1: Let $f(x) = x^{20} - x$. We know from algebra that this function has roots at $x = 0$ and $x = 1$. However, if we plot the graph of f for $-4 \leq x \leq 4$, the plot will look like the sketch in Figure 1.a. It is hard to tell where the roots might be by looking at this plot. We do get a hint on how to restrict our attention. Clearly, any roots are between -2 and 2. We plot the graph for $-2 \leq x \leq 2$ in Figure 1.b. Here our picture is improving. We guess that we can restrict our attention to $-1.5 \leq x \leq 1.5$ and we obtain the plot in Figure 1.1.c. Finally, we plot the function over the interval $[-0.1, 0.1]$. We can keep "zooming in" on the graph to estimate the zeros of the function as closely as the accuracy of our calculator or computer allows.

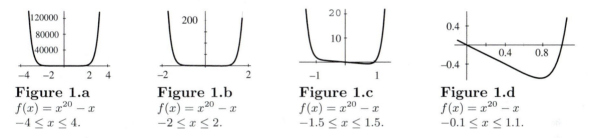

Figure 1.a   **Figure 1.b**   **Figure 1.c**   **Figure 1.d**
$f(x) = x^{20} - x$ $f(x) = x^{20} - x$ $f(x) = x^{20} - x$ $f(x) = x^{20} - x$
$-4 \leq x \leq 4.$ $-2 \leq x \leq 2.$ $-1.5 \leq x \leq 1.5.$ $-0.1 \leq x \leq 1.1.$

We can sometimes get a better start by restricting the range of the output. In Figure 2, we plot f over $[-4, 4]$ as we did in Figure 1.a, except that we restrict the range to $-0.1 \leq y \leq 0.1$. The result is a rather strange looking graph, which appears to be two nearly vertical lines. Those lines cross the x–axis at the roots of the function.

The process that we went through in our example is called *zooming in*. We can get closeup views by restricting the sizes of the domain and range of the function being plotted. Similarly, we say that we are *zooming out* if we make the plotted range and domain larger. ■

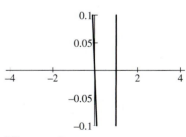

Figure 2. *That part of $f(x) = x^{20} - x$ that lies between the lines $y = -0.1$ and $y = 0.1$.*

A function f is said to be *increasing* on the interval $[a, b]$ if, as x moves from left to right on the interval, $f(x)$ goes uphill. Similarly, f is *decreasing* on the interval $[a, b]$ if, as x moves from left to right on the interval, $f(x)$ goes downhill.

EXAMPLE 2: The function illustrated in Figure 3 is increasing on the interval $[0, 1]$ and it is decreasing elsewhere. ■

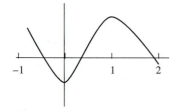

Figure 3. *A function that is increasing on* $[0,1]$ *and decreasing elsewhere.*

EXAMPLE 3: Let $f(x) = x + 2\sin(x)$. As x moves to the right from $x = 0$, find the first number b where f stops increasing and starts decreasing.

SOLUTION: Since the sine function goes through a complete cycle on the interval $[0, 2\pi]$, we first plot f over that interval. (See Figure 4.a.) We see from that plot that we can restrict our attention to the interval $[1.5, 3]$. This gives us the plot in Figure 4.b. In Figure 4.c, we restrict the plot to $[2, 2.3]$ where we see that $2.05 < b < 2.15$. We estimate $b \approx 2.1$. This calculation has an error of no more than 0.05. We can improve this estimate by continuing to zoom in on b. ∎

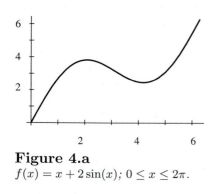

Figure 4.a
$f(x) = x + 2\sin(x);\ 0 \le x \le 2\pi.$

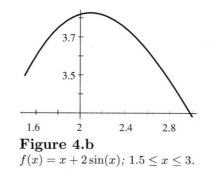

Figure 4.b
$f(x) = x + 2\sin(x);\ 1.5 \le x \le 3.$

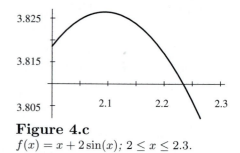

Figure 4.c
$f(x) = x + 2\sin(x);\ 2 \le x \le 2.3.$

⊞ EXERCISES 2.2

When using your calculator to evaluate trigonometric functions, be sure that it is in radian mode. Unless specifically stated otherwise, angles will be measured in radians.

In Exercises 1–9, plot the function over the given domain.

1. $\sin(x), \quad 0 \le x \le 2\pi.$

2. $\sin(x) + \cos(x), \quad -\dfrac{\pi}{2} \le x \le 2\pi.$

3. $\cos(x) - \dfrac{x}{3}, \quad 0 \le x \le 2\pi.$

4. $\cos(5x), \quad -\dfrac{\pi}{5} \le x \le \dfrac{2\pi}{5}.$

5. $\cos(10x) - \dfrac{x}{3}, \quad 0 \le x \le 2\pi.$

6. $\sqrt{x^2 + 1}, \quad 0 \le x \le 2.$

7. $|\cos(x)|, \quad 0 \le x \le 4\pi.$

8. $|\cos(x)| + x, \quad 0 \le x \le 6\pi.$

9. $|\cos(x)| + |x|, \quad 0 \le x \le 6\pi.$

10. Plot the following functions over the given domain.

   a. $\sin(x)\cos(10x), \quad 0 \le x \le 2\pi.$

   b. $\sin(x)\cos(30x), \quad 0 \le x \le 2\pi.$

   c. $\sin(x)\cos(100x), \quad 0 \le x \le 2\pi.$

While your calculator or computer can handle $\sin(x)\cos(nx)$ for small n, n need not be very large before even the best computer monitor fails to have enough pixels to draw a good graph. You can zoom in for a better local picture. However, you then lose the overall shape of the curve.

11. A common substitution used to simplify problems in engineering and physics is x for $\sin(x)$ "for small values of x." Plot the graph of $|\sin(x) - x|$ over the given interval and find an approximation for the maximum error on that interval.

   a. $-\dfrac{\pi}{2} \leq x \leq \dfrac{\pi}{2}$.   b. $-\dfrac{\pi}{4} \leq x \leq \dfrac{\pi}{4}$.

   c. $-\dfrac{\pi}{8} \leq x \leq \dfrac{\pi}{8}$.   d. $-\dfrac{\pi}{16} \leq x \leq \dfrac{\pi}{16}$.

12. $p(x) = x - \dfrac{x^3}{6}$ is also used to approximate $\sin(x)$ "for small values of x." Plot the graph of $|\sin(x) - p(x)|$ over the given interval and find an approximation for the maximum error on that interval. Is this approximation for $\sin(x)$ more accurate than using x as we did in Exercise 11?

   a. $-\dfrac{\pi}{2} \leq x \leq \dfrac{\pi}{2}$.   b. $-\dfrac{\pi}{4} \leq x \leq \dfrac{\pi}{4}$.

   c. $-\dfrac{\pi}{8} \leq x \leq \dfrac{\pi}{8}$.   d. $-\dfrac{\pi}{16} \leq x \leq \dfrac{\pi}{16}$.

In Exercises 13–15, plot the graph of $f(x)$ over the given interval. Zoom in to approximate the first point greater than x_0 where f stops increasing and starts decreasing with an error less than 0.01.

13. $f(x) = \dfrac{11}{14} - \dfrac{13}{7}x + \dfrac{1}{14}x^2 + x^3$,
    $-2 \leq x \leq 2$, $x_0 = -2$.

14. $f(x) = -\dfrac{55}{84}x + \dfrac{38}{21}x^2 + \dfrac{251}{84}x^3 + x^4$
    $-2 \leq x \leq 1$, $x_0 = -1$.

15. $f(x) = \sin(2x)\left(-\dfrac{55}{84} + \dfrac{38}{21}x + \dfrac{251}{84}x^2 + x^3\right)$
    $-2 \leq x \leq 1$, $x_0 = -2$.

In Exercises 16–18, approximate the first point greater than x_0, with an error less than 0.01, where f stops decreasing and starts increasing.

16. $-\dfrac{55}{84}x + \dfrac{38}{21}x^2 + \dfrac{251}{84}x^3 + x^4$ $x_0 = -2$.

17. $-\dfrac{55}{84}x + \dfrac{38}{21}x^2 + \dfrac{251}{84}x^3 + x^4$ $x_0 = -0.5$.

18. $\sin(2x)\left(-\dfrac{55}{84} + \dfrac{38}{21}x + \dfrac{251}{84}x^2 + x^3\right)$ $x_0 = -0.5$.

19. Plot $f(x) = \cos(x) + \dfrac{x}{3}$, $0 \leq x \leq 2\pi$.

   a. Zoom in on the appropriate intervals until you can estimate numbers a and b with an error of less than 0.001 such that $[a, b]$ is the interval over which $f(x)$ is negative.

   b. Approximate the numbers a and b with an error of less than 0.01 such that f is increasing on $[0, a]$, decreasing on $[a, b]$ and increasing on $[b, 2\pi]$.

20. Let $f(x) = \sin(x) + \dfrac{x^2}{6}$. Plot f over the interval $[0, \pi]$. Approximate the numbers a and b with an error of less than 0.01 such that $[a, b]$ is the interval over which f is decreasing.

21. Find the first integer n so that $n\sin(x) + x$ is not always increasing on $[0, \pi]$.

22. Find the first integer n so that $\sin(x) + \dfrac{x^2}{n}$ is not always increasing on $[0, \pi]$.

2.3   Functions from $\mathbb{R}$ into $\mathbb{R}^n$

So far, we have considered only functions where the domain and range are subsets of the real line, although there is nothing in the definition of a function that requires that this be the case. We want to consider what happens when the domain or range is a subset of $\mathbb{R}^n$ rather than a subset of $\mathbb{R}$. We will postpone the investigation of functions with domain in $\mathbb{R}^n$ until they become useful to us later. In this section, we study functions that are defined on subsets of the real line, but have their images in $\mathbb{R}^n$. We will refer to such functions as *vector-valued functions*. These functions are particularly useful when describing the motion of an object in space.

Suppose that a particle is moving in $\mathbb{R}^3$ as in Figure 1.a. If we

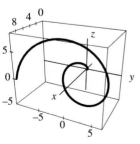

Figure 1.a
$\vec{r}(t) = (t, t\cos(t), t\sin(t))$.

let $\vec{r}(t)$ denote the position of the particle at time t, then we have defined a function from $\mathbb{R}$ into $\mathbb{R}^3$, where we are thinking of $\mathbb{R}$ as a time line. If $\mathbb{R}^3$ is xyz–space, then at each time t, the position of the particle $\vec{r}(t)$ has an x–coordinate, a y–coordinate, and a z–coordinate, denoted by $x(t)$, $y(t)$ and by $z(t)$ respectively. At time t we have: $\vec{r}(t) = (x(t), y(t), z(t))$.

In a similar fashion, if the motion of the particle is restricted to $\mathbb{R}^2$, and if $\vec{r}(t)$ denotes the position of the particle at time t, then $\vec{r}$ defines a function from $\mathbb{R}$ into $\mathbb{R}^2$. If we denote the coordinates of $\mathbb{R}^2$ by x and y, at time t we have $\vec{r}(t) = (x(t), y(t))$. (See Figure 1.b.)

There are two basic types of problems that we model with vector-valued functions:

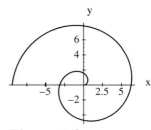

Figure 1.b
$\vec{r}(t) = (t\cos(t), t\sin(t))$.

The first type of problem involves using a function from a subset of $\mathbb{R}$ into $\mathbb{R}^n$ as a description of a particle moving in $\mathbb{R}^n$. In this type of problem, we might interpret $\mathbb{R}$ as the time line and $\mathbb{R}^n$ is the space in which the particle is moving. Thus, if A is a subset of $\mathbb{R}$ and $\vec{r}\colon A \to \mathbb{R}^n$, then $\vec{r}(t)$ denotes the position of the particle at time t, and the image of $\vec{r}$ is the path of motion. This concept is consistent with notions such as velocity, acceleration, force and momentum. In this book the domain in such problems is most often an interval of time.

The second type of problem uses a vector-valued function to describe how a straight piece of wire might be bent and stretched out of shape. In this application the domain of the function is the position of the original piece of wire, and the image of the function is the position of the final bent wire. A point t in the domain represents the position of a cross-section in the original wire and $\vec{r}(t)$ is the position of the same cross-section after the wire is bent. We will not encounter this application until Chapter 11, where we begin to develop tools to calculate quantities such as the energy of an object moving in space.

If $\vec{r}$ is a function from $\mathbb{R}$ into $\mathbb{R}^3$ defined by

$$\vec{r}(t) = \begin{pmatrix} x(t) \\ y(t) \\ z(t) \end{pmatrix},$$

then x, y and z are functions from $\mathbb{R}$ into $\mathbb{R}$, called *coordinate functions*. We refer to $\vec{r}$ as a *parametric representation*, or a *parametrization*, and refer to t as the *parameter*.

EXAMPLE 1: Let $\vec{r}(t) = (t, \cos(t), \sin(t))$. Then

$x(t) = t$ is the x–coordinate function.

$y(t) = \cos(t)$ is the y–coordinate function.

$z(t) = \sin(t)$ is the z–coordinate function.   ■

The first functions from $\mathbb{R}$ into $\mathbb{R}^n$ that we consider are parametric representations for lines in $\mathbb{R}^n$.

Parametrizations of Lines

Let (a, b) be a vector in the plane, and let (x_0, y_0) be a position in the plane. If we define the function $\vec{r}$ from $\mathbb{R}$ into $\mathbb{R}^2$ by

$$\vec{r}(t) = (x_0 + ta, y_0 + tb) = (x_0, y_0) + t(a, b),$$

then the image of $\vec{r}$ is the set of all points (x, y) satisfying

$$x = x_0 + ta, \quad \text{and} \tag{1}$$

$$y = y_0 + tb. \tag{2}$$

If $a \neq 0$, then we can solve Equation (1) for t and substitute the result into Equation (2) to obtain:

$$y = y_0 + (x - x_0)\left(\frac{b}{a}\right)$$

or

$$y - y_0 = \left(\frac{b}{a}\right)(x - x_0).$$

The image of $\vec{r}$ is a line with slope $\frac{b}{a}$ and contains the point (x_0, y_0).

Geometrically, for a given time t, the vector $\vec{r}(t) = (x_0, y_0) + t(a, b)$ is a point in the plane obtained by adding a multiple of (a, b) to (x_0, y_0). Note that (a, b) is a vector that points in the direction of any line in the plane with slope (b/a). (What happens if $a = 0$?) We may generalize this idea to $\mathbb{R}^n$. If $\vec{r}_0$ is a position in $\mathbb{R}^n$ and $\vec{d}$ is a nonzero vector in $\mathbb{R}^n$, then

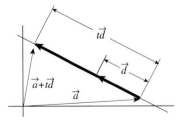

Figure 2. *Parametrization of a straight line.*

$$\vec{r}(t) = \vec{r}_0 + t\vec{d}$$

is a parametrization for a straight line, which contains the point $\vec{r}_0$ and has direction $\vec{d}$. The vector $\vec{d}$ is called a *direction vector* for the line. See Figure 2.

EXAMPLE 2: Find a function $\vec{r}(t)$ from $\mathbb{R}$ into $\mathbb{R}^3$ so that the image of $\vec{r}(t)$ contains $(1, -1, 2)$ and has $(0, -1, 2)$ as a direction vector.

SOLUTION: We simply apply the equation $\vec{r}(t) = \vec{r}_0 + t\vec{d}$, where $\vec{r}_0 = (1, -1, 2)$ and $\vec{d} = (0, -1, 2)$:

$$\vec{r}(t) = \begin{pmatrix} 1 \\ -1 \\ 2 \end{pmatrix} + t \begin{pmatrix} 0 \\ -1 \\ 2 \end{pmatrix} = \begin{pmatrix} 1 \\ -1 - t \\ 2 + 2t \end{pmatrix}. \qquad \blacksquare$$

EXAMPLE 3: Find a function $\vec{r}(t)$ from $\mathbb{R}$ into $\mathbb{R}^3$ so that $\vec{r}(t)$ is a parametrization of the line that contains $\vec{p} = (1, 1, 2)$ and $\vec{q} = (-1, 2, 1)$.

SOLUTION: We parametrize the line by

$$\vec{r}(t) = \vec{r}_0 + t\vec{d}. \tag{3}$$

With $\vec{r}$ as the position of a particle moving in space, because the timing of the motion is not specified in the problem, the choice is ours. We assume that the particle is at $\vec{p} = (1, 1, 2)$ at time $t = 0$ and at $\vec{q} = (-1, 2, 1)$ at time $t = 1$. Using the values in Equation (3) gives us two equations with two unknowns, $\vec{d}$ and $\vec{r}_0$. We have:

$$\vec{r}(0) = (1, 1, 2), \text{ which gives } (1, 1, 2) = \vec{r}_0$$

and

$$\vec{r}(1) = (-1, 2, 1), \text{ which gives } (-1, 2, 1) = \vec{d} + \vec{r}_0 = \vec{d} + (1, 1, 2).$$

Thus

$$\vec{d} = (-1, 2, 1) - (1, 1, 2) = (-2, 1, -1)$$

so that

$$\vec{r}(t) = \begin{pmatrix} -2 \\ 1 \\ -1 \end{pmatrix} t + \begin{pmatrix} 1 \\ 1 \\ 2 \end{pmatrix} = \begin{pmatrix} -2t + 1 \\ t + 1 \\ -t + 2 \end{pmatrix}$$

is a parametrization for the line. $\qquad \blacksquare$

EXAMPLE 4: Let $\vec{p}$ and $\vec{q}$ be as in the previous example. Find a function $\vec{r}(t)$ from $\mathbb{R}$ into $\mathbb{R}^3$ so that $\vec{r}(t)$ is a parametrization of the line such that $\vec{r}(1) = \vec{p}$ and $\vec{r}(2) = \vec{q}$.

SOLUTION: Let $\vec{r}(t)$ be of the form $\vec{r}(t) = \vec{r}_0 + t\vec{d}$. Evaluating $\vec{r}(t)$ at both $t = 1$ and $t = 2$ gives us two equations with two unknowns

$\vec{d}$ and $\vec{r}_0$:

$$\vec{r}(1) = \vec{r}_0 + \vec{d} = \begin{pmatrix} 1 \\ 1 \\ 2 \end{pmatrix} \quad \text{and} \quad \vec{r}(2) = \vec{r}_0 + 2\vec{d} = \begin{pmatrix} -1 \\ 2 \\ 1 \end{pmatrix}.$$

We solve this system of equations to obtain $\vec{d} = (-2, 1, -1)$ and $\vec{r}_0 = (3, 0, 3)$. Thus

$$\vec{r}(t) = \begin{pmatrix} 3 - 2t \\ t \\ 3 - t \end{pmatrix}$$

is the desired parametrization for the line.   ■

It is common knowledge that if two planes intersect and they are not coplaner, then their intersection is a line.

EXAMPLE 5: Find a parametrization for the common part of the planes

$$2x + 3y + z \;\; = \;\; 1 \qquad\qquad (4)$$

$$x + y - z \;\; = \;\; 2. \qquad\qquad (5)$$

SOLUTION: We use Equations (4) and (5) to write two of the variables in terms of the third. We eliminate z by adding Equations (4) and (5) to obtain $3x + 4y = 3$ or $x = 1 - 4y/3$. Subtracting Equation (5) from Equation (4) twice yields $z = -1 - y/3$. We have x and z in terms of y and the function

$$\vec{r}(y) = \begin{pmatrix} 1 - \frac{4y}{3} \\ y \\ -1 - \frac{y}{3} \end{pmatrix}$$

parametrizes our line.   ■

Definition: Parallel, Perpendicular and Skew Lines

Suppose that L_1 and L_2 are lines with direction vectors $\vec{d}_1$ and $\vec{d}_2$, respectively. Then:

(a) L_1 and L_2 are *parallel* if and only if there is a number t so that

$$\vec{d}_1 = t\vec{d}_2 \qquad (\vec{d}_1 \text{ is a multiple of } \vec{d}_2).$$

(b) L_1 and L_2 are *perpendicular* or *orthogonal* if and only if:

(i) $\vec{d}_1 \cdot \vec{d}_2 = 0$ ($\vec{d}_1$ and $\vec{d}_2$ are perpendicular); and

(ii) L_1 and L_2 intersect.

(c) L_1 and L_2 are *skew* if and only if:

　　(i) L_1 and L_2 do not intersect; and

　　(ii) L_1 and L_2 are not parallel.

Geometrically, two lines are skew if they do not lie in a common plane.

EXAMPLE 6: Let L_1 be parametrized by

$$\vec{r}_1(t) = \begin{pmatrix} t \\ 1 + 3t \\ -2 + 6t \end{pmatrix}.$$

Then $(0, 1, -2)$ is a point of L_1, and $(1, 3, 6)$ is a direction vector for L_1. The line L_2, parametrized by

$$\vec{r}_2(t) = \begin{pmatrix} t \\ 3t \\ 6t \end{pmatrix},$$

is parallel to L_1, but contains the origin. The vector $\vec{d} = (1, 3, 6)$ is a direction vector for both lines L_1 and L_2. Let $\vec{d}_3 = (-3, -1, 1)$. Since

$$\vec{d} \cdot \vec{d}_3 = 0,$$

$\vec{d}$ and $\vec{d}_3$ are orthogonal. If we let L_3 be the line parametrized by

$$\vec{r}_3(t) = \begin{pmatrix} -3t \\ -t \\ t \end{pmatrix},$$

then L_2 and L_3 are perpendicular. We suspect that L_1 and L_3 are skew. They are clearly not parallel. To see that they do not intersect, we must show that if t_1 and t_2 are real numbers, then $\vec{r}_1(t_1) \neq \vec{r}_3(t_2)$. Notice that we must show this for all numbers t_1 and t_2. (Why can't we just show that $\vec{r}_1(t) \neq \vec{r}_2(t)$ for all t?) We proceed by finding numbers t_1 and t_2 such that

$$x_1(t_1) = x_3(t_2) \quad \text{and} \quad y_1(t_1) = y_3(t_2) \tag{6}$$

and then show that

$$z_1(t_1) \neq z_3(t_2).$$

To solve Equation (6), we proceed as follows:

$$x_1(t_1) = t_1 \quad \text{and} \quad x_3(t_2) = -3t_2$$

$$y_1(t_1) = 1 + 3t_1 \quad \text{and} \quad y_3(t_2) = -t_2.$$

Setting $x_1(t_1) = x_3(t_2)$ and $y_1(t_1) = y_3(t_2)$, we obtain two equations in two unknowns:

$$t_1 = -3t_2; \quad \text{and} \tag{7}$$

$$1 + 3t_1 = -t_2. \tag{8}$$

Substituting Equation (7) in Equation (8) for t_1 we obtain:

$$1 + 3(-3t_2) = -t_2, \quad \text{or}$$

$$t_2 = \frac{1}{8} \quad \text{and} \quad t_1 = -\frac{3}{8}.$$

This yields:

$$z_1(t_1) = -2 + 6t_1 = -2 - \frac{9}{4} = -\frac{17}{4}, \quad \text{and}$$

$$z_3(t_2) = t_2 = \frac{1}{8}.$$

Since $z_1(t_1) \neq z_3(t_2)$ we have shown that the lines L_1 and L_3 do not intersect. ∎

We can visualize two intersecting lines as two intersecting highways, one running east to west and the other south to north. A truck's position on one highway is given by $\vec{r}_1(t)$ and a car's position on the other highway is given by $\vec{r}_2(t)$. The truck crosses the intersection at time t_1 while the car crosses at time t_2. Both drivers would hope that $t_1 \neq t_2$.

Torque and the Line of Action of a Force

Suppose that $\vec{F}$ is a force acting on a mass at position $\vec{A}$. Recall from Section 1.5 that the torque about the position $\vec{p}_0$ induced by $\vec{F}$ is given by $(\vec{A} - \vec{p}_0) \times \vec{F}$. Let $\mathcal{L}$ be the line parametrized by $\vec{r}(\tau) = \vec{F}\tau + \vec{A}$ as illustrated in Figure 3. Then $\mathcal{L}$ is called the *line of action of the force $\vec{F}$*.

Now, suppose that $\vec{B} = \tau_0 \vec{F} + \vec{A}$ is an arbitrary point on $\mathcal{L}$. The torque about $\vec{p}_0$ induced by $\vec{F}$ acting at point $\vec{B}$ is given by $(\vec{B} - \vec{p}_0) \times \vec{F}$. Now,

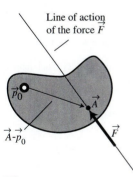

Line of action of the force $\vec{F}$

Figure 3. *The line of action of the force $\vec{F}$ acting on a mass at position $\vec{A}$.*

$$\begin{aligned}
(\vec{B} - \vec{p}_0) \times \vec{F} &= \left((\tau_0 \vec{F} + \vec{A}) - \vec{p}_0 \right) \times \vec{F} \\
&= \tau_0 \vec{F} \times \vec{F} + (\vec{A} - \vec{p}_0) \times \vec{F} \\
&= (\vec{A} - \vec{p}_0) \times \vec{F} \ (\text{since } \vec{F} \times \vec{F} = \vec{0}.)
\end{aligned}$$

We have established a fact that is very useful in mechanics.

> **Theorem 1** *If $\vec{p_0}$ is a point, if the force $\vec{F}$ is a direction vector for the line $\mathcal{L}$, and if $\vec{A}$ and $\vec{B}$ are any two points on $\mathcal{L}$, then $(\vec{B} - \vec{p_0}) \times \vec{F} = (\vec{A} - \vec{p_0}) \times \vec{F}$. That is, the force can be applied at any point on its line of action without changing the torque.*

EXERCISES 2.3

In Exercises 1–6, find a function $\vec{r}(t)$ from $\mathbb{R}$ into $\mathbb{R}^3$ that parametrizes the line satisfying the given conditions.

1. Contains $(1, 2, -1)$ and has direction vector $(1, -2, 3)$.

2. Contains $(1, 2, -1)$ and has direction vector $(1, 1, 1)$.

3. Contains $(0, 0, 0)$ and is parallel to the line of Exercise 1.

4. Contains $(1, 1, 1)$ and $(0, 2, 1)$.

5. Contains $(0, 0, 0)$ and is parallel to the line of Exercise 2.

6. Contains $(0, 0, 0)$ and is parallel to the line of Exercise 4.

In Exercises 7–10, find a function $\vec{r}(t)$ from $\mathbb{R}$ into $\mathbb{R}^2$ so that the image of $\vec{r}(t)$ is the line L (in the plane) satisfying the given conditions.

7. L contains $(1, 2)$ and has direction vector $(0, 1)$.

8. L contains $(1, 3)$ and $(-1, 2)$.

9. L contains $(1, 5)$ and is parallel to the line $y = 2x + 6$.

10. L contains $(1, 5)$ and is perpendicular to the line $y = 2x + 6$.

In Exercise 11–13, parametrize the line of action of the force $\vec{F}$ acting on a mass at $\vec{r_0}$.

11. $\vec{F} = (2, 1, -3)$N and $\vec{r_0} = (1, -1, 3)$m.

12. $\vec{F} = (0, -6, 1)$N and $\vec{r_0} = (1, 0, -2)$m.

13. $\vec{F} = (-4, 1, 2)$N and $\vec{r_0} = (-5, -1, 2)$m.

14. Identify the functions that have the same image as $\vec{r}(t) = (2t + 3, 4t - 1, 8t - 6)$.

    a. $\vec{r_1}(t) = (t + 3, 2t - 1, 4t - 6)$

    b. $\vec{r_2}(t) = (2t - 3, 4t + 1, 8t + 6)$

    c. $\vec{r_3}(t) = (t + 4, 2t + 1, 4t - 2)$

In Exercises 15–19, find a direction vector for the line determined by the given function.

15. $\vec{r}(t) = (2t, 3t)$.

16. $\vec{r}(t) = (1, 2 + 3t)$.

17. $\vec{r}(t) = (2t, 1 + 3t, -t)$.

18. $\vec{r}(t) = (2 + 3t, 1 - 6t, 24 + 5t)$.

19. $\vec{r}(t) = (\pi t, 1 - t, \sqrt{2} + t)$.

In Exercises 20–23, determine whether the given planes intersect in a line. If the intersection is a line, find a parametrization for the line.

20. $2x - y + z = 1, \ x + y + z = 3$.

21. $x - y - z = 1, \ x + y + 3z = 0$.

22. $2x - y + z = 1, \ 4x - 2y + 2z = 3$.

23. $x - y + z = 0, \ x + y + z = 2$.

In Exercises 24–27, L_1 is the line parametrized by $\vec{r_1}(t) = (3t, 1 + 6t, 3 - t)$. Determine whether L_1 intersects the line L satisfying the given conditions.

24. L contains the points $(0, 0, 0)$ and $(0, 1, 3)$.

25. L is parametrized by $\vec{r}(t) = (6t, 1 - 3t, 3 + 6t)$.

26. L is parametrized by $\vec{r}(t) = (3 + t, 7 - 6t, 2 + t)$.

27. L is parametrized by $\vec{r}(t) = (6t, 1 + 5t, 3 - t)$.

As above, L_1 is the line parametrized by $\vec{r}_1(t) = (3t, 1+6t, 3-t)$. In Exercises 28–33, determine whether L_1 is perpendicular to the line L satisfying the given conditions.

28. L has direction vector $(-1, 1, 3)$ and contains the point $(0, 1, 3)$.

29. L has direction vector $(-1, 1, 3)$ and contains the point $(0, 0, 0)$.

30. L has direction vector $(1, -1, -3)$ and contains the point $(0, 1, 3)$.

31. L is parametrized by $\vec{r}(t) = (-t, 1, 3+3t)$.

32. L is parametrized by $\vec{r}(t) = (t, 1, 3-3t)$.

33. L is parametrized by $\vec{r}(t) = (t, 6, 3-3t)$.

34. Let $\mathcal{L}$ be the line of action of the force $\vec{F} = (1, 2, 3)$ Newtons acting at position $\vec{A} = (-1, 1, 1)$ meters.

   a. Show that the points $\vec{B} = (0, 3, 4)$ meters and $\vec{C} = (-2, -1, -2)$ meters are on $\mathcal{L}$.

   b. Verify Theorem 1 by showing that the torque acting on a mass about the origin generated by $\vec{F}$ is the same vector if it acts at each of the positions $\vec{A}$, $\vec{B}$ or $\vec{C}$.

35. Suppose that L is the line parametrized by $\vec{r}(t) = t(a, b, c)$. Let $\vec{x}$ and $\vec{y}$ be points in L, and let u and v be real numbers. Show that $u\vec{x} + v\vec{y}$ is in L.

2.4   The Wrapping Function and Other Functions

In the previous section, our consideration of functions from $\mathbb{R}$ into $\mathbb{R}^n$ was restricted to those that describe straight lines. We continue our introduction of functions from $\mathbb{R}$ into $\mathbb{R}^n$ with some more examples.

EXAMPLE 1: (The Wrapping Function) Suppose that a particle is rotating about the origin in a circular path of radius 1 unit at a constant speed of 1 rotation every 2π seconds in a counterclockwise direction. Note that since the circumference is 2π, our particle travels 2π units in 2π seconds, and so has speed 1 unit/second. We assume that the particle is at $(1, 0)$ at time $t = 0$. Let $\vec{W}(t)$ denote the position of the particle at time t, and let θ denote the angle that the ray drawn emanating from the origin and containing $\vec{W}(t)$ makes with the x–axis. Since the particle is moving with speed 1 unit/second, $\theta = t$ is the radian measure of θ. Thus, the x–coordinate at time t, which is $x(t)$, is $\cos(t)$ and the y–coordinate at time t, which is $y(t)$, is $\sin(t)$. See Figure 1.

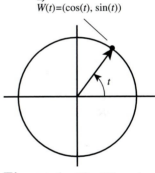

$\vec{W}(t) = (\cos(t), \sin(t))$

Figure 1. *The Wrapping Function.*

Thus we see that $\vec{W}(t) = (\cos t, \sin t)$ is a function from $\mathbb{R}$ into $\mathbb{R}^2$, which has as its image the unit circle. The function $\vec{W}(t)$ is called the *wrapping function*. The wrapping function will be of recurring interest to us since it is a fundamental building block in the modeling of circular motion. ∎

EXAMPLE 2: We can parametrize ellipses using a variant of the wrapping function. The function

$$E(t) = \begin{pmatrix} A\cos(t) \\ B\sin(t) \end{pmatrix}$$

parametrizes the ellipse $x^2/A^2 + y^2/B^2 = 1$. To see this, notice that (x, y) is in E if and only if there is some t so that $x = A\cos(t)$ and $y = B\sin(t)$. If this is the case, then the point (x, y) is also on the specified ellipse, since

$$\frac{x^2}{A^2} + \frac{y^2}{B^2} = \frac{A^2\cos^2(t)}{A^2} + \frac{B^2\sin^2(t)}{B^2} = 1$$

as desired. In Figures 2.a and 2.b, we assume that $0 \le t \le 2\pi$ to illustrate the geometry of our parametrization. ∎

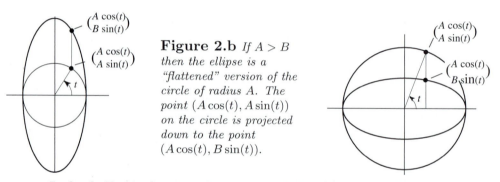

Figure 2.a *If $A < B$ then the ellipse is a "stretched" version of the circle of radius A. The point $(A\cos(t), A\sin(t))$ on the circle is projected up to the point $(A\cos(t), B\sin(t))$.*

Figure 2.b *If $A > B$ then the ellipse is a "flattened" version of the circle of radius A. The point $(A\cos(t), A\sin(t))$ on the circle is projected down to the point $(A\cos(t), B\sin(t))$.*

In both Examples 1 and 2, we used the trigonometric identity $\sin^2 t + \cos^2 t = 1$ to parametrize an ellipse and a circle. The identity $\sec^2(t) - \tan^2(t) = 1$ can parametrize hyperbolas with equations of the form $\dfrac{x^2}{a^2} - \dfrac{y^2}{b^2} = 1$ by letting $x = a\sec(t)$ and $y = b\tan(t)$.

EXAMPLE 3: Parametrize the hyperbolas $\mathcal{H}_1$ with equation $x^2 - \frac{y^2}{4} = 1$ and $\mathcal{H}_2$ with equation $\frac{y^2}{4} - x^2 = 1$.

SOLUTION: Because both graphs are in two pieces, it takes two functions to parametrize each of them completely. $\mathcal{H}_1$ opens to the right and to the left while $\mathcal{H}_2$ opens up and down. See Figure 3.

For $\mathcal{H}_1$, let $x(t) = \sec(t)$ and $y(t) = 2\tan(t)$. Let $\vec{r}_1(t) = (x(t), y(t))$, $-\frac{\pi}{2} < t < \frac{\pi}{2}$ and let $\vec{r}_2(t) = (x(t), y(t))$, $\frac{\pi}{2} < t < \frac{3\pi}{2}$. The function $\vec{r}_1$ parametrizes the "right half" of the hyperbola while $\vec{r}_2$ parametrizes the "left half."

For $\mathcal{H}_2$, let $x(t) = \tan(t)$ and $y(t) = 2\sec(t)$. Let $\vec{s}_1(t) = (x(t), y(t))$, $-\frac{\pi}{2} < t < \frac{\pi}{2}$ and let $\vec{s}_2(t) = (x(t), y(t))$, $\frac{\pi}{2} < t < \frac{3\pi}{2}$. The function $\vec{s}_1$ parametrizes the "top half" of the hyperbola while

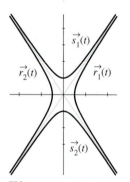

Figure 3. *The hyperbolas $\mathcal{H}_1$ and $\mathcal{H}_2$. $\mathcal{H}_1$ is parametrized with $\vec{r}_1$ and $\vec{r}_2$. $\mathcal{H}_2$ is parametrized with $\vec{s}_1$ and $\vec{s}_2$.*

$\vec{s}_2$ parametrizes the "bottom half."   ■

Parametrizing Graphs of Functions

When we parametrize a curve, we are focusing on the image of
the function rather than its graph. Recall that the image is the set of
all second coordinates (or outputs) of the function, while the graph
is the set of all ordered pairs in the function.

EXAMPLE 4: Consider $y = x^2$. This describes a function, and
the image of that function is the set of nonnegative real numbers.
$y(x) = x^2$ is a parametrization of the set of nonnegative real numbers.
The graph of $y(x) = x^2$ is a parabola.   ■

EXAMPLE 5: The graph of $\vec{r}(t) = (t\cos(t), t\sin(t))$ is the set
of all points in $\mathbb{R}^3$ of the form $(t, \vec{r}(t)) = (t, t\cos(t), t\sin(t))$, while
its image is the set of all ordered pairs of the form $(t\cos(t), t\sin(t))$.
Thus, the image of the function is in $\mathbb{R}^2$, while its graph is in $\mathbb{R}^3$.
The image of $\vec{r}$ is the set parametrized by $\vec{r}$. See Figures 3.a and 3.b.   ■

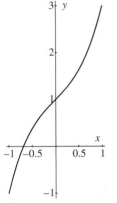

Figure 4.a *The graph of*
$f(x) = x^3 - x + 1$ *and the*
image of $\vec{h}(t) = (t, f(t))$.

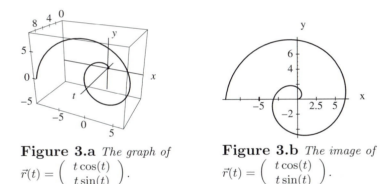

Figure 3.a *The graph of*
$\vec{r}(t) = \begin{pmatrix} t\cos(t) \\ t\sin(t) \end{pmatrix}$.

Figure 3.b *The image of*
$\vec{r}(t) = \begin{pmatrix} t\cos(t) \\ t\sin(t) \end{pmatrix}$.

If the domain of a function f is a subset of the reals, then $\vec{f}(t)$ is
a parametrization of the *image* of f. We can parametrize the graph
of a function by $\vec{h}(t) = (t, f(t))$.

EXAMPLE 6: $\vec{h}(t) = (t, t^2)$ is a parametrization of the parabola
of Example 4, and $\vec{h}(t) = (t, t\cos(t), t\sin(t))$ is a parametrization of
the graph of the function in Example 5.   ■

EXAMPLE 7: Figure 4.a provides a sketch of the graph of $f(x) =$
$x^3 - x + 1$. A parametrization for the graph of f is $\vec{h}(t) = (t, f(t))$.
The image of $\vec{h}$ is the graph of f. A parametrization for the *image* of
$\vec{h}$ is $\vec{r}(t) = (t, t, f(t))$. The image of $\vec{r}$, illustrated in Figure 4.b, is the
graph of $\vec{h}$. Now, the *graph* of $\vec{r}$ is parametrized by $\vec{s}(t) = (t, t, t, f(t))$.
Notice that it takes four variables to describe $\vec{r}$ and its image is not
easy to visualize.   ■

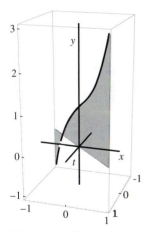

Figure 4.b *The graph of*
$\vec{h}(t) = (t, f(t))$ *and the image*
of $\vec{r}(t) = (t, t, f(t))$.

EXAMPLE 8: Parametrize the graph $\mathcal{G}$ of the equation $x^2 + 2x = 3y^3 + y^2 + y$.

SOLUTION: We start by solving for x in terms of y. Proceed by completing the square in x.

$$
\begin{aligned}
x^2 + 2x &= 3y^3 + y^2 + y \\
x^2 + 2x + 1 &= 3y^3 + y^2 + y + 1 \\
(x+1)^2 &= 3y^3 + y^2 + y + 1 \\
(x+1) &= \pm\sqrt{3y^3 + y^2 + y + 1}
\end{aligned}
$$

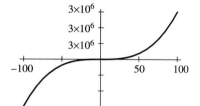

Figure 5.a *The graph of g over* $-100 \le y \le 100$.

We get x as two functions of y:

$$
\begin{aligned}
x &= f_1(y) = \sqrt{3y^3 + y^2 + y + 1} - 1; \text{ and} \\
x &= f_2(y) = -\sqrt{3y^3 + y^2 + y + 1} - 1.
\end{aligned}
$$

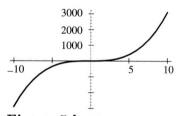

Figure 5.b *The graph of g over* $-10 \le y \le 10$.

The union of the graphs of f_1 and f_2 is $\mathcal{G}$. The domain of both of these functions is the set of all y such that $3y^3 + y^2 + y + 1 \ge 0$. Factoring doesn't seem to be easy; so we solve the inequality graphically. Let $g(y) = 3y^3 + y^2 + y + 1$. Figure 5.a sketches g over the interval $[-100, 100]$.

We zoom in on the origin to find the roots of g by restricting the domain of the graph to $[-10, 10]$ as illustrated in Figure 5.b. From there, we zoom in on the interval $[-3, 3]$ (Figure 5.c). Continuing this process, we determine that $g(y) \ge 0$ for $y \ge a$, where $a \approx -0.635$. The domain of f_1 and f_2 is the set $y \ge a$, where $a \approx -0.635$. With the tools we have, we need two functions to parametrize $\mathcal{G}$: $\vec{r}_1(t) = (f_1(t), t)$ and $\vec{r}_2(t) = (f_2(t), t)$. Note that $\vec{r}_1(a) = \vec{r}_2(a)$, so that even though we used two functions to parametrize $\mathcal{G}$, the graph $\mathcal{G}$ is connected. The image of $\vec{r}_1$ is that part of $\mathcal{G}$ to the right of $\vec{r}_1(a)$, and the image of $\vec{r}_2$ is that part of $\mathcal{G}$ to the left of $\vec{r}_1(a)$ as illustrated in Figure 6. It is noteworthy that although we obtained our parametrizations for $\mathcal{G}$ by writing x as two functions of y, an inspection of our graph in Figure 6 shows us that $\mathcal{G}$ is the graph of a function of x. Since we know that $\mathcal{G}$ is a function of x, it is not easy to write down that function. ∎

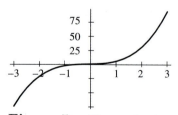

Figure 5.c *The graph of g over* $-3 \le y \le 3$.

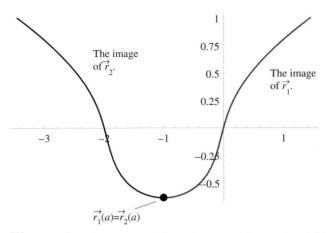

Figure 6. *The image of $\vec{r}_1$ is that part of $\mathcal{G}$ to the right of $\vec{r}_1(a)$ and the image of $\vec{r}_2$ is that part of $\mathcal{G}$ to the left of $\vec{r}_1(a)$.*

We refer to Example 7 when using calculators or computers to graph functions where x is a function of y. If you want technology to graph $x = f(y)$ and the coordinate axes positioned in the usual way with the x–axis horizontal and the y–axis vertical, then you can parametrize the graph of f by $\vec{r}(t) = (f(t), t)$ and draw the image of $\vec{r}$.

EXERCISES 2.4

1. Let $\vec{W}$ be the wrapping function. Find the coordinates of the following points:

   a. $\vec{W}(0)$.   b. $\vec{W}\left(\dfrac{\pi}{3}\right)$.

   c. $\vec{W}\left(\dfrac{\pi}{6}\right)$.   d. $\vec{W}\left(-\dfrac{3\pi}{4}\right)$.

   e. $\vec{W}\left(\dfrac{\pi}{2}\right)$.   f. $\vec{W}\left(\dfrac{100\pi}{3}\right)$.

2. Let $\vec{r}$ be defined by $x(t) = 3\sin(\pi t), y(t) = \cos(2\pi t)$. Compute the coordinates of the following points:

   a. $\vec{r}\left(\dfrac{1}{2}\right)$.   b. $\vec{r}(2)$.

   c. $\vec{r}\left(\dfrac{3}{4}\right)$.   d. $\vec{r}\left(\dfrac{20}{3}\right)$.

3. Let $\vec{f}$ be defined by $x(t) = 3t^2, y(t) = 4\sin(\pi t/6)$, and $z(t) = 2\cos(\pi t)$. Find the coordinates of the following points:

   a. $\vec{f}(0)$.   b. $\vec{f}(1)$.   c. $\vec{f}(3)$.   d. $\vec{f}(30)$.

4. Sketch the graphs of $x = cy^2$ for $c = 1, 2$, and 3 on the same coordinate system.

5. Sketch the graphs of $x + y^2 = c$ for $c = 1, 2$, and 3 on the same coordinate system.

In Exercises 6–16, parametrize the graph of the following equations and sketch the graph.

6. $x^2 + y^2 = 9$.

7. $x^2 + 9y^2 = 1$.

8. $4x^2 + 16y^2 = 16$.

9. $4x^2 + 9y^2 = 36$.

10. $4x^2 - 16y^2 = 1$.

11. $4y^2 - 9x^2 = 36$

12. $4x^2 - 16y^2 = 0$. Hint: Use two parametrizing functions.

13. $4y^2 - 9x^2 = 0$,

14. $4x + 16y^2 - y = 16$.

15. $3x - y^2 + y + 6 = 0$,

16. $4x^2 + y^2 - 4y = 0$. Hint: Complete the square in y.

17. Without the aid of a calculator, match the parametrization with the figure sketching its image.

    a. $\vec{f}(t) = (3t + 2, 6t - 5)$.

    b. $\vec{h}(t) = (2\cos 3t, 2\sin 3t)$.

    c. $\vec{h}(t) = (2\cos t, 4\sin t)$.

    d. $\vec{r}(t) = (4\cos 2t, 2\sin 2t)$.

    e. $\vec{r}(t) = (t, 4t^2)$.

    f. $\vec{r}(t) = (4t^2, t)$.

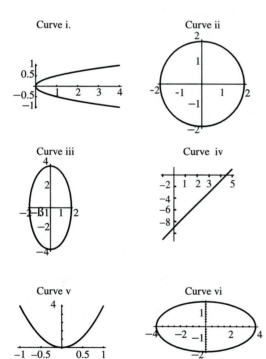

Figure 7. *Curves for Exercise 17.*

Without the use of a calculator, sketch the images of the functions in Exercises 18–22.

18. $\vec{h}(s) = (s, s^2 + 4s + 4)$, $-4 \le s \le 2$.

19. $\vec{v}(r) = (r^2 + 4r + 4, r)$, $-4 \le r \le 2$.

20. $\vec{s}(t) = (t, \sin t)$, $-2\pi \le t \le 2\pi$.

21. $\vec{s}(t) = (\sin t, t)$, $-2\pi \le t \le 2\pi$.

22. $\vec{r}(t) = (t, -t)$, $-2 \le t \le 2\pi$.

23. Parametrize the graphs of $x^2 + 4y^2 = c$ for $c = 1, 4$, and 16 and sketch the graphs on the same coordinate axis.

In Exercises 24 and 25, draw the Graphs of $\mathcal{H}_1$, $\mathcal{H}_2$ and $\mathcal{L}$ on the same coordinate axes.

24. $\mathcal{H}_1$ is the graph of $x^2 - 2y^2 = 1$, $\mathcal{H}_2$ is the graph of $2y^2 - x^2 = 1$, and $\mathcal{L}$ is the graph of $2y^2 - x^2 = 0$.

25. $\mathcal{H}_1$ is the graph of $4x^2 - y^2 = 16$, $\mathcal{H}_2$ is the graph of $y^2 - 4x^2 = 16$, and $\mathcal{L}$ is the graph of $y^2 - 4x^2 = 0$.

26. Draw the graph of $4x^2 - y^2 = c_i$ for $c_i = -4$, $-2, 0, 2$, and 4 on the same axes.

2.5   Sketching Parametrized Curves

In the previous sections, most of the curves we sketched were familiar objects such as circles, lines, ellipses, and hyperbolas. In many cases, we must look at the coordinate functions themselves to learn about the parametrized curve.

EXAMPLE 1:   The image of $\vec{r}(t)$ defined by $x(t) = [[t]]$ and $y(t) = t$ is sketched in Figure 1. The procedure in sketching was to look at places where the coordinate functions change values. In this case, x changes at integer values of t. ∎

Figure 1.

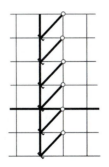

When $0 \le t < 1$, $x(t) = 0$ and $0 \le y(t) < 1$.

When $1 \le t < 2$, $x(t) = 1$ and $1 \le y(t) < 2$.

When $2 \le t < 3$, $x(t) = 2$ and $2 \le y(t) < 3$.

And so on ...!

EXAMPLE 2:   Let $\vec{r}(t) = (t - [[t]], t)$. The image of $\vec{r}$ is sketched in Figure 2. As in Example 1, interesting things happen to $\vec{r}$ when t takes on integer values. ∎

Figure 2.

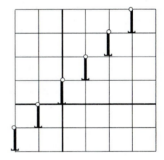

When $0 \le t < 1$, $x(t) = t$ and $y(t) = t$.

When $1 \le t < 2$, $x(t) = t - 1$ and $y(t) = t$.

When $2 \le t < 3$, $x(t) = t - 2$ and $y(t) = t$.

And so on ...!

In more complicated curves, there are some relationships between the coordinate functions of the parametrization and the parametrized curve that can help in visualizing the behavior of the curve. Suppose that we have a curve in xy–space parametrized by $\vec{r}(t) = (x(t), y(t))$, and that the coordinate axes are drawn in the usual way, with the x–axis horizontal and the y–axis vertical. If $x(t)$ is increasing, then $\vec{r}(t)$ moves to the right as t increases. If $x(t)$ is decreasing, then $\vec{r}(t)$

must move to the left as t increases. Similarly, if $y(t)$ is increasing, then $\vec{r}(t)$ is moving uphill as time increases. Some examples should help clarify this. ∎

EXAMPLE 3: A particle's position is given by $\vec{r}(t) = (x(t), y(t))$ as it moves along the curve in Figure 3 in the direction indicated by the arrows. As it passes through $\vec{p}_1$ at time t_1, it is moving to the right and it is going up. We would expect $x(t)$ and $y(t)$ to be increasing. At t_2, the particle is moving down and to the right. $x(t)$ is increasing and $y(t)$ is decreasing. At t_3, the particle is moving to the right, but $y(t)$ stops decreasing and starts increasing and the particle stops going down and starts going up. At t_4, the particle is moving to the left and down. At that time, both x and y are decreasing. ∎

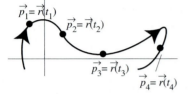

Figure 3. *The curve for Example 3.*

EXAMPLE 4: A particle is moving on the same curve as in Example 3, but its direction is reversed as illustrated in Figure 4. Let $\vec{r}(t) = (u(t), v(t))$ be the parametrization that gives its position as a function of time. As it passes through $\vec{p}_1$ at time t_4, it is moving to the left and it is going down. We would expect $u(t)$ and $v(t)$ to be decreasing. At t_3, the particle is moving up and to the left. $u(t)$ is decreasing and $v(t)$ is increasing. At t_2, the particle is moving to the left, but $v(t)$ stops decreasing and starts increasing and the particle stops going down and starts going up. At t_1, the particle is moving to the right and up. At that time, both u and v are increasing. ∎

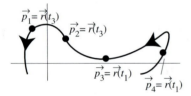

Figure 4. *The curve for Example 4.*

EXAMPLE 5: Let $\mathcal{C}$ be the curve parametrized by $\vec{r}(t) = (x(t), y(t))$, where $x(t) = t^2$ and $y(t) = t^3 - t^2$, for $-0.5 \le t \le 1$. The coordinates x and y are sketched in Figures 5.a and 5.b. The function x is decreasing when $-0.5 \le t \le 0$, so $\vec{r}(t)$ is moving to the left. The function $y(t)$ is increasing as t goes from -0.5 to 0, so $\vec{r}(t)$ is going uphill. Both $x(-0.5)$ and $y(-0.5)$ are negative and $\vec{r}(0) = \vec{0}$. Thus, $\vec{r}$ describes a particle moving from $\vec{A}$ to $\vec{B}$ as time goes from -0.5 to 0. The image of $\vec{r}$ as t goes from -0.5 to 0 is shaded in Figures 5.a, b and c.

If we zoom in on the graph of y, we find that y stops going down and starts going up at $t_0 \approx 0.666$. On the interval $[0, t_0]$, $x(t)$ is increasing and $y(t)$ is decreasing, so that on that interval, $\vec{r}(t)$ is going down and moving to the right. See Figures 5.d, e and f.

As t goes from t_0 to 1.0, x and y are increasing and $\vec{r}(t)$ is moving to the right and up. ∎

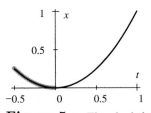

Figure 5.a *The shaded region of the graph of $x(t)$ is the region where $-0.5 \leq t \leq 0$.*

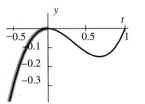

Figure 5.b *The shaded region of the graph of $y(t)$ is the region where $-0.5 \leq t \leq 0$.*

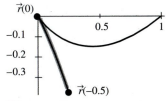

Figure 5.c *The shaded region of the image of $\vec{r}(t)$ is the region where $-0.5 \leq t \leq 0$.*

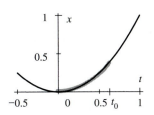

Figure 5.d *The shaded region of the graph of $x(t)$ is the region where $0 \leq t \leq t_0$.*

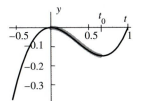

Figure 5.e *The shaded region of the graph of $y(t)$ is the region where $0 \leq t \leq t_0$.*

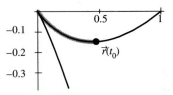

Figure 5.f *The shaded region of the image of $\vec{r}(t)$ is the region where $0 \leq t \leq t_0$.*

EXERCISES 2.5

In Exercises 1–4, sketch the image of the given function.

1. $\vec{R}(t) = (t, [[t]])$.

2. $\vec{u}(t) = (t, [[t]] - t)$.

3. $\vec{r}(t) = ([[t]] - t, [[t]] - t)$.

4. $\vec{r}(t) = ([[t]], [[t]])$.

The functions $\vec{r}_1(t) = (x_1(t), y_1(t))$ and $\vec{r}_2(t) = (x_2(t), y_2(t))$ are parametrizations of the curve sketched in Figures 5.a and b. The shaded arrows indicate the direction of motion along the curve given by the parametrizations.

5. In Figure 6.a, at each of the times t_1, t_2, t_3, t_4 and t_5, determine whether x_1 is increasing or decreasing and whether y_1 increasing or decreasing.

6. In Figure 6.b, at each of the times t_1, t_2, t_3, t_4 and t_5, determine whether x_2 is increasing or decreasing and whether y_2 increasing or decreasing.

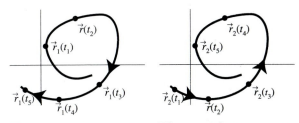

Figure 6.a *Exercise 5.*   **Figure 6.b** *Exercise 6.*

In Exercises 7–10, you are given the graphs of x and y.

a. Find approximate values for $\vec{r}(t)$, for $t = -4$, -2, 0, 2, and 4.

b. Determine the intervals on which $\vec{r}$ is moving to the right and the intervals on which $\vec{r}$ is moving to the left.

c. Determine the intervals on which $\vec{r}$ is moving up and the intervals on which $\vec{r}$ is moving down.

d. Sketch the image of $\vec{r}(t) = (x(t), y(t))$.

7.

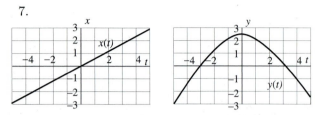

Figure 7. *The graphs of x and y for Exercise 7.*

8.

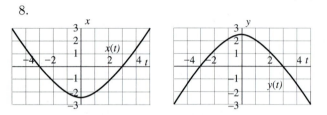

Figure 8. *The graphs of x and y for Exercise 8.*

9.

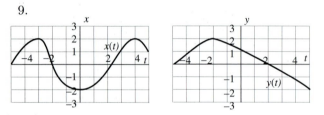

Figure 9. *The graphs of x and y for Exercise 9.*

10.

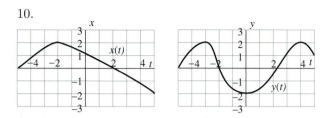

Figure 10. *The graphs of x and y for Exercise 10.*

Use your graphing calculator in Exercises 11–15 to sketch x and y over the given domain and

a. Determine the intervals on which $\vec{r}$ is moving to the right and the intervals on which $\vec{r}$ is moving to the left.

b. Determine the intervals on which $\vec{r}$ is moving up and the intervals on which $\vec{r}$ is moving down.

c. Sketch the image of $\vec{r}(t) = (x(t), y(t))$.

11. $x(t) = t^2$, $y(t) = t^3$, $-2 \le t \le 2$.

12. $x(t) = t^3$, $y(t) = t^2$, $-2 \le t \le 2$.

13. $x(t) = t^2$, $y(t) = t^2$, $-2 \le t \le 2$.

14. $x(t) = t^2$, $y(t) = t^3 - t$, $-2 \le t \le 2$.

15. $x(t) = t^3 - t$, $y(t) = t^2$, $-2 \le t \le 2$.

2.6   Compositions of Functions

> **Definition:** Composition of Functions
>
> ───────────────────
>
> Suppose that f and g are functions and that the domain of g contains some points in the image of f. Then $g \circ f$ is the function defined by $g \circ f(x) = g(f(x))$. We call this the *composite* (or *composition*) of g and f.

The domain of $g \circ f$ is the set of all x in the domain of f such that $f(x)$ is in the domain of g. Thus, the domain of $g \circ f$ is a subset of the domain of f, but it excludes all of the points in the domain of f whose images are not in the domain of g.

Recall the notion of a function machine. The idea is that if x is in the domain of f, then we can compute $f(x)$. In order to compute

$g(f(x))$, it has to be the case that $y = f(x)$ will fit into the g machine; that is, $y = f(x)$ is in the domain of g. Thus, in order to construct $g \circ f(x) = g(f(x))$, it must be true that x is in the domain of f (i.e., x fits into the f function machine); and then, $f(x)$ must be in the domain of g (i.e., $f(x)$ fits into the g function machine).

Figure 1. *The composition of functions machines.*

EXAMPLE 1: If $f(x) = x^2$ and $g(x) = x - 2$, then we can represent $g(f(7))$ as $7 \to \boxed{\text{f}} \to 49 \to \boxed{\text{g}} \to 47$.   ∎

EXAMPLE 2: If $f(x) = \sin x$ and $g(x) = 2\pi x$, then $f \circ g(x) = \sin(2\pi x)$ and $g \circ f(x) = 2\pi \sin(x)$.   ∎

EXAMPLE 3: Let $f(x) = \sin x$ and $g(x) = 2x+3$. Then $f \circ g(x) = \sin(2x + 3)$ and $g \circ f(x) = 2\sin(x) + 3$.   ∎

EXAMPLE 4: If $f(x) = 2x^2 + 3x + 6$ and $g(3) = 2$, then $f \circ g(3) = 8 + 6 + 6 = 20$.   ∎

EXAMPLE 5: Let $f(x) = 1/(1 - x)$. The domain of f is the set of all numbers except $x = 1$.

(a) If $g(x) = x + 1$, then $f \circ g = -1/x$ and the domain of $f \circ g$ is the set of all numbers except $x = 0$.

(b) If $g(x) = 2x - 1$, then $f \circ g = 1/(2 - 2x)$ and the domain of $f \circ g$ is the set of all numbers except $x = 1$.

(c) If $g(x) = \sin x$, then the domain of $f \circ g$ is the set of all numbers except $x = n\pi/2, n = 1, 5, 9, 13\dots$ or $-3, -7, -11, \dots$.

EXAMPLE 6: Let $\vec{r}(t)$ be defined by $x(t) = 1/t$, $y(t) = 1/[(t - 2)(t + 3)]$, $z(t) = \csc(t)$. The domain of $\vec{r}$ is the set of all t except

$t = 0, 2, -3$, and $n\pi$ for integer n. If $g(t) = \pi t$, then the domain of $\vec{r} \circ g$ is the set of all t except those values of t for which $\pi t = 0, 2, -3$, or is of the form $n\pi$; that is, $t = 0, t = 2/\pi, t = -3/\pi$, or $t = n$, where n is an integer.   ∎

EXAMPLE 7: Let u be the function defined by $u(t) = 3t^2 + 4t, 0 < t < 25$. Then the graph of u is a parabola with a restricted domain (if $t \leq 0$ or $t \geq 25$, then t does not fit in the u function machine). Let g be defined by $g(t) = 5t - 5$. Now g is defined for all real numbers. The composite $u \circ g$ will be defined for all numbers t for which $g(t)$ is in the domain of u; that is, the number t is in the domain of $u \circ g$ if and only if

$$
\begin{aligned}
0 &< g(t) < 25, \\
0 &< 5t - 5 < 25, \\
5 &< 5t < 30, \\
1 &< t < 6.
\end{aligned}
$$

Thus the domain of $u \circ g$ is the interval $(1, 6)$.   ∎

Shifting, Speeding Up or Slowing Down Functions

If f is a function from $\mathbb{R}$ to $\mathbb{R}$ and g is of the form $g(t) = t + a$, then the image of $f \circ g$ is the same as the image of f, but $f \circ g$ describes things happening $|a|$ units of time earlier or later, according to whether a is positive or negative. If $g(t) = t + a$, we say that $f \circ g$ shifts f by a units. Thus if a is positive, the graph of f shifts to the left by a units and if a is negative, the graph of f shifts to the right by $|a|$ units. If $g(t)$ is of the form $g(t) = mt + b = m(t + b/m)$, then $f \circ g$ speedsup (or slows down) the particle motion by a factor of m and shifts the motion b/m units of time. In Figures 2.a-f, we use the graph of the sine function to illustrate these ideas.   ∎

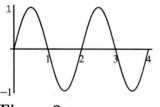

Figure 2.a
$\sin(\pi x)$, $0 \leq x \leq 4$. We *have two cycles of the sine function.*

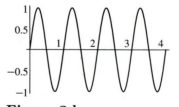

Figure 2.b
$\sin(2\pi x)$, $0 \leq x \leq 4$. We *have 4 cycles of the sine function.*

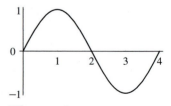

Figure 2.c
$\sin(\frac{\pi}{2}x)$, $0 \leq x \leq 4$. We *have 1 cycle of the sine function.*

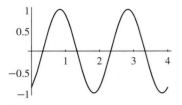

Figure 2.d
$\sin(\pi(x - 1/3))$,
$0 \le x \le 4$. *The graph*
of Figure 2.a is shifted
to the right 1/3 unit.

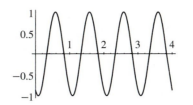

Figure 2.e
$\sin(2\pi(x - 1/3))$,
$0 \le x \le 4$. *The graph*
in Figure 2.b is shifted
to the right 1/3 unit.

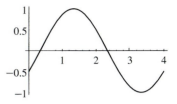

Figure 2.f
$\sin(\frac{\pi}{2}(x - 1/3))$,
$0 \le x \le 4$. *The graph*
in Figure 2.c is shifted
to the right 1/3 unit.

Composites and Images of Functions

We have been focusing on the effects of composites and graphs of functions. We now turn our attention to composites and images of functions. As an example, consider the wrapping function $\vec{W}(t)$ from $\mathbb{R}$ into $\mathbb{R}^2$. Then $\vec{W}(t) = (\cos t, \sin t)$ describes a particle moving around a unit circle at a rate of one rotation every 2π units of time. Now, let g be the function defined by $g(t) = 2\pi t$. Then $\vec{W} \circ g(t) = (\cos 2\pi t, \sin 2\pi t)$, which describes a particle moving around the circle one rotation every unit of time. Taking the composite of $\vec{W}$ and g has the effect of speeding things up by the factor of 2π. In general, if k is a real number, g is the function from $\mathbb{R}$ into $\mathbb{R}$ defined by $g(t) = kt$, and if $\vec{r}$ is a function from $\mathbb{R}$ into $\mathbb{R}^n$, then $\vec{r} \circ g$ is a parametrization for the same set as is $\vec{r}$, but $\vec{r} \circ g$ traces out the path k times faster than does $\vec{r}$. Notice that if $0 < k < 1$, then $\vec{r} \circ g$ slows down the action. If $k < 0$, then $\vec{r} \circ g$ traces its path in a direction opposite to that of $\vec{r}$.

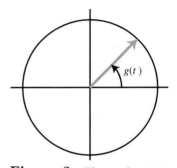

Figure 3. *The angle at time t is measured by g(t).*

EXAMPLE 8: Let $\vec{W}$ be the wrapping function, $g(t) = 2\pi t, h(t) = -t, s(t) = 4\pi t$, $u(t) = \frac{1}{2}\pi t$, and $v(t) = 2\pi t + \frac{\pi}{3}$. Then:

- $\vec{W}$ describes a particle moving counterclockwise around a circle once every 2π units of time.

- $\vec{W} \circ g$ describes a particle moving counterclockwise around a circle once every unit of time.

- $\vec{W} \circ h$ describes a particle moving clockwise around a circle once every 2π units of time.

- $\vec{W} \circ s$ describes a particle moving counterclockwise around a circle twice every unit of time.

- $\vec{W} \circ u$ describes a particle moving counterclockwise around a circle once every four units of time.

- $\vec{W} \circ v$ describes a particle moving counterclockwise around a circle once every unit of time, so that at time $t = 0$ the particle is at $(\cos(\pi/3), \sin(\pi/3)) = (1/2, \sqrt{3}/2)$.

Notice that values in the image of $\vec{W}$ are in $\mathbb{R}^2$, so they are not in the domain of g, h, u, nor v. Hence none of the composites $g \circ \vec{W}$, $h \circ \vec{W}$, $s \circ \vec{W}$, $u \circ \vec{W}$, and $v \circ \vec{W}$ $h \circ \vec{W}$ are defined.   ∎

As was mentioned in the beginning of this section, in order for $f \circ g$ to be defined at the number x, it is necessary that x be in the domain of g and that $g(x)$ be in the domain of f. If $\vec{r}$ is a function from $\mathbb{R}$ into $\mathbb{R}^n$, then $\vec{r}$ parametrizes a path in $\mathbb{R}^n$. This path is the image of $\vec{r}$ and it might be thought of as the track on which an object moves. We can use functions from $\mathbb{R}$ into $\mathbb{R}$ and composites to control how an object might move on the path. In Example 7, we used g to control how a particle moves around the circle parametrized by the wrapping function. If $\vec{r}(t) = \vec{v}t + \vec{r}_0$, then the path is a straight line and $g(t)$ can be used to describe how a particle might move on that line.

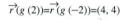

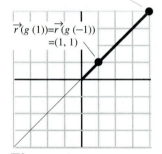

Figure 4.a

The image of $\vec{r}(t^2)$.

EXAMPLE 9: Let $\vec{r}(t) = (t, t)$. The image of $\vec{r}$ is the line $y = x$. If $g(t) = t^2$, then the image of $\vec{r}(g(t))$ is just that part of the line $y = x$ in the first quadrant as illustrated in Figure 4.a. If $g(t) = \sin(t)$, then the image of $\vec{r}(g(t))$ is just that part of the line $y = x$ satisfying $-1 \le x \le 1$. See Figure 4.b.   ∎

Inverses of Functions

Definition: The Inverse of a Function

A function f is the *inverse* of the function g if and only if $f \circ g(x) = x$ for all x in the domain of g and $g \circ f(x) = x$ for all x in the domain of f.

If the function f has an inverse, then we will denote that inverse by f^{-1}.

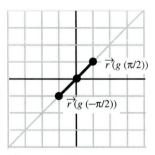

Figure 4.b

The image of $\vec{r}(\sin(t))$.

NOTE: The domain of g must be the image of f and the domain of f must be the image of g. The function f^{-1} is the inverse of f and has the following behavior:

$$y = f(x) \quad \text{if and only if} \quad x = f^{-1}(y).$$

EXAMPLE 10: Let f be defined by $f(x) = 3x + 2$. If g is the inverse of f, then:

$$f \circ g(x) = 3g(x) + 2 = x,$$

or

$$g(x) = \frac{x - 2}{3}. \qquad \blacksquare$$

EXAMPLE 11: If we use the procedure from the preceding example for $f(x) = x^2 + 1$, we get into trouble. Let us suppose that g is the inverse of f. Then the inverse g must satisfy:

$$f \circ g(x) = x,$$

or

$$[g(x)]^2 + 1 = x.$$

Solving for $g(x)$, we obtain

$$g(x) = \pm\sqrt{x - 1}.$$

Thus, there are two choices for g and no way to decide which one we want! This is enough to tell us that f does not have an inverse. $\blacksquare$

The following theorem tells us when a function does have an inverse.

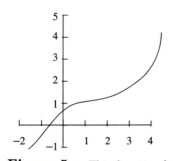

Figure 5.a *This function has an inverse. No horizontal line passes through the graph at more than one point.*

> **Theorem 1** *The function f has an inverse if and only if whenever a and b are distinct elements of the domain of f, then $f(a) \neq f(b)$.*[4]
>
> ---
> [4]Such functions are called *one–to–one functions.*

The proof of this theorem is left as an exercise. In the case of a function from a subset of $\mathbb{R}$ into $\mathbb{R}$, Theorem 1 has a nice geometric interpretation. If f is a function from a subset of $\mathbb{R}$ into $\mathbb{R}$, then f has an inverse if and only if it is true that every horizontal line L passes through the graph of f no more than once. The "horizontal line test" is illustrated in Figures 5.a and 5.b.

EXAMPLE 12: Returning to Example 11, we see that although $f(x) = x^2 + 1$ fails to have an inverse, the function g where $g(x) = x^2 + 1$, $x \geq 0$ does have an inverse. We have picked a subset of the graph of g that intersects each horizontal line at most once. In this case, $g^{-1}(x) = \sqrt{x - 1}$. See Figures 6.a-6.c. $\blacksquare$

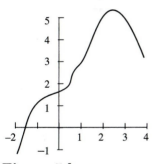

Figure 5.b *This function does not have an inverse. The line $y = 4$ passes through the graph at more than one point.*

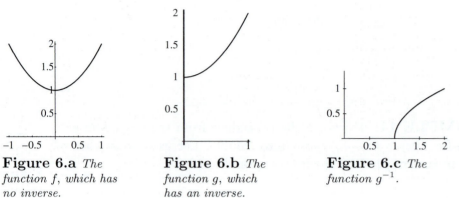

Figure 6.a *The function f, which has no inverse.*

Figure 6.b *The function g, which has an inverse.*

Figure 6.c *The function g^{-1}.*

According to the definition of the inverse of a function, if f^{-1} is the inverse of f, then $f^{-1} \circ f(x) = x$. Thus, if $(x, f(x))$ is a point on the graph of f, then $(f(x), f^{-1} \circ f(x)) = (f(x), x)$ is a point on the graph of f^{-1}. So the point (a, b) is on the graph of f if and only if the point (b, a) is on the graph of f^{-1}. Geometrically, (a, b) is just the mirror image of (b, a) reflected over the diagonal, $y = x$, as in Figure 7.

We now have a reasonably easy method of sketching the inverse of a function whose domain and range are subsets of $\mathbb{R}$, provided that we have a sketch of the graph itself. We simply draw the mirror image of the function reflected across the diagonal $y = x$ as in Figure 8, where we display the function g from Example 11 and its inverse drawn on the same graph.

Figure 7. The point (b, a) is the mirror image of (a,b) through the line $y = x$.

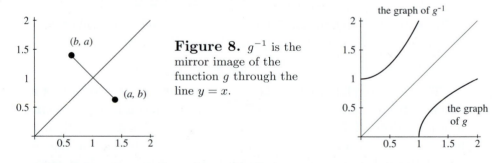

Figure 8. g^{-1} is the mirror image of the function g through the line $y = x$.

EXAMPLE 13: The following data are known.

x	0	0.25	0.5	0.75	1	1.25	1.5	1.75
$f(x)$	0	0.101	0.388	0.787	1.262	1.798	2.382	3.006
x	2	2.25	2.5	2.75	3	3.25	3.5	3.75
$f(x)$	3.666	4.356	5.074	5.817	6.582	7.368	8.173	8.995

We can translate the given information into information about

f^{-1} as follows.

x	0	0.101	0.388	0.787	1.262	1.798	2.382	3.006
$f^{-1}(x)$	0	0.25	0.5	0.75	1	1.25	1.5	1.75
x	3.666	4.356	5.074	5.817	6.582	7.368	8.173	8.995
$f^{-1}(x)$	2	2.25	2.5	2.75	3	3.25	3.5	3.75

The graphs of f and its inverse are sketched in Figure 9. ∎

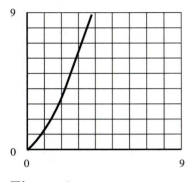

Figure 9.a *The graph of f.*

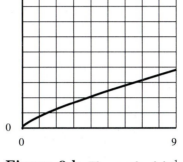

Figure 9.b *The graph of f^{-1}.*

EXAMPLE 14: The sine function does not have an inverse, since the horizontal line $y = 1$ passes through the graph of the sine function in infinitely many places. To resolve this problem, we restrict the domain of the sine function so that the resulting function has an inverse. (See Figure 10.)

We let Sin be the function defined by $\mathrm{Sin}(x) = \sin(x), -\frac{\pi}{2} \leq x \leq \frac{\pi}{2}$. The Sin function has an inverse Sin^{-1}, which is also called the Arcsine function. Thus,

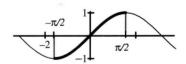

Figure 10. *A portion of the sine function with an inverse.*

$$\sin(\mathrm{Sin}^{-1}(x)) = \sin(\mathrm{Arcsin}(x)) = x.$$

We can use the mirror image technique to sketch the Arcsine function. The points $(-\pi/2, -1) \approx (-1.571, -1)$, $(-\pi/4, -\sqrt{2}/2) \approx (-0.785, -0.707)$, $(0, 0)$, $(\pi/4, \sqrt{2}/2) \approx (0.785, 0.707)$, and $(\pi/2, 1) \approx (1.571, 1)$ are points on the graph of $\mathrm{Sin}(x)$. Thus we plot the points $(-1, -1.571)$, $(-0.707, -0.785)$, $(0, 0)$, $(0.707, 0.785)$, and $(1, 1.571)$ as points in the graph of $\mathrm{Arcsin}(x)$ and fill in the graph so that it looks like the reflection of the $\mathrm{Sin}(x)$ function as shown in Figure 11. ∎

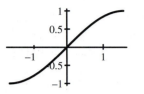

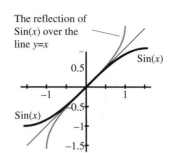

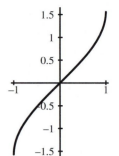

Figure 11.a *The graph of Sin(x).*

Figure 11.b *The graph of Sin(x) and its reflection over the diagonal.*

Figure 11.c *The graph of Arcsin(x).*

By similar restrictions of the domain, we can define the functions Cos(x), Tan(x), Cot(x), Sec(x), and Csc(x). We are then able to construct inverses for these functions.

EXAMPLE 15: We choose Sec(x) to be the restriction of the secant function to an appropriate domain. We select the "largest" subset of the domain of the secant function that has zero as the left end point and such that no horizontal line passes through the graph of the secant function more than once over that set. See Figure 12.a. The domain of Sec(x) is the set $0 \leq x < \pi/2$ or $\pi/2 < x \leq \pi$, which may be written $[0, \pi/2) \cup (\pi/2, \pi]$. The image of Sec(x) is $y \leq -1$ or $y \geq 1$, which may be written $(-\infty, -1] \cup [1, \infty)$. The function Sec(x) has an inverse Arcsec(x). The graph of Arcsec(x) can be obtained by reflecting the graph of Sec(x) over the line $y = x$. Arcsec(x) is sketched in Figure 12.b. The domain of Arcsec(x) is the image of Sec(x), which is $(-\infty, -1] \cup [1, \infty)$. The image of the Arcsec(x) is the domain of Sec(x), which is $[0, \pi/2) \cup (\pi/2, \pi]$. ∎

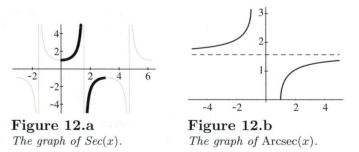

Figure 12.a
The graph of Sec(x).

Figure 12.b
The graph of Arcsec(x).

For the other trigonometric functions, see Exercises 42–46 at the end of this section.

We often want to simplify the composition of a trigonometric function with an inverse trigonometric function. If both the trigonometric function and the inverse trigonometric function are of the same type (e.g., $\sin(\text{Arcsin}(x))$), then the simplification is immediate. However, there are occasions when we have a trigonometric function composed with an inverse of a different trigonometric function. We will encounter this situation extensively in the chapter on techniques of integration.

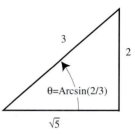

Figure 13.
$\theta = (\text{Arcsin}(2/3))$.

EXAMPLE 16: Simplify the composition $\cos(\text{Arcsin}(2/3))$.

SOLUTION: One way to solve this is to represent $\text{Arcsin}(2/3)$ by the angle θ of the right triangle having hypotenuse of length 3, opposite leg of length 2 and, by the Pythagorean Theorem, adjacent leg of length $\sqrt{3^2 - 2^2} = \sqrt{5}$. (See Figure 13.) Now, $\cos(\text{Arcsin}(2/3)) = \cos\theta = \sqrt{5}/3$. ■

Drawing Inverse Functions With a Graphing Calculator

We have seen that sketching the inverse of a function f is often most easily done by drawing first f and then its reflection over the diagonal $y = x$. On a calculator, parametrize the graph of f with $\vec{r}(t) = (t, f(t))$ and then draw $\vec{r}_1(t) = (f(t), t)$. Indeed, if $\vec{r}(t) = (x(t), y(t))$ parametrizes the curve $\mathcal{C}$, then $\vec{r}_1(t) = (y(t), x(t))$ will parametrize the reflection of $\mathcal{C}$ over the diagonal.

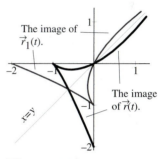

Figure 14. *The images of* $(x(t), y(t))$ *and* $(y(t), x(t))$.

EXAMPLE 17: In Figure 14, the curve parametrized by $\vec{r}(t) = (t^2, t^3 - t^2)$ and its reflection over the diagonal parametrized by $\vec{r}_1(t) = (t^3 - t^2, t^2)$ are drawn on the same coordinate system. ■

EXERCISES 2.6

In Exercises 1–6, find $f \circ g(2)$ (or $\vec{f} \circ g(2)$) where f, $\vec{f}$, and g are the given functions.

1. $f(x) = \sin x$, $g(2) = \frac{\pi}{3}$.

2. $f(x) = \cos x$, $g(2) = \frac{\pi}{3}$.

3. $f(x) = \sin(3x)$, $g(2) = \frac{\pi}{3}$.

4. $\vec{f}(x) = (\sin x, \cos x, \sin(3x) + \cos(3x))$, $g(2) = \frac{\pi}{3}$.

5. $f(x) = x + 1$, $g(x) = x^2 - 1$.

6. $\vec{f}(x) = (x^2, x + 1, 3)$, $g(x) = x^2 + 3$.

In Exercises 7–10, determine $f \circ g(x)$ (or $\vec{f} \circ g(x)$) where f, $\vec{f}$, and g are the given functions.

7. $f(x) = x^2 + 2x + 1$, $g(x) = 2x + 1$.

8. $f(x) = x^3 + 6x^2 + 2x + 2$, $g(x) = x + 2$.

9. $\vec{f}(x) = (x^2 + 2x + 1, x^2 - 2)$, $g(x) = x^2 - 2x + 3$.

10. $\vec{f}(x) = (x + 1, 2x + 3, \frac{x}{3})$, $g(x) = x^5 + 2x$.

In Exercises 11–16, determine the domain of $f \circ g$.

11. $f(x) = \dfrac{1}{(x - 1)}$ and $g(x) = x^2$.

12. $f(x) = \dfrac{1}{x}$ and $x^2 - x - 2$.

13. $f(x) = \sqrt{x - 1}$ and $g(x) = x^2$.

14. $f(x) = \dfrac{1}{\sqrt{x - 1}}$ and $g(x) = x^2$.

15. $f(x) = \sqrt{x - 1}$ and $g(x) = \sin(x)$.

16. $f(x) = \sqrt{x - 1}$ and $g(x) = x^2 - x$.

Describe in a sentence or two the motion defined by $\vec{W} \circ g$ in Exercises 17–23.

17. $g(t) = 6\pi t$.

18. $g(t) = -6\pi t$.

19. $f(t) = \omega t$.

20. $g(t) = \pi \sin(t)$.

21. $g(t) = \sin(t) + \frac{\pi}{2}$.

22. $g(t) = \pi \sin(\omega t)$.

23. $g(t) = A \sin(\omega t)$.

The curve $\mathcal{C}$ (illustrated in Figure 15) is parametrized by $\vec{r}(t)$, $-1.25 \le t \le 2.5$. In Exercises 24–29, determine whether $\vec{r} \circ g$ is defined, and if it is, determine the domain and sketch the image of $\vec{r} \circ g$.

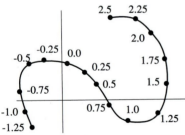

Figure 15. *The curve $\mathcal{C}$ of Exercises 24–29. Values for t are given next to the corresponding point $\vec{r}(t)$.*

24. $g(t) = t^2$.   25. $g(t) = t^2$.

26. $g(t) = t^2$.   27. $g(t) = \sin t$.

28. $g(t) = 1.25 \sin t$.   29. $g(t) = 0.5 \sin 2t$.

Let $\vec{r}(t) = t(1, 2, -1) + (-1, 0, -1)$. In Exercises 30–32, describe the motion defined by $\vec{r} \circ g$.

30. $g(t) = 6\sin(t)$.   31. $g(t) = \pi \sin(\omega t)$.

32. $g(t) = A \sin(\omega t)$.

Let $\vec{r}(t) = (t, t^2)$. In Exercises 33–35, describe the motion defined by $\vec{r} \circ g$.

33. $g(t) = 6\sin(t)$.

34. $g(t) = \pi \cos(\omega t)$.

35. $g(t) = A \cos\left(\omega t + \frac{\pi}{4}\right)$.

Let F be the largest subset of f so that 0 is in the domain of F and so that F has an inverse. In Exercises 36–39, determine the domain of F and F^{-1}, the inverse of F.

36. $f(x) = 2x + 1$.

37. $f(x) = x^2 + 2x + 1$.

38. $f(x) = -x - 1$.

39. $f(x) = x^3$.

40. Explain why we know that each of the functions sketched in Figure 16 has an inverse. For each of the functions, find the domain and image of the function and its inverse. Sketch the inverse function.

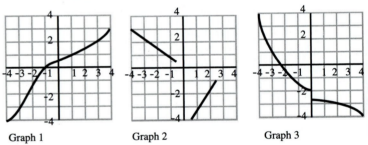

Graph 1   Graph 2   Graph 3

Figure 16. *Graphs for Exercise 40.*

41. Given that f has an inverse, its domain is $[2, 5]$ and its image is $[-1, 3]$, what are the domain and image of f^{-1}?

For each of the functions f in Exercises 42–45, find the largest interval I or segment I, so that:

a. *The origin is either a point of the interval or the left end point of I.*

b. *The set $\{(x, f(x)) \mid x \text{ is in } I \text{ and } f(x) \text{ is defined}\}$ has an inverse.*

42. $f(x) = \cos x$.   43. $f(x) = \tan x$.

44. $f(x) = \cot x$.   45. $f(x) = \csc x$.

Definition: Cos x, Tan x, Cot x, Sec x and Csc x

We define Cos x, Tan x, Cot x and Csc x to be the subsets of $\cos x$, $\tan x$, $\cot x$, $\sec x$ and $\csc x$, respectively, with domain the set I of the appropriate part of the above problem.

Arccos(x) is the inverse of Cos(x).

Arctan(x) is the inverse of Tan(x) etc.

46. Sketch the graph of Arccos(x), Arctan(x), Arccot(x) and Arccsc(x).

In Exercises 47–49, find $f \circ g(2)$ where f and g are the given functions.

47. $f(x) = \tan(x), g(2) = \text{Arctan}(36)$.

48. $f(x) = \cos(x), g(2) = \text{Arcsin}\left(\frac{1}{5}\right)$.

49. $f(x) = \sin(x), g(2) = \text{Arccos}\left(\frac{1}{5}\right)$.

50. $f(x) = \cos(x), g(2) = \text{Arctan}\left(\frac{1}{5}\right)$.

In Exercises 51–54, determine $f \circ g(x)$ (or $\vec{f} \circ g(x)$) where f, $\vec{f}$, and g are the given functions.

51. $f(x) = \tan(x)$, $g(x) = \text{Arcsin}(x)$.

52. $\vec{f}(x) = (\tan(x), \sin(x))$, $g(x) = \text{Arcsin}(x)$.

53. $\vec{f}(x) = (\cos(x), \tan(x), \sec(x))$,
    $g(x) = \text{Arccos}(x)$.

54. $\vec{f}(x) = (\cos(x), \tan(x), \sec(x))$,
    $g(x) = \text{Arctan}(x)$.

In Exercises 55–60, sketch the graph of $\sin \circ g$, $\cos \circ g$, $\tan \circ g$, and $\sec \circ g$.

55. $g(x) = x$.   56. $g(x) = 3x$.

57. $g(x) = 3\pi x$.   58. $g(x) = 3\pi x + \pi$.

59. $g(x) = \dfrac{\pi x}{2}$.   60. $g(x) = \dfrac{\pi x}{2} + \pi$.

Let f be the function sketched in Figure 17. In Exercises 61–65, sketch the graph of $f \circ g$.

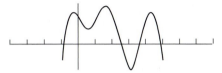

Figure 17. *Illustration for Exercises 61-65.*

61. $g(x) = x + 2$.   62. $g(x) = 2x$.

63. $g(x) = 2x - 2$.   64. $g(x) = \dfrac{x}{2}$.

65. $g(x) = \dfrac{x}{3} - 2$.

Let g be the greatest integer function $g(x) = [[x]]$ and let a be a sequence. Then $f(x) = a \circ g(x) = a[[x]]$ is a function with domain the set of all $x \geq N$, where N is the initial integer in the domain of a. The function f is called the step function *determined by the sequence a. For Exercises 66–69, sketch the step function determined by the given sequence.*

66. $a(n) = \dfrac{1}{n+3}$, $n \geq -2$.

67. $a(n) = 1 + (-1)^n$, $n \geq 0$.

68. $s(n)$ is the sequence of partial sums of $a(n) = 1 + (-1)^n$, $n \geq 0$.

2.7 Building New Functions

Many problems involved with motion are quite complicated when taken as a whole but can often be broken down into several simple pieces. The simple pieces can be modeled, and then the models can be easily reassembled into one model describing the complicated motion. In the previous section, we developed one of our basic tools: composites of functions. In this section, we introduce some additional modeling tools.

EXAMPLE 1: A particle p_1 moves along a straight line. A second particle p_2 moves around a circle centered at p_1. We want to parametrize the motion of p_2. The essence of the solution to this problem is that if $\vec{r}_1$ is the position of p_1 when viewed from the origin and $\vec{r}_2$ is the position of p_2 relative to p_1, then $\vec{r}_1 + \vec{r}_2$ gives the position of p_2 when viewed from (relative to) the origin. See Figure 1.

■

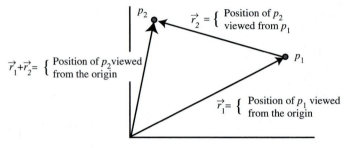

Figure 1. *The position of p_2 relative to the origin is its position relative to the position of p_1 plus the position of p_1 relative to the origin.*

EXAMPLE 2: Suppose that the position of p_1 relative to the origin is given by $\vec{r}_1(t) = (2t+1, -2t)$ and the position of p_2 viewed from p_1 is given by $\vec{r}_2(t) = (\cos(2\pi t), \sin(2\pi t))$. Then the position of p_2 relative to the origin is

$$\vec{r}(t) = \vec{r}_1(t) + \vec{r}_2(t) = \begin{pmatrix} 2t + 1 + \cos(2\pi t) \\ -2t + \sin(2\pi t) \end{pmatrix}.$$

See Figures 2.a-2.d below.

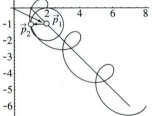

Figure 2.a $\vec{r}(0.5)$.

Figure 2.b $\vec{r}(1)$.

Figure 2.c $\vec{r}(1.5)$.

Figure 2.d $\vec{r}(2)$.

EXAMPLE 3: Suppose that a particle p_1 rotates about the origin in a circle of radius one, five rotations per second, and a particle p_2 rotates about p_1 in a circle with radius one, four rotations per second. At time $t = 0$, the particle p_1 is at $(0,1)$ and p_2 is at $(0,2)$. Let

$$\vec{r}_1(t) = \begin{pmatrix} \cos\left(10\pi t + \frac{\pi}{2}\right) \\ \sin\left(10\pi t + \frac{\pi}{2}\right) \end{pmatrix} \quad \text{and} \quad \vec{r}_2(t) = \begin{pmatrix} \cos\left(8\pi t + \frac{\pi}{2}\right) \\ \sin\left(8\pi t + \frac{\pi}{2}\right) \end{pmatrix}.$$

Then

$$(\vec{r}_1 + \vec{r}_2)(t) = \begin{pmatrix} \cos\left(10\pi t + \frac{\pi}{2}\right) + \cos\left(8\pi t + \frac{\pi}{2}\right) \\ \sin\left(10\pi t + \frac{\pi}{2}\right) + \sin\left(8\pi t + \frac{\pi}{2}\right) \end{pmatrix}$$

describes the motion of p_2, relative to the coordinate axes. Figure 3.a displays the path of p_2. Figures 3.b–3.k provide an animation of the particle p_2 as time progresses from 0 seconds to 0.55 seconds. ∎

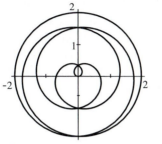

Figure   3.a *The path of p_2.*

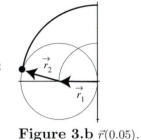

Figure 3.b $\vec{r}(0.05)$.

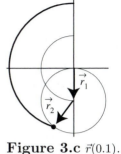

Figure 3.c $\vec{r}(0.1)$.

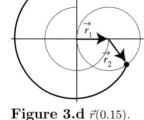

Figure 3.d $\vec{r}(0.15)$.

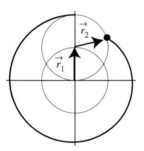

Figure 3.e $\vec{r}(0.2)$.

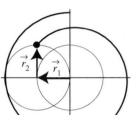

Figure 3.f $\vec{r}(0.25)$.

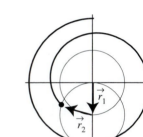

Figure 3.g $\vec{r}(0.3)$.

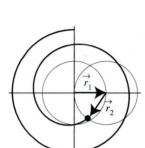

Figure 3.h $\vec{r}(0.35)$.

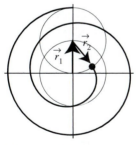

Figure 3.i $\vec{r}(0.4)$.

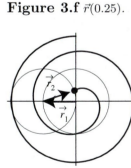

Figure 3.j $\vec{r}(0.45)$.

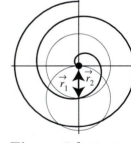

Figure 3.k $\vec{r}(0.5)$.

Figure 3.l $\vec{r}(0.55)$.

We can add functions to unscramble complicated functions.

EXAMPLE 4: The function $\vec{r}(t) = \begin{pmatrix} 2t + 1 + \cos(t) \\ -t + 3 + \sin(t) \end{pmatrix}$ can be written as the sum of two functions $\vec{r}_1(t) = \begin{pmatrix} 2t + 1 \\ -t + 3 \end{pmatrix}$ and $\vec{r}_2(t) = \begin{pmatrix} \cos(t) \\ \sin(t) \end{pmatrix}$. The function $\vec{r}_1$ models a point $\vec{p}_1$ moving along a straight line and the function $\vec{r}_2$ describes circular motion.

$\vec{r}(t) = \vec{r}_1(t) + \vec{r}_2(t)$ describes a point circling about $\vec{p}_1$ as $\vec{p}_1$ moves along a straight line. ■

We have been building new functions by adding two functions together. Another useful tool is to multiply a vector valued function by a scalar valued function. If $\rho(t)$ is a real valued function and $\vec{r}(t)$ is a vector valued function, then $\rho(t)\vec{r}(t)$ will scale the vector $\vec{r}(t)$ by the factor $\rho(t)$. This is particularly interesting when the vector valued function is the wrapping function.

EXAMPLE 5: Let $\rho(t) = t$. Sketch the image of $\rho(t)\vec{W}(t)$.

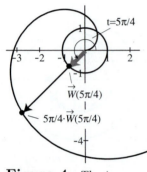

SOLUTION: We proceed by sketching the image of the wrapping function and then we sketch the result of scaling $\vec{W}(t)$ by $\rho(t)$. See Figure 4. ■

EXAMPLE 6: Let $\rho(t) = 2\sin(3t)$. Sketch the image of $\rho(t)\vec{W}(t)$.

SOLUTION: We proceed as in the previous example. We sketch the image of $\rho(t)\vec{W}(t)$ over a full rotation of the wrapping function. That is, we restrict our attention to $0 \leq t \leq 2\pi$. By sketching $\rho(t)$, we get a feel for how $\vec{W}(t)$ is scaled by ρ. Notice that $\rho(t) = 0$ at $t = 0$, $\pi/3$, $2\pi/3, \ldots$ and ρ is either $+2$ or -2 at other multiples of $\pi/6$. See Figures 5.a–5.f. ■

Figure 4. *The image of $\rho(t)\vec{W}(t)$, where $\rho(t) = t$.*

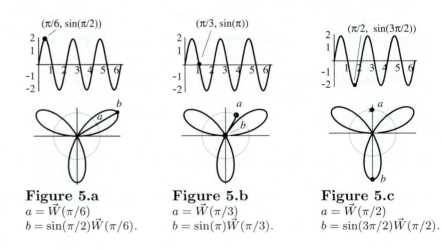

Figure 5.a
$a = \vec{W}(\pi/6)$
$b = \sin(\pi/2)\vec{W}(\pi/6)$.

Figure 5.b
$a = \vec{W}(\pi/3)$
$b = \sin(\pi)\vec{W}(\pi/3)$.

Figure 5.c
$a = \vec{W}(\pi/2)$
$b = \sin(3\pi/2)\vec{W}(\pi/2)$.

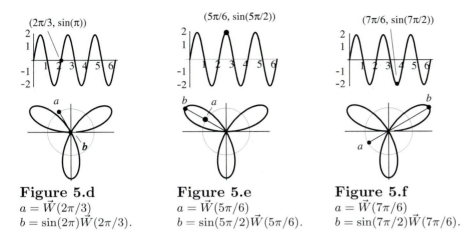

Figure 5.d
$a = \vec{W}(2\pi/3)$
$b = \sin(2\pi)\vec{W}(2\pi/3)$.

Figure 5.e
$a = \vec{W}(5\pi/6)$
$b = \sin(5\pi/2)\vec{W}(5\pi/6)$.

Figure 5.f
$a = \vec{W}(7\pi/6)$
$b = \sin(7\pi/2)\vec{W}(7\pi/6)$.

EXAMPLE 7: Example 6 models a particle oscillating back and forth on a track as the track rotates about its center. Suppose that the center of that track is moving along the line parametrized by $\vec{r}_1(t) = (2t, t)$. The particle's position, relative to the track's center, is given by $\vec{r}_2(t) = 2\sin(3t)\vec{W}(t)$. The position of the particle relative to the origin is

$$\vec{r}_1(t) + \vec{r}_2(t) = \begin{pmatrix} 2t \\ t \end{pmatrix} + 2\sin(3t)\begin{pmatrix} \cos(t) \\ \sin(t) \end{pmatrix}.$$

The particle's rather strange looking trajectory is sketched in Figure 6. ∎

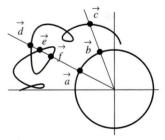

Figure 6. *The image of*
$\vec{r}_1(t) + \vec{r}_2(t) =$

$$\begin{pmatrix} 2t \\ t \end{pmatrix} + 2\sin(3t)\begin{pmatrix} \cos(t) \\ \sin(t) \end{pmatrix}.$$

EXAMPLE 8: Let $\vec{r}$ be a function from $\mathbb{R}$ into $\mathbb{R}^n$ such that $\|\vec{r}(t)\|$ is never zero. Let $f(t)$ be defined by $f(t) = 1/\|\vec{r}(t)\|$. (See Figure 7.) Then $f\vec{r}$ is a function from $\mathbb{R}$ into $\mathbb{R}^n$ such that at each time t, $f\vec{r}(t) = \vec{r}(t)/\|\vec{r}(t)\|$ is a vector with length one, which points in the same direction as $\vec{r}(t)$. In Figure 7, $\vec{c} = \vec{r}(t_1)$, $\vec{d} = \vec{r}(t_2)$, $\vec{e} = \vec{r}(t_4)$, and $\vec{f} = \vec{r}(t_3)$, while $\vec{b} = f\vec{r}(t_1)$ and $\vec{a} = f\vec{r}(t_2) = f\vec{r}(t_3) = f\vec{r}(t_4)$. ∎

In this section, we have focused on adding vector valued functions and scaling vector valued functions with real valued functions. Of course, there will be times when you want to take the dot product and the cross product of vector valued functions. We will postpone a serious look at the dot and cross product of functions until we have use for them. There is some common shorthand notation for the types of constructions we have been using.

Let ρ be a real valued function and let $\vec{f}$ and $\vec{g}$ be vector valued functions.

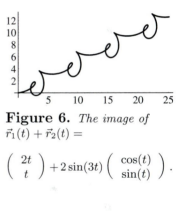

Figure 7.
$\vec{r}(t)$ *and* $\vec{r}(t)/\|\vec{r}(t)\|$.

$\rho\vec{f}$ is the function defined by $[\rho\vec{f}](x) = \rho(x)\vec{f}(x)$.

$\vec{f} + \vec{g}$ is the function defined by $[\vec{f} + \vec{g}](x) = \vec{f}(x) + \vec{g}(x)$.

$\vec{f} \cdot \vec{g}$ is the function defined by $[\vec{f} \cdot \vec{g}](x) = \vec{f}(x) \cdot \vec{g}(x)$.

$\vec{f} \times \vec{g}$ is the function defined by $[\vec{f} \times \vec{g}](x) = \vec{f}(x) \times \vec{g}(x)$.   ■

EXERCISES 2.7

In Exercises 1 and 2, compute $\vec{f} + \vec{g}$ and $\vec{f} \cdot \vec{g}$ at the indicated points.

1. $\vec{f}(t) = (\cos t, \sin t)$, $\vec{g}(t) = (\cos t, \sin t)$, $t = \pi$, $t = 3\pi/2$, $t = 3\pi/4$, $t = 5\pi/3$.

2. $\vec{f}(t) = (\sin t)\hat{\imath} + (2t/\pi)\hat{\jmath} + ((1/2)\cos^2 t)\hat{k}$, $\vec{g}(t) = (\cos t)\hat{\imath} + (\sin t)\hat{\jmath} + (t/\pi)\hat{k}$, $t = \pi/6$, $t = 5\pi/4$, $t = 10\pi/3$.

3. If $f(x) = 3x + 1$ and $\vec{g}(x) = (2x, \cos x)$, compute $f\vec{g}$ at $x = \pi$, $x = \pi/2$, $x = 10\pi/6$.

4. Find an expression for $f(t) = \|\vec{W}(t)\|$.

In Exercises 5–10, find a function $\vec{h}$ such that, for each x in the domain of $\vec{h}$, the function $\vec{h}(x)$ is a unit vector that points in the same direction as $\vec{g}$.

5. $\vec{g}(x) = (x^2 + 1, 3x, x^3 + 2)$.

6. $\vec{g}(x) = (\sin x, \cos x)$.

7. $\vec{g}(x) = (\cos x, \sin x, x)$.

8. $\vec{g}(x) = (x \cos x, x \sin x)$.

9. $\vec{g}(x) = (a \cos x)\hat{\imath} + (b \sin x)\hat{\jmath}$.

10. $\vec{g}(x) = (a \cos x)\hat{\imath} + (b \sin x)\hat{\jmath} + x\hat{k}$.

In Exercises 11–14, the position of an object p_1 relative to the origin is given by $\vec{r}_1(t)$. A second object p_2 has position relative to p_1 given by $\vec{r}_2(t)$. Find the position of p_2 relative to the origin and describe in words the motion of p_2 as seen from the origin.

11. $\vec{r}_1(t) = (-5t + 2, 3t)$, $\vec{r}_2(t) = (3t, 1 - t)$.

12. $\vec{r}_1(t) = (-5t, 3t)$, $\vec{r}_2(t) = 3(\cos(t), \sin(t))$.

13. $\vec{r}_1(t) = 5(\cos(t), \sin(t))$, $\vec{r}_2(t) = 3(\cos(t), \sin(t))$.

14. $\vec{r}_1(t) = (-5t, 3t)$, $\vec{r}_2(t) = (\cos(t), \cos(t))$.

Let $\vec{r}_1(t) = (\alpha t, \beta t)$, $\vec{r}_2(t) = \rho \vec{W}(\omega t)$ and $\vec{r}(t) = \vec{r}_1(t) + \vec{r}_2(t)$. In Exercises 15–18, explain in words the difference between $\vec{r}(t)$ in the given two cases.

15. Case 1: $\alpha = 1$, $\beta = 0$, and $\omega = 1$
    Case 2: $\alpha = 0$, $\beta = 1$, and $\omega = 1$.

16. Case 1: $\alpha = 1$, $\beta = -1$, and $\omega = 1$
    Case 2: $\alpha = 1$, $\beta = 1$, and $\omega = 1$.

17. Case 1: $\alpha = 1$, $\beta = 1$, and $\omega = 1$
    Case 2: $\alpha = 1$, $\beta = 1$, and $\omega = 2\pi$.

18. Case 1: $\alpha = 1$, $\beta = 1$, and $\omega = 2\pi$
    Case 2: $\alpha = 1$, $\beta = 1$, and $\omega = -2\pi$.

In Exercises 19–22, the position of an object p_1 relative to the origin is given by $\vec{r}_1(t)$. A second object p_2 has position relative to the origin given by $\vec{r}_2(t)$. Find the position of p_2 relative to p_1.

19. $\vec{r}_1(t) = (-5t + 2, 3t)$, $\vec{r}_2(t) = (3t, 1 - t)$.

20. $\vec{r}_1(t) = (-5t, 3t)$, $\vec{r}_2(t) = 3(\cos(t), \sin(t))$.

21. $\vec{r}_1(t) = 5(\cos(t), \sin(t))$,
    $\vec{r}_2(t) = 3(\cos(t), \sin(t))$.

22. $\vec{r}_1(t) = (-5t, 3t)$, $\vec{r}_2(t) = (\cos(t), \cos(t))$.

23. Let $\vec{v}_1 = (3, 1)$ ft and let $\vec{v}_2 = (1, 0)$ ft. A particle p_1 moves uniformly along the line passing through $\vec{v}_1$ and $\vec{v}_2$ so that at time $t = 0$ min, p_1 is at $\vec{v}_2$, and at time $t = 1$, p_1 is at $\vec{v}_1$. A particle p_2 rotates counterclockwise about p_1 in a circle of radius 2 ft, at a rate of 3 rotations/min. At $t = 0$, p_2 is at $(3, 0)$ ft. Find a function $\vec{r}$ from $\mathbb{R}$ into $\mathbb{R}^2$ so that $\vec{r}(t)$ will give the position of p_2 at time t.

24. Same as Exercise 23 except that p_2 is rotating clockwise.

Without the aid of a calculator or computer, sketch the image of $f\vec{W}$ in Exercises 25-28.

25. $f(t) = -t$.

26. $f(t) = \sin(2t)$.

27. $f(t) = \cos(2t)$.

28. $f(t) = \cos(3t)$.

29. Use the parametric plot feature to plot the images of $\sin(nt)\vec{W}(t)$, $0 \leq t \leq 2\pi$ for $n = 1, 2, \ldots, 7$. What seems to be the correlation between n and the number of "leaves" on the image. Plot the image of $\sin(10t)$, and $\sin(11t)$ to verify your conjecture.

30. Use the parametric plot feature to plot the images of $\cos(nt)\vec{W}(t)$, $0 \leq t \leq 2\pi$ for $n = 1, 2, \ldots, 7$. What seems to be the correlation between n and the number of "leaves" on the image. Plot the image of $\cos(10t)$, and $\cos(11t)$ to verify your conjecture.

31. Figure 8 gives a sketch of the image of $\cos(25t)\vec{W}(t)$. Find all the values for t_1, t_2, t_3 and t_4 so that $\cos(25t_i)\vec{W}(t_i) = \vec{a}_i$ and $0 \leq t_i \leq 2\pi$.

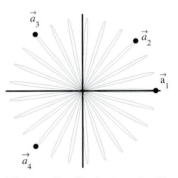

Figure 8. *Illustration for Exercise 31.*

32. **Rolling Wheel Problem 1:** A wheel of radius 1 ft rolls along a horizontal plane at a rate of 20 rotations/min with the center of the wheel moving in the negative x–direction. There is a tack in the wheel that, at time $t = 0$ min, is at the bottom of the wheel.

   a. Let $\theta(t)$ denote the angle between the negative y–axis and the line segment from the center of the wheel to the tack measured in radians as illustrated in Figure 9. Find $\theta(t)$.

   b. Let $\vec{r}_1(t)$ denote the position of the center of the wheel relative to the original wheel center. Find $\vec{r}_1(t)$. Hint: The length of $\vec{r}_1(t)$ must be the length of the arc swept out by $\theta(t)$.

   c. Let $\vec{r}_2(t)$ give the position of the tack relative to the center of the wheel. Find $\vec{r}_2(t)$.

   d. Find a function $\vec{r}_3(t)$ that gives the position of the tack relative to the given coordinate axes.

      i. The axes are chosen so that at time $t = 0$ min, the origin is at the wheel's center.

      ii. The axes are chosen so that at time $t = 0$ min, the origin is at the bottom of the wheel.

      iii. The axes are chosen so that at time $t = 0$ min, the origin is 10 ft above the wheel's center.

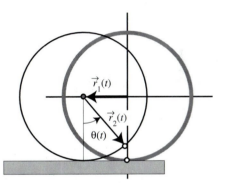

Figure 9. *The rolling wheel.*

33. **Ant on the Wheel Problem 1:** Suppose that there is an ant on the wheel of Exercise 32. At time $t = 0$ min he is at the tack. Walking at a steady pace directly toward the wheel's center, he can reach the wheel's center in 1 min. Find functions $\vec{r}_a, \vec{r}_b$ and $\vec{r}_c$ to describe the motion of the ant relative to the axes of Parts i, ii and iii of Exercise 32. The domain of the parametrization is $[0, 1]$.

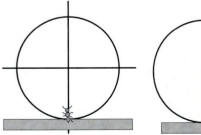

Figure 10.a *The initial position of the ant.*

Figure 10.b *The position of the ant at time t.*

34. **Rotating Pendulum Problem 1:**   A ball moves back and forth along a track of length $2L$.   The equation of motion relative to the center of the track is given by $r(t) = L\cos(\omega t)$. This track is attached to a disc centered at the origin with radius R.   The track is located so that its center is at the point $(a, 0)$, and the ball is at the end of the track at the point $(a + L, 0)$ at time $t = 0$.   See Figure 11.a.   The disc is set in motion so that it spins at a rate of k revolutions per unit time in a counterclockwise direction, as illustrated in Figure 11.b. Let $\theta(t)$ be the angle that the track makes with the horizontal.   Find $\theta(t)$ and describe the position of the ball as a function of time.

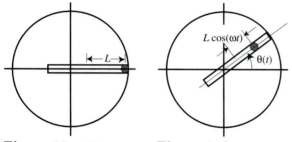
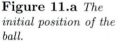

Figure 11.a *The initial position of the ball.*

Figure 11.b *The position of the ball at time t.*

35. **Wheel in a Cylinder Problem 1:** A wheel of radius r rolls around the interior of a cylinder of radius R. (Of course, $r < R$.) Assume that the center of the cylinder is at the origin and, at time $t = 0$, the point of tangency is at the point $(R, 0)$ as in Figure 12.a. Let P denote the original point of tangency on the wheel. We will investigate the motion of this point P on the wheel. Let $\vec{v}_1(t)$ denote the vector emanating from the origin to the center of the wheel and let $\theta(t)$ denote the angle $\vec{v}_1(t)$ makes with the x–axis at time t.   Let $\vec{v}_2(t)$ be the vector that emanates at the center of the wheel and terminates at P, and let $\phi(t)$ be the angle that $\vec{v}_2(t)$ makes with the horizontal at time t. See Figure 12.b.

   a. Show that $\phi(t) = -\left(\frac{R-r}{r}\right)\theta(t)$.

   b. Describe the position of P as a function of time.

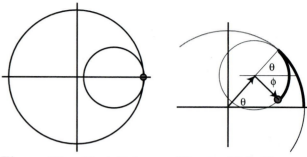

Figure 12.a *The initial position of P.*

Figure 12.b *The position of P at time t.*

36. **Piston Problem 1:**

   Let D be a disc of radius R centered at the origin. One end of a rod of length $L \geq R$ is attached to the point $(R, 0)$ on the disc. The other end of the rod is constrained to lie on the positive y–axis and it is attached to a piston which moves inside an air tight chamber as in Figure 13. As the disc rotates, the piston will move up and down in its chamber. Let $\vec{v}$ be the vector drawn emanating from the origin and terminating at the point p where the rod is attached to the cylinder. Let θ be the angle that $\vec{v}$ makes with the horizontal, and let y denote the distance from the origin to the end of the rod that is attached to the piston.

   a. Express y as a function of θ.

   b. Assume that when the piston is at its topmost position that it completely fills its chamber. Assuming that the chamber has a radius of ρ units, express the enclosed chamber volume as a function of θ.

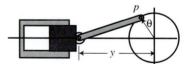

Figure 13. *The piston from Exercise 36.*

Chapter 3

Limits, Continuity and Derivatives

3.1   Average Velocity and Average Rate of Change

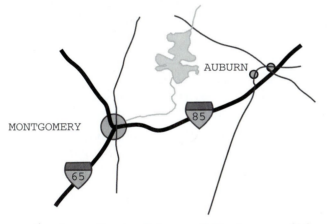

Figure 1. *A map showing Auburn and Montgomery, Alabama.*

Average Velocity

In Figure 1, we display a map featuring a portion of Alabama. If we were to drive from Auburn to Montgomery in an hour, and if our odometer showed that we had driven 55 miles, then we would know that we had averaged 55 miles/hour. The idea of average speed is a natural concept familiar to all of us. Average speed is simply distance traveled divided by time elapsed. In physics and engineering, there are two notions closely related to distance and average speed: *displacement* and *average velocity*. Both of these are vector quantities. The displacement in our trip from Auburn to Montgomery would be

91

the vector represented by the directed line segment emanating from our starting point in Auburn and ending in Montgomery. The length would be the distance "as the crow flies." The units of displacement are the same as the units of distance; i.e., miles, feet, meters, etc. The average velocity is displacement divided by time elapsed.

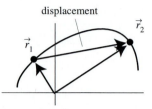

Figure 2.a
The displacement.

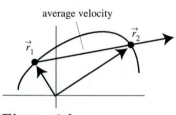

Figure 2.b
The displacement scaled by $\frac{1}{change\ in\ time}$.

Definition: Average Velocity

Suppose that a particle is moving in $\mathbb{R}^n$, and that it is at $\vec{r}_1$ at time t_1 and at $\vec{r}_2$ at time t_2.

The *displacement*[1] in moving from $\vec{r}_1$ to $\vec{r}_2$ is the *vector* $\vec{r}_2 - \vec{r}_1$.

The *average velocity*[2] is:

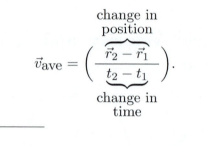

$$\vec{v}_{\text{ave}} = \left(\frac{\overbrace{\vec{r}_2 - \vec{r}_1}^{\text{change in position}}}{\underbrace{t_2 - t_1}_{\text{change in time}}} \right).$$

---

[1] See Figure 2.a
[2] See Figure 2.b

As Figures 2.a and 2.b show, the average velocity has the same direction as the displacement.

The average velocity can also be computed in terms of the coordinate functions for $\vec{r}$, that is, if $\vec{r}_1 = (x_1, y_1)$ and $\vec{r}_2 = (x_2, y_2)$, then the average velocity can be expressed as:

$$\vec{v}_{\text{ave}} = \left(\underbrace{\overbrace{\frac{x_2 - x_1}{t_2 - t_1}}^{\text{Change in } x}}_{\text{Change in } t}, \underbrace{\overbrace{\frac{y_2 - y_1}{t_2 - t_1}}^{\text{Change in } y}}_{\text{Change in } t} \right).$$

It is important to observe that both displacement and average velocity are vector quantities, and neither displacement nor average velocity depends upon the path of the particle or the total distance traveled in getting from $\vec{r}_1$ to $\vec{r}_2$. The following examples illustrate these concepts.

EXAMPLE 1: If a particle moves from $(3, 2, 1)$ ft to $(-1, 3, 4)$ ft

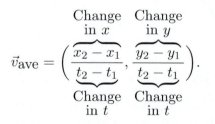

in 2 seconds, then the displacement would be:

$$\vec{d} = (-1, 3, 4) - (3, 2, 1) \text{ ft} = (-4, 1, 3) \text{ ft},$$

and the average velocity would be:

$$\vec{v} = \frac{\vec{d}}{2} \text{ ft/sec} = \left(-2, \frac{1}{2}, \frac{3}{2}\right) \text{ ft/sec.} \qquad \blacksquare$$

EXAMPLE 2: Suppose that a particle's motion is described by

$$\vec{r}(t) = (\cos 2\pi t, \sin 2\pi t) \text{ m.}$$

It is rotating on a circle with radius one m, one rotation/sec. Since the circle's circumference is 2π m, the average speed of the particle, over any time interval, would be 2π m/sec. The following figures illustrate that things are not so intuitive when it comes to average velocity. $\blacksquare$

Figure 3.a.
Let $t_1 = 0$ sec and $t_2 = \frac{1}{4}$ sec. The displacement on this time interval is

$$\vec{d} = \vec{r}(t_2) - \vec{r}(t_1) = (0, 1) - (1, 0) = (-1, 1) \text{ m.}$$

The average velocity is

$$\vec{v}_{ave} = \frac{\vec{d}}{t_2 - t_1} = \frac{(-1, 1)}{1/4} = (-4, 4) \text{ m/s.}$$

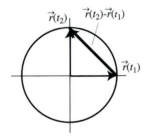

Figure 3.b.
Let $t_1 = 0$ sec and $t_2 = \frac{1}{2}$ sec. The displacement on this interval is

$$\vec{d} = \vec{r}\left(\frac{1}{2}\right) - \vec{r}(0) = (-1, 0) - (1, 0) = (-2, 0) \text{ m.}$$

The average velocity is

$$\vec{v}_{ave} = \frac{\vec{d}}{t_2 - t_1} = \frac{(-2, 0)}{1/2} = (-4, 0) \text{ m/s.}$$

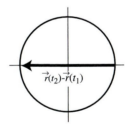

Figure 3.c.
Let $t_1 = 0$ sec and $t_2 = \frac{3}{4}$ sec. The displacement on this time interval is

$$\vec{d} = \vec{r}(t_2) - \vec{r}(t_1) = (0, -1) - (1, 0) = (-1, -1) \text{ m.}$$

The average velocity is

$$\vec{v}_{ave} = \frac{\vec{d}}{t_2 - t_1} = \frac{(-1, -1)}{3/4} = \left(-\frac{4}{3}, \frac{4}{3}\right) \text{ m/s.}$$

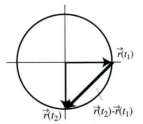

Figure 3.d.

Let $t_1 = 0$ sec and $t_2 = 1$ sec. The displacement on this interval is

$$\vec{d} = \vec{r}(1) - \vec{r}(0) = (1,0) - (1,0) = (0,0) \ m.$$

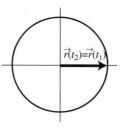

The average velocity is

$$\vec{v}_{ave} = \frac{\vec{d}}{t_2 - t_1} = \frac{(0,0)}{1} = (0,0) \ m/s.$$

Average Rate of Change

Average velocity is a special case of the more general concept of *average rate of change*. If A is a subset of $\mathbb{R}$ and $\vec{f}: A \to \mathbb{R}^n$ is a function, then we can compute the change in $\vec{f}(t)$ relative to the change in t.

Definition: Average Rate of Change

Suppose that $\vec{f}: \mathbb{R} \to \mathbb{R}^n$ is a function from the real numbers to $\mathbb{R}^n$ and that t_1 and t_2 are two different numbers. Then the *average rate of change* of $\vec{f}(t)$ from t_1 to t_2 is given by:

$$\frac{\vec{f}(t_2) - \vec{f}(t_1)}{t_2 - t_1}.$$

Again, we note in the case of $f: \mathbb{R}^3 \to \mathbb{R}^3$, we can express $\vec{f}(t)$ in terms of coordinate functions $(x(t), y(t), z(t))$. The rate of change can be expressed as

$$\left(\underbrace{\frac{\overbrace{x(t_2) - x(t_1)}^{\substack{\text{change} \\ \text{in } x}}}{t_2 - t_1}, \frac{\overbrace{y(t_2) - y(t_1)}^{\substack{\text{change} \\ \text{in } y}}}{t_2 - t_1}, \frac{\overbrace{z(t_2) - z(t_1)}^{\substack{\text{change} \\ \text{in } z}}}{t_2 - t_1}}_{\text{change in time}} \right).$$

Average velocity, which is just the average rate of change of position, has provided us with some examples of vector-valued functions. We have seen that the average rate of change of a vector-valued function can be computed by evaluating the rates of change of its coordinate functions, which are real-valued functions. As the next few examples show, there are many applications of the notion of rates of change of real-valued functions.

EXAMPLE 3: A circle of radius r is expanding. What is the average rate of change of the circumference of the circle from $r_1 = 1$ m to $r_2 = 2$ m?

SOLUTION: The circumference C is a function of the radius, given by $C(r) = 2\pi r$. Thus the average rate of change is

$$\frac{C(r_2) - C(r_1)}{r_2 - r_1} = \frac{2\pi(2) - 2\pi(1) \text{ m}}{1 \text{ m}} = 2\pi.$$

Notice that the rate of change of the circumference of the circle due to a change in its radius is a unit free or dimensionless quantity. ■

EXAMPLE 4: A sphere of radius r is expanding. What is the average rate of change of the volume of the sphere from $r_1 = 1$ m to $r_2 = 2$ m? From $r_1 = 1$ m to $r_3 = 1.5$ m? From $_1r = 1$ m to $r_4 = 1.1$ m?

SOLUTION: The volume V is a function of the radius, given by $V(r) = \frac{4}{3}\pi r^3$. Thus the average rate of change of V from r_1 to r_2 is:

$$\frac{V(r_2) - V(r_1)}{r_2 - r_1} = \frac{\frac{4}{3}\pi(2)^3 - \frac{4}{3}\pi(1)^3 \text{ m}^3}{1 \text{ m}} = \frac{28\pi}{3} \text{ m}^2.$$

For r_1 to r_3:

$$\frac{V(r_3) - V(r_1)}{r_2 - r_1} = \frac{\frac{4}{3}\pi(1.5)^3 - \frac{4}{3}\pi(1)^3 \text{ m}^3}{0.5 \text{ m}} = \frac{19\pi}{3} \text{ m}^2.$$

For r_1 to r_4:

$$\frac{V(r_4) - V(r_1)}{r_4 - r_1} = \frac{\frac{4}{3}\pi(1.1)^3 - \frac{4}{3}\pi(1)^3 \text{ m}^3}{0.1 \text{ m}} = \frac{331\pi}{75} \text{ m}^2. \qquad ■$$

EXAMPLE 5: What is the average rate of change of the volume V of a sphere as the surface area S changes from $S_1 = 2\pi$ m^2 to $S_2 = 3\pi$ m^2?

SOLUTION: We know that the volume of a sphere is given by $V(r) = \frac{4}{3}\pi r^3$ and that the surface area is given by $S(r) = 4\pi r^2$. Now, solve this latter equation for r in terms of S to get

$$r = \sqrt{\frac{S}{4\pi}} = \frac{1}{2}\sqrt{\frac{S}{\pi}}.$$

Substitute this expression into the one for V to obtain $V(S) = \dfrac{S^{3/2}}{6\sqrt{\pi}}$.

Thus the average rate of change is:

$$\frac{V(3\pi) - V(2\pi)}{3\pi - 2\pi} = \frac{\frac{(3\pi)^{3/2}}{6\sqrt{\pi}} - \frac{(2\pi)^{3/2}}{6\sqrt{\pi}} \text{ m}^3}{3\pi - 2\pi \text{ m}^2}$$

$$= \frac{3^{3/2} - 2^{3/2}}{6} \text{ m} \approx 0.395 \text{ m.} \qquad \blacksquare$$

EXAMPLE 6: Let $y(x) = \sqrt{x}$. What is the average rate of change of y as x changes from $x = 1$ to $x = 2$? From $x = 1$ to $x = 1.5$? From $x = 1$ to $x = 1.1$?

SOLUTION: The average rate of change from $x = 1$ to $x = 2$ is given by

$$\frac{y(2) - y(1)}{2 - 1} = \frac{\sqrt{2} - \sqrt{1}}{1} = \sqrt{2} - 1 \approx 0.414.$$

For $x = 1$ to $x = 1.5$:

$$\frac{y(1.5) - y(1)}{1.5 - 1} = \frac{\sqrt{1.5} - \sqrt{1}}{0.5} \approx 0.449.$$

For $x = 1$ to $x = 1.1$:

$$\frac{y(1.1) - y(1)}{1.1 - 1} = \frac{\sqrt{1.1} - \sqrt{1}}{0.1} \approx 0.488. \qquad \blacksquare$$

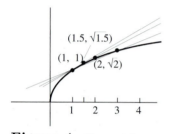

Figure 4. *If x_1 and x_2 are close together, then the average rate of change of a real-valued function as the x–coordinate changes from x_1 to x_2 approximates the slope of the line tangent to the graph of the function.*

In Example 6, the average rate of change of y as x changes from $x = 1$ to $x = 1.5$ is simply the slope of the line containing the points $(1, y(1))$ and $(1, y(1.5))$. Similarly, the average rate of change of y as x changes from $x = 1$ to $x = 1.1$ is simply the slope of the line containing the points $(1, y(1))$ and $(1, y(1.1))$. (See Figure 4.) This illustrates the fact that the rate of change for a function from a subset of R^1 to R^1 is, in fact, the slope of the secant line of the graph of the function through the two points (x_1, y_1) and (x_2, y_2). Notice also that as x_2 gets close to x_1, this slope gets close to the slope of the tangent line at x_1.

EXAMPLE 7: Continuing with Example 6, let $x_1 = 1$ and $x_2 = 1.01$. The rate of change is given by:

$$\frac{y(1.01) - y(1)}{1.01 - 1} = \frac{\sqrt{1.01} - \sqrt{1}}{0.01} \approx 0.499.$$

It appears that as x_2 gets close to $x = 1$, the average rate of change (the slope of the secant line) gets close to the value 0.5. Is the slope of the tangent line at $x = 1$ equal to 0.5? In Section 3.2 we show that this is indeed the case.

■

EXERCISES 3.1

1. Figure 5.a displays the path of a particle as a function of time (measured in sec).

   a. Find the displacement vectors from $t = 1$ sec to $t = 2$ sec and from $t = 1$ sec to $t = 1.5$ sec.

   b. Find the average velocity vectors from $t = 1$ sec to $t = 2$ sec and from $t = 1$ sec to $t = 1.5$ sec.

2. Figure 5.b displays the path of a particle as a function of time (measured in sec).

   a. Find the displacement vectors from $t = 1$ sec to $t = 2$ sec and from $t = 1$ sec to $t = 1.5$ sec.

   b. Find the average velocity vectors from $t = 1$ sec to $t = 2$ sec and from $t = 1$ sec to $t = 1.5$ sec.

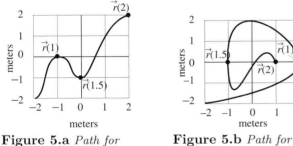

Figure 5.a *Path for Exercise 1.*   **Figure 5.b** *Path for Exercise 2.*

In Exercises 3–6, a man drives his car east for 100 mi and then turns around and drives west for 50 mi. It takes 2 hr for the entire trip.

3. What is the total distance traveled?

4. What is his displacement?

5. What is his average velocity?

6. What is his average speed?

In Exercises 7–10, a particle moves in a straight line from $(1, 3, -1)$ ft to $(0, 0, 0)$ ft in 2 sec and then it moves in a straight line to $(1, 1, 1)$ ft in 3 sec.

7. What is the total distance traveled?

8. What is the particle's displacement?

9. What is the average velocity?

10. What is the average speed?

In Exercises 11–13, a particle moves along the x–axis with its position at time t given by $x(t) = 4t^2 - 6t$ meters.

11. During the time interval from $t = 0$ sec to $t = 2$ sec, what is the displacement?

12. During the time interval from $t = 0$ sec to $t = 2$ sec, what is the average velocity?

13. What is the average velocity for the particle for the time interval starting at $t = 0$ sec and ending at $t = \frac{3}{2}$ sec?

In Exercises 14–16, $\vec{r}(t) = (4 + t, t - 2t^2)$ m. Find the average velocity for the time intervals.

14. 0 to 4 sec.

15. 1 to 2 sec.

16. 3 to 5 sec.

In Exercises 17–20, a particle moves with constant speed in a circle with a radius of 5 m in a counter-clockwise direction. It makes 1 revolution every 16 sec.

17. What is the average speed for 1/4 revolution?

18. What is the average velocity for 1/4 revolution?

19. What is the average speed for 5/4 revolutions?

20. What is the average velocity for 5/4 revolutions?

In Exercises 21–24, a right triangle expands so that the base b is always twice as long as the height. Find the average rate of change of the area relative to the change in the length of the base for the given change of b.

21. b changes from 2 in to 3 in.

22. b changes from 2 in to 2.5 in.

23. b changes from 2 in to 2.1 in.

24. b changes from 2 in to 2.01 in.

25. Particle A moves along the x–axis with its position given by $A(t) = (3t, 0)$ m. Particle B is moving along the y–axis with its position given by $B(t) = (0, 4t)$ m. Find the average rate of change of the distance between A and B relative to time when t changes from $t = 1$ sec to $t = 1.5$ sec.

In Exercises 26–28, a right circular cone is expanding in such a way that the height h always remains twice the length of the radius r. Find the average rate of change of the volume of the cone for the given change in r. The volume of a right circular cone is given by $V = \dfrac{\pi r^2 h}{3}$.

26. r changes from 3 ft to 3.5 ft.

27. r changes from 3 ft to 3.1 ft.

28. r changes from 3 ft to 3.01 ft.

In Exercises 29–34, a certain species of bacteria, under suitable conditions, is observed to have a population of

$$P(t) = \frac{1000 \cdot 3^t}{9 + 3^t} \text{ per square mm,}$$

where t is measured in hr. Find the average rate of change of the population for the given time interval.

29. t changes from $t_0 = 1$ to $t_1 = 2$ hr.

30. t changes from $t_0 = 1$ to $t_1 = 1.5$ hr.

31. t changes from $t_0 = 1$ to $t_1 = 1.1$ hr.

32. t changes from $t_0 = 5$ to $t_1 = 6$ hr.

33. t changes from $t_0 = 5$ to $t_1 = 5.5$ hr.

34. t changes from $t_0 = 5$ to $t_1 = 5.1$ hr.

35. Graph the lines passing through $(t_0, P(t_0))$ and $(t_1, P(t_1))$ for Exercises 29–31 against the curve $P(t)$. (Use your calculator to obtain a sketch of P.)

36. Graph the lines passing through $(t_0, P(t_0))$ and $(t_1, P(t_1))$ for Exercises 32–34 against the curve $P(t)$. (Use your calculator to obtain a sketch of P.)

3.2   Limits: An Intuitive Approach

EXAMPLE 1: In archery, the goal is not necessarily to hit the very center of the target, but simply to hit the bull's-eye. You probably shoot for the center of the target, but there can be some error in your aim since your goal is merely to hit inside a circle containing the center. How much error you can allow and still hit the bulls-eye depends only on the diameter of the bull's-eye. Make the bull's-eye smaller, and your aim must be better.

EXAMPLE 2: A machinist is to make a wheel with a circumference of 4π m. The wheel, then, must have a radius of 2 m. While it is unlikely that he can be exact in his work, he can machine the wheel to a radius of 2 m with an error within a specified tolerance.

Both of these examples illustrate an idea that is fundamental to the study of calculus: the notion of limits. We will study limits more closely and use them extensively in the sections to come. In this section, we give an intuitive introduction to the notion. In Section 6 we give a more formal treatment. An in-depth study of limits more properly belongs in an advanced calculus course.

Definition: Limits, an Intuitive Definition

Suppose that

(a) A is a subset of $\mathbb{R}$,

(b) f is a function from A into $\mathbb{R}$,

(c) z is a number (not necessarily in A),

(d) there are numbers a and b so that the segments $a < x < z$ and $z < x < b$ are subsets of A; and

(e) $f(x)$ is close to the number L whenever x is close to z, but $\neq z$. Moreover, the error in the approximation of L by f can be made as small as we need by being sure that x is chosen "close enough" to z.

Then we say that

$$L \text{ is the limit of } f(x) \text{ as } x \text{ approaches } z,$$

and we write

$$L = \lim_{x \to z} f(x).$$

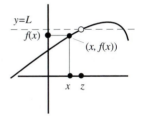

Figure 1. *As x gets close to z, $f(x)$ gets close to L and $(x, f(x))$ gets close to the line $y = L$.*

The idea is that we can control how close $f(x)$ is to L by controlling how close x is to z (being careful that $x \neq z$) as in Figure 1. In Example 2, the machinist controls the error in the circumference of the wheel by controlling the accuracy of the length of the radius of the wheel. In this example, $f(x) = 2\pi x$ and $z = 2$.

EXAMPLE 3: Let f be defined by

$$f(x) = \begin{cases} -1 & \text{if } x \leq 1 \\ x & \text{if } x > 1 \end{cases}$$

The function is sketched in Figure 2. As x gets close to 1 from the left, $f(x) = -1$, but as x gets close to 1 from the right, $f(x)$ gets close to 1. Thus, there is no single number that is well-approximated by $f(x)$ as x gets close to the number 1. In this case, $\lim_{x \to 1} f(x)$ does not exist. ∎

EXAMPLE 4: Let g be defined by

$$g(x) = \begin{cases} x^2 & \text{if } x \neq 3 \\ 1 & \text{if } x = 3. \end{cases}$$

The function g is sketched in Figure 3. Notice that $\lim_{x \to 3} g(x) = 9$ even though $g(3) = 1$. ∎

EXAMPLE 5: Let $y(x) = \sqrt{x}$. Let

$$f(h) = \frac{y(1+h) - y(1)}{h}.$$

Find $\lim_{h \to 0} f(h)$.

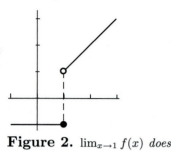

Figure 2. $\lim_{x \to 1} f(x)$ *does not exist.*

SOLUTION: In Examples 6 and 7 of Section 3.1, we evaluated $f(1)$, $f(0.5)$, $f(0.1)$, and $f(0.01)$. The function $f(h)$ is the average rate of change of y as x changes from $x = 1$ to $x = 1 + h$. At the end of the examples, we speculated that $f(h)$ is getting close to 0.5 as h gets close to 0. That is, we guessed that perhaps $\lim_{h \to 0} f(h) = 0.5$. The difficulty in evaluating $\lim_{h \to 0} f(h)$ is that we are looking at a quotient where both the numerator $f(h)$ and the denominator h are getting close to 0. We can do some algebra to get a better look at what is happening to the ratio.

In a standard high school algebra course, you learned to rationalize the denominator. That is, if you were given an expression like $\frac{1}{\sqrt{2}}$ you would multiply both the numerator and denominator by $\sqrt{2}$ to obtain $\frac{\sqrt{2}}{2}$. Similarly, if the expression were

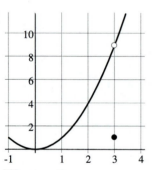

Figure 3.
$\lim_{x \to 3} g(x) \neq g(3).$

$$\frac{1}{1 - \sqrt{3}},$$

you would multiply the numerator and denominator by $1 + \sqrt{3}$ to obtain:

$$\frac{1}{1 - \sqrt{3}} = \left(\frac{1}{1 - \sqrt{3}}\right)\left(\frac{1 + \sqrt{3}}{1 + \sqrt{3}}\right) = \frac{1 + \sqrt{3}}{1 - 3} = \frac{1 + \sqrt{3}}{-2}.$$

In the present example, it is useful to "rationalize the numerator" instead of the denominator.

$$\lim_{h \to 0} f(h) = \lim_{h \to 0} \frac{y(1+h) - y(1)}{h} = \lim_{h \to 0} \frac{\sqrt{1+h} - 1}{h}$$

$$= \lim_{h \to 0} \left(\frac{\sqrt{1+h}-1}{h} \right) \left(\frac{\sqrt{1+h}+1}{\sqrt{1+h}+1} \right)$$

$$= \lim_{h \to 0} \frac{(\sqrt{1+h})^2 - 1}{h(1 + \sqrt{1+h})}$$

$$= \lim_{h \to 0} \frac{h}{h(1 + \sqrt{1+h})} = \lim_{h \to 0} \frac{1}{1 + \sqrt{1+h}}.$$

Finally, we are no longer dividing a number that is close to 0 by another number that is close to 0 and we can now evaluate the limit.

$$\lim_{h \to 0} \frac{\sqrt{1+h}-1}{h} = \lim_{h \to 0} \frac{1}{1 + \sqrt{1+h}} = \frac{1}{2}. \qquad \blacksquare$$

EXERCISES 3.2

In Exercises 1–12, compute
$\lim_{h \to 0} \left(f(x_0 + h) - f(x_0) \right) / h.$

1. $f(x) = 3x^2, \quad x_0 = 1.$

2. $f(x) = ax^2 + x, \quad x_0 = 3.$

3. $f(x) = (2x^2 + x)^2, \quad x_0 = -2.$

4. $f(x) = x^{-2}, \quad x_0 = 3.$

5. $f(x) = (2x^2 + 3x + 1)^{-1}, \quad x_0 = 0.$

6. $f(x) = x^4, \quad x_0 = -1.$

7. $f(x) = x^{-3}, \quad x_0 = -1.$

8. $f(x) = x^5, \quad x_0 = -3.$

9. $f(x) = \sqrt{x+1}, \quad x_0 = 3.$

10. $f(x) = x^{3/2}, \quad x_0 = 9.$

11. $f(x) = \sqrt{x^2 + 1}, \quad x_0 = 0.$

12. $f(x) = x^{-1/2}, \quad x_0 = 4.$

In Exercises 13–16, evaluate $f(0.1)$, $f(0.01)$, $f(0.001)$, and $f(0.0001)$. Based on this data, what would you guess $\lim_{x \to 0} f(x)$ is?

13. $f(x) = \dfrac{\sin x}{x}.$

14. $f(x) = \dfrac{\sin x}{\sqrt{x}}.$

15. $f(x) = \dfrac{1 - \cos x}{x}.$

16. $f(x) = \dfrac{1 - \cos x}{x^2}.$

17. Let $f(x) = \dfrac{\ln(x^x)}{\ln(x)}.$

(On your calculator there should be keys marked ln or $\log_e$.) Pick numbers closer and closer to 1 and compute this function. What is your guess for $\lim_{x \to 1} f(x)$? The reasons for this behavior will be explained in a later chapter.

3.3 Instantaneous Rate of Change: The Derivative

Instantaneous Velocity

Suppose that we drive our car from Auburn to Montgomery and we average 55 mi/hr. During the trip, our speedometer varies; at one time it may read 60 mi/hr, while at another time it may read 50 mi/hr (as this writer's car does when going uphill!). At any time in the trip, the speedometer reading is, most likely, not equal to the average speed, but it is related to it. The quantity that the speedometer is attempting to measure is called *instantaneous speed*. A notion that is similarly related to average velocity is *instantaneous velocity*. Let $\vec{r}(t)$ denote the position of a particle moving in space. If h is a number, then

$$\vec{v}_{\text{ave}} = \frac{\vec{r}(t+h) - \vec{r}(t)}{h} \tag{1}$$

is the average velocity over the time period from t to $t+h$. See Figures 1.a and b.

If there is a vector-valued function defined by

$$\vec{v}(t) = \lim_{h \to 0} \frac{\vec{r}(t+h) - \vec{r}(t)}{h},$$

then $\vec{v}(t)$ is called the *instantaneous velocity* of the particle. It is often denoted by $\vec{r}'(t)$ or $d\vec{r}/dt$ where the $d\vec{r}$ is meant to suggest the change in $\vec{r}$, versus the change dt in t . In Figure 1.b, we illustrate our approximations for $\vec{v}(t)$ (at time $t = 0$) for values of h of 1, 0.5, 0.25 and 0.1. We emphasize that $\vec{v}(t)$ is a vector.

We can determine the units of measurement of velocity by looking at its definition. Velocity is the limit of a ratio $(\vec{r}(t+h) - \vec{r}(t))/h$. The numerator of this ratio $(\vec{r}(t+h) - \vec{r}(t))$ has units of length. The denominator h is a change in time. Thus the ratio must be measured in (units of length)/(units of time).

Notice in Figure 1.b that as h gets small, the vector $\vec{v}_{\text{ave}}$ given by Equation (1) is a vector that gets close to a direction vector for the line tangent to the path of the particle at the position $\vec{r}(t)$.

Because of our familiarity with speedometers on cars, for example, the notion of instantaneous velocity is much closer to our intuitive notion of velocity than that of average velocity. From this point on **"velocity" will mean instantaneous velocity unless specifically stated otherwise**. The *instantaneous speed* of the particle is defined to be the magnitude of the velocity.

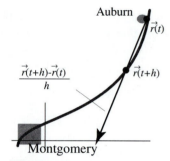

Figure 1.a *Average velocity for the time period from t to $t+h$.*

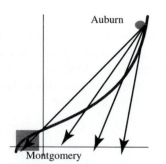

Figure 1.b *Approximations for $\vec{v}(t)$ (at time $t = 0$) for values of h of 1, 0.5, 0.25, and 0.1.*

The instantaneous speed at time $t = \|\vec{v}(t)\| = v(t)$.

We will have to postpone the full development of the idea of rate of change of distance traveled, but it is true that instantaneous speed measures the rate of change of distance. This is the quantity measured by the speedometer.

The important facts about velocity we have at this point are given in the following table.

Facts about velocity

Let $\vec{r}(t)$ give the position of an object at time t.

- The velocity $\vec{v}(t) = \vec{r}'(t)$ is a vector that is tangent to the object's path at $\vec{r}(t)$.

- The units of measurement of velocity are (units of length)/(units of time).

- The speed of the object is $\|\vec{v}(t)\|$, which is also measured in (units of length)/(units of time).

We first consider an object moving along a straight line. If the path is parametrized by $\vec{r} = \vec{v}t + \vec{b}$ then this motion is very easy to describe as the following theorem shows.

Theorem 1 *Let $\vec{r}(t) = \vec{v}t + \vec{b}$ be a parametrization of the straight line with direction vector $\vec{v}$ and containing the vector $\vec{b}$. Then $\vec{r}$ describes a particle moving in a straight line with constant velocity $\vec{v}$ at position $\vec{b}$ at time $t = 0$.*

Proof: The velocity of the particle is approximated by

$$\frac{\vec{r}(t+h) - \vec{r}(t)}{h} = \frac{\vec{v}(t+h) + \vec{b} - (\vec{v}t + \vec{b})}{h} = \vec{v}. \qquad \blacksquare$$

EXAMPLE 1: A particle is moving in a straight line with constant velocity $\vec{v} = (2, -1, 4)$ m/s. It is at the position $\vec{b} = (1, 4, -2)$ m at time $t = 0$ sec. From Theorem 1, the parametrization for the path of the particle may be written:

$$\vec{r}(t) = \vec{v}t + \vec{b} = \begin{pmatrix} 2 \\ -1 \\ 4 \end{pmatrix} t + \begin{pmatrix} 1 \\ 4 \\ -2 \end{pmatrix} = \begin{pmatrix} 2t + 1 \\ -t + 4 \\ 4t - 2 \end{pmatrix} \text{ m.} \qquad \blacksquare$$

EXAMPLE 2: As in Example 1, a particle is moving in a straight line with constant velocity $\vec{v} = (2, -1, 4)$ m/s but it is at the position $(1, 4, -2)$ m at time $t = 1$ s. The parametrization for the particle's motion will be of the form $\vec{r}(t) = \vec{v}t + \vec{b}$. The velocity is $(2, -1, 4)$ m/s, but we have to work a little to find $\vec{b}$.

$$\vec{r}(t) = \vec{v}t + \vec{b} = (2, -1, 4)t + \vec{b} \text{ m.}$$

$$\vec{r}(1) \;\;=\;\; \vec{v} + \vec{b} = (2, -1, 4) + \vec{b} = (1, 4, -2) \text{ m}.$$

Solving for $\vec{b}$ we obtain:

$$\vec{b} = (1, 4, -2) - (2, -1, 4) = (-1, 5, -6) \text{ m};$$

and so

$$\vec{r}(t) = (2, -1, 4)t + (-1, 5, -6) \text{ m}. \qquad \blacksquare$$

Theorem 1 and Examples 1 and 2 demonstrate that straight line motion with constant velocity is relatively simple. However, things get more complicated for the general case. Let $\vec{r}$ parametrize the motion of a particle in $\mathbb{R}^2$. Then $\vec{r}(t) = (x(t), y(t))$ and

$$\vec{v}(t) \;\;=\;\; \lim_{h \to 0} \frac{\vec{r}(t+h) - \vec{r}(t)}{h} = \lim_{h \to 0} \begin{pmatrix} \frac{x(t+h) - x(t)}{h} \\ \frac{y(t+h) - y(t)}{h} \end{pmatrix}$$

$$=\;\; \begin{pmatrix} \lim_{h \to 0} \frac{x(t+h) - x(t)}{h} \\ \lim_{h \to 0} \frac{y(t+h) - y(t)}{h} \end{pmatrix}.$$

EXAMPLE 3: Find $\vec{v}(t)$ if $\vec{r}(t) = \left(t, t^2\right)$ m.

SOLUTION: $x(t) = t$ m and $y(t) = t^2$ m.

$$\lim_{h \to 0} \frac{\vec{r}(t+h) - \vec{r}(t)}{h}$$

$$=\;\; \begin{pmatrix} \lim_{h \to 0} \frac{x(t+h) - x(t)}{h} \\ \lim_{h \to 0} \frac{y(t+h) - y(t)}{h} \end{pmatrix} = \begin{pmatrix} \lim_{h \to 0} \frac{(t+h) - t}{h} \\ \lim_{h \to 0} \frac{(t+h)^2 - t^2}{h} \end{pmatrix}$$

$$=\;\; \begin{pmatrix} \lim_{h \to 0} \frac{h}{h} \\ \lim_{h \to 0} \frac{t^2 + 2th + h^2 - t^2}{h} \end{pmatrix} = \begin{pmatrix} 1 \\ \lim_{h \to 0} 2t + h \end{pmatrix}$$

$$=\;\; \begin{pmatrix} 1 \\ 2t \end{pmatrix} \text{ m/s}.$$

Thus, $\vec{v}(t) = (1, 2t)$ m/s. $\qquad \blacksquare$

EXAMPLE 4: Let $\vec{r}$ be as in Example 3. To find the velocity at $t = 1$, we evaluate $\vec{v}$ at $t = 1$:

$$\vec{v}(1) = (1, 2). \qquad \blacksquare$$

Instantaneous Rate of Change

Just as average velocity is a special case of rate of change, so also is instantaneous velocity a special case of instantaneous rate of change.

Definition: The Derivative

Let A be a subset of $\mathbb{R}$ and $\vec{f}: A \to \mathbb{R}^n$. If there is a function $\vec{f}'$ defined by

$$\lim_{h \to 0} \frac{\vec{f}(t+h) - \vec{f}(t)}{h},$$

then $\vec{f}'(t)$ is the *instantaneous rate of change*, or *derivative*, of $\vec{f}$ at t.

We use other symbols to denote the derivative of f, namely, df/dt and Df. These are all common notations and we use them freely. The computation of the derivative is called *differentiation*; we say we *differentiate* f to obtain f'. The physical/geometric intuition we want for the derivative of a function is that it is the (instantaneous) rate of change of the *image* of the function.

EXAMPLE 5: We already know that if A is a subset of $\mathbb{R}$ and $\vec{f}: A \to \mathbb{R}^n$, then $\vec{f}'(t)$ equals the velocity at time t, which is the instantaneous rate of change of position. Let $\vec{f}(t) = (x(t), y(t), z(t))$. We have

$$\vec{f}'(t) = \lim_{h \to 0} \frac{\vec{f}(t+h) - \vec{f}(t)}{h}$$

$$= \lim_{h \to 0} \begin{pmatrix} \frac{x(t+h)-x(t)}{h} \\ \frac{y(t+h)-y(t)}{h} \\ \frac{z(t+h)-z(t)}{h} \end{pmatrix} = \begin{pmatrix} \lim_{h \to 0} \frac{x(t+h)-x(t)}{h} \\ \lim_{h \to 0} \frac{y(t+h)-y(t)}{h} \\ \lim_{h \to 0} \frac{z(t+h)-z(t)}{h} \end{pmatrix}$$

$$= \begin{pmatrix} x'(t) \\ y'(t) \\ z'(t) \end{pmatrix} = \begin{pmatrix} \text{rate of change of the } x\text{--coordinate function} \\ \text{rate of change of the } y\text{--coordinate function} \\ \text{rate of change of the } z\text{--coordinate function} \end{pmatrix}.$$

This is a particularly nice result because it means that if we can find the derivative, or rate of change, of real-valued functions, then we can find the derivative of vector-valued functions. ∎

EXAMPLE 6: In Examples 11 and 14 we will see that if $x(t) = 2t^3$ then $x'(t) = 6t^2$, and if $y(t) = \sqrt{t}$ then $y'(t) = \frac{1}{2\sqrt{t}}$. Thus, if $\vec{r}(t) = (2t^3, \sqrt{t})$, then $\vec{r}'(t) = (6t^2, \frac{1}{2\sqrt{t}})$. ∎

We have, for the most part, been focusing on the derivative as applied to the rate of change of position of a particle moving in space. The next few examples illustrate that derivatives of real-valued functions are important in their own right.

EXAMPLE 7: As in Example 3 of Section 3.1, let $C(r) = 2\pi r$, which is the circumference of a circle of radius r. Find the derivative of $C(r)$ at $r = 1$ m.

SOLUTION: $C(r) = 2\pi r$. The average rate of change from r to $r + h$ is given by:
$$\frac{2\pi(r + h) - 2\pi r}{h} = \frac{2\pi h}{h} \frac{\text{m}}{\text{m}} = 2\pi.$$

Notice that the average rate of change does not depend upon either r or h. Thus we see that for any r, $C'(r)$ is simply 2π, the slope of line $C(r) = 2\pi r$. ∎

EXAMPLE 8: Let $V(r) = \frac{4}{3}\pi r^3$, which is the volume of a sphere of radius r. Find the derivative of $V(r)$ at $r = 1$ m.

SOLUTION: The average rate of change from $r = 1$ to $r = 1 + h$ is given by:

$$\frac{\frac{4}{3}\pi(1 + h)^3 \text{ m}^3 - \frac{4}{3}\pi(1)^3 \text{ m}^3}{h \text{ m}} = \frac{4}{3}\pi\left(\frac{(1 + h)^3 - 1}{h}\right) h \text{ m}^2$$

$$= \frac{4}{3}\pi\left(\frac{1 + 3h + 3h^2 + h^3 - 1}{h}\right) \text{ m}^2$$

$$= \frac{4}{3}\pi\left(3 + 3h + h^2\right) \text{ m}^2.$$

Now, as h gets close to 0 , both $3h$ and h^2 get close to 0. Thus

$$V'(1) = \lim_{h \to 0} \frac{\frac{4}{3}\pi(1 + h)^3 - \frac{4}{3}\pi(1)^3}{h} \text{ m}^2$$

$$= \lim_{h \to 0} \frac{4}{3}\pi\left(3 + 3h + h^2\right) \text{ m}^2 = 4\pi \text{ m}^2. \quad ∎$$

Slopes of Tangent Lines

Up to now we have considered the derivative as the instantaneous rate of change. However, there is also a geometric view of the derivative. Just as the slope of a straight line gives us geometric insight into the "steepness" of a line, the derivative gives us geometric insight into how "steep" a curve is at a particular point on it. We begin by discussing tangent lines to graphs.

If f is a function from a subset of $\mathbb{R}$ into $\mathbb{R}$, then the line **L** is tangent to f at x if:

1. **L** is the *only line* that is "flat against f" at $(x, f(x))$; and

2. **L** contains the point $(x, f(x))$.

It is important to note that x must be in the domain of f in order for there to be a tangent line at x.

EXAMPLE 9: Consider the graph of the function sketched in Figure 2.

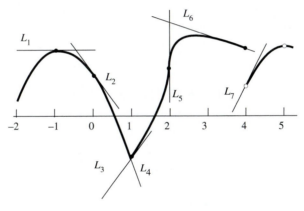

Figure 2. *An example of lines tangent to a graph and points where no tangent line is defined.*

1. At $x = -1$, the line L_1 is tangent to f. It is horizontal and thus has slope 0.

2. At $x = 0$, the line L_2 is tangent to f. The slope of L_2 is approximately -1.

3. At $x = 1$, there are two lines "flat" against f, L_3 and L_4, so there is no tangent line to f at $x = 1$.

4. At $x = 2$, the vertical line L_5 is "flat" against f, so the slope of the tangent to f at $x = 2$ is undefined.

5. At $x = 4$, the lines L_6 and L_7 both appear to be "flat" against f, so there is no tangent line at $x = 4$.

6. At $x = 5$, there is no line tangent to f since 5 is not in the domain of f.

An important observation from this example: There are no tangent lines to the graph of f at points where there are "corners" on the graph (as at $x = 1$ in the graph of Example 9), where there are "breaks" in the graph (as at $x = 4$ in the same graph), or where the function is undefined (as at $x = 5$ in the same graph). ■

Recall that

$$\frac{f(x + h) - f(x)}{h}$$

is the slope of the secant line to the graph of a function $f\colon A \to \mathbb{R}$ containing the points $(x, f(x))$ and $(x+h, f(x+h))$. If f is "relatively smooth," then as h gets small the slope of this secant line must approximate the slope of the line tangent to f at x. Since $f'(x)$ is also approximated by $\frac{f(x+h)-f(x)}{h}$, for small h, the derivative $f'(x)$ is the slope of the line tangent to f at x. See Figure 3.

If we refer back to Example 9, we expect that if the derivative of f exists at x, then at that point the graph of f has neither a corner there nor a break.

Furthermore, it is implicit that x must be in the domain of f in order for the derivative of f to be defined at x.

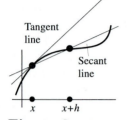

Figure 3. *The secant line containing $(x, f(x))$ and $(x + h, f(x + h))$ approximates the slope of the line tangent to the graph of f at $(x, f(x))$.*

EXAMPLE 10: Let $f(x) = 2x$. We already know that the slope of the graph of $y = 2x$ is 2, so the slope of the line tangent to f at any point x should be 2. The derivative f' is

$$\lim_{h \to 0} \frac{f(x + h) - f(x)}{h} = \lim_{h \to 0} \frac{2(x + h) - 2x}{h} = \lim_{h \to 0} \frac{2h}{h} = 2. \qquad ■$$

EXAMPLE 11: Let $f(x) = 2x^3$. Then $f'(x)$ is given by

$$
\begin{aligned}
f'(x) &= \lim_{h \to 0} \frac{2(x + h)^3 - 2x^3}{h} \\[2mm]
&= \lim_{h \to 0} \frac{2(x^3 + 3x^2h + 3xh^2 + h^3) - 2x^3}{h} \\[2mm]
&= \lim_{h \to 0} \frac{2(3x^2h + 3xh^2 + h^3)}{h} \\[2mm]
&= \lim_{h \to 0} 2(3x^2 + 3xh + h^2) = 6x^2. \qquad ■
\end{aligned}
$$

EXAMPLE 12: Let $f(x) = 2x^3$.

(a) Find the slope of the line tangent to f at $x = 3$.

(b) Find the equations for all the lines that are tangent to f and have slope 24.

SOLUTION: In Example 11, we found that $f'(x) = 6x^2$. For part (a) we merely need to evaluate this derivative at $x = 3$. Thus the slope of the line tangent to f at $x = 3$ is $6(3)^2 = 54$. For part (b) we must find all x such that $f'(x) = 24$, that is, $6x^2 = 24$, which has as solutions $x = 2$ and $x = -2$. Thus, there are two lines that satisfy the conditions given in the problem. See Figure 4.

Figure 4. $f'(x) = 24$ at $x = 2$ and $x = -2$.

1. l_1 is the line containing $(2, f(2)) = (2, 16)$ with slope $f'(2) = 24$. Thus the equation for l_1 is $y = m(x - x_1) + y_1 = 24(x - 2) + 16 = 24x - 32$.

2. l_2 is the line containing $(-2, f(-2)) = (-2, -16)$ with slope $f'(-2) = 24$. Thus the equation for l_2 is given by $y = m(x - x_1) + y_1 = 24(x + 2) - 16 = 24x + 32$. ∎

EXAMPLE 13: Find the equation for the line containing the point $(1, 2)$ and tangent to the parabola $y = x^2 + x$.

SOLUTION: We first note that the point $(1, 2)$ is on the parabola. The problem would be more complicated if the point were not on the parabola (cf. Exercises 3.3.54, 3.3.56, 3.3.57 and 3.3.58). We know that the derivative of $y = x^2$ is $2x$ and that the derivative of $y = x$ is 1. Thus the derivative of $y = x^2 + x$ is given by $y' = 2x + 1$. Since the derivative at a point is the slope of the tangent line there, it must be the case that our line has slope $2(1) + 1 = 3$. The equation of a straight line is given by

$$y - y_1 = m(x - x_1)$$

where (x_1, y_1) is a point on the line and m is the slope. Since the tangent line passes through the point $(1, 2)$, we have that

$$y - 2 = 3(x - 1)$$

is the desired equation. ∎

EXAMPLE 14: Let $f(x) = \sqrt{x}$. Then $f'(x) = \lim_{h \to 0} \frac{\sqrt{x+h} - \sqrt{x}}{h}$. Multiplying the numerator and the denominator by $(\sqrt{x + h} + \sqrt{x})$ we obtain:

$$\lim_{h\to 0}\frac{(\sqrt{x+h}-\sqrt{x})}{h} = \lim_{h\to 0}\frac{(\sqrt{x+h}-\sqrt{x})(\sqrt{x+h}+\sqrt{x})}{h(\sqrt{x+h}+\sqrt{x})}$$

$$= \lim_{h\to 0}\frac{(\sqrt{x+h}\sqrt{x+h}-\sqrt{x}\sqrt{x})}{h(\sqrt{x+h}+\sqrt{x})}$$

$$= \lim_{h\to 0}\frac{h}{h(\sqrt{x+h}+\sqrt{x})}$$

$$= \lim_{h\to 0}\frac{1}{(\sqrt{x+h}+\sqrt{x})}=\frac{1}{2\sqrt{x}}. \qquad\blacksquare$$

Figure 5.a $f(x)=|x|$

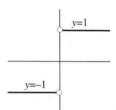

Figure 5.b $f'(x)=\dfrac{|x|}{x}$

EXAMPLE 15: Let $f(x)=\sqrt{x}$. Find the slope of the line tangent to f at $x=1, x=4$, and $x=8$.

SOLUTION: $f'(x)$ is the slope of the line tangent to f at x. From Example 14, $f'(x)=\frac{1}{2\sqrt{x}}$. Thus $f'(1)=\frac{1}{2}, f'(4)=\frac{1}{4}$, and $f'(8)=\frac{1}{4\sqrt{2}}$. $\blacksquare$

EXAMPLE 16: Let $f(x)=\sqrt{x}$. Determine where the line tangent to f has slope 1.

SOLUTION: We must find the value for x that satisfies the condition $f'(x)=1$. The equation $\frac{1}{2\sqrt{x}}=1$ is equivalent to $\sqrt{x}=\frac{1}{2}$, or $x=\frac{1}{4}$. Why isn't $x=-\frac{1}{4}$ a solution? $\blacksquare$

EXAMPLE 17: Let $f(x)=|x|$. Find $f'(x)$.

SOLUTION: Recall that

$$|x|=\left\{\begin{array}{ll} x, & \text{if } x\geq 0; \\ -x, & \text{if } x<0. \end{array}\right. \quad \text{See Figure 5.a.}$$

Therefore, if $x>0$, then $f(x)=x$, and $f'(x)=1$, and if $x<0$, then $f(x)=-x$, and $f'(x)=-1$. The derivative of $|x|$ is not defined at $x=0$. Note that there is a corner on the graph of $|x|$ at $x=0$. A nice way to write this is $f'(x)=|x|/x$. Thus

$$f'(x)=\left\{\begin{array}{ll} 1, & \text{if } x>0; \\ -1, & \text{if } x<0. \end{array}\right. =\frac{|x|}{x}. \text{ See Figure 5.b.} \qquad \blacksquare$$

EXAMPLE 18: Find $g'(x)$, where $g(x)=|x^2-4|$.

SOLUTION: The appropriate approach to this example is to break it down into the cases where (x^2-4) is positive and where (x^2-4)

is negative.

$$|x^2 - 4| = \begin{cases} x^2 - 4 & \text{when } x^2 - 4 \geq 0 \\ -(x^2 - 4) = 4 - x^2 & \text{when } x^2 - 4 < 0. \end{cases}$$

The graph of g is sketched in Figure 6.a. Since the function changes the way it is expressed at $x = \pm 2$, we consider these points as special cases. Before looking at these special points, we look at the two cases where $x^2 - 4 \neq 0$:

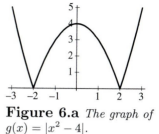

Figure 6.a *The graph of $g(x) = |x^2 - 4|$.*

For the first case:

$$(x^2 - 4) > 0 \text{ when } x > 2 \text{ or when } x < -2.$$

In this case, $g'(x)$ is given by:

$$\begin{aligned} g'(x) &= \lim_{h \to 0} \frac{((x+h)^2 - 4) - (x^2 - 4)}{h} \\ &= \lim_{h \to 0} \frac{(x^2 + 2xh + h^2 - 4 - x^2 + 4)}{h} \\ &= \lim_{h \to 0} 2x + h = 2x. \end{aligned}$$

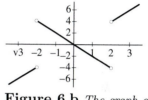

Figure 6.b *The graph of $g'(x)$.*

Thus

$$g'(x) = 2x \text{ when } x > 2 \text{ or } x < -2.$$

Now, for the other case:

$$(x^2 - 4) < 0 \text{ when } -2 < x < 2.$$

Going through the same procedures as above, we find that

$$g'(x) = -2x \text{ when } -2 < x < 2.$$

Thus

$$g'(x) = \begin{cases} 2x & \text{when } x > 2 \text{ or when } x < -2 \\ -2x & \text{when } -2 < x < 2 \end{cases}.$$

Finally, we consider the special cases: $x = \pm 2$. At $x = 2$, the expression

$$\frac{g(2 + h) - g(2)}{h}$$

is quite different according to whether $h > 0$ or $h < 0$. If $h > 0$, then

$$\frac{g(2 + h) - g(2)}{h} = \frac{((2 + h)^2 - 4) - (2^2 - 4)}{h} = 4 + h \approx 4.$$

If $h < 0$, then

$$\frac{g(2+h) - g(2)}{h} = \frac{\left(4 - (2+h)^2\right) - \left(4 - 2^2\right)}{h} = -4 - h \approx -4.$$

Thus there is no number that is well approximated by $\frac{g(2+h) - g(2)}{h}$ for all h close to 0. We conclude that 2 is not in the domain of g'. In exactly the same way, -2 is not in the domain of g'.

Notice that the graph of $g(x)$ has corners at $x = \pm 2$. Figure 6.b gives a sketch of g'.   ∎

EXAMPLE 19: In Examples 3 and 11 we found $f'(t)$ for $f(t) = 2t^3$, $f(t) = t^2$ and $f(t) = t$. Let

$$\vec{r}(t) = \begin{pmatrix} 2t^3 \\ t^2 \\ t \end{pmatrix}.$$

Find $\vec{v}(t)$.

SOLUTION: Writing $\vec{r}(t)$ in terms of its coordinate functions $(x(t), y(t), z(t))$ we have that $x(t) = 2t^3$, $y(t) = t^2$, and $z(t) = t$. We know that $x'(t) = 6t^2$, $y'(t) = 2t$, and $z'(t) = 1$. Thus

$$\vec{v}(t) = \vec{r}'(t) = \begin{pmatrix} x'(t) \\ y'(t) \\ z'(t) \end{pmatrix} = \begin{pmatrix} 6t^2 \\ 2t \\ 1 \end{pmatrix}.$$ ∎

EXAMPLE 20: Parametrize the line tangent to the image of

$$\vec{r}(t) = \begin{pmatrix} 2t^3 \\ t^2 \\ t \end{pmatrix}$$

at $t = 1$.

SOLUTION: Recall that if $\vec{r}$ is a vector-valued function, then $\vec{r}'(t)$ is a direction vector for the line tangent to the image of $\vec{r}$. Using the work done in the previous example, we have that

$$\vec{r}'(1) = \begin{pmatrix} x'(1) \\ y'(1) \\ z'(1) \end{pmatrix} = \begin{pmatrix} 6 \\ 2 \\ 1 \end{pmatrix}$$

is a direction vector for the line. The point $\vec{r}_0 = \vec{r}(1) = (2, 1, 1)$ is on the line. Thus the line tangent to $\vec{r}(t)$ is

$$\vec{r}'t + r_0 = \begin{pmatrix} 6 \\ 2 \\ 1 \end{pmatrix} t + \begin{pmatrix} 2 \\ 1 \\ 1 \end{pmatrix} = \begin{pmatrix} 6t + 2 \\ 2t + 1 \\ t + 1 \end{pmatrix}.$$ ∎

Of particular importance in the exercises below are Exercises 16 and 17 where you are asked to provide an argument for the following theorem.

Theorem 2 *Let* $\vec{W}(t) = (\cos(t), \sin(t))$. *Then*

- $\vec{W}'(t) = (-\sin(t), \cos(t))$.

- $D\cos(t) = -\sin(t)$.

- $D\sin(t) = \cos(t)$.

The derivative formulas for the sine and cosine functions are important and should be memorized. These formulas were developed using radian measure for the angle t. When we do calculus on the trigonometric functions, the angles are assumed to be measured in radians. The formulas we have developed do not work otherwise.

EXERCISES 3.3

Exercises 1–3 refer to the function f sketched in Figure 7.

1. For what values of x is $f'(x) = 0$? (Where are the tangent lines horizontal?)

2. For what values of x is $f'(x) > 0$? (Where do the tangent lines have positive slope?)

3. For what values of x is $f'(x) < 0$? (Where do the tangent lines have negative slope?)

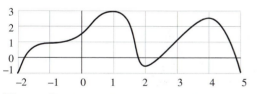

Figure 7. *Sketch of the function f for Exercises 1–3.*

Exercises 4–7 refer to the function f sketched in Figure 8.

4. For what values of x is the derivative of f NOT defined?

5. For what values of x is $f'(x) = 0$?

6. For what values of x is $f'(x) > 0$?

7. For what values of x is $f'(x) < 0$?

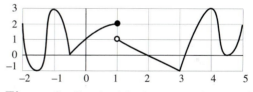

Figure 8. *Sketch of the function f for Exercises 4–7.*

An object's position at time t sec is given by $\vec{r}(t) = (4 + t, t - 2t^2)$ m. Find the velocity at the times given in Exercises 8–10.

8.   0 sec.   9.   1 sec.   10.   5 sec.

In Exercises 11–13, a particle moves counterclockwise with constant speed in a circle centered at the origin with a radius of 5 m. It makes 1 revolution every 16 sec.

11. What is the particle's speed?

12. What is the instantaneous velocity when the particle is on the positive x–axis? (Hint: the velocity vector is tangent to the circle when drawn emanating from the particle, and the magnitude is the speed.)

13. What is the instantaneous velocity when the particle is on the positive y–axis?

Exercises 14–16 refer to the wrapping function $\vec{W}(t) = (\cos(t), \sin(t))$ which describes a particle moving counterclockwise around the unit circle, 1 rotation every 2π units of time.

14. What is the particle's speed?

15. Argue that either $\vec{W}'(t) = (-\sin(t), \cos(t))$ or $\vec{W}'(t) = (\sin(t), -\cos(t))$. Hint: A tangent vector to a circle must be perpendicular to the radius drawn to the point of tangency. Refer to Section 1.4.

16. Check $\vec{W}'(0)$ to argue that the proper choice for $\vec{W}'(t)$ is $(-\sin(t), \cos(t))$.

Find the derivative of the functions in Exercises 17–32.

17. $f(x) = 3x^2$.

18. $f(x) = ax^2 + x$.

19. $f(x) = (2x^2 + x)^2$.

20. $f(x) = x^{-2}$.

21. $f(x) = (2x^2 + 3x + 1)^{-1}$.

22. $f(x) = x^4$.

23. $f(x) = x^{-3}$.

24. $f(x) = x^5$.

25. $f(x) = \sqrt{x + 1}$.

26. $f(x) = x^{3/2}$.

27. $f(x) = \sqrt{x^2 + 1}$.

28. $f(x) = x^{-1/2}$.

29. $f(x) = (x^2 + 1)^{-1/2}$.

30. $f(x) = |x + 1|$.

31. $f(x) = |x^2 + 1|$.

32. $f(x) = |x^2 - 1|$.

In Exercises 33–39, find an equation of the line that is tangent to g at the given point.

33. $g(x) = x^3$, $x = 0$.

34. $g(x) = x^{-2}$, $x = -1$.

35. $g(x) = x^3 + x^2$, $x = 1$.

36. $g(x) = \sqrt{x + 1}$, $x = 3$.

37. $g(x) = (x + 1)^{-1/2}$, $x = 3$.

38. $g(x) = \sin(x)$, $x = \frac{\pi}{4}$.

39. $g(x) = \cos(x)$, $x = \frac{\pi}{3}$.

In Exercises 40–46, find $\vec{r}'$.

40. $\vec{r}(t) = (3t, t + 1, 2t)$.

41. $\vec{r}(t) = (1, 3t, 2)$.

42. $\vec{r}(t) = \left(t^2, 5t, \frac{1}{t}\right)$.

43. $\vec{r}(t) = (\sqrt{t}, t)$.

44. $\vec{r}(t) = \left(\frac{1}{t+1}, 3t^2 + t\right)$.

45. $\vec{r}(t) = (2t^3, t, \sqrt{t + 2})$.

46. $\vec{r}(t) = (t, \sin(t), \cos(t))$.

In Exercises 47–53, find a parametrization for the line tangent to the path of $\vec{r}$ at $t = 1$.

47. $\vec{r}(t) = (3t, t + 1, 2t)$.

48. $\vec{r}(t) = (1, 3t, 2)$.

49. $\vec{r}(t) = \left(t^2, 5t, \frac{1}{t}\right)$.

50. $\vec{r}(t) = (\sqrt{t}, t)$.

51. $\vec{r}(t) = \left(\frac{1}{t+1}, 3t^2 + t\right)$.

52. $\vec{r}(t) = (2t^3, t, \sqrt{t + 2})$.

53. $\vec{r}(t) = (t, \sin(t), \cos(t))$.

54. Find the equations for the lines containing the point $(1, 0)$ and tangent to the parabola $y = x^2$.

55. Find an equation for the line containing the point $(1, 1)$ and tangent to the parabola $y = x^2$.

56. Find the equations for the lines containing the point $(2, 1)$ and tangent to the parabola $y = x^2$.

57. Find the equations for the two lines containing the point $(-1, 0)$ and tangent to the parabola $x^2 = 2y$.

58. Find the equations for the two lines containing the point $(0, -1)$ and tangent to the parabola $y^2 = 2x$.

3.4 Linear Approximations of Functions

Using the Tangent Line to Approximate a Function

Suppose that a function f is differentiable at x_0. If we let $x = x_0 + h$, then an approximation of the derivative of f at x_0 is given by

$$f'(x_0) \approx \frac{f(x_0 + h) - f(x_0)}{h} = \frac{f(x) - f(x_0)}{x - x_0}.$$

Thus $f(x) \approx f(x_0) + f'(x_0)(x - x_0)$. We have that the tangent line $y(x) = f(x_0) + f'(x_0)(x - x_0)$ is a good approximation for $f(x)$, provided that x_0 is close to x. See Figure 1. How close x must be to x_0 and how good an approximation we obtain will be quantified later.

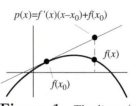

Figure 1. *The line $p(x)$ approximates $f(x)$ for x close to x_0.*

Definition: Linear Approximations

The function $p(x) = f(x_0) + f'(x_0)(x - x_0)$ is referred to in several ways:

(a) *linear approximation of f centered at x_0;*

(b) *linearization of f centered at x_0;*

(c) *first order approximation of f centered at x_0; and*

(d) *first degree Taylor polynomial for f centered at x_0.*

If c is a number, then $p(c) = f(x_0) + f'(x_0)(c - x_0)$ is referred to as the *first order approximation* for $f(c)$ (centered at x_0).

EXAMPLE 1: Find the first order approximation for $\sqrt{1.1}$ centered at $x_0 = 1$.

SOLUTION: Let $f(x) = \sqrt{x}$. Then by Example 14 of Section 3.3, $f'(x) = \frac{1}{2\sqrt{x}}$. We have that $f'(x_0) = f'(1) = \frac{1}{2}$ and $p(x) = \frac{1}{2}(x - 1) + 1 = \frac{x+1}{2}$ approximates $f(x) = \sqrt{x}$ for values of x close to $x_0 = 1$. Thus, $p(1.1) = \frac{2.1}{2} = 1.05$ is our approximation. This author's calculator value for $\sqrt{1.1}$ appeared as 1.048808848. ∎

EXAMPLE 2: Use a first degree Taylor polynomial to approximate the volume of a sphere of radius 1.1 m.

SOLUTION: The volume of a sphere of radius r is $V(r) = \frac{4}{3}\pi r^3$. The number 1.1 m is close to 1 m, so we use the first degree Taylor

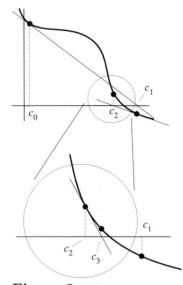

Figure 2. *The point c_0 is the initial point and c_1 and c_2 are the first two points obtained using Newton's method. By zooming in 300%, we can graphically obtain c_3.*

polynomial centered at $x_0 = 1$ m. In Example 8 of Section 3.3, we showed that $V'(1) = 4\pi$ m^2. $V(1) = \frac{4}{3}\pi$ m^3. Thus $p(r) = \frac{4}{3}\pi + 4\pi(r - 1)$ is the first degree Taylor polynomial centered at $r = 1$ m. Thus $V(1.1) \approx V(1) + V'(1)\left((1.1) - 1\right) = \frac{4}{3}\pi + 0.4\pi$ m^3. ∎

Newton's Method

A recurring problem in nearly every field that uses mathematical modeling is that of solving an equation for an unknown. Newton's method for approximating solutions for equations of the form $f(x) = 0$ is easily programmed on a computer. Suppose that we have an initial guess for the solution of $f(x) = 0$, say $x = c_0$. We approximate $f(x)$ using the linear approximation, $y(x) = f(c_0) + f'(c_0)(x - c_0)$ at $x = c_0$, and then let c_1 be the solution to $y(x) = 0$. The number c_1 is our new approximation for a solution to $f(x) = 0$. We repeat the process using c_1 instead of c_0, to obtain c_2, etc. (See Figure 2.)

Starting with c_0, let c_1 be the solution of $f(c_0) + f'(c_0)(x - c_0) = 0$, or

$$c_1 = c_0 - \frac{f(c_0)}{f'(c_0)}.$$

Let c_2 be the solution of $f(c_1) + f'(c_1)(x - c_1) = 0$, or

$$c_2 = c_1 - \frac{f(c_1)}{f'(c_1)}.$$

Having obtained $c_0, c_1, \ldots, c_n$, let c_{n+1} be the solution of $f(c_n) + f'(c_n)(x - c_n) = 0$, or

$$c_{n+1} = c_n - \frac{f(c_n)}{f'(c_n)}.$$

EXAMPLE 3: We want to approximate the root of $y = x^3 + 3x^2 - 5$, using $c_0 = 1$ as our initial guess. The following table lists the values of c_n, $y(c_n)$, $y'(c_n)$, and the resulting c_{n+1}.

n	c_n	$y(c_n)$	$y'(c_n)$	c_{n+1}
0	1	-1	9	1.11111
1	1.11111	0.075446	10.3704	1.10384
2	1.10384	0.000376	10.2784	1.10380
3	1.10380	0.000006	10.2780	1.10380

In just three steps we have reached a point in this *algorithm*[3] so that:

---

[3] Algorithm is a mathematical term that means procedure or method.

(a) $y(c_3) = 0.000006$, which approximates 0 with an accuracy of 5 decimal places; and

(b) The difference between c_3 and c_4 is less than 10^{-5}.

Statements (a) and (b) suggest tests to determine whether our approximation for a root of a function f is accurate enough. Let E denote a number that represents our acceptable error. At the nth step, we can apply one of two tests to decide if our root is acceptable:

Test 1: If $|f(c_n)| < E$, then stop.

Test 2: If $|c_n - c_{n+1}| < E$, then stop.

In practice, if we are asked to find the root of a function, there are two things that might be sought:

(a) A value for x such that $f(x)$ is close to 0; or

(b) A value for x that is close to a number x_0 that is a root of f.

Although, theoretically, if f is continuous, condition (b) implies condition (a), condition (b) is hard to verify since we don't know the value of x_0. In practice we only know if c_n is close to c_{n+1}.

Both tests mentioned above have drawbacks. If $f'(c_n)$ is close to zero, then $f(x)$ may be close to 0, but c_n may not be close to the root x_0. In this situation, using Test 1 alone would make us stop prematurely. (See Figure 3.a.) If $|f'(c_n)|$ is large, then c_n may be close to c_{n+1}, yet c_n may not be close to the root x_0. Test 2 would have us stop too soon in this case. (See Figure 3.b.) Thus we usually employ both tests (and occasionally other esoteric tests) to ensure convergence to a root.

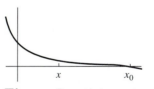

Figure 3.a $f(x)$ *can be close to 0 even if x is not a good approximation of the root of f.*

Word of warning: Newton's method is far from foolproof. A fundamental problem is that there is no assurance that we will ever get close to a root of the function. If $f'(x)$ is defined on an interval containing the root in its interior, is not zero or undefined on this interval, and if c_0 is "close enough" to the root of f, then it can be shown that the algorithm will work. (In practice, it is usually not trivial to decide what "close enough" means.) If our function does not satisfy these requirements, it is possible for the points c_n either to "cycle" periodically around the root or to spiral away from the root, rather than converging to the root as desired.

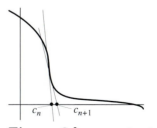

Figure 3.b c_n *can be close to c_{n+1} without being close to the root.*

Another problem arises when Test 2 fails, as discussed above. Just because c_n is close to c_{n+1}, it does not follow that c_n is close to a root of f. In fact, if we were using this algorithm on a computer, we might well reach an n such that c_n is so close to c_{n+1} that the computer cannot tell the difference between them, yet c_n is a long way from the desired root. In this case, we need to use another algorithm.

A third problem may occur when the function has more than one root. We may be searching for a specific root, and Newton's method may converge to a different root. Examples of which Newton's method fails to work are postponed until we get some tools for differentiating functions.

In spite of these problems, Newton's method is widely used in practice because it usually works, and when it does, it usually works fast. When employing this algorithm, the user needs to be aware that problems in its use may lie deeper than his or her ability as a programmer.

EXAMPLE 4: Use Newton's method to approximate $\sqrt{2}$.

SOLUTION: Let $f(x) = x^2 - 2$. We want to solve the equation $f(x) = 0$. Let $c_0 = 1$. The following table lists the values of $c_n, y(c_n), y'(c_n)$, and c_{n+1} for $n = 0, 1, \ldots, 5$.

n	c_n	$y(c_n)$	$y'(c_n)$	c_{n+1}
0	1.0	-1.0	2	1.5
1	1.5	0.25	3	1.41666
2	1.41666	0.006944	2.83333	1.414209
3	1.414209	-0.0000124	2.82842	1.414213
4	1.414213	0.0000005	2.828427	1.414213

By Test 1, we have an accuracy of 10^{-6}. By Test 2, c_n and c_{n+1} differ by less than 10^{-6}. The value for $\sqrt{2}$ with an accuracy of nine digits is 1.41421356.

We give a summary of Newton's method.

Newton's Method

Assume that A is a subset of $\mathbb{R}$, that $f : A \to \mathbb{R}$ is continuously differentiable, and that we want to approximate a root c of f.
Step I: Set the acceptable tolerances. That is, establish a number $E > 0$, which will be our acceptable error, and decide which test for error will be used.
Step II: Find a point c_0 that is an initial approximation for the root.

Step III: Let $c_1 = c_0 - \frac{f(c_0)}{f'(c_0)}$.

Step IV: Apply your test for error. If your answer has the desired accuracy, go to the next step. Otherwise set $c_0 = c_1$ and return to Step III.

Step V: Output your answer and stop.

EXERCISES 3.4

1. Sketch the linear approximation $p(x)$ for each of the following functions at $x = x_0$ and indicate on your graph $f(x_0 + 0.5)$ and $p(x_0 + 0.5)$.

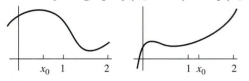

Figure 4. *Graphs for Exercise 1.*

In Exercises 2–5, select an appropriate x_0 and find the first order approximation for the square root of the number x, then compare your answer with your calculator values.

   2. $x = 5$ 3. $x = 9.2$ 4. $x = 99$ 5. $x = 0.0069$

In Exercises 6 and 7, a right triangle expands so that the base b is always twice as long as the height.

6. Find the instantaneous rate of change of the area relative to the change in the length of the base when $b = 2$ ft.

7. Find the first order approximation, centered at $b = 2$ ft, for the area when $b = 2.1$ ft.

In Exercises 8 and 9, a particle A moves along the $x-$ axis with its position given by $A(t) = (3t, 0)$ m. Particle B moves along the y–axis with its position given by $B(t) = (0, 4t)$ m.

8. Find the instantaneous rate of change of the distance between A and B relative to time when $t = 1$ sec.

9. Find the first order approximation, centered at $t = 1$ sec, for the distance between A and B at time $t = 0.9$ sec.

In Exercises 10 and 11, a right circular cone expands in such a way that the height always remains twice the length of the radius, r.

10. Find the instantaneous rate of change of the volume of the cone when $r = 3$ ft.

11. Find the first order approximation, centered at $r = 3$ ft, for the volume when $r = 2.9$ ft.

We can linearize a vector-valued function by linearizing each of its coordinate functions. This process simply gives a parametrization for the tangent line. The first order approximation for vector-valued functions takes on the same form as the first order approximation for a real-valued function:

$$\vec{r}(t) = \begin{pmatrix} x(t) \\ y(t) \\ z(t) \end{pmatrix} \approx \vec{p}(t)$$

$$= \begin{pmatrix} x(t_0) + x'(t_0)(t - t_0) \\ y(t_0) + y'(t_0)(t - t_0) \\ z(t_0) + z'(t_0)(t - t_0) \end{pmatrix}$$

$$= \vec{r}(t_0) + (t - t_0)\vec{r}'(t_0).$$

The function

$$\vec{p}(t) = \vec{r}(t_0) + (t - t_0)\vec{r}'(t_0)$$

is called

- the *first order approximation for $\vec{r}$ centered at t_0,*

- the *linear approximation for $\vec{r}$ centered at t_0,* or

- the *first degree Taylor polynomial for $\vec{r}$ centered at t_0.*

12. Find the linear approximation for $\vec{r}(t) = \left(t^2, \sqrt{1 - t}, \frac{1}{t}\right)$ centered at $t_0 = -1$.

Approximating the derivative numerically: *By definition, if f has a derivative at $x = x_0$ and if h is a small number, then $f'(x_0)$ is approximated by $\frac{f(x_0+h)-f(x_0)}{h}$. We may actually use this formula to numerically approximate the derivative of a function. We call this direct method of approximating the derivative Method 1. We expect to do better in our approximation if we use h and $-h$ to approximate the derivative, and then average the two results. We call this second approach Method 2. The formula for Method 2 is:*

$$f'(x_0)$$
$$\approx \frac{1}{2}\left(\frac{f(x_0+h)-f(x_0)}{h} + \frac{f(x_0-h)-f(x_0)}{-h}\right)$$
$$= \frac{f(x_0+h)-f(x_0-h)}{2h}.$$

13. Let $f(x) = x^3$. We know that $f'(1) = 3$. Use both Method 1 and Method 2 to approximate $f'(1)$ for the given values of h:

    a. $h = 0.1$.   b. $h = 0.01$.   c. $h = 0.001$.

14. Does Method 2 actually seem to give more accurate answers for a given h?

15. Let $f(x) = x^4$. Use both methods to approximate $f'(1)$ for the given values of h:

    a. $h = 0.1$.   b. $h = 0.01$.   c. $h = 0.001$.

16. Let $f(x) = \sin x$, with x measured in radians. Divide the interval $[0, 2\pi]$ into 10 equal intervals with the points $x_0 = 0, x_1 = \frac{\pi}{5}, x_2 = \frac{2\pi}{5}, \ldots, x_{10} = 2\pi$. Use Method 2 and $h = 0.01$ to approximate f' at each of these points. Use these values to obtain a rough sketch of the graph of f' over the interval $[0, 2\pi]$. Does the graph of f' look familiar?

In Exercises 17–20, use Newton's method to approximate the square root of the given number to an accuracy of 10^{-4} for both tests. Compare your results with those of Exercises 2–5.

17. $x = 5$.   18. $x = 9.2$.

19. $x = 99$.   20. $x = 0.0069$.

In Exercises 21–23, $f(x) = \frac{1}{x}$.

21. Find the linear approximation $p(x)$ of f centered at $x_0 = 1$.

22. Graph the error $|f(x) - p(x)|$ for $1 \leq x \leq 5$.

23. If you wanted to use $p(x)$ as an approximation, but needed to ensure that the error was less than 0.2, what would be the allowable range of x?

3.5   More on Limits

In Section 2 we introduced the notion of limits and developed the idea just enough to define the derivative. We now come back to explore the idea more extensively.

One Sided Limits

Let's take another look at Examples 17 and 18 of section 3.3.

EXAMPLE 1: In Example 3.3.17, we saw that if $f(x) = |x|$, then

$$f'(x) = \begin{cases} 1 & \text{when } x > 0 \\ -1 & \text{when } x < 0 \end{cases}.$$

Thus, $\lim_{x \to 0} f'(x)$ is undefined. However, if we let x get close to 0, being careful to keep $x > 0$, then $f'(x) = 1$. Similarly, if we let x get close to 0, being careful to keep $x < 0$, then $f'(x) = -1$. ∎

EXAMPLE 2: In Example 3.3.18, we saw that if $g(x) = |x^2 - 4|$, then

$$g'(x) = \begin{cases} 2x & \text{when } x > 2 \text{ or when } x < -2 \\ -2x & \text{when } -2 < x < 2 \end{cases}$$

Thus, neither $\lim_{x \to 2} g'(x)$ nor $\lim_{x \to -2} g'(x)$ is defined. However,

> if we let x get close to 2, being careful to keep $x > 2$, then $g'(x)$ gets close to 4.

> If we let x get close to 2, being careful to keep $x < 2$, then $g'(x)$ gets close to -4.

> If we let x get close to -2, being careful to keep $x > -2$, then $g'(x)$ gets close to -4.

> Finally, if we let x get close to -2, being careful to keep $x < -2$, then $g'(x)$ gets close to 4. ∎

In both of the above examples, the functions f' and g' are well-behaved even at points where the limits fail to exist. It is useful to introduce some vocabulary to describe the behavior illustrated here.

Definition: One Sided Limits

Suppose that A is a subset of $\mathbb{R}$, f is a function from A into $\mathbb{R}$, and z is a number (not necessarily in A). If

(a) there is a number a so that the segment $a < x < z$ is a subset of A, and

(b) $f(x)$ well approximates the number L if $x < z$ is close to z,

then we will say that *L is the limit of $f(x)$ as x approaches z from the left* and we write

$$L = \lim_{x \to z^-} f(x).$$

If

(a) there is a number b so that the segment $z < x < b$ is a subset of A, and

(b) $f(x)$ well-approximates the number L if $x > z$ is close to z,

then we will say that L *is the limit of* $f(x)$ *as* x *approaches* z *from the right* and we write

$$L = \lim_{x \to z^+} f(x).$$

Figures 1.a and 1.b illustrate the concept of one-sided limits.

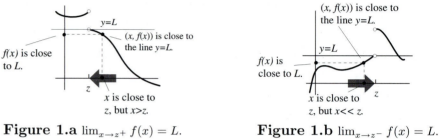

Figure 1.a $\lim_{x \to z^+} f(x) = L.$ **Figure 1.b** $\lim_{x \to z^-} f(x) = L.$

It should be clear that

$$\lim_{x \to z} f(x) = L$$

if and only if

$$\lim_{x \to z^+} f(x) = \lim_{x \to z^-} f(x) = L.$$

EXAMPLE 3: From Example 2,

$$g'(x) = \begin{cases} 2x & \text{when } x > 2 \text{ or when } x < -2 \\ -2x & \text{when } -2 < x < 2 \end{cases}$$

We have

$$\lim_{x \to 2^+} g'(x) = 4 \qquad\qquad \lim_{x \to 2^-} g'(x) = -4$$

$$\lim_{x \to -2^+} g'(x) = -4 \qquad\qquad \lim_{x \to -2^-} g'(x) = 4. \qquad \blacksquare$$

In order to dispel any idea that right and left limits must exist, we give the following example.

EXAMPLE 4: Let $f(x) = \sin\left(\frac{1}{x}\right)$. Then as x moves toward 0, f will oscillate back and forth between 1 and -1. See Figure 2. The limit as x approaches 0 is not defined from either side! $\blacksquare$

Figure 2. *The graph of*
$f(x) = \sin\left(\frac{1}{x}\right)$.

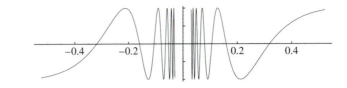

Continuous Functions

In Example 4 of Section 3.2, we saw a function with $\lim_{x \to 3} g(x) = 9$, but $g(3) = 1$. Admittedly, this function is a bit contrived. If the function behaved as it should, we would expect $g(3) = \lim_{x \to 3} g(x)$. That is, since the limit exists and since 3 is in the domain of g, our impulse is to evaluate the limit by direct substitution! Points in the domains of functions that have the property that $\lim_{x \to x_0} g(x) = g(x_0)$ seem more natural to us. We give them a name.

Definition: Points of Continuity and Points of Discontinuity

If f is a function from a subset of $\mathbb{R}$ into $\mathbb{R}$, then:

The function f is *continuous at the point a* if $\lim_{x \to a} f(x) = f(a)$.[4]

If the function f is continuous at each point of its domain, then f is *continuous*.

If a is in the domain of f, and f is not continuous at a, then a is called a *point of discontinuity of f*, and f is *discontinuous at a*.[5]

---

[4]It is implicit in the definition that if f is continuous at a, then a is in the domain of f.

[5]The function f can be discontinuous only at points in the domain of f. Note that if f has any discontinuity, then f is not continuous.

EXAMPLE 5: The function f, sketched in Figure 3, illustrates two types of discontinuity and a couple of points that are often mistaken for points of discontinuity. At $x = -1$, there is a true break or jump; in fact, $\lim_{x \to -1} f(x)$ is undefined. f is not continuous at $x = -1$. We have a limit at $x = -3$, but -3 is not in the domain of f and we have neither continuity nor discontinuity there. At $x = 1$, $\lim_{x \to 1} f(x) = \frac{1}{2}$, but $f(1) = -1$. The point $x = 1$ is a point of discontinuity. Even though there is a break in f at $x = 2$, that break is not in the domain of f; and so, f is neither continuous nor discontinuous at $x = 2$. ∎

Figure 3. *The graph has four breaks but only two points of discontinuity.*

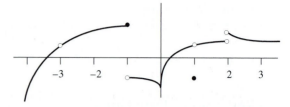

We have already pointed out that if f has a derivative at a point a, then there is no break there. We would expect that if f has a derivative at a point, then f is continuous there. This is the case.

Theorem 1 *Suppose that A is a subset of $\mathbb{R}$ and $f\colon A \to \mathbb{R}$ has a derivative at $x = a$. Then f is continuous at $x = a$.*

Proof: We have that $\lim_{x \to a} \left(\frac{f(x) - f(a)}{x - a} \right) = f'(a)$. Thus if $|x - a|$ is small, then

$$\left(\frac{f(x) - f(a)}{x - a} \right) \approx f'(a) \quad \text{or}$$

$$f(x) - f(a) \approx (x - a)f'(a). \qquad \blacksquare$$

Since $|x - a|$ is close to 0, $f(x) - f(a) \approx 0$, which implies that $f(x) \approx f(a)$.

Finally, since $\lim_{x \to a} \frac{f(x) - f(a)}{x - a}$ is defined, we know that f is defined in a region surrounding $x = a$. $\qquad \blacksquare$

Building Continuous Functions From Continuous Pieces

The following theorem allows us to break a difficult limit problem into easier component pieces.

Theorem 2 *Suppose that f and g are functions with the same domain, c is a number, $\lim_{x \to a} f(x) = L$, and $\lim_{x \to a} g(x) = K$. Then*

   (a) $\lim_{x \to a} cf(x) = cL,$

   (b) $\lim_{x \to a} (fg)(x) = \lim_{x \to a} f(x)g(x) = LK,$

(c) $\lim_{x \to a}(f + g)(x) = L + K$, and

(d) If $L \neq 0$, then $\lim_{x \to a}(g/f)(x) = \lim_{x \to a} g(x)/f(x) = K/L$.

We have not given a careful enough definition of limits to prove this theorem. However, the idea behind the proof is reasonably straightforward. For example, in Part (b), if x is close to a, then $f(x)$ is close to L and $g(x)$ is close to K. Thus $f(x)g(x)$ is close to LK. ∎

Corollary 1 *If f and g are functions that are continuous at the point a, and c is a number, then*

(a) *cf is continuous at a.*

(b) *$f + g$ is continuous at a.*

(c) *fg is continuous at a.*

(d) *if $f(a) \neq 0$, then $\dfrac{g}{f}$ is continuous at $x = a$.*

The utility of this corollary is demonstrated by the following example. Starting with the fact that $f(x) = x$ is continuous, it follows from Corollary 1(a) that $g(x) = 10x$ is continuous. From Corollary 1(c), x^2 is continuous and therefore so is x^3; etc. Thus, from Corollary 1(b), $10x + x^2 + x^3$ is continuous. Finally, from Corollary 1(d), $\frac{10x+x^2+x^3}{x-1}$ is continuous everywhere except at $x = 1$. Using the same idea, we can prove the more general results for polynomials and rational functions.

Theorem 3 *If f is a polynomial (that is, $f(x) = a_0 + a_1 x + a_2 x^2 + \ldots + a_n x^n$), then f is continuous.*

Recall that a function f is called a rational function if there are polynomials P_1 and P_2 such that $f = P_1/P_2$. For some examples, let

$$f(x) = \frac{x}{x^2 + 1},$$

$$g(x) = \frac{x^2 + 3x + 2}{x^3 + 6x + 1},$$

$$h(x) = \frac{1}{x^3 + 6x + 1}.$$

However, the following is not a rational function:

$$u(x) = \frac{\sqrt{x}}{x^2 + 1}.$$

Corollary 2 *Every rational function is continuous at each point of its domain.*

Proof: Suppose that $f(x) = \frac{P(x)}{Q(x)}$. If x is in the domain of f, then $Q(x) \neq 0$. Since $Q(x) \neq 0$, Corollary 2 tells us that $\frac{P}{Q}$ is continuous at the number x.   ■

EXAMPLE 6: The domain of the function

$$f(x) = \frac{x^3 + 6x^2 - 2x + 45}{(x - 1)(x + 2)}$$

is the set of real numbers except $x = 1$ and $x = -2$. The function f is clearly continuous on its domain.   ■

Limits of Composites

Suppose that $\lim_{x \to L} f(x) = B$ and $\lim_{x \to a} g(x) = L$. Consider what happens to $f \circ g(x) = f(g(x))$ as x approaches a. As x moves toward a, we have that $g(x)$ is close to L, and $f(g(x))$ is close to B. This leads us to the following theorem.

Theorem 4 *If $f(x)$ and $g(x)$ are continuous functions, and if $\lim_{x \to L} f(x) = B$ and $\lim_{x \to a} g(x) = L$, then $\lim_{x \to a} f \circ g(x) = B$ and $f \circ g$ is continuous.*

EXAMPLE 7: Let $f(x) = x^2 + 6x^5 + 10$ and let $g(x) = x^{10} + 1$. Then

$$\lim_{x \to 0} g(x) = 1, \text{ and } \lim_{x \to 1} f(x) = 17.$$

Thus, $\lim_{x \to 0} f \circ g(x) = 17$.

$$\lim_{x \to 0} f(x) = 10, \text{ and } \lim_{x \to 10} g(x) = 10^{10} + 1.$$

Hence $\lim_{x \to 0} g \circ f(x) = 10^{10} + 1$.   ■

The Intermediate Value Theorem

Suppose that A is a subset of $\mathbb{R}$, $f : A \rightarrow \mathbb{R}$ is continuous, and the interval $[a, b]$ is a subset of A. Suppose further that y_0 is a number between $f(a)$ and $f(b)$. Then the line $y = y_0$ separates the points $(a, f(a))$ and $(b, f(b))$ in the plane (see Figure 4).

The only way that f can fail to intersect the line $y = y_0$ is for f to have a break. But, since f is continuous at each point of $[a, b]$, f must cross the line $y = y_0$. This demonstrates the following very important theorem.

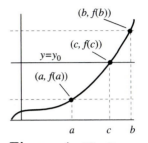

Figure 4. *The line $y = y_0$ separates the points $(a, f(a))$ and $(b, f(b))$ in the plane.*

> **Theorem 5 (The Intermediate Value Theorem)** *Suppose that A is a subset of $\mathbb{R}$ containing the interval $[a, b]$, and that f is a continuous function from A into $\mathbb{R}$. If y_0 is a number between $f(a)$ and $f(b)$, then there is at least one number c between a and b so that $f(c) = y_0$.*

A rigorous proof of Theorem 5 is beyond the scope of this text.

The following corollary is an easy consequence of the Intermediate Value Theorem. It is one of many applications.

> **Corollary 3** *Suppose that A is a subset of $\mathbb{R}$ containing the interval $[a, b]$, and that f is a continuous function from A into $\mathbb{R}$. If, for every x between a and b, $f(x) \neq 0$, then $f(x)$ is either positive at each x in (a, b), or $f(x)$ is negative at each x in (a, b).*

It is clear that if f is a real-valued function, then the only places that f can change signs are at points where $f(x) = 0$, points of discontinuity, and points that are not in the domain of f. As the following example illustrates, this application of the Intermediate Value Theorem will be useful in Chapter 5 when we will be concerned with graphing functions.

EXAMPLE 8: Let $f(x) = (x + 1)/[(x - 1)(x + 2)]$. Determine where f is positive and where f is negative.

SOLUTION: The domain of f is the union of $(-\infty, -2), (-2, 1)$ and $(1, \infty)$. Since f is a rational function, f is continuous. The function $f(x) = 0$ only at $x = -1$. Thus, f can change signs only at -2, -1 and 1; and so, if $f(x)$ is positive (negative) for any value in $(-\infty, -2)$, it will be positive (negative) everywhere on $(-\infty, -2)$. Similarly, f cannot change signs on the segments $(-2, -1)$, $(-1, 1)$ and $(1, \infty)$.

We now pick values in each interval to test whether the function is positive or negative on that interval.

$f(-3) < 0$, so f is negative on $(-\infty, -2)$,

$f(-3/2) > 0$, so f is positive on $(-2, -1)$,

$f(0) < 0$, so f is negative on $(-1, 1)$; and

$f(2) > 0$, so f is positive on $(1, \infty)$. ■

EXAMPLE 9: Solve the inequality $\dfrac{x+1}{(x-1)(x+2)} \geq 0$.

SOLUTION: We have already done the work! By the previous example we know that

$$\frac{x+1}{(x-1)(x+2)} > 0 \quad \text{if} \quad -2 < x < -1 \quad \text{or} \quad x > 1$$

and

$$\frac{x+1}{(x-1)(x+2)} = 0 \quad \text{at} \quad x = -1.$$

Thus

$$\frac{x+1}{(x-1)(x+2)} \geq 0 \quad \text{if} \quad -2 < x \leq -1 \quad \text{or} \quad x > 1. \quad ■$$

The methods applied in the above examples are fine if you can find where the function crosses the x–axis. But solving for 0 can be a problem! We have Newton's method as a tool. The Intermediate Value Theorem provides another method for approximating a root, as we illustrate in the following example.

EXAMPLE 10: Let $f(x) = 2x^3 - x^2 + 4x + 5$. Approximate a root of f.

SOLUTION: Since f is a polynomial, we know it is continuous. We first do a little fishing using a calculator to evaluate f at some values:

$$f(-5) = -290 \quad f(-4) = -155 \quad f(-3) = -70$$
$$f(-2) = -23 \quad f(-1) = -2 \quad f(0) = 5.$$

Our choice to start evaluating f at $x = -5$ was somewhat arbitrary. We discovered that f changes signs on the interval $[-1, 0]$. The Intermediate Value Theorem tells us that there is a root for f between -1 and 0. We now begin a general procedure.

-0.5 is the midpoint of $[-1, 0]$. -0.5 approximates our root with an error of less than $\frac{1}{2}$.	$f(-0.5) = 2.5$	$\Rightarrow$	there is a root in $[-1, -0.5]$.
-0.75 is the midpoint of $[-1, -0.5]$. -0.75 approximates our root with an error of less than $\frac{1}{4}$.	$f(-0.75) = 0.549$	$\Rightarrow$	there is a root in $[-1, -0.75]$.
-0.875 is the midpoint of $[-1, -0.75]$. -0.875 approximates our root with an error of less than $\frac{1}{8}$.	$f(-0.875) = -0.605$	$\Rightarrow$	there is a root in $[-0.875, -0.75]$.
-0.8125 is the midpoint of $[-0.875, -0.75]$. -0.8125 approximates our root with an error of less than $\frac{1}{16}$.	$f(-0.8125) = 0.017$	$\Rightarrow$	there is a root in $[-0.875, -0.8125]$.
$\vdots$	$\vdots$	$\vdots$	$\vdots$

We have employed a general algorithm, which we now outline.

The Interval-Halving Method[6]

Assume that A is a subset of $\mathbb{R}$, $f : A \to \mathbb{R}$ is continuous, and that we want to approximate a root of f.

Step I: Set the acceptable tolerances. That is, establish a number $E > 0$ that will be our acceptable error.

Step II: Find numbers a and b such that $[a, b]$ is a subset of the domain of f and such that $f(a)f(b) < 0$. The condition that $f(a)f(b) < 0$ tells us that f changes sign on $[a, b]$.

Step III: Let $c = \frac{a+b}{2}$. The number c is our current approximation of the root. The error in this approximation is no more than $\frac{|b-a|}{2}$. Check to see if $\frac{|b-a|}{2} < E$. If the answer is yes, then output c as your answer and quit. Otherwise continue to Step IV.

---

[6]Also known as the Bisection Method.

Step IV: Is $f(a)f(c) < 0$? If the answer is yes, then the root is between a and c. In this case, set $b = c$, leave a unchanged, and then return to Step III. Otherwise, we know that the root is between c and b. In this case, set $a = c$, leave b, unchanged and return to step III.

In order for the Interval-Halving method to work, the graph of the function must cross the x–axis. Thus, it could not be used to find the root of something like $f(x) = \left(x - \frac{2}{3}\right)^2$. However, though Newton's Method does not work for $f(x) = x^{1/3}$, the Interval-Halving Method finds the root.

Limits of functions from $\mathbb{R}$ into $\mathbb{R}^n$

Let A be a subset of $\mathbb{R}$, and let $\vec{f}(t) = (x(t), y(t), z(t))$ be a function from A into $\mathbb{R}^3$. Let $\vec{L} = (L_x, L_y, L_z)$. Since the distance from $(x(t), y(t), z(t))$ to $\vec{L} = (L_x, L_y, L_z)$ is $\|\vec{f}(t) - \vec{L}\| = \sqrt{(x(t) - L_x)^2 + (y(t) - L_y)^2 + (z(t) - L_z)^2}$, $\vec{f}(t)$ is close to $\vec{L}$ if and only if $x(t)$ is close to L_x, $y(t)$ is close to L_y, and $z(t)$ is close to L_z. We have

$$\lim_{t \to a} \vec{f}(t) = \lim_{t \to a} \begin{pmatrix} x(t) \\ y(t) \\ z(t) \end{pmatrix} = \begin{pmatrix} \lim_{t \to a} x(t) \\ \lim_{t \to a} y(t) \\ \lim_{t \to a} z(t) \end{pmatrix} = \begin{pmatrix} L_x \\ L_y \\ L_z \end{pmatrix} = \vec{L}.$$

Thus, $\lim_{t \to a} \vec{f}(t)$ exists if and only if the limits of each of the coordinate functions exist and $\vec{f}$ is continuous at $t = a$ if and only if each of the coordinate functions is continuous at $t = a$. It is apparent that we have vector versions of our limit and continuity theorems.

Theorem 6 *Suppose that $\vec{f}$ and $\vec{g}$ are vector-valued functions, h is a real-valued function with the same domain, and c is a number. Suppose further that $\lim_{x \to a} \vec{f}(x) = \vec{L}$, $\lim_{x \to a} \vec{g}(x) = \vec{K}$, and $\lim_{x \to a} h(x) = M$. Then*

(a) $\lim_{x \to a} c\vec{f}(x) = c\vec{L}$;

(b) $\lim_{x \to a} (h\vec{g})(x) = \lim_{x \to a} h(x)\vec{g}(x) = M\vec{K}$;

(c) $\lim_{x \to a} (\vec{f} + \vec{g})(x) = \vec{L} + \vec{K}$;

(d) If $M \neq 0$, then $\lim_{x \to a} \left(\frac{1}{h} \vec{g} \right)(x) = \lim_{x \to a} \frac{1}{h(x)} \vec{g}(x) = \vec{K}/M$;

(e) $\lim_{x \to a} (\vec{f} \cdot \vec{g})(x) = \lim_{x \to a} \vec{f}(x) \cdot \vec{g}(x) = \vec{L} \cdot \vec{K}$; and

(f) $\lim_{x \to a} \|\vec{f}(x)\| = \|\vec{L}\|$.

Proof: We prove only Part (e):

Expand $\vec{f}$ and $\vec{g}$:

$$\vec{f}(t) = (f_x(t), f_y(t), f_z(t)) \text{ and } \vec{g}(t) = (g_x(t), g_y(t), g_z(t)).$$

Then

$$\vec{f}(t) \cdot \vec{g}(t) = (f_x(t)g_x(t) + f_y(t)g_y(t) + f_z(t)g_y(t)).$$

By Theorem 2(b),

$$\lim_{x \to a} f_x(t)g_x(t) = L_x K_x,$$

$$\lim_{x \to a} f_y(t)g_y(t) = L_y K_y, \text{ and}$$

$$\lim_{x \to a} f_z(t)g_y(t) = L_z K_z.$$

Then by Theorem 2(c),

$$\lim_{x \to a} (f_x(t)g_x(t) + f_y(t)g_y(t) + f_z(t)g_y(t))$$

$$= L_x K_x + L_y K_y + L_z K_z = \vec{L} \cdot \vec{K}. \qquad \blacksquare$$

Of course, a similar theorem is also true for one-sided limits.

Theorem 7 *If $\vec{f}$, $\vec{g}$, and h are functions with a common domain and are continuous at the point a, and c is a number, then*

(a) $c\vec{f}$ *is continuous at a.*

(b) $\vec{f} + \vec{g}$ *is continuous at a.*

(c) $h\vec{g}$ *is continuous at a.*

(d) *If $h(a) \neq 0$, then $\frac{1}{h}\vec{f}$ is continuous at a.*

(e) $\vec{f} \cdot \vec{g}$ *is continuous at a.*

(f) $\|\vec{f}\|$ *is continuous at a.*

The following theorem is an application of Theorem 6 and Theorem 7.

Theorem 8 *If A is a subset of the real numbers, and $\vec{f}$ is a function from A into $\mathbb{R}^n$ such that $\vec{f}(t)$ is never $\vec{0}$, then the function $\vec{h}(t) = \left(\frac{1}{\|\vec{f}(t)\|}\right) \vec{f}(t)$ is a continuous function such that*

(a) *$\vec{h}(t)$ is always a unit vector.*

(b) *$\vec{h}(t)$ points in the same direction as $\vec{f}(t)$.*

Proof: We already know that if $\vec{v} \neq \vec{0}$, then $\left(\frac{1}{\|\vec{v}\|}\right) \vec{v}$ is a unit vector that points in the same direction as does $\vec{v}$. By Theorem 7.f, $\|\vec{f}(t)\|$ is a continuous real-valued function. Thus by Theorem 6.d, $\frac{1}{\|\vec{f}(t)\|}$ is continuous and, finally, by Theorem 7.c, $\left(\frac{1}{\|\vec{f}(t)\|}\right) \vec{f}(t)$ is continuous. ∎

EXAMPLE 11: If $\vec{r}(t) = (1 + t^2, t, t^3)$, then $\vec{r}$ is continuous and $\|\vec{r}(t)\|$ is never 0. Thus the function

$$\vec{g}(t) = \frac{\vec{r}(t)}{\|\vec{r}(t)\|} = \frac{1}{\sqrt{1 + 3t^2 + t^4 + t^6}} \begin{pmatrix} 1 + t^2 \\ t \\ t^3 \end{pmatrix}$$

is a continuous vector-valued function such that, for each t, $\vec{g}(t)$ is a unit vector that points in the same direction as does $\vec{r}(t)$. ∎

EXAMPLE 12: Let $\vec{r}(t) = (t, t, t)$. Then $\frac{1}{\sqrt{3t^2}}(t, t, t)$ is a unit vector that points in the same direction as $\vec{r}(t)$ everywhere except at $t = 0$. ∎

We also have limit and continuity properties for the cross product if we are working in $\mathbb{R}^2$ or $\mathbb{R}^3$.

Theorem 9 *Suppose that $\vec{f}$ and $\vec{g}$ are funtions from a connected subset of $\mathbb{R}^1$ to $\mathbb{R}^2$ (or $\mathbb{R}^3$).*

(a) *If $\lim_{x \to a} \vec{f}(x) = \vec{L}$ and $\lim_{x \to a} \vec{g}(x) = \vec{K}$, then $\lim_{x \to a}(\vec{f}(x) \times \vec{g}(x)) = \vec{L} \times \vec{K}$.*

(b) *If $\vec{f}(x)$ and $\vec{g}(x)$ are continuous at the point a, then $\vec{f}(x) \times \vec{g}(x)$ is continuous at a.*

EXERCISES 3.5

1. Determine the points of discontinuity in the function sketched in Figure 5.

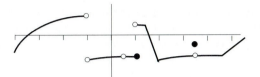

Figure 5. *Function for Exercise 1.*

2. Determine the points of discontinuity in the function sketched in Figure 6.

Figure 6. *Function for Exercise 2.*

In Exercises 3–10, determine the indicated limits if they exist. If the limit fails to exist, explain why.

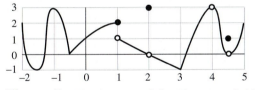

Figure 7. *The function f for Exercises 3-10.*

3. $\lim_{x \to 4.5^-} f(x)$ 4. $\lim_{x \to 4.5^+} f(x)$

5. $\lim_{x \to 4.5} f(x)$ 6. $\lim_{x \to 1^-} f(x)$

7. $\lim_{x \to 1^+} f(x)$ 8. $\lim_{x \to 2} f(x)$

9. $\lim_{x \to 1} f(x)$ 10. $\lim_{x \to 4} f(x)$

In Exercises 11–17, determine the indicated limits if they exist. If the limit fails to exist, explain why.

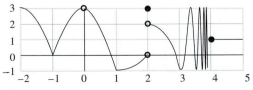

Figure 8. *The function f for Exercises 11–17.*

11. $\lim_{x \to 0} f(x)$ 12. $\lim_{x \to 2^-} f(x)$

13. $\lim_{x \to 2^+} f(x)$ 14. $\lim_{x \to 2} f(x)$

15. $\lim_{x \to 4^+} f(x)$ 16. $\lim_{x \to 4^-} f(x)$

17. $\lim_{x \to 4} f(x)$

18. Determine at which points of discontinuity of the function in Exercise 1 has a limit.

19. Determine at which points of discontinuity of the function in Exercise 2 has a limit.

Determine the points of discontinuity for the functions in Exercises 20–28.

20. $f(x) = |x|$.

21. $f(x) = \dfrac{x^2 + 2x + 3}{x^2 + x + 1}$.

22. $f(x) = \dfrac{x^3 + 2x^2 + x}{x^2 + 3x + 2}$.

23. $f(x) = [[x]]$.

24. $f(x) = [[x^2]]$.

25. $f(x) = \dfrac{|x|}{x}$.

26. $f(x) = \dfrac{|x^2 - 1|}{x^2 - 1}$.

27. $\vec{f}(t) = ([[t]], [[t]])$.

28. $\vec{f}(t) = ([[t]], [[t^2]])$.

Solve the inequalities in Exercises 29–31.

29. $(x - 1)(x + 1)(x - 4) \le 0$.

30. $\dfrac{(x + 3)(x - 4)}{x} > 0$.

31. $\dfrac{x^3}{(x - 2)(x + 10)} \ge 0$.

Compute the limits in Exercises 32–41.

32. $\lim_{x \to 1} x^6 + 6x^2 + 10x$.

33. $\lim_{x \to 0} \dfrac{x^2 + 2x}{x^3 + 6x}$.

34. $\lim_{x \to -1} \dfrac{x^3 + 2x^2 + x}{x^2 + 3x + 2}$.

35. $\lim\limits_{h \to 0} \dfrac{f(x+h) - f(x)}{h}$, where $f(x) = x$.

36. $\lim\limits_{h \to 0} \dfrac{f(x+h) - f(x)}{h}$, where $f(x) = (x+1)^{-1/2}$.

37. $\lim\limits_{h \to 0} \dfrac{g(t+h) - g(t)}{h}$, where $g(t) = \dfrac{t}{t^2 + 1}$.

38. $\lim_{h \to 1^+} [[h]]$.

39. $\lim_{h \to 1^-} [[h]]$.

40. $\lim_{t \to \sqrt{2}^-} \left([[t]], [[t^2]] \right)$.

41. $\lim_{t \to \sqrt{2}^+} \left([[t]], [[t^2]] \right)$.

For the given sequence a and integer n in Exercises 42–44, let $f(x) = a([[x]])$. Compute $\lim_{x \to n^+} f(x)$ and $\lim_{x \to n^-} f(x)$.

42. $a(i) = i$, $i \geq 1$. Let $n = 2$.

43. $a(i) = \dfrac{2^i}{i!}$, $i \geq 0$. Let $n = 4$.

44. $a(0) = 1$ and $a(i+1) = (-1)(1 + a^i(i))$. Let $n = 3$.

Determine the points of discontinuity of the functions given in Exercises 45–47.

45. $f(t) = t - [[t]]$.

46. $g(t) = t - [[6t]]$.

47. $f \circ g$ and $g \circ f$ where $f(x) = [[x]]$ and $g(x) = x^2$.

In Exercises 48–51, show that the function f has a root in the interval $[a, b]$. Using a calculator, do four steps of the Interval-Halving method to approximate a root of the function on the interval $[a, b]$, and determine a bound on the error of your estimate.

48. $f(x) = x^3 + 2x^2 + 3x - 5$,　$a = 0$ and $b = 1$.

49. $f(x) = x^3 + 2x^2 + \pi x - \sqrt{2}$,　$a = -1$ and $b = 1$.

50. $f(x) = \sin(x) - \cos(x)$,　$a = 0$ and $b = \frac{\pi}{2}$. (Assume that the sine and cosine functions are continuous.)

51. $f(x) = \operatorname{Arctan}(x) - \sqrt{1.1}$,　$a = 0$ and $b = 3$. (Assume that the Arctangent function is continuous.)

52. Let $f(x) = \dfrac{1}{x^2 - 1}$.　$f(0) = -1$ and $f(2) = \frac{1}{3}$. Must there be a root for f in the interval $[0, 2]$?

3.6　Limits: A Formal Approach

EXAMPLE 1: Suppose that we are machining a wheel to drive a belt in a widget. It is important that the circumference of the wheel be 1 m, with an error of no more than 0.001 m. Assume that we can machine the wheel so that the radius r of the circle is as accurate as necessary, but we cannot expect it to be exact. How much error can we allow in our measurement of r and still be sure that the circumference of our wheel will be within its tolerances?

SOLUTION: Let $C(r)$ denote the circumference of the circle with radius r. Then $C(r) = 2\pi r$ m. We want our radius to approximate the value that will give $C(r) = 1$ m; that is, we want to approximate $r_0 = \frac{1}{2\pi}$. We can machine our wheel so that r is as close to $r_0 = \frac{1}{2\pi}$ as we please; so we need only determine just how close r must be to $\frac{1}{2\pi}$ in order that the error in the wheel's circumference is less than

0.001 m. We must choose r so that

$$|C(r) - 1| < 0.001.$$

Equivalently, we write

$$-0.001 < C(r) - 1 < 0.001.$$

Dividing by 2π yields:

$$-\frac{0.001}{2\pi} < \frac{2\pi r - 1}{2\pi} < \frac{0.001}{2\pi},$$

or

$$-\frac{0.001}{2\pi} < r - \frac{1}{2\pi} < \frac{0.001}{2\pi},$$

which is the same as

$$\left| r - \frac{1}{2\pi} \right| < \frac{0.001}{2\pi}.$$

Hence to ensure that the error in the circumference is less than 0.001, we must measure r with an error of no more than $\frac{0.001}{2\pi}$.

In Figure 1, we sketch the graph of C as a function of r. The vertical shaded region in the blowup circle is the region $\left| r - \frac{1}{2\pi} \right| < \frac{0.001}{2\pi}$. The horizontal shaded region contains the set of points with y–coordinates with the acceptable error.   ■

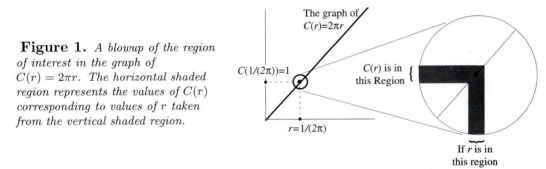

Figure 1. *A blowup of the region of interest in the graph of $C(r) = 2\pi r$. The horizontal shaded region represents the values of $C(r)$ corresponding to values of r taken from the vertical shaded region.*

The graph of $C(r)=2\pi r$

$C(1/(2\pi))=1$

$C(r)$ is in this Region

$r=1/(2\pi)$

If r is in this region

EXAMPLE 2: Suppose that we want to measure the area A of a square and that we can allow an error for the measurement of A of no more than ϵ. Find a number δ so that if we measure the length of a side with an error of no more than δ, then our calculation of A will be within our tolerances.

SOLUTION: Let r_0 denote the actual length of a side of the square and let r denote our measurement of r_0. The actual area has the form $A = r_0^2$ and our calculation of A is r^2. We want

$$|r_0^2 - r^2| < \epsilon,$$

which is equivalent to

$$|(r_0 - r)(r_0 + r)| < \epsilon.$$

Since r_0 and r are lengths, they are positive. Thus we must have

$$|r_0 - r|(r_0 + r) < \epsilon.$$

The term $(r + r_0)$ complicates the solution of this problem. We will put a bound on this term by using the following procedure. Let δ_1 be any positive number and let $k = r_0 + \delta_1$. If we measure carefully enough, we can be sure that

$$|r - r_0| < \delta_1 \text{ or } -\delta_1 < r - r_0 < \delta_1.$$

But this implies that if $|r - r_0| < \delta_1$, then $r < r_0 + \delta_1 = k$. Since r_0 is less than k, we have that $r + r_0 < 2k$, and

$$|r^2 - r_0^2| = |r_0 - r|(r_0 + r) < |r_0 - r|(2k).$$

We have bounded the troublesome term $(r_0 + r)$. Now let $\delta_2 = \epsilon/2k$, and let δ be the minimum of δ_1 and δ_2. If $|r_0 - r| < \delta$, then

$$
\begin{aligned}
|r_0^2 - r^2| &= |r_0 - r|(r_0 + r) \\
&< \left(\frac{\epsilon}{2k}\right)(2k) = \epsilon.
\end{aligned}
$$

Thus if we measure a side of the square with an error of no more than δ, then our computation of A will have an error of no more than ϵ. In Figure 2, we show what we have done. ∎

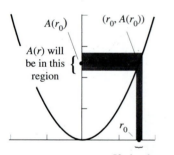

Figure 2. *The horizontal shaded region represents acceptable values of $A(r)$. As long as r is taken from the vertical shaded region, $A(r)$ remains safely within the horizontal region.*

EXAMPLE 3: Suppose that we have a sphere and that we can measure the radius of the sphere to any given degree of accuracy, but not exactly. Suppose further that we want to know the volume of the sphere to within 0.001 m³. The volume of a sphere is given by $V(r) = (4/3)\pi r^3$. How much error can there be in our measurement of the radius of the sphere if we are to be sure that the error in our computation of the volume of the sphere is less than 0.001 m³?

SOLUTION: Let r_0 denote the true radius of the sphere and let r denote our measurement of the radius. The true volume of the sphere is $V(r_0)$, and our approximation of the volume of the sphere is $V(r)$. The error is $\epsilon = |V(r_0) - V(r)|$. Again, we must have $\epsilon < 0.001$.

$$\epsilon \;=\; |V(r_0) - V(r)| = \left|\frac{4}{3}\pi(r_0^3 - r^3)\right| = \frac{4}{3}\pi|r_0^3 - r^3|$$

$$= \;\frac{4}{3}\pi|(r_0 - r)(r_0^2 + r_0 r + r^2)| = \frac{4}{3}\pi(r_0^2 + r_0 r + r^2)|r_0 - r|.$$

Note that the last equation holds, since both r and r_0 are positive numbers. As in Example 2, we need to put a bound of the term $(r_0^2 + r_0 r + r^2)$. Since r is a length that we are measuring, we can find a number k so that $k > r_0$, and we can be careful enough in our measurement of r so that we know that $r < k$. Let δ denote the maximum error in our measurement of r, so $|r_0 - r| < \delta$. Then

$$\epsilon \;=\; \frac{4}{3}\pi(r_0^2 + r_0 r + r^2)|r_0 - r| < \frac{4}{3}\pi(3k^2)|r_0 - r|$$

$$< \;\frac{4}{3}\pi(3k^2)\delta = 4\pi k^2 \delta.$$

It follows that if δ, the error in our measurement, is less than $\frac{0.001}{4\pi k^2}$, then we have

$$\epsilon < 4\pi k^2 \delta < 4\pi k^2 \left(\frac{0.001}{4\pi k^2}\right) = 0.001. \qquad\blacksquare$$

In the previous sections, we considered $\lim_{x\to a} f(x) = L$ in terms of L being the number that is approximated by f when x is close to a. This is an intuitive statement that does not well define the notion of a limit. In the formal definition of the limit of a function, the number ϵ should be thought of as the acceptable error in our approximation of L by $f(x)$, and δ should be thought of as the accuracy required in our measurement x of the number a.

We are now ready to formalize the notion of limit, and the student is advised to compare this definition with the intuitive notion presented in Section 3.2.

Definition: Limits, a Formal Definition

Suppose that f is a function from a subset of $\mathbb{R}$ into $\mathbb{R}$. The statement that the number L is the *limit of $f(x)$ as x goes to a* means that:

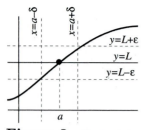

Figure 3. *Geometric interpretation of Condition (b).*

(a) a is a number (not necessarily in the domain of f).

(b) If $\epsilon > 0$, then there is a number $\delta > 0$ such that if $0 < |a - x| < \delta$, then

    (i) x is in the domain of f, and

    (ii) $|L - f(x)| < \epsilon$.

Condition (b)(i) says that while a may not be in the domain of f, a is surrounded by the domain of f. Condition (b)(ii) states that if $\epsilon > 0$, then there is a number δ so that if $x \neq a$ is a measurement of a with an error of less than δ, then $f(x)$ is a value that approximates L with an error of less than ϵ.

If we look at the graph of the function f, we see that there is a nice geometric meaning of Condition (b):

(b'). If $\epsilon > 0$, then there is a number $\delta > 0$ such that if $(x, f(x))$, $x \neq a$ lies between the vertical lines $x = a - \delta$ and $x = a + \delta$, then $(x, f(x))$ is between the horizontal lines $y = L - \epsilon$ and $y = L + \epsilon$. See Figure 3.

This interpretation permits us to use our graphing calculators to *verify* that L is a limit.

EXAMPLE 4: We expect that $f(x) = \sin(x)\cos(x)$ is continuous. That is, we would expect that $\lim_{x \to \pi/2} = f\left(\frac{\pi}{4}\right) = \frac{1}{2}$. Find a number δ such that if $|x - \frac{\pi}{2}| < \delta$, then $|f(x) - f\left(\frac{\pi}{4}\right)| < 0.001$.

SOLUTION: We use our calculator to graph f in a trial and error approach. We make an initial guess for δ. Then we draw the graph of f and the lines $y = \frac{1}{2} - 0.001$ and $y = \frac{1}{2} + 0.001$ on the same axes, restricting the domain to $\frac{\pi}{4} - \delta < x < \frac{\pi}{4} + \delta$. In this example, we first choose $\delta = 0.1$ and plot the graph which gives the graph in Figure 4.a. We see that 0.1 is too large, so we try $\delta = 0.05$ to obtain Figure 4.b. Inspection of this graph leads us to believe that $\delta = 0.03$ will likely work. We generate the graph in Figure 4.c and discover that this choice for δ does work. ∎

In problems such as these, there is no unique choice for δ. If we find one choice that works, then any smaller choice for delta will also be a solution.

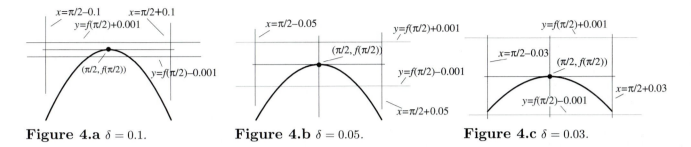

Figure 4.a $\delta = 0.1$.   **Figure 4.b** $\delta = 0.05$.   **Figure 4.c** $\delta = 0.03$.

EXAMPLE 5: We have argued that $D\sin(x) = \cos(x)$. Thus if $f(x) = \sin(x)$, then $f'\left(\frac{\pi}{3}\right) = \cos\left(\frac{\pi}{3}\right) = \frac{1}{2}$. By the definition of the derivative, we know that

$$\lim_{h\to 0}\frac{\sin\left(\frac{\pi}{3}+h\right)-\sin\left(\frac{\pi}{3}\right)}{h} = \lim_{h\to 0}\frac{\sin\left(\frac{\pi}{3}+h\right)-\frac{1}{\sqrt{2}}}{h} = \frac{1}{2}.$$

Find a number δ so that if $0 < |h| < \delta$, then

$$\left|\frac{\sin\left(\frac{\pi}{3}+h\right)-\frac{1}{\sqrt{2}}}{h} - \frac{1}{2}\right| < 0.001.$$

SOLUTION: Let $g(h) = \dfrac{\sin\left(\frac{\pi}{3}+h\right)-\frac{1}{\sqrt{2}}}{h} - 1/2$. We want to find a number δ such that if $0 < |h| < \delta$, then $|g(h)| < 0.001$. Notice the condition that $0 < |h|$. This condition is imposed because 0 is not in the domain of g. We start as we did in Example 4. We choose a reasonable δ and look at the graph of g and the lines $y = 0.001$ and $y = -0.001$ on the same coordinate system, keeping $|x| < \delta$. Most graphing calculators finesse the fact that 0 is not in the domain of g by filling in the limiting value. In Figure 5.a, we try $\delta = 0.01$. Inspection of Figure 5.a leads us to try $\delta = 0.002$. We see in Figure 5.b that this choice for δ works. That is, if $0 < |h| < 0.002$, then $\left|\dfrac{\sin\left(\frac{\pi}{3}+h\right)-\frac{1}{\sqrt{2}}}{h} = \dfrac{1}{2}\right| < 0.001$. ∎

Figure 5.a.
g(h) restricted to the
domain $0 < |h| < 0.01$.

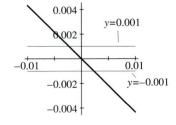

Figure 5.b.
g(h) restricted to the
domain $0 < |h| < 0.002$.

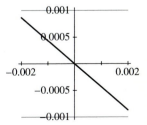

While we can use graphing calculators to check continuity and limits, it is usually the case that we must resort to algebra to prove that a limit is actually what we expect it to be. In Example 4, we have found a δ that works for $\epsilon = 0.001$. If we start with a smaller ϵ, then we will have to find a new δ. In order to prove that $\lim_{x \to x_0} = L$, we must have a method for finding δ as a function of ϵ.

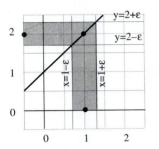

Figure 6. *If x is between the vertical lines $x = 1 + \epsilon$ and $x = 1 - \epsilon$, then $f(x)$ is between the horizontal lines $y = 2 + \epsilon$ and $y = 2 - \epsilon$.*

EXAMPLE 6: Let $f(x) = x + 1$. Show that $\lim_{x \to 1} f(x) = 2$.

SOLUTION: Let $\epsilon > 0$. We must find a number δ so that if $|x-1| < \delta$, then $|f(x)-2| < \epsilon$, which is equivalent to $|(x+1)-2| < \epsilon$ or $|x-1| < \epsilon$. Thus we may let $\delta = \epsilon$, and if x is a measurement of 1 with an error less than $\delta = \epsilon$, then $f(x)$ approximates 2 with an error of less than ϵ. See Figure 6. ∎

EXAMPLE 7: Show that $\lim_{x \to 1} (3x + 1) = 4$.

SOLUTION: Let ϵ be a positive number. We must find a number $\delta > 0$ so that if $|x-1| < \delta$, then $|(3x+1) - 4| < \epsilon$. This is equivalent to showing that

$$|3x - 3| < \epsilon,$$

which is the same as

$$3|x - 1| < \epsilon,$$

or

$$|x - 1| < \frac{\epsilon}{3}.$$

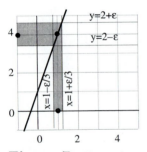

Figure 7. *If x is between the vertical lines $x = 1 + \frac{\epsilon}{3}$ and $x = 1 - \frac{\epsilon}{3}$, then $f(x)$ is between the horizontal lines $y = 2 + \epsilon$ and $y = 2 - \epsilon$.*

Hence if we let $\delta = \frac{\epsilon}{3}$, then if $|x-1| < \delta$, we have that $|(3x+1)-4| < \epsilon$. See Figure 7. ∎

EXAMPLE 8: Show that $\lim_{x \to 2} x^2 = 4$.

SOLUTION: Let $\epsilon > 0$. We want to find a number $\delta > 0$ so that if $|x - 2| < \delta$, then $|x^2 - 4| < \epsilon$. Now

$$|x^2 - 4| = |x - 2||x + 2|.$$

We need to bound $|x + 2|$, which we accomplish using a strategy similar to that used in Examples 2 and 3. We first note that we can measure x so that $|x - 2| < \delta_1$ for some positive number δ_1. We choose, for example, $\delta_1 = 1$. Now we solve the inequality $|x - 2| < 1$ in order to obtain the bound on $|x + 2|$.

$$-1 < x - 2 < 1 \text{ or } 1 < x < 3.$$

Thus if $|x - 2| < 1$, then $|x + 2| < 5$, which gives

$$
\begin{aligned}
|x^2 - 4| &= |x - 2||x + 2| \\
&< 5|x - 2|.
\end{aligned}
$$

Let $\delta_2 = \epsilon/5$, and let δ be the minimum of δ_1 and δ_2. If $|x - 2| < \delta$, then

$$
\begin{aligned}
|x^2 - 4| &< 5|x - 2| \\
&< 5\delta \\
&\leq 5\left(\frac{\epsilon}{5}\right) = \epsilon. \text{ See Figure 8.} \quad \blacksquare
\end{aligned}
$$

We close this section with a theorem that will be useful later on.

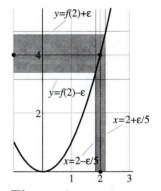

Figure 8. *If x is between the vertical lines $x = 2 + \frac{\epsilon}{5}$ and $x = 2 - \frac{\epsilon}{5}$, then $f(x)$ is between the horizontal lines $y = 4 + \epsilon$ and $y = 4 - \epsilon$.*

Theorem 1 *If f is a function from a subset of $\mathbb{R}$ into $\mathbb{R}$ which is continuous at x_0, and if $f(x_0) > 0$, then there is a number $\delta > 0$ so that if $|x - x_0| < \delta$, then $f(x) > 0$. If $f(x_0) < 0$, then there is a number $\delta > 0$ so that if $|x - x_0| < \delta$, then $f(x) < 0$.*

Proof: We prove only the case that $f(x_0) > 0$. Let $\epsilon = f(x_0)$. Since f is continuous at x_0, $\lim_{x \to x_0} f(x) = f(x_0)$; so there is a number $\delta > 0$ such that if $|x - x_0| < \delta$, then $|f(x) - f(x_0)| < \epsilon = f(x_0)$. This is equivalent to

$$
-f(x_0) < f(x) - f(x_0) < f(x_0),
$$

or

$$
0 < f(x) < 2f(x_0).
$$

Thus

$$
0 < f(x) \text{ if } |x - x_0| < \delta.
$$

The argument for the case that $f(x_0) < 0$ is similar. $\blacksquare$

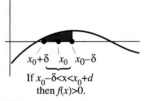

Figure 9. *If a continuous function is positive at x_0, then it is positive in a segment centered at x_0.*

EXERCISES 3.6

1. The radius of the Earth is approximately 6,370 km. Assume that the Earth is a perfectly round sphere. Put a belt along the equator, completely encircling the Earth. If you wanted to expand the belt so that it is extended 1 m above the earth at all points, how much belt would you need to add? How much belt would you need to add if you did this same experiment on the moon (radius $\approx$ 1,740 km)?

In Exercises 2–5, find a number δ so that if $|x - x_0| < \delta$, then $|f(x) - f(x_0)| < 0.0001$.

2. $f(x) = 2x + 1$, $x_0 = 2$.

3. $f(x) = x/3 + 2$, $x_0 = 1$.

4. $f(x) = x^2$, $x_0 = 1$.

5. $f(x) = 3x^2 + x - 1$, $x_0 = -1$.

Let $f(x) = x^2$. We have already shown that $f'(x) = 2x$. In Exercises 6–9, algebraically, determine how small h must be in order to guarantee that $|f'(x_0) - [f(x_0 + h) - f(x_0)]/h| < \epsilon$.

6. $x_0 = 0$, $\epsilon = 0.1$.

7. $x_0 = 0$, $\epsilon = 0.01$.

8. $x_0 = 3$, $\epsilon = 0.1$.

9. $x_0 = 3$, $\epsilon = 0.01$.

10. Let $f(x) = x^2$. Find δ as a function of ϵ so that if $0 < |h| < \delta$, then $\left| 2x - \frac{f(x+h) - f(x)}{h} \right| < \epsilon$.

In Exercises 11–17, compute $f'(x)$ and determine how small h must be in order to guarantee that $\left| f'(x_0) - \frac{f(x_0 + h) - f(x_0)}{h} \right| < \epsilon$.

11. $f(x) = x^3$, $x_0 = 3$, $\epsilon = 0.1$.

12. $f(x) = x^3$, $x_0 = 4$, $\epsilon = 0.001$.

13. $f(x) = \sqrt{x}$, $x_0 = 4$, $\epsilon = 0.1$.

14. $f(x) = \sqrt{x}$, $x_0 = 4$, $\epsilon = 0.01$.

15. $f(x) = \sqrt{x}$, $x_0 = 4$, $\epsilon = 0.0001$.

16. $f(x) = \sin(x)$, $x_0 = \frac{2\pi}{3}$, $\epsilon = 0.0001$.

17. $f(x) = \cos(x)$, $x_0 = \frac{5\pi}{6}$, $\epsilon = 0.0001$.

18. Let $f(x) = \tan(x)$. Find a number δ such that if
$$0 < |h| < \delta, \text{ then } 3.9 < \left| \frac{f\left(\frac{\pi}{3} + h\right) - f\left(\frac{\pi}{3}\right)}{h} \right| <$$
4.1.

In Exercises 19 and 20, $x_0 = 1$ and
$$f(x) = \begin{cases} 0 & \text{if } x \leq 1 \\ 1 & \text{if } x > 1 \end{cases}.$$

19. Show that if $\delta > 0$, then there is an h_1 such that $|h_1| < D$ and $\dfrac{f(x_0 + h_1) - f(x_0)}{h_1} > 1000$ and there is an h_2 such that $|h_2| < \delta$ and $\dfrac{f(x_0 + h_2) - f(x_0)}{h_2} < \dfrac{1}{1000}$.

20. Show that if $N > 0$ and $\delta > 0$, then there are h_1 and h_2 with $|h_1| < \delta$ and $|h_2| < \delta$ such that $\dfrac{f(x_0 + h_1) - f(x_0)}{h_1} > N$ and $\dfrac{f(x_0 + h_2) - f(x_0)}{h_2} < \dfrac{1}{N}$.

In Exercises 21–23, $x_0 = 0$ and
$$f(x) = \begin{cases} \sin\left(\frac{1}{x}\right) & \text{if } x \neq 0 \\ 0 & \text{if } x = 0 \end{cases}.$$

21. Show that if $\delta > 0$, then there is an h_1 such that $|h_1| < \delta$, yet $\dfrac{f(x_0 + h_1) - f(x_0)}{h_1} > 1000$, and there is an h_2 such that $|h_2| < \delta$ and $\dfrac{f(x_0 + h_2) - f(x_0)}{h_2} < \dfrac{1}{1000}$.

22. Show that if $N > 0$ and $\delta > 0$, then there are h_1 and h_2 with $|h_1| < D$ and $|h_2| < \delta$ such that $\dfrac{f(x_0 + h_1) - f(x_0)}{h_1} > N$ and $\dfrac{f(x_0 + h_2) - f(x_0)}{h_2} < \dfrac{1}{N}$.

23. Argue that the function f is not differentiable at x_0.

24. Show that $x \sin\left(\frac{1}{x}\right)$ lies between the graphs of $y = x$ and $y = -x$. Argue that since $\lim_{x \to 0} x = 0$, it follows that $\lim_{x \to 0} x \sin\left(\frac{1}{x}\right) = 0$. (Note: This principle is sometimes called the Sandwich Theorem.)

25. Prove that the following two statements are equivalent:

    i) $a = b$

    ii) If ϵ is any positive number, then $|a - b| < \epsilon$.

Chapter 4

Differentiation Rules

4.1   The Sum and Product Rules and Higher Derivatives

We have encountered the derivative of a function when looking for tangents to graphs of functions, and when introducing velocity. When we compute the slope of a function $y = f(x)$, we are computing the rate of change of y as x changes (the rise over the run). When we compute the derivative of a function from a subset of $\mathbb{R}$ into $\mathbb{R}^n$, we are computing the rate of change of position. As we have pointed out before, if

$$\vec{r}(t) = \begin{pmatrix} x(t) \\ y(t) \\ z(t) \end{pmatrix},$$

then

$$\vec{r}'(t) = \lim_{h \to 0} \frac{\vec{r}(t+h) - \vec{r}(t)}{h} = \begin{pmatrix} \lim_{h \to 0} \frac{x(t+h)-x(t)}{h} \\ \lim_{h \to 0} \frac{y(t+h)-y(t)}{h} \\ \lim_{h \to 0} \frac{z(t+h)-z(t)}{h} \end{pmatrix} = \begin{pmatrix} x'(t) \\ y'(t) \\ z'(t) \end{pmatrix}.$$

Thus, if we are able to find the derivative of real-valued functions, then we can also find the derivative of vector-valued functions. For this reason, we will concentrate on differentiating real-valued functions.

In Exercise 3.3.15 and 3.3.16, you showed that $D \sin(x) = \cos(x)$ and $D \cos(x) = -\sin(x)$. In this chapter, we introduce other rules to help us in differentiating functions. Each of these rules allows us to break down a function into simpler components which, hopefully, we can differentiate.

Theorem 1 *Suppose that c is a positive number and that f and g are real-valued functions so that f' and g' are defined at x. Then*

(a) $(cf)'(x) = cf'(x).$

(b) $(f + g)'(x) = f'(x) + g'(x)$ *(Sum rule).*

(c) $(fg)'(x) = f'(x)g(x) + f(x)g'(x)$ *(Product rule).*

Proof:

(a) By the definition of the derivative,

$$
\begin{aligned}
(cf)'(x) &= \lim_{h \to 0} \frac{cf(x + h) - cf(x)}{h} \\
&= \lim_{h \to 0} \frac{c(f(x + h) - f(x))}{h} \\
&= c \lim_{h \to 0} \frac{f(x + h) - f(x)}{h} \\
&= cf'(x).
\end{aligned}
$$

(b) This is left as an exercise.

(c) Again, by the definition of the derivative,

$$
\begin{aligned}
(fg)'(x) &= \lim_{h \to 0} \frac{fg(x + h) - fg(x)}{h} \\
&= \lim_{h \to 0} \frac{f(x + h)g(x + h) - f(x)g(x)}{h} \\
&= \lim_{h \to 0} \frac{f(x + h)g(x + h) - f(x)g(x + h) + f(x)g(x + h) - f(x)g(x)}{h} \\
&= \lim_{h \to 0} \frac{f(x + h)g(x + h) - f(x)g(x + h)}{h} \\
&\quad + \lim_{h \to 0} \frac{f(x)g(x + h) - f(x)g(x)}{h} \\
&= \lim_{h \to 0} g(x + h) \frac{f(x + h) - f(x)}{h} + \lim_{h \to 0} f(x) \frac{g(x + h) - g(x)}{h}.
\end{aligned}
$$

By Theorem 1 of Section 3.5, g is continuous at x since $g'(x)$ exists. Thus, $\lim_{h \to 0} g(x + h) = g(x)$. We have

$$
\begin{aligned}
(fg)'(x) &= g(x) \lim_{h \to 0} \frac{f(x + h) - f(x)}{h} + f(x) \lim_{h \to 0} \frac{g(x + h) - g(x)}{h} \\
&= g(x)f'(x) + f(x)g'(x). \qquad \blacksquare
\end{aligned}
$$

EXAMPLE 1: We know that $D(x^2) = 2x$ and $D(\sin(x)) = \cos(x)$. Thus

$$D(x^2 + \sin(x)) = 2x + \cos(x).$$

$$D(x^2 \sin(x)) = \sin(x)D(x^2) + x^2 D(\sin(x)) = 2x \sin(x) + x^2 \cos(x).$$

There is an obvious counterpart of Theorem 1 for vector-valued functions.

Theorem 2 *If h is a real-valued function and $\vec{f}$ and $\vec{g}$ are vector-valued functions defined on a common subset of $\mathbb{R}$ and differentiable at t, and if c is a number, then*

(a) $D(c\vec{f})(t) = cD\vec{f}(t).$

(b) $D(\vec{f} + \vec{g})(t) = \vec{f}'(t) + \vec{g}'(t).$

(c) $D(\vec{f} \cdot \vec{g})(t) = \vec{f}(t) \cdot \vec{g}'(t) + \vec{f}'(t) \cdot \vec{g}(t)$ *(where $\vec{f} \cdot \vec{g}(t)$ is the dot (or inner) product of $\vec{f}(t)$ and $\vec{g}(t)$).*

(d) $D(\vec{f} \times \vec{g})(t) = \vec{f}'(t) \times \vec{g}(t) + \vec{f}(t) \times \vec{g}'(t).$

(e) $D(h\vec{f})(t) = h(t)\vec{f}'(t) + h'(t)\vec{f}(t).$

We give an argument for parts (c) and (d).

Part (c): Assume that $\vec{f}$ and $\vec{g}$ are functions from a subset A, of $\mathbb{R}$, into $\mathbb{R}^2$. Suppose that

$$\vec{f}(t) = (f_x(t), f_y(t)) \text{ and } \vec{g}(t) = (g_x(t), g_y(t)).$$

Then

$$
\begin{aligned}
D(\vec{f} \cdot \vec{g})(t) &= D(f_x(t)g_x(t) + f_y(t)g_y(t)) \\
&= D(f_x(t)g_x(t)) + D(f_y(t)g_y(t)) \\
&= f_x(t)g_x'(t) + f_x'(t)g_x(t) + f_y(t)g_y'(t) + f_y'(t)g_y(t)
\end{aligned}
$$

$$= \quad [f_x(t)g'_x(t) + f_y(t)g'_y(t)] + [f'_x(t)g_x(t) + f'_y(t)g_y(t)]$$

$$= \quad \vec{f} \cdot \vec{g}'(t) + \vec{f}' \cdot \vec{g}(t). \qquad \blacksquare$$

Part (d): We employ the same notation that we used in Part (c), but we suppress the variable t.

$$D\left(\vec{f} \times \vec{g}\right) = D(f_y g_z - f_z g_y, f_z g_x - f_x g_z, f_x g_y - f_y g_x)$$

$$= \quad ((f'_y g_z + f_y g'_z - f'_z g_y - f_z g'_y), (f'_z g_x + f_z g'_x - f'_x g_z - f_x g'_z),$$
$$(f'_x g_y + f_x g'_y - f'_y g_x - f_y g'_x))$$

$$= \quad ((f'_y g_z - f'_z g_y), (f'_z g_x - f'_x g_z), (f'_x g_y - f'_y g_x))$$
$$+ ((f_y g'_z - f_z g'_y), (f_z g'_x - f_x g'_z), (f_x g'_y - f_y g'_x))$$

$$= \quad \vec{f}' \times \vec{g} + \vec{f} \times \vec{g}'. \qquad \blacksquare$$

The next theorem gives us an explicit calculation for the derivative of a monomial.

Theorem 3 (Power Rule) *If* n *is a positive integer, then* $Dx^n = nx^{n-1}$.

Proof: We have already proved the cases where $n = 1, 2$, and 3 in earlier sections. We now show that if the theorem is true for $n = k$, then it is true for $n = k + 1$. The validity of the theorem will follows by induction.[1]
Assume that $Dx^k = kx^{k-1}$.

Then

$$Dx^{k+1} = D(x^k x) = xDx^k + x^k Dx \text{ (by the Product Rule)}$$

$$= x(kx^{k-1}) + x^k$$

$$= kx^k + x^k$$

$$= (k+1)x^k. \qquad \blacksquare$$

EXAMPLE 2: Let $f(x) = x^{100} + \pi x^{510}$.
Then

$$\frac{df(x)}{dx} = f'(x) = D(x^{100}) + D(\pi x^{510}) = 100x^{99} + 510\pi x^{509}. \qquad \blacksquare$$

---

[1]See Appendix A for a short introduction to mathematical induction.

EXAMPLE 3: Given that $D(\sqrt{x^2 + 2x + 1}) = \dfrac{x + 1}{\sqrt{x^2 + 2x + 1}}$,
compute df/dx, where $f(x) = (x^{100} + 1)\sqrt{x^2 + 2x + 1}$.

SOLUTION: By the product rule,

$$
\begin{aligned}
f'(x) &= (x^{100} + 1)D\left(\sqrt{x^2 + 2x + 1}\right) \\
&\quad + \left(\sqrt{x^2 + 2x + 1}\right)D(x^{100} + 1) \\
&= \frac{(x^{100} + 1)(x + 1)}{\sqrt{x^2 + 2x + 1}} + \sqrt{x^2 + 2x + 1}\,(100x^{99}). \quad \blacksquare
\end{aligned}
$$

EXAMPLE 4: Compute $d\vec{r}(t)/dt$, where $\vec{r}(t) = [(x^{32} + x^{999}),\ (x^{5261} - x^{91})]$.

SOLUTION:

$$
\begin{aligned}
\frac{d\vec{r}(t)}{dt} &= (D(x^{32} + x^{999}),\ D(x^{5261} - x^{91})) \\
&= (32x^{31} + 999x^{998},\ 5261x^{5260} - 91x^{90}). \quad \blacksquare
\end{aligned}
$$

The above examples illuminate the power of the tools provided by the theorems of this chapter. With our new tools, differentiating the functions given in Examples 1–3 is reduced to nearly trivial "cookbook" techniques. Computing these derivatives using limits would be a tedious task indeed!

EXAMPLE 5: Find an equation for the line tangent to the graph of the function $f(x) = (2x^{100} + x^{99} + 25x + 2)$ at $x = 1$.

SOLUTION: The slope of the line is $f'(1)$, and a point on the line is $(1, f(1)) = (1, 30)$.

$$f'(x) = 200x^{99} + 99x^{98} + 25.$$

$$f'(1) = 324.$$

Using the point-slope form of the equation of a line,

$$y - y_0 = m(x - x_0)$$

$$y - 30 = 324(x - 1) \text{ or } y = 324x - 294. \quad \blacksquare$$

EXAMPLE 6: Let $\vec{r}(t) = [(2t^{25} + 1), (2t^{30} + t), t^5]$. Parametrize the line tangent to the image of $\vec{r}$ at $t = 1$.

SOLUTION: $\vec{r}(1) = (3, 3, 1)$ is a point on the line. The derivative $\vec{r}\,'(1)$ is a direction vector for the line.

$$\vec{r}\,'(t) = (50t^{24}, 60t^{29} + 1, 5t^4)$$

$$\vec{r}\,'(1) = (50, 61, 5).$$

A parametrization for the line is given by

$$\vec{s}(t) = \vec{v}t + \vec{b} = (50, 61, 5)t + (3, 3, 1)$$

$$= (50t + 3, 61t + 3, 5t + 1). \qquad \blacksquare$$

Higher Derivatives

If f is a differentiable function from a subset of $\mathbb{R}$ into $\mathbb{R}$, then f', df/dx, and Df are all symbols that denote the derivative of f. The derivative f' is a new function; if it is also differentiable, we can take the derivative of f'. The derivative of f' is called the second derivative of f. It is denoted by f'', d^2f/dx^2 and D^2f.

EXAMPLE 7: Find the second derivative of $f(x) = x^3 + 2x^2 - 3x + 2$.

SOLUTION:

$$f'(x) = 3x^2 + 4x - 3.$$

$$f''(x) = 6x + 4. \qquad \blacksquare$$

As long as the function f is sufficiently well-behaved, there is nothing to prevent us from taking the derivative over and over again. If we take the derivative of f'', we obtain the third derivative of f, which will be denoted by f''', d^3f/dx^3, and D^3f. If we have the kth derivative of f, then the $(k + 1)^{\text{th}}$ derivative is obtained by taking the derivative of the kth derivative of f. It should be clear that the symbols d^kf/dx^k and D^kf denote the kth derivative of f. We also write $f^{(k)}$ to denote the kth derivative of f rather than trying to attach k primes to f. Thus, $f^{(3)}(x) = f'''(x)$, $f^{(4)}(x) = f''''(x)$, etc.

EXAMPLE 8: Let $f(x) = x^5 - 2x^2 + 3$.

$$f'(x) = 5x^4 - 4x.$$

$$f''(x) = 20x^3 - 4.$$

$$f^{(3)}(x) = 60x^2. \qquad \blacksquare$$

A practical application of higher derivatives is the acceleration of a particle.

Definition: Acceleration

The *acceleration* of a particle is the instantaneous rate of change of velocity relative to the change in time.

The units for acceleration are (units of length)/(units of time)^2.

It follows from the definition that the acceleration $\vec{a}(t)$ is the second derivative of the position function $\vec{r}$. Notice that just as $\vec{r}$ and $\vec{v}$ are vector quantities, so is $\vec{a}$. If

$$\vec{r}(t) = \begin{pmatrix} x(t) \\ y(t) \\ z(t) \end{pmatrix} \quad \text{and} \quad \vec{v}(t) = \vec{r}'(t) = \begin{pmatrix} x'(t) \\ y'(t) \\ z'(t) \end{pmatrix},$$

then

$$\vec{a}(t) = \vec{v}'(t) = \vec{r}''(t) = \begin{pmatrix} x''(t) \\ y''(t) \\ z''(t) \end{pmatrix}.$$

EXAMPLE 9: A particle moves along a path given by the following parametrization:

$$\vec{r}(t) = \begin{pmatrix} t^2 + 3t - 10 \\ t^5 \\ t \end{pmatrix} \text{ m.}$$

Find the velocity and acceleration of the particle.

SOLUTION:

$$\vec{v}(t) \quad = \quad \vec{r}'(t) = \begin{pmatrix} 2t + 3 \\ 5t^4 \\ 1 \end{pmatrix} \text{ m/s.}$$

$$\vec{a}(t) \quad = \quad \vec{v}'(t) = \begin{pmatrix} 2 \\ 20t^3 \\ 0 \end{pmatrix} \text{ m/s}^2. \qquad \blacksquare$$

Some Related Rates Applications

Recall that $x'(t)$ is the rate that the image of x is changing as t changes. This rate leads to some quite useful applications of the derivative.

EXAMPLE 10: Suppose that $h(t) = u(t)v(t)$. Suppose further that when $t = t_0$, $u(t_0) = 2$ units, $v(t_0) = -2$ units, u is increasing at a rate of 3 units/sec, and v is decreasing at a rate of 5 units/sec. What is the rate of change of h at time $t = t_0$?

SOLUTION: At time $t = t_0$, we are given that $u(t_0) = 2$, $v(t_0) = -2$, $u'(t_0) = 3$, $v'(t_0) = -5$. By the product rule, $h'(t) = u(t)v'(t) + u'(t)v(t)$. Thus $h'(t_0) = (2)(-5) + (3)(-2) = -16$ units^2/sec.   ∎

EXAMPLE 11: Suppose that the radius of a circle is changing with time. At time t_0, the radius is 3 m, and it is decreasing at a rate of 2 m/sec. At what rate is the area of the circle changing at time t_0?

SOLUTION: Let $r(t)$ denote the radius of the circle at time t. The area function is given by $A(t) = \pi r^2(t) = \pi r(t)r(t)$. Then $A'(t) = \pi(r(t)r'(t) + r'(t)r(t)) = 2\pi r(t)r'(t)$. Thus at time t_0, we have $A'(t_0) = 2\pi r(t_0)r'(t_0) = 2\pi(3)(-2) = -12\pi$ m^2/sec.   ∎

EXAMPLE 12: The base of a triangle is increasing at a rate of 3 in/sec, and its altitude is decreasing at a rate of 2 in/sec. Find the rate of change of the area of the triangle when the altitude is 1 in and the base is 2 in.

SOLUTION: This problem is an application of the product rule. Let $a(t)$ denote the height of the altitude at time t, and let $b(t)$ denote the length of the base at time t. Then,

$$A(t) = \frac{1}{2}a(t)b(t)$$

gives the area of the triangle at time t, and

$$\frac{dA(t)}{dt} = \frac{1}{2}(a(t)b'(t) + b(t)a'(t)).$$

We are given that at a certain time, say t_0, $a(t_0) = 1$ and $b(t_0) = 2$. We are also given that $a'(t) = -2$ and $b'(t) = 3$. Thus

$$A'(t_0) = \frac{1}{2}[(1)(3) + (2)(-2)] = \frac{1}{2}(-1) = -\frac{1}{2} \text{ in}^2/\text{sec}.$$

EXERCISES 4.1

In Exercises 1 and 2, $f(2) = 3$, $f'(2) = -1$, $g(2) =$
-4 and $g'(2) = 2$.

1. Find $D(f(x) + g(x))$ at $x = 2$.

2. Find $D(f(x)g(x))$ at $x = 2$.

In Exercises 3–11, compute the derivative of f.

3. $f(x) = 2x^2 + 16x$.

4. $f(x) = 16x^6 + 5x^{100} + 2$.

5. $f(x) = 10x^{16} + 20x + 2 + \sin(x) + \cos(x)$.

6. $f(x) = x^3 + \sqrt{x+1}$.

7. $f(x) = (x^{100} + 2x^5 + \pi x^{32})(x^6 - 6x^2 + x + 22)$.

8. $f(x) = (x^{21} + \frac{1}{2}x^5)\sin(x)$.

9. $f(x) = (2x^5 + x^{31} + 5x^2 + 2)(6x^2 + 2x^3 + x^4 + 2x + 1)$.

10. $f(x) = \sin(x)\cos(x)$.

11. $f(x) = (x^7 + \pi x^4)(x+1)^{-1/2}$.

Compute the derivative of the functions in Exercises 12–21.

12. $f(x) = 2.135^{51}$.

13. $g(y) = \pi^{999}$.

14. $y(x) = (20x^2 + 15x + 2)\sin(x)\cos(x)$.

15. $g(x) = (2x^{31} + 5x^{20} + 2x + 3)^2$.

16. $s(t) = (5t^2 + 2t^3 + t + 1)^2$.

17. $s(t) = (5t^2 + 2t^3 + t + 1)^3$.

18. $g(\theta) = \sin^3(\theta)$.

19. $\vec{r}(t) = (2t^3 + 3t^2 + t, t^{100} - 5t^2)$.

20. $\vec{s}(x) = (\pi x^6 - 2x + 2, \ 5x^{21} + x^{51} + 25x^4, \ x^3 + x - \pi)$.

21. $\vec{f}(x) = ((\pi + 2)^2, \ 2x^3 + x + 1, \ 999)$.

Find the second and third derivatives of the functions in Exercises 22–28.

22. $f(x) = x^3 - 2x^2 + 3$.

23. $g(x) = ax^3 + bx^2 + c$.

24. $\rho(t) = \sin(t)(t - 1)$.

25. $q(t) = \sin(t)\cos(t)$.

26. $g(t) = t\sin(t)\cos(t)$.

27. $\lambda(t) = \sin^2(t)$.

28. $w(t) = t^n\cos(t)$.

In Exercises 29–31, find an equation for the tangent line to the function at the given point.

29. $f(x) = x^{25} + 2x^3 + 3$, $x_0 = 1$.

30. $f(x) = \sin(x)\cos(x)$, $x_0 = \frac{\pi}{4}$.

31. $r(x) = \sin^3(x)$, $x_0 = \frac{\pi}{3}$.

In Exercises 32–34, find $d(\vec{f} \cdot \vec{g})/dt$ and $d(\vec{f} \times \vec{g})/dt$.

32. $\vec{f}(t) = (1, t, \sin(t))$, $g(t) = (t, 0, \cos(t))$.

33. $\vec{f}(t) = (t^2, 0, t)$, $g(t) = (\cos(t), 0, t)$.

34. $\vec{f}(t) = (t^2, 1, t^4)$, $g(t) = (2t, 2t^3, -1)$.

In Exercises 35–37, $g(1) = 3$, $g'(1) = -2$,
$\vec{r}(1) = (1, 0, 2)$, $\vec{r}'(1) = (-1, 1, 0)$, $\vec{s}(1) = (1, 1, 1)$, and
$\vec{s}'(1) = (0, -1, 2)$.

35. Find $D(g(x)\vec{r}(x))$ at $x = 1$.

36. Find $D(\vec{r}(x) \cdot \vec{s}(x))$ at $x = 1$.

37. Find $D(\vec{r}(x) \times \vec{s}(x))$ at $x = 1$.

In Exercises 38–40, $g(2) = -2$, $g'(2) = 3$,
$\vec{r}(2) = (2, -1, 0)$, $\vec{r}'(2) = (0, -3, 1)$, $\vec{s}(2) = (-2, 4, 0)$,
and $\vec{s}'(2) = (1, 0, -2)$.

38. Find $D(g(x)\vec{r}(x))$ at $x = 2$.

39. Find $D(\vec{r}(x) \cdot \vec{s}(x))$ at $x = 2$.

40. Find $D(\vec{r}(x) \times \vec{s}(x))$ at $x = 2$.

In Exercises 41–44, parametrize the line tangent to the image of the function at the given point.

41. $\vec{r}(t) = (2t^3 + 3t + 1, t^6 + 5t + 6, 2t^{100} + 5t)$, $t_0 = 1$.

42. $\vec{s}(t) = (6t^2, t^4 + t^2 + t, t^3 + t + 3)$, $t_0 = 2$.

43. $\vec{a}(t) = ((\pi + 2)^{25}, 3t + 1)$, $t_0 = 20$.

44. $\vec{r}(t) = t\vec{W}(t)$, $t_0 = \frac{\pi}{3}$.

In Exercises 45–48, determine where the tangent line to the function is horizontal, and give the equation of the line(s).

45. $f(x) = x^3 - x$.

46. $f(x) = \frac{1}{3}x^3 + x^2 + x + 2$.

47. $f(x) = \frac{1}{5}x^5 + \frac{2}{3}x^3 + x$.

48. $f(x) = x + \sin(x)$.

In Exercises 49–52, find equations for all the lines tangent to the function that have slope 1.

49. $f(x) = x^3 - x$.

50. $f(x) = \frac{1}{3}x^3 + x^2 + x + 2$.

51. $f(x) = \frac{1}{5}x^5 + \frac{2}{3}x^3 + x$.

52. $f(x) = x + \sin(x)$.

In Exercises 53 and 54, $\vec{r}(t)$ gives the position of a particle in space as a function of time. Compute the acceleration for the parametrized motion. Time is measured in sec.

53. $\vec{r}(t) = (t^3, t^2 - 2)$ m.

54. $\vec{r}(t) = (10t^3, t^4, t^2)$ m.

In Exercises 55–57, $\vec{r}(t)$ gives the position of a particle in space as a function of time. Find the unit vectors in the directions of the velocity vector and the acceleration vector. Time is measured in sec.

55. $\vec{r}(t) = (t^2, t, -t^2)$ m.

56. $\vec{r}(t) = (t^2, \cos(t), \sin(t))$ m.

57. $\vec{r}(t) = (3\cos(t), t, -3\sin(t))$ m.

In Exercises 58–61, use both Newton's method and the Interval-Halving method to approximate the root on the given interval with an accuracy of at least 10^{-4}.

58. $f(x) = x^3 - x - 1$, $0 \le x \le 1$.

59. $f(x) = x^2 + x - 1$, $-2 \le x \le -1$.

60. $f(x) = x^4 + x - 3$, $1 \le x \le 2$.

61. $f(x) = x + \cos(x)$, $0 \le x \le \pi$.

In Exercises 62–65, compute $d^3 f/dx^3$.

62. $f(x) = 6x^5 - 2x^3 + 5$.

63. $f(x) = x^{10} - 2x^2$.

64. $f(t) = t^3 + t^2 + t + 2$.

65. $f(\theta) = \sin(\theta)$

The area of the ellipse $\frac{x^2}{A^2} + \frac{y^2}{B^2} = 1$ is given by πAB. In Exercises 66 and 67, consider both A and B to be functions of time.

66. If $A(t) = t^3 + 2t + 3$ and $B(t) = t^4 + t^8$, find the rate at which the area of the ellipse $\frac{x^2}{A^2(t)} + \frac{y^2}{B^2(t)} = 1$ is changing.

67. At time $t = t_0$, $A(t_0) = 3$ m, $B(t_0) = 1$ m, the area of the ellipse $\frac{x^2}{A^2(t)} + \frac{y^2}{B^2(t)} = 1$ is decreasing by 2 m^2/sec and $A(t)$ is increasing by 4 m/sec. What is the rate of change of $B(t)$ at $t = t_0$?

68. At $t = t_0$, one side of a rectangle is 3 m long, and its length is increasing at a rate of 2 m/sec. The other side has constant length of 5 m. What is the rate of change of the area of the rectangle?

In Exercises 69–72, $\vec{r}(t) = \left(t^3 + t^2 + t, t^2, 5t^{10}\right)$ ft denote the position of a particle at time t sec.

69. Find the velocity of the particle.

70. Find the rate that the velocity is changing.

71. Let $\vec{v} = (1, 2, -1)$.

    a. Find the derivative of $\text{proj}_{\vec{v}} \vec{r}(t)$.

    b. Find the derivative of $\text{orth}_{\vec{v}} \vec{r}(t)$.

72. Find the rate of change of the projection of $\vec{r}(t)$ onto the plane $2x - 3y + z = 1$.

Definition: The Momentum of a Point Mass

If $M(t)$ denotes the mass of a particle at time t and $\vec{r}(t)$ denotes its position at time t, then $\vec{p}(t) = (d/dt)\,(M(t)\vec{r}(t))$ is called the *momentum* of the particle (at time t).

---

73. At time t sec, the mass of a particle is given by $M(t) = t^3 + t^2$ kg, and its position is $\vec{r}(t) = (t^3, 6t^4, t)$ m. Find its momentum and the rate that the momentum is changing.

Definition: The Momentum of a System of Point Masses

The *momentum* of a system of point masses is simply the sum of the momenta of the individual particles. That is, if $P_1, P_2, \ldots, P_n$ is a system of "point" masses such that

P_1 has mass $m_1(t)$ and position $\vec{r}_1(t)$,
P_2 has mass $m_2(t)$ and position $\vec{r}_2(t)$,
$\vdots$ $\vdots$ $\vdots$
P_n has mass $m_n(t)$ and position $\vec{r}_n(t)$,
then

$$\vec{p}(t) = (d/dt)\,(m_1(t)\vec{r}_1(t) + m_2(t)\vec{r}_2(t) + \ldots + m_n(t)\vec{r}_n(t)).$$

Exercises 74–76 refer to the system $\{P_1, P_2, P_3\}$ where, at time t sec,

P_1 *is a $2t^2$ kg mass and is located at $(t, t^2, -t)$ m;*

P_2 *is a t^4 kg mass and is located at $(t^3 + 1, -2t^2, t)$ m; and*

P_3 *is a 1 kg mass and is located at $(t^2, t, 4 - t^2)$ m.*

74. Find the center of mass of the system as a function of time.

75. Find the momentum of the system as a function of time.

Definition: The Kinetic Energy of a Point Mass

If M denotes the mass of a particle and $\vec{v}(t)$ is its velocity at time t, then $\frac{1}{2}M\|\vec{v}(t)\|^2$ is called the *kinetic energy* of the particle. In the metric system (m, kgs, sec), kinetic energy is measured in *Joules*. The *kinetic energy of a system of point masses* is the sum of the kinetic energies of the individual masses. That is, if $P_1, P_2, \ldots, P_n$ is a system of "point" masses such that:

P_1 has mass m_1 and velocity $\vec{v}_1(t)$,
P_2 has mass m_2 and velocity $\vec{v}_2(t)$,
$\vdots$ $\vdots$
P_n has mass m_n and velocity $\vec{v}_n(t)$,
then

Total Kinetic Energy $=$
$\frac{1}{2}m_1\|\vec{v}_1(t)\|^2 + \frac{1}{2}m_2\|\vec{v}_2(t)\|^2 + \ldots$
$+ \frac{1}{2}m_n\|\vec{v}_n(t)\|^2$ Joules.

---

76. Find the kinetic energy of the system.

77. Suppose that the function f can be differentiated n times for all n.

　　a. Find the first four derivatives of $xf(x)$.

　　b. A pattern should have developed in Part a. Write down a formula for the nth derivative of $xf(x)$.

　　c. Assuming that your formula is correct for $n = k$, show that your formula is correct for $n = k + 1$.

In Exercises 78–80, you have used induction to prove that your formula works for all n.

78. Let h be a small number. Then $f''(x_0) \approx$ $\dfrac{f'(x_0 + h) - f'(x_0 - h)}{2h}$ (Method 2 from the exercises in Section 3.4 applied to f'). Use the same method on each of $f'(x_0 + h)$ and $f'(x_0 - h)$, and then substitute $h/2$ for h to obtain

$$f''(x_0) \approx \frac{f(x_0 - h) - 2f(x_0) + f(x_0 + h)}{h^2}.$$

(1)

79. Use Equation (1) from Exercise 78 to estimate $f''(x_0)$, where $f(x) = \tan(x)$, $x_0 = \frac{\pi}{4}$, $h = 0.01$.

80. Use Equation (1) from Exercise 78 to estimate $f''(x_0)$, where $f(x) = \sec^2(x)$, $x_0 = \frac{\pi}{4}$, $h = 0.01$.

4.2   The Quotient Rule

The product and sum rules provided us with the tools necessary to differentiate polynomials of any degree. The quotient rule enables us to differentiate rational functions with reasonable ease. We begin with a special case.

Theorem 1 *If f is a real-valued function that has a derivative at x, and $f(x) \neq 0$, then*

$$D\left(\frac{1}{f(x)}\right) = -\frac{f'(x)}{[f(x)]^2}.$$

Proof (assuming $D(1/f)$ exists): Differentiating both sides of $f(x)/f(x) = 1$:

$$D\left(f(x)\frac{1}{f(x)}\right) = D(1) = 0.$$

Applying the product rule to the left-hand side, we obtain

$$D\left(f(x)\frac{1}{f(x)}\right) = f'(x)\frac{1}{f(x)} + f(x)D\left(\frac{1}{f(x)}\right) = 0,$$

or

$$D\left(\frac{1}{f(x)}\right) = -\frac{f'(x)}{[f(x)]^2}. \qquad \blacksquare$$

EXAMPLE 1: Let $g(x) = \frac{1}{x^{100}+2x^3}$. To compute dg/dx, observe that g is of the form $\frac{1}{f}$, where $f(x) = x^{100} + 2x^3$. Differentiating f we obtain $f'(x) = 100x^{99} + 6x^2$. Employing Theorem 1, we get

$$g'(x) = -\frac{f'(x)}{[f(x)]^2} = \frac{-(100x^{99} + 6x^2)}{(x^{100} + 2x^3)^2}. \qquad \blacksquare$$

Theorem 2 *If n is an integer (positive or negative), then $Dx^n = nx^{n-1}$.*

Proof: We have already shown that if $n \geq 0$ then $Dx^n = nx^{n-1}$. If $n < 0$, let $k = -n$. Then $x^n = \frac{1}{x^k}$. By Theorem 1,

$$D(x^n) = D\left(\frac{1}{x^k}\right) = \frac{-kx^{k-1}}{x^{2k}} = -\frac{k}{x^{k+1}} = nx^{n-1}. \qquad \blacksquare$$

EXAMPLE 2: $D(x^{-100} - x^{-53} + x^{25}) = -100x^{-101} + 53x^{-54} + 25x^{24}$. ∎

We now state and prove the Quotient Rule.

Theorem 3 (Quotient Rule) *If f and g are real-valued functions that are differentiable at x, and $f(x) \neq 0$, then*

$$D\left(\frac{g(x)}{f(x)}\right) = \frac{f(x)g'(x) - g(x)f'(x)}{[f(x)]^2}.$$

Proof: We apply Theorem 1 and the Product Rule.

$$
\begin{aligned}
D\left(\frac{g(x)}{f(x)}\right) &= g'(x)\frac{1}{f(x)} + g(x)D\left(\frac{1}{f(x)}\right) \quad \text{(Product Rule)} \\[2mm]
&= \frac{g'(x)}{f(x)} + g(x)\left(\frac{-f'(x)}{[f(x)]^2}\right) \quad \text{(Theorem 1)} \\[2mm]
&= \frac{f(x)g'(x) - g(x)f'(x)}{[f(x)]^2}. \quad\blacksquare
\end{aligned}
$$

EXAMPLE 3:

$$
\begin{aligned}
D\left(\frac{x^3 + 5x^2}{x^2 + 2x}\right) &= \frac{(x^2 + 2x)D(x^3 + 5x^2) - (x^3 + 5x^2)D(x^2 + 2x)}{(x^2 + 2x)^2} \\[2mm]
&= \frac{(x^2 + 2x)(3x^2 + 10x) - (x^3 + 5x^2)(2x + 2)}{(x^2 + 2x)^2}. \quad\blacksquare
\end{aligned}
$$

The quotient rule allows us to determine the derivatives of the other trigonometric functions.

EXAMPLE 4: $D\sec(x) = D[1/\cos(x)] = \sin(x)/\cos^2(x) = \tan(x)\sec(x)$. ∎

The derivation of the derivatives of the other trigonometric functions is left for the exercises. We provide a list of the derivatives of the trigonometric functions here.

Theorem 4

$D\sin(x) = \cos(x)$	$D\tan(x) = \sec^2(x)$
$D\sec(x) = \tan(x)\sec(x)$	$D\cos(x) = -\sin(x)$
$D\cot(x) = -\csc^2(x)$	$D\csc(x) = -\cot(x)\csc(x)$

We now have the following formaulas for differentiating functions. These formulas should be committed to memory.

Differentiation Formulas

$$D(cf(x)) = cf'(x)$$

$$D(f(x) + g(x)) = f'(x) + g'(x) \quad \text{(Sum Rule)}$$

$$D(f(x)g(x)) = f'(x)g(x) + f(x)g'(x) \quad \text{(Product Rule)}$$

$$D(1/f(x)) = -\frac{f'(x)}{[f(x)]^2}$$

$$D(g(x)/f(x)) = \frac{f(x)g'(x) - g(x)f'(x)}{[f(x)]^2} \quad \text{(Quotient Rule)}$$

$$D(x^n) = nx^{n-1} \textbf{ provided that } n \text{ is an integer} \quad \text{(Power Rule)}$$

$$D(c\vec{f})(t) = cD\vec{f}(t)$$

$$D(\vec{f} + \vec{g})(t) = \vec{f}\,'(t) + \vec{g}\,'(t)$$

$$D(\vec{f} \cdot \vec{g})(t) = \vec{f}(t) \cdot \vec{g}\,'(t) + \vec{f}\,'(t) \cdot \vec{g}(t)$$

$$D(h\vec{f})(t) = h(t)\vec{f}\,'(t) + h'(t)\vec{f}(t)$$

$$D(\vec{f} \times \vec{g})(t) = \vec{f}(t) \times \vec{g}\,'(t) + \vec{f}\,'(t) \times \vec{g}(t)$$

$D \sin(x) = \cos(x)$	$D \tan(x) = \sec^2(x)$
$D \sec(x) = \tan(x)\sec(x)$	$D \cos(x) = -\sin(x)$
$D \cot(x) = -\csc^2(x)$	$D \csc(x) = -\cot(x)\csc(x)$

EXERCISES 4.2

Compute the derivative of each of the functions in Exercises 1–17.

1. $f(x) = x^{-101} - 10x^{-99} + x^2$.

2. $f(x) = \tan(x) + \sec(x) + \cot(x) + \csc(x)$.

3. $f(x) = \pi x^5 + x^{-2} + 1/(x^3 + 1)$.

4. $f(x) = \dfrac{1}{x^{24} + x^{-13} - 3x^2 + 2}$.

5. $f(x) = \dfrac{1}{\pi + 2}$.

6. $g(x) = \dfrac{\pi + 0.9999}{x^{-23} + x^{51} + 6x^6}$.

7. $h(t) = (1 + \sqrt{2})(t^{100} + t^{-5} + t^{-31} + 2.5)$.

8. $u(s) = \dfrac{s^2 + 2s^3 - \pi s}{s^3 - 2s}$.

9. $f(t) = \dfrac{t^3 + 3t}{5t^{-6} + t^6}$.

10. $f(\theta) = \tan(\theta)\theta^2$.

11. $f(x) = \dfrac{\sec(x)}{x^3 + 2x}$.

12. $f(t) = \dfrac{\cot(x)}{\sin(x) + x}$.

13. $f(x) = \dfrac{(x^2 + 2x + 3 - \sqrt{2})\tan(x)}{x^2 - \pi x}$.

14. $h(x) = \dfrac{\cos(x) + 1}{(x^3 + 3x - 4)\sin(x)}$.

15. $f(x) = \dfrac{kx^2 + dx}{ax^{-2} + bx + c}$.

16. $f(t) = \dfrac{x^3 - 2t}{x^2 + t^{-2}}$. **Caution, what is the variable? !!!**

17. $\vec{r}(t) = \left(\dfrac{t^{-35} + 1}{\tan(x)}, \dfrac{\sec(x)}{t^5 - 1} \right)$.

Derive the formulas in Exercises 18–21.

18. $D\tan(x) = \sec^2(x)$.

19. $D\cot(x) = -\csc^2(x)$.

20. $D\sec(x) = \sec(x)\tan(x)$.

21. $D\csc(x) = -\csc(x)\cot(x)$.

In Exercises 22–27, find $\frac{d^2}{dx^2} f(x)$.

22. $f(x) = \sec(x)$.   23. $f(x) = \dfrac{\sin(x)}{x}$.

24. $f(x) = \dfrac{1}{x+1}$.   25. $f(x) = \dfrac{\sin(x)}{x}$.

26. $f(x) = \dfrac{x-1}{x+1}$.   27. $f(x) = \dfrac{x^2}{x^2+1}$.

In Exercises 28–32, find an equation for the line tangent to f at the given point.

28. $f(x) = 3x + x^{100} + \dfrac{x^2 + 2x}{x - 2}$, $x_0 = 1$.

29. $f(x) = \left[\dfrac{x^2 + 3x}{x^5 + 1} \right]^2$, $x_0 = 0$.

30. $f(\theta) = \sec(\theta) + \tan(\theta)$, $\theta_0 = \dfrac{\pi}{4}$.

31. $f(\theta) = \csc(\theta) + \cot(\theta)$, $\theta_0 = \dfrac{2\pi}{3}$.

32. $f(\theta) = \dfrac{\sin(\theta)}{1 + \tan(\theta)}$, $\theta_0 = \dfrac{5\pi}{4}$.

In Exercises 33–36, parametrize the line tangent to the image of the function at the given point.

33. $\vec{r}(t) = \left(t^{-5}, t^3 + t^{-12}, t^3 \right)$ at $t = 1$.

34. $\vec{r}(t) = \left(\dfrac{t^3}{t+1}, \dfrac{t-2}{t^3-12}, t^3 \right)$ at $t = 0$.

35. $\vec{w}(\theta) = (\sec(\theta), \tan(\theta), \csc(t))$ at $\theta = -\frac{\pi}{4}$.

36. $\vec{s}(u) = (\sin(u)\cos(u), \cot(u), u\csc(u))$ at $u = \dfrac{5\pi}{4}$.

37. $\vec{r}(t) = (t^3, t^{-4} + t, t)$ m gives the position of an object at time t sec. Find the unit vector in the direction of the velocity at $t = 1$ sec.

38. $\vec{r}(t) = \left(\dfrac{t^3}{t+1}, \dfrac{t^2-2}{t-12}, t^{13} \right)$ m gives the position of an object at time t sec. Find the unit vector in the direction of the velocity at $t = 0$ sec.

In Exercises 39–43, $f(t_0) = 3$, $g(t_0) = -1$, $h(t_0) = 3$, $f'(t_0) = -2$, $g'(t_0) = 4$, and $h'(t_0) = -6$. Find $s'(t_0)$.

39. $s(t) = \dfrac{g(t)}{h(t)}$.

40. $s(t) = \dfrac{g(t)}{h(t) + f(t)}$.

41. $s(t) = \dfrac{g(t) + h(t)}{f(t)}$.

42. $s(t) = \dfrac{f(t)g(t)}{h(t)}$.

43. $s(t) = \dfrac{4f(t)h(t) - (2g(t)/h(t))}{f(t) - g(t)}$.

In Exercises 44–46 determine where the slope of the line tangent to f is positive and where the slope of the tangent line is negative.

44. $f(x) = \dfrac{x}{x+1}$.   45. $f(x) = \dfrac{x-1}{x+1}$.

46. $f(x) = \dfrac{x^2-1}{x^2+1}$.

In Exercises 47–50, determine where the rate of change of f is positive.

47. $f(x) = \dfrac{x+1}{x^2-1}$.   48. $f(x) = \sec(x)$.

49. $f(x) = \cot(x)$.   50. $f(x) = \sin(x)\cos(x)$.

In Exercises 51 and 52, determine the values for x where the slope of the line tangent to f is m.

51. $f(x) = \dfrac{x}{x+1}$, $m = \dfrac{1}{2}$.

52. $f(x) = \dfrac{x-1}{x+1}$, $m = 2$.

53. Find equations for the lines tangent to $f(x) = \dfrac{x-1}{x+1}$ with slope $m = 2$.

54. Show that the rate of change of $\dfrac{x^2}{1+x}$ is always positive.

In Exercises 55–58, graph $f(x)$ and $f'(x)$, $a < x < b$ on the same axes.

  a. Estimate the intervals or segments where f is increasing and where f is decreasing.

  b. Estimate the intervals where f' is positive and where f' is negative.

  c. What is the relationship between the intervals in Part a and Part b?

55. $f(x) = \dfrac{-1 + \frac{4x}{3}}{1 + x^2}$, $a = -2$, $b = 2$.

56. $\dfrac{-1 + \frac{4x^3}{5}}{-2 + \cos(x)}$, $a = -2$, $b = 2$.

57. $f(x) = \dfrac{\sin(x)}{-2 + \cos(x)}$, $a = -2\pi$, $b = 2\pi$.

58. $\dfrac{x}{-1 + x^2}$, $a = -3$, $b = 3$.

59. Use both Newton's method and the Interval-Halving method to approximate the root of $\frac{x^4 + x - 1}{x + 2}$ on the interval $[0, 1]$ with an error of less than 10^{-4}.

60. Usually, Newton's method is the method of choice in finding roots. When it works, it usually is much faster than the Interval-Halving method. Also, the Interval-Halving method requires that the function cross the axis. Do four steps of the Interval-Halving method to approximate the root of $f(x) = x^{31}$ on the interval $[-1, 2]$. Now do four steps of Newton's method starting with $x_0 = 2$. Why does Newton's method close in on the root more slowly than the Interval-Halving method does?

4.3   The Chain Rule

When we introduced composite functions in Section 2.6, we argued that the function g in the composition $f \circ g$ can be used to control the timing of the motion on the image of f. The Chain Rule explains exactly how this works.

Theorem 1 (Chain Rule) *If $f \circ g$ is defined at x, f is differentiable at $g(x)$, and if g is differentiable at x, then $D[f(g(x))] = f'(g(x))g'(x)$.*

Proof:

$$
\begin{aligned}
D(f(g(x))) &= \lim_{h \to 0} \frac{f(g(x+h)) - f(g(x))}{h} \\[2mm]
&= \lim_{h \to 0} \left(\frac{f(g(x+h)) - f(g(x))}{h} \right) \left(\frac{g(x+h) - g(x)}{g(x+h) - g(x)} \right) \\[2mm]
&= \lim_{h \to 0} \left(\frac{f(g(x+h)) - f(g(x))}{g(x+h) - g(x)} \right) \left(\frac{g(x+h) - g(x)}{h} \right)
\end{aligned}
$$

$$= f'(g(x))g'(x).\qquad\blacksquare$$

Note: $f'(g(x))$ is the derivative of f evaluated at $g(x)$, not the derivative of $f(g(x))$. This argument assumes that if h is "sufficiently small," then $[g(x+h)-g(x)]\neq 0$. The case that $g(x+h)=g(x)$ for arbitrarily small h is left to another course.

EXAMPLE 1: Let $f(x)=x^{100}$ and $g(x)=x^2+6x+5$. Then $f(g(x))=(x^2+6x+5)^{100}$. To compute the derivative of $f(g(x))$ directly would be cumbersome indeed. However, Theorem 1 makes the task easy: $f'(x)=100x^{99}$ and $g'(x)=2x+6$. Thus $f'(g(x))g'(x)=100(x^2+6x+5)^{99}(2x+6)$. $\blacksquare$

> **Theorem 2** If r is a rational number (that is, $r=\frac{p}{q}$, where p and q are integers), then $Dx^r=rx^{r-1}$.

Proof (Assuming that Dx^r exists): Let $f(x)=x^q$ and $g(x)=x^{p/q}$. Then $f(g(x))=(x^{p/q})^q=x^p$. By Theorem 1:

$$f'(g(x))g'(x) = px^{p-1};$$

$$q(x^{p/q})^{q-1}g'(x) = px^{p-1};$$

$$q(x^{p/q})^{q-1}Dx^{p/q} = px^{p-1},$$

and

$$Dx^{p/q} = \frac{px^{p-1}}{q(x^{p/q})^{q-1}} = \left(\frac{p}{q}\right)x^{(p/q)-1}.\qquad\blacksquare$$

EXAMPLE 2: $Dx^{0.234}=0.234x^{-0.766}$. $\blacksquare$

EXAMPLE 3: Compute the derivative of $(x^3+x^{5/2}+6)^{55/3}$.

SOLUTION: $(x^3+x^{5/2}+6)^{55/3}$ is of the form $f(g(x))$, where $f(x)=x^{55/3}$ and $g(x)=x^3+x^{5/2}+6$. This yields:

$$f'(x) = \frac{55}{3}x^{52/3};$$

$$g'(x) = 3x^2+\frac{5}{2}x^{3/2};$$

$$f'(g(x)) = \frac{55}{3}(x^3+x^{5/2}+6)^{52/3},$$

and

$$f'(g(x))g'(x) = \frac{55}{3}(x^3+x^{5/2}+6)^{52/3}\left(3x^2+\frac{5}{2}x^{3/2}\right).\qquad\blacksquare$$

EXAMPLE 4: Compute the velocity of a particle whose position is defined by:

$$\vec{r}(t) = \begin{pmatrix} (2t^{1/2} + 5t^3 + 6)^{5/2} \\ (t^{1/3} + 3)^{1/4} \\ (t^{-7/6} + 2)^{25/24} \end{pmatrix}.$$

SOLUTION: $\vec{v}(t) = \vec{r}\,'(t)$.

$$D(2t^{1/2} + 5t^3 + 6)^{5/2} = \frac{5}{2}(2t^{1/2} + 5t^3 + 6)^{3/2}(t^{-1/2} + 15t^2).$$

$$D(t^{1/3} + 3)^{1/4} = \frac{1}{4}(t^{1/3} + 3)^{-3/4}\left(\frac{1}{3}t^{-2/3}\right) = \frac{1}{12}t^{-2/3}(t^{1/3} + 3)^{-3/4}.$$

$$D(t^{-7/6} + 2)^{25/24} = \frac{25}{24}(t^{-7/6} + 2)^{1/24}\left(-\frac{7}{6}t^{-13/6}\right)$$

$$= -\frac{175}{144}t^{-13/6}(t^{-7/6} + 2)^{1/24}.$$

Thus

$$\vec{v}(t) = \begin{pmatrix} \frac{5}{2}(2t^{1/2} + 5t^3 + 6)^{3/2}(t^{-1/2} + 15t^2) \\ \frac{1}{12}t^{-2/3}(t^{1/3} + 3)^{-3/4} \\ -\frac{175}{144}t^{-13/6}(t^{-7/6} + 2)^{1/24} \end{pmatrix}. \quad\blacksquare$$

In Example 18 of Section 3.3 we constructed a function for which the derivative was defined everywhere except at two points. That function was the composition of a polynomial and the absolute value function. The next example shows that even for quite simple functions there may be points where the derivative is not defined.

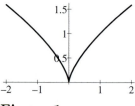

Figure 1.
The graph of $y = x^{2/3}$

EXAMPLE 5: Let $y = x^{2/3}$. Then $y' = \frac{2}{3}x^{-1/3}$. The derivative is not defined at $x = 0$. See Figure 1. $\quad\blacksquare$

EXAMPLE 6: The function $y = x^{4/3}$ has domain the set of real numbers, and so does the derivative, $y' = \frac{4}{3}x^{1/3}$. However, the second derivative, $y'' = \frac{4}{9}x^{-2/3}$, has as its domain the set of real numbers with the exception of $x = 0$. $\quad\blacksquare$

EXAMPLE 7:

$$D\left[\csc\left(x^{3/2} - 2x + 3\right)\right]^{5/6}$$

$$= \frac{5}{6} \left[\csc \left(x^{3/2} - 2x + 3 \right) \right]^{-1/6} D \csc \left(x^{3/2} - 2x + 3 \right)$$

$$= \frac{5}{6} \left[\csc \left(x^{3/2} - 2x + 3 \right) \right]^{-1/6} (-1) \cot \left(x^{3/2} - 2x + 3 \right)$$

$$\times \csc \left(x^{3/2} - 2x + 3 \right) D \left(x^{3/2} - 2x + 3 \right)$$

$$= -\frac{5}{6} \left[\csc \left(x^{3/2} - 2x + 3 \right) \right]^{-1/6} \cot \left(x^{3/2} - 2x + 3 \right)$$

$$\times \csc \left(x^{3/2} - 2x + 3 \right) \times \left(\frac{3}{2} x^{1/2} - 2 \right)$$

$$= -\frac{5}{6} \left[\csc \left(x^{3/2} - 2x + 3 \right) \right]^{5/6} \cot \left(x^{3/2} - 2x + 3 \right) \left(\frac{3}{2} x^{1/2} - 2 \right).$$

EXAMPLE 8: Consider a particle moving on a circle of radius R. Then the position of the particle at time t is given by

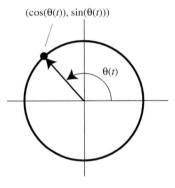

$(\cos(\theta(t)), \sin(\theta(t)))$

$$\vec{r}(t) = \begin{pmatrix} R \cos(\theta(t)) \\ R \sin(\theta(t)) \end{pmatrix} = R\vec{W}(\theta(t)),$$

where $\theta(t)$ denotes the angle the position vector makes with the x–axis. (See Figure 2.) Recall That $\vec{W}$ is the wrapping function from Section 2.3.

Figure 2. $\vec{r}(t) = R\vec{W}(\theta(t)).$

$$\vec{r}'(t) = R\vec{W}'(\theta(t))\theta'(t) = R\frac{d\vec{W}}{d\theta}\frac{d\theta}{dt} = R \begin{pmatrix} -\sin(\theta(t)) \\ \cos(\theta(t)) \end{pmatrix} \frac{d\theta}{dt}.$$

$|\theta'(t)|$ is called the *angular speed* of the particle. Since $\vec{W}'$ is a unit vector, $R|\theta'(t)|$ is the speed of the particle. ∎

EXAMPLE 9: A particle is moving in a 5 ft circle about the origin. When the position vector makes a $\frac{\pi}{6}$ radian angle with the x–axis, $\theta'(t) = 20$ radians/sec. Find the velocity of the particle at that point.

SOLUTION:

$$\vec{v}(t) = R\theta'(t)\vec{W}'(\theta(t)) \;=\; (5)(20)\vec{W}'(\theta(t))$$

$$= \; (100)(-\sin(\theta(t)), \cos(\theta(t)))$$

$$= (100)\left(-\frac{1}{2}, \frac{\sqrt{3}}{2}\right) \text{ ft/sec}$$

$$= (-50, 50\sqrt{3}) \text{ ft/sec.} \qquad \blacksquare$$

More Related Rates

The Chain Rule has a nice intuitive interpretation in terms of rates of change:

$$\underbrace{\frac{d}{dt}f(g(t))}_{\substack{\text{The rate that} \\ f(g(t)) \text{ is chang-} \\ \text{ing as } t \text{ changes.}}} = \underbrace{f'(g(t))}_{\substack{\text{The rate that } f \\ \text{is changing at} \\ g(t) \text{ as } g(t) \\ \text{changes.}}} \qquad \underbrace{g'(t)}_{\substack{\text{The rate that } g \\ \text{is changing as } t \\ \text{changes.}}}$$

EXAMPLE 10: Let $\vec{r}$ be a function describing the position of a particle moving in space. Recall that in Section 2.6, we emphasized that if we composed a function $\vec{r}$ with a linear function $g(t) = mt+b$, then the number m changes the speed of the motion being described. We can use the chain rule here: $\frac{d}{dt}\vec{r}(g(t)) = \vec{r}'(g(t))g'(t) = m\vec{r}'(g(t))$. Thus one of the effects of composing $\vec{r}$ with g is to change the speed by the factor m. $\qquad \blacksquare$

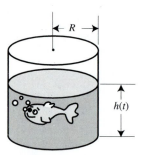

Figure 3. *A cylindrical tank of radius R. The depth of water at time t is h(t).*

EXAMPLE 11: Water is being pumped into a cylindrical tank at a rate of 3 m³/sec. If the tank is standing on its circular base and the base has a radius of R m, what is the rate of change of the depth of the water?

SOLUTION: As shown in Figure 3, let $h(t)$ denote the depth of water at time t. Then the volume of water in the tank at time t is given by

$$V(h(t)) = \pi R^2 h(t).$$

The rate of change of the volume of water is 3 m³/sec. Thus:

$$\frac{d}{dt}V(h(t)) = 3.$$

The chain rule gives us that

$$3 = \frac{d}{dt}V(h(t)) = \frac{dV}{dh}\frac{dh}{dt} = \pi R^2 \frac{dh}{dt}.$$

Solving for dh/dt yields

$$\frac{dh}{dt} = \frac{3}{\pi R^2} \text{ m/sec.} \qquad \blacksquare$$

EXAMPLE 12: Water is being pumped out of a conical tank at a rate of 1 m^3/sec. The radius of the base of the tank is 4 m and the height of the tank is 8 m as in Figure 4. Find the rate that the depth is decreasing when the water level is 2 m from the top.

SOLUTION: Let $r(t)$ denote the radius of the surface of the water, and let $h(t)$ denote the depth of the water at time t. By similar triangles, we have:

$$\frac{r(t)}{h(t)} = \frac{4}{8} = \frac{1}{2}.$$

Thus

$$r(t) = \frac{h(t)}{2}.$$

Since the volume of a right cone is $V = \left(\frac{1}{3}\right)\pi r^2 h$,

$$V(t) = \frac{1}{3}\pi[r(t)]^2 h(t) = \frac{1}{12}\pi[h(t)]^3$$

is the volume of water in the tank at time t. We are given that $dV/dt = -1$ (we use -1 since $V(t)$ is decreasing). Thus

$$V'(t) = \frac{1}{4}\pi[h(t)]^2 h'(t) = -1.$$

At the time t when $h(t) = 8 - 2 = 6$, we have

$$-1 = \frac{1}{4}\pi(6)^2 h'(t) = 9\pi h'(t).$$

Therefore, $h'(t) = -\frac{1}{9\pi}$ m/sec when the surface of the water is 2 m from the top. ∎

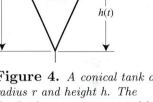

Figure 4. *A conical tank of radius r and height h. The depth of water at time t is $h(t)$.*

EXERCISES 4.3

Compute the derivatives of each of the functions given in Exercises 1–34

1. $f(x) = 3x^{4/5} + x^{-22/7} + 9x^{-3/2} + \pi^{2/3}$.

2. $f(x) = (x^2 + 5x + 6)^{4/3}$.

3. $f(x) = (x^{-5/2} + x^{-3/8})^{-3/5}$.

4. $f(x) = (x^3 + 4x^{99/4} + \sqrt{x})^{92/3}$.

5. $f(x) = (x^{-1/7} + 2x^{5/2} - 4x)^{-2/5}(x^{-2/9} - x^3)$.

6. $f(x) = (x^{8/3} + x^{5/8})^{9.8}$.

7. $f(x) = (x^{0.76} + x^{1.6})^{3.2}$.

8. $f(x) = [(x^{2/3} + 1)^{2/3} + 1]^{2/3}$.

9. $f(x) = [[(x^{2/3} + 1)^{2/3} + 1]^{2/3} + 1]^{2/3}$.

10. $f(x) = (x^2 + 6x^{-1/7})^3/(x^3 - 2x^{3/2})$.

11. $f(x) = (x^{-1/2} + \pi x^{3/2})^3(x^{5/2} - 6x^{91/3})^{1/6}$.

12. $f(x) = [(x^2 + 6x^{-1/7})^3/(x^3 - 2x^{3/2})]^3$.

13. $f(x) = [(x^{-1/2} + \pi x^{3/2})^3(x^{5/2} - 6x^{91/3})^{1/6}]^{3/4}$.

14. $f(x) = \cos(x + 1)$.

15. $f(x) = \sin(x^2 + 1)$.

16. $f(x) = \cos(3x + \pi)$.

17. $f(x) = \cos(2\sqrt{x} + x^2 + 1)$.

18. $f(x) = \tan(x^2 - 6x + 2)$.

19. $f(x) = \tan(\sqrt{x^2 + 1})$.

20. $f(x) = \sec(\pi x^{0.53} - 2x)$.

21. $f(x) = \cot[(2x + 3)^{0.2}]$.

22. $f(x) = \cot^3(x^2 + 2x + 1)$.

23. $f(x) = \cot^3[(x^2 + 2x + 1)^2]$.

24. $f(x) = \csc(x^2 + 2x + 3)$.

25. $f(x) = \csc[(x^{1/2} + \pi^2)^2]$.

26. $f(x) = \sin(x^{22} + 3x - 5) + \cos(x^{1.2}) - \sec(6x^2)$.

27. $f(x) = \cos[(x^{1/2} - x^3 + 1)(3x^3 + 2x^{-5/2} + \pi)]$.

28. $f(x) = \tan^3(x^2 + 2x - 1)$.

29. $f(x) = \cot^{1/2}(5x^{1.2} + 6x - 3)$.

30. $f(x) = \csc^3((2x^3 + 5x^2 + 6x + 1)^4)$.

31. $f(x) = \dfrac{\sin(x^2 + 2x + 3)}{\csc(x^2 + 2x + 3)}$.

32. $f(x) = x^2 - 3x^{-5/4}\cos\left(\dfrac{x^3 + 2}{x^2 + 1}\right)$.

33. $f(x) = (\cos^2 x + \tan(x^2 + 2) - \sec^3(3x) + 6\cot^2(\pi x))^{3/4}$.

34. $f(x) = \left(\dfrac{\tan^3(x^5 + 5x)}{\cot^4(x + 3)}\right)^4$.

35. Given that $\theta(2) = \frac{\pi}{3}$ and that $\theta'(2) = -4$, find $\frac{d}{dt}\vec{W}(\theta(t))$ at $t = 2$.

36. Given that $g(2) = 4$, $f(4) = 2$ and $f'(4) = -5$, find $(g \circ f(t))'$ at $t = 4$.

In Exercises 37–39, find an equation for the line tangent to f at the given point.

37. $f(x) = \left(\dfrac{3x^3 + 6x}{2x + 1}\right)^3$ at $x = 0$.

38. $f(x) = \left(\dfrac{2x^2 + 2x}{5x + 1}\right)^{3/2}$ at $x = 1$.

39. $f(x) = \sin^2\left(\dfrac{\pi(x^2 + x)}{3}\right)$ at $x = 1$.

In Exercises 40–42, determine where the given function has a horizontal tangent line.

40. $f(x) = (x^2 + 2x)^3$.

41. $h(t) = \left(\dfrac{3t + 1}{t - 2}\right)^{1/2}$.

42. $u(t) = \left(\dfrac{t^2}{6 - t}\right)^3$.

In Exercises 43–46, find the point(s) in the domain of the following functions where the derivative fails to exist.

43. $f(x) = (x - 1)^{1/3}$.   44. $f(x) = \sqrt{|x|}$.

45. $f(x) = (x^2 - 4)^{2/5}$ 46. $f(x) = \sqrt{|x^2|}$

47. Do the first five steps of Newton's method to approximate the root of $f(x) = x^{1/3}$. Is the method getting close to the root?

48. Define $f(x) = \sqrt{|x|}$. Starting with $x_0 = 1$, do five steps of Newton's method to approximate the root of f. If you change the starting point what happens?

49. A particle is moving along an ellipse given by $\vec{r}(t) = (a\cos(\omega t), b\sin(\omega t))$ for constants $a, b > 0$.

    a. Find the velocity of the particle.

    b. Find the speed of the particle.

    c. Find the unit vector in the direction of the velocity.

50. A particle is moving along the circle of radius R, with its position at time t given by

$$\vec{r}(t) = R\left(\cos(\omega\sin(t)), \sin(\omega\sin(t))\right).$$

    a. Describe the motion of the particle.

    b. Find the velocity of the particle.

    c. Find the speed of the particle.

51. Let n be a positive integer.

    a. Show that if $n = 1$ and $1 \leq i \leq n$, then $\left(\frac{d^i}{dx^i}\right) x^n = \frac{n!}{(n-i)!} x^{n-i}$.

    b. Show that if $n = 2$ and $1 \leq i \leq n$, then $\left(\frac{d^i}{dx^i}\right) x^n = \frac{n!}{(n-i)!} x^{n-i}$

    c. What is $\left(\frac{d^3}{dx^3}\right) x^2$?

    d. Show that if $n = 3$ and $1 \leq i \leq n$, then $\left(\frac{d^i}{dx^i}\right) x^n = \frac{n!}{(n-i)!} x^{n-i}$.

    e. What is $\left(\frac{d^4}{dx^4}\right) x^3$?

    f. A pattern seems to be developing. At this point, it is reasonable to conjecture that if n is an integer and $1 \leq i \leq n$, then

    $$\left(\frac{d^i}{dx^i}\right) x^n = \frac{n!}{(n-i)!} x^{n-i}. \qquad (2)$$

    Assuming that Equation (2) is true for $n = k$, show that it is true for $n = k + 1$.

    g. What is true about $\left(\frac{d^{n+1}}{dx^{n+1}}\right) x^n$?

52. Let r be any rational number that is not a positive integer, and let $f(x) = x^r$.

    a. Derive formulas for the first five derivatives of f.

    b. A pattern should have developed that will lead to a formula for the nth derivative of f. What is this formula?

    c. From the first part of this exercise, you should know that your formula works for the first derivative of f. Assume that your formula is true for the kth derivative of f, and show that your formula works for the $(k+1)^{\text{th}}$ derivative of f.

53. Use Newton's method to approximate a root for $f(x) = \sin(x) - \cos(2x)$, $0 \leq x \leq \frac{\pi}{2}$, with an accuracy of at least 10^{-4}.

54. A particle moves along the circle of radius R, with its position at time t sec given by

$$\vec{r}(t) = R\left(\cos\left(\pi\frac{t}{t+1}\right), \sin\left(\pi\frac{t}{t+1}\right)\right),$$

where $t \geq 0$. Distance is measured in m.

    a. Describe the motion of the particle.

    b. Find the velocity of the particle.

    c. Find the speed of the particle.

    d. As t gets very large, what is happening to the particle's speed?

55. A particle is traveling on the spiral parametrized by $\vec{r}(t) = (\cos t, \sin t, t)$, with a constant speed of 5m/sec. Find the velocity of the particle when it is at the point $\left(\frac{\sqrt{2}}{2}, \frac{\sqrt{2}}{2}, \frac{\pi}{4}\right)$ m. (Warning: we are not assuming that $\vec{r}(t)$ is the position of the particle at time t.)

56. The dimensions of a right circular cylinder are changing with time. When the radius is 1 cm and the height is 3 cm, the radius is getting larger at a rate of 3 cm/sec and the height is getting smaller at a rate of 2 cm/sec. At what rate is the volume of the cylinder changing? At what rate is the surface area (including the top and bottom) changing?

57. The base of an isosceles triangle is a constant length of 3 m and its area is changing at a rate of 3 m²/sec. At what rate is the altitude changing when the area is 6 m²? What is the rate of change of the perimeter when the area is 6 m²?

58. One leg of a right triangle is fixed at 2 m as other leg is increasing at a rate of 2 m/sec. What is the rate of change of the length of the hypotenuse

    a. when the leg of variable length is 1 m long?

    b. when the hypotenuse is 3 m long?

    c. as a function of the length of the variable leg?

59. One leg of a right triangle is fixed at 2 m as the hypotenuse is decreasing at 3 m/sec. What is the rate of change of the other leg as a function of the length of the hypotenuse?

60. One leg of a right triangle is increasing at a rate of 2 ft/sec when it is 6 ft long while the hypotenuse is constant. What is the rate of change of the other leg?

61. A 100-ft cable passes over a pulley 25 ft above ground as illustrated in Figure 5. One end of the cable is pulled horizontally away from a point directly below the pulley at a rate of 5 ft/sec while the other end is free to be pulled through the pulley. How fast is the free end rising when it is 5 ft off the ground?

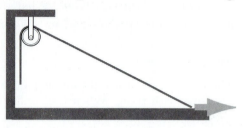

Figure 5. *Illustration for Exercise 61*

62. Gas is pumped into a spherical balloon at the rate of 4π m^3/sec. What is the rate of change of the radius of the balloon when the radius of the balloon is 4 m?

63. Sand is poured onto a conical pile at a rate of 5 ft^3/min. If the diameter of the base of the pile is always twice the height of the pile, find the rate of change of the height of the pile when the pile is 10 ft high.

64. The sand of Exercise 63 is poured into a cylindrical tank with a diameter of 10 ft. What is the rate of change of the height of the sand when the height is 10 ft?

65. A particle moves along the line $y = 2x + 4$ so that its x–coordinate is changing at a rate of -2 ft/sec. Find the velocity of the particle.

66. A particle moves along the graph of $y = x^2$. When $x = -1$, the x–coordinate is changing at a rate of -5 m/sec. What is the velocity of the particle then?

67. A particle moves along the graph of $y = x^2$. The x–coordinate at time t is given by $x(t) = \sin t$ feet. Find the velocity of the particle. Find the rate of change of the distance between the particle and the point $(0, 1)$.

68. Consider the particle from Exercise 67. How is the distance from the particle to the origin changing when $x = -1$?

69. A particle P moves along the x–axis at a speed of 2 ft/sec, while a particle Q moves along the line $y = 2x$ so that it is always directly above (or below) P. Find the speed of Q. (All motion is in the xy–plane.)

70. When a particle P is at $(1, 2, 1)$ m, it has a velocity of $(-1, 0, 2)$ m/sec. At the same time a particle Q is at $(3, -1, 2)$ m and has a velocity of $(2, 1, 0)$ m/sec. Find the rate at which the distance between P and Q is changing at that point in time.

71. **Rolling Wheel Problem 2:** In Rolling Wheel Problem 1 (Exercise 2.5.8), we located the tack at time t with the function

$$\vec{r}(t) = \begin{pmatrix} -40t + \sin(40\pi t) \\ -\cos(40\pi t) \end{pmatrix}.$$

a. Find the velocity and the acceleration of the tack as a function of time.

b. Find the projection of the acceleration in the direction of the velocity as a function of time.

c. Find the unit vector in the direction of the velocity.

72. **Ant on the Wheel Problem 2:** In the Ant on the Wheel Problem 1 (Exercise 2.5.9), we showed that the ant's position is located by

$$\vec{r}(t) = \begin{pmatrix} -40t + (1 - t)\sin(40\pi t) \\ -(1 - t)\cos(40\pi t) \end{pmatrix}.$$

Find the velocity and the acceleration of the ant as a function of time.

73. **Rotating Pendulum Problem 2:** In Rotating Pendulum Problem 1 (Exercise 2.5.10), we located the ball as a function of time by

$$\vec{r}(t) = (a + L\cos(\omega t)) \begin{pmatrix} \cos(2\pi k t) \\ \sin(2\pi k t) \end{pmatrix}.$$

Find the velocity of the ball as a function of time.

74. **Wheel in a Cylinder Problem 2:** In Wheel in a Cylinder Problem 1 (Exercise 2.5.11), we showed that the original point of tangency can be located by

$$\vec{r}(t) = \begin{pmatrix} (R-r)\cos(\theta(t)) + r\cos\left(\frac{R-r}{r}\theta(t)\right) \\ (R-r)\sin(\theta(t)) - r\sin\left(\frac{R-r}{r}\theta(t)\right) \end{pmatrix}.$$

Find the velocity of the original point of tangency as a function of time.

75. **Piston Problem 2:**

In Piston Problem 1 (Exercise 2.5.12), we showed that the enclosed volume is given by

$$V(t) = \pi\rho^2(L + R$$
$$- \left(R\sin(\theta(t)) + \sqrt{L^2 - R^2\cos^2(\theta(t))}\right)).$$

a. Assume that $L > R$. For the given function θ, find the rate that the volume of the chamber is changing.
   i. $\theta(t) = \omega t$.
   ii. $\theta(t) = t^2$.
   iii. $\theta(t) = R\sin(t)$.

b. For $L > R$, find $dV/d\theta$. What is the derivative in the limiting case where $L = R$? (Hint: $\sqrt{\sin^2(\theta)} = |\sin(\theta)|$.)

c. For $L > R$, find $dV^2/d\theta^2$. What is the second derivative in the limiting case where $L = R$?

In Exercises 76–80, find dy/dt at $t = t_0$.

76. $y(t) = (x(t) + 1)^2$, $x(t_0) = 3$, $x'(t_0) = -2$.

77. $y(t) = \dfrac{x(t) + 1}{x(t) - 1}$, $x(t_0) = 3$, $x'(t_0) = -2$.

78. $y(t) = \dfrac{x^2(t) + 1}{x^3(t) - x(t)}$, $x(t_0) = 2$, $x'(t_0) = -3$.

79. $y(t) = \sin(x(t))\tan^2(x(t))$, $x(t_0) = \frac{\pi}{3}$, $x'(t_0) = 4$.

80. $y(t) = (x^2(t) - 1)\sqrt{x^3(t) + x(t)}$, $x(t_0) = 2$, $x'(t_0) = -5$.

81. A parallelepiped is spanned by three vectors, $\vec{A}(t), \vec{B}(t)$ and $\vec{C}(t)$, which are functions of time. Find the change in the volume of the parallelepiped as a function of time given that $\vec{A}(t) = (t^2, t, 1)$, $\vec{B}(t) = (0, 3t, t^2)$, and $\vec{C}(t) = (0, 0, t^3)$.

4.4 Implicit Differentiation

Suppose that we have a graph of an equation of the form $f(x, y) = c$. For some familiar examples we call on the conic sections:

(a) $ax^2 - y = 0$ and $ay^2 - x = 0$ (parabolas in standard position);

(b) $\frac{x^2}{a^2} + \frac{y^2}{b^2} = 1$ (ellipses in standard position); and

(c) $\frac{x^2}{a^2} - \frac{y^2}{b^2} = 1$ and $\frac{y^2}{a^2} - \frac{x^2}{b^2} = 1$ (hyperbolas in standard position).

All parabolas can be expressed as graphs of functions. Either y can be solved in terms of x or vice-versa. In the case of an ellipse, the graph can be broken into two pieces, each of which is a function of x:

$$y = b\sqrt{1 - \frac{x^2}{a^2}} \text{ and } y = -b\sqrt{1 - \frac{x^2}{a^2}}.$$

Hyperbolas can similarly be broken into two functions of x.

It is not always so easy to solve for y in terms of x or x in terms of y. For example, what would you do with $x + y + \cos(xy) = 0$? This example is a case of a function being *implicitly defined*.

Definition: Implicitly Defined Functions

The equation $f(x, y) = c$ *implicitly defines y as a function of x near the point (x_0, y_0)* if:

(a) (x_0, y_0) is in the graph of the equation; and

(b) there is a disc centered at $(x_0, y_0)^a$ so that the intersection of the interior of the disc and the graph of $f(x, y) = c$ is a function (of x).

---

^{a}The interior of a disc centered at a point (a, b) is often called a *neighborhood* of the point.

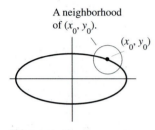

A neighborhood of (x_0, y_0).

(x_0, y_0)

Figure 1. *An equation of an ellipse in standard position implicitly defines y as a function of x at every point on the ellipse except where y crosses the x-axis.*

EXAMPLE 1: The ellipse $\frac{x^2}{a^2} + \frac{y^2}{b^2} = 1$, illustrated in Figure 1, implicitly defines y as a function of x at all (x_0, y_0) on the ellipse, except at the points $(a, 0)$ and $(-a, 0)$, where x is implicitly defined as a function of y. ∎

The graph of any function $y = g(x)$ can be represented implicitly by letting $f(x, y) = y - g(x) = 0$.

EXAMPLE 2: The graph of the function $y = x^2$ can be represented by $f(x, y) = y - x^2$. Then the solution set for $f(x, y) = 0$ is precisely the graph of the parabola $y = x^2$. ∎

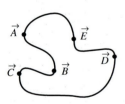

$\vec{A}$ $\vec{E}$

$\vec{C}$ $\vec{B}$ $\vec{D}$

Figure 2. *The graph implicitly defines y as a function of x at every point on the curve except at the points $\vec{A}$, $\vec{B}$, $\vec{C}$, and $\vec{D}$. It does implicitly define y at $\vec{E}$ even though the tangent line is vertical there. At each of these points, x is implicitly defined by y.*

EXAMPLE 3: The graph in Figure 2 illustrates an important fact: Let (a, b) be a point on the graph of $f(x, y) = c$. If the graph of $f(x, y) = c$ has a tangent line at each of its points and if (a, b) is a point on the graph with a *nonvertical* tangent line, then $f(x, y) = c$ implicitly defines y as a function of x in a neighborhood of (a, b). At points where the tangent line is vertical, x can be written as a function of y. ∎

The next example illustrates the fact that even for nicely defined equations of the form $f(x, y) = c$, tangent lines may not exist at all points on the graph.

EXAMPLE 4: Let $\mathcal{E}$ be the graph of $x^2 - y^2 = 0$. Then $\mathcal{E}$ consists of two straight lines through the origin, one having slope 1 and the other having slope -1, as illustrated in Figure 3. $\mathcal{E}$ implicitly defines y as a function of x at each of its points except at the origin. There the graph does not have a well defined tangent line.   ∎

In later chapters, we will be able to state some powerful theorems that tell us when graphs of equations of the form $f(x, y) = c$ actually define y as a function of x. For the present, let us assume that the equation does define y as a *differentiable* function of x in a neighborhood of a point (x, y) on the graph. With this assumption, we can compute the derivative of y.

EXAMPLE 5: Find the slope of the tangent line to the circle

$$x^2 + y^2 = 1 \tag{3}$$

at the point $\left(\frac{1}{\sqrt{2}}, \frac{1}{\sqrt{2}}\right)$.

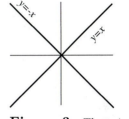

Figure 3. *There is no unique line tangent to $\mathcal{E}$ at the origin*

SOLUTION: If we assume that y is a function of x near the indicated point, then Equation (3) becomes:

$$x^2 + y^2(x) = 1. \tag{4}$$

Differentiating both sides of Equation (4), we obtain

$$2x + 2y(x)y'(x) = 0,$$

which we may solve for $y'(x)$ to yield:

$$y'(x) = -\frac{x}{y(x)}. \tag{5}$$

We want to solve for y' when $x = \frac{1}{\sqrt{2}}$ and $y(x) = \frac{1}{\sqrt{2}}$. We simply substitute these values into Equation (5) to obtain

$$y' = -\frac{\frac{1}{\sqrt{2}}}{\frac{1}{\sqrt{2}}} = -1.$$

It is important to notice that, in this example, we can't solve for y' when $y = 0$. These are the points $(1, 0)$ and $(-1, 0)$. At these points, the tangent lines are vertical, and y cannot be expressed as a function of x on any disc centered at either of these points.   ∎

EXAMPLE 6: Returning to Example 4, the graph of $x^2 - y^2 = 0$ is a pair of lines $y = x$ and $y = -x$. The graph of $x^2 - y^2 = 0$ implicitly defines the function $y = x$ near the point $(1, 1)$, for example.

Similarly, it implicitly defines $y = -x$ near the point $(1, -1)$. Since there is no well defined way to choose between the two lines at the origin $(0, 0)$, the graph of $x^2 - y^2 = 0$ does not implicitly define a function there. Use implicit differentiation to find y' for $x^2 - y^2 = 0$.

SOLUTION: Assume that y is a function of x, and take the derivative (with respect to x) of both sides of $x^2 - y^2 = 0$. This yields:

$$2x - 2yy' = 0.$$

Thus

$$y' = \frac{x}{y}.$$

We can find y' at every point of the graph of $x^2 - y^2 = 0$ except where $y = 0$. This is completely consistent with our observation that there is no tangent line to the graph at $(0, 0)$. ∎

We can also use implicit differentiation to find the second derivative.

EXAMPLE 7: In Example 5 we found the slope of a line tangent to the unit circle at a particular point. The derivative was given by $y' = -\frac{x}{y}$. In order to find the second derivative of y, we use the quotient rule and substitute in our expression for y' as follows.

$$
\begin{aligned}
y'' &= -\frac{yx' - xy'}{y^2} = -\frac{y - x\left(-\frac{x}{y}\right)}{y^2} \\
&= -\frac{\frac{y^2 + x^2}{y}}{y^2} = -\frac{y^2 + x^2}{y^3} \\
&= -y^{-3} \quad \text{(since } y^2 + x^2 = 1\text{)}
\end{aligned}
$$

∎

EXAMPLE 8: Figure 4 is a sketch of the graph of

$$x^2 y^2 - \left((1 + x)^2 \left(9 - x^2\right)\right) = 0.$$

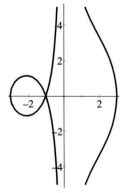

Figure 4. *The graph of the equation in Example 8.*

Inspection of the graph indicates that y is not defined implicitly at $(-3, 0)$, $(-1, 0)$ or $(3, 0)$. x is implicitly defined at $(-3, 0)$ and $(3, 0)$, but not at $(-1, 0)$. Implicit differentiation with respect to x yields

$$2xy^2 + 2xyy' - 2(1 + x)(9 - x^2) + (2 + x)^2 x = 0.$$

Evaluating this equation at each of the points $(-3, 0)$, $(-1, 0)$, and $(3, 0)$ yields $0 = 0$, and we are unable to solve for y'. If we solve for x' implicitly, we get $x' = 0$ at each of $(-3, 0)$ and $(3, 0)$. At $(-1, 0)$ we have $0 = 0$, and we cannot find x' there, which is consistent with the observation that neither x nor y is defined at $(-1, 0)$.

EXERCISES 4.4

In Exercises 1–6, use implicit differentiation to find y'.

1. $x^3 y + xy - xy^3 = 0$.

2. $\dfrac{x}{y} + (1 + x^2 y^3)^{-1/2} = x^4$.

3. $x = \left(\dfrac{x+y}{x-y}\right)^2$.

4. $x^2 y^{1/2} - y^{3/2} = 7$.

5. $\dfrac{x^2}{a^2} + \dfrac{y^2}{b^2} = 1$, where $a, b > 0$.

6. $\dfrac{x^2}{a^2} - \dfrac{y^2}{b^2} = 1$, where $a, b > 0$.

In Exercises 7–9, find the second derivative of y.

7. $xy^2 - y = 7$.

8. $\dfrac{x^2}{a^2} + \dfrac{y^2}{b^2} = 1$, where $a, b > 0$.

9. $\dfrac{x^2}{a^2} - \dfrac{y^2}{b^2} = 1$, where $a, b > 0$.

In Exercises 10–14, find the lines tangent and normal to the curve at the indicated points.

10. $xy + y^2 = x^3 - 1$ at $(1, 0)$.

11. $x^3 + xy + y^3 = 11$ at $(1, 2)$.

12. $y^2 - x^2 = 5$ at $(2, 3)$.

13. $x^2 - y^2 - 5\sqrt{x^2 + y^2} = 0$ at $(4, 3)$.

14. $x^2 + 2xy + y^2 + 2x - y = 26$ at $(2, 3)$.

The curves in Exercises 15–19 are graphs of the form $f(x, y) = c$.

a. Find the points on the curve where y is not implicitly defined by x.

b. Find the points on the curve where x is not implicitly defined by y.

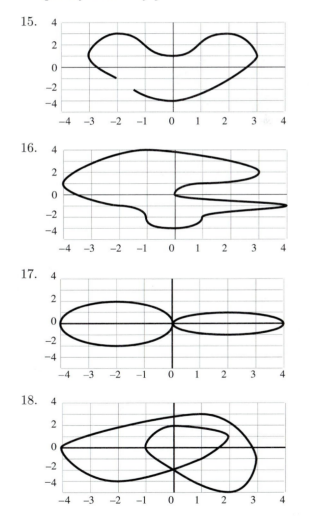

15.

16.

17.

18.

19.
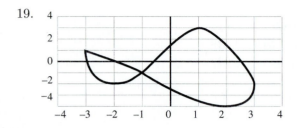

Exercises 20–22 refer to Figure 5, which displays the graph of $x^{2/3} + y^{2/3} = 1$.

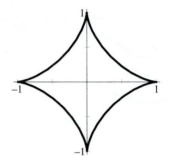

Figure 5. *The graph of $x^{2/3} + y^{2/3} = 1$.*

20. Explain why y is *not* implicitly defined by x at $(1,0)$ nor at $(-1,0)$.

21. Explain why x is *not* implicitly defined by y at $(0,1)$ nor at $(0,-1)$.

22. Use implicit differentiation to show that $y' = -\left(\frac{y}{x}\right)^{1/3}$. This implies that $y' = 0$ at $(1,0)$. But, in Part a, you argued that y is not a function of x in a neighborhood of $(1,0)$. y is not defined implicitly by x. Is it reasonable for $y'(x)$ to be defined when it is not a function of x? Where are you dividing by zero?

4.5   Higher Taylor Polynomials

We know how to construct a linear approximation for a function. We can employ the tools of calculus to build a polynomial of any order that approximates a given function. But first we use derivatives to prove a theorem about polynomials.

Theorem 1 *Suppose that*

$$a_n x^n + a_{n-1} x^{n-1} + \cdots + a_1 x + a_0 = b_n x^n + b_{n-1} x^{n-1} + \cdots + b_1 x + b_0,$$

for all x.
Then $a_0 = b_0$, $a_1 = b_1$, $\cdots$, $a_n = b_n$.

Proof: Let $a(x) = a_n x^n + a_{n-1} x^{n-1} + \cdots + a_1 x + a_0$ and $b_n x^n + b_{n-1} x^{n-1} + \cdots + b_1 x + b_0$.

$$a'(x) = na_nx^{n-1} + (n-1)a_{(n-1)}x^{n-2} + \cdots + 2a_2x + a_1$$

$$b'(x) = nb_nx^{n-1} + (n-1)b_{(n-1)}x^{n-2} + \cdots + 2b_2x + b_1$$

$$a''(x) = n(n-1)a_nx^{n-2} + (n-1)(n-2)a_{n-1}x^{n-3} + \cdots + (3)(2)a_3x + 2a_2$$

$$b''(x) = n(n-1)b_nx^{n-2} + (n-1)(n-2)b_{n-1}x^{n-3} + \cdots + (3)(2)b_3x + 2b_2$$

$$a'''(x) = n(n-1)(n-2)a_nx^{n-3} + (n-1)(n-2)(n-3)a_{n-1}x^{n-4} + \cdots + (4)(3)(2)a_4x + (3)(2)a_3$$

$$b'''(x) = n(n-1)(n-2)b_nx^{n-3} + (n-1)(n-2)(n-3)b_{n-1}x^{n-4} + \cdots + (4)(3)(2)b_4x + (3)(2)b_3$$

$$\vdots \qquad \vdots \qquad \qquad \vdots$$

$$a^{(n)}(x) = n(n-1)(n-2)\ldots(1)a_n = n!a_n$$

$$b^{(n)}(x) = n(n-1)(n-2)\ldots(1)b_n = n!b_n$$

Then

$$a(0) = b(0) \quad \Rightarrow \quad a_n = b_n.$$

$$a'(0) = b'(0) \quad \Rightarrow \quad a_{n-1} = b_{n-1}.$$

$$\vdots \qquad \qquad \vdots \qquad \qquad \vdots$$

$$a^{(n)}(0) = b^{(n)}(0) \quad \Rightarrow \quad n!a_n = n!b_n, \text{ or } a_n = b_n. \qquad \blacksquare$$

EXAMPLE 1: Suppose that $a_3x^3 + a_2x^2 + a_1x + a_0 = 2x^3 - 6x$. Show that $a_3 = 2$, $a_2 = 0$, $a_1 = -6$, and $a_0 = 0$.

SOLUTION: This is, of course, a consequence of Theorem 1. However, let us work the problem directly by emulating the proof of Theorem 1.

If we set $x = 0$ on both sides of

$$a_3x^3 + a_2x^2 + a_1x + a_0 = 2x^3 - 6x, \qquad (6)$$

we obtain $a_0 = 0$. Now, differentiate both sides of Equation (6) to obtain

$$3a_3x^2 + 2a_2x + a_1 = 6x^2 - 6. \qquad (7)$$

Let $x = 0$ in Equation (7) to obtain $a_1 = -6$. Now differentiate both sides of Equation (7) to get

$$6a_3x + 2a_2 = 12x \qquad (8)$$

Evaluating Equation (8) at $x = 0$ yields $a_2 = 0$. Differentiating one more time yields $a_3 = 2$. ∎

If f has a derivative at $x = x_0$, then the graph of $p(x) = f(x_0) + f'(x_0)(x - x_0)$ is the line tangent to f at x_0. It is the first order polynomial that "best fits" f in the sense that $p(x_0) = f(x_0)$ and $p'(x_0) = f'(x_0)$. Can we find a second order polynomial p_2 that "best fits" f in the same sense, so that $f(x_0) = p_2(x_0), f'(x_0) = p_2'(x_0)$, and $f''(x_0) = p_2''(x_0)$? The answer is yes, provided that $f''(x_0)$ is defined.

The procedure is as follows. We write p_2 as

$$p_2(x) = a_0 + a_1(x - x_0) + a_2(x - x_0)^2 \tag{9}$$

We evaluate Equation (9) at $x = x_0$, which gives $p_2(x_0) = a_0$. But we require that $p_2(x_0) = f(x_0)$, so a_0 must equal $f(x_0)$. Differentiating Equation (9) yields

$$p_2'(x) = a_1 + 2a_2(x - x_0). \tag{10}$$

Evaluating Equation (10) at x_0, and setting the result equal to $f'(x_0)$, we have $a_1 = f'(x_0)$. Finally, evaluating $p_2''(x) = 2a_2$ at x_0, we get $a_2 = \frac{f''(x_0)}{2}$. So we see that

$$p_2(x) = f(x_0) + f'(x_0)(x - x_0) + \frac{f''(x_0)}{2}(x - x_0)^2.$$

Just as

$$p_1(x) = f(x_0) + f'(x_0)(x - x_0)$$

is called the first order Taylor Polynomial for f centered at x_0, the polynomial $p_2(x)$, given above, is called the *second order Taylor Polynomial for f centered at x_0*. ∎

EXAMPLE 2: Find the first and second order Taylor polynomials for $f(x) = x^6 - 2x^2 - 4$ at $x_0 = 1$. Use both approximating polynomials to estimate $f(1.1)$, and compare these estimates with the true value.

SOLUTION: $f'(x) = 6x^5 - 4x$ and $f''(x) = 30x^4 - 4$. So $f(1) = -5$, $f'(1) = 2$, and $f''(1) = 26$. Hence

$$p_1(x) = -5 + 2(x - 1),$$

and

$$p_2(x) = -5 + 2(x - 1) + \frac{26}{2}(x - 1)^2.$$

Now, $f(1.1) = -4.648439$, $p_1(1.1) = -4.8$, and $p_2(1.1) = -4.67$. It is not too surprising that p_2 gives a more accurate approximation than does p_1. ∎

In Figures 1.a-1.c, we zoom in on the graphs of f, p_1 and p_2 near $x_0 = 1$ to see their relationships. Inspection of the graphs in Figure 1.a does not give a clear picture of what is happening. We expect that p_2 would give a more accurate approximation to f than does p_1. In Figure 1.b, we zoom in closer to $x_0 = 1$ by graphing the functions over the interval $[x_0 - 0.5, x_0 + 0.5]$. p_2 clearly does a better job of approximating f than does p_1 on this interval. Restricting our graphs to the interval $[0.9, 1.1]$, it becomes difficult to see the difference between f and p_2.

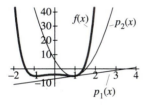

Figure 1.a *The first and second order Taylor polynomial approximations of f on the interval $(-2, 4)$.*

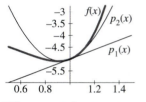

Figure 1.b *The first and second order Taylor polynomial approximations of f on the interval $(0.5, 1.5)$.*

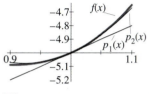

Figure 1.c *The first and second order Taylor polynomial approximations of f on the interval $(0.9, 1.1)$.*

EXAMPLE 3: Let $f(x)$, $p_1(x)$, and $p_2(x)$ be as in Example 2 and let $\text{Error}_n(x) = |f(x) - p_n(x)|$. The Error function provides a tool to measure the accuracy of the approximating functions p_1 and p_2. In Figures 2.a and 2.b, we sketch $\text{Error}_1(x)$ and $\text{Error}_2(x)$ over the interval $[0.9, 1.1]$. ∎

We have introduced first and second order Taylor polynomials. Higher order Taylor polynomials are similarly defined.

Definition: Taylor and Maclaurin Polynomials

Suppose that A is a subset of $\mathbb{R}$, and that f is a function from A into $\mathbb{R}$. Suppose further that the nth derivative of f exists at $x = x_0$. The polynomial

$$p_n(x) = a_0 + a_1(x - x_0) + a_2(x - x_0)^2 + \cdots + a_n(x - x_0)^n$$

where

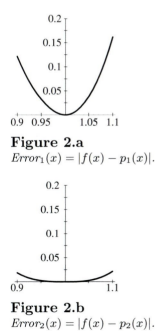

Figure 2.a
$\text{Error}_1(x) = |f(x) - p_1(x)|$.

Figure 2.b
$\text{Error}_2(x) = |f(x) - p_2(x)|$.

$$a_i = \frac{f^{(i)}(x_0)}{n!}, \text{ for } 0 \leq i \leq n$$

is called the nth order *Taylor Polynomial for f centered at $x = x_0$.* If $x_0 = 0$, then p_n is called the *Maclaurin polynomial* for f.

EXAMPLE 4: If we expand the Taylor polynomial

$$p_n(x) = a_0 + a_1(x - x_0) + \ldots + a_n(x - x_0)^n$$

for $f(x) = x^6 - 2x^2 - 4a$ at $x_0 = 1$, then from Example 2, we have that $a_0 = -5$, $a_1 = 2$, and $a_2 = 13$. Starting with $f''(x) = 30x^4 - 4$ from Example 2, we see that $f'''(x) = 120x^3$ and $a_3 = \frac{120}{3!} = 20$. $f^{(4)}(x) = 360x^2$ and $a_4 = \frac{360}{4!} = 15$ and $f^{(5)}(x) = 720$ so $a_5 = \frac{720}{5!} = 6$. Thus

$$p_3(x) = -5 + 2(x - 1) + \frac{26}{2}(x - 1)^2 + 20(x - 1)^3,$$

$$p_4(x) = -5 + 2(x - 1) + \frac{26}{2}(x - 1)^2 + 20(x - 1)^3 + 15(x - 1)^4$$

and

$$p_5(x) = -5 + 2(x - 1) + \frac{26}{2}(x - 1)^2 + 20(x - 1)^3$$
$$+ 15(x - 1)^4 + 6(x - 1)^5.$$

In Figures 3.a-c, we compare the graphs of p_3, p_4 and p_5 with f over the interval $[-2, 2]$. An inspection of the error functions in Figures 4.a-c reveals that as n gets larger, p_n better approximates f over a larger interval. ∎

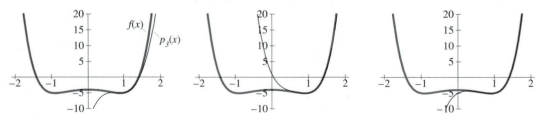

Figure 3.a p_3 *and* f.   **Figure 3.b** p_4 *and* f.   **Figure 3.c** p_5 *and* f.

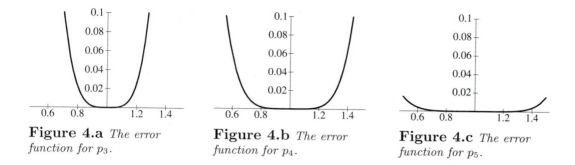

Figure 4.a *The error function for p_3.*

Figure 4.b *The error function for p_4.*

Figure 4.c *The error function for p_5.*

EXAMPLE 5: Compare the graphs of the second, third, fourth, and fifth order Maclaurin polynomial approximations for f to the graph of f, where $f(x) = x/(1+x)$.

1. Find a number δ such that if $|x - x_0| < \delta$, then the error in p_3, $\mathrm{Error}_3(x) = |p_2(x) - f(x)| < 0.01$.

2. Find a number δ such that if $|x - x_0| < \delta$, then the error in p_5, $\mathrm{Error}_5(x) = |p_5(x) - f(x)| < 0.01$.

SOLUTION:

$$
\begin{aligned}
f(x) &= \frac{x}{1+x} & &\Rightarrow a_0 = f(0) = 0. \\
f'(x) &= \frac{(1+x)-x}{(1+x)^2} = \frac{1}{(1+x)^2} & &\Rightarrow a_1 = f'(0) = 1 \\
f''(x) &= -\frac{2(1+x)}{(1+x)^4} = -\frac{2}{(1+x)^3} & &\Rightarrow a_2 = \frac{f''(0)}{2} = -1 \\
f'''(x) &= \frac{6(1+x)^2}{(1+x)^6} = \frac{6}{(1+x)^4} & &\Rightarrow a_3 = \frac{f''(0)}{3!} = 1 \\
f^{(4)}(x) &= -\frac{24(1+x)^3}{(1+x)^8} = -\frac{24}{(1+x)^5} & &\Rightarrow a_4 = \frac{f^{(4)}(0)}{4!} = -1 \\
f^{(5)}(x) &= \frac{120(1+x)^4}{(1+x)^{10}} = -\frac{120}{(1+x)^6} & &\Rightarrow a_4 = \frac{f^{(5)}(0)}{5!} = 1
\end{aligned}
$$

Therefore $p_1(x) = x$, $p_2(x) = x - x^2$, $p_3(x) = x - x^2 + x^3$, $p_4(x) = x - x^2 + x^3 - x^4$, and $p_5(x) = x - x^2 + x^3 - x^4 + x^5$. In Figures 5.a-d, we compare sketches of the Maclaurin polynomials p_2, p_3, p_4, and p_5 with the function f.

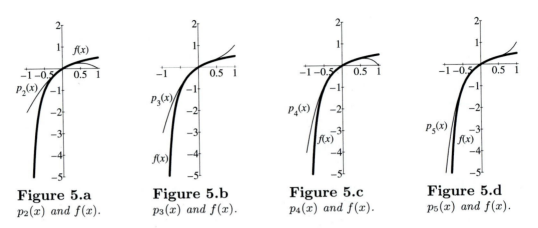

Figure 5.a
$p_2(x)$ and $f(x)$.

Figure 5.b
$p_3(x)$ and $f(x)$.

Figure 5.c
$p_4(x)$ and $f(x)$.

Figure 5.d
$p_5(x)$ and $f(x)$.

Now, to find the δ asked for in Part 1, we look at the error function $\text{Error}_3(x)$, $-1 < x < 1$, restricting the range of the function to $[0, 0.01]$ in Figure 6.a. Zooming in on the domain $-0.25 < x < 0.25$ allows us to obtain Figure 6.b. We see that if we let $\delta = 0.25$, then $\text{Error}_3(x) < 0.01$ when $-\delta < x < \delta$.

To find the δ asked for in Part 2, we graph the error function $\text{Error}_5(x)$, $-1 < x < 1$, again restricting the range of the function to $[0, 0.01]$. See Figure 6.c. Zooming in on the domain $-0.5 < x < 0.5$, we obtain Figure 6.d. At this point, we observe that $\text{Error}_5(x)$ is decreasing on $[-0.5, 0]]$ and increasing on $[0, 0.5]$. Thus, if we choose δ such that $\text{Error}_5(x) - \delta \leq 0.01$ and $\text{Error}_5(x) + \delta \leq 0.01$, then we know that $\text{Error}_5(x) \leq 0.01$ on the interval $[-\delta, \delta]$. We try letting $\delta = 0.05$. However, $\text{Error}_5(-0.5) \approx 0.03125$, which is too large, so $\delta < 0.5$. Trying $\delta = 0.4$, we obtain $\text{Error}_5(-0.4) \approx 0.00682667 < 0.01$. Since $\text{Error}_5(0.4) \approx 0.00292571 < 0.01$, we have answered Part 2 with $\delta = 0.4$. ∎

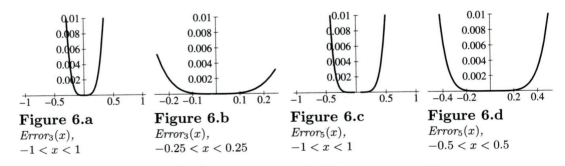

Figure 6.a
$Error_3(x)$,
$-1 < x < 1$

Figure 6.b
$Error_3(x)$,
$-0.25 < x < 0.25$

Figure 6.c
$Error_5(x)$,
$-1 < x < 1$

Figure 6.d
$Error_5(x)$,
$-0.5 < x < 0.5$

EXAMPLE 6: Derive a formula for $\frac{d^n}{dx^n}\cos(x)$.

SOLUTION: Let $f(x) = \cos(x)$. We know that $f'(x) = -\sin(x)$,

$f''(x) = -\cos(x)$, $f'''(x) = \sin(x)$, and $f^{(4)}(x) = \cos(x)$. In order to see a trend, it's helpful to take a couple more derivatives: $f^{(5)}(x) = -\sin(x)$ and $f^{(6)}(x) = -\cos(x)$. It seems that when n is an even integer $f^{(n)}(x)$ is either $\cos(x)$ or $-\cos(x)$. When n is an odd integer, $f^{(n)}(x)$ is $\sin(x)$ or $-\sin(x)$. Our solution appears to have the form

$$\frac{d^n}{dx^n}\cos x = \begin{cases} (-1)^{??}\cos(x) & \text{if } n = 2i \\ (-1)^{??}\sin(x) & \text{if } n = 2i+1. \end{cases}$$

The question is, what is "??" in the two cases? We consider them separately.

Case 1. The integer n is even. That is, $n = 2i$. Exploring a little we see that when $i = 0$, we have $f^{(0)}(x) = (+1)\cos x$. When $i = 1$, $f^{(2i)}(x) = f''(x) = (-1)\cos(x)$, and when $i = 2$, $f^{(2i)}(x) = f^{(4)}(x) = \cos(x)$. This suggests that "??"$= (-1)^i$. Our "educated" guess for $f^{(2i)}(x)$ is $f^{(2i)}(x) = (-1)^i\cos(x)$. A quick check shows that this works when $i = 3$.

Case 2. The integer n is odd. That is, $n = 2i + 1$. When $i = 0$, we have $f'(x) = (-1)\sin x$. When $i = 1$, $f^{(2i+1)}(x) = f'''(x) = (+1)\sin(x)$, and when $i = 2$, $f^{(2i+1)}(x) = f^{(5)}(x) = -\sin(x)$. We again hypothesize that "??"$= i$. Thus our guess for the nth derivative of $\cos x$ is

$$\frac{d^n}{dx^n}\cos x = \begin{cases} (-1)^i\cos(x) & \text{if } n = 2i \\ (-1)^i\sin(x) & \text{if } n = 2i+1. \end{cases}$$

We know that our formula works for the first few integers. We can use induction to show that it works for all n by first showing that it works for all even integers, and then for all odd integers.

For the even integers, assume that we know the formula works for $n = 2k$. To show that it works for $n = 2(k+1)$. We simply take two derivatives of $\frac{d^{2k}}{dx^{2k}}\cos x = (-1)^k\cos x$ as follows.

$$\begin{aligned} \frac{d^{2k+1}}{dx^{2k+1}}\cos x &= \left(\frac{d}{dx}\right)\frac{d^{2k}}{dx^{2k}}\cos x = \frac{d}{dx}(-1)^k\cos x \\ &= (-1)(-1)^k\sin x = (-1)^{k+1}\sin x \end{aligned}$$

and

$$\begin{aligned} \frac{d^{2k+2}}{dx^{2k+2}}\cos x &= \left(\frac{d}{dx}\right)\frac{d^{2k+1}}{dx^{2k+1}}\cos x = \frac{d}{dx}(-1)^{k+1}\sin x \\ &= (-1)^{k+1}\cos x. \end{aligned}$$

For the odd integers, assume that we know the formula works for $n = 2k + 1$. We need to show that it works for $n = 2(k + 1) + 1$. As above, we take two derivatives of $\frac{d^{2k+1}}{dx^{2k+1}} \cos x = (-1)^k \sin x$. The details are left to the student. ∎

EXAMPLE 7:   Find the nth order Maclaurin polynomial for $\cos(x)$.

SOLUTION: We have

$$
\begin{aligned}
f(x) &= \cos(x) \Longrightarrow a_0 = f(0) = 1 \\[2mm]
f'(x) &= -\sin(x) \Longrightarrow a_1 = f'(0) = 0 \\[2mm]
f''(x) &= -\cos(x) \Longrightarrow a_2 = \frac{f''(0)}{2} = -\frac{1}{2} \\[2mm]
f'''(x) &= \sin(x) \Longrightarrow a_3 = \frac{f'''(0)}{3!} = 0 \\[2mm]
f^{(4)}(x) &= \cos(x) \Longrightarrow a_4 = \frac{f^{(4)}(0)}{4!} = \frac{1}{4!} \\[2mm]
&\qquad\qquad \vdots \qquad\qquad\qquad \vdots \\[2mm]
f^{(2i)}(x) &= (-1)^i \cos(x) \Longrightarrow a_{2i} = \frac{(-1)^i}{(2i)!}
\end{aligned}
$$

and

$$
f^{(2i+1)}(x) = (-1)^i \sin(x) \Longrightarrow a_{2i+1} = 0.
$$

Thus

$$
p_{2i}(x) = 1 - \frac{1}{2}x^2 + \frac{1}{24}x^4 + \cdots + \frac{(-1)^i}{(2i)!}x^{2i}.
$$

Notice that since $\sin(0) = 0$, the $(2i + 1)^{\text{st}}$ order Maclaurin polynomial for $\cos(x)$ is the same as the $(2i)^{\text{th}}$ order polynomial. ∎

We can also use Taylor polynomials to approximate solutions of implicit equations, which we demonstrate in the following example.

EXAMPLE 8:   Find the first, second, and third order Taylor polynomials approximating the graph of

$$
x^2 + 2xy - y^3 + 11 = 0 \tag{11}
$$

near the point $(-1, 2)$.

SOLUTION: We assume that y is a function of x near $(-1, 2)$. We will center the polynomial for y at $x = -1$ since that is where we have some information. Differentiating Equation (11) implicitly, we obtain

$$2x + 2y + 2xy' - 3y^2 y' = 0. \tag{12}$$

We can solve Equation (12) for y' to get $y' = \frac{2x+2y}{3y^2-2x}$, and substituting $x = -1$ and $y = 2$ shows us that $y'(1) = \frac{1}{7}$. Assuming that y'' exists, we can differentiate Equation (12) implicitly to get

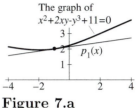

The graph of
$x^2+2xy-y^3+11=0$

Figure 7.a

p_1 compared to the graph of $x^2 + 2xy - y^3 + 11 = 0$ near the point $(-1, 2)$.

$$2+2y'+2y'+2xy''-6yy'^2-3y^2y'' = 2+4y'+2xy''-6yy'^2-3y^2y'' = 0. \tag{13}$$

We solve Equation (13) for y'' in terms of x, y and y' to get

$$y'' = \frac{2 + 4y' - 6yy'^2}{3y^2 - 2x}. \tag{14}$$

Substituting $x = -1$, $y = 2$, and $y' = \frac{1}{7}$, into Equation (14), we obtain $y''(1) = \frac{57}{343}$. We could use the quotient rule and implicit differentiation on Equation (14) to find y'''; however, Equation (13) is easier to work with. Implicitly differentiating Equation (13), we obtain:

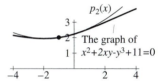

Figure 7.b

p_2 compared to the graph of $x^2 + 2xy - y^3 + 11 = 0$ near the point $(-1, 2)$.

$$4y'' + 2y'' + 2xy''' - 12yy'y'' - 6y'^3 - 6yy'y'' - 3y^2y''' =$$

$$6y'' + 2xy''' - 18yy'y'' - 6y'^3 - 3y^2y''' = 0. \tag{15}$$

We solve Equation (15) for y''' in terms of x, y, y', and y''.

$$y''' = \frac{6y'' - 18yy'y'' - 6y'^3}{3y^2 - 2x}. \tag{16}$$

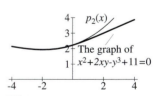

Figure 7.c

p_3 compared to the graph of $x^2 + 2xy - y^3 + 11 = 0$ near the point $(-1, 2)$.

We now let $x = 1$, $y = 2$, $y' = \frac{1}{7}$, and $y'' = \frac{57}{343}$ in Equation (16) to get $y'''(1) = \frac{663}{16807}$. Our Taylor polynomials will be centered at $x_0 = -1$.

$$
\begin{aligned}
p_1(x) &= y(1) + y'(1)(x-1) = 2 + \tfrac{1}{7}(x+1) \\
p_2(x) &= y(1) + y'(1)(x-1) + \tfrac{1}{2}y''(x-1)^2 \\
&= 2 + \tfrac{1}{7}(x+1) + \tfrac{57}{686}(x+1)^2 \\
p_3(x) &= 2 + \tfrac{1}{7}(x+1) + \tfrac{57}{686}(x+1)^2 + \tfrac{221}{33614}(x+1)^3
\end{aligned}
$$

The polynomials p_1, p_2, and p_3 are compared to the graph of $x^2 + 2xy - y^3 + 3 = 0$ near $(1, 2)$. ∎

Approximating Solutions to Differential Equations

In the last example, we used the fact that we had an equation involving x, y, and y' to solve for y'' in terms of x, y, and y'. We then used that result to find y''' in terms of x, y, y', and y''. Except for the tedium, there is nothing to keep us from finding as many derivatives as we need in this way. We are of course making the (not terribly brash) assumption that these derivatives do exist! It often turns out that when we try to model a real problem, we wind up with an equation that involves derivatives of y called a differential equation. We can use this equation and implicit differentiation to find derivatives of y. This in turn can be used to write down Taylor polynomials to approximate the true function y.

EXAMPLE 9: It has been experimentally determined that the rate of radioactive decay of a mass is proportional to the current radioactive content of the mass. Let $y(t)$ denote the radioactive content of the mass at time t. The rate of change of y is negative since it represents a loss of radioactive content. The constant k depends on the material and the scale of time. The scale of time might be in years. Thus we have that

$$y' = -ky. \tag{17}$$

Successive differentiations of Equation (17) show us that $y^{(n)} = (-1)^n k^n y$. Assume that we start our clock at the instant we measure the current value for y. Let λ denote this initial measurement. Thus we know $y(0) = \lambda$. Hence, we also know that $y^{(n)}(0) = (-1)^n k^n \lambda$. Although we don't know what $y(x)$ is exactly, we do now have approximating polynomials. The nth order Maclaurin polynomial for y is ∎

$$p_n(x) = \lambda - k\lambda x + \frac{k^2 \lambda}{2} x^2 - \frac{k^3 \lambda}{3!} x^3 + \ldots + \frac{(-1)^n k^n \lambda}{n!} x^n.$$

EXERCISES 4.5

In Exercises 1-3, find the first and second order Taylor polynomials for the function at x_0 and use all three to estimate $f(x_0 + 0.01)$. Does p_2 seem to consistently do better than p_1?

1. $f(x) = x^{25} + 2x^3 + 3$, $\quad x_0 = 1$.

2. $f(x) = \sin(x)\cos(x)$, $\quad x_0 = \pi/4$.

3. $r(x) = \sin^3(x)$, $\quad x_0 = 0$.

4. Let $f(x) = \dfrac{1 + x^2}{x^3 - 2x}$.

    a. Find $p_1(x)$, the first order Taylor polynomial for f at $x_0 = 1$.

    b. Use $p_1(x)$ to approximate $f(1.1)$.

    c. Compare the graphs of $p_1(x)$ and $f(x)$ over the interval $[0, 2]$.

    d. Evaluate $\text{Error}_1(0.9)$ and $\text{Error}_1(1.1)$.

    e. Sketch the error function $\text{Error}_1(x)$ over the interval $[0.9, 1.1]$, and find an number κ such that $p_1(x)$ approximates $f(x)$ with an error of no more than κ on the interval $[0.9, 1.1]$.

5. Let $p_1(x)$ denotes the first order Taylor polynomial for $f(x) = \dfrac{\sin\left(\frac{\pi x}{3}\right)}{x^3 - 2x}$ at $x_0 = 1$.

    a. Find $p_1(x)$.

    b. Use $p_1(x)$ to approximate $f(1.1)$.

    c. Compare the graphs of $p_1(x)$ and $f(x)$ over the interval $[0, 2]$.

    d. Evaluate $\text{Error}_1(0.9)$ and $\text{Error}_1(1.1)$.

    e. Sketch the error function $\text{Error}_1(x)$ over the interval $[0.9, 1.1]$, and find an number κ such that $p_1(x)$ approximates $f(x)$ with an error of no more than κ on the interval $[0.9, 1.1]$.

6. Find the 5th Maclaurin polynomial for $\sin(x)$.

7. Find the 5th Maclaurin polynomial for $\cos(x)$.

8. Let $p_n(x)$ denote the nth order Maclaurin polynomial for $\cos(x)$.

    a. Find $p_8(x)$.

    b. Compare the graph of $p_n(x)$ with the graph of $\cos(x)$ over the interval $[-2\pi, 2\pi]$ for $n = 2, 4, 6$, and 8.

    c. Find the first integer n so that $p_n(x)$ approximates $\cos(x)$ over the interval $[-4, 4]$ with an error of less than 0.01.

9. Let $p_n(x)$ denote the nth order Maclaurin polynomial for $\sin(x)$.

    a. Find $p_9(x)$

    b. Compare the graph of $p_n(x)$ with the graph of $\sin(x)$ over the interval $[-2\pi, 2\pi]$ for $n = 1, 3, 7$, and 9.

    c. Compare the derivative of $p_9(x)$ with the eighth order Maclaurin polynomial for $\cos(x)$.

10. Let $p_n(x)$ denote the nth order Taylor polynomial for $f(x) = \frac{1}{x}$ at $x_0 = 1$.

    a. Find $p_1(x)$, $p_2(x)$, $p_3(x)$, and $p_4(x)$

    b. Find an expression for $p_n(x)$.

    c. Evaluate $p_n(2)$, for $n = 1, 2, 3$, and 4.

    d. Evaluate $p_n(2)$, for $n = 25$, and 26.

    e. Does $p_n(x)$ seem to approximate $\frac{1}{x}$ at $x = 2$?

    f. Evaluate $p_n(2)$, for $0 < n \leq 25$. Does $p_n(x)$ seem to approximate $\frac{1}{x}$ at $x = 2$ as n gets large?

    g. Compare $p_n(0.9)$ with $f(1.9)$, for $n = 1, 2, \ldots, 25$. Does $p_n(x)$ seem to approximate $\frac{1}{x}$ at $x = 0.9$ as n gets large?

    h. Compare $p_n(0.9)$ with $f(1.9)$, for $n = 1, 2, \ldots, 100$. Does $p_n(x)$ seem to approximate $\frac{1}{x}$ at $x = 1.9$ as n gets large?

Chapter 5

The Geometry of Functions and Curves

5.1   Horizontal and Vertical Asymptotes

In this chapter, we use the tools of calculus we have developed to obtain information about the geometry of graphs of functions and of parametrized paths. In this section, we extend our notion of limits to define horizontal and vertical asymptotes.

EXAMPLE 1:   The graph of $f(x) = \frac{1}{x}$ is sketched in Figure 1. Although neither $\lim_{x \to 0^+} f(x)$ nor $\lim_{x \to 0^-} f(x)$ is defined, the graph of f is reasonably well behaved near the origin. As x "gets close to 0" while staying positive, $f(x)$ gets to be a very large positive number. Similarly, as x "gets close to 0" while staying negative, $f(x)$ is a negative number with very large magnitude. In fact, if $a > 0$ is close to zero then the graph of $f(x)$, for $0 < x < a$ looks quite like a ray on the positive y–axis. In the same way, if $a < 0$ is close to zero, then the graph of $f(x)$, for $a < x < 0$, looks quite like a ray on the negative y–axis. ■

It is this type of behavior that we capture in the following definitions.

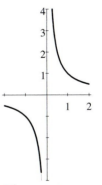

Figure 1. *The graph of* $f(x) = \frac{1}{x}$.

Definition: Vertical Asymptotes

(a) The limit of f as x approaches the number a from the right is ∞ if, whenever E is a number, there is a number D such that if

---

185

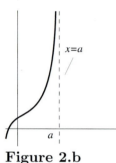

Figure 2.a

$\lim_{x \to a^+} f(x) = \infty.$

Figure 2.b

$\lim_{x \to a^-} f(x) = \infty.$

$a < x < a + D$, then x is in the domain of f and $f(x) > E$.

If the limit of f as x approaches the number a from the right is ∞, we write

$$\lim_{x \to a^+} f(x) = \infty,$$

and we will say that ∞ is the *right hand limit of f at a*.

(b) The limit of f as x approaches the number a from the left is ∞ if, whenever E is a number, there is a number D such that if

$a - D < x < a$, then x is in the domain of f and $f(x) > E$.

If the limit of f as x approaches the number a from the left is ∞, we write

$$\lim_{x \to a^-} f(x) = \infty,$$

and we say that ∞ is the *left hand limit of f at a*.

We also have the analogous definitions for limits to $-\infty$. If the right hand or left hand limit of f at a is $\pm\infty$, then the line $x = a$ is called a *vertical asymptote* of the function f. (See Figures 2.a and 2.b.)

EXAMPLE 2: Let $f(x) = \dfrac{1}{(x-2)^3}$. Then $\lim_{x \to 2^+} f(x) = \infty$, and $\lim_{x \to 2^-} f(x) = -\infty$. The line $x = 2$ is a vertical asymptote for the function f. ∎

EXAMPLE 3: $\lim_{x \to \pi/2^-} \tan(x) = \infty$, and $\lim_{x \to \pi/2^+} \tan(x) = -\infty$. The line $x = \frac{\pi}{2}$ is an asymptote for $\tan(x)$. ∎

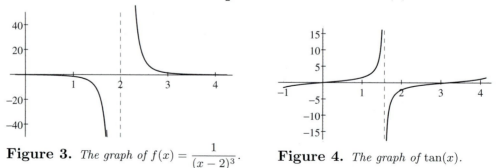

Figure 3. *The graph of* $f(x) = \dfrac{1}{(x-2)^3}$.   **Figure 4.** *The graph of* $\tan(x)$.

The idea of a *horizontal asymptote* is very similar to the idea of a vertical asymptote. An example will help explain.

EXAMPLE 4: We return to the function in Example 1. Let $f(x) = \frac{1}{x}$. We can make $\frac{1}{x}$ be as close to 0 as we please by making the magnitude of x large enough. We say that $\lim_{x \to \infty} f(x) = 0$ and $\lim_{x \to -\infty} f(x) = 0$. Notice, that if a is a very large positive number, then the graph of $f(x)$, for $x > a$, looks very much like the ray $x > a$ on the x–axis. Similarly, if a is a very large negative number, then the graph of $f(x)$, for $x < a$, looks very much like the ray $x < a$ on the x–axis.   ■

Definition: Horizontal Asymptotes

(a) The limit of f as x goes to ∞ is L

    (i) if the domain is unbounded to the right[1] and

    (ii) if we can make $f(x)$ as close to L as we please by choosing x large enough.

We state Part (ii) formally by saying if E is a positive number (E is the acceptable error), then there is a number K such that if $x > K$ is in the domain of f, then $|f(x) - L| < E$.

As expected, we use the notation

$$\lim_{x \to \infty} f(x) = L$$

to state that the limit of f as x goes to ∞ is L. See Figure 5.

(b) The limit of f as x goes to $-\infty$ is L

    (i) if the domain of f is unbounded to the left[2] and

    (ii) if we can make $f(x)$ as close to L as we please by choosing $x < 0$ in the domain of f with $|x|$ large enough. More formally, if E is a positive number (E is the acceptable error), then there is a number K such that if $x < K$ is in the domain of f, then $|f(x) - L| < E$.

---

[1] A set is unbounded to the right if we can find points in the domain of f as large as we please. More formally, if N is a number, then we can find a point x in the set such that $x > N$.

[2] A set is unbounded to the left if we can find negative numbers in the set with magnitude as large as we please. More formally, if N is an number, then we can find a point x in the set such that $x < N$.

In this case we write

$$\lim_{x \to -\infty} f(x) = L.$$

If either of Conditions (a) or (b) are satisfied, then the line $y = L$ is called a *horizontal asymptote of f*.

Figure 5. *The line $y = L$ is a horizontal asymptote.*

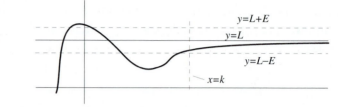

The x–axis is the only horizontal asymptote of $f(x) = 1/x$. The function $g(x) = x$ has no horizontal asymptote.

EXAMPLE 5:   Let $f(x) = \frac{|x|}{x}$. Then $\lim_{x\to\infty} f(x) = 1$ and $\lim_{x\to-\infty} f(x) = -1$. The lines $y = 1$ and $y = -1$ are both horizontal asymptotes of f. See Figure 6. ■

EXAMPLE 6:   The lines $y = \frac{\pi}{2}$ and $y = -\frac{\pi}{2}$ are horizontal asymptotes of $f(x) = \text{Arctan}(x)$. See Figure 7 ■

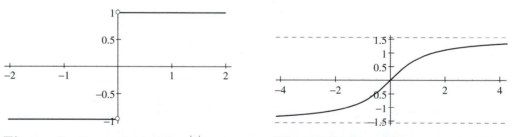

Figure 6. *The graph of $f(x) = \frac{|x|}{x}$.*   **Figure 7.** *The graph of $f(x) = \arctan(x)$.*

Question: Is it possible for a function to have more than two horizontal asymptotes?

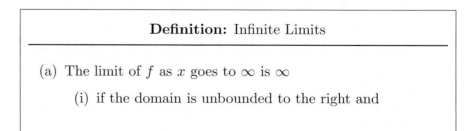

Definition: Infinite Limits

(a) The limit of f as x goes to ∞ is ∞

    (i) if the domain is unbounded to the right and

(a) *(Continued)*

    (ii) if we can make $f(x)$ as large as we please by choosing x large enough. Formally, if N is a number, then there is a number K such that if $x > K$ is in the domain of f, then $f(x) > N$. In this case we write

$$\lim_{x \to \infty} f(x) = \infty.$$

Intuitively, as x moves along the x–axis in the positive direction and $y = N$ is any horizontal line, eventually $f(x)$ is above and stays above that line.

(b) The limit of f as x goes to $-\infty$ is ∞

    (i) if the domain of f is unbounded to the left and

    (ii) if we can make $f(x)$ as large as we please by choosing $x < 0$ in the domain of f with $|x|$ large enough. Formally, if N is a number, then there is a number K such that if $x < K$ is in the domain of f, then $f(x) > N$.

In this case we write

$$\lim_{x \to -\infty} f(x) = \infty.$$

Intuitively, as x moves along the x–axis in the negative direction and $y = N$ is any horizontal line, eventually $f(x)$ is above and stays above that line.

The statements

$$\lim_{x \to \infty} f(x) = -\infty \text{ and } \lim_{x \to -\infty} f(x) = -\infty$$

can be defined similarly. Intuitively, $\lim_{x \to \infty} f(x) = -\infty$ means that as x moves along the x–axis in the positive direction, then given any horizontal line $y = N$, eventually $f(x)$ is below and stays below that line. Similarly $\lim_{x \to -\infty} f(x) = -\infty$ means that as x moves along the x–axis in the negative direction, then given any horizontal line $y = N$, eventually $f(x)$ is below and stays below that line.

EXAMPLE 7: Let $f(x) = x^3$. Then $\lim_{x \to \infty} f(x) = \infty$ and $\lim_{x \to -\infty} f(x) = -\infty$. ∎

EXAMPLE 8: Let $f(x) = \dfrac{2x^2 + 3x + 25}{x^2 - 2x - 10^5}$. In this case, it is not immediately clear what happens to f as x gets large. A standard technique for rational functions is to divide both the numerator and denominator by the highest power of x that occurs in the denominator.

$$\begin{aligned} f(x) & = \frac{2x^2 + 3x + 25}{x^2 - 2x - 10^5} \\[2mm] & = \frac{2 + \frac{3}{x} + \frac{25}{x^2}}{1 - \frac{2}{x} - \frac{10^5}{x^2}} \end{aligned}$$

As x gets large, the terms $\frac{3}{x}, \frac{25}{x^2}, \frac{2}{x}$ and $\frac{10^5}{x^2}$ each approach 0. Thus

$$\begin{aligned} \lim_{x \to \infty} f(x) & = \lim_{x \to \infty} \frac{2x^2 + 3x + 25}{x^2 - 2x - 10^5} \\[3mm] & = \lim_{x \to \infty} \frac{2 + \frac{3}{x} + \frac{25}{x^2}}{1 - \frac{2}{x} - \frac{10^5}{x^2}} = 2. \end{aligned}$$

Similarly, $\lim_{x \to -\infty} f(x) = 2$.   ∎

EXAMPLE 9: One might be tempted to assume that if $f(x)$ does not get close to a real number as x approaches ∞, then either $\lim_{x \to \infty}(f) = -\infty$ or $\lim_{x \to \infty} f(x) = \infty$. This is not the case. The function $\sin x$ alternates between -1 and 1 as x gets large; therefore, $\sin x$ has no limit at infinity.   ∎

We list some rules for limits that are helpful.

Theorem 1 (Some Rules for Limits at Infinity) *Let f and g be real-valued functions with common domain such that $\lim_{x \to \infty} f(x) = A$ and $\lim_{x \to \infty} g(x) = B$. Let c be a real number. Then*

$$\lim_{x \to \infty} (cf)(x) = cA,$$

$$\lim_{x \to \infty} (f + g)(x) = A + B,$$

$$\lim_{x \to \infty} (fg)(x) = AB, \text{ and}$$

$$\lim_{x \to \infty} (f/g)(x) = \frac{A}{B}, \text{ provided that } B \neq 0.$$

Of course we have similar rules if $\lim_{x \to -\infty} f(x) = A$ and $\lim_{x \to -\infty} g(x) = B$.

EXAMPLE 10: Let $f(x) = \dfrac{x^2 + 2x + 1}{3x^2 - 1}$ and $g(x) = \dfrac{2x^3 + 2x + 1}{5x^3 - x}$.

Then
$$\lim_{x \to \infty} f(x) = \frac{1}{3} \text{ and } \lim_{x \to \infty} g(x) = \frac{2}{5}.$$

Thus

$$\lim_{x \to \infty} \left(\frac{x^2 + 2x + 1}{3x^2 - 1} + \frac{2x^3 + 2x + 1}{5x^3 - x} \right)$$

$$= \lim_{x \to \infty} \frac{x^2 + 2x + 1}{3x^2 - 1} + \lim_{x \to \infty} \frac{2x^3 + 2x + 1}{5x^3 - x} = \frac{1}{3} + \frac{2}{5} = \frac{11}{15}.$$

And

$$\lim_{x \to \infty} \left(\left(\frac{x^2 + 2x + 1}{3x^2 - 1} \right) \left(\frac{2x^3 + 2x + 1}{5x^3 - x} \right) \right)$$

$$= \left(\lim_{x \to \infty} \frac{x^2 + 2x + 1}{3x^2 - 1} \right) \left(\lim_{x \to \infty} \frac{2x^3 + 2x + 1}{5x^3 - x} \right)$$

$$= \left(\frac{1}{3} \right) \left(\frac{2}{5} \right) = \frac{2}{15}. \qquad \blacksquare$$

We have the expected limits of composite functions.

Theorem 2 *Suppose that $g(x)$ is continuous at $x = L$.*

If $\displaystyle\lim_{x \to \infty} f(x) = L$, *then* $\displaystyle\lim_{x \to \infty} g(f(x)) = g(L)$.

If $\displaystyle\lim_{x \to -\infty} f(x) = L$, *then* $\displaystyle\lim_{x \to -\infty} g(f(x)) = g(L)$

EXAMPLE 11: Compute $\displaystyle\lim_{x \to \infty} \sqrt{4 + \frac{1}{x}}$.

SOLUTION: Let $g(x) = \sqrt{x}$ and $f(x) = 4 + \frac{1}{x}$. Then $\lim_{x \to \infty} f(x) = 4$. Thus, by Theorem 2, we have that $\lim_{x \to \infty} \sqrt{4 + \frac{1}{x}} = \lim_{x \to \infty} g(f(x)) = g(4) = 2.$ $\qquad \blacksquare$

EXAMPLE 12: Compute $\lim_{x \to \infty} \sin \left(\frac{1}{x} \right)$.

SOLUTION: Let $g(x) = \sin x$ and $f(x) = \frac{1}{x}$. Then $\lim_{x \to \infty} f(x) = 0$. Thus, by Theorem 2

$$\lim_{x \to \infty} \sin \left(\frac{1}{x} \right) = \lim_{x \to \infty} g(f(x)) = g(0) = 0. \qquad \blacksquare$$

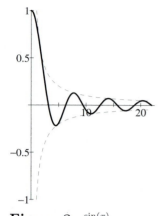

Figure 8. $\frac{\sin(x)}{x}$
is sandwiched between
$\frac{1}{x}$ *and* $-\frac{1}{x}$.

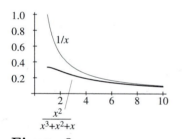

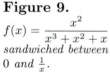

Figure 9.
$f(x) = \dfrac{x^2}{x^3 + x^2 + x}$ *is
sandwiched between*
0 *and* $\frac{1}{x}$.

Theorem 3 (The Sandwich Theorem) *Suppose that f, g, and h are functions such that:*

  (a) the domain of g is a subset of the domains of both f and h and that the domain of g is unbounded to the right;

  (b) either $f(x) \le g(x) \le h(x)$ for all x in the domain of g or $h(x) \le g(x) \le f(x)$ for all x in the domain of g; and

  (c) $\lim_{x \to \infty} f(x) = \lim_{x \to \infty} h(x) = L$.

Then $\lim_{x \to \infty} g(x) = L$.

The Sandwich Theorem does exactly what its name implies. If the function g is sandwiched between f and h, and if f and h have a common limit, then g must share that limit. The analogous theorem for limits as $x \to -\infty$ is also true.

EXAMPLE 13: $\lim_{x \to \infty} \frac{\sin x}{x} = 0$ since $-\frac{1}{x} \le \frac{\sin x}{x} \le \frac{1}{x}$ (as long as $x > 0$). See Figure 8. ∎

EXAMPLE 14: Let $f(x) = \dfrac{x^2}{x^3 + x^2 + x}$. Show that $\lim_{x \to \infty} f(x) = 0$.

SOLUTION: Notice that if $x > 0$, then $f(x) > 0$. We divide both the numerator and the denominator by x^3 to obtain

$$f(x) = \frac{\frac{1}{x}}{1 + \frac{1}{x} + \frac{1}{x^2}} < \frac{1}{x}.$$

Thus, as long as x is positive, $f(x)$ is sandwiched between $\frac{1}{x}$ and 0 as illustrated in Figure 9. Since $\lim_{x \to \infty} \frac{1}{x} = 0$, $\lim_{x \to \infty} f(x) = 0$. ∎

Limits of Sequences

Limits of sequences are of special importance. Recall from Chapter 1 that a sequence is a function with domain a terminal set of integers. That is, the function a is a sequence provided that there is a number N such that the domain of a is the set of all integers $n \ge N$. It seems reasonable to call N the *initial number in the domain of a*. Limits of sequences are of particular interest.

EXAMPLE 15: Let $a(n) = \frac{1}{n}$. Then

$$\lim_{n \to \infty} a(n) = \lim_{n \to \infty} \frac{1}{n} = 0.$$

∎

We state the following theorem but postpone its proof until a later section.

Theorem 4 *If* $-1 < r < 1$, *then* $\lim_{n \to \infty} r^n = 0$.

EXAMPLE 16: Let $a(n) = \left(\frac{1}{2}\right)^n$. Then $r = \frac{1}{2}$, and by Theorem 4, $\lim_{n \to \infty} \left(\frac{1}{2}\right)^n = 0$. If we list the first few terms, $\frac{1}{2}$, $\frac{1}{4}$, $\frac{1}{8}$, $\frac{1}{16}$, $\cdots$, we see that this agrees with our intuition. ∎

EXAMPLE 17: Evaluate $\lim_{n \to \infty} \frac{2^n}{n!}$.

SOLUTION: First, we expand the first few terms to see what is happening. $a(0) = 1$, $a(1) = 2$, $a(2) = \frac{(2)(2)}{(1)(2)}$, $a(3) = \frac{(2)(2)(2)}{(1)(2)(3)}$, $a(4) = \frac{(2)(2)(2)(2)}{(1)(2)(3)(4)}$, $\ldots$. We have the following information:

$$a(3) \quad = \quad \frac{(2)(2)(2)}{(1)(2)(3)} \quad = \quad 2\left(\frac{2}{3}\right)$$

$$a(4) \quad = \quad \frac{(2)(2)(2)(2)}{(1)(2)(3)(4)} \quad < \quad 2\left(\frac{2}{3}\right)^2$$

$$\vdots \qquad\qquad \vdots \qquad\qquad \vdots$$

$$a(n) \quad = \quad \frac{(2)(2)(2)\cdots(2)}{(1)(2)(3)\cdots(n)} \quad < \quad 2\left(\frac{2}{3}\right)^{n-2}.$$

By Theorem 4, $\lim_{n \to \infty} \left(\frac{2}{3}\right)^{n-2} = 0$. Thus, $2^n/n!$ is sandwiched between 0 and $2\left(\frac{2}{3}\right)^{n-2}$, which both converge to 0. It follows that $\lim_{n \to \infty} \frac{2^n}{n!} = 0$. ∎

In Exercise 38 you are asked to extend the argument for Example 17 to prove the following theorem.

Theorem 5 *If c is a number, then* $\lim_{n \to \infty} \frac{c^n}{n!} = 0$.

EXERCISES 5.1

Compute the limits in Exercises 1–4.

1. $\displaystyle \lim_{x \to 2^+} \frac{1}{x-2}$.

2. $\displaystyle \lim_{x \to 2^-} \frac{1}{x-2}$.

3. $\displaystyle \lim_{x \to 2^+} \frac{x^2-2}{(x-2)^3}$.

4. $\displaystyle \lim_{x \to 2^-} \frac{x^2-2}{(x-2)^3}$.

In Exercises 5–8, $\lim_{x \to 0^-} f(x) = \infty$ and $\lim_{x \to 0^-} g(x) = -\infty$. What can be said about the following:

5. $\displaystyle \lim_{x \to 0^-} f(x)g(x)$.

6. $\displaystyle \lim_{x \to 0^-} [f(x) + g(x)]$.

7. $\displaystyle \lim_{x \to 0^-} [f(x) - g(x)]$.

8. $\displaystyle \lim_{x \to 0^-} f(x)/g(x)$.

Compute the limits in Exercises 9–17.

9. $\lim\limits_{x \to \infty} \dfrac{x^3 + x^2 - 2}{4x^3 + \pi x^2 - 2}.$

10. $\lim\limits_{x \to -\infty} \dfrac{x^3 + x^2 - 2}{4x^3 + \pi x^2 - 2}.$

11. $\lim\limits_{x \to \infty} \dfrac{x^4 + 2x^2 - 10^{25}}{10^{-25}x^5 + 2x^2 - 10^{25}}.$

12. $\lim\limits_{x \to \infty} \dfrac{x^{10} - 2x^4 + 5{,}000x - 6}{5x^{10} - (\text{the national debt})}.$

13. $\lim\limits_{x \to \infty} \dfrac{1}{x} \cos x.$

14. $\lim\limits_{x \to -\infty} \dfrac{1}{x}(\sin x + x).$

15. $\lim\limits_{x \to \infty} 2^{-x}.$

16. $\lim\limits_{x \to \infty} \operatorname{Arctan} x.$

17. $\lim\limits_{x \to -\infty} \operatorname{Arctan} x.$

Find the horizontal and vertical asymptotes in Exercises 18–20.

18. $f(x) = \dfrac{(x-2)(x+3)}{(x+5)(6x-18)}.$

19. $g(x) = \tan x.$

20. $h(t) = \operatorname{Arctan} t.$

Find the limit, as $n \to \infty$, of the sequences in Exercises 21–28.

21. $a_n = \dfrac{n^3 + n^2 - 2}{4n^3 + \pi n^2 - 2}.$

22. $a_n = \dfrac{-n^3 + n^2 - 2}{-4n^3 + \pi n^2 - 2}.$

23. $a_n = \dfrac{n^4 + 2n^2 - 10^{25}}{10^{-25}n^5 + 2n^2 - 10^{25}}.$

24. $a_n = \dfrac{1}{n} \sin n.$

25. $a_n = \dfrac{1}{-n}(\sin(-n) - n).$

26. $a_n = 2^{-n}.$

27. $a_n = \operatorname{Arctan} n.$

28. $a_n = \operatorname{Arctan}(-n).$

In Exercises 29–32, $\lim_{x \to \infty} f(x) = \infty$ and $\lim_{x \to \infty} g(x) = -\infty$. What can be said about the following:

29. $\lim\limits_{x \to \infty} f(x)g(x).$

30. $\lim\limits_{x \to \infty} [f(x) + g(x)].$

31. $\lim\limits_{x \to \infty} [f(x) - g(x)].$

32. $\lim\limits_{x \to \infty} \dfrac{f(x)}{g(x)}.$

In Exercises 33–37, $\lim_{x \to \infty} f(x) = \infty$ and $\lim_{x \to \infty} g(x) = 0$. What can be said about the following:

33. $\lim\limits_{x \to \infty} f(x)g(x).$

34. $\lim\limits_{x \to \infty} [f(x) + g(x)].$

35. $\lim\limits_{x \to \infty} [f(x) - g(x)].$

36. $\lim\limits_{x \to \infty} \dfrac{f(x)}{g(x)}.$

37. $\lim\limits_{x \to \infty} \dfrac{g(x)}{f(x)}.$

38. Prove Theorem 5.

Find the limit, as $n \to \infty$, of the following sequences in Exercises 38–43.

38. $a_n = \dfrac{n-1}{n}.$

39. $a_n = \dfrac{n!}{(n+1)!}.$

40. $a_n = \dfrac{\sin n}{n}.$

41. $a_n = 3.5 + (0.999)^n.$

42. $a_n = \dfrac{n^2 + 5n - 7}{3n^2 - 6n + 1}.$

43. $a_n = \dfrac{n!}{n^n}.$

5.2   Increasing and Decreasing Functions

Without the tools of calculus, usually the only techniques available for sketching a function are either to use graphing calculators or to use brute force: plot a few points of the graph and then connect the points with a smooth curve (or surface). A graph produced by brute force will not accurately reflect any twists and turns the graph may

take between the few selected points. On the other hand, if the scale
is too large or too small, interesting portions of the graph may be
missed. The derivative of a function offers some sophisticated analyt-
ical tools, which will enable us to improve our sketches considerably.
In order to describe the "shape" of the graph of a real-valued function and to pick out technically important parts of its graph, we need
the tools of calculus. We first formalize the notions of increasing and
decreasing functions, which were introduced in Chapter 2.

Definition: Increasing and Decreasing Functions

Let f be a function and let S be a subset in the domain of f.

The function f is *increasing* on S if whenever $a < b$ are
numbers in S, then $f(a) < f(b)$.

The function f is *decreasing* on S if whenever $a < b$ are in
S, then $f(a) > f(b)$.

See Figures 1. a–c.

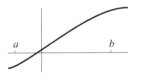

Figure 1.a *A function that
is increasing on* (a, b).

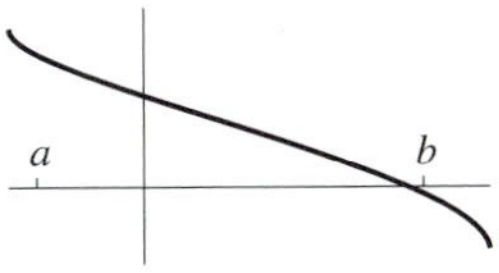

Figure 1.b *A function that
is decreasing on* (a, b).

Figure 1.c *A function that
is neither increasing nor
decreasing on* (a, b).

Thus, if we consider a "point" moving along the graph from left
to right, f is increasing if the point moves uphill and f is decreasing
if the point moves downhill. See Figure 2.

It is often useful to divide the domain of a function into segments so that the function is increasing on some of the segments and
decreasing on the others. The following theorem is our workhorse.

Theorem 1 *Suppose that $f'(x) > 0$ for $a < x < b$. Then f
is increasing on the interval (a, b). Similarly, if $f'(x) < 0$ for
$a < x < b$, then f is decreasing on the interval (a, b).*

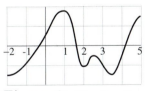

Figure 2. *The function
is increasing on the segments*
$(-2, 1), \left(2, \frac{5}{2}\right),$ *and* $\left(\frac{7}{2}, 5\right)$ *and
it is decreasing on the segments*
$(1, 2)$ *and* $\left(\frac{5}{2}, \frac{7}{2}\right)$.

We will postpone the proof of Theorem 1 until the end of the
section; however, **notice that Theorem 1 gives us no information if $f'(x) = 0$. Nor does it give us information at points
where f is not differentiable.**

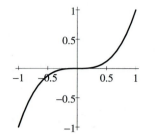

Figure 3.a $y = x^3$.

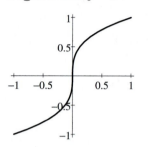

Figure 3.b $y = x^{1/3}$.

EXAMPLE 1: The function $y = x^3$ is clearly increasing on its domain (the real line), yet $y'(0) = 0$. (See Figure 3.a.) The function $y = x^{1/3}$ is increasing but it is not differentiable at $x = 0$. (See Figure 3.b.)   ∎

Theorem 2 *Suppose that f is differentiable and that f' is continuous.*

If $f'(x_0) > 0$, then f is increasing on a segment containing x_0.

If $f'(x_0) < 0$, then f is decreasing on a segment containing x_0.

Proof: Suppose that $f'(x_0) > 0$. Since f' is continuous, by Theorem 1 of Section 3.6, there is a number $\delta > 0$ such that if $|x - x_0| < \delta$, then $f'(x) > 0$. Let $a = x_0 - \delta$ and $b = x_0 + \delta$. Then (a, b) is a segment containing x_0 such that $f'(x) > 0$ for all x in (a, b). By Theorem 1, f is increasing on (a, b).

The argument for the case that $f'(x_0) < 0$ is similar.   ∎

Theorem 2 leads us to the idea of a function increasing or decreasing at a point.

Definition: Increasing/Decreasing at a Point

Let x_0 be a point in the domain of the function f.

f is increasing at x_0 if there is a segment containing x_0 in its interior such that f is increasing on that segment.

f is decreasing at x_0 if there is a segment containing x_0 in its interior such that f is decreasing on that segment.

By Theorem 2, we know that if f' is continuous and $f'(x) > 0$, then f is increasing at x. Similarly, if f' is continuous and $f'(x) < 0$, then f is decreasing at x. Like Theorem 1, Theorem 2 gives us no information if $f'(x_0) = 0$.

EXAMPLE 2: The graph of $f(x) = x^3$ is increasing at $x_0 = 0$ even though $f'(0) = 0$. The graph of $g(x) = x^2$ is a parabola and $g'(0) = 0$. The function g is neither increasing nor decreasing at $x_0 = 0$.   ∎

EXAMPLE 3: Let $f(x) = \frac{x^3}{3} + x^2 + 2x + 3$. Determine where f is increasing and where f is decreasing.

SOLUTION: The derivative $f'(x) = x^2 + 2x + 2$. We determine where f' is positive (hence, where f is increasing) and where f' is negative (hence, where f is decreasing). Since $f'(x) = (x+1)^2 + 1$, we see that f' is greater than 0 everywhere. Thus f is increasing on the entire real line. This agrees with the sketch of f provided in Figure 4. ∎

There is a generic procedure to find where a function is increasing and where it is decreasing.

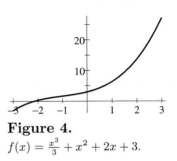

Figure 4.

$f(x) = \frac{x^3}{3} + x^2 + 2x + 3$.

Steps for Finding Where a Function f is Increasing and Decreasing

1. Find all points where either $f'(x) = 0$ or where f' is undefined.

2. Consider each segment of the form (a, b), $(-\infty, a)$, or (b, ∞), where

   - a and b are numbers found in Step 1, and

   - there are no numbers from Step 1 in the interior of the segment.

   We know that f' does not change sign on any of the segments found in this step.

3. From each segment from Step 2, select a convenient sample point c and evaluate $f'(c)$.

   - If $f'(c) > 0$, then we know that $f'(x) > 0$ for every x in the segment, and f is increasing on the segment.

   - If $f'(c) < 0$, then we know that $f'(x) < 0$ for every x in the segment, and f is decreasing on the segment.

EXAMPLE 4: Determine where $f(x) = \frac{x^6}{6} - \frac{5x^4}{4} + 2x^2 + 6$ is increasing and where it is decreasing.

SOLUTION: The derivative of f is $f'(x) = x^5 - 5x^3 + 4x = x(x^2 - 4)(x^2 - 1)$. The functions f and f' are sketched in Figures 5.a and 5.b. The function f' is defined on all the real numbers, so the endpoints of the segments from Step 2 are the numbers where $f'(x) = 0$, namely, $x = 0, \pm 1$ and ± 2. We must examine the behavior of f

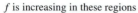

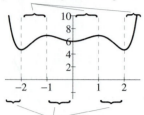

f is decreasing in these regions

Figure 5.a *The graph of*
$$f(x) = \frac{x^6}{6} - \frac{5x^4}{4} + 2x^2 + 6.$$

f′ is positive in these regions

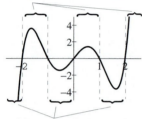

f′ is negative in these regions

Figure 5.b *The graph of*
$f'(x) = x^5 - 5x^3 + 4x.$

on the intervals $(-\infty, -2), (-2, -1), (-1, 0), (0, 1), (1, 2),$ and $(2, \infty)$. Following Step 3, we evaluate f' at a point in each interval.

$f'(-3) < 0$. Thus f' is negative on the segment $(-\infty, -2)$, and f is decreasing there.

$f'(-1.5) > 0$. Thus f' is positive on the segment $(-2, -1)$, and f is increasing there.

$f'(-0.5) < 0$. Thus f' is negative on the segment $(-1, 0)$, and f is decreasing there.

$f'(0.5) > 0$. Thus f' is positive on the segment $(0, 1)$, and f is increasing there.

$f'(1.5) < 0$. Thus f' is negative on the segment $(1, 2)$, and f is decreasing there.

$f'(3) > 0$. Thus f' is positive on the segment $(2, \infty)$, and f is increasing there.

See Figures 5.a and 5.b.

EXAMPLE 5: Sketch the graph of $f(x) = \dfrac{x^2 + 1}{x - 1}$.

SOLUTION: First, note that f has a vertical asymptote at $x = 1$. (See Figure 6.a.)

We now employ the derivative of f to determine where f is increasing and where f is decreasing.

$$f'(x) = \frac{(x - 1)2x - (x^2 + 1)}{(x - 1)^2}$$

$$= \frac{x^2 - 2x - 1}{(x - 1)^2}.$$

This derivative is zero at $x = 1 + \sqrt{2}$ and $x = 1 - \sqrt{2}$, and f' is not defined at $x = 1$. Thus the only values where f' can change signs are $x = 1 + \sqrt{2}, x = 1 - \sqrt{2}$, and $x = 1$. Note that f' is continuous on each of the segments $(-\infty, 1 - \sqrt{2}), (1 - \sqrt{2}, 1), (1, 1 + \sqrt{2})$, and $(1 + \sqrt{2}, \infty)$. Since f' is not equal to zero on any of these segments, f' cannot change sign on any of them. To determine where f' is positive (hence, where f is increasing) and where f' is negative (hence, where f is decreasing), we choose a number in each interval and compute the derivative at that number.

$f'(-2) > 0$. Thus f' is positive on the segment $(-\infty, 1 - \sqrt{2})$, and f is increasing there.

$f'(0) < 0$. Thus f' is negative on the segment $(1 - \sqrt{2}, 1)$, and f is decreasing there.

$f'(2) < 0$. Thus f' is negative on the segment $(1, 1 + \sqrt{2})$, and f is decreasing there.

$f'(3) > 0$. Thus f' is positive on the segment $(1, \infty)$, and f is increasing there.

See Figure 6.a.

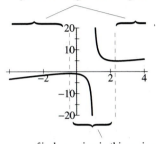
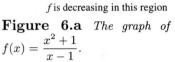

f is increasing in these regions

f is decreasing in this region

Figure 6.a *The graph of* $f(x) = \dfrac{x^2 + 1}{x - 1}$.

The next step in graphing f is to determine (if possible) where f is positive and where f is negative. Notice that f is never zero (the numerator is always positive), so the only place that f can change signs is at $x = 1$ (where f is not defined). Since $f(0) = -1$, f is negative everywhere on the segment $(-\infty, 1)$. Since $f(2) = 5$, f is positive everywhere on the segment $(1, \infty)$. See Figure 6.b.

Finally, we select some points to plot. Interesting things seem to happen at $x = 1 - \sqrt{2} \approx -0.414$, where f changes from an increasing function to a decreasing function, at $x = 1 + \sqrt{2} \approx 2.414$, where f changes from decreasing to increasing, and at $x = 1$, where f is not defined. $f(-0.414) \approx -0.828$ and $f(2.414) \approx 4.828$. We pick some other convenient numbers to plot:

x	0	2	3	-1	-2	-3
$f(x)$	-1	5	5	-1	$-\frac{5}{3}$	$-\frac{5}{2}$

■

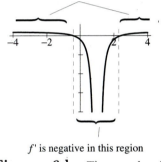
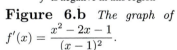

f' is positive in these regions

f' is negative in this region

Figure 6.b *The graph of* $f'(x) = \dfrac{x^2 - 2x - 1}{(x - 1)^2}$.

Critical Points, Relative and Absolute Extrema

Thus far, our strategy in graphing functions has been to find the intervals on which the function is increasing and those on which it is decreasing. We then plot the points where the function changes from increasing to decreasing and decreasing to increasing. These points are precisely where the derivative f' equals zero or is undefined. Notice that if a function is increasing and then at some point starts to decrease, that point denotes the "top of a hill," or cusp. Similarly, if the function is decreasing and at some point starts to increase, then that point denotes the "bottom of a valley," or cusp. At each of these points, the tangent line will be horizontal, vertical, or undefined. The following definition makes these concepts precise.

Definition: Critical Points, Relative and Absolute Extrema

- Let c be a point in the domain of f. The number c is a *critical point* of f if either $f'(c) = 0$ or c is not in the domain of f'.

- The function f attains a *relative maximum* at c if there is a segment (a, b) containing c and lying in the domain of f such that if x is in (a, b), then $f(x) \leq f(c)$.

- The function f attains a *relative minimum*[3] at c if there is a segment (a, b) containing c and lying in the domain of f such that if x is in (a, b), then $f(x) \geq f(c)$.

- The function f attains a *relative extremum* at c if f attains either a relative maximum or a relative minimum there.[4]

- Suppose that the set A is a subset of the domain of f. The number M is the *(absolute) maximum* for f on the set A if M is in the image of f and $f(x) \leq M$ for all x in A. [5]

- The number m is the *(absolute) minimum* for f on the set A if m is in the image of f and $f(x) \geq m$ for all x in A.[6]

- The maximum and minimum values for f are called the *(absolute) extrema*.

---

[3]The plural forms for maximum, minimum, and extremum are maxima, minima, and extrema.

[4]Notice that if f attains a relative extremum at c, then c is an interior point in the domain of f.

[5]The point c does not have to be in the interior of the domain of f in order for f to attain an absolute maximum there. To find an absolute maximum, we must evaluate f at each point in its domain that is not an interior point as well as check out the relative maxima.

[6]As with an absolute maximum, f can achieve its absolute minimum at c without c being in the interior of the domain of f.

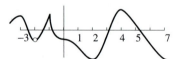

Figure 7. *The function f attains relative maxima at $x = -3, -1$, and 4, and it attains a relative minimum at $x = 2$. The values $x = -3, -1, 0, 2,$ and 4 are the critical points. The value -2 is not a critical point since f is not defined at -2. The number 0 is a critical point, but f does not attain a relative extremum there. It is not clear from this figure whether f does or does not have any absolute extrema.*

Notice that if f is constant on a segment, then f attains both a relative maximum and a relative minimum at each point in that segment! Figure 7 illustrates the ideas defined above.

Notice also that the critical point "0" in Figure 7 illustrates the important fact that a function can have a critical point without attaining a relative extremum there. However, by Theorem 3, if f has a derivative at c and if $f(c)$ is an extremum for f, then $f'(c) = 0$.

Theorem 3 *If f is differentiable at c and f attains a relative extremum at c, then $f'(c) = 0$. Recall that if f is differentiable at c, then c is in the interior of the domain of f.*

Proof: Suppose that the function f is differentiable[7] at c and that f attains a relative maximum at c. Then there are numbers a and b with $a < c < b$ so that if x is in (a, c) or (c, b), then $f(x) \le f(c)$. Thus, if x is in (a, c), then

$$\frac{f(x) - f(c)}{x - c} \ge 0 \text{ (since } x < c\text{).}$$

It follows that

$$\lim_{x \to c} \frac{f(x) - f(c)}{x - c} = f'(c) \ge 0. \tag{1}$$

But, if x is in (c, b), then

$$\frac{f(x) - f(c)}{x - c} \le 0 \text{ (since } x > c\text{).}$$

So it also follows that

$$\lim_{x \to c} \frac{f(x) - f(c)}{x - c} = f'(c) \le 0. \tag{2}$$

The only way that both Equations (1) and (2) can hold at the same time is for $f'(c) = 0$. In a similar fashion, we can argue that if f attains a relative minimum at c, then $f'(c) = 0$. ∎

Theorem 3 gives us an important tool for finding relative extrema of a differentiable function.

Steps for Finding Relative Extrema

---

1. Find all critical points.

2. For each critical point c, determine if it yields a relative maximum or a relative minimum:

---

[7]Since f is differentiable at c, we know that there are numbers a and b such that c is in (a, b) and (a, b) is a subset of the domain of f.

2. *Continued*

(a) If there is a segment (a, b) containing c so that $f'(x) \geq 0$ on (a, c) and $f'(x) \leq 0$ on (c, b), then f attains a relative maximum at c.

(b) If there is a segment (a, b) containing c so that $f'(x) \geq 0$ on (a, c) and $f'(x) \leq 0$ on (c, b), then f attains a relative maximum at c.

(c) If there is a segment (a, b) containing c so that $f'(x) \leq 0$ on (a, c) and $f'(x) \geq 0$ on (c, b), then f attains a relative minimum at c.

WARNING: It is possible for $f'(c) = 0$ without f attaining a relative extremum there. (See the comment in Figure 7 and Exercise 30.) Figures 8.a and 8.b show the relative extrema from Examples 4 and 5.

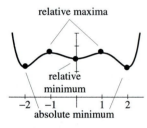

Figure 8.a *From Example 4, $f(x) = \frac{x^6}{6} - 5\frac{x^4}{4} + 2x^2 + 6$ has critical points at -2, -1, 0, 1, and 2. It attains relative extrema at each of these points, and it attains its absolute minimum at $x = -2$ and $x = 2$. It does not have an absolute maximum.*

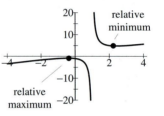

Figure 8.b *From Example 5, $f(x) = \dfrac{x^2 + 1}{x - 1}$ has critical points at $1 - \sqrt{2}$ and $1 + \sqrt{2}$. It attains relative extrema at each of these points. It does not have absolute extrema.*

Theorem 3 told us that if a differentiable function f has a relative extremum at c and c is an interior point in the domain of f, then $f'(c) = 0$. Is it true that all differentiable functions have extrema? Not necessarily! As we see in Figure 8.a, the graph of $f(x) = \frac{x^6}{6} - \frac{5x^4}{4} + 2x^2 + 6$ does have an absolute minimum but no absolute maximum. The graph of $f \colon \mathbb{R} \to \mathbb{R}$ given by $f(x) = x$ has no absolute extrema at all. However, as the next theorem states, if f is continuous on a closed interval, then f must have extrema on that interval.

Theorem 4 *If f is continuous on the interval $[a, b]$, then there are numbers c and d in $[a, b]$ such that $f(c)$ is the maximum value for f in $[a, b]$ and $f(d)$ is the minimum value for f in $[a, b]$.*

The proof of this theorem requires the completeness axiom, and it is left for a more advanced course.

EXAMPLE 6: Let $f(x) = x^2$. Find the maximum and minimum values for f on the interval $[-1, 2]$.

SOLUTION: This example is sufficiently simple that we do not need to resort to the techniques of calculus. The graph of f is the parabola sketched in Figure 9. $f(0) = 0$ is clearly its minimum, and $f(2) = 4$ is its maximum. ∎

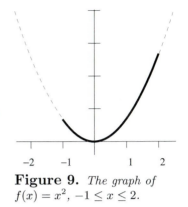

Figure 9. *The graph of* $f(x) = x^2$, $-1 \le x \le 2$.

Example 6 illustrates the fact that if f is differentiable on the interval $[a, b]$, then there are two places to look for absolute extrema: the values in $[a, b]$ where f attains its relative extrema, and the endpoints of the interval, a and b.

Steps for Finding Absolute Extrema of a Differentiable Function

1. Find all critical points.

2. List the endpoints of the domain (if any).

3. Evaluate the function at the critical points and endpoints.

4. If the function has an absolute maximum, then the largest value is the absolute maximum. If the function has an absolute minimum, then the smallest value from Step 3 is the absolute minimum.

EXAMPLE 7: The function f of Example 5 does not attain any absolute minima on the interval $[-1, 1]$. Note that this does not violate Theorem 4, since $[-1, 1]$ is not a subset of the domain of f. ∎

The following table outlines the general procedure for using the tools we have discussed so far to graph real valued functions.

Steps for Graphing Functions

1. Find the vertical and horizontal asymptotes.

2. Determine where the function is positive and where it is negative.

3. Determine where the function is increasing and where it is decreasing.

4. Find any relative and absolute extrema and plot these values.

5. If necessary, plot a few more points.

6. Now sketch the graph.

EXAMPLE 8: Let $f(x) = \cos^2 x + \sin x$, $0 \le x \le \pi$. Determine where f is positive and where f is negative, where f is increasing and where f is decreasing, and locate the relative and absolute extrema. Sketch the graph of f.

SOLUTION: To determine where f is positive and where f is negative, we first determine where f can change sign. Since f is continuous, this will be precisely where $f(x) = 0$. Thus we must solve the equation

$$\cos^2 x + \sin x = 0,$$

which is equivalent to

$$\sin^2 x - \sin x - 1 = 0.$$

The solutions for this quadratic equation are

$$\sin x = \frac{1 \pm \sqrt{5}}{2} \approx 0.5 \pm 1.12.$$

Since $\sin x$ is never greater than 1, the only possibility is

$$\sin x \approx -0.62.$$

But x was restricted to the interval $[0, \pi]$, where $\sin x$ is positive. Thus $f(x)$ is not 0 in the interval $[0, \pi]$, so f does not change sign. Since $f\left(\frac{\pi}{2}\right) = 1$, we see that f is always positive on $[0, \pi]$.

To determine where f is increasing and where f is decreasing, we determine where f' is positive and where f' is negative.

$$
\begin{aligned}
f'(x) &= (2\cos x)(-\sin x) + \cos x \\
&= (1 - 2\sin x)\cos x.
\end{aligned}
$$

Thus, $f'(x) = 0$ at $x = \frac{\pi}{2}$, $x = \frac{\pi}{6}$, and $x = \frac{5\pi}{6}$. In order to determine where f is increasing and where f is decreasing, we need only determine the sign of f' on the segments $\left(0, \frac{\pi}{6}\right)$, $\left(\frac{\pi}{6}, \frac{\pi}{2}\right)$, $\left(\frac{\pi}{2}, \frac{5\pi}{6}\right)$, and $\left(\frac{5\pi}{6}, \pi\right)$.

For the segment $\left(0, \frac{\pi}{6}\right)$: $\cos x$ is positive here and $\sin x < \frac{1}{2}$. Thus $(1 - 2\sin x)$ is positive, and f' is positive on the segment $\left(0, \frac{\pi}{6}\right)$. Thus f is increasing on $\left(0, \frac{\pi}{6}\right)$.

For the segment $\left(\frac{\pi}{6}, \frac{\pi}{2}\right)$: $f'\left(\frac{\pi}{4}\right) = \left(\frac{1}{\sqrt{2}}\right)\left(1 - \frac{2}{\sqrt{2}}\right) < 0$. Thus $f' < 0$ and f is decreasing on $\left(\frac{\pi}{6}, \frac{\pi}{2}\right)$.

For the segment $\left(\frac{\pi}{2}, \frac{5\pi}{6}\right)$: $f' > 0$, so f is increasing on $\left(\frac{\pi}{2}, \frac{5\pi}{6}\right)$.

For the segment $\left(\frac{5\pi}{6}, \pi\right)$: f' is negative on $\left(\frac{5\pi}{6}, \pi\right)$, so f is decreasing on $\left(\frac{5\pi}{6}, \pi\right)$.

We shade in these regions in Figure 10.a. With this information, it is apparent where f attains relative extrema.

$f'\left(\frac{\pi}{6}\right) = 0$, f is increasing on $\left(0, \frac{\pi}{6}\right)$ and decreasing on $\left(\frac{\pi}{6}, \frac{\pi}{2}\right)$. Therefore, f attains a relative maximum at $\frac{\pi}{6}$.

$f'\left(\frac{\pi}{2}\right) = 0$, f is decreasing on $\left(\frac{\pi}{6}, \frac{\pi}{2}\right)$ and increasing on $\left(\frac{\pi}{2}, \frac{5\pi}{6}\right)$. Therefore, f attains a relative minimum at $\frac{\pi}{2}$.

$f'\left(\frac{5\pi}{6}\right) = 0$, f is increasing on $\left(\frac{\pi}{2}, \frac{5\pi}{6}\right)$ and decreasing on $\left(\frac{5\pi}{6}, \pi\right)$. Therefore, f attains a relative maximum at $\frac{5\pi}{6}$.

Since $f(0) = f\left(\frac{\pi}{2}\right) = f(\pi) = 1$, the minimum for f is 1. Similarly, $f\left(\frac{\pi}{6}\right) = f\left(\frac{5\pi}{6}\right) = \frac{3}{4} + \frac{1}{2} = \frac{5}{4}$, so $\frac{5}{4}$ is the maximum for f.

To obtain the graph of f (see Figure 10.b), we plot the relative extrema of f and the endpoints of f, and use the information about where f is increasing and where f is decreasing to fill in the graph. ∎

EXAMPLE 9: In Figure 11, we have a sketch of $f'(x)$. From the figure, we see that the critical points for f are the end points of the domain $x = -3$ and $x = 6$ and the points where $f'(x) = 0$, $x = -2$, $x = 1$, $x = 3$ and $x = 5$. f is decreasing on the segments $(-3, -2)$ and $(1, 5)$. f is increasing on the segments $(-2, 1)$ and $(5, 6)$. Since f is decreasing on $(-3, -2)$, $f(-3)$ is a local maximum. The function f is decreasing to the left of $x = -2$ and increasing to the right of $x = -2$. Therefore, $f(-2)$ is a local minimum. Similarly, $f(5)$ is a local minimum. Since f is increasing to the left of $x = 1$ and decreasing to the right of $x = 1$, $f(1)$ is a local maximum. Although

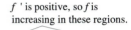

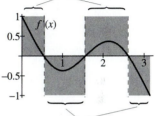

Figure 10.a *Regions where $f'(x)$ is positive and where it is negative indicate where the graph is increasing and where it is decreasing.*

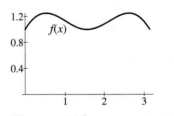

Figure 10.b *The graph of f.*

$f'(3) = 0$, f is decreasing both to the right and to the left of $x = 3$. Therefore, f is decreasing at $x = 3$ and f attains neither a local minimum or maximum there. ∎

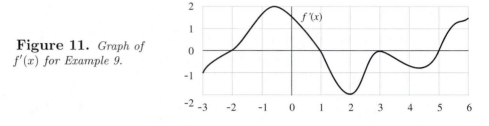

Figure 11. *Graph of* $f'(x)$ *for Example 9.*

The Mean Value Theorem and a Proof of Theorem 1

The Intermediate Value Theorem gives us information about function values between two points on a graph. Similarly, the following theorem gives us information about the derivative between two points on a graph.

Theorem 5 (Rolle's Theorem) *Let f be a function from a subset of $\mathbb{R}$ into $\mathbb{R}$ that is continuous on the interval $[a, b]$ and such that f' is defined at each point of (a, b). If $f(a) = 0 = f(b)$, then there is a point c with $a < c < b$ such that $f'(c) = 0$. See Figure 12.*

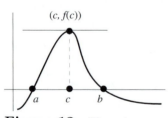

Figure 12. *There is a horizontal tangent line at* $(c, f(c))$.

Proof: Consider the graph of f. Either $f(x) = 0$ for all x in the interval $[a, b]$, or not. In the first case, $f'(x) = 0$ for all x in (a, b), so we can choose any such value for c. In the second case, where f is zero at the endpoints but takes on nonzero values on (a, b), we know that f must attain either a maximum or minimum on (a, b). We can use Theorem 4 to pick a point c that yields a maximum or minimum for f, and note that Theorem 3 shows that $f'(c) = 0$. ∎

The essence of the above theorem is that since the slope of the line from $(a, f(a))$ to $(b, f(b))$ is zero, then there must exist some point c between a and b such that the graph of f at c has a horizontal tangent line. We generalize this notion in the following theorem.

Theorem 6 (The Mean Value Theorem) *If f is a function from a subset of $\mathbb{R}$ into $\mathbb{R}$ that is continuous on the interval $[a, b]$, and if f' is defined at each point of (a, b), then there is a point c of (a, b) so that $f'(c) = \dfrac{f(b) - f(a)}{b - a}$.*

Proof: We construct an auxiliary function that translates f vertically by shifting the point $(a, f(a))$ to $(a, 0)$ and then linearly shifts the resulting graph so that $(b, f(b))$ moves to the point $(b, 0)$. Define

$$h(x) = f(x) - f(a) - \frac{f(b) - f(a)}{b - a}(x - a).$$

The function h is continuous on $[a, b]$, and h' is defined at each point of (a, b). Notice also that $h(a) = 0 = h(b)$. Thus we can employ Rolle's Theorem to conclude that there must be a point c between a and b such that $h'(c) = 0$. However,

$$h'(x) = f'(x) - \frac{f(b) - f(a)}{b - a}.$$

Thus

$$h'(c) = 0 = f'(c) - \frac{f(b) - f(a)}{b - a}.$$

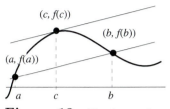

Figure 13. *The tangent line at $(c, f(c))$ has the same slope as the line passing through $(a, f(a))$ and $(b, f(b))$.*

Solving this equation for $f'(c)$ yields $f'(c) = \frac{f(b) - f(a)}{b - a}$, as desired. ∎

Geometrically, the Mean Value Theorem states that there is a point c between a and b so that the slope of the line tangent to f at $(c, f(c))$ is the same as the slope of the line passing through $(a, f(a))$ and $(b, f(b))$. See Figure 13.

We now return to the proof of Theorem 1 that was promised at the beginning of this section. In Theorem 2 we proved that a function was increasing or decreasing at a point by examining the sign of the derivative at that point. We now show that if a function f is differentiable on an interval (a, b) and $f'(x) > 0$ for all x in (a, b), then f is increasing on (a, b). Similarly, if $f'(x) < 0$ for all x in (a, b), then f is decreasing on (a, b).

We now prove Theorem 1.

Theorem 1 *Let f be a differentiable function on an interval (a, b) such that $f'(x) \neq 0$ for all x in (a, b). Then f is increasing (respectively decreasing) on (a, b) if $f'(x) > 0$ (respectively $f'(x) < 0$) for all x in (a, b).*

Proof: We prove the case where $f'(x) > 0$ and leave the other case to the reader. Let x and y be in (a, b), where $x < y$. By the Mean Value Theorem, we have that:

$$\frac{f(y) - f(x)}{y - x} = f'(c)$$

for some c with $x < c < y$. Since $f'(c) > 0$, this implies that:

$$\frac{f(y) - f(x)}{y - x} > 0.$$

Since $y - x > 0$, we see that

$$f(y) - f(x) > 0,$$

which is equivalent to

$$f(y) > f(x).$$

Thus f is increasing on (a, b).   ∎

EXERCISES 5.2

In Exercises 1–4, you are given the graph of $f'(x)$.

  a. Locate the critical points of f.

  b. Determine the intervals where f is increasing and where f is decreasing.

  c. Locate the points in the domain where f attains relative extrema.

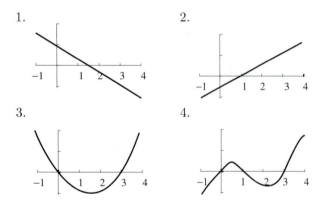

1.

2.

3.

4.

In Exercises 5–24 find all vertical and horizontal asymptotes, determine where the function is increasing and where it is decreasing, and find all relative extrema. Sketch the graph.

  5. $f(x) = x^2 + 1$.

  6. $f(x) = \frac{x^3}{3} + x^2 + x$.

  7. $g(x) = x^2 - 1$.

  8. $r(x) = (x + 1)(x - 1)(x + 2)$.

  9. $f(x) = \dfrac{x + 6}{x - 1}$.

 10. $g(x) = \dfrac{1 - x^4}{x}$.

 11. $s(x) = \dfrac{x}{1 - x^4}$.

12. $f(x) = \dfrac{1 - x}{1 - x^2}$.

13. $f(x) = \dfrac{(6 - x)(1 + x)}{x - 1}$.

14. $f(x) = \dfrac{(6 - x)(1 + x)}{x^2 - 1}$.

15. $f(x) = \dfrac{x^2 + 2x + 1}{x - 1}$.

16. $f(x) = \sin x$.

17. $f(x) = \tan x$.

18. $f(x) = x + \sin x$.

19. $f(x) = x - \cos x$.

20. $f(x) = 3x + \sin x$.

21. $f(x) = \sin x + \cos x$.

22. $f(x) = (\sin x + \cos x)^2$.

23. $f(x) = (\sin x + \cos x)^2 - 1$.

24. $f(x) = \sqrt{3} \sin x + \cos x$.

In Exercises 25–29, determine whether the function attains absolute extrema on the interval $[\alpha, \beta]$, and if so, find them.

25. $f(x) = x^2 + 1$. Refer to your sketch for Exercise 5.

  a. $[\alpha, \beta] = [-2, 2]$ b. $[\alpha, \beta] = [0, 4]$

26. $f(x) = \frac{x^3}{3} + x^2 + x$. Refer to your sketch for Exercise 6.

  a. $[\alpha, \beta] = [-2, 2]$ b. $[\alpha, \beta] = [0, 4]$

27. $g(x) = \dfrac{1 - x^4}{x}$. Refer to your sketch for Exercise 10.

  a. $[\alpha, \beta] = [-2, 2]$ b. $[\alpha, \beta] = [1, 4]$

28. $f(x) = \dfrac{(6-x)(1+x)}{x-1}$. Refer to your sketch for Exercise 13.
    a. $[\alpha, \beta] = [-2, 2]$ b. $[\alpha, \beta] = [-4, 0]$

29. $f(x) = x - \cos x$. Refer to your sketch for Exercise 19.
    a. $[\alpha, \beta] = [0, \pi]$ b. $[\alpha, \beta] = [-\pi, 2\pi]$

30. Find a function f so that $f'(0) = 0$, but $f(0)$ is not a relative extremum.

In Exercises 31–33, verify the Mean Value Theorem for the functions on the indicated interval $[a, b]$; i.e., find a value c in $[a, b]$ so that $f'(c) = \dfrac{f(b) - f(a)}{b - a}$.

31. $f(x) = x^3 + x$; $a = 0, b = 2$.

32. $f(x) = \sin x$; $a = 0, b = \pi$.

33. $f(x) = x^3 + 2x^2 + 5x + 1$; $a = -1, b = 1$.

34. Let $x(t) = t \sin t$.

    a. Argue that $x(t)$ has exactly one critical point in each segment $\left(-\frac{\pi}{2} + n\pi, \frac{\pi}{2} + n\pi\right)$.

    b. Use Newton's method to approximate the critical point in each of the segments $\left(-\frac{\pi}{2} + n\pi, \frac{\pi}{2} + n\pi\right)$, $n = 1$, 2, and 3 with an accuracy of four decimals.

    c. Sketch the graph of $x(t)$.

35. Let $x(t) = t \cos t$.

    a. Argue that $x(t)$ has exactly one critical point in each segment $\left(-\frac{\pi}{2} + n\pi, \frac{\pi}{2} + n\pi\right)$.

    b. Use Newton's method to approximate the critical point in each of the segments $\left(-\frac{\pi}{2} + n\pi, \frac{\pi}{2} + n\pi\right)$, $n = 1$, 2, and 3 with an accuracy of four decimals.

    c. Sketch the graph of $x(t)$.

36. Match each function sketched in the first column of graphs with its derivative in the second column.

Graph of function a.

Graph of derivative a.

Graph of function b.

Graph of derivative b.

Graph of function c.

Graph of derivative c.

5.3 Increasing and Decreasing Curves in the Plane

Suppose that we have a parametrized curve or path $\mathcal{C}$ in the plane; that is, $\mathcal{C}$ is the image of a vector-valued function defined on a subset of $\mathbb{R}$. The task in this section is to use knowledge of the parametrization of $\mathcal{C}$ to determine the geometry of the curve. An analogy might be to think of $\mathcal{C}$ as a railroad track. A parametrization $\vec{r}$ for $\mathcal{C}$ might give the position of a train on the track as a function of time. The geometry of the track will not depend on variations in speed as the train is moving on the track. However, we can learn about the track by watching the train move.

First of all, recall that a parametrized curve $\mathcal{C}$ need not be a function of x. However, we can employ the ideas developed in the previous section by breaking $\mathcal{C}$ into pieces, each of which is a function of x. We call on the ellipse for an example.

EXAMPLE 1: Assume that a and b are positive numbers. Let $\mathcal{C}$ be the graph of $\frac{x^2}{a^2} + \frac{y^2}{b^2} = 1$. While $\mathcal{C}$ is not a function, it is the union of two functions, the top half and the bottom half:

$$f_1(x) = b\sqrt{1 - x^2/a^2} \quad \text{and} \quad f_2(x) = -b\sqrt{1 - x^2/a^2}.$$

It is straightforward to see that f_1 is increasing when $-a \leq x < 0$, and it is decreasing when $0 < x \leq a$. In the same way, f_2 is decreasing when $-a \leq x < 0$, and it is increasing when $0 < x \leq a$. Neither f_1 nor f_2 is increasing or decreasing at $x = 0$.

A parametrization for $\mathcal{C}$ is

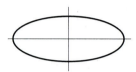

Figure   1.a   *The   ellipse* $x^2/a^2 + y^2/b^2 = 1.$

$$\vec{r}(\theta) = \begin{pmatrix} x(\theta) \\ y(\theta) \end{pmatrix} = \begin{pmatrix} a\cos(\theta) \\ b\sin(\theta) \end{pmatrix}.$$

Notice that $x'(\theta) = -a\sin(\theta) = 0$ at $\theta = 0, \pm\pi, \pm 2\pi, \dots$. If n is an integer, then the image of $\vec{r}(\theta)$, $n\pi \leq \theta \leq (n+1)\pi$ is a function (of x). If n is even, then the image of $\vec{r}(\theta)$, $n\pi \leq \theta \leq (n+1)\pi$ is the top half of the ellipse, and if n is odd, then the image of $\vec{r}(\theta)$, $n\pi \leq \theta \leq (n+1)\pi$ is the bottom half of the ellipse. The parts of $\mathcal{C}$ that are functions correspond exactly to the images of the segments of the time line where $x'(\theta) \neq 0$. We have a general theorem.

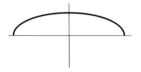

Figure 1.b *The top half of the ellipse.*

> **Theorem 1** *Let $\vec{r}(t) = (x(t), y(t))$ be a function that is differentiable on the interval (a, b). If $x'(t) \neq 0$ for $a < t < b$, then the image of $\vec{r}(t)$, $a \leq t \leq b$, is a function (of x). That is, no vertical line passes through the image of $\vec{r}(t)$, $a \leq t \leq b$ at more than one point.*

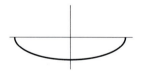

Figure 1.c *The bottom half of the ellipse.*

Proof: Since $x'(t) \neq 0$ for $a < t < b$, x is either an increasing function of t on $[a, b]$ or a decreasing function on $[a, b]$. In either case, the function $x(t)$ is one-to-one and has an inverse on $[a, b]$. Let us call that inverse T. Recall that for x to have an inverse T we must have $T(x(t)) = t$ and $x(T(c)) = c$. Let t be in $[a, b]$, and let $c = x(t)$ so that $T(c) = t$. Then $(x(t), y(t)) = (c, y(T(c)))$ is the only point in the image of $\vec{r}(t)$, $a \leq t \leq b$, with first coordinate c.   ∎

EXAMPLE 2: For an example that is a bit more complicated than the ellipse of Example 1, suppose that a wheel with radius 1, lying in the xy–plane, is rotating counterclockwise about its center at a rate of one radian/sec (1 rotation every 2π sec). Suppose further that the center of the wheel is moving in a straight line away from the origin at a rate of $\frac{1}{2}$ units/sec. (Recall Section 2.7.) Place the coordinate system so that the center is moving along the x–axis, and let $\mathcal{C}$ be the path of a particle on the rim of the wheel. Then a parametrization for $\mathcal{C}$ would be of the form

$$\vec{r}(t) = \left(\frac{t}{2} + \cos t, \sin t\right).$$

$\mathcal{C}$ is sketched in Figure 2.

We see that $x'(t) = \frac{1}{2} - \sin(t) = 0$ when $\sin(t) = \frac{1}{2}$, or when $t = \frac{\pi}{6} \pm 2n\pi$ or $t = \frac{5\pi}{6} \pm 2n\pi$. Thus $\mathcal{C}$ is a function on the intervals

$$\left[\frac{5\pi}{6} - 2\pi, \frac{\pi}{6}\right], \quad \left[\frac{\pi}{6}, \frac{5\pi}{6}\right],$$
$$\left[\frac{5\pi}{6}, \frac{\pi}{6} + 2\pi\right], \quad \left[\frac{\pi}{6} + 2\pi, \frac{5\pi}{6} + 2\pi\right],$$
$$\left[\frac{5\pi}{6} + 2\pi, \frac{\pi}{6} + 4\pi\right], \quad \left[\frac{\pi}{6} + 4\pi, \frac{5\pi}{6} + 4\pi\right],$$
$$\vdots \qquad\qquad \vdots$$

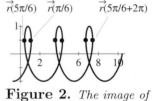

$\vec{r}(5\pi/6)$ $\vec{r}(\pi/6)$ $\vec{r}(5\pi/6+2\pi)$

Figure 2. *The image of* $\vec{r}(t) = \left(\frac{t}{2} + \cos t, \sin t\right).$

■

Theorem 5.2.1 can be rephrased to say that if the slope of the line tangent to the graph of a function is positive at every point in the segment (a, b), then the function is increasing on $[a, b]$. Similarly, if the slope of the line tangent to the graph of a function is negative at every point in the segment (a, b), then the function is decreasing on $[a, b]$. We can extend this technique for parametrized curves in $\mathbb{R}^2$. Let

$$\vec{r}(t) = (x(t), y(t))$$

be a vector valued function with image a curve $\mathcal{C}$ in the plane. Suppose that $x'(t) \neq 0$ on the interval $[a, b]$. Then we know from Theorem 1 that the part of $\mathcal{C}$ that is the image of $\vec{r}(t) = (x(t), y(t))$, $a \le t \le b$ is a function of the variable x. Furthermore, from Theorem 5.2.1, this function is increasing if its tangent line has positive slope at every point in (a, b).

Suppose that $\mathcal{C}$ is parametrized by a continuously differentiable function[8] $\vec{r}(t) = (x(t), y(t))$. If $x'(t_0) \neq 0$, then there is a segment (a, b) containing t_0 such that $x'(t) \neq 0$ for any $a < t < b$. Let $\tilde{\mathcal{C}}$ be the image of $\vec{r}(t) = (x(t), y(t))$, $a \le t \le b$. Then $\vec{r}'(t) = (x'(t), y'(t))$ is a direction vector for the line tangent to $\tilde{\mathcal{C}}$ at time t. Thus, as illustrated in Figure 3, $m(t) = \frac{y'(t)}{x'(t)}$ is the slope of the line tangent

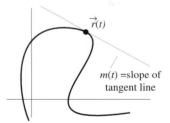

$\vec{r}(t)$

$m(t) =$ slope of tangent line

Figure 3. $m(t) = \frac{y'(t)}{x'(t)}$ *is the slope of the line tangent to the curve at the point* $\vec{r}(t)$.

---

[8]A function is continuously differentiable if it has a derivative at each point in its domain and the derivative is continuous.

to $\tilde{\mathcal{C}}$ at $(x(t), y(t))$. We have the following theorem.

Theorem 2 *Suppose that $\vec{r}(t) = (x(t), y(t))$ is continuously differentiable on $[a, b]$ and that $x'(t) \neq 0$ on $[a, b]$. Let $m(t) = y'(t)/x'(t)$.*

> *If $m(t) > 0$ for $a < t < b$, then $\mathcal{C}$ is increasing on the image of $[a, b]$.*

> *If $m(t) < 0$ for $a < t < b$, then $\mathcal{C}$ is decreasing on the image of $[a, b]$.*

Notice that the idea of a curve increasing or decreasing depends only on the slope of the tangent line. Thus, while we use a parametrization to determine where the curve is increasing and where it is decreasing, these regions on the curve must be independent of the particular parametrization. More formally, we have the following theorem.

Theorem 3 *Suppose that $\vec{r}_1(t) = (x_1(t), y_1(t))$ and $\vec{r}_2 = (x_2(t), y_2(t))$ are differentiable parametrizations for the same curve, and that $\vec{r}_1(t_1) = \vec{r}_2(t_2)$. If $x_1'(t_1) \neq 0$ and $x_2'(t_2) \neq 0$, then $m_1(t_1) = m_2(t_2)$, where $m_1(t) = \dfrac{y_1'(t)}{x_1'(t)}$ and $m_2(t) = \dfrac{y_2'(t)}{x_2'(t)}$.*

Many authors denote $m(t)$ by $\frac{dy}{dx}$ to reflect that $m(t)$ is the rate that y is changing as x changes (the rise over the run) at time t. Their formulation looks like:

$$m(t) = \frac{dy}{dx} = \frac{\frac{dy}{dt}}{\frac{dx}{dt}}.$$

If $\vec{r}(t) = (x(t), y(t))$ is continuously differentiable, then $m(t)$ is continuous[9], and $m(t)$ can change sign only at times when $m(t) = 0$ or when $m(t)$ is undefined. If (a, b) is a subset of the domain of $\vec{r}$, them m can change sign on (a, b) only at points where either $y'(t) = 0$ or $x'(t) = 0$. Points where $x'(t) = y'(t) = 0$ are problematic. If we think of the parametrization as locating a particle at time t, and if $\vec{r}'(t_0) = \vec{0}$, then the velocity at t_0 is $\vec{0}$. At this time, the particle has stopped. It can move off in any direction it chooses. Thus $\vec{r}(t_0)$ is a position where the image of $\vec{r}$ can have a corner. We summarize these statements in the following theorem.

---

[9]If $x'(t) = 0$, then t is not in the domain of m.

Theorem 4 *Let $\vec{r}(t)$ be a continuously differentiable parametrization for the curve C.*

If $\vec{r}\,'(t) \neq 0$ and $x'(t) = 0$, then the tangent to the image of $\vec{r}$ is vertical, and $m(t)$ is not defined there.

If $x'(t) \neq 0$ and $y'(t) = 0$, then the tangent is horizontal there.

$m(t)$ can change sign only at points where either $x'(t) = 0$, $y'(t) = 0$, or points where $\vec{r}$ is not defined.

We now present the procedure for determining where a curve in the plane is increasing and where it is decreasing.

Steps for Finding where a Curve C is Increasing and where it is Decreasing

Let $\vec{r} = (x(t), y(t))$ be a parametrization for C with a continuous derivative.

1. Find all points where either $x'(t) = 0$, $y'(t) = 0$, or where $\vec{r}$ is undefined. These are the values where m is undefined or where $m = 0$.

2. List the segments of the form (a, b) where

   - a and b are numbers found in Step 1, and
   - there are no numbers from Step 1 between a and b.

3. We know that

   - m does not change sign on any of the segments found in the previous step.
   - the image of $\vec{r}$ is a function on each of these segments.

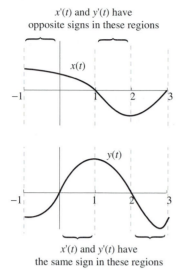

x'(t) and y'(t) have
opposite signs in these regions

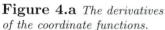

x'(t) and y'(t) have
the same sign in these regions

Figure 4.a *The derivatives of the coordinate functions.*

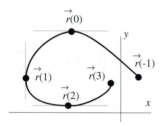

Figure 4.b *A sketch of the curve C.*

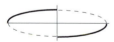

Figure 5.a $\mathcal{C}$ *is increasing in the second and fourth quadrants.*

Figure 5.b $\mathcal{C}$ *is decreasing in the first and third quadrants.*

4. For each segment (a, b) from Step 2, select a convenient sample point c between a and b and evaluate $m(c)$.

- If $m(c) > 0$, then we know that $m(t) > 0$ for every t in (a, b), and $\mathcal{C}$ is increasing on the image of (a, b).
- If $m(c) < 0$, then we know that $m(t) < 0$ for every t in (a, b), and $\mathcal{C}$ is decreasing on the image of (a, b).

EXAMPLE 3: The curve C is parametrized by $\vec{r}(t) = (x(t), y(t))$, $-1 \leq t \leq 3$. The derivatives of the coordinate functions $x'(t)$ and $y'(t)$ are sketched in Figure 4.a.

$x'(t) = 0$ at $t = 1$ and $t = 3$. $y'(t) = 0$ at $t = 0$ and $t = 2$. The curve C has vertical tangents at $\vec{r}(1)$ and $\vec{r}(3)$. It will have horizontal tangents at $\vec{r}(0)$ and $\vec{r}(2)$. $x'(t) > 0$ and $y'(t) < 0$ on $(-1, 0)$, so $m(t) < 0$ and C is decreasing between $\vec{r}(-1)$ and $\vec{r}(0)$. $x'(t)$ and $y'(t)$ both positive on the segment $0 < t < 1$ so that $m(t) > 0$ and C is increasing there. The coordinate functions have opposite signs on the segment $1 < t < 2$, so C is decreasing there. Finally, we see that $x'(t)$ and $y'(t)$ have opposite signs on $2 < t < 3$, which means that C is decreasing between $\vec{r}(2)$ and $\vec{r}(3)$.

EXAMPLE 4: Let $\vec{r}(t) = (t, t^2)$. Notice that the image of $\vec{r}(t)$ is the same as the graph of $f(x) = x^2$.

$$m(t) = \frac{y'(t)}{x'(t)} = \frac{2t}{1} = 2t.$$

Thus the image of $\vec{r}(t)$ is increasing when $t > 0$ and decreasing when $t < 0$. This corresponds to our usual interpretation of increasing and decreasing real-valued functions in the plane. In fact, the notion of increasing and decreasing functions is merely a special case of increasing and decreasing parametrized curves, since any function $f: \mathbb{R} \to \mathbb{R}$ can be represented as the image of the vector function $\vec{r}(t) = (t, f(t))$. ∎

EXAMPLE 5: Let $\mathcal{C}$ be the ellipse parametrized by $\vec{r}(t) = (a \cos t, b \sin t)$ from Example 1. Determine where $\mathcal{C}$ is increasing and where it is decreasing.

SOLUTION: $\vec{r}'(t) = (-a \sin t, b \cos t)$.

$$x'(t) = 0 \quad \text{at} \quad t = n\pi \text{ for } n = 0, \pm 1, \pm 2, \ldots$$
$$y'(t) = 0 \quad \text{at} \quad t = \frac{(2n+1)\pi}{2} \text{ for } n = 0, \pm 1, \pm 2, \ldots$$

The resulting segments are $\left(0, \frac{\pi}{2}\right)$, $\left(\frac{\pi}{2}, \pi\right)$, $\left(\pi, \frac{3\pi}{2}\right)$, ...

$m\left(\frac{\pi}{4}\right) = -\frac{a}{b} < 0$; therefore, $m(t) < 0$ for t in $\left(0, \frac{\pi}{2}\right)$, and $\mathcal{C}$ is decreasing there. This is the part of $\mathcal{C}$ in the first quadrant.

$m\left(\frac{3\pi}{4}\right) = \frac{a}{b} > 0$; therefore, $m(t) > 0$ for t in $\left(\frac{\pi}{2}, \pi\right)$, and $\mathcal{C}$ is increasing there. This is the part of $\mathcal{C}$ in the second quadrant.

$m\left(\frac{5\pi}{4}\right) = -\frac{a}{b} < 0$; therefore, $m(t) < 0$ for t in $\left(\pi, \frac{3\pi}{2}\right)$, and $\mathcal{C}$ is decreasing there. This is the part of $\mathcal{C}$ in the third quadrant.

$m\left(\frac{7\pi}{4}\right) = \frac{a}{b} > 0$; therefore, $m(t) > 0$ for t in $\left(\pi, \frac{3\pi}{2}\right)$, and $\mathcal{C}$ is increasing there. This is the part of $\mathcal{C}$ in the fourth quadrant. ∎

EXAMPLE 6: Returning to the curve in Example 2, let $\mathcal{C}$ be the image of

$$\vec{r}(t) = \left(\frac{t}{2} + \cos t, \sin t\right).$$

Determine where $\mathcal{C}$ is increasing and where it is decreasing.

SOLUTION: In Example 2, we determined that $x'(t) = 0$ when $t = \frac{\pi}{6} \pm 2n\pi$ or $t = \frac{5\pi}{6} \pm 2n\pi$. We know that $y'(t) = 0$ when t is an odd multiple of $\frac{\pi}{2}$; that is, $y'(t) = 0$ when $t = \frac{(2n+1)\pi}{2}$ for $n = 0, \pm 1, \pm 2, \ldots$.

The segments of interest are $\left(-\frac{\pi}{2}, \frac{\pi}{6}\right)$, $\left(\frac{\pi}{6}, \frac{\pi}{2}\right)$, $\left(\frac{\pi}{2}, \frac{5\pi}{6}\right)$, $\left(\frac{5\pi}{6}, \frac{3\pi}{2}\right)$, ...

$x'\left(\frac{\pi}{7}\right) \approx 0.06$ $y'\left(\frac{\pi}{7}\right) \approx 0.9$, so $m(t) > 0$, and $\mathcal{C}$ increases on $\left(-\frac{\pi}{2}, \frac{\pi}{6}\right)$.

$x'\left(\frac{\pi}{3}\right) \approx -0.366$ $y'\left(\frac{\pi}{3}\right) = 0.5$, so $m(t) < 0$, and $\mathcal{C}$ decreases on $\left(\frac{\pi}{6}, \frac{\pi}{2}\right)$.

$x'\left(\frac{2\pi}{3}\right) \approx -0.366$ $y'\left(\frac{2\pi}{3}\right) = -0.5$, so $m(t) > 0$, and $\mathcal{C}$ increases on $\left(\frac{\pi}{2}, \frac{5\pi}{6}\right)$.

$x'(\pi) = 0.5$ $y'(\pi) = -1$, so $m(t) < 0$, and $\mathcal{C}$ decreases on $\left(\frac{5\pi}{6}, \frac{3\pi}{2}\right)$.

And so on.

The information that we have gained by determining where $m(t) > 0$ and where $m(t) < 0$ is completely consistent with the sketch of $\mathcal{C}$ in Figure 6. If $m(t) > 0$, then as you move from left to right on $\mathcal{C}$, you go uphill. If $m(t) < 0$, then as you move from left to right on $\mathcal{C}$, you go downhill. ∎

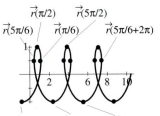

Figure 6. *Points where $\mathcal{C}$ can change from increasing to decreasing.*

We can also use the techniques from Section 5.2 to find relative extrema for parametrized functions into $\mathbb{R}^2$.

EXAMPLE 7: Let $\vec{r}$ be the function defined in Example 2. Recall that $m(t)$ is the slope of the line tangent to $\mathcal{C}$ at $\vec{r}(t)$. Thus $m(t) = 0$ or is undefined at $t = \frac{\pi}{2} + n\pi$, $\frac{\pi}{6} + 2n\pi$, and $\frac{5\pi}{6} + 2n\pi$, where n is any integer. Since $\vec{r}$ is increasing on $\left(\frac{\pi}{2} + 2m\pi, \frac{5\pi}{6} + 2m\pi\right)$ and decreasing on $\left(\frac{\pi}{6} + 2m\pi, \frac{\pi}{2} + 2m\pi\right)$ we see that $\vec{r}$ attains relative maxima at $t = \frac{\pi}{2} + 2m\pi$. Similarly, we see that $\vec{r}$ attains relative minima at $t = \frac{3\pi}{2} + 2m\pi$. ∎

Polar Equations

Functions that model something moving on a diameter of a wheel that is rotating provide a tool that has proved quite useful. We made use of such models in the Ant on the Wheel and Rotating Pendulum problems. These functions merit further attention.

If ρ is a real-valued function of θ, then

$$\vec{P}(\theta) = \rho(\theta)\vec{W}(\theta) = \rho(\theta)\begin{pmatrix} \cos(\theta) \\ \sin(\theta) \end{pmatrix}$$

locates a point $|\rho(\theta)|$ units from the origin that lies on a line making an angle θ with the positive x–axis. The angle θ is the *polar angle* of the point. The function ρ is called a *polar function* for the curve $\mathcal{C}$ parametrized by $\vec{P}$. The equation $r = \rho(\theta)$ is called a *polar equation* for $\mathcal{C}$. See Figure 7.

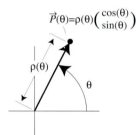

Figure 7. $\vec{P}(\theta) = \rho(\theta)\vec{W}(\theta)$.

EXAMPLE 8: Let $\mathcal{C}$ be the curve illustratied in Figure 8. It is parametrized by $\vec{P}(\theta) = \sin(2\theta)\begin{pmatrix} \cos(\theta) \\ \sin(\theta) \end{pmatrix}$. The equation $r = \rho(\theta) = \sin(2\theta)$ is a *polar equation* for $\mathcal{C}$. ∎

It can be quite difficult to analyze curves in the plane with polar equations using the techniques we have developed in this section, even when the polar equations seem simple. For example, it is difficult to determine where the curve $\mathcal{C}$ of Example 8 is increasing and where it is decreasing. We can, however, obtain a great deal of information about curves in the plane with polar equations by using the techniques of Section 5.2 to get information about the polar function itself. ∎

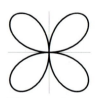

Figure 8. *The curve C from Example 8.*

EXAMPLE 9: Let us return to the curve $\mathcal{C}$ of Example 8,

$$\vec{P}(\theta) = \sin(2\theta)\begin{pmatrix} \cos(\theta) \\ \sin(\theta) \end{pmatrix}.$$

The polar function is $\rho(\theta) = \sin(2\theta)$. In Figure 8, we sketch the curve $\mathcal{C}$. Below, we describe how we arrived at this picture.

We proceed by determining where the function ρ is positive, where it is negative, where it is increasing, and where it is decreasing.

$\sin(2\theta)$ is positive and increasing on $\left(0, \frac{\pi}{4}\right)$. This implies that as the polar ray rotates counterclockwise, it is increasing in length.

$\sin(2\theta)$ is positive and decreasing on $\left(\frac{\pi}{4}, \frac{\pi}{2}\right)$. This implies that as the polar ray rotates counterclockwise, it is decreasing in length.

$\sin(2\theta)$ is negative and decreasing on $\left(\frac{\pi}{2}, \frac{3\pi}{4}\right)$. This implies that as the polar ray rotates counterclockwise, it is increasing in length.

$\sin(2\theta)$ is negative and increasing on $\left(\frac{3\pi}{4}, \pi\right)$. This implies that as the polar ray rotates counterclockwise, it is decreasing in length.

And so on. See Figures 9.a–d.

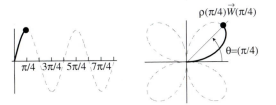

Figure 9.a *The graph of* $\sin(2\theta)$ *and the image of* $\rho(\theta)\vec{W}(\theta)$ *for* $0 \leq \theta \leq \frac{\pi}{4}$.

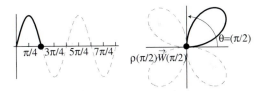

Figure 9.b *The graph of* $\sin(2\theta)$ *and the image of* $\rho(\theta)\vec{W}(\theta)$ *for* $0 \leq \theta \leq \frac{\pi}{2}$.

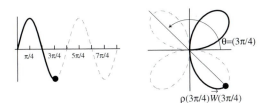

Figure 9.c *The graph of* $\sin(2\theta)$ *and the image of* $\rho(\theta)\vec{W}(\theta)$ *for* $0 \leq \theta \leq \frac{3\pi}{4}$.

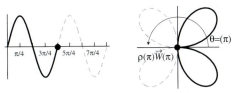

Figure 9.d *The graph of* $\sin(2\theta)$ *and the image of* $\rho(\theta)\vec{W}(\theta)$ *for* $0 \leq \theta \leq \pi$.

EXAMPLE 10: Let $\mathcal{C}$ be the image of $\vec{P}(\theta) = (1 + \sin(2\theta)) \begin{pmatrix} \cos(\theta) \\ \sin(\theta) \end{pmatrix}$.

In Figure 10.a, we sketch the polar function $\rho(\theta) = (1 + \sin(2\theta))$.

$1 + \sin(2\theta)$ is positive and increasing on $\left(0, \frac{\pi}{4}\right)$. This implies that as the polar ray rotates counterclockwise, it is increasing in length.

$1 + \sin(2\theta)$ is positive and decreasing on $\left(\frac{\pi}{4}, \frac{3\pi}{4}\right)$. This implies that as the polar ray rotates counterclockwise, it is decreasing in length.

$1 + \sin(2\theta)$ is positive and increasing on $\left(\frac{3\pi}{4}, \frac{5\pi}{4}\right)$. This implies that as the polar ray rotates counterclockwise, it is increasing in length.

$1 + \sin(2\theta)$ is positive and decreasing on $\left(\frac{5\pi}{4}, \frac{7\pi}{4}\right)$. This implies that as the polar ray rotates counterclockwise, it is decreasing in length.

And so on.

The polar function attains its maximum at $\theta = \frac{\pi}{4},\ \frac{5\pi}{4}\ \frac{9\pi}{4},\ \dots,$ and it attains its minimum at $\theta = \frac{3\pi}{4},\ \frac{7\pi}{4}\ \frac{11\pi}{4},\ \dots.$ The maximum value for ρ is 2, and its minimum value is 0. This information is used to sketch the curve in Figure 10.b. In Figures 10.c–h we display sketches of three more examples. ∎

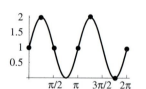

Figure 10.a
The graph of
$\rho(\theta) = 1 + \sin(2\theta).$

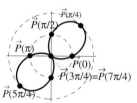

Figure 10.b
The image of
$(1 + \sin(2\theta))\, \vec{W}(\theta).$

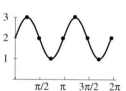

Figure 10.c
The graph of
$\rho(\theta) = 2 + \sin(2\theta).$

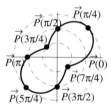

Figure 10.db
The image of
$(2 + \sin(2\theta))\, \vec{W}(\theta).$

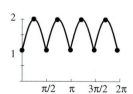

Figure 10.e
The graph of
$\rho(\theta) = 1 + |\sin(2\theta)|.$

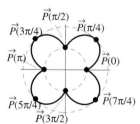

Figure 10.f
The image of
$(1 + |\sin(2\theta)|)\, \vec{W}(\theta).$

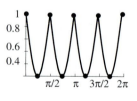

Figure 10.g
The graph of
$\rho(\theta) = 1 - |\sin(2\theta)|.$

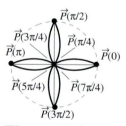

Figure 10.h
The image of
$(1 - |\sin(2\theta)|)\, \vec{W}(\theta).$

EXERCISES 5.3

1. Suppose that $\vec{r}(t) = (x(t),\ y(t))$ is a parametrization for the curve C illustrated in Figure 11.a such that $\vec{r}'(t) \neq 0$.

   a. Locate all points $\vec{v}$ on the curve such that if $r(t_0) = \vec{v}$, then $x'(t_0) = 0$.

   b. Locate all points $\vec{v}$ on the curve such that if $r(t_0) = \vec{v}$, $y'(t_0) = 0$.

2. Repeat Parts a and b of Exercise 1 for the curve in Figure 11.b.

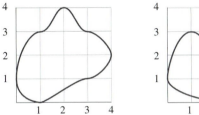 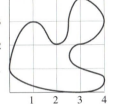

Figure 11.a *Curve for Exercise 1.*   **Figure 11.b** *Curve for Exercise 2.*

In Exercises 3–6, $\vec{r}(t) = (x(t), y(t))$, $a \leq t \leq b$, models a particle moving along a path. Given $\vec{r}'(c)$, determine if the particle is moving to the left or to the right, up or down, and if its path is increasing or decreasing at the time $t = c$.

3. $\vec{r}'(c) = (1, -2)$.   4. $\vec{r}'(c) = (-1, -2)$.

5. $\vec{r}'(c) = (-1, 2)$.   6. $\vec{r}'(c) = (1, 2)$.

7. A curve C, sketched in Figure 12, is parameterized by a function $\vec{r}(t) = (x(t), y(t))$, $0 \leq t \leq 8$. Determine the intervals where x is increasing, x is decreasing, y is increasing, and y is decreasing.

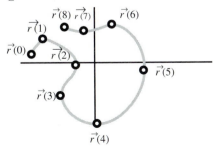

Figure 12. *Curve for Exercise 7.*

8. The curve in Figure 13.a is parametrized by $\vec{r}(t) = (x(t),\ y(t))$, with motion around the curve indicated by the arrows so that $\vec{r}'(t) \neq \vec{0}$.

   a. For each t_i, determine whether $x'(t_i) > 0$, $x'(t_i) < 0$, or $x'(t_i) = 0$.

   b. For each t_i, determine whether $y'(t_i) > 0$, $y'(t_i) < 0$, or $y'(t_i) = 0$.

9. The curve in Figure 13.b is parametrized by $\vec{r}(t) = (x(t),\ y(t))$, with motion around the curve indicated by the arrows so that $\vec{r}'(t) \neq \vec{0}$.

   a. For each t_i, determine whether $x'(t_i) > 0$, $x'(t_i) < 0$, or $x'(t_i) = 0$.

   b. For each t_i, determine whether $y'(t_i) > 0$, $y'(t_i) < 0$, or $y'(t_i) = 0$.

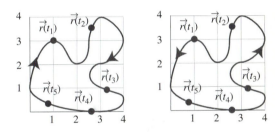

Figure 13.a *Curve for Exercise 8.*   **Figure 13.b** *Curve for Exercise 9.*

In Exercises 10–12, you are given sketches of $x'(t)$ and $y'(t)$.

   a. Determine the intervals in the domain of $\vec{r}$ where the curve is a function of x.

   b. Find the values of t where the the curve has horizontal tangents and the values where the curve has vertical tangents.

   c. Determine the intervals where the curve is increasing and the intervals where the curve is decreasing.

10. 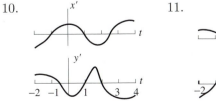   11.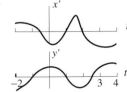

12.

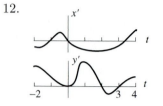

In Exercises 13–15, find an equation for the line tangent to the parametrized curve at $\vec{r}(t_0)$.

13. $\vec{r}(t) = (\cos t, \sin t), \quad t_0 = \frac{\pi}{3}.$

14. $\vec{r}(t) = (t\cos(2\pi t), t\sin(2\pi t)), \quad t_0 = \frac{1}{3}.$

15. $\vec{r}(t) = \sin(4\pi t)\,(\cos(2\pi t), \sin(2\pi t)), \quad t_0 = \frac{1}{6}.$

For the parametrized curves in Exercises 16–23,

a. Determine the intervals in the domain of $\vec{r}$ where the curve is a function of x.

b. Find the values of t where the the curve has horizontal tangents and the values where the curve has vertical tangents.

c. Determine the intervals where the curve is increasing and the intervals where the curve is decreasing.

d. Plot some points and sketch the curve.

16. $\vec{r}(t) = (\cos t, \sin t).$

17. $\vec{r}(t) = (t^2, t).$

18. $\vec{r}(t) = (t, t^2).$

19. $\vec{r}(t) = (t^2 + 6t + 2, t).$

20. $\vec{r}(t) = (t^2 + 6t + 2, t^2).$

21. $\vec{r}(t) = (t^2 + 6t + 2, t^3).$

22. $\vec{r}(t) = \vec{a}t + \vec{b}, \quad \vec{a} = (a_1, a_2), a_1 \neq 0.$

23. $\vec{r}(t) = (t + \cos t, 1 - \sin t).$

In Exercises 24–27, C is the curve parametrized by $\vec{r}(t) = (\sin at, \sin bt).$

a. Determine the values of t where $\vec{r}(t) = \vec{0}.$

b. Calculate $m(t)$ for each value of t from Part a.

c. Find the intervals where C is a function of x.

d. Find the intervals where the curve is increasing and the intervals where the curve is decreasing.

e. Match the curve C with one of the curves sketched in Figures 14.a–c.

24. $a = 1, \ b = 2.$ 25. $a = 1, \ b = 3.$

26. $a = 2, \ b = 3.$ 27. $a = 1, \ b = 4.$

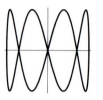

Figure 14.a

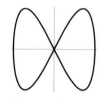

Figure 14.b

Figure 14.c

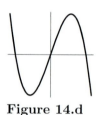

Figure 14.d

Figure 14.e

Figure 14.f

28. Let C be the curve parametrized by $\vec{r}(t) = \left(\frac{3t}{1+t^3}, \frac{3t^2}{1+t^3}\right).$

a. Find the domain of $\vec{r}.$

b. Find the intervals where C is a function of x.

c. Determine the intervals where $m(t)$ is positive and where $m(t)$ is negative.

d. Plot some points of C and sketch its graph.

e. Show that $\vec{r}$ is a parametrization of the graph of $x^3 + y^3 = 3xy$. This curve is called the *folium of Descartes*.

29. **Wheel in a Cylinder Problem 3:** Let C be the curve parametrized by $\vec{r}(t) = \begin{pmatrix} \cos^3(t) \\ \sin^3(t) \end{pmatrix}$. C is called a *hypocycloid of four cusps*.

   a. Determine segments where C is a function.

   b. Determine where C is increasing and where it is decreasing.

   c. Show that C is the graph of the equation $x^{2/3} + y^{2/3} = 1$.

   d. Show that

   $$\vec{g}(t) = \begin{pmatrix} \frac{3}{4}\cos t + \frac{1}{4}\cos(3t) \\ \frac{3}{4}\sin t - \frac{1}{4}\sin(3t) \end{pmatrix}$$

   also parametrizes C. Notice that this is the parametrization for the Wheel Rolling Inside a Cylinder of Exercise 2.7.17, with $R = 1$ and $r = 0.25$.

Match the curves sketched in Figures 15.a–l with the polar functions given in Exercises 30–41.

30. $\rho(\theta) = \sin(3\theta)$.

31. $\rho(\theta) = \sin(4\theta)$.

32. $\rho(\theta) = \cos(\theta)$.

33. $\rho(\theta) = \cos(2\theta)$.

34. $\rho(\theta) = \cos(3\theta)$.

35. $\rho(\theta) = \cos(4\theta)$.

36. $\rho(\theta) = 1 + \cos(\theta)$.

37. $\rho(\theta) = 1 + \cos(2\theta)$.

38. $\rho(\theta) = 1 - \cos(\theta)$.

39. $\rho(\theta) = 1 + 2\cos(\theta)$.

40. $\rho(\theta) = 1 + 2\cos(2\theta)$.

41. $\rho(\theta) = 1 + |\cos(2\theta)|$.

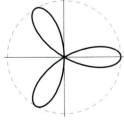

Figure 15.a

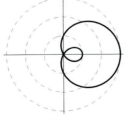

Figure 15.b

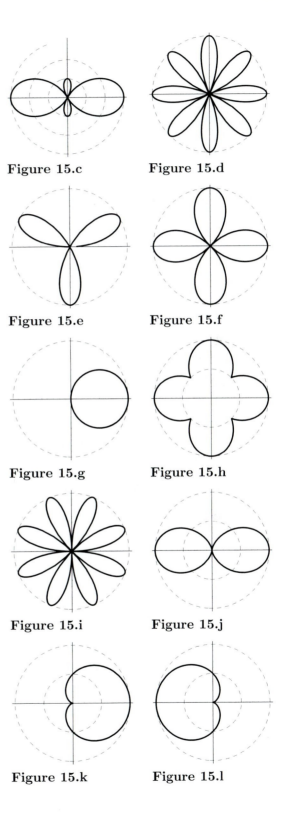

Figure 15.c   Figure 15.d

Figure 15.e   Figure 15.f

Figure 15.g   Figure 15.h

Figure 15.i   Figure 15.j

Figure 15.k   Figure 15.l

42. The curve $\mathcal{C}$ is parametrized by $\vec{P}(t) = \rho(t)\vec{W}(t)$, where $\rho(t) = \sin(3t)$.

    a. Find the slope of the line tangent to $\mathcal{C}$ at $\vec{P}\left(\frac{\pi}{4}\right)$.

    b. Find an equation for the line tangent to $\mathcal{C}$ at $\vec{P}\left(\frac{\pi}{4}\right)$.

    c. Find a parametrization for the line tangent to $\mathcal{C}$ at $\vec{P}\left(\frac{\pi}{4}\right)$.

43. **Ant on the Wheel Problem 3:** In the Ant on the Wheel Problem 1 (Exercise 2.7.33), we showed that an ant walking toward the center of a wheel rotating with a frequency of f rotations/sec can be located in time by the parametrization

$$\vec{r}(t) = \begin{pmatrix} (R - t)\sin(2\pi ft) \\ -(R - t)\cos(2\pi ft) \end{pmatrix}.$$

Let $R = 1$ ft and $f = 3$. Figures 16.a–b show the graphs of $x(t)$ and $x'(t)$.

Figure 16.a *The graph of $x(t)$.*   **Figure 16.b** *The graph of $x'(t)$.*

The zeros of $x(t)$ are $a_1 = 0$, $a_2 = 0.1667$, $a_3 = 0.3333$, $a_4 = 0.5$, $a_5 = 0.6667$, $a_6 = 0.8333$ and $a_7 = 1$. The zeros of $x'(t)$ are $b_1 = 0.0803$, $b_2 = 0.2463$, $b_3 = 0.4119$, $b_4 = 0.5767$, $b_5 = 0.7394$, $b_6 = 0.8924$ and $b_7 = 1$. Figures 17.a–b show the graphs of $y(t)$ and $y'(t)$. The zeros of $y(t)$ are given by $c_1 = 0.0833$, $c_2 = 0.25$, $c_3 = 0.4167$, $c_4 = 0.5833$, $c_5 = 0.75$, $c_6 = 0.9167$ and $c_7 = 1$. The zeros of $y'(t)$ are $d_1 = 0.1633$, $d_2 = 0.3292$, $d_3 = 0.4944$, $d_4 = 0.6585$, $d_5 = 0.8183$ and $d_6 = 0.9544$. Using this information:

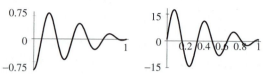

Figure 17.a *The graph of $y(t)$.*   **Figure 17.b** *The graph of $y'(t)$.*

    a. Determine the time intervals where the ant's path is a function of x.

    b. Determine the time intervals where the ant's path is increasing and the intervals where it is decreasing.

    c. Show that the ant's path can be parametrized as a function of its polar angle by $\vec{r}(\theta) = \left(R - \frac{\theta}{2\pi f}\right)\vec{W}\left(\theta - \frac{\pi}{2}\right)$.

    d. Find the values for θ where the polar function $\rho(\theta) = \left(R - \frac{\theta}{2\pi f}\right)$ is increasing and the values where $\rho(\theta)$ is decreasing.

44. **Ant on the Wheel Problem 4:** In the previous exercise you were given the zeros of $x(t), x'(t), y(t)$, and $y'(t)$ when $R = 1$ and $f = 2$. Find the zeros of these functions when $R = 1$ and $f = 2$. (One method is to use the Interval-Halving Method to find the zeros.) Then use this information to:

    a. Determine the time intervals where the ant's path is a function of x.

    b. Determine the time intervals where the ant's path is increasing and the intervals where it is decreasing.

45. **Rotating Pendulum Problem 3:** In the Rotating Pendulum Problem 1 (Exercise 2.7.34), we showed that a ball rolling back and forth on a track on a rotating disc can be located by

$$\vec{r}(t) = (a + L\cos(\omega t))\begin{pmatrix} \cos(2\pi kt) \\ \sin(2\pi kt) \end{pmatrix}.$$

Let $\rho(t) = (a + L\cos(\omega t))$ and let $\theta(t) = 2\pi kt$. Let $k = \frac{1}{4}$, $\omega = 2\pi$, $L = 1$, and $a = 0$.

a. Express t as a function of θ.

b. Express ρ as a function of θ. Note that ρ is now a true polar function with θ as the polar angle. We can now reparametrize the ball's path as a function of its polar angle. That is, the ball is located by $\vec{P}(\theta) = \rho(\theta)\vec{W}(\theta)$.

c. Sketch the graph of ρ (as a function of θ.)

d. Sketch the path of the ball.

5.4   Concavity

Though knowing where f is increasing and decreasing gives us a reasonable sketch of f, it does not describe the "smoothness" of f at all. If f' is a constant, then we know that f is a straight line. Except for this special case, we do not know how f bends. Figures 1.a–c indicate functions that have the same y–coordinates at the endpoints and that are all increasing. Yet there is a big difference in the behavior of these functions. We say that the first function is *concave up* whereas the second is *concave down*. The third function is neither concave up nor concave down.

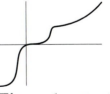

Figure 1.a *This function is increasing and concave up.*

Figure 1.b *This function is increasing and concave down.*

Figure 1.c *This function is increasing but neither concave up nor concave down.*

The geometric realization of the notion of concavity is exactly what the name indicates. Consider taking a tangent line and moving the point of tangency from left to right on a segment (a, b), as in Figure 2. If f' is increasing on (a, b), then the tangent line will rotate in a counterclockwise fashion (see Figure 2.a), and the graph must "cup up" on (a, b). If f' is decreasing on (a, b), then the tangent line will rotate in a clockwise fashion, and the graph must "cup down" (as in Figure 2.b).

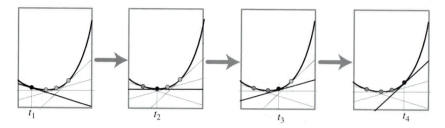

Figure 2.a *The tangent line swings counterclockwise as x moves from left to right.*

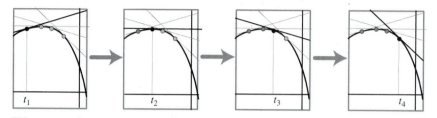

Figure 2.b *The tangent line swings clockwise as x moves from left to right.*

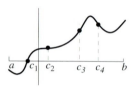

Figure 3. *Concavity changes at c_1, c_2, c_3 and c_4.*

The function sketched in Figure 3 is concave up on (a, c_1), (c_2, c_3), and (c_4, b) and it is concave down on (c_1, c_2) and (c_3, c_4). The points where $x = c_1$, c_2, c_3, and c_4 are points where the concavity changes.

The previous discussion motivates the following definition.

Definition: Concavity of Functions

A function f is *concave up* on the segment (a, b) if f' is increasing on (a, b), and f is *concave down* on (a, b) if f' is decreasing on (a, b). If f changes concavity at $x = c$, then f has an *inflection point* at c.

By applying the methods of Section 5.2 to f', we have a tool for determining the concavity of a function f.

Theorem 1 *If $f'' < 0$ on (a, b), then f' is decreasing, and f is concave down on (a, b). If $f'' > 0$ on (a, b), then f' is increasing, and f is concave up on (a, b). If c is a point of inflection for f and $f''(c)$ exists, then $f''(c) = 0$.*

Example 4 llustrates the fact that $f''(c)$ need not be defined if c is an inflection point.

It is also quite easy to see that if $x = c$ is a critical point and f is concave up at c, then $f(c)$ is a relative minimum for f. Similarly, if $x = c$ is a critical point and f is concave down at c, then $f(c)$ is a relative maximum for f. See Figure 4. We formalize this observation in the Second Derivative Test.

Theorem 2 (The Second Derivative Test) *Suppose that* $f'(c) = 0$.

If $f''(c) < 0$, then f attains a relative maximum at c.

If $f''(c) > 0$, then f attains a relative minimum at c.

If $f''(c) = 0$, then the test gives us no information.

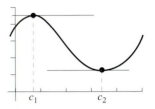

Figure 4. $f'(c_1) = f'(c_2) = 0$. *f is concave down at c_1 and it attains a relative maximum there. f is concave up at c_2 and it attains a relative minimum there.*

EXAMPLE 1: Let $f(x) = x^3$. Since $f'(x) = 3x^2$, the function f is always increasing. The second derivative $f''(x) = 6x$. So the point $x = 0$ is the only inflection point. Thus f is concave down on $(-\infty, 0)$ and concave up on $(0, \infty)$. (See Figure 5.a.) ■

EXAMPLE 2: The function $y = x^2 + 2x$ is an equation of a parabola with vertex at $(-1, -1)$, which opens in the positive y direction. The derivative is given by $y'(x) = 2x + 2$. Thus $y'(x) = 0$ at $x = -1$. Since $y''(x) = 2 > 0$, the curve is always concave up, and y attains its minimum at $x = -1$. This agrees with the fact that we are dealing with a parabola. (See Figure 5.b.) ■

EXAMPLE 3: Determine the concavity of the function $y = \tan x$ and locate the inflection points in the segment $\left(-\frac{\pi}{2}, \frac{\pi}{2}\right)$. (See Figure 5.c.)

SOLUTION: $D \tan x = \sec^2 x$ and $D^2 \tan x = 2 \sec^2(x) \tan(x)$. Now $\sec^2 x > 0$ for all x in the interval $\left(-\frac{\pi}{2}, \frac{\pi}{2}\right)$. We also see that $\tan x < 0$ on $\left(-\frac{\pi}{2}, 0\right)$ and $\tan x > 0$ on $\left(0, \frac{\pi}{2}\right)$. We conclude that $\tan x$ is concave down on $\left(-\frac{\pi}{2}, 0\right)$ and concave up on $\left(0, \frac{\pi}{2}\right)$. The point $x = 0$ is the only inflection point for $\tan x$ on $\left(-\frac{\pi}{2}, \frac{\pi}{2}\right)$. ■

EXAMPLE 4: Let $f(x) = x^{1/3}$. Then f has a vertical tangent line at $x = 0$. However, $f''(x) = -\frac{2}{9}x^{-5/3} < 0$ on $(-\infty, 0)$ and $f''(x) > 0$ on $(0, \infty)$. Thus 0 is an inflection point of f even though 0 is not in the domain of f''. (See Figure 5.d.) ■

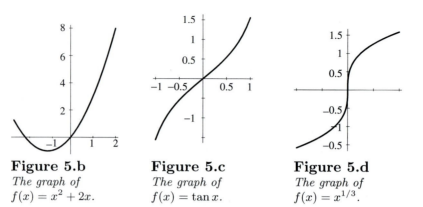

Figure 5.a
The graph of
$f(x) = x^3.$

Figure 5.b
The graph of
$f(x) = x^2 + 2x.$

Figure 5.c
The graph of
$f(x) = \tan x.$

Figure 5.d
The graph of
$f(x) = x^{1/3}.$

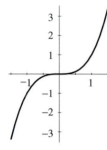

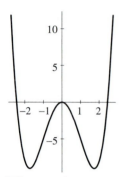

Figure 6. *The graph of*
$f(x) = x^4 - 6x^2.$

EXAMPLE 5: Let $f(x) = x^4 - 6x^2$. Determine

(a) where f is positive and where f is negative;

(b) where f is increasing and where f is decreasing; and

(c) where f is concave up and where f is concave down.

Sketch the function.

SOLUTION:

(a) $f(x) = x^2(x^2 - 6) = x^2(x - \sqrt{6})(x + \sqrt{6})$. Thus f can change sign at $x = 0$, $x = \sqrt{6}$ and $x = -\sqrt{6}$.

f is positive on $(-\infty, -\sqrt{6})$ since $f(-3)$ is positive.

f is negative on $(-\sqrt{6}, 0)$ since $f(-2)$ is negative.

f is negative on $(0, \sqrt{6})$ since $f(2)$ is negative.

f is positive on $(\sqrt{6}, \infty)$ since $f(3)$ is positive.

As x approaches ∞, $f(x)$ also approaches ∞, and as x approaches $-\infty$, $f(x)$ approaches ∞ (since the powers of x are even).

(b) $f'(x) = 4x^3 - 12x = 4x(x^2 - 3) = 4x(x + \sqrt{3})(x - \sqrt{3})$. Thus f' can change sign at $x = 0$, $x = \sqrt{3}$, and $x = -\sqrt{3}$.

f is decreasing on $(-\infty, -\sqrt{3})$ since $f'(-2)$ is negative.

f is increasing on $(-\sqrt{3}, 0)$ since $f'(-1)$ is positive.

f is decreasing on $(0, \sqrt{3})$ since $f'(1)$ is negative.

f is increasing on $(\sqrt{3}, \infty)$ since $f'(2)$ is positive.

(c) $f''(x) = 12x^2 - 12 = 12(x + 1)(x - 1)$. Thus f'' can change sign at $x = 1$ and $x = -1$.

f is concave up on $(-\infty, -1)$ since $f''(-2)$ is positive.

f is concave down on $(-1, 1)$ since $f''(0)$ is negative.

f is concave up on $(1, \infty)$ since $f''(2)$ is positive. ■

We outline the general procedure for employing the tools of calculus in graphing real valued functions.

Graphing Functions

Step I: Determine where the function is positive and where it is negative.

Step II: Locate the critical points, points of inflection, and relative extrema.

Step III: Determine where the function is increasing and where it is decreasing.

Step IV: Determine the concavity of the function.

Step V: Plot a few points. Usually we try to plot at least the points obtained in Step IV.

Step VI: Sketch the graph.

The above procedure is a rule of thumb only. Sometimes it is not reasonable to get all of the information in all of the steps.

EXAMPLE 6: Let $f(x) = x^3 - 2x^2 + 5x - \pi$. Unless we had a calculator or computer with an equation solver at our disposal, we would skip Step I. On the other hand, it is easy to determine where f is increasing and where it is decreasing, and it is easier still to determine the concavity of f. With this information, we can get an accurate sketch of f by plotting a few points. ■

EXAMPLE 7: Going back to Example 5, if the only goal were to find the relative extrema of $f(x) = x^4 - 6x^2$, we would only have to determine where $f'(x) = 0$ and check the concavity at those points. ■

EXAMPLE 8: Find the absolute extrema for the function $f(x) = x + \frac{1}{x}$ on the set $0 < x < \infty$.

SOLUTION: (See Figure 7.) $f'(x) = 1 - x^{-2}$, so the only critical point is at $x = 1$ (note that x must be positive). The second derivative

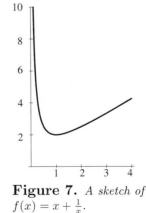

Figure 7. *A sketch of* $f(x) = x + \frac{1}{x}$.

$f''(x) = 2x^{-3}$, so when we evaluate the second derivative at the critical point, we obtain $f''(1) = 2 > 0$. Thus f has a relative minimum at $x = 1$. Since $f \to \infty$ as both $x \to 0$ and as $x \to \infty$, we see that f has an absolute minimum of 2 at $x = 1$ and that f does not have an absolute maximum. ∎

Concavity of Parametrized Curves in the Plane

We can employ the techniques of this section to determine where parametrized curves in the plane are concave up and where they are concave down. Let $\vec{r}(t) = (x(t), y(t))$ parametrize the curve $\mathcal{C}$. Then $m(t) = \frac{y'(t)}{x'(t)}$ is the slope of the line tangent to $\mathcal{C}$ at the point $(x(t), y(t))$. Let $\vec{m}(t) = (x(t), m(t))$. Then $\mathcal{C}$ will be concave up at points where the image of $\vec{m}$ is increasing, and $\mathcal{C}$ will be concave down at points where the image of $\vec{m}$ is decreasing. In the previous section, we learned how to determine where $\vec{m}(t)$ is increasing and where $\vec{m}(t)$ is decreasing.

We look at $\frac{m'(t)}{x'(t)}$, which is the rate that m is changing as x changes at the point $\vec{r}(t)$. That is, $\frac{m'(t)}{x'(t)}$ is the rate of change of the slope of the tangent line at $\vec{r}(t)$. If $\frac{m'(t)}{x'(t)} > 0$, then the image of $\vec{m}(t)$ is increasing. This tells us that as we move from left to right on $\mathcal{C}$ at the point $\vec{r}(t)$, the slope of the tangent line is increasing and $\mathcal{C}$ is concave up. Similarly, if $\frac{m'(t)}{x'(t)} < 0$, then $\mathcal{C}$ is concave down at $\vec{r}(t)$. Computationally, since $m(t) = \frac{y'(t)}{x'(t)}$, we have

$$\frac{m'(t)}{x'(t)} = \frac{\frac{d}{dt}\left(\frac{y'(t)}{x'(t)}\right)}{x'(t)} = \frac{x'(t)y''(t) - y'(t)x''(t)}{(x'(t))^3}. \qquad (3)$$

Suppose that $\mathcal{C}$ is a curve in the plane parametrized by $\vec{r}$. Suppose further that $\vec{r}$ is continuously differentiable. If t_0 is a time such that $x'(t_0) \neq 0$, then there is a neighborhood (a, b) of t_0 such that $x'(t) \neq 0$ for $a < t < b$. Thus the part of $\mathcal{C}$ which is the image of (a, b) can be written as a function of x, say $y = f(x)$. Since $\frac{m'(t)}{x'(t)}$ is the rate that the slope of the tangent line is changing as x changes, it must be the case that $\frac{d^2 y}{dx^2} = \frac{m'(t)}{x'(t)}$. We follow convention by using $\frac{d^2 y}{dx^2}$ to denote $\frac{m'(t)}{x'(t)}$. It should be remembered that this describes the concavity of $\mathcal{C}$ only on an image of a segment of the domain of $\vec{r}$.

EXAMPLE 9: Let $\mathcal{C}$ be the image of the function with polar equation $r = \theta$, illustrated in Figure 8a. That is, let $\vec{P}(\theta) = \theta \vec{W}(\theta)$. In Figure 8.b, we zoom in by restricting θ to $-\frac{5\pi}{6} \leq \theta \leq \frac{5\pi}{6}$. Observe that $\vec{P}\left(-\frac{\pi}{2}\right) = \vec{P}\left(\frac{\pi}{2}\right) = \left(0, \frac{\pi}{2}\right)$. There is no well-defined tangent line there. However, as we see in Figures 8.c and 8.d, those parts of $\mathcal{C}$ that are the images of $\left(-\frac{5\pi}{6}, -\frac{\pi}{3}\right)$ and $\left(\frac{\pi}{3}, \frac{5\pi}{6}\right)$ are functions of x that are concave down.

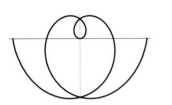

Figure 8.a *The image of $\theta\vec{W}(\theta)$,* $-\pi \le \theta \le \pi$.

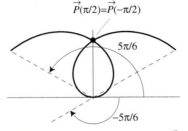

Figure 8.b *The image of $\theta\vec{W}(\theta)$,* $-\frac{5\pi}{6} \le \theta \le \frac{\pi}{6}$.

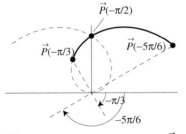

Figure 8.c *The image of $\theta\vec{W}(\theta)$,* $-\frac{5\pi}{6} \le \theta \le 0$.

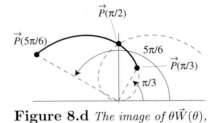

Figure 8.d *The image of $\theta\vec{W}(\theta)$,* $0 \le \theta \le \frac{\pi}{6}$.

$x'(\theta) = \cos(\theta) - \theta\sin(\theta)$, and $x''(\theta) = -2\sin(\theta) - \theta\cos(\theta)$.
$y'(\theta) = \sin(\theta) + \theta\cos(\theta)$, and $y''(\theta) = 2\cos(\theta) - \theta\sin(\theta)$.
Thus $x'y'' = 2\cos^2(\theta) - 3\theta\sin(\theta)\cos(\theta) + \theta^2\sin^2(\theta)$, and
$y'x'' = -2\sin^2(\theta) - 3\theta\sin(\theta)\cos(\theta) - \theta^2\cos^2(\theta)$.
So the rate that the slope of the tangent line is changing at $\theta = \frac{\pi}{2}$ is given by

$$\left.\frac{d^2y}{dx^2}\right|_{\theta=\pi/2} = \frac{m'(\pi/2)}{x'(\pi/2)} = \left.\frac{2 + \theta^2}{(\cos(\theta) - \theta\sin(\theta))^3}\right|_{\theta=\pi/2} = -\frac{16 + 2\pi^2}{\pi^3}.$$

Similarly,

$$\left.\frac{d^2y}{dx^2}\right|_{\theta=-\pi/2} = -\frac{16 + 2\pi^2}{\pi^3}. \qquad\blacksquare$$

EXAMPLE 10: In Example 6 of Section 5.3, we determined where the graph of $\vec{r}(t) = \left(\frac{t}{2} + \cos t, \sin t\right)$ was increasing and decreasing. We showed that $dx/dt = \frac{1}{2} - \sin t$ and $m(t) = \dfrac{(\cos t)}{\left(\frac{1}{2} - \sin t\right)}$.
By employing Equation (3), we see that

$$\frac{d^2y}{dx^2} = \frac{m'(t)}{x'(t)} = \frac{\frac{d}{dt}\frac{\cos t}{1/2-\sin t}}{1/2 - \sin t}$$

$$= \frac{-\frac{1}{2}\sin t + 1}{\left(\frac{1}{2} - \sin t\right)^3}.$$

Since $-\frac{1}{2}\sin t + 1$ is always positive, we know that $\frac{d^2y}{dx^2}$ can change sign only when $\left(\frac{1}{2} - \sin t\right) = 0$. Thus $\frac{m'(t)}{x'(t)}$ changes sign when $t = \frac{\pi}{6} + 2n\pi$ and when $t = \frac{5\pi}{6} + 2n\pi$. On the intervals $\left(\frac{\pi}{6} + 2n\pi, \frac{5\pi}{6} + 2n\pi\right)$ the graph of $\vec{r}$ is concave down and on the intervals $\left(\frac{5\pi}{6} + 2n\pi, \frac{\pi}{6} + 2(n+1)\pi\right)$, the graph of $\vec{r}$ is concave up. This information is consistent with the sketch of Figure 6 in Section 5.3.

EXERCISES 5.4

Figures 9.a–9.d are reference figures for Exercises 1–12.

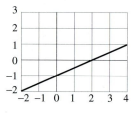

Figure 9.a

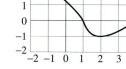

Figure 9.b

Figure 9.c

Figure 9.d

In Exercises 1–4,

   a. Approximate the location of the inflection points of the function f.

   b. Determine where f is concave up and where f is concave down.

1. The graph of f is the function sketched in Figure 9.a.

2. The graph of f is the function sketched in Figure 9.b.

3. The graph of f is the function sketched in Figure 9.c.

4. The graph of f is the function sketched in Figure 9.d.

In Exercises 5–8,

   a. Determine where f is increasing and where f is decreasing.

   b. Approximate the location of the local extrema of f.

   c. The inflection points of f.

   d. Determine where f is concave up and where f is concave down.

5. The graph of derivative of f is sketched in Figure 9.a.

6. The graph of derivative of f is sketched in Figure 9.b.

7. The graph of derivative of f is sketched in Figure 9.c.

8. The graph of derivative of f is sketched in Figure 9.d.

In Exercises 9–12,

a. Approximate the location of the inflection points of f.

b. Determine where f is concave up and where f is concave down.

9. The graph of the second derivative of f is sketched in Figure 9.a.

10. The graph of the second derivative of f is sketched in Figure 9.b.

11. The graph of the second derivative of f is sketched in Figure 9.c.

12. The graph of the second derivative of f is sketched in Figure 9.d.

13. Given that $f(-1) = 2$, $f(0) = 3$, and $f(1) = 5$, use the Mean Value Theorem to show that there are numbers c_1 and c_2 such that $-1 < c_1 < 0$, $0 < c_2 < 1$, and $f'(c_1) < f'(c_2)$. Explain why f is not concave down on the interval $[-1, 1]$.

In Exercises 14–21, use the second derivative test to find the relative extrema of the given function.

14. $y = x^2 - 2x - 6$.

15. $f(x) = x^3 - 6x^2 + 9x + 1$.

16. $g(x) = 4x^3 - x^5$.

17. $f(\beta) = \beta - \cos \beta$.

18. $f(t) = \dfrac{t^3 + 1}{t^3 - 1}$.

19. $f(x) = \dfrac{(x+1)^2}{x^2 + 1}$.

20. $h(t) = \dfrac{t^2}{t + 1}$.

21. $r(t) = \dfrac{1}{t^2 + 4}$.

In Exercises 22–41, find all relative maxima, relative minima, and points of inflection; then sketch the function.

22. $f(x) = x^3 + 2x^2 - 4x + 1$.

23. $f(x) = x^3 + 3x^2 + 6x + 8$.

24. $g(x) = (x + 6)^2(x - 2)^2$.

25. $h(x) = (x + 6)^2(x - 2)^3$.

26. $r(t) = t^3 + \dfrac{24}{t}$.

27. $h(t) = t^3 - \dfrac{24}{t}$.

28. $f(x) = \dfrac{x^2 - 4}{x}$.

29. $f(x) = \dfrac{x^2 - 4}{x + 2}$.

30. $g(t) = \dfrac{t^2 - 4}{t + 1}$.

31. $f(t) = \dfrac{t^2 - 4}{t + 4}$.

32. $f(x) = \dfrac{x}{x^2 + 1}$.

33. $f(t) = \dfrac{t^2 + 1}{t^2 - 1}$.

34. $f(t) = \sin t + \cos t$.

35. $g(\beta) = \sin^2 \beta$.

36. $f(x) = 2\sqrt{3} \sin x + \cos(2x)$.

37. $f(x) = \sin x + \sqrt{3} \cos x$.

38. $f(t) = \sin(t) \cos(t)$.

39. $f(t) = |t^2 - 1|$.

40. $f(t) = |\sin t|$.

41. $f(x) = |x^3|$.

In Exercises 42–47, investigate the function for absolute extrema.

42. $f(x) = x - 2$.

43. $f(x) = \dfrac{x}{x^2 + 1}$.

44. $f(x) = \sin x$.

45. $f(x) = \dfrac{x}{x^2 - 1}$.

46. $f(x) = \dfrac{x}{x^2 - 1}$, $-4 \le x \le 2$.

47. $f(x) = \dfrac{x}{x^2 - 4}$, $-1 \le x \le 1$.

48. Let $f(x) = \dfrac{ax + b}{cx + d}$ with $ad - bc \ne 0$.

    a. Show that f has no extrema.

    b. Show that f is either always increasing or always decreasing.

    c. Find all inflection points.

49. Let $f(x) = \dfrac{x^2 - a^2}{x - b}$.

    a. Show that if $b \le a$, then f has no extrema.

    b. Show that if $b > a$, then f has exactly one relative maximum and one relative minimum.

50. Let $\mathcal{C}$ be the curve parametrized by $\vec{r}(t) = (x(t), y(t))$, where $x(t) = \dfrac{1 - t^2}{1 + t^2}$ and $y(t) = \dfrac{2t}{1 + t^2}$.

    a. Find a number b so that the images of $(-\infty, b)$ and (b, ∞) are functions.

    b. Determine where $\mathcal{C}$ is increasing and decreasing.

    c. Determine the concavity of $\mathcal{C}$.

    d. Show that $\mathcal{C}$ is a subset of the unit circle centered at the origin.

    e. Are there any points on the unit circle that are not in $\mathcal{C}$?

51. Match the function in the first column with its derivative in the second column and its second derivative in the third column.

Function a.   Derivative a.   2nd derivative a.

Function b.   Derivative b.   2nd derivative b.

Function c.   Derivative c.   2nd derivative c.

Function d.   Derivative d.   2nd derivative d.

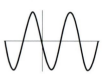

Function e.   Derivative e.   2nd derivative e.

In Exercises 52–55, calculate $\dfrac{d^2 y}{dx^2}$ for the image of $\vec{r}(t)$ at $\vec{r}(t_0)$.

52. $\vec{r}(t) = (t^3, t^2 - 4)$, $t_0 = 2$.

53. $\vec{r}(t) = (t \cos t, t \sin(t))$, $t_0 = \frac{\pi}{4}$.

54. $\vec{r}(t) = (x(t), y(t))$, where $x(t) = \dfrac{1 - t^2}{1 + t^2}$ and $y(t) = \dfrac{2t}{1 + t^2}$, $t_0 = -2$.

55. $\vec{r}(t) = (x(t), y(t))$, where $x(t) = \dfrac{3t}{1 + t^3}$ and $y(t) = \dfrac{3t^2}{1 + t^3}$, $t_0 = 2$.

56. Let $\mathcal{C}$ be the curve parametrized by $\vec{r}(t) = (t^4, t^3)$.

    a. Find a number b so that the images of $(-\infty, b)$ and (b, ∞) are functions.

    b. Determine where $\mathcal{C}$ is increasing and decreasing.

    c. Determine the concavity of $\mathcal{C}$.

    d. Sketch $\mathcal{C}$.

In Exercises 57–60, determine where the parametrized curve is concave up and where it is concave down. Sketch the curve.

57. $\vec{r}(t) = (\cos t, \sin t)$.

58. $\vec{r}(t) = (t^2, t^3)$.

59. $\vec{r}(t) = \vec{a}t + \vec{b}, \quad \vec{a} = (a_1, a_2), a_1 \neq 0, \vec{b} = (b_1, b_2)$.

60. $\vec{r}(t) = (t + \cos t, 1 - \sin t)$.

61. Let $f(t) = \sin\left(\frac{1}{t}\right)$.

    a. Find the critical points for $f(t)$.

    b. Use Newton's method to find the inflection point in each of the intervals $\left(\frac{2}{3\pi + 2n\pi}, \frac{2}{\pi + 2n\pi}\right)$, for $n = 1$, 2, and 3. Use a point "slightly" smaller than the midpoint of the given segment for the starting point.

5.5   Tangential and Normal Components of Acceleration

A fundamental idea in understanding the geometry of parametrized curves in space is to find the normal and tangential components of the acceleration of a particle moving along the curve. That is, we decompose the acceleration vector into two parts. One part is parallel to the tangent vector and the other part is normal, or perpendicular, to the tangent vector. We then use this information to fit the curve with a circle that describes how the curve is turning at a point.

First, however, we need to address more carefully a point that we have been assuming to be true. Suppose that $\vec{r}(t)$, $t_0 \leq t$ describes the position of a particle moving in space. We have been claiming that $\|\vec{v}(t)\|$ is the speed of the particle. Now the speed is supposed to be the rate of change of distance traveled. We define

$$s(t) = \text{total distance traveled from time } t_0 \text{ to time } t.$$

The assumption is that if $h > 0$, then

$$s(t + h) - s(t) \text{ approximates } \|\vec{r}(t + h) - \vec{r}(t)\| \text{ (see Figure 1)}$$

well enough that

$$s'(t) = \lim_{h \to 0^+} \frac{s(t + h) - s(t)}{h} = \lim_{h \to 0^+} \left\| \frac{\vec{r}(t + h) - \vec{r}(t)}{h} \right\| = \|\vec{v}(t)\|.$$

Recall that if the projection of $\vec{a}$ in the direction of $\vec{v}$ points in the same direction as $\vec{v}$, then $\vec{a} \cdot \vec{v}$ is positive. If the projection of $\vec{a}$ in

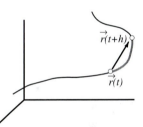

Figure 1. $s(t + h) - s(t)$ *approximates the length of the displacement $\vec{r}(t + h) - \vec{r}(t)$.*

the direction of $\vec{v}$ is opposite the direction of $\vec{v}$, then $\vec{a} \cdot \vec{v}$ is negative. If $\vec{a} \cdot \vec{v} = 0$, then the acceleration is changing only the direction of motion, not the speed. In fact, we have the following theorem.

Theorem 1 *If $\vec{v}(t)$ is the velocity of a particle at time t, then*

$$\frac{d\|\vec{v}(t)\|}{dt} = \frac{\vec{a}(t) \cdot \vec{v}(t)}{\|\vec{v}(t)\|}.$$

Since $\|\vec{v}(t)\|$ is the speed of the particle, $d\|\vec{v}(t)\|/dt$ is the rate of change of speed of the particle. Thus if $\vec{a}(t) \cdot \vec{v}(t)$ is positive, then the particle is speeding up at time t, and if $\vec{a}(t) \cdot \vec{v}(t)$ is negative, then the particle is slowing down at time t. If the particle is moving at constant speed, then $\vec{a}(t) \cdot \vec{v}(t) = 0$.

Proof: Recall that

$$\|\vec{v}(t)\|^2 = \vec{v}(t) \cdot \vec{v}(t).$$

If we take the derivative of both sides, then we obtain

$$2\|\vec{v}(t)\|\frac{d}{dt}\|\vec{v}(t)\| = \vec{v}(t) \cdot \vec{v}'(t) + \vec{v}'(t) \cdot \vec{v}(t) = 2\vec{v}(t) \cdot \vec{v}'(t).$$

Solving for $\dfrac{d}{dt}\|\vec{v}(t)\|$, we obtain

$$\frac{d}{dt}\|\vec{v}(t)\| = \frac{\vec{a}(t) \cdot \vec{v}(t)}{\|\vec{v}(t)\|}. \qquad \blacksquare$$

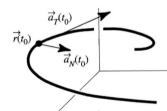

Figure 2. *A sketch of $\vec{a}_T(t_0)$ and $\vec{a}_N(t_0)$.*

Definition: Tangential and Normal Acceleration

The *tangential acceleration* $\vec{a}_T$ is defined as the projection of $\vec{a}$ in the direction of $\vec{v}$, that is: $\vec{a}_T = \dfrac{\vec{a} \cdot \vec{v}}{v^2}\vec{v} = \dfrac{\vec{a} \cdot \vec{v}}{\|\vec{v}\|}\left(\dfrac{\vec{v}}{\|\vec{v}\|}\right).$
The *normal acceleration* $\vec{a}_N$ is $\vec{a}_N = \vec{a} - \vec{a}_T$. Thus $\vec{a}_N = \mathrm{orth}_{\vec{v}}\vec{a}$.

Note: According to Theorem 1, if the particle is moving at constant speed, then $\vec{a}_T(t) = \vec{0}$.

EXAMPLE 1: When a particle is at $\vec{r}_0 = (-1, -1, -1)$m, its velocity is $\vec{v} = (1, -2, 3)$m/s and its acceleration is $\vec{a} = (-2, 3, 0)$m/s^2. Find the normal and tangential components of its acceleration and determine if it is speeding up or slowing down.

SOLUTION:

$$\vec{a}_T = \mathrm{proj}_{\vec{v}}\vec{a} = \frac{\vec{a} \cdot \vec{v}}{\vec{v} \cdot \vec{v}}\vec{v}$$

$$= \frac{(-2,3,0)\mathrm{m/s^2} \cdot (1,-2,3)\mathrm{m/s}}{(1,-2,3)\mathrm{m/s} \cdot (1,-2,3)\mathrm{m/s}}(1,-2,3)\mathrm{m/s}$$

$$= -\frac{4}{7}(1,-2,3)\mathrm{m/s^2}.$$

$$\vec{a}_N = \vec{a} - \vec{a}_T$$

$$= (-2,3,0)\mathrm{m/s^2} - \left(-\frac{4}{7},\frac{8}{7},-\frac{12}{7}\right)\mathrm{m/s^2}$$

$$= \left(-\frac{10}{7},\frac{13}{7},\frac{12}{7}\right)\mathrm{m/s^2}.$$

Notice that $\vec{a} \cdot \vec{v} = (-2,3,0) \cdot (1,-2,3) = -8 < 0$. Therefore, the particle is slowing down. ∎

EXAMPLE 2: Consider the function $\vec{r}(t) = t\vec{W}(t) = (t\cos t, t\sin t)$. Then

$$\vec{v}(t) = \vec{r}'(t) = (\cos t - t\sin t, \sin t + t\cos t),$$

and

$$\vec{a}(t) = \vec{v}'(t) = (-2\sin t - t\cos t, 2\cos t - t\sin t).$$

Some simple computations show that $\vec{a}(t) \cdot \vec{v}(t) = t$ and $\|\vec{v}(t)\| = (1 + t^2)^{1/2}$; and so,

$$\frac{d}{dt}\|\vec{v}(t)\| = \frac{\vec{v}(t) \cdot \vec{a}(t)}{\|\vec{v}(t)\|} = \frac{t}{(1+t^2)^{1/2}}$$

and

$$\vec{a}_T(t) = \frac{t}{(1+t^2)}(\cos t - t\sin t, \sin t + t\cos t).$$ ∎

EXAMPLE 3: Let C be the graph of the function $f(x)$, $a \le x \le b$. Find $\vec{a}_N$, $\vec{a}_T$, $\|\vec{a}_N\|$, and $\|\vec{a}_T\|$. Assume that C is parametrized by $\vec{r}(x) = (x, f(x))$.

SOLUTION:

$$\vec{v}(x) = (1, f'(x)), \quad \vec{a}(x) = (0, f''(x)),$$

$$\vec{a}_T(x) = \frac{(0, f''(x)) \cdot (1, f'(x))}{1 + (f'(x))^2}(1, f'(x)) = \frac{f'(x)f''(x)}{1 + (f'(x))^2}(1, f'(x))$$

$$\vec{a}_N(x) = \vec{a}(x) - \vec{a}_T(x)$$

$$= \left(-\frac{f'(x)f''(x)}{1 + (f'(x))^2}, f''(x) - \frac{(f'(x))^2 f''(x)}{1 + (f'(x))^2}\right)$$

$$= \frac{f''(x)}{1 + (f'(x))^2}(-f'(x), 1),$$

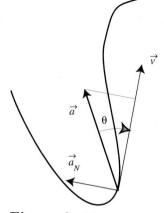

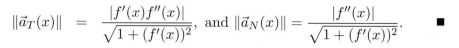

$$\|\vec{a}_T(x)\| = \frac{|f'(x)f''(x)|}{\sqrt{1+(f'(x))^2}}, \text{ and } \|\vec{a}_N(x)\| = \frac{|f''(x)|}{\sqrt{1+(f'(x))^2}}. \qquad \blacksquare$$

It is often the case that you need only $\|\vec{a}_N\|$ and $\|\vec{a}_T\|$. Since $\vec{a}_T$ is the projection of $\vec{a}$ onto $\vec{v}$, we have a relatively simple expression for $\|\vec{a}_T\|$:

$$\|\vec{a}_T\| = \frac{|\vec{v} \cdot \vec{a}|}{\|\vec{v}\|}. \qquad (4)$$

It turns out that we have an equally simple expression for $\|\vec{a}_N\|$. Consider the illustration in Figure 3.

$$\|\vec{a}_N\| = \|\vec{a}\| \sin\theta = \frac{\|\vec{a}\|\,\|\vec{v}\|}{\|\vec{v}\|} \sin\theta = \frac{\|\vec{v} \times \vec{a}\|}{\|\vec{v}\|}. \qquad (5)$$

Figure 3. *The vectors $\vec{v}$, $\vec{a}$ and $\vec{a}_N$ drawn emanating from $\vec{r}$.*

Notice that if the curve C lies in $\mathbb{R}^2$, then it can be parametrized by $\vec{r}(t) = (x(t), y(t))$, and the equation for $\|\vec{a}_N\|$ takes on a simple form.

$$\|\vec{a}_N\| = \frac{|x''y' - y''x'|}{\|\vec{r}'\|}. \qquad (6)$$

To recapitulate:

$\vec{a}(t) = \vec{v}'(t) = \vec{r}''(t)$ is the acceleration. Its magnitude is measured in (unit of length)/(unit of time)^2.

$v'(t) = d\|\vec{v}(t)\|/dt = (\vec{a}(t) \cdot \vec{v}(t))/\|\vec{v}(t)\|$ is the rate of change of the speed of the particle and it is measured in (unit of length)/(unit of time)^2. If $v'(t) > 0$, the motion is speeding up. If $v'(t) < 0$, the motion is slowing down.

$\vec{a}_T(t) = \dfrac{\vec{a}(t) \cdot \vec{v}(t)}{\vec{v}(t) \cdot \vec{v}(t)} \vec{v}(t)$ is the tangential acceleration. It points in (or opposite to) the direction of motion, and its magnitude is the magnitude of the rate of change of speed, $|v'(t)|$. It is measured in (unit of length) /(unit of time)^2.

$$\|\vec{a}_T(t)\| = \frac{|\vec{a}(t) \cdot \vec{v}(t)|}{\|\vec{v}(t)\|}.$$

$\vec{a}_N(t) = \vec{a}(t) - \vec{a}_T(t)$ is the normal component of the acceleration. It is normal to the direction of motion and is measured in (unit of length)/(unit of time)^2. Physically, it is that part of the acceleration that is changing the direction of motion.

$$\|\vec{a}_N(t)\| = \frac{\|\vec{a}(t) \times \vec{v}(t)\|}{\|\vec{v}(t)\|}.$$

EXAMPLE 4: Find $\|\vec{a}_N(t)\|$ for the ellipse parametrized by $\vec{r}(t) = (A\cos t, B\sin t)$.

SOLUTION:

$$\vec{r}\,'(t) = (-A\sin t, B\cos t) \text{ and } \vec{r}\,''(t) = (-A\cos t, -B\sin t).$$

Using Equation 6, we have

$$
\|\vec{a}_N(t)\| = \frac{|-AB\cos^2(t) - AB\sin^2(t)|}{\sqrt{A^2\sin^2(t) + B^2\cos^2(t)}}
$$

$$
= \frac{|AB|}{\sqrt{A^2\sin^2(t) + B^2\cos^2(t)}}. \qquad \blacksquare
$$

The Tangent, Normal, and Binormal Vectors

Assume that $\vec{a}_N(t) \neq \vec{0}$ and $\vec{v}(t) \neq \vec{0}$. Let $\hat{t}(t) = \frac{\vec{v}(t)}{\|\vec{v}(t)\|}$, $\hat{n}(t) = \frac{\vec{a}_N(t)}{\|\vec{a}_N(t)\|}$ and $\hat{b}(t) = \hat{t}(t) \times \hat{n}(t)$. See Figure 4. Since $\hat{t}(t)$ and $\hat{n}(t)$ are perpendicular, $\hat{t}(t)$, $\hat{n}(t)$, and $\hat{b}(t)$ are orthogonal unit vectors. We present the following definition.

Definition: The Tangent, Normal, and Binormal Vectors

If $\vec{a}_N(t) \neq \vec{0}$ and $\vec{v}(t) \neq \vec{0}$, then

$\hat{t}(t) = \vec{v}(t)/\|\vec{v}(t)\|$ is called the *(unit) tangent vector* at $\vec{r}(t)$.

$\hat{n}(t) = \vec{a}_N(t)/\|a_N(t)\|$ is called the *(unit) normal vector* at $\vec{r}(t)$.

$\hat{b}(t) = \hat{t}(t) \times \hat{n}(t)$ is called the *(unit) binormal vector* at $\vec{r}(t)$.

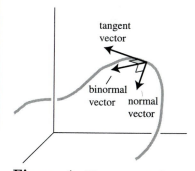

Figure 4. *The tangential, normal, and binormal vectors.*

One of the purposes of these unit vectors is to give a local coordinate system for each point on the path using intrinsic properties of the curve and the direction of motion along the curve. Imagine that you are driving a car on a curvy road. The velocity vector is pointed straight ahead, the normal component of acceleration is pointed to the inside of the turn, and the binormal vector is pointed either up (if the inside of the turn is to your left) or down (if the inside of the turn is to your right).

Those three directions may be used to fashion a coordinate system with the origin at your position. This coordinate system is time dependent; it changes as time changes. (What happens if you are traveling in a straight line?)

EXAMPLE 5: From Example 1, a particle is at $\vec{r}_0 = (-1, -1, -1)$ m, its velocity is $\vec{v} = (1, -2, 3)$m/s, and its acceleration is $\vec{a} = (-2, 3, 0)$m/s^2. We found that $\vec{a}_N = \left(-\frac{10}{7}, \frac{13}{7}, \frac{12}{7}\right)$ m/s^2. Find the unit vectors $\hat{t}$, $\hat{n}$, and $\hat{b}$.

SOLUTION:

$$\hat{t} = \frac{\vec{v}}{\|\vec{v}\|} = \frac{1}{\sqrt{14}}(1, -2, 3).$$

$$\hat{n} = \frac{\vec{a}_N}{\|\vec{a}_N\|} = \frac{7}{\sqrt{413}}\left(-\frac{10}{7}, \frac{13}{7}, \frac{12}{7}\right) \text{ m/s}^2 = \frac{1}{\sqrt{413}}(-10, 13, 12).$$

We use the result from Exercise 20 where you are asked to show that

$$\hat{b}(t) = \frac{\vec{v}(t) \times \vec{a}(t)}{\|\vec{v}(t) \times \vec{a}(t)\|}.$$

Thus

$$\hat{b} = \frac{\vec{v} \times \vec{a}}{\|\vec{v} \times \vec{a}\|} = \frac{1}{\sqrt{118}}(-9, -6, -1). \qquad \blacksquare$$

EXAMPLE 6: Find the binormal vector for the function $\vec{r}(t) = (\cos t, \sin t, t)$.

SOLUTION: The velocity vector is $\vec{v}(t) = (-\sin t, \cos t, 1)$ and $\vec{a}(t) = (-\cos t, -\sin t, 0)$. Thus

$$\hat{b}(t) = \frac{\vec{v}(t) \times \vec{a}(t)}{\|\vec{v}(t) \times \vec{a}(t)\|} = \frac{(\sin t, -\cos t, 1)}{\sqrt{2}}. \qquad \blacksquare$$

EXERCISES 5.5

1. In each of the diagrams, a particle is moving with velocity $\vec{v}$ and acceleration $\vec{a}$. Determine whether the particle is speeding up or slowing down and whether it is turning to the right or to the left.

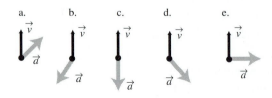

The curves in Exercises 2 and 3 are paths of a particle moving at constant speed. Sketch a vector that points in the direction of the acceleration at the points $\vec{p}_1$, $\vec{p}_2$, and $\vec{p}_3$.

2.   3.

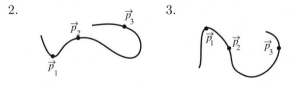

In Exercises 4–7, you are given $\vec{a}(t_0)$ and $\vec{v}(t_0)$. Find $\vec{a}_T(t_0)$, $\vec{a}_N(t_0)$, $\hat{t}(t_0)$, $\hat{n}(t_0)$, and $\hat{b}(t_0)$.

4. $\vec{a}(t_0) = (1, -1, 0)$, $\vec{v}(t_0) = (0, 1, 2)$.

5. $\vec{a}(t_0) = (2, -1, 1)$, $\vec{v}(t_0) = (0, 1, 0)$.

6. $\vec{a}(t_0) = (0, -3, -2)$, $\vec{v}(t_0) = (2, 1, 0)$.

7. $\vec{a}(t_0) = (2, -1, 4)$, $\vec{v}(t_0) = (-2, 1, 3)$.

In Exercises 8–12, find the normal and tangential components of acceleration at the given point. Assume that the parametrization is given by $\vec{r}(t) = (t, f(t))$.

8. $f(t) = t^2$, $t_0 = 2$.   9. $f(t) = t^3$, $t_0 = 2$.

10. $f(t) = t^3$, $t_0 = 0$.   11. $f(t) = \sin t$, $t_0 = 0$.

12. $f(t) = \sin t$, $t_0 = \frac{\pi}{4}$.

In Exercises 13–16, C is the curve parametrized by $\vec{r}$. Find $\vec{a}_T$ and $\vec{a}_N$ at the given point.

13. $\vec{r}(t) = (\cos t, \sin t)$, $\vec{r}(\pi/3)$.

14. $\vec{r}(t) = (2\cos t, \sin t)$, $\vec{r}(0)$.

15. $\vec{r}(t) = (t\cos t, t\sin t)$, $\vec{r}(0)$.

16. $\vec{r}(t) = (t, \cos t, \sin t)$, $\vec{r}(0)$.

In Exercises 17–19, find the vectors $\hat{t}(t), \hat{n}(t)$ and $\hat{b}(t)$.

17. $\vec{v}(t) = (t^2, t^3, t)$, $t_0 = 1$.

18. $\vec{v}(t) = (t^2 + t, 2t + 1, 2t + 2)$, $t_0 = 0$.

19. $\vec{v}(t) = (t^2 - t, t - 1, t^3)$, $t_0 = 1$.

20. Show that $\vec{v}(t) \times \vec{a}(t) = \vec{v}(t) \times \vec{a}_N(t)$ and that
$$\hat{b}(t) = \frac{\vec{v}(t) \times \vec{a}(t)}{\|\vec{v}(t) \times \vec{a}(t)\|}.$$

21. The paths in Figures 5.a and 5.b are parametrized to model particles moving in the direction indicated by the arrows at a constant speed. Sketch a representation of $\vec{v}(t)$ and $\vec{a}(t)$, and determine the direction of the binormal vector $\hat{b}(t)$ at the points $\vec{P}_1$, $\vec{P}_2$, and $\vec{P}_3$.

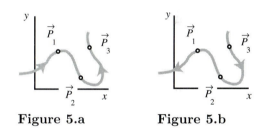

Figure 5.a   Figure 5.b

5.6   Circular Motion and Curvature

Circular Motion

A particularly important case of parametrized motion is the description of circular motion. Suppose that a particle is moving around a circle of radius R centered at the origin. Then the motion will be of the form

$$\vec{r}(t) = R\vec{W}(\theta(t)).$$

As illustrated in Figure 1, $\theta(t)$ is the angle swept out at time t. Differentiating $\vec{r}(t)$, we obtain

$$\vec{v}(t) = \vec{r}'(t) = R\theta'(t)\vec{W}'(\theta(t)). \qquad (7)$$

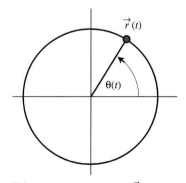

Figure 1. $\vec{r}(t) = R\vec{W}(\theta(t))$.

The rotation is in the xy–coordinate plane. The vector $\hat{k}$ is perpendicular to the plane of rotation, so it points along the axis of rotation. We define $\vec{\omega}(t) = \theta'(t)\hat{k}$. $\vec{\omega}(t)$ is called *angular velocity* of the motion, and $\|\vec{\omega}(t)\| = |\theta'(t)|$ is called *angular speed* of the motion. See

Figure 2. Now,

$$\begin{aligned}
\vec{\omega}(t) \times \vec{r}(t) &= R\left(0, 0, \theta'(t)\right) \times (\cos(t), \sin(t), 0) \\
&= R\theta'(t)\left(-\sin(t), \cos(t), 0\right) \\
&= \vec{r}'(t).
\end{aligned}$$

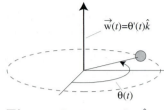

Figure 2. $\vec{\omega}(t) = \theta'(t)\hat{k}$.

Thus we have that if a particle is rotating about a point in the plane, then its velocity is

$$\vec{v}(t) = \vec{\omega}(t) \times \vec{r}(t), \text{ where } \vec{\omega}(t) = \theta'(t)\hat{k}. \tag{8}$$

Since $\vec{\omega}(t)$ and $\vec{r}(t) = R\vec{W}(t)$ are perpendicular, we know that $\|\vec{\omega}(t) \times \vec{r}(t)\| = \|\vec{\omega}(t)\| \, \|\vec{r}(t)\| = R\|\vec{\omega}(t)\|$. The speed is given by

$$\|\vec{v}(t)\| = R|\theta'(t)|. \tag{9}$$

Differentiating both sides of Equation (8) yields

$$\vec{a}(t) = \vec{\omega}\,'(t) \times \vec{r}(t) + \vec{\omega}(t) \times \vec{r}'(t).$$

Substituting $\vec{r}'(t) = \vec{\omega}(t) \times \vec{r}(t)$, we have

$$\vec{a}(t) = \vec{\omega}\,'(t) \times \vec{r}(t) + \vec{\omega}(t) \times (\vec{\omega}(t) \times \vec{r}(t)). \tag{10}$$

We denote $\vec{\omega}\,'(t) = \theta''(t)\hat{k}$ by $\vec{\alpha}(t)$. $\vec{\alpha}(t)$ is called the *angular acceleration*. With this notation, Equation (10) becomes

$$\vec{a}(t) = \vec{\alpha}(t) \times \vec{r}(t) + \vec{\omega}(t) \times (\vec{\omega}(t) \times \vec{r}(t)). \tag{11}$$

In your science and engineering courses, you are likely to see Equations (8) and (11) written with the parameter t suppressed as below:

$$\vec{v} = \vec{\omega} \times \vec{r} \text{ and } \vec{a} = \vec{\alpha} \times \vec{r} + \vec{\omega} \times (\vec{\omega} \times \vec{r}).$$

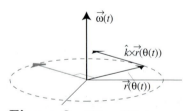

Figure 3. $\vec{\alpha}(t) \times \vec{r}(t)$ *is tangent to the circle of motion.*

We gain some useful information by inspecting the two parts in the sum in Equation (11). First, consider

$$\vec{\alpha}(t) \times \vec{r}(t) = \theta''(t)\hat{k} \times \vec{r}(t).$$

Recall that $\vec{r}(t)$ points radially away from the center of rotation and hence is perpendicular to $\hat{k}$. Thus $\hat{k} \times \vec{r}(t)$ (and therefore $(\vec{\alpha}(t) \times \vec{r}(t))$) is tangent to the circle of motion. See Figure 3.

Now consider $\vec{\omega} \times (\vec{\omega} \times \vec{r})$. $\vec{\omega} \times \vec{r}$ is tangent to the circle of motion. It follows that $\vec{\omega} \times (\vec{\omega} \times \vec{r})$ is normal to the circle of motion.

We may conclude that $\vec{\alpha}(t) \times \vec{r}(t)$ is the tangential component of the acceleration and that $\vec{\omega} \times (\vec{\omega} \times \vec{r})$ is the normal component of the acceleration. Thus

$$\vec{a}(t) = \vec{a}_N(t) + \vec{a}_T(t),$$

where

$$\vec{a}_T(t) = \vec{\alpha}(t) \times \vec{r}(t) \text{ and } \vec{a}_N(t) = \vec{\omega}(t) \times (\vec{\omega}(t) \times \vec{r}(t)) . \qquad (12)$$

Often in applications, the items of interest are $\|\vec{a}_N(t)\|$ and $\|\vec{a}_T(t)\|$.

$$
\begin{aligned}
\|a_T(t)\| &= \|\vec{\alpha}(t) \times \vec{r}(t)\| \\
&= \|\theta''(t)\hat{k} \times R\vec{W}(\theta(t))\| \\
&= \|R\theta''(t)\hat{k} \times \vec{W}(\theta(t))\| \\
&= R|\theta''(t)| \, \|\hat{k} \times \vec{W}(\theta(t))\| \\
&= R|\theta''(t)|. \\
\|a_N(t)\| &= \|\vec{\omega}(t) \times (\vec{\omega}(t) \times \vec{r}(t))\,\| \\
&= \left\| \theta'(t)\hat{k} \times \left(\theta'(t)\hat{k} \times R\vec{W}(\theta(t)) \right) \right\| \\
&= \left\| R(\theta'(t))^2 \left(\hat{k} \times \left(\hat{k} \times \vec{W}(\theta(t)) \right) \right) \right\| \\
&= R(\theta'(t))^2 = R\|\vec{\omega}(t)\|^2.
\end{aligned}
$$

Since $\|\vec{v}(t)\| = |\theta'(t)|R$, we have

$$\|\vec{a}_N(t)\| = \frac{\|\vec{v}(t)\|^2}{R} = \frac{\vec{v}(t) \cdot \vec{v}(t)}{R}. \qquad (13)$$

All of the equations for circular motion in the xy–plane extend directly to circular motion in an arbitrary plane. In the general case,

$\vec{\omega}(t)$ is a vector that is perpendicular to the plane of rotation (it points in the direction of the axis of rotation), and $\|\vec{\omega}(t)\|$ is the angular speed.

$\vec{\alpha}(t)$, the angular acceleration, is a scaler multiple of $\vec{\omega}(t)$.

The relationship between the direction of the angular velocity vector and the rotation follows the right hand rule. If you point the right hand thumb in the direction of the angular velocity vector, then your fingers will point in the direction of rotation. See Figure 4.

We have the following equations for circular motion.

Figure 4. *The right hand rule for circular motion.*

Equations for Circular Motion

Given that a particle is rotating in a plane about a point $\vec{C}$ and $\vec{r}(t)$ is the radius vector pointing from $\vec{C}$ to the particle's position, then

$$
\begin{aligned}
\vec{v}(t) &= \vec{\omega}(t) \times \vec{r}(t), \\
\vec{a}_T(t) &= \vec{\alpha}(t) \times \vec{r}(t), \\
\vec{a}_N(t) &= \vec{\omega}(t) \times (\vec{\omega}(t) \times \vec{r}(t)) = \vec{\omega}(t) \times \vec{v}(t), \\
\|\vec{a}_T(t)\| &= \|\vec{r}(t)\| \, \|\vec{\alpha}(t)\|,
\end{aligned}
$$

$$\|\vec{a}_N(t)\| = \|\vec{r}(t)\| \|\vec{\omega}(t)\|^2 = \|\vec{\omega}(t)\| \|\vec{v}(t)\|, \text{ and}$$

$$\|\vec{a}_N(t)\| = \frac{\vec{v}(t) \cdot \vec{v}(t)}{\|\vec{r}(t)\|}.$$

EXAMPLE 1: A particle is moving around a circle of radius 3 m with angular velocity (at a given time) $(1, -2, 3)$ radians/sec. Its angular acceleration is $\left(-\frac{1}{2}, 2, -\frac{3}{2}\right)$ rad/s^2.

Its angular speed is $\|(1, -2, 3)\|$ radians/sec $= \sqrt{14}$ radians/sec.

The particle's speed is $R\|\vec{\omega}\| = 3\sqrt{14}$ m/s.

$\|\vec{a}_N\| = R\|\vec{\omega}\|^2 = 42$ m/s^2.

$\|a_T\| = R|\vec{\alpha}| = 3 \times \sqrt{1/4 + 4 + 9/4} = 3\sqrt{13/2}$ m/s^2. ∎

EXAMPLE 2: A particle is moving around a circle with center $\vec{C} = \hat{\imath} + 2\hat{\jmath}$ m. At time t_0, the particle is at $\vec{r}(t_0) = -\hat{\imath} + 3\hat{\jmath}$ m, its angular velocity is $-2\hat{k}$ rad/s, and its angular acceleration $3\hat{k}$ rad/s^2. Find its velocity and the normal and tangential components of its acceleration at time t_0.

SOLUTION: $\vec{R} = -2\hat{\imath} + \hat{\jmath}$ m is the radius vector pointing from the center of rotation to the particle's position.

$$\begin{aligned}
\vec{v} &= \vec{\omega} \times \vec{R} = (0, 0, -2) \times (-2, 1, 0) = (2, 4, 0) \text{ m/s.} \\
\vec{a}_N &= \vec{\omega} \times \vec{v} = (0, 0, -2) \times (2, 4, 0) = (8, -4, 0) \text{ m/s}^2. \\
\vec{a}_T &= \vec{\alpha} \times \vec{r} = (0, 0, 3) \times (-2, 1, 0) = (-3, 6, 0) \text{ m/s}^2.
\end{aligned}$$ ∎

EXAMPLE 3: Suppose that a particle moves around a circle of radius ρ with constant angular speed $\|\vec{\omega}\|$. Notice that in this case the speed is a constant $v = \|\vec{\omega}\|\rho$, and there is no tangential component to its acceleration. The magnitude of $\vec{a}(t) = \vec{a}_N(t)$ is simply $\frac{v^2}{\rho}$ and its angular speed is $\|\vec{\omega}\| = \frac{v}{\rho}$. ∎

EXAMPLE 4: Suppose that a car is moving with a constant speed of 25 m/sec around a circular track with a radius of 100 m. Then the acceleration is normal to the track, and its magnitude is $(25\frac{m}{s})^2/100$ m $= 6.25\frac{m}{s^2}$. ∎

EXAMPLE 5: A particle's position is given by

$$\vec{s}(t) = \rho\vec{W}\left(\frac{v}{\rho}t\right),$$

where ρ and v are constants. Then its angular speed is $\|\vec{\omega}\| = \frac{v}{\rho}$ and, according to Equation (9), its speed is $\|\vec{\omega}\|\rho = \rho\frac{v}{\rho} = v$. It follows from Equation (13) that the magnitude of its acceleration is $a = \frac{v^2}{\rho}$. ∎

Curvature

We use the results of Example 5 to measure the way a path is curving at a point on the path. Suppose that $\vec{r}(t)$ parametrizes the curve C and that $\vec{r}_0 = \vec{r}(t_0)$ is a point on the curve. Assume that $\vec{r}'(t) \neq 0$ on the domain of $\vec{r}$.[10] $\vec{a}_N(t_0)$ is that part of the acceleration that is causing the path to curve at $\vec{r}_0$. Let $v = \|\vec{v}(t_0)\|$ and

$$\rho = \frac{\|\vec{v}(t_0)\|^2}{\|\vec{a}_N(t_0)\|}. \tag{14}$$

Consider the circle S parametrized by

$$\vec{s}(t) = \rho\vec{W}\left(\frac{vt}{\rho}\right) = \frac{1}{\kappa}\vec{W}\left(\frac{vt}{\rho}\right), \qquad \text{where } \kappa = \frac{1}{\rho}. \tag{15}$$

$\vec{s}$ models a point traveling around a circle of radius ρ with speed v and with acceleration of magnitude $\frac{v^2}{\rho}$. In the event that $\vec{a}_N(t) = 0$, we define $\rho = \infty$ and $\kappa = 0$. In order to see the relationship between $\vec{r}(t)$ and $\vec{s}(t)$, suppose that $\vec{r}(t)$ denotes the position of a particle p_1 on the curve C at time t and $\vec{s}(t)$ denotes the position of another particle p_2 on the circle at the same time. Then the speed of p_1 at time t_0 is the same as the speed of p_2, and the magnitude of the normal acceleration $\|a_N(t_0)\|$ of p_1 at time t_0 is the same as the (centripetal) acceleration of p_2. As illustrated in Figure 5, the circle parametrized by $\vec{s}(t)$ does not change just because the velocity of p_1 on the curve changes.

Suppose now that C is a curve parameterized by $\vec{r}(\tau)$. We want to relax the condition that the parameter τ is time. In this more general situation, we cannot assume that $\vec{r}'$ is velocity and $\vec{r}''$ is acceleration. If we let $\vec{r}''_N(\tau)$ denote the component of $\vec{r}''(t)$ that is normal to C at $\vec{r}(\tau_0)$, then Equation (14) can be reformulated by letting

$$\rho = \frac{\|\vec{r}'(\tau_0)\|^2}{\|\vec{r}''_N(\tau_0)\|}.$$

---

[10]This assumes that the particle does not stop as it moves along the curve.

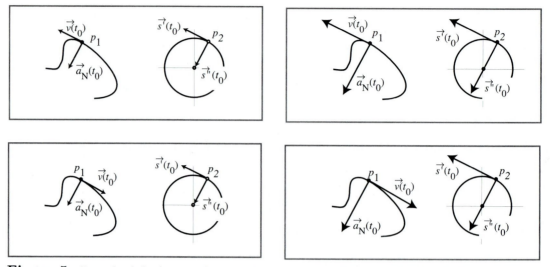

Figure 5. *In each of the frames, the particle p_1 moves on the curve C with velocity $\vec{v}_0$. The particle p_2 moves on the circle of radius ρ with the same speed.*

Equation (15) can be rewritten as

$$\vec{s}(\tau) = \rho \vec{W} \left(\frac{\|\vec{r}'(\tau_0)\| \tau}{\rho} \right).$$

Of course, $\vec{r}''_T$ and $\vec{r}''_N$ will be defined as their counterparts a_T and a_N. That is,

$$\vec{r}''_T(\tau) = \mathrm{proj}_{\vec{r}'(\tau)} \vec{r}''(\tau) = \left(\frac{\vec{r}'(\tau) \cdot \vec{r}''(\tau)}{\vec{r}'(\tau) \cdot \vec{r}'(\tau)} \right) \vec{r}'(\tau),$$

$$\vec{r}''_N(\tau) = \mathrm{orth}_{\vec{r}'(\tau)} \vec{r}''(\tau) = \vec{r}''(\tau) - \vec{r}''_T(\tau),$$

$$\|\vec{r}''_T(\tau)\| = \frac{|\vec{r}'(\tau) \cdot \vec{r}''(\tau)|}{\|\vec{r}'(\tau)\|} \quad \text{and} \quad \|\vec{r}''_N(\tau)\| = \frac{\|\vec{r}'(\tau) \times \vec{r}''(\tau)\|}{\|\vec{r}'(\tau)\|}. \quad (16)$$

Definition: Curvature and Radius of Curvature

Suppose that $\vec{r}$ parametrizes a curve C, $\vec{r}_0 = \vec{r}(\tau_0) \neq 0$ and $\vec{r}''$ is continuous at τ_0.

$$\rho(\vec{r}_0) = \begin{cases} \dfrac{\|\vec{r}'(\tau_0)\|^2}{\|\vec{r}''_N(\tau_0)\|}, & \text{if } \vec{r}''_N(\tau_0) \neq 0 \\ \infty, & \text{otherwise} \end{cases}$$

is called the *radius of curvature* of the curve C at $\vec{r}(t_0)$.

$$\kappa(\vec{r}_0) = \frac{\|\vec{r}''_N(\tau_0)\|}{\|\vec{r}'(\tau_0)\|^2} = \begin{cases} \dfrac{1}{\rho(t_0)}, & \text{if } \rho(\tau_0) \neq 0 \\ 0, & \text{if } \rho(\tau_0) = \infty \end{cases}$$

is called the *curvature* of C at $\vec{r}(\tau_0)$.

We can use Equation (16) to obtain ρ and κ without actually computing $\vec{r}''_N$. Suppressing τ, we obtain the following working equations.

Equations for Curvature and Radius of Curvature

$$\rho = \frac{(\vec{r}' \cdot \vec{r}')^{3/2}}{\|\vec{r}' \times \vec{r}''\|} \quad \text{and} \quad \kappa = \frac{\|\vec{r}' \times \vec{r}''\|}{(\vec{r}' \cdot \vec{r}')^{3/2}}.$$

To complete the picture, let P denote the plane containing the vectors $\vec{r}''_T(\tau_0)$ and $\vec{r}''_N(\tau_0)$, when drawn from the point $\vec{r}_0 = \vec{r}(\tau_0)$. The plane P is called the *osculating plane* for the curve C at $\vec{r}_0$. Now, let S_0 be a copy of the circle parametrized by $\vec{s}(\tau)$ drawn in the osculating plane with center at the point

$$\vec{r}_0 + \rho \frac{\vec{r}''_N(t_0)}{\|\vec{r}''_N(t_0)\|}.$$

The circle S_0 is called the *osculating circle* or the *circle of curvature* for C at $\vec{r}(t_0)$, and the center of S_0 is called the *center of curvature*. **The radius of curvature, the curvature, and the osculating circle and plane depend only on the geometry of the curve, and they are independent of the parametrization used to describe the curve.** That is, we will get the same curvature at a point on the curve regardless of the parametrization used to describe the curve. The term osculating is derived from the Latin word *osculum*, which means "kiss." The osculating circle touches the curve as though it were kissing it. See Figure 6.

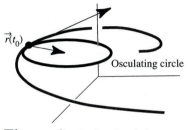

Figure 6. *A sketch of the osculating circle at a point on the curve.*

EXAMPLE 6: Given that $\vec{r}_0 = \vec{r}(t_0) = (1, 1, 1)$, $\vec{r}'(t_0) = (-1, 0, 1)$, and $\vec{r}''(t_0) = (0, -1, -1)$, find $\rho(\vec{r}_0)$ and the center of curvature at ρ_0.

SOLUTION: $\vec{r}' \cdot \vec{r}' = 2$ and $\|\vec{r}' \times \vec{r}''\| = \sqrt{3}$. Thus

$$\rho(\vec{r}_0) = \frac{(\vec{r}' \cdot \vec{r}')^{3/2}}{\|\vec{r}' \times \vec{r}''\|} = \sqrt{\frac{8}{3}}.$$

To find the center of curvture, we first find $\vec{r}''_T$ and $\vec{r}''_N$.

$$\vec{r}''_T = \frac{\vec{r}' \cdot \vec{r}''}{\vec{r}' \cdot \vec{r}'} \vec{r}' = \left(\frac{1}{2}, 0, -\frac{1}{2}\right)$$

and

$$\vec{r}''_N = \vec{r}'' - \vec{r}''_T = \left(-\frac{1}{2}, -1, -\frac{1}{2}\right).$$

The center of curvature is

$$
\begin{aligned}
\vec{r}_0 + \rho \frac{\vec{r}''_N}{\|\vec{r}''\|} &= (1,1,1) + \sqrt{\frac{8}{3}} \frac{1}{\sqrt{2}} \left(-\frac{1}{2}, -1, -\frac{1}{2} \right) \\
&= (1,1,1) + \left(\frac{2}{\sqrt{3}} \right) \left(-\frac{1}{2}, -1, -\frac{1}{2} \right) \\
&= \left(1 - \frac{1}{\sqrt{3}}, 1 - \frac{2}{\sqrt{3}}, 1 - \frac{1}{\sqrt{3}} \right) \qquad \blacksquare
\end{aligned}
$$

EXAMPLE 7: Let f be a real valued function, and let C be the curve parametrized by $\vec{r}(x) = (x, f(x))$. Then $\vec{r}'(x) = (1, f'(x))$, $\vec{r}''(x) = (0, f''(x))$, and the radius of curvature at $(x, f(x))$ is

$$
\begin{aligned}
\rho(\vec{r}(x)) &= \frac{(\vec{r}'(x) \cdot \vec{r}'(x))^{3/2}}{\|\vec{r}'(x) \times \vec{r}''(x)\|} = \frac{\left(1 + (f'(x))^2\right)^{3/2}}{|f''(x)|}. \\
\kappa(\vec{r}(x)) &= \frac{|f''(x)|}{(1 + (f'(x))^2)^{3/2}}. \qquad \blacksquare
\end{aligned}
$$

In the example below, we see that the curvature is large when the "rate of turn" of the curve is large, and the curvature is nearly zero when the rate of turn in the curve is small.

EXAMPLE 8: Let $f(x) = x^3 - 5x^2 + x + 5$.

(a) Compare the graph of f to the graph of its curvature over the interval $[-1, 5]$.

(b) Find the curvature and the center of the osculating circle at $(3.25, f(3.25))$ and at $(2.8, f(2.8))$.

(c) Sketch the osculating circle against the graph of f at $(3.25, f(3.25))$ and at $(2.8, f(2.8))$.

SOLUTION:

(a) We use the equation from Example 7. Let $\vec{r}(x) = (x, f(x))$.

$$
\begin{aligned}
\kappa(\vec{r}(x)) &= \frac{|f''(x)|}{(1 + (f'(x))^2)^{3/2}} \\
&= \frac{|6x - 10|}{(9x^4 - 60x^3 + 106x^2 - 20x + 2)^{3/2}}
\end{aligned}
$$

In Figure 7 we sketch the graph of f and the curvature of f on the same axis. Observe that the magnitude of $\kappa(\vec{r}(x))$ is large where f is curving sharply and it approaches 0 when f flattens out.

(b) $\rho(\vec{r}(3.25)) = 1/\kappa(\vec{r}(3.25)) = 2.537$. The center of the osculating circle is

$$(3.25, f(3.25)) + \rho(3.25)\frac{\vec{a}_N(3.25)}{\|\vec{a}_N(3.25)\|} \approx (1.116, -9.008).$$

$\rho(\vec{r}(2.8)) = 1/\kappa(\vec{r}(2.8)) = 5.204$. The center of the osculating circle is

$$(2.8, f(2.8)) + \rho(2.8)\frac{\vec{a}_N(2.8)}{\|\vec{a}_N(2.8)\|} \approx (7.629, -8.06).$$

(c) We generate sketches of

$$\vec{s}(t) = (1.116, -9.008) + 2.537\vec{W}(t), \ 0 \le t \le 2\pi$$

in Figure 8.a, and

$$\vec{s}(t) = (7.629, -8.06) + 5.204\vec{W}(t), \ 0 \le t \le 2\pi$$

in Figure 8.b.   ∎

EXAMPLE 9: Let $\mathcal{E}$ be the ellipse parametrized by $\vec{r}(t) = (\cos t, 2\sin t)$, and let $\vec{c}(t)$ be the center of curvature of $\mathcal{E}$ at $\vec{r}(t)$.

(a) Sketch the curves parametrized by $\vec{c}(t)$ and $\vec{r}(t)$.

(b) Sketch the osculating circle for $\mathcal{E}$ at $\vec{r}\left(\frac{\pi}{3}\right)$.

(c) Sketch the osculating circle for $\mathcal{E}$ at $\vec{r}\left(\frac{\pi}{2}\right)$.

SOLUTION:

$$\vec{r}'(t) = (-\sin(t), 2\cos(t)), \ \text{and} \ \vec{r}'' = (-\cos(t), -2\sin(t)).$$

$$\vec{r}''_T(t) = \frac{\vec{r}''(t) \cdot \vec{r}'(t)}{\vec{r}'(t) \cdot \vec{r}'(t)}\vec{r}'(t)$$

$$= \frac{-3\cos(t)\sin(t)}{\sin^2(t) + 4\cos^2(t)}(-\sin(t), 2\cos(t)),$$

$$\vec{r}''_N(t) = \vec{r}''(t) - \vec{r}''_T(t)$$

$$= (-\cos t - A\sin t, -2\sin t + 2A\cos t),$$

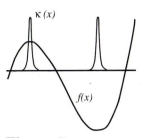

Figure 7. *Sketches of the graph of the function* $f(x) = x^3 - 5x^2 + x + 5$ *and the curvature function* $\kappa(x)$.

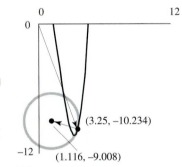

Figure 8.a *The osculating circle at* $(3.25, f(3.25))$.

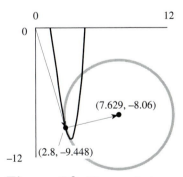

Figure 8.b *The osculating circle at* $(2.8, f(2.8))$.

$$\text{where } A = \frac{3\cos t \sin t}{\sin^2 t + 4\cos^2 t},$$

$$\|\vec{r}''_N(t)\| = \sqrt{1 + 3\sin^2 t + A^2 + 3A^2\cos^2 t - 6A\cos t \sin t},$$

$$\|\vec{r}'(t)\| = \sqrt{\sin^2(t) + 4\cos^2(t)},$$

and

$$\vec{c}(t) = \vec{r}(t) + \rho \frac{\vec{r}''_N(t)}{\|\vec{r}''_N(t)\|}$$

$$= (-\sin t, 2\cos t) + B(-\cos t - A\sin t, -2\sin t + 2A\cos t),$$

where

$$B = \frac{\sin^2 t + 4\cos^2 t}{1 + 3\sin^2 + A^2 + 3A^2\cos^2 t - 6A\cos t \sin t}. \qquad \blacksquare$$

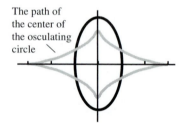

Figure 9.a *The curves parametrized by $\vec{c}(t)$ and $\vec{r}(t)$.*

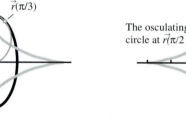

Figure 9.b *The osculating circle for $\mathcal{E}$ at $\vec{r}\left(\frac{\pi}{3}\right)$.*

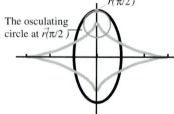

Figure 9.c *The osculating circle for $\mathcal{E}$ at $\vec{r}\left(\frac{\pi}{2}\right)$.*

EXERCISES 5.6

1. Explain the difference between angular speed and angular acceleration.

2. If two particles move around a circle with the same angular speed, must they have the same velocity?

3. A particle moves around a circle of radius 3 in. At time t_0, its angular speed is 10 radians/sec and the magnitude of its angular acceleration is 2 radians/sec². Find the particle's speed, $\|\vec{a}_N\|$, and $\|\vec{a}_T\|$.

4. A particle moves around a circle of radius 10 m with a constant speed of 5 m/s. Find its angular speed and the magnitude of its normal acceleration.

5. A particle's position is given by $\vec{r}(t) = \vec{W}(\theta(t))$, where $\theta(t) = \sin(t)$. Find $\|\vec{a}_N(t)\|$.

6. At a given time, a particle's velocity is $(-1, 1, 2)$m/s and its acceleration is $(1, 0, 2)$ m/s². Find the radius of curvature and the curvature at that point in the particle's trajectory.

7. A particle moves in a circular path. At a fixed time t_0, its radius vector is $(-2, 0, 1)$ m, $\vec{\omega}(t_0) = (1, -1, 2)$ rad/s and $\vec{\alpha}(t_0) = (-2, 2, -4)$rad/s². Find $\vec{v}(t_0)$, $\vec{a}_N(t_0)$, $\vec{a}_T(t_0)$ and $\vec{a}(t_0)$.

In Exercises 8–11, a particle moves around a circle of radius R with speed v and angular speed ω.

8. Find R given that $v = 3$ m/s and $\omega = 6$ rad/s.

9. Find ω given that $R = 3$ m and $v = 6$ m/s.

10. Find $\|\vec{a}_N\|$ given that $R = 3$ m and $v = 6$ m/s.

11. Find $\|\vec{a}_N\|$ given that $R = 3$ m and $\omega = 6$ rad/s.

In Exercises 12–20, a particle rotates in the plane $\mathcal{P}$ in a counterclockwise direction when viewed from the position p_0 with an angular speed of 3π rad/s. Find $\vec{\omega}$.

12. $\mathcal{P}$ is the xy–coordinate plane, $p_0 = \hat{k}$.

13. $\mathcal{P}$ is the xy–coordinate plane, $p_0 = -\hat{k}$.

14. $\mathcal{P}$ is the xy–coordinate plane, $p_0 = (1, 1, 2)$.

15. $\mathcal{P}$ is the xy–coordinate plane, $p_0 = (1, 1, -2)$.

16. $\mathcal{P}$ is the yz–coordinate plane, $p_0 = (1, 1, -2)$.

17. $\mathcal{P}$ is the yz–coordinate plane, $p_0 = (-1, 1, -2)$.

18. $\mathcal{P}$ is the xz–coordinate plane, $p_0 = (-1, 1, -2)$.

19. $\mathcal{P}$ has equation $x + y + z = 0$, $p_0 = (1, 1, 1)$.

20. $\mathcal{P}$ has equation $x + y + z = 0$, $p_0 = (-1, -1, -1)$.

In Exercises 21–24, explain why the given values are NOT consistent with circular motion about the origin.

21. At a given time, the position vector is $(1, 1, 1)$ and the velocity vector is $(-1, 0, 2)$.

22. At a given time, the position vector is $(1, 1, 1)$ and $\vec{a}_T = (-1, 0, 2)$.

23. At a given time, the position vector is $(1, 1, 1)$ and $\vec{\omega} = (-1, 0, 2)$.

24. At a given time $\vec{\omega} = (1, 1, 1)$ and $\vec{\alpha} = (-1, 0, 2)$.

25. A particle moves around a circle of radius 3 m. At a given instant, its velocity is $(3, 2, 1)$ m/s. Find $\|\vec{a}_N\|$.

26. A particle moves around a circle. At a given instant, its velocity is $(3, 2, 1)$ m/s and $\|\vec{a}_N\| = 5$ m/s^2. Find the radius of the circle.

27. A particle moves around a circle. At a given instant, its position is $(1, 2, 2)$ m, its velocity is $(3, 2, 1)$m/s and $\vec{a}_N = (1, -1, -1)$m/s^2. Find the radius and center of the circle.

28. A particle moves along a path. At a given instant, its velocity is $(3, 2, 1)$ m/s, and the radius of curvature of its position is 3 m. Find $\|\vec{a}_N\|$.

29. A particle moves along a path. At a given instant, its position is $\vec{r}_0$, its velocity is $(3, 2, 1)$m/s, and $\|\vec{a}_N\| = 5$m/s^2. Find the radius of curvature of the path of the particle at $\vec{r}_0$.

30. At a given instant, a particle's position is $(1, 2, 2)$ m, its velocity is $(3, 2, 1)$ m/s, and $\vec{a}_N = (1, -1, -1)$m/s^2. Find the radius and center of curvature of the particle's path at $(1, 2, 2)$ m.

31. At a given instant, a particle's position is $(1, 1, 1)$ m, its velocity is $(3, 2, 1)$ m/s and its acceleration is $(-1, 0, 2)$ m/s^2. Find the radius and center of curvature of the particle's path at $(1, 1, 1)$ m.

In Exercises 32–36, find the curvature and the center of curvature for the graph of the function at the given point. Assume that the parametrization is given by $\vec{r}(x) = (x, f(x))$.

32. $f(x) = x^2$, $x_0 = 2$.   33. $f(x) = x^3$, $x_0 = 2$.

34. $f(x) = x^3$, $x_0 = 0$.   35. $f(x) = \sin x$, $x_0 = 0$.

36. $f(x) = \sin x$, $x_0 = \frac{\pi}{4}$.

37. Show that if the second derivative of f exists at x_0 and x_0 is an inflection point for f, then the curvature of the graph of f at x_0 is zero or undefined.

In Exercises 38–42, let C be the curve parametrized by $\vec{r}$. Find $\vec{a}_T$, $\vec{a}_N$, the curvature, and the center of curvature for C at the given point.

38. $\vec{r}(t) = (\cos t, \sin t)$, $\vec{r}\left(\frac{\pi}{3}\right)$.

39. $\vec{r}(t) = (2\cos t, \sin t)$, $\vec{r}(0)$.

40. $\vec{r}(t) = (t\cos t, t\sin t)$, $\vec{r}(0)$.

41. $\vec{r}(t) = (t, \cos t, \sin t)$, $\vec{r}(0)$.

42. $\vec{r}(t) = (t^2, t^3 + t)$, $\vec{r}(1)$.

43. A car goes around a curve at a speed of 25mi/hr. The normal component $\vec{a}_N$ of its acceleration has magnitude of 10mi/hr^2. What is the curvature of the highway?

In Exercises 44–46, $\kappa(x)$ denotes the curvature for the graph of $f(x)$ at $(x, f(x))$. Sketch the graphs of $\kappa(x)$, $a \leq x \leq b$; and $f(x)$, $a \leq x \leq b$, on the same coordinate system.

44. $f(x) = x^2$, $a = -2$, $b = 2$.

45. $f(x) = \sin(x)$, $a = -2\pi$, $b = 2\pi$.

46. $f(x) = \sin(x) + \cos(x)$, $a = -2\pi$, $b = 2\pi$.

47. Show that the curvature for the ellipse $\frac{x^2}{a^2} + \frac{y^2}{b^2} = 1$, $a^2 > b^2$ has a minimum at $(0, \pm b)$ and a maximum at $(\pm a, 0)$. Suggestion: Use the parametrization $r(t) = (a \cos(t), b \sin(t))$.

48. Show that the curvature for the parabola $y = ax^2$ attains its maximum at the parabola's vertex.

49. Let C be the curve parametrized by $\vec{r}(t) = (t^2, t^3 + t)$, $-2 \leq t \leq 2$.

a. Use your calculator to draw a graph of C.

b. Determine the value of t where it appears from the graph that the curvature is a maximum.

c. Show that the value for t obtained in Part b is a local maximum for $\kappa(t)$.

d. Use your calculator to draw a graph of $\kappa(t)$.

50. The function $f(x)$ in Figure 10 is constructed from sections of circles and straight lines. Sketch the graph of the curvature $\kappa(x)$.

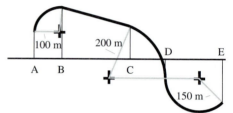

Figure 10. *Illustration for Exercise 50.*

5.7   Applications of Maxima and Minima

In this section, we consider some practical applications of finding extrema. A process for finding extrema is often referred to as an optimization method.

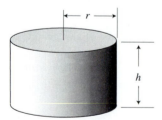

Figure 1. *A cylindrical can of height h and radius r.*

EXAMPLE 1: A production engineer wishes to make a tin can using exactly 50 square centimeters of tin. The can must have the shape of a right cylinder with a top and a bottom. Determine the radius and height the can must have to maximize the volume.

SOLUTION: The first thing to do is to write down the pertinent information. The surface of the can will be the area of the side together with the area of the top and bottom. Let h denote the height of the can, and let r denote its radius. (See Figure 1.) Then the total surface area of the can in terms of r and h is:

$$2\pi r^2 + 2\pi rh = 50. \qquad (17)$$

The volume of the can, which is the quantity that we want to maximize, is given by

$$V = \pi r^2 h. \qquad (18)$$

Now, our problem is that the volume V in Equation (18) is expressed as a function of two variables r and h. In order to employ the tools we have developed in the previous sections, we must express V in terms of only one variable. We can do this by using Equation (17) to write h in terms of r:

$$h = \frac{50 - 2\pi r^2}{2\pi r} = \frac{25}{\pi r} - r. \tag{19}$$

Combining Equation (19) and Equation (18) yields:

$$V(r) = \pi r^2 h = \pi r^2 \left(\frac{25}{\pi r} - r \right) = 25r - \pi r^3.$$

All that remains is to determine for what values of r the function V attains its maximum. To accomplish this, we first find the critical points for V:

$$V'(r) = 25 - 3\pi r^2.$$

$V'(r) = 0$ at $r = \pm \frac{5}{\sqrt{3\pi}}$. Since r is positive, $r = \frac{5}{\sqrt{3\pi}}$ is the only critical point. To check that V attains its maximum there, we employ the second derivative test.

$$V''(r) = -6\pi r$$

is negative for all $r > 0$, so V must attain its maximum at $r = 5/\sqrt{3\pi}$. Now that we have the optimal radius of the can, Equation (19) will yield the height of the can:

$$h = \frac{25}{\pi r} - r = \frac{5\sqrt{3\pi}}{\pi} - \frac{5}{\sqrt{3\pi}} = \frac{10}{\sqrt{3\pi}}. \qquad \blacksquare$$

The steps in solving the problem of Example 1 are generally applicable to optimization problems.

Steps for Solving Optimization Problems

Step 1. Write the given information in equation form. In Example 1, the quantities that could vary were the height and the radius.

Step 2. Write the quantity to be maximized or minimized in functional form. In Example 1, the function to be maximized was V, and it was a function of two variables, r and h.

Step 3. Use the equations from Step 1 to express all of the variables in the function to be optimized as functions of one of the variables. In Example 1, we wrote h as a function of r.

Step 4. Use Step 3 to express the function to be optimized as a function of one variable. This was Equation (4) in our first example.

Step 5. Employ the techniques of calculus to find the maximum or minimum of the function from Step 4.

EXAMPLE 2: A farmer wishes to enclose a rectangular field. Since one side of the field is bounded by a river, he needs to fence only three sides. Assuming that the river is straight, determine the largest rectangular area he can enclose with 5000 yd of fencing. (See Figure 2.)

SOLUTION:

Step 1. Let x denote the depth of the region and y the width so that

$$2x + y = 5000. \tag{20}$$

Step 2. The area is given by

$$A = xy. \tag{21}$$

Step 3. Solving for y in Equation (20) gives

$$y = 5000 - 2x. \tag{22}$$

Step 4. Combining Equation (22) and Equation (21), we obtain:

$$A(x) = x(5000 - 2x) = 5000x - 2x^2.$$

Step 5. $A'(x) = 5000 - 4x$. $A' = 0$ at $x = 1250$. $A''(x) = -4$, so A attains its maximum at $x = 1250$, and the maximum area that can be enclosed is $A(1250) = 3{,}125{,}000$ sq yd. ∎

Figure 2. *The enclosed field.*

In some cases you need to check the endpoints of your interval to determine the absolute extrema.

EXAMPLE 3: We want to construct a surface (such as a dining room table) that is a rectangle with four semicircular discs attached along the sides of the table. (See Figure 3.) The perimeter of the surface should be 20 ft. Find the dimensions of the table which:

(a) maximize the area of the table.

(b) minimize the area of the table.

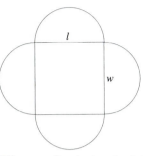

Figure 3. *A sketch of the table.*

SOLUTION: Let l and w denote, respectively, the length and width of the rectangle. Then the perimeter of the table is given by:

$$
\begin{aligned}
20 &= 2\pi\left(\frac{w}{2}\right) + 2\pi\left(\frac{l}{2}\right) \\
&= \pi w + \pi l. \quad\quad (23)
\end{aligned}
$$

The area of the rectangle is then wl ft^2. The area of the four semicircular discs is $\pi\left(\frac{w}{2}\right)^2 + \pi\left(\frac{l}{2}\right)^2$ ft^2. Thus the total area is given by

$$
A = wl + \pi\left(\frac{w}{2}\right)^2 + \pi\left(\frac{l}{2}\right)^2. \quad\quad (24)
$$

We solve the Perimeter Equation (23) for one of the variables and substitute the resulting expression into the Area Equation (24). Solving for l in Equation (23), we obtain:

$$
l = \frac{20 - \pi w}{\pi}. \quad\quad (25)
$$

Substituting Equation (25) into the Area Equation (24) gives us:

$$
\begin{aligned}
A &= w\left(\frac{20 - \pi w}{\pi}\right) + \pi\left(\frac{w}{2}\right)^2 + \pi\left(\frac{20 - \pi w}{2\pi}\right)^2 \\
&= \frac{20w}{\pi} - w^2 + \frac{\pi w^2}{4} + \pi\left(\frac{400 - 40\pi w + \pi^2 w^2}{4\pi^2}\right) \\
&= \frac{20w}{\pi} - w^2 + \frac{\pi w^2}{4} + \frac{100}{\pi} - 10w + \frac{w^2\pi}{4} \\
&= \frac{20w}{\pi} - w^2 + \frac{\pi w^2}{2} + \frac{100}{\pi} - 10w.
\end{aligned}
$$

We now take the derivative of A with respect to w to get:

$$A' = \frac{20}{\pi} - 2w + \pi w - 10.$$

Setting A' equal to zero we find that $w = \dfrac{10 - \frac{20}{\pi}}{\pi - 2} \approx 3.183$ ft. Thus $l = \dfrac{20 - \pi w}{\pi} \approx 3.183$ft. So our rectangle is in fact a square. If we check the second derivative we find that $A'' = -2 + \pi > 0$. Thus the value of $w \approx 3.183$ ft yields a minimum area. Substituting these values of w and l into the area equation gives us a minimum area of approximately 26 sq ft. We know that this is the absolute minimum since A is a quadratic. Otherwise, we would need to check the endpoints of our domain. This solves part (b). How do we find the solution to part (a)? Notice that both the width and length are nonnegative numbers. Thus $w \geq 0$ and $l \geq 0$. In our perimeter equation, since $l \geq 0$ we see that:

$$
\begin{aligned}
\frac{20 - \pi w}{\pi} &\geq 0 \\
20 &\geq \pi w \\
\frac{20}{\pi} &\geq w.
\end{aligned}
$$

Thus the variable w is restricted to the interval $0 \leq w \leq \frac{20}{\pi}$. If $w = 0$, then $l = \frac{20}{\pi}$, and if $w = \frac{20}{\pi}$, then $l = 0$. In either case we get a "degenerate" rectangle, that is, a straight line. Our table is then a circular disc with area $\pi \left(\frac{20}{\pi}\right)^2 \approx 127.32$ square feet. ■

EXAMPLE 4: Dirty Sam is lost in the desert. At one point, a search party is 5 mi directly west of Sam. The search party is proceeding East at 4 mi/hr, while Sam is crawling north at 1 mi/hr. How close does the search party come to finding Sam?

SOLUTION: Place the coordinate axes so that the origin is at Sam's initial location with the y–axis pointing north and the positive x–axis pointing east. Let $\vec{r}(t)$ denote the position of the search party at time t, and let $\vec{s}(t)$ denote the position of Sam at time t. Then

$$\vec{r}(t) = (-5 + 4t, 0) \quad \text{and} \quad \vec{s}(t) = (0, t).$$

The distance between Sam and the search party is

$$d(t) = \sqrt{(4t - 5)^2 + t^2}.$$

We want to minimize d.

$$d'(t) \;=\; \frac{1}{2}((4t-5)^2 + t^2)^{-1/2}(8(4t-5)+2t)$$

$$\quad\;\;\;\, =\; ((4t-5)^2 + t^2)^{-1/2}(17t-20).$$

So

$$d'(t) = 0 \text{ at } t = \frac{20}{17}.$$

The second derivative test may be used to demonstrate that d is minimized at $t = \frac{20}{17}$.

$$d\left(\frac{20}{17}\right) \approx 1.213 \text{ mi.}$$

The rescue party comes within 1.213 mi of Sam.   ■

Note: This problem could also be solved in a slightly different manner by employing the technique of implicit differentiation from Section 4.4. Instead of differentiating d, we could, to simplify the differentiation, use d^2 instead. That is:

$$d^2 = (4t-5)^2 + t^2.$$

Differentiating this implicitly yields:

$$2dd' = 2(4t-5)4 + 2t.$$

Thus

$$d' = \frac{34t-40}{2d} = \frac{17t-20}{d}.$$

The critical point is $t = \frac{20}{17}$ as before.   ■

EXAMPLE 5: There are 70 Spiffy Chili Patty hamburger stands in Southern California. On the average, each stand sells 5000 chili burgers a day. The daily operating cost for a stand is \$400, and a hamburger is sold for \$0.50 above the cost of the materials. Each new burger stand reduces the average number of burgers sold by 1%. How many new stands should be built in order to maximize profits?

SOLUTION: Let x denote the number of new stands built. The total daily operating cost will be

$$C(x) = \$400(70 + x) = \$(28{,}000 + 400x).$$

The daily revenue is

$$R(x) = \$0.50(5000)(1 - 0.01x)(70 + x).$$

The daily profit is

$$
\begin{aligned}
P(x) &= R(x) - C(x) \\
&= \$0.5(5000)(1 - .01x)(70 + x) - \$(28{,}000 + 400x) \\
&= \$(2500(70 + .3x - .01x^2) - 28{,}000 - 400x) \\
&= \$(147{,}000 + 350x - 25x^2).
\end{aligned}
$$

The first derivative is

$$
P'(x) = 350 - 50x.
$$

The second derivative is

$$
P''(x) = -50.
$$

$P'(x) = 0$ when $x = 7$. Since $P''(x)$ is negative, 7 is the optimal number of stands for Spiffy Chili Patty to build.   ∎

EXERCISES 5.7

In Exercises 1–6, the sum of two nonnegative numbers is 230. Find the numbers that satisfy the given conditions.

1. The sum of their squares is a maximum.

2. The sum of their squares is a minimum.

3. Their product is a maximum.

4. Their product is a minimum.

5. The product of their squares is a maximum.

6. The product of one and the square of the other is a maximum.

7. Two particles, P and Q, move in $\mathbb{R}^3$. Let $\vec{p}(t) = (3t, t, -t)$ be the position of P at time t, and let $\vec{q}(t) = (t - 1, 2t + 1, 2t)$ be the position of Q. At what time is the distance between P and Q a minimum?

8. Find the dimensions of the rectangle with maximum area that can be inscribed in a circle of radius 5.

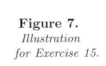

Figure 4. *An inscribed rectangle.*

9. A rectangular box with a square top and bottom is to be made using 3 sq ft of cardboard. Find the dimensions of the box with maximum volume.

10. A rectangular box is to be constructed with a square top and bottom. The material to be used for the bottom costs $0.15/sq ft while the material for the sides and top cost only $0.10/sq ft. Find the dimensions of the box with a volume of 5 cu ft that costs the least to build.

11. A tin can is made with thicker tin on the top and bottom than on the side. The tin used for the top and bottom costs 20% more than the material for the side. Find the dimensions of the can that minimize cost if the can must have a volume of 5 cu in.

12. A 3 ft by 4 ft piece of cardboard has a square cut out of each corner, and the edges are then folded up to make an open box. Find the dimensions of the box with maximum volume.

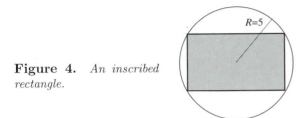

Figure 5. *Illustration for Exercise 12.*

13. Find the dimensions of the rectangle of maximum area that can be inscribed in the region bounded between the x–axis and the parabola $y = x^2 - 4$.

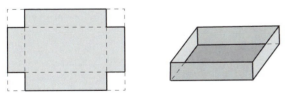

Figure 6.
Illustration for Exercise 13.

14. Find the dimensions of the rectangle of maximum area that can be inscribed in the ellipse $\frac{x^2}{4} + \frac{y^2}{9} = 1$ with its sides parallel to the axes.

15. Find the dimensions of the trapezoid of maximum area that can be inscribed in a circle of radius R with its base a diameter of the circle.

Figure 7.
Illustration for Exercise 15.

16. Find the dimensions of the isosceles triangle of minimum area that contains a circle of radius R.

Figure 8.
Illustration for Exercise 16.

17. Find the dimensions of the right circular cone of minimum volume that circumscribes a hemisphere of radius R where the base of the hemisphere lies on the base of the cone.

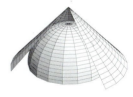

Figure 9.
Illustration for Exercise 17.

18. Find the dimensions of the right circular cylinder of maximum volume that can be inscribed in a sphere of radius R.

19. If a widget manufacturer produces 80 widgets a day, he can sell them all at $65 each. For each $2 increase in unit price, he sells 1 less widget. Each widget costs $10 to produce. What price optimizes his profit?

20. Several inmates plan to break out of prison. They are housed in a 100 ft high building without windows. The building is surrounded by a 15 ft tall fence, 10 ft from the building. Help from the outside is to provide a ladder that is to pass over the fence and lean against the building. The inmates will lower themselves by a rope

from the top of the building to the top of the ladder. What is the shortest ladder that will suffice?

21. A woman is in a race that involves running and cycling. She is 5 mi from the nearest point P on a straight road, and the finish line is 25 mi from P on the road. She must run to the road and then cycle to the finish line. She can run 10 mi/hr, and she can cycle 20 mi/hr. She is allowed to place her bicycle at a point on the road of her choice. Assuming that she can run in a straight line to any point on the road, what is the best place to leave her bicycle?

22. A hallway goes completely around a rectangular building. On one side of the building, the hallway is 8 ft wide, while on the other sides, it is 10 ft wide. What is the longest ladder that can be carried around the hall while keeping it parallel to the floor?

5.8   The Remainder Theorem for Taylor Polynomials

Recall that

$$p_k(x) = a_0 + a_1(x - x_0) + a_2(x - x_0)^2 + \ldots + a_k(x - x_0)^k,$$

where

$$a_i = \left(\frac{f^{(i)}(x_0)}{i!} \right)$$

is the Taylor polynomial for f at x_0. We asserted that $p_k(x) \approx f(x)$. Our knowledge of maximizing functions allows us to quantify the error in these approximations.

EXAMPLE 1: Before stating Taylor's Theorem, let us recall that the $(2k + 1)^{\text{st}}$–degree Maclaurin polynomial for $\sin(x)$ is given by

$$p_k(x) = x - \frac{x^3}{3!} + \frac{x^5}{5!} - \ldots + (-1)^k \frac{x^{2k+1}}{(2k + 1)!}.$$

In Figure 1, we compare the graphs of $p_1, \ldots, p_{15}, p_{21}$, and p_{25} with the graph of $\sin(x)$.

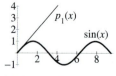

Figure 1.a
$\sin(x)$ and $p_1(x)$.

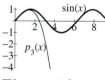

Figure 1.b
$\sin(x)$ and $p_3(x)$.

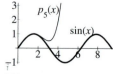

Figure 1.c
$\sin(x)$ and $p_5(x)$.

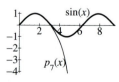

Figure 1.d
$\sin(x)$ and $p_7(x)$.

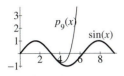

Figure 1.e
$\sin(x)$ *and* $p_9(x)$.

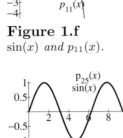

Figure 1.f
$\sin(x)$ *and* $p_{11}(x)$.

Figure 1.g
$\sin(x)$ *and* $p_{13}(x)$.

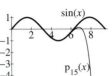

Figure 1.h
$\sin(x)$ *and* $p_{15}(x)$.

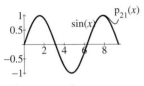

Figure 1.i
$\sin(x)$ *and* $p_{21}(x)$.

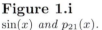

Figure 1.j
$\sin(x)$ *and* $p_{25}(x)$.

In Figure 2, we sketch the error functions $\mathrm{error}_n(x) = |\sin(x) - p_n(x)|$ for $n = 3,\ 7,\ 11,\ 15,$ and 25.

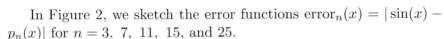

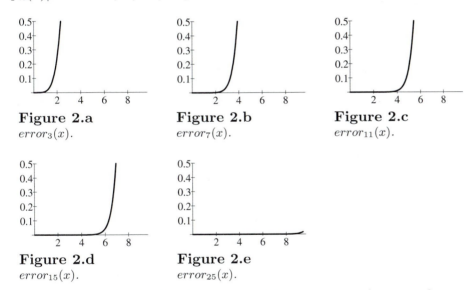

Figure 2.a
$error_3(x)$.

Figure 2.b
$error_7(x)$.

Figure 2.c
$error_{11}(x)$.

Figure 2.d
$error_{15}(x)$.

Figure 2.e
$error_{25}(x)$.

The graphs sketched in Figures 1 and 2 point to two facts, neither of which should be too surprising.

(a) The closer x is to x_0, the better the approximation obtained from $p_k(x)$; and

(b) The higher the degree of the approximating polynomial, the better the approximation.

These facts suggest the following two questions:

Given a k, how close must x be to x_0 to ensure that the $k^{\text{th}}-$

degree approximating polynomial approximates $f(x)$ within allowable tolerances?

Given an x, how large must k be in order to ensure that the kth-degree approximating polynomial approximates $f(x)$ within allowable tolerances?

The answers to both of these questions can be found in the following theorem.

Theorem 1 (Taylor's Theorem) *Suppose that $f^{(n+1)}(t)$ is continuous on $x_0 \leq t \leq x$ (or $x \leq t \leq x_0$). Then there is a number c between x_0 and x such that*

$$f(x) = a_0 + a_1(x - x_0) + a_2(x - x_0)^2 + \cdots + a_n(x - x_0)^n + R_n(x),$$

where

$$R_n(x) = \frac{f^{(n+1)}(c)}{(n+1)!}(x - x_0)^{n+1}$$

and

$$a_k = \frac{f^{(k)}(x_0)}{k!}.$$

We provide a proof for the case $n = 1$ of Theorem 1 at the end of this section. The argument for the general case is more complicated but similar.

An inspection of the error function, $R_n(x) = \left[\frac{1}{(n+1)!}\right] f^{(n+1)}(c)$ $(x - x_0)^{n+1}$, verifies the observations made from the table of Example 1. The larger $|x - x_0|$, the greater the term $(x - x_0)^{n+1}$ in $R_n(x)$; and the larger the integer n, the smaller the term $\frac{1}{(n+1)!}$ in $R_n(x)$. The following theorem is our tool to estimate the error for Taylor polynomials.

Theorem 2 *Suppose that $E > 0$ and that $f^{(n+1)}(t)$ is continuous for all t satisfying $x_0 - E \leq t \leq x_0 + E$. Suppose further that $a_k = \frac{f^{(k)}(x_0)}{k!}$ for $0 \leq k \leq n$, and that M is greater than or equal to the maximum value for $|f^{(n+1)}(t)|$ on the interval $x_0 - E \leq t \leq x_0 + E$. Then*

$$p_n(x) = a_0 + a_1(x - x_0) + a_2(x - x_0)^2 + \cdots + a_n(x - x_0)^n$$

approximates $f(x)$ with an error of no more than $\left[\frac{M}{(n+1)!}\right] E^{n+1}$.

Thus the job of quantifying the error in Taylor's polynomial approximations boils down to the task of finding an upper bound M for the maximum of a real valued function $|f^{(n+1)}(x)|$.

EXAMPLE 2: The 5th–degree Maclaurin polynomial for $f(x) = \sin x$ is given by

$$p_5(x) = x - \frac{x^3}{3!} + \frac{x^5}{5!}.$$

Indeed, since $\frac{d^6}{dx^6} \sin x \Big|_{x=0} = 0$, it follows that p_5 is actually the 6th– degree Maclaurin polynomial for $\sin x$. We also see that $\frac{d^7}{dx^7} \sin x = \cos x$. The absolute maximum for $|\cos(x)|$ is 1. Thus, if $E > 0$, the number $M = 1$ is an upper bound for $|\frac{d^7}{dx^7} \sin x|$ for all x in the interval $[-E, E]$. We now know that $p_5(x)$ approximates $\sin x$ with an error of no more than $\frac{1}{7!}E^7$. Thus, if $E = \frac{\pi}{4}$, then the error $|R_5(x)|$ in $p_5(x)$ is less than or equal to $\frac{(\pi/4)^7}{7!} < \frac{1}{7!} = \frac{1}{5040} \approx 0.00019841 < 10^{-3}$. Figure 3 compares the estimated error with $\text{error}_n(x)$.

On the other hand, if we let $E = \pi$, and all we know is that x is in the interval $[-\pi, \pi]$, then we can bound the error by

$$|R_5(x)| \leq \frac{1}{7!}(3.1416)^7 = \frac{1}{5040}(3.1416)^7 \approx 0.59927434.$$

See Figure 4.

From the table in Example 1, $p_5(\pi) = 0.5240439$. We know that $\sin(\pi) = 0$. Thus the true value of our error was 0.5240439. ∎

We see that if we want to approximate $\sin\left(\frac{\pi}{4}\right)$, then we do not get a good approximation for our error if we choose $E = \pi$. In general, we want E to be as small as reasonable for the given problem.

EXAMPLE 3: We want to use a 3rd–degree polynomial to approximate $f(x) = x^{-1/2}$ on the interval $[1.5, 2.5]$. We proceed by expanding the Taylor polynomial centered at $x_0 = 2$. Find a bound for the error that is valid for all x in the interval $[1.5, 2.5]$.

SOLUTION:

$$f(x) = x^{-1/2} \quad \Rightarrow \quad a_0 = f(2) = \frac{1}{\sqrt{2}}.$$

$$f'(x) = \frac{-1}{2x^{3/2}} \quad \Rightarrow \quad a_1 = f'(2) = \frac{-1}{4\sqrt{2}}.$$

$$f''(x) = \frac{3}{4x^{5/2}} \quad \Rightarrow \quad a_2 = \frac{f''(2)}{2!} = \frac{3}{16(2!)\sqrt{2}} = \frac{3}{32\sqrt{2}}.$$

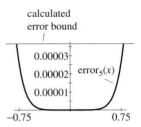

calculated
error bound

$\text{error}_5(x)$

Figure 3. *The true error compared to the estimated error bound over the interval* $\left[-\frac{\pi}{4}, \frac{\pi}{4}\right]$.

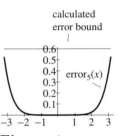

calculated
error bound

$\text{error}_5(x)$

Figure 4. *The true error compared to the estimated error bound over the interval* $[-\pi, \pi]$.

$$f'''(x) = \frac{-15}{8x^{7/2}} \quad \Rightarrow \quad a_3 = f'''(2) = \frac{-15}{64(3!)\sqrt{2}} = \frac{-5}{128\sqrt{2}}.$$

Thus

$$p_3(x) = \frac{1}{\sqrt{2}} - \frac{1}{4\sqrt{2}}(x-2) + \frac{3}{32\sqrt{2}}(x-2)^2 - \frac{5}{128\sqrt{2}}(x-2)^3.$$

Now

$$f^{(4)}(x) = \frac{105}{16}x^{-9/2}$$

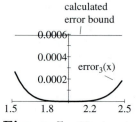

Figure 5. *The true error compared to the estimated error bound over the interval* $[1.5, 2.5]$.

is always positive. Since $f^{(5)}(x)$ is negative, $f^{(4)}(x)$ is decreasing. Thus $|f^{(4)}(x)| = f^{(4)}(x)$ attains its maximum on the interval $[1.5, 2.5]$ at $x = 1.5$. Let $M = 1.06 > \frac{105}{16}(1.5)^{-9/2} = f^{(4)}(1.5)$. We have that

$$|R_4(x)| < \frac{1.06}{4!}(0.5)^4 \approx 0.00276$$

for all x in $[1.5, 2.5]$. See Figure 5.

The maximum error for x in the interval $[1.9, 2.1]$ will be much smaller. First, f will attain its maximum on the interval $[1.9, 2.1]$ at $x = 1.9$. Thus let $M = 0.367 > \frac{105}{16}(1.9)^{-9/2}$. We have that

$$|R_4(x)| < \frac{0.367}{4!}(.1)^4 \approx 1.5292 \times 10^{-6}$$

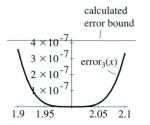

Figure 6. *The true error compared to the estimated error bound over the interval* $[1.9, 2.1]$.

for all x in $[1.9, 2.1]$. See Figure 6.

EXAMPLE 4: Find the third order Taylor polynomial for $f(x) = 1/(1+x^2)$ centered at $x_0 = 1$, and find a bound for the error on the interval $[0, 2]$.

SOLUTION:

$$\begin{aligned}
f(x) &= \frac{1}{1+x^2}, & f(1) &= \frac{1}{2}, & a_0 &= \tfrac{1}{2} \\
f'(x) &= \frac{-2x}{(1+x^2)^2}, & f'(1) &= \frac{1}{2}, & a_1 &= \tfrac{1}{2} \\
f''(x) &= \frac{6x^2-2}{(1+x^2)^3}, & f''(1) &= \frac{1}{2}, & a_2 &= \tfrac{1}{4} \\
f'''(x) &= 24x\frac{1-x^2}{(1+x^2)^4}, & f'''(1) &= 0, & a_2 &= 0
\end{aligned}$$

We have

$$p_3(x) = \frac{1}{2} + \frac{1}{2}(x-1) + \frac{1}{4}(x-1)^2.$$

To get a bound on our error, we need to bound

$$f^{(4)}(x) = 24\frac{(1 - 10x^2 + 5x^4)}{(1+x^2)^5}$$

over the interval $[0, 2]$. To find out where $f^{(4)}$ attains its maximum on this interval would be tedious at best. We resort to a graphing calculator or computer to sketch $f^{(4)}$ in Figure 7.

An inspection of the graph indicates that $f^{(4)}$ attains its maximum on the interval $[0, 2]$ at $x = 0$. We let $M = f^{(4)}(0) = 24$. In this case, $E = 1$ and a bound for the error is $\left(\frac{M}{24}\right)1^4 = 1$. ■

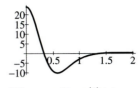

Figure 7. $f^{(4)}(x)$.

Numerical Differentiation

In the exercises for Section 3.4, we introduced two methods for approximating the derivative of a function:

Method 1: $f'(x) \approx \dfrac{f(x + h) - f(x)}{h}$ and

Method 2: $f'(x) \approx \dfrac{f(x + h) - f(x - h)}{2h}$.

We asserted that Method 2 is better than Method 1. One of the most important applications of Taylor's Remainder Theorem is that it can be used to quantify the errors in numerical methods. In this section, we employ the remainder theorem to quantify our assertion that Method 2 generally gives better results than does Method 1. Later, we will use Taylor's theorem to quantify the error in numerical methods for calculating integrals.

Taylor's Remainder Theorem states that there are numbers μ_1 between x and $x + h$, μ_2 between x and $x + h$, and μ_3 between $x - h$ and x such that

$$f(x + h) = f(x) + f'(x)h + \frac{1}{2}f''(\mu_1)h^2 \qquad (26)$$

$$f(x + h) = f(x) + f'(x)h + \frac{1}{2}f''(x)h^2 + \frac{1}{6}f'''(\mu_2)h^3, \quad (27)$$

and

$$f(x - h) = f(x) - f'(x)h + \frac{1}{2}f''(x)h^2 - \frac{1}{6}f'''(\mu_3)h^3. \qquad (28)$$

Of course, Equation (26) requires that f'' be continuous on $[x, x + h]$ and Equations (27) and (28) require that f''' be continuous on $[x - h, x + h]$.

In order to quantify the error in Method 1, let M be the maximum of $|f''(x)|$ on $[x, x + h]$ and solve Equation (26) for f' to obtain

$$f'(x) = \frac{f(x + h) - f(x)}{h} + E(h) \text{ where } |E(h)| \leq \frac{1}{2}Mh.$$

Thus we have that the error from Method 1 is proportional to h.

In order to quantify the error in Method 2, let M be the maximum of $|f'''(x)|$ on $[x-h, \ x+h]$ and subtract Equation (28) from Equation (27) to obtain

$$f(x + h) - f(x - h) = 2f'(x)h + \frac{h^3}{6}\left(f'''(\mu_2) + f'''(\mu_3)\right). \qquad (29)$$

By the Intermediate Value Theorem, there is a μ between μ_2 and μ_3 such that $2\mu = \mu_2 + \mu_3$. We can rewrite Equation (29) as

$$f'(x) = \frac{f(x + h) - f(x - h)}{2h} - \frac{h^2}{3}f'''(\mu). \qquad (30)$$

Since Method 1 has an error proportional to h, and Method 2 has an error proportional to h^2, Method 2 is the method of choice. We express the working version of Equation (30) as a theorem.

Theorem 3 *Suppose that $h > 0$, f''' is continuous on $[x-h, \ x+h]$, and M is the maximum of $|f'''|$ on $[x - h, \ x + h]$. Then*

$$f'(x) = \frac{f(x + h) - f(x - h)}{2h} + E(h), \ \ where \ |E(h)| \leq \frac{h^2}{3}M.$$

EXAMPLE 5: Approximate the derivative of $\text{Arcsin}(x)$ at $x = 0.5$ using $h = 0.1$ and $h = 0.01$.

SOLUTION: Using a calculator for $h = 0.1$, we obtain

$$\text{Arcsin}(0.4) \ = \ 0.41152$$

$$\text{Arcsin}(0.6) \ = \ 0.64530$$

$$\frac{\text{Arcsin}(0.51) - \text{Arcsin}(0.49)}{0.2} \ = \ 1.1599.$$

Again using a calculator for $h = 0.01$, we have

$$\text{Arcsin}(0.49) \ = \ 0.51209$$

$$\text{Arcsin}(0.51) \ = \ 0.53518$$

$$\frac{\text{Arcsin}(0.51) - \text{Arcsin}(0.49)}{0.02} \ = \ 1.1545.$$

In Section 7.6 we will learn how to calculate the derivative of Arcsin(x). The answer is (within calculator accuracy) 1.15470054. When we used $h = 0.1$, our answer was accurate to the second decimal. When we used $h = 0.01$, our answer was accurate to the third decimal. Try $h = 0.0001$ and see how the accuracy improves.   ■

We can use the same techniques to approximate the second derivative of a function.

The fourth order Taylor polynomial for f yields:

$$f(x + h) = f(x) + f'(x)h + \frac{f''(x)h^2}{2} + \frac{f'''(x)h^3}{6}$$

$$+ \frac{f^{(4)}(\mu_1)h^4}{24} \tag{31}$$

$$f(x - h) = f(x) - f'(x)h + \frac{f''(x)h^2}{2} - \frac{f'''(x)h^3}{6}$$

$$+ \frac{f^{(4)}(\mu_2)h^4}{24} \tag{32}$$

We add Equations (31) and (32) to obtain

$$f(x + h) + f(x - h) = 2f(x) + f''(x)h^2$$

$$+ \frac{\left(f^{(4)}(\mu_1) + f^{(4)}(\mu_2)\right)h^4}{24} \tag{33}$$

Since there is a μ between μ_1 and μ_2 such that $2f^{(4)}(\mu) = f^{(4)}(\mu_1) + f^{(4)}(\mu_2)$, we can solve Equation (33) to obtain

$$f''(x) = \frac{f(x + h) + f(x - h) - 2f(x)}{h^2} - \frac{f^{(4)}(\mu)h^2}{12}.$$

We have the following theorem.

Theorem 4 *Suppose that $h > 0$, $f^{(4)}$ is continuous on $[x - h, x + h]$ and M is the maximum of $|f^{(4)}|$ on $[x - h, x + h]$. Then*

$$f''(x) = \frac{f(x + h) + f(x - h) - 2f(x)}{h^2} + E(h),$$

where $|E(h)| \leq \dfrac{Mh^2}{12}$.

EXAMPLE 6: Approximate the second derivative of Arcsin(x) at $x = 0.5$ using $h = 0.01$.

SOLUTION: We have the following data

$$\text{Arcsin}(0.49) = 0.51208975$$

$$\text{Arcsin}(0.5) = 0.52359877$$

$$\text{Arcsin}(0.51) = 0.53518479$$

$$\frac{\text{Arcsin}(0.49) + \text{Arcsin}(0.51) - 2\,\text{Arcsin}(0.5)}{0.0001} = 0.7699201.$$

Again, in Section 7.6, we will learn that the true answer is (within calculator accuracy) 0.7698004.   ∎

In "real life," most applications of numerical differentiation arise because it is impractical to calculate the derivatives directly. Thus you will certainly not be taking higher derivatives to find the bound in your error! If you are using Theorems 1 or 2 to approximate the first or second derivative of a function, you do know that your error is proportional to h^2. Thus you can expect to improve your accuracy drastically by decreasing the size of h. If $h = 0.1$, then the accuracy of your answer will be proportional to 0.01, while $h = 0.01$ gives an accuracy proportional to 0.0001. Beware, however, since making h smaller works only until computer round-off error begins to degrade your answer.

An Argument for Taylor's Theorem for $n = 1$

Recall that the Mean Value Theorem states that if f is a function from a subset of the real numbers into the real numbers, which is continuous on the interval $[a, b]$, and if f' is defined on the open interval (a, b), then there is a number c between a and b such that the slope of the line tangent to f at c is equal to the slope of the line passing through the points $(a, f(a))$ and $(b, f(b))$. That is, $f'(c) = [f(b) - f(a)]/(b - a)$. See Figure 8.a.

We now generalize this theorem for parametrized curves in $\mathbb{R}^2$. It is known as Cauchy's Formula or the Extended Mean Value Theorem.

Theorem 5 (Cauchy's Formula) *Suppose   that C is   a parametrized curve in the plane. Let $\vec{r}(t) = (x(t), y(t))$ be a differentiable parametrization for C, and let $[a, b]$ be a subset of the domain of $\vec{r}$. If $\vec{r}(a) \neq \vec{r}(b)$, there is a number c between a and b such that $\vec{r}'(c)$ is a direction vector for the line passing* *through the points $\vec{r}(a)$ and $\vec{r}(b)$. In particular, there is a number c between a and b such that*

$$y'(c)[x(b) - x(a)] = x'(c)[y(b) - y(a)].$$

See Figure 8.b.

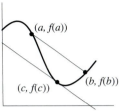

Figure 8.a *There is a number c such that $f'(c)$ is the slope of the line passing through $(a, f(a))$ and $(b, f(b))$.*

Proof: We use the Mean Value Theorem to prove the extended version. First, observe that the vectors (a, b) and $(b, -a)$ are orthogonal since $(a, b) \cdot (b, -a) = 0$. Thus $\vec{r}'(c) = (x'(c), y'(c))$ points in the direction of the vector $\vec{r}(b) - \vec{r}(a) = (x(b) - x(a), y(b) - y(a))$ if and only if $(y'(c), -x'(c)) \cdot [\vec{r}(b) - \vec{r}(a)] = 0$. Let f be the function defined by

$$f(t) = y(t)[x(b) - x(a)] - x(t)[y(b) - y(a)] - [x(b)y(a) - y(b)x(a)],$$

so that

$$\begin{aligned} f'(t) &= y'(t)[x(b) - x(a)] - x'(t)[y(b) - y(a)] \\ &= (y'(t), -x'(t)) \cdot [\vec{r}(b) - \vec{r}(a)]. \end{aligned}$$

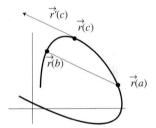

Figure 8.b *There is a number c such that $\vec{r}'(c)$ has the same direction as the line passing through $\vec{r}(a)$ and $\vec{r}(b)$.*

Notice that f is a function with range a subset of the real line, f is continuous and differentiable on $[a, b]$, and $f(a) = f(b) = 0$. Thus, by the Mean Value Theorem, there is a number c between a and b such that $f'(c) = 0$, and the number c satisfies the conditions of the theorem. ∎

Corollary 1 *Suppose that $f(t)$ and $g(t)$ are continuous, $[a, b]$ is a subset of the domains of f and g, the functions f and g are differentiable on (a, b), and g' is not 0 on (a, b). Then there is a number c such that $\frac{f'(c)}{g'(c)} = \frac{f(b) - f(a)}{g(b) - g(a)}$.*

Proof: Since $g'(c)$ is not 0 on (a, b), it follows from the Mean Value Theorem that $g(b) - g(a) \neq 0$. Let $\vec{r}(t) = (f(t), g(t))$. Then, by Cauchy's Formula, there is a number c such that $g'(c)[f(b) - f(a)] - f'(c)[g(b) - g(a)] = 0$. Thus we have that $\frac{f'(c)}{g'(c)} = \frac{f(b) - f(a)}{g(b) - g(a)}$. ∎

Finally, we are in a position to prove the remainder theorem for first order Taylor polynomials. The argument for $n > 1$ is similar but more complicated. Let $p_1(x) = f(x_0) + f'(x_0)(x - x_0)$ and let $R_1(x)$ be defined by

$$f(x) = f(x_0) + f'(x_0)(x - x_0) + R_1(x),$$

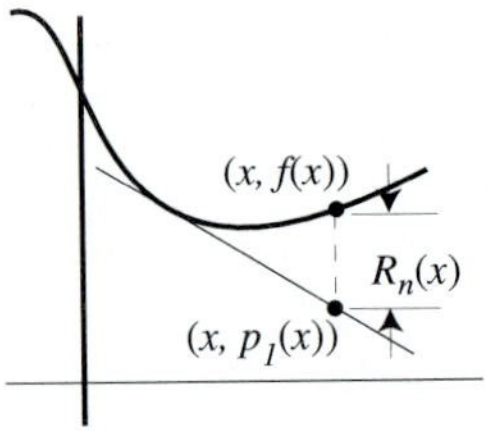

as illustrated in Figure 9.

Now,

Figure 9. *The remainder*
$R_1(x)$.

$$R_1'(x) = f'(x) - f'(x_0) \text{ and } R_1''(x) = f''(x) \qquad (34)$$

and

$$R_1(x_0) = 0 \text{ and } R_1'(x_0) = 0.$$

Consider

$$\frac{R_1(x) - R_1(x_0)}{(x - x_0)^2} = \frac{R_1(x)}{(x - x_0)^2}. \qquad (35)$$

We let $h(x) = (x - x_0)^2$, and note that the left side of Equation (35) is of the form $[R_1(x) - R_1(x_0)]/[h(x) - h(x_0)]$. By Corollary 1, there is a t between x_0 and x such that

$$\frac{R_1'(t)}{h'(t)} = \frac{R_1(x) - R_1(x_0)}{h(x) - h(x_0)}.$$

But using Equation (34) we see that

$$\frac{R_1'(t)}{h'(t)} = \frac{f'(t) - f'(x_0)}{2(t - x_0)}.$$

A similar argument shows that there is a number c between x_0 and t such that

$$\frac{f'(t) - f'(x_0)}{2(t - x_0)} = \frac{f''(c)}{2}.$$

Thus we have shown that there is a c between x_0 and x such that

$$\frac{R_1(x)}{(x - x_0)^2} = \frac{f''(c)}{2}.$$

This completes the argument for this case of Taylor's Theorem. ■

EXERCISES 5.8

In Exercises 1–7, find a bound for the error for the nth-degree Taylor polynomial for f at x_0 that is valid in the interval $[x_0 - E, x_0 + E]$.

1. $f(x) = 2 + 3x + 4x^2 + x^3$, $x_0 = 2$, $n = 3$, and $E = 256$.

2. $f(x) = \cos(x)$, $x_0 = 0$, $n = 4$, and $E = 0.1$.

3. $f(x) = \dfrac{1}{1 + x}$, $x_0 = 0$, $n =, 4$ and $E = 0.1$.

4. $f(x) = x^{3/4}$, $x_0 = 1$, $n = 4$, and $E = 0.1$.

5. $f(x) = \sin(x) + \cos(x)$, $x_0 = 0$, $n = 6$, and $E = \pi$.

6. $f(x) = \dfrac{1}{1 - x^2}$, $x_0 = 0$, $n = 2$, and $E = 0.1$. Look at the graph of $f^{(3)}$ to obtain M.

7. $f(x) = \tan(x)$, $x_0 = 0$, $n = 2$, and $E = 0.1$.

8. Let $p_n(x)$ denote the nth order Maclaurin polynomial for $\cos(x)$. Compare the error given in Theorem 2 with the actual error over the interval $[-\pi, \pi]$ for $n = 2, 4, 6$, and 8.

9. Let $p_n(x)$ denote the nth order Maclaurin polynomial for $\sin(x)$. Compare the error given in Theorem 2 with the actual error over the interval $[-\pi, \pi]$ for $n = 1, 3, 5$, and 7.

10. Find the fourth order Taylor polynomial for $f(x) = x\sin(x)$ centered at $x_0 = \frac{\pi}{2}$. Compare the error given in Theorem 2 with the actual error over the interval $[0, \pi]$.

11. Find an integer n such that the nth-degree Maclaurin polynomial for $\cos(x)$ will have an accuracy of at least 1.0×10^{-10} on the interval $[-1, 1]$.

12. Find an integer n such that the nth-degree Maclaurin polynomial for $\cos(x)$ will have an accuracy of at least 1.0×10^{-10} on the interval $[-\pi, \pi]$.

In Exercises 13 and 14, use Theorems 3 and 4 of this section to approximate the first and second derivative of the function f at the given point. Compare your answer to the derivatives taken directly.

13. $f(x) = \sin(x)$, $x = \frac{\pi}{6}$, $h = 0.01$, and $h = 0.00000001$.

14. $f(x) = \sec^6(x)$, $x = \frac{\pi}{3}$, $h = 0.01$, and $h = 0.00000001$.

15. Approximate the derivative of $f(x) = \text{Arcsin}(x)$ at $x = -0.9, -0.8, \ldots, 0.8, 0.9$ using $h = 0.01$. Sketch the graph of the derivative of $\text{Arcsin}(x)$ and the graph of $\frac{1}{\sqrt{1-x^2}}$ on the same page and on the same coordinate system. What is your conclusion based on this information?

16. Approximate the derivative of $f(x) = \text{Arccos}(x)$ at $x = -0.9, -0.8, \ldots, 0.8, 0.9$ using $h = 0.01$. Sketch the graph of the derivative of $\text{Arccos}(x)$ and the graph of the derivative of $\text{Arcsin}(x)$ on the same page and on the same coordinate system. What is your conclusion based on this information?

17. Your calculator has a built-in ln or $\log_e$ function key. Use this key to approximate the derivative of $f(x) = \ln(x)$ at $x = 0.1, 0.2, \ldots, 0.9, 1.0$ using $h = 0.01$. Sketch the graph of the derivative of $\ln(x)$ and the graph of the function $\frac{1}{x}$ on the same page and on the same coordinate system. What is your conclusion based on this information?

18. The following information was gathered by observing a function f.

x_i	0	0.1	0.2	0.3
$f(x_i)$	1	1.10517	1.2214	1.34986
x_i	0.4	0.5	0.6	0.7
$f(x_i)$	1.49182	1.64872	1.82212	2.01375
x_i	0.8	0.9	1.0	1.1
$f(x_i)$	2.22554	2.4596	2.71828	3.00417
x_i	1.2	1.3	1.4	1.5
$f(x_i)$	3.32012	3.6693	4.0552	4.48169

Use this data to plot f' and f''. Is there a relationship between f, f', and f''?

19. Let $f(x) = \frac{1}{x+1}$.

   a. Find the $2^{\text{nd}}, 3^{\text{rd}}$, and 4th degree Maclaurin polynomials for $f(x)$.

   b. Find a bound on the remainder function $R(x)$ on the interval $[0, 5]$ for each of the polynomials in Part a.

   c. Compare your answer to Part b with the actual error function, that is, $f(x) - p_2(x)$, $f(x) - p_3(x)$ and $f(x) - p_4(x)$. Sketch both the answers to Part b and these error functions.

Chapter 6

Antiderivatives

6.1 Antiderivatives and the Integral

Suppose that a particle is moving in space, and we know its velocity as a function of time t. Using our knowledge of the particle's velocity, can we work backward and find the position of the particle as a function of time? That is, given the derivative of a function, is there a reasonable way to construct the original function? This is not a trivial question; however, in this chapter, we will show that the answer is yes (sort of), if the function is continuous, and if it is, then there exists a function F such that $F'(x) = f(x)$. We also discover that there is quite a difference in knowing that such a function F exists and actually finding it. The definition of an *antiderivative* can simplify this discussion.

Definition: Antiderivative

If $F'(x) = f(x)$, then F is called an *antiderivative* of f.

EXAMPLE 1: Suppose that $\vec{a}(t)$ is the acceleration acting on a particle. Then the velocity $\vec{v}(t)$ of the particle is an antiderivative of the function $\vec{a}$, and the position function $\vec{r}(t)$ of the particle is an antiderivative of $\vec{v}$. ∎

Indeed, if a function has an antiderivative, then it has many antiderivatives. Fortunately, these antiderivatives are related. As the next theorem states, if two functions have the same derivative on an

271

interval, then they differ at most by a constant on that interval.

> **Theorem 1** *Suppose that f and g are functions from a subset of $\mathbb{R}$ into $\mathbb{R}$; S is a set of the form (a, b) where a and b are real numbers, or ∞, or $-\infty$; and $f'(x) = g'(x)$ for all x in S. Then there is a number c so that $f(x) = g(x) + c$ for all x in S.*

Proof: Let $h(x) = f(x) - g(x)$. We want to show that there is a number c such that $h(x) = c$ for all x. Since $f'(x) = g'(x)$, we know that $h'(x) = f'(x) - g'(x) = 0$. Let d be any point of S. Let x be any point of S other than d. The Mean Value Theorem tells us that there is a point e between d and x so that

$$h'(e) = \frac{h(d) - h(x)}{d - x}.$$

But $h'(e) = 0$. Thus $h(d) - h(x) = 0$, or $h(d) = h(x)$ for all x in S. Let $c = h(d)$. Then $h(x) = c$ for all x in S. ∎

Theorem 1 leads us to the following definition.

> ### Definition: The Integral
>
> $\int f(x)\, dx$ denotes the set of all functions with f as their derivative. $\int f(x)\, dx$ is called the *integral* of f.

Suppose that f has an antiderivative on a set S as described in Theorem 1. By Theorem 1, any two members of $\int f(x)\, dx$ differ on S by at most a constant. If F is any function such that $F'(x) = f(x)$, then we write

$$\int f(x)\, dx = F(x) + C.$$

This is shorthand for the statement that $\int f(x)\, dx$ is the set of all functions with f as their derivative. We get a unique solution for each choice of C. This notation makes sense only if the domain of f is a segment, ray, or the real line. Otherwise f can have antiderivatives that differ by something more than a constant function (see Exercise 54). When we compute $\int f(x)\, dx$, we say that we are *integrating f*.

EXAMPLE 2: Since $Dx^3 = 3x^2$, $\int 3x^2\, dx = x^3 + C$. Figure 1 displays members of $\int 3x^2\, dx$ obtained by choosing different values for C. ∎

EXAMPLE 3: Since $D(-\cos x) = \sin x$, $\int \sin(x)\, dx = -\cos x + C$. Figure 2 displays members of $\int \sin(x)\, dx$ obtained by choosing different values for C. ∎

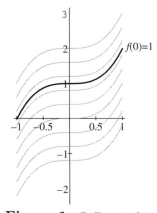

Figure 1. *Different choices for C yield distinct members of $\int 3x^2 dx$.*

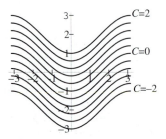

Figure 2. *Different choices for C yield distinct members of $\int \sin(x) dx$.*

EXAMPLE 4: Find a function f such that $f'(x) = 3x^2$ and $f(0) = 1$.

SOLUTION: f must be a member of $\int 3x^2 \, dx$, so $f(x)$ is of the form $x^3 + C$. Since we know that $f(0) = 1$, we are in a position to calculate C:

$$f(0) = 0^3 + C = 1.$$

Hence

$$C = 1$$

and

$$f(x) = x^3 + 1.$$

As Figure 3 illustrates, a sketch of our desired member of $\int 3x^2 \, dx$ can be obtained by vertically shifting the graph of any member of $\int 3x^2 \, dx$ into position. ∎

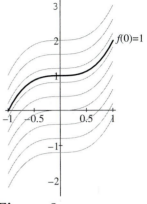

Figure 3. *The member f of $\int 3x^2 dx$ such that $f(0) = 1$.*

The integral is a little confusing in that it is a set of functions. If f is a function that is not the derivative of any function, then $\int f(x) \, dx$ is the empty set. The equation $\int f(x) \, dx = \int g(x) \, dx$ means that $\int f(x) \, dx$ and $\int g(x) \, dx$ contain exactly the same functions.

Theorem 2 *Suppose that S is a set, f and g are functions from S into $\mathbb{R}$, and c is a constant. Then*

(a) $\int (f + g)(x) \, dx = \int f(x) \, dx + \int g(x) \, dx$; *and*

(b) $\int cf(x) \, dx = c \int f(x) \, dx$.

The statements in Theorem 2 require some translation:

$\int f(x) \, dx + \int g(x) \, dx$ is the set of functions obtained by adding members of $\int f(x) \, dx$ to members of $\int g(x) \, dx$. Thus, if the derivative of F is f and the derivative of G is g, then $F + G$ is in $\int f(x) \, dx + \int g(x) \, dx$.

The notation $c \int f(x) \, dx$ stands for the set obtained by multiplying members of $\int f(x) \, dx$ by the constant c.

From the above translation of the statements in Theorem 2, we see that Theorem 2 is simply a consequence of the Sum and Constant Theorems for derivatives: $D(f + g)(x) = Df(x) + Dg(x)$ and $D(cf(x)) = cDf(x)$.

EXAMPLE 5: In Examples 2 and 3, we showed that

$$\int 3x^2 \, dx = x^3 + C \text{ and } \int \sin(x) \, dx = -\cos(x) + C.$$

Thus

$$\int (3x^2 + \sin(x)) \; dx \;\; = \;\; \int 3x^2 \; dx + \int \sin(x) \; dx$$

$$= \;\; x^3 - \cos(x) + C. \qquad \blacksquare$$

EXAMPLE 6: Find a function f so that $f'(t) = \cos(t) + \sin(t)$, and such that $f\left(\frac{\pi}{2}\right) = 0$.

SOLUTION: f is a member of $\int (\cos(t) + \sin(t)) \; dt = \sin(t) - \cos(t) + C$. Thus there is a number C such that

$$f(t) = \sin(t) - \cos(t) + C.$$

Since $f\left(\frac{\pi}{2}\right) = 0$,

$$f\left(\frac{\pi}{2}\right) = \sin\left(\frac{\pi}{2}\right) - \cos\left(\frac{\pi}{2}\right) + C = 1 - 0 + C = 0.$$

Therefore,

$$C = -1 \text{ and } f(t) = \sin(t) - \cos(t) - 1. \qquad \blacksquare$$

EXAMPLE 7: Find a function f such that $f''(x) = \cos(x)$, $f'\left(\frac{\pi}{2}\right) = 0$, and $f\left(\frac{\pi}{2}\right) = 1$.

SOLUTION: First, we know that f' is in $\int \cos(x) \; dx = \sin(x) + C$. Thus there is a C_1 such that

$$f'(x) = \sin(x) + C_1.$$

Since $f'\left(\frac{\pi}{2}\right) = 0$,

$$f'\left(\frac{\pi}{2}\right) = \sin\left(\frac{\pi}{2}\right) + C_1 = 1 + C_1 = 0.$$

Therefore

$$C_1 = -1 \text{ and } f'(x) = \sin(x) - 1.$$

We now know that $f(x)$ is in $\int f'(x) \; dx = \int (\sin(x) - 1) \; dx = -\cos(x) - x + C$. Hence, there is a number C_2 such that

$$f(x) = -\cos(x) - x + C_2.$$

Since $f\left(\frac{\pi}{2}\right) = 1$,

$$f\left(\frac{\pi}{2}\right) = -\cos\left(\frac{\pi}{2}\right) - \frac{\pi}{2} + C_2 = -\frac{\pi}{2} + C_2 = 1,$$

$$C_2 = 1 + \frac{\pi}{2},$$

and

$$f(x) = -\cos(x) - x + 1 + \frac{\pi}{2}. \qquad \blacksquare$$

The power rule for differentiation can be translated into the following power rule for antiderivatives.

Theorem 3 *If r is a rational number and $r \neq -1$, then*

$$\int x^r \, dx = \frac{x^{r+1}}{r+1} + C.$$

Proof:

$$D\left(\frac{x^{r+1}}{r+1}\right) = x^r. \qquad \blacksquare$$

EXAMPLE 8:

$$\int (x^6 + 5x^{3/2} + x^{21/6}) \, dx = \int x^6 \, dx + 5 \int x^{3/2} \, dx + \int x^{21/6} \, dx$$

$$= \frac{x^7}{7} + \frac{5x^{5/2}}{\frac{5}{2}} + \frac{x^{27/6}}{\frac{27}{6}} + C = \frac{x^7}{7} + 2x^{5/2} + \frac{6x^{27/6}}{27} + C. \qquad \blacksquare$$

Recall that if $\vec{f}$ is a function from a subset of $\mathbb{R}$ into $\mathbb{R}^3$, then we compute the derivative of $\vec{f}$ by differentiating the coordinate functions. If

$$\vec{f}(t) = (x(t), y(t), z(t)), \text{ then } \vec{f}'(t) = (x'(t), y'(t), z'(t)).$$

If $\vec{F}(x) = (F_x(t), F_y(t), F_z(t))$ and $\vec{G}(x) = (G_x(t), G_y(t), G_z(t))$ have the same derivative, then there are constants $C_x, C_y,$ and C_z such that:

$$\begin{aligned} F_x(t) &= G_x(t) + C_x, \\ F_y(t) &= G_y(t) + C_y, \end{aligned}$$

and

$$F_z(t) = G_z(t) + C_z.$$

Thus, letting $\vec{C} = (C_x, C_y, C_z)$, we have:

$$\vec{F}(x) = \vec{G}(x) + \vec{C}.$$

Similarly, if $\vec{F}$ and $\vec{G}$ are functions from a subset S of $\mathbb{R}$ into $\mathbb{R}^2$ with the same derivative, and S is a segment, ray, or the whole line as in Theorem 1, then there is a constant vector $\vec{C}$ so that $\vec{F}(x) = \vec{G}(x) + \vec{C}$. In general, if $\vec{f}$ is a function from S into $\mathbb{R}^n$, where S is a segment, ray, or the real line, then the set of antiderivatives of $\vec{f}$, denoted by $\int \vec{f}(t)\, dt$, will be of the form $\vec{F}(t) + \vec{C}$, where $\vec{F}$ is any antiderivative of $\vec{f}$.

EXAMPLE 9: Let $\vec{f}(t) = (t^3, 4t^{5/2}, \pi t)$. Then

$$\int \vec{f}(t)\, dt = \left(\frac{t^4}{4} + C_x,\ \frac{8t^{7/2}}{7} + C_y,\ \frac{\pi t^2}{2} + C_z \right). \qquad \blacksquare$$

EXAMPLE 10: Suppose that the acceleration acting on a particle is given by $\vec{a}(t) = (t, t^2, t^3)$ m/sec^2. At time $t = 0$ the particle is observed to have a velocity of $(1, 1, 1)$ m/sec, and it is at position $(-1, 2, 0)$ m. Find its velocity and position as functions of time.

SOLUTION: The particle's velocity is an antiderivative of $\vec{a}$. Thus there is a vector $\vec{V}_0 = (C_x, C_y, C_z)$ so that

$$\vec{v}(t) = \begin{pmatrix} \frac{t^2}{2} + C_x \\ \frac{t^3}{3} + C_y \\ \frac{t^4}{4} + C_z \end{pmatrix}.$$

Letting $t = 0$, we obtain $(C_x, C_y, C_z) = (1, 1, 1)$ which gives

$$\vec{v}(t) = \begin{pmatrix} \frac{t^2}{2} + 1 \\ \frac{t^3}{3} + 1 \\ \frac{t^4}{4} + 1 \end{pmatrix}.$$

The particle's position function is an antiderivative of $\vec{v}$, so

$$\vec{r}(t) = \begin{pmatrix} \frac{t^3}{6} + t + x_0 \\ \frac{t^4}{12} + t + y_0 \\ \frac{t^5}{20} + t + z_0 \end{pmatrix}.$$

Again, letting $t = 0$, we obtain $(x_0, y_0, z_0) = (-1, 2, 0)$, and

$$\vec{r}(t) = \begin{pmatrix} \frac{t^3}{6} + t - 1 \\ \frac{t^4}{12} + t + 2 \\ \frac{t^5}{20} + t \end{pmatrix}. \qquad \blacksquare$$

Direction Fields

No matter how sophisticated you become in using the tools of integration, in "real world" applications you encounter functions that are often very difficult, if not impossible, to integrate directly. There are many numerical methods to generate such integrals, some of which we will discuss in later chapters. We have, however, the tools from Chapter 5 to help us visualize solutions to $F'(x) = f(x)$. You can generate a graph of $f(x)$ using your calculator or computer and then determine where f is positive and where it is negative. You can use that information to determine where $F(x)$ is increasing and where it is decreasing. You can also obtain information about the concavity of $F(x)$. Another approach that is useful is to generate a sketch of a *direction field*. Suppose that $F(x)$ is the particular solution satisfying $F(x_0) = y_0$. Then (x_0, y_0) is a point in the graph of F, and the slope of F at x_0 is $F'(x_0) = f(x_0)$. The direction field associated with the equation $F'(x) = f(x)$ is a function defined on a subset of $\mathbb{R}^2$ that assigns $f(x_0)$ to the point (x_0, y_0). Keep in mind that $f(x_0)$ is the slope of the line tangent $F(x_0)$ at x_0. A picture will be helpful. We start with a rectangular grid of points in the plane. We sketch the direction field by drawing a short line interval centered at each point (x, y) of the grid with slope $f(x)$.

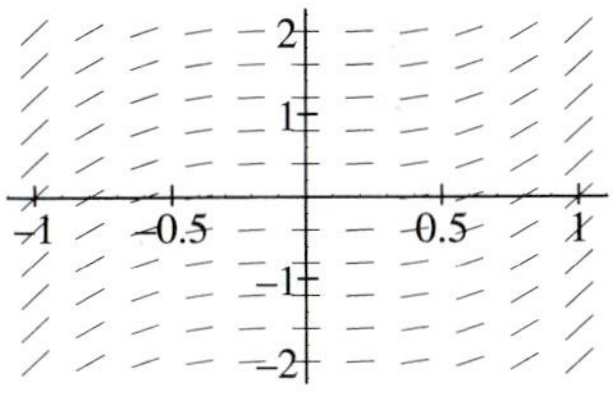

Figure 4. *The direction field for $F'(x) = 3x^2$.*

EXAMPLE 11: Sketch the direction field for $F'(x) = 3x^2$ over the rectangle $-1 \le x \le 1$, $-2 \le y \le 2$.

SOLUTION: The number of points in the grid is somewhat arbitrary. We need enough points to provide us with the information but not so many that the picture is too cluttered to be helpful. In the example, we chose the grid of points to have 11 columns and 11 rows. Figure 4 is the resulting sketch of the direction field for this problem.

With the picture of the direction field, it's fairly easy to see how the solutions to $F'(x) = f(x)$ must look. In Figure 5, we display the solution $F'(x) = 3x^2$ that satisfies the condition that $F(0) = 1$.   ∎

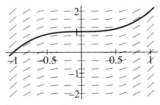

Figure 5. *The direction field for $F'(x) = 3x^2$ and the solution satisfying $F(0) = 1$.*

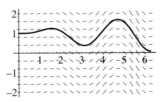

Figure 6. *The direction field for $\frac{3x^2 \sin(2x)}{(2+x)^2}$ and the solution satisfying $F(0) = 1$.*

EXAMPLE 12: In Figure 6, we sketch the direction field for
$$F'(x) = \frac{3x^2 \sin(2x)}{(2+x)^2} \text{ and display the particular solution for } F(0) = 1.$$

■

EXERCISES 6.1

In Exercises 1–37, compute the antiderivative of the function and check your answer by differentiation.

1. x^2.

2. $x^{1/3}$.

3. $x^2 + x^{1/3}$.

4. $x^{1/4} + 6x - 3$.

5. $5x^{100} - 2x^{-6} + 10$.

6. $x(x^2 + 2x - 3.)$

7. $5t^{5/2} - 6t - 2 + \frac{\pi}{3}$.

8. $x^2 - 2\sin x$.

9. $\cos(2x)$.

10. $\sin(5x)$.

11. $x^3 - \cos(0.312)$.

12. $\sin(3x - 2)$.

13. $\cos(5x - \pi)$.

14. $\sin\left(3x - \frac{\pi}{2}\right)$.

15. $(t^2 - 1)(2t + 3)$.

16. $\frac{1}{x^2}$.

17. $\frac{x+2}{\sqrt{x}}$.

18. $\frac{x^2 - 1}{x+1}$.

19. $\begin{pmatrix} x^3 + 3x + 2 \\ 5x^2 - x - 2 \\ 6x - 3 \end{pmatrix}$.

20. $\begin{pmatrix} \sin(2x) \\ \cos(3x) + 2 \\ 256.3 \end{pmatrix}$.

21. $\begin{pmatrix} 5x - x^{3/2} \\ x - (x+1)^2 \\ \sqrt{x+1} \end{pmatrix}$.

22. $\int \sin(t) - 3\cos(t) + t^3 - t^2 + 6t^{-1/2} \, dt$.

23. $\int \frac{x^3}{2} - 5x + 3(x^2 - 1) \, dx$.

24. $\int \cos\left(\frac{\pi}{25}\right) \sin(x) \, dx$.

25. $\int a^2 x^3 - 2b \, dx$.

26. $\int [(3cx + 5x) + d] \, dx$.

27. $\int 3tx + 5 \, dx$.

28. $\int 3tx + 5 \, dt$.

29. $\int (2 - t^2)(5x + 3) \, dx$.

30. $\int (2 - t^2)(5x + 3) \, dt$.

31. $\int (2 - t^2)(5x + 3) \, ds$.

32. $\int (9x^3 - 2zx + yz) \, dz$.

33. $\int (x - 1)^2 \, dx$.

34. $\int (x - 1)^3 \, dx$.

35. $\int (x - 1)^{10} \, dx$.

36. $\int \vec{f}(x) \, dx$, where
$$\vec{f}(x) = \begin{pmatrix} x^{3/2} - \sin(x) \\ \cos(2x) - 100x^{99} \\ (x - 1)^{77/3} \end{pmatrix}.$$

37. $\int \vec{g}(t) \, dt$, where
$$\vec{g}(t) = \begin{pmatrix} (t^2 - 2t + 3)(t + 1) \\ \sin(t - 1) \\ \cos(t + a) \end{pmatrix}.$$

Verify the following identities in Exercises 38–42.

38. $\int \sec(x)\tan(x) \, dx = \sec(x) + C$.

39. $\int \sec^2 x \, dx = \tan x + C$.

40. $\int \csc^2 x \, dx = -\cot x + C$.

41. $\int \csc(x)\cot(x) \, dx = -\csc x + C$.

42. $\int \csc^2 x \, dx = -\cot x + C$.

Find the function f satisfying the given conditions in Exercises 43–51.

43. $f'(x) = 3x^2 + 5x; f(1) = 2$.

44. $f'(x) = 5x^4 - 2x + 3; f(0) = 3$.

45. $f'(t) = t^{-3/2} + 3t^{-2} + 5; f(1) = -1$.

46. $f''(x) = 3x^2 - 2x + 3; f(0) = 1, f'(0) = 3$.

47. $f''(s) = 5s^{7/2} - 3s - 6; f(1) = 2, f'(1) = 1$.

48. $f'''(x) = \cos(2x); f''\left(\frac{\pi}{2}\right) = 1, f'\left(\frac{\pi}{2}\right) = 0,$ $f\left(\frac{\pi}{2}\right) = 1$.

49. $f''(x) = \sqrt{x - 1}; f'(2) = 1, f(2) = 0$.

50. $\vec{f}'(x) = \begin{pmatrix} 2x+3 \\ 5x^2 - 2x - 2 \\ -4x^3 - 3x^2 + 5 \end{pmatrix}$;

$\vec{f}(0) = \begin{pmatrix} 1 \\ -2 \\ 0 \end{pmatrix}$.

51. $\vec{f}'(x) = \begin{pmatrix} \sin(2x) \\ \cos(4x) \\ 25 \end{pmatrix}$; $\vec{f}\left(\frac{\pi}{4}\right) = \begin{pmatrix} 0 \\ 3 \\ -1 \end{pmatrix}$.

Below is a sketch of a member of $\int f(x)\,dx$. In Exercises 52–54, sketch the graph of the member of $\int f(x)\,dx$ satisfying the given condition.

52. $g(0) = 1$.   53. $g(1) = 1$.   54. $g(1) = -1$.

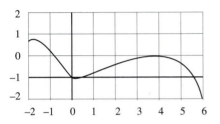

Figure 7. *Figure for Exercises 52–54.*

55. Let $f(x) = x^{-2}$. Find functions F and G such that $F'(x) = G'(x) = f(x)$ for all $x \neq 0$ but such that

$$F(x) - G(x) = \begin{cases} 1 & \text{if } x > 0 \\ 26 & \text{if } x < 0. \end{cases}$$

In Exercises 56–60, $L(x)$ is a member of $\int \frac{1}{x}\,dx$, $x > 0$.

56. Show that L is an increasing function.

57. Show that L is concave down.

58. Assume that $L(1) = 0$. Use a second order Taylor polynomial centered at $x_0 = 1$ to approximate $L(0.9)$. Then use that approximation for L to approximate the second order Taylor polynomial for L centered at $x_0 = 0.9$. Now use that polynomial to approximate $L(0.8)$. Continue this process to get an approximate value for L at $x = 0.1$.

59. As in Exercise 58, assume that $L(1) = 0$. Use a second order Taylor polynomial centered at $x_0 = 1$ to approximate $L(1.1)$. Then use that approximation for L to approximate the second order Taylor polynomial for L centered at $x_0 = 1.1$. Now use that polynomial to approximate $L(1.2)$. Continue this process to get an approximate value for L at $x = 2$.

60. Use the data from the previous two exercises to sketch the graph of $L(x)$.

Figure 8 displays the direction field for $f'(x) = \frac{1}{x}$. In Exercises 61–63, use the information in the figure to sketch the graph of the solution that satisfies the given condition.

61. $f(1) = 1$.   62. $f(1) = 0$.   63. $f(1) = -1$.

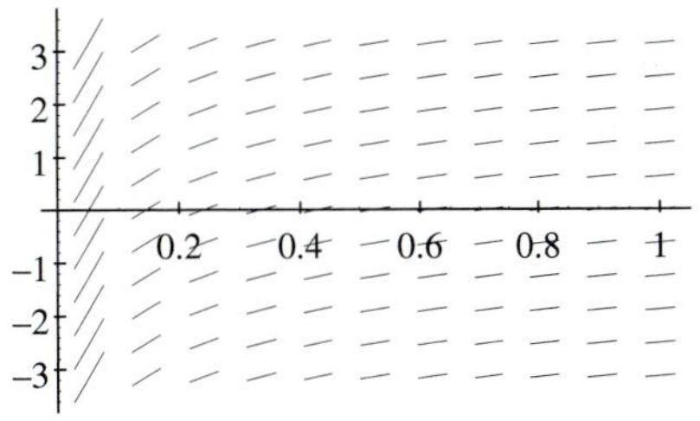

Figure 8. *The direction field for $f'(x) = \frac{1}{x}$.*

Figure 9 displays the direction field for $f'(x) = \frac{1}{x^2+1}$. Use the information in the figure to sketch the graph of the solution that satisfies the conditions given in Exercises 64–66.

64. $f(0.5) = -2$.   65. $f(0.5) = 0$. 66. $f(0.5) = 1.5$.

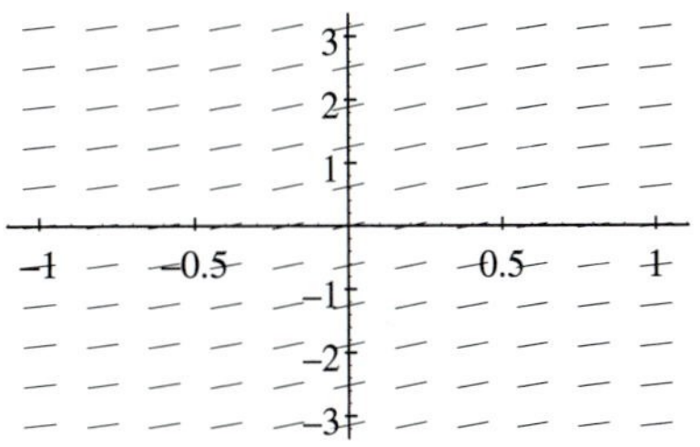

Figure 9. *The direction field for $f'(x) = \frac{1}{x^2+1}$.*

Figure 10 displays the direction field for $f'(x) =$ $\frac{1}{\sqrt{1-x^2}}$. *Use the information in the figure to sketch the graph of the solution that satisfies the conditions given in Exercises 67–69.*

67. $f(0.5) = -1.5$. 68. $f(0.5) = 0$. 69. $f(0.5) = 1.5$.

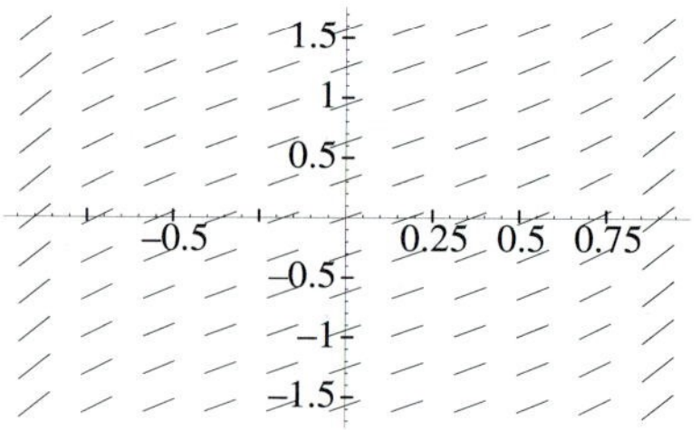

Figure 10. *The direction field for* $f'(x) = \dfrac{1}{\sqrt{1-x^2}}$.

6.2   The Chain Rule in Reverse

Thinking of the chain rule in reverse gives us a particularly useful tool for computing integrals. Since

$$Df(g(x)) = f'(g(x))g'(x),$$

we know that

$$\int f'(g(x))g'(x) \ dx = f(g(x)) + C. \tag{1}$$

An alternate form of Equation (1), which may be easier to remember, is

$$\int F(g(x))g'(x) \ dx = \int F(u) \ du\big|_{u=g(x)}. \tag{2}$$

The use of Equations (1) and (2) is something of an art, which must be developed with a lot of practice.

EXAMPLE 1: Compute $\int (x^2 + 2x)^5 (2x + 2) \ dx$.

SOLUTION: We let $g(x) = x^2 + 2x$ and see that $g'(x) = 2x + 2$. So, if $u = g(x)$, then

$$\int \underbrace{(x^2 + 2x)^5}_{u} \underbrace{(2x + 2) \ dx}_{du} = \int u^5 \ du\big|_{u=x^2+2x}$$

$$= \frac{u^6}{6} + C\big|_{u=x^2+2x} = \frac{(x^2 + 2x)^6}{6} + C. \qquad \blacksquare$$

EXAMPLE 2: $\int (x^2 + 2x)^5(x+1)\ dx$ is not quite in the form $\int u^5\ du$, since $D(x^2 + 2x) = 2x + 2$. However,

$$\int (x^2 + 2x)^5(x+1)\ dx = \frac{1}{2}\int 2(x^2 + 2x)^5(x+1)\ dx$$

$$= \frac{1}{2}\int (x^2 + 2x)^5(2x + 2)\ dx = \frac{1}{2}\int u^5\ du|_{u=x^2+2x}$$

$$= \left(\frac{1}{2}\frac{u^6}{6} + C\right)|_{u=x^2+2x} = \frac{(x^2 + 2x)^6}{12} + C. \qquad \blacksquare$$

EXAMPLE 3:

$$\int (x+1)\sin(x^2 + 2x)\ dx = \frac{1}{2}\int (2x + 2)\sin(x^2 + 2x)\ dx$$

$$= \frac{1}{2}\int \sin(u)du|_{u=x^2+2x} = \frac{1}{2}(-\cos(u)) + C|_{u=x^2+2x}$$

$$= -\frac{\cos(x^2 + 2x)}{2} + C. \qquad \blacksquare$$

EXAMPLE 4:

$$\int \sin^2(x)\cos(x)\ dx = \int u^2\ du|_{u=\sin(x)}$$

$$= \frac{u^3}{3} + C|_{u=\sin(x)} = \frac{\sin^3(x)}{3} + C. \qquad \blacksquare$$

EXAMPLE 5:

$$\int (x+1)\sin^2(x^2 + 2x)\cos(x^2 + 2x)\ dx$$

$$= \frac{1}{2}\int u^2\ du|_{u=\sin(x^2+2x)}$$

$$= \frac{1}{2}\frac{u^3}{3} + C|_{u=\sin(x^2+2x)} = \frac{\sin^3(x^2 + 2x)}{6} + C. \qquad \blacksquare$$

EXAMPLE 6: Find a function f such that $f'(x) = (x^2+1)\sqrt{x^3 + 3x}$ and $f(1) = 0$.

SOLUTION: f must be a member of $\int (x^2 + 1)\sqrt{x^3 + 3x}\ dx$.

$$\int (x^2 + 1)\sqrt{x^3 + 3x}\ dx = \frac{1}{3}\int \sqrt{u}\ du|_{u=x^3+3x}$$

$$= \quad \frac{1}{3}\frac{2u^{3/2}}{3} + C\big|_{u=x^3+3x} = \frac{2(x^3 + 3x)^{3/2}}{9} + C.$$

$$f(x) = \frac{2(x^3 + 3x)^{3/2}}{9} + C.$$

Since $f(1) = 0$,

$$f(1) = 0 = \frac{2(4)^{3/2}}{9} + C = \frac{16}{9} + C.$$

$$C = -\frac{16}{9}.$$

And so

$$f(x) = \frac{2(x^3 + 3x)^{3/2}}{9} - \frac{16}{9}. \qquad \blacksquare$$

Separable Differential Equations

Suppose that we have an equation of the form

$$f(y)y' = g(x)$$

which we want to solve for y. If we assume that y is a function of x, then we have

$$f(y(x))y'(x) = g(x). \qquad (3)$$

If we can find the antiderivative of both f and g, then we can find the antiderivative of both sides of Equation (3), and the antiderivatives of the two sides will differ by at most a constant.

EXAMPLE 7: To solve the differential equation $y^2 y' = x$, we translate it into $[y(x)]^2 y'(x) = x$. Integrating both sides (with respect to x), we obtain:

$$\int [y(x)]^2 y'(x) \, dx = \int x \, dx + C,$$

where we have combined the constants from both sides in C.

Now we use the chain rule in reverse to obtain:

$$\int u^2 \, du\big|_{u=y(x)} = \int x \, dx + C$$

$$\frac{[y(x)]^3}{3} = \frac{x^2}{2} + C,$$

or simply

$$\frac{y^3}{3} = \frac{x^2}{2} + C.$$

Thus we have the family of solutions

$$y = \left(\frac{3x^2}{2} + C\right)^{1/3}.$$

Note that we replaced the term $3C$ by C. We are able to do this since $3C$ is also an arbitrary constant. ∎

EXAMPLE 8: Solve $yy' = (y^2 + 1)^2 x^3$.

SOLUTION: We start by getting the equation into the form $f(y)y' = g(x)$:

$$\left[\frac{y}{(y^2 + 1)^2}\right] y' = x^3, \quad \text{or} \quad \left[\frac{y(x)}{(y^2(x) + 1)^2}\right] y'(x) = x^3.$$

Now we may integrate both sides with respect to x. To find an antiderivative of the left-hand side, we let $u = (y^2(x) + 1)$. Then $du = 2y(x)y'(x)\,dx$. Thus:

$$\int \left[\frac{y(x)}{(y^2(x) + 1)^2}\right] y'(x)\,dx = \frac{1}{2} \int u^{-2}\,du\big|_{u=(y^2(x)+1)}$$

$$= -\frac{1}{2}u^{-1}\big|_{u=(y^2(x)+1)} = -\frac{1}{2y^2 + 2}.$$

We omit the constant of integration since we only want one antiderivative. The solution to our original problem is:

$$-\frac{1}{2y^2 + 2} = \frac{x^4}{4} + C, \quad \text{or}$$

$$-\frac{1}{y^2 + 1} = \frac{x^4}{2} + k, \quad \text{where } (k = 2C). \quad \blacksquare$$

Notice that when we solve a differential equation of the form $f(y)y' = g(x)$, we obtain a whole family of solutions (because of the constant C). If we are given the value of y for a particular value of x, then we can solve for C.

EXAMPLE 9: Solve the Equation

$$y^2 y' = 1, \ y(0) = 1. \tag{4}$$

SOLUTION: Integrating both sides of the equation, we obtain $y^3(x) = x + C$. Setting $y(0) = 1$, we obtain $y^3(1) = 1 = C$, which yields

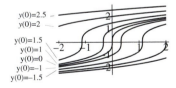

Figure 1. *Solutions for* $y^2 y' = 1$ *satisfying* $y(0) = -2.5$, $y(0) = -2$, $y(0) = -1.5$, $y(0) = -1$, $y(0) = 0$, $y(0) = 1$, $y(0) = 1.5$, $y(0) = 2$, *and* $y(0) = 2.5$.

$y^3(x) = x + 1$. In this case, we may solve for y explicitly to obtain $y(x) = (x+1)^{1/3}$. In Figure 1, we illustrate solutions for $y^2 y' = x$ for several choices for $y(0)$. ∎

Notice that in each of the solutions for Equation (4) pictured in Figure 1, the tangent line is vertical when $y(x) = 0$. This is consistent with our original equation. If we solve $y^2 y' = 1$ for y', we obtain $y'(x) = \frac{1}{y^2(x)}$. If y is any solution for Equation (4) and if $y(x_0) = 0$, then the slope of the tangent line to the graph at that point is undefined. In general, if $f(y_0) = 0$, then a solution for Equation (4) satisfying the condition that $f(y(x_0)) = 0$ may not exist, and if it does, it will have a vertical tangent line at that point. Attempts to solve separable equations of the form of Equation (4) satisfying the condition that $y(x_0) = y_0$ and $f(y_0) = 0$ must be handled on a case by case basis. However, we have the following theorem.

Theorem 1 *If $f(y)$ is continuous in a neighborhood of y_0, $g(x)$ is continuous in a neighborhood of x_0, and $f(y_0) \neq 0$, then there is a neighborhood $\mathcal{N}$ of x_0 and exactly one solution to the equation $f(y)y' = g(x)$, valid on $\mathcal{N}$ satisfying the condition that $y(x_0) = y_0$.*

The following example illustrates the problems that may arise in trying to obtain solutions to Equation (4) satisfying $y(x_0) = y_0$ where $f(y_0) = 0$.

EXAMPLE 10: The Equation $yy' = x$ has as its solution $\frac{y^2}{2} = \frac{x^2}{2} + C$, which may be rewritten as

$$y^2 - x^2 = k \text{ where } k = 2C. \tag{5}$$

As Figure 2 illustrates, if $k \neq 0$, then the solution is a hyperbola. If you are given that $y(x_0) = y_0$, then there are three possibilities.

I. $y_0^2 > x_0^2$. In this case, $k > 0$ and $y^2 - x^2 = k$ is an equation of a hyperbola that opens vertically. If $y_0 > 0$, then the desired solution is $y(x) = \sqrt{k + x^2}$ (the top branch of the hyperbola in Figure 3). If $y_0 < 0$, then the desired solution is $y(x) = -\sqrt{k + x^2}$ (the bottom branch of the hyperbola in Figure 3). In this case, it is not possible for $y_0 = 0$.

II. $y_0^2 < x_0^2$. In this case, $k < 0$ and $x^2 - y^2 = -k$ is an equation of a hyperbola that opens horizontally. If $y_0 > 0$, then the desired solution is $y(x) = \sqrt{k + x^2}$ (the top branch of the hyperbola

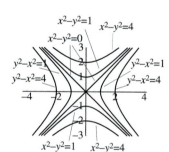

Figure 2. *Solutions to* $yy' = x$.

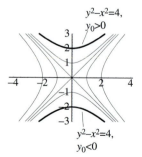

Figure 3. *Solutions for* $yy' = x$, $y_0^2 > x_0^2$.

as illustrated in Figure 4.a). If $y_0 < 0$ and $x_0 > 0$, then the desired solution is $y(x) = -\sqrt{k + x^2}$ (the bottom branch of the hyperbola). See Figure 4.b. In this case, it is not possible for $y_0 = 0$.

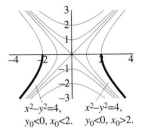

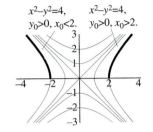

Figure 4.a *Solutions for* $yy' = x$, $y_0^2 < x_0^2$, $y_0 > 0$.

Figure 4.b *Solutions for* $yy' = x$, $y_0^2 < x_0^2$, $y_0 < 0$.

III. $y_0^2 = x_0^2$. In this case, $k = 0$ and $x^2 - y^2 = 0$ is an equation of a pair of lines. If $y_0 > 0$ and $x_0 > 0$, then the desired solution is $y(x) = x$, $x \geq 0$ as illustrated in Figure 5.a. If $y_0 > 0$ and $x_0 < 0$, then the desired solution is $y(x) = -x$, $x \geq 0$. See Figure 5.b. The cases where $y_0 < 0$ and $x_0 > 0$ and $y_0 < 0$ and $x_0 < 0$ are illustrated in Figures 5.c and 5.d.

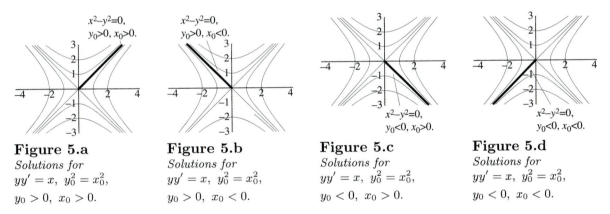

Figure 5.a
Solutions for
$yy' = x$, $y_0^2 = x_0^2$,
$y_0 > 0$, $x_0 > 0$.

Figure 5.b
Solutions for
$yy' = x$, $y_0^2 = x_0^2$,
$y_0 > 0$, $x_0 < 0$.

Figure 5.c
Solutions for
$yy' = x$, $y_0^2 = x_0^2$,
$y_0 < 0$, $x_0 > 0$.

Figure 5.d
Solutions for
$yy' = x$, $y_0^2 = x_0^2$,
$y_0 < 0$, $x_0 < 0$.

A word about notation: A differential equation of the form $f(y)y' = g(x)$ can be written $f(y)\left(\frac{dy}{dx}\right) = g(x)$. In many books, this will be written

$$f(y)dy = g(x)dx.$$

What dy and dx mean in this context is not clear, but we can easily translate this notation back into $f(y)y' = g(x)$. Later we will give a precise meaning to this notation. At this point, however, the notation $f(y)dy = g(x)dx$ is a good trick that might make it easier to remember how to attack the problem.

EXAMPLE 11: We write the equation $yy' = (y^2 + 1)^2 x^3$ as

$$ydy = (y^2 + 1)^2 x^3 \, dx,$$

or

$$\frac{y}{(y^2+1)^2} dy = x^3 dx.$$

$$\int \frac{y}{(y^2+1)^2} \, dy = \int x^3 \, dx + C \quad \text{(the constants of integration for both sides are combined in } C\text{).}$$

$$-\frac{1}{2y^2 + 2} = \frac{x^4}{4} + C$$

$$\frac{1}{y^2 + 1} = \frac{x^4}{2} + 2C$$

$$= \frac{x^4 + 4C}{2},$$

or

$$y(x) = \pm\sqrt{\frac{2}{x^4 + 4C} - 1}.$$

Thus

$$y(x) = \pm\sqrt{\frac{2}{x^4 + 4C} - 1}, \quad y(x_0) = 0.$$

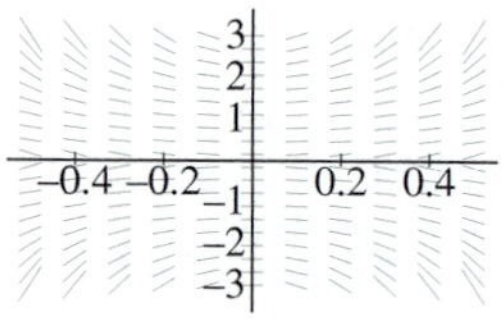

Figure 6.a *The direction field for the equation* $yy' = (y^2 + 1)^2 x^3$.

In our original equation, $f(y) = \frac{y}{y^2+1}$, so we would expect to have trouble if we want $y_0 = 0$. This is indeed the case. For any x_0, we can find a C so that

$$0 = \pm\sqrt{\frac{2}{x_0^4 + 4C} - 1}.$$

However, since there is nothing to help choose between the positive and negative roots, there are two solutions for our problem.

Figure 6.b *The direction field for the equation* $yy' = (y^2 + 1)^2 x^3$ *and several particular solutions.*

Direction Fields

There is a direction field associated with an equation of the form $f(y)y' = g(x)$. Just as in the case of antiderivatives, the direction field is a function that assigns to each point (x, y) in its domain the value $y'(x, y) = \frac{g(x)}{f(y)}$. The function $y'(x, y)$ is the slope of the solution to $f(y)y' = g(x)$ that contains (x, y) in its graph. Note that in order to find the slope function at (x, y), $\frac{g(x)}{f(y)}$ must be defined at (x, y).

Thus, as Examples 9, 10 and 11 illustrate, we must consider places where $f(y) = 0$ on a case by case basis.

EXAMPLE 12: Figure 6.a shows the direction field for the equation $yy' = (y^2 + 1)^2 x^3$ from Example 11. In Figure 6.b, particular solutions are shown for $y(0) = \pm 2$, $y(0) = \pm 1$, and $y = \pm 0.5$.

EXERCISES 6.2

In Exercises 1–5, $\int f(x)dx = \sqrt{x^2 + x} + C$. Find the given integral.

1. $\int 2x f(x^2 + 1)\ dx$.

2. $\int x^2 f(x^3 + 1)\ dx$.

3. $\int \cos(x) f(\sin(x))\ dx$.

4. $\int \cos(3x) f(\sin(3x))\ dx$.

5. $\int \sec(3x)\tan(3x) f(\sec(3x))\ dx$.

Evaluate the integrals in Exercises 6–30.

6. $\int (x^2 + x)^6 (2x + 1)\ dx$.

7. $\int (x^2 + x + 6)^{25}(2x + 1)\ dx$.

8. $\int x^3 + x^{1/2} - \frac{1}{2}x\ dx$.

9. $\int \left[x^3 + x^{1/2} - \frac{1}{2}x \right]^{1/3} \left[3x^2 + \frac{1}{2\sqrt{x}} - \frac{1}{2} \right]\ dx$.

10. $\int \dfrac{x}{(x^2 + 1)^2}\ dx$.

11. $\int \dfrac{x}{x^4 + 2x^2 + 1}\ dx$.

12. $\int \dfrac{4x^3 + 4x}{(x^4 + 2x^2 + 1)^2}\ dx$.

13. $\int \dfrac{x}{(x^4 + 2x^2 + 1)^5}\ dx$.

14. $\int (x^3 + x)(x^4 + 2x^2 + 1)^{2/3}\ dx$.

15. $\int x\sin(x^2)\ dx$.

16. $\int x\sin(x^2 + 1)\ dx$.

17. $\int (x + 1)\cos((x + 1)^2)\ dx$.

18. $\int \sin(x)\cos(x)\ dx$.

19. $\int \sin^2(x)\cos(x)\ dx$.

20. $\int \sin^{1/2}(x + 5)\cos(x + 5)\ dx$.

21. $\int [\sin(x) + 1]\cos(x)\ dx$.

22. $\int \tan(x + 1)\cos(x + 1)\ dx$.

23. $\int \dfrac{\tan(x + 1)}{\cos(x + 1)}\ dx$.

24. $\int x\sin(x^2 + 1)\cos^{5/2}(x^2 + 1)\ dx$.

25. $\int \dfrac{1}{\cos^2(x - 3)}\ dx$.

26. $\int \dfrac{\cos(t)}{\sqrt{1 - \sin^2(t)}}\ dt$.

27. $\int t^2 \sin(t^3 + 6)\ dt$.

28. $\int (x^3 - 1)\sec^2(x^4 - 4x)\ dx$.

29. $\int (3t^5 + 1)\sec\left(\frac{t^6}{2} + t\right)\tan\left(\frac{t^6}{2} + t\right)\ dt$.

30. $\int \sec^2(t)\tan(t)\ dt$.

In 31–38, find the function f that satisfies the given conditions.

31. $f'(x) = x\sin(x^2)$; $f(0) = 32$.

32. $f'(x) = \left(x^2 - \frac{1}{3}\right)(x^3 - x)^{5/3}$; $f(1) = 2$.

33. $f'(t) = t\sec^2(t^2)$; $f(0) = 1$.

34. $f'(x) = \dfrac{x^3}{(x^4 - 7)^{3/2}}$; $f(2) = 31$.

35. $f'(w) = w\cos(w^2 - 4)\sin(w^2 - 4)$; $f(2) = \pi$.

36. $f'(\beta) = \cos(\beta)(1 - \sin^2(\beta))$; $f\left(\frac{\pi}{2}\right) = 1$.

37. $f'(\beta) = \cos^3(\beta)$; $f(\pi) = 0$.

38. $f'(t) = \sin^5(t)$; $f(0) = 0$.

Solve the differential equations in Exercises 39–47

39. $(y + 1)y' = 2x$.

40. $\sin(y)y' = \cos(x)$.

41. $y \sin(y^2 + 2)y' = \sqrt{x + 2}$.

42. $(2y + 1)y' = (y^2 + y)^6 x(x^2 + 1)^{1/4}$.

43. $[\cos(y)/x]y' = \sqrt{\sin(y)}$.

44. $(x^3 + x)^3 \, dy = (3x^2 + 1) \, dx$.

45. $\cos^3(x)y \, dy = \sin(x) \sec(y^2) \, dx$.

46. $x^2 y' = y^2, y(1) = 2$.

47. $(x + 1)^3 y' = (y + 2)^{1/4}, y(0) = 1$.

48. Sketch the direction field for $(y^2 + 1)y' = 2x$ on the grid below.

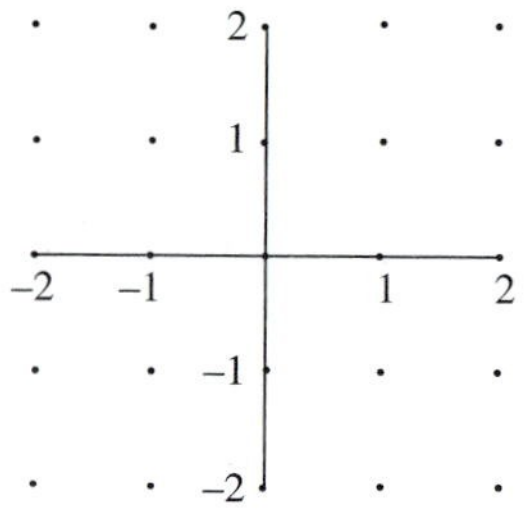

49. In Figure 7, we have the direction field for a differential equation of the form $y' = \frac{f(x)}{f(y)}$. Sketch the solutions that satisfy the given conditions.
   a. $y(0) = -1$.   b. $y(0) = 0$.   c. $y(0) = 1$.

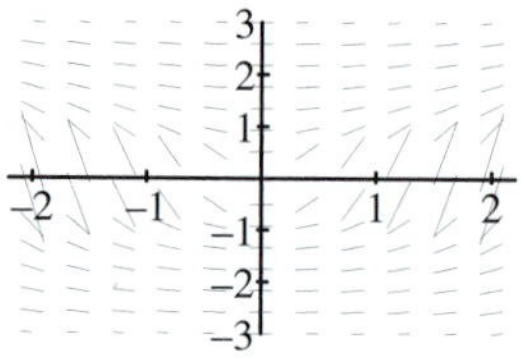

Figure 8. *Direction field for Exercise 49.*

6.3   Acceleration, Velocity, and Position

Using the techniques of the previous sections, we can sometimes determine the position of a particle as a function of time knowing velocity (as a function of time) and the position of the particle at some given time.

EXAMPLE 1: Suppose that a particle is moving in 3–space so that at time $t = 2$ sec the particle is at $(3, 1, 2)$, with distance measured in feet, and the velocity of the particle at time t is given by $\vec{v}(t) = (t, t, -t)$ ft/sec. Find a function $\vec{r}$ such that $\vec{r}(t)$ denotes the position of the particle at time t.

SOLUTION: $\vec{r}'(t) = \vec{v}(t) = (t, t, -t)$. Let $x(t), y(t)$, and $z(t)$ denote the coordinate functions of $\vec{r}$. Then

$$x'(t) = t \text{ implies that } x(t) = \frac{t^2}{2} + C_x,$$

$$y'(t) = t \text{ implies that } y(t) = \frac{t^2}{2} + C_y,$$

and

$$z'(t) = -t \text{ implies that } z(t) = -\frac{t^2}{2} + C_z.$$

Thus

$$\vec{r}(t) = \begin{pmatrix} \frac{t^2}{2} + C_x \\ \frac{t^2}{2} + C_y \\ -\frac{t^2}{2} + C_z \end{pmatrix}.$$

Since $\vec{r}(2) = (3, 1, 2)$,

$$\vec{r}(2) = \begin{pmatrix} \frac{4}{2} + C_x \\ \frac{4}{2} + C_y \\ -\frac{4}{2} + C_z \end{pmatrix} = \begin{pmatrix} 2 + C_x \\ 2 + C_y \\ -2 + C_z \end{pmatrix} = \begin{pmatrix} 3 \\ 1 \\ 2 \end{pmatrix} \text{ ft.}$$

We have that $C_x = 1$, $C_y = -1$, $C_z = 4$, so

$$\vec{r}(t) = \begin{pmatrix} \frac{t^2}{2} + 1 \\ \frac{t^2}{2} - 1 \\ -\frac{t^2}{2} + 4 \end{pmatrix} \text{ ft.} \qquad \blacksquare$$

EXAMPLE 2: Suppose that the velocity of a particle is given by

$$\vec{v}(t) = \begin{pmatrix} t(t^2 + 1)^{100} \\ (t + 1)(t^2 + 2t)^3 \\ t \end{pmatrix},$$

and the initial position is given by $\vec{r}(0) = (3, -5, 2)$, where distance is measured in meters and time in seconds. Integrating the first coordinate function gives us

$$
\begin{aligned}
\int t(t^2 + 1)^{100} \, dt &= \frac{1}{2} \int u^{100} \, du \Big|_{u=t^2+1} \\
&= \frac{u^{101}}{202} + C_x \Big|_{u=t^2+1} \\
&= \frac{(t^2 + 1)^{101}}{202} + C_x.
\end{aligned}
$$

Similarly, for the second and third coordinates, we have

$$\int (t + 1)(t^2 + 2t)^3 \, dt = \frac{(t^2 + 2t)^4}{8} + C_y,$$

and

$$\int t \, dt = \frac{t^2}{2} + C_z,$$

hence

$$\vec{r}(t) = \begin{pmatrix} \frac{(t^2+1)^{101}}{202} + C_x \\ \frac{(t^2+2t)^4}{8} + C_y \\ \frac{t^2}{2} + C_z \end{pmatrix}.$$

Since

$$\vec{r}(0) = \begin{pmatrix} \frac{1}{202} + C_x \\ C_y \\ C_z \end{pmatrix} = \begin{pmatrix} 3 \\ -5 \\ 2 \end{pmatrix},$$

$C_x = \frac{605}{202}, C_y = -5$, and $C_z = 2$.

Thus

$$\vec{r}(t) = \begin{pmatrix} \frac{(t^2+1)^{101}}{202} + \frac{605}{202} \\ \frac{(t^2+2t)^4}{8} - 5 \\ \frac{t^2}{2} + 2 \end{pmatrix}. \qquad \blacksquare$$

The motion in each of these examples involved a fairly complicated velocity function. In the case that the velocity is a constant vector, then the following theorem shows us that the equation of motion is quite simple.

Theorem 1 *If a particle is moving with constant velocity $\vec{v}$, then its path is a straight line and its position is given by $\vec{r}(t) = \vec{v}t + \vec{r}_0$, where $\vec{r}_0$ is the particle's position at $t = 0$.*

The proof of this theorem is an easy exercise.

If we know the acceleration $\vec{a}(t)$ of a particle, then $\vec{v}(t)$ is an antiderivative of $\vec{a}$. Thus, if we know $\vec{a}(t)$ and if we know $\vec{v}(t)$ at some initial time t_0, then we can determine $\vec{v}(t)$ for all time, provided that we can compute the antiderivative of $\vec{a}$. Hence, if we can also compute the antiderivative of $\vec{v}$, and if we know $\vec{r}(t)$ for some time t_1 in the domain of $\vec{v}$, then we know $\vec{r}(t)$ for all time. (t_0 and t_1 need not be the same, although in practice they are usually chosen to be equal.)

EXAMPLE 3: Suppose that $\vec{a}(t) = (t, 1, -t), \vec{v}(0) = (1, 1, 1)$, and

$\vec{r}(0) = (2, 1, -1)$. Then

$$\vec{v}(t) = \int \begin{pmatrix} t \\ 1 \\ -t \end{pmatrix} dt = \begin{pmatrix} \frac{t^2}{2} + C_x \\ t + C_y \\ -\frac{t^2}{2} + C_z \end{pmatrix}.$$

Since $\vec{v}(0) = (C_x, C_y, C_z) = (1, 1, 1)$,

$$\vec{v}(t) = \begin{pmatrix} \frac{t^2}{2} + 1 \\ t + 1 \\ -\frac{t^2}{2} + 1 \end{pmatrix}.$$

Now

$$\vec{r}(t) = \int \begin{pmatrix} \frac{t^2}{2} + 1 \\ t + 1 \\ -\frac{t^2}{2} + 1 \end{pmatrix} dt = \begin{pmatrix} \frac{t^3}{6} + t + K_x \\ \frac{t^2}{2} + t + K_y \\ -\frac{t^3}{6} + t + K_z \end{pmatrix}.$$

Since $\vec{r}(0) = (K_x, K_y, K_z) = (2, 1, -1)$,

$$\vec{r}(t) = \begin{pmatrix} \frac{t^3}{6} + t + 2 \\ \frac{t^2}{2} + t + 1 \\ -\frac{t^3}{6} + t - 1 \end{pmatrix}. \qquad \blacksquare$$

EXAMPLE 4: Let $\vec{r}_1(t) = t\vec{e} + \vec{r}_0$ be a parametrization of a line L in $\mathbb{R}^n$. If p is a particle moving along L, then there is a function g from $\mathbb{R}$ into $\mathbb{R}$ such that $\vec{r}(t) = \vec{r}_1(g(t)) = g(t)\vec{e} + \vec{r}_0$. In this case,

$$\vec{v}(t) = \vec{r}'(t) = g'(t)\vec{e},$$

and

$$\vec{a}(t) = \vec{v}'(t) = g''(t)\vec{e}.$$

Since both $\vec{v}$ and $\vec{a}$ point in the direction of either $\vec{e}$ or $-\vec{e}$, we see that $\vec{a}$ and $\vec{v}$ have the same or opposite directions, so $\vec{a}_T(t) = \vec{a}(t)$ and $\vec{a}_N = \vec{0}$. $\blacksquare$

   Example 4 shows that if a particle is moving in a straight line, then $\vec{a}_N(t) = \vec{0}$. The converse is also true. Thus we have the following theorem.

Theorem 2 *The path of a particle in motion lies in a straight line if and only if the normal component of its acceleration is zero.*

Motion Under Constant Acceleration

   In Exercise 16, you are asked to show that if the acceleration is a constant $\vec{a}$, then

$$\vec{v}(t) = \vec{a}t + \vec{v}_0, \tag{6}$$

and

$$\vec{r}(t) = \frac{1}{2}\vec{a}t^2 + t\vec{v}_0 + \vec{r}_0, \tag{7}$$

where $\vec{v}_0$ is the velocity at time $t = 0$, and $\vec{r}_0$ is the position at time $t = 0$. This simple case is important because if we are interested in the path of a projectile near the surface of the earth, and if the distances traveled and the projectile are both small compared to the radius of the earth, then we can often assume that the acceleration due to gravity is a constant. This acceleration is directed toward the center of the earth, and we denote its magnitude by g.

$$g \approx 9.8 \text{ m/sec}^2 \approx 32 \text{ ft/sec}^2.$$

When we discuss projectile problems, we assume that gravity is the only thing that changes the motion of the projectile. Factors such as air resistance are critical in high velocity projectile problems but lead to complications that we will not attack until Section 8.2.

Returning to Equations (6) and (7), if $\vec{a} = (a_x, a_y, a_z)$, and we consider the coordinate functions separately, we obtain the following:

x–motion

(a) $v_x(t) = v_x(0) + a_x t$

(b) $x(t) = \frac{1}{2}a_x t^2 + v_x(0)t + x(0)$

y–motion

(a) $v_y(t) = v_y(0) + a_y t$

(b) $y(t) = \frac{1}{2}a_y t^2 + v_y(0)t + y(0)$

z–motion

(a) $v_z(t) = v_z(0) + a_z t$

(b) $z(t) = \frac{1}{2}a_z t^2 + v_z(0)t + z(0)$

In practical problems with constant acceleration, we can often simplify things by carefully choosing our coordinate axes. We assume that $\vec{a}$ is not the zero vector. If $\vec{v}(0)$ points in the same or opposite direction as $\vec{a}$, then $\vec{a} = k\vec{v}(0)$, for some scalar k, and the vector form of Equation (7) becomes

$$\vec{r}(t) = \frac{1}{2}k\vec{v}(0)t^2 + \vec{v}(0)t + \vec{r}(0)$$

$$= \left(\frac{1}{2}kt^2 + t\right)\vec{v}(0) + \vec{r}(0).$$

This parametrizes the path of a particle traveling in the line with direction $\vec{v}(0)$ and containing $\vec{r}(0)$.

If $\vec{v}(0)$ does not point in the same or opposite direction as $\vec{a}$, then we may place our coordinate axes so that the particle is at the origin at time $t = 0$ and so that the xy–coordinate plane contains $\vec{v}(0)$ and $\vec{a}(0)$ if they are drawn emanating from the origin. This gives us $\vec{v}(0) = (v_x(0), v_y(0), 0)$ and $\vec{a} = (a_x, a_y, 0)$. Then Equation (7) shows us that $z(t) = 0$ for all t. Thus motion with constant acceleration is constrained to a plane. We, whenever reasonable, assume that motion with constant acceleration takes place in the xy–plane. Unless otherwise stated, we assume that $\vec{v}(0) = (v_x(0), v_y(0))$ and $\vec{a} = (a_x, a_y)$. Carrying this process one step further, we position our axes so that $\vec{a}$ points in the (positive or negative) y–direction. This gives us $\vec{a} = (0, a_y)$. Using this and $\vec{r}(0) = (0, 0)$ in Equation (7), we see that

$$x(t) = v_x(0)t \quad \text{and} \quad y(t) = v_y(0)t + \frac{1}{2}a_y t^2.$$

Solving for t in terms of $x(t)$ and substituting this into the expression for $y(t)$, we obtain

$$y(t) = \frac{v_y(0)}{v_x(0)}x(t) + \frac{1}{2}\frac{a_y}{v_x^2(0)}x^2(t).$$

Thus the point (x, y) lies in the path of the particle if and only if

$$y = \frac{v_y(0)}{v_x(0)}x + \frac{1}{2}\frac{a_y}{v_x^2(0)}x^2.$$

But this is an equation of a parabola. We have the following important result.

> **Theorem 3** *The path of a particle under constant acceleration will be a point, a subset of a straight line, or a parabola.*

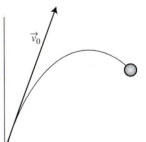

Figure 1. *The trajectory of a projectile with initial velocity $\vec{v}_0$.*

EXAMPLE 5: A ball is dropped from a height of 10 m. How long does it take to reach the ground, and how fast is it moving at that point?

SOLUTION: We first choose a convenient orientation and position for our coordinate axes. Let us put the origin at ground level with the y–axis pointing up so that the ball is at position $(0, 10)$ m when it is dropped. Its acceleration is pointing down, so $\vec{a}(t) = (0, -9.8)$ m/sec^2. The ball is not moving the instant we drop it, so $\vec{v}(0) = (0, 0)$. Now, $\vec{v}(t)$ is an antiderivative of $\vec{a}$. Thus $\vec{v}(t) = (0, -9.8)t + \vec{C}$. Since $\vec{v}(0) = \vec{0}$, we see that $\vec{C} = \vec{0}$, so

$$\vec{v}(t) = (0, -9.8t).$$

Similarly, $\vec{r}(t) = (0, -9.8t^2/2) + \vec{K}$.

$$\vec{r}(0) = (0, 10) = \vec{K},$$

which implies that

$$\vec{r}(t) = \left(0, -\frac{9.8t^2}{2} + 10\right).$$

The ball strikes the ground when $\vec{r}(t) = \left(0, -\frac{9.8t^2}{2} + 10\right) = (0, 0)$. Solving $-\frac{9.8t^2}{2} + 10 = 0$ yields $t = \sqrt{\frac{20}{9.8}}$. Thus it will take $\sqrt{\frac{20}{9.8}}$ sec for the ball to get to the ground. Its velocity at time $t = \sqrt{\frac{20}{9.8}}$ is $\vec{v}\left(\sqrt{\frac{20}{9.8}}\right) = \left(0, -9.8\sqrt{\frac{20}{9.8}}\right)$ m/sec. Its speed is $9.8\sqrt{\frac{20}{9.8}}$ m/sec.   ∎

EXAMPLE 6: A ball is pitched up with a speed of 98 m/sec, from a height of 1470 m above the ground.

(a) How high above the ground does the ball go?

(b) How long is the ball in the air?

(c) What is the velocity when it strikes the ground?

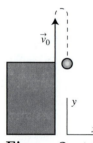

Figure 2. *A ball is pitched up with a speed of 98 m/sec from a height of 1470 m above the ground.*

SOLUTION: Place the coordinate system so that the origin is at the point where the ball will strike the ground and with the positive y–axis pointing up. (See Figure 2.) Let t_1 be the time the ball reaches its maximum height, and t_2 be the time that the ball hits the ground. $y(0) = 1470$ m, $v_y(0) = 98$ m/sec, and $a_y = -9.8$ m/sec^2.

(a) To determine where $y(t)$ is a maximum, we solve $y'(t) = v_y(t) = 0$ for t. Equation (6) shows us that $v_y(t_1) = 0 = -(9.8)t_1 + 98$, or $t_1 = 10$. Using Equation (7), we find that

$$y(10) = -\left(\frac{1}{2}\right)(9.8)(10)^2 + (10)(98) + 1470 = 1960 \text{ m}.$$

(b) We want to find t_2 such that $y(t_2) = 0$. Using Equation (7) we obtain:

$$y(t_2) = -\left(\frac{1}{2}\right)(9.8)(t_2)^2 + t_2(98) + 1470 = 0.$$

Taking the positive solution for t_2 gives $t_2 = 30$ sec.

(c) We now use Equation (6) to find that $v_y(30) = -(9.8)(30) + 98 = -196$ m/sec. ■

EXAMPLE 7: A shell is fired from a cannon at ground level, with a speed of 300 ft/sec. The cannon is aimed 37° above the horizontal.

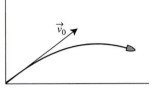

Figure 3. *The cannon shell's trajectory.*

(a) What is the maximum height of the shell?

(b) How long is the shell in the air?

(c) What is the range of the shell (i.e., how far from the cannon does the shell hit the ground)?

(d) What is its velocity upon impact with the ground?

SOLUTION: We place the coordinate axes with the origin at the position from which the shell is fired so that the motion is in the first quadrant. (See Figure 3.)

Recall that $\cos(37°) \approx \frac{4}{5}$ and $\sin(37°) \approx \frac{3}{5}$. Thus, $v_x(0) = \left(\frac{4}{5}\right)300 = 240$ and $v_y(0) = \left(\frac{3}{5}\right)300 = 180$. Let t_1 denote the time that the shell reaches its maximum height, and let t_2 be the time that the shell hits the ground.

(a) $y(t)$ attains its maximum when $v_y(t) = 0$ (when $y'(t) = 0$). We employ Equation (6) and solve for $v_y(t_1) = 0$:

$$v_y(0) + ta_y(t_1) = 0,$$
$$180 - 32t_1 = 0,$$

or

$$t_1 = \frac{45}{8}.$$

We now evaluate Equation (7) at $t_1 = \frac{45}{8}$ to obtain

$$y(t_1) = \left(\frac{1}{2}\right)(-32)\left(\frac{45}{8}\right)^2 + 180\left(\frac{45}{8}\right) = 506.25 \text{ ft.}$$

(b) Using Equation (7), we may solve $y(t) = 0$ for t to get t_2 :

$$y(t_2) = \left(\frac{1}{2}\right)(-32)(t_2)^2 + t_2(180) = 0.$$

$$t_2(180 - 16t_2) = 0.$$

Since $t = 0$ is the time the shell is fired, the desired solution must be $t_2 = 180/16 = 11.25$ sec.

(c) Equation (7) gives us that $x(t_2) = 2700$ ft.

(d) Equation (6) shows that

$$
\begin{aligned}
\vec{v}(t_2) &= (0, -32)t_2 + (240, 180) \\
&= (0, -32)(11.25) + (240, 180) \\
&= (0, -360) + (240, 180) \\
&= (240, -180) \text{ ft/sec.}
\end{aligned}
$$

∎

EXERCISES 6.3

In Exercises 1–12, find $\vec{r}(t)$:

1. $\vec{v}(t) = \begin{pmatrix} 3t^2 + 2t + 1 \\ 5t + 2 \\ t^{1/2} - 3t + 5 \end{pmatrix}$ m/sec,

   $\vec{r}(1) = (1, 2, -1)$ m.

2. $\vec{v}(t) = \begin{pmatrix} \cos(\pi t) \\ \sin(\pi t) \\ \sec^2\left(\frac{\pi t}{2}\right) \end{pmatrix}$ m/sec,

   $\vec{r}\left(\frac{1}{2}\right) = (0, 2, 1)$ m.

3. $\vec{v}(t) = \begin{pmatrix} (t + \pi)\sin(t^2 + 2\pi t) \\ t^2 \cos(t^3) \\ \cos(t)\sin(t) \end{pmatrix}$ m/sec,

   $\vec{r}(0) = (1, -1, 0)$ m.

4. $\vec{v}(t) = \begin{pmatrix} t(t^2 + 1)^{-3/4} \\ \csc^2(t)\tan^2(t) \end{pmatrix}$ m/sec

   $\vec{r}\left(\frac{\pi}{4}\right) = (0, 3)$ m.

5. $\vec{v}(t) = (2, -1, 2)$ m/sec, $\vec{r}(1) = (4, -1, 2)$ m.

6. $\vec{a}(t) = (0, 0, 0)$ m/sec^2,
   $\vec{r}(0) = (0, 0, 0)$ m,
   $\vec{v}(0) = (0, 0, 0)$ m/sec.

7. $\vec{a}(t) = (0, 0, 0)$ m/sec^2,
   $\vec{r}(0) = (0, 0, 0)$ m,
   $\vec{v}(0) = (1, 2, 3)$ m/sec.

8. $\vec{a}(t) = (2, 1, -1)$ m/sec^2,
   $\vec{r}(0) = (1, 2, 7)$ m,
   $\vec{v}(1) = (4, -3, 7)$ m/sec.

9. $\vec{a}(t) = (0, -32, 0)$ m/sec^2,
   $\vec{r}(2) = (22, 50, 110)$ m,
   $\vec{v}(2) = (225, 32, 100)$ m/sec.

10. $\vec{a}(t) = \begin{pmatrix} t^2 \\ 2t + 3 \\ \sqrt{t + 1} \end{pmatrix}$ m/sec^2,

    $\vec{r}(0) = (1, 1, 1)$ m,
    $\vec{v}(0) = (1, 0, -1)$ m/sec.

11. $\vec{a}(t) = \begin{pmatrix} t^2 \\ 2t + 3 \\ \sqrt{t + 1} \end{pmatrix}$ m/sec^2,

    $\vec{r}(-1) = (1, 1, 1)$ m,
    $\vec{v}(1) = (1, 0, -1)$ m/sec.

12. $\vec{a}(t) = \begin{pmatrix} \sin t \\ \cos t \\ \sin(2t) \end{pmatrix}$ m/sec^2,

$\vec{r}\left(\frac{\pi}{2}\right) = (0,0,0)$ m,

$\vec{v}\left(\frac{\pi}{2}\right) = (-1,2,1)$ m/sec.

In Exercises 13–15, $\vec{r}(t)$ describes the motion of a particle that travels in a straight line with $\vec{a}(t) = 0$ m/sec^2 and that moves from $(1,2,3)$ m (at $t = 0$ sec) to $(-1,5,2)$ m in the given time.

13. 1 sec.   14. $\frac{1}{2}$ sec.   15. 3 sec.

16. Show that if a particle's acceleration is constant, then its position can be parametrized by $\vec{r}(t) = \frac{1}{2}\vec{a}t^2 + \vec{v}_0 t + \vec{r}_0$, where $\vec{a}$ is the constant acceleration of the particle, $\vec{v}_0$ is the velocity of the particle at time $t = 0$, and $\vec{r}_0$ is the position at $t = 0$.

17. Given $\vec{a}(t) = (3,0)$ m/sec^2, $\vec{v}(0) = (5,0)$ m/sec, and $\vec{r}(0) = (2,0)$ m, find $\vec{v}(t)$ and $\vec{r}(t)$. Where is the particle and what is its velocity at $t = 2$ sec?

18. Given $\vec{a}(t) = (3,0)$ m/sec^2, $\vec{v}(0) = (5,2)$ m/sec, and $\vec{r}(0) = (2,0)$ m, find $\vec{v}(t)$ and $\vec{r}(t)$. Where is the particle and what is its velocity at $t = 2$ sec?

19. Given $\vec{a}(t) = (0,2)$ m/sec^2, $\vec{v}(0) = (5,2)$ m/sec, and $\vec{r}(0) = (2,0)$ m, find $\vec{v}(t)$ and $\vec{r}(t)$. Where is the particle and what is its velocity at $t = 2$ sec?

20. Given $\vec{a}(t) = (3,2)$ m/sec^2, $\vec{v}(0) = (5,2)$ m/sec, and $\vec{r}(0) = (2,0)$ m, find $\vec{v}(t)$ and $\vec{r}(t)$. Where is the particle and what is its velocity at $t = 2$ sec?

21. Given $\vec{a}(t) = (2\hat{\imath} - 6\hat{\jmath} + \hat{k})$ m/sec^2, $\vec{v}(0) = 2\hat{\imath}$ m/sec, and $\vec{r}(0) = -2\hat{k}$ m, find $\vec{v}(t)$ and $\vec{r}(t)$. Where is the particle and what is its velocity at $t = 2$ sec?

22. Given $\vec{a}(t) = (0, 2t^2)$ m/sec^2, $\vec{v}(0) = (1,2)$ m/sec, and $\vec{r}(0) = (0,1)$ m, derive expressions for $\vec{v}(t)$ and $\vec{r}(t)$.

In Exercises 23 and 24, a car traveling along a straight line and initially moving at 50 ft/sec stops in 10 sec. Assume that the acceleration is constant while the car is stopping.

23. What is the acceleration?

24. How far does the car travel while stopping?

25. The x and y coordinates of a particle moving in a plane are given as functions of time by

$$x(t) = (3 + 4t^5) \text{ m}$$

and

$$y(t) = (2t + t^2) \text{ m}.$$

What are the velocity and acceleration at $t = 2$ sec?

26. A car traveling at 50 ft/sec stops in a distance of 175 ft. Assume the acceleration is constant and find:

a. how long it takes for the car to stop;

b. the magnitude of the acceleration while stopping.

27. A ball is dropped and takes 5 sec to strike the ground. How high (in m) above the ground is the point of release?

28. A ball is pitched up with a speed of 49 m/sec and takes 15 sec to reach the ground. How high above the ground is the point of release?

29. A stone is released from rest, and another is pitched up from ground level with a speed of 30 m/sec. The two collide when the stone released from rest has fallen one third of the distance to the ground. How much time elapses between the release and collision? How high above the ground was the first stone released?

30. An object released from rest falls two thirds of the total distance in the last second.

a. How long does it fall?

b. From what height is it released?

31. A projectile is fired with a speed $v(0)$ at an angle θ above the horizontal.

a. Show that the range is $R = \dfrac{v^2(0)\sin(2\theta)}{g}$.

b. Show that the total time of flight is $T = \dfrac{2v(0)\sin(\theta)}{g}$.

32. A piece of pipe is positioned so that it points directly toward a target hanging from the ceiling. A ball is shot through the pipe toward the target. The target is released from the ceiling the instant the ball leaves the end of the pipe. Show that the ball will strike the target.

33. An object is thrown horizontally with a speed of 50 m/sec from the top of a 40 m building.

    a. How long does it take before the object reaches the ground?

    b. What is its speed as it reaches the ground?

34. Rework Exercise 33 for the object thrown *above* the horizontal at an angle of 37°.

35. Rework Exercise 33 for the object thrown *below* the horizontal at an angle of 37°.

36. For the situation described in Exercise 34, determine the maximum height above ground reached by the object and the time to reach this height.

[PROJECT]: *In Exercises 37–41, a Patriot missile is trying to intercept an enemy missile. The Patriot has an onboard computer designed to guide it to its target. It knows the trajectory of the target. It can accelerate in any direction by pointing its thrusters, but the acceleration has a constant magnitude A. The method employed is called* proportional navigation. *Proportional navigation works by trying to point the acceleration directly toward the target. Every h seconds the Patriot can update the direction of its acceleration. We assume that the Patriot's internal clock turns on when the homing device is turned on so that $t_0 = 0$. Let $t_1 = h$, $t_2 = 2h$, and in general let $t_{i+1} = t_i + h = (i+1)h$. Let $\vec{T}(t) = (T_x(t), T_y(t), T_z(t))$ denote the position of the target at time t, and let $\vec{r}(t) = (x(t), y(t), z(t))$ denote the position of the Patriot at time t. Then its acceleration on the time interval $[t_i, t_{i+1}]$ will be $\vec{a}_{i+1} = A\vec{d}_i/\|\vec{d}_i\|$, where*

$$\vec{d}_i = \begin{pmatrix} T_x(t_i) - x(t_i) \\ T_y(t_i) - y(t_i) \\ T_z(t_i) - z(t_i) \end{pmatrix}.$$

Let $\vec{r}_i = \vec{r}(t_i)$ and $\vec{v}_i = \vec{v}(t_i) = \vec{r}'(t_i)$. Then, on the time interval $[t_i, t_{i+1}]$, we have that the equations of motion of the Patriot are given by

$$\vec{v}(t) = \vec{a}_{i+1}(t - t_i) + \vec{v}_i, \text{ which gives us that}$$
$$\vec{v}_{i+1} = \vec{a}_{i+1}h + \vec{v}_i, \text{ and}$$
$$\vec{r}(t) = \frac{1}{2}\vec{a}_{i+1}(t - t_i)^2 + \vec{v}_i(t - t_i) + r_i, \text{ so}$$
$$\vec{r}_{i+1} = \frac{1}{2}\vec{a}_{i+1}h^2 + \vec{v}_ih + r_i.$$

Let dist(t) denote the distance between the Patriot and the threatening missile. Assume that the Patriot can change its acceleration 10 times per sec and the magnitude of its acceleration is 10 m/s². The incoming missile's position is given by

$$\vec{T}(t) = \begin{pmatrix} 0 \\ 120{,}000 - 400t \\ 60{,}000 - 200t \end{pmatrix} \text{ m.}$$

The Patriot's initial position and velocity are

$$\vec{r}_0 = \begin{pmatrix} 0 \\ 800 \\ 2{,}000 \end{pmatrix} \text{ m, and } \vec{v}_0 = \begin{pmatrix} 100 \\ 800 \\ 3{,}000 \end{pmatrix} \text{ m/s.}$$

37. Find dist(t_i) for each i until the threatening missile would hit the ground or until the Patriot hits the missile.

38. Plot the graph of the dist function.

39. What is the minimum value for dist(t_i)?

40. Does the Patriot perform better if it can update its acceleration 20 times a sec?

41. How would the Patriot's performance be affected if the magnitude of its acceleration were 5 m/s²?

6.4 Antiderivatives and Area

Antiderivatives can be used in attacking a wide range of problems. In this section, we use the antiderivative to find areas. Suppose that f is a positive valued function that is continuous on $[a, b]$ and that we want to find the area bounded between the graph of f, the x–axis, and the lines $x = a$ and $x = b$, as illustrated in Figure 1. The idea is to define a function A so that $A(x)$ denotes the area bounded between the graph over the interval $[a, x]$. We have represented $A(x)$ in Figure 1. If we can find A, then $A(b)$ will be the area between the lines $x = a$ and $x = b$.

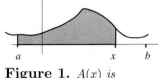

Figure 1. $A(x)$ is represented by the shaded area.

Nearly all that we know about A is that $A(a) = 0$. If we can find the derivative of A, then possibly we can find A. To this end, consider

$$A'(x) = \lim_{h \to 0} \frac{A(x + h) - A(x)}{h}.$$

Let $R(h)$ be the area of the rectangle with width h and height $f(x)$ as in Figure 2. Then $R(h) = hf(x) \approx A(x + h) - A(x)$.

$$A'(x) \approx \frac{A(x + h) - A(x)}{h} \approx f(x).$$

We conclude that $A'(x) = f(x)$.

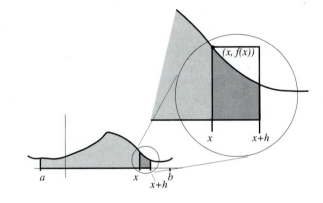

Figure 2. $A(x + h) - A(x)$ is approximated by the area of the rectangle of height $f(x)$ and width h.

EXAMPLE 1: Find the area of the region bounded between the graph of $f(x) = \sin(x)$, the x–axis, and the lines $x = 0$ and $x = \pi$.

SOLUTION: Let $A(x)$ denote the area of the region bounded between the graph of $\sin(x)$ and the x–axis over the interval $[0, x]$. Then $A'(x) = \sin(x)$, which gives us that $A(x) = -\cos(x) + C$. Since there is no area between 0 and 0, $A(0) = 0$. We use this information to find C. $0 = -\cos(0) + C$, which implies that $C = 1$ and that $A(x) = 1 - \cos(x)$. Thus $A(\pi) = 2$ is the area between $x = 0$ and $x = \pi$. ∎

EXAMPLE 2: Find the area of the region bounded between the graphs of $y = x$ and $y = x^2$.

SOLUTION: Define $A(x)$ to be the area bounded between the graphs of $f(x) = x$ and $f(x) = x^2$ over the interval $[0, x]$. Then, as illustrated in Figure 3, $A(x + h) - A(x) \approx (x - x^2)h$. Therefore,

$$A'(x) \approx \frac{A(x+h) - A(x)}{h} \approx x - x^2.$$

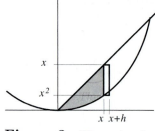

Figure 3. *The rectangle with height $x - x^2$ and width h approximates $A(x + h) - A(x)$.*

We conclude that

$$A'(x) = x - x^2, \quad \text{or} \quad A(x) = \frac{x^2}{2} - \frac{x^3}{3} + C.$$

By our definition of A, it is true that $A(0) = 0$. It follows that $C = 0$ so that $A(x) = \frac{x^2}{2} - \frac{x^3}{3}$. The desired area is $A(1) = \frac{1}{2} - \frac{1}{3} = \frac{1}{6}$. ∎

Area Swept Out By a Vector-Valued Function

Kepler's Second Law states that as a planet revolves about the sun, the vector emanating from the sun and terminating at the planet sweeps out equal areas in equal times.

Suppose that $\vec{r}(t)$, $a \leq t \leq b$ defines a path. As time passes, the position vector sweeps out a surface as illustrated in Figure 4.a. Let $A(t)$ denote the area swept out by the position vector from time a to time t. See Figure 4.b. Then $A(a) = 0$ and $A(b)$ is the total area of the surface.[1] Although it is not obvious how to find A directly, we can find $A'(t)$. By the definition of the derivative,

$$A'(t) = \lim_{h \to 0} \frac{A(t+h) - A(t)}{h}.$$

As shown in Figure 4.c, $A(t + h) - A(t)$ can be approximated by the area of the triangle T with vertices $\vec{0}$, $\vec{r}(t)$, and $\vec{r}(t + h)$. The vectors $(\vec{r}(t + h) - \vec{r}(t))$ and $\vec{r}(t)$ are adjacent edges of T.

---

[1] If $\vec{r}(t)$ is the position of a planet relative to the sun, then Kepler's Second Law is equivalent to stating that $A'(t) = c$.

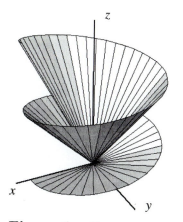

Figure 4.a *The area swept out by the curve parametrized by $\vec{r}(t) = (\cos(2\pi t), \sin(2\pi t), t)$, $0 \leq t \leq 2$.*

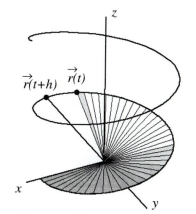

Figure 4.b *The shaded region is $A(t)$.*

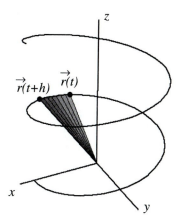

Figure 4.c $A(t+h) - A(t)$ *is approximated by the area of the triangle with vertices $\vec{0}$, $\vec{r}(t)$, and $\vec{r}(t+h)$.*

Thus

$$
\begin{aligned}
\text{The area of T} &= \frac{1}{2}\|\vec{r}(t) \times (\vec{r}(t+h) - \vec{r}(t))\| \\
A'(t) &\approx \frac{1}{2h}\|\vec{r}(t) \times (\vec{r}(t+h) - \vec{r}(t))\| \\
&= \frac{1}{2}\left\|\vec{r}(t) \times \frac{\vec{r}(t+h) - \vec{r}(t)}{h}\right\| \\
&\approx \frac{1}{2}\|\vec{r}(t) \times \vec{r}'(t)\|.
\end{aligned}
$$

EXAMPLE 3: Let C be the curve parametrized by

$$
\vec{r}(t) = \begin{pmatrix} \cos(2\pi t) \\ \sin(2\pi t) \\ t \end{pmatrix}, \quad 0 \leq t \leq 2,
$$

illustrated in Figure 4.a. According to the discussion above,

$$
A'(t) = \frac{1}{2}\|\vec{r}(t) \times \vec{r}'(t)\| = \sqrt{t^2 + 1} \text{ and } A(0) = 0.
$$

We will not have the tools to calculate A until Chapter 10, but we do know the rate at which A is changing.   ∎

The computations appear simpler in the case where the curve lies in the plane. Recall from Chapter 1 that if $\vec{a} = (a_x, a_y)$ and $\vec{b} = (b_x, b_y)$ are vectors in the plane, then the area of the rectangle determined by $\vec{a}$ and $\vec{b}$ is the absolute value of $\begin{vmatrix} a_x & a_y \\ b_x & b_y \end{vmatrix}$.

EXAMPLE 4: Find the area swept out by

$$\vec{r}(t) = t\,(\cos t, \sin t), \ 0 \le t \le 2\pi.$$

See Figure 5.

SOLUTION: We know that $A'(0) = 0$ and that

$$
\begin{aligned}
A'(t) &= \frac{1}{2}\|\vec{r}(t) \times \vec{r}'(t)\| \\[2mm]
&= \frac{1}{2}\ \text{the absolute value of}\ \begin{vmatrix} t\cos t & t\sin t \\ \cos t - t\sin t & \sin t + t\cos t \end{vmatrix} \\[2mm]
&= \frac{1}{2}\left|\left(t\cos t\sin t + t^2\cos^2 t\right) - \left(t\sin t\cos t - t^2\sin^2 t\right)\right| \\[2mm]
&= \frac{1}{2}t^2.
\end{aligned}
$$

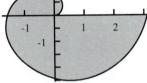

Figure 5. *The area swept out by* $\vec{r}(t) = t(\cos t, \sin t)$, $0 \le t \le 2\pi$.

Thus $A(t) = \frac{t^3}{6} + C$. Since $A(0) = 0$, $C = 0$, and $A(t) = \frac{t^3}{6}$. The total area is $A(2\pi) = \frac{8\pi^3}{6} = \frac{4\pi^3}{3}$. ∎

EXERCISES 6.4

In Exercises 1–5, find the area bounded between $f(x)$, $a \le x \le b$, *and the* x–*axis.*

1. $f(x) = x$, $a = 0$, $b = 2$.

2. $f(x) = x^3$, $a = 1$, $b = 3$.

3. $f(x) = \sin(x)$, $a = 0$, $b = \pi$.

4. $f(x) = \cos(x)$, $a = -\frac{\pi}{2}$, $b = \frac{\pi}{2}$.

5. $f(x) = \sin x + \cos x$, $a = 0$, $b = \frac{\pi}{2}$.

6. Find the area bounded between the graphs of $y = x^2$ and $y = 4$.

7. Find the area bounded between the graphs of $y = x$ and $y = x^4$.

8. Find the area bounded between the graphs of $y = x^3$ and $y = x^2$.

9. Show that if $a > 0$, then the area bounded between the graph of x^r and the x–axis from $x = 0$ to $x = a$ is $\dfrac{a^{r+1}}{r+1}$.

10. Find the area of the region swept out by $\vec{r}(t) = (\cos t, 3\sin t)$, $0 \le t \le 2\pi$.

11. Find the area of the region swept out by $\vec{r}(t) = (4\cos t, 2\sin t)$, $0 \le t \le 2\pi$.

12. Use the fact that $\vec{r}(t) = (a\cos t, b\sin t)$, $0 \le t \le 2\pi$ parametrizes the ellipse $\frac{x^2}{a^2} + \frac{y^2}{b^2} = 1$ to show that the area of the ellipse $\frac{x^2}{a^2} + \frac{y^2}{b^2} = 1$ is $\frac{ab\pi}{2}$.

13. Find the area of the region swept out by $\vec{r}(t) = t^2(\cos t, \sin t)$, $0 \le t \le 2\pi$.

14. Find the area of the region swept out by $\vec{r}(t) = t^2(\cos t, \sin t)$, $\pi/2 \le t \le 2\pi$.

15. Find the area of the region swept out by $\vec{r}(t) = (2\pi - t)(\cos t, \sin t)$, $0 \le t \le 2\pi$.

16. Show that if ρ is a real valued function and $A(t)$ is the area swept out by $\vec{r}(t) = \rho(t)\begin{pmatrix} \cos(t) \\ \sin(t) \end{pmatrix}$, $a \le t \le b$, then $A'(t) = \rho^2(t)$ and $A(a) = 0$.

17. Let $A(t)$ denote the area of the region swept out by $\vec{r}(t) = (t, t\cos t, t\sin t)$, $a \le t$. Find $A'(t)$.

18. Let T be the triangle with vertices $(0,0)$, $(a,0)$, and (c,d), where a, b, and c are positive numbers as illustrated in Figure 6. Then a is the length of the base of the triangle and d is its height. Let $\vec{r}(t)$ parametrize the line segment from $(a,0)$ to (c,d) such that $\vec{r}(0) = (a,0)$ and $\vec{r}(1) = (c,d)$. Then the region swept out by $\vec{r}(t)$, $0 \le t \le 1$ is the triangle T. Show that the area swept out is $\frac{1}{2}ad$.

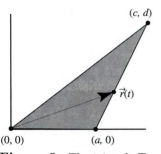

Figure 6. *The triangle T.*

[PROJECT] Kepler's Laws 1: *In Exercises 19–24, we derive Kepler's Second Law from Newton's Law of Gravitation.*

19. Show that if $\vec{r}(t)$ is a differentiable function from an interval into $\mathbb{R}^3$ and $\vec{r}(t) \ne 0$, then
$$\frac{d}{dt}\|\vec{r}(t)\| = \frac{\vec{r}(t) \cdot \vec{r}'(t)}{\|\vec{r}(t)\|}.$$

20. From Exercise 19 show that the rate of change of $\|\vec{r}(t)\|$ is zero if $\vec{r}'(t) = 0$ or $\vec{r}(t)$ and $\vec{r}''(t)$ are orthogonal.

21. Let $\vec{v}(t) = \vec{r}'(t)$ and $\vec{a}(t) = \vec{r}''(t) = \vec{v}'(t)$. Show that $\dfrac{d}{dt}(\vec{r}(t) \times \vec{r}'(t)) = \vec{r}(t) \times \vec{a}(t)$.

22. Explain why Exercise 21 shows that if $\vec{r}$ and $\vec{r}''$ point in the same or opposite directions, then the rate of change of area swept by $\vec{r}$ is constant.

23. Newton's Law of Gravitation states that two objects O_1 and O_2 of masses M_1 and M_2 exert an attractive force with magnitude $\dfrac{GM_1M_2}{d^2}$, where d is the distance between the objects. The number G is an experimentally determined constant of proportionality. Thus, if O_1 is at position $\vec{r}_1$ and O_2 is at $\vec{r}_2$, then the resulting gravitational force on O_2 due to O_1 is $\dfrac{GM_1M_2}{\|\vec{r}_2 - \vec{r}_1\|^3}(\vec{r}_2 - \vec{r}_1)$. Argue that if we place the origin of our coordinate axis at the center of mass of the sun and if $\vec{r}(t)$ denotes the position of the earth's mass center at time t, then the acceleration of the earth due to the gravitational force due to the sun is of the form $\vec{a}(t) = g(t)\vec{r}(t)$.

24. Show that if $\vec{r}''(t) = g(t)\vec{r}(t)$, then $\dfrac{d}{dt}(\vec{r}(t) \times \vec{r}''(t)) = 0$ and hence, the rate of change of area swept by $\vec{r}$ is constant. That is, $\vec{r}$ sweeps out equal areas in equal times.

6.5   Area and Riemann Sums

In the previous section, we used antiderivatives to find areas of regions bounded between graphs of functions. In this section, we attack the same problem with a different approach. We approximate the area of the region by decomposing it into a collection of rectangles and then adding up the areas of the rectangles. It seems that we are introducing notions completely divorced from antiderivatives and integrals. However, the chapter ends with a beautiful theorem that ties things together.

EXAMPLE 1: Find an approximation for the area of the region bounded by the graph of $y = x^2$, the x-axis, and the line $x = 1$.

SOLUTION: (See Figure 1.) We begin by dividing the interval $[0, 1]$ into n subintervals of equal length: $[0, \frac{1}{n}], [\frac{1}{n}, \frac{2}{n}], \ldots, [\frac{n-1}{n}, 1]$. Construct rectangles with base given by the aforementioned subintervals and height the value of the function $y = x^2$ evaluated at the right endpoint of the subinterval.

Let R_1 be the rectangle with base $[0, \frac{1}{n}]$ and height $(\frac{1}{n})^2$.

Let R_2 be the rectangle with base $[\frac{1}{n}, \frac{2}{n}]$ and height $(\frac{2}{n})^2$.

$$\vdots \qquad\qquad \vdots \qquad\qquad \vdots$$

Let R_n be the rectangle with base $[\frac{n-1}{n}, \frac{n}{n}]$ and height $(\frac{n}{n})^2$.

We see in Figure 1 that the area of the region is approximated by adding up the areas of these rectangles. Thus the area is approximated by

$$
\begin{aligned}
\text{Area} \;&\approx\; \text{Area}(R_1) + \text{Area}(R_2) + \cdots + \text{Area}(R_n) \\
&= \left(\frac{1}{n}\right)\left(\frac{1}{n}\right)^2 + \left(\frac{1}{n}\right)\left(\frac{2}{n}\right)^2 + \cdots + \left(\frac{1}{n}\right)\left(\frac{n}{n}\right)^2 \\
&= \left(\frac{1}{n}\right)^3 \left(1^2 + 2^2 + \cdots + n^2\right).
\end{aligned}
$$

■

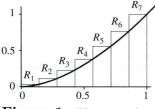

Figure 1. *The sum of the areas of the rectangles approximates the area of the bounded region.*

EXAMPLE 2: Approximate the area of the region bounded by the graphs of $y = x^2$ and $y = x^3$.

SOLUTION: First, we sketch the region in Figure 2. We then construct a collection of rectangles such that the sum of their areas approximates the area of the region. Observing that the two graphs $y = x^2$ and $y = x^3$ intersect at $x = 0$ and $x = 1$, we divide the interval $[0, 1]$ into n intervals of equal length: $[0, \frac{1}{n}], [\frac{1}{n}, \frac{2}{n}], \ldots, [\frac{n-1}{n}, 1]$.

Let R_1 be the rectangle with base $[0, \frac{1}{n}]$ and height $(\frac{1}{n})^2 - (\frac{1}{n})^3$.

Let R_2 be the rectangle with base $[\frac{1}{n}, \frac{2}{n}]$ and height $(\frac{2}{n})^2 - (\frac{2}{n})^3$.

$$\vdots \qquad\qquad\qquad \vdots \qquad\qquad\qquad \vdots$$

Let R_n be the rectangle with base $[\frac{n-1}{n}, \frac{n}{n}]$ and height $(\frac{n}{n})^2 - (\frac{n}{n})^3 = 0$.

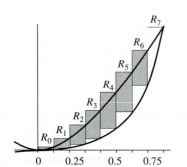

Figure 2. *As in Figure 1, the sum of the areas of the rectangles approximates the area of the bounded region.*

This region is approximated by the union of these rectangles. (See Figure 2.) Therefore, the area of the region is approximated by

$$
\begin{aligned}
\text{Area} \quad &\approx \quad \text{Area}(R_1) + \text{Area}(R_2) + \cdots + \text{Area}(R_n) \\[2mm]
&= \left(\frac{1}{n}\right)\left[\left(\frac{1}{n}\right)^2 - \left(\frac{1}{n}\right)^3\right] + \left(\frac{1}{n}\right)\left[\left(\frac{2}{n}\right)^2 - \left(\frac{2}{n}\right)^3\right] + \cdots \\[2mm]
&\quad + \left(\frac{1}{n}\right)\left[\left(\frac{n}{n}\right)^2 - \left(\frac{n}{n}\right)^3\right] \\[2mm]
&= \left(\frac{1}{n^3}\right)\left[1^2 + 2^2 + 3^2 + \cdots + n^2\right] \\[2mm]
&\quad - \left(\frac{1}{n^4}\right)\left[1^3 + 2^3 + 3^3 + \cdots + n^3\right]. \qquad\blacksquare
\end{aligned}
$$

The methods we have been using to approximate areas lead us to the more general notion of a *Riemann sum*. In order to pursue this notion, we will need some new vocabulary.

The notation $\sum_{i=1}^{n} a_i$ is shorthand for the sum $a_1 + a_2 + \cdots + a_n$ and is read "the sum from i equals 1 to n of a_i." In Example 1 the expression $1^2 + 2^2 + \cdots + n^2$ can be written as $\sum_{i=1}^{n} i^2$. Similarly, in Example 2, the expression $1^3 + 2^3 + 3^3 + \cdots + n^3$ can be written as $\sum_{i=1}^{n} i^3$. This notation can be used for a variety of sums. If we want the sum of the numbers $(-2)^5 + (-1)^5 + 0^5 + 1^5 + 2^5$, we express this as $\sum_{i=-2}^{2} i^5$. The subscript of the $\sum$ symbol tells us where to start and the superscript tells us where to end. The letter in the subscript is called the *index*. The value of this letter is incremented (increased) by one until we have reached the superscript value. The Greek symbol $\sum$ represents the Roman letter S, which is the first letter of the word "sum." We refer to this notation as *summation notation* .

EXAMPLE 3: We express the sum $2 + 4 + 6 + 8$ in summation notation as $\sum_{i=1}^{4} 2i$.

EXAMPLE 4: Let $f(x) = x^2 + 3x$ and let $x_i = \frac{i}{10}$ for $i = 0, 1, \ldots, 10$. Then $\sum_{i=0}^{10} f(x_i) = \left(\left(\frac{0}{10}\right)^2 + 3\left(\frac{0}{10}\right)\right) + \left(\left(\frac{1}{10}\right)^2 + 3\left(\frac{1}{10}\right)\right) + \cdots + \left(\left(\frac{10}{10}\right)^2 + 3\left(\frac{10}{10}\right)\right).$ $\blacksquare$

There are a number of properties of sums that can be expressed quite easily in summation notation.

Properties of Sums

(a) If c is a constant, then $\sum_{i=m}^{n} ca_i = c\sum_{i=m}^{n} a_i$.

(b) $\sum_{i=m}^{n}(a_i \pm b_i) = \sum_{i=m}^{n} a_i \pm \sum_{i=m}^{n} b_i$.

(c) $\sum_{i=m}^{n} a_i = \sum_{i=m}^{s} a_i + \sum_{i=s+1}^{n} a_i$ for $m \leq s < n$.

In some cases it is possible to express a sum in closed form; that is, we can express it as an algebraic formula.

EXAMPLE 5: Show that $\displaystyle\sum_{i=1}^{n} i = \frac{n(n+1)}{2}$.

SOLUTION: We use induction on n (see Appendix A). For $n = 1$ it is obvious that $\sum_{i=1}^{1} i = \frac{1(1+1)}{2} = 1$. Assume that $\sum_{i=1}^{n} i = \frac{n(n+1)}{2}$ is true for n. We now show that this is true for $n + 1$.

$$
\begin{aligned}
\sum_{i=1}^{n+1} i &= \sum_{i=1}^{n} i + \sum_{i=n+1}^{n+1} i \\
&= \frac{n(n+1)}{2} + (n+1) \\
&= \frac{(n+1)((n+1)+1)}{2}.
\end{aligned}
$$

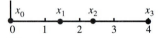

Figure 3.a *A partition $\mathcal{P}_1$ consisting of four points that divide the interval $[0,4]$ into three nonoverlapping subintervals.*

The mathematician Carl Friedrich Gauss invented a way of computing this sum when he was a schoolboy. In one column write the numbers 1 to n. In the next column write the numbers n down to 1. Add the two columns. Each term in this last column adds up to $n + 1$, so the total of the terms in the third column is $n(n + 1)$. But this is the sum of columns one and two. Thus the sum of the numbers from 1 to n is $\frac{n(n+1)}{2}$. ■

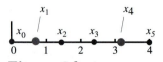

Figure 3.b *A refinement $\mathcal{P}_2$ of $\mathcal{P}_1$ consisting of six points that divide the interval $[0,4]$ into five nonoverlapping subintervals.*

We now want to formalize the ideas in Examples 1 and 2 for computing approximations for areas. The first idea was that of dividing an interval into a collection of nonoverlapping subintervals. To generalize this procedure, we introduce the notions of partition and refinement.

Definition: Partitions and Refinements

Let $[a, b]$ be an interval. The ordered set $\mathcal{P} = \{x_0, x_1, \ldots, x_n\}$ is a *partition* of $[a, b]$ provided that $a = x_0 \leq x_1 \leq x_2 \leq \ldots \leq x_n = b$. Thus, we have that $[x_0, x_1], [x_1, x_2], \ldots, [x_{n-1}, x_n]$ is a collection of nonoverlapping intervals that "fill up" the interval $[a, b]$.

If $\mathcal{P}_1$ and $\mathcal{P}_2$ are partitions of $[a, b]$, then $\mathcal{P}_2$ is called a *refinement* of $\mathcal{P}_1$ if $\mathcal{P}_1$ is a subset of $\mathcal{P}_2$. Thus $\mathcal{P}_2$ *refines* $\mathcal{P}_1$ if $\mathcal{P}_2$ divides $[a, b]$ more finely than does $\mathcal{P}_1$. See Figures 3.a and 3.b.

The second idea was that of approximating the area by adding up the areas of a collection of rectangles $R_1, R_2, \ldots, R_n$, where R_i has as its base the interval between x_{i-1} and x_i, and the height of R_i is determined by evaluating a function at a point in the interval $[x_{i-1}, x_i]$. We formalize this procedure in the following definition.

Definition: Selections and Riemann Sums

If $\mathcal{P} = \{x_0, x_1, \ldots, x_n\}$ is a partition of the interval $[a, b]$, and $\mathcal{S} = \{s_1, s_2, \ldots, s_n\}$ is a set of numbers such that, for each $i \leq n$, $x_{i-1} \leq s_i \leq x_i$, then $\mathcal{S}$ is called a *selection* from $\mathcal{P}$. If f is a function, $[a, b]$ is a subset of the domain of f, $\mathcal{P} = \{x_0, x_1, \ldots x_n\}$ is a partition of $[a, b]$, and $\mathcal{S} = \{s_1, s_2, \ldots s_n\}$ is a selection from $\mathcal{P}$, then

$$\sum_{i=1}^{n} f(s_i)(x_i - x_{i-1})$$

is called the *Riemann sum for f determined by $\mathcal{P}$ and $\mathcal{S}$.*

Note: We are not necessarily computing areas since $f(s_i)(x_i - x_{i-1})$ is negative if $f(s_i)$ is negative!

EXAMPLE 6: Let $f(x) = x^2$ and let $\mathcal{P} = \left\{1, \frac{5}{4}, \frac{3}{2}, \frac{13}{8}, 2\right\}$ be a partition of $[1, 2]$. Each of $\mathcal{S}_1 = \left\{\frac{5}{4}, \frac{3}{2}, \frac{13}{8}, 2\right\}$, $\mathcal{S}_2 = \left\{1, \frac{5}{4}, \frac{3}{2}, \frac{13}{8}\right\}$, and $\mathcal{S}_3 = \left\{\frac{9}{8}, \frac{5}{4}, \frac{25}{16}, \frac{7}{4}\right\}$ is a selection from $\mathcal{P}$. The set $\left\{\frac{9}{8}, \frac{25}{16}, \frac{5}{4}, \frac{7}{4}\right\}$ is not a selection from $\mathcal{P}$. Why?

The Riemann sum for f determined by $\mathcal{P}$ and $\mathcal{S}_1$ is:

$$\left(\frac{5}{4}\right)^2 \left(\frac{5}{4} - 1\right) + \left(\frac{3}{2}\right)^2 \left(\frac{3}{2} - \frac{5}{4}\right) + \left(\frac{13}{8}\right)^2 \left(\frac{13}{8} - \frac{3}{2}\right) + (2)^2 \left(2 - \frac{13}{8}\right)$$

$$= \frac{555}{128}.$$

The Riemann sum for f determined by $\mathcal{P}$ and $\mathcal{S}_2$ is:

$$(1)^2\left(\frac{5}{4}-1\right)+\left(\frac{5}{4}\right)^2\left(\frac{3}{2}-\frac{5}{4}\right)+\left(\frac{3}{2}\right)^2\left(\frac{13}{8}-\frac{3}{2}\right)+\left(\frac{13}{8}\right)^2\left(2-\frac{13}{8}\right)$$

$$=\frac{979}{512}.$$

The Riemann sum for f determined by $\mathcal{P}$ and $\mathcal{S}_3$ is:

$$\left(\frac{9}{8}\right)^2\left(\frac{5}{4}-1\right)+\left(\frac{5}{4}\right)^2\left(\frac{3}{2}-\frac{5}{4}\right)+\left(\frac{25}{16}\right)^2\left(\frac{13}{8}-\frac{3}{2}\right)+\left(\frac{7}{4}\right)^2\left(2-\frac{13}{8}\right)$$

$$=\frac{4425}{2048}. \qquad\qquad \blacksquare$$

The Riemann sum is a particularly strong and useful method for approximating certain types of physical quantities. The general problem will be: given a particular quantity, find a function f and an interval $[a,b]$ such that if $\mathcal{P}$ is a partition of $[a,b]$ and $\mathcal{S}$ is a selection from $\mathcal{P}$, then the Riemann sum for f determined by $\mathcal{P}$ and $\mathcal{S}$ is an approximation for the given quantity. It is "usually true" that the finer the partition, the better the approximation. This leads us to the following "Rule of Thumb."

"Rule of Thumb for Riemann Sums"

(a) I is an interval $[a,b]$ and f is a real valued function defined on I.

(b) Q is a physical quantity (such as area or volume) that can be approximated by:

   (i) Breaking up I into "very small" nonoverlapping intervals $[x_{i-1},x_i]$ $i=1,\ldots,n$.

   (ii) Choosing an arbitrary "selection point" s_i from each $[x_{i-1},x_i]$.

   (iii) Calculating the sum $\sum f(s_i)\text{length}[x_{i-1},x_i]$ to approximate Q.

Some examples follow.

EXAMPLE 7: Let $g(x)=x^2+2x+1$. Let A be the area of the region bounded by the graph of g, the x–axis, the line $x=1$, and the line $x=2$.

(a) Find a function f so that if $\mathcal{P} = \{x_0, x_1, \ldots, x_n\}$ is a partition of $[1, 2]$, and $\mathcal{S} = \{s_1, s_2, \ldots s_n\}$ is a selection from $\mathcal{P}$, then

$$\sum_{i=1}^{n} f(s_i)(x_i - x_{i-1}) \approx A.$$

(b) Compute the particular Riemann sum for f where the partition divides the interval $[1, 2]$ into four intervals of equal length, and, for each i, s_i is the midpoint between x_{i-1} and x_i.

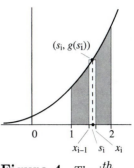

(s_i, $g(s_i)$)

x_{i-1} s_i x_i

Figure 4. *The ith rectangle.*

SOLUTION: For Part (a) we observe that g is a parabola with vertex $(-1, 0)$. Figure 4 is a sketch of g.

Let $\mathcal{P} = \{x_0, x_1, \ldots, x_n\}$ be a partition of $[1, 2]$, and let $\mathcal{S} = \{s_1, s_2, \ldots, s_n\}$ be a selection from $\mathcal{P}$. Now, for each i with $1 \le i \le n$, let R_i be the rectangle with base the interval $[x_{i-1}, x_i]$ and height $g(s_i)$ as illustrated in Figure 4. Thus

$$A \approx \sum_{i=1}^{n} \text{Area}(R_i) = \sum_{i=1}^{n} g(s_i)(x_i - x_{i-1}).$$

And so we may let $f = g$.

To solve Part (b) we observe that if $\mathcal{P}$ divides $[1, 2]$ into four intervals of equal length, then $x_0 = 1$, $x_1 = \frac{5}{4}$, $x_2 = \frac{3}{2}$, $x_3 = \frac{7}{4}$, and $x_4 = 2$.

If we choose each s_i as the midpoint of $[x_{i-1}, x_i]$, then $s_1 = \frac{9}{8}$, $s_2 = \frac{11}{8}$, $s_3 = \frac{13}{8}$, and $s_4 = \frac{15}{8}$. Thus this particular Riemann sum is

$$f(s_1)(x_1 - x_0) + f(s_2)(x_2 - x_1) + f(s_3)(x_3 - x_2) + f(s_4)(x_4 - x_3)$$

$$= \left[\left(\frac{9}{8} \right)^2 + 2 \left(\frac{9}{8} \right) + 1 \right] \left(\frac{1}{4} \right) + \left[\left(\frac{11}{8} \right)^2 + 2 \left(\frac{11}{8} \right) + 1 \right] \left(\frac{1}{4} \right) +$$

$$\left[\left(\frac{13}{8} \right)^2 + 2 \left(\frac{13}{8} \right) + 1 \right] \left(\frac{1}{4} \right) + \left[\left(\frac{15}{8} \right)^2 + 2 \left(\frac{15}{8} \right) + 1 \right] \left(\frac{1}{4} \right)$$

$$= 6\frac{21}{64}. \qquad \blacksquare$$

x_{i-1} s_i x_i

(s_i, $g(s_i)$)

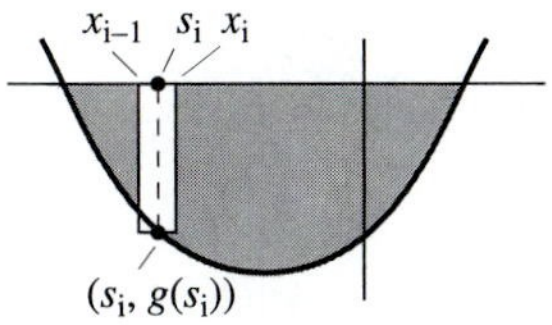

Figure 5. *The graph of $g(x) = x^2 + 2x - 3$, and the ith rectangle.*

EXAMPLE 8: Let $g(x) = x^2 + 2x - 3$. Find a function f and an interval $[a, b]$ so that if $\mathcal{P}$ is a partition of $[a, b]$, and $\mathcal{S}$ is a selection from $\mathcal{P}$, then the Riemann sum for f determined by $\mathcal{P}$ and $\mathcal{S}$ approximates the area bounded by the graph of g and the x–axis.

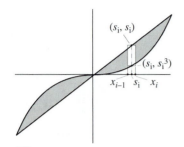

Figure 6. *The region bounded by the graphs of $g(x) = x^3$ and $h(x) = x$, and the ith rectangle.*

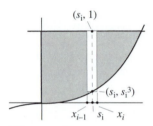

Figure 7. *The region bounded by the graph of $y = x^3$, the y-axis, and the line $y = 2$; and the ith rectangle.*

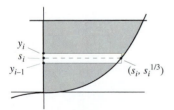

Figure 8. *The region bounded by the graph of $y = x^3$, the y-axis, and the line $y = 2$; and the ith rectangle, where we are partitioning the interval $[0, 2]$ of the y-axis.*

SOLUTION: First we sketch g in Figure 5, which displays the area we wish to approximate. The graph g crosses the x–axis at $x = -3$ and at $x = 1$. Let $\mathcal{P} = \{x_0, x_1, \ldots x_n\}$ be a partition of the interval $[-3, 1]$, and let $\mathcal{S} = \{s_1, s_2, \ldots s_n\}$ be a selection from $\mathcal{P}$. Consider the rectangle R_i with base $[x_{i-1}, x_i]$ as shown in Figure 5. The height of R_i must be $|g(s_i)| = -g(s_i)$. Thus the area of R_i is $-g(s_i)(x_i - x_{i-1})$. The desired area is approximated by $\sum_{i=1}^{n} -g(s_i)(x_i - x_{i-1})$, so, if we let $f(x) = -g(x)$, any Riemann sum for f over $[-3, 1]$ approximates the desired area.   ∎

EXAMPLE 9: Find a function f and an interval $[a, b]$ so that if $\mathcal{P}$ is a partition of $[a, b]$, and $\mathcal{S}$ is a selection from $\mathcal{P}$, then the Riemann sum for f determined by $\mathcal{P}$ and $\mathcal{S}$ approximates the area of the region bounded by the graphs of $g(x) = x^3$ and $h(x) = x$.

SOLUTION: In Figure 6 we sketch the desired region. Note that $(-1, -1), (0, 0)$, and $(1, 1)$ are the points of intersection of g and h. Hence $[a, b]$ should be the interval $[-1, 1]$. Clearly if we let $\mathcal{P} = \{x_0, x_1, \ldots x_n\}$ be a partition of $[-1, 1]$ and $\mathcal{S} = \{s_1, s_2, \ldots s_n\}$ be a selection from $\mathcal{P}$, and if, for each i with $1 \le i \le n$, we let R_i be the rectangle with base the interval $[x_{i-1}, x_i]$ and height $|s_i - s_i{}^3|$, then $\sum_{i=1}^{n} \text{Area}(R_i) = \sum_{i=1}^{n} |s_i - s_i{}^3|(x_i - x_{i-1})$ will approximate the area of the given region. Thus, if we let $f(x) = |x - x^3|$, any Riemann sum for f over $[-1, 1]$ is an approximation for the area of the given region.   ∎

It is not required that the function f in our Riemann sum be a function of x. As the next example illustrates, we can, in some cases, use rectangles that are based on the y–axis and f will then be a function of y. When we are in the fortunate position of having a choice of which axis to use, our choice is usually determined by the ease in evaluating one Riemann sum over the other.

EXAMPLE 10: Find a function f and an interval $[a, b]$ so that any Riemann sum for f over $[a, b]$ approximates the area of the region bounded by the graph of $y = x^3$, the y–axis, and the line $y = 2$.

SOLUTION: In Figure 7 we sketch the desired region. Note that our points of intersection are at $(0, 0), (0, 2)$, and $(\sqrt[3]{2}, 2)$.

We solve this problem using two methods.

Solution 1: In this approach, we partition the interval $[0, \sqrt[3]{2}]$ of the x–axis with $\mathcal{P} = \{x_0, x_1, \ldots x_n\}$. Let $\mathcal{S} = \{s_1, s_2, \ldots, s_n\}$ be a selection from $\mathcal{P}$. If, for each $i \le n$, we let R_i be the area of the rectangle of width $(x_i - x_{i-1})$ and height $2 - s_i{}^3$ as illustrated in

Figure 7, then

$$\sum_{i=1}^{n} R_i = \sum_{i=1}^{n} (2 - s_i^{\,3})(x_i - x_{i-1})$$

is an approximation for the area of the given region. Thus, if we let $f(x) = 2 - x^3$, then any Riemann sum of f over $[0, \sqrt[3]{2}]$ approximates the area of the given region.

Solution 2: In this approach, we divide the interval $[0, 2]$ of the y–axis with a partition $\mathcal{P} = \{y_0, y_1, \ldots y_n\}$. Let $\mathcal{S} = \{s_0, s_1, \ldots s_n\}$ be a selection from $\mathcal{P}$. In Figure 8, we see that if we let R_i be the rectangle with width $s_i^{\,1/3}$ and height $y_i - y_{i-1}$, then

$$\sum_{i=1}^{n} R_i = \sum_{i=1}^{n} s_i^{\,1/3} (y_i - y_{i-1})$$

approximates the desired area. Thus, if we let $f(y) = y^{1/3}$, then any Riemann sum for f over $[0, 2]$ approximates the desired area. ■

We now summarize the geometric information we have captured via the Riemann sum.

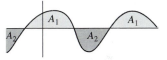

Figure 9. *The regions A_1 and A_2.*

Geometric Interpretation of the Riemann Sum

Suppose that f is a function and $[a, b]$ is a subset of the domain of f.

(a) If f is positive over the interval, then any Riemann sum for f over $[a, b]$ approximates the area bounded by the graph of f, the x–axis, and the lines $x = a$ and $x = b$. If f is negative over the interval $[a, b]$ then any Riemann sum for f over $[a, b]$ approximates the negative of the area bounded by the graph of f, the x–axis, and the lines $x = a$ and $x = b$. In general, if, as in Figure 9, we let A_1 denote the area bounded by the graph of f, the x–axis, and the lines $x = a$ and $x = b$ where f is positive, and if we let A_2 denote the area bounded by the graph of f, the x–axis, and the lines $x = a$ and $x = b$ where f is negative, then any Riemann sum for f over the interval $[a, b]$ approximates the number $A_1 - A_2$.

(b) It is a theorem that **if the function f is continuous**, we can control the accuracy of our approximation. In this case, by partitioning the interval into smaller and smaller pieces, we can ensure that the Riemann sums approach a limit.

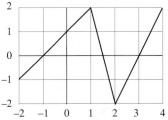

Figure 10.a *The function for Example 11.*

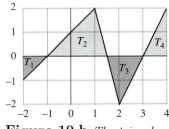

Figure 10.b *The triangles T_1, T_2, T_3, and T_4.*

EXAMPLE 11: Find the number approximated by the function f illustrated in Figure 10.a.

SOLUTION: From Figure 10.b, we see that the number approximated by the function is the sum of the areas of triangle T_2 and T_4 minus the sum of the areas of triangles T_1 and T_3, which is $\frac{1}{2}(2.5)(2) + \frac{1}{2}(1)(2) - \left(\frac{1}{2}(1)(1) + \frac{1}{2}(1.5)(2)\right) = 1.5.$ ■

EXERCISES 6.5

In Exercises 1–3, calculate the Riemann sum for the function f and partition illustrated to the right for the given selection.

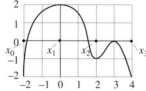

1. $s_i = x_i,\ 1 \le i \le 3.$

2. $s_i = x_{i-1},\ 1 \le i \le 3.$

3. s_i is the midpoint of $[x_{i-1}, x_i],\ 1 \le i \le 3.$

In Exercises 4–8, let $f(x) = 2x^2 + 3x + 1$. Compute the Riemann sum determined by $\mathcal{P}$ and $\mathcal{S}$.

4. $\mathcal{P} = \{0, \frac{1}{4}, \frac{1}{2}, \frac{3}{4}, 1\}, \mathcal{S} = \{0, \frac{1}{4}, \frac{1}{2}, \frac{3}{4}\}.$

5. $\mathcal{P} = \{0, \frac{1}{4}, \frac{1}{2}, \frac{3}{4}, 1\}, \mathcal{S} = \{\frac{1}{4}, \frac{1}{2}, \frac{3}{4}, 1\}.$

6. $\mathcal{P} = \{0, \frac{1}{4}, \frac{1}{2}, \frac{3}{4}, 1\}, s_i$ is the midpoint of $[x_{i-1}, x_i]$ for $i = 1, 2, 3, 4.$

7. $\mathcal{P} = \{-1, -\frac{1}{2}, 0, \frac{1}{5}, \frac{1}{2}, 1\}, s_i$ is the number where f attains its maximum on the interval $[x_{i-1}, x_i]$ for $i = 1, 2, 3, 4, 5.$

8. $\mathcal{P} = \{-1, -\frac{1}{2}, 0, \frac{1}{5}, \frac{1}{2}, 1\}, s_i$ is the number where f attains its minimum on the interval $[x_{i-1}, x_i]$, for $i = 1, 2, 3, 4, 5.$

In Exercises 9–11, let $\mathcal{P}$ be the partition that divides $[a, b]$ into five intervals of equal length. Find a selection $\mathcal{S}$ from $\mathcal{P}$ so that if R is the Riemann sum for f determined by $\mathcal{P}$ and $\mathcal{S}$, then the choice of any other selection yields a Riemann sum greater than or equal to R. Compute the sum for f determined by $\mathcal{S}$ and $\mathcal{P}$.

9. $f(x) = 2x^2 + 3x + 1, [a, b] = [0, 5].$

10. $f(x) = x^3, [a, b] = [-2, 3].$

11. $f(x) = 2x^2 + 3x + 1, [a, b] = [-1, 4].$

In Exercises 12–14, let $\mathcal{P}$ be the partition that divides $[a, b]$ into five intervals of equal length. Find a selection $\mathcal{S}$ from $\mathcal{P}$ so that if R is the Riemann sum for f determined by $\mathcal{P}$ and $\mathcal{S}$, then the choice of any other selection yields a Riemann sum less than or equal to R. Compute the sum for f determined by $\mathcal{S}$ and $\mathcal{P}$.

12. $f(x) = 2x^2 + 3x + 1, [a, b] = [0, 5].$

13. $f(x) = x^3, [a, b] = [-2, 3].$

14. $f(x) = 2x^2 + 3x + 1, [a, b] = [-1, 4].$

In Exercises 15 and 16, find a function f and an interval $[a, b]$ so that the Riemann sums for f over $[a, b]$ approximate the area of R. Sketch R.

15. R is the region bounded by the graph of $y = 2x^2 + 3x + 1$, the x–axis, and the lines $x = 1$ and $x = 2.$

16. R is the region bounded by the graph of $y = 2x^2 + 3x + 1$, the x–axis, and the lines $x = -1$ and $x = 1.$

Each of the functions below are made of straight lines and half circles. Find the number that is approximated by the Riemann sums for the function.

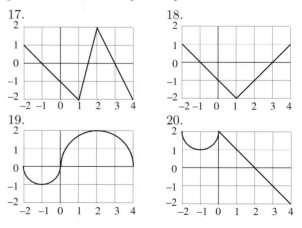

21. **[PROJECT]**: *Let f be a function that is continuous on the interval* $[a, b]$.

    a. For each integer n, let $\mathcal{P}_n = \{x_0, x_1, \cdots, x_{2^n}\}$ be the partition of $[a, b]$ that divides $[a, b]$ into 2^n subintervals of equal length.

        i. Find $\mathcal{P}_1$, $\mathcal{P}_2$, and $\mathcal{P}_3$ if $[a, b] = [0, 2]$.

        ii. Find $\mathcal{P}_1$, $\mathcal{P}_2$, and $\mathcal{P}_3$ if $[a, b] = [-2, 2]$.

        iii. Write down the first three terms of $\mathcal{P}_n$ for the interval $[a, b]$.

    b. Let $\mathcal{S}_{*n} = \{s_{*1}, s_{*2}, \cdots, s_{*(2^n)}\}$, be the selection from $\mathcal{P}_n$ such that f attains its minimum value over $[x_{i-1}, x_i]$ at s_{*i}.

        i. Explain why we know that s_{*i} exists.

        ii. Find $\mathcal{S}_{*n}$ for $n = 1, 2, 3$ $f(x) = x^2$ and $[a, b] = [0, 2]$.

        iii. Find $\mathcal{S}_{*n}$ for $n = 1, 2, 3$ $f(x) = x^2$ and $[a, b] = [-2, 2]$.

    c. Let L_n be the Riemann sum for the function f determined by $\mathcal{P}_n$ and $\mathcal{S}_{*n}$. That is,

$$L_n = \sum_{i=1}^{2^n} f(s_{*1})(x_i - x_{i-1}).$$

        i. Find L_1, L_2, and L_3 for $f(x) = x^2$ and $[a, b] = [0, 2]$.

        ii. Find L_1, L_2, and L_3 for $f(x) = x^2$ and $[a, b] = [-2, 2]$.

        iii. Argue that for any continuous function on $[a, b]$, $L_1 \leq L_2 \leq L_3$.

21.   c. iv. Argue that for any continuous function on $[a, b]$, L_1, L_2, L_3, $\cdots$ is a nondecreasing sequence. That is, argue that $L_i \leq L_{i+1}$ for all i.

    d. Let $\mathcal{S}_n^* = \{s_1^*, s_2^*, \cdots, s_{2^n}^*\}$, be the selection from $\mathcal{P}_n$ such that f attains its maximum value over $[x_{i-1}, x_i]$ at s_i^*.

        i. Find $\mathcal{S}_n^*$ for $n = 1, 2, 3$ $f(x) = x^2$, and $[a, b] = [0, 2]$.

        ii. Find $\mathcal{S}_n^*$ for $n = 1, 2, 3$ $f(x) = x^2$, and $[a, b] = [-2, 2]$.

    e. Let U_n be the Riemann sum for the function f determined by $\mathcal{P}_n$ and $\mathcal{S}_n^*$. That is,

$$U_n = \sum_{i=1}^{2^n} f(s_i^*)(x_i - x_{i-1}).$$

        i. Find U_1, U_2, and U_3 for $f(x) = x^2$ and $[a, b] = [0, 2]$.

        ii. Find U_1, U_2, and U_3 for $f(x) = x^2$ and $[a, b] = [-2, 2]$.

        iii. Argue that for any continuous function on $[a, b]$, $U_1 \geq U_2 \geq U_3$.

        iv. Argue that for any continuous function on $[a, b]$, U_1, U_2, U_3, $\cdots$ is a nonincreasing sequence. That is, argue that $U_i \geq U_{i+1}$ for all i.

        v. Argue that $L_n \leq U_n$ for all n.

    f. Argue that if f is nondecreasing on $[a, b]$, then $\lim_{n \to \infty} (U_n - L_n) = 0$.

6.6   The Definite Integral

In the previous section, we used the concept of Riemann sum to approximate the areas of various regions. We also asserted that if f is a continuous function, and if $[a, b]$ is a subset of the domain of f, then there was a particular number so that the Riemann sums for f over $[a, b]$ well approximate that number. That is, as we further and further refine the partitions, the sums approach a common limit. It is this number that we call the *definite integral of f over* $[a, b]$.

Suppose that f is a function and $[a, b]$ is an interval such that the Riemann sums for f over $[a, b]$ approximate a particular number, say L. Clearly, if $\mathcal{P}_1$ and $\mathcal{P}_2$ are partitions of $[a, b]$ and $\mathcal{P}_1$ is a subset of $\mathcal{P}_2$, then the points of $\mathcal{P}_2$ are closer together than the points of $\mathcal{P}_1$; thus the Riemann sums for f determined by $\mathcal{P}_2$ should be better approximations to the number L than are the Riemann sums determined by $\mathcal{P}_1$.[2] It is this idea that we use to formalize the notion of the definite integral.

For completeness we reiterate a definition from the previous section.

Definition: Refinement of a Partition

Let $\mathcal{P}_1$ and $\mathcal{P}_2$ be two partitions of the interval $[a, b]$. We say that the partition $\mathcal{P}_2$ *refines* the partition $\mathcal{P}_1$ if $\mathcal{P}_1$ is a subset of $\mathcal{P}_2$. Thus, if $\mathcal{P}_2$ refines $\mathcal{P}_1$, then $\mathcal{P}_2$ divides the interval $[a, b]$ more finely than does $\mathcal{P}_1$.

We now can define the notion of a limit of Riemann sums.

Definition: Limit of Riemann Sums

We say that the number L is the *limit of the Riemann sums for the function f over the interval $[a, b]$* if L is the only number that is approximated by the Riemann sums for f over $[a, b]$. That is, L is the limit of the collection of Riemann sums for f over $[a, b]$ if it is true that if E is a positive number, then there is a partition $\mathcal{P}$ for $[a, b]$ such that if $\mathcal{P}'$ is a refinement for $\mathcal{P}$ and $\sum_{i=1}^{n} f(s_i)(x_i - x_{i-1})$ is a Riemann sum for f determined by $\mathcal{P}'$, then

$$\left| L - \sum_{i=1}^{n} f(s_i)(x_i - x_{i-1}) \right| < E,$$

for all selections $\mathcal{S}$ of $\mathcal{P}'$.

Notation: If L is the limit of the set of Riemann sums of f over $[a, b]$, then L is denoted by $\int_a^b f(x)\, dx$. The notation $\int_a^b f(x)\, dx$ is called the *integral of f from a to b*; the numbers a and b is called the *limits of integration*.

To recall the comment at the end of the last section, if f is a continuous function and $[a, b]$ is a subset of the domain of f, then $\int_a^b f(x)\, dx = A_1 - A_2$, where A_1 denotes the area between the graph of f, the x–axis, and the lines $x = a$ and $x = b$ where f is positive,

---

[2]However, it should be noted that these approximations also depend heavily upon the selections $\mathcal{S}_1$ and $\mathcal{S}_2$. Even though the points of $\mathcal{P}_2$ may be closer together than the points of $\mathcal{P}_1$, the choice of selections $\mathcal{S}_1$ and $\mathcal{S}_2$ can cause the Riemann sum determined by $\mathcal{P}_1$ and $\mathcal{S}_1$ to be a closer approximation of the area than the Riemann sum determined by $\mathcal{P}_2$ and $\mathcal{S}_2$.

and where A_2 denotes the area between the graph of f, the x–axis, and the lines $x = a$ and $x = b$ where f is negative (recall Figure 9 of Section 6.5).

(a) From Example 7 of Section 6.5, $\int_1^2 (x^2 + 2x + 1) \, dx$ is the area of the region bounded by the graph of $g(x) = x^2 + 2x + 1$, the x–axis, and the lines $x = 1$ and $x = 2$.

(b) From Example 8 of Section 6.5, $\int_{-3}^1 (x^2 + 2x - 3) \, dx$ is the area of the region bounded by the graph of $g(x) = x^2 + 2x - 3$ and the x–axis.

(c) From Example 9 of Section 6.5, we see that $\int_{-1}^1 |x - x^3| \, dx = \int_{-1}^0 x^3 - x \, dx + \int_0^1 x - x^3 \, dx$ is the area of the region bounded by the graphs of $y = x^3$ and $y = x$.

(d) From Example 10 of Section 6.5, we see that both $\int_0^1 y^{1/3} \, dy$ and $\int_0^{\sqrt 2} (2 - x^3) \, dx$ denote the area of the region dbounded by the graphs of $y = x^3, y = 2$, and $x = 0$.

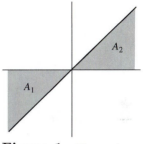

Figure 1. *The regions A_1 and A_2 for $f(x) = x$, $-1 \leq x \leq 1$.*

The definite integral is perhaps one of the most useful notions the student will encounter. A large portion of the remainder of the text will be devoted to applying and computing the definite integrals of various functions. One important observation about our notation: in the integral notation, the variable used makes no difference; for example, if f is any function, then $\int_a^b f(x) \, dx$, $\int_a^b f(t) \, dt$, $\int_a^b f(y) \, dy$, and $\int_a^b f(z) \, dz$ all represent the same number. In the expression $\int_a^b \square \, dt$ (where $\square$ represents a function), t is the variable for integration. Likewise in the expression $\int_a^b \square \, dz$, the variable for integration is z. Thus, if we have the expression $\int_a^b f(x) \, dt$, the function $f(x)$ is treated as a constant. Thus $\int_a^b f(x) \, dt = f(x) \int_a^b \, dt$.

It should be pointed out here that the geometric interpretation of the integral makes a few (very few, unfortunately) integrals very easy to compute. However, once we have mastered the techniques for solving integrals, we will have at our disposal a method for solving many geometric problems.

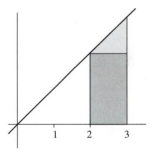

Figure 2. *The region bounded by the graph of $f(x) = x$ and the x–axis on the interval $[2, 3]$.*

EXAMPLE 1: Compute $\int_{-1}^1 x \, dx$.

SOLUTION: In Figure 1, we see that $\int_{-1}^1 x \, dx = A_1 - A_2$, where A_1 and A_2 are areas of triangles of equal size. Thus $\int_{-1}^1 x \, dx = 0$. ∎

EXAMPLE 2: Compute $\int_2^3 x \, dx$.

SOLUTION: In Figure 2, we see that $\int_2^3 x\,dx$ is the sum of the areas of two regular figures: the triangle with height 1 and base 1 and the rectangle of height 2 and base 1. Thus we have $\int_2^3 x\,dx = \frac{1}{2} + 2 = \frac{5}{2}$. ∎

EXAMPLE 3: Returning to Example 11 of the previous section, let f be the function illustrated in Figure 3a. Then $\int_{-2}^4 f(x)\,dx =$ Area(T_2) + Area(T_4) − Area(T_1) − Area(T_3) = 1.5.

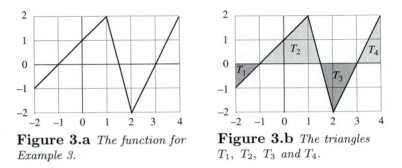

Figure 3.a *The function for Example 3.*   **Figure 3.b** *The triangles T_1, T_2, T_3 and T_4.*

We now list some facts about the definite integral, which we will be using extensively throughout the course.

Theorem 1 *If f is continuous on the interval $[a, b]$, then $\int_a^b f(x)\,dx$ exists.*

The proof of Theorem 1 is best left to a course in advanced calculus.

Important Note: The essence of the proof is that if f is continuous on $[a, b]$ and if U_n and L_n are defined as in Exercise 7 of the previous section, then $\lim\limits_{n\to\infty} U_n = \lim\limits_{n\to\infty} L_n$ and that limit is $\displaystyle\int_a^b f(x)\,dx$.

Theorem 2 *If $a \le b \le c$, then*

(a) *if $\int_a^c f(x)\,dx$ exists, then each of $\int_a^b f(x)\,dx$ and $\int_b^c f(x)\,dx$ exist;*

(b) *if $\int_a^b f(x)\,dx$ and $\int_b^c f(x)\,dx$ exist, then $\int_a^c f(x)\,dx = \int_a^b f(x)\,dx + \int_b^c f(x)\,dx$; and*

(c) *if k is any number, then $k\int_a^b f(x)\,dx = \int_a^b kf(x)\,dx$.*

A rigorous proof of Theorem 2 requires a careful application of the definition of the limit of Riemann sums. We give only the basic idea behind the argument for Part (b). The arguments for Parts (a) and (c) have the same flavor. Let $\mathcal{P} = \{x_0, x_1, \ldots, x_n\}$ be a partition of the interval $[a, b]$, and let $\mathcal{S} = \{s_1, \ldots, s_n\}$ be a selection for $\mathcal{P}$ so that $\sum_{i=1}^{n} f(s_i)(x_i - x_{i-1})$ approximates $\int_a^b f(x)\,dx$. Let $\mathcal{P}' = \{x_0', x_1', \ldots, x_m'\}$ be a partition of the interval $[b, c]$, and let $\mathcal{S}' = \{s_1', \ldots, s_m'\}$ be a selection for $\mathcal{P}'$ so that $\sum_{i=1}^{m} f(s_i')(x_i' - x_{i-1}')$ approximates $\int_b^c f(x)\,dx$.

Then $\mathcal{P}'' = \{x_0, x_1, \ldots, x_n = x_0', x_1', \ldots, x_m'\}$ is a partition of the interval $[a, c]$, and $\mathcal{S}'' = \{s_1, \ldots, s_n = s_0', s_1', \ldots, s_m'\}$ is a selection for $\mathcal{P}''$. Furthermore, the Riemann sum for f determined by $\mathcal{P}''$ and $\mathcal{S}''$ is an approximation for $\int_a^c f(x)\,dx$.

We have

$$\int_a^c f(x)\,dx \approx \sum_{i=1}^{n} f(s_i)(x_i - x_{i-1}) + \sum_{i=1}^{m} f(s_i')(x_i' - x_{i-1}')$$

$$\approx \int_a^b f(x)\,dx + \int_b^c f(x)\,dx. \qquad \blacksquare$$

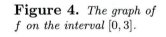

Figure 4. *The graph of f on the interval $[0, 3]$.*

Theorem 3 *If $\int_a^b f(x)\,dx$ and $\int_a^b g(x)\,dx$ exist, then $\int_a^b (f + g)(x)\,dx$ exists, and $\int_a^b (f + g)(x)\,dx = \int_a^b f(x)\,dx + \int_a^b g(x)\,dx$.*

Sketch of proof: Let $\mathcal{P} = \{x_0, x_1, \ldots, x_n\}$ be a partition of the interval $[a, b]$ and let $\mathcal{S} = \{s_1, \ldots, s_n\}$ be a selection for $\mathcal{P}$. Then

$$\int_a^b (f + g)(x)\,dx \approx \sum_{i=1}^{n} (f + g)(s_i)(x_i - x_{i-1})$$

$$= \sum_{i=1}^{n} f(s_i)(x_i - x_{i-1}) + \sum_{i=1}^{n} g(s_i)(x_i - x_{i-1})$$

$$\approx \int_a^b f(x)\,dx + \int_a^b g(x)\,dx. \qquad \blacksquare$$

EXAMPLE 4: Let a be the sequence defined by $a(n) = (-1)^n n$, and let $f(x) = a[[x]]$. Compute $\int_0^3 f(x)\,dx$. The graph of f is sketched in Figure 4.

SOLUTION: Using the geometry of the problem, we see that

$$\int_0^1 f(x)\,dx = 0, \quad \int_1^2 f(x)\,dx = -1, \quad \text{and} \quad \int_2^3 f(x)\,dx = 2.$$

Thus

$$\int_0^3 f(x)\,dx \;=\; \int_0^1 f(x)\,dx + \int_1^2 f(x)\,dx + \int_2^3 f(x)\,dx$$

$$= \; 0 - 1 + 2 = 1.$$

EXERCISES 6.6

Compute $\int_{-2}^4 f(x)\,dx$, where f is sketched below.

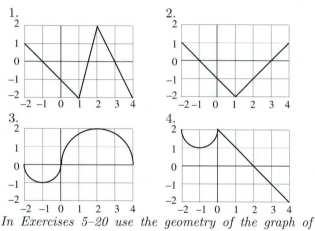

In Exercises 5–20 use the geometry of the graph of the functions to compute the following integrals.

5. $\int_{-1}^1 x\,dx$.

6. $\int_{-1}^0 2x\,dx$.

7. $\int_{-3}^4 3s\,ds$.

8. $\int_{-2}^1 (3x+1)\,dx$.

9. $\int_{-1}^2 (3t+5)\,dt$.

10. $\int_{-1}^1 x^3\,dx$.

11. $\int_0^{2\pi} \cos x\,dx$.

12. $\int_0^{2\pi} \sin x\,dx$.

13. $\int_0^\pi \sin 2x\,dx$.

14. $\int_0^\pi \cos x\,dx$.

15. $\int_0^\pi (t + \cos t)\,dt$.

16. $\int_{-1}^1 (s + s^3)\,ds$.

17. $\int_{-1}^1 |x|\,dx$.

18. $\int_1^4 [[x]]\,dx$.

19. $\int_{-1}^1 (|x| + x)\,dx$.

20. $\int_{-1}^1 \sqrt{1 - x^2}\,dx$.

In Exercises 21 and 22, express as a definite integral the number that is approximated by the Riemann sum.

21. $\displaystyle\sum_{i=1}^n (s_i)^3 \cos(s_i)(x_i - x_{i-1})$.
    The partitioned interval is $[-1, 3]$.

22. $\displaystyle\sum_{i=1}^n \sqrt{(s_i)^2 + 4}\,(x_i - x_{i-1})$.
    The partitioned interval is $[0, 5]$.

23. Given that $\displaystyle\int_a^b f(x)\,dx = 3$ and $\displaystyle\int_a^b g(x)\,dx = -5$, find $\displaystyle\int_a^b f(x) + g(x)\,dx$.

24. Given that $\displaystyle\int_1^2 f(x)\,dx = 3$ and $\displaystyle\int_2^4 f(x)\,dx = -5$, find $\displaystyle\int_1^4 f(x)\,dx$.

25. Given that $\displaystyle\int_1^4 f(x)\,dx = 3$ and $\displaystyle\int_2^4 f(x)\,dx = -5$, find $\displaystyle\int_1^2 f(x)\,dx$.

26. Given that $\int_{-1}^{2} x^2 \, dx = 3$, compute $\int_{-1}^{2} (x^2 + 3x + 5) \, dx$.

27. Given that $\int_{-\pi/2}^{\pi/2} \cos x \, dx = 2$, compute $\int_{0}^{\pi/2} \cos x \, dx$, $\int_{-\pi/2}^{\pi} \cos x \, dx$, and $\int_{0}^{\pi} \sin x \, dx$.

28. Show that $\int_{0}^{a} mx \, dx = \frac{ma^2}{2}$.

29. Show that $\int_{0}^{a} mx + 1 \, dx = \frac{ma^2}{2} + a$.

30. Show that $\int_{a}^{b} mx \, dx = \frac{mb^2}{2} - \frac{ma^2}{2}$. (Assume that $0 \le a < b$.)

In Exercises 31 and 32 let a be the sequence defined by $a(n) = (-1)^n n$ and let $f(x) = a[[x]]$ as in Example 4. Compute the following integrals.

31. $\int_{0}^{4} f(x) \, dx$.

32. $\int_{0}^{3.5} f(x) \, dx$.

33. Let $g(x)$ be a continuous increasing function on the interval $[a, b]$, which has an inverse g^{-1}. Show that

$$\int_{a}^{b} g(x) \, dx + \int_{g(a)}^{g(b)} g^{-1}(y) \, dy = bg(b) - ag(a),$$

and explain geometrically why this is true.

6.7   Volumes

In this section, we use Riemann sums to approximate volumes of solids, and we demonstrate how these volumes can be expressed as definite integrals.

We begin by considering two approaches for approximating the volume of a simple cone. The student may be familiar with the type of toy that consists of a collection of thin discs of different radii that can be stacked according to size, with the largest at the bottom, so that the resulting shape is an approximation of a cone (see Figure 1.a). Alternately, we could use concentric cylindrical shells of varying heights to approximate the volume of the cone (see Figure 1.b).

Figure 1.a *Approximating the volume with thin discs.*

Figure 1.b *Approximating the volume with thin shells.*

We expand on these simple ideas to compute volumes of various solids. Given a solid, our job is to find a function f and an interval $[a, b]$ so that $\int_{a}^{b} f(t) \, dt$ is the volume of that solid.

EXAMPLE 1:   Find a function f and an interval $[a, b]$ so that

$\int_a^b f(x)\, dx$ is the volume of the right circular cone of radius 1 ft and height 2 ft illustrated in Figure 2.

SOLUTION: We use two techniques to solve this problem.

The Disc Method: In this method we approximate the solid with a collection of thin discs. We first place the coordinate plane in a convenient position: the origin is the center of the base of the cone so that the y–axis coincides with the axis of the cone as in Figure 3.

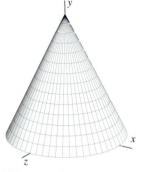

Figure 2. *The right circular cone of Example 1.*

Let $\mathcal{P}$ be a partition of the interval $[0, 2]$ of the y–axis, and let $\mathcal{S}$ be a selection from $\mathcal{P}$. If we let C_i be the volume of the disc of thickness $(y_i - y_{i-1})$ and radius $(2 - s_i)/2$, then $\sum_{i=1}^{n} C_i$ approximates the volume of the cone.

Figure 3. *The construction of the discs.*

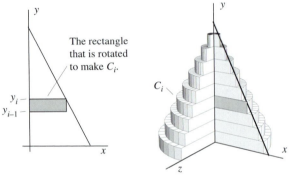

We have

$$\sum_{i=1}^{n} C_i = \sum_{i=1}^{n} \pi \left(\frac{2 - s_i}{2} \right)^2 (y_i - y_{i-1}). \qquad (8)$$

But, if we let f be the function defined by

$$f(y) = \pi \left(\frac{2 - y}{2} \right)^2,$$

then the sum in Equation (8) is a Riemann sum for f over $[0, 2]$. Thus the volume of the cone is given by

$$\int_0^2 f(y)\, dy = \int_0^2 \pi \left(\frac{2 - y}{2} \right)^2 dy. \qquad \blacksquare$$

The Shell Method: In this method we approximate the solid cone with a collection of cylindrical shells. Let $\mathcal{P} = \{x_0, x_1, \ldots, x_n\}$ be a partition of $[0, 1]$, and let $\mathcal{S} = \{s_1, s_2, \ldots s_n\}$ be a selection from $\mathcal{P}$.

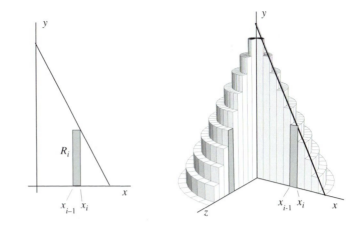

Figure 4. *The construction of the shells.*

If we rotate the rectangle R_i of Figure 4 about the y–axis, we obtain the shell with volume S_i. Clearly, the volume of the cone is approximated by $\sum_{i=1}^{n} S_i$. To determine the volume of the shell S_i, we notice that the volume of the cylinder of radius x_i and height $-2s_i + 2$ is $\pi x_i^2(-2s_i + 2)$ and the volume of the cylinder of radius x_{i-1} and height $-2s_i + 2$ is $\pi x_{i-1}^2(-2s_i + 2)$. We see in Figure 5 that the volume of the shell S_i is the difference between these two volumes, which is:

$$
\begin{aligned}
S_i &= \pi x_i^2(-2s_i + 2) - \pi x_{i-1}^2(-2s_i + 2) \\
&= \pi(-2s_i + 2)(x_i^2 - x_{i-1}^2).
\end{aligned}
$$

Figure 5. *The shell obtained by rotating R_i about the y–axis.*

Thus the volume of the cone is approximated by

$$
\sum_{i=1}^{n} S_i = \sum_{i=1}^{n} \pi(-2s_i + 2)(x_i^2 - x_{i-1}^2). \tag{9}
$$

The problem now is to express the sum in Equation (9) as a Riemann sum. We can rewrite Equation (9) as follows:

$$
\sum_{i=1}^{n} S_i = \sum_{i=1}^{n} \pi(-2s_i + 2)(x_i + x_{i-1})(x_i - x_{i-1}). \tag{10}
$$

In order to simplify the above expression, we choose our selection from $\mathcal{P}$ a little more carefully. Notice that if we choose s_i to be the midpoint of the interval $[x_{i-1}, x_i]$, then $2s_i = x_i + x_{i-1}$. With this choice of selection, Equation (10) becomes

$$
\sum_{i=1}^{n} S_i = \sum_{i=1}^{n} \pi 2s_i(-2s_i + 2)(x_i - x_{i-1})
$$

$$= 4\pi \sum_{i=1}^{n} (-s_i^2 + s_i)(x_i - x_{i-1}). \qquad (11)$$

But the sum in Equation (11) is a Riemann sum for $f(x) = x - x^2$ over the interval $[0, 1]$, so $4\pi \int_0^1 (x - x^2)\, dx$ represents the desired volume. ∎

EXAMPLE 2: The region in the plane bounded by the graph of $y = x^2$ and the line $y = x$ is rotated about the y–axis. Find a function f and an interval $[a, b]$ so that $\int_a^b f(x)\, dx$ is the volume of the resulting solid shown in Figure 6.

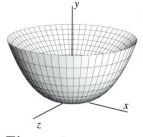

Figure 6. *The solid in Example 2.*

SOLUTION: As in the previous example, we employ two methods.
The Disc Method: The word disc is a misnomer here, for the shape looks more like that of a washer, as the drawing in Figure 7 illustrates.

Let $\mathcal{P} = \{y_0, y_1, \ldots, y_n\}$ be a partition for the interval $[0, 1]$ of the y–axis, and let $\mathcal{S} = \{s_1, s_2, \ldots, s_n\}$ be a selection from $\mathcal{P}$. Let W_i be the volume obtained by rotating the rectangle R_i illustrated in Figure 7 about the y–axis so that W_i is the volume of a thin washer. The volume of the given solid will be approximated by $\sum_{i=1}^{n} W_i$. Now,

$$
\begin{aligned}
W_i &= \pi(\sqrt{s_i})^2 (y_i - y_{i-1}) - \pi s_i^2 (y_i - y_{i-1}) \\
&= \pi(s_i - s_i^2)(y_i - y_{i-1}).
\end{aligned}
$$

Thus the desired volume is approximated by

$$\sum_{i=1}^{n} \pi(s_i - s_i^2)(y_i - y_{i-1}). \qquad (12)$$

But Equation (12) is a Riemann sum for $f(y) = \pi(y - y^2)$, so the sum in Equation (12) approximates both $\int_0^1 \pi(y - y^2)\, dy$ and the desired volume V. Thus $V = \int_0^1 \pi(y - y^2)\, dy$. ∎

Figure 7. *The ith rectangle R_i is rotated about the y–axis to build the ith disc.*

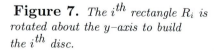

The Shell Method: Here we approximate the given solid with a collection of cylindrical shells. Let $\mathcal{P} = \{x_0, x_1, \ldots, x_n\}$ be a partition of the interval $[0, 1]$ on the x–axis, and let $\mathcal{S} = \{s_1, s_2, \ldots, s_n\}$ be a selection from $\mathcal{P}$. For each i, let S_i be the volume of the shell obtained by rotating the rectangle R_i of Figure 8 about the y–axis.

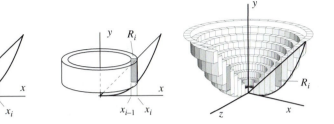

Figure 8. *The ith rectangle R_i is rotated about the y–axis to build the ith shell.*

Now

$$
\begin{aligned}
S_i &= \pi x_i^2(s_i - s_i^2) - \pi x_{i-1}^2(s_i - s_i^2) \\
&= \pi(s_i - s_i^2)(x_i^2 - x_{i-1}^2).
\end{aligned}
$$

Thus the volume V is approximated by the sum of the S_i.

$$
V \approx \sum_{i=1}^{n} \pi(s_i - s_i^2)(x_i^2 - x_{i-1}^2).
$$

Our task is to express $\sum_{i=1}^{n} \pi(s_i - s_i^2)(x_i^2 - x_{i-1}^2)$ as a Riemann sum. Observe that

$$
x_i^2 - x_{i-1}^2 = (x_i + x_{i-1})(x_i - x_{i-1}).
$$

If we choose s_i to be the midpoint of $[x_{i-1}, x_i]$, then $2s_i = (x_i + x_{i-1})$. With this choice of s_i, our sum becomes

$$
\begin{aligned}
V &\approx \sum_{i=1}^{n} \pi(s_i - s_i^2)(x_i + x_{i-1})(x_i - x_{i-1}) \\
&= \sum_{i=1}^{n} 2\pi s_i(s_i - s_i^2)(x_i - x_{i-1}) \\
&\approx \int_0^1 2\pi(x^2 - x^3)\, dx. \qquad\blacksquare
\end{aligned}
$$

The same techniques may be used to find volumes obtained by rotating regions about the x–axis.

EXAMPLE 3: Let R be the region bounded between the graph of $y = x^3$, the x–axis, and the line $x = 1$, and let S be the solid obtained by rotating R about the x–axis. See Figure 9.a. Express the volume V of S as a definite integral.

SOLUTION: We use discs. Let $\{x_0, x_1, \ldots, x_n\}$ partition the interval $[0, 1]$ on the x–axis and let $\{x_1, s_2, \ldots, s_n\}$ be a selection from that partition. For each i, let R_i be the rectangle or width $x_i - x_{i-1}$ and height s_i^3 as illustrated in Figure 9.b. R_i is rotated about the x–axis to obtain a disc D_i of radius s_i^3 and width $x_i - x_{i-1}$ as shown in Figure 9c. The volume of D_i is $V_i = \pi(s_i^3)^2(x_i - x_{i-1})$. The sum of the V_i's approximates the volume of the solid. Thus

$$V \approx \sum_{i=1}^{n} V_i = \sum_{i=1}^{n} \pi(s_i^3)^2(x_i - x_{i-1}) \approx \int_0^1 \pi x^6 \, dx. \qquad \blacksquare$$

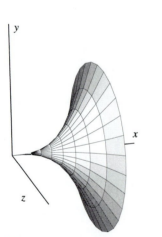

Figure 9.a *The solid of Example 3*

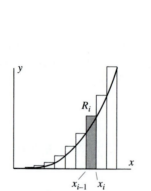

Figure 9.b *The region to be rotated and the ith rectangle.*

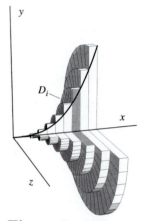

Figure 9.c *The rectangles are rotated about the x–axis to approximate the solid.*

We need only a slight modification of our technique to handle the following example, where the "slices" are squares rather than discs.

EXAMPLE 4: Find the volume of the right pyramid with a 1 ft $\times$1 ft square base and an altitude of 2 ft. (See Figure 10.a.)

SOLUTION: We place our coordinate system so that:

(a) the center of the base is at the origin;

(b) the vertex is on the positive y–axis; and

(c) the sides intersect the x–axis at $x = \pm\frac{1}{2}$. (See Figure 10b.)

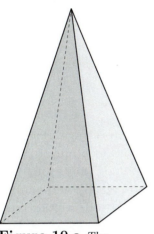

Figure 10.a *The pyramid of Example 4.*

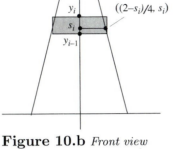

Figure 10.b *Front view of the pyramid and the ith thin box.*

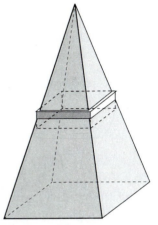

Figure 10.c *The pyramid and the ith thin box.*

With this placement one of the sides intersects the xy–plane in the line $y = 2 - 4x$.

Now let $\mathcal{P} = \{y_0, y_1, \ldots, y_n\}$ be a partition for the interval $[0, 2]$ of the y–axis, and let $\mathcal{S} = \{s_1, s_2, \ldots s_n\}$ be a selection from $\mathcal{P}$. Denote by B_i the volume of the box drawn in Figure 10.c. Then

$$B_i = \left(\frac{2 - s_i}{2}\right)^2 (y_i - y_{i-1})$$

so that

$$\text{Volume} \approx \sum_{i=1}^{n} B_i = \sum_{i=1}^{n} \left(\frac{2 - s_i}{2}\right)^2 (y_i - y_{i-1}). \qquad (13)$$

But the sum in Equation (13) is also a Riemann sum for $f(y) = ((2 - y)/2)^2$ over the interval $[0, 2]$. Thus we have that

$$V = \int_0^2 \left(\frac{2 - y}{2}\right)^2 dy. \qquad \blacksquare$$

EXERCISES 6.7

In Exercises 1–10, the region R is rotated about the x–axis. Express the resulting volume as a definite integral or a sum of definite integrals.

1. R is the region bounded by the graphs of $y = x^2, y = 0$, and $x = 2$.

2. R is the region bounded by the graphs of $y = x^2, y = 0, x = -2$, and $x = 2$.

3. R is the region bounded by the graph of $y = -x^2 + 1$ and the x–axis.

4. R is the region bounded by the graphs of $y = x, y = 0, x = -1$, and $x = 2$.

5. R is the region bounded by the graphs of $y = x^3, y = 0, x = -1$, and $x = 1$.

6. R is the region bounded by the graphs of $y = x^3, x = 0, y = -1$, and $y = 1$.

7. R is the region bounded by the graph of $y = x^3 - 2x^2 - x + 2$ and the x–axis.

8. R is the region bounded by the graphs of $y = \sin x, y = 0, x = 0$, and $x = \pi$.

9. R is the region bounded by the graphs of $y = x^3$ and $y = x$.

10. R is the region bounded by the graphs of $y = x^3$ and $y = x^2$.

For Exercises 11–20, the regions of Exercises 1–10 are rotated about the y–axis. Express the resulting volume as a definite integral.

In Exercises 21–23, the region R is rotated about the line L. Express the resulting volume as a definite integral.

21. R is the region of Exercise 1. L is the line $y = -2$.

22. R is the region of Exercise 4. L is the line $x = 1$.

23. R is the region of Exercise 10. L is the line $y = 1$.

If B is a planar region and $\vec{v}$ is a point in 3–space, then the cone with base B and vertex $\vec{v}$ is the set of all points from the line segments from $\vec{v}$ to each point in B, together with $\vec{v}$ and the points of B. In Exercises 24–26, express the volumes of the following cones as a definite integral.

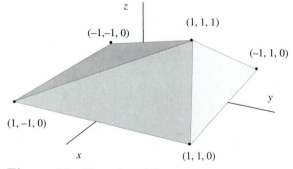

Figure 11. *The solid of Exercise 24.*

24. B is the square in the xy–plane with vertices $(1, 1, 0), (-1, 1, 0), (1, -1, 0)$, and $(-1, -1, 0)$ and $\vec{v} = (1, 1, 1)$.

25. B is the half disc $x^2 + y^2; \leq 1, z = 0, x \geq 0$ and $\vec{v} = (0, 0, 1)$.

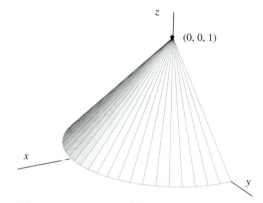

Figure 12. *The solid of Exercise 25.*

26. B is the disc $x^2 + y^2 \leq 1, z = 0$, and $\vec{v} = (1, 1, 1)$.

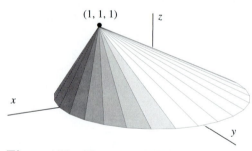

Figure 13. *The solid of Exercise 26.*

In Exercises 27–29, express the volume of the solid S as a definite integral.

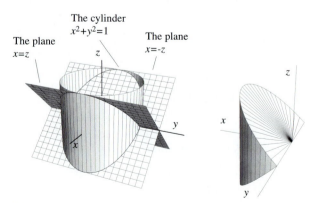

Figure 14. *The solid of Exercise 27.*

27. S is the set of points in 3–space lying in the cylinder $x^2 + y^2 \leq 1$ below the plane $x = z$, and above the plane $x = -z$.

28. S is the set of points in $\{(x, y, z) \mid 1 \geq y \geq x^2, z \geq 0\}$, which are below the plane that contains the points $(1, 1, 0), (-1, 1, 0)$, and $(0, 0, 1)$.

29. S is the set of points $\{(x, y, z) \mid y \geq x^2, x^2 + y^2 \leq 1$ and $-1 \leq z \leq 1\}$.

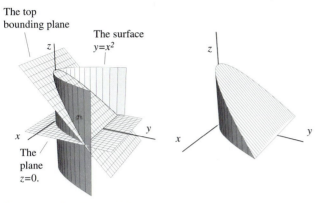

Figure 15. *The solid of Exercise 28.*

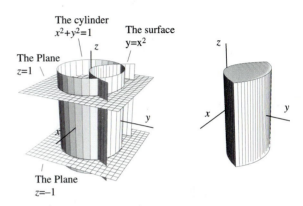

Figure 16. *The solid of Exercise 29.*

6.8   The Fundamental Theorem of Calculus

In the previous sections we have demonstrated the versatility of the definite integral. With the computers that we have available today, we can use Riemann sums to approximate any integral to an amazing degree of accuracy. Yet the power of the integral will not be completely realized until we develop some method for computing areas exactly. It will probably surprise the student that nearly 2200 years ago Archimedes used techniques very similar to the ones we have been using. Unfortunately, these ad hoc techniques were not very useful. It wasn't until the mid–1600's that Newton and Leibniz were able to exploit the surprising relationship between the integral and the derivative. Both Newton and Leibniz independently invented the calculus to aid in the understanding of physical problems. Newton developed a theory of gravitation, and Leibniz computed the volumes of revolution. However, the credit for the discovery of the relationship between the integral and the derivative must be shared with Newton's teacher, the English mathematician Isaac Barrow. In this section we present Barrow's beautiful theorem, which is likely the most important discovery in the history of mathematics, as it is used in science.

   Before proceeding, we must clear up a few "loose ends." We have defined $\int_a^b f(x)\ dx$ only in the case $a < b$. We need to take care of the other possibilities:

Definitions

(a) If $a = b$, then define $\int_a^b f(x)\ dx = 0$.

(b) If $b < a$, then define $\int_a^b f(x)\ dx = -\int_b^a f(x)\ dx$.

Each of these definitions is motivated by the Riemann sum. For the first, we note that given any partition $\{x_0, x_1, \ldots, x_n\}$ of $[a, b]$ where $b = a$, we have that each term $x_i - x_{i-1} = 0$ in the Riemann sum for $i = 1, \ldots n$. For the second definition, note that any partition $\{x_0, x_1, \ldots, x_n\}$ of $[a, b]$ gives rise to a partition $\{y_0, y_1, \ldots, y_n\}$ of $[b, a]$ where $y_i = x_{n-i}$ (write the partition in the reverse order). Thus for the Riemann sum which approximates $\int_a^b f(x)\ dx$ we have the terms $x_i - x_{i-1}$ whereas in the Riemann sum which approximates $\int_b^a f(x)\ dx$ we have the terms $y_i - y_{i-1} = x_{n-i} - x_{n-i+1} = -(x_{j+1} - x_j)$ for $j = n - i$.

We can now proceed to the Fundamental Theorem, which reveals the relationship between our work with definite integrals and Riemann sums and our earlier work with antiderivatives.

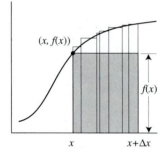

Figure 1.a
$S > f(x)\Delta x$.

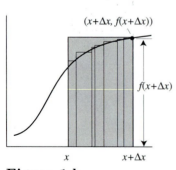

Figure 1.b
$S < f(x + \Delta x)\Delta x$.

Theorem 1 (The Fundamental Theorem of Calculus) *Let f be a function that is continuous on the interval $[a, b]$.*

(a) *If F is defined by $F(x) = \int_a^x f(t)\ dt$ for each x in $[a, b]$, then $F'(x) = f(x)$.*

(b) *If G is any function so that $G'(x) = f(x)$ for every x in $[a, b]$, then $\int_a^b f(x)\ dx = G(b) - G(a)$.*

Proof of Part (a): We consider a simple case here that is easy to visualize geometrically. Suppose that f is increasing on $[a, b]$. We know that $F(x + \Delta x) - F(x) = \int_x^{x+\Delta x} f(t)\ dt$. If $\Delta x > 0$ and if S is any Riemann sum for f over the interval $[x, x + \Delta x]$, then $S > f(x)\Delta x$ (see Figure 1.a) and $S < f(x + \Delta x)\Delta x$ (see Figure 1.b.) Thus

$$f(x) < \frac{F(x + \Delta x) - F(x)}{\Delta x} < f(x + \Delta x). \tag{14}$$

Since f is continuous, we have that $\lim_{\Delta x \to 0} f(x + \Delta x) = f(x)$.

Thus it follows from Equation (14) that

$$F'(x) = \lim_{\Delta x \to 0} \frac{F(x + \Delta x) - F(x)}{\Delta x} = f(x).$$

Proof of Part (b): Let G be any member of $\int f(x)\, dx$ (i.e., $G' = f$). If F is defined as in the proof of part (a), we know that there is a number C so that $F(x) = G(x) + C$. By the definition of F, we have $\int_a^b f(x)\, dx = F(b) - F(a)$. Thus we obtain:

$$\int_a^b f(x)\, dx = (G(b) + C) - (G(a) + C) = G(b) - G(a). \qquad \blacksquare$$

Notation: If F is a function, then we use the notation

$$F(x)\Big|_a^b = F(b) - F(a).$$

EXAMPLE 1: Compute $\int_1^2 x^2\, dx$.

SOLUTION: $\frac{x^3}{3}$ is an antiderivative of x^2. Thus, by the Fundamental Theorem of Calculus,

$$\int_1^2 x^2\, dx = \frac{x^3}{3}\Big|_1^2 = \frac{8}{3} - \frac{1}{3} = \frac{7}{3}. \qquad \blacksquare$$

EXAMPLE 2: Evaluate $\int_0^{-\pi/2} D(x \sin x)\, dx$.

SOLUTION: An antiderivative of $D(x \sin x)$ is $x \sin x$. Thus

$$\begin{aligned}
\int_0^{-\pi/2} D(x \sin x)\, dx &= x \sin x \Big|_0^{-\pi/2} \\
&= -\frac{\pi}{2} \sin\left(-\frac{\pi}{2}\right) - 0 \sin 0 \\
&= -\frac{\pi}{2}(-1) - 0 = \frac{\pi}{2}. \qquad \blacksquare
\end{aligned}$$

EXAMPLE 3: Evaluate $D \int_\pi^x t \sin t\, dt$.

SOLUTION: By Part (a) of Theorem 1, $D \int_\pi^x t \sin t\, dt = x \sin x$. $\quad\blacksquare$

EXAMPLE 4: Let $f(x) = \int_a^x g(t)\, dt$. Then $f(h(x)) = \int_a^{h(x)} g(t)\, dt$. By the Chain Rule,

$$D[f(h(x))] = f'(h(x))h'(x).$$

By the Fundamental Theorem of Calculus,

$$f'(h(x)) = g(h(x)).$$

Thus

$$D \int_a^{h(x)} g(t) \ dt = g(h(x))h'(x). \qquad \blacksquare$$

EXAMPLE 5: Evaluate $D \int_1^{x^3} \frac{1}{t} \ dt$

SOLUTION: $D \int_1^{x^3} \frac{1}{t} \ dt = \left(\frac{1}{x^3} \right) 3x^2.$ $\qquad \blacksquare$

EXAMPLE 6: Evaluate $D \int_{x^2}^{x^3} t \sin t \ dt.$

SOLUTION: First we judiciously apply Part (b) of Theorem 2 in Section 6.6 to rewrite our integral.

$$
\begin{aligned}
D \int_{x^2}^{x^3} t \sin t \ dt &= D \left(\int_{x^2}^{a} t \sin t \ dt + \int_{a}^{x^3} t \sin t \ dt \right) \\
&= D \left(- \int_{a}^{x^2} t \sin t \ dt \right) + D \int_{a}^{x^3} t \sin t \ dt \\
&= -x^2 \sin(x^2) 2x + x^3 \sin(x^3) 3x^2. \qquad \blacksquare
\end{aligned}
$$

Most integrals do not have straightforward solutions. However, in many cases we can employ the techniques of Section 6.2 to compute the definite integral. Recall that

$$\int f'(g(x))g'(x) \ dx = f(g(x)) + C.$$

Thus

$$\int_a^b f'(g(x))g'(x) \ dx = f(g(x))\big|_a^b = f(g(b)) - f(g(a)),$$

or

$$\int_a^b f(g(x))g'(x) \ dx = \int_{g(a)}^{g(b)} f(u) \ du.$$

In plain words, we can use the chain rule to make substitutions in integration, but we must be careful in accounting for the limits of integration. We illustrate this technique in the following examples.

EXAMPLE 7: Compute $\int_0^{\pi/2} \sin^2 x \cos x \ dx.$

SOLUTION: We want to find a function f and a function g so that:

(a) $\sin^2 x \cos x = f(g(x))g'(x)$; and

(b) we can compute the antiderivative of $f(x)$.

The choices for f and g in this example are clear! If we let $g(x) = \sin x$ and $f(x) = x^2$, then $\sin^2 x = f(g(x))$ and $g'(x) = \cos x$. Thus

$$\int_0^{\pi/2} \sin^2 x \cos x \; dx = \int_{\sin 0}^{\sin(\pi/2)} u^2 \; du$$

$$= \left. \frac{u^3}{3} \right|_{\sin 0}^{\sin(\pi/2)} = \frac{1}{3}. \qquad \blacksquare$$

EXAMPLE 8: Compute $\int_0^{\sqrt{\pi}} x \sin(x^2 + \pi) \; dx$.

SOLUTION: Again, the choices for f and g are clear. We let $f(x) = \sin x$ and $u = g(x) = x^2 + \pi$ so that

$$\int_0^{\sqrt{\pi}} x \sin(x^2 + \pi) \; dx = \frac{1}{2} \int_0^{\sqrt{\pi}} \sin(x^2 + \pi) 2x \; dx$$

$$= \frac{1}{2} \int_{g(0)}^{g(\sqrt{\pi})} \sin u \; du = \frac{1}{2} \int_\pi^{\pi + \pi} \sin u \; du$$

$$= \frac{1}{2} \left(-\cos u \big|_\pi^{2\pi} \right) = -1. \qquad \blacksquare$$

EXAMPLE 9: Compute $\int_0^{\pi/4} \sec^3 x \tan x \; dx$.

SOLUTION: Let $g(x) = \sec x$ and $f(x) = x^2$. Then $g'(x) = \sec x \tan x$ and

$$\int_0^{\pi/4} \sec^3 x \tan x \; dx = \int_{\sec 0}^{\sec \pi/4} u^2 \; du$$

$$= \left. \frac{u^3}{3} \right|_1^{\sqrt{2}} = \frac{2\sqrt{2} - 1}{3}. \qquad \blacksquare$$

The notation we have been using in Examples 7–9 is awkward, to say the least. We can simplify the process by thinking of "change of variables." If we have $\int_a^b f(g(x))g'(x) \; dx$, we let $u = g(x)$ and $du = g'(x) \; dx$. When $x = a$, we have that $u = g(a)$, and when $x = b$, we see that $u = g(b)$. We now return to Examples 7–9 using this technique.

EXAMPLE 10: Compute $\int_0^{\pi/2} \sin^2 x \cos x \, dx$.

SOLUTION: Let $u = \sin x$. Then $du = \cos x \, dx$. When $x = 0$, $u = 0$; and when $x = \frac{\pi}{2}$, $u = 1$. Thus

$$\int_0^{\pi/2} \sin^2 x \cos x \, dx = \int_0^1 u^2 \, du = \frac{u^3}{3}\Big|_0^1 = \frac{1}{3}. \qquad \blacksquare$$

EXAMPLE 11: Compute $\int_0^{\sqrt{\pi}} x \sin(x^2 + \pi) \, dx$.

SOLUTION: Let $u = x^2 + \pi$. Then $du = 2x \, dx$. When $x = 0$, $u = \pi$; and when $x = \sqrt{\pi}$, $u = 2\pi$. Thus

$$
\begin{aligned}
\int_0^{\sqrt{\pi}} x \sin(x^2 + \pi) \, dx &= \frac{1}{2} \int_0^{\sqrt{\pi}} \underbrace{\sin(x^2 + \pi)}_{\sin u} \; \underbrace{2x \, dx}_{du} \\
&= \frac{1}{2} \int_\pi^{2\pi} \sin u \, du \\
&= \frac{1}{2}\left(-\cos u\big|_\pi^{2\pi}\right) = -1. \qquad \blacksquare
\end{aligned}
$$

Note that once we change the variables from x to u, we do not have to do any more conversions!

EXAMPLE 12: Compute $\int_0^{\pi/4} \sec^3 x \tan x \, dx$.

SOLUTION: Let $u = \sec x$. Then $du = \sec x \tan x \, dx$. When $x = 0$, $u = 1$; and when $x = \frac{\pi}{4}$, $u = \sqrt{2}$. Thus

$$
\begin{aligned}
\int_0^{\pi/4} \sec^3 x \tan x \, dx &= \int_1^{\sqrt{2}} u^2 \, du \\
&= \frac{u^3}{3}\Big|_1^{\sqrt{2}} = \frac{2\sqrt{2} - 1}{3}. \qquad \blacksquare
\end{aligned}
$$

EXERCISES 6.8

In Exercises 1–27, use the theorems and techniques of this section to evaluate the given expression.

1. $\displaystyle\int_0^2 x^2\,dx.$

2. $\displaystyle\int_{-1}^2 x^3\,dx.$

3. $\displaystyle\int_{-1}^1 t^3 + t^2\,dt.$

4. $\displaystyle\int_{-1}^1 (x+1)(x^3 + x^2)\,dx.$

5. $\displaystyle\int_{-1}^5 (x^3 + 6x^2 + 1)\,dx.$

6. $\displaystyle\int_0^{\pi/4} \cos x + \sin x\,dx.$

7. $\displaystyle\int_0^{\pi/4} \sec\theta\tan\theta\,d\theta.$

8. $\displaystyle\int_1^2 \frac{x^{2/3} + 1}{3x^2}\,dx.$

9. $\displaystyle\int_0^{\pi/4} \cos(2\theta)\,d\theta.$

10. $\displaystyle\int_0^1 \frac{2x}{\sqrt{x^2 + 1}}\,dx.$

11. $\displaystyle\int_0^1 \frac{t}{\sqrt{t^2 + 1}}\,dt.$

12. $\displaystyle\int_0^1 \frac{z^2 + 1}{\sqrt{z^3 + 3z + 1}}\,dz.$

13. $\displaystyle\int_0^3 (z^3 + 3z + 1)^{-1/2}(z^2 + 1)\,dz.$

14. $\displaystyle\int_1^9 \frac{t - 1}{\sqrt{t}}\,dt.$

15. $\displaystyle\int_0^1 \frac{x}{\sqrt{x^2 + 1}}\,dx.$

16. $\displaystyle\int_0^1 \frac{x}{x^4 + 2x^2 + 1}\,dx.$

17. $\displaystyle\int_{\pi/4}^{\pi/2} \frac{1}{\sec\theta}\,d\theta.$

18. $\displaystyle\int_0^{\sqrt{\pi/2}} \theta\sin(9\theta^2)\cos(9\theta^2)\,d\theta.$

19. $\displaystyle\int_0^{\pi/2} \frac{\cos\theta}{\sqrt{1 - \sin^2\theta}}\,d\theta.$

20. $\displaystyle\int_0^{\pi/2} \frac{\sin\theta\cos\theta}{\sqrt{1 - \sin^2\theta}}\,d\theta.$

21. $\displaystyle\int_0^{\pi/2} \cos^3 t\,dt.$

22. $\displaystyle\int_{-1}^1 \sqrt{|x|}\,dx.$

23. $\displaystyle D\int_0^x \frac{1}{1 + t^2}\,dt.$

24. $\displaystyle D\int_0^{x^2+1} \frac{1}{1 + t^2}\,dt.$

25. $\displaystyle\int_0^1 D\frac{1}{1 + t^2}\,dt.$

26. $\displaystyle D\int_{x^2}^3 \sec(t^3)\,dt.$

27. $\displaystyle D\int_{\sin x}^{\cos x} \sqrt{1 - t^3}\,dt.$

In Exercises 28–32, find the area bounded by the graph of f and the x–axis over the given interval.

28. $f(x) = x^2,\ -1 \le x \le 1.$

29. $f(x) = x^3,\ -1 \le x \le 1.$

30. $f(x) = \dfrac{x}{(x^2 + 1)^2},\ 0 \le x \le 1.$

31. $f(x) = \cos x + x,\ 0 \le x \le \pi.$

32. $f(x) = \cos x + x,\ -\pi \le x \le \pi.$

In Exercises 33 and 34, find the area of the region bounded by the graph of f and the x–axis.

33. $f(x) = x^2 - 1.$

34. $f(x) = x^3 - x$.

In Exercises 35–38, find the area of the region bounded by the graphs of f and g.

35. $f(x) = x^2$, $g(x) = x$.

36. $f(x) = x^3$, $g(x) = x$.

37. $f(x) = x^3 - x$, $g(x) = x$.

38. $f(x) = x^3$, $g(x) = x^2$.

In Exercises 39–42, find the volume obtained by rotating the region R about the x–axis.

39. R is the region bounded by the graph of $f(x) = x^2$, the x–axis, and the line $x = 1$.

40. R is the region bounded by the graph of $f(x) = x^3 + x$, the x–axis, and the line $x = 2$.

41. R is the region bounded by the graphs of $f(x) = \sqrt{x}$ and $y = x$.

42. R is the region bounded by the graphs of $f(x) = x^3$ and $g(x) = x^2$.

43. Find the volume obtained by rotating the region of Exercise 39 about the y–axis.

44. Find the volume obtained by rotating the region of Exercise 41 about the y–axis.

45. Find the volume obtained by rotating the region of Exercise 42 about the y–axis.

46. Find the volume obtained by rotating the region R of Exercise 39 about the line $x = 1$.

47. Find the volume obtained by rotating the region R of Exercise 40 about the line $y = -3$.

48. Find the volume obtained by rotating the region R of Exercise 42 about the line $y = 5$.

In Exercises 49 and 50, approximate $\int_1^2 f(x)\,dx$ using the Riemann sums determined by partitioning the interval $[1, 2]$ into ten intervals of equal length and using the given selection points where

a. s_i is the left endpoint of $[x_i, x_{i+1}]$.

b. s_i is the right endpoint of $[x_i, x_{i+1}]$.

c. s_i is the point of $[x_i, x_{i+1}]$ such that $f(s_i) = \dfrac{f(x_i) + f(x_{i+1})}{2}$. (Use the Intermediate Value Theorem to argue that there is such a point.)

49. $f(x) = \frac{1}{x}$.

50. $f(x) = \sqrt{x^2 - 1}$.

Chapter 7

Some Transcendental Functions

We have been working with some fairly complicated functions. However, all of them have been built using polynomials, various roots of polynomials, and trigonometric functions. Functions that can be defined by adding, dividing, and taking roots and powers of such functions (except for the trigonometric functions) are called *algebraic*.[1] Two examples are

$$f(x) = \sqrt{\frac{x^2 + x + 1}{x^3 + 2x^2 - 3}} \text{ and } f(x) = \sqrt{\frac{x^2 + x + 1}{x^3 + 2x^2 - 3} - (x^2 + 2)^{3/4}}.$$

A function is *transcendental* if it is not algebraic. It can be shown that the trigonometric functions are transcendental. Other transcendental functions that are of interest to us are antiderivatives of functions that can be built from algebraic and trigonometric functions.

7.1 Antiderivatives Revisited

The Fundamental Theorem of Calculus tells us that if f is continuous on the segment (a, b), then f has an antiderivative. Indeed, if c is in (a, b) and we define $F(x) = \int_c^x f(t)\ dt$, then we know that F is defined on (a, b) and that F is the antiderivative of f such that $F(c) = 0$. Even though we may not be able to find F exactly, we have more information about F than is immediately obvious. For

---

[1] More generally, a function $f(x)$ is algebraic if there are polynomials $p_0(x), p_1(x), \ldots, p_n(x)$ such that $f(x)$ satisfies the equation $p_0(x)\,(f(x))^0 + p_1(x)\,(f(x))^1 + \ldots + p_n(x)\,(f(x))^n = 0$.

335

example, we know that $F'(x) = f(x)$. This is enough to determine where F is increasing and where F is decreasing. In the same way, if f is differentiable on (a, b), then $F''(x) = f'(x)$, so we can determine the concavity of F. This is enough information to obtain a rough sketch of F.

EXAMPLE 1: Let $f(x) = \frac{1}{x}$. Sketch the graph of $F(x) = \int_1^x f(t)\, dt$.

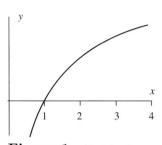

Figure 1. *Sketch of $F(x)$, using only the fact that $F(1) = 0$ and knowledge of the slope and concavity.*

SOLUTION: Since $f(x)$ is continuous on $(0, \infty)$, we know that $F(x)$ is defined on the same set. Also, we know that $F(1) = 0$. Since $F'(x) = \frac{1}{x}$ and $\frac{1}{x} > 0$ on $(0, \infty)$, F is increasing there. Going one step further, $F''(x) = f'(x) = -\frac{1}{x^2}$. Thus $F''(x) < 0$ and F is always concave down. We can now obtain the rough sketch of F in Figure 1. ∎

EXAMPLE 2: Let $f(x) = \sqrt{1 - x^2}$. Sketch the graph of $F(x) = \int_0^x f(t)\, dt$.

SOLUTION: The domain of f is $(-1, 1)$. Since f is continuous on its domain, we know that F is defined there. Since f is positive, F is an increasing function. $F''(x) = f'(x) = -\frac{x}{\sqrt{1 - x^2}}$. Thus $F''(x) > 0$ on $(-1, 0)$, which tells us that F is concave up there. Similarly, $F''(x) < 0$ on $(0, 1)$, which tells us that F is concave down there. We can use this information to obtain the sketch in Figure 2. ∎

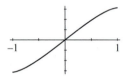

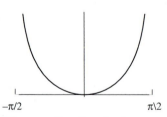

Figure 2. *Sketch of $F(x) = \int_0^x \sqrt{1 - t^2}\, dt$, using only the fact that $F(0) = 0$ and knowledge of the slope and concavity.*

EXAMPLE 3: Sketch the graph of $F(x) = \int_0^x \tan(t)\, dt$.

SOLUTION: We know that $\tan(x)$ is continuous on $(-\pi/2, \pi/2)$, so F will be defined on that segment. We know that $F'(x) = \tan(x)$. Thus $F'(x) > 0$ on $(0, \frac{\pi}{2})$ and F is increasing there. Similarly, $F'(x) < 0$ on $(-\frac{\pi}{2}, 0)$ and F is decreasing there. $F''(x) = \sec^2(x) > 0$, so F is concave up.

We obtain the sketch of F using this information in Figure 3. ∎

Euler's Method

The sketches in Figures 1, 2, and 3 are about as good as we can do without being able to plot some actual values for F. Finding values for F is usually not a trivial matter. We now introduce a process that gives us a better picture of F by allowing us to obtain rough approximations for F at selected points.

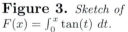

$-\pi/2$ $\pi\backslash 2$

Figure 3. *Sketch of $F(x) = \int_0^x \tan(t)\, dt$.*

EXAMPLE 4: Let $f(x) = \frac{1}{x}$ as in Example 1. We know that $F(1) = 0$. Using the linear approximation $p_1(x)$ for F centered at

$x_0 = 1$, we have

$$F(x) \approx p_1(x) = f(x_0)(x - x_0) = x - 1.$$

Thus $F(1.5) \approx 0.5$. We then use this approximation for $F(1.5)$ to obtain an *approximation* for the linear approximation for F centered at 1.5.

$$F(x) \approx f(1.5)(x - 1.5) + F(1.5) \approx \left(\frac{1}{1.5}\right)(x - 1.5) + 0.5.$$

This gives us an approximation for $F(2)$.

$$F(2) \approx \left(\frac{1}{1.5}\right)(2 - 1.5) + 0.5 \approx 0.833.$$

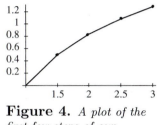

Figure 4. *A plot of the first few steps of our approximations of F.*

We then use this approximation to *approximate* the linear approximation for F centered at $x_0 = 2$. Continuing this process, we obtain approximations for F at $x = 2.5$, $x = 3$, etc. See Figure 4.

In Table 1 below, we tabulate the results of this process and compare them to the true values of F rounded to three decimal places. Figure 5 compares the graph we obtain using our approximated values with a sketch using true values for F. ∎

Table 1: *The values for our approximation of $F(x_i)$ compared with the true values of $F(x_i)$ rounded to three decimal places.*

x_i	Approximated values for $F(x_i)$	$F(x_i)$ rounded to 3 decimal places
1.0	0	0
1.5	0.5	0.405
2.0	0.833	0.693
2.5	1.083	0.916
3.0	1.283	1.099
3.5	1.45	1.253
4.0	1.593	1.385
4.5	1.718	1.504
5.0	1.829	1.609
5.5	1.929	1.705
6.0	2.02	1.792

Figure 5. *The graph using values for our approximation of $F(x_i)$ compared with a graph using the true values of $F(x_i)$ rounded to three decimal places.*

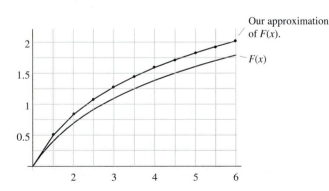

Our approximation of $F(x)$.

$F(x)$

The process used above to approximate values of F is called *Euler's method*. The general procedure is as follows: we know that $F(c) = \int_c^c f(t)\ dt = 0$. Let $x_0 = c$ and $y_0 = 0$. Pick a small number h. In Example 4, we chose $h = 0.5$. Having values for x_i and y_i, let $y_{i+1} = F'(x_i)h + y_i = f(x_i)h + y_i$, and let $x_{i+1} = x_i + h$. We use the pair (x_i, y_i) to approximate $(x_i, F(x_i))$. This process allows us to obtain (x_1, y_1) from (x_0, y_0), (x_2, y_2) from (x_1, y_1), etc. The simplicity of this algorithm makes it very appealing. However, this approximation method can be very inaccurate if h is chosen to be too large or if too many steps are used. The errors can compound very quickly, as we will see in the next example. On the other hand, if we choose h to be too small, then we soon introduce roundoff errors due to the limits on computer accuracy. There are much better methods that are the subject of great interest in numerical analysis. Indeed, this is an area of much current research in mathematics.

In Example 4, we approximated values for $F(x) = \int_1^x \frac{1}{t}\ dt$ only for $x > 1$. We can use Euler's method to approximate F for values of $x < 1$ by taking h to be negative.

EXAMPLE 5: Let $F(x) = \int_1^x \frac{1}{t}\ dt$. Sketch F on the interval $0.25 \le x \le 1$.

SOLUTION: We know from Example 1 that F is increasing and that it is concave down. We now want to approximate a few values of F. Let $h = -0.25$. With this choice of h, we use Euler's method to approximate $F(0.75)$, $F(0.5)$, and $F(0.25)$. Table 2 tabulates the values of y_0, y_1, y_2, and y_3, as well as the true values for $F(1)$, $F(0.75), F(0.5)$, and $F(0.25)$ rounded to three decimal places.

x_i	Approximated values for $F(x_i)$	$F(x_i)$ rounded to 3 decimal places
1.00	0	0
0.75	-0.25	-0.288
0.50	-0.583	-0.693
0.25	-1.083	-1.386

Table 2: *The values for our approximation of $F(x_i)$ compared with the true values of $F(x_i)$ rounded to three decimal places.*

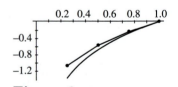

Figure 6. *The sketch of F using Euler's approximations compared with the sketch using true values.*

Figure 6 compares the graph obtained by using our approximations with the graph obtained using true values for F. Notice that the error in our approximation of $F(0.25)$ is nearly 22% of the true value for $F(0.25)$. ∎

EXAMPLE 6: Let us look at $F(x) = \int_1^x \frac{1}{t}\, dt$ again, this time using a smaller value of h. If we let $h = -0.05$, then it takes 15 steps to approximate $F(0.25)$, and we obtain $F(0.25) \approx -1.31441$. The true value for $F(0.25)$ rounded to five decimal places is -1.38629. The error in our approximation of $F(0.25)$ is now less than 5% of the true value. So we see that using a smaller value for h did improve our approximation.

Now suppose that we want to approximate $F(0.05)$ using $h = -0.05$. This will take 19 steps. Our approximation is -2.59774, while the true value of $F(0.05) = -2.99573$ rounded to 5 decimal places. Our error is back up to 13% of the true value! Even using our smaller step size, Euler's method produces substantial errors as we move away from our starting point at $x = 1$. Even though each individual step now has a smaller error, we need more steps. Many small errors add up to one large error!

Thus we see that Euler's method gives us only rough estimates of F. Nonetheless, it is often a good tool for gaining basic information about the nature of the graph of F. ∎

EXERCISES 7.1

In Exercises 1–10 state the domain and sketch the graph of $F(x) = \int_c^x f(t)\, dt$.

1. $f(x) = \dfrac{1}{1+x}$, $c = 0$.

2. $f(x) = \dfrac{1}{1+x^2}$, $c = 0$.

3. $f(x) = \dfrac{x}{1+x}$, $c = 1$.

4. $f(x) = \dfrac{1}{\sqrt{1-x^2}}$, $c = 0$.

5. $f(x) = \dfrac{1}{\sqrt{x^2-1}}$, $c = 2$.

6. $f(x) = \dfrac{1}{x\sqrt{1-x^2}}$, $c = 0$.

7. $f(x) = \dfrac{1}{x\sqrt{x^2-1}}$, $c = 2$.

8. $f(x) = \dfrac{1}{x\sqrt{x^2-1}}$, $c = -2$.

9. $f(x) = \dfrac{x^2}{\sqrt{x^2-1}}$, $c = 2$.

10. $f(x) = \dfrac{1}{|x|\sqrt{x^2-1}}$, $c = -2$.

11. Do 10 steps using Euler's method with $h = 0.1$ to obtain approximate values for $F(0.1)$, $F(0.2)$, ..., $F(1)$, where $F(x) = \int_0^x \frac{1}{1+t^2}\, dt$. Compare your estimate for $F(1)$ with the true value of $F(1) = \frac{\pi}{4}$.

12. Do 10 steps using Euler's method with $h = -0.1$ to obtain approximate values for $F(-0.1)$, $F(-0.2)$, ..., $F(-1)$, where $F(x) = \int_0^x \frac{1}{1+t^2}\, dt$. Compare your estimate for $F(-1)$ with the true value of $F(-1) = -\frac{\pi}{4}$.

13. Let $F(x) = \int_0^x \frac{1}{\sqrt{1-t^2}}\, dt$.

    a. What should h be if we are to use 10 steps of Euler's Method to approximate $F\left(\frac{1}{2}\right)$?

    b. Use the h from Part a.2 to obtain an approximation of $F\left(\frac{1}{2}\right)$.

    c. Given that $F\left(\frac{1}{2}\right) = \frac{\pi}{6}$, what is your error?

    d. Explain why you cannot expect to be successful in using Euler's Method to approximate $F(2)$.

7.2   Numerical Methods

In the last section we used Euler's method to approximate the antiderivatives of some functions. The problem with this method is that the approximations can drift away from the real antiderivative rather rapidly. Over the next few chapters, we will spend a lot of time learning how to calculate the antiderative for a large variety of functions. It is an unfortunate fact that no matter how many techniques of integration we learn or how sophisticated our calculator, there will always be functions that we cannot integrate directly — even when we know that the integral exists! The best we can do with such functions is to approximate the integral numerically.

If $f(x)$ is a continuous real valued function, then the Fundamental Theorem of Calculus tells us that we can write its antiderivative as

$$F(x) = \int_{x_0}^{x} f(t) \, dt + C.$$

For simplicity's sake, assume that $F(x_0) = 0$ so that we may take $C = 0$. Let h be a number. We want to find a way to approximate $F(x_0 + h)$, $F(x_0 + 2h)$, $F(x_0 + 3h), \ldots, F(x_0 + nh)$ that will be more accurate than Euler's method. In this section, we present two popular numerical integration algorithms: the Trapezoidal Rule and Simpson's Rule. Both of these rules use the fact that

$$F(x_0 + ih) = \int_{x_0}^{x_0 + ih} f(t) \, dt$$

$$= \int_{x_0}^{x_0 + h} f(t) \, dt + \int_{x_0 + h}^{x_0 + 2h} f(t) \, dt + \cdots + \int_{x_0 + (i-1)h}^{x_0 + ih} f(t) \, dt.$$

Each of the methods provides a method for approximating $\int_{x}^{x+h} f(x) \, dx$.

The Trapezoidal Rule

Let $x_i = x_0 + ih, \quad 0 \leq i \leq n$. Then

$$F(x_i) = \int_{x_0}^{x_i} f(t) \, dt$$

$$= \int_{x_0}^{x_1} f(t) \, dt + \int_{x_1}^{x_2} f(t) \, dt + \cdots + \int_{x_{i-1}}^{x_i} f(t) \, dt.$$

The definition of the definite integral suggests that we may take an arbitrary selection point s_j from the segment $[x_{j-1}, x_j]$ for each

$1 \le j \le i$ and use

$$f(s_j)(x_j - x_{j-1}) = hf(s_j)$$

to approximate

$$\int_{x_{j-1}}^{x_j} f(t) \, dt.$$

The Trapezoidal Rule tells us to choose the selection point s_j to be a point in $[x_{j-1}, x_j]$ such that $f(s_j) = \frac{f(x_{j-1}) + f(x_j)}{2}$. (The Intermediate Value Theorem assures us that there is such a point.)

In order to see what is happening geometrically, assume that $f(x) > 0$ on the interval $[x_{j-1}, x_j]$. If we use an arbitrary selection point s_j, then $f(s_j)h$ approximates $\int_{x_{j-1}}^{x_j} f(x) \, dx$ with the area of a rectangle as illustrated in Figure 1.a. By choosing the selection point s_j to be a point in $[x_{j-1}, x_j]$ such that $f(s_j) = \frac{f(x_{j-1}) + f(x_j)}{2}$, then $f(s_j)h$ approximates $\int_{x_{j-1}}^{x_j} f(x) \, dx$ with the area of a trapezoid as illustrated in Figure 1.b.

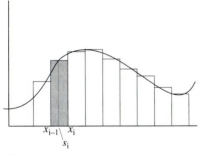
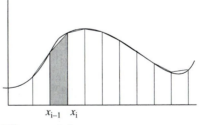

Figure 1.a *The rectangle with area $f(s_2)h$.*

Figure 1.b *The trapezoid with area $\left(\frac{f(x_1) + f(x_2)}{2} \right) h$.*

Intuitively, we would expect that $f(s_j)h = \left(\frac{f(x_{j-1}) + f(x_j)}{2} \right) h$ would provide a better approximation to

$$\int_{x_{j-1}}^{x_j} f(x) \, dx$$

than we would obtain using an arbitrary selection point. We now show that the error in this approximation is proportional to h^3.

Theorem 1 (The Trapezoidal Rule) *Suppose that f is a real valued function such that f'' is continuous on the interval $[x, x+h]$, and let M denote the maximum of $|f''|$ on $[x, x+h]$. Then*

$$\int_x^{x+h} f(t) \, dt = \frac{f(x) + f(x+h)}{2} h + E(h) \text{ where } |E(h)| \le \frac{M}{2} h^3.$$

The proof of Theorem 1 will require the following theorem.

Theorem 2 *Suppose that $f^{(n)}(x)$ is continuous in a neighborhood (a, b) of x_0 and that $F(x) = \int_{x_0}^x f(t) \, dt$, $a < x < b$. Then there is a number μ between x and x_0 such that*

$$
\begin{aligned}
F(x) \;=\; & f(x_0)(x - x_0) + \frac{f'(x_0)}{2}(x - x_0)^2 + \frac{f''(x_0)}{3!}(x - x_0)^3 + \\
& \cdots + \frac{f^{n-1}(x_0)}{n!}(x - x_0)^n + \frac{f^n(\mu)}{(n+1)!}(x - x_0)^{n+1}.
\end{aligned}
$$

Proof: This is immediate from Taylor's Theorem since $F(x_0) = 0$, and by the Fundamental Theorem of Calculus, we have that $F^{(i+1)}(x) = f^{(i)}(x)$. ∎

Proof of Theorem 1: By Theorem 2, there is a μ between x and $x + h$ such that

$$\int_x^{x+h} f(x)dx = f(x)h + \frac{f'(x)}{2}h^2 + \frac{f''(\mu)}{6}h^3. \tag{1}$$

By Taylor's Remainder Theorem there is a μ_1 between x and $x + h$ such that

$$f(x + h) = f(x) + f'(x)h + \frac{1}{2}f''(\mu_1)h^2.$$

Solving for $f'(x)$, we have

$$f'(x) = \frac{f(x+h) - f(x)}{h} - \frac{f''(\mu_1)}{2}h.$$

Thus, we may write Equation (1) as

$$\int_x^{x+h} f(x)dx$$

$$= f(x)h + \left(\frac{\left(\frac{f(x+h)-f(x)}{h}\right) - \left(\frac{f''(\mu_1)}{2}\right)h}{2}\right)h^2 + \frac{f''(\mu)}{6}h^3$$

$$= f(x)h + \frac{h}{2}\left(f(x+h) - f(x)\right) + \left(\frac{-f''(\mu_1)}{4} + \frac{f''(\mu)}{6}\right)h^3$$

$$= f(x)h + \frac{h}{2}\left(f(x+h) - f(x)\right) + E(h), \text{ where}$$

$$|E(h)| = \left|\left(\frac{f''(\mu)}{6} - \frac{f''(\mu_1)}{4}\right)h^3\right| \leq \left(\left|\frac{f''(\mu)}{6}\right| + \left|\frac{f''(\mu_1)}{4}\right|\right)h^3$$

$$\leq \left(\left|\frac{f''(\mu)}{4}\right| + \left|\frac{f''(\mu_1)}{4}\right|\right)h^3$$

$$= \left|\frac{f''(\mu_2)}{2}h^3\right|,$$

for some μ_2 between x and $x + h$.   ∎

EXAMPLE 1: Sketch the graph of $F(x) = \int_0^x \frac{dt}{t^2+1}$ on the interval $[0, 1]$.

SOLUTION: Let $h = \frac{1}{10}$. We estimate $F(h) = F(0.1) = \int_0^{0.1} \frac{dt}{t^2+1}$, $F(2h) = F(0.2) = \int_0^{0.2} \frac{dt}{t^2+1}$, ..., $F(10h) = F(1) = \int_0^1 \frac{dt}{t^2+1}$. We carry our arithmetic to a six decimal place accuracy.

$$F(0.1) \approx \frac{h}{2}\left(f(0) + f(0.1)\right) = (0.05)\left(1 + \frac{1}{1.01}\right) = 0.099505$$

$$F(0.2) \approx 0.099505 + \frac{h}{2}\left(f(0.1) + f(0.2)\right) = 0.197087$$

$$F(0.3) \approx 0.197087 + \frac{h}{2}\left(f(0.2) + f(0.3)\right) = 0.291035$$

$$F(0.4) \approx 0.291035 + \frac{h}{2}\left(f(0.3) + f(0.4)\right) = 0.380010$$

$$F(0.5) \approx 0.380010 + \frac{h}{2}\left(f(0.4) + f(0.5)\right) = 0.463114$$

$$F(0.6) \approx 0.463114 + \frac{h}{2}\left(f(0.5) + f(0.6)\right) = 0.539878$$

$$F(0.7) \approx 0.539878 + \frac{h}{2}\left(f(0.6) + f(0.7)\right) = 0.610200$$

$$F(0.8) \approx 0.610200 + \frac{h}{2}\left(f(0.7) + f(0.8)\right) = 0.674245$$

$$F(0.9) \approx 0.674245 + \frac{h}{2}\left(f(0.8) + f(0.9)\right) = 0.732357$$

$$F(1.0) \approx 0.732357 + \frac{h}{2}\left(f(0.9) + f(1.0)\right) = 0.784981$$

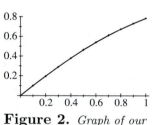

Figure 2. *Graph of our approximation to*

$$\int_0^x \frac{1}{(t^2 + 1)}\, dt.$$

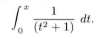

We plot the graph in Figure 2. ∎

We might expect that the error would accumulate rapidly in the scheme described here. Fortunately, it is not so bad. Suppose we want to estimate $\int_a^b f(x)\, dx$. Let $h = \frac{b-a}{N}$ and let $x_i = a + ih$ for $0 \le i \le N$, so that $\{x_0, x_1, \dots, x_N\}$ partitions $[a, b]$ into N subintervals of equal length. We now use the Trapezoidal Rule to estimate $\int_{x_{j-1}}^{x_j} f(x)\, dx$ and sum these terms to approximate $\int_a^b f(x)\, dx$. That is,

$$\int_a^b f(x)\, dx \approx \frac{h}{2} \sum_{i=1}^{N} \left(f(x_{i-1}) + f(x_i)\right).$$

For each i, this estimate introduces an error with a magnitude of no more than $\frac{Mh^3}{2}$ (where M is the maximum value for $|f''(x)|$ on $[a, b]$). Since there are N subintervals, we have introduced a total error of no more than $\frac{NMh^3}{2}$. But $N = \frac{b-a}{h}$, so that $\frac{NMh^3}{2} = \frac{b-a}{h}\frac{Mh^3}{2} = \frac{(b-a)Mh^2}{2}$. Thus

$$\int_a^b f(x)\, dx = \frac{h}{2} \sum_{i=1}^{N} [f(x_{i-1}) + f(x_i)] + E(h) \tag{2}$$

where

$$|E(h)| \le \frac{(b-a)Mh^2}{2}.$$

EXAMPLE 2: Let $f(x) = \frac{1}{x^2+1}$ as in Example 1. Then

$$f''(x) = \frac{3x^2 - 1}{(x^2 + 1)^3} \quad \text{and} \quad f'''(x) = \frac{24x(1 - x^2)}{(x^2 + 1)^4}.$$

Thus $f''(x)$ is increasing on $[0, 1]$ and it attains its maximum $M = \frac{1}{4}$ at $x = 1$. Hence the total error in our approximation for $F(x)$ in Example 1 is never greater than $\left(\frac{1}{8}\right)(0.1)^2 = 0.00125$. Indeed, in the next few sections we will learn how to integrate $f(x) = \frac{1}{x^2+1}$ directly, and it will turn out that $\int_0^1 \frac{1}{x^2+1}\, dx = \frac{\pi}{4} = 0.785398$ (carried out to six decimal places). Our estimate for $F(1) = \int_0^1 \frac{1}{1+x^2}\, dx$ is accurate to three decimal places. This is consistent with our theoretical error bound. ∎

If you are entering a field that seriously uses mathematics as a modeling tool, then you will more than likely use Equation (2) to numerically approximate definite integrals. There are certain considerations that are important in "real life" applications of the Trapezoidal Rule. The first is that even with the availability of inexpensive computing power, run time on a computer can quickly become an issue. Thus, you will want to reduce the number of operations in a numerical computation as much as possible. In this case, we observe that when we use Equation (2), we are adding up all of the terms, except for $f(x_0)$ and $f(x_N)$, twice. We can cut the number of additions in Equation (2) nearly in half by rewriting it. The resulting formulation is called the *Composite Trapezoidal Rule*.

Theorem 3 The Composite Trapezoidal Rule: *Suppose that f'' is continuous on $[a, b]$, M is the maximum of $|f''|$ on $[a, b]$, N is an integer, and $h = (b-a)/N$. For each $0 \leq i \leq N$ let $x_i = a + ih$. Then*

$$\int_a^b f(x) \, dx = h \left(\frac{f(a) + f(b)}{2} + \sum_{i=1}^{N-1} f(x_i) \right) + E(h)$$

where

$$|E(h)| \leq \frac{(b-a)Mh^2}{2}.$$

The second reality you can expect when you employ the Trapezoidal Rule is that you will seldom be able to actually obtain f'', much less find the maximum for $|f''(x)|$. The important thing is that you are aware that your approximation will have an error that is proportional to the square of h.

A third consideration in applying the Trapezoidal Rule is that while we can theoretically get whatever accuracy we need just by letting h be small enough, the computer itself imposes a limit on how small we may let h be without generating unacceptable computer roundoff error. For this reason, we have numerical methods for approximating definite integrals that have errors that are proportional to h^n for just about any n we choose. Of course the number of calculations for each step goes up as n goes up. We will introduce the Composite Simpson's Rule, which has an error proportional to h^4, and leave more accurate methods to a course in numerical analysis.

Simpson's Rule

We go back to the problem of graphing $F(x) = \int_{x_0}^{x} f(t)\, dt$. As with the Trapezoidal Rule, we let $h > 0$ and let $x_i = x_0 + ih$, $0 \le i \le 2n$. We deviate a little from what we did previously by integrating over subintervals of length $2h$.

$$
\begin{aligned}
F(x_{2i}) &= \int_{x_0}^{x_{2i}} f(t)\, dt \\
&= \int_{x_0}^{x_2} f(t)\, dt + \int_{x_2}^{x_4} f(t)\, dt + \cdots + \int_{x_{2i-2}}^{x_{2i}} f(t)\, dt.
\end{aligned}
$$

This time, we approximate $f(x)$ on $[x_{2(i-1)}, x_{2i}]$ with a "well fitting" parabola, and integrate the parabola to approximate $\int_{x_{2i-2}}^{x_{2i}} f(t)\, dt$. See Figure 3.a.

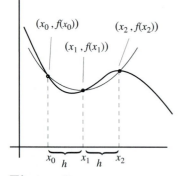

Figure 3.a
Approximating part of the function with a parabola.

Let $x_1 = x_0 + h$ and $x_2 = x_0 + 2h$, and let $y_0 = f(x_0)$, $y_1 = f(x_1)$ and $y_2 = f(x_2)$. Let $\mathcal{P}$ be the parabola $g(x) = Ax^2 + Bx + C$, such that $g(x_0) = f(x_0)$, $g(x_1) = f(x_1)$, and $g(x_2) = f(x_2)$. Let $\mathcal{P}' = \mathcal{P} - (x_1, 0)$ be the translation of $\mathcal{P}$ by $-(x_1, 0)$ so that $\mathcal{P}'$ will be the parabola with a vertical axis that passes through the points $(-h, y_0)$, $(0, y_1)$, and (h, y_2) as in Figure 3.b. $\mathcal{P}'$ is the graph of a function $p(x) = ax^2 + bx + c$. By the geometry of the parabolas,

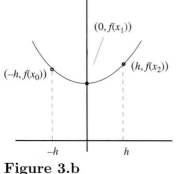

Figure 3.b
The parabola translated by $(-x_1, 0)$.

$$
\begin{aligned}
\int_{x_0}^{x_2} g(x)\, dx &= \int_{-h}^{h} p(x)\, dx \\
&= \int_{-h}^{h} ax^2 + bx + c\, dx \\
&= \frac{2ah^3}{3} + 2ch.
\end{aligned}
$$

Now we know that $p(-h) = y_0$, $p(0) = y_1$, and $p(h) = y_2$, so we have three equations with three unknowns:

$$
\begin{aligned}
c &= y_1 \\
ah^2 - bh + c &= y_0 \\
ah^2 + bh + c &= y_2.
\end{aligned}
$$

Thus

$$c = y_1 \text{ and } a = \frac{y_0 + y_2 - 2y_1}{2h^2}.$$

We now have that

$$\int_{x_0}^{x_0+2h} f(t) \ dt \ \approx \ \int_{x_0}^{x_2} g(x) \ dx = \frac{2ah^3}{3} + 2ch$$

$$= \ \frac{h}{3} \left(f(x_0) + f(x_2) + 4f(x_1) \right).$$

Theorem 4 (Simpson's Rule) *Suppose that $h > 0$ is a number, $f^{(4)}(x)$ is continuous in a neighborhood of $[x_0 - h, x_0 + h]$, and M is the maximum of $\left| f^{(4)}(x) \right|$ on $[x_0 - h, x_0 + h]$. Then*

$$\int_{x_0-h}^{x_0+h} f(t) \ dt = \frac{h}{3} \left(f(x_0 - h) + f(x_0 + h) + 4f(x_0) \right) + E(h)$$

where

$$E(h) \le \frac{h^5 M}{30}.$$

Proof: Let $F(x) = \int_{x_0}^{x_0+x} f(t) \ dt$. Calling on Theorem 2, we have

$$\int_{x_0}^{x_0+h} f(t) \ dt = F(h)$$

$$= \ f(x_0)h + \frac{f'(x_0)}{2}h^2 + \frac{f''(x_0)}{3!}h^3 + \frac{f'''(x_0)}{4!}h^4 + \frac{f^{(4)}(\mu_1)}{5!}h^5,$$

and

$$\int_{x_0-h}^{x_0} f(t) \ dt = -F(-h)$$

$$= \ f(x_0)h - \frac{f'(x_0)}{2}h^2 + \frac{f''(x_0)}{3!}h^3 - \frac{f'''(x_0)}{4!}h^4 + \frac{f^{(4)}(\mu_2)}{5!}h^5.$$

Adding these equations gives us

$$\int_{x_0-h}^{x_0+h} f(x) \ dx = 2hf(x_0) + \frac{2h^3}{3!}f''(x_0) + \frac{h^5}{5!} \left(f^{(4)}(\mu_1) + f^{(4)}(\mu_2) \right).$$

By the Intermediate Value Theorem, there is a μ_3 between $x_0 - h$ and $x_0 + h$ such that

$$\left(f^{(4)}(\mu_1) + f^{(4)}(\mu_2) \right) = 2f^{(4)}(\mu_3).$$

Therefore,

$$\int_{x_0-h}^{x_0+h} f(x)\, dx = 2hf(x_0) + \frac{2h^3}{3!}f''(x_0) + \frac{2h^5}{5!}f^{(4)}(\mu_3). \qquad (3)$$

By Theorem 5.8.4, we know that

$$f''(x_0) = \frac{f(x_0 + h) - 2f(x_0) + f(x_0 - h)}{h^2} - \frac{h^2}{12}f^{(4)}(\mu_4)$$

for some μ_4 in $[x_0 - h, x_0 + h]$.

Substituting the above into Equation (3) we obtain

$$\int_{x_0-h}^{x_0+h} f(x)\, dx$$

$$= \; 2hf(x_0) + \frac{2h^3}{3!}\left(\frac{f(x_0 + h) - 2f(x_0) + f(x_0 - h)}{h^2} - \frac{h^2}{12}f^{(4)}(\mu_4) \right)$$

$$+ \frac{2h^5}{5!}f^{(4)}(\mu_3)$$

$$= \; 2hf(x_0) + \frac{h}{3}\left(f(x_0 + h) - 2f(x_0) + f(x_0 - h) \right) - \frac{h^5}{36}f^{(4)}(\mu_4)$$

$$+ \frac{h^5}{60}f^{(4)}(\mu_3).$$

We have that

$$\int_{x_0-h}^{x_0+h} f(x)\, dx = 2hf(x_0) + \frac{h}{3}\left(f(x_0 + h) - 2f(x_0) + f(x_0 - h) \right) + E(h),$$

where

$$|E(h)| \; = \; \left| \frac{h^5}{60}f^{(4)}(\mu_3) - \frac{h^5}{36}f^{(4)}(\mu_4) \right|$$

$$\leq \; \left| \frac{h^5}{36}f^{(4)}(\mu_3) \right| + \left| \frac{h^5}{36}f^{(4)}(\mu_4) \right|$$

$$\leq \; \left| \frac{h^5}{18}M \right|. \qquad\qquad \blacksquare$$

We put the pieces together to get the Composite Simpson's Rule.

Theorem 5 (The Composite Simpson's Rule) *Suppose that $f^{(4)}(x)$ is continuous on $[a, b]$, M is the maximum of $\left|f^{(4)}(x)\right|$ on $[a, b]$, N is an integer, and $h = \frac{b-a}{2N}$. For each $0 \leq i \leq 2N$, let $x_i = a + ih$. Then*

$$\int_a^b f(x)\, dx =$$

$$\frac{h}{3}\left(f(a) + f(b) + 2\sum_{i=1}^{N-1} f(x_{2i}) + 4\sum_{i=1}^{N} f(x_{2i-1})\right) + E(h),$$

where

$$|E(h)| \leq \frac{h^4(b-a)M}{36}.$$

Proof: By Theorem 4, for every $1 \leq i \leq N$ we have that

$$\int_{x_{2(i-1)}}^{x_{2i}} f(x)\, dx = \frac{h}{3}\left(f(x_{2(i-1)}) + 4f(x_{2i-1}) + f(x_{2i})\right) + E_i(h),$$

where

$$|E_i(h)| \leq \frac{h^5 M}{18}.$$

Thus

$$
\begin{aligned}
\int_a^b f(x)\, dx &= \sum_{i=1}^{N} \int_{x_{2(i-1)}}^{x_{2i}} f(x)\, dx \\
&= \sum_{i=1}^{N} \left\{\frac{h}{3}\left[f(x_{2(i-1)}) + 4f(x_{2i-1}) + f(x_{2i})\right] + E_i(h)\right\} \\
&= \frac{h}{3}\sum_{i=1}^{N}\left[f(x_{2(i-1)}) + 4f(x_{2i-1}) + f(x_{2i})\right] + \sum_{i=1}^{N} E_i(h) \\
&= \frac{h}{3}\sum_{i=1}^{N}\left[f(x_{2(i-1)}) + 4f(x_{2i-1}) + f(x_{2i})\right] + E(h),
\end{aligned}
$$

where

$$|E(h)| = \left| \sum_{i=1}^{N} E_i(h) \right| \le \sum_{i=1}^{N} |E_i(h)| \le N\left(\frac{h^5 M}{18} \right).$$

However, since $h = (b-a)/(2N)$, we have that $N = (b-a)/(2h)$. Thus

$$|E(h)| \le N\left(\frac{h^5 M}{18} \right) = \frac{(b-a)h^4 M}{36}. \qquad \blacksquare$$

EXAMPLE 3: Use the Composite Simpson's Rule to approximate $\int_0^1 \frac{dx}{1+x^2}$ with $N = 5$.

SOLUTION: Let $h = \frac{b-a}{2N} = 0.1$. We have that $f^{(4)}(x) = \frac{24(5x^4 - 10x^2 + 1)}{(1+x^2)^5}$, and we find that $M = 24$. Thus a bound for our error is

$$E = \frac{(0.1)^4(24)}{36} = \left(\frac{2}{3} \right) \times 10^{-4}.$$

$$\int_0^1 \frac{dx}{1+x^2} \approx \frac{0.1}{3}\left(f(0) + f(1) + 2\sum_{i=1}^{4} f(x_{2i}) + 4\sum_{i=1}^{5} f(x_{2i-1}) \right)$$

$$= \frac{0.1}{3}\left[1 + 0.5 + 2(0.961538 + 0.862069 + 0.735294 + 0.609756) \right.$$

$$\left. + 4(0.990099 + 0.917431 + 0.8 + 0.671141 + 0.552486) \right]$$

$$= 0.785398.$$

We know from Example 2 that $\int_0^1 \frac{dx}{1+x^2} = \frac{\pi}{4}$. Our answer matches our calculator value to six decimals.

EXERCISES 7.2

In Exercises 1–3, calculate $\int_a^b f(x)\, dx$ and approximate the integral using the Composite Trapezoidal Rule and the Composite Simpson's Rule. With each method, use $n = 4$, then $n = 8$. Carry our your answer to four decimal places.

1. $f(x) = x^3$, $a = 0$ and $b = 2$.

2. $f(x) = \sin^3 x \cos x$, $a = 0$ and $b = \frac{\pi}{2}$.

3. $f(x) = \sec^2 x$, $a = 0$ and $b = \frac{\pi}{4}$.

4. Let $F(x) = \int_1^x \frac{dt}{t}$, and let $G(x) = \int_1^x F(t)\, dt$.

   a. Do 10 steps of the Trapezoidal Rule with $h = 0.1$ to approximate $F(1.1)$, $F(1.2)$, ..., $F(2)$.

   b. Use the data from Part a. and the Trapezoidal Rule to approximate $G(1)$, $G(1.1)$, $G(1.2)$, ..., $G(2)$.

   c. Sketch the graphs of F and G.

5. The following data was observed from a function f.

i	0	1	2	3	4
x_i	0	0.2	0.4	0.6	0.8
$f(x_i)$	1	1.221	1.491	1.822	2.225

i	5	6	7	8	9
x_i	1.0	1.2	1.4	1.6	1.8
$f(x_i)$	2.718	3.320	4.055	4.953	6.0496

i	10	11	12	13	14
x_i	2.0	2.2	2.4	2.6	2.8
$f(x_i)$	7.389	9.025	11.023	13.463	16.444

a. Let $F(x) = \int_0^x f(t)\ dt$. Use Simpson's Rule to approximate $F(0.4)$, $F(0.8)$, $F(1.2)$, $F(1.6)$, $F(2.0)$, $F(2.4)$, $F(2.8)$.

b. Can we use Simpson's rule to approximate $F(0.2)$?

c. Use Method 2 (Exercises in Section 3.4) to approximate $f'(x)$ at $x = 0.2, 0.4, \ldots, 2.6$.

d. Compare the graphs of F, f and f'.

e. Let $L(x) = \int_1^x \frac{dt}{t}$ (this is the function F in Exercise 4). Use Simpson's rule with $N = 10$ to approximate $L(f(1.1)), L(f(1.2)), \ldots L(f(2))$.

6. Let $AT(x) = \int_0^x \frac{dt}{1+t^2}$. For $x = 1, 2, 3$ and 4 :

a. Use Simpson's rule with $N = 10$ to approximate $AT(x)$.

b. Evaluate $\tan(AT(x))$.

c. What do you suppose the function $AT(x)$ is?

7. Let $AS(x) = \int_0^x \frac{dt}{\sqrt{1-t^2}}$. For $x = 0.25, 0.5$ and 0.75 :

a. Use Simpson's rule with $N = 10$ to approximate $AS(x)$.

b. Evaluate $\sin(AS(x))$.

c. Use Simpson's rule with $N = 10$ to approximate $AS(\sin(x))$.

d. What do you suppose the function $AS(x)$ is?

7.3   The ln function

We have learned how to find the anti-derivatives of x^n for all rational n except $n = -1$. So it is natural to turn our attention to $\int \frac{1}{x}\ dx$. The Fundamental Theorem of Calculus assures us that the integral of $\frac{1}{x}$ exists over any interval that does not contain 0, so we define

$$\ln(x) = \int_1^x \frac{1}{t}\ dt \text{ for } x > 0.$$

By the Fundamental Theorem of Calculus, $D\ln(x) = \frac{1}{x}$. This is the function F that you sketched in Problem 4.a of the previous section and it is the function we sketched using Euler's method in Section 7.1. As it stands, the ln function probably does not seem very useful. That perspective will change as we explore this function over the next few sections.

EXAMPLE 1: $\int \frac{\ln(x)}{x}\ dx = \int u\ du|_{u=\ln x} = \frac{u^2}{2}+C|_{u=\ln x} = \frac{\ln^2(x)}{2}+C.$ ∎

So far, we have only found the anti-derivative of $f(x) = \frac{1}{x}$ for the case when x is positive. The next theorem will complete the picture for us.

Theorem 1 $D \ln|x| = \frac{1}{x}$ and $\int \frac{1}{x} \, dx = \ln|x| + C$.

The proof of Theorem 1 is left as an exercise.

The Fundamental Theorem of Calculus tells us that $\int_a^b \frac{1}{x} \, dx = \ln|b| - \ln|a|$, **but only in the case that 0 is not between a and b.**

EXAMPLE 2: Compute $\int_{-1}^0 \frac{x^2}{x^3-1} \, dx$.

SOLUTION: Let $u(x) = x^3 - 1$. Then $u'(x) = 3x^2$. Thus,

$$
\begin{aligned}
\int_{-1}^0 \frac{x^2}{x^3 - 1} \, dx &= \frac{1}{3} \int_{-1}^0 \frac{1}{x^3 - 1} 3x^2 \, dx \\[2mm]
&= \frac{1}{3} \int_{u=-2}^{u=-1} \frac{du}{u} \\[2mm]
&= \frac{1}{3} \ln|u| \Big|_{-2}^{-1} \\[2mm]
&= \frac{1}{3} (\ln(1) - \ln(2)) = -\frac{1}{3} \ln(2). \qquad \blacksquare
\end{aligned}
$$

The following properties give us insight into the ln function. The first three properties were established in Example 1 of the Section 7.1.

Property L_1: The domain of ln is $(0, \infty)$.

Property L_2: $\ln(1) = 0$.

Property L_3: ln is an increasing function and it is concave down.

Property L_4: If b is a positive number, then there is a number x such that $\ln(x) = b$.

Proof: Let n be an integer such that $\frac{n}{2} > b$. We will show below that $\ln(2^n) > \frac{n}{2} > b$. Since ln is differentiable, it is continuous. Thus, by the Intermediate Value Theorem, it follows that there is an x between 1 and 2^n such that $\ln(x) = b$. We now prove that $\ln(2^n) > \frac{n}{2}$.

$\ln(2) > \frac{1}{2}$: Let R_2 be the rectangle of height $\frac{1}{2}$ and with base the interval $[1, 2]$. Then R_2 lies entirely under the graph of $f(x) = \frac{1}{x}$. Thus $\ln(2) = \int_1^2 \frac{1}{t}\, dt$ is greater than the area of R_2, which is $\frac{1}{2}$.

$\ln(2^2) > 2/2$: Let R_3 be the rectangle of height $\frac{1}{4}$ and base the interval $[2, 4]$. The area of R_3 is $\frac{1}{2}$. Since R_3 lies entirely under the graph of $f(x) = \frac{1}{x}$, the area of R_3 is less than $\int_2^4 \frac{1}{t}\, dt$. Thus

$$
\begin{aligned}
\ln(2^2) &= \int_1^4 \frac{1}{t}\, dt \\[2mm]
&= \int_1^2 \frac{1}{t}\, dt + \int_2^4 \frac{1}{t}\, dt \\[2mm]
&> \text{Area}(R_2) + \text{Area}(R_3) \\[2mm]
&= \frac{1}{2} + \frac{1}{2} = 1.
\end{aligned}
$$

In a similar manner, we can show that $\int_4^8 \frac{1}{t}\, dt > \frac{1}{2}$ so that

$$
\begin{aligned}
\ln(2^3) &= \int_1^8 \frac{1}{t}\, dt \\[2mm]
&= \int_1^2 \frac{1}{t}\, dt + \int_1^2 \frac{1}{t}\, dt + \int_4^8 \frac{1}{t}\, dt \\[2mm]
&> \frac{1}{2} + \frac{1}{2} + \frac{1}{2} = \frac{3}{2}.
\end{aligned}
$$

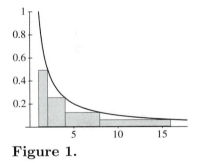

Figure 1.

The case for $n = 4$.

In general, $\int_{2^{n-1}}^{2^n} \frac{1}{t}\, dt > \frac{1}{2}$, so that

$$
\ln(2^n) = \int_1^{2^n} \frac{1}{t} = \int_1^2 \frac{1}{t}\, dt + \int_2^4 \frac{1}{t}\, dt + \cdots + \int_{2^{n-1}}^{2^n} \frac{1}{t}\, dt
$$

$$
> \frac{1}{2} + \frac{1}{2} + \cdots + \frac{1}{2} = \frac{n}{2}. \qquad \blacksquare
$$

The case for $n = 4$ is sketched in Figure 1.

Property L_5: If b is a negative number, then there is a number a between 0 and 1 such that $\ln(a) = b$.

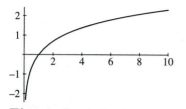

Figure 2. *A sketch of the* ln *function.*

The argument for L_5 is similar to the proof of L_4 and it is left as an exercise.

Properties L_1–L_5 provide us with enough information to obtain a rough sketch of the ln function. (See Figure 2.)

Property L_6: If a and b are numbers, then $\ln(ab) = \ln(a) + \ln(b)$.

Proof:

$$D(\ln(a) + \ln(x)) = D\ln(a) + D\ln(x) = 0 + \frac{1}{x} = \frac{1}{x},$$

and

$$D(\ln(ax)) = \frac{1}{ax}a = \frac{1}{x}.$$

Thus, the functions $\ln(a) + \ln(x)$ and $\ln(ax)$ have the same derivative. By Theorem 1 of Section 6.1, there is a number C so that

$$\ln(a) + \ln(x) = \ln(ax) + C. \tag{4}$$

Evaluating Equation (4) at $x = 1$ yields $\ln(a) = \ln(a) + C$. Thus

$$C = 0 \text{ and } \ln(ax) = \ln(a) + \ln(x). \qquad\blacksquare$$

Property L_7: If r is a rational number, then $\ln(x^r) = r\ln(x)$.

Proof:

$$D\left[\ln\left(x^r\right)\right] = \frac{rx^{r-1}}{x^r} = \frac{r}{x} \quad \text{and} \quad D[r\ln(x)] = \frac{r}{x}.$$

Thus there is a constant C such that

$$\ln\left(x^r\right) = r\ln(x) + C. \tag{5}$$

Evaluating Equation (5) at $x = 1$ shows us that $C = 0$. $\qquad\blacksquare$

EXERCISES 7.3

In Exercises 1–6 find the domain of the function.

1. $\ln(x-3)$.

2. $\ln(1-x)$.

3. $(\ln x)^5/x$.

4. $\ln(x^2-1)$.

5. $\ln(1-x^2)$.

6. $\ln(x^2+x)$.

In Exercises 7–10 compute the derivative.

7. $\ln(\sqrt{x}) + \frac{1}{2}\ln(x) + \ln(x^5) + 5\ln(x)$.

8. $\ln|\sin x| + \sin(\ln(x))$.

9. $\ln|\cos x| + \ln|\sec(x)|$.

10. $\ln\left(\frac{x}{\ln x}\right)$.

11. Find the equation of the line tangent to the graph of the function $y = (\sin \pi x)\ln x$ at $x = 2$. (Use your calculator value for $\ln(2)$.)

12. Find a function $\vec{s}$ from $\mathbb{R}$ into $\mathbb{R}^3$ so that the trace of $\vec{s}$ is the line tangent to the trace of $\vec{r}(t) = \left(t^2 \ln t, \ln t, \frac{1}{t}\right)$ at $t = 1$.

In Exercises 13–23 compute the anti-derivatives of the following functions.

13. $\dfrac{1}{x+1}$.

14. $\dfrac{x+1}{x^2+2x+25}$.

15. $\dfrac{(\ln x)^5}{x}$.

16. $\dfrac{\ln(x^5)}{x}$.

17. $(\cot t)(\ln|\sin t|)$.

18. $\cos x \csc x$.

19. $\dfrac{\cos x}{\sin^3 x}$.

20. $\tan x$.

21. $\cot x$.

22. $\cos x \tan x$.

23. $\sec x$. Hint: multiply by $\dfrac{\sec x + \tan x}{\sec x + \tan x}$.

24. Find the 4th–order Maclaurin polynomial for $f(x) = \ln(x+1)$.

25. Find the 4th–order Maclaurin polynomial for $f(x) = \ln(ax+1)$. (Assume that $a > 0$.)

26. Find the nth–order Maclaurin polynomial for the functions in Exercise 24.

27. Find the nth–order Maclaurin polynomial for the functions in Exercise 25.

28. Use the 4th–degree Taylor polynomial about $x_0 = 1$ to approximate $\ln(1.5)$ and find a bound for the error in this approximation. Compare your answer with your calculator value for $\ln(1.5)$.

29. Recall that $\log_a(b) = c$ if and only if $a^c = b$.

    a. Use your calculator to compare $\frac{\log_{10}(n)}{\ln(n)}$ with $\frac{1}{\ln(10)}$ for $n = 2, 3, 4$, and 6.

    b. If you have a graphing calculator, use it to sketch the graph of $f(x) = \frac{\log_{10}(x)}{\ln(x)}$ over the interval $[2,6]$.

    c. Find $\log_2(n)$, for $n = 1, 2, 4, 8, 16$, and 32.

    d. What seems to be true about the ratios $\frac{\log_2(n)}{\ln(n)}$, for $n = 2, 4, 8$, and 16? (Hint: use the fact that $\ln(a^r) = r\ln(a)$.)

    e. Show that $\dfrac{\log_3(9)}{\ln(9)} = \dfrac{1}{\ln(3)}$, $\dfrac{\log_{10}(100)}{\ln(100)} = \dfrac{1}{\ln(10)}$, and $\dfrac{\log_5(25)}{\ln(25)} = \dfrac{1}{\ln(5)}$.

    f. Show that $\log_a(a^r) = \dfrac{\ln(a^r)}{\ln(a)}$.

30. Let $f(x) = 1 - \ln(x)$.

    a. Use a graphing calculator to sketch the graph of f over the interval $[1,4]$. Apparently, f has a root in that interval. Check $f(1) \times f(4)$ to verify this observation.

    b. Find the first integer n such that f changes signs on the interval $[n, n+1]$.

    c. Use $x_0 = \frac{2n+1}{2}$ (the midpoint of the interval $[n, n+1]$) as the starting point and do just enough steps of Newton's method to find a number A such that $|f(A)| < 10^{-5}$. Use your calculator to evaluate $\ln(A)$. Compare A with your calculator value for $e^1 = e$.

31. Let $f(x) = 1 + \ln(x)$.

   a. If you have a graphing calculator, use it to sketch the graph of f over the interval $\left[\frac{1}{4}, 1\right]$. Apparently, f has a root in that interval. Check $f\left(\frac{1}{4}\right) \times f(1)$ to verify this observation.

   b. Find the first integer n such that f changes signs on the interval $\left[\frac{1}{n+1}, \frac{1}{n}\right]$.

   c. Use $x_0 = \left(\frac{1}{n+1} + \frac{1}{n}\right)/2$ (the midpoint of the interval $[n, n+1]$) as the starting point and do just enough steps of Newton's method to find a number B such that $|f(B)| < 10^{-5}$. Use your calculator to evaluate $\ln(B)$.

   d. Compare B with $\frac{1}{A}$, where A is the answer to Problem 30.c.

32. Let $f(x) = 1 - \ln(x) - \ln^2(x)$.

   a. Use your graphing calculator to approximate the root of f that lies between $x = 1$ and $x = 5$.

   b. Starting with your guess for the root in Part a, do enough steps of Newton's method to find a number C so that $|f(C)| < 10^{-5}$.

In Exercises 33–35 use differentiation to verify the integration formulas.

33. $\displaystyle\int \frac{dx}{x^2 + 3x + 2} = \ln\left|\frac{x+1}{x+2}\right| + C.$

34. $\displaystyle\int \frac{dx}{6x^2 + 7x + 2} = \frac{1}{36}\ln\left|\frac{x + 1/2}{x + 2/3}\right| + C.$

35. $\displaystyle\int \frac{dx}{(cx + a)(dx + b)} = \frac{1}{ad - bc}\ln\left|\frac{dx + b}{cx + a}\right| + C$

Use Exercise 35 to evaluate the integrals in Exercises 36–41.

36. $\displaystyle\int \frac{dx}{(x-1)(x+3)}.$

37. $\displaystyle\int \frac{dx}{(x-2)(2x+3)}.$

38. $\displaystyle\int \frac{dx}{(3x-1)(x+1)}.$

39. $\displaystyle\int \frac{dx}{x^2 + 3x - 4}.$

40. $\displaystyle\int \frac{dx}{2x^2 + 6x - 8}.$

41. $\displaystyle\int \frac{dx}{6x^2 + 7x + 2}.$

42. Let $f(x) = \dfrac{\ln(x)}{x}$

   a. What is the domain of f?

   b. Show that the number A of Exercise 30.c approximates a critical point of f.

   c. Determine where f is increasing and where f is decreasing.

   d. Show that f'' has a root on the interval $[1, 5]$.

   e. Find the first integer n such that f'' changes signs on the interval $[n, n+1]$.

   f. Starting with the information from Exercise 42.e, do just enough steps of the interval halving method to find a number C such that $f''(C) < 10^{-5}$.

   g. Determine where f is concave up and where f is concave down.

   h. Use the above information to sketch the graph of f. Compare your graph with the one generated by your calculator.

43. Let $f(x) = \dfrac{x}{\ln(x)}$.

   a. What is the domain of f?

   b. Show that the number A of Exercise 10 approximates a critical point of f. Where else can f' change sign?

   c. Determine where f is increasing and where f is decreasing.

   d. Show that f'' has a root on the interval $[2, 8]$. Where else can the concavity of f change?

43.   e. Find the first integer $n \geq 2$ such that f'' changes signs on the interval $[n, n+1]$.

   f. Starting with the information from Exercise 43.e, do just enough steps of the interval halving method to find a number C such that $f''(C) < 10^{-5}$.

   g. Determine where f is concave up and where f is concave down.

   h. Use this information to sketch the graph of f. Compare your graph with the one generated by your calculator.

Use Simpson's Rule (with $2n = 10$) to approximate the integrals in Exercises 44 and 45.

44. $\int_1^3 \ln(x)\, dx$.

45. $\int_1^3 x \ln^2(x)\, dx$.

7.4   The function e^x

Since ln is a strictly increasing function, ln has an inverse function, which we now define.

Definition: The Function E

$E(x)$ is defined to be the inverse function of $\ln(x)$. Thus $E(\ln(x)) = \ln(E(x)) = x$ as illustrated in Figure 1.

Since the expressions $y = E(x)$ and $x = \ln(y)$ are equivalent, we obtain a graph of E by reflecting the graph of $y = \ln(x)$ over the diagonal $y = x$.

Property E_1: $D(E(x)) = E(x)$.

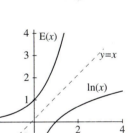

Figure 1. *The graph of $\ln(x)$ and $E(x)$.*

Proof: Starting with the definition of E, we have

$$\ln(E(x)) = x.$$

Taking the derivative of both sides (assuming that E is differentiable), we obtain

$$\ln'(E(x))E'(x) = 1.$$

Since $\ln'(x) = \frac{1}{x}$, we have

$$\frac{1}{E(x)}E'(x) = 1, \text{ or } E'(x) = E(x). \qquad \blacksquare$$

Thus E has the remarkable property that E is its own derivative! Note that our assumption that E is differentiable implicitly assumes that E is continuous.

Property E_2: $E(0) = 1$ since $E(0) = E(\ln(1)) = 1$.

> **Property E₃:** E is positive, increasing and concave up.

Proof: The fact that E is positive follows from the fact that the range of E is the domain of ln (since E is the inverse of ln). Thus E' and E'' are also positive since $E = E' = E''$. ∎

> **Property E₄:** The domain of E is $\mathbb{R}$ since the domain of E is the image of ln.

> **Property E₅:** $E(a + b) = E(a)E(b)$.

Proof: Let x and y be numbers such that $\ln(x) = a$ and $\ln(y) = b$. Then

$$E(a + b) = E(\ln(x) + \ln(y)) = E(\ln(xy)) = xy,$$

and

$$E(a)E(b) = E(\ln(x))E(\ln(y)) = xy.$$ ∎

> **Property E₆:** If r is a rational number, then $E(rx) = (E(x))^r$.

Proof: $(E(x))^r = E(\ln((E(x))^r)) = E(r\ln(E(x))) = E(rx).$ ∎

We now define Euler's number.

> **Definition:** Euler's Number
>
> ---
>
> e is defined as that number such that $\ln(e) = 1$. Equivalently, $e = E(1)$.

e is the number we approximated in Exercise 10.c of the previous section. The number e is an irrational number which is approximated by

$$e \approx 2.7183.$$

You can get your calculator to approximate e by asking for e^1.

By Property E₆, if r is a rational number, then $e^r = (E(1))^r = E(r)$. Thus E is an exponential function. Since E is defined for all real numbers and it is continuous, we may define

$$e^x = E(x) \text{ for all numbers } x.$$

The functions $\ln(x)$ and e^x were sketched in Figure 1. Some of the important facts we have developed about the functions e^x and $\ln x$ are listed below.

Summary of Properties of e^x and $\ln x$.

(a) $De^x = e^x$ and $D\ln(x) = \frac{1}{x}$.

(b) $\ln(e^x) = x$ and $e^{\ln(x)} = x$.

(c) The domain of $\ln(x)$ is $(0, \infty)$ and the image of $\ln(x)$ is $(-\infty, \infty)$.

(d) The domain of e^x is $(-\infty, \infty)$ and the image of e^x is $(0, \infty)$.

(e) $\ln(x)$ is increasing and concave down.

(f) e^x is positive, increasing, and concave up.

(g) $\ln(ab) = \ln(a) + \ln(b)$ and $\ln(x^r) = r\ln(x)$.

(h) $e^{x+y} = e^x e^y$ and $e^{xy} = (e^x)^y$.

Property (b) shows us that ln is the logarithm to the base e. Thus

$$\ln(x) = \log_e(x).$$

We call ln the *natural logarithm.*

EXAMPLE 1: Given that $\ln(3) \approx 1.10$ and $\ln(4) \approx 1.39$, it follows from Property (g) that $\ln(12) = \ln(3 \times 4) = \ln(3) + \ln(4) \approx 2.49$. ■

EXAMPLE 2: Using the chain rule, $De^{3x} = e^{3x}D(3x) = 3e^{3x}$. ■

EXAMPLE 3: $\int e^{2x}\, dx = \frac{1}{2}\int e^{2x}2\, dx = \frac{1}{2}\int e^u\, du|_{u=2x} = \frac{1}{2}e^{2x} + C$. ■

EXAMPLE 4: $\int_0^{\pi/2} e^{\sin x}\cos x\, dx = \int_{\sin 0}^{\sin \pi/2} e^u\, du = e^{\sin \pi/2} - e^{\sin 0} = e - 1$. ■

EXAMPLE 5: Solve the differential equation $(x-1)y' = y$.

SOLUTION: Writing the equation in the form $f(y)y' = g(x)$ gives us that

$$\frac{y'}{y} = \frac{1}{x-1}.$$

Integrating both sides with respect to x, we find that

$$\ln|y| = \ln|x - 1| + C,$$

or

$$e^{\ln|y|} = e^{\ln|x-1|+C} = e^{\ln|x-1|}e^C.$$

This is equivalent to

$$|y| = e^C|x - 1|.$$

We can let $k = e^C$ or $-e^C$ according to the initial conditions so that k will absorb the complications of the absolute value signs, leaving us with

$$y = k(x - 1). \qquad \blacksquare$$

EXAMPLE 6: Find the solution to the differential equation of Example 5 given that $y(2) = -1$.

SOLUTION: Starting with $y = k(x - 1)$, we see that $y(2) = -1 = k(2 - 1) = k$. Thus $k = -1$, and the solution is $y = 1 - x$. $\qquad \blacksquare$

Approximating the number e:

Let n be a positive integer and consider the integral

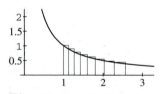

Figure 2.a *A Riemann sum with value greater than* $\int_1^{(1+\frac{1}{n})^n} \frac{1}{x}\, dx.$

$$\int_1^{(1+\frac{1}{n})^n} \frac{1}{t}\, dt = \ln\left(1 + \frac{1}{n}\right)^n.$$

We partition the interval $\left[1, \left(1 + \frac{1}{n}\right)^n\right]$ with

$$x_0 = 1, x_1 = \left(1 + \frac{1}{n}\right), x_2 = \left(1 + \frac{1}{n}\right)^2, \ldots, x_n = \left(1 + \frac{1}{n}\right)^n.$$

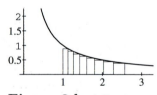

Figure 2.b *A Riemann sum with value less than* $\int_1^{(1+\frac{1}{n})^n} \frac{1}{x}\, dx.$

Notice that

$$x_i - x_{i-1} = \left(1 + \frac{1}{n}\right)^i - \left(1 + \frac{1}{n}\right)^{i-1} = \left(1 + \frac{1}{n}\right)^{i-1}\frac{1}{n}.$$

If we choose the left end point in $[x_i, x_{i-1}]$ as our selection point s_i, (see Figure 2.a), we have that

$$s_i = \left(1 + \frac{1}{n}\right)^{i-1};$$

and

$$\int_1^{(1+\frac{1}{n})^n} \frac{1}{x}\, dx \;\leq\; \sum_{i=1}^{n} \frac{1}{s_i}(x_i - x_{i-1})$$

$$= \sum_{i=1}^{n} \frac{1}{\left(1+\frac{1}{n}\right)^{i-1}} \left(1+\frac{1}{n}\right)^{i-1} \frac{1}{n}$$

$$= \underbrace{\frac{1}{n} + \frac{1}{n} + \cdots + \frac{1}{n}}_{n \text{ times}} = 1.$$

However, if, in the interval $[x_{i-1}, x_i]$ we choose the right end point for the selection point s_i (see Figure 2.b), we have that

$$s_i = \left(1+\frac{1}{n}\right)^{i};$$

and

$$\int_1^{(1+\frac{1}{n})^n} \frac{1}{x}\, dx \;\geq\; \sum_{i=1}^{n} \frac{1}{s_i}(x_i - x_{i-1})$$

$$= \sum_{i=1}^{n} \frac{1}{\left(1+\frac{1}{n}\right)^{i}} \left(1+\frac{1}{n}\right)^{i-1} \frac{1}{n}$$

$$= \sum_{i=1}^{n} \left(\frac{1}{1+\frac{1}{n}}\right) \frac{1}{n}$$

$$= \sum_{i=1}^{n} \frac{1}{1+n}$$

$$= \underbrace{\frac{1}{n+1} + \frac{1}{n+1} + \cdots + \frac{1}{n+1}}_{n \text{ times}} = \frac{n}{n+1}.$$

Thus for any integer n, we have that

$$\frac{n}{1+n} \;\leq\; \int_1^{(1+\frac{1}{n})^n} \frac{1}{x}\, dx \leq 1.$$

This implies that

$$\frac{n}{1+n} \;\leq\; \ln\left(1+\frac{1}{n}\right)^{n} \leq 1,$$

which is equivalent to

$$e^{\frac{n}{1+n}} \le \left(1 + \frac{1}{n}\right)^n \le e^1.$$

Since, as n gets large, $\frac{n}{1+n}$ gets close to 1 (i.e., $\lim_{n \to \infty} \frac{n}{1+n} = 1$), it follows that

$$e^1 \le \lim_{n \to \infty} \left(1 + \frac{1}{n}\right)^n \le e^1.$$

Thus we see that   ■

$$\lim_{n \to \infty} \left(1 + \frac{1}{n}\right)^n = e.$$

Later, we will develop a method for approximating e^x using a Maclaurin series. This will give us another way to approximate e.

EXERCISES 7.4

1. Sketch the graph of $y = e^{-x}$.

In Exercises 2–4 solve for x.

2. $e^{2x} - e^x = 0$.

3. $e^{2x} - 2e^x + 1 = 0$.

4. $\ln^2 x + 2\ln x + 1 = 0$.

In Exercises 5–12 compute the derivative.

5. e^{2x+1}.

6. $e^{(x^2)}$.

7. $(e^x)^2$.

8. $e^{\pi \ln(2x)}$.

9. $e^x \ln|\sin x|$.

10. $\dfrac{\ln|\cos x|}{e^{x^3}}$.

11. $\ln\left(\dfrac{e^x}{\ln x}\right)$.

12. $\vec{r}(x) = \left(e^{\sin x}, \tan(\ln x), \dfrac{e^{\csc x}}{x^2 + 1}\right)$.

13. Find the equation of the line tangent to the graph of the function $y = e^{x^2+2x+1}$ at $x = -1$.

14. Find a function $\vec{s}$ from $\mathbb{R}$ into $\mathbb{R}^3$ so that the trace of $\vec{s}$ is the line tangent to the trace of $\vec{r}(t) = \left(e^t, \ln t, \frac{1}{t}\right)$ at $t = 1$.

In Exercises 15–20 compute the anti-derivative.

15. e^{4x}.

16. e^{x+1}.

17. $xe^{(x^2)}$.

18. $e^{5\ln x}$.

19. $\dfrac{e^{2t}}{1 + e^{2t}}$.

20. $\dfrac{e^{2t}}{\sqrt{1 - e^{2t}}}$.

21.

Definition: Hyperbolic Trigonometric Functions

The *hyperbolic sine* of x, denoted by $\sinh x$, is defined by $\sinh x = \frac{e^x - e^{-x}}{2}$. The *hyperbolic cosine* of x, denoted by $\cosh x$, is defined by $\cosh x = \frac{e^x + e^{-x}}{2}$. The *hyperbolic tangent* of x, denoted by $\tanh x$, is defined by $\tanh x = \frac{\sinh x}{\cosh x}$.

Prove that the following identities hold:

a. $D\sinh x = \cosh x$.

b. $D\cosh x = \sinh x$.

c. $D\tanh x = \dfrac{1}{\cosh^2 x}$.

d. $\sinh(x + y) = \sinh x \cosh y + \cosh x \sinh y$.

22. Determine where $\sinh x$ is increasing, decreasing, concave up, and concave down. Sketch the graph of $y = \sinh x$.

23. Sketch the graph of $y = \cosh x$.

In Exercises 24 and 25 find the relative extrema and the points of inflection.

24. $y(x) = xe^x$.

25. $y(x) = x^2 e^{-x}$.

In Exercises 26–28 evaluate the integrals.

26. $\int_0^2 e^x \, dx$.

27. $\int_{-1}^1 e^{-x} \, dx$.

28. $\int_0^1 \sinh x \, dx$.

29. Evaluate $\left(1 + \frac{1}{n}\right)^n$ for $n = 1, 10, 100$, and 1000. Compare your results with the value for e in your calculator.

In Exercises 30–33 solve the differential equations.

30. $y' = xy$, $y(0) = 2$.

31. $yy' = (x^2 + 1)(y^2 + 1)$, $y(1) = -2$.

32. $y' = y$.

33. $y' = x \tan(y + 2)$.

In Exercises 34 and 35 find the 4th–order Maclaurin polynomial.

34. $f(x) = e^x$.

35. $f(x) = e^{ax}$.

36. Find the nth–order Maclaurin polynomial for the functions in Exercise 34.

37. Find the nth–order Maclaurin polynomial for the functions in Exercise 35.

38. Find a bound for the error in the 4th–order Maclaurin polynomial approximation for $e = e^1$. Compare your answer with your calculator's value for e.

39. Let $f(x) = e^x \ln(x)$.

a. Use your calculator to graph f' over the interval $[0.01, 3]$. Where do you expect f to be increasing? Where do you expect it to be decreasing?

b. Show that f'' has a root between $x_0 = 0.1$ and $x_1 = 1$.

c. Use the interval halving method to find a root of f'' with an error of no more than 10^{-5}. (Use x_0 and x_1 from Part b to get started.) Is this an inflection point?

40. Let $f(x) = e^x + \ln(x)$.

a. Show that f is increasing. Why does this show that f has at most one root?

b. Show that f has a root between $x_1 = 0.1$ and $x_2 = 1$.

c. Find the first integer n such that f has a root between $a = \frac{n}{10}$ and $b = \frac{n+1}{10}$.

d. Using $x_0 = \frac{a+b}{2}$ as a starting point, do just enough steps of Newton's method to find a number c such that $|f(c)| < 10^{-5}$.

e. Show that f'' is an increasing function. Why does this function show that f has at most one inflection point?

f. Show that f'' has a root between $x_1 = 0.1$ and $x_2 = 1$.

g. Find the first integer n such that f'' has a root between $a = \frac{n}{10}$ and $b = \frac{n+1}{10}$.

h. Using $x_0 = \frac{a+b}{2}$ as a starting point, do just enough steps on Newton's method to find a number c such that $|f''(c)| < 10^{-5}$.

i. Sketch the graph of f.

41. Recall that $\log_a(b) = c$ if and only if $a^c = b$.

a. Find $\log_2(n)$, for $n = 1, 2, 4, 8, 16$, and 32.

b. What seems to be true about the ratios $\frac{\log_2(n)}{\ln(n)}$?

In Exercises 42 and 43 use differentiation to verify the formulas.

42. $\int xe^x\ dx = xe^x - e^x + C.$

43. $\int x^2 e^x\ dx = x^2 e^x - 2xe^x + 2e^x + C.$

In Exercises 44–48 use the integration formulas of Exercises 42 and 43 to compute the integrals.

44. $\int (x+1)\,e^x\ dx.$

45. $\int (x^2 + x + 1)\,e^x\ dx.$

46. $\int \sin(x)\cos(x)\,e^{\cos(x)}\ dx.$

47. $\int \ln(x)\dfrac{e^{\ln x}}{x}\ dx.$

48. $\int \sin^2(x)\cos(x)e^{\sin(x)}\ dx.$

In Exercises 49–51 use Simpson's rule (with $N = 5$) to estimate the integrals.

49. $\int_{-1}^{1} e^{x^2}\ dx.$

50. $\int_{1}^{4} e^{1/x}\ dx.$

51. $\int_{0}^{2} \ln(x^2 + x + 10)\ dx.$

52. Sketch the graph of the curve that is parametrized by
$$\vec{r}(t) = e^\theta \begin{pmatrix} \cos(\theta) \\ \sin(\theta) \end{pmatrix},\ 0 \le \theta \le 2\pi.$$

7.5   Exponents and Logarithms

If a is a positive number and $r = \frac{m}{n}$ is a rational number (where we let n be positive), then we can define a^r to be the nth root of a^m. As yet, however, if x is an irrational number, a^x has not been defined. For example, what does 2^π mean? We use the natural logarithm function and the exponential function to define the function $y = a^x$. Recall that if r is a rational number, $\ln a^r = r \ln a$. Thus by exponentiating both sides, we see that $a^r = e^{r \ln a}$. We use this observation to define a^x for any number x.

Definition: a^x

If a is a positive number, then a^x is defined to be $e^{x \ln a}$ for all x.

Why is it that we require that a be positive?

The following theorem shows that the laws of exponents that we learned in high school algebra actually hold.

Theorem 1 *If a is positive, then*

(a) $a^{x+y} = a^x a^y$,

(b) $(a^x)^y = a^{xy}$,

(c) $a^0 = 1$,

(d) $a^1 = a$,

(e) $a^{-x} = \dfrac{1}{a^x}.$

Proof: We prove (a) and leave the remaining proofs to the reader.

$$a^{x+y} = e^{(x+y)\ln a} = e^{x\ln a + y\ln a} = e^{x\ln a}e^{y\ln a} = a^x a^y. \qquad \blacksquare$$

We now turn to the techniques for differentiating exponential functions.

EXAMPLE 1: Compute the derivative of 3^x.

SOLUTION: We use the definition of 3^x.

$$D3^x = De^{x\ln 3} = e^{x\ln 3}D(x\ln 3) = 3^x\ln 3. \qquad \blacksquare$$

EXAMPLE 2: Compute the derivative of a^x for $a > 0$.

SOLUTION:

$$Da^x = De^{x\ln a} = e^{x\ln a}D(x\ln a) = a^x\ln a. \qquad \blacksquare$$

We use the same technique to integrate a^x.

EXAMPLE 3: Compute the anti-derivative of a^x.

$$
\begin{aligned}
\int a^x\, dx &= \int e^{x\ln a}\, dx = \frac{1}{\ln a}\int e^{x\ln a}\ln a\, dx \\[2mm]
&= \frac{1}{\ln a}\int e^u\, du\Big|_{u=x\ln a} = \frac{1}{\ln a}e^{x\ln a} + C \\[2mm]
&= \frac{a^x}{\ln a} + C. \qquad \blacksquare
\end{aligned}
$$

We can extend this technique quite easily to find the derivatives of more complicated exponential functions that have the form $y = x^{f(x)}$

EXAMPLE 4: Compute the derivative of x^x.

SOLUTION: $Dx^x = De^{x\ln x} = e^{x\ln x}D(x\ln x) = x^x(\ln x + 1).$ $\qquad \blacksquare$

We can also compute these derivatives with the aid of Implicit Differentiation (Section 4.4). This method is often referred to as *logarithmic differentiation.*

EXAMPLE 5: Compute the derivative of $x^{\tan x}$.

SOLUTION: Taking the natural logarithm of both sides of the equation $y = x^{\tan x}$ yields $\ln y = \ln(x^{\tan x}) = \tan x\ln x$. If we differentiate both sides of this expression, we obtain

$$\frac{y'}{y} = \sec^2 x\ln x + \frac{\tan x}{x}.$$

Now multiply both sides by y and replace y with $x^{\tan x}$.

$$y' = y\sec^2 x \ln x + y\frac{\tan x}{x} = x^{\tan x}\sec^2 x \ln x + x^{\tan x}\frac{\tan x}{x}. \quad \blacksquare$$

In the exercises, we see that a^x is either increasing or decreasing according to whether a is greater than 1 or a is less than 1 (if $a = 1$, then the graph of $y = a^x$ is a horizontal line). In either case, the function $y = a^x$ has an inverse. We thus have the following definition.

Definition: $\log_a x$

The function $y = \log_a x$ is that function so that

$$\log_a(a^x) = x = a^{\log_a x}.$$

Or, equivalently,

$$y = \log_a x \text{ if and only if } a^y = x.$$

In the exercises, we see that $\log_a x$, often referred to as the logarithm to the base a of x, behaves exactly as we want it to, i.e., $\log_a xy = \log_a x + \log_a y$ and that $\log_a x^y = y\log_a x$. The student undoubtedly has seen the concept of logarithm before and is probably most familiar with the logarithms to the base 10. In the next theorem, we show that it is quite simple to translate from a logarithm base a to the ln function.

Theorem 2

$$\log_a x = \frac{\ln x}{\ln a}.$$

Proof:

$$e^{\ln x} = x = a^{\log_a x}.$$

Taking ln of both sides, we obtain:

$$\ln\left(e^{\ln x}\right) = \ln\left(a^{\log_a x}\right)$$

$$(\ln x)(\ln e) = (\log_a x)(\ln a)$$

$$(\ln x) = (\log_a x)(\ln a) \quad \text{(recall that } \ln e = 1\text{).}$$

Thus

$$\log_a x = \frac{\ln x}{\ln a}.$$ ■

As a Corollary we can translate a logarithm from one base to another.

Corollary 1

$$\log_b x = \frac{\log_a x}{\log_a b}.$$

Proof:

$$\log_b x = \left(\frac{\ln x}{\ln b}\right) = \left(\frac{\left(\frac{\ln x}{\ln a}\right)}{\left(\frac{\ln b}{\ln a}\right)}\right) = \left(\frac{\log_a x}{\log_a b}\right).$$ ■

We now investigate the derivative of the logarithm function.

Theorem 3 $D \log_a x = \dfrac{1}{x \ln a}.$

Proof: We apply Theorem 2.

$$D \log_a x = D \frac{\ln x}{\ln a} = \frac{1}{x \ln a}.$$ ■

The anti-derivative of $\log_a x$ will have to await another section.

EXERCISES 7.5

In Exercises 1–8 compute the derivative.

1. 2^x.

2. $2^{(x^2)}$.

3. $(2^x)^2$.

4. $\log_{10} x^2$.

5. $\log_3(x^3 + 2x)$.

6. x^{2x}.

7. $x^{(x+\sin x)}$.

8. $\sin x^{\cos x}$.

In Exercises 9–12 evaluate the integrals.

9. $\int 2^x \, dx$.

10. $\int (\sin x) 2^{\cos x} \, dx$.

11. $\int \dfrac{\log_2 x}{x} \, dx$.

12. $\int \dfrac{(\log_2 x)^5}{x} \, dx$.

In Exercises 13–15 evaluate the logarithms.

13. $\log_2 8$.

14. $\log_{\sqrt{2}} 4$.

15. $\log_{(1/4)} \frac{1}{2}$.

16. For which values of a is $y(x) = \log_a x$ an increasing function? For which values of a is $y(x) = \log_a x$ a decreasing function?

17. Sketch the graph of $\log_{(1/2)} x$ and the graph of $\log_2 x$.

18. Prove the following (using the fact that $\log_a x = \frac{\ln x}{\ln a}$):

   a. $\log_a xy = \log_a x + \log_a y$.

   b. $\log_a x^y = y \log_a x$.

   c. $\log_a \frac{x}{y} = \log_a x - \log_a y$.

19. Determine where $f(x) = \log_x 2$ is increasing and where it is decreasing.

20. Determine where $f(x) = \log_x \frac{1}{2}$ is increasing and where it is decreasing.

21. What does the graph of $\log_x x$ look like?

22. What is the domain of $\log_{\sin x} x$?

23. Prove the identity $\log_{e^x} x = \dfrac{\ln x}{x}$.

7.6   Euler's Formula (Optional)

If the equation $ax^2 + bx + c = 0$ does not factor easily, then we use the quadratic formula,

$$x = \frac{-b \pm \sqrt{b^2 - 4ac}}{2a},$$

which works fine as long as $b^2 - 4ac \geq 0$. In order to handle the case where $b^2 - 4ac < 0$, we introduced the *imaginary number* $i = \sqrt{-1}$. If $b^2 - 4ac < 0$, then $ax^2 + bx + c = 0$ has solutions of the form $x = \alpha \pm i\beta$.

EXAMPLE 1: The solutions to $x^2 + x + 1 = 0$ are $x = \frac{1}{2}(-1 \pm \sqrt{-3}) = -\frac{1}{2} \pm \frac{i}{2}\sqrt{3}$. ∎

Elementary Properties of Complex Numbers

(a) We denote by i the complex number $\sqrt{-1}$. Thus, $i^2 = -1, i^3 = -i$, and $i^4 = 1$.

(b) If a and b are real numbers, then $z = a + bi$ is a complex number where a is called the *real part of z* and b is called the *imaginary part of z*.

(c) Geometrically, $a + bi$ is identified with the point (a, b) in the xy–plane. The *distance* from $a + bi$ to $c + di$ is given by $\sqrt{(a - c)^2 + (b - d)^2}$.

(d) Complex numbers are added and subtracted just as vectors. $(a + bi) + (c + di) = (a + c) + (b + d)i$ and $(a + bi) - (c + di) = (a + c) - (b + d)i$.

(e) Unlike vectors, two complex numbers can be multiplied in a natural way: $(a + bi)(c + di) = ac + bci + adi + (bi)(di) = (ac + bd(i^2)) + (ad + bc)i = (ac - bd) + (ad + bc)i$.

(f) $a - bi$ is called the *conjugate* of $a + bi$. It is easily verified that $(a + bi)(a - bi) = a^2 + b^2$ If the complex number z is a solution to $ax^2 + bx + c = 0$, then so is its conjugate.

EXAMPLE 2:

(a) $2 + 3i$ is identified with the point $(2, 3)$ in the plane.

(b) The real number 4 is the complex number $4 + 0i$ which is identified with the point $(4, 0)$ in the plane.

(c) The number i is identified with the point $(0, 1)$ in the plane.

(d) The distance between $2 + 3i$ and $-5 + 6i$ is given by
$$\sqrt{(2 - (-5))^2 + (3 - 6)^2} = \sqrt{7^2 + 3^2} = \sqrt{49 + 9} = \sqrt{58}.$$

(e) $(2 + 3i) + (-5 + 6i) = -3 + 9i$ and $(2 + 3i)(-5 + 6i) = -10 + 12i - 15i - 18 = -28 - 3i$.

(f) $1 + i$ and its conjugate $1 - i$ are the solutions to $x^2 - 2x + 2 = 0$. ∎

There is a very useful and important relationship between complex numbers and e^x called Euler's formula.

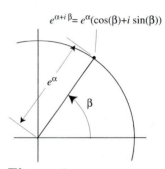

Figure 1.
$e^{\alpha + i\beta} = e^\alpha (\cos \beta + i \sin \beta)$
is identified with the point
$e^\alpha (\cos \beta, \sin \beta)$.

Euler's Formula

$$e^{i\theta} = \cos \theta + i \sin \theta$$

In the problem set, we lead you through a set of exercises designed to show that this formula makes sense. Geometrically, since $e^{i\theta} = \cos \theta + i \sin \theta$ is identified with the point in the plane $(\cos \theta, \sin \theta)$, $y = e^{i\theta}$ is just another way to express the wrapping function.

Let $z = \alpha + i\beta$. Then

$$e^z = e^{\alpha + i\beta} = e^\alpha e^{i\beta} = e^\alpha (\cos \beta + i \sin \beta),$$

which is identified with the vector $e^\alpha (\cos \beta, \sin \beta)$ in the plane as illustrated in Figure 1.

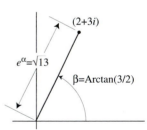

Figure 2. $z = 2 + 3i$

EXAMPLE 3: Write $e^{2 + i\pi/4}$ as $\alpha + i\beta$.

SOLUTION: $e^{2 + i\pi/4} = e^2 \left(\cos \left(\frac{\pi}{4} \right) + i \sin \left(\frac{\pi}{4} \right) \right) = e^2 \frac{\sqrt{2}}{2} + ie^2 \frac{\sqrt{2}}{2}$. ∎

EXAMPLE 4: Write $z = 2 + 3i$ as $e^{\alpha + i\beta}$.

SOLUTION: As in Figure 2, $e^\alpha = $ the magnitude of $(2, 3) = \sqrt{13}$. Therefore, $\alpha = \ln(e^\alpha) = \ln(\sqrt{13}) \approx 1.282$. The angle β is $\arctan \left(\frac{3}{2} \right) \approx 0.983$. We now have $z = 2 + 3i \approx e^{1.282 + 0.983i}$. ∎

EXERCISES 7.6

1. Find (a) i^5, and (b) i^6.

For Exercises 2–5 find (i) $z_1 + z_2$, (ii) $z_1 z_2$, (iii) the distance between z_1 and z_2, and (iv) the product of z_1 and its conjugate.

2. $z_1 = 3 - 2i$, $z_2 = -4 + i$.

3. $z_1 = 5 - i$, $z_2 = 3 - 2i$.

4. $z_1 = e^{1+i\pi/3}$, $z_2 = e^{2-i\pi/4}$.

5. $z_1 = e^{-2-i\pi/6}$, $z_2 = e^{1-i3\pi/4}$.

For Exercises 6–9 express z as $e^{\alpha+i\beta}$.

6. $z = 1 - i$.

7. $z = 1 + i$.

8. $z = 2 - i$.

9. $z = 2i + \sqrt{3}$.

In Exercises 10–12 express e^z as $\alpha + i\beta$.

10. $z = 3 + i\frac{\pi}{4}$.

11. $z = 2 - i\frac{3\pi}{4}$.

12. $z = -1 + i\frac{2\pi}{3}$.

13. Let $f(x) = e^x$.

   a. Find the Maclaurin polynomial of order $2m$ for $f(x)$.

   b. Use the result of Part a. to find the Maclaurin polynomial of order $2m$ for $f(3x) = e^{3x}$.

   c. Use the result of Part a. to find the Maclaurin polynomial of order $2m$ for $f(cx) = e^{cx}$.

   d. Use the result of Part a. to find the Maclaurin polynomial of order $2m$ for $f(ix) = e^{ix}$.

   e. Find the Maclaurin polynomial of order $2m$ for $g(x) = \sin x$.

   f. Find the Maclaurin polynomial of order $2m$ for $g(x) = \cos x$.

Observe from Parts d., e., and f. that the Maclaurin polynomial of order $2m$ for e^{ix} can be factored so that it is precisely $A + iB$, where A is the Maclaurin polynomial of order $2m$ for $\cos x$, and B is the Maclaurin polynomial of order $2m$ for $\sin x$.

7.7   Inverse Trigonometric Functions

Recall from Theorem 1 of Section 2.6 that a function f from $\mathbb{R}$ to $\mathbb{R}$ has an inverse if and only if, whenever a and b are in the domain of f and $a \neq b$, then $f(a) \neq f(b)$. This is equivalent to saying that each horizontal line in the plane passes through at most one point of the graph of f. Clearly, none of the trigonometric functions have inverses. We can, however, define the following functions, called the *principal trigonometric functions,* which do have inverses.

Definition: Principal Trigonometric Functions

$$\text{Sin } x = \sin x \text{ for } -\frac{\pi}{2} \leq x \leq \frac{\pi}{2}.$$
$$\text{Cos } x = \cos x \text{ for } 0 \leq x \leq \pi.$$
$$\text{Tan } x = \tan x \text{ for } -\frac{\pi}{2} < x < \frac{\pi}{2}.$$

$$\begin{aligned}
\text{Cot } x &= \cot x \text{ for } 0 < x < \pi. \\
\text{Sec } x &= \sec x \text{ for } 0 \le x < \frac{\pi}{2} \text{ or } \frac{\pi}{2} < x \le \pi. \\
\text{Csc } x &= \csc x \text{ for } -\frac{\pi}{2} \le x < 0 \text{ or } 0 < x \le \frac{\pi}{2}.
\end{aligned}$$

We denote these inverses as $\text{Arcsin } x$, $\text{Arccos } x$, $\text{Arctan } x$, $\text{Arccot } x$, $\text{Arcsec } x$, and $\text{Arccsc } x$. Some texts use $\sin^{-1} x$ or $\text{Sin}^{-1} x$ to denote the $\text{Arcsin } x$, for example. However, we wish to avoid confusing this with $\dfrac{1}{\sin x}$.

The sketches of the graphs of the inverse trigonometric functions can be obtained by reflecting the principal trigonometric functions across the diagonal $y = x$. In Figure 1, we have sketches for the graphs of $\text{Arcsin } x$, $\text{Arctan } x$, and $\text{Arcsec } x$. We leave the sketches of the graphs of $\text{Arccos } x$, $\text{Arccot } x$, and $\text{Arccsc } x$ as an exercise.

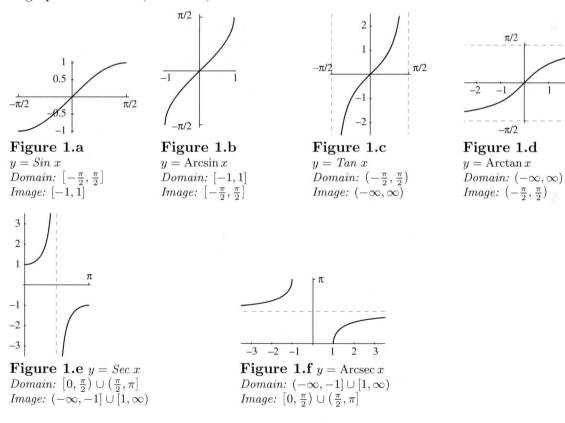

Figure 1.a
$y = Sin\ x$
Domain: $\left[-\frac{\pi}{2}, \frac{\pi}{2}\right]$
Image: $[-1, 1]$

Figure 1.b
$y = \text{Arcsin } x$
Domain: $[-1, 1]$
Image: $\left[-\frac{\pi}{2}, \frac{\pi}{2}\right]$

Figure 1.c
$y = Tan\ x$
Domain: $\left(-\frac{\pi}{2}, \frac{\pi}{2}\right)$
Image: $(-\infty, \infty)$

Figure 1.d
$y = \text{Arctan } x$
Domain: $(-\infty, \infty)$
Image: $\left(-\frac{\pi}{2}, \frac{\pi}{2}\right)$

Figure 1.e $y = Sec\ x$
Domain: $\left[0, \frac{\pi}{2}\right) \cup \left(\frac{\pi}{2}, \pi\right]$
Image: $(-\infty, -1] \cup [1, \infty)$

Figure 1.f $y = \text{Arcsec } x$
Domain: $(-\infty, -1] \cup [1, \infty)$
Image: $\left[0, \frac{\pi}{2}\right) \cup \left(\frac{\pi}{2}, \pi\right]$

EXAMPLE 1: $\text{Arccos } \frac{1}{2} = \frac{\pi}{3}.$ ■

EXAMPLE 2: Compute $\sin(\text{Arccos}\,x)$.

SOLUTION: Let $\theta = \text{Arccos}\,x$. Then $\cos\theta = x$. Draw a right triangle, as in Figure 2, having one angle of measure θ which has adjacent side of length $|x|$ and hypotenuse of length 1. Clearly, for this triangle, $\cos\theta = |x|/1$. We now use the Pythagorean theorem to find the length of the opposite side which is $\sqrt{1-x^2}$. Thus, up to sign, $\sin(\text{Arccos}\,x) = \sqrt{1-x^2}/1 = \sqrt{1-x^2}$. Since the range of the $\text{Arccos}\,x$ is between 0 and π and $\sin\theta$ is positive over this domain we see that $\sin(\text{Arccos}\,x) = \sqrt{1-x^2}$. ∎

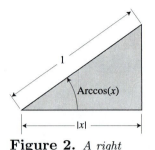

Figure 2. *A right triangle with adjacent side of length* $|x|$ *and hypotenuse of length* 1.

EXAMPLE 3: Compute $\cos(\text{Arcsec}\,x)$ and $\tan(\text{Arcsec}\,x)$.

SOLUTION: Let $\theta = \text{Arcsec}\,x$. Then $\sec\theta = x$ which is equivalent to $\cos\theta = \frac{1}{x}$. For the second expression we again draw a right triangle, as in Figure 3, having one angle of measure θ which has adjacent side of length 1 and hypotenuse of length $|x|$. Using the Pythagorean theorem we find that the length of the opposite side is $\sqrt{x^2-1}$. By inspection of Figure 3 we see that $\tan(\text{Arcsec}\,x) = \pm\sqrt{x^2-1}/1$. If $x \geq 1$ then the $\text{Arcsec}\,x$ is between 0 and $\frac{\pi}{2}$. For these values tan is positive. However, if $x \leq -1$ then the $\text{Arcsec}\,x$ is between $\frac{\pi}{2}$ and π. For these values tan is negative. Thus

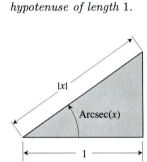

Figure 3. *A right triangle with adjacent side of length 1 and hypotenuse of length* $|x|$.

$$\tan(\text{Arcsec}\,x) = \begin{cases} \sqrt{x^2-1} & \text{if } x \geq 1 \\ -\sqrt{x^2-1} & \text{if } x \leq -1. \end{cases}$$

∎

EXERCISES 7.7

1. Sketch the graphs of $\text{Cos}\,\theta$, $\text{Arccos}\,x$, $\text{Cot}\,\theta$, $\text{Arccot}\,x$, $\text{Csc}\,\theta$, and $\text{Arccsc}\,x$.

2. Simplify the following expressions:

   a. $\cos(\text{Arcsin}\,x)$, $\tan(\text{Arcsin}\,x)$, $\cot(\text{Arcsin}\,x)$ $\sec(\text{Arcsin}\,x)$, and $\csc(\text{Arcsin}\,x)$.

   b. $\sin(\text{Arctan}\,x)$, $\cos(\text{Arctan}\,x)$, $\cot(\text{Arctan}\,x)$, $\sec(\text{Arctan}\,x)$, and $\csc(\text{Arctan}\,x)$.

   c. $\sin(\text{Arccot}\,x)$, $\cos(\text{Arccot}\,x)$, $\tan(\text{Arccot}\,x)$, $\cot(\text{Arccot}\,x)$, $\sec(\text{Arccot}\,x)$, and $\csc(\text{Arccot}\,x)$.

2.   d. $\sin(\text{Arccsc}\,x)$, $\cos(\text{Arccsc}\,x)$, $\tan(\text{Arccsc}\,x)$, $\cot(\text{Arccsc}\,x)$, $\sec(\text{Arccsc}\,x)$, and $\csc(\text{Arccsc}\,x)$.

In Exercises 3–8 simplify the expressions.

3. $\text{Arccos}\,\frac{1}{2}$.

4. $\text{Arcsin}\,\frac{1}{2}$.

5. $\text{Arctan}\,1$.

6. $\text{Cos}\left(\text{Arctan}\,\frac{3}{4}\right)$.

7. $\sec\left(\text{Arccos}\,\frac{4}{5}\right)$.

8. $\csc(\text{Arcsec}\,\sqrt{2})$.

7.8   Derivatives of Inverse Trigonometric Functions

To compute the derivative of $\operatorname{Arcsin} x$, let $f(x) = \operatorname{Sin} x$ so that $f^{-1}(x) = \operatorname{Arcsin}(x)$.

$$D(\sin(\operatorname{Arcsin} x)) = Dx.$$

Using the Chain Rule, we get

$$\cos(\operatorname{Arcsin} x)D(\operatorname{Arcsin} x) \;=\; 1$$

$$D(\operatorname{Arcsin} x) \;=\; \frac{1}{\cos(\operatorname{Arcsin} x)}$$

$$\;=\; \frac{1}{\sqrt{1-x^2}}. \qquad\blacksquare$$

To compute the derivative of $\operatorname{Arcsec} x$, we let $f(x) = \operatorname{Sec} x$. The inverse function is $f^{-1}(x) = \operatorname{Arcsec}(x)$.

$$D(\sec(\operatorname{Arcsec} x)) \;=\; Dx$$

$$\sec(\operatorname{Arcsec} x)\tan(\operatorname{Arcsec} x)D(\operatorname{Arcsec} x) \;=\; 1$$

$$x\tan(\operatorname{Arcsec} x)D(\operatorname{Arcsec} x) \;=\; 1$$

$$D(\operatorname{Arcsec} x) \;=\; \frac{1}{x\tan(\operatorname{Arcsec} x)}.$$

By Example 3 of Section 7.7,

$$\tan(\operatorname{Arcsec} x) = \begin{cases} \sqrt{x^2-1} & \text{if } x \geq 1 \\ -\sqrt{x^2-1} & \text{if } x \leq -1. \end{cases}$$

Thus

$$D(\operatorname{Arcsec} x) \;=\; \begin{cases} \frac{1}{x\sqrt{x^2-1}} & \text{if } x \geq 1 \\ \frac{1}{-x\sqrt{x^2-1}} & \text{if } x \leq -1 \end{cases}$$

$$\;=\; \frac{1}{|x|\sqrt{x^2-1}}. \qquad\blacksquare$$

In the exercises, the student is asked to compute the derivatives of the remaining inverses of the trigonometric functions which we compile in Theorem 1.

Theorem 1

$$D \operatorname{Arcsin} x \;=\; \frac{1}{\sqrt{1 - x^2}}.$$

$$D \operatorname{Arccos} x \;=\; -\frac{1}{\sqrt{1 - x^2}}.$$

$$D \operatorname{Arctan} x \;=\; \frac{1}{1 + x^2}.$$

$$D \operatorname{Arccot} x \;=\; -\frac{1}{1 + x^2}.$$

$$D \operatorname{Arcsec} x \;=\; \frac{1}{|x|\sqrt{x^2 - 1}}.$$

$$D \operatorname{Arccsc} x \;=\; -\frac{1}{|x|\sqrt{x^2 - 1}}.$$

Once we have the formulas for computing derivatives of the inverse trigonometric functions, you should have little trouble using them. The difficulty you will likely encounter is simply remembering them. These formulas should be committed to memory post haste! The most important formulas in Theorem 1 are those for $\operatorname{Arcsin} x$, $\operatorname{Arctan} x$, and $\operatorname{Arcsec} x$ (the others are merely the negatives of these).

Of course, we use Theorem 1 to compute anti-derivatives, which, as usual, will be more difficult. From Theorem 1 we have:

Theorem 2

$$\int \frac{1}{\sqrt{1 - x^2}}\, dx \;=\; \operatorname{Arcsin} x + C$$

$$\int \frac{1}{1 + x^2}\, dx \;=\; \operatorname{Arctan} x + C$$

$$\int \frac{1}{x\sqrt{x^2 - 1}}\, dx \;=\; \operatorname{Arcsec} |x| + C$$

The first two parts of Theorem 2 are simply restatements of the corresponding parts from Theorem 1. It is left to the student to verify the last part (Exercise 36).

The following examples illustrate the diversity of functions which can be integrated using Theorem 2.

EXAMPLE 1:

$$\int \frac{1}{\sqrt{1-4x^2}}\,dx = \frac{1}{2}\int \frac{1}{\sqrt{1-4x^2}}2\,dx$$

$$= \frac{1}{2}\int \frac{1}{\sqrt{1-u^2}}\,du\Big|_{u=2x}$$

$$= \frac{1}{2}\operatorname{Arcsin} u + C\Big|_{u=2x}$$

$$= \frac{1}{2}\operatorname{Arcsin}(2x) + C. \qquad\blacksquare$$

EXAMPLE 2:

$$\int_{-1}^{1} \frac{1}{\sqrt{1-x^2}}\,dx = \operatorname{Arcsin} x\Big|_{-1}^{1}$$

$$= \frac{\pi}{2} - \left(-\frac{\pi}{2}\right) = \pi. \qquad\blacksquare$$

EXAMPLE 3:

$$\int \frac{1}{1+(1+x)^2}\,dx = \int \frac{1}{1+u^2}\,du\Big|_{u=1+x}$$

$$= \operatorname{Arctan} u + C\Big|_{u=1+x}$$

$$= \operatorname{Arctan}(1+x) + C. \qquad\blacksquare$$

EXAMPLE 4:

$$\int \frac{1}{x\sqrt{x^2-a^2}}\,dx = \int \frac{1}{ax\sqrt{\left(\frac{x}{a}\right)^2-1}}\,dx$$

$$= \frac{1}{a}\int \frac{1}{u\sqrt{u^2-1}}\,du\Big|_{u=\frac{x}{a}}$$

$$= \frac{1}{a}\operatorname{Arcsec}|u| + C\Big|_{u=\frac{x}{a}}$$

$$= \frac{1}{a}\operatorname{Arcsec}\left|\frac{x}{a}\right| + C. \qquad\blacksquare$$

EXAMPLE 5:
$$\int \frac{1}{\sqrt{2x-x^2}}\,dx.$$

We proceed by completing the square in the denominator:

$$2x - x^2 = -(x^2 - 2x + 1) + 1$$

$$= 1 - (x-1)^2$$

so that

$$\int \frac{1}{\sqrt{2x - x^2}} \, dx \;=\; \int \frac{1}{\sqrt{1 - (x - 1)^2}} \, dx$$

$$=\; \int \frac{1}{\sqrt{1 - u^2}} \, du\Big|_{u = x - 1}$$

$$=\; \operatorname{Arcsin} u + C\big|_{u = x - 1}$$

$$=\; \operatorname{Arcsin}(x - 1) + C. \qquad \blacksquare$$

EXERCISES 7.8

In Exercises 1–22 evaluate the integrals.

1. $\displaystyle\int \frac{1}{4x^2 + 1} \, dx.$

2. $\displaystyle\int \frac{1}{x\sqrt{4x^2 - 1}} \, dx.$

3. $\displaystyle\int \frac{x}{4x^2 + 1} \, dx.$

4. $\displaystyle\int \frac{1}{\sqrt{1 - 9x^2}} \, dx.$

5. $\displaystyle\int \frac{1}{\sqrt{9 - 4x^2}} \, dx.$

6. $\displaystyle\int \frac{x}{\sqrt{3 - 2x^2}} \, dx.$

7. $\displaystyle\int \frac{1}{3 + 2x^2} \, dx.$

8. $\displaystyle\int \frac{1 + x}{3 + 2x^2} \, dx.$

9. $\displaystyle\int \frac{1}{a^2 + b^2 x^2} \, dx.$

10. $\displaystyle\int \frac{1}{x\sqrt{x^2 - 9}} \, dx.$

11. $\displaystyle\int \frac{1}{x\sqrt{a^2 x^2 - b^2}} \, dx.$

12. $\displaystyle\int \frac{x}{\sqrt{x^2 - 9}} \, dx.$

13. $\displaystyle\int \frac{x}{\sqrt{a^2 - b^2 x^2}} \, dx.$

14. $\displaystyle\int \frac{e^x}{1 + e^{2x}} \, dx.$

15. $\displaystyle\int \frac{e^x}{1 + e^x} \, dx.$

16. $\displaystyle\int \frac{1}{\sqrt{e^{2x} - 1}} \, dx.$

17. $\displaystyle\int \frac{e^{2x}}{1 + e^{2x}} \, dx.$

18. $\displaystyle\int \frac{1}{x}(1 - \ln^2 x) \, dx.$

19. $\displaystyle\int \frac{1}{x \ln x \sqrt{\ln^2 x - 1}} \, dx.$

20. $\displaystyle\int \frac{1}{x^2 + 2x + 1} \, dx.$

21. $\displaystyle\int \frac{1}{x^2 + 2x + 2} \, dx.$

22. $\displaystyle\int \frac{1}{(x - 1)\sqrt{x^2 - 2x}} \, dx.$

In Exercises 23–30 compute the derivatives.

23. $\operatorname{Arcsin}(x^2 + 2x + 3).$

24. $\operatorname{Arctan}\sqrt{x^5 - 2x + 5}.$

25. $\operatorname{Arccos}(\ln(x)).$

26. $\operatorname{Arcsec}(2^x).$

27. $\log_3(\operatorname{Arccot}(x)).$

28. $\left[\operatorname{Arccsc}(x^2 + x + \pi)\right]^e.$

29. $\operatorname{Arcsin}(x^3 + x^\pi + \pi^x).$

30. $\sec(\mathrm{Arcsec}(\pi^x))$.

In Exercises 31–34 solve the differential equations.

31. $\cos(x)y' = 1 + y^2$.

32. $y' = x\sqrt{1 - 4x^2}$.

33. $(1 + 4x^2)y' = y\sqrt{y^2 - 16}$.

34. $(x^2 + 2)y' = (y^2 + 2)x$.

35. Verify $D\,\mathrm{Arcsec}\,|x| = \dfrac{1}{x\sqrt{x^2 - 1}}$. Hint: for $x \neq 0$, $D|x| = \frac{|x|}{x}$.

36. Verify the derivative formulas for the inverse trigonometric functions that have not been done in the text.

7.9   Tables of Integrals

We have discussed the chain rule in reverse (often called the method of substitution) for computing integrals. There are numerous other methods, which we will see in Chapter 10. However, our goal is to model physical processes. To this end we include here a list of integral tables so that we can move beyond techniques of integration and apply these results. Many such tables contain over 700 integrals. Indeed, if you have access to a computer package that does symbolic integration, it may seem that you need need neither integral tables nor methods of integration. It is the case that sometimes tables are more helpful than a symbolic integration package. While we haven't found anything in a table of integrals that our symbolic package can't handle properly, it quite often happens that the answer given by the computer package is in a format that is hard to use. For example, Mathematica$^{©}$ tells us that

$$\int \frac{dx}{x^2 - a^2} = -\frac{1}{a}\mathrm{Arctanh}\left(\frac{x}{a}\right).$$

You might correctly guess that $\mathrm{Arctanh}(x)$ is the inverse of the tanh function defined in Exercise 7.4.21. However, it is not trivial to get an explicit expression for $\mathrm{Arctanh}(x)$. Most integration tables will give

$$\int \frac{dx}{x^2 - a^2} = -\frac{1}{2a}\left(\ln|x - a| - \ln|x + a|\right).$$

Which solution would you prefer?

The integration techniques we take up in Chapter 10 are important for two reasons. First, it is very unlikely that the integral that you arrive at in your research or work will appear in any given table. Usually, you must apply some method to transform your integral into one that is in a table. Second, the techniques we introduce are commonly used in the derivation of formulas used in the applications of calculus.

We list below some of the more common integrals grouped into categories. This list is by no means exhaustive. Two good sources for tables of integrals are:

(a) **CRC Handbook of Chemistry and Physics**, published by the Chemical Rubber Company; and,

(b) **Handbook of Mathematical Functions with Formulas, Graphs and Mathematical Tables**, edited by Milton Abramowitz and Irene A. Stegun, National Bureau of Standards, Applied Math Series, 55

Note that in the following the "$+ C$"–term is understood.

Elementary Functions

1. $\int cf(x)\, dx = c \int f(x)\, dx$

2. $\int f(x) \pm g(x)\, dx = \int f(x)\, dx \pm \int g(x)\, dx$

3. $\int x^n\, dx = \dfrac{x^{n+1}}{n+1}, \quad n \neq -1$

4. $\int x^{-1}\, dx = \ln|x|$

5. $\int \dfrac{f'(x)}{f(x)}\, dx = \ln|f(x)|$

Trigonometric Functions

6. $\int \sin ax\, dx = -\dfrac{\cos ax}{a}$

7. $\int \cos ax\, dx = \dfrac{\sin ax}{a}$

8. $\int \tan ax\, dx = \dfrac{\ln|\sec ax|}{a}$

9. $\int \cot ax\, dx = \dfrac{\ln|\sin ax|}{a}$

10. $\int \sec ax\, dx = \dfrac{\ln|\sec ax + \tan ax|}{a}$

11. $\int \csc ax\, dx = -\dfrac{\ln|\csc ax + \cot ax|}{a}$

12. $\int \sin^2 ax\, dx = \dfrac{x}{2} - \dfrac{\sin 2ax}{4a}$

13. $\displaystyle\int \cos^2 ax\ dx = \frac{x}{2} + \frac{\sin 2ax}{4a}$

14. $\displaystyle\int \tan^2 ax\ dx = \frac{\tan ax}{a} - x$

15. $\displaystyle\int \cot^2 ax\ dx = -\frac{\cot ax}{a} - x$

16. $\displaystyle\int \sec^2 ax\ dx = \frac{\tan ax}{a}$

17. $\displaystyle\int \csc^2 ax\ dx = -\frac{\cot ax}{a}$

18. $\displaystyle\int \sin^n ax\ dx = -\frac{\cos ax \sin^{n-1}}{na} + \frac{n-1}{n} \int \sin^{n-2} ax\ dx$

19. $\displaystyle\int \cos^n ax\ dx = \frac{\sin ax \cos^{n-1}}{na} + \frac{n-1}{n} \int \cos^{n-2} ax\ dx$

20. $\displaystyle\int \sec^n ax\ dx = \frac{\tan ax \sec^{n-2} ax}{a(n-1)} + \frac{n-2}{n-1} \int \sec^{n-2} ax\ dx$

21. $\displaystyle\int \csc^n ax\ dx = -\frac{\cot ax \csc^{n-2} ax}{a(n-1)} + \frac{n-2}{n-1} \int \csc^{n-2} ax\ dx$

22. $\displaystyle\int \tan^n ax\ dx = \frac{\tan^{n-1} ax}{a(n-1)} - \int \tan^{n-2} ax\ dx$

23. $\displaystyle\int \cot^n ax\ dx = -\frac{\cot^{n-1} ax}{a(n-1)} - \int \cot^{n-2} ax\ dx$

24. $\displaystyle\int x^n \sin ax\ dx = -\frac{x^n \cos ax}{a} + \frac{n}{a} \int x^{n-1} \cos ax\ dx$

25. $\displaystyle\int x^n \cos ax\ dx = \frac{x^n \sin ax}{a} - \frac{n}{a} \int x^{n-1} \sin ax\ dx$

26. $\displaystyle\int \sin ax \cos bx\ dx = -\frac{\cos(a+b)x}{2(a+b)} - \frac{\cos(a-b)x}{2(a-b)},\ a^2 \neq b^2$

27. $\displaystyle\int \sin ax \sin bx\ dx = \frac{\sin(a-b)x}{2(a-b)} + \frac{\sin(a+b)x}{2(a+b)},\ a^2 \neq b^2$

28. $\displaystyle\int \cos ax \cos bx\ dx = \frac{\sin(a-b)x}{2(a-b)} + \frac{\sin(a+b)x}{2(a+b)},\ a^2 \neq b^2$

29. $\displaystyle\int \sin^n ax \cos^m ax \; dx = \frac{\sin^{n+1} ax \cos^{m-1} ax}{a(m+n)}$
$\displaystyle + \frac{m-1}{m+n} \int \sin^n ax \cos^{m-2} ax \; dx, \; m \neq -n$

30. $\displaystyle\int \text{Arcsin } ax \; dx = x\text{Arcsin } ax + \frac{\sqrt{1-a^2x^2}}{a}$

31. $\displaystyle\int \text{Arccos } ax \; dx = x\text{Arccos}(ax) - \frac{\sqrt{1-a^2x^2}}{a}$

32. $\displaystyle\int \text{Arctan}(ax) \; dx = x\text{Arctan}(ax) - \frac{\ln(a+a^2x^2)}{2a}$

33. $\displaystyle\int \text{Arccot } ax \; dx = x\text{Arccot } ax + \frac{\ln(a+a^2x^2)}{2a}$

34. $\displaystyle\int \text{Arcsec } ax \; dx = x\text{Arcsec } ax - \frac{\ln(ax+\sqrt{a^2x^2-1})}{a}$

35. $\displaystyle\int \text{Arccsc } ax \; dx = x\text{Arccsc } ax + \frac{\ln(ax+\sqrt{a^2x^2-1})}{a}$

Exponential and Logarithmic Functions

36. $\displaystyle\int e^{ax} \; dx = \frac{e^{ax}}{a}$

37. $\displaystyle\int xe^{ax} \; dx = (ax-1)\frac{e^{ax}}{a}$

38. $\displaystyle\int x^n e^{ax} \; dx = \frac{x^n e^{ax}}{a} - \frac{n}{a} \int x^{n-1}e^{ax} \; dx$

39. $\displaystyle\int b^{ax} \; dx = \frac{b^{ax}}{a \ln b}$

40. $\displaystyle\int x^n b^{ax} \; dx = \frac{x^n b^{ax}}{a \ln b} - \frac{n}{a \ln b} \int x^{n-1}b^{ax} \; dx, \; b > 0, b \neq 1$

41. $\displaystyle\int e^{ax} \sin bx \; dx = \frac{e^{ax}}{a^2+b^2}(a \sin bx - b \cos bx)$

42. $\displaystyle\int e^{ax} \cos bx \; dx = \frac{e^{ax}}{a^2+b^2}(a \cos bx + b \sin bx)$

43. $\int \ln ax \; dx = x \ln ax - x$

44. $\displaystyle\int \frac{\ln ax}{x}\ dx = \frac{1}{2}(\ln ax)^2$

45. $\displaystyle\int x^n \ln ax\ dx = \frac{x^{n+1}}{n+1}\ln ax - \frac{x^{n+1}}{(n+1)^2},\ \ n \neq -1$

46. $\displaystyle\int \frac{dx}{x \ln ax}\ dx = \ln|\ln ax|$

Hyperbolic Trigonometric Functions

47. $\displaystyle\int \sinh ax\ dx = \frac{\cosh ax}{a}$

48. $\displaystyle\int \cosh ax\ dx = \frac{\sinh ax}{a}$

49. $\displaystyle\int \tanh ax\ dx = \frac{\ln|\cosh ax|}{a}$

50. $\displaystyle\int \coth ax\ dx = \frac{\ln|\sinh ax|}{a}$

51. $\displaystyle\int \text{sech}\ ax\ dx = \frac{1}{a}\text{Arcsin}(\tanh ax)$

52. $\displaystyle\int \text{csch}\ ax\ dx = \frac{1}{a}\ln\left|\tanh \frac{ax}{2}\right|$

53. $\displaystyle\int \sinh^2 ax\ dx = \frac{\sinh 2ax}{4a} - \frac{x}{2}$

54. $\displaystyle\int \cosh^2 ax\ dx = \frac{\sinh 2ax}{4a} + \frac{x}{2}$

55. $\displaystyle\int \tanh^2 ax\ dx = x - \frac{\tanh ax}{a}$

56. $\displaystyle\int \coth^2 ax\ dx = x - \frac{\coth ax}{a}$

57. $\displaystyle\int \text{sech}^2 ax\ dx = \frac{\tanh ax}{a}$

58. $\displaystyle\int \text{csch}^2 ax\ dx = -\frac{\coth ax}{a}$

59. $\displaystyle\int \sinh^n ax\ dx = \frac{\sinh^{n-1} ax \cosh ax}{na} + \frac{n-1}{n}\int \sinh^{n-2} ax\ dx$
$n \neq 0$

60. $\int \cosh^n ax \; dx = \dfrac{\cosh^{n-1} ax \sinh ax}{na} + \dfrac{n-1}{n} \int \cosh^{n-2} ax \; dx$
$n \neq 0$

61. $\int \tanh^n ax \; dx = - \dfrac{\tanh^{n-1} ax}{(n-1)a} + \int \tanh^{n-2} ax \; dx, \quad n \neq 1$

62. $\int \coth^n ax \; dx = - \dfrac{\coth^{n-1} ax}{(n-1)a} + \int \coth^{n-2} ax \; dx, \quad n \neq 1$

63. $\int \operatorname{sech}^n ax \; dx = \dfrac{\operatorname{sech}^{n-2} ax \tanh ax}{(n-1)a} + \dfrac{n-2}{n-1} \int \operatorname{sech}^{n-2} ax \; dx$
$n \neq 1$

64. $\int \operatorname{csch}^n ax \; dx = - \dfrac{\operatorname{csch}^{n-2} ax \coth ax}{(n-1)a} - \dfrac{n-2}{n-1} \int \operatorname{csch}^{n-2} ax \; dx$
$n \neq 1$

65. $\int e^{ax} \sinh bx \; dx = \dfrac{e^{ax}}{2} \left(\dfrac{e^{bx}}{a+b} - \dfrac{e^{-bx}}{a-b} \right), \quad a^2 \neq b^2$

66. $\int e^{ax} \cosh bx \; dx = \dfrac{e^{ax}}{2} \left(\dfrac{e^{bx}}{a+b} + \dfrac{e^{-bx}}{a-b} \right), \quad a^2 \neq b^2$

$$\sqrt{ax+b}$$

67. $\int \dfrac{\sqrt{ax+b}}{x} \; dx = 2\sqrt{ax+b} + b \int \dfrac{dx}{x\sqrt{ax+b}}$

68. $\int \dfrac{dx}{x\sqrt{ax+b}} = \dfrac{2}{\sqrt{-b}} \operatorname{Arctan} \sqrt{\dfrac{ax+b}{-b}}, \quad \text{if } b < 0$

69. $\int \dfrac{dx}{x\sqrt{ax+b}} = \dfrac{1}{\sqrt{b}} \ln \left| \dfrac{\sqrt{ax+b} - \sqrt{b}}{\sqrt{ax+b} + \sqrt{b}} \right|, \quad \text{if } b > 0$

70. $\int \dfrac{\sqrt{ax+b}}{x^2} \; dx = - \dfrac{\sqrt{ax+b}}{x} + \dfrac{a}{2} \int \dfrac{dx}{x\sqrt{ax+b}}$

71. $\int \dfrac{dx}{x^2\sqrt{ax+b}} = - \dfrac{\sqrt{ax+b}}{bx} - \dfrac{a}{2b} \int \dfrac{dx}{x\sqrt{ax+b}}$

$$\sqrt{a^2+x^2}$$

72. $\int \sqrt{a^2+x^2} \; dx = \dfrac{x}{a}\sqrt{a^2+x^2} + \dfrac{a^2}{2} \ln \left| x + \sqrt{a^2+x^2} \right|$

73. $\displaystyle \int \frac{\sqrt{a^2 + x^2}}{x}\, dx = \sqrt{a^2 + x^2} - a \ln \left| \frac{a + \sqrt{a^2 + x^2}}{x} \right|$

74. $\displaystyle \int \frac{dx}{x\sqrt{a^2 + x^2}} = \frac{1}{a} \ln \left| \frac{x}{\sqrt{a^2 + x^2} + a} \right|$

75. $\displaystyle \int \frac{dx}{\sqrt{a^2 + x^2}} = \ln \left| x + \sqrt{a^2 + x^2} \right|$

$$\sqrt{a^2 - x^2}$$

76. $\displaystyle \int \frac{dx}{\sqrt{a^2 - x^2}} = \text{Arcsin } \frac{x}{a}$

77. $\displaystyle \int \sqrt{a^2 - x^2}\, dx = \frac{x}{2}\sqrt{a^2 - x^2} + \frac{a^2}{2} \text{ Arcsin } \frac{x}{a}$

78. $\displaystyle \int x^2 \sqrt{a^2 - x^2}\, dx = \frac{a^4}{8} \text{ Arcsin } \frac{x}{a} - \frac{1}{8} x \sqrt{a^2 - x^2}(a^2 - 2x^2)$

79. $\displaystyle \int \frac{\sqrt{a^2 - x^2}}{x}\, dx = \sqrt{a^2 - x^2} - a \ln \left| \frac{a + \sqrt{a^2 - x^2}}{x} \right|$

80. $\displaystyle \int \frac{\sqrt{a^2 - x^2}}{x^2}\, dx = -\text{Arcsin } \frac{x}{a} - \frac{\sqrt{a^2 - x^2}}{x}$

81. $\displaystyle \int \frac{dx}{x\sqrt{a^2 - x^2}} = -\frac{1}{a} \ln \left| \frac{a + \sqrt{a^2 - x^2}}{x} \right|$

82. $\displaystyle \int \frac{dx}{x^2 \sqrt{a^2 - x^2}} = -\frac{\sqrt{a^2 - x^2}}{a^2 x}$

$$\sqrt{x^2 - a^2}$$

83. $\displaystyle \int \frac{dx}{\sqrt{x^2 - a^2}} = \text{Arccosh } \frac{x}{a}$

84. $\displaystyle \int \sqrt{x^2 - a^2}\, dx = \frac{x}{2}\sqrt{x^2 - a^2} - \frac{a^2}{2} \text{ Arccosh } \frac{x}{a}$

85. $\displaystyle \int \frac{\sqrt{x^2 - a^2}}{x}\, dx = \sqrt{x^2 - a^2} - a \text{ Arcsec } \left| \frac{x}{a} \right|$

86. $\displaystyle \int \frac{\sqrt{x^2 - a^2}}{x^2}\, dx = \text{Arccosh } \frac{x}{a} - \frac{\sqrt{x^2 - a^2}}{x}$

87. $\displaystyle\int \frac{x^2}{\sqrt{x^2 - a^2}} \, dx = \frac{a^2}{2} \operatorname{Arccosh} \frac{x}{a} + \frac{x}{2}\sqrt{x^2 - a^2}$

88. $\displaystyle\int \frac{dx}{x\sqrt{x^2 - a^2}} \, dx = \frac{1}{a} \operatorname{Arcsec} \left|\frac{x}{a}\right|$

$$\sqrt{2ax - x^2}$$

89. $\displaystyle\int \frac{dx}{\sqrt{2ax - x^2}} = \operatorname{Arcsin}\left(\frac{x - a}{a}\right)$

90. $\displaystyle\int \sqrt{2ax - x^2} \, dx = \frac{x - a}{2}\sqrt{2ax - x^2} + \frac{a^2}{2} \operatorname{Arcsin}\left(\frac{x - a}{a}\right)$

91. $\displaystyle\int \frac{\sqrt{2ax - x^2}}{x} \, dx = \sqrt{2ax - x^2} + a \operatorname{Arcsin}\left(\frac{x - a}{a}\right)$

92. $\displaystyle\int \frac{x}{\sqrt{2ax - x^2}} \, dx = a \operatorname{Arcsin}\left(\frac{x - a}{a}\right)$

93. $\displaystyle\int \frac{dx}{x\sqrt{2ax - x^2}} = -\frac{1}{a}\sqrt{\frac{2a - x}{x}}$

Miscellaneous

94. $\displaystyle\int \sqrt{a^2 + b^2 x^2} \, dx = \frac{a^2}{2b}\left(\frac{bx\sqrt{a^2 + b^2 x^2}}{a^2} + \ln\left|\frac{\sqrt{a^2 + b^2 x^2} + bx}{a}\right|\right)$

95. $\displaystyle\int \frac{dx}{x^2 - a^2} = -\frac{1}{2a}\left(\ln|x - a| - \ln|x + a|\right)$

EXAMPLE 1: Compute $\displaystyle\int \frac{d\theta}{\tan\theta\sqrt{\sin^2\theta + 1}}$.

SOLUTION: If we express $\tan\theta$ as $\frac{\sin\theta}{\cos\theta}$ we have

$$\int \frac{\cos\theta \, d\theta}{\sin\theta\sqrt{\sin^2\theta + 1}}.$$

Letting $x = \sin\theta$, we obtain

$$\int \frac{\cos\theta \, d\theta}{\sin\theta\sqrt{\sin^2\theta + 1}} = \int \frac{dx}{x\sqrt{x^2 + 1}} + C.$$

Using formula 74, we see that

$$\int \frac{dx}{x\sqrt{x^2+1}} = \ln\left|\frac{x}{\sqrt{x^2+1}+1}\right| + C.$$

Thus

$$\int \frac{d\theta}{\tan\theta\sqrt{\sin^2\theta+1}} = \ln\left|\frac{\sin\theta}{\sqrt{\sin^2\theta+1}+1}\right| + C. \qquad \blacksquare$$

EXAMPLE 2: Compute $\displaystyle\int \frac{dx}{\sqrt{e^x+1}}$.

SOLUTION: If we let $u = e^x$, then $du = e^x dx$ or, equivalently, $\frac{du}{u} = dx$. Thus,

$$\int \frac{dx}{\sqrt{e^x+1}} = \int \frac{du}{u\sqrt{u+1}} + C.$$

Using formula 69, we see that

$$\int \frac{du}{u\sqrt{u+1}} = \ln\left|\frac{\sqrt{u+1}-1}{\sqrt{u+1}+1}\right| + C.$$

So

$$\int \frac{dx}{\sqrt{e^x+1}} = \ln\left|\frac{\sqrt{e^x+1}-1}{\sqrt{e^x+1}+1}\right| + C. \qquad \blacksquare$$

EXERCISES 7.9

In Exercises 1–17 evaluate the integrals.

1. $\int x^2 e^{2x}\ dx$.

2. $\displaystyle\int \frac{x}{\sqrt{2x+3}}\ dx$.

3. $\int x\ln(5x)\ dx$.

4. $\int \sin^3(2x)\cos^4(2x)\ dx$.

5. $\int \sqrt{2e^{3x}-e^{4x}}\ dx$.

6. $\displaystyle\int \frac{\tan x}{\sqrt{9-\sec^2 x}}\ dx$.

7. $\int e^x \operatorname{Arccsc}(3e^x)\ dx$.

8. $\displaystyle\int \frac{\ln(3\sqrt{x})}{x}\ dx$.

9. $\int \sqrt{4-e^{2x}}\ dx$.

10. $\int \sqrt{4+e^{2x}}\ dx$.

11. $\int \sec^5\theta\tan\theta\ln(2\sec\theta)\ d\theta$.

12. $\int \sin(2x)e^{3\sin x}\ dx$.

13. $\int (x+1)^2 e^x\ dx$.

14. $\int (x+1)^2\sin(3x)\ dx$.

15. $\displaystyle\int \frac{e^x}{\sqrt{1+e^{2x}}}\ dx$.

16. $\int \sqrt{1+(x+1)^2}\ dx$.

17. $\int \sqrt{x^2+2x+3}\ dx$.
    (Hint: Complete the square.)

In a later chapter, we will show that if $f(x)$ is a differentiable function, then the length of the graph of f over the interval $[a,b]$ is given by $\int_a^b \sqrt{1+(f'(x))^2}\ dx$. In Exercises 18–20 find the length of the graphs of the parabolas.

18. $f(x) = x^2 \quad 1 \le x \le 2$.

19. $f(x) = x^2 + 2x \quad 0 \le x \le 1$.

20. $f(x) = x^2 - 4x + 1 \quad -1 \le x \le 1$.

Chapter 8

Applications of Separation of Variables

8.1 Separation of Variables and Exponential Growth

In this Chapter we exploit separation of variables to solve differential equations of the form $f(y)y' = f(x)$. We solved this type of equation in Section 6.2 by assuming that y was a function of x and integrating both sides with respect to x. We now apply this method to some equations whose solutions involve the transcendental functions we developed in Chapter 7.

EXAMPLE 1: Solve $y' = 3xy$, $y(0) = 2$.

SOLUTION: We rewrite the equation as $y'(x)/y(x) = 3x$. Integrating both sides with respect to x, we have

$$\ln|y(x)| = \frac{3x^2}{2} + C$$

$$e^{\ln|y(x)|} = e^{\frac{3x^2}{2}+C}$$

$$|y(x)| = e^C e^{\frac{3x^2}{2}}$$

$$y(x) = \pm e^C e^{\frac{3x^2}{2}}.$$

If we let $A = \pm e^C$, we have

$$y(x) = Ae^{3x^2/2}.$$

387

Now we use the fact that $y(0) = 2$ to get $A = 2$. The solution is

$$y(x) = 2e^{3x^2/2}.$$ ∎

EXAMPLE 2: Solve $(2x + 1)y' = y - 3$, $y(1) = 2$.

SOLUTION: Separating variables, we have

$$\frac{y'(x)}{y(x) - 3} = \frac{1}{2x + 1}.$$ (1)

Integrating both sides yields

$$\ln |y(x) - 3| = \frac{1}{2} \ln |2x + 1| + C$$

$$e^{\ln |y(x)-3|} = e^{\frac{1}{2} \ln |2x+1| + C}$$

$$= e^C e^{\ln \sqrt{|2x+1|}}$$

$$|y(x) - 3| = e^C \sqrt{|2x + 1|}.$$

Letting $A = \pm e^C$, we have

$$y(x) - 3 = A\sqrt{|2x + 1|}, \text{ or } y(x) = 3 + A\sqrt{|2x + 1|}.$$

Since $y(1) = 2$, we have that

$$2 = y(1) = 3 + A\sqrt{3}, \text{ or } A = -\frac{1}{\sqrt{3}}.$$

Our solution is

$$y(x) = 3 - \frac{\sqrt{|2x + 1|}}{\sqrt{3}}.$$

We cannot divide by zero in Equation (1); hence, our domain cannot contain $x = -\frac{1}{2}$. Our initial condition implies that $x = 1$ is in the domain. Thus the set $x > -\frac{1}{2}$ is our solution domain. This permits us to drop the absolute value sign, and our solution is

$$y(x) = 3 - \frac{1}{\sqrt{3}}\sqrt{(2x + 1)}, \; x > -\frac{1}{2}.$$

Why don't we have to worry about $y = 3$ in Equation (1)?   ∎

EXAMPLE 3: Solve $(2x + 1)y' = y - 3$, $y(-1) = 2$.

SOLUTION: We start out exactly as above to find that the solution is of the form

$$y(x) = 3 + A\sqrt{|2x + 1|}.$$

As before, we cannot divide by zero, so our domain cannot contain $x = -\frac{1}{2}$. This time our initial condition implies that $x = -1$ is in the domain. Thus, the set $x < -\frac{1}{2}$ is our solution domain. In this region, $|2x + 1| = -2x - 1$, and we have

$$y(x) = 3 + A\sqrt{-(2x + 1)}, \ x < -\frac{1}{2}.$$

Using our initial condition, we obtain

$$2 = y(-1) = 3 + A\sqrt{1} = 3 + A.$$

This gives us $A = -1$, and our solution is

$$y(x) = 3 - \sqrt{-(2x + 1)}, \ x < -\frac{1}{2}.$$

Exponential Rates of Change

Suppose that $y(t)$ measures a quantity such as the speed of an object, the number of bacteria in a culture, the amount of a radioactive substance remaining in an object, or the amount of money in a savings account. We are interested in quantities that change at a rate proportional to the current quantity. That is, we are interested in quantities that satisfy an equation of the form

$$y'(t) = ky(t). \tag{2}$$

Such quantities are said to satisfy the *exponential rate of change model*. Equations of this form represent a simple, yet important, class of separable differential equations.

We assume that we have a measurement for y at time t_0. The measured quantity $y(t_0) = y_0$ is called the *initial condition*.

EXAMPLE 4: Recall Example 9 in Section 4.5. If $y(t)$ denotes the radioactive content of a mass at time t, then the rate of radioactive decay is proportional to the current radioactive content. The number k in Equation (2) depends on the material and the scale of time. Since the radioactive content is decreasing, the rate of change is negative. In this case, the constant k of Equation (2) is negative. ∎

With the techniques available in Chapter 4, we were forced to use a Maclaurin polynomial to approximate our solution. We can do much better now. Separating variables, Equation (2) can be written in the form

$$\frac{y'(t)}{y(t)} = k.$$

Integrating both sides yields

$$\ln |(y(t)| = kt + C \quad \text{or} \quad e^{\ln |(y(t)|} = e^{kt+C} = e^{kt}e^{C}.$$

We now have

$$|y(t)| = e^{kt}e^{C} \quad \text{or} \quad y(t) = (\pm e^{C})e^{kt}.$$

This can be rewritten as

$$y(t) = Ae^{kt}.$$

When $t = 0$, $y(0) = A = y_0$. Thus

$$y(t) = y_0 e^{kt}. \tag{3}$$

EXAMPLE 5: Under laboratory conditions, a particular type of bacterial culture is known to double its size each day. If $y(t)$ denotes the number of organisms in the culture at time t, then y is known to satisfy the exponential rate of change model. At the start of an experiment there are 2000 organisms in the culture. What is the count 15 hr later?

SOLUTION: We are given that y satisfies the equation

$$y(t) = Ae^{kt}, \ y(0) = 2000, \ \text{and} \ y(24) = 4000.$$

Substituting $y(0) = 2000$ into $y(t) = Ae^{kt}$, we learn that $A = 2000$, so we have that $y(t) = 2000e^{kt}$. Using the fact that $y(24) = 4000$, we get

$$4000 = 2000e^{24k} \quad \text{or} \quad \frac{\ln(2)}{24} = k.$$

This gives us that

$$y(t) = 2000e^{t\ln(2)/24}.$$

Thus

$$y(15) = 2000e^{(\ln(2))(15)/24} \approx 3084.$$

EXAMPLE 6: The *half-life* of a radioactive material is the time required for half of the radioactive content to decay. In Example 4 we showed that the rate of change of radioactive content is exponential. That is, $y(t) = y_0 e^{kt}$, where y_0 is the radioactive content at time $t = 0$. The half-life of a radioactive material is the solution to

$$y(t) = \frac{y(0)}{2} = \frac{y_0}{2} \quad \text{or} \quad \frac{y_0}{2} = y_0 e^{kt}.$$

Solving for t, we get

$$\text{half-life} = \frac{\ln\left(\frac{1}{2}\right)}{k} = -\frac{\ln(2)}{k}.$$ ∎

EXAMPLE 7: Carbon 14 has a half-life of approximately 5700 yr. The remains of a bison were found to have lost 15% of their original Carbon 14 content. Approximately how long ago did the bison die?

SOLUTION: We know that if we take $t = 0$ as the time of the bison's death, then the carbon content is given by Equation (3) with

$$k = -\frac{\ln(2)}{\text{half-life}} = -\frac{0.693147}{5700} = -1.22 \times 10^{-4}.$$

We want to solve the equation

$$0.85y_0 = y_0 e^{-1.22t \times 10^{-4}}.$$

Thus

$$t = \frac{\ln 0.85}{-1.22 \times 10^{-4}} \approx 1{,}334 \text{ yr.}$$ ∎

Working From Experimental Data

Suppose that we have been observing the growth of an organism, and at times $t_0, t_1, \ldots t_n$, we have measured the mass of the organism. If we let $M(t)$ denote the mass at time t, then we have $M(t_0) = M_0$, $M(t_1) = M_1$, ... , $M(t_n) = M_n$. We want to see if the growth is exponential. That is, can we find numbers A and k so that $M(t) = Ae^{kt}$? One approach would be to look at the ratios

$$\frac{M(t_1)}{M(t_0)} = e^{k(t_1-t_0)}, \ \frac{M(t_2)}{M(t_0)} = e^{k(t_2-t_0)}, \ \ldots, \ \frac{M(t_n)}{M(t_0)} = e^{k(t_n-t_0)}.$$

Taking the logarithm of both sides of each of these equations, we get

$$\ln\frac{M(t_1)}{M(t_0)} = k(t_1 - t_0), \ \ln\frac{M(t_2)}{M(t_0)} = k(t_2 - t_0), \ \ldots,$$

$$\ln\frac{M(t_n)}{M(t_0)} = k(t_n - t_0).$$

Since these masses are laboratory measurements, they are not exact. However, the $n+1$ measurements give us n estimates for the number k. If all of these estimates approximate a single number, then it seems likely that the growth is exponential. A reasonable choice for the number k would be the average of the k's obtained above.

EXAMPLE 8: We are observing the growth of an organism. Let $M(t)$ denote the mass of the organism. We have measured the mass at times t_0, t_1, ... t_8 to obtain the following data:

i	0	1	2	3	4	5	6	7	8
t_i	0	1	3	4	5	8	12	14	15
$M(t_i)$	2.5	3.73	8.3	12.3	18.5	61.3	303	676	1007
$\frac{M(t_i)}{M(t_0)}$		1.492	3.32	4.92	7.4	24.52	121.2	270.4	402.8
$\ln\left(M(t_i)/M(t_0)\right)$		0.4	1.2	1.593	2.001	3.199	4.797	5.6	5.998
$\frac{\ln(M(t_i)/M(t_0))}{t_i - t_0}$		0.4001	0.4	0.3983	0.4003	0.3999	0.3998	0.4	0.3999

Each of the numbers in the last row represents an estimate for k. Averaging, we take $k = 0.3998$. We model the mass by $M(t) = Ae^{0.3998t}$. To choose A, we again use our measurements. Since $A = M(t)/e^{0.3998t}$, we have nine measurements to work with, and we should use all of our data. We build a new chart.

i	0	1	2	3	4	5	6	7	8
t_i	0	1	3	4	5	8	12	14	15
$M(t_i)$	2.5	3.73	8.3	12.3	18.5	61.3	303	676	1007
$e^{0.3998t_i}$	1	1.4915	3.3181	4.9491	7.3817	24.493	121.22	269.67	402.22
$M(t_i)/e^{0.3998t_i}$	2.5	2.5008	2.5014	2.4853	2.5062	2.5027	2.4996	2.5068	2.5036

Each of the numbers in the last row is an estimate for A, and the average of these values is 2.5007. Taking $A = 2.5$, the mass in 20 days would be

$$M(20) \approx 2.5e^{(0.3998)(20)} \approx 7452. \qquad \blacksquare$$

EXERCISES 8.1

In Exercises 1–8, solve the following differential equation and give the solution domain.

1. $y'(t) = 2y(t)$, $y(0) = 10$.

2. $y'(t) = 2y(t)$, $y(0) = -10$.

3. $y'(t) = 2y(t) + 3$, $y(0) = 10$.

4. $y'(t) = 4y(t) - 6$, $y(0) = -2$.

5. $x^2 y' = 4y$, $y(1) = 3$.

6. $e^x y' = e^y$, $y(2) = 0$.

7. $(x - 1)y' = y$, $y(0) = -2$.

8. $(2x + 4)y' = y - 1$, $y(0) = 5$.

9. Explain why we cannot use separation of variables to solve

$$(x - 2)y' = y - 3, \quad y(0) = 3.$$

In Exercises 10 and 11, use the given conditions to find k and solve the equation.

10. $y'(t) = ky(t)$, $y(0) = 2$, and $y(2) = 4$.

11. $y'(t) = ky(t)$, $y(0) = 4$, and $y(2) = 2$.

12. Compare the graphs of the solution for $y'(t) = 2y(t)$, $y(0) = C$ for $C = -1$, 0, 1, 2, and 3.

13. Compare the graphs of the solution for $y'(t) = ky(t)$, $y(0) = 1$ for $k = -1$, 0, 1, 2, and 3.

In Exercises 14 and 15, a bacterial culture is known to grow exponentially with the constant of proportionality $k = 3$. Time is measured in hr.

14. How long will it take for the culture's population to double?

15. If the original population is 100, what will the population be in 4 hr?

In Exercises 16 and 17, a bacterial culture is known to grow exponentially.

16. If the culture's population doubles every hr, what is the constant k?

17. If the original population is 100, what will the population be in 4 hr?

18. The remains of the bison of Example 7 were found to have lost 25% of their original Carbon 14 content. Approximately how long ago did the bison die?

19. If the half-life of a radioactive material is 100 yr, what percent of the original material would be left in 6 mo?

20. If a material decays by 25% in 1 yr, what is its half-life?

In Exercises 21–24, $P(t)$ denotes the amount of money invested in a savings account at time t, measured in years. The account is said be compounded continuously if $P(t)$ satisfies the equation $P'(t) = IP(t)$. (The rate of growth of the principle is proportional to the principle.) The number I, converted to a decimal, is the annual interest rate. A thousand dollars is put into the account.

21. If the annual interest rate is 3% ($I = 0.03$), how much money will be in the account at the end of 5 yr?

22. If the annual interest rate is 10%, how much money will be in the account at the end of 5 yr?

23. What must the interest rate be if you want to double your money in 10 yr?

24. Suppose the annual interest rate is 3% ($I = 0.03$) and an additional $1,000 is put into the account at the end of the first year. How much money will be in the account at the end of 5 yr?

The intensity of light traveling in a medium is known to decay exponentially as a function of depth. In Exercises 25 and 26, the intensity of light 10 m below the surface of a lake is half the intensity at the surface.

25. What is the constant of proportionality?

26. If I_0 is the intensity at the surface, what is the intensity at a depth of 5 m?

The population of a town is growing exponentially. If $P(t)$ is the population of the town at time t, then the growth of the town satisfies the equation $P'(t) = kP(t)$. The census is taken every five years with the following results:

Year	1970	1975	1980
Population	30,000	38,521	49,463
Year	1985	1990	1995
Population	63,510	81,548	104,700

Use this data in Exercises 27–30.

27. Find k, the annual rate of growth of the town.

28. What will be the population in 2010?

29. What was the population in 1960?

30. How long does it take for the population to double?

In Exercises 31–34, use the following data obtained from a culture which grows exponentially:

Time after culture taken (in hr)	1	2	4	7
Population count	135	182	332	817

31. Find the constant of proportionality.

32. What was the population when the culture was taken?

33. What will the population count be 24 hr after the culture was taken?

34. How long does it take for the population to double?

Use the following data in Exercises 35–38, which were obtained from an exponentially decaying culture count:

Time after culture taken (in days)	1	3	4	6
Population count	1482	813	602	331

35. Find the constant of proportionality.

36. What was the population when the culture was taken?

37. What will the population count be 24 days after the culture was taken?

38. How long does it take for the population to decrease to $\frac{1}{2}$ of its original size?

8.2   Equations of the Form $y' = ky + b$

Suppose that $y(t)$ measures a quantity as a function of time, that we have a measurement for y at time t_0, and that the rate of change of y satisfies the equation

$$y'(t) = ky(t) + b, \tag{4}$$

where k and b are constants.

We use separation of variables and rewrite Equation (4) as

$$\frac{y'(t)}{ky(t) + b} = 1. \tag{5}$$

Integrating both sides of Equation (5) with respect to t, we obtain

$$\frac{1}{k} \ln |ky(t) + b| = t + C.$$

Solving for $y(t)$, we get

$$|ky(t) + b| = e^{k(t+C)},$$

or

$$y(t) = \frac{1}{k} \left(-b \pm e^{k(t+C)} \right) = \frac{1}{k} \left(-b \pm e^{kt} e^{kC} \right).$$

If we let $A = \pm \frac{e^{kC}}{k}$, then our solution to Equation (4) is

$$y(t) = Ae^{kt} - \frac{b}{k}. \tag{6}$$

The method for solving Equation (4) is a straightforward application of separation of variables. We recommend that you solve equations of this type by using the method rather than trying to remember the solution given by Equation (6). Note that the number A can be positive or negative according to the initial condition $y_0 = y(t_0)$. ∎

EXAMPLE 1: Solve $y'(x) = 2y(x) + 3, \; y(2) = 1.$

SOLUTION: Rewriting the equation so that the variables are separated, we obtain

$$\frac{y'(x)}{2y(x) + 3} = 1.$$

Integrating both sides, we have

$$\frac{1}{2} \ln |2y(x) + 3| = x + C \quad \text{or} \quad y = \frac{1}{2}\left(-3 \pm e^{2x} e^{2C}\right) = \frac{1}{2}(-3 + Ae^{2x}).$$

The initial condition $y(2) = 1$ gives us that

$$1 = \frac{1}{2}(-3 + Ae^4) \quad \text{or} \quad A = 5e^{-4}.$$

Our solution is

$$y(x) = \frac{1}{2}\left(-3 + 5e^{-4}e^{2x}\right) = \frac{1}{2}\left(-3 + 5e^{2x-4}\right). \qquad ∎$$

LR Series Circuits

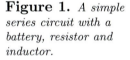

Resistor

Battery

Inductor

Figure 1. *A simple series circuit with a battery, resistor and inductor.*

Suppose that we have a simple circuit that is a closed loop composed of a battery, a resistor, and an inductor as illustrated in Figure 1.

Such a circuit is called an *LR series circuit*. The arrows indicate the direction of the flow of charge. The *inductance* (of the inductor) is measured in henrys (denoted by H), the *resistance* (of the resistor) is measured in ohms (denoted by Ω) and the *electromotive force* (of the battery) is measured in volts (denoted by V). The voltage generated by the battery is constant. If $q(t)$ denotes the charge density at a cross section, then the *current* at that cross section is defined to be $i(t) = q'(t)$. Current is measured in amperes (A), and electric charge is measured in coulombs.

If we pick two points a and b in the wire, then we denote by $E(a, b)$ the change in *electric potential* as the charge moves from point a to point b in the circuit. E is measured in volts. We assume that $E(a, b) = 0$ if there is no inductor or resistor between a and b. If there is only a resistor between a and b, then the voltage will drop by Ri volts, where R is the measured resistance. If there is only an inductor between a and b, then the voltage will drop by Ldi/dt volts, where L is the inductance. If there is only a battery between

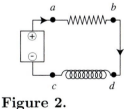

a b

c d

Figure 2.
$E(a, b) = -Ri,$
$E(b, d) = 0,$
$E(d, c) = -Ldi/dt$ *and*
$E(c, a) = V.$

a and b, then $E(a, b) = V$, where V is the voltage of the battery. See Figure 2. **Kirchhoff's Law** states that as the current goes all the way around the circuit, the total change in electric potential is 0.

Kirchhoff's Law tells us that

$$E(a, a) = V - Ri - L\,di/dt = 0. \tag{7}$$

If $L = 0$, then $i(t) = \frac{V}{R}$. Otherwise, Equation (7) can be written as

$$Li'(t) = V - Ri(t).$$

Separating variables, we have that

$$L\frac{i'(t)}{V - Ri(t)} = 1.$$

Integrating both sides gives us that

$$-\frac{L}{R}\ln|V - Ri(t)| = t + c \quad \text{or} \quad \ln|V - Ri(t)| = -t\frac{R}{L} + k.$$

Exponentiating both sides, we have that

$$V - Ri(t) = Ae^{-tR/L}, \quad \text{where} \quad A = \pm e^k.$$

We start our clock at the instant the battery is connected to the circuit. This gives us the initial condition $i(0) = 0$. We use this to solve for A, obtaining $A = V$. The equation we arrive at is

$$i(t) = \frac{1}{R}\left(V - Ve^{-tR/L}\right) = \frac{V}{R}\left(1 - e^{-tR/L}\right). \qquad \blacksquare$$

Newton's Law of Cooling

Newton's law of cooling states that if an object with temperature T_0 is placed in a medium of uniform temperature, then the rate of change of the temperature of the object (as a function of time) is proportional to the difference between the temperature of the object and the temperature of the medium.

EXAMPLE 2: After a potato has been heated to $250°F$, it is removed from the oven and placed in a room that is held at $70°$. The potato has cooled to $200°F$ in 5 min. What will its temperature be in 15 min?

SOLUTION: The rate of change of the temperature $T(t)$ of the potato is given by

$$T'(t) = k(T(t) - 70), \ T(0) = 250, \ \text{and} \ T(5) = 200.$$

Separating the variables, we get

$$\frac{T'(t)}{T(t) - 70} = k$$

$$\ln|T(t) - 70| = kt + C$$

$$T(t) = 70 \pm e^{kt+C}$$

$$= 70 + Ae^{kt}, \quad \text{where } A = \pm e^C.$$

Using $T(0) = 250$, we have $250 = 70 + A$, or $A = 180$. Thus

$$T(t) = 70 + 180e^{kt}.$$

We use $T(5) = 200$ to find k:

$$200 = 70 + 180e^{5k}$$

$$130 = 180e^{5k}$$

$$5k = \ln\left(\frac{13}{18}\right)$$

$$k = \frac{\ln\left(\frac{13}{18}\right)}{5}$$

$$\approx -0.065.$$

The temperature of the potato at time t minutes after the potato is taken from the oven is

$$T(t) = 70 + 180e^{-0.065t}.$$

We are asked to find $T(15)$.

$$T(15) = 70 + 180e^{(-0.065)(15)} \approx 142.5°. \qquad \blacksquare$$

Projectile Motion With Air Resistance

EXAMPLE 3: One model of a projectile moving in a medium of constant density will experience a resistive force that is proportional to its speed. The constant of proportionality depends on the shape of the object and the medium in which it is traveling. Suppose that an object with mass m is moving vertically. The coefficient of air resistance is k. The force due to gravity is $-mg$. We have the following equation describing the motion:

$$ma(t) = -mg - kv(t) \quad \text{or} \quad v'(t) = -g - \frac{k}{m}v(t).$$

If we let $\rho = \frac{k}{m}$, then this equation can be rewritten as

$$-\frac{v'(t)}{g + \rho v(t)} = 1.$$

Integrating both sides with respect to time, we have

$$-\frac{1}{\rho} \ln |g + \rho v(t))| = t + C$$

or

$$g + \rho v(t) = \pm e^{-\rho(t+C)} = Ae^{-\rho t}.$$

Solving for v, we obtain

$$v(t) = \frac{1}{\rho}\left(Ae^{-\rho t} - g\right), \quad \text{where } \rho = \frac{k}{m}. \tag{8}$$

Since $\rho > 0$, as t gets large the number $Ae^{-\rho t}$ gets small. This implies that the velocity approaches $-\frac{g}{\rho}$. The quantity $\lim_{t \to \infty} v(t) = \lim_{t \to \infty} \frac{1}{\rho}\left(Ae^{-\rho t} - g\right) = -\frac{g}{\rho}$ is called the *terminal velocity* of the object.   ∎

EXAMPLE 4: A certain 100 kg mass is known to have a terminal velocity of -150 m/sec. Find the constant k of air resistance.

SOLUTION: $-\frac{9.8}{\rho} = -\frac{9.8}{k/100} = -\frac{980}{k} = -150$. Thus, $k = \frac{980}{150} \approx 6.53$.   ∎

EXAMPLE 5: A 0.1 kg ball is thrown straight up with an initial speed of 50 m/sec. The constant of resistance due to the atmosphere is known to be $k = 0.02$.

  (a) How high will the ball go?

  (b) How long will it take to return to the ground?

  (c) How high would the ball go if there were no air resistance?

  (d) How long would it take to reach the ground if there were no air resistance?

SOLUTION: $\rho = \frac{k}{m} = \frac{0.02}{0.1} = 0.2$. Using Equation (8) we have

$$v(t) = \frac{1}{0.2}\left(Ae^{-0.2t} - 9.8\right) = 5(Ae^{-0.2t} - 9.8).$$

Since $v(0) = 50 = 5(A - 9.8)$, we have that $A = 19.8$. Thus

$$v(t) = 5\left(19.8e^{-0.2t} - 9.8\right) \text{ m/sec.}$$

(a) The ball will reach its maximum height when
$$v(t) = 5\left(19.8e^{-0.2t} - 9.8\right) = 0 \text{ m/sec.}$$

$$
\begin{aligned}
5\left(19.8e^{-0.2t} - 9.8\right) &= 0 \\
19.8e^{-0.2t} &= 9.8 \\
e^{-0.2t} &= \frac{9.8}{19.8} \\
t &= \frac{\ln\left(\frac{9.8}{19.8}\right)}{-0.2} \approx 3.516 \text{ s.}
\end{aligned}
$$

Now, let $y(t)$ denote the height of the ball at time t. Then

$$
\begin{aligned}
y(t) &= \int 5\left(19.8e^{-0.2t} - 9.8\right) dt \\
&= 5\left(-99e^{-0.2t} - 9.8t\right) + C.
\end{aligned}
$$

We use $y(0) = 0$ to obtain $C = 495$, so

$$y(t) = -495e^{-0.2t} - 49t + 495 \text{ m.} \tag{9}$$

Our answer is $y(3.516) = 77.69$ m.

(b) In order to determine when the ball hits the ground, we use Equation (9) to find the time t so that $y(t) = 0$. Unfortunately, we cannot solve this in closed form. First, we obtain a sketch of the graph of y to get an approximation for our root and then use one of our numerical methods such as Newton's method, or a numerical equation solving feature in a calculator or software package. In Figure 3, we see that t_0 is between 8 sec and 8.5 sec. The NSolve command in Mathematica gives $t = 8.10476$. Notice that when air resistance is considered, the ball takes more time to come down than to go up!

(c) Neglecting air resistance, the equations of motion are

$$v(t) = -9.8t + 50 \text{ m/sec}, \quad \text{and} \quad y(t) = -4.9t^2 + 50t \text{ m}.$$

The ball will reach its maximum height when $v(t) = 0$ m/sec, or when $t = \frac{50}{9.8} \approx 5.1$ sec, at which time $y = 127.6$ m.

(d) To find when the ball reaches the ground, we solve $y(t) = -4.9t^2 + 50t = 0$ m, to obtain $t = \frac{50}{4.9} \approx 10.2$ sec.

In Figure 4, we display sketches of y in both cases. Air resistance makes a substantial difference! ■

We now consider a more general case.

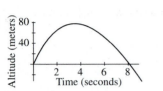

Figure 3. *A graph of y over the interval $[0, 9]$.*

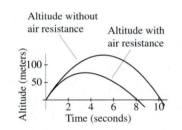

Figure 4. *Altitude as a function of time with and without air resistance.*

EXAMPLE 6: An object moves in a medium in 3–space with initial velocity $\vec{v}_0 = (v_x(0), v_y(0), v_z(0))$. A constant external force $\vec{f} = (f_x, f_y, f_z)$ acts on the object, and the medium provides a resistive force proportional to the object's speed. The forces acting on the object are sketched in Figure 5. Find the velocity of the object as a function of time.

SOLUTION: Let k denote the constant of proportionality for the resistive force, and let $\vec{v}(t) = (v_x(t), v_y(t), v_z(t))$ denote the velocity at time t. The resistive force acts in a direction opposite the velocity, so a unit vector pointing in the direction of the resistive force is $-(1/\|\vec{v}(t)\|)\vec{v}(t)$. The resistive force will be

$$\vec{F}_r(t) = -k\|\vec{v}(t)\|\frac{\vec{v}(t)}{\|\vec{v}(t)\|} = -k \begin{pmatrix} v_x(t) \\ v_y(t) \\ v_z(t) \end{pmatrix}.$$

The force equation for the object is

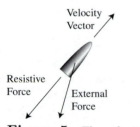

Figure 5. *The velocity vector and the forces acting on the object.*

$$m\vec{a}(t) = m\vec{v}'(t) = m \begin{pmatrix} v_x'(t) \\ v_y'(t) \\ v_z'(t) \end{pmatrix} = -k \begin{pmatrix} v_x(t) \\ v_y(t) \\ v_z(t) \end{pmatrix} + \begin{pmatrix} f_x \\ f_y \\ f_z \end{pmatrix}$$

$$= \begin{pmatrix} -kv_x(t) + f_x \\ -kv_y(t) + f_y \\ -kv_z(t) + f_z \end{pmatrix}.$$

This leaves us with three equations that can be solved independently:

$$v_x'(t) = -\rho v_x(t) + \frac{f_x}{m},$$

$$v_y'(t) = -\rho v_y(t) + \frac{f_y}{m},$$

$$v_z'(t) = -\rho v_z(t) + \frac{f_z}{m}, \quad \text{where } \rho = \frac{k}{m}. \qquad \blacksquare$$

EXAMPLE 7: A 10 kg projectile moves with an initial velocity of $\vec{v}_0 = (-50, 10, 0)$ m/sec. Its air resistance coefficient is $k = 2$, and the external force $f = (0, 0, -98)$ N is due to gravity. Find the velocity of the projectile as a function of time. What is its terminal velocity?

SOLUTION: $\rho = \frac{k}{m} = 0.2$, so we have the equations

$$v_x'(t) = -\rho v_x(t) + \frac{f_x}{m} = -0.2 v_x(t).$$

$$v_y'(t) = -\rho v_y(t) + \frac{f_y}{m} = -0.2 v_y(t).$$

$$v_z'(t) = -\rho v_z(t) + \frac{f_z}{m} = -0.2 v_z(t) - 9.8.$$

To find $v_x(t)$ we have

$$\frac{v_x'(t)}{v_x(t)} = -0.2 \quad \text{or} \quad v_x(t) = A_x e^{-0.2t}.$$

Since $v_x(0) = -50$, we have $A_x = -50$ and

$$v_x(t) = -50 e^{-0.2t} \text{m/sec}.$$

In exactly the same way, we find

$$v_y(t) = 10 e^{-0.2t} \text{m/sec}.$$

Finding $v_z(t)$ requires a little more work. We have

$$v_z'(t) = -0.2 v_z(t) - 9.8.$$

Separating variables, we obtain

$$-\frac{v_z'(t)}{0.2 v_z(t) + 9.8} = 1.$$

Integrating, we get

$$\ln |0.2 v_z(t) + 9.8| = -0.2t + C$$

or

$$v_z(t) = 5(A_z e^{-0.2t} - 9.8).$$

Using $v_z(0) = 0$, we have $A_z = 9.8$, so

$$v_z(t) = 49(e^{-0.2t} - 1) \text{ m/sec.}$$

The terminal velocity is

$$\lim_{t \to \infty} \vec{v}(t) = \lim_{t \to \infty} \begin{pmatrix} -50e^{-0.2t} \\ 10e^{-0.2t} \\ 49(e^{-0.2t} - 1) \end{pmatrix} = \begin{pmatrix} 0 \\ 0 \\ -49 \end{pmatrix} \text{ m/sec.} \qquad \blacksquare$$

EXAMPLE 8: The projectile of the previous example has as its initial position $\vec{r}_0 = (500, 1000, 2500)$ m. Find its position as a function of time.

SOLUTION: We know from the previous problem that the projectile's velocity is given by

$$\vec{v}(t) = \begin{pmatrix} -50e^{-0.2t} \\ 10e^{-0.2t} \\ 49(e^{-0.2t} - 1) \end{pmatrix}.$$

$$\vec{r}(t) = \int \vec{v}(t)\, dt = \int \begin{pmatrix} -50e^{-0.2t} \\ 10e^{-0.2t} \\ 49(e^{-0.2t} - 1) \end{pmatrix} dt$$

$$= \begin{pmatrix} 250e^{-0.2t} + C_x \\ -50e^{-0.2t} + C_y \\ 49(-5e^{-0.2t} - t) + C_z \end{pmatrix}.$$

Our initial conditions give us that

$$\vec{r}(0) = \begin{pmatrix} 250 + C_x \\ -50 + C_y \\ -245 + C_z \end{pmatrix} = \begin{pmatrix} 500 \\ 1000 \\ 2500 \end{pmatrix},$$

so

$$\begin{pmatrix} C_x \\ C_y \\ C_z \end{pmatrix} = \begin{pmatrix} 250 \\ 1050 \\ 2745 \end{pmatrix}.$$

Our position function is

$$\vec{r}(t) = \begin{pmatrix} 250e^{-0.2t} + 250 \\ -50e^{-0.2t} + 1050 \\ 49(-5e^{-0.2t} - t) + 2745 \end{pmatrix}. \qquad \blacksquare$$

EXAMPLE 9: The 0.1 kg ball of Example 5 is thrown at an angle of 60 deg with the horizontal on a level field, with an initial speed of 50 m/sec. See Figure 6. How high will the ball go, and how far away will it land?

SOLUTION: As in Example 5, $k = 0.02$ and $\rho = 0.2$. Let $\vec{v}(t) = (v_x(t), v_y(t))$ be the velocity function, and let $\vec{r}(t) = (x(t), y(t))$ be the position function. We have the initial conditions

$$\vec{v}(0) = \begin{pmatrix} v_x(0) \\ v_y(0) \end{pmatrix} = \begin{pmatrix} 50\cos\left(\frac{\pi}{3}\right) \\ 50\sin\left(\frac{\pi}{3}\right) \end{pmatrix} = \begin{pmatrix} 25 \\ 43.3 \end{pmatrix}$$

and

$$\vec{r}(0) = \begin{pmatrix} 0 \\ 0 \end{pmatrix}.$$

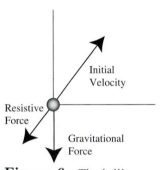

The equations of motion are

$$v'_x(t) = -\rho v_x(t) + f_x/m = -0.2v_x(t).$$

$$v'_y(t) = -\rho v_y(t) + f_y/m = -0.2v_y(t) - 9.8.$$

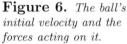

Figure 6. *The ball's initial velocity and the forces acting on it.*

Proceeding exactly as we did in the previous example, we get

$$v_x(t) = A_x e^{-0.2t} \text{ and } v_y(t) = \frac{1}{0.2}(A_y e^{-0.2t} - 9.8) = 5A_y e^{-0.2t} - 49.$$

Using our initial conditions, we get $A_x = 25$ and $A_y = 18.46$ so that

$$v_x(t) = 25e^{-0.2t} \text{ and } v_y(t) = 92.3e^{-0.2t} - 49.$$

The ball will reach its maximum altitude when $v_y(t) = 0$. Thus we solve

$$0 = 92.3e^{-0.2t} - 49$$

$$e^{-0.2t} = 49/92.3$$

to get

$$t = -5\ln\left(\frac{49}{92.3}\right) \approx 3.166.$$

We now integrate the velocity function to get the position function:

$$x(t) = \int 25e^{-0.2t}\,dt = -125e^{-0.2t} + C_x.$$

Using $x(0) = 0$ we find that $C_x = 125$ m. We have

$$x(t) = 125(1 - e^{-0.2t}).$$

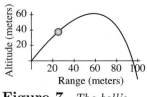

Figure 7. *The ball's trajectory.*

Note that, even if the ball were going over a cliff and t could become quite large, its range would be less than 125 m.

Turning to the y–coordinate of our position function, we have

$$y(t) = \int 92.3e^{-0.2t} - 49\, dt = -461.5e^{-0.2t} - 49t + C_y.$$

Setting $y(0) = 0$ gives us $C_y = 461.5$. Therefore,

$$y(t) = -461.5e^{-0.2t} - 49t + 461.5.$$

The ball's trajectory is sketched in Figure 7.

We are now ready to determine how high the ball will go, and where it will land. We have seen that the ball reaches its maximum altitude at $t = 3.166$. Thus the maximum height is

$$y(3.166) = -461.5e^{(-0.2)(3.166)} - 49(3.166) + 461.5 = 61.36 \text{ m}.$$

To find the range, we must first find the time $t \neq 0$ when $y(t) = 0$. (Why can't we assume that it takes the same amount of time to get down as it took to get to the top of its trajectory?)

We cannot solve the equation $y(t) = 0$ in closed form, but we can use Newton's method or an equation solver. In Figure 8, we sketch the graph of $y(t)$ over the time interval $[0, 8]$.

The graph of Figure 8 indicates $t_0 \approx 7$ sec. After three iterations of Newton's method, we have $t = 7.176286$. If we do one more iteration, t doesn't change in the first six decimal places and $y(7.176286) = -7 \times 10^{-13}$. Notice how much longer the ball takes to come down than it took going up. Why?

At time $t = 7.176$, $x(t) = 95.24$ m.   ■

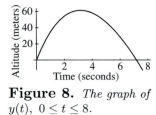

Figure 8. *The graph of $y(t)$, $0 \le t \le 8$.*

EXERCISES 8.2

1. An LR circuit has a 12 volt battery, a 10 ohm resistor, and a 0.01 henry inductor. Assuming that $i(0) = 0$, find $i(t)$.

In Exercises 2 and 3, $q(t)$ denotes the charge on a capacitor at time t. If we have a circuit with a battery of voltage V, a resistor of resistance R and a capacitor with capacitance C (C is measured in farads (F)), then the charge on the capacitor satisfies the equation $Rq'(t) + \frac{q(t)}{C} = V$. Solve for $q(t)$ in terms of R, C, and V.

2. Assume that $q(0) = 0$ coulombs.

3. Assume that $q(0) = 2$ coulombs.

4. A metal ball is heated to 300°F and then placed in a room held at 72°F. Two minutes later the ball has cooled to 156°. What will its temperature be 5 min after it is allowed to start cooling?

5. An object's temperature is initially 32°F. It is placed in a room that is held at 72°F, and two hr later its temperature is 50°F. How long will it take the object to warm up to 60°?

6. An object has been heated, and at time $t = 0$ it is placed in a room that is held at a temperature of $70°$F. At time $t = 5$ its temperature is $258°$, and at $t = 10$, the temperature of the object is $224°$. Given that its temperature function T satisfies the equation $T'(t) = k(T(t) - 70)$, what was the object's original temperature?

In Exercises 7–13, a 2 kg ball is shot straight up with an initial speed of 50 m/sec. The constant of resistance due to the atmosphere is known to be $k = 0.02$.

7. Find the velocity function $v(t)$.

8. Find the altitude function $y(t)$.

9. Determine how high the ball will go.

10. How long does it take the ball to return to the ground?

11. Use your calculator to sketch the graph of $y(t)$ from the time the ball is shot up until the time it reaches the ground.

12. Do Exercises 7–10 assuming there is no air resistance. Plot the altitude function (assuming no air resistance) and the function $y(t)$ on the same axis.

13. Plot the function $y(t)$ of Exercise 8 and the altitude function given in Equation 9 for the 0.1 kg ball of Example 5 on the same axis. Explain the differences in the graphs of the functions.

In Exercises 14–16, a 5 kg projectile moves with an initial velocity of $\vec{v}_0 = (-30, 10, 10)$ m/sec. Its air resistance coefficient is $k = 0.03$, and the external force $f = (0, 0, -49)$ N is due to gravity.

14. Find the projectile's velocity as a function of time.

15. What is its terminal velocity?

16. If the projectile's initial position is $(200, 1500, 2500)$ find its position as a function of time.

In Exercises 17–22, a 0.2 kg ball is thrown at an angle of 60 degrees with the horizontal on a level field with an initial speed of 50 m/sec. The constant of air resistance is $k = 0.01$.

17. Find the ball's velocity as a function of time.

18. Find its position as a function of time. (Place the coordinate axis with its origin at the point where the ball is thrown.)

19. How high will the ball go?

20. How far away will it land?

21. Draw the trajectory of the ball.

22. Repeat Exercises 17–21 neglecting air resistance. Compare the trajectory obtained by neglecting air resistance with the trajectory of Exercise 21.

23. **[PROJECT]:** *A ball is shot straight up with an initial speed of 50 m/sec. The constant of resistance due to the atmosphere is known to be $k = 0.2$. Let $T(m)$ denote the time when the velocity of the ball of mass m kg is zero and let $Y(m)$ denote the maximum altitude of a ball of mass m kg.*

   a. Show that $T(m)$ is the root of the equation $\left(\frac{10}{m} + 9.8\right) e^{-0.2t/m} = 9.8$.

   b. Find $T(m)$.

   c. Find $Y(m)$.

   d. Calculate $Y(m)$ for $m = \frac{n}{10}$, $n = 1, 2, \ldots$ until you find the first integer n so that $Y\left(\frac{n}{10}\right) > 25$. If you have done this correctly, $Y\left(\frac{n-1}{10}\right) \leq 25$.

   e. Letting $a = \frac{n-1}{10}$ and $b = \frac{n}{10}$, use the interval halving method to find a number m such that $|Y(m) - 25| < 10^{-5}$.

8.3   The Logistic Equation

In Chapter 10, we will learn that

$$\int \frac{dx}{ax + bx^2} = \frac{1}{a}\left(\ln|x| - \ln|a + bx|\right) + C. \tag{10}$$

Until we derive this formula, it is easy to verify by differentiation. In the meantime, it is useful in the more complicated growth problems that we wish to study here.

EXAMPLE 1: Use the Integration Formula (10) to calculate

$$\int \frac{\sin x}{4\cos x + 2\cos^2 x}\,dx.$$

SOLUTION:

$$\int \frac{\sin x}{4\cos x + 2\cos^2 x}\,dx \;=\; -\int \frac{du}{4u + 2u^2}\bigg|_{u=\cos x}$$

$$=\; -\frac{1}{4}\left(\ln|u| - \ln|4 + 2u|\right) + C\bigg|_{u=\cos x}$$

$$=\; -\frac{1}{4}\left(\ln|\cos x| - \ln|4 + 2\cos x|\right) + C.\ \blacksquare$$

EXAMPLE 2: Solve the differential equation $y'(x) = 5y(x) - 2y^2(x)$, $y(0) = -3$.

SOLUTION: Separating the variables, we obtain

$$1 = \frac{y'(x)}{5y(x) - 2y^2(x)}. \tag{11}$$

Integrating both sides with respect to x:

$$x + C \;=\; \int \frac{y'(x)}{5y(x) - 2y^2(x)}\,dx = \int \frac{du}{5u - 2u^2}\bigg|_{u=y(x)}$$

$$=\; \frac{1}{5}\left(\ln|u| - \ln|5 - 2u|\right)\bigg|_{u=y(x)} = \frac{1}{5}\ln\left(\frac{|y(x)|}{|5 - 2y(x)|}\right),$$

or

$$5x + k = \ln\left(\frac{|y(x)|}{|5 - 2y(x)|}\right).$$

Exponentiating both sides, we have

$$e^{5x+k} = \left| \frac{y(x)}{5 - 2y(x)} \right|.$$

Letting $A = \pm e^k$, we obtain

$$Ae^{5x} = \frac{y(x)}{5 - 2y(x)}.$$

We now use $y(0) = -3$ to find $A = -\frac{3}{11}$, so

$$-\frac{3}{11}e^{5x} = \frac{y(x)}{5 - 2y(x)}.$$

We finish the problem by solving for y. First, we cross multiply to obtain

$$11y(x) = -3e^{5x}\left(5 - 2y(x)\right) = -15e^{5x} + 6e^{5x}y(x).$$

$$y(x)(11 - 6e^{5x}) = -15e^{5x},$$

or

$$y(x) = -\frac{15e^{5x}}{11 - 6e^{5x}},$$

which can be rewritten as

$$y(x) = \frac{15}{6 - 11e^{-5x}}.$$

We should be careful about the domain of our solution. Our method of separation of variables works only as long as the denominator in Equation (11) is not zero. It is easily verified that our solution $y(x)$ is never zero, and $5 - 2y(x)$ is never zero. The only place where $y(x)$ is not defined is when $11e^{-5x} - 6 = 0$, which occurs when $x = -\frac{\ln(6/11)}{5} \approx 0.1212$. The domain for y will be the largest segment containing zero but not including $x = 0.1212$. Thus the domain of y is $(-\infty, 0.1212)$. ■

The exponential growth function models population growth quite well as long as there are no deaths due to disease. In a population where such deaths do occur, we might assume that the death rate due to disease is proportional to the number of two person interactions. If $P(t)$ is the population at time t, then the number of possible two person interactions is $\frac{P(t)(P(t)-1)}{2}$. The equation of population growth would be

$$\begin{aligned} P'(t) &= k_1 P(t) - k_2 P(t)\left(P(t) - 1\right) \\ &= (k_1 + k_2)P(t) - k_2 P^2(t) \\ &= c_1 P(t) - c_2 P^2(t). \end{aligned}$$

The *logistic equation* is defined as

$$P'(t) = c_1 P(t) - c_2 P^2(t). \tag{12}$$

Notice that Example 2 was just a special case of Equation (12), where $c_1 = 5$ and $c_2 = 2$.

Since we are considering population growth, we may assume that c_1, c_2, and $P(0)$ are all positive.

We now solve Equation (12). Proceeding exactly as in Example 2, we use separation of variables to obtain

$$\frac{P'(t)}{c_1 P(t) - c_2 P^2(t)} = 1. \tag{13}$$

Using the Integration Formula (10), we have:

$$
\begin{aligned}
t + C &= \int \frac{P'(t)}{c_1 P(t) - c_2 P^2(t)}\, dt \\
&= \int \frac{du}{c_1 u - c_2 u^2}\bigg|_{u = P(t)} \\
&= \frac{1}{c_1} \left(\ln|u| - \ln|c_1 - c_2 u|\right)\big|_{u = P(t)} \\
&= \frac{1}{c_1} \left(\ln|P(t)| - \ln|c_1 - c_2 P(t)|\right).
\end{aligned}
$$

Thus

$$\frac{1}{c_1} \left(\ln|P(t)| - \ln|c_1 - c_2 P(t)|\right)| = t + C.$$

This can be rewritten as

$$\ln\left(\frac{|P(t)|}{|c_1 - c_2 P(t)|}\right) = c_1 t + K.$$

Exponentiating both sides, we have

$$\left|\frac{P(t)}{c_1 - c_2 P(t)}\right| = e^{c_1 t + K}.$$

Letting $A = \pm e^K$,

$$\frac{P(t)}{c_1 - c_2 P(t)} = A e^{c_1 t}. \tag{14}$$

$$P(t) = A e^{c_1 t} \left(c_1 - c_2 P(t)\right).$$

Solving for $P(t)$, we obtain

$$P(t) = \frac{c_1 A}{Ac_2 + e^{-c_1 t}}. \tag{15}$$

If we are given the initial condition $P(0)$, then we can use Equation (14) to obtain A.

$$A = \frac{P(0)}{c_1 - P(0)c_2}. \tag{16}$$

Substituting this value for A into Equation (15), with some algebra, we obtain

$$P(t) = \frac{c_1 P(0)}{P(0)c_2 + (c_1 - P(0)c_2)\, e^{-c_1 t}}. \tag{17}$$

We now consider the domain of our solution $P(t)$. First of all, note that $P(t)$ is never zero. We need to determine if the denominator is ever zero. To this end, we set

$$0 = P(0)c_2 + (c_1 - P(0)c_2)\, e^{-c_1 t},$$

or

$$e^{-c_1 t} = -\frac{P(0)c_2}{c_1 - c_2 P(0)} = \frac{P(0)}{P(0) - \frac{c_1}{c_2}}. \tag{18}$$

If $P(0) < \frac{c_1}{c_2}$, then

$$\frac{P(0)}{P(0) - \frac{c_1}{c_2}} < 0.$$

In this case there is no solution to Equation (18). Hence, the denominator of (18) is never zero, and the domain of P will be $(-\infty, \infty)$.

If $P(0) > \frac{c_1}{c_2}$, we can solve Equation (18) for t. In this case, let

$$T = -\frac{1}{c_1} \ln\left(\frac{P(0)}{P(0) - \frac{c_1}{c_2}} \right). \tag{19}$$

Recall that we are assuming that 0 is in the domain of our solution, and that c_1, c_2, and $P(0)$ are all positive. Therefore, since $P(0) > \frac{c_1}{c_2}$,

$$\frac{P(0)}{P(0) - \frac{c_1}{c_2}} > 1 \text{ and } T < 0.$$

The domain for P is (T, ∞).

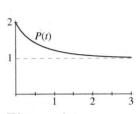

Figure 1.a
An example where $P(0) > \frac{c_1}{c_2}$. In this case P decreases toward the equilibrium population.

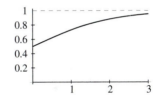

Figure 1.b
An example where $P(0) < \frac{c_1}{c_2}$. In this case P increases toward the equilibrium population.

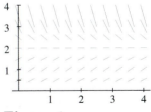

Figure 1.c
The direction field for the case where $c_1 = 2$ and $c_2 = 1$.

We have shown that if $P(0) > \frac{c_1}{c_2}$, then the domain of P is (T, ∞). If $P(0) < \frac{c_1}{c_2}$, then the domain of P is $(-\infty, \infty)$.

In either case, no matter what $P(0)$ is, we have from Equation (15) that

$$\lim_{t \to \infty} P(t) = \lim_{t \to \infty} \frac{c_1 A}{Ac_2 + e^{-c_1 t}} = \frac{c_1}{c_2}.$$

The number $\frac{c_1}{c_2}$ is called the *equilibrium population*. If $P(0)$ is greater than $\frac{c_1}{c_2}$, then $P(t)$ will decrease toward $\frac{c_1}{c_2}$. If $P(0)$ is less than $\frac{c_1}{c_2}$, then $P(t)$ will increase toward $\frac{c_1}{c_2}$. See Figures 1.a-c.

Why is it important to know that T in Equation (19) is negative in the above discussion? What is $\lim_{t \to T^+} P(t)$?

EXAMPLE 3: A new pond is stocked with 50 bass. The population of bass is known to obey the logistic equation $P'(t) = c_1 P(t) - c_2 P^2(t)$, where $c_1 = 0.4$ and $c_2 = 0.001$ and time is measured in yr. Find

(a) the equilibrium population, and

(b) the expected population in five yr.

SOLUTION: The equilibrium population is $\frac{c_1}{c_2} = 400$. Using Equation (15), we have

$$P(t) = \frac{c_1 A}{c_2 A + e^{-c_1 t}} = \frac{0.4A}{0.001A + e^{-0.4t}}.$$

This models the expected population t yr after the pond is stocked. We use Equation (16) to obtain $A = \frac{50}{0.4 - 0.05} = 142.857$, so

$$P(t) = \frac{(0.4)(142.857)}{(0.001)(142.857) + e^{-0.4t}} = \frac{57.143}{0.142857 + e^{-0.4t}},$$

and

$$P(5) = 205.41.$$

In Figure 2, we graph the bass population over 15 yr as a function of time. ∎

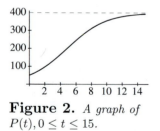

Figure 2. *A graph of* $P(t), 0 \le t \le 15.$

EXAMPLE 4: Suppose we know that a bacterial culture is growing according to the logistic equation $P'(t) = c_1 P(t) - c_2 P^2(t)$, where $c_1 = 0.4$ and $c_2 = 0.001$ and time is measured in days. The population at time $t = 0$ is 600 individuals. Since $\frac{c_1}{c_2} = 400$, we know that the culture is overpopulated, and that the population is shrinking. Find the domain for $P(t)$, and determine how long ago the population count was 1000.

SOLUTION: Rather than starting with Equation (19), we use Equations (15) and (16) to obtain $P(t)$ and work from there.

$$P(t) = \frac{c_1 A}{c_2 A + e^{-c_1 t}} = \frac{0.4A}{0.001A + 0.4e^{-0.4t}}$$

$$\text{where} \quad A = \frac{600}{0.4 - 0.6} = -3000.$$

This gives us that

$$P(t) = \frac{(0.4)(-3000)}{(0.001)(-3000) + e^{-0.4t}} = \frac{-1200}{-3 + e^{-0.4t}}.$$

The lower bound for the domain of $P(t)$ will be the time T when the denominator is zero. Thus T is the solution for

$$0 = -3 + e^{-0.4T},$$

or

$$T = -\frac{1}{0.4}\ln(3) \approx -2.75.$$

The domain for $P(t) = (-2.75, \infty)$.

We want to find the time t such that $P(t) = 1000$.

$$1000 = \frac{-1200}{-3 + e^{-0.4t}}$$

$$-3000 + 1000e^{-0.4t} = -1200$$

$$e^{-0.4t} = 1.8$$

$$t = -\frac{1}{0.4}\ln(1.8) \approx -1.47.$$

The culture count was 1000 approximately 1.47 days ago. ∎

Working With Data

EXAMPLE 5: Suppose that we stock a pond with a new species of bass. We believe that the population growth will follow the logistic growth equation, and we want to find c_1 and c_2. We initially stock the pond with 50 fish. In five yr, we drain the pond and count 205 fish. We refill and restock the pond, and in 10 yr we count 355 fish. Note that the data we are using here are very nearly the values we obtained for $P(5)$ and $P(10)$ in Example 3. Hence, we expect our solutions for c_1 and c_2 to be close to 0.4 and 0.001. We start with Equation (15) and let

$$a = c_1 A \quad \text{and} \quad b = c_2 A \tag{20}$$

so that we have

$$P(t) = \frac{a}{e^{-c_1 t} + b}. \tag{21}$$

Letting $t = 0$, we have

$$P(0) = \frac{a}{1+b} \quad \text{or} \quad a = P(0)(1+b). \tag{22}$$

Substituting Equation (22) into Equation (21), we have that

$$P(t) = \frac{P(0)(1+b)}{e^{-c_1 t} + b}.$$

Solving for b, we obtain

$$b = \frac{P(0) - P(t)e^{-c_1 t}}{P(t) - P(0)}. \tag{23}$$

By using two values of t, we obtain two equations:

$$b = \frac{P(0) - P(t_1)e^{-c_1 t_1}}{P(t_1) - P(0)},$$

and

$$b = \frac{P(0) - P(t_2)e^{-c_1 t_2}}{P(t_2) - P(0)}.$$

Subtracting the second equation from the first, we eliminate b.

$$0 = \frac{P(0) - P(t_1)e^{-c_1 t_1}}{P(t_1) - P(0)} - \frac{P(0) - P(t_2)e^{-c_1 t_2}}{P(t_2) - P(0)}.$$

This can be rewritten as

$$\begin{aligned}
0 &= \left(\frac{P(t_2)}{P(t_2) - P(0)}\right) e^{-c_1 t_2} - \left(\frac{P(t_1)}{P(t_1) - P(0)}\right) e^{-c_1 t_1} \\
&+ \left(\frac{P(0)}{P(t_1) - P(0)} - \frac{P(0)}{P(t_2) - P(0)}\right).
\end{aligned} \tag{24}$$

If $t_2 = 2t_1$, then we can use the quadratic formula to solve for e^{c_1}. We continue with our example to show how this works. We have that

$t_1 = 5$ and $t_2 = 10 = 2t_1$. We also know that $P(0) = 50$, $P(5) = 205$, and $P(10) = 355$. Putting these values into Equation (24) we get

$$
\begin{aligned}
0 &= \frac{355}{305}e^{-10c_1} - \frac{205}{155}e^{-5c_1} + \frac{50}{155} - \frac{50}{305} \\
&= \frac{355}{305}\left(e^{-5c_1}\right)^2 - \frac{205}{155}e^{-5c_1} + \frac{50}{155} - \frac{50}{305}.
\end{aligned}
$$

Using the quadratic formula to solve for e^{-5c_1}, we obtain

$$
e^{-5c_1} \approx 0.136302 \quad \text{or} \quad c_1 = -\frac{\ln(0.136302)}{5} \approx 0.399.
$$

Using Equation (23) to solve for b, we get

$$
b = \frac{P(0) - P(5)e^{(-5)(0.399)}}{P(5) - P(0)} \approx 0.1423.
$$

We use Equation (22) to find a and Equation (20) to find A.

$$
a = p(0)(1 + b) \approx 50(1.1423) = 57.115 \quad \text{and} \quad A = \frac{a}{c_1} \approx 143.1.
$$

Finally, we use Equation (20) again to find c_2.

$$
c_2 = \frac{b}{A} \approx 9.9 \times 10^{-4}.
$$

Let us reflect on what we have done here. We began with a population that we assumed to have logistic growth, but where the constants in the logistic equation were unknown. Using population measurements for various times, we were able to find the constants c_1 and c_2. This permits us to model this bass population with the logistic equation

$$
P'(t) = c_1 P(t) - c_2 P^2(t),
$$

with the constants we found above.

Since we obtained the data for this problem by rounding off the results from Example 3, our c_1 and c_2 were vere close to the constants in that example. If we had used exact data from the function $P(t)$ of Example 3, and if there were no roundoff error in our computations, we would have found that $c_1 = 0.4$ and $c_2 = 0.001$.   ∎

EXERCISES 8.3

In Exercises 1–5 use Integration Formula (10) to evaluate the integrals.

1. $\int \dfrac{dx}{2x + 3x^2}$.

2. $\int \dfrac{dx}{-x - 5x^2}$.

3. $\int \dfrac{e^x\,dx}{2e^x + 3e^{2x}}$.

4. $\int \dfrac{\sec^2(x)\,dx}{\tan(x) + 3\tan^2(x)}$.

5. $\int \dfrac{1 + \tan^2(x)\,dx}{\tan(x) + 3\tan^2(x)}$.

6. Explain why you can't evaluate $\int_{-1}^{0.5} \frac{dx}{x+x^2}$.

7. Explain why you can't evaluate $\int_{-1}^{1} \frac{dx}{3x-2x^2}$.

In Exercises 8–13, solve the separable differential equations.

8. $y'(x) = 2y(x) + 3y^2(x)$, $\quad y(0) = 1$.

9. $y'(x) = 2y(x) - 3y^2(x)$, $\quad y(0) = -3$.

10. $y'(x) = 2y(x) + 3y^2(x)$, $\quad y(0) = 0$.

11. $y'(t) = c_1 y(t) - c_2 y^2(t)$, $c_1 = 0.2, c_2 = 0.01$, and $y(0) = 25$.

12. $y'(t) = c_1 y(t) + c_2 y^2(t)$, $c_1 = 0.2, c_2 = 0.01$, and $y(0) = 25$.

13. $y'(t) = c_1 y(t) + c_2 y^2(t)$, $c_1 = 0.2, c_2 = 0.01$, and $y(0) = -25$.

14. What are the domains for each of your solutions in Exercises 4–13?

In Exercises 15–18, suppose that a culture's initial count is 1200 and its growth is logistic. The growth is measured in hr. We know that $c_1 = 0.2$.

15. What must c_2 be in order for the population to be at equilibrium?

16. If $c_2 = 0.001$, then

   a. What is the equilibrium population?

   b. What will the population be in 3 hr?

   c. Is the population increasing or decreasing?

   d. What is the domain of $P(t)$?

17. If $c_2 = 0.0001$, then

   a. What is the equilibrium population?

   b. What will the population be in 3 hr?

   c. Is the population increasing or decreasing?

   d. What is the domain of $P(t)$?

18. Compare the graphs for the solutions for Exercises 16 and 17 over the interval $0 \leq t \leq 10$.

19. Suppose that the cricket population in a field obeys the logistic equation with $c_1 = 0.3$ and $c_2 = 0.002$. Compare the graphs of $P(t)$ for $P(0) = 10$, $P(0) = 100$, and $P(0) = 1000$.

20. Find the derivative of $P(t)$ as given in Equation (17), and show that if $P(0) > \frac{c_1}{c_2}$, then P is a decreasing function, and if $P(0) < \frac{c_1}{c_2}$, then P is an increasing function.

21. In the derivation of Equation (12), we had $c_1 = k_1 + k_2$ and $c_2 = k_2$, where k_1 can be taken as the birth rate minus the death rate due to old age, and k_2 is the death rate due to disease. A rare ape populates an island. The birth rate is 0.2, the death rate due to old age is $\frac{1}{40}$, and the death rate due to disease is $\frac{1}{600}$, with time measured in yr.

   a. What is the equilibrium population?

   b. Suppose that the population had reached one less than the equilibrium population when a storm killed half of the apes. How long will it take for the population to reach its original number?

   c. Due to a period of extraordinarily good weather, the population grew to 50 apes above the equilibrium population, and then the weather patterns returned to normal. How long will it take for the population to reach one more than the equilibrium population?

22. A bacterial population obeys the logistic equation. The population count is initially 50. After 5 hr, the population is 421, and after 10 hr, the population is 1076. Find the function that gives the population as a function of time, in hr, and find the equilibrium population.

23. The population of a bacterial culture obeys the logistic equation. The population count is initially 200. After 1 hr, the population is 178, and after 2 hr, the population is 171. Find the function that gives the population as a function of time, in hr, and find the equilibrium population.

Chapter 9

L'Hôpital's Rule, Improper Integrals, and Series

9.1 L'Hôpital's Rule $\left(\frac{0}{0}\text{ and }\frac{\infty}{\infty}\right)$

If $\lim_{x \to a} f(x) = 0$ and $\lim_{x \to a} g(x) = 0$, discerning the limit of $\frac{f(x)}{g(x)}$ as x approaches a can be quite difficult.

EXAMPLE 1:

(a) $\displaystyle\lim_{x \to 0^+} \frac{x}{x^2} = \infty.$

(b) $\displaystyle\lim_{x \to 0^-} \frac{x}{x^2} = -\infty.$

(c) $\displaystyle\lim_{x \to 0^+} \frac{x}{x} = \lim_{x \to 0^-} \frac{x}{x} = \lim_{x \to 0} \frac{x}{x} = 1.$

(d) $\displaystyle\lim_{x \to 0} \frac{\sin x}{x} = 1.$

(e) $\displaystyle\lim_{x \to 0} \frac{\sin(3x)}{x} = 3.$ ∎

Fortunately, l'Hôpital gave us some techniques to evaluate these limits.[1]

Suppose that $\lim_{x \to a^+} f(x) = 0$, $\lim_{x \to a^+} g(x) = 0$, there is a number c such that f and g are differentiable on (a, c), and $g'(x)$ is

---

[1] Though these techniques are called l'Hôpital's rules, they were not discovered by l'Hôpital. He learned these techniques from his teacher, Johann Bernoulli. However, l'Hôpital published these results in a small tract, and they have been known as l'Hôpital's rules ever since.

not 0 anywhere in (a, c). Assume further that $f(a) = g(a) = 0$. For a given x in (a, c), by the Extended Mean Value Theorem, there is a t between a and x such that

$$\frac{f(x)}{g(x)} = \frac{f(x) - f(a)}{g(x) - g(a)} = \frac{f'(t)}{g'(t)}.$$

Thus, if $\lim_{t \to a^+} \frac{f'(t)}{g'(t)} = L$, then $\lim_{x \to a^+} \frac{f(x)}{g(x)} = L$. The limit L may be a real number, $+\infty$, or $-\infty$. We have the following theorem, known as l'Hôpital's Rule.

Theorem 1 (l'Hôpital's Rule $\frac{0}{0}$) *Suppose that*

(a) $\displaystyle\lim_{x \to a^+} f(x) = 0;$

(b) $\displaystyle\lim_{x \to a^+} g(x) = 0;$

(c) *there is a number c such that, if x is in (a, c), then $g(x) \neq 0$; and*

(d) $\displaystyle\lim_{x \to a^+} \frac{f'(x)}{g'(x)} = L$, *where L is a real number, $+\infty$, or $-\infty$.*

Then

$$\lim_{x \to a^+} \frac{f(x)}{g(x)} = L.$$

A similar rule holds for $\lim_{x \to a^-} \frac{f(x)}{g(x)}$, and, of course, for $\lim_{x \to a} \frac{f(x)}{g(x)}$.

EXAMPLE 2: Evaluate $\lim_{x \to 1} \dfrac{\ln x}{x - 1}$.

SOLUTION: Note that $\lim_{x \to 1} \ln x = 0$ and $\lim_{x \to 1} x - 1 = 0$. Hence l'Hôpital's rule applies, and we have

$$\lim_{x \to 1} \frac{\ln x}{x - 1} = \lim_{x \to 1} \frac{\frac{1}{x}}{1} = 1. \qquad \blacksquare$$

EXAMPLE 3: Evaluate $\lim_{x \to 0} \dfrac{x - \sin x}{x^2}$.

SOLUTION: $\lim_{x \to 0}(x - \sin x) = 0$ and $\lim_{x \to 0} x^2 = 0$, and l'Hôpital's rule yields

$$\lim_{x \to 0} \frac{x - \sin x}{x^2} = \lim_{x \to 0} \frac{1 - \cos x}{2x}.$$

However, the term on the right is also of the form $\frac{0}{0}$. We may use l'Hôpital's rule again to obtain

$$\lim_{x \to 0} \frac{x - \sin x}{x^2} = \lim_{x \to 0} \frac{1 - \cos x}{2x} = \lim_{x \to 0} \frac{\sin x}{2} = 0.$$

One also encounters difficulties in ascertaining the limit of $\frac{f(x)}{g(x)}$ if $f(x)$ and $g(x)$ are each approaching ∞. There is, however, a l'Hôpital rule for this case.

Theorem 2 (l'Hôpital's Rule $\frac{\infty}{\infty}$) *Suppose that*

(a) $\lim\limits_{x \to a^+} f(x) = +\infty \ \ or \ -\infty$;

(b) $\lim\limits_{x \to a^+} g(x) = +\infty \ \ or \ -\infty$; *and*

(c) $\lim\limits_{x \to a^+} \dfrac{f'(x)}{g'(x)} = L$, *where L is a real number, $+\infty$, or $-\infty$.*

Then
$$\lim_{x \to a^+} \frac{f(x)}{g(x)} = L.$$

As in Theorem 1, a similar rule holds for $\lim_{x \to a^-} \frac{f(x)}{g(x)}$, and for $\lim_{x \to a} \frac{f(x)}{g(x)}$. Unfortunately, the proof of Theorem 2 should be left to an advanced calculus course or a course in introductory analysis.

Suppose that we know that $\lim_{x \to \infty} f(x) = 0$ and that $\lim_{x \to \infty} g(x) = 0$. We can make a subsitution $t = 1/x$ to obtain

$$\lim_{x \to \infty} \frac{f(x)}{g(x)} \ = \ \lim_{t \to 0^+} \frac{f\left(\frac{1}{t}\right)}{g\left(\frac{1}{t}\right)} = \lim_{t \to 0^+} \frac{-t^{-2} f'\left(\frac{1}{t}\right)}{-t^{-2} g'\left(\frac{1}{t}\right)}$$

$$= \ \lim_{t \to 0^+} \frac{f'\left(\frac{1}{t}\right)}{g'\left(\frac{1}{t}\right)} = \lim_{x \to \infty} \frac{f'(x)}{g'(x)}.$$

We have proved the following theorem.

Theorem 3 (l'Hôpital's Rule $\frac{0}{0}$) *Suppose that*

(a) $\lim\limits_{x \to \infty} f(x) = 0$;

(b) $\lim\limits_{x \to \infty} g(x) = 0$;

(c) *There is a number c such that if $x > c$, then $g(x) \neq 0$; and*

(d) $\displaystyle \lim_{x \to \infty} \frac{f'(x)}{g'(x)} = L$, *where L is a real number, $+\infty$, or $-\infty$.*

Then

$$\lim_{x \to \infty} \frac{f(x)}{g(x)} = L.$$

A similar rule holds for $\lim_{x \to -\infty} \frac{f(x)}{g(x)}$. We also have the following variation of Theorem 2.

Theorem 4 (l'Hôpital's Rule $\frac{\infty}{\infty}$) *Suppose that*

(a) $\displaystyle \lim_{x \to \infty} f(x) = +\infty$ *or* $-\infty$;

(b) $\displaystyle \lim_{x \to \infty} g(x) = +\infty$ *or* $-\infty$;

(c) $\displaystyle \lim_{x \to \infty} \frac{f'(x)}{g'(x)} = L$, *where L is a real number, $+\infty$, or $-\infty$.*

Then

$$\lim_{x \to \infty} \frac{f(x)}{g(x)} = L.$$

As in Theorem 3, a similar rule holds for $\lim_{x \to -\infty} \frac{f(x)}{g(x)}$.

EXAMPLE 4: Evaluate $\displaystyle \lim_{x \to \infty} \frac{\ln x}{x}$.

SOLUTION: Since $\lim_{x \to \infty} \ln x = \lim_{x \to \infty} x = \infty$, we may apply Theorem 4.

$$\lim_{x \to \infty} \frac{\ln x}{x} = \lim_{x \to \infty} \frac{\frac{1}{x}}{1} = 0. \qquad \blacksquare$$

EXAMPLE 5: Evaluate $\displaystyle \lim_{x \to \infty} \frac{x^{-2}}{x^{-3}}$.

SOLUTION: Since $\lim_{x \to \infty} x^{-2} = \lim_{x \to \infty} x^{-3} = 0$, we may try to apply Theorem 3.

$$\lim_{x \to \infty} \frac{x^{-2}}{x^{-3}} = \lim_{x \to \infty} \frac{-2x^{-3}}{-3x^{-4}},$$

which is again of the form $\frac{0}{0}$. However, it is clear that if we keep repeating this process, we continue to obtain a rational function of the form $\frac{0}{0}$. We would have saved ourselves a lot of trouble if we had initially observed that $\frac{x^{-2}}{x^{-3}} = x$ so that

$$\lim_{x\to\infty} \frac{x^{-2}}{x^{-3}} = \lim_{x\to\infty} x = \infty. \qquad \blacksquare$$

Limits of the form $0 \cdot \infty$.

If $\lim_{x\to a} f(x) = 0$ and $\lim_{x\to a} g(x) = \infty$, it is not at all clear what is happening to $\lim_{x\to a} f(x)g(x)$. For example, if $a = 0$, $f(x) = x^2$, and $g(x) = \frac{1}{x}$, then

$$\lim_{x\to 0^+} f(x)g(x) = \lim_{x\to 0} x = 0.$$

On the other hand, if $a = 0$, $f(x) = x$ and $g(x) = \frac{1}{x^2}$, then

$$\lim_{x\to 0^+} f(x)g(x) = \lim_{x\to 0} \frac{1}{x} = \infty.$$

If $\lim_{x\to a} f(x) = 0$ and $\lim_{x\to a} g(x) = \infty$, then we can rewrite the limit as

$$\lim_{x\to a} f(x)g(x) = \lim_{x\to a} \frac{g(x)}{\frac{1}{f(x)}},$$

which is of the form $\frac{\infty}{\infty}$. Or

$$\lim_{x\to a} f(x)g(x) = \lim_{x\to a} \frac{f(x)}{\frac{1}{g(x)}},$$

which is of the form $\frac{0}{0}$.

EXAMPLE 6: Compute $\lim_{x\to 0^+} x \ln x$.

SOLUTION: Though the expression is not of the form $\frac{0}{0}$ or $\frac{\infty}{\infty}$, we may rewrite $\lim_{x\to 0^+} x \ln x$ as $\lim_{x\to 0^+} \frac{\ln x}{\frac{1}{x}}$, which is of the form $\frac{-\infty}{\infty}$. Thus

$$\lim_{x\to 0^+} \frac{\ln x}{\frac{1}{x}} = \lim_{x\to 0^+} \frac{\frac{1}{x}}{-x^{-2}} = \lim_{x\to 0^+} -x = 0. \qquad \blacksquare$$

Limits of the form (1^∞), (0^∞), (0^0) or (∞^0).

We know that $x^0 = 1$ for all x, except for $x = 0$. It's also true that $0^x = 0$ for $x \neq 0$. Also, if x is large and y is close to zero, then what is x^y? Exponentials give us another set of indeterminate forms. More formally, suppose that $\lim_{x\to a} f(x) = 0$, $\lim_{x\to a} g(x) = 0$, $\lim_{x\to a} h(x) = 1$ and $\lim_{x\to a} y(x) = \infty$. Then all of the following limits are indeterminate.

$$\lim_{x \to a} [f(x)]^{g(x)} \qquad\qquad \lim_{x \to a} [y(x)]^{g(x)}$$

$$\lim_{x \to a} [g(x)]^{y(x)} \qquad\qquad \lim_{x \to a} [h(x)]^{y(x)}$$

We handle these cases by going back to the fundamentals and rewriting the limit using logarithms.

EXAMPLE 7: Suppose that $\lim\limits_{x \to a} f(x) = 0$ and $\lim\limits_{x \to a} g(x) = 0$.

$$\lim_{x \to a} f(x)^{g(x)} = \lim_{x \to a} e^{(g(x) \ln(f(x)))} e^{\lim_{x \to a} (g(x) \ln(f(x)))}.$$

Now

$$\lim_{x \to a} g(x) \ln(f(x))$$

is of the form $(0 \cdot \infty)$ which we know how to attack.   ∎

EXAMPLE 8: To find $\lim_{x \to 0^+} (1 + x)^{1/x}$, we write $(1 + x)^{1/x}$ as $e^{(1/x) \ln(1+x)}$ and compute $\lim_{x \to 0^+} \left(\frac{1}{x}\right) \ln(1 + x) = \lim_{x \to 0^+} \frac{\ln(1+x)}{x}$, which is of the form $\frac{0}{0}$. Employing l'Hôpital's rule, we obtain

$$\lim_{x \to 0^+} \frac{\ln(1 + x)}{x} = \lim_{x \to 0^+} \frac{1}{x + 1} = 1.$$

Thus

$$\lim_{x \to 0^+} (1 + x)^{1/x} = \lim_{x \to 0^+} e^{(1/x) \ln(1+x)} = e^1 = e.$$ ∎

EXERCISES 9.1

In Exercises 1–27, evaluate the limits.

1. $\lim\limits_{x \to 2} \dfrac{x^2 - 4}{x - 2}$.

2. $\lim\limits_{x \to 0} \dfrac{e^x - 1}{\sin(2x)}$.

3. $\lim\limits_{x \to 2^+} \dfrac{\sqrt{x - 2}}{x - 2}$.

4. $\lim\limits_{x \to 2} \dfrac{x^{1/4} - 2^{1/4}}{x - 2}$.

5. $\lim\limits_{x \to 2^+} \dfrac{(x - 2)^{1/4}}{x - 2}$.

6. $\lim\limits_{x \to 2} \dfrac{x - 2}{x^{1/4} - 2^{1/4}}$.

7. $\lim\limits_{x \to \infty} \dfrac{x - 2}{x^{1/4} - 2^{1/4}}$.

8. $\lim\limits_{x \to \infty} \dfrac{x^{1/4} - 2^{1/4}}{x - 2}$.

9. $\lim\limits_{x \to \infty} \dfrac{\sin x}{x}$.

10. $\lim\limits_{x \to \infty} \dfrac{e^x}{\ln x}$.

11. $\lim\limits_{x \to \infty} \dfrac{\ln x}{e^x}$.

12. $\lim\limits_{x \to \infty} \dfrac{\ln x}{1/x}$.

13. $\lim\limits_{x \to \infty} \dfrac{e^x - e^{-x}}{e^x + e^{-x}}$.

14. $\lim\limits_{x \to \infty} \dfrac{e^x - e^{-x}}{\ln x}$.

15. $\lim\limits_{x \to 0^+} \dfrac{\ln x}{1/x^2}$.

16. $\lim\limits_{x \to 0^+} \dfrac{1/x}{\ln x}$.

17. $\lim\limits_{x \to 0^+} \dfrac{x + 1 - e^x}{x^2}$.

18. $\lim\limits_{x \to 0} \dfrac{1 - \cos x}{\sin x}$.

19. $\lim\limits_{x \to 0} \dfrac{1 + \cos x}{\sin x}$.

20. $\lim\limits_{x \to \infty} \dfrac{2^x}{3^x}$.

21. $\lim\limits_{x \to \pi/2} (\cos x) e^{\pi/2 - x}$.

22. $\lim\limits_{x \to 0^+} (\sin x) \ln(\sin x)$.

23. $\lim\limits_{x \to 0^+} (1/x) \ln(1 + x)$.

24. $\lim\limits_{x \to 0^+} (\tan x) \ln x$.

25. $\lim\limits_{x \to 0^+} (e^{1/x}) \ln(1 + x)$.

26. $\lim\limits_{x \to \infty} (e^{1/x}) \ln(1 + x)$.

27. $\lim\limits_{x \to \pi^+} \csc(x \cos x + \pi)$ (Notice that it is important to decide if $(x \cos x + \pi)$ is positive or negative!).

28. Show that $\lim\limits_{x \to \infty} \left(\dfrac{bx^2}{x - 1} - bx \right) = b$ (Hint: Add and then take the limit).

29. Given that $\lim\limits_{x \to a} f(x) = \infty$ and $\lim\limits_{x \to a} g(x) = \infty$, what do we know about $\lim\limits_{x \to a} (f(x) - g(x))$?

In Exercises 30–43 compute the limits.

30. $\lim\limits_{x \to 1} \left[\dfrac{1}{\ln x} - \dfrac{x}{\ln x} \right]$.

31. $\lim\limits_{x \to 0} \left[\dfrac{1}{x} - \dfrac{1}{\sin x} \right]$.

32. $\lim\limits_{x \to \infty} \left[\dfrac{1}{x(e^x - 1)} - \dfrac{1}{3x} \right]$.

33. $\lim_{x \to \infty} [\sec x - \tan x]$.

34. $\lim_{x \to 0} x^{\sin x}$ (Hint: $x^{\sin x} = e^{(\sin x)\ln x}$.).

35. $\lim_{x \to 0} \sin^x x$.

36. $\lim\limits_{x \to \infty} \left(1 + \dfrac{1}{x} \right)^x$.

37. $\lim_{x \to 0^+} (1 + x)^{2x}$.

38. $\lim_{x \to \infty} (1 + x)^{2/x}$.

39. $\lim\limits_{x \to 0^+} \left(1 + \dfrac{1}{x} \right)^x$.

40. $\lim\limits_{x \to 0^+} \left(1 + \dfrac{a}{x} \right)^x$.

41. $\lim_{x \to \pi/2^-} (\sin x)^{\tan x}$.

42. $\lim_{x \to \infty} (a + ax)^{b/x}$.

43. $\lim_{x \to 0^-} (\sin(x))^{\sec(x)}$.

9.2   Improper Integrals

The Fundamental Theorem of Calculus not only tells us when functions have antiderivatives, but also gives us an avenue for computing the definite integral over intervals, provided that the function is continuous. In this section we use one sided limits to handle definite integrals of functions that have some isolated points of discontinuity. We also extend the concept of the definite integral to functions over unbounded intervals (e.g. $[0, \infty)$).

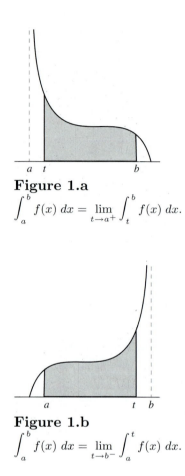

Figure 1.a
$$\int_a^b f(x)\, dx = \lim_{t \to a^+} \int_t^b f(x)\, dx.$$

Figure 1.b
$$\int_a^b f(x)\, dx = \lim_{t \to b^-} \int_a^t f(x)\, dx.$$

Definition: Definite Integrals over Points of Discontinuity

(a) If f is continuous on the interval $a < x \leq b$ but discontinuous at a, and if $\lim_{t \to a^+} \int_t^b f(x)\, dx$ exists and is finite, then we define

$$\int_a^b f(x)\, dx = \lim_{t \to a^+} \int_t^b f(x)\, dx.   \text{(See Figure 1.a.)}$$

(b) If f is continuous on the interval $a \leq x < b$ but discontinuous at b, and if $\lim_{t \to b^-} \int_a^t f(x)\, dx$ exists and is finite, then we define

$$\int_a^b f(x)\, dx = \lim_{t \to b^-} \int_a^t f(x)\, dx.   \text{(See Figure 1.b.)}$$

(c) If f is continuous everywhere on the interval $a \leq x \leq b$ except at the number c, then we define

$$\int_a^b f(x)\, dx = \lim_{t \to c^-} \int_a^t f(x)\, dx + \lim_{t \to c^+} \int_t^b f(x)\, dx,$$

provided that these limits are finite.

The integrals defined above are called *improper integrals*. If a definite integral of any of the above types exists, then it *converges* to its value. If it does not converge, it *diverges*.

EXAMPLE 1: Evaluate $\int_0^1 \frac{1}{\sqrt{x}}\, dx$ and $\int_0^1 \frac{1}{x}\, dx$.

SOLUTION: $\frac{1}{\sqrt{x}}$ is continuous on $(0, 1]$, so we can apply part (a) of the above definition.

$$\int_0^1 \frac{1}{\sqrt{x}}\, dx = \lim_{t \to 0^+} \int_t^1 \frac{1}{\sqrt{x}}\, dx = \lim_{t \to 0^+} \frac{\sqrt{x}}{1/2}\Big|_t^1$$

$$= \lim_{t \to 0^+} \left(2\sqrt{1} - 2\sqrt{t} \right) = 2.$$

Thus $\int_0^1 \frac{1}{\sqrt{x}}\, dx$ converges. The same definition applies to $\int_0^1 \frac{1}{x}\, dx$.

It follows that

$$\int_0^1 \frac{1}{x}\,dx = \lim_{t\to 0^+} \int_t^1 \frac{1}{x}\,dx = \lim_{t\to 0^+} \ln x \Big|_t^1$$

$$= \lim_{t\to 0^+} (\ln 1 - \ln t) = \infty.$$

Thus $\displaystyle\int_0^1 \frac{1}{x}\,dx$ diverges.   ■

EXAMPLE 2:

$$\int_{-1}^0 \frac{|x|}{x}\,dx = \lim_{t\to 0^-} \int_{-1}^t \frac{|x|}{x}\,dx = \lim_{t\to 0^-} \int_{-1}^t -1\,dx = \lim_{t\to 0^-} [t-1] = -1.$$

$$\int_0^2 \frac{|x|}{x}\,dx = \lim_{t\to 0^+} \int_t^2 \frac{|x|}{x}\,dx = \lim_{t\to 0^+} \int_t^2 1\,dx = \lim_{t\to 0^+} [2-t] = 2.$$

And so,

$$\int_{-1}^2 \frac{|x|}{x}\,dx = \int_{-1}^0 \frac{|x|}{x}\,dx + \int_0^2 \frac{|x|}{x}\,dx = 1.$$

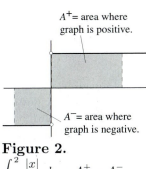

A^+ = area where graph is positive.

A^- = area where graph is negative.

Figure 2.
$\int_{-1}^2 \frac{|x|}{x}\,dx = A^+ - A^-.$

This agrees with the geometric intepretation of the integral. (See Figure 2.)   ■

Having seen how to handle definite integrals over points of discontinuity, we now turn to integrals over unbounded intervals, which we deal with in a similar fashion.

Definition: The Definite Integral Over Unbounded Intervals

(a) If f is continuous on the interval $a \le x < \infty$, and if $\lim_{t\to\infty} \int_a^t f(x)\,dx$ exists and is finite, then we define

$$\int_a^\infty f(x)\,dx = \lim_{t\to\infty} \int_a^t f(x)\,dx.$$

(See Figure 3a.)

(b) If f is continuous on the interval $-\infty < x \le b$, and if $\lim_{t\to-\infty} \int_t^b f(x)\,dx$ exists and is finite, then we define

$$\int_{-\infty}^b f(x)\,dx = \lim_{t\to-\infty} \int_t^b f(x)\,dx.$$

(See Figure 3b.)

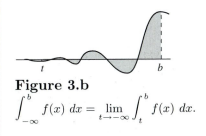

Figure 3.a
$\int_a^\infty f(x)\,dx = \lim_{t\to\infty} \int_a^t f(x)\,dx.$

Figure 3.b
$\int_{-\infty}^b f(x)\,dx = \lim_{t\to-\infty} \int_t^b f(x)\,dx.$

(c) If $\int_{-\infty}^{a} f(x)\,dx$ exists and $\int_{a}^{\infty} f(x)\,dx$ exists, then we define

$$\int_{-\infty}^{\infty} f(x)\,dx = \int_{-\infty}^{a} f(x)\,dx + \int_{a}^{\infty} f(x)\,dx.$$

The integrals defined above are also called *improper integrals*. If an integral of any of the above types exists, then it *converges* to its value. If it does not converge, it *diverges*.

EXAMPLE 3:

$$\int_{1}^{\infty} x^{-2}\,dx = \lim_{t\to\infty} \int_{1}^{t} x^{-2}\,dx = \lim_{t\to\infty} -x^{-1}\Big|_{1}^{t}$$

$$= \lim_{t\to\infty} \left(-\frac{1}{t} + \frac{1}{1}\right) = 1.$$

Thus $\int_{1}^{\infty} x^{-2}\,dx$ converges. However,

$$\int_{1}^{\infty} x^{-1}\,dx = \lim_{t\to\infty} \int_{1}^{t} x^{-1}\,dx = \lim_{t\to\infty} \ln x\Big|_{1}^{t}$$

$$= \lim_{t\to\infty} (\ln t - \ln 1) = \infty.$$

Thus $\int_{1}^{\infty} x^{-1}\,dx$ diverges.   ∎

EXAMPLE 4: Since $\int_{0}^{t} \sin x\,dx$ ranges back and forth between 0 and 2 as t gets large, $\lim_{t\to\infty} \int_{0}^{t} \sin x\,dx$ does not exist, and hence $\int_{0}^{\infty} \sin x\,dx$ is not defined.   ∎

According to our definition, in order to calculate $\int_{-\infty}^{\infty} f(x)\,dx$, we pick an arbitrary number c and calculate $\int_{-\infty}^{c} f(x)\,dx$ and $\int_{c}^{\infty} f(x)\,dx$ independently and then add the results.

EXAMPLE 5: Calculate $\int_{-\infty}^{\infty} \dfrac{dx}{1+x^2}$.

SOLUTION: For convenience, we choose $c = 0$.

$$\int_{0}^{\infty} \frac{dx}{1+x^2} = \lim_{a\to\infty} \int_{0}^{a} \frac{dx}{1+x^2}$$

$$= \lim_{a\to\infty} (\arctan a - 0) = \frac{\pi}{2}.$$

$$\int_{-\infty}^{0} \frac{dx}{1+x^2} = \lim_{a\to-\infty} \int_{a}^{0} \frac{dx}{1+x^2}$$

$$= \lim_{a\to-\infty} (0 - \arctan a) = \frac{\pi}{2}.$$

$$\int_{-\infty}^{\infty} \frac{dx}{1+x^2} = \int_{-\infty}^{0} \frac{dx}{1+x^2} + \int_{0}^{\infty} \frac{dx}{1+x^2}$$

$$= \frac{\pi}{2} + \frac{\pi}{2} = \pi. \qquad \blacksquare$$

There are a few theorems that will be useful to us in the next section, but first we need a fundamental property of the real line, called the Completeness Axiom.

The Completeness Axiom

If $a(n)$ is a nondecreasing sequence of numbers, and U is a number such that for all n, $a(n) \leq U$, $a(n)$ converges. [2]

---

[2] Note that this does not say that the sequence converges to U. The number U is simply an *upper bound* for a.

Theorem 1 *Suppose that f and g are integrable on $[a, \infty)$, and $f(x) \geq 0$ and $g(x) \geq 0$ for all x in $[a, \infty)$. Then*

  (a) *If $f(x) \leq g(x)$ for all x in $[a, \infty)$, and $\int_{a}^{\infty} g(x)\, dx$ converges, then $\int_{a}^{\infty} f(x)\, dx$ converges.*

  (b) *If $f(x) \geq g(x)$ for all x in $[a, \infty)$, and $\int_{a}^{\infty} g(x)\, dx$ diverges, then $\int_{a}^{\infty} f(x)\, dx$ diverges.*

Proof:

Part (a): Let $G(x) = \int_{a}^{x} g(t)\, dt$, $F(x) = \int_{a}^{x} f(t)\, dt$, and $b(n) = F(n)$ for integers $n \geq a$. Then $b(n)$ is a nondecreasing sequence. (Why?) We have $b(n) \leq G(n) \leq \int_{a}^{\infty} g(x)\, dx$ for all $n \geq a$. It follows from the Completeness Axiom that

$$\lim_{n\to\infty} b(n) = \lim_{x\to\infty} F(x) = \int_{a}^{\infty} f(x)\, dx$$

converges.

Part (b): Let F and G be defined as above. Since $\lim_{x\to\infty} G(x) = \infty$, and since $F(x) \geq G(x)$ for all x in $[a, \infty)$, $\lim_{x\to\infty} F(x) = \infty$. $\blacksquare$

EXAMPLE 6: We already know that $\int_1^\infty x^{-2}\,dx$ converges. Since $e^{-x} \leq 1$ for $x \geq 1$, it follows that $e^{-x}x^{-2} \leq x^{-2}$ for $x \geq 1$. Thus we know that $\int_1^\infty e^{-x}x^{-2}\,dx$ converges. Of course, this does not tell us the value to which it converges! ∎

EXAMPLE 7: $\int_1^\infty x^{-1/2}\,dx$ diverges. Since $e^x \geq 1$ for $x \geq 1$, it follows that $e^x x^{-1/2} \geq x^{-1/2}$ for $x \geq 1$. Thus we know that $\int_1^\infty e^x x^{-1/2}\,dx$ diverges. ∎

Theorem 2 *Suppose that f is integrable on the interval $[a, \infty)$. If $\int_a^\infty |f(x)|dx$ converges, then $\int_a^\infty f(x)\,dx$ converges.*

Proof: Let

$$f_+(x) = \begin{cases} f(x) \text{ if } f(x) \geq 0 \\ 0 \text{ otherwise} \end{cases} \quad \text{and} \quad f_-(x) = \begin{cases} f(x) \text{ if } f(x) \leq 0 \\ 0 \text{ otherwise} \end{cases}.$$

We argue that $\int_a^\infty f_-(x)\,dx$ and $\int_a^\infty f_+(x)\,dx$ exist. It follows that $\int_a^\infty f(x)\,dx = \int_a^\infty f_-(x)\,dx + \int_a^\infty f_-(x)\,dx$ exists.

Since $-f_-(x) = |f_-(x)| \leq |f(x)|$ for all $a \leq x < \infty$, it follows from Theorem 1 that $\int_a^\infty f_-(x)\,dx = -\int_a^\infty -f_-(x)\,dx$ exists.

Similarly, since $f_+(x) \leq |f(x)|$ for all $a \leq x < \infty$, it follows from Theorem 1 that $\int_a^\infty f_+(x)\,dx$ exists. ∎

EXAMPLE 8: Show that $\int_1^\infty x^{-2}\sin(x)\,dx$ converges.

SOLUTION: $|\sin(x)| \leq 1$, so by Theorem 1, $\int_1^\infty x^{-2}|\sin(x)|\,dx$ converges. It follows from Theorem 2 that $\int_1^\infty x^{-2}\sin(x)\,dx$ converges. ∎

EXERCISES 9.2

In Exercises 1–20, determine whether the integrals converge and evaluate those that converge.

1. $\int_1^\infty \dfrac{1}{x^{4/3}}\,dx.$

2. $\int_1^\infty \dfrac{1}{x^{3/2}}\,dx.$

3. $\int_{-\infty}^1 \dfrac{dx}{(x-2)^5}.$

4. $\int_{-\infty}^0 \dfrac{dx}{x-1}.$

5. $\int_0^\infty e^x\,dx.$

6. $\int_0^\infty \cos x\,dx.$

7. $\int_{-\infty}^\infty \dfrac{1}{x^2+1}\,dx.$

8. $\int_{-\infty}^\infty \dfrac{x^2}{x^4+x^2}\,dx.$

9. $\int_0^1 \dfrac{dx}{(x-1)^{2/3}}.$

10. $\int_0^2 \dfrac{dx}{(x-1)^{2/3}}.$

11. $\int_0^1 \dfrac{dx}{(x-1)^2}$.

12. $\int_{-3}^3 \dfrac{x\,dx}{x^2-9}$.

13. $\int_0^\infty \dfrac{\sin x}{1+\cos^2 x}\,dx$.

14. $\int_1^\infty \dfrac{dx}{(x+1)^{1/2}}\,dx$.

15. $\int_0^1 \dfrac{2\sqrt{x}}{\sqrt{x}}\,dx$.

16. $\int_0^3 \dfrac{dx}{(3-x)^{2/3}}$.

17. $\int_{-\pi/2}^0 \tan^2 x\,dx$.

18. $\int_0^1 \dfrac{1}{x^2}\cos\left(\dfrac{1}{x}\right)\,dx$.

19. $\int_0^1 \sec(\pi x)\,dx$.

20. $\int_1^\infty \ln x\,dx$.

In Exercises 21–25, use Theorems 1 and 2 to determine the convergence of the integrals.

21. $\int_1^\infty e^x x^{4/3}\,dx$.

22. $\int_1^\infty x^{-3}\,dx$.

23. $\int_1^\infty \cos^2(x)x^{-3}\,dx$.

24. $\int_1^\infty \cos(x)\sin(x)x^{-3}\,dx$.

25. $\int_1^\infty [[x^{-3}]]\,dx$.   (Recall that $[[x]]$ is the greatest integer less than or equal to x.).

In Exercises 26–29, let R be the region "bounded" by the graph of f, the x–axis, and the lines $x = 0$ and $x = 1$. Determine whether the area of R is finite, and if it is, calculate it.

26. $f(x) = x^{-1/2}$ 27. $f(x) = x^{-1/3}$

28. $f(x) = x^{-0.51}$ 29. $f(x) = x^{-0.49}$

30. Let S denote the solid obtained by rotating the region R in Exercise 26 about the x–axis. Determine whether the volume of S is finite, and if it is, compute it.

31. Let S denote the solid obtained by rotating the region R in Exercise 27 about the x–axis. Determine whether the volume of S is finite, and if it is, compute it.

32. Let S denote the solid obtained by rotating the region R in Exercise 28 about the x–axis. Determine whether the volume of S is finite, and if it is, compute it.

33. Let S denote the solid obtained by rotating the region R in Exercise 29 about the x–axis. Determine whether the volume of S is finite, and if it is, compute it.

34. Find the area of the region bounded by the graph of $f(x) = (1+x^2)^{-1}$ and the x–axis.

35. Show that if $p > 1$, then $\int_1^\infty x^{-p}\,dx$ converges.

36. Show that if $0 \le p < 1$, then $\int_0^1 x^{-p}\,dx$ converges.

37. Show that if $p \ge 1$, then $\int_0^1 x^{-p}\,dx$ diverges.

38. Show that if $0 \le p \le 1$, then $\int_1^\infty x^{-p}\,dx$ diverges.

In Exercises 39–43, show that $\lim_{a\to\infty}\int_{-a}^a f(x)\,dx = 0$ while $\int_{-\infty}^\infty f(x)\,dx$ is undefined. (Note that $\lim_{a\to\infty}\int_{-a}^a f(x)\,dx$ is not necessarily equal to $\int_{-\infty}^\infty f(x)\,dx$.)

39. $f(x) = \sin(x)$.

40. $f(x) = \sin(x)\cos(x)$.

41. $f(x) = \sin(2x)\cos(2x)$.

42. $f(x) = \sin(nx)\cos(nx)$.

43. $f(x) = \dfrac{x}{1+x^2}$.

44. Let f be continuous on $[a,b]$ and let $\{U_n\}$ and $\{L_n\}$ be defined as in Exercise 7 of Section 6.5. Show that both $\{U_n\}$ and $\{L_n\}$ converge.

9.3   Series

Suppose that we wish to add up the terms of a sequence $a(i)$ for $i \geq N$. Without some tools, doing so would be like trying to add up the grains of sand on Fort Walton Beach. We can handle such "infinite sums" with an approach similar to that which we used for improper integrals. As with improper integrals, often the best we can do is to show that the sum exists, without calculating its value.

Let $a(i)$ for $i \geq N$ be a sequence, and for each $n \geq 1$, let $s(n)$ denote the sum of the first n terms of a; that is, s is the sequence defined by

$$s(n) = a(N) + a(N+1) + \cdots + a(N+n-1) = \sum_{i=N}^{N+n-1} a(i).$$

The sequence s is called the *sequence of partial sums of a*.

Notice that

$$\sum_{i=N}^{N+n-1} a(i) = \int_{N}^{N+n} a([[x]])dx.$$

In Figure 1, we illustrate $a([[x]])$ for $a(i) = \frac{1}{i}$ with $i \geq 1$.

The *series* constructed from the sequence a is the formal expression

$$\sum_{i=N}^{\infty} a(i) = a(N) + a(N+1) + a(N+2) + a(N+3) + \cdots.$$

If the sequence of partial sums of a converges, then the limit of the sequence of partial sums of a is denoted by $\sum_{i=N}^{\infty} a(i)$, and the infinite series is said to *converge*. A series that does not converge is said to *diverge*.

We emphasize that, although we are adding to our vocabulary, the ideas here are the same as those of the last section.

$$\sum_{i=N}^{\infty} a(i) \text{ is just the improper integral } \int_{N}^{\infty} a([[x]]) \, dx.$$

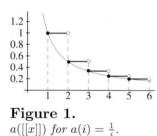

Figure 1.
$a([[x]])$ *for* $a(i) = \frac{1}{i}$.

EXAMPLE 1: Show that $\displaystyle\sum_{i=1}^{\infty} i^{-2}$ converges.

SOLUTION: We know that

$$\sum_{i=1}^{\infty} \frac{1}{i^2} = \int_{1}^{\infty} \frac{1}{[[x]]^2} \, dx.$$

But $\frac{1}{[[x]]^2} \leq 1/(x-1)^2$ for all $x > 1$. Thus, by Theorem 1 of the previous section, $\int_1^\infty \frac{1}{[[x]]^2} \, dx$ converges since $\int_2^\infty \frac{1}{(x-1)^2} \, dx$ converges. ∎

EXAMPLE 2: Show that $\displaystyle\sum_{i=1}^\infty i^{-(1/2)}$ diverges.

SOLUTION: We know that

$$\int_1^\infty \frac{1}{x^{1/2}} dx$$

diverges by using Theorem 1 of Section 9.2 with Exercise 9.2.1.14. Also,

$$f(x) = a([[x]]) = \frac{1}{[[x]]^{1/2}} \geq \frac{1}{x^{1/2}} \text{ for } x \geq 1.$$

Again by Theorem 1 of the previous section,

$$\sum_{i=1}^\infty \frac{1}{i^{1/2}} = \int_1^\infty \frac{1}{[[x]]^{1/2}} dx$$

diverges, since

$$\int_1^\infty \frac{1}{x^{1/2}} dx$$

diverges. ∎

The Integral Test

The above examples point to an important test for convergence.

Theorem 1 (The Integral Test) *Suppose that $a(i)$ for $i \geq N$ is a sequence of nonnegative numbers.*

(a) *If f is a nonincreasing function that is integrable on $[N, \infty)$ such that $a(i) \leq f(i)$ for all $i \geq N$, then*

$$\sum_{i=N}^\infty a(i) \text{ converges, provided that } \int_N^\infty f(x) \, dx \text{ converges.}$$

(b) *If f is a nonincreasing function that is integrable on $[N, \infty)$ such that $a(i) \geq f(i)$ for all $i \geq N$, then*

$$\sum_{i=N}^\infty a(i) \text{ diverges, provided that } \int_N^\infty f(x) \, dx \text{ diverges.}$$

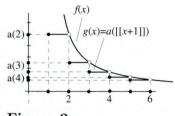

Figure 2.a
$a([[x+1]]) \leq f(x)$.

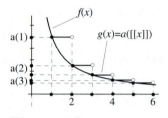

Figure 2.b $a([[x]]) \geq f(x)$.

Proof: The proof closely parallels the solutions to Examples 1 and 2. For simplicity, we assume $N = 1$.

Part (a): Let $g(x) = a([[x + 1]])$, $x \geq 1$ as in Figure 2.a. Then $g(x) \leq f(x)$ for $x \geq 1$. Now,

$$\int_1^2 g(x)\, dx \ = \ a(2) \leq \int_1^2 f(x)\, dx,$$

$$\int_2^3 g(x)\, dx \ = \ a(3) \leq \int_2^3 f(x)\, dx,$$

$$\int_3^4 g(x)\, dx \ = \ a(4) \leq \int_3^4 f(x)\, dx,$$

$$\int_5^6 g(x)\, dx \ = \ a(5) \leq \int_5^6 f(x)\, dx,$$

and so on. Thus, by Theorem 1 of Section 9.2,

$$\sum_{i=1}^{\infty} a(i) = a(1) + \sum_{i=2}^{\infty} a(i) = a(1) + \int_1^{\infty} g(x)\, dx$$

converges, since

$$\int_1^{\infty} f(x)\, dx \text{ converges.}$$

Part (b): Let $g(x) = a([[x]])$, $x \geq 1$. Then, as illustrated in Figure 2.b, $g(x) \geq f(x)$ for $x \geq 1$.

$$\int_1^2 g(x)\, dx \ = \ a(1) \geq \int_1^2 f(x)\, dx,$$

$$\int_2^3 g(x)\, dx \ = \ a(3) \geq \int_2^3 f(x)\, dx,$$

$$\int_3^4 g(x)\, dx \ = \ a(4) \geq \int_3^4 f(x)\, dx,$$

$$\int_5^6 g(x)\, dx \ = \ a(5) \geq \int_5^6 f(x)\, dx,$$

and so on. Again, by Theorem 1 of Section 9.2,

$$\sum_{i=1}^{\infty} a(i) = \int_1^{\infty} g(x)\, dx \text{ converges, since } \int_1^{\infty} f(x)\, dx \text{ converges.}$$

The p–Series Test

As a consequence of the Integral Test, we have the following theorem.

Theorem 2 (The p–Series Test) *The series $\sum_{n=1}^{\infty} \left(\frac{1}{n}\right)^p$ diverges if $p \leq 1$ and converges if $p > 1$.*

Proof: Assume that $p \neq 1$, and let $f(x) = x^{-p}$.

$$\int_1^{\infty} x^{-p} \, dx = \lim_{t \to \infty} \frac{x^{1-p}}{1-p}\Big|_1^t$$

$$= \lim_{t \to \infty} \left[\frac{t^{1-p}}{1-p} - \frac{1}{1-p}\right]$$

$$= \begin{cases} \infty & \text{if } p < 1 \\ \frac{1}{p-1} & \text{if } p > 1 \end{cases} .$$

By the Integral Test, the p–series converges if $p > 1$ and diverges if $p < 1$. If $p = 1$, then

$$\int_1^{\infty} x^{-1} \, dx = \lim_{t \to \infty} \ln x \Big|_1^t = \infty.$$

Again, by the Integral Test, $\sum_{i=1}^{\infty} \frac{1}{i} = \infty$.
Note: The series $\sum_{i=1}^{\infty} \frac{1}{i}$ is called the *harmonic series*.   ∎

EXAMPLE 3: By the p–Series Test, $\sum_{i=1}^{\infty} i^{-1.5}$ converges.   ∎

The Comparison Test

The following theorem, which is similar in nature to the Integral Test, proves quite useful in determining convergence and divergence of series.

Theorem 3 (The Comparison Test) *Suppose that $a(i)$, for $i \geq N$, and $b(i)$, for $i \geq N$, are sequences of nonnegative numbers such that $a(i) \leq b(i)$.*

$$\text{If } \sum_{i=N}^{\infty} b(i) \text{ converges, then so must } \sum_{i=N}^{\infty} a(i).$$

It follows immediately that

$$\text{If } \sum_{i=N}^{\infty} a(i) \text{ diverges, then so must } \sum_{i=N}^{\infty} b(i).$$

Proof: Let $f(x) = a([[x]])$ and $g(x) = b([[x]])$. Then

$$\sum_{i=N}^{\infty} a(i) = \int_{N}^{\infty} f(x)\,dx \quad \text{and} \quad \sum_{i=N}^{\infty} b(i) = \int_{N}^{\infty} g(x)\,dx.$$

Since $f(x) \le g(x)$, it follows from Theorem 1 that if $\int_{N}^{\infty} f(x)\,dx$ diverges, then so must $\int_{N}^{\infty} g(x)\,dx$. Similarly, if $\int_{N}^{\infty} g(x)\,dx$ converges, then so must $\int_{N}^{\infty} f(x)\,dx$.

EXAMPLE 4: We know that $\sum_{i=1}^{\infty} \frac{1}{i}$ diverges. It follows from the comparison test that $\sum_{i=2}^{\infty} \frac{1}{i-1}$ must also diverge. ∎

The Limit Comparison Test

With some effort we could show that $\sum_{i=2}^{\infty} \frac{1}{i^2-1}$ converges by comparing it to $\sum_{i=2}^{\infty} \frac{1}{i^2}$. However, there is another test that is helpful here. The observation to make is that $\frac{1}{i^2-1}$ and $\frac{1}{i^2}$ get large at about the same rate. There is not much difference between $\frac{1}{10^{10}}$ and $\frac{1}{10^{10}-1}$. In general, suppose that a and b are sequences of positive terms, and we know that $\sum_{i=1}^{\infty} b(i)$ converges. The Comparison Test tells us that if for large n, $a(n) \le b(n)$, then $\sum_{i=1}^{\infty} b(i)$ converges. We can paraphrase this to say that if a converges to 0 faster than b, then the convergence of b gives us the convergence of a. The Limit Comparison Test extends this idea to say that if the sequences a and b converge to 0 at "the same rate," then either both $\sum_{i=1}^{\infty} b(i)$ and $\sum_{i=1}^{\infty} a(i)$ converge or they both diverge. We formalize this idea in the following theorem.

Theorem 4 (The Limit Comparison Test) *Suppose that a and b are sequences of positive numbers and that $0 < L < \infty$ is a number such that $\lim_{n \to \infty} \frac{a(n)}{b(n)} = L$. Then either both $\sum_{i=1}^{\infty} b(i)$ and $\sum_{i=1}^{\infty} a(i)$ converge or they both diverge.*

Proof: Let $A = \frac{L}{2}$ and $B = \frac{3L}{2}$. Then $A < L < B$. Since $\lim_{n \to \infty} \frac{a(n)}{b(n)} = L$, there is an integer N such that if $n > N$, then $A < \frac{a(n)}{b(n)} < B$. We have that $a(n) < Bb(n)$ and $a(n) > Ab(n)$.

If $\sum_{i=1}^{\infty} b(i)$ converges, then so must $\sum_{i=1}^{\infty} Bb(i)$. Since $a(n) < Bb(n)$ for $n > N$, it follows from the Comparison Test that $\sum_{i=1}^{\infty} a(i)$ converges.

If $\sum_{i=1}^{\infty} b(i)$ diverges, then so must $\sum_{i=1}^{\infty} Ab(i)$. Since $a(n) > Ab(n)$ for $n > N$, it follows from the Comparison Test that $\sum_{i=1}^{\infty} a(i)$ diverges.

EXAMPLE 5: Determine the convergence of

$$\sum_{i=1}^{\infty} \frac{i+1}{2i^3 - 1}.$$

SOLUTION: The fact that the highest power of i in the numerator is 1 while the highest power of i in the denominator is 3 suggests using the Limit Comparison Test to compare $\frac{i+1}{2i^3-1}$ with $\frac{1}{i^2}$. We know that $\sum_{i=1}^{\infty} \frac{1}{i^2}$ converges since it is a p–series with $p > 1$.

$$\lim_{i \to \infty} \frac{1/i^2}{(i+1)/(2i^3 - 1)} = \lim_{i \to \infty} \frac{2i^3 - 1}{i^3 + 1} = 2.$$

It follows from the Limit Comparison Test that $\displaystyle\sum_{i=1}^{\infty} \frac{i+1}{2i^3 - 1}$ converges. ■

Geometric Series

Given an arbitrary sequence, it is almost impossible to determine the value to which the sequence of partial sums converges. However, if we have a geometric sequence, then the following theorem provides us with the answer.

Theorem 5 (Geometric Series) *Let $a(n) = cr^n$, where $c \neq 0$. If $0 \leq r < 1$, then the series $\sum_{n=1}^{\infty} cr^n = \frac{cr}{1-r}$, and if $r \geq 1$, then $\sum_{n=1}^{\infty} cr^n$ diverges.*

Proof: Let $s(n) = cr + cr^2 + cr^3 + \cdots + cr^n$. Clearly, if $r \geq 1$ then

$\lim_{n \to \infty} s(n) = \pm\infty$. Assume that $0 \le r < 1$. Then

$$
\begin{aligned}
rs(n) &= cr^2 + cr^3 + \cdots cr^{n+1}; \text{ and} \\
rs(n) - s(n) &= c\left(r^{n+1} - r\right).
\end{aligned}
$$

Solving for $s(n)$, we obtain

$$
s(n) = \frac{c(r^{n+1} - r)}{r - 1}.
$$

Since $0 \le r < 1$, we see that $\lim_{n \to \infty} \frac{r^{n+1}}{r-1} = 0$. Thus

$$
\sum_{n=1}^{\infty} cr^n = \lim_{n \to \infty} s(n) = \lim_{n \to \infty} c\left(\frac{r^{n+1}}{r - 1} + \frac{r}{1 - r}\right) = \frac{cr}{1 - r}. \qquad \blacksquare
$$

EXAMPLE 6: Use Theorem 5 to compute $\sum_{i=1}^{\infty} \left(\frac{1}{2^i}\right)$.

SOLUTION: The series $\sum_{i=1}^{\infty} \frac{1}{2^i}$ is of the form $\sum_{i=1}^{\infty} r^i$, where $r = \frac{1}{2}$. Thus

$$
\sum_{i=1}^{\infty} \frac{1}{2^i} = \frac{\frac{1}{2}}{1 - \left(\frac{1}{2}\right)} = \frac{\frac{1}{2}}{\frac{1}{2}} = 1.
$$

EXAMPLE 7: It is well known that any number with a repeating decimal representation is rational. Show that $0.31313131\ldots$ is a rational number. (That is, $0.31313131\ldots$ can be written in the form $\frac{p}{q}$, where p and q are integers.)

SOLUTION: Let $r = \frac{1}{100}$. Then

$$
0.31313131\ldots = \sum_{n=1}^{\infty} 31r^i = 31\frac{r}{1 - r} = 31\frac{\frac{1}{100}}{1 - \frac{1}{100}} = \frac{31}{99}. \qquad \blacksquare
$$

EXAMPLE 8: Let a be the sequence defined by $a(n) = \frac{1}{n!}$. To show that

$$
\sum_{i=1}^{\infty} \frac{1}{i!} = 1 + \frac{1}{2} + \frac{1}{3 \cdot 2} + \cdots + \frac{1}{n!} + \cdots
$$

converges, let b be the sequence defined by $b(n) = 2^{n-1}$. The series

$$
\sum_{i=1}^{\infty} \frac{1}{2^{i-1}} = 1 + \frac{1}{2} + \frac{1}{4} + \cdots + \frac{1}{2^{n-1}} + \cdots
$$

converges since it is a geometric series. The terms $a(n)$ and $b(n)$ are nonnegative for all n. If we can show that $a(n) \leq b(n)$ for all n, we can use the Comparison Test. It is easy to see that

$$\underbrace{n \cdot (n-1) \cdot (n-2) \cdots 2}_{n-1 \text{ terms}} \geq \underbrace{2 \cdot 2 \cdot 2 \cdots 2}_{n-1 \text{ terms}}$$

for all $n > 1$, hence $\frac{1}{n!} \leq \frac{1}{2^{n-1}}$. We also know that $a(1) = 1 = b(1)$. Thus $a(n) \leq b(n)$ for all n, and by the Comparison Test, the series $\sum_{i=1}^{\infty} \frac{1}{i!}$ converges.

Note that while we know that the series $\sum_{i=1}^{\infty} \frac{1}{i!}$ converges, we do not know its value! ∎

The nth Term Test for Divergence

The following theorem, which we state without proof, is different from the previous convergence theroems in that it is a tool that can be used to show divergence but not convergence.

Theorem 6 (The nth Term Test for Divergence)
If $\sum_{i=1}^{\infty} a(i)$ converges, then $\lim_{i \to \infty} a(i) = 0$.

This says that a series cannot possibly converge unless its terms approach 0. Hence if we can show that $\lim_{i \to \infty} a(i) \neq 0$, then we know that $\sum_{i=1}^{\infty} a(i)$ fails to converge. If you start with a series that requires some effort to determine convergence, then this test is one to use first. It will not show convergence, but it is usually a quick test to see if the series *might* diverge.

EXAMPLE 9: Show that $\displaystyle\sum_{n=1}^{\infty} \frac{n^2}{n^2 + n + 1}$ fails to converge.

SOLUTION: Employing l'Hôpital's rule, we have

$$\lim_{n \to \infty} \frac{n^2}{n^2 + n + 1} = \lim_{x \to \infty} \frac{x^2}{x^2 + x + 1} = \lim_{x \to \infty} \frac{2x}{2x + 1} = 1.$$

Thus we know that $\displaystyle\sum_{n=1}^{\infty} \frac{n^2}{n^2 + n + 1}$ fails to converge.   ∎

EXERCISES 9.3

In Exercises 1–7, determine which of the geometric series converge. If the series converges, determine its limit.

1. $\sum_{n=1}^{\infty} \left(\frac{3}{4} \right)^n$.

2. $\sum_{n=0}^{\infty} \left(\frac{3}{4} \right)^n$.

3. $\sum_{n=1}^{\infty} \left(\frac{4}{5} \right)^n$.

4. $\sum_{n=3}^{\infty} (4/5)^n$.

5. $\sum_{n=1}^{\infty} \left(\frac{4}{3} \right)^n$.

6. $\sum_{n=1}^{\infty} (0.999)^n$.

7. $\sum_{n=1}^{\infty} (1.0001)^n$.

In Excercises 8–11, use Theorem 6 to show that the series fail to converge.

8. $\sum_{i=1}^{\infty} \frac{i}{i + 32}$.

9. $\sum_{n=1}^{\infty} \frac{n^2}{3n^2 - 24n + \pi}$.

10. $\sum_{n=1}^{\infty} (-1)^n$.

11. $\sum_{n=1}^{\infty} \frac{n}{\ln(n + 1)}$.

12. Explain why Theorem 6 does not imply that the harmonic series converges.

In Exercises 13–16, determine which of the p–series converge.

13. $\sum_{i=1}^{\infty} i^{-0.99}$.

14. $\sum_{n=2}^{\infty} n^{-2}$.

15. $\sum_{n=1}^{\infty} \left(\frac{1}{n} \right)^{3/4}$.

16. $\sum_{n=1}^{\infty} \left(\frac{1}{n} \right)^{4/3}$.

In Exercises 17–32, determine the convergence of the series.

17. $\sum_{n=0}^{\infty} \frac{1}{(n + 1)^2}$.

18. $\sum_{n=0}^{\infty} \frac{1}{(n + 1)^{4/3}}$.

19. $\sum_{n=1}^{\infty} \frac{1}{(n + 1)^{3/4}}$.

20. $\sum_{n=1}^{\infty} \frac{\ln n}{n}$.

21. $\sum_{n=0}^{\infty} \left(\frac{1}{2} \right)^{n+1}$.

22. $\sum_{n=2}^{\infty} \frac{n}{\ln n}$.

23. $\sum_{n=1}^{\infty} \frac{\ln n}{n!}$.

24. $\sum_{n=0}^{\infty} \frac{n}{3n + 1}$.

25. $\sum_{n=1}^{\infty} \frac{1}{n^{3/2}}$.

26. $\sum_{n=1}^{\infty} \frac{10^5}{n + n^2}$.

27. $\sum_{j=0}^{\infty} \frac{1}{(j + 2) \ln(j + 2)}$.

28. $\sum_{i=1}^{\infty} \frac{\text{Arctan } i}{i^2 + 1}$.

29. $\displaystyle\sum_{i=2}^{\infty} \frac{\operatorname{Arctan} i}{i^2 - 1}.$

30. $\displaystyle\sum_{i=1}^{\infty} \frac{i}{(i+1)i^3}.$

31. $\displaystyle\sum_{n=1}^{\infty} \frac{n^n}{n!}.$

32. $\displaystyle\sum_{i=1}^{\infty} \frac{i^2(\operatorname{Arctan} i)}{i^2 + 1}.$

In Exercises 33–37, express the numbers in the form $\frac{p}{q}$, where p and q are integers.

33. $0.99999999\ldots$

34. $0.33333333\ldots$

35. $0.66666666\ldots$

36. $5.3333333\ldots$

37. $6.012121212\ldots$

9.4   Alternating Series and Absolute Convergence

An *alternating series* is one whose terms alternate from positive to negative. If

$$\sum_{n=1}^{\infty} a(n) = a(1) + a(2) + a(3) + \cdots$$

is a series of positive terms, then

$$\sum_{n=1}^{\infty} (-1)^n a(n) = -a(1) + a(2) - a(3) + \cdots$$

and

$$\sum_{n=1}^{\infty} (-1)^{n+1} a(n) = a(1) - a(2) + a(3) - \cdots$$

are examples of alternating series.

Theorem 1 (The Alternating Series Test) *If a is a decreasing sequence of positive numbers (that is, $a(1) > a(2) > a(3) > \cdots > 0$), then $\sum_{n=1}^{\infty}(-1)^{n+1}a(n)$ and $\sum_{n=1}^{\infty}(-1)^n a(n)$ converge if and only if $\lim_{n\to\infty} a(n) = 0$.*

Proof: By Theorem 6 of Section 9.3 we know that if $\sum_{n=1}^{\infty}(-1)^{n+1}a(n)$ converges, then $\lim_{n\to\infty}(-1)^{n+1}a(n) = 0$. Clearly, $\lim_{n\to\infty}(-1)^{n+1}a(n) = 0$ implies that $\lim_{n\to\infty} a(n) = 0$. Now suppose that $\lim_{n\to\infty} a(n) = 0$. Let $s(n)$ denote the nth partial sum of the sequence given by $(-1)^{n+1}a(n)$, so that $s(n) = a(1) - a(2) + a(3) - a(4) + \cdots + (-1)^{n+1}a(n)$.

Then

$$
\begin{aligned}
s(2n+1) &= [a(1) - a(2)] + [a(3) - a(4)] + \cdots + [a(2n-1) - a(2n)] \\
&\quad + a(2n+1), \text{ and} \quad (1)
\end{aligned}
$$

$$
\begin{aligned}
s(2n+1) &= a(1) - [a(2) - a(3)] - [a(4) - a(5)] - \cdots \\
&\quad - [a(2n) - a(2n+1)]. \quad (2)
\end{aligned}
$$

The terms $[a(i) - a(i+1)]$, for $i = 1, 2, \ldots, 2n$, and $a(2n+1)$ are positive; thus from Equation (1) it follows that $s(2n+1) > 0$. From Equation (2) we see that $s(2n+1) < a(1)$. Since the sequence $s(2n+1)$ is an increasing sequence bounded above by $a(1)$, the Completeness Axiom shows us that $\lim_{n\to\infty} s(2n+1)$ is a real number. A similar argument shows that $\lim_{n\to\infty} s(2n)$ is also a real number. Since $s(2n+1) - s(2n) = a(2n+1)$, we see that $\lim_{n\to\infty} s(2n+1) - \lim_{n\to\infty} s(2n) = \lim_{n\to\infty} a(2n+1) = 0$. Thus

$$
\lim_{n\to\infty} s(2n+1) = \lim_{n\to\infty} s(2n) = \sum_{n=1}^{\infty} (-1)^{n+1} a(n). \qquad \blacksquare
$$

EXAMPLE 1: Show that $\sum_{n=1}^{\infty} (-1)^n \frac{1}{n}$ converges.

SOLUTION: Since $-\sum_{n=1}^{\infty} (-1)^n \frac{1}{n} = \sum_{n=1}^{\infty} (-1)^{n+1} \frac{1}{n}$, we see that $\sum_{n=1}^{\infty} (-1)^n \frac{1}{n}$ will converge if $\sum_{n=1}^{\infty} (-1)^{n+1} \frac{1}{n}$ converges. Clearly the sequence a defined by $a(n) = \frac{1}{n}$ is decreasing, and $\lim_{n\to\infty} \frac{1}{n} = 0$. Thus, by the alternating series test, $\sum_{n=1}^{\infty} (-1)^n \frac{1}{n}$ converges. $\blacksquare$

The alternating series test is the only convergence test we have that applies to a series whose terms are not all positive. Of course, not every series that has some negative terms is an alternating series. For a general series having some terms positive and some negative, it is useful to consider taking absolute values of each term. A series $\sum_{n=1}^{\infty} a(n)$ *converges absolutely* if $\sum_{n=1}^{\infty} |a(n)|$ converges. Example 1 is an example of a series that converges, but not absolutely, since $\sum_{n=1}^{\infty} \left| (-1)^n \left(\frac{1}{n} \right) \right|$ is the harmonic series. Such a series *converges conditionally*.

Theorem 2 *If $\sum_{n=1}^{\infty} a(n)$ converges absolutely, then it converges.*

Proof: Define the sequences c and d by

$$
c(n) = \begin{cases} a(n) & \text{if } a(n) \geq 0 \\ 0 & \text{if } a(n) < 0 \end{cases}
\qquad
d(n) = \begin{cases} |a(n)| & \text{if } a(n) \leq 0 \\ 0 & \text{if } a(n) > 0 \end{cases}
$$

Let $s_a(n) = \sum_{i=1}^{n} |a(i)|$ and $s_c(n) = \sum_{i=1}^{n} c(i)$. The sequence s_c is nondecreasing, and, for each n, it is clear that $s_c(n) \leq s_a(n)$. It follows from the Comparison Test that $\sum_{n=1}^{\infty} c(n)$ converges. Similarly,

$\sum_{n=1}^{\infty} d(n)$ converges. Since $\sum_{n=1}^{\infty} a(n) = \sum_{n=1}^{\infty} c(n) - \sum_{n=1}^{\infty} d(n)$, we see that $\sum_{n=1}^{\infty} a(n)$ converges. ∎

EXERCISES 9.4

In Exercises 1–18, test the following series for convergence and absolute convergence.

1. $\displaystyle\sum_{n=1}^{\infty} \frac{(-1)^n}{n^2}$.

2. $\displaystyle\sum_{n=1}^{\infty} (-1)^{2n} \left(\frac{1}{n}\right)$.

3. $\displaystyle\sum_{n=1}^{\infty} \cos(n\pi)/(n+1)$.

4. $\displaystyle\sum_{n=1}^{\infty} \frac{(-1)^n}{(n+1)\ln(n+1)}$.

5. $\displaystyle\sum_{n=1}^{\infty} \frac{\cos(n\pi)\ln(n+1)}{(n+1)}$.

6. $\displaystyle\sum_{n=1}^{\infty} \cos(2n\pi)n^{-2/3}$.

7. $\displaystyle\sum_{n=1}^{\infty} \frac{(-1)^{n-1}}{(n+1)^{4/3}}$.

8. $\displaystyle\sum_{n=1}^{\infty} \frac{\cos(n\pi)n}{(n+1)^{3/2}}$.

9. $\displaystyle\sum_{n=1}^{\infty} \sin(n)n^{-4/3}$.

10. $\displaystyle\sum_{n=1}^{\infty} (-1)^n \frac{n+1}{n+6}$.

11. $\displaystyle\sum_{n=1}^{\infty} (-1)^n \frac{n^2+1}{n+6}$.

12. $\displaystyle\sum_{n=1}^{\infty} (-1)^n \frac{n+1}{n^2+6}$.

13. $\displaystyle\sum_{n=1}^{\infty} (-1)^n n \sin(1/n)$.

14. $\displaystyle\sum_{n=1}^{\infty} (-1)^n \left(\frac{n}{n+6}\right)^n$.

15. $\displaystyle\sum_{n=1}^{\infty} (-1)^n \frac{2^{3n}}{5}$.

16. $\displaystyle\sum_{n=1}^{\infty} (-1)^n \frac{2^{3n}}{15}$.

17. $\displaystyle\sum_{n=1}^{\infty} (-1)^n \left(\frac{n^n}{n!}\right)$.

18. $\displaystyle\sum_{k=1}^{\infty} (-1)^k \left(\frac{\ln k}{k}\right)$.

19. Explain why $-1 < \displaystyle\sum_{k=1}^{\infty} (-1)^k \left(\frac{1}{k}\right)$.

20. Explain why $-0.5 > \displaystyle\sum_{k=1}^{\infty} (-1)^k \left(\frac{1}{k}\right)$.

9.5   The Ratio Test and Power Series

Theorem 1 (The Ratio Test) *Suppose that $\sum_{n=1}^{\infty} a(n)$ is a series of positive terms. Let $r = \lim_{n\to\infty} \frac{a(n+1)}{a(n)}$.*

(a) *If $r < 1$, then $\sum_{n=1}^{\infty} a(n)$ converges.*

(b) *If $r > 1$, then $\sum_{n=1}^{\infty} a(n)$ diverges.*

(c) *If $r = 1$, then the Ratio Test provides no information about the convergence of $\sum_{n=1}^{\infty} a(n)$.*

Proof:

Part (a): Suppose that $r < 1$. Let t be a number such that $r < t < 1$. Since $r = \lim_{n\to\infty} \frac{a(n+1)}{a(n)}$, there is a number N such that if $n \geq N$ then $\frac{a(n+1)}{a(n)} < t$. We then have

$$\frac{a(N+1)}{a(N)} < t \quad \Rightarrow \quad a(N+1) < a(N)t$$

$$\frac{a(N+2)}{a(N+1)} < t \quad \Rightarrow \quad a(N+2) < a(N+1)t \Rightarrow a(N+2) < a(N)t^2$$

$$\frac{a(N+3)}{a(N+2)} < t \quad \Rightarrow \quad a(N+3) < a(N+2)t \Rightarrow a(N+3) < a(N)t^3$$

$$\vdots \qquad\qquad\qquad \vdots \qquad\qquad\qquad \vdots$$

$$\frac{a(N+m)}{a(N+m-1)} < t \quad \Rightarrow \quad a(N+m) < a(N+m-1)t \Rightarrow a(N+m) < a(N)t^m$$

$$\vdots \qquad\qquad\qquad \vdots \qquad\qquad\qquad \vdots$$

(where "$\Rightarrow$" stands for "implies").

By the above we see that

$$\sum_{n=1}^{\infty} a(n) \;=\; \sum_{n=1}^{N} a(n) + \sum_{n=N+1}^{\infty} a(n)$$

$$=\; \sum_{n=1}^{N} a(n) + \sum_{m=1}^{\infty} a(N+m)$$

$$< \sum_{n=1}^{N} a(n) + \sum_{m=1}^{\infty} a(N)t^m.$$

Since the series $\sum_{n=1}^{N} a(n)$ is finite, and the series $\sum_{m=1}^{\infty} a(N)t^m$ is a geometric series with $t < 1$, it follows that $\sum_{n=1}^{\infty} a(n)$ converges.

Part (b): Suppose that $r > 1$. Let t be a number such that $r > t > 1$. Since $r = \lim_{n\to\infty} \frac{a(n+1)}{a(n)}$, there is a number N such that if $n \geq N$ then $\frac{a(n+1)}{a(n)} > t$. We then have, as before,

$$\frac{a(N+1)}{a(N)} > 1 \quad \Rightarrow \quad a(N+1) > a(N)$$

$$\frac{a(N+2)}{a(N+1)} > 1 \quad \Rightarrow \quad a(N+2) > a(N+1) \Rightarrow a(N+2) > a(N)$$

$$\frac{a(N+3)}{a(N+2)} > 1 \quad \Rightarrow \quad a(N+3) > a(N+2) \Rightarrow a(N+3) > a(N)$$

$$\vdots \qquad\qquad \vdots \qquad\qquad \vdots$$

$$\frac{a(N+m+1)}{a(N+m)} > 1 \quad \Rightarrow \quad a(N+m+1) > a(N+m)$$

$$\Rightarrow a(N+m+1) > a(N)$$

$$\vdots \qquad\qquad \vdots \qquad\qquad \vdots$$

It follows that $\lim_{n\to\infty} a(n) = \lim_{m\to\infty} a(N+m) \geq a(N)$. Since $a(N) \neq 0$, by Theorem 6 of Section 9.3, the series $\sum_{m=1}^{\infty} a(n)$ diverges.

Part (c): Consider the p–series $\sum_{n=1}^{\infty} \left(\frac{1}{n}\right)^p$.

$$\lim_{n\to\infty} \frac{\frac{1}{(n+1)^p}}{\frac{1}{n^p}} = \lim_{n\to\infty} \left(\frac{n}{n+1}\right)^p = 1.$$

However, we know that the p–series converges if $p > 1$ and diverges if $p \leq 1$. ∎

Useful Observation Suppose that $\sum_{n=1}^{\infty} a(n)$ is any series (we are not requiring that the terms are positive!). If we use the Ratio Test to check for absolute convergence and find that $\lim_{n\to\infty} |a(n+1)|/|a(n)| > 1$, then our argument for Part (b) shows that the sequence $|a(n)|$ does not converge to 0. It follows that $a(n)$ does not converge to 0. By Theorem 6 of Section 9.3, $\sum_{n=1}^{\infty} a(n)$ does not converge. ***Thus, if the ratio test shows that the series does not converge absolutely, then it follows that the series is divergent.***

EXAMPLE 1: Test $\sum_{n=1}^{\infty} frac2^n n!$ for convergence.

SOLUTION: We employ the Ratio Test

$$\lim_{n \to \infty} \frac{2^{n+1}/(n+1)!}{2^n/n!} = \lim_{n \to \infty} \frac{(2)2^n/[(n+1)n!]}{2^n/n!} = \lim_{n \to \infty} \frac{2}{n+1} = 0.$$

Thus, by the Ratio Test, $\sum_{n=1}^{\infty} \dfrac{2^n}{n!}$ converges.   ∎

If you need to test a series for convergence, the following signpost may help.

Signpost for Series

Given a series $\sum_{n=0}^{\infty} a_n$:

1. Is $\lim_{n \to \infty} a_n = 0$? If not, then the series diverges.

2. If Step 1 is true, then check for the convergence of $\sum_{n=0}^{\infty} |a_n|$.

   (a) The Comparison Test is usually used with a p–Series or Geometric Series.

   (b) The Ratio Test is used if a_n contains factorials or powers of n.

   (c) The Integral Test is used if $|a|$ is a decreasing sequence and you can integrate the function f which has $f(n) = a_n$.

If the series $\sum_{n=0}^{\infty} |a_n|$ converges, then the series $\sum_{n=0}^{\infty} a_n$ converges absolutely.

---

3. If the series in Step 2 does not converge and all of the terms a_n are positive, then $\displaystyle\sum_{n=0}^{\infty} a_n$ diverges.

4. If instead, the series in Step 2 diverges, but the series $\displaystyle\sum_{n=0}^{\infty} a_n$ is an alternating series with $|a_{n+1}| < |a_n|$ for all n and $\displaystyle\lim_{n\to\infty} a_n = 0$, then $\displaystyle\sum_{n=0}^{\infty} a_n$ converges conditionally.

The signpost for series is layed out as a flow chart below.

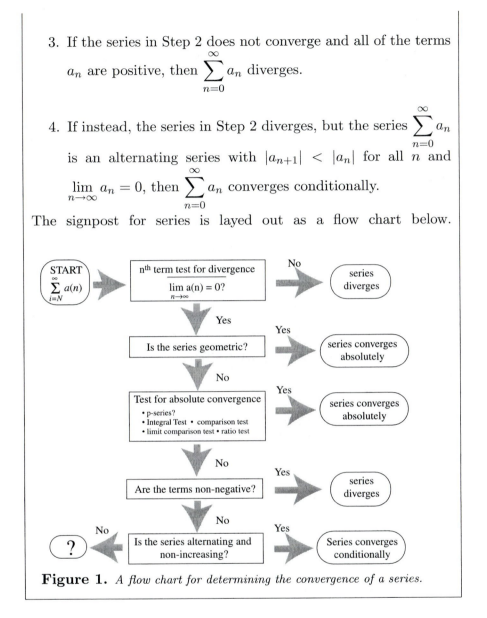

Figure 1. *A flow chart for determining the convergence of a series.*

Power Series and Radius of Convergence

Recall that if f is a function and all of its derivatives exist at the number c, then the nth Taylor polynomial for f at c is

$$P_n(x) = \sum_{k=0}^{n} a(k)(x-c)^k, \text{ where } a(k) = \frac{f^{(k)}(c)}{k!}.$$

Formally we can let n approach ∞ and obtain

$$\lim_{n\to\infty} P_n(x) = \sum_{k=0}^{\infty} a(k)(x-c)^k.$$

Our interest here is twofold:

(a) For what values of x does $\displaystyle\sum_{k=0}^{\infty} a(k)(x-c)^k$ converge?

(b) Given that $\displaystyle\sum_{k=0}^{\infty} a(k)(x-c)^k$ converges, does it converge to $f(x)$?

We have a reasonable collection of tools to study (a), and this is what we will concentrate on here. Question (b) is somewhat harder to answer. Recall, however, that part of Taylor's Theorem gives us an error function for $P_n(x)$. That is, for each x,

$$|f(x) - P_n(x)| < \frac{f^{(n+1)}(\mu)}{(n+1)!}(x-c)^{n+1},$$

for some μ between c and x.

A series of the form $\sum_{k=0}^{\infty} a(k)(x-c)^k$ is a *power series*. For a given x, we use the Ratio Test to determine when $\sum_{k=0}^{\infty} a(k)(x-c)^k$ is absolutely convergent.

EXAMPLE 2: Determine the values for x so that $\displaystyle\sum_{k=0}^{\infty} x^k$ converges.

SOLUTION: We first employ the Ratio Test on the series $\displaystyle\sum_{k=0}^{\infty} |x^k|$.

$$\lim_{k\to\infty} \frac{|x^{k+1}|}{|x^k|} = \lim_{k\to\infty} |x| = |x|.$$

The Ratio Test tells us that $\sum_{k=0}^{\infty} x^k$ converges absolutely for all $|x| < 1$. According to the useful observation preceding Theorem 1, $\sum_{k=0}^{\infty} x^k$ diverges for all $|x| > 1$. The cases $x = \pm 1$ must be handled separately. When $x = 1$, the series is $\sum_{k=0}^{\infty} 1^k$, and when $x = -1$, the series is $\sum_{k=0}^{\infty} (-1)^k$. In either case, $\lim_{n\to\infty} a(n) \neq 0$. We know that these two series diverge. Thus $\sum_{k=0}^{\infty} x^k$ converges if and only if $|x| < 1$.   ■

Rules for Determining Convergence of a Power Series

If the power series $\sum_{k=0}^{\infty} a(k)(x - c)^k$ passes the Ratio Test for absolute convergence at any value $x \neq c$, then either

(a) There is a number R such that:

   (i) If $|x - c| < R$, then $\sum_{k=0}^{\infty} a(k)(x - c)^k$ converges;

   (ii) If $|x - c| > R$, then $\sum_{k=0}^{\infty} a(k)(x - c)^k$ diverges;

   (iii) If $|x - c| = R$, then another test must be used to determine if $\sum_{k=0}^{\infty} a(k)(x - c)^k$ converges; or,

(b) $\sum_{k=0}^{\infty} a(k)(x - c)^k$ converges for all x.

If Case (a) holds, the number R is called the *radius of convergence* of the series. If Case (b) holds, then we will say that the radius of convergence is infinite and we will write $R = \infty$.

In Example 2, the radius of convergence is 1.

Definition: Set of Convergence

The *set of convergence* of the series $\sum_{k=0}^{\infty} a(k)(x - c)^k$ is the set to which x belongs if and only if $\sum_{k=0}^{\infty} a(k)(x - c)^k$ converges.

The set of convergence in Example 2 is the set $(-1, 1) = \{x \mid -1 < x < 1\}$.

EXAMPLE 3: The nth degree Maclaurin polynomial for e^x is

$$P_n(x) = 1 + x + \frac{x^2}{2} + \frac{x^3}{2 \cdot 3} + \cdots + \frac{x^n}{n!}.$$

By Taylor's Theorem, there is a number t between 0 and x such that the error in $P_n(x)$ is $R(t) = \frac{t^{n+1}}{(n+1)!}$. Since t is is between 0 and x, $|t|^{n+1} \leq |x^{n+1}|$. It follows that $|R(x)| \leq \frac{|x|^{n+1}}{(n+1)!}$. Clearly, $\lim_{n \to \infty} \frac{|x|^{n+1}}{(n+1!)} = 0$ for all x. It follows that $\sum_{n=0}^{\infty} \frac{x^n}{n!}$ converges to e^x for all x. Thus the radius of convergence for $\sum_{n=0}^{\infty} \frac{x^n}{n!}$ is infinite. Let us see if this agrees with what we get when we use the Ratio Test:

$$\lim_{n \to \infty} \frac{|x^{n+1}/(n+1)!|}{|x^n/n!|} = \lim_{n \to \infty} \left| \frac{x}{n+1} \right| = 0$$

for every x. Thus, by the Ratio Test, $R = \infty$. The set of convergence for $\sum_{n=0}^{\infty} \frac{x^n}{n!}$ is the real line. ∎

EXAMPLE 4: Find the radius of convergence and the set of convergence for $\displaystyle\sum_{n=0}^{\infty} \frac{x^n}{n}$.

SOLUTION: We will use the Ratio Test to determine the radius of convergence for $\sum_{n=0}^{\infty} \frac{x^n}{n}$.

$$\lim_{n \to \infty} \frac{\left| x^{n+1}/(n+1) \right|}{\left| x^n/n \right|} = \lim_{n \to \infty} |x| \frac{n}{n+1} = |x|.$$

Thus $R = 1$.

The set of convergence contains the set $(-1, 1) = \{x \mid -1 < x < 1\}$. We need to check the endpoints individually. If $x = 1$, we obtain the series $\sum_{n=1}^{\infty} \frac{1}{n}$, which is the harmonic series and is divergent. If $x = -1$, we obtain the series $\sum_{n=1}^{\infty} (-1)^n \left(\frac{1}{n} \right)$, which is a convergent alternating series, since $\lim_{n \to \infty} \frac{1}{n} = 0$. Hence the set of convergence is $[-1, 1) = \{x \mid -1 \leq x < 1\}$. ∎

EXAMPLE 5: Find the radius and set of convergence for $\displaystyle\sum_{n=1}^{\infty} \frac{(x-3)^n}{n}$.

SOLUTION: Using the Ratio Test we obtain

$$\lim_{n \to \infty} \frac{\left| (x-3)^{n+1}/(n+1) \right|}{\left| (x-3)^n/n \right|} = \lim_{n \to \infty} |x-3| \frac{n}{n+1} = |x-3|.$$

Thus $\sum_{n=1}^{\infty} \frac{(x-3)^n}{n}$ converges if $|x-3| < 1$ and diverges if $|x-3| > 1$. The radius of convergence R is 1. The set of convergence contains the set of all x such that $|x-3| < 1$, or $2 < x < 4$. If $x < 2$ or $x > 4$, then x is not in the set of convergence. If $x = 2$, then we have the series $\sum_{n=1}^{\infty} \frac{(-1)^n}{n}$, which is a convergent alternating series. If $x = 4$, then we have the series $\sum_{n=1}^{\infty} \frac{1}{n}$, which is the harmonic series and diverges. The set of convergence is $[2, 4) = \{x \mid 2 \leq x < 4\}$. ∎

EXERCISES 9.5

Test the series in Exercises 1–10 for convergence using the Ratio Test. If the Ratio Test fails to determine whether the series converges, use some other test.

1. $\displaystyle\sum_{n=0}^{\infty} \frac{n^2}{n!}$.

2. $\displaystyle\sum_{n=0}^{\infty} \frac{n^2}{2^n}$.

3. $\displaystyle\sum_{n=1}^{\infty} \frac{n!}{n^2}$.

4. $\displaystyle\sum_{n=0}^{\infty} \frac{2^n}{3^{n+1}}$.

5. $\displaystyle\sum_{n=0}^{\infty} \frac{3^n}{2^{n+3}}$.

6. $\displaystyle\sum_{n=1}^{\infty} \frac{2^n}{n^2}$.

7. $\displaystyle\sum_{n=1}^{\infty} \frac{(0.5)^n}{n^2}$.

8. $\displaystyle\sum_{n=1}^{\infty} \frac{2 \cdot 4 \cdots (2n)}{(2n)!}$.

9. $\displaystyle\sum_{n=1}^{\infty} \frac{n!}{1 \cdot 3 \cdot 5 \cdots (2n-1)}$.

10. $\displaystyle\sum_{n=1}^{\infty} \frac{(-1)^n}{2n}$.

Find the radius and set of convergence of the following power series in Exercises 11–25.

11. $\displaystyle\sum_{n=1}^{\infty} \frac{(-1)^n}{n} x^n$.

12. $\displaystyle\sum_{n=1}^{\infty} n^{-1/2} x^n$.

13. $\displaystyle\sum_{n=1}^{\infty} (-1)^n (x-1)^n$.

14. $\displaystyle\sum_{n=1}^{\infty} \frac{(x+2)^n}{n}$.

15. $\displaystyle\sum_{n=0}^{\infty} \frac{(x-3)^n}{n!}$.

16. $\displaystyle\sum_{n=0}^{\infty} \frac{n(x-2)^{n-1}}{2^n}$.

17. $\displaystyle\sum_{n=1}^{\infty} \frac{(x-5)^n}{\sqrt{n}}$.

18. $\displaystyle\sum_{n=0}^{\infty} (-2)^n (x+\pi)^n$.

19. $\displaystyle\sum_{n=1}^{\infty} \frac{(2x-5)^n}{n}$.

20. $\displaystyle\sum_{n=0}^{\infty} (3x+\pi)^n$.

21. $\displaystyle\sum_{n=1}^{\infty} \frac{x^n}{1 \cdot 3 \cdots (2n-1)}$.

22. $\displaystyle\sum_{n=1}^{\infty} \frac{nx^n}{1 \cdot 3 \cdots (2n-1)}$.

23. $\displaystyle\sum_{n=1}^{\infty} \frac{n!x^n}{1 \cdot 3 \cdots (2n-1)}$.

24. $\displaystyle\sum_{n=1}^{\infty} \frac{[1 \cdot 3 \cdots (2n-1)]x^n}{n}$.

25. $\displaystyle\sum_{n=1}^{\infty} \frac{[2 \cdot 4 \cdots (2n)]x^n}{1 \cdot 3 \cdots (2n-1)}$.

26. Given that $x = 3$ is *not* in the set of convergence of $\displaystyle\sum_{n=1}^{\infty} a(n)x^n$, can $\displaystyle\sum_{n=1}^{\infty} a(n)x^n$ converge at $x = 4$? Explain your answer.

27. Given that $x = 3$ is *not* in the set of convergence of $\sum_{n=1}^{\infty} a(n)x^n$, can $\sum_{n=1}^{\infty} a(n)x^n$ converge at $x = -3$? Explain your answer.

28. Given that $x = 3$ is in the set of convergence of $\sum_{n=1}^{\infty} a(n)x^n$, *must* $\sum_{n=1}^{\infty} a(n)x^n$ converge at $x = -3$? Explain your answer.

29. Given that $x = 3$ is *not* in the set of convergence of $\sum_{n=1}^{\infty} a(n)(x - 4)^n$, can $\sum_{n=1}^{\infty} a(n)x^n$ converge at $x = 6$? Explain your answer.

30. Given that $x = 3$ is *not* in the set of convergence of $\sum_{n=1}^{\infty} a(n)(x - 4)^n$, can $\sum_{n=1}^{\infty} a(n)x^n$ converge at $x = 5$? Explain your answer.

31. Find the set of convergence of $\sum_{n=0}^{\infty} \dfrac{1}{x^n}$.

32. Find the set of convergence of $\sum_{n=0}^{\infty} \dfrac{1}{(x - 1)^n}$.

9.6　Power Series of Functions

Recall from Secton 4.5 that if the nth derivative of f exists at $x = c$, then the nth order Taylor polynomial for f centered at c is

$$\sum_{i=0}^{n} a_i(x - c)^i, \quad \text{where } a_i = \frac{f^{(i)}(c)}{i!}.$$

In Section 5.8, we defined the remainder $R_n(x)$ by $R_n(x) = f(x) - p_n(x)$. Clearly,

$$\sum_{i=0}^{\infty} a_i(x - c)^i = \lim_{n \to \infty} \sum_{i=0}^{i} a_n(x - c)^i = f(x)$$

if and only if $\lim_{n \to \infty} R_n(x) = 0$.

The notions of Taylor's polynomials and Maclaurin's polynomials extend naturally to the ideas of Taylor's series and Maclaurin's series. If $f(x)$ is a function such that $f^{(n)}(c)$ is defined for all $n > 0$, and if $a(n) = \frac{f^{(n)}(c)}{n!}$, then $\sum_{n=0}^{\infty} a(n)(x - c)^n$ is called the *Taylor series* for f at $x = c$. If $c = 0$, then the Taylor series for f at c is called the *Maclaurin series* for f. If the function f has a Taylor series at $x = c$, and if the series converges to f in a neighborhood of c, then f is said to be *analytic* at c. Thus, if all the derivatives of a function exist at a point c, then the Taylor's series for f at c is a power series.

The Maclaurin series for $\sin(x)$ is $\sum_{n=0}^{\infty} \dfrac{(-1^n)}{(2n + 1)!} x^{2n+1}$,

the Maclaurin series for $\cos(x)$ is $\sum_{n=0}^{\infty} \dfrac{(-1^n)}{(2n)!} x^{2n}$, and the Maclaurin series for e^x is $\displaystyle\sum_{n=0}^{\infty} \frac{1}{n!} x^n$.

In Section 9.5, we developed tools for determining the set of convergence for a power series $\sum_{n=0}^{\infty} a(n)(x-c)^n$. We found that if the set of convergence contains at least one point other than c, then either the set of convergence is the real line or there is a number R such that the set of convergence is the set of points $c - R < x < c + R$, together with perhaps one or both of the endpoints $c - R$ and $c + R$. For each x in the set of convergence of $\sum_{n=0}^{\infty} a(n)(x-c)^n$, let

$$g(x) = \sum_{n=0}^{\infty} a(n)(x-c)^n.$$

Then $g(x)$ is a function. Its domain is exactly the set of convergence of $\sum_{n=0}^{\infty} a(n)\,(x-c)^n$.

In Example 3 of the previous section, we used Taylor's Theorem to show that the MacLaurin series for e^x converges for all x. We also used the Ratio Test to show that the set of convergence is $(-\infty, \infty)$.

There are examples of functions whose Taylor series converge to completely different functions. However, in this text, we restrict our attention to functions that have the property that a Taylor series for the function will converge to the function at every point in the set of convergence.

You can often obtain the Taylor polynomial for composites in a natural way.

EXAMPLE 1:

(a) $\displaystyle\cos(2x) = \sum_{n=0}^{\infty} \frac{(-1)^n}{(2n)!} (2x)^{2n} = \sum_{n=0}^{\infty} \frac{(-1)^n \left(2^{2n}\right)}{(2n)!} (x)^{2n}.$

(b) $\displaystyle\cos(x^2) = \sum_{n=0}^{\infty} \frac{(-1)^n}{(2n)!} \left(x^2\right)^{2n} = \sum_{n=0}^{\infty} \frac{(-1)^n}{(2n)!} (x)^{4n}.$

(c) $\displaystyle x\cos(x) = x \sum_{n=0}^{\infty} \frac{(-1)^n}{(2n)!} (x)^{2n} = \sum_{n=0}^{\infty} \frac{(-1)^n}{(2n)!} x \cdot x^{2n}$

$\displaystyle \quad\quad\quad = \sum_{n=0}^{\infty} \frac{(-1)^n}{(2n)!} x^{2n+1}.$

(d) $\displaystyle\cos(2x+1) = \sum_{n=0}^{\infty} \frac{(-1^n)}{(2n)!} (2x+1)^{2n}$

$$= \sum_{n=0}^{\infty} \frac{(-1)^n}{(2n)!} \left(2\left(x + \frac{1}{2} \right) \right)^{2n}$$

$$= \sum_{n=0}^{\infty} \frac{(-1)^n 2^{2n}}{(2n)!} \left(x + \frac{1}{2} \right)^{2n}. \qquad \blacksquare$$

Notice in Part (d) that we changed the center of the Taylor Series. In the exercises, you are asked to show that if

$$f(x) = \sum_{i=0}^{\infty} a(n)(x - c)^n,$$

then

$$f(\alpha x + \beta) = \sum_{i=0}^{\infty} a(n)\alpha^n \left(x - \frac{c - \beta}{\alpha} \right)^n.$$

EXAMPLE 2: Let

$$f(x) = \sum_{i=0}^{\infty} \frac{1}{3n}(x - 2)^n.$$

Then

$$f(4x - 3) = \sum_{i=0}^{\infty} \frac{1}{3n}((4x - 3) - 2)^n = \sum_{i=0}^{\infty} \frac{4^n}{3n} \left(x - \frac{5}{4} \right)^n.$$

We started with the series centered at $c = 2$ for $f(x)$ and obtained the series for $f(4x - 3)$ centered at $c = \frac{5}{4}$. $\qquad \blacksquare$

We can differentiate and integrate a Taylor polynomial for a function f term by term to obtain a Taylor polynomial approximation for the derivative of f or the integral of f. The following theorem provides us with rules for doing Calculus on Taylor series.

Theorem 1 *Facts about $f(x) = \displaystyle\sum_{n=0}^{\infty} a(n)(x - c)^n$:*

(a) *f is differentiable for all x satisfying $|x - c| < R$, and $f'(x) = \sum_{n=0}^{\infty} na(n)(x - c)^{n-1}$. The radius of convergence of the series for f' is the same as the radius of convergence of the series for $f(x)$. (Thus a power series may be differentiated term by term.)*

(b) $\displaystyle\sum_{n=0}^{\infty} \frac{a(n)(x-c)^{n+1}}{n+1}$ *is an antiderivative of f. (Thus a power series can be integrated term by term.) The radius of convergence of the series for $\int f(x)\,dx$ is the same as the radius of convergence of the series for $f(x)$.*

(c) *Direct computation shows that* $a(n) = \dfrac{f^{(n)}(c)}{n!}$. *Thus* $\sum_{n=0}^{k} a(n)(x-c)^n$ *is the* k^{th} *Taylor polynomial for f near $x = c$.*

(d) *If $\sum_{n=0}^{\infty} b(n)(x-c)^n$ converges to f near c, then $b(n) = a(n)$ for every n. Thus this power series for f at $x = c$ is the only power series at $x = c$ that converges to f near c.*

EXAMPLE 3: Let $f(x) = \dfrac{1}{1+x}$. Letting $a(n) = \dfrac{f^{(n)}(0)}{n!}$, we get

$$\frac{1}{1+x} = 1 - x + x^2 - x^3 + x^4 - \cdots = \sum_{n=0}^{\infty} (-1)^n x^n. \qquad (3)$$

Since $D\left(\frac{1}{1+x}\right) = -\frac{1}{(1+x)^2}$, the Maclaurin series for $-\frac{1}{(1+x)^2}$ is obtained by differentiating Equation (3) term by term.

$$-\frac{1}{(1+x)^2} = -1 + 2x - 3x^2 + 4x^3 + \cdots = \sum_{n=0}^{\infty} n(-1)^n x^{n-1}$$

$$= \sum_{n=1}^{\infty} n(-1)^n x^{n-1}.$$

Multiplying by -1, we obtain

$$\frac{1}{(1+x)^2} = 1 - 2x + 3x^2 - 4x^3 + \cdots = \sum_{n=0}^{\infty} n(-1)^{n-1} x^{n-1}$$

$$= \sum_{n=1}^{\infty} n(-1)^{n-1} x^{n-1}.$$

Integrating Equation (3) term by term, we get

$$\ln(1+x) = C + x - \frac{x^2}{2} + \frac{x^3}{3} - \frac{x^4}{4} + \cdots$$

$$= C + \sum_{n=1}^{\infty} (-1)^{n+1} \frac{x^n}{n}.$$

Since $\ln(1) = 0$, the constant of integration is zero, and

$$\ln(1 + x) = x - \frac{x^2}{2} + \frac{x^3}{3} - \frac{x^4}{4} + \cdots = \sum_{n=1}^{\infty} (-1)^{n+1} \frac{x^n}{n}. \qquad \blacksquare$$

EXAMPLE 4: Find the Maclaurin series for $g(x) = \frac{1}{1+x^2}$.

SOLUTION: We already have a series expansion for $f(x) = \frac{1}{1+x}$ from Example 3. Notice that $g(x) = f(x^2)$.

$$f(x) = \frac{1}{1+x} = 1 - x + x^2 - x^3 + x^4 - \cdots = \sum_{n=0}^{\infty} (-1)^n x^n, \text{ so}$$

$$g(x) = f(x^2) = 1 - \left(x^2\right) + \left(x^2\right)^2 - \left(x^2\right)^3 + \left(x^2\right)^4 - \cdots$$

$$= \sum_{n=0}^{\infty} (-1)^n x^{2n}. \qquad \blacksquare$$

EXAMPLE 5: Find the Maclaurin series for $\text{Arctan}(x)$.

SOLUTION:

$$\text{Arctan}(x) = \int \frac{1}{1+x^2} \, dx$$

$$= \int 1 - x^2 + x^4 - x^6 + x^8 - \cdots \, dx$$

$$= C + x - \frac{x^3}{3} + \frac{x^5}{5} - \frac{x^7}{7} + \cdots$$

$$= C + \sum_{n=1}^{\infty} (-1)^{n+1} \frac{x^{2n-1}}{2n-1}.$$

Since $\text{Arctan}(0) = 0$, the constant of integration is zero, and the desired series expansion is

$$\text{Arctan}(x) = \sum_{n=1}^{\infty} (-1)^{n+1} \frac{x^{2n-1}}{2n-1}. \qquad \blacksquare$$

EXAMPLE 6: Find the Maclaurin series for $x \, \text{Arctan}(x)$.

SOLUTION: By Example 3, $\text{Arctan}(x) = x - \frac{x^3}{3} + \frac{x^5}{5} - \frac{x^7}{7} + \cdots$.
Multiplying by x leads to

$$
\begin{aligned}
x\,\text{Arctan}(x) &= x\left[x - \frac{x^3}{3} + \frac{x^5}{5} - \frac{x^7}{7} + \cdots\right] \\
&= x^2 - \frac{x^4}{3} + \frac{x^6}{5} - \frac{x^8}{7} + \cdots \\
&= \sum_{n=1}^{\infty}(-1)^{n+1}\frac{x^{2n}}{2n-1}.
\end{aligned}
$$

∎

EXERCISES 9.6

In Exercises 1–11, find the Maclaurin series expansion for the given function.

1. $f(x) = \dfrac{1}{1+x}$.

2. $f(x) = \dfrac{x}{1-x}$.

3. $f(x) = \dfrac{1}{1-x^2}$.

4. $f(x) = x^2\,\text{Arctan}(x)$.

5. $g(x) = \ln(1-x)$.

6. $f(x) = \sin(2x)$.

7. $f(x) = \cos\left(\frac{x}{2}\right)$.

8. $f(x) = x^2\cos(x^2)$.

9. $f(x) = (x+1)\sin(x^3)$.

10. $f(x) = (x^2 + x)e^{-x^2}$.

11. $g(x) = \sin^2(x)$ $\left(\text{Hint: } \sin^2(x) = \frac{1}{2}(1 - \cos(2x))\right)$.

12. Show that $x = 3$ is not in the domain of
$$f(x) = \sum_{i=0}^{\infty} x^n.$$

13. Show that $x = 3$ is not in the domain of
$$f(x) = \sum_{i=0}^{\infty} \frac{(x-5)^n}{3n+2}.$$

14. Find the domain of the function in Exercise 12.

15. Find the domain of the function in Exercise 13.

16. Given that $x = 3$ is *not* is in the domain of
$$f(x) = \sum_{n=1}^{\infty} a(n)x^n,\text{ can } x = 4 \text{ be in the domain}$$
of f? Explain your answer.

17. Given that $x = 3$ is *not* in the domain of
$$g(x) = \sum_{n=1}^{\infty} a(n)x^n,\text{ can } x = -3 \text{ be in the domain}$$
of g? Explain your answer.

18. Given that $x = 3$ is in the domain of
$$f(x) = \sum_{n=1}^{\infty} a(n)x^n,\ \textit{must } x = -3 \text{ be in the do-}$$
main of f? Explain your answer.

19. Given that $x = 3$ is *not* in the domain of
$$f(x) = \sum_{n=1}^{\infty} a(n)(x-4)^n,\text{ can } x = 6 \text{ be in the}$$
domain of f? Explain your answer.

20. Can $x = 5$ be in the domain of $f(x) = \sum_{n=1}^{\infty} a(n)(x-4)^n$ if $x = 3$ is *not* in the domain of f? Explain your answer.

In Exercises 21–23, direct substitution of $g(x)$ into $f(x) = \sum_{n=1}^{\infty} \frac{n^2}{n!}(x-4)^n$ gives a series expansion of $f(g(x))$ of the form $\sum_{n=1}^{\infty} A(n)(x-C)^n$.

21. Find $\{A(n)\}$ and C if $g(x) = 4x - 2$.

22. Find $\{A(n)\}$ and C if $g(x) = 2x + 5$.

23. Find $\{A(n)\}$ and C if $g(x) = \alpha x + \beta$.

In Exercises 24–27, direct substitution of $g(x) = 3x+2$ into $f(x) = \sum_{n=1}^{\infty} a(n)(x-c)^n$ gives a series expansion of $f(g(x))$ of the form $f(g(x)) = \sum_{n=1}^{\infty} A(n)(x-C)^n$. The domain of $f(x)$ is $(-2, 4]$.

24. What is the center of convergence of $\sum_{n=1}^{\infty} a(n)(x-c)^n$?

25. What is C (the center of convergence of $\sum_{n=1}^{\infty} A(n)(x-C)^n$)?

26. What is the domain of $f(g(x))$?

27. What is the domain of $f(g(x))$?

28. The derivative of $f(x) = \sum_{n=0}^{\infty} a(n)(x-c)^n$ is $f'(x) = \sum_{n=1}^{\infty} na(n)(x-c)^{n-1}$. Show that the radius of convergence for $\sum_{n=0}^{\infty} a(n)(x-c)^n$ is the same as the radius of convergence for its derivative.

9.7   Radius of Convergence for Rational Functions

We start this section with an example.

EXAMPLE 1: Recall that from Example 3 of the previous section that the Maclaurin series for $f(x) = \frac{1}{1+x}$ is given by

$$\frac{1}{1+x} = 1 - x + x^2 - x^3 + x^4 - \cdots = \sum_{n=0}^{\infty} (-1)^n x^n.$$

The radius of convergence of this series is 1. Notice that f is undefined at $x = 1$, and that the distance from $x = 0$ to $x = 1$ is also 1. The significance of the point $x = 0$ is that the Maclaurin series is centered at 0 (i.e., $c = 0$). The fact that the radius of convergence equals the distance from $x = 0$ to the closest point where f is undefined is more than a coincidence!   ∎

Before proceding, we pause to review complex numbers. Recall that, by definition, $i = \sqrt{-1}$. The roots of the equation $x^2 + 1 = 0$ are $\pm i$. Application of the quadratic formula shows that $x = -2 \pm i$ are the roots of $x^2 + 4x + 5 = 0$. Numbers of the form $\alpha + \beta i$ are complex numbers; α is called the real part and β is the imaginary part of the complex number. We add and multiply two complex numbers just as we would multiply two polynomials. Thus $(a + bi) + (c + di) =$

$(a + c) + (b + d)i$, and $(a + bi)(c + di) = ac + (ad + bc)i + (bi)(di) = (ac - bd) + (bc + bd)i$.

EXAMPLE 2: $(2 + 3i)(1 - 4i) = (2 + 12) + (-8 + 3)i = 14 - 5i.$ ■

The complex number $a + bi$ is identified with the point (a, b) in the *complex plane*. Rather than having an x–axis and a y–axis, we have a real axis and an imaginary axis. See Figure 1. Of importance in this section is that the distance between two points in the complex plane is defined just as the distance between two points in the xy–plane. Thus the distance between $a_1 + b_1 i$ and $a_2 + b_2 i$ is $\sqrt{(a_1 - a_2)^2 + (b_1 - b_2)^2}$.

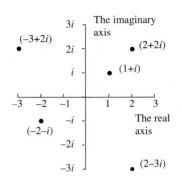

Figure 1. *The complex plane.*

EXAMPLE 3:

- The distance between $1 + 2i$ and $2 + 3i$ is $\sqrt{(1 - 2)^2 + (2 - 3)^2} = \sqrt{2}$.

- The distance between $1 + 2i$ and i is $\sqrt{(1 - 0)^2 + (2 - 1)^2} = \sqrt{2}$.

- The distance between 1 and i is $\sqrt{(1 - 0)^2 + (0 - 1)^2} = \sqrt{2}$.

- The distance between 0 and i is $\sqrt{(0 - 0)^2 + (0 - 1)^2} = \sqrt{2}$.

EXAMPLE 4: Let $f(x) = \frac{1}{1+x^2}$. In Example 4 of Section 9.6, we saw that the Maclaurin series for f is given by

$$\frac{1}{1 + x^2} = 1 - x^2 + x^4 - x^6 + \cdots = \sum_{n=1}^{\infty} (-1)^n x^{2n}.$$

The radius of convergence of this series is also 1, even though $\frac{1}{1+x^2}$ is defined everywhere (or is it?). Notice that the complex number $i = \sqrt{-1}$ is a solution of $1 + x^2 = 0$, and that in the complex plane i is 1 unit in distance from 0. ■

The relationship between the radius of convergence of the Taylor series for $\frac{1}{1+x^2}$ centered at $c = 0$, and the distance between c and the roots of $1 + x^2$ is just one example that illustrates the following theorem.

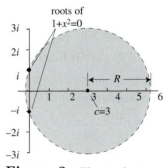

Figure 2. *The circle in the complex plane centered at $c = 3$ and containing the solutions to the equation $1 + x^2 = 0$.*

Theorem 1 *Suppose that f is a function whose power series about $x = c$ has a radius of convergence of R_f, and $P(x) = a_0 + a_1 x + \cdots a_n x^n$ is a polynomial function with (complex) roots $A_1, A_2, \ldots A_n$. If $P(c) \neq 0$, then $\frac{f(x)}{P(x)}$ has a power series at $x = c$, and the radius of convergence of the series for $\frac{f(x)}{P(x)}$ at $x = c$ is at least as large as the minimum of R_f and the distances from $(c, 0)$ to the $A_i's$.*

EXAMPLE 5: Find a lower bound for the radius of convergence of the power series expansion of $\dfrac{\sin(x)}{1+x^2}$ about $c = 3$.

SOLUTION: The radius of convergence for the power series for $\sin x$ is ∞. The roots for $1 + x^2$ are $\pm i = (0, 1), (0, -1)$. The distance from $(3, 0)$ to $(0, 1)$ is $\sqrt{10}$, and the distance from $(3, 0)$ to $(0, -1)$ is $\sqrt{10}$. See Figure 2. The radius of convergence for the power series of $\dfrac{\sin(x)}{1+x^2}$ is at least $\min\{\sqrt{10}, \infty\} = \sqrt{10}$. ■

EXAMPLE 6: Find a lower bound for the radius of convergence of the power series expansion of $\dfrac{\sin(x)}{(x^2+1)(x^2+x+1)}$ about $c = 3$.

SOLUTION: As in Example 5, the radius of convergence of the series expansion for $\sin(x)$ is ∞, and the roots for $x^2 + 1$ are $\pm i = (0, \pm 1)$. The roots for $x^2 + x + 1$ are $\left(\frac{1}{2}\right)\left(-1 \pm i\sqrt{3}\right) = \left(-\frac{1}{2}, \pm\frac{\sqrt{3}}{2}\right)$. See Figure 3.

(a) The distance from $(3, 0)$ to $(0, 1)$ is $\sqrt{10}$.

(b) The distance from $(3, 0)$ to $(0, -1)$ is $\sqrt{10}$.

(c) The distance from $(3, 0)$ to $\left(-\frac{1}{2}, \frac{\sqrt{3}}{2}\right)$ is $\sqrt{13}$.

(d) The distance from $(3, 0)$ to $\left(-\frac{1}{2}, -\frac{\sqrt{3}}{2}\right)$ is $\sqrt{13}$.

The radius of convergence of the series for $\dfrac{\sin(x)}{(x^2+1)(x^2+x+1)}$ is at least $\min(\sqrt{10}, \sqrt{13}, \infty) = \sqrt{10}$. ■

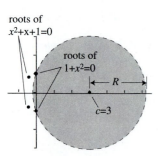

Figure 3. *The smallest circle in the complex plane centered at $c = 3$, which contains no roots of $1 + x^2$ or $x^2 + x + 1$ in its interior.*

EXERCISES 9.7

In Exercises 1–4, find the distance between the given pair of complex numbers.

1. $1 + 3i$ and $-2 + i$.   2. $3i$ and $2 + i$.

3. 2 and $-i$.   4. 4 and $\sqrt{-9}$.

In Exercises 5–9, use the techniques of this section to find a lower bound for the radius of convergence for the Taylor Series expansion of the following functions near the given point.

5. $f(x) = \dfrac{x}{x^2+1}$, at $c = 2$.

6. $f(x) = \dfrac{x}{(x^2+4)(x^2+x+1)}$, at $c = 1$.

7. $f(x) = \dfrac{e^x}{(x^2+4)(x^2+x+1)}$, at $c = 1$.

8. $f(x) = \dfrac{\ln(1-x)}{(x^2+4)(x^2+x+1)}$, at $c = 0$.

9. $f(x) = \dfrac{\text{Arctan}(x)}{(x^2+4)(x^2+x+1)}$, at $c = 0$.

Chapter 10

Techniques of Integration

In this chapter, we discuss various methods for evaluating integrals. So far the tools at our disposal are the chain rule in reverse and tables of integrals. As mentioned in Chapter 7, integral tables (or the computer equivalents such as Maple or Mathematica) are extremely valuable tools for evaluating integrals, yet familiarity with the use of such tools is insufficient. Some integrals you encounter will not be listed in integral tables. More importantly, however, is that sometimes converting an integral into an equivalent integral makes it easier to evaluate, or it allows you to gain insight into properties of the solution.

Many integrals do not even have "closed form" solutions, that is, solutions that can be written as a finite number of terms. In these cases a numerical approximation of the definite integral is all that can be obtained.

10.1 Integration By Parts

"Integration by parts" is a method often useful in evaluating integrals of products of functions. We start with the product rule for differentiation,

$$D[f(x)g(x)] = f'(x)g(x) + f(x)g'(x).$$

The above equation can be rewritten as

$$f(x)g'(x) = D[f(x)g(x)] - g(x)f'(x). \qquad (1)$$

By taking the antiderivative of both sides of Equation (1), we obtain:

$$\int f(x)g'(x)\ dx = f(x)g(x) - \int g(x)f'(x)\ dx. \qquad (2)$$

459

In terms of a definite integral, Equation (2) has the following counterpart.

$$\int_a^b f(x)g'(x)\ dx = f(x)g(x)\Big|_a^b - \int_a^b g(x)f'(x)\ dx \qquad (3)$$

In order to use this method, we need to determine how to transform our given integral into one that fits Equation (2) or (3). We need to decide which part of our integrand will assume the part of $f(x)$, and which part will be the term $g'(x)$. The following "steps" provide us with a rule of thumb.

Steps for Using Integration by Parts

Given an integral of the form

$$\int h(x)\ dx,$$

choose functions f and g' so that:

(a) $f(x)g'(x) = h(x)$;

(b) it is "relatively" easy to integrate $g'(x)$, obtaining $g(x)$; and

(c) it is easier to find the antiderivative of gf' than it is to find the antiderivative of fg'.

In the following examples, we illustrate the usefulness of Equations (2) and (3).

EXAMPLE 1: Compute $\int_0^1 xe^x\ dx$.

SOLUTION: Reiterating the above steps, we choose f and g' so that:

(a) $f(x)g'(x) = xe^x$;

(b) it is relatively easy to find g; and

(c) it is easier to find the antiderivative of gf' than it is to find the antiderivative of fg'.

This process is something of an art, which is learned by practice. In this case we let $f(x) = x$ and $g'(x) = e^x$. Then Equation (3) gives us

$$\int_0^1 \underbrace{xe^x}_{fg'}\ dx = \underbrace{xe^x}_{fg}\Big|_0^1 - \int_0^1 \underbrace{e^x}_{gf'}\ dx$$

$$= \ e - e^x \Big|_0^1 = e - e + e^0 = e^0 = 1. \qquad \blacksquare$$

EXAMPLE 2: Compute $\int_0^\pi x^2 \cos x \, dx$.

SOLUTION: Let $f(x) = x^2$ and $g'(x) = \cos x$. Then $f'(x) = 2x$ and $g(x) = \sin x$. We have

$$\int_0^\pi \underbrace{x^2}_{f} \underbrace{\cos x}_{g'} \, dx = \underbrace{x^2}_{f} \underbrace{\sin x}_{g} \Big|_0^\pi - \int_0^\pi \underbrace{(\sin x)}_{g} \underbrace{(2x)}_{f'} \, dx.$$

In order to compute $\int_0^\pi 2x \sin x \, dx$, we see that $2x \sin x$ is of the form $f(x)g'(x)$, where, in this case, $f(x) = 2x$ and $g'(x) = \sin x$; so

$$\int_0^\pi \underbrace{2x}_{f} \underbrace{\sin x}_{g'} \, dx = \underbrace{(2x)}_{f} \underbrace{(-\cos x)}_{g} \Big|_0^\pi - \int_0^\pi \underbrace{(-\cos x)}_{g} \underbrace{(2)}_{f'} \, dx$$

$$= \ -2x \cos x + 2 \sin x \Big|_0^\pi = 2\pi.$$

Thus $\int_0^\pi x^2 \cos x \, dx = x^2 \sin x \Big|_0^\pi - 2\pi = -2\pi.$ $\blacksquare$

EXAMPLE 3: Compute $\int \sin^2 x \, dx$.

SOLUTION: We already know how to handle this problem by using the identity $\sin^2 x = \frac{1 - \cos 2x}{2}$. The technique of integration by parts gives us another approach:

$$\int \sin^2 x \, dx = \int \underbrace{\sin x}_{f} \underbrace{\sin x}_{g'} \, dx$$

$$= \ \underbrace{(\sin x)}_{f} \underbrace{(-\cos x)}_{g} - \int \underbrace{(-\cos x)}_{g} \underbrace{(\cos x)}_{f'} \, dx$$

$$= \ -\sin x \cos x + \int \cos^2 x \, dx$$

$$= \ -\sin x \cos x + \int (1 - \sin^2 x) \, dx$$

$$= \ -\sin x \cos x + x - \int \sin^2 x \, dx.$$

It appears that we have merely arrived at an expression that contains our original problem. However, the integral on the right is preceded by a negative sign. Adding $\int \sin^2 x \, dx$ to both sides of the equation and solving for $\int \sin^2 x \, dx$, we obtain

$$\int \sin^2 x \, dx = \frac{-\sin x \cos x + x}{2} + C,$$

where the $+C$ indicates our arbitrary constant.   ■

We can use the notation that we developed in Section 6.2 to facilitate our work. If we let $u = f(x)$ and $dv = g'(x)dx$, then integration by parts can be expressed as

$$\int u \, dv = uv - \int v \, du.$$

We must emphasize that this is merely a notational convenience.

EXAMPLE 4: Compute $\int \sin(2x) \cos x \, dx$.

SOLUTION: Let $u = \sin(2x)$ and $dv = \cos x$. Then $u' = 2\cos(2x)$ and $v = \sin x$ so that we have

$$\int \sin(2x) \cos x \, dx = \sin(2x) \sin x - \int 2 \sin x \cos(2x) \, dx.$$

$\int 2 \sin x \cos(2x) \, dx$ is an integral of a product, and we proceed by using integration by parts. Note that if we choose $u = \sin x$, then we will merely undo the previous integration by parts and end up with the statement $\int \sin(2x) \cos x \, dx = \int \sin(2x) \cos x \, dx$ (try it!). Let $u = \cos(2x)$ and $dv = \sin x$ so that $u' = -2 \sin(2x)$ and $v = -\cos x$. Thus

$$-\int 2 \sin x \cos(2x) \, dx$$

$$= -2 \left((\cos(2x))(-\cos x) - 2 \int \cos x \sin(2x) \, dx \right)$$

$$= 2 \cos(2x) \cos x + 4 \int \cos x \sin(2x) \, dx.$$

Thus

$$\int \sin(2x) \cos x \, dx$$

$$= \sin(2x) \sin x + 2 \cos(2x) \cos x + 4 \int \cos x \sin(2x) \, dx.$$

Solving for $\int \sin(2x) \cos x \, dx$, we have

$$\int \sin(2x) \cos x \, dx = -\frac{1}{3} \left(\sin(2x) \sin x + 2 \cos(2x) \cos x \right) + C. \quad \blacksquare$$

EXAMPLE 5: Show that if $n > 1$, then

$$\int \sec^n x \, dx = \frac{1}{n-1} \left(\sec^{n-2} x \tan x + (n-2) \int \sec^{n-2} x \, dx \right).$$

SOLUTION: Since $\sec^n x = \sec^{n-2} x \sec^2 x$, we can let $u = \sec^{n-2} x$ and $dv = \sec^2 x$. Thus

$$
\begin{aligned}
u' &= (n-2) \sec^{n-3} x \sec x \tan x \\
&= (n-2) \sec^{n-2} x \tan x,
\end{aligned}
$$

and

$$v = \tan x.$$

Using integration by parts, we obtain

$$\int \sec^n x \, dx = \sec^{n-2} x \tan x - \int (n-2) \sec^{n-2} x \tan^2 x \, dx.$$

Since $\tan^2 x = \sec^2 x - 1$, we have

$$\int \sec^n x \, dx = \sec^{n-2} x \tan x - (n-2) \int \sec^{n-2} x (\sec^2 x - 1) \, dx$$

$$= \sec^{n-2} x \tan x - (n-2) \int \sec^n x \, dx + (n-2) \int \sec^{n-2} x \, dx.$$

Adding $(n-2) \int \sec^n x \, dx$ to both sides, we get

$$(n-1) \int \sec^n x \, dx = \sec^{n-2} x \tan x + (n-2) \int \sec^{n-2} x \, dx,$$

or

$$\int \sec^n x \, dx$$

$$= \frac{1}{n-1} \left(\sec^{n-2} x \tan x + (n-2) \int \sec^{n-2} x \, dx \right).$$

Remarks: It is often difficult for the student to discern what method should be tried to evaluate a given integral. We have discussed two methods thus far: the chain rule in reverse, and integration by parts. In the next two sections we develop more techniques, which may increase the student's confusion about which technique is appropriate for a particular problem. We leave some signposts, which we hope will guide the student through this morass of techniques.

Signposts for Integration by Parts

(a) Integration by parts is usually used when the integrand is a product of polynomials, exponential, or trigonometric functions, such as:
$\int (x^3 + x)e^x \, dx, \int x \sin x \, dx$, or $\int e^x \cos x \, dx$.

(b) If $\int f(x)g(x) \, dx$ is an integral with $f(x)$ a polynomial and $g(x)$ an exponential or trigonometric function, the best strategy for integration by parts is usually to differentiate f and integrate g. The reason for this is that after taking a sufficient number of derivatives of the polynomial f, we arrive at a constant. However, don't be fooled by an integral of the type $\int xe^{x^2} \, dx$, which can be solved quite easily using the chain rule in reverse.

(c) On occasion we are presented with an integral of the form $\int f(x) \, dx$, where using the chain rule in reverse does not solve the problem. In such cases, we can sometimes use integration by parts, where we differentiate $f(x)$ and integrate 1. Examples of this technique are seen in Exercises 4 and 20 below. In other cases, we may need to factor $f(x)$ to apply integration by parts, such as: $\int \sin^6 x \, dx = \int \sin^5 x \sin x \, dx$.

We do not claim that these are the only signposts for the use of integration by parts, merely that when an integrand is of one of the previous types, then integration by parts is a reasonable technique to try.

EXERCISES 10.1

Evaluate the following integrals.

1. $\int_0^1 x^2 e^x \, dx.$

2. $\int \cos x \sin x \, dx.$

3. $\int x^2 e^{2x} \, dx.$

4. $\int x e^x \cos x \, dx.$

5. $\int x e^{(x^2)} \, dx.$

6. $\int x \cos x \, dx.$

7. $\int \ln |x| \, dx.$

8. $\int x \sec^2 x \, dx.$

9. $\int x \ln |x| \, dx.$

10. $\int \text{Arctan}\, x \, dx.$

11. $\int \sqrt{x} \ln |x| \, dx.$

12. $\int \text{Arcsin}\, x \, dx.$

13. $\int x^{-1} \ln |x| \, dx.$

14. $\int \ln |x^2 - 1| \, dx.$

15. $\int (\sin x) \ln |\cos x| \, dx.$

16. $\int \sec^3 \theta \, d\theta.$

17. $\int x^3 \sin x \, dx.$

18. $\int \sec^5 \theta \, d\theta.$

19. $\int_0^{\frac{1}{2}} \frac{x}{\sqrt{1 - x^2}} \, dx.$

20. $\int \csc^3 \theta \, d\theta.$

21. $\int_0^1 \sqrt{1 - x^2} \, dx.$

22. $\int \sin^4 x \, dx.$

23. $\int_0^{\frac{1}{2}} \frac{x^3}{\sqrt{1 - x^2}} \, dx.$

24. $\int \sin^4 3x \, dx.$

25. $\int_0^{\frac{1}{2}} x \sqrt{1 - x^2} \, dx.$

26. $\int \cos^5 x \, dx.$

27. $\int_0^{\frac{1}{2}} x^3 \sqrt{1 - x^2} \, dx.$

28. $\int \cos^4 3x \, dx.$

29. $\int e^x \cos x \, dx.$

30. **The Gamma Function:** Define
    $\Gamma(t) = \int_0^\infty x^{t-1} e^{-x} \, dx$ for $t \geq 1$.

    a. Compute $\Gamma(1)$ and $\Gamma(2)$.

    b. Using integration by parts, show that
       $\Gamma(t + 1) = t\Gamma(t)$.

    c. Using induction, show that $\Gamma(n + 1) = n!$
       for integers $n \geq 0$.

31. Sketch the graph of the curve that is
    parametrized by
    $$\vec{r}(\theta) = \sin(4\theta) \begin{pmatrix} \cos(\theta) \\ \sin(\theta) \end{pmatrix}, \quad 0 \leq \theta \leq \frac{\pi}{4}.$$
    Use the results of Exercise 6.4.16 to find the area
    of this region.

32. Find the area of the region swept out by
    $$\vec{r}(\theta) = \sin(n\theta) \begin{pmatrix} \cos(\theta) \\ \sin(\theta) \end{pmatrix}, \quad 0 \leq \theta \leq \frac{\pi}{n}.$$

33. Find the area of the region swept out by
    $$\vec{r}(\theta) = \cos(n\theta) \begin{pmatrix} \cos(\theta) \\ \sin(\theta) \end{pmatrix}, \quad 0 \leq \theta \leq \frac{\pi}{n}.$$

34. Find the area of the region swept out by
    $$\vec{r}(\theta) = \cos(n\theta) \begin{pmatrix} \cos(\theta) \\ \sin(\theta) \end{pmatrix}, \quad -\frac{\pi}{2n} \leq \theta \leq \frac{\pi}{2n}.$$

10.2   Trigonometric Substitutions

One of the most useful tools we have for computing integrals is the chain rule in reverse. Recall that

$$\int_a^b f(g(x))g'(x)\, dx = \int_{g(a)}^{g(b)} f(u)\, du, \tag{4}$$

and in the case of antiderivatives:

$$\int f(g(x))g'(x)\, dx = \int f(u)\, du \Big|_{u=g(x)}. \tag{5}$$

If, in Equations (4) and (5), $u = g(x)$ has an inverse, then Equations (4) and (5) can be rewritten in the following forms:

$$\int_{g^{-1}(c)}^{g^{-1}(d)} f(g(x))g'(x)\, dx = \int_c^d f(u)\, du,$$

and

$$\int f(g(x))g'(x)\, dx \Big|_{x=g^{-1}(u)} = \int f(u)\, du.$$

Equivalently, we may write

$$\int_{g^{-1}(c)}^{g^{-1}(d)} f(g(u))g'(u)\, du = \int_c^d f(x)\, dx, \tag{6}$$

and

$$\int f(g(u))g'(u)\, du \Big|_{u=g^{-1}(x)} = \int f(x)\, dx. \tag{7}$$

Although Equations (6) and (7) are equivalent to Equations (4) and (5) respectively, using them is a little more sophisticated than just doing the chain rule in reverse. Equations (6) and (7) suggest substituting a more complicated expression for x in the hopes that it will give some insight into the antiderivative of the original expression.

One useful set of substitutions is based upon the trigonometric identities $\cos^2 x + \sin^2 x = 1$ and $\tan^2 x + 1 = \sec^2 x$.

EXAMPLE 1: Compute $\int \sqrt{25 - x^2}\, dx$.

SOLUTION: We let $x = 5\sin(u) = g(u)$. Then $dx = g'(u)du = 5\cos(u)du$, and $u = \text{Arcsin}\left(\frac{x}{5}\right)$. Employing Equation (7), we have:

$$
\begin{aligned}
\int \sqrt{25 - x^2}\, dx &= \int \sqrt{25 - 25\sin^2(u)}\; 5\cos(u)\, du \Big|_{u = \text{Arcsin}(x/5)} \\
&= \int 25\cos^2(u)\, du \Big|_{u = \text{Arcsin}(x/5)} \\
&= \frac{25}{2}(\sin(u)\cos(u) + u) + C \Big|_{u = \text{Arcsin}(x/5)}.
\end{aligned}
$$

Since $\sin\left(\text{Arcsin}\left(\frac{x}{5}\right)\right) = \frac{x}{5}$, and $\cos\left(\text{Arcsin}\left(\frac{x}{5}\right)\right) = \sqrt{1 - x^2/25}$, we see that

$$
\int \sqrt{25 - x^2}\, dx = \frac{25}{2}\left[\frac{x}{5}\sqrt{1 - \frac{x^2}{25}} + \text{Arcsin}\left(\frac{x}{5}\right)\right] + C. \quad \blacksquare
$$

EXAMPLE 2: Compute $\int_0^1 \sqrt{1 + 4t^2}\, dt$. The substitution $t = \frac{\tan u}{2} = g(u)$ eliminates the radical in the integrand. With this choice of $g(u)$, $dt = g'(u)du = \frac{\sec^2 u}{2}du$, and $u = g^{-1}(t) = \text{Arctan}(2t)$. The endpoints become $g^{-1}(0) = 0$ (when $t = 0, u = 0$), and $g^{-1}(1) = \text{Arctan}(2)$ (when $t = 1, u = \text{Arctan}(2)$). Thus

$$
\begin{aligned}
\int_0^1 \sqrt{1 + 4t^2}\, dt &= \int_0^{\text{Arctan}(2)} \sqrt{1 + \tan^2(u)}\, \frac{\sec^2(u)}{2}\, du \\
&= \frac{1}{2}\int_0^{\text{Arctan}(2)} \sec^3(u)\, du.
\end{aligned}
$$

From Exercise 16 in Section 10.1, we have

$$
\int \sec^3(u)\, du = \frac{1}{2}\left[\sec(u)\tan(u) + \ln|\sec(u) + \tan(u)|\right] + C.
$$

Thus

$$
\begin{aligned}
\int_0^1 \sqrt{1 + 4t^2}\, dt &= \frac{1}{4}\left[\sec(u)\tan(u) + \ln|\sec(u) + \tan(u)|\right]\Big|_0^{\text{Arctan}(2)} \\
&= \frac{1}{4}\left[2\sqrt{5} + \ln(\sqrt{5} + 2)\right]. \quad \blacksquare
\end{aligned}
$$

In general, if all else fails and the integrand involves powers of $(x^2 + 1), (x^2 - 1)$, or $(1 - x^2)$, then trigonometric substitution is something to try. The identities to remember are: $\sin^2 x + \cos^2 x = 1$ and $\tan^2 x + 1 = \sec^2 x$.

Signposts for Using Trigonometric Substitutions

If the integrand involves $x^2 + 1$, try $x = \tan u$.

If the integrand involves $x^2 - 1$, try $x = \sec u$.

If the integrand involves $1 - x^2$, try $x = \sin u$.

Finally, there are a few more identities that are helpful in evaluating integrals involving $(x^2 + 1)$, $(x^2 - 1)$ or $(1 - x^2)$. Recall that we have defined the hyperbolic functions:[1]

$$\sinh(x) \quad = \quad \frac{e^x - e^{-x}}{2};$$

$$\cosh(x) \quad = \quad \frac{e^x + e^{-x}}{2};$$

$$\tanh(x) \quad = \quad \frac{\sinh(x)}{\cosh(x)} = \frac{e^x - e^{-x}}{e^x + e^{-x}};$$

$$\coth(x) \quad = \quad \frac{1}{\tanh(x)} = \frac{e^x + e^{-x}}{e^x - e^{-x}};$$

$$\operatorname{sech}(x) \quad = \quad \frac{1}{\cosh(x)} = \frac{2}{e^x + e^{-x}}; \text{ and,}$$

$$\operatorname{csch}(x) \quad = \quad \frac{1}{\sinh(x)} = \frac{2}{e^x - e^{-x}}.$$

We also have the following identities which parallel those for the trigonometric functions: $\cosh^2(x) - \sinh^2(x) = 1$ and $1 - \tanh^2(x) = \operatorname{sech}^2(x)$.

We can restrict the domain of each of the hyperbolic functions so that we obtain one-to-one functions. Thus, with the appropriate domain restrictions, their inverse functions exist. As with trigonometric functions, we represent the inverses by adding "Arc" to the function names. The following theorem shows that we can explicitly represent these inverses in terms of the logarithmic function.

---

[1] The standard pronunciation of $\sinh x$ is "cinch of x". The function $\cosh x$ rhymes with "gosh" and $\tanh x$ rhymes with "branch."

Theorem 1 *The inverses of the hyperbolic trigonometric functions are*

$$\text{Arcsinh}(x) = \ln(x + \sqrt{x^2 + 1});$$

$$\text{Arccosh}(x) = \ln(x + \sqrt{x^2 - 1}), x \geq 1;$$

$$\text{Arctanh}(x) = \frac{1}{2} \ln\left(\frac{1 + x}{1 - x}\right), |x| < 1;$$

$$\text{Arccoth}(x) = \frac{1}{2} \ln\left(\frac{x + 1}{x - 1}\right) = \text{Arctanh}\left(\frac{1}{x}\right), |x| > 1;$$

$$\text{Arcsech}(x) = \ln\left(\frac{1 + \sqrt{1 - x^2}}{x}\right) = \text{Arccosh}\left(\frac{1}{x}\right), 0 < x \leq 1;$$

and

$$\text{Arccsch}(x) = \ln\left(\frac{1}{x} + \frac{\sqrt{1 + x^2}}{|x|}\right) = \text{Arcsinh}\left(\frac{1}{x}\right), x \neq 0.$$

Proof: To prove that $\text{Arcsinh}(x) = \ln(x + \sqrt{x^2 + 1})$, we solve the equation $x = \sinh(y)$ for y.

$$\sinh(y) = \frac{e^y - e^{-y}}{2} = x$$
$$e^y - e^{-y} - 2x = 0$$
$$e^{2y} - 1 - 2xe^y = 0.$$

If we let $Z = e^y$, then this last equation becomes

$$Z^2 - 2xZ - 1 = 0.$$

Using the quadratic formula, we obtain

$$Z = \frac{2x \pm \sqrt{4x^2 + 4}}{2}$$
$$= x \pm \sqrt{x^2 + 1}.$$

Taking the natural logarithm of both sides yields

$$y = \ln\left(x \pm \sqrt{x^2 + 1}\right).$$

However, $\sqrt{x^2 + 1} > x$, so we must choose the plus sign, which leaves us with

$$y = \ln\left(x + \sqrt{x^2 + 1}\right).$$

The reader should check that this is indeed the required inverse. We leave the remaining arguments as exercises. ∎

Having constructed these new inverse hyperbolic functions, we now investigate their derivatives.

Theorem 2 *The derivatives of the inverse hyperbolic trigonometric functions are*

$$\frac{d}{dx}\operatorname{Arcsinh}(x) = \frac{1}{\sqrt{x^2+1}}$$

$$\frac{d}{dx}\operatorname{Arccosh}(x) = \frac{1}{\sqrt{x^2-1}}$$

$$\frac{d}{dx}\operatorname{Arctanh}(x) = \frac{1}{1-x^2}$$

$$\frac{d}{dx}\operatorname{Arccoth}(x) = \frac{1}{1-x^2}$$

$$\frac{d}{dx}\operatorname{Arcsech}(x) = -\frac{1}{x\sqrt{1-x^2}}$$

$$\frac{d}{dx}\operatorname{Arccsch}(x) = -\frac{1}{|x|\sqrt{1+x^2}}$$

These formulas can be derived directly by differentiating the given functions using the identities of Theorem 1.

We also have the associated integral formulas.

Theorem 3

$$\int \frac{1}{\sqrt{x^2+1}}\,dx = \operatorname{Arcsinh} x.$$

$$\int \frac{1}{\sqrt{x^2-1}}\,dx = \operatorname{Arccosh} x.$$

$$\int \frac{1}{1-x^2}\,dx = \operatorname{Arctanh} x.$$

$$\int \frac{1}{1 - x^2} \, dx = \text{Arccoth}\, x.$$

$$\int \frac{1}{x\sqrt{1 - x^2}} \, dx = -\text{Arcsech}\, x.$$

$$\int \frac{1}{|x|\sqrt{1 + x^2}} \, dx = -\text{Arccsch}\, x.$$

Each of the integral formulas in Theorem 3 can be obtained by using trigonometric substitutions, but knowledge of the integral formulas in Theorem 3 can make life easier, as we demonstrate in the next exercise.

EXAMPLE 3: Compute $\displaystyle\int \frac{1}{\sqrt{9x^2 + 25}} \, dx$.

SOLUTION: We first rewrite our integral as

$$\int \frac{1}{\sqrt{9x^2 + 25}} \, dx = \frac{1}{5} \int \frac{1}{\sqrt{\left(\frac{3x}{5}\right)^2 + 1}} \, dx.$$

We will solve this problem using two methods.

Trigonometric Substitution Method: Let $x = \frac{5}{3}\tan(u)$ so that $dx = \frac{5}{3}\sec^2(u)du$. Then we have

$$\int \frac{1}{\sqrt{9x^2 + 25}} \, dx = \frac{1}{5} \int \frac{5}{3} \frac{\sec^2 u}{\sqrt{\tan^2 u + 1}} \, du = \frac{1}{3} \int \sec u \, du$$

$$= \frac{1}{3} \ln|\sec u + \tan u| + C$$

$$= \frac{1}{3} \ln \left| \sqrt{1 + \left(\frac{3x}{5}\right)^2} + \frac{3x}{5} \right| + C.$$

Hyperbolic Function Method: Let $x = \left(\frac{5}{3}\right) u$, so that $dx = \left(\frac{5}{3}\right) du$. Then we have

$$\int \frac{1}{\sqrt{9x^2 + 25}} \, dx = \frac{1}{5} \int \frac{\frac{5}{3}}{\sqrt{u^2 + 1}} \, du = \frac{1}{3} \int \frac{1}{\sqrt{u^2 + 1}} \, du$$

$$= \frac{1}{3} \text{Arcsinh}\, u + C = \frac{1}{3} \text{Arcsinh} \left(\frac{3x}{5}\right) + C.$$

Theorem 1 shows that the two solutions are the same. The solution obtained by the second method is considerably less difficult. ∎

EXAMPLE 4: Compute $\int \dfrac{1}{x^2 + 4x - 5}\, dx$.

SOLUTION: We first complete the square in the denominator of the integrand.

$$\int \frac{1}{x^2 + 4x - 5}\, dx = \int \frac{1}{x^2 + 4x + 4 - 9}\, dx = \int \frac{1}{(x+2)^2 - 9}\, dx$$

$$= \frac{1}{9}\int \frac{1}{\left(\frac{x+2}{3}\right)^2 - 1}\, dx = -\frac{1}{3}\int \frac{1}{1 - u^2}\, du\Big|_{u = \frac{x+2}{3}}$$

$$= -\frac{1}{3}\,\mathrm{Arctanh}\left(\frac{x+2}{3}\right) + C.$$

We could also have solved this integral using trigonometric substitutions. The only difference would have appeared at the last equality. If we let $u = \sin(\theta)$, we have that $du = \cos(\theta)$, and

$$\int \frac{1}{x^2 + 4x - 5}\, dx$$

$$= -\frac{1}{3}\int \frac{1}{1 - u^2}\, du\Big|_{u = \frac{x+2}{3}} = -\frac{1}{3}\int \frac{1}{1 - \sin^2(\theta)}\cos(\theta)\, d\theta$$

$$= -\frac{1}{3}\int \sec(\theta)\, d\theta = -\frac{1}{3}\ln(\sec(\theta) + \tan(\theta)) + C$$

$$= -\frac{1}{3}\ln\left(\frac{1}{\sqrt{1 - u^2}} + \frac{u}{\sqrt{1 - u^2}}\right) + C$$

$$= -\frac{1}{3}\ln\left(\frac{1}{\sqrt{1 - \left(\frac{x+2}{3}\right)^2}} + \frac{\left(\frac{x+2}{3}\right)}{\sqrt{1 - \left(\frac{x+2}{3}\right)^2}}\right) + C. \qquad ∎$$

EXERCISES 10.2

In Exercises 1–24, compute the given integrals.

1. $\displaystyle\int \frac{dx}{x^2\sqrt{4 + x^2}}$.

2. $\displaystyle\int \sqrt{25 - x^2}\, dx$.

3. $\displaystyle\int \sqrt{25 - 9x^2}\, dx$.

4. $\displaystyle\int \sqrt{x^2 + 9}\, dx$.

5. $\displaystyle\int \frac{x^2}{(9 - x^2)^{3/2}}\, dx$.

6. $\displaystyle\int \frac{\sqrt{x^2 - a^2}}{x}\, dx$.

7. $\displaystyle\int \sqrt{b^2 x^2 + a^2}\, dx$.

8. $\displaystyle\int \frac{dx}{x\sqrt{b^2x^2 - a^2}}$.

9. $\displaystyle\int \frac{dx}{\sqrt{16x^2 - 4}}$.

10. $\displaystyle\int \frac{1}{(x^2 - 4x + 5)^2} \, dx$.

11. $\displaystyle\int \frac{x\,dx}{(x^2 - 4x + 3)^2}$.

12. $\displaystyle\int x\,\mathrm{Arcsin}(x) \, dx$.

13. $\displaystyle\int \frac{x^2}{\sqrt{x^2 - 9}} \, dx$.

14. $\displaystyle\int \frac{9 - x^2}{x} \, dx$.

15. $\displaystyle\int x\sqrt{25 - 9x^2} \, dx$.

16. $\displaystyle\int x\sqrt{x^2 + 9} \, dx$.

17. $\displaystyle\int \frac{x^2 + 25}{x} \, dx$.

18. $\displaystyle\int \frac{\sqrt{a^2 - x^2}}{x^2} \, dx$.

19. $\displaystyle\int \frac{dx}{x\sqrt{a^2 - b^2x^2}}$.

20. $\displaystyle\int \frac{dx}{\sqrt{16 - 4x^2}}$.

21. $\displaystyle\int \frac{dx}{\sqrt{4x^2 - 24x + 27}}$.

22. $\displaystyle\int \frac{dx}{x^2 - 4x + 3} \, dx$.

23. $\displaystyle\int \frac{e^x}{e^{2x} - 1} \, dx$.

24. $\displaystyle\int \sqrt{\frac{x+1}{x}} \, dx$.

Hint: Multiply by $\sqrt{\dfrac{x+1}{x+1}}$

25. Show that the area inside the ellipse $\dfrac{x^2}{a^2} + \dfrac{y^2}{b^2} = 1$ is πab.

26. Find the volume inside the ellipsoid $\dfrac{x^2}{a^2} + \dfrac{y^2}{b^2} + \dfrac{z^2}{c^2} = 1$.

27. Prove the last five identities in Theorem 1.

28. Find the area of the region swept out by $\vec{r}(\theta) = (\theta, \cos(\theta), \sin(\theta))$, $0 \le \theta \le 2\pi$. (See Section 6.4.)

29. Find the area of the region swept out by $\vec{r}(\theta) = (\theta, \theta\cos(\theta), \theta\sin(\theta))$, $0 \le \theta \le 2\pi$. (See Section 6.4.)

10.3 Rational Functions

We now consider how to evaluate the integral of a rational function $F(x) = \frac{Q(x)}{P(x)}$, where Q and P are polynomials. If the degree of the numerator is not less than the degree of the denominator, the first step in integrating a rational function $F(x) = \frac{Q(x)}{P(x)}$ is to divide the denominator into the numerator. Then we have that $F(x) = G(x) + \frac{R(x)}{P(x)}$, where G and R are polynomials, and the degree of R is less than the degree of P.

EXAMPLE 1: Using long division, we can show that

$$\frac{x^5 + x^3 - 2x^2 + 1}{x^2 + 1} = x^3 - 2 + \frac{3}{x^2 + 1}.$$

Thus

$$\int \frac{x^5 + x^3 - 2x^2 + 1}{x^2 + 1} \, dx = \int x^3 - 2 + \frac{3}{x^2 + 1} \, dx$$

$$= \frac{x^4}{4} - 2x + 3\operatorname{Arctan}(x) + C. \quad \blacksquare$$

EXAMPLE 2: By long division, $\dfrac{x^3}{x^2 - x - 2} = x + 1 + \dfrac{3x + 2}{x^2 - x - 2}.$

Thus

$$\int \frac{x^3}{x^2 - x - 2} \, dx = \int x \, dx + \int \, dx + \int \frac{3x + 2}{x^2 - x - 2} \, dx$$

$$= \frac{x^2}{2} + x + \int \frac{3x + 2}{x^2 - x - 2} \, dx. \qquad (8)$$

Unfortunately, $\displaystyle\int \frac{3x + 2}{x^2 - x - 2} \, dx$ does not integrate easily in its present form. We can, however, express the integrand in a form that we can integrate:

$$\frac{3x + 2}{x^2 - x - 2} = \frac{8}{3(x - 2)} + \frac{1}{3(x + 1)}. \qquad (9)$$

(Convince yourself that these expressions are equal.)

Thus

$$\int \frac{3x + 2}{x^2 - x - 2} \, dx = \int \frac{8}{3(x - 2)} \, dx + \int \frac{1}{3(x + 1)} \, dx$$

$$= \frac{8}{3} \ln |x - 2| + \frac{1}{3} \ln |x + 1| + C.$$

Combining this with Equation (8), we have

$$\int \frac{x^3}{x^2 - x - 2} \, dx = \frac{x^2}{2} + x + \frac{8}{3} \ln |x - 2| + \frac{1}{3} \ln |x + 1| + C. \quad \blacksquare$$

It was not a coincidence that we were able to express $\frac{3x+2}{x^2-x-2}$ in a more tractable form. The key is recognizing that the denominator of the left side of Equation (9) factors into $(x - 2)$ and $(x + 1)$, which are the denominators of the right side.

Suppose that we want the antiderivative of the rational function $\frac{Q(x)}{P(x)}$. Any polynomial can be written as the product of linear and irreducible quadratic factors. If we can find such factors for our denominator $P(x)$, then we can evaluate the integral $\int \frac{Q(x)}{P(x)} \, dx$ by expressing the rational function in terms of its *partial fractions*, which is essentially "undoing" a common denominator.

EXAMPLE 3: Consider $f(x) = \dfrac{1}{(x+1)(x-2)(x+3)}$. The expression $(x+1)(x-2)(x+3)$ is the common denominator of anything of the form $\frac{A}{x+1} + \frac{B}{x-2} + \frac{C}{x+3}$. We try to find $A, B,$ and C so that

$$\frac{1}{(x+1)(x-2)(x+3)} = \frac{A}{x+1} + \frac{B}{x-2} + \frac{C}{x+3}. \qquad (10)$$

If we add together the terms in the right-hand side of Equation (10), we obtain

$$\frac{1}{(x+1)(x-2)(x+3)}$$
$$= \frac{A(x-2)(x+3) + B(x+1)(x+3) + C(x+1)(x-2)}{(x+1)(x-2)(x+3)}. (11)$$

In order for Equation (11) to hold, the numerators must be equal. Thus

$$1 = A(x-2)(x+3) + B(x+1)(x+3) + C(x+1)(x-2). \qquad (12)$$

We must now solve for $A, B,$ and C. There are two standard ways to do this, both of which are illustrated below.

Method 1: Since Equation (12) must hold for all x, we simply choose three convenient values for x (e.g., the roots for the denominator on the left side of Equation (10)).

Let $x = 2$. Using Equation (12), we have

$$1 = A(0)(5) + B(3)(5) + C(3)(0), \text{ or } B = \frac{1}{15}.$$

Let $x = -3$. Using Equation (12), we have

$$1 = A(-5)(0) + B(-2)(0) + C(-2)(-5), \text{ or } C = \frac{1}{10}.$$

Let $x = -1$. Using Equation (12), we have

$$1 = A(-3)(2) + B(0)(2) + C(0)(-3), \text{ or } A = -\frac{1}{6}.$$

Method 2: First, we perform the multiplications indicated in Equation (12) to obtain

$$1 = (x^2 + x - 6)A + (x^2 + 4x + 3)B + (x^2 - x - 2)C,$$

or (collecting like terms of x)

$$1 = (A + B + C)x^2 + (A + 4B - C)x + (-6A + 3B - 2C). \quad (13)$$

We can equate coefficients of like powers of x in both sides of Equation (13) to obtain:

$$0 \;=\; A + B + C. \tag{14}$$

$$0 \;=\; A + 4B - C. \tag{15}$$

$$1 \;=\; -6A + 3B - 2C. \tag{16}$$

Solving Equations (14), (15), and (16) yields the same values for A, B, and C as were obtained in Method 1. ∎

There are two other types of rational functions that we must consider. First, how do we decompose rational functions where the denominator has roots of multiplicity greater than one (e.g., $(x + 1)^3$)? Second, what is the partial fraction decomposition involving quadratics which cannot be factored into linear terms (e.g., $x^2 + 1$)? The next theorem gives us the answers to both of these questions.

Theorem 1 *Suppose that $F(x) = \frac{R(x)}{P(x)}$, where $R(x)$ and $P(x)$ are polynomials, and the degree of R is less than the degree of P.*

(a) *If $n \geq 1$ is an integer, and $G(x)$ is a polynomial such that $P(x) = (ax+b)^n G(x)$, then there are constants $A_1, A_2, \ldots A_n$ and a polynomial $H(x)$, with degree less than the degree of G, such that*

$$F(x) = \frac{A_1}{ax + b} + \frac{A_2}{(ax + b)^2} + \cdots + \frac{A_n}{(ax + b)^n} + \frac{H(x)}{G(x)}.$$

(b) *If* $n \geq 1$ *is an integer and* $G(x)$ *is a polynomial such that* $P(x) = (ax^2 + bx + c)^n G(x)$, *then there are constants* $A_1, A_2, \ldots, A_n$ *and* $B_1, B_2, \ldots B_n$, *and a polynomial* $H(x)$ *such that*

$$F(x) = \frac{A_1 x + B_1}{ax^2 + bx + c} + \frac{A_2 x + B_2}{(ax^2 + bx + c)^2} + \cdots$$
$$+ \frac{A_n x + B_n}{(ax^2 + bx + c)^n} + \frac{H(x)}{G(x)}.$$

EXAMPLE 4: Write $\dfrac{x^2 + 1}{(x-1)^2(x+1)}$ as a sum of its partial fractions.

SOLUTION: We want to find A, B, and C such that

$$\frac{x^2 + 1}{(x-1)^2(x+1)} = \frac{A}{x-1} + \frac{B}{(x-1)^2} + \frac{C}{x+1}. \qquad (17)$$

Multiplying both sides of Equation (17) by $(x-1)^2(x+1)$, we obtain

$$x^2 + 1 = A(x-1)(x+1) + B(x+1) + C(x-1)^2. \qquad (18)$$

We solve Equation (18) using a minor modification of Method 1. We first set $x = 1$, which gives us $B = 1$, and then set $x = -1$, which gives us $C = \frac{1}{2}$. Since we do not have any more roots for the denominator of Equation (17), we pick any other value of x. We choose $x = 0$. This yields $1 = -A + B + C$. Solving this equation, we obtain $A = \frac{1}{2}$. Thus

$$\frac{x^2 + 1}{(x-1)^2(x+1)} = \frac{\frac{1}{2}}{x-1} + \frac{1}{(x-1)^2} + \frac{\frac{1}{2}}{x+1}. \qquad \blacksquare$$

EXAMPLE 5: From Example 4,

$$\int \frac{x^2 + 1}{(x-1)^2(x+1)}\, dx = \int \frac{\frac{1}{2}}{x-1} + \frac{1}{(x-1)^2} + \frac{\frac{1}{2}}{x+1}\, dx$$

$$= \int \frac{1}{2(x-1)}\, dx + \int \frac{1}{(x-1)^2}\, dx + \int \frac{1}{2(x+1)}\, dx$$

$$= \frac{1}{2}\ln|x-1| - \frac{1}{x-1} + \frac{1}{2}\ln|x+1| + C$$

$$= \frac{1}{2}\ln|(x-1)(x+1)| - \frac{1}{x-1} + C$$

$$= \frac{1}{2} \ln \left| x^2 - 1 \right| - \frac{1}{x - 1} + C. \qquad \blacksquare$$

With more complicated rational functions, we can combine Methods 1 and 2 to find the partial fraction decomposition, as the next example illustrates.

EXAMPLE 6: Find the partial fraction decomposition for

$$\frac{2x^4 - x^3 + 4x^2 - x}{(x^2 + 1)^2 (x - 1)}.$$

SOLUTION: We first note that the degree of the numerator is 4 and the degree of the denominator is 5. Since $x^2 + 1$ cannot be factored, we use part (b) of Theorem 1 to find A, B, C, D, and E so that:

$$\frac{2x^4 - x^3 + 4x^2 - x}{(x^2 + 1)^2 (x - 1)} = \frac{Ax + B}{x^2 + 1} + \frac{Cx + D}{(x^2 + 1)^2} + \frac{E}{x - 1}. \qquad (19)$$

Multiply both sides of Equation (19) by the denominator of the left side to obtain:

$$\begin{aligned}
2x^4 - x^3 &+ 4x^2 - x \\
&= (Ax + B)(x^2 + 1)(x - 1) + (Cx + D)(x - 1) + E(x^2 + 1)^2 \\
&= (A + E)x^4 + (B - A)x^3 + (A - B + C + 2E)x^2 \\
&\quad + (B - A - C + D)x - D - B + E.
\end{aligned} \qquad (20)$$

The only real root of the denominator of the left side of Equation (19) is $x = 1$. Substituting this into Equation (20), we obtain $4 = 4E$, or $E = 1$. Notice that the coefficient for the x^4 term on the right side of Equation (20) is $A + E$. Thus $A + E = 2$. Since $E = 1$, we see that $A = 1$. The coefficient for the x^3 term on the right side of Equation (20) is $B - A$. Thus $B - A = -1$. Since $A = 1$, we see that $B = 0$. We now pick two other values for x to solve for C and D. We choose $x = 0$, which yields $0 = B(1)(-1) + D(-1) + E(1)$. Since $B = 0$ and $E = 1$, we have $0 = -D + 1$. Thus $D = 1$. We choose $x = -1$, which yields $8 = (-A + B)(2)(-2) + (-C + D)(-2) + E(4)$. Substituting in the known values for A, B, D, and E, we obtain $C = 1$. Thus

$$\frac{2x^4 - x^3 + 4x^2 - x}{(x^2 + 1)^2 (x - 1)} = \frac{x}{x^2 + 1} + \frac{x + 1}{(x^2 + 1)^2} + \frac{1}{x - 1}. \qquad \blacksquare$$

EXAMPLE 7: Calculate $\displaystyle \int \frac{x^5 + x^4 + x^3 + 2x^2 - 1}{(x^2 + 1)^2 (x - 1)} \, dx.$

SOLUTION: Since the degree of the numerator is not less than the degree of the denominator, we first divide to obtain:

$$\int \frac{x^5 + x^4 + x^3 + 2x^2 - 1}{(x^2 + 1)^2(x - 1)} \, dx = \int 1 + \frac{2x^4 - x^3 + 4x^2 - x}{(x^2 + 1)^2(x - 1)} \, dx.$$

We now apply the partial fraction decomposition of Example 6 to obtain

$$\int 1 + \frac{x}{x^2 + 1} + \frac{x + 1}{(x^2 + 1)^2} + \frac{1}{x - 1} \, dx = x + \frac{1}{2} \ln(x^2 + 1) -$$

$$\frac{1}{2(x^2 + 1)} + \frac{1}{2} \left[\text{Arctan}(x) + \frac{x}{x^2 + 1} \right] + \ln|x - 1| + C. \quad \blacksquare$$

EXERCISES 10.3

Find the antiderivatives of the following functions.

1. $\dfrac{3x - 1}{x^2 - 1}$.

2. $\dfrac{x}{x^2 - 1}$.

3. $\dfrac{6x^2 + x - 1}{x^3 - x}$.

4. $\dfrac{x^2 + 4x + 2}{x^2 + x}$.

5. $\dfrac{x^2 + 2}{(x - 1)(x + 2)(x - 3)}$.

6. $\dfrac{x^2 - x + 2}{(2x - 1)(x - 1)(3x - 2)}$.

7. $\dfrac{x^4}{(x^2 - x)(x + 1)}$.

8. $\dfrac{1}{x^3 + x^2}$.

9. $\dfrac{2x^2 + 5x + 2}{x^3 + 2x^2 + x}$.

10. $\dfrac{x^2 + 2x - 3}{(x - 1)^2(2x + 1)^2}$.

11. $\dfrac{x^2 + x}{(x - 1)(x^2 + 1)}$.

12. $\dfrac{3x^2}{(x^2 + x + 1)(x - 1)}$.

13. $\dfrac{4}{(x + 2)(x^2 + 4)}$.

14. $\dfrac{x}{(x^2 + 4)^2}$.

15. $\dfrac{1}{x^3 + x^2 + x + 1}$.

16. $\dfrac{x - 1}{x^4 + x^2}$.

17. $\dfrac{x^4 + 2x^3 + 5x^2 + 5x + 2}{(x^2 + x + 1)^2(x + 1)}$.

18. $\dfrac{e^x}{e^{2x} - 1}$.

10.4 Integration Factors

In many applications, such as the mechanics of springs or pendulums, we model the problem by setting the force proportional to a function involving only the position of the object. Thus we have an equation of the form $mx''(t) = G(x(t))$. If we divide both sides of this by m, we have

$$x''(t) = F(x(t)). \tag{21}$$

In order to solve Equation (21), we use a technique called an *integrating factor*. Multiply both sides of Equation (21) by $x'(t)$ and integrate with respect to t. Then we have

$$\int x''(t)x'(t)\ dt = \int F(x(t))x'(t)\ dt. \tag{22}$$

Notice that the right hand side of Equation (22) can be written as $\int F(x)\ dx$. If we let $u = x'(t)$ in the left hand side of Equation (22) we get $\int u\ du$. Thus we have

$$\int u\ du = \int F(x)\ dx$$

$$\frac{u^2}{2} = \int F(x)\ dx$$

Solving for u and replacing u with $x'(t)$ we obtain

$$x'(t) = \pm\sqrt{2\int F(x)\ dx}, \tag{23}$$

where the $\pm$ sign is determined by the initial conditions on $x'(t)$. We solve Equation (23) by dividing both sides of Equation (23) by the right hand side and integrating with respect to t.

EXAMPLE 1: Solve the equation $x'' = -Kx, x(0) = 0, x'(0) = v_0 > 0$, and $K > 0$.

SOLUTION: Letting $F(x) = -Kx$ in Equation (23) we have:

$$x'(t) = \pm\sqrt{2\int -Kx\ dx}$$

$$= \pm\sqrt{-Kx^2(t) + C}.$$

Since $x'(0) > 0$, we must use the positive solution. Setting $t = 0$, we get $v_0 = \sqrt{C}$. Thus $C = v_0^2$. We now have

$$\frac{x'(t)}{\sqrt{v_0^2 - Kx^2(t)}} = 1.$$

Integrating both sides with respect to t, and using the substitution $x = \left(\frac{v_0}{\sqrt{K}}\right)\sin\theta$, we get

$$\frac{1}{\sqrt{K}}\operatorname{Arcsin}\left(\frac{x(t)}{v_0}\right) = t + C_1,$$

or

$$x(t) = v_0\sin(t\sqrt{K} + C) \quad \text{(where } C = C_1\sqrt{K}\text{)}.$$

Since $x(0) = 0$, we see that $C = 0$ and $x(t) = v_0\sin(t\sqrt{K})$. A simple check shows that this solution is valid even when $x'(t)$ is negative or zero. ∎

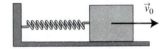

Figure 1. *A block is attached to a spring, and it is resting on a frictionless plane.*

EXAMPLE 2: Suppose that a block of mass m kg is attached to a spring with spring constant $k > 0$, and it is resting on a frictionless plane. Suppose further that the coordinate axes are positioned so that the block is at the origin when the system is at rest. The block is then put in motion so that:

(a) at time $t = 0$, the block is at the origin;

(b) all of the motion is on the x–axis; and

(c) $\vec{v}(0) = v_0$ (as in Figure 1).

The equation of motion is $m\vec{a} = -kx$, or $x''(t) = -\left(\frac{k}{m}\right)x(t)$. This is of the form $x''(t) = -Kx(t)$. From Example 1, we have

$$x(t) = v_0\sin\left(t\sqrt{\frac{k}{m}}\right).$$ ∎

Pendulum Motion

Suppose that a ball of mass m is suspended by a (massless) wire of length L, and the ball is allowed to swing as in Figure 2.

Let $\theta(t)$ denote the angle that the wire makes with the vertical, and let $s(t) = L\theta(t)$ so that $s(t)$ is the arc determined by $\theta(t)$. Then $s''(t)$ is the tangential part of the acceleration of the ball, which is $-g\sin(\theta(t))$. Since $s(t) = L\theta(t)$, we have the following differential equation:

$$s''(t) = L\theta''(t) = -g\sin(\theta(t)),$$

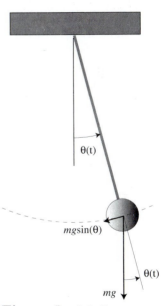

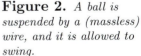

Figure 2. *A ball is suspended by a (massless) wire, and it is allowed to swing.*

which we write as

$$\theta''(t) = -\frac{g}{L}\sin(\theta(t)).$$

Assume that $\theta(0) = 0$ and that $\theta'(0) > 0$. Letting $F(\theta) = \left(-\frac{g}{L}\right)\sin(\theta)$, we can apply Equation (23) to obtain:

$$
\begin{aligned}
\theta'(t) &= \sqrt{2\int F(\theta)\,d\theta + C} \\[2mm]
&= \sqrt{2\int -\frac{g}{L}\sin(\theta)\,d\theta + C} \\[2mm]
&= \sqrt{\frac{2g}{L}\cos(\theta(t)) + C}.
\end{aligned}
\tag{24}
$$

Letting $t = 0$ we find that $C = [\theta'(0)]^2 - \frac{2g}{L}$, thus giving us

$$\theta'(t) = \sqrt{\frac{2g}{L}(\cos(\theta(t)) - 1) + [\theta'(0)]^2}\tag{25}$$

Solving Equation (25) for $\theta(t)$ is not at all trivial. A trick that is often used in applications is to recall that the second order Maclaurin polynomial for $\sin x$ is $p_2(x) = x$. The error is less than $\left|\frac{x^3}{3!}\right|$. Thus, if x is "small," then x is a good approximation for $\sin x$. So $F(\theta) = -\frac{g\theta}{L}$ will be a good approximation for small values of θ. Our equation of motion is then given by

$$\theta''(t) = -\frac{mg}{L}\theta(t).\tag{26}$$

Equation (26) was solved in Example 1:

$$\theta(t) = \theta'(0)\sin\left(t\sqrt{\frac{g}{L}}\right).$$

Notice that the motion is periodic and that the period depends only upon g (which is a constant) and the length of the wire L, not upon the mass of the ball! (This was Galileo's observation of a swinging lantern in a church in Pisa when he was seventeen.)

EXERCISES 10.4

In Exercises 1–5, find $x'(t)$.

1. $x''(t) = -[x(t)]^{-2}$, where $x(0) = 2$, $x'(0) = 1$.

2. $x''(t) = -[x(t)]^{-2}$, where $x(0) = 2$, $x'(0) = -1$.

3. $x''(t) = [x(t)]^{-2}$, where $x(0) = 2$, $x'(0) = 1$.

4. $x''(t) = [x(t)]^{-2}$, where $x(0) = 2$, $x'(0) = -1$.

5. $x''(t) = 3x + 1$, where $x(0) = 5$, $x'(0) = 2$.

In Exercises 6–8, find $x(t)$.

6. $x''(t) = 2[x(t)]^3$, where $x(0) = 2$, $x'(0) = 2$.

7. $x''(t) = 5[x(t)]^5$, where $x(0) = 2$, $x'(0) = 8$.

8. $x''(t) = \dfrac{3[x(t)]^{1/2}}{4}$, where $x(0) = 1$, $x'(0) = 1$.

Chapter 11

Work, Energy, and the Line Integral

11.1 Work

Our understanding of the definite integral is now sufficiently deep to allow us to introduce the concepts of work and energy. We begin with the simplest case: a constant force doing work on a particle that moves in a constant direction.

Definition: Work

Suppose that $\vec{F}$ is a constant force acting on a mass as it moves in a constant direction through a displacement $\vec{D}$. Then

$$W = \vec{F} \cdot \vec{D} \qquad (1)$$

is called the *work done on the mass by* $\vec{F}$ *during the displacement* $\vec{D}$.

It must be emphasized here that our definition assumes a constant force and a constant direction of motion. The metric units of work are called Joules after the nineteenth–century British scientist James Joule. One Joule (J) is equal to one Newton–meter. In the British system of units, the unit of work is the *foot-pound*.

EXAMPLE 1: A 10N force acts on a 2 kg mass as it moves for 10 meters along a straight line. Find the work done if the angle between the force and direction of motion is: (a) 0°; (b) 30°; (c) 60°; and (d) 90°.

SOLUTION: For any angle θ we write the work done by $\vec{F}$ as

$$W = \vec{F} \cdot \vec{D} = \|\vec{F}\| \|\vec{D}\| \cos\theta,$$

or

$$W = (10)(10)\cos\theta = 100\cos\theta.$$

(a) $\theta = 0°$ implies that $W = 100\cos(0°) = 100$ J.

(b) $\theta = 30°$ implies that $W = 100\cos(30°) = 50\sqrt{3}$ J.

(c) $\theta = 60°$ implies that $W = 100\cos(60°) = 50$ J.

(d) $\theta = 90°$ implies that $W = 100\cos(90°) = 0$ J.   ■

Our definition of work applies when the force and direction of motion are constant. Of course, in real world problems, these requirements may often not be satisfied. The definite integral can be employed to extend the idea of work to quite complicated situations.

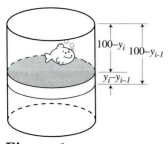

Figure 1. C_i *is the cylinder of height* $y_i - y_{i-1}$.

EXAMPLE 2:
A cylindrical tank 100 m high with a radius of 25 m is full of water. Find a function f and an interval $[a, b]$ so that $\int_a^b f(y)\, dy$ is the amount of work required to pump the water out the top of the tank. Compute the work done.

SOLUTION: The first step is to obtain a sketch of the problem. We place the coordinate axes so that the origin is at the base of the tank and so that the y–axis is the axis of the cylinder as in Figure 1.

Since the water at the bottom of the tank has farther to be lifted, it is clear that more work is done on the water that is at the bottom of the tank than is done on the water at the top. Let $\mathcal{P} = \{y_0, y_1, \ldots, y_n\}$ be a partition of the interval $[0, 100]$ on the y–axis. Let $\mathcal{S} = \{s_1, s_2, \ldots, s_n\}$ be a selection from $\mathcal{P}$. For each i, let C_i be the cylinder of height $y_i - y_{i-1}$ as illustrated in Figure 1. For each i, we let $W(C_i)$ be the work done in pumping the water in the C_i^{th} cylinder out of the tank. If y_i and y_{i-1} are very close together so that C_i is very thin, a good approximation for $W(C_i)$ is given by[1]

$$
\begin{aligned}
W(C_i) &\approx \text{Force} \times \text{Displacement} \\[6pt]
&= \text{Mass} \times \text{Acceleration} \times \text{Displacement} \\[6pt]
&= \text{Volume} \times \text{Density} \times \text{Acceleration} \times \text{Displacement} \\[6pt]
&= \pi(25)^2(y_i - y_{i-1}) \times (10)^3 \times 9.8 \times (100 - s_i).
\end{aligned}
$$

---

[1] Water has a density of 10^3 kg/m^3.

Thus the total work is approximated by

$$W \approx \sum_{i=1}^{n} W(C_i)$$

$$= \sum_{i=1}^{n} \pi(25)^2(10)^3(9.8)(100 - s_i)(y_i - y_{i-1}).$$

But this approximation for W is also an approximation for

$$\int_0^{100} \pi(25)^2(10)^3(9.8)(100 - y)\, dy.$$

We conclude that

$$W = \int_0^{100} \pi(25)^2(10)^3(9.8)(100 - y)\, dy$$

$$= \int_0^{100} 7125\pi(10)^3(100 - y)\, dy$$

$$= 7125\pi(10)^3 \left(100y - \frac{y^2}{2}\right)\Big|_0^{100}$$

$$= 3.5625\pi 10^{10}\ \text{J.} \qquad \blacksquare$$

EXAMPLE 3: It has been established experimentally that the magnitude of the force required to stretch a spring is proportional to the distance that the spring has been stretched; i.e. $F = ks$, where s is the distance the spring has been stretched. Suppose that we have a spring with a spring constant k of 2 N/m. Determine the work required to stretch the spring 3 cm.

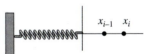

Figure 2. W_i denotes the work done in pulling the end of the spring from x_{i-1} to x_i.

SOLUTION: First, since in the metric system the unit for work is expressed in terms of m, we convert 3 cm to 0.03 m. In Figure 2, we place our coordinate axis so that the origin is at the end of the relaxed spring.

Let $\mathcal{P} = \{x_0, x_1, \ldots, x_n\}$ be a partition for the interval $[0, 0.03]$ and let W_i denote the work done in pulling the end of the spring from x_{i-1} to x_i. We choose two selections, $\mathcal{S}_1$ and $\mathcal{S}_2$, from $\mathcal{P}$. Since the force is an increasing function along the interval $[x_{i-1}, x_i]$, it is obvious that

$$2x_{i-1}(x_i - x_{i-1}) < W_i < 2x_i(x_i - x_{i-1}),$$

and

$$\sum_{i=i}^{n} 2x_{i-1}(x_i - x_{i-1}) < \sum_{i=i}^{n} W_i < \sum_{i=i}^{n} 2x_i(x_i - x_{i-1}).$$

The sum on the left is a Riemann sum with selection $S_1 = \{x_0, x_1, \ldots, x_{n-1}\}$, whereas the sum on the right is a Riemann sum with selection $S_2 = \{x_1, x_2, \ldots, x_n\}$. Each of these Riemann sums is an approximation for $\int_0^{0.03} 2x \, dx$, so the sum in the middle of our inequality must also approximate this integral. We have

$$\text{Work} = \int_0^{0.03} 2x \, dx = x^2 \Big|_0^{0.03} = 0.0009 \text{ J.} \qquad \blacksquare$$

EXERCISES 11.1

1. Find the work done by $\vec{F} = (1, 2, -1)$ N with a displacement of $(-1, 0, 3)$ m.

In Exercises 2 and 3, find the work done by $\vec{F} = (1, -3, -2)$ N on a 2 kg mass as it moves from $\vec{A}$ to $\vec{B}$ along a straight line.

2. $\vec{A} = (1, 2, 3)$ m and $\vec{B} = (1, -1, 2)$ m.

3. $\vec{A} = (-1, 0, 1)$ m and $\vec{B} = (1, 6, -3)$ m.

4. In Exercise 2, find the work done by $\vec{F}$ as the mass goes along a straight line from $\vec{B}$ to $\vec{A}$.

In Exercises 5–9, you are given the magnitude of the frictional force $\vec{F}(x)$ opposing the motion of a 2 kg block as it slides 1 m on a table along a straight line from the position $\vec{A}$. Let x denote the distance from $\vec{A}$ measured in m and let $a = 1$ m. Find the work done.

5. $F(x) = \dfrac{1}{1 + \left(\frac{x}{a}\right)^2}$ N.

6. $F(x) = \dfrac{1}{1 + \frac{x}{a}}$ N.

7. $F(x) = e^{x/a}$ N.

8. $F(x) = \ln\left(1 + \frac{x}{a}\right)$ N.

9. A force of 25 N is required to stretch a spring of natural length 0.8 m to a length of 0.85 m. Compute the work done (against the spring) in stretching the spring from its natural length to a length of 1.0 m.

10. A 2 kg block, resting on a level plane, is attached to a relaxed spring with spring constant $k = 5$ N/m, and then the block is moved 10 cm directly away from the fixed end of the spring. The magnitude of the frictional force opposing the motion of the block is 0.5 N. Determine the work done against gravity in moving the block.

11. Suppose that the magnitude of the frictional force opposing the motion of the block of Exercise 10 is given by $F(x) = e^{x/a}$ N, where x denotes the distance of the block from its initial position. Let $a = 1$ m. Determine the work done in moving the block.

12. A cylindrical tank 15 m high with a radius of 5 m is full of water. Find the work done in pumping the water out the top of the tank.

13. Find the work done in pumping the water out of the tank in Exercise 12 to a level 10 m above the top of the tank.

14. The tank in Exercise 12 is full of sandy water. The sand has settled some so that the density of the sandy water is $(\mu + e^{kx})$ kg/m^3, where x denotes the distance from the top of the tank. Compute the work done in pumping the water out the top of the tank.

15. Compute the work done by the gravitational force if the water is drained out the bottom of the tank in Exercise 14.

16. A conical tank has base radius 3 m and height 10 m. The base is situated 10 m above the water line of a reservoir of water as in Figure 3a. Compute the work done filling the tank with water from the reservoir.

17. The tank of Exercise 16 is oriented with the vertex of the cone below the base as in Figure 3.b. Compute the work done filling the tank with water from the reservoir.

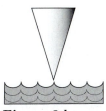

Figure 3.a
The cone's vertex points up.

Figure 3.b
The cone's vertex points down.

18. A spherical tank with radius R is full of water. Compute the work done in pumping the water out the top of the tank.

19. A parabolic tank is full of water. The tank is described by $x^2 + y^2 = z + 1$, $z \le 0$. See Figure 4. Calculate the work done in pumping the water out the top of the tank.

20. A tank, built in the shape of the ellipsoid $\frac{x^2}{a^2} + \frac{y^2}{b^2} + z^2 = 1$, is full of water. See Figure 5. Find the work done in pumping the water out the top of the tank.

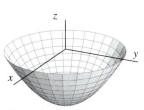
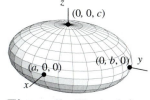

Figure 4. *The tank for Exercise 19.*

Figure 5. *The tank for Exercise 20.*

21. A chain of length 1 m and density 1 kg/m is hanging by end **A** from the top of a table. See Figures 6.

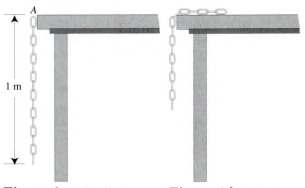

Figure 6.a *The chain hanging from end A.*

Figure 6.b *The chain as it is dragged onto the table.*

22. The magnitude of the frictional force opposing the motion of the chain in Exercise 21 is $\frac{1}{4}$ the weight of that portion of the chain on the table. Find the work done in pulling the end **A** across the table until the chain is entirely on the table given that there is no friction.

23. Solve Exercise 21 with the condition that the mass density of the chain at a distance x m from end **A** is given by e^{kx} kg/m.

24. The magnitude of the gravitational force between the earth and an object is inversely proportional to the square of the distance between the mass and the center of the earth, and directly proportional to the product of the mass of the earth and the mass of the object. Assuming that the earth is a sphere with radius R m, find the work done in sending a rocket of mass m to an altitude h above the surface of the earth.

Assume that the path of flight is a straight line directly away from the center of the earth.

In Exercises 25–27, the magnitude of the force between two electrically charged particles with charges q_1 and q_2 is equal to $\frac{kq_1q_2}{r^2}$ N, where r is the distance between the particles. If the particles have like charges, i.e., q_1 and q_2 are either both positive or both negative, then the electrical force pushes the particles apart. If the particles have opposite charges, then the force is one of attraction.

25. Suppose that a particle P_1 with charge q_1 is at a fixed location, and that the coordinate axes are positioned so that P_1 is at the position $(0, 1)$ m. Find the work done on a particle P_2 with charge q_2 by the electrical force between P_1 and P_2 as P_2 moves along a straight line from $(0, 0)$ m to $(0, -1)$ m if $q_1 > 0$ and $q_2 < 0$.

26. Find the work done in Exercise 25 if both charges are positive.

27. Find the work done in Exercise 25 if P_2 moves along a straight line from $(0, 0)$ m to $(1, 0)$ m.

11.2   The Work–Energy Theorem

Suppose that $\vec{F}$ is a function from $\mathbb{R}$ into $\mathbb{R}^3$ and that M is a mass, so that at time t, $\vec{F}(t)$ is the total (or net, or resultant) force acting on M. In this section, we extend the notion of work so that we can speak of the work done by the variable force $\vec{F}$ on M over some time interval.

Let t_I and t_F denote the initial time and the final time of our time interval. Let $\vec{r}(t)$ denote the position of M at time t, so

$$\vec{a}(t) = \frac{d^2\vec{r}(t)}{dt^2} = \frac{\vec{F}(t)}{M}.$$

We break the total work done on M into small pieces of work, and then we try to approximate each piece of work done. The sum of the approximations of the pieces of work gives a good approximation of the total work done. The most obvious way to break up the total work done is to partition the time interval $[t_I, t_F]$ as $\{t_0, t_1, \ldots, t_n\}$. The assumption we make is that we can make the intervals $[t_{i-1}, t_i]$ small enough so that we can, with little error, treat $\vec{F}$ as though it were constant on the interval $[t_{i-1}, t_i]$, and so that the polygonal path

$$[\vec{r}(t_0), \vec{r}(t_1)] \cup [\vec{r}(t_1), \vec{r}(t_2)] \cup \cdots \cup [\vec{r}(t_{n-1}), \vec{r}(t_n)]$$

closely approximates the path of the mass M. (See Figure 1.)

Hopefully, if $\mathbb{S} = \{s_1, s_2, \ldots, s_n\}$ is a selection from the partition $\{t_0, t_1, \ldots, t_n\}$, then the work W done by $\vec{F}$ on M over the time interval $[t_I, t_F]$ is well approximated by

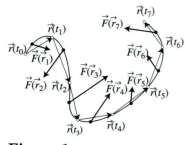

Figure 1.
$\vec{F}$ is evaluated at the selection points on the path.

$$W \approx \sum_{i=1}^{n} \vec{F}(s_i) \cdot (\vec{r}(t_i) - \vec{r}(t_{i-1})). \qquad (2)$$

Se attempt to express Equation (2) as a Riemann sum so that we can evaluate W with a definite integral. Observe that

$$\vec{F}(s_i) \cdot \frac{\vec{r}(t_i) - \vec{r}(t_{i-1})}{t_i - t_{i-1}}$$

is an approximation of

$$\vec{F}(s_i) \cdot \vec{r}\,'(s_i).$$

Thus

$$
\begin{aligned}
\text{Work} \quad &\approx \quad \sum_{i=1}^{n} \vec{F}(s_i) \cdot (\vec{r}(t_i) - \vec{r}(t_{i-1})) \\
&= \quad \sum_{i=1}^{n} \vec{F}(s_i) \cdot \frac{\vec{r}(t_i) - \vec{r}(t_{i-1})}{t_i - t_{i-1}}(t_i - t_{i-1}) \\
&\approx \quad \sum_{i=1}^{n} \vec{F}(s_i) \cdot \vec{r}\,'(s_i)(t_i - t_{i-1}) \\
&\approx \quad \int_{t_I}^{t_F} \vec{F}(t) \cdot \vec{r}\,'(t) \; dt.
\end{aligned}
$$

Thus we conclude that

$$\text{Work} \; = \; \int_{t_I}^{t_F} \vec{F}(t) \cdot \vec{r}\,'(t) \; dt.$$

Since $\vec{r}\,'(t) = \vec{v}(t)$ and $\vec{F}(t) = M\vec{a}(t) = M\frac{d}{dt}\vec{v}(t)$, the above can be rewritten as

$$\text{Work} \; = \; \int_{t_I}^{t_F} M\vec{v}\,'(t) \cdot \vec{v}(t) \; dt. \qquad (3)$$

Recall that

$$\frac{dv^2(t)}{dt} = \frac{d(\vec{v}(t) \cdot \vec{v}(t))}{dt} = 2\vec{v}(t) \cdot \frac{d\vec{v}(t)}{dt}.$$

Substituting this expression into Equation (3), we see that

$$\text{Work} \; = \; \frac{1}{2} \int_{t_I}^{t_F} M\frac{d}{dt}(v^2(t)) \; dt.$$

However,

$$\frac{1}{2} \int_{t_I}^{t_F} M\frac{d}{dt}(v^2(t)) \; dt = \frac{1}{2}Mv^2(t)\Big|_{t_I}^{t_F} = \frac{1}{2}Mv^2(t_F) - \frac{1}{2}Mv^2(t_I).$$

So it must be the case that

$$\text{Work} \; = \; \frac{1}{2}Mv^2(t_F) - \frac{1}{2}Mv^2(t_I). \qquad (4)$$

Equation (4) leads us to the following definition.

Definition: Kinetic Energy

If a mass M moves with speed v, then $\frac{1}{2}Mv^2$ is called the *kinetic energy* of M.

Of course, the idea of kinetic energy is designed with Equation (4) in mind. We summarize the above discussion in the following theorem.

Theorem 1 (Work–Energy Theorem) *If the net or resultant force $\vec{F}(t)$ acts on a mass M over the time interval $[t_I, t_F]$, then the work done on M is equal to the change in kinetic energy of the mass during the motion.*

$$Work = \frac{1}{2}Mv^2(t_F) - \frac{1}{2}Mv^2(t_I).$$

The units for kinetic energy are the same as the units for work.

EXAMPLE 1: A 2 kg mass slides 10 m on a horizontal frictionless surface. It starts from rest and is pushed by a 10 N force. What is the speed of the block at the end of the 10 m?

SOLUTION: By the Work–Energy Theorem, the work done by a 10 N force equals the change in kinetic energy. We use the notation KE_I, KE_F to denote the initial kinetic energy and the final kinetic energy, respectively. We also use the notation $W_{I \to F}$ to denote the work done from time t_I to time t_F, and we use V_I to denote the speed at time t_I and V_F to denote the speed a time t_F.

$$KE_I = 0 \quad \text{and} \quad KE_F = \frac{1}{2}MV_F^2.$$

$$W_{I \to F} = \vec{F} \cdot \vec{D} = (10)(10) = 100J.$$

Using the Work–Energy Theorem, we have

$$100 = \frac{1}{2}MV_F^2 - \frac{1}{2}MV_I^2$$

$$= \frac{1}{2}(2)V_F^2 - 0.$$

Thus   ■

$$V_F = 10 \text{ m/s}.$$

If a block slides along a surface, it is reasonable to assume that no matter how slick the surface, there is a drag caused by friction. We assume that the frictional force is a vector pointing opposite to the direction of motion. If the surface is horizontal, then the magnitude of the frictional force is proportional to the weight of the block, that is, $\vec{f} = \mu M g$. The constant μ is called the *coefficient of friction.*

EXAMPLE 2: The block of the preceding problem is pushed as before, but now the coefficient of friction is $\frac{1}{4}$. Again, find the speed of the block after it has moved 10 m.

SOLUTION: The force due to friction is $\vec{f} = \mu M g = \frac{1}{4}(2)(9.8) = 4.9$ N. The total force is $\vec{F} = 10 - \vec{f} = 5.1$ N.

$$W_{I \to F} = \vec{F} \cdot \vec{D} = (5.1)(10) = 51 \text{ J}.$$

Using the Work–Energy Theorem, we have

$$\begin{aligned} 51 &= \frac{1}{2}MV_F^2 - \frac{1}{2}MV_I^2 \\ &= \frac{1}{2}(2)V_F^2 - 0. \end{aligned}$$

Thus

$$V_F = \sqrt{51} \text{ m/s.} \qquad \blacksquare$$

EXAMPLE 3: A 50 N force is applied in the "up" direction to a 2 kg mass initially at rest. After moving up a distance of 10 m, what is the speed of the mass?

SOLUTION: As before, $KE_I = 0$ and $KE_F = \frac{1}{2}MV_F^2$. The resultant force is given by $\vec{F} = 50 - Mg = 50 - (2)(9.8) = 30.4$ N.

$$W_{I \to F} = \vec{F} \cdot \vec{D} = (30.4)(10) = 304 \text{ J}.$$

Using the Work–Energy Theorem, we have

$$\begin{aligned} 304 &= \frac{1}{2}MV_F^2 - \frac{1}{2}MV_I^2 \\ &= \frac{1}{2}(2)V_F^2 - 0. \end{aligned}$$

Thus

$$V_F = \sqrt{304} \text{ m/s.} \qquad \blacksquare$$

EXAMPLE 4: A 2 kg block slides along a straight track. At position **A** of the track, the block has a speed of 1 m/s. The frictional force between the block and the track is $\frac{1}{1+x^2}$ N, where x is the distance from position **A** measured in m.

(a) Determine how much work the frictional force will do on the block in stopping it.

(b) Determine how far the block will slide before stopping.

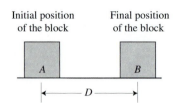

Figure 2. *The block stops at a distance D, at point* **B**.

SOLUTION: Figure 2 shows the setup for the example. The work done by the frictional force as the block slides from **A** to **B** is given by

$$W = -\int_0^D \frac{1}{1+x^2}\, dx.$$

By the Work–Energy Theorem, this is also equal to the change in kinetic energy. Thus

$$\begin{aligned} W &= \frac{1}{2}MV_F^2 - \frac{1}{2}MV_I^2 \\ &= 0 - \frac{1}{2}(2)(1)^2 = -1 \text{ J.} \end{aligned}$$

Therefore, the solution to Part (a) is $W = -1$ J.

For Part (b),

$$\begin{aligned} W &= -\int_0^D \frac{1}{1+x^2}\, dx \\ &= -\left. \text{Arctan}\, x \right|_0^D = -\text{Arctan}\, D. \end{aligned}$$

From Part (a) we know that $W = -1$ J, so

$$D = \tan(1) \text{ m} \approx 1.57 \text{ m.} \qquad \blacksquare$$

EXAMPLE 5: The magnitude of the gravitational force between the sun and a rocket in space is $F(x) = GMm/r^2$, where r is the distance between the center of mass of the rocket and the center of mass of the sun, M is the mass of the sun, m is the mass of the rocket, and G is the gravitational constant. Suppose that the rocket is moving directly away from the sun. If the rocket is initially at a distance R from the center of the sun, how fast must the rocket move away from the sun to escape the sun's gravitational "field,"

Initial position of the block

Final position of the block

A

B

D

assuming that no force other than the sun's gravitational pull acts on the rocket?

SOLUTION: Since the motion is along a straight line, we place the x–axis so that the center of the sun is at the origin and the rocket is moving in the positive x direction. If we let W denote the work done *against* the gravitational pull of the sun as the rocket's distance to the sun "goes to infinity," then by the Work–Energy Theorem, we want the initial velocity v_0 to be such that

$$\frac{1}{2}mv_0^2 = W \text{ or } v_0 = \sqrt{\frac{2W}{m}}.$$

We are given that the magnitude of the gravitational force acting on the rocket at position x is $F(x) = GMm/x^2$. Thus

$$W = \int_R^\infty \frac{GMm}{x^2}\, dx = GMm \lim_{r\to\infty} -\frac{1}{x}\bigg|_R^r = \frac{GMm}{R}.$$

Thus, $v_0 = \sqrt{2GM/R}$ is the speed necessary for a rocket to escape the sun's gravitational field.   ∎

EXERCISES 11.2

1. A 10 kg block slides on a table where the frictional force opposing the motion is 5 N. Use the Work–Energy Theorem to determine how far the block slides before coming to rest, if its initial speed is 2 m/sec.

In Exercises 2 and 3, a 3 kg ball is released from rest and allowed to fall under the influence of gravity. Use the Work–Energy Theorem to determine the speed of the ball after it has fallen.

2. 50 m, and

3. 100 m.

4. Determine what the initial velocity of the block in Example 4 must be for the block to travel 2 m before stopping.

In Exercises 5 and 6, a 2 kg block, resting on a level plane, is attached to a relaxed spring with spring constant $k = 10$ N/m. The block is then moved 10 cm directly away from the fixed end of the spring. Use the Work–Energy Theorem to determine how far the block will travel before stopping.

5. Assume that there is no friction between the plane and the block.

6. The frictional force opposing the motion is 0.25 N.

In Exercises 7–9, a 2 kg block moves in a straight line away from a position on a level plane. $f(x)$ denotes the frictional force opposing the motion of the block, where x is the distance from the initial position. If v_0 is the initial velocity of the block and $a = 1$ m, determine the distance that the block will move before coming to rest from the frictional force.

7. $f(x) = e^{x/a}$ N.

8. $f(x) = \dfrac{1}{\frac{x}{a} + 1}$ N.

9. $f(x) = \dfrac{\left(\frac{x}{a}\right)^2}{\left(\left(\frac{x}{a}\right)^2 - 1\right)}$ N.

In Exercises 10 and 11, one end of a chain of length 1 m and density 1 kg/m is hanging off a table, so that half of the chain is on the table and half is not. The chain is then allowed to slide off the table. Determine the velocity of the chain the instant that it leaves the table.

10. Assume that the table top is frictionless.

11. The frictional force opposing the chain's sliding is $\frac{1}{4}$ the weight of that portion of the chain that is on the table.

12. Find the velocity of the chain of Exercise 10 if there is a 1 kg mass attached to the end of the chain that is hanging off the table.

13. Find the velocity of the chain of Exercise 11 if there is a 1 kg mass attached to the end of the chain that is hanging off the table.

14. Find the velocity of the chain of Exercise 10 if there is a 0.5 kg mass attached to the end of the chain that is initially on the table.

15. Find the velocity of the chain of Exercise 11 if there is a 0.5 kg mass attached to the end of the chain that is initially on the table.

16. Determine the initial speed of the rocket of Exercise 24 of the previous section if the rocket is to reach an altitude h above the surface of the earth.

17. Find a function h so that $h(v_0)$ gives the altitude achieved by the rocket in Exercise 16 if the rocket has an initial speed of v_0.

11.3   Fundamental Curves

In the next few sections we are interested in computing the work done on a mass M as M moves through space. To keep things manageable, we introduce the idea of a fundamental curve.

Definition: Fundamental Curve

Let C be a subset of $\mathbb{R}^n$. We say that C is a *fundamental curve* provided that there is a parametrization $\vec{r}$ for C such that:

(a) The domain of $\vec{r}$ is an interval $[a, b]$;

(b) $\vec{r}$ is one-to-one on $[a, b]$, except possibly $\vec{r}(a) = \vec{r}(b)$;

(c) $\vec{r}'$ is continuous on $[a, b]$; and

(d) $\vec{r}' \neq \vec{0}$ on (a, b).

If $\vec{r}(a) = \vec{r}(b)$, then $\vec{r}$ is called a *simple closed curve*.

There are two important physical interpretations of a fundamental curve.

(a) The first is to think of $[a, b]$ as being the time interval during which a particle moves from one end of C to the other. Condition (b) states that the curve does not loop around on itself except perhaps at the endpoints. Equivalently, a particle moving along C passes through any point of C only once, except perhaps the starting and ending points.

(b) The second is to think of $[a, b]$ as a straight piece of wire that is stretched and bent to fit C. Condition (b) states that the wire is bent so that it does not meet itself, except perhaps the ends are soldered together. Conditions (c) and (d) ensure that the wire is not broken nor bent in a way that it has sharp corners. Conditions (c) and (d) state that a particle can move along C with a smooth motion without stopping.

EXAMPLE 1: Show that the semicircle $C = \{(x, y, 0) \mid y = \sqrt{1 - x^2}\}$ is a fundamental curve.

SOLUTION: To show that C is a fundamental curve, we must find a parametrization for C. There are many choices; the most obvious to try is

$$\vec{r}(t) = (t, \sqrt{1 - t^2}, 0), -1 \le t \le 1.$$

However, $\vec{r}'(t)$ is not defined at the points $t = -1$ and $t = 1$, which are in the domain of $\vec{r}(t)$. Thus $\vec{r}(t)$ does not satisfy our requirements for a parametrization of a fundamental curve. Another candidate for a parametrization, and one that works, is

$$\vec{r}(t) = (\cos t, \sin t, 0), \ 0 \le t \le \pi.$$

It is easily seen that this choice for $\vec{r}$ satisfies the conditions for a fundamental curve. ∎

If we let $\vec{A} \ne \vec{B}$ be the endpoints for a fundamental curve C, and if $\vec{r}(t)$ is a parametrization for C with domain $[a, b]$, then we say that $\vec{r}(t)$ "goes from" $\vec{A}$ to $\vec{B}$ if $\vec{r}(a) = \vec{A}$ and $\vec{r}(b) = \vec{B}$. It is intuitively correct to think of $\vec{r}$ as representing the position of a particle as it moves along the curve C from $\vec{A}$ to $\vec{B}$. Similarly, we will say that a parametrization $\vec{r}(t)$ for C "goes from" $\vec{B}$ to $\vec{A}$ if $\vec{r}(a) = \vec{B}$ and $\vec{r}(b) = \vec{A}$.

EXAMPLE 2: Let C be the straight line segment connecting the points $\vec{A} = (1, 2, 0)$ and $\vec{B} = (2, 0, 1)$. Find a parametrization for C that goes from $\vec{A}$ to $\vec{B}$, and a parametrization for C that goes from $\vec{B}$ to $\vec{A}$.

SOLUTION: We have done this before when we parametrized straight lines. Recall that a parametrization for a straight line is of the form $\vec{r}(t) = t\vec{v} + \vec{b}$. If we choose $\vec{v}$ and $\vec{b}$ so that $\vec{r}_1(0) = \vec{A}$ and $\vec{r}_1(1) = \vec{B}$, we obtain

$$
\begin{aligned}
\vec{r}_1(t) &= t\vec{B} + (1-t)\vec{A},\ 0 \le t \le 1 \\
&= t(2,0,1) + (1-t)(1,2,0) \\
&= (t+1, 2-2t, t),\ \text{for } 0 \le t \le 1
\end{aligned}
$$

as a parametrization for C that goes from $\vec{A}$ to $\vec{B}$.

If we choose $\vec{v}$ and $\vec{b}$ so that $\vec{r}_2(0) = \vec{B}$ and $\vec{r}_2(1) = \vec{A}$, we obtain

$$
\begin{aligned}
\vec{r}_2(t) &= t\vec{A} + (1-t)\vec{B},\ 0 \le t \le 1 \\
&= t(1,2,0) + (1-t)(2,0,1) \\
&= (2-t, 2t, 1-t),\ \text{for } 0 \le t \le 1
\end{aligned}
$$

as a parametrization for C that goes from $\vec{B}$ to $\vec{A}$.   ∎

EXAMPLE 3: Find a parametrization for the upper half of the ellipse in the xy–plane given by $E = \{(x,y,0) \mid \frac{x^2}{9} + \frac{y^2}{4} = 1\}$ that goes from $\vec{A} = (3,0,0)$ to $\vec{B} = (-3,0,0)$. See Figure 1.

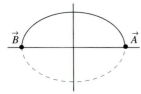

Figure 1. *The upper half of the ellipse E.*

SOLUTION: A parametrization for E is given by $\vec{r}(t) = (3\cos t, 2\sin t, 0)$, for all $0 \le t \le \pi$. To see this, first observe that if $x = 3\cos t$ and $y = 2\sin t$, then $\frac{x^2}{9} + \frac{y^2}{4} = \cos^2 t + \sin^2 t = 1$. To complete the argument that the image of $\vec{r}(t)$ is E, it must be shown that if $(x,y,0)$ is in E, then there is a t in $[0,\pi]$ so that $\vec{r}(t) = (x,y,0)$. We leave this as an exercise for the student.   ∎

Notice that if C is a fundamental curve and $\vec{r}(t)$ is a parametrization for C, then $\vec{r}\,'(t)$ is a vector that is parallel to the line tangent to C at $\vec{r}(t)$. See Figure 2.

Thus, if C is a fundamental curve, a parametrization $\vec{r}_1(t)$ for the line tangent to $\vec{r}(t_0)$ is given by:

$$\vec{r}_1(t) = t\vec{r}\,'(t_0) + \vec{r}(t_0).$$

EXAMPLE 4: Let H be the portion of the helix that is parameterized by

$$\vec{r}(t) = \left(\cos t, \sin t, \frac{t}{\pi}\right),\ 0 \le t \le 2\pi.$$

See Figure 2. Find a parametrization for the line tangent to H at $\vec{p} = \left(0, 1, \frac{1}{2}\right).$

SOLUTION: $\vec{r}\left(\frac{\pi}{2}\right) = \vec{p}$, and $\vec{d} = \vec{r}'\left(\frac{\pi}{2}\right) = \left(-1, 0, \frac{1}{\pi}\right)$ is a direction vector for the line. We may use a parametrization of the form $\vec{r}_1(t) = t\vec{d} + \vec{p}$. Thus

$$\vec{r}_1(t) = t\left(-1, 0, \frac{1}{\pi}\right) + \left(0, 1, \frac{1}{2}\right) = \left(-t, 1, \frac{1}{\pi}t + \frac{1}{2}\right).$$ ∎

If f is a function from a subset of $\mathbb{R}$ into $\mathbb{R}$, and if f is differentiable on $[a, b]$, then the function $\vec{r}$ from $\mathbb{R}$ into $\mathbb{R}^2$ defined by $\vec{r}(t) = (t, f(t))$ is a parametrization for that portion of the graph of f with domain $[a, b]$.

Figure 2. $\vec{r}'(t)$ *drawn emanating from* $\vec{r}(t)$ *at* $\vec{p}$.

EXAMPLE 5: $\vec{r}(t) = (t, \sin t), 0 \le t \le \pi$, is a parametrization for the graph of $\sin t$ for t between 0 and π. ∎

EXAMPLE 6: $\vec{r}(t) = (\cos t, \sin t), \ 0 \le t \le 2\pi$ is a parametrization of the unit circle that describes a particle moving in a counterclockwise fashion, while $\vec{r}(t) = (\cos t, -\sin t), \ 0 \le t \le 2\pi$ describes a particle moving around the unit circle in a clockwise fashion. ∎

EXERCISES 11.3

In Exercises 1–4, find parametrizations $\vec{s}$ and $\vec{r}$ with domain $[0, 1]$ for the straight line segment with endpoints $\vec{A}$ and $\vec{B}$, so that $\vec{s}$ goes from $\vec{A}$ to $\vec{B}$ and $\vec{r}$ goes from $\vec{B}$ to $\vec{A}$.

1. $\vec{A} = (1, 0, 2), \ \vec{B} = (-1, 2, 0)$.

2. $\vec{A} = (-1, 2, 3), \ \vec{B} = (1, 0, 2)$.

3. $\vec{A} = (1, 3), \ \vec{B} = (0, -1)$.

4. $\vec{A} = (-5, 3, 2), \ \vec{B} = (1, -1, 4)$.

5. Find a parametrization $\vec{r}$ that goes from $\vec{A} = (-1, 2, 3)$ to $\vec{B} = (-1, 2, 0)$ with domain the time interval $[1, 2]$.

6. Find a parametrization $\vec{r}$ that goes from $\vec{A} = (1, 0, 2)$ to $\vec{B} = (2, 0, 1)$ with domain the time interval $\left[\frac{1}{2}, 1\right]$.

7. Find a parametrization $\vec{r}$ that goes from $\vec{A} = (-5, 3, 2)$ to $\vec{B} = (1, -1, 4)$ with domain the time interval $[3, 4]$.

8. Let T be the triangle with vertices $\vec{A} = (0, 0, 0), \vec{B} = (1, 0, 1)$ and $C = (1, 1, 1)$. Let C_1 be the side from $\vec{A}$ to $\vec{B}$, let C_2 be the side from $\vec{B}$ to $\vec{C}$, and let C_3 be the side from $\vec{C}$ to $\vec{A}$. Find a parametrization $\vec{r}_1$ for C_1 with domain $[0, 1]$, a parametrization $\vec{r}_2$ for C_2 with domain the interval $[1, 2]$, and a parametrization $\vec{r}_3$ for C_3 with domain the interval $[2, 3]$. Note: if we define the function

$$\vec{r}(t) = \begin{cases} \vec{r}_1(t) & \text{if } 0 \le t \le 1 \\ \vec{r}_2(t) & \text{if } 1 \le t \le 2 \\ \vec{r}_3(t) & \text{if } 2 \le t \le 3, \end{cases}$$

then $\vec{r}$ describes a particle moving around T in the time interval $[0, 3]$.

9. A mass M moves through space. The position function for M is $\vec{r}(t) = (3t^2, \pi \cos(\pi t), e^t)$. At time $t = 1$, all forces acting on M are relaxed, so from that time on, the velocity of M is constant. Find a parametrization for the motion of M for the time interval $[1, 2]$.

11.4   Line Integrals of Type I and Arc Length

In the next four sections we discuss two types of line integrals. Let us consider two examples in order to motivate the type of line integral we study in this section. The reason for the use of the term "line integral" will become clear from context.

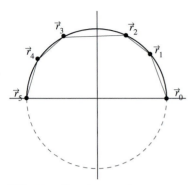

Figure 1. *A typical partition for C.*

EXAMPLE 1: As we all know, the number π is defined to be half the length of the circumference of the unit circle. We also know that π is an irrational number. One method of obtaining an approximation for the number π is the following: let $\vec{r}_0, \vec{r}_1, \ldots, \vec{r}_n$ be points on the upper half of the unit circle $C = \{(x, y) \mid x^2 + y^2 = 1, y \geq 0\}$ such that, if a particle moves around C from the point $(1, 0)$ to the point $(-1, 0)$, it encounters $\vec{r}_{i-1}$ before $\vec{r}_i$ for each i. If we add up the lengths of the straight line segments $[\vec{r}_0, \vec{r}_1], [\vec{r}_1, \vec{r}_2], \ldots, [\vec{r}_{n-1}, \vec{r}_n]$, we should have an approximation of π. Clearly, the more finely we partition C, the better our approximation will be. Thus

$$\pi \approx \sum_{i=1}^{n} \|\vec{r}_i - \vec{r}_{i-1}\|.$$

In Figure 1 we illustrate a case where $n = 5$. ∎

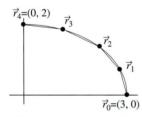

Figure 2. *A partition for E used to approximate the work done in moving the block along the path E.*

EXAMPLE 2: Suppose that, as a block moves on a table, the magnitude of the frictional force between the block and the table is given by $f(x, y) = 3xy$ N. The coordinate system is set up so that the units of distance are measured in m. Suppose further that the block is moving around the elliptical path $E = \{(x, y) \mid x^2/9 + y^2/4 = 1\}$ counterclockwise, from $\vec{A} = (3, 0)$ to $\vec{B} = (0, 2)$. Approximate the work done against the frictional force.

SOLUTION: Let $\{\vec{r}_0, \vec{r}_1, \ldots, \vec{r}_n\}$ be a set of points that partitions E so that $\vec{r}_0 = (3, 0)$ and $\vec{r}_n = (0, 2)$, and so that, for each i, the block passes through $\vec{r}_{i-1}$ before encountering $\vec{r}_i$ as in Figure 2.

If we compute the work done as the block moves along each of the straight line segments $[\vec{r}_0, \vec{r}_1], [\vec{r}_1, \vec{r}_2], \ldots, [\vec{r}_{n-1}, \vec{r}_n]$, then the total work done moving along these segments should approximate the work done moving along the path E. If $\vec{r}_{i-1}$ is sufficiently close to $\vec{r}_i$, then the magnitude of the frictional force along the interval $[\vec{r}_{i-1}, \vec{r}_i]$ can be approximated by $3x_i y_i$ where $\vec{r}_i = (x_i, y_i)$. The force we exert to overcome friction is always in the direction of motion, and thus the work done against friction on the interval $[\vec{r}_{i-1}, \vec{r}_i]$ is approximated by

$$f(\vec{r}_i)\|\vec{r}_i - \vec{r}_{i-1}\| = 3x_i y_i \|\vec{r}_i - \vec{r}_{i-1}\|.$$

The total work done is approximated by the sum

$$W \approx \sum_{i=1}^{n} 3x_i y_i \|\vec{r}_i - \vec{r}_{i-1}\|.$$ ∎

These two examples provide us with a natural way for generalizing the concepts of Riemann Sum and the integral. We first need to describe the concept of a partition for a fundamental curve.

Definition: Partition and Selection for Fundamental Curves

Let C be a fundamental curve with endpoints $\vec{A}$ and $\vec{B}$ in $\mathbb{R}^n$. A *partition* $\mathcal{P}$ of C from $\vec{A}$ to $\vec{B}$ is a set of points, $\{\vec{r}_0, \vec{r}_1, \ldots, \vec{r}_n\}$, of C, such that if a particle moves from $\vec{A}$ to $\vec{B}$ along C, it encounters $\vec{r}_{i-1}$ before encountering $\vec{r}_i$ for each i. Thus, if $\vec{g}$ is a parametrization for C that goes from $\vec{A}$ to $\vec{B}$, then $\vec{g}^{-1}(\vec{r}_{i-1}) \leq \vec{g}^{-1}(\vec{r}_i)$ for all i (remember that g is one-to-one and onto its image, so the inverse exists). We say that $\{\vec{s}_0, \vec{s}_1, \ldots, \vec{s}_n\}$ is a *selection* from $\{\vec{r}_0, \vec{r}_1, \ldots, \vec{r}_n\}$ if $\vec{s}_i$ is a point of C that is between $\vec{r}_{i-1}$ and $\vec{r}_i$.

We now need to define our generalization of Riemann Sum.

Definition: C–sum for f from $\vec{A}$ to $\vec{B}$

If f is a function from $\mathbb{R}^n$ into $\mathbb{R}$, C is a fundamental curve with endpoints $\vec{A}$ and $\vec{B}$, the set $\mathcal{P}$ is a partition of C from $\vec{A}$ to $\vec{B}$, and $\mathcal{S}$ is a selection from $\mathcal{P}$, then the sum

$$\sum_{i=1}^{n} f(\vec{s}_i) \|\vec{r}_i - \vec{r}_{i-1}\|$$

is called a *C–sum for f from $\vec{A}$ to $\vec{B}$*.

Let C be the semicircle $\{(x, y) \mid x^2 + y^2 = 1, y \geq 0\}$. We saw in Example 1 that if $f(\vec{r}) = 1$ for all $\vec{r}$, then π is approximated by the set of all C–sums for f from $\vec{A} = (1, 0)$ to $\vec{B} = (-1, 0)$. In Example 2, we saw that if E is the set of points on the top quarter of the ellipse $\{(x, y) \mid \frac{x^2}{9} + \frac{y^2}{4} = 1, x \geq 0, y \geq 0\}$ and $f(\vec{r}) = 3xy$, then the work done in moving the block from $\vec{A} = (3, 0)$ to $\vec{B} = (0, 2)$ is approximated by the set of E–sums for f from $\vec{A}$ to $\vec{B}$.

Having defined the concepts of partition for a fundamental curve and C–sum for f from $\vec{A}$ to $\vec{B}$, we can now define the limit of such sums and the line integral (of type I).

Definition: Limit of C–sums

We say that L is the *limit of the set of C–sums for f from $\vec{A}$ to $\vec{B}$* if it is true that for any positive number k there is a partition $\mathcal{P}$ of C so that if $\mathcal{P}'$ is any refinement of $\mathcal{P}$ (i.e., $\mathcal{P}$ is a subset of $\mathcal{P}'$) where $\mathcal{P}' = \{\vec{r}_0, \vec{r}_1, \ldots, \vec{r}_n\}$, and if $\mathcal{S} = \{\vec{s}_1, \vec{s}_2, \ldots, \vec{s}_n\}$ is any selection from $\mathcal{P}'$, then

$$\left| L - \sum_{i=1}^{n} f(\vec{s}_i) \|\vec{r}_i - \vec{r}_{i-1}\| \right| \leq k.$$

Definition: Line Integral of Type I

If L is the limit of the set of C–sums for f from $\vec{A}$ to $\vec{B}$, then L is denoted by $\int_{\vec{A}_C}^{\vec{B}} f \, dr$. We call $\int_{\vec{A}_C}^{\vec{B}} f \, dr$ a *line integral (of type I)*.

The following theorem tells us when we can be certain that the line integral exists, and how to compute it.

Theorem 1 *If C is a fundamental curve with endpoints $\vec{A}$ and $\vec{B}$, f is a continuous function from $\mathbb{R}^n$ into $\mathbb{R}$, and $\vec{r}$ is a parametrization for C with domain $[a, b]$ such that $\vec{r}(a) = \vec{A}$ and $\vec{r}(b) = \vec{B}$, then*

$$\int_{\vec{A}_C}^{\vec{B}} f \, dr \text{ exists and is equal to } \int_a^b f(\vec{r}(t)) \, \|\vec{r}'(t)\| \, dt.$$

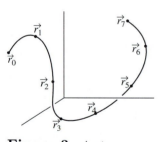

Figure 3. *A given partition and selection for the curve.*

To argue that this is true, let $\mathcal{P} = \{\vec{r}_0, \vec{r}_1, \ldots, \vec{r}_n\}$ be a partition of C. For each i, let $t_i = \vec{r}^{-1}(\vec{r}_i)$ (remember, $\vec{r}(t)$ is one-to-one and onto its image, so the inverse exists), and choose our selection $\mathcal{S}$ so that $\vec{s}_i = \vec{r}_i$ for each i. Then $\sum_{i=1}^{n} f(\vec{r}_i) \|\vec{r}_i - \vec{r}_{i-1}\|$ is an approximation of $\int_{\vec{A}_C}^{\vec{B}} f \, dr$. (See Figure 3.)

We have

$$\int_{\vec{A}_C}^{\vec{B}} f(\vec{r}) \, dr \quad \approx \quad \sum_{i=1}^{n} f(\vec{r}(t_i)) \|\vec{r}(t_i) - \vec{r}(t_{i-1})\|$$

$$= \sum_{i=1}^{n} f(\vec{r}(t_i)) \frac{\|\vec{r}(t_i) - \vec{r}(t_{i-1})\|}{t_i - t_{i-1}} (t_i - t_{i-1})$$

$$\approx \sum_{i=1}^{n} f(\vec{r}(t_i)) \|\vec{r}'(t_i)\| (t_i - t_{i-1})$$

$$\approx \int_{a}^{b} f(\vec{r}(t)) \|\vec{r}'(t)\| \, dt. \qquad \blacksquare$$

EXAMPLE 3: Let H be the portion of the helix that is the image of the function $\vec{r}(t) = (\cos t, \sin t, t), 0 \le t \le 2\pi$. Find the length of H.

SOLUTION: Let $\mathcal{P} = \{\vec{r}_0, \vec{r}_1, \ldots, \vec{r}_n\}$ be a partition of H from $\vec{r}(0) = (1, 0, 0)$ to $\vec{r}(2\pi) = (1, 0, 2\pi)$. Then the length of H is approximated by

$$L \approx \sum_{i=1}^{n} \|r_i - r_{i-1}\|.$$

But $\sum_{i=1}^{n} \|r_i - r_{i-1}\|$ is an approximation of $\int_{A_C}^{\vec{B}} dr$. By Theorem 1, we have

$$int_{A_C}^{\vec{B}} \, dr = \int_{0}^{2\pi} \|\vec{r}'(t)\| \, dt = \int_{0}^{2\pi} \sqrt{\cos^2 t + \sin^2 t + 1} \, dt$$

$$= \int_{0}^{2\pi} \sqrt{2} \, dt = 2\pi\sqrt{2}. \qquad \blacksquare$$

EXAMPLE 4: Compute the work done in Example 2.

SOLUTION: We saw in Example 2 that the work done on the block is well approximated by

$$W \approx \sum_{i=1}^{n} 3x_i y_i \|\vec{r}_i - \vec{r}_{i-1}\| \approx \int_{(3,0)_E}^{(0,2)} f \, dr, \quad \text{where } f(x,y) = 3xy.$$

To compute W, we must first find a parametrization for E. One such parametrization is $\vec{r}(t) = (3\cos t, 2\sin t), 0 \le t \le \frac{\pi}{2}$. Then

$$W = \int_{(3,0)_E}^{(0,2)} f \, dr = \int_{0}^{\pi/2} f(\vec{r}(t)) \|\vec{r}'(t)\| \, dt$$

$$= \int_{0}^{\pi/2} 3(3\cos t)(2\sin t)\sqrt{9\sin^2 t + 4\cos^2 t} \, dt$$

$$= 18 \int_0^{\pi/2} \cos t \sin t \sqrt{9 - 9\cos^2 t + 4\cos^2 t} \ dt$$

$$= 18 \int_0^{\pi/2} \cos t \sin t \sqrt{9 - 5\cos^2 t} \ dt.$$

If we let $g(t) = 9 - 5\cos^2 t$, then $g'(t) = 10\cos t \sin t$. Thus

$$W = \frac{18}{10} \int_{g(0)}^{g(\pi/2)} \sqrt{g(t)}\, g'(t) \ dt = \frac{9}{5} \int_4^9 \sqrt{u} \ du$$

$$= \left(\frac{9}{5}\right) \frac{2u^{3/2}}{3} \Big|_4^9 = 22.8 \text{ J.} \qquad \blacksquare$$

EXAMPLE 5: Find the length of the graph of $f(x) = x^2$ from $x = 0$ to $x = 1$.

SOLUTION: A parametrization for f is given by $\vec{r}(x) = (x, x^2), 0 \le x \le 1$. Then $\vec{r}'(x) = (1, 2x)$, and $\|\vec{r}'(x)\| = \sqrt{1 + 4x^2}$. Thus

$$L = \int_0^1 \sqrt{1 + 4x^2} \ dx$$

$$= \frac{1}{4} \left(2x\sqrt{1 + 4x^2} + \ln\left|2x + \sqrt{1 + 4x^2}\right|\right)\Big|_0^1 \quad \text{(from the tables)}$$

$$= \frac{1}{4}\left((2\sqrt{5} + \ln(2 + \sqrt{5})) - (0 + \ln 1)\right)$$

$$= \frac{\sqrt{5}}{2} + \frac{1}{4}\ln(2 + \sqrt{5}). \qquad \blacksquare$$

In all of the previous examples there is a common idea: think of C as a piece of wire that is stretched and bent into a shape by the parametrization $\vec{r}$. A physical quantity is estimated by a C–sum and the corresponding Riemann sum.

$\displaystyle\sum$	$f(\vec{s}_i)$	$\|\vec{r}_i - \vec{r}_{i-1}\|$
Sum over all the pieces of curve	f evaluated at a point in piece of curve	Approximation of the length of the part of C bounded by r_{i-1} and r_i

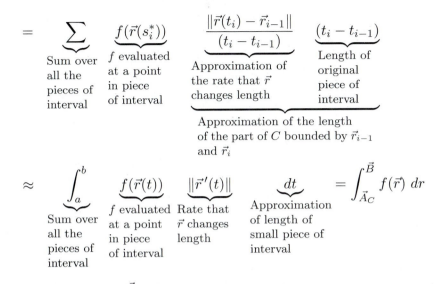

$$= \underbrace{\sum}_{\substack{\text{Sum over} \\ \text{all the} \\ \text{pieces of} \\ \text{interval}}} \underbrace{f(\vec{r}(s_i^*))}_{\substack{f \text{ evaluated} \\ \text{at a point} \\ \text{in piece} \\ \text{of interval}}} \underbrace{\frac{\|\vec{r}(t_i) - \vec{r}_{i-1}\|}{(t_i - t_{i-1})}}_{\substack{\text{Approximation of} \\ \text{the rate that } \vec{r} \\ \text{changes length}}} \underbrace{(t_i - t_{i-1})}_{\substack{\text{Length of} \\ \text{original} \\ \text{piece of} \\ \text{interval}}}$$

Approximation of the length
of the part of C bounded by $\vec{r}_{i-1}$
and $\vec{r}_i$

$$\approx \underbrace{\int_a^b}_{\substack{\text{Sum over} \\ \text{all the} \\ \text{pieces of} \\ \text{interval}}} \underbrace{f(\vec{r}(t))}_{\substack{f \text{ evaluated} \\ \text{at a point} \\ \text{in piece} \\ \text{of interval}}} \underbrace{\|\vec{r}'(t)\|}_{\substack{\text{Rate that} \\ \vec{r} \text{ changes} \\ \text{length}}} \underbrace{dt}_{\substack{\text{Approximation} \\ \text{of length of} \\ \text{small piece of} \\ \text{interval}}} = \int_{\vec{A}_C}^{\vec{B}} f(\vec{r}) \, dr$$

Recalling that $\int_{\vec{A}_C}^{\vec{B}} f(\vec{r}) \, dr$ is the limit of the C–sums, it is helpful to think of dr as being a short length of wire and $f(\vec{r})$ as being a weighting factor evaluated at a point in dr. At the same time, think of $\|\vec{r}'(t)\|$ as the rate that $\vec{r}$ changes length and dt as the length of a small segment in $[a, b]$ (the domain of $\vec{r}$) so that $\|\vec{r}'(t)\|dt = dr$ is a length of the piece of wire in C.

A second useful interpretation of the components in $\int_{\vec{A}_C}^{\vec{B}} f(\vec{r}) \, dr = \int_a^b f(\vec{r}(t)) \, \|\vec{r}'(t)\| \, dt$ is to think of $\|\vec{r}'(t)\|$ as the speed of a particle moving along C and dt as a small segment of time so that $\|\vec{r}'(t)\|dt$ is the distance traveled over the segment of time dt.

Speed and Velocity Revisited (*Optional*)

Let $\vec{r}(t)$, $a \le t \le b$, represent the position of a particle at time t. We define $S(t)$ to be the distance the particle has traveled at time t:

$$S(t) = \int_a^t \|\vec{r}'(x)\| \, dx, a \le t \le b.$$

Then $\frac{dS(t)}{dt}$ should be the *instantaneous speed* of the particle at time t. Fortunately, this agrees with what we have been doing. We have referred to the speed of a particle as the magnitude of its velocity. The Fundamental Theorem of Calculus tells us that

$$\frac{dS(t)}{dt} = \frac{d}{dt} \int_a^t \|\vec{r}'(x)\| \, dx = \|\vec{r}'(t)\|.$$

In our definition of a fundamental curve, we did not allow $\|\vec{r}'(t)\|$ to be 0. Thus $\frac{dS(t)}{dt}$ is positive and S is an increasing function, so $S(t)$ has an inverse $T(s)$. That is, there is a function T such that

$T(S(t)) = t$ and $S(T(s)) = s$. The function T picks out the time t at which the particle has traveled a distance s; i.e., the particle will have traveled a distance s at time $T(s)$. The domain of T is the interval $0 \le s \le$ length of C. If we assume that T has a derivative, then we can compute T' using the chain rule.

$$
\begin{aligned}
t &= T(S(t)) \\
1 &= \frac{dT(S(t))}{dt} = \frac{dT(S(t))}{dS(t)}\frac{dS(t)}{dt} \\
\frac{dT(s)}{ds} &= \frac{1}{dS(t)/dt} = \frac{1}{S'(T(s))} = \frac{1}{\|\vec{r}'(T(s))\|}. \tag{5}
\end{aligned}
$$

EXAMPLE 6: We all are familiar with the mile markers along interstate highways. If you travel east on an interstate highway, when you cross a state line, the mileage marker is set at zero. The 150 mile marker indicates that you are at a point on the highway 150 miles from the western state border. You are provided with a parametrization of the highway in terms of arc length. If you set a stop watch at zero as you cross the state line (going east), and you check the time elapsed when you reach the 150 mile marker, then your watch will give you the value for $T(150)$. On the other hand, if your watch indicates that 2.2 hours have elapsed when you reach the 150 mile marker, then $S(2.2) = 150$. Your speedometer reading as you cross the 150 mile marker is a measurement of $\frac{dS(150)}{dt}$. ∎

EXAMPLE 7: Find $dT(s)/ds$ for the function $\vec{r}(t) = (1, t^2)$ for $0 \le t \le 2$.

SOLUTION: The derivative of $\vec{r}$ is $\vec{r}'(x) = (0, 2x)$. So

$$
S(t) = \int_0^t \|(0, 2x)\|\ dx = \int_0^t |2x|\ dx = t^2.
$$

Thus $T(s) = \sqrt{s}$. Using Equation (5) we have that $\frac{dT(s)}{ds} = \frac{1}{\|(0,2\sqrt{s})\|} = \frac{1}{2\sqrt{s}}$. If we compute $\frac{dT(s)}{ds}$ directly we also obtain $\frac{1}{2}s^{-1/2}$. ∎

EXERCISES 11.4

1. In Figure 4, you are given a partition $\{\vec{r}_0, \vec{r}_1, \ldots, \vec{r}_{10}\}$. Calculate the C–sum for the curve determined by the partition that approximates the length of C.

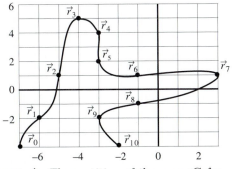

Figure 4. *The partition of the curve C for Exercises 1–3.*

2. Let $\mathcal{P}$ be the partition $\{\vec{r}_0, \vec{r}_1, \ldots, \vec{r}_{10}\}$ of the curve in Figure 4. For each i, let $\vec{s}_i{}^* = \vec{r}_i$. Then $\mathcal{S} = \{\vec{s}_1{}^*, \vec{s}_2{}^*, \ldots, \vec{s}_n{}^*\}$ is a selection from $\mathcal{P}$. Given that $f(x, y) = xy + 1$, evaluate the C–sum for f determined by $\mathcal{P}$ and $\mathcal{S}$.

3. Let $\mathcal{P}$ be the partition $\{\vec{r}_0, \vec{r}_1, \ldots, \vec{r}_{10}\}$ of the curve in Figure 4. For each i, let $\vec{s}_i{}^* = \vec{r}_{i-1}$. Then $\mathcal{S} = \{\vec{s}_1{}^*, \vec{s}_2{}^*, \ldots, \vec{s}_n{}^*\}$ is a selection from $\mathcal{P}$. Given that $f(x, y) = xy + 1$, evaluate the C–sum for f determined by $\mathcal{P}$ and $\mathcal{S}$.

In Exercises 4–9, find the length of the curves parameterized by $\vec{r}$.

4. $\vec{r}(t) = (5t^2, 2t^3),\ 0 \le t \le 1$.

5. $\vec{r}(t) = (t^2, \cos(t^2), \sin(t^2)),\ 0 \le t \le \sqrt{2\pi}$.

6. $\vec{r}(t) = (t, \ln(\cos t)),\ 0 \le t \le 1$.

7. $\vec{r}(t) = (e^t, \cos(e^t), \sin(e^t)),\ 0 \le t \le \ln 2$.

8. $\vec{r}(t) = (t \cos t, t \sin t),\ 0 \le t \le 1$.

9. $\vec{r}(t) = (t, t \cos t, t \sin t),\ 0 \le t \le 1$.

Find the length of the graph of the functions given in Exercises 10–12.

10. $f(x) = \sqrt{1 - x^2},\ 0 \le x \le \frac{1}{2}$.

11. $f(x) = x^{3/2},\ 0 \le x \le 1$.

12. $f(x) = \frac{1}{2}(e^x + e^{-x}),\ 0 \le x \le 1$.

Let C be the straight line segment connecting the points $\vec{A} = (1, 0, 1)$ and $\vec{B} = (1, 2, -1)$. In Exercises 13–18, calculate $\int_{\vec{A}_C}^{\vec{B}} f\ dr$.

13. $f(x, y, z) = x + y + z$.

14. $f(x, y, z) = e^x + e^y + e^z$.

15. $f(x, y, z) = x^{-2} + y + z^2$.

16. $f(x, y, z) = e^{x+y+z}$.

17. $f(x, y, z) = xyz$.

18. $f(x, y, z) = \dfrac{xz}{y + 1}$.

In Exercises 19 and 20, let C be the straight line segment connecting the points $\vec{A} = (1, 0, 1)$ and $\vec{B} = (1, 2, -1)$. Explain why you cannot evaluate $\int_{\vec{A}_C}^{\vec{B}} f\ dr$.

19. $f(x, y, z) = \dfrac{xy}{z}$.

20. $f(x, y, z) = \dfrac{z}{xy}$.

Let C be the circular arc from the point $\vec{A} = (2, 0)$ to $\vec{B} = (0, 2)$, parametrized by $\vec{r}(t) = (2 \cos t, 2 \sin t), 0 \le t \le \frac{\pi}{2}$. In Exercises 21–23, calculate $\int_{\vec{A}_C}^{\vec{B}} f\ dr$.

21. $f(x, y) = x + y$

22. $f(x, y) = e^x + e^y$

23. $f(x, y) = xy$

11.5   Center of Mass and Moment of Inertia

In this section we employ the techniques of Section 11.4 to find the mass and center of mass of a "one-dimensional" piece of material such as a piece of wire. We also find the kinetic energy of the mass as it rotates about an axis. We first need the concept of density.

Let M be a piece of material such as an extremely thin wire that can be represented as a fundamental curve $\vec{r}(t)$. The *density function* ρ is a function from the position $\vec{r}(t)$ of the wire into $\mathbb{R}$ such that if ΔL is a very short length of M containing the point $\vec{r}$, then $\rho(\vec{r})\Delta L$ approximates the mass of ΔL. That is, $\rho(\vec{r})$ is the derivative of the total mass with respect to the arc length. More formally:

$$\rho(\vec{r}) = \lim_{\substack{\text{length of piece} \\ \text{of } M \text{ contain-} \\ \text{ing } \vec{r} \to 0}} \frac{\text{mass of piece of } M \text{ containing } \vec{r}}{\text{length of piece of } M \text{ containing } \vec{r}}.$$

From the definition, we see that if ρ is a density function for the wire M, and if $\vec{A}$ and $\vec{B}$ are the endpoints of M, then using our "rule of thumb" from Section 6.5,

$$\int_{\vec{A}_M}^{\vec{B}} \underbrace{\rho(\vec{r})}_{\text{density}} \underbrace{dr}_{\text{length}}$$

should be the mass of the wire, where $\rho(\vec{r})dr$ may be interpreted as the mass of a small piece of wire.

Definition: Center of Mass for Point Masses

Let $m_1,\ m_2, \ldots, m_n$ be point masses located at (x_1, y_1, z_1), $(x_2, y_2, z_3), \ldots, (x_n, y_n, z_n)$. Let

$$X = \frac{\sum_{i=1}^{n} x_i m_i}{\sum_{i=1}^{n} m_i}, \qquad Y = \frac{\sum_{i=1}^{n} y_i m_i}{\sum_{i=1}^{n} m_i}, \text{ and } Z = \frac{\sum_{i=1}^{n} z_i m_i}{\sum_{i=1}^{n} m_i}.$$

The point $\vec{C} = (X, Y, Z)$ is called the *center of mass* for the collection of masses.

Integrals can be used to extend this concept to configurations that are not the sum of a finite number of point masses. Let C be a piece of wire represented as a fundamental curve $\vec{r}$ with endpoints $\vec{A}$ and $\vec{B}$. Let $\rho(x, y, z)$ be a density function defined on C. Let $\mathcal{P} = \{\vec{r}_0, \vec{r}_1, \ldots, \vec{r}_n\}$ be a partition of C, and let $\mathcal{S} = \{\vec{s}_1, \vec{s}_2, \ldots, \vec{s}_n\}$, where $s_i = (x_i, y_i, z_i)$, be a selection from $\mathcal{P}$.

Let X be the number approximated by

$$\frac{\sum_{i=1}^{n} x_i \rho(\vec{s}_i)\|\vec{r}_i - \vec{r}_{i-1}\|}{\text{total mass of } C}. \quad \text{Then} \quad X = \frac{\int_{\vec{A}_C}^{\vec{B}} x\rho(\vec{r})\, dr}{\int_{\vec{A}_C}^{\vec{B}} \rho(\vec{r})\, dr}.$$

Similarly, let

$$Y = \frac{\int_{\vec{A}_C}^{\vec{B}} y\rho(\vec{r})\, dr}{\int_{\vec{A}_C}^{\vec{B}} \rho(\vec{r})\, dr}, \quad \text{and} \quad Z = \frac{\int_{\vec{A}_C}^{\vec{B}} z\rho(\vec{r})\, dr}{\int_{\vec{A}_C}^{\vec{B}} \rho(\vec{r})\, dr}.$$

We have the following definition.

Definition: Center of Mass

Let C be a piece of wire represented as a fundamental curve $\vec{r}$ with endpoints $\vec{A}$ and $\vec{B}$. Let $\rho(x, y, z)$ be a density function defined on C, and let

$$X = \frac{\int_{\vec{A}_C}^{\vec{B}} x\rho(\vec{r})\, dr}{\int_{\vec{A}_C}^{\vec{B}} \rho(\vec{r})\, dr} \qquad Y = \frac{\int_{\vec{A}_C}^{\vec{B}} y\rho(\vec{r})\, dr}{\int_{\vec{A}_C}^{\vec{B}} \rho(\vec{r})\, dr}, \quad \text{and } Z = \frac{\int_{\vec{A}_C}^{\vec{B}} z\rho(\vec{r})\, dr}{\int_{\vec{A}_C}^{\vec{B}} \rho(\vec{r})\, dr}.$$

Then the point (X, Y, Z) is called the *center of mass* of C.

Physically, if you place your finger under the center of mass of a piece of material, then the system will be perfectly balanced. Observe that there is no reason to expect the center of mass of a piece of material to be on that piece. Indeed, as the next example shows, the center of mass of a piece of wire with a constant density function bent into the shape of a circle will be the center of the circle.

EXAMPLE 1: Let C be a piece of wire with constant mass density m, bent into a circle of radius r. Find the center of mass of C.

SOLUTION: We place the origin at the center of the circle so that a parametrization for C is given by $\vec{r}(\theta) = r(\cos(\theta), \sin(\theta))$, for $0 \leq \theta \leq 2\pi$. Then

$$X = \frac{\int_{(r,0)_C}^{(r,2\pi)} mx\, dr}{2\pi r m} = \frac{\int_0^{2\pi} mr\cos(\theta)r\, d\theta}{2\pi r m} = \frac{mr \int_0^{2\pi} \cos(\theta)\, d\theta}{2\pi m} = 0,$$

and

$$Y = \frac{\int_{(r,0)_C}^{(r,2\pi)} my\, dr}{2\pi r m} = \frac{\int_0^{2\pi} mr\sin(\theta)r\, d\theta}{2\pi r m} = \frac{mr \int_0^{2\pi} \sin(\theta)\, d\theta}{2\pi m} = 0. \blacksquare$$

EXAMPLE 2: Let C be the line segment from $\vec{A} = (1,1,1)$ to $\vec{B} = (2,2,2)$, representing a straight piece of wire. Suppose that the mass density at a point $\vec{r}$ of C is the square of its distance from the origin. Find the center of mass of C.

SOLUTION: A parametrization for C is $\vec{r}(t) = (t+1, t+1, t+1), 0 \leq t \leq 1$. The density function is $\rho(t) = (t+1)^2 + (t+1)^2 + (t+1)^2 = 3(t^2 + 2t + 1)$. Then,

$$
X = \frac{\int_{\vec{A}_C}^{\vec{B}} x\rho(\vec{r})\ dr}{\int_{\vec{A}_C}^{\vec{B}} \rho(\vec{r})\ dr} = \frac{\int_0^1 (t+1)3(t+1)^2\sqrt{3}\ dt}{\int_0^1 3(t+1)^2\sqrt{3}\ dt}
$$

$$
= \frac{3\sqrt{3}\frac{(t+1)^4}{4}\big|_0^1}{3\sqrt{3}\frac{(t+1)^3}{3}\big|_0^1} = \frac{\frac{15}{4}}{\frac{7}{3}} = \frac{45}{28}.
$$

Similarly, we have that $Y = \frac{45}{28} = Z$. Thus the center of mass is the point $\left(\frac{45}{28}, \frac{45}{28}, \frac{45}{28}\right)$. ∎

The last application we consider in this section is the kinetic energy of a rotating body. First, we consider a point mass m rotating about the z–axis, at a rate of ω radians/second, as if it were on the end of a string of length r. We assume that it stays in the xy–plane, so a parametrization for its trajectory is given by $\vec{r}(t) = r(\cos(\omega t), \sin(\omega t))$. Its velocity is $\vec{r}'(t) = wr(-\sin(\omega t), \cos(\omega t))$, and its speed is $v = \omega r$. The kinetic energy of the particle is $\frac{1}{2}mv^2 = \frac{1}{2}m\omega^2 r^2$.

Consider a finite system of point masses as in the discussion on center of mass. Let $m_1, m_2, \ldots, m_n$ denote the point masses located at (x_1, y_1, z_1), (x_2, y_2, z_3), $\ldots, (x_n, y_n, z_n)$. Our assumption is that the system is held rigid (as if tied together by a massless wire) and allowed to rotate about the x–axis at a rate of ω radians/second. The radius of rotation of the ith point mass is $\sqrt{y_i^2 + z_i^2}$. Thus its speed is $\omega\sqrt{y_i^2 + z_i^2}$, and its kinetic energy is $\frac{1}{2}mv^2 = \frac{1}{2}m_i\omega^2(y_i^2 + z_i^2)$. To find the total kinetic energy of the system, we simply add up the kinetic energy of all of the point masses. Thus the total kinetic energy is $\sum_{i=1}^n \frac{1}{2}m_i\omega^2(y_i^2 + z_i^2) = \frac{1}{2}\omega^2 \sum_{i=1}^n m_i(y_i^2 + z_i^2)$.

Finally, suppose that we have a one-dimensional piece of material that can be described as a fundamental curve C with endpoints $\vec{A}$ and $\vec{B}$, and that $\rho(\vec{r})$ is a mass density function for C. Recall our interpretation of dr as a small section of the material, and $\vec{r} = (x, y, z)$ as a point in dr. If C is rotating about the x–axis at a rate of ω

radians/second, $\sqrt{y^2 + z^2}$ is the radius of rotation of dr, so its speed is $\omega\sqrt{y^2 + z^2}$. Its mass is given by $\rho(\vec{r})\,dr$. The kinetic energy of the system rotating about the x–axis is then given by

$$\int_{\vec{A}_C}^{\vec{B}} \frac{1}{2}\omega^2(y^2 + z^2)\rho(\vec{r})\,dr = \frac{1}{2}\omega^2 \int_{\vec{A}_C}^{\vec{B}} (y^2 + z^2)\rho(\vec{r})\,dr.$$

Similarly, if we have the same system rotating about the y–axis at a rate of ω radians/sec, the kinetic energy is given by

$$\frac{1}{2}\omega^2 \int_{\vec{A}_C}^{\vec{B}} (x^2 + z^2)\rho(\vec{r})\,dr,$$

and if it is rotating about the z–axis, its kinetic energy is

$$\frac{1}{2}\omega^2 \int_{\vec{A}_C}^{\vec{B}} (x^2 + y^2)\rho(\vec{r})\,dr.$$

EXAMPLE 3: Let C be the graph of $y = x^2$, $0 \le x \le 1$ in the xy–plane. Assume that the mass density of C is a constant, m. Find its kinetic energy if it is rotating at a rate of 10 radians/sec about (a) the x–axis, (b) the y–axis, and (c) the z–axis.

SOLUTION: A parametrization for C is $\vec{r}(t) = (t, t^2)$, $0 \le t \le 1$. The derivative is given by $\vec{r}\,'(t) = (1, 2t)$, and the norm of the derivative is given by $\|\vec{r}\,'(t)\| = \sqrt{1 + 4t^2}$.

(a) $\begin{aligned}[t] \text{KE}_x &= \frac{1}{2}\omega^2 \int_{\vec{A}_C}^{\vec{B}} (y^2 + z^2)\rho(\vec{r})\,dr = \frac{1}{2}\omega^2 \int_{\vec{A}_C}^{\vec{B}} (y^2 + z^2)m\,dr \\[2mm] &= \frac{1}{2}m\omega^2 \int_0^1 \left((t^2)^2 + 0^2\right)\sqrt{1 + 4t^2}\,dt \\[2mm] &= \frac{1}{2}m\omega^2 \int_0^1 t^4\sqrt{1 + 4t^2}\,dt \text{ Joules.} \end{aligned}$

(b) $\begin{aligned}[t] \text{KE}_y &= \frac{1}{2}\omega^2 \int_{\vec{A}_C}^{\vec{B}} (x^2 + z^2)\rho(\vec{r})\,dr = \frac{1}{2}\omega^2 \int_{\vec{A}_C}^{\vec{B}} (x^2 + z^2)m\,dr \\[2mm] &= \frac{1}{2}m\omega^2 \int_0^1 (t^2 + 0^2)\sqrt{1 + 4t^2}\,dt \\[2mm] &= \frac{1}{2}m\omega^2 \int_0^1 t^2\sqrt{1 + 4t^2}\,dt \text{ Joules.} \end{aligned}$

$$\text{(c)} \quad \text{KE}_z \;=\; \frac{1}{2}\omega^2 \int_{\vec{A}_C}^{\vec{B}} (x^2 + y^2)\rho(\vec{r})\; dr = \frac{1}{2}\omega^2 \int_{\vec{A}_C}^{\vec{B}} (x^2 + y^2)m\; dr$$

$$\;=\; \frac{1}{2}m\omega^2 \int_0^1 \left(t^2 + (t^2)^2\right)\sqrt{1 + 4t^2}\; dt \;\text{Joules.}$$

Note that in each of the above examples, the term $\frac{1}{2}\omega^2$ can be factored out. The remaining integral is called the *moment of inertia* about the appropriate axis. Thus

$$I_x \;=\; \int_{\vec{A}_C}^{\vec{B}} (y^2 + z^2)\rho(\vec{r})\; dr \quad \text{is called the moment of inertia about the } x\text{–axis,}$$

$$I_y \;=\; \int_{\vec{A}_C}^{\vec{B}} (x^2 + z^2)\rho(\vec{r})\; dr \quad \text{is called the moment of inertia about the } y\text{–axis,}$$

$$I_z \;=\; \int_{\vec{A}_C}^{\vec{B}} (x^2 + y^2)\rho(\vec{r})\; dr \quad \text{is called the moment of inertia about the } z\text{–axis.}$$

If we know the moment of inertia of a piece of material about its given axis, and if we know its rate of rotation about that axis, then we can compute the kinetic energy of the material without any calculus. For example, if a mass with $I_x = 25$ is rotating about the x–axis with angular speed $\omega = 2\pi$ radians/sec, then the total kinetic energy of the system is $\text{KE} = \frac{1}{2}I_x\omega^2 = \frac{1}{2} \cdot 25 \cdot (2\pi)^2 = 50\pi^2$ Joules.

EXERCISES 11.5

In Exercises 1–3 let L be the line segment with endpoints $\vec{A} = (1, 2, 3)$ m and $\vec{B} = (2, -1, 3)$ m. For the density function ρ, find the segment's mass, center of mass, and moment of inertia about each of the coordinate axes.

1. $\rho(x, y, z) = x$ kg/m.

2. $\rho(x, y, z) = y^2$ kg/m.

3. $\rho(x, y, z) = x + y + z$ kg/m.

4. Find the kinetic energy of the line L of Exercise 1 if it is rotating about the x–axis at a rate of 3 rotations/sec.

5. Find the kinetic energy of the line L of Exercise 1 if it is rotating about the y–axis at a rate of 3 rotations/sec.

6. Let C be a circular hoop of radius R with constant mass density $\rho = m$. Find the moment of inertia about a diameter line segment.

7. Let C be the part of the unit circle $x^2 + y^2 = 1$ lying in the upper half plane. Given a constant mass density $\rho(x, y) = m$, find the center of mass of C.

8. Find the moment of inertia about the x–axis of the semicircle C of Exercise 7.

In Exercises 9 and 10, let C be a piece of wire bent into the shape of the graph of $y = x^2 + 1$, $0 \le x \le 1$ (x and y are measured in m). Set up the integrals for the center of mass and the moment of inertia about the y-axis, given the density function ρ.

9. $\rho(x, y) = c$

10. $\rho(x, y) = x + y$

In Exercises 11 and 12, let C be the spiral parametrized by $\vec{r}(t) = (t, \sin t, \cos t)$ for $0 \le t \le 2\pi$. Its mass density is the constant m.

11. Find the center of mass of C.

12. What is the kinetic energy of C if it is rotating about the x-axis at a rate of 25 radians/sec?

11.6   Vector Fields

EXAMPLE 1: If M and m are masses, then the gravitational force between them is proportional to the product of the masses, and inversely proportional to the square of the distance between them. If we place our coordinate axes so that the origin is at the center (of mass) of M, and if $\vec{r}$ is the position vector of m, then the magnitude of the gravitational force due to M on m is $GMm/\|\vec{r}\|^2$, where G is a constant. The direction of the force is toward M, or in the direction $-\vec{r}$. Notice that $-\vec{r}/\|\vec{r}\|$ is a unit vector that points in the direction of the force. Thus

$$\vec{F}(\vec{r}) = \left(\frac{GMm}{\|\vec{r}\|^2}\right) \left(\frac{-\vec{r}}{\|\vec{r}\|}\right) = -\left(\frac{GMm}{\|\vec{r}\|^3}\right) \vec{r} = -\frac{GMm}{r^3} \vec{r}$$

is a function from $\mathbb{R}^3$ into $\mathbb{R}^3$ that assigns to the position $\vec{r}$ ($\vec{r} \ne \vec{0}$) the force that would be exerted on m by M if m were placed at position $\vec{r}$. ∎

Example 1 is just one instance of what is called a vector field. If D is a subset of $\mathbb{R}^n$, then a function $\vec{f}$ from D into $\mathbb{R}^n$ is called a *vector field*. The intuitive interpretation of a vector field is that it assigns to each position $\vec{r}$ in the domain D a directed line segment $\vec{f}(\vec{r})$. If $\vec{f}$ is a vector field with domain D such that $\vec{f}(\vec{r})$ denotes the force that would be exerted on a mass m if it were at position $\vec{r}$, then $\vec{f}$ is called a *force field*. Notice that a force field is not a force but rather a function that assigns to each position in its domain the force that would act on a mass if it were at that position. A force field is just one type of vector field. Another example of a vector field is illustrated below.

EXAMPLE 2: Let D denote a river, and let $\vec{f}$ be the function that assigns to each position in D the velocity of a water particle at that position in D. (See Figure 2.) ∎

Figures 3.a–3.d we provide examples of vector fields.

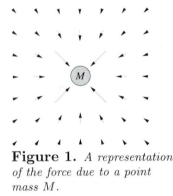

Figure 1. *A representation of the force due to a point mass M.*

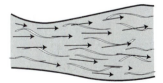

Figure 2. *The vector field, which gives the velocity of each particle in a river.*

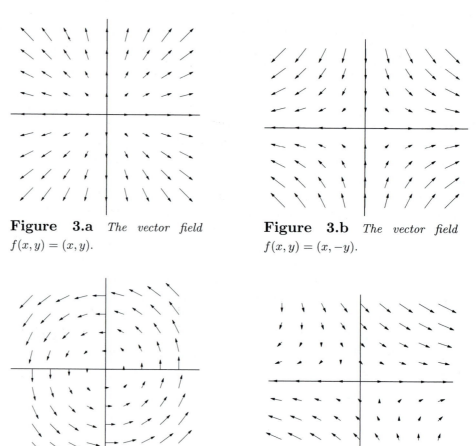

Figure 3.a *The vector field*
$f(x, y) = (x, y)$.

Figure 3.b *The vector field*
$f(x, y) = (x, -y)$.

Figure 3.c *The vector field*
$f(x, y) = (-y, x)$.

Figure 3.d *The vector field*
$f(x, y) = (x + y, -y)$.

Coulomb's Law provides us with a rich source of examples of vector fields. If Q is a point mass with charge[2] q_0 at position $\vec{r}_0 = (x_0, y_0, z_0)$, then the *electric field* due to Q at $\vec{r} = (x, y, z)$ is defined by

$$\vec{E}(\vec{r}) = \frac{kq_0}{\|\vec{r} - \vec{r}_0\|^3}(\vec{r} - \vec{r}_0).$$

Thus the magnitude of $\vec{E}(\vec{r})$ is inversely proportional to the square of the distance from $\vec{r}$ to the charged particle Q. If the charge on Q is positive, then $\vec{E}$ points radially away from Q. If the charge on Q is negative, then $\vec{E}$ points directly toward Q. The unit of measure of an electric field is Newtons/Coulomb (abbreviated N/C.) If the mass M is a positively charged mass, then Figure 1 provides a visualization

---

[2]The unit of measure of a charged particle is a *Coulomb*. A charge on a mass can be negative or positive.

of the electric field.

The importance of an electric field is that if $\vec{E}$ is the electric field due to a charged particle and if another particle Q_1 with charge q_1 is placed in space at position $\vec{r}$, then the force acting on q_1 is $q_1 \vec{E}(\vec{r})$. See Figure 4.

EXAMPLE 3: We can obtain the electric field generated by two charged particles by adding the fields generated by each particle separately.

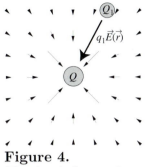

Figure 4.
The electric force acting on a charged particle placed in the electric field due to Q.

EXAMPLE 4: Suppose that a particle of charge q_1 is placed at $\vec{A}$ and a particle of charge q_2 is placed at $\vec{B}$. Then the electric field at position $\vec{r} = (x, y, z)$ due to the charge at $\vec{A}$ is

$$\vec{E}_1(\vec{r}) = \frac{kq_1}{\|\vec{r} - \vec{A}\|^3}(\vec{r} - \vec{A}),$$

and the electric field at position $\vec{r}$ due to the charge at $\vec{B}$ is

$$\vec{E}_2(\vec{r}) = \frac{kq_2}{\|\vec{r} - \vec{B}\|^3}(\vec{r} - \vec{B}).$$

To obtain the electric field due to both charges, we simply add $\vec{E}_1$ and $\vec{E}_2$.

$$\vec{E}_1 + \vec{E}_2 = \frac{k}{\|\vec{r} - \vec{A}\|^3}(\vec{r} - \vec{A}) + \frac{k}{\|\vec{r} - \vec{B}\|^3}(\vec{r} - \vec{B}).$$

In Figure 5.a, we display the field generated by two particles with charges of equal magnitude but opposite signs. Figure 5.b illustrates the field generated by two equally charged particles.   ■

We can use the idea of adding vector fields to find the electric field due to a charged wire. Suppose that C is a wire in space with end points $\vec{A}$ and $\vec{B}$ and that $\rho(\vec{r})$ gives the charge per unit length at the position $\vec{r}$ on the wire. (The units on ρ are Coulombs per unit length.) If $\{\vec{r}_0, \vec{r}_1, \ldots, \vec{r}_n\}$ is a partition of C, and $\{\vec{s}_1{}^*, \vec{s}_2{}^*, \ldots \vec{s}_n{}^*\}$ is a selection for the partition, as in Figure 5, then an approximation of the total charge on C is given by

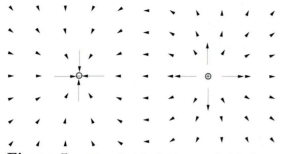

Figure 5.a *The electric field generated by two particles with charges of equal magnitude but opposite signs.*

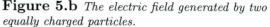

Figure 5.b *The electric field generated by two equally charged particles.*

$$\sum_{i=1}^{n} \rho(\vec{s_i}^*)\|\vec{r_i} - \vec{r}_{i-1}\|.$$

It follows that the charge on C is

$$\int_{C_A}^{B} r\, dr.$$

To calculate the electric field due to the charged wire at the point $\vec{p} = (x, y, z)$, we start with a partition $\{\vec{r}_0, \vec{r}_1, \ldots, \vec{r}_n\}$ of C and a selection $\{\vec{s}_1^*, \vec{s}_2^*, \ldots, \vec{s}_n^*\}$ for the partition. The electric field at the point $\vec{r}$ due to the section of the wire from $\vec{r}_{i-1}$ to $\vec{r}_i$ is approximated by

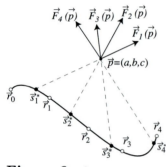

Figure 6. *A partition and selection of the charged wire C.*

$$\vec{E}_i(\vec{p}) = \frac{k\rho(\vec{s_i}^*)\|r_i - \vec{r}_{i-1}\|}{\|\vec{p} - \vec{s_i}^*\|^3}(\vec{p} - \vec{s_i}^*).$$

Thus the electric field at $\vec{r}$ is approximated by

$$\sum_{i=1}^{n} \vec{E}_i(\vec{p}) = \sum_{i=1}^{n} \frac{k\rho(\vec{s_i}^*)}{\|\vec{p} - \vec{s_i}^*\|^3}(\vec{p} - \vec{s_i}^*)\|\vec{r_i} - \vec{r}_{i-1}\|$$

as in Figure 6. Thus the electric field at $\vec{p}$ due to the charged wire is

$$\vec{E}(\vec{p}) = \int_{A_C}^{\vec{B}} \frac{k\rho(\vec{r})}{\|\vec{p} - \vec{r}\|^3}(\vec{p} - \vec{r})\, dr. \tag{6}$$

The expression $\vec{p} - \vec{r}$ in Equation (6) is a vector, and so the result of the integral is a vector quantity. If $\vec{p} = (a, b, c)$ and $\vec{r} = (x, y, z)$, then the vector field can be broken into its components:

$$\vec{E}(\vec{p}) = \begin{pmatrix} E_x(\vec{p}) \\ E_y(\vec{p}) \\ E_z(\vec{p}) \end{pmatrix} = \begin{pmatrix} \int_{A_C}^{\vec{B}} \frac{k\rho(\vec{r})}{\|\vec{p}-\vec{r}\|^3}(a - x)\, dr \\ \int_{A_C}^{\vec{B}} \frac{k\rho(\vec{r})}{\|\vec{p}-\vec{r}\|^3}(b - y)\, dr \\ \int_{A_C}^{\vec{B}} \frac{k\rho(\vec{r})}{\|\vec{p}-\vec{r}\|^3}(c - z)\, dr \end{pmatrix}.$$

EXAMPLE 5: A coordinate system is placed so that a straight 2 m wire with a constant charge density ρ is centered at the origin with end points on the x–axis. Find the electric field induced by the wire at the point $\vec{p} = (-1, -1, -1)$.

SOLUTION: The endpoints of the wire are at $(-1, 0, 0)$ m and $(1, 0, 0)$ m. A parametrization for the wire is $\vec{r}(t) = (2t - 1, 0, 0)$, $0 \le t \le 1$. The derivative $\vec{r}'(t) = (2, 0, 0)$ and $\|\vec{r}'(t)\| = 2$. The point $\vec{p}$ of Equation (6) is $(-1, -1, -1)$.

$$
\begin{aligned}
E_x(-1, -1, -1) &= \int_{\vec{A}_C}^{\vec{B}} \frac{k\rho}{\|(-1, -1, -1) - \vec{r}\|^3}(-1 - x)\, dr \\
&= k\rho \int_0^1 \frac{-2t}{(4t^2 + 2)^{3/2}}(2)\, dt \\
&= -\frac{k\rho}{2} \int_2^6 u^{-3/2}\, du = \frac{k}{\rho}\left(\frac{1}{\sqrt{6}} - \frac{1}{\sqrt{2}}\right). \\
E_y(-1, -1, -1) &= \int_{\vec{A}_C}^{\vec{B}} \frac{k\rho}{\|(-1, -1, -1) - \vec{r}\|^3}(-1 - y)\, dr \\
&= -k\rho \int_0^1 \frac{1}{(4t^2 + 2)^{3/2}}(2)\, dt = -\frac{k\rho}{\sqrt{6}}.
\end{aligned}
$$

You are asked to find $E_z(-1, -1, -1)$ in Exercise 13.   ■

EXERCISES 11.6

In Exercises 1–6, plot the vector field by drawing $\vec{F}(x, y)$ emanating from each point on the 3×4 grid of points in Figure 7.

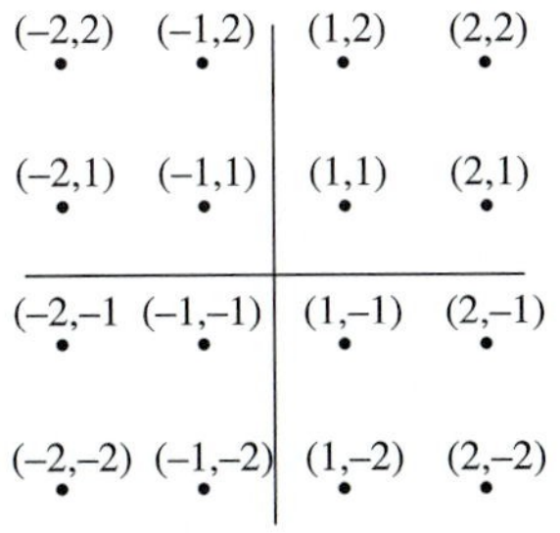

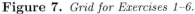

(−2,2) (−1,2) (1,2) (2,2)

(−2,1) (−1,1) (1,1) (2,1)

(−2,−1) (−1,−1) (1,−1) (2,−1)

(−2,−2) (−1,−2) (1,−2) (2,−2)

Figure 7. *Grid for Exercises 1–6.*

1. $\vec{F}(x, y) = (-x, -y)$.
2. $\vec{F}(x, y) = (-x, y)$.
3. $\vec{F}(x, y) = (x, x^2)$.
4. $\vec{F}(x, y) = (y, -x)$.
5. $\vec{F}(x, y) = (x^2, x^2)$.
6. $\vec{F}(x, y) = \dfrac{1}{\sqrt{x^2 + y^2}}(-y, x)$.

In Exercises 7–10, a 2 C charge is placed at the origin. Determine the electric field at the given position.

7. $(1, 1, 1)$ m.

8. $(-1, 4, -2)$ m.

9. $(5, 0, -3)$ m.

In Exercises 10–12, a 2 C charge is placed at the origin and a 3 C charge is placed at $(2, 3, -1)$ m.

10. Determine the electric field at $(1, 2, 1)$ m.

11. Determine the electric field at $(1, 1, 1)$ m.

12. Express the electric field as a function of position.

13. Show that $E_z(-1, -1, -1)$ of Example 5 is $-\frac{k\rho}{\sqrt{6}}$.

14. Give a geometric reason for the fact that in Example 5, $E_z(-1, -1, -1) = E_y(-1, -1, -1)$.

15. Give a geometric reason for the fact that in Example 5, $E_x(-1, -1, -1)$ is negative.

For Exercises 16–21, a coordinate system is placed so that a straight 2 m wire with a constant charge density ρ is centered at the origin with endpoints on the x–axis (as in Example 5).

16. Find the electric field induced by the wire at the point $\vec{p} = (1, 1, 1)$. Express your answer as a definite integral and then use a numerical integrator to calculate the integral.

17. Use a geometric argument to explain why, in Exercise 10, $\vec{E}_x(1, 1, 1)$ is positive.

18. Without doing any calculations, would you expect $\vec{E}_x(2, 3, 4)$ to be positive? Explain your answer geometrically.

19. Without doing any calculations, would you expect $\vec{E}_y(2, 3, 4) = E_z(2, 3, 4)$? Explain your answer geometrically.

20. Let C be a circle with axis the x–axis (the x–axis is perpendicular to the plane containing the circle) as illustrated in Figure 8. Give a geometric reason for the fact $\|\vec{E}(\vec{r})\|$ is constant on C.

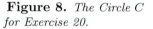

Figure 8. *The Circle C for Exercise 20.*

21. Show by computation that $\vec{E}_x(0, c, d) = \vec{0}$. Give a geometric reason for this answer.

22. A wire with constant charge density ρ is shaped as a circle of radius 1 m. It is centered at the origin and lies in the xy–coordinate plane. See Figure 9. Calculate the electric field induced by the wire at the point $(0, 0, c)$.

Figure 9. *Illustration for Exercises 22 and 23.*

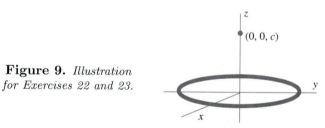

23. Give a geometric reason that, in Exercise 22, $\vec{E}_x(0, 0, c) = \vec{0}$.

11.7   Line Integrals of Type II and Work

Suppose that a mass M is moving through space and there is a force field $\vec{F}$ acting on that particle. Then $\vec{F}$ does work on M. This section is devoted to developing techniques for computing the work done on a mass as it moves through a force field.

Assume that the mass is moving from point $\vec{A}$ to point $\vec{B}$ along the fundamental curve C. Let $\mathcal{P} = \{\vec{r}_0, \vec{r}_1, \ldots, \vec{r}_n\}$ be a partition of C from $\vec{A}$ to $\vec{B}$. If we calculate the work that would be done on

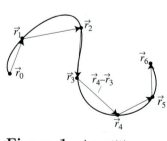

Figure 1. *A partition of the curve C.*

the particle by $\vec{F}$ if it were moving from $\vec{A}$ to $\vec{B}$ along the chain of straight line intervals $[\vec{r}_0, \vec{r}_1], [\vec{r}_1, \vec{r}_2], \ldots, [\vec{r}_{n-1}, \vec{r}_n]$, then we should have an approximation to the work done by moving the particle along the curve C. If the partition $\mathcal{P}$ is fine enough, then the sum $\sum \vec{F}(\vec{r}_i) \cdot (\vec{r}_{i+1} - \vec{r}_i)$ will approximate the work done along this polygonal path. See Figure 1.

This is the motivation for the introduction of the line integral of type II. ∎

Definition: C-sum for $\vec{F}$ from $\vec{A}$ to $\vec{B}$

Suppose that

(a) C is a fundamental curve with endpoints $\vec{A}$ and $\vec{B}$;

(b) $\vec{F}$ is a function from D into $\mathbb{R}^n$, where D is a subset of $\mathbb{R}^n$ which contains C;

(c) $\mathcal{P} = \{\vec{r}_0, \vec{r}_1, \ldots, \vec{r}_n\}$ is a partition of C from $\vec{A}$ to $\vec{B}$; and

(d) $\mathcal{S} = \{\vec{s}_1, \vec{s}_2, \ldots, \vec{s}_n\}$ is a selection from $\mathcal{P}$.

Then $\sum_{i=1}^{n} \vec{F}(\vec{s}_i) \cdot (\vec{r}_i - \vec{r}_{i-1})$ is called a *C–sum for $\vec{F}$ from $\vec{A}$ to $\vec{B}$*. See Figure 2.

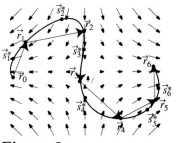

Figure 2. *The partitioned curve C lying in a vector field.*

If the set of C–sums for $\vec{F}$ from $\vec{A}$ to $\vec{B}$ has a limit, then that limit is denoted by $\int_{A_C}^{\vec{B}} \vec{F} \cdot d\vec{r}$, and the limit is called a *line integral (of type II)*. The integral $\int_{A_C}^{\vec{B}} \vec{F} \cdot d\vec{r}$ can be thought of intuitively as the work done on a particle as it moves from $\vec{A}$ to $\vec{B}$ along a path C under the influence of the force field $\vec{F}$.

It often happens that students confuse line integrals of Type I with line integrals of Type II. Though they may be denoted in a similar fashion, and, as the following theorem shows, computed similarly, they are quite different. The line integral of type I consists of a real valued function over the incremental *lengths* of the path, $f(\vec{r})\|\vec{r}'(t)\|\, dt$, whereas the line integral of type II consists of the dot product of a vector valued function over the incremental path itself, $\vec{F}(\vec{r}(t)) \cdot \vec{r}'(t)\, dt$. The line integral of type I is usually associated with calculations such as arc length, mass, or center of mass. The line integral of type II is most often associated with the calculation of work in the presence of a vector field.

Theorem 1 *If D is a subset of $\mathbb{R}^n$, $\vec{F}$ is a continuous function from D into $\mathbb{R}^n$, C is a fundamental curve contained in D with endpoints $\vec{A}$ and $\vec{B}$, and $\vec{r}$ is a parametrization for C with domain $[a, b]$; then $\int_{\vec{A}_C}^{\vec{B}} \vec{F} \cdot d\vec{r}$ exists and is equal to*

$$\int_a^b \vec{F}(\vec{r}(t)) \cdot \vec{r}'(t) \, dt.$$

The argument here is similar to that for Theorem 1 of Section 11.4. Let $\mathcal{P} = \{\vec{r}_0, \vec{r}_1, \ldots, \vec{r}_n\}$ be a partition of C, and let $\{\vec{s}_1, \vec{s}_2, \ldots, \vec{s}_n\}$ be a selection from $\mathcal{P}$. For each i, let $t_i = \vec{r}^{-1}(\vec{r}_i)$ and $s_i^* = \vec{r}^{-1}(\vec{s}_i)$ so that $\mathcal{T} = \{t_0, t_1, \ldots, t_n\}$ is a partition of $[a, b]$ and $\{s_1^*, s_2^*, \ldots, s_n^*\}$ is a selection from $\mathcal{T}$.

The vector $\vec{r}_i - \vec{r}_{i-1}$ approximates the displacement along C from $\vec{r}_{i-1}$ to $\vec{r}_i$. The point $\vec{s}_i$ is on C between $\vec{r}_{i-1}$ and $\vec{r}_i$. Thus $\vec{F}(\vec{s}_i)$ approximates the average force along C from $\vec{r}_{i-1}$ to $\vec{r}_i$. The total work done from $\vec{r}_{i-1}$ to $\vec{r}_i$ by $\vec{F}$ along C is given by $\vec{F}(\vec{s}_i) \cdot (\vec{r}_i - \vec{r}_{i-1})$.

Thus we have

$$
\begin{aligned}
\int_{\vec{A}_C}^{\vec{B}} \vec{F} \cdot d\vec{r} \ &\approx \ \sum_{i=1}^n \vec{F}(\vec{s}_i) \cdot (\vec{r}_i - \vec{r}_{i-1}) \\
&= \ \sum_{i=1}^n \vec{F}(\vec{r}(s_i^*)) \cdot (\vec{r}(t_i) - \vec{r}(t_{i-1})) \\
&= \ \sum_{i=1}^n \vec{F}(\vec{r}(s_i^*)) \cdot \frac{\vec{r}(t_i) - \vec{r}(t_{i-1})}{t_i - t_{i-1}} (t_i - t_{i-1}) \\
&\approx \ \sum_{i=1}^n \vec{F}(\vec{r}(s_i^*)) \cdot \vec{r}'(s_i^*)(t_i - t_{i-1}) \\
&\approx \ \int_a^b \vec{F}(\vec{r}(t)) \cdot \vec{r}'(t) \, dt.
\end{aligned}
$$

Theorem 1 reduces a line integral of Type II to the more familiar and computable definite integral.

Comment: Notice that we can also translate a line integral of Type II into a line integral of Type I since

$$\int_{\vec{A}_C}^{\vec{B}} \vec{F} \cdot d\vec{r} = \int_a^b \vec{F}(\vec{r}(t)) \cdot \vec{r}'(t)\, dt$$

$$= \int_a^b \vec{F}(\vec{r}(t)) \cdot \frac{\vec{r}'(t)}{\|\vec{r}'(t)\|} \|\vec{r}'(t)\|\, dt$$

$$= \int_a^b \frac{\vec{F} \cdot \vec{r}'}{\|\vec{r}'\|}\, dr.$$

EXAMPLE 1: Find the work done on a particle moving from $\vec{A} = (0,0,0)$ to $\vec{B} = (1,1,1)$ along C by a force field $\vec{F}$ defined by

$$\vec{F}(x,y,z) = (yz, 1 + xz, z + xy),$$

where

(a) C is the straight line segment $[\vec{A}, \vec{B}]$,

(b) C is the image of $\vec{r}$, where $\vec{r}(t) = (t^2, t, t^3), 0 \le t \le 1$,

(c) C is the chain of straight line segments $[\vec{A}, \vec{P}], [\vec{P}, \vec{B}]$, where $\vec{P} = (1,1,0)$. (Our distances are measured in meters.)

SOLUTION:

(a) A parametrization for C is $\vec{r}(t) = (t,t,t), 0 \le t \le 1$. We then have that $\vec{r}'(t) = (1,1,1)$, $\vec{F}(\vec{r}(t)) = \vec{F}(t,t,t) = (t^2, (1+t^2), (t+t^2))$, and $\vec{F}(\vec{r}(t)) \cdot \vec{r}'(t) = t^2 + (1 + t^2) + (t + t^2) = 1 + t + 3t^2$. Thus

$$\int_{\vec{A}_C}^{\vec{B}} \vec{F} \cdot d\vec{r} = \int_0^1 (1 + t + 3t^2)\, dt = t + \frac{t^2}{2} + t^3 \Big|_0^1 = \frac{5}{2} \text{ Joules.}$$

(b) $\vec{F}(\vec{r}(t)) = \vec{F}(t^2, t, t^3) = (t^4, 1 + t^5, 2t^3)$. Then $\vec{r}'(t) = (2t, 1, 3t^2)$ and $\vec{F}(\vec{r}(t)) \cdot \vec{r}'(t) = 1 + 9t^5$. Thus

$$\int_{\vec{A}_C}^{\vec{B}} \vec{F} \cdot d\vec{r} = \int_0^1 (1 + 9t^5)\, dt = \frac{5}{2} \text{ Joules.}$$

(c) Let W_1 be the work done along $[\vec{A}, \vec{P}]$. Then $\vec{r}(t) = (t, t, 0)$ is a parametrization for $[\vec{A}, \vec{P}]$. The derivative of this parametrization is $\vec{r}'(t) = (1, 1, 0)$, and we see that $\vec{F}(\vec{r}(t)) = (0, 1, t^2)$. Thus $\vec{F}(\vec{r}(t)) \cdot \vec{r}'(t) = 1$. We have

$$W_1 = \int_{\vec{A}_{[\vec{A}, \vec{P}]}}^{\vec{B}} \vec{F} \cdot d\vec{r} = \int_0^1 dt = t \Big|_0^1 = 1.$$

Let W_2 be the work done along $[\vec{P}, \vec{B}]$. Then $\vec{r}(t) = (1, 1, t)$ is a parametrization for $[\vec{P}, \vec{B}]$. We have $\vec{F}(\vec{r}(t)) = (t, 1+t, 1+t)$ and $\vec{r}'(t) = (0, 0, 1)$. Thus $\vec{F}(\vec{r}(t)) \cdot \vec{r}'(t) = 1 + t$ and

$$W_2 = \int_{\vec{P}_{[\vec{P}, \vec{B}]}}^{\vec{B}} \vec{F} \cdot d\vec{r} = \int_0^1 (1 + t) \, dt = t + \frac{t^2}{2} \Big|_0^1 = \frac{3}{2}.$$

The total work $W = W_1 + W_2 = \frac{5}{2}$ Joules.   ■

Suppose that $\vec{F}(\vec{v}) = (F_x(\vec{v}), F_y(\vec{v}), F_z(\vec{v}))$ is a function from $\mathbb{R}^3$ into $\mathbb{R}^3$, and $\vec{r}(t) = (x(t), y(t), z(t))$ is a parametrization for the fundamental curve C with domain $[a, b]$. It is sometimes useful, from both the physical and the computational standpoints, to express the integral $\int_{\vec{A}_C}^{\vec{B}} \vec{F} \cdot d\vec{r}$ as a sum of its component parts. Since $\vec{r}'(t) = (x'(t), y'(t), z'(t))$, we see that

$$\int_{\vec{A}_C}^{\vec{B}} \vec{F} \cdot d\vec{r} = \int_a^b \vec{F}(\vec{r}(t)) \cdot \vec{r}'(t) \, dt$$

$$= \int_a^b (F_x(\vec{r}(t)), F_y(\vec{r}(t)), F_z(\vec{r}(t))) \cdot (x'(t), y'(t), z'(t)) \, dt$$

$$= \int_a^b F_x(\vec{r}(t)) x'(t) \, dt + \int_a^b F_y(\vec{r}(t)) y'(t) \, dt + \int_a^b F_z(\vec{r}(t)) z'(t) \, dt.$$

This inspires the following notation.

Notation

The student may encounter the following notation:

$\int_a^b F_x(\vec{r}(t))x'(t)\,dt$ is denoted by $\int_{\vec{A}_C}^{\vec{B}} F_x\,dx$.

$\int_a^b F_y(\vec{r}(t))y'(t)\,dt$ is denoted by $\int_{\vec{A}_C}^{\vec{B}} F_y\,dy$.

$\int_a^b F_z(\vec{r}(t))z'(t)\,dt$ is denoted by $\int_{\vec{A}_C}^{\vec{B}} F_z\,dz$.

Thus we have that $\int_{\vec{A}_C}^{\vec{B}} \vec{F}\cdot d\vec{r}$ is written as

$\int_{\vec{A}_C}^{\vec{B}} F_x\,dx + F_y\,dy + F_z\,dz$.

EXAMPLE 2:   Suppose, as in Example 1, that $\vec{F}(x,y,z) = (yz, 1+xz, z+xy)$, where C is the straight line segment with endpoints $\vec{A} = (0,0,0)$ and $\vec{B} = (1,1,1)$, and $\vec{r}(t) = (t,t,t)$. Then $F_x(x,y,z) = yz$, $F_y(x,y,z) = 1+xz$, and $F_z(x,y,z) = z+xy$. We also have that $x(t) = t, y(t) = t$ and $z(t) = t$. Thus

$$
\begin{aligned}
\int_{\vec{A}_C}^{\vec{B}} F_x\,dx &= \int_0^1 F_x(\vec{r}(t))x'(t)\,dt \\
&= \int_0^1 y(t)z(t)\,dt = \int_0^1 t^2\,dt = \frac{t^3}{3}\Big|_0^1 = \frac{1}{3}.
\end{aligned}
$$

$$
\begin{aligned}
\int_{\vec{A}_C}^{\vec{B}} F_y\,dy &= \int_0^1 F_y(\vec{r}(t))y'(t)\,dt \\
&= \int_0^1 1 + x(t)z(t)\,dt = \int_0^1 (1+t^2)\,dt = \frac{4}{3}.
\end{aligned}
$$

$$
\begin{aligned}
\int_{\vec{A}_C}^{\vec{B}} F_z\,dz &= \int_0^1 F_z(\vec{r}(t))z'(t)\,dt \\
&= \int_0^1 (z(t) + x(t)y(t))\,dt = \int_0^1 (t+t^2)\,dt = \frac{5}{6}.
\end{aligned}
$$

Note that these values sum to $\frac{5}{2}$, our solution in Example 1.

EXAMPLE 3:   Compute $\int_{\vec{A}_C}^{\vec{B}}(xy\,dx - y\,dy + dz)$, where C is the straight line segment joining $\vec{A} = (0,0,0)$ and $\vec{B} = (2,-2,1)$.

SOLUTION: We have $F_x(x, y, z) = xy$, $F_y(x, y, z) = -y$, and $F_z(x, y, z) = 1$. A parametrization $\vec{r}$ for C from $\vec{A}$ to $\vec{B}$ is defined by $\vec{r}(t) = (2t, -2t, t)$; thus $x(t) = 2t$, $y(t) = -2t$, and $z(t) = t$. We have

$$\int_{\vec{A}_C}^{\vec{B}} (xy\,dx - y\,dy + dz)$$

$$= \int_0^1 x(t)y(t)x'(t)\,dt + \int_0^1 y(t)y'(t)\,dy + \int_0^1 z'(t)\,dt$$

$$= \int_0^1 (2t)(-2t)(2)\,dt + \int_0^1 (-2t)(-2)\,dt + \int_0^1 dt$$

$$= -8\int_0^1 t^2\,dt + 4\int_0^1 t\,dt + \int_0^1 dt = \frac{1}{3}. \qquad \blacksquare$$

Before concluding this section, it might be helpful to see where the integrals of the form $\int_{\vec{A}_C}^{\vec{B}} F_x\,dx$, $\int_{\vec{A}_C}^{\vec{B}} F_y\,dy$, and $\int_{\vec{A}_C}^{\vec{B}} F_z\,dz$ come from in terms of the approximating C-sums for the original line integral $\int_{\vec{A}_C}^{\vec{B}} \vec{F} \cdot d\vec{r}$. Suppose that we have the curve C with endpoints $\vec{A}$ and $\vec{B}$, and $\vec{F}(\vec{v}) = (F_x(\vec{v}), F_y(\vec{v}), F_z(\vec{v}))$. Let $P = \{\vec{r}_0, \vec{r}_1, \ldots, \vec{r}_n\} = \{(x_0, y_0, z_0), (x_1, y_1, z_1), \ldots, (x_n, y_n, z_n)\}$ be a partition of C from $\vec{A}$ to $\vec{B}$, and let $S = \{\vec{s}_1, \vec{s}_2, \ldots, \vec{s}_n\}$ be a selection from P. Then the corresponding C-sum for $\vec{F}$ is

$$\sum_{i=1}^n \vec{F}(\vec{s}_i) \cdot (\vec{r}_i - \vec{r}_{i-1})$$

$$= \sum_{i=1}^n (F_x(\vec{s}_i), F_y(\vec{s}_i), F_z(\vec{s}_i)) \cdot ((x_i - x_{i-1}), (y_i - y_{i-1}), (z_i - z_{i-1}))$$

$$= \sum_{i=1}^n (F_x(\vec{s}_i)(x_i - x_{i-1}) + F_y(\vec{s}_i)(y_i - y_{i-1}) + F_z(\vec{s}_i)(z_i - z_{i-1}))$$

$$= \sum_{i=1}^n F_x(\vec{s}_i)(x_i - x_{i-1}) + \sum_{i=1}^n F_y(\vec{s}_i)(y_i - y_{i-1}) + \sum_{i=1}^n F_z(\vec{s}_i)(z_i - z_{i-1}).$$

Thus

$\int_{\vec{A}_C}^{\vec{B}} F_x\,dx$ is the limit of the sums of the form

$\sum_{i=1}^n F_x(\vec{s}_i)(x_i - x_{i-1})$;

$\int_{\vec{A}_C}^{\vec{B}} F_y\,dy$ is the limit of the sums of the form

$\sum_{i=1}^n F_y(\vec{s}_i)(y_i - y_{i-1})$; and

$\int_{\vec{A}_C}^{\vec{B}} F_z \, dz$ is the limit of the sums of the form

$\sum_{i=1}^{n} F_z(\vec{s}_i)(z_i - z_{i-1})$.

From this derivation, we see that we can use different parametrizations for each of these integrals.

Note: on page 522 we introduced notation so that

$$\int_{\vec{A}_C}^{\vec{B}} \vec{F} \cdot d\vec{r} = \int_{\vec{A}_C}^{\vec{B}} F_x \, dx + \int_{\vec{A}_C}^{\vec{B}} F_y \, dy + \int_{\vec{A}_C}^{\vec{B}} F_z \, dz.$$

The terms on the right side of the above equation are the expressions one most often encounters in physics and engineering courses. They are usually computed directly rather than using $\int_a^b \vec{F}(\vec{r}) \cdot \vec{r}'(t) \, dt$. This is **not** to say that they are computed more easily by dealing with them directly; the point is that one encounters texts where the integrals are computed using techniques that seem different but are actually the same. What these texts are doing is parametrizing C by the variable x for $\int_{\vec{A}_C}^{\vec{B}} F_x \, dx$, parametrizing C by y for $\int_{\vec{A}_C}^{\vec{B}} F_y \, dy$ and parametrizing C by z for $\int_{\vec{A}_C}^{\vec{B}} F_z \, dz$.

EXAMPLE 4: Compute the work done by $\vec{F}(x,y,z) = (x,y,z)$ in going around the square with vertices $\vec{A} = (0,0,0)$, $\vec{B} = (1,0,0)$, $\vec{C} = (1,1,0)$, and $\vec{D} = (0,1,0)$ in a counterclockwise fashion. See Figure 3.

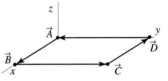

Figure 3. *The path from $\vec{A}$ to $\vec{B}$, to $\vec{C}$, to $\vec{D}$, and finally to $\vec{A}$.*

SOLUTION: Parametrize the segments $[\vec{A}, \vec{B}]$ and $[\vec{C}, \vec{D}]$ by x, and parametrize the segments $[\vec{B}, \vec{C}]$ and $[\vec{D}, \vec{A}]$ by y:

$\vec{r}_1(t) = (x,0,0)$, $0 \le x \le 1$ parametrizes $[\vec{A}, \vec{B}]$.

$\vec{r}_2(t) = (1,y,0)$, $0 \le y \le 1$ parametrizes $[\vec{B}, \vec{C}]$.

$\vec{r}_3(t) = (1-x,1,0)$, $0 \le x \le 1$ parametrizes $[\vec{C}, \vec{D}]$.

$\vec{r}_4(t) = (0,1-y,0)$, $0 \le y \le 1$ parametrizes $[\vec{D}, \vec{A}]$.

Thus

$$\int_{[\vec{A},\vec{B}]} \vec{F} \cdot d\vec{r} = \int_0^1 (x,0,0) \cdot (1,0,0) \, dx = \int_0^1 x \, dx = \frac{1}{2}.$$

$$\int_{[\vec{B},\vec{C}]} \vec{F} \cdot d\vec{r} = \int_0^1 (1,y,0) \cdot (0,1,0) \, dy = \int_0^1 y \, dy = \frac{1}{2}.$$

$$\int_{[\vec{C},\vec{D}]} \vec{F} \cdot d\vec{r} = \int_0^1 (1-x,1,0) \cdot (-1,0,0) \, dx = \int_0^1 x - 1 \, dx$$

$$= \frac{x^2}{2} - x\big|_0^1 = -\frac{1}{2}.$$

$$\int_{[\vec{D},\vec{A}]} \vec{F} \cdot d\vec{r} = \int_0^1 (0, 1-y, 0) \cdot (0, -1, 0)\ dy = \int_0^1 y - 1\ dy$$

$$= \frac{y^2}{2} - y\big|_0^1 = -\frac{1}{2}.$$

And

$$\text{Total Work} = \frac{1}{2} + \frac{1}{2} - \frac{1}{2} - \frac{1}{2} = 0.$$ ∎

EXERCISES 11.7

In Exercises 1 and 2, calculate $\int_{\vec{A}_C}^{\vec{B}} (x, y, z) \cdot d\vec{r}$ for the given curve.

1. C is the straight line segment from $(0, 1, 0)$ to $(2, 2, -1)$.

2. C is the helix parametrized by $\vec{r}(t) = (t, \cos(2\pi t), \sin(2\pi t))$, $0 \le t \le 1$.

3. Evaluate $\int_{(1,0,0)_C}^{(2,1,1)} (x+z)dx + y\ dy + (x+y+z)dz$,

    where C is a straight line segment.

In Exercises 4–7, calculate $\int_{\vec{A}_C}^{\vec{B}} \frac{\vec{r}}{\|\vec{r}\|^3} \cdot d\vec{r}$ over the given curve.

4. C is the straight line segment from $(1, 1, 1)$ to $(2, 2, 2)$.

5. C is the straight line segment from $(1, 0, 0)$ to $(0, 1, 1)$.

6. C is parametrized by $\vec{r}(t) = (at + x_0, bt + y_0, ct + z_0)$, $0 \le t \le 1$.

7. C is the straight line segment from $\vec{r}_0 = (x_0, y_0, z_0)$ to $\vec{r}_1 = (x_1, y_1, z_1)$.

In Exercises 8–10, calculate the C–sum that approximates $\int_{\vec{A}_C}^{\vec{B}} \vec{F} \cdot d\vec{r}$ determined by the partition and selection in Figure 4.

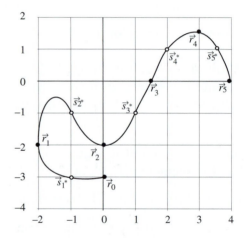

Figure 4. *The selection $\{\vec{r}_0, \vec{r}_1, \cdots, \vec{r}_5\}$ and selection $\{\vec{s}_1{}^*, \cdots, \vec{s}_5{}^*\}$ for Exercises 8–10.*

8. $\vec{F}(x, y) = (x, -y)$.

9. $\vec{F}(x, y) = (x, xy)$.

10. $\vec{F}(x, y) = (-y, x)$.

11. Calculate $\int_{\vec{A}_C}^{\vec{B}} (x + y + z)dx$ where C is the straight line interval from $(-1, 1, 0)$ to $(1, -2, 3)$.

12. Calculate $\int_{\vec{A}_C}^{\vec{B}} (x + y + z)dy$ where C is the straight line interval from $(-1, 1, 0)$ to $(1, -2, 3)$.

13. Calculate $\int_{A_C}^{\vec{B}}(x + y + z)dz$, where C is the straight line interval from $(-1, 1, 0)$ to $(1, -2, 3)$.

14. Calculate $\int_{A_C}^{\vec{B}}(x^2 + y + z)dx$, where C is the arc of the circle of radius 4, centered at the origin, and lying in the xy–coordinate plane going from $(4, 0, 0)$ to $(-4, 0, 0)$ as illustrated in Figure 5.

15. Calculate $\int_{A_C}^{\vec{B}}(x^2 + y + z)dy$, where C is the arc of the circle of radius 4, centered on the z–axis, and lying in the plane $z = 1$ going from $(4, 0, 1)$ to $(-4, 0, 1)$ as illustrated in Figure 6.

16. Calculate $\int_{A_C}^{\vec{B}}(x^2 + y + z)dz$, where C is the arc of the circle of radius 4, centered at the origin, and lying in the xy–coordinate plane going from $(4, 0, 0)$ to $(-4, 0, 0)$ as illustrated in Figure 5.

17. Calculate $\int_{A_C}^{\vec{B}}(x^2 + y + z)dz$ where C is the arc of the circle of radius 4, centered on the z–axis, and lying in the plane $z = 1$ going from $(4, 0, 1)$ to $(-4, 0, 1)$ as illustrated in Figure 6.

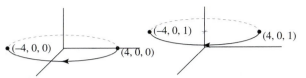

Figure 5. *The arc for Exercises 14 and 16.*   **Figure 6.** *The arc for Exercises 15 and 17.*

18. A vector field $\vec{F}$ points radially away from the origin, and C is a curve from $\vec{A}$ to $\vec{B}$ lying entirely on the surface of a ball of radius R centered at the origin. Explain why $\int_{A_C}^{\vec{B}} \vec{F} \cdot d\vec{r} = 0$.

19. A vector field $\vec{F}$ points radially away from the x–axis. That is, $\vec{F}(\vec{r})$ points away from the x–axis and is perpendicular to the x–axis. C is a curve from $\vec{A}$ to $\vec{B}$ lying entirely on the surface of cylinder of radius R with axis the x–axis. Explain why $\int_{A_C}^{\vec{B}} \vec{F} \cdot d\vec{r} = 0$.

11.8   Partial Derivatives

In modeling situations where we have more than one input variable, we often allow one input variable to vary while holding the other variables constant, and then analyze the resulting behavior. This helps us to understand the effect that each input variable has on our model. Formally, given a function f from $\mathbb{R}^n$ into $\mathbb{R}$ and a point $\vec{r} = (x_1, x_2, \ldots, x_n)$, we are interested in the rate of change of f if we hold all of the variables $x_1, x_2, \ldots, x_n$ fixed except for one, say x_i. We call this the *partial derivative of f with respect to x_i at $\vec{r}$,* denoted by $\frac{\partial f}{\partial x_i}$.

EXAMPLE 1: Let $f(x, y, z) = x(\sin y)(\cos z)$.

$$
\begin{aligned}
\frac{\partial f(x, y, z)}{\partial x} &= \lim_{\Delta x \to 0} \frac{(x + \Delta x)(\sin y)(\cos z) - x(\sin y)(\cos z)}{\Delta x} \\
&= \lim_{\Delta x \to 0} \frac{(\sin y)(\cos z)((x + \Delta x) - x)}{\Delta x} \\
&= \sin y \cos z \lim_{\Delta x \to 0} \frac{\Delta x}{\Delta x} = (\sin y)(\cos z).
\end{aligned}
$$

$$
\frac{\partial f(x, y, z)}{\partial y} = \lim_{\Delta y \to 0} \frac{x(\sin(y + \Delta y))(\cos z) - x(\sin y)(\cos z)}{\Delta y}
$$

$$= \lim_{\Delta y \to 0} \frac{x(\cos z)(\sin(y + \Delta y) - \sin(y))}{\Delta y}$$

$$= x(\cos z) \lim_{\Delta y \to 0} \frac{\sin(y + \Delta y) - \sin y}{\Delta y} = x(\cos z)(\cos y).$$

$$\frac{\partial f(x, y, z)}{\partial z} = \lim_{\Delta z \to 0} \frac{x(\sin y)(\cos(z + \Delta z)) - x(\sin y)(\cos z)}{\Delta z}$$

$$= \lim_{\Delta z \to 0} \frac{x(\sin y)(\cos(z + \Delta z) - \cos z))}{\Delta z}$$

$$= x(\sin y) \lim_{\Delta z \to 0} \frac{\cos(z + \Delta z) - \cos z}{\Delta z} = x(\sin y)(-\sin z).$$

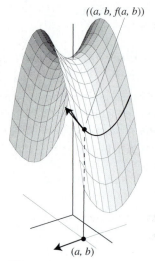

Figure 1.a *As you move away from (a, b) in the positive x–direction, $f(x, y)$ increases.*

Example 1 suggests that we can compute $\frac{\partial f}{\partial x}$ by differentiating f, treating the variables y and z as constants. Similarly, we compute $\frac{\partial f}{\partial y}$ (respectively, $\frac{\partial f}{\partial z}$) by differentiating f, treating the variables x and z (respectively, x and y) as constants.

EXAMPLE 2: Geometrically, we might think of $f(x, y)$ as the altitude when you are at position (x, y). If the rate of change of f as x changes at the point (a, b) ($\frac{\partial f(a,b)}{\partial x}$) is positive, then as you move through (a, b) in the positive x–direction you will be going uphill. Conversely, if you are going uphill as you move through (a, b) in the positive x–direction, then $\frac{\partial f(a,b)}{\partial x} \geq 0$. See Figure 1.a. As illustrated in Figure 1.b, we would expect that $\frac{\partial f(a,b)}{\partial y} \leq 0$ at the point (a, b).   ∎

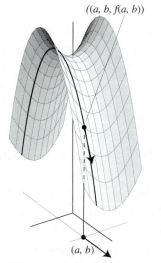

Figure 1.b *As you move away from (a, b) in the positive y–direction, $f(x, y)$ decreases.*

EXAMPLE 3: Compute $\frac{\partial f}{\partial x}$, $\frac{\partial f}{\partial y}$, and $\frac{\partial f}{\partial z}$, where

(a) $f(x, y, z) = \cos(xy) + \sin z$.

(b) $f(x, y, z) = \cos\left(\frac{xy}{z}\right) + \text{Arctan}(xy)$.

SOLUTION:

(a) To compute $\frac{\partial f}{\partial x}$, we simply take the derivative of f, treating y and z as constants. Thus $\frac{\partial f}{\partial x} = -y \sin(xy)$. To compute $\frac{\partial f}{\partial y}$, we take the derivative of f, treating x and z as constants. Thus $\frac{\partial f}{\partial y} = -x \sin(xy)$. To compute $\frac{\partial f}{\partial z}$, we take the derivative of f, treating x and y as constants. Thus $\frac{\partial f}{\partial z} = \cos z$.

(b) $\frac{\partial f(x,y,z)}{\partial x} = -\left(\frac{y}{z}\right) \sin\left(\frac{xy}{z}\right) + \frac{y}{1+(xy)^2}$.

$\frac{\partial f(x,y,z)}{\partial y} = -\left(\frac{x}{z}\right) \sin\left(\frac{xy}{z}\right) + \frac{x}{1+(xy)^2}$.

$\frac{\partial f(x,y,z)}{\partial z} = \left(\frac{xy}{z^2}\right) \sin\left(\frac{xy}{z}\right)$.   ∎

A thorough study of partial derivatives and their uses will await another section. However, there is one more definition that we find helpful.

Definition: Gradient of f

Suppose that f is a function from $\mathbb{R}^3$ into $\mathbb{R}$ and that each of the partial derivatives exists. Then the *gradient of f*, denoted by grad f or ∇f (pronounced "del f"), is defined by:

$$\nabla f(\vec{r}) = \frac{\partial f(\vec{r})}{\partial x}\hat{i} + \frac{\partial f(\vec{r})}{\partial y}\hat{j} + \frac{\partial f(\vec{r})}{\partial z}\hat{k}.$$

Thus the gradient of f is a function from $\mathbb{R}^3$ into $\mathbb{R}^3$.

Note: The gradient can be defined analogously for any function from $\mathbb{R}^n$ into $\mathbb{R}$ as long as the partial derivatives exist.

EXAMPLE 4: Compute ∇f, where f is defined by $f(x, y, z) = xyz$.

SOLUTION: $\frac{\partial f(x,y,z)}{\partial x} = yz$, $\frac{\partial f(x,y,z)}{\partial y} = xz$, and $\frac{\partial f(x,y,z)}{\partial z} = xy$. Thus $\nabla f(x, y, z) = yz\hat{i} + xz\hat{j} + xy\hat{k}$. ∎

EXERCISES 11.8

Compute $\frac{\partial f}{\partial x}, \frac{\partial f}{\partial y}$, and $\frac{\partial f}{\partial z}$ for the functions in Exercises 1–3.

1. $f(x, y, z) = x^{yz}$.

2. $f(x, y, z) = Kx/\sqrt{x^2 + y^2 + z^2}$.

3. $f(\vec{r}) = \sin \|\vec{r}\|$, for $\vec{r} \neq \vec{0}, \vec{r} = (x, y, z)$.

Compute the gradient of U in Exercises 4–10.

4. $U(x, y, z) = x \sin z + \tan z$.

5. $U(\vec{r}) = \frac{K}{\|\vec{r}\|}$ for $\vec{r} \neq \vec{0}, \vec{r} = (x, y, z)$.

6. $U(x, y) = \cos^2 x + \text{Arcsin}(xy)$.

7. $U(x, y, z) = \ln(xyz)$.

8. $U(x, y, z) = \log_3(xyz)$.

9. $U(x, y, z) = \text{Arcsin}(xyz)$.

10. $U(\vec{r}) = \text{Arctan} \|\vec{r}\|$, $\vec{r} = (x, y, z)$.

In Exercises 11 and 12, compute $\int_{\vec{A}_C}^{\vec{B}} \nabla f \cdot d\vec{r}$, where C is the straight line segment $\vec{A}\vec{B}$.

11. $\vec{A} = (1, 2, -1), \vec{B} = (0, 1, 0)$, and $f(x, y, z) = xy + yz + xyz$.

12. $\vec{A} = (0, 1, 1), \vec{B} = (-1, 0, 0)$, and $f(x, y, z) = x^2 + y^2 + z^2$.

13. Given that $U(\vec{r}) = \frac{K}{\|\vec{r}\|}$, find functions V_1 and V_2 so that $\nabla V_1 = \nabla V_2 = \nabla U$, but $V_1(1, 1, 1) = 10$ and $V_2(0, 1, -1) = 10, \vec{r} = (x, y, z)$.

Exercises 14–17 refer to the function graphed in Figure 2.a.

14. Would you expect $\frac{\partial f(a,b)}{\partial x}$ to be positive or negative? Explain.

15. Would you expect $\frac{\partial f(a,b)}{\partial y}$ to be positive or negative? Explain.

16. Would you expect $\frac{\partial f(c,d)}{\partial x}$ to be positive or negative? Explain.

17. Would you expect $\frac{\partial f(c,d)}{\partial y}$ to be positive or negative? Explain.

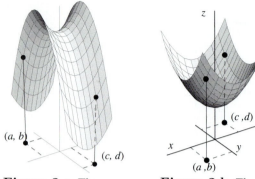

Figure 2.a *The graph of the function f for Exercises 14–17.*

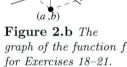

Figure 2.b *The graph of the function f for Exercises 18–21.*

Exercises 18–21 refer to the function graphed in Figure 2.b.

18. Would you expect $\frac{\partial f(a,b)}{\partial x}$ to be positive or negative? Explain.

19. Would you expect $\frac{\partial f(a,b)}{\partial y}$ to be positive or negative? Explain.

20. Would you expect $\frac{\partial f(c,d)}{\partial x}$ to be positive or negative? Explain.

21. Would you expect $\frac{\partial f(c,d)}{\partial y}$ to be positive or negative? Explain.

22. The function graphed in Figure 2.a is $f(x,y) = x^2 - y^2 + 5$. Find $\nabla f(x,y)$ and determine where $\nabla f(x,y) = \vec{0}$. Refering to Figure 2.a, explain geometrically why this is true.

23. The function graphed in Figure 2.b is $f(x,y) = x^2 + y^2 + 1$. Find $\nabla f(x,y)$ and determine where $\nabla f(x,y) = \vec{0}$. Refering to Figure 2.b, explain geometrically why this is true.

24. Locate the points on the graph of f in Figure 3 where you would expect ∇f to be zero.

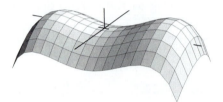

Figure 3. *The graph of the function in Exercise 24.*

11.9   Potential Functions and the Gradient

Suppose that $\vec{r}$ is a function from a subset of $\mathbb{R}$ into $\mathbb{R}^n$ and that g is a function from a subset of $\mathbb{R}^n$ (that contains the image of $\vec{r}$) into $\mathbb{R}$, so that the composite $f(t) = g(\vec{r}(t))$ is a function from $\mathbb{R}$ into $\mathbb{R}$. (See Figure 1.)

As long as everything is "smooth enough," we can differentiate f in the standard way. Let $x(t), y(t)$ and $z(t)$ denote the coordinate functions of $\vec{r}$. For the sake of simplicity, we will assume that $x'(t) \neq 0, y'(t) \neq 0$, and $z'(t) \neq 0$.

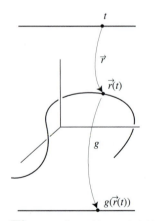

Figure 1. $f(t) = g(\vec{r}(t))$.

$$\begin{aligned} f'(t) &= \lim_{h \to 0} \frac{f(t+h) - f(t)}{h} = \lim_{h \to 0} \frac{g(\vec{r}(t+h)) - g(\vec{r}(t))}{h} \\ &= \lim_{h \to 0} \frac{g(x(t+h), y(t+h), z(t+h)) - g(x(t), y(t), z(t))}{h} \end{aligned}$$

We need to put the quotient in a form that allows us to find the limit. Since $[-g(x(t+h), y(t+h), z(t))] + [g(x(t+h), y(t+h), z(t)) -$

$g(x(t+h), y(t), z(t))] + [g(x(t+h), y(t), z(t))] = 0$, we can introduce this expression into our numerator without changing its value. This maneuver ultimately allows us to write the quotient in a more manageable way. We have

$$\frac{g(x(t+h), y(t+h), z(t+h)) - g(x(t), y(t), z(t))}{h}$$

$$= \frac{g(x(t+h), y(t+h), z(t+h)) - g(x(t+h), y(t+h), z(t))}{h}$$

$$+ \frac{g(x(t+h), y(t+h), z(t)) - g(x(t+h), y(t), z(t))}{h}$$

$$+ \frac{g(x(t+h), y(t), z(t)) - g(x(t), y(t), z(t))}{h}.$$

Thus

$$\frac{g(x(t+h), y(t+h), z(t+h)) - g(x(t), y(t), z(t))}{h} =$$

$$\left(\frac{g(x(t+h), y(t+h), z(t+h)) - g(x(t+h), y(t+h), z(t))}{z(t+h) - z(t)} \right) \left(\frac{z(t+h) - z(t)}{h} \right)$$

$$+ \left(\frac{g(x(t+h), y(t+h), z(t)) - g(x(t+h), y(t), z(t))}{y(t+h) - y(t)} \right) \left(\frac{y(t+h) - y(t)}{h} \right)$$

$$+ \left(\frac{g(x(t+h), y(t), z(t)) - g(x(t), y(t), z(t))}{x(t+h) - x(t)} \right) \left(\frac{x(t+h) - x(t)}{h} \right)$$

$$\approx \frac{\partial g(x(t+h), y(t+h), z(t))}{\partial z} z'(t) + \frac{\partial g(x(t+h), y(t), z(t))}{\partial y} y'(t)$$

$$+ \frac{\partial g(x(t), y(t), z(t))}{\partial x} x'(t)$$

$$\approx \frac{\partial g(x(t), y(t), z(t))}{\partial z} z'(t) + \frac{\partial g(x(t), y(t), z(t))}{\partial y} y'(t)$$

$$+ \frac{\partial g(x(t), y(t), z(t))}{\partial x} x'(t)$$

$$= \frac{\partial g(\vec{r}(t))}{\partial z} z'(t) + \frac{\partial g(\vec{r}(t))}{\partial y} y'(t) + \frac{\partial g(\vec{r}(t))}{\partial x} x'(t)$$

$$= \nabla g(\vec{r}(t)) \cdot \vec{r}'(t).$$

We have proven the following theorem.

Theorem 1 *Suppose that*

(*a*) *D is an open[3] subset of $\mathbb{R}^n$,*

(*b*) *f is a function from D into $\mathbb{R}$ such that the partial derivatives of f are continuous; and*

(*c*) *$\vec{r}$ is a differentiable function from a subset of $\mathbb{R}$ into D.*

Then

$$\frac{df(\vec{r}(t))}{dt} = \nabla f(\vec{r}(t)) \cdot \vec{r}'(t). \tag{7}$$

---

[3] Recall that a subset U of $\mathbb{R}^n$ is open if every point of U is the center of a ball lying entirely in U.

It is correct to think of ∇f as the derivative of f. With this in mind, Equation (7) is the *chain rule*. If $\vec{r}(t) = (x(t), y(t), z(t))$, then the procedure can be rewritten in several ways, which we compile in the following box.

The Chain Rule

$$
\begin{aligned}
\frac{df(\vec{r}(t))}{dt} &= \nabla f(\vec{r}(t)) \cdot \vec{r}'(t) \\[2mm]
&= \left(\frac{\partial f}{\partial x}, \frac{\partial f}{\partial y}, \frac{\partial f}{\partial z} \right) \Big|_{\vec{r}(t)} \cdot (x'(t), y'(t), z'(t)) \\[2mm]
&= \frac{\partial f}{\partial x}\frac{dx}{dt} + \frac{\partial f}{\partial y}\frac{dy}{dt} + \frac{\partial f}{\partial z}\frac{dz}{dt}.
\end{aligned}
$$

EXAMPLE 1: Let $\vec{r}(t) = (\cos(\pi t), \sin(\pi t))$, and let $g(x, y) = x^2 + y^2$. Then g, evaluated at $\vec{r}_0$, is simply the square of the norm of $\vec{r}_0$, and the function $\vec{r}$ describes motion on the unit circle. As we would expect,

$$g(\vec{r}(t)) = 1, \quad \text{and} \quad \frac{dg(\vec{r}(t))}{dt} = 0.$$

Now, since $\nabla g(x, y) = (2x, 2y)$, we see that $\nabla g(\vec{r}(t)) = (2\cos(\pi t), 2\sin(\pi t))$. We also know that $\vec{r}'(t) = (-\pi \sin(\pi t), \pi \cos(\pi t))$. Thus $\nabla g(\vec{r}(t)) \cdot \vec{r}'(t) = 0$. ∎

EXAMPLE 2: Suppose that the temperature in a room is given by $T(x, y, z) = e^{x+y+z}$, and that at time $t = 0$ a fly is at position $(1, 0, -1)$ and flying with velocity $\vec{v}(0) = (1, -1, 1)$. At what rate is the fly feeling the temperature change?

SOLUTION: Let $\vec{r}(t)$ be the function that gives the position of the fly at time t. Then $\vec{r}(0) = (1, 0, -1)$ and $\vec{r}\,'(0) = (1, -1, 1)$. Thus

$$\left. \frac{dT(\vec{r}(t))}{dt} \right|_{t=0} = \nabla T(\vec{r}(0)) \cdot \vec{r}\,'(0) = (e^0, e^0, e^0) \cdot (1, -1, 1) = 1. \quad \blacksquare$$

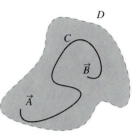

Figure 2. *The curve C lies in the set D.*

Potential Functions and Conservative Forces

Let D be an open subset of $\mathbb{R}^n$, let f be a function from D into $\mathbb{R}$ with continuous partial derivatives, and let C be a fundamental curve from $\vec{A}$ to $\vec{B}$ lying entirely in D as in Figure 2. We will also let $\vec{r}$ be a parametrization for C with domain $[a, b]$. Then

$$
\begin{aligned}
\int_{\vec{A}_C}^{\vec{B}} \nabla f \cdot \, d\vec{r} &= \int_a^b \nabla f(\vec{r}(t)) \cdot \vec{r}\,'(t) \, dt \\[2mm]
&= \int_a^b \frac{df(\vec{r}(t))}{dt} \, dt \ \text{(by Theorem 1)} \\[2mm]
&= f(\vec{r}(b)) - f(\vec{r}(a)) = f(\vec{B}) - f(\vec{A}).
\end{aligned}
$$

We have proven the following theorem.

Theorem 2 *Let D be an open subset of $\mathbb{R}^n$, let f be a function from D into $\mathbb{R}$ with continuous partial derivatives, and let C be a fundamental curve from $\vec{A}$ to $\vec{B}$ lying entirely in D. Let $\vec{r}$ be a parametrization for C with domain $[a, b]$. Then*

$$\int_{\vec{A}_C}^{\vec{B}} \nabla f \cdot \, d\vec{r} = f(\vec{B}) - f(\vec{A}).$$

Theorem 2 tells us a great deal. In particular, if the vector field F is known to be the gradient of a function f and if C is **any** path from $\vec{A}$ to $\vec{B}$ that lies in the domain of F, then $\int_{\vec{A}_C}^{\vec{B}} \nabla f \cdot d\vec{r} = f(\vec{B}) - f(\vec{A})$. Thus, if ∇f is a force field, then the work that ∇f does on a mass as it moves from position $\vec{A}$ to position $\vec{B}$ is simply $f(\vec{B}) - f(\vec{A})$, **regardless of the path**, as long as the mass remains in the domain of f.

EXAMPLE 3: Exercise 11 of the previous section asks that we compute $\int_{\vec{A}_C}^{\vec{B}} \nabla f \cdot d\vec{r}$, where C is the line segment connecting $\vec{A} = (1, 2, -1)$ and $\vec{B} = (0, 1, 0)$ and $f(x, y, z) = xy + yz + xyz$. The solution is now trivial:

$$\int_{\vec{A}_C}^{\vec{B}} \nabla f \cdot d\vec{r} = f(\vec{B}) - f(\vec{A}) = 0 - [2 - 2 - 2] = 2.$$

Theorem 1 of Section 6.1 stated that if f and g were two functions from $\mathbb{R}$ into $\mathbb{R}$ such that $f'(x) = g'(x)$ for all x, then f and g differed by a constant. We now state the analogous theorem for functions of several variables.

Theorem 3 *Let D be an open subset of $\mathbb{R}^n$ such that every pair of points in D can be connected with a fundamental curve lying entirely in D. If f and g are functions from D into $\mathbb{R}$ with continuous partial derivatives such that $\nabla f = \nabla g$, then there is a constant C such that $f(\vec{r}) = g(\vec{r}) + C$.*

Proof: Let $\vec{A}$ be a point in D. By Theorem 2, if $\vec{r}$ is in D, then

$$f(\vec{r}) - f(\vec{A}) = g(\vec{r}) - g(\vec{A}),$$

and so, if we let $C = f(\vec{A}) - g(\vec{A})$, then $f(\vec{r}) = g(\vec{r}) + C$.

Definition: Potential Functions and Conservative Forces

Let D be an open subset of $\mathbb{R}^n$ such that every pair of points in D can be connected by a fundamental curve lying entirely in D, and let f be a function from D into $\mathbb{R}$ with continuous partial derivatives. Then f is called a *potential function*. A continuous vector field with domain D that is the gradient of some potential function is often called a *conservative force field*.

Potential functions are so named because of their relation to the notion of *potential energy*. Suppose that $\vec{F} = \nabla f$, and $\vec{r}_0$ is a point in D. Then

$$U_{\vec{r}_0}(\vec{r}) = -\int_{\vec{r}_0}^{\vec{r}} \vec{F} \cdot d\vec{r} = -[f(\vec{r}) - f(\vec{r}_0)]$$

is called the *potential (energy)* of $\vec{F}$ relative to $\vec{r}_0$, and it represents the work that $\vec{F}$ performs on a mass as it moves back to $\vec{r}_0$ from $\vec{r}$.

The following theorem ties things together.

Theorem 4 *Suppose that $U_{\vec{r}_0}$ is the potential of $\vec{F}$ relative to $\vec{r}_0$. Then $\vec{U}_{\vec{r}_0}$ is that potential function for $\vec{F}$ such that $U_{\vec{r}_0}(\vec{r}_0) = 0$. Furthermore, if ϕ is any potential fucntion for $\vec{F}$, then $U_{\vec{r}_0}(\vec{r}) = \phi(\vec{r}) - \phi(\vec{r}_0)$.*

The idea of potential energy makes sense only in terms of a conservative force field. If $U(\vec{r})$ is a potential energy function for a force field $\vec{F}$, then $\nabla U(\vec{r}) = -\vec{F}(\vec{r})$, and $U(\vec{A}) - U(\vec{B})$ measures the work done **against** $\vec{F}$ as a mass moves from $\vec{B}$ to $\vec{A}$ (while staying in the domain of $\vec{F}$).

EXAMPLE 4: Let $U(\vec{r}) = \frac{k}{\|\vec{r}\|}$. In Exercise 5 of the previous section, it is shown that $\nabla U(\vec{r}) = -\frac{k\vec{r}}{\|\vec{r}\|^3}$. This implies that U is a potential function for $\vec{F}(\vec{r}) = \frac{k\vec{r}}{\|\vec{r}\|^3}$. Indeed, every potential function for $\vec{F}(\vec{r}) = \frac{k\vec{r}}{\|\vec{r}\|^3}$ is of the form $\phi(\vec{r}) = \frac{k}{\|\vec{r}\|} + C$. The potential function for $\vec{F}$ relative to $\vec{r}_0 = (1, 1, 1)$ is $\phi(\vec{r}) = \frac{k}{\|\vec{r}\|} - \frac{k}{\sqrt{3}}$.   ∎

EXAMPLE 5: Given that $\vec{F}(\vec{r}) = (x, y, z)$ is conservative, find a function $U(\vec{r})$ such that $-\nabla U = \vec{F}$.

SOLUTION: We solve this problem using two different methods. The first method entails the solution of a partial differential equation, whereas the second method appeals to Theorem 2.

Method 1: We want U to satisfy

$$-\nabla U(x, y, z) = \begin{pmatrix} -\frac{\partial U(x,y,z)}{\partial x} \\ -\frac{\partial U(x,y,z)}{\partial y} \\ -\frac{\partial U(x,y,z)}{\partial z} \end{pmatrix} = \begin{pmatrix} x \\ y \\ z \end{pmatrix}$$

so that

$$\frac{\partial U(x, y, z)}{\partial x} = -x \tag{8}$$

$$\frac{\partial U(x, y, z)}{\partial y} = -y \tag{9}$$

$$\frac{\partial U(x, y, z)}{\partial z} = -z. \tag{10}$$

From Equation (8) we obtain

$$U(x,y,z) = -\frac{x^2}{2} + \text{terms whose partial derivative with respect to } x \text{ is } 0.$$

$$= -\frac{x^2}{2} + \text{a function involving only } y \text{ and } z,$$

since, if $f(y,z)$ is a differentiable function, $\partial f(y,z)/\partial x = 0$.
Thus

$$U(x,y,z) = -\frac{x^2}{2} + f(y,z). \tag{11}$$

We now use Equation (9) to obtain more information about $f(y,z)$. Since $\frac{\partial U(x,y,z)}{\partial y} = -y$, we see that

$$-y = \frac{\partial U(x,y,z)}{\partial y} = 0 + \frac{\partial f(y,z)}{\partial y}.$$

Thus it must follow that

$$f(y,z) = -\frac{y^2}{2} + \text{terms whose partial derivative with respect to } y \text{ is } 0.$$

$$= -\frac{y^2}{2} + \text{a function involving only } z.$$

Thus

$$f(y,z) = -\frac{y^2}{2} + g(z). \tag{12}$$

Combining Equation (12) with Equation (11), we have

$$U(x,y,z) = -\frac{x^2}{2} - \frac{y^2}{2} + g(z). \tag{13}$$

We finally use Equation (10). Since $\frac{\partial U(x,y,z)}{\partial z} = -z$, we see that (using Equation (13))

$$-z = \frac{\partial U(x,y,z)}{\partial z} = 0 + 0 + \frac{dg(z)}{dz}.$$

Notice that since $g(z)$ is a function of a single variable, the derivative of g is $\frac{dg(z)}{dz}$. Thus

$$g(z) = -\frac{z^2}{2} + C. \tag{14}$$

Combining Equation (14) with Equation (13), we have

$$U(x,y,z) = -\frac{x^2}{2} - \frac{y^2}{2} - \frac{z^2}{2} + C.$$

Method 2: We can obtain a function U such that $\nabla U = -\vec{F}$ by using Theorem 2. For each $\vec{v} = (x, y, z)$, let C be the straight line segment going from the origin $\vec{0}$ to the point $\vec{v}$. We parametrize C by $\vec{r}(t) = t\vec{v} = (tx, ty, tz), 0 \le t \le 1$. We then define

$$
\begin{aligned}
U(x, y, z) &= -\int_{\vec{0}_C}^{\vec{v}} \vec{F} \cdot d\vec{r} \\
&= -\int_0^1 (tx, ty, tz) \cdot (x, y, z) \, dt \\
&= -\frac{x^2}{2} - \frac{y^2}{2} - \frac{z^2}{2}. \qquad \blacksquare
\end{aligned}
$$

Lest the reader think that every vector field is conservative, we present the following example.

EXAMPLE 6: Let $\vec{F}(x, y) = \left(-\dfrac{y}{x^2 + y^2}, \dfrac{x}{x^2 + y^2}\right)$. In Exercise 18, you are asked to show that

$$
\int_{(1,0)_{C_1}}^{(-1,0)} \vec{F} \cdot d\vec{r} \neq \int_{(1,0)_{C_2}}^{(-1,0)} \vec{F} \cdot d\vec{r},
$$

where C_1 is the upper half of the unit circle and C_2 is the lower half of the unit circle. $\qquad \blacksquare$

EXERCISES 11.9

In Exercises 1–3, $-\nabla U$ is a conservative force field acting on a mass M. Compute the work done by $-\nabla U$ as M moves from $\vec{A}$ to $\vec{B}$.

1. $U(x, y, z) = xyz, \vec{A} = (1, 3, 1)$ and $\vec{B} = \vec{0}$.

2. $U(x, y, z) = xy \cos z, \vec{A} = (-1, 2, \pi)$ and $\vec{B} = \left(1, 3, \frac{\pi}{2}\right)$.

3. $U(x, y, z) = \frac{1}{\|(x,y,z)\|}$, $(x, y, z) \neq \vec{0}$, $\vec{A} = (1, 1, 1)$, and $\vec{B} = (0, 0, 1)$.

Find the potential energy function, relative to the origin, associated with $\vec{F}$ for Exercises 4–8.

4. $\vec{F}(x, y) = (\cos x, \sin y)$.

5. $\vec{F}(x, y, z) = \left(x^2, y^2, z^2\right)$.

6. $\vec{F}(x, y, z) = (yz, xz, xy)$.

7. $\vec{F}(x, y) = \left(y^2 + 2xy, 2xy + x^2\right)$.

8. $\vec{F}(x, y, z) = \big(yz \cos(xyz) + x^2, xz \cos(xyz) + y^2, xy \cos(xyz) + z^2\big)$.

In Exercises 9–14 , find $\frac{d}{dt}\phi(\vec{r}(t))$ at $t = t_0$.

9. $\phi(x, y) = x + y + xy$, $\vec{r}(t_0) = (-2, 1)$ and $\vec{r}'(t_0) = (4, -2)$.

10. $\phi(x, y) = xy + \sin(xy)$, $\vec{r}(t_0) = \left(\frac{\pi}{3}, 3\right)$ and $\vec{r}'(t_0) = (1, -3)$.

11. $\phi(x, y, z) = x + xy + xyz$, $\vec{r}(t_0) = (3, -1, 1)$ and $\vec{r}'(t_0) = (1, -2, -1)$.

12. $\phi(x, y, z) = e^{xy}z$, $\vec{r}(t_0) = (1, -1, 2)$ and $\vec{r}'(t_0) = (1, 0, 4)$.

13. $\phi(\vec{r}) = \|\vec{r}\|$, $\vec{r}(t_0) = (1, -1, 2)$ and $\vec{r}'(t_0) = (1, 0, 4)$.

14. $\phi(\vec{r}) = \frac{1}{\|\vec{r}\|}$, $\vec{r}(t_0) = (1, -1, 2)$ and $\vec{r}'(t_0) = (1, 0, 4)$.

In Exercises 15–17, $f(x, y, z) = x^2 y^3 z$. Find the rate that f is changing at $(1, 1, 1)$ as a particle moves with velocity $\vec{v}$.

15. $\vec{v} = (1, 1, 1)$.

16. $\vec{v} = (-1, -1, -1)$.

17. $\vec{v} = (3, 2, 1)$.

18. Let $\vec{F}(x, y) = \left(-\dfrac{y}{x^2 + y^2}, \dfrac{x}{x^2 + y^2} \right)$. Show that

$$\int_{(1,0)_{C_1}}^{(-1,0)} \vec{F} \cdot d\vec{r} \neq \int_{(1,0)_{C_2}}^{(-1,0)} \vec{F} \cdot d\vec{r},$$

where C_1 is the upper half of the unit circle and C_2 is the lower half of the unit circle.

19. Let $f(x, y, z) = x^2 y^3 z$. If a particle is to move through position $(1, 1, 1)$ with a speed of 1 m/s, what direction should it move to maximize the rate of change of f?

11.10   The Gradient and Directional Derivatives

Suppose that D is an open subset of $\mathbb{R}^n$, and f is a function from D into $\mathbb{R}$ with continuous partial derivatives. We would like to have a rate of change of f at a point. When $n = 1$ (when f is a function defined on the real line), there are only two possible directions, positive and negative. In this case the derivative gives us valuable information in terms of whether the function is increasing or decreasing. However, when $n > 1$, we have an infinite number of directions from which to choose. The rate that f changes at a point $\vec{r}_0$ depends on the direction and speed we are moving through $\vec{r}_0$. In order to standardize the idea of rate of change of f, we assume that we are moving with speed one so that our discussion of the rate of change of f will depend on direction alone. Let $\hat{v}$ be a *unit vector*, and let $\vec{r}(t) = t\vec{v} + \vec{r}_0$. Then $\vec{r}(t)$ describes a particle moving through space with speed 1 m/s and direction $\hat{v}$, so that it is at $\vec{r}_0$ at time $t = 0$. Let $g(t) = f(\vec{r}(t))$. Then $g'(t)\big|_{t=0}$ denotes the rate that the particle "feels f change" as it moves through $\vec{A}$ with speed 1 m/s in the direction of $\hat{v}$. See Figure 1.

Now

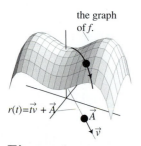

the graph of f.

$r(t) = t\vec{v} + \vec{A}$

$\vec{A}$

$\vec{v}$

Figure 1. *The line parametrized by $\vec{r}(t) = \vec{v}t + \vec{A}$ and its image in the graph of the function f.*

$$
\begin{aligned}
g'(t)\big|_{t=0} &= \frac{d}{dt} f(\vec{r}(t)) \Big|_{t=0} \\
&= \nabla f(\vec{r}(t)) \cdot \vec{r}\,'(t) \Big|_{t=0} \\
&= \nabla f(\vec{r}_0) \cdot \hat{v}.
\end{aligned}
$$

Definition: Directional Derivatives

Let $\hat{v}$ be a unit vector. Then

$$\frac{\partial f}{\partial \hat{v}}(\vec{r}) = \nabla f(\vec{r}) \cdot \hat{v}$$

is called the *derivative of f in the direction of $\hat{v}$ at $\vec{r}$.*

EXAMPLE 1: Suppose that Luke Skywalker has found himself in a hot spot in outer space and his heat shield is rapidly evaporating. He quickly calculates that, relative to a central coordinate system, $T(x, y, z) = e^{xy} + \ln z$ gives the temperature at position (x, y, z), and he is at position $(1, 1, 1)$. What rate of change of T will Luke experience if he moves in the direction $\left(\frac{1}{\sqrt{2}}, \frac{1}{\sqrt{2}}, 0\right)$?

SOLUTION: We are asked to find $\frac{\partial}{\partial \hat{v}} T(x, y, z) = \frac{\partial}{\partial \hat{v}}(e^{xy} + \ln z)$, where $\hat{v} = (1/\sqrt{2}, 1/\sqrt{2}, 0)$. Since $\nabla T(x, y, z) = \left(ye^{xy}, xe^{xy}, \frac{1}{z}\right)$, $\nabla T(1, 1, 1) = (e, e, 1)$. We simply calculate

$$
\begin{aligned}
\frac{\partial T(x, y, z)}{\partial \hat{v}} &= \nabla T(1, 1, 1) \cdot \left(\frac{1}{\sqrt{2}}, \frac{1}{\sqrt{2}}, 0\right) \\
&= (e, e, 1) \cdot \left(\frac{1}{\sqrt{2}}, \frac{1}{\sqrt{2}}, 0\right) \\
&= \frac{e}{\sqrt{2}} + \frac{e}{\sqrt{2}} + 0 = \sqrt{2}e.
\end{aligned}
$$

This directional derivative is positive, so this direction would be a poor choice for Skywalker, making him even hotter than he was already! ∎

Since $\nabla f(\vec{A}) \cdot \hat{v} = \|\nabla f(\vec{A})\| \|\hat{v}\| \cos(\theta) = \|\nabla f(\vec{A})\| \cos(\theta)$, where θ is the angle between $\nabla f(\vec{A})$ and $\hat{v}$, the derivative of f in the direction of $\hat{v}$ will be a maximum if $\hat{v}$ points in the same direction as does $\nabla f(\vec{A})$. This yields a very important interpretation of the gradient of a function:

The Geometry of the Gradient

- $\nabla f(\vec{A})$ is a vector that points in the direction of **maximum** increase of f, and

- the magnitude of the gradient of f at the point $\vec{A}$ is the rate of increase of change of f if you move from $\vec{A}$ in the direction of $\nabla f(\vec{A})$.

EXAMPLE 2:   We return to Luke Skywalker in his hot spot. What direction must he move to ensure that he will cool down as quickly as possible?

SOLUTION: Since $\nabla T(x, y, z) = \left(ye^{xy}, xe^{xy}, \frac{1}{z}\right)$, the gradient ∇ $T(1, 1, 1) = (e, e, 1)$ points in the direction of maximum increase of T. He wants to move in the direction of maximum decrease of the temperature T, which would be $-(e, e, 1) = (-e, -e, -1)$.   ■

If f is a potential function, then ∇f is called the *gradient field* for f. If the domain of the potential function f is a subset of $\mathbb{R}^2$, then ∇f is a vector-valued function defined on the domain of f, and a graphical representation of ∇f can be obtained by drawing the vector $\nabla f(x, y)$ emanating from the point (x, y) at several places in the domain of f.

EXAMPLE 3:   In the figures below, we sketch the graph of $f(x, y) = \sin(xy)$, $-\pi \le x \le \pi$, $-\pi \le y \le \pi$, and its gradient field.

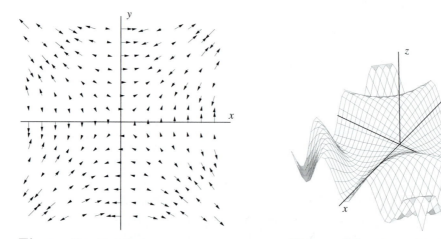

Figure 2.a $\nabla \sin(xy)$.   **Figure 2.b** *The graph of* $\sin(xy)$.

Notice that the lengths of the vectors are scaled in the sketch of the gradient field so that they do not overlap.   ■

Suppose that $U(x, y)$ represents the altitude at position (x, y). Suppose also that $\vec{F} = -\nabla U$. Then $\vec{F}(x, y)$ will point in the direction of steepest descent. Figure 3.a displays a picture of Pickens Nose Mountain viewed from the Appalachian Trail just north of the Georgia border. In Figure 3.b, we have sketched $\vec{F} = -\nabla U$. If it were raining, then the water would flow in the direction of the vectors illustrated in Figure 3.b. The longer the vectors, the steeper the descent and the faster the flow of water.

Level Surfaces and Curves

Another notion that is helpful in visualizing real valued functions defined on subsets of $\mathbb{R}^n$ is that of a level curve or level surface.

Definition: Level (Equipotential) Surfaces and Curves

Suppose that f is a potential function, c is a number, and E is the set of points satisfying $f(\vec{r}) = c$.

If the domain of f is a subset of $\mathbb{R}^2$, then E is called an *equipotential curve*, a *level curve*, or a *contour*.

If the domain of f is a subset of $\mathbb{R}^3$, then E is called an *equipotential surface* or a *level surface*.

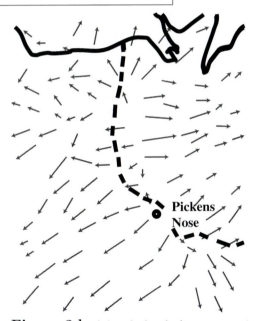

Figure 3.a *A view of Pickens Nose Mountain taken from the Appalachian Trail.*

Figure 3.b *A hand–sketched representation of -1 times the altitude gradient field.*

EXAMPLE 4: The idea of a level surface or equipotential surface is familiar to anyone who has done any backpacking. The curves on a contour map are simply the level surfaces for the function $f(x, y) =$ the altitude at position (x, y) on the map. In figure 4.a, we intersect a plane, parallel to the surface of the earth, 4000 ft above sea level, with a computer generated picture of Pickens Nose Mountain (locally, we are thinking of a flat earth). In Figure 4.b, we have attempted to sketch paths on Pickens Nose Mountain that are of constant altitude.

When these curves are projected onto a map, they provide a contour map. Figure 4.c is a *contour map* of the region. If $U(x, y)$ represents the altitude at the longitude x and latitude y, then the *contours* drawn on the map represent level altitude curves. There is a 40 ft altitude difference between positions on adjacent contours. ∎

EXAMPLE 5: In Figure 5.a, we show the plane $z = 0.5$ intersecting the graph of $f(x, y) = \sin(xy)$, $-\pi \le x \le \pi$ and $-\pi \le y \le \pi$. Figure 5.b shows the contour $\sin(xy) = 0.5$. Figure 5.c shows a contour map of $\sin(xy)$.

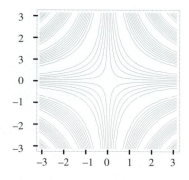

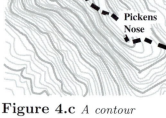

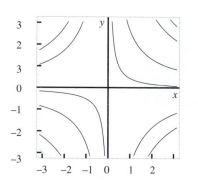

Figure 4.a *A level plane intersecting a computer generated picture of Pickens Nose.*

Figure 4.b *Equipotential curves sketched on Pickens Nose Mountain.*

Figure 4.c *A contour map of the region.*

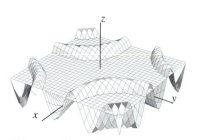

Figure 5.a *The plane $z = 0.5$ intersecting the graph of $\sin(xy)$.*

Figure 5.b *The equipotential curve that corresponds to the plane $z = 0.5$ intersecting the graph of $\sin(xy)$.*

Figure 5.c *A contour map of the function $\sin(xy)$.*

EXAMPLE 6: The contour lines drawn on a weather map are the equipotential surfaces for the function $g(x, y) =$ the barometric pressure at position (x, y) on the map.   ∎

EXAMPLE 7: Let $f(\vec{r}) = \|\vec{r}\|$ for $\vec{r}$ in $\mathbb{R}^3$. The equipotential surfaces are spheres centered at the origin as illustrated in Figure 6. If $f(\vec{r}) = \|\vec{r}\|$ for $\vec{r}$ in $\mathbb{R}^2$, then the equipotential surfaces are circles centered at the origin.   ∎

EXAMPLE 8: If U is a potential energy function for a force field $-\nabla U$, and if S is the equipotential surface $U(\vec{r}) = c$, then the work done by $-\nabla U$ on a mass as it moves from one point on S to another is 0.   ∎

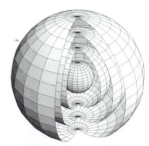

Figure 6. *The level curves $f(\vec{r}) = 1$, $f(\vec{r}) = 2$, $f(\vec{r}) = 3$, and $f(\vec{r}) = 4$.*

There is a strong relationship between the equipotential surfaces and the gradient. Consider a potential function U mapping a subset of $\mathbb{R}^2$ into $\mathbb{R}$. We assume that ∇U is continuous at $\vec{r}_0$ and that $\nabla U \neq 0$. Let C be the level curve defined by $U(\vec{r}) = U(\vec{r}_0)$. There is a neighborhood O containing $\vec{r}_0$ such that we can parametrize that part of C lying in O. Let $\vec{r}(t)$ be such a parametrization with $\vec{r}(t_0) = \vec{r}_0$. Since C lies in a level curve, $U(\vec{r}(t))$ is a constant function, and

$$\frac{dU(\vec{r}(t))}{dt} = 0.$$

However, we also have that

$$\left.\frac{dU(\vec{r}(t))}{dt}\right|_{t=t_0} = \nabla U(\vec{r}_0) \cdot \vec{r}\,'(t_0),$$

so that $\vec{r}\,'(t_0)$ is orthogonal to $\nabla U(\vec{r}_0)$. Since $\vec{r}\,'(t_0)$ is tangent to C at $\vec{r}_0$, it follows that $\nabla U(\vec{r}_0)$ is normal to C at $\vec{r}_0$. We have argued that the gradient vector at a point is orthogonal to the level curve containing the point.

Using the same argument, we can show that if U is a potential function defined on a subset of $\mathbb{R}^3$, then $\nabla U(\vec{r}_0)$ is orthogonal to the level surface $U(r) = U(r_0)$.

In Figure 7.a, we display the gradient field sketched in Figure 3.b overlayed on the contour map from Figure 4.c. In Figure 7.b, we display our sketch of $\nabla \sin(xy)$ overlayed on a contour map of $\sin(x, y)$.   ∎

Figure 7.a *The vector field from Figure 3.b overlayed on the contour map of Pickens Nose Mountain.*

Tangent Planes to Level Surfaces and Tangent Lines to Level Curves

Let U be a potential function defined on a subset of $\mathbb{R}^2$. Since the vector $\vec{n} = \nabla U(\vec{r}_0)$ is normal to the level curve $U(\vec{r}) = U(\vec{r}_0)$

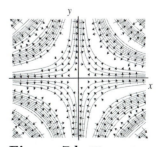

Figure 7.b *The vector field $\nabla \sin(xy)$ overlayed on a contour map of the graph of $\sin(xy)$.*

and it is also normal to the line $\vec{n} \cdot \vec{r} = \vec{n} \cdot \vec{r}_0$, we may conclude that the line $\vec{n} \cdot \vec{r} = \vec{n} \cdot \vec{r}_0$ is tangent to the curve $U(r) = U(\vec{r}_0)$. Similarly, if the domain of U is a subset of $\mathbb{R}^3$, then the plane $\vec{n} \cdot \vec{r} = \vec{n} \cdot \vec{r}_0$ is tangent to the surface $U(\vec{r}) = U(\vec{r}_0)$. See Figures 8.a–d. ■

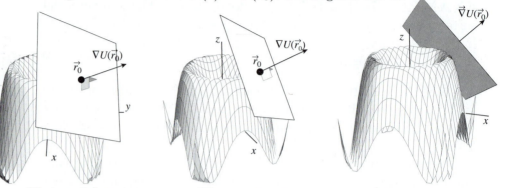

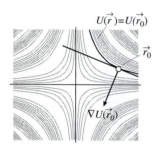

Figure 8.a $\nabla U(\vec{r}_0)$ is a vector that is normal to the level curve S at $\vec{r}_0$.

Figure 8.b $\nabla U(\vec{r}_0)$ is a vector that is normal to the level surface S at $\vec{r}_0$, viewed from position $(2, 0, 1)$.

Figure 8.c $\nabla U(\vec{r}_0)$ is a vector that is normal to the level surface S at $\vec{r}_0$, viewed from position $(2, -1, 1)$.

Figure 8.d $\nabla U(\vec{r}_0)$ is a vector that is normal to the level surface S at $\vec{r}_0$, viewed from position $(1, -3.1)$.

EXAMPLE 9: Let $U(x, y, z) = xyz$. Then $\vec{a} = (2, -1, 2)$ is a point on the surface S with equation $U(x, y, z) = -4$. The gradient $\nabla U(x, y, z) = (yz, xz, xy)$, and $\nabla U(2, -1, 2) = (-2, 4, -2)$ is a vector that is normal to S at $\vec{a}$. Thus an equation for the plane tangent to S at $\vec{a}$ is

$$(-2, 4, -2) \cdot [(x, y, z) - (2, -1, 2)] = 0, \text{ or}$$

$$-2x + 4y - 2z + 12 = 0. \qquad ■$$

EXAMPLE 10: Let $U(x, y) = \sin(x + y)$. Then $\nabla U(x, y) = (\cos(x + y), \cos(x + y))$. The point $\vec{a} = (0, 0)$ is on the level curve S given by $U(x, y) = 0$. The gradient at $\vec{a}$ is $\nabla U(0, 0) = (1, 1)$. Thus the line L with equation

$$\nabla U(0, 0) \cdot (x, y) = 0 \quad \text{or} \quad x + y = 0$$

is tangent to S at $(0, 0)$. ■

EXERCISES 11.10

In Exercises 1–5, find the derivative of the given function in the direction of $\frac{\vec{v}}{\|\vec{v}\|}$ at the point $\vec{a}$.

1. $g(x, y, z) = \dfrac{x^3 y}{z + 1}$, $\vec{a} = (1, -1, 0), \vec{v} = (1, -1, 2)$.

2. $f(x, y) = \dfrac{x + y}{xy}$, $\vec{a} = (1, 1), \vec{v} = (-1, -1)$.

3. $u(x, y) = \cos x \sin y$, $\vec{a} = \left(\pi, \frac{\pi}{2}\right), \vec{v} = (2, -3)$.

4. $u(x, y, z) = \dfrac{x + y}{e^z}$, $\vec{a} = (1, 1, 0), \vec{v} = (1, 1, 1)$.

5. $u(x,y) = x^y, \vec{a} = (1,1), \vec{v} = (1,1)$.

6. Find a unit vector that points in the direction of maximum increase for the function g of Exercise 1 at the given point $\vec{a}$.

7. Find a unit vector that points in the direction of maximum decrease for the function g of Exercise 1 at the given point $\vec{a}$.

8. Find a unit vector that points in the direction of maximum increase for the function f of Exercise 2 at the given point $\vec{a}$.

9. Find a unit vector that points in the direction of maximum decrease for the function f of Exercise 2 at the given point $\vec{a}$.

In Exercises 10–13, find an equation for the plane tangent to the given surface at the indicated point.

10. $x^2 + y^2 + z^2 = 14$, at $\vec{a} = (1,2,3)$.

11. $e^{xyz} = 1$, at $\vec{a} = (1,1,0)$.

12. $\ln\left(\frac{xy}{z}\right) = 1$, at $\vec{a} = (e,e,e)$.

13. $2^{x+y+z} = 4$, at $\vec{a} = (1,1,0)$.

In Exercises 14–16, find an equation for the line tangent to the level curve at the indicated point.

14. $f(x,y) = \dfrac{x+y}{xy}$, $\vec{a} = (1,1)$.

15. $f(x,y) = \cos x \sin y$, $\vec{a} = \left(\pi, \frac{\pi}{2}\right)$.

16. $f(x,y) = x^y$, $\vec{a} = (1,1)$.

17. Describe the level curves for $f(x,y) = \dfrac{x^2}{4} - y^2$.

18. Describe the level surfaces for $f(x,y,z) = \dfrac{x^2}{4} + \dfrac{y^2}{9} + z^2$.

Figure 9 is a contour map for a function f. Make a copy of the figure for Exercises 19–23.

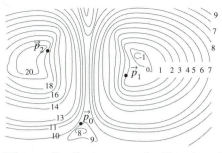

Figure 9. *Contour map for Exercises 19–23*

19. Draw a vector emanating from $\vec{p}_0$ that points in the direction of maximum increase of f at $\vec{p}_0$.

20. Draw a vector emanating from $\vec{p}_1$ that points in the direction of maximum increase of f at $\vec{p}_1$.

21. Draw a vector emanating from $\vec{p}_1$ that points in the direction of maximum increase of f at $\vec{p}_2$.

22. Locate likely points where f attains its local maximia.

23. Locate likely points where f attains its local minimia.

24. Plot the vector field ∇f by drawing an estimate of a $\nabla f(\vec{r})$ emanating from each point $\vec{r}$ in the grid in Figure 10.

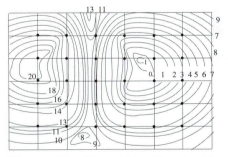

Figure 10. *The grid for Exercise 24.*

In Exercises 25–28, you are given the gradient field for a function. Sketch the level curves for the function passing through the points $\vec{a}$, $\vec{b}$, $\vec{c}$, and $\vec{d}$.

25.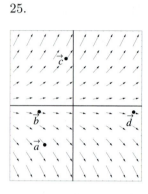

26.

27.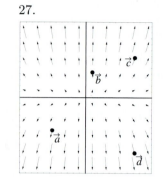

28.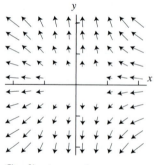

29. Match the graphs in the first column with the contour maps in the second column and the gradient maps in the third column.

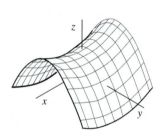

Graph 1.

Contour map 1.

Gradient map 1.

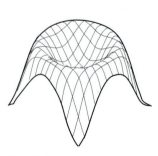

Graph 2.

Contour map 2.

Gradient map 2.

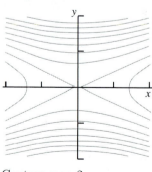

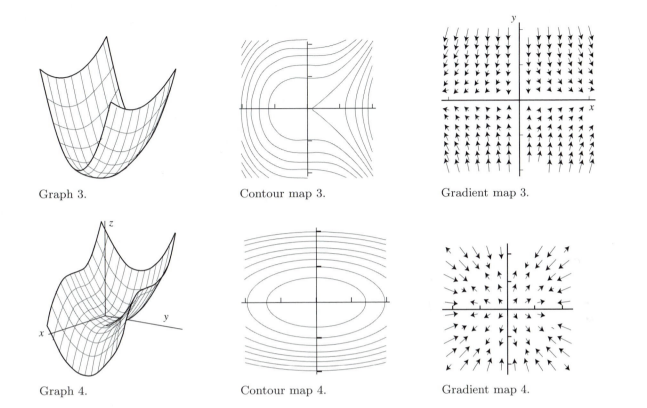

Graph 3.　　　　　Contour map 3.　　　　　Gradient map 3.

Graph 4.　　　　　Contour map 4.　　　　　Gradient map 4.

11.11　The Curl and Iterated Partial Derivatives

Definition: $\mathrm{Curl}(\vec{F})$

Suppose that $\vec{F}$ is a vector field in $\mathbb{R}^3$ (i.e., $\vec{F}$ is a function from a subset of xyz–space into xyz–space). We define

$$\mathrm{Curl}(\vec{F}) = \begin{pmatrix} \frac{\partial F_z}{\partial y} - \frac{\partial F_y}{\partial z} \\ \frac{\partial F_x}{\partial z} - \frac{\partial F_z}{\partial x} \\ \frac{\partial F_y}{\partial x} - \frac{\partial F_x}{\partial y} \end{pmatrix}.$$

　　A memory device that is helpful here is to think of $\vec{\nabla} = \left(\frac{\partial}{\partial x}, \frac{\partial}{\partial y}, \frac{\partial}{\partial z} \right)$, and to recall that $\mathrm{Curl}(\vec{F})$ is simply the result of formally computing the cross product $\vec{\nabla} \times \vec{F}$.

$$\vec{\nabla} \times \vec{F} = \left(\left| \begin{array}{cc} \frac{\partial}{\partial y} & \frac{\partial}{\partial z} \\ F_y & F_z \end{array} \right|, -\left| \begin{array}{cc} \frac{\partial}{\partial x} & \frac{\partial}{\partial z} \\ F_x & F_z \end{array} \right|, \left| \begin{array}{cc} \frac{\partial}{\partial x} & \frac{\partial}{\partial y} \\ F_x & F_y \end{array} \right| \right)$$

$$= \left(\frac{\partial F_z}{\partial y} - \frac{\partial F_y}{\partial z}, \frac{\partial F_x}{\partial z} - \frac{\partial F_z}{\partial x}, \frac{\partial F_y}{\partial x} - \frac{\partial F_x}{\partial y} \right)$$

$$= \text{Curl}(\vec{F}).$$

EXAMPLE 1: Let $\vec{F}(x, y, z) = (x + y, y + z, xyz)$. Then

$$\text{Curl}(\vec{F}(x, y, z)) = \begin{pmatrix} xz - 1 \\ -yz \\ -1 \end{pmatrix}. \qquad \blacksquare$$

If $\vec{F}$ is a vector field defined in $\mathbb{R}^2$, then we can represent $\vec{F}(x, y) = (F_x(x, y), F_y(x, y))$ as a vector field in $\mathbb{R}^3$ by $\vec{F}(x, y, z) = (F_x(x, y), F_y(x, y), 0)$. Since $\frac{\partial F_z}{\partial y} = \frac{\partial F_y}{\partial z} = \frac{\partial F_x}{\partial z} = \frac{\partial F_z}{\partial x} = 0$, we see that $\text{Curl}(\vec{F})$ points in the z direction. Thus

$$\text{Curl}(\vec{F}) = \left(0, 0, \frac{\partial F_y}{\partial x} - \frac{\partial F_x}{\partial y} \right).$$

EXAMPLE 2: Let $\vec{F}(x, y) = (xy, e^{xy})$. Then $\text{Curl}(\vec{F}) = (0, 0, ye^{xy} - x)$. $\blacksquare$

In order to carry our study of conservative forces further, we need to take partial derivatives of partial derivatives of partial derivatives, etc.

Notation: If f is a function from xyz–space into $\mathbb{R}$, then

$$\frac{\partial^2 f(x, y, z)}{\partial x \, \partial y} = \frac{\partial}{\partial x} \left[\frac{\partial f(x, y, z)}{\partial y} \right].$$

You first evaluate the partial derivative with respect to y, and then take the partial derivative with respect to x. Similarly, we can find a third partial derivative such as

$$\frac{\partial^3 f(x, y, z)}{\partial x^2 \, \partial y} = \frac{\partial}{\partial x} \left[\frac{\partial}{\partial x} \left[\frac{\partial f(x, y, z)}{\partial y} \right] \right].$$

EXAMPLE 3: Let $f(x, y, z) = xy^2 z^3$. Then

$$\frac{\partial^2 f(x,y,z)}{\partial x^2} = 0 \qquad \frac{\partial^2 f(x,y,z)}{\partial x\,\partial y} = 2yz^3 \qquad \frac{\partial^2 f(x,y,z)}{\partial z\,\partial y} = 6xyz^2$$

$$\frac{\partial^2 f(x,y,z)}{\partial z^2} = 6xy^2 z \qquad \frac{\partial^2 f(x,y,z)}{\partial y\,\partial x} = 2yz^3 \qquad \frac{\partial^3 f(x,y,z)}{\partial x\,\partial y\,\partial z} = 6yz^2$$

$$\frac{\partial^3 f(x,y,z)}{\partial z^3} = 6xy^2.$$

■

EXAMPLE 4: Let $f(u,v) = \cos(u)\sin(uv)$. Then

$$\frac{\partial f}{\partial u} = -\sin(u)\sin(uv) + v\cos(u)\cos(uv)$$

$$\frac{\partial^2 f}{\partial u^2} = -\cos(u)\sin(uv) - v\sin(u)\cos(uv) - v\sin(u)\cos(uv)$$

$$-v^2\cos(u)\sin(uv)$$

$$\frac{\partial^2 f}{\partial v\,\partial u} = -u\sin(u)\cos(uv) + \cos(u)\cos(uv)$$

$$-uv\cos(u)\sin(uv).$$

Theorem 1

(a) *Suppose that f is a function from a subset of xy–space into $\mathbb{R}$ such that all of the second partial derivatives of f are continuous. Then $\frac{\partial f^2}{\partial x\,\partial y} = \frac{\partial f^2}{\partial y\,\partial x}$.*

(b) *Suppose that f is a function from a subset of xyz–space into $\mathbb{R}$ such that all of the second partial derivatives of f are continuous. Then $\frac{\partial f^2}{\partial x\,\partial y} = \frac{\partial f^2}{\partial y\,\partial x}$, $\frac{\partial f^2}{\partial x\,\partial z} = \frac{\partial f^2}{\partial z\,\partial x}$, and $\frac{\partial f^2}{\partial z\,\partial y} = \frac{\partial f^2}{\partial y\,\partial z}$.*

(c) *Suppose that f is a function from a subset of xyz–space into $\mathbb{R}$ such that all of the third partial derivatives of f are continuous. Then $\frac{\partial f^3}{\partial x\,\partial y\,\partial z} = \frac{\partial f^3}{\partial z\,\partial y\,\partial x} = \frac{\partial f^3}{\partial z\,\partial x\,\partial y}$, etc.*

Proof: We will argue only part (a). Consider

$$\lim_{\Delta x \to 0, \Delta y \to 0} \frac{f(x + \Delta x, y + \Delta y) - f(x + \Delta x, y) - f(x, y + \Delta y) + f(x, y)}{\Delta x \, \Delta y}$$

$$= \lim_{\Delta x \to 0, \Delta y \to 0} \frac{f(x + \Delta x, y + \Delta y) - f(x + \Delta x, y)}{\Delta x \, \Delta y} - \frac{f(x, y + \Delta y) - f(x, y)}{\Delta x \, \Delta y}$$

$$= \lim_{\Delta x \to 0} \frac{\frac{\partial f(x + \Delta x, y)}{\partial y} - \frac{\partial f(x, y)}{\partial y}}{\Delta x} = \frac{\partial}{\partial x} \frac{\partial f(x, y)}{\partial y} = \frac{\partial^2 f(x, y)}{\partial x \, \partial y}.$$

In exactly the same way we can show that

$$\lim_{\Delta x \to 0, \Delta y \to 0} \frac{f(x + \Delta x, y + \Delta y) - f(x + \Delta x, y) - f(x, y + \Delta y) + f(x, y)}{\Delta x \, \Delta y}$$

$$= \frac{\partial}{\partial y} \frac{\partial f(x, y)}{\partial x} = \frac{\partial^2 f(x, y)}{\partial y \, \partial x}. \qquad \blacksquare$$

Recall that a vector field $\vec{F}$ is *conservative* if there is a potential function f such that $\vec{F}(x, y, z) = \vec{\nabla} f(x, y, z) = \left(\frac{\partial f(x,y,z)}{\partial x}, \frac{\partial f(x,y,z)}{\partial y}, \frac{\partial f(x,y,z)}{\partial z} \right)$. We have the following theorem.

Theorem 2 *If $\vec{F}$ is a conservative force field, and if the partial derivatives of the coordinate functions of $\vec{F}$ are continuous, then $Curl(\vec{F}) = 0$.*

Proof: Let f be a function such that $\vec{\nabla} f = F$. Then

$$\mathrm{Curl}(\vec{F}) = \begin{pmatrix} \frac{\partial F_z}{\partial y} - \frac{\partial F_y}{\partial z} \\[2mm] \frac{\partial F_x}{\partial z} - \frac{\partial F_z}{\partial x} \\[2mm] \frac{\partial F_y}{\partial x} - \frac{\partial F_x}{\partial y} \end{pmatrix} = \begin{pmatrix} \frac{\partial^2 f}{\partial y \, \partial z} - \frac{\partial^2 f}{\partial z \, \partial y} \\[2mm] \frac{\partial^2 f}{\partial z \, \partial x} - \frac{\partial^2 f}{\partial x \, \partial z} \\[2mm] \frac{\partial^2 f}{\partial x \, \partial y} - \frac{\partial^2 f}{\partial y \, \partial x} \end{pmatrix} = \vec{0}. \qquad \blacksquare$$

Note: Theorem 2 does **not** state that if $\mathrm{Curl}(\vec{F}) = \vec{0}$ then $\vec{F}$ is a conservative force field. However, under suitable conditions on the domain of $\vec{F}$, the force field is conservative. We will investigate this matter in Section 15.3.

EXAMPLE 5: Show that $\vec{F}(x, y, z) = (y, xz, xy)$ is not conservative.

SOLUTION: We will show that $\text{Curl}(\vec{F}) \neq 0$.

$$\text{Curl}(\vec{F}) = \begin{pmatrix} \frac{\partial F_z}{\partial y} - \frac{\partial F_y}{\partial z} \\ \frac{\partial F_x}{\partial z} - \frac{\partial F_z}{\partial x} \\ \frac{\partial F_y}{\partial x} - \frac{\partial F_x}{\partial y} \end{pmatrix} = \begin{pmatrix} x - x \\ 0 - y \\ z - 1 \end{pmatrix} = \begin{pmatrix} 0 \\ -y \\ z - 1 \end{pmatrix} \neq \vec{0}.$$ ∎

Note: Theorem 2 shows that if f is a function from a subset of $\mathbb{R}^n$ into $\mathbb{R}$ with continuous second partial derivatives, then $\text{Curl}(\text{grad}(f)) = \vec{\nabla} \times \vec{\nabla} f = 0$.

EXERCISES 11.11

In Exercises 1–4, compute $\frac{\partial^2 f}{\partial x^2}$ and $\frac{\partial^2 f}{\partial x\, \partial y}$.

1. $f(x, y) = x \sin(xy)$.

2. $f(x, y) = e^{xy}$.

3. $f(x, y) = y^x$.

4. $f(x, y, z) = 3^{xyz}$.

In Exercises 5–9, compute $\text{Curl}(\vec{F})$.

5. $\vec{F}(x, y) = (e^{xy}, y \sin(x) \cos(y))$.

6. $\vec{F}(x, y) = \left(\dfrac{y}{\sqrt{x^2 + y^2}}, \dfrac{x}{\sqrt{x^2 + y^2}} \right)$.

7. $\vec{F}(x, y, z) = (xy \sin(y), xze^{xy}, xyz)$.

8. $\vec{F}(\vec{r}) = \dfrac{k\vec{r}}{\|\vec{r}\|^3}$.

9. $\vec{F}(x, y, z) = (3x^2 y, x^3 + y^3, 0)$.

In Exercises 10 and 11, compute $\vec{\nabla} \cdot (\vec{\nabla} \times \vec{F})$.

10. $\vec{F}(x, y, z) = (xy, xy + xz, xyz)$.

11. $\vec{F}(x, y, z) = (x \sin(xy), y \cos(xy), e^{xyz})$.

12. Show that if $\vec{F}$ is a vector field with continuous second partial derivatives, then $\vec{\nabla} \cdot (\vec{\nabla} \times \vec{F}) = 0$.

*In Exercises 13–17, determine if the force field is **not** conservative according to Theorem 2.*

13. $\vec{F}(x, y, z) = (e^{xy}, y \sin(x) \cos(y), 0)$.

14. $\vec{F}(x, y, z) = \left(z, \dfrac{y}{\sqrt{x^2 + y^2}}, \dfrac{x}{\sqrt{x^2 + y^2}} \right)$.

15. $\vec{F}(x, y, z) = (xy \sin(y), xze^{xy}, xyz)$.

16. $\vec{F}(\vec{r}) = \dfrac{k\vec{r}}{\|\vec{r}\|^3}$,   where $\vec{r} = (x, y, z)$.

17. $\vec{F}(x, y, z) = (3x^2 y, x^3 + y^3, 0)$.

In Exercises 18–21,

$$h(x, y) = \begin{cases} xy \dfrac{y^2 - x^2}{x^2 + y^2}, & \text{if } (x, y) \neq \vec{0} \\ 0, & \text{if } (x, y) = \vec{0} \end{cases}$$

See Figure 1.

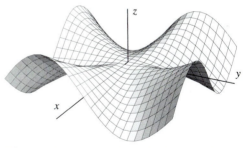

Figure 1. *The graph of the function h.*

18. Calculate $\frac{\partial h(x,y)}{\partial y}$, $\frac{\partial h(x,y)}{\partial x}$, $\frac{\partial^2 h(x,y)}{\partial x\, \partial y}$ and $\frac{\partial^2 h(x,y)}{\partial y\, \partial x}$ for $(x, y) \neq \vec{0}$.

19. Calculate $\frac{\partial h}{\partial y}$ at $(0, 0)$ by computing $\lim_{\Delta y \to 0} \frac{h(0, 0+\Delta y) - h(0,0)}{\Delta y}$, and then compute $\frac{\partial^2 h}{\partial x\, \partial y}$ at $(0, 0)$ by computing $\lim_{\Delta x \to 0} \left[\frac{\partial h(0+\Delta x, 0)}{\partial y} - \frac{\partial h(0,0)}{\partial y} \right] / \Delta x$.

20. Calculate $\frac{\partial h}{\partial x}$ at $(0,0)$ by computing
$\lim_{\Delta x \to 0} \frac{h(0+\Delta x, 0) - h(0,0)}{\Delta x}$, and then compute
$\frac{\partial^2 h}{\partial x \, \partial y}$ at $(0,0)$ by computing
$\lim_{\Delta y \to 0} \left[\frac{\partial h(0, 0+\Delta y)}{\partial x} - \frac{\partial h(0,0)}{\partial x} \right] / \Delta y$.

21. Does the fact that $\frac{\partial^2 h}{\partial x \, \partial y}$ at $(0,0)$ does not equal $\frac{\partial^2 h}{\partial y \, \partial x}$ at $(0,0)$ violate Theorem 1?

Chapter 12

Optimization of Functions From $\mathbb{R}^n$ into $\mathbb{R}$

12.1   Tests for Local Extrema

Once again Luke Skywalker is in a hot spot, and he has a map that gives the temperature as a function of position. He now wants to really cool off, by going to the coolest place in space. The temperature function is $T(x, y, z) = x^2 + y^2 + z^2$. It is easy to see that his destination is $(0, 0, 0)$. The problem is more complicated than it seems at first glance however, because the temperature function is expected to change as time progresses. He receives an updated function every few months, and he needs a technique for finding the coolest spot. What he needs, then, is a general method for minimizing a real valued function defined on $\mathbb{R}^3$. We first find a method for choosing points in space that are likely candidates for minimizing (or maximizing) such real-valued functions. Recall that when we tried to minimize real valued functions defined on subsets of the real line, we picked out points where the derivative was 0. Since the derivative of a real valued function T defined on $\mathbb{R}^n$ is the gradient of T, we might try looking at points where $\vec{\nabla} T = \vec{0}$. We begin with the definition for maxima and minima for functions from $\mathbb{R}^n$.

Definition: Local Extrema

Suppose that T is a real valued function defined on a subset U of $\mathbb{R}^n$, and $\vec{r}_0$ is a point in U such that every ball[1] centered at $\vec{r}_0$ contains a point of U distinct from $\vec{r}_0$.

(a) If there is a ball B centered at $\vec{r}_0$ such that $T(\vec{r}_0) \leq T(\vec{r})$ for all $\vec{r}$ lying in B and U, then T is said to attain a *relative minimum* or a *local minimum* at $\vec{r}_0$, and $T(\vec{r}_0)$ is called a *relative* or *local minimum*.

(b) If there is a ball B centered at $\vec{r}_0$ such that $T(\vec{r}_0) \geq T(\vec{r})$ for all $\vec{r}$ lying in B and U, then T is said to attain a *relative maximum* or a *local maximum* at $\vec{r}_0$, and $T(\vec{r}_0)$ is called a *relative* or *local maximum*.

(c) The relative maxima and minima are called *relative* or *local extrema*.

---

[1]A set B is a *ball* in $\mathbb{R}^n$ centered at $\vec{r}_0$, if there is a number $R > 0$ such that B is the set of all points $\vec{r}$ such that $\text{dist}(\vec{r}, \vec{r}_0) < R$.

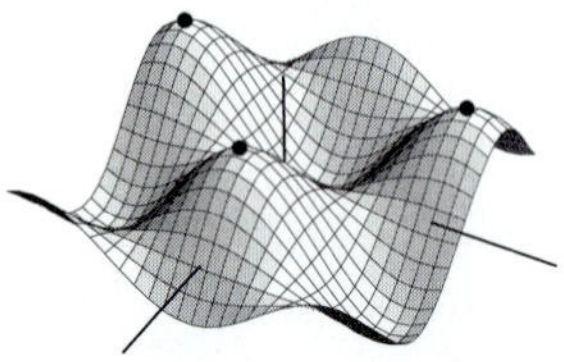

A simple example of a function f from $\mathbb{R}^2$ into $\mathbb{R}$ is the function $z = f(x, y)$, which represents the altitude of a point on the earth's surface at latitude x and longitude y. With this model, the local maxima are hill tops, while local minima are bottoms of lakes. Figures 1.a and 1.b illustrate this idea.

We now present our main theorem for finding extrema for real valued functions from $\mathbb{R}^n$.

Theorem 1 (The First Derivative Test) *Suppose that T is a differentiable function from an open subset of $\mathbb{R}^n$ into $\mathbb{R}$. If T attains a local extremum at $\vec{r}_0$, then $\vec{\nabla} T|_{\vec{r}_0} = 0$.*

Proof: Let $\vec{r}(t) = \vec{v}t + \vec{r}_0$. Then $T \circ \vec{r}(t) = T(\vec{r}(t))$ is a function from a subset of $\mathbb{R}$ into $\mathbb{R}$. If T attains a local extremum at $\vec{r}_0$, then $T(\vec{r}(t))$ attains a local extremum at $t = 0$. Thus, from Chapter 5, we know that $\frac{d}{dt} T(\vec{r}(t))|_{t=0} = 0$. This, together with the chain rule, gives us

$$\frac{d}{dt} T(\vec{r}(t))|_{t=0} = \vec{\nabla} T|_{\vec{r}_0} \cdot \vec{r}\,'(t)|_{t=0} = \vec{\nabla} T|_{\vec{r}_0} \cdot \vec{v} = 0,$$

regardless of the choice of $\vec{v}$. It follows that $\vec{\nabla} T|_{\vec{r}_0} = \vec{0}$. ∎

Notice that Theorem 1 reinforces the idea that the gradient of a

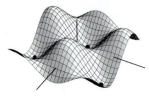

Figure 1.a *The relative maxima for* $f(x, y) = \sin x \cos y$.

Figure 1.b *The relative minima for* $f(x, y) = \sin x \cos y$.

function points in the direction of maximum increase of the function, and that -1 times the gradient points in the direction of maximum decrease of the function. If a function has a local maximum at $\vec{r}_0$, then there is no direction to move from $\vec{r}_0$ that will increase the function locally.

Following the terminology of Chapter 5, we say that $\vec{r}_0$ is a *critical point* of f if $\vec{\nabla} f|_{\vec{r}_0} = \vec{0}$.

EXAMPLE 1: Figure 2 reveals that $f(x, y) = x^2 + xy + y^2 + x + y$ has at least one relative minimum. We can find where this occurs by finding the critical points.

$$\vec{\nabla} f(x, y) = (2x + y + 1, x + 2y + 1).$$

Setting this equal to $(0, 0)$ we have a pair of equations

$$2x + y + 1 = 0 \text{ and } x + 2y + 1 = 0.$$

We solve these simultaneously to obtain the critical point $\left(-\frac{1}{3}, -\frac{1}{3}\right)$. ∎

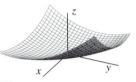

Figure 2. *The graph of* $f(x, y) = x^2 + xy + y^2 + x + y$.

Just as a function f from $\mathbb{R}$ into $\mathbb{R}$ need not reach a local extremum at a point x_0 when $f'(x_0) = 0$, a function f from a subset of $\mathbb{R}^n$ into $\mathbb{R}$ might have $\vec{\nabla} f|_{\vec{r}_0} = \vec{0}$ without attaining an extremum there.

EXAMPLE 2: Let $f(x, y) = x^2 - y^2$. The graph of f is displayed in Figure 3. The gradient $\vec{\nabla} f(x, y) = (2x, -2y)$. Thus $\vec{\nabla} f(x, y) = \vec{0}$ at $(0, 0)$, so $(0, 0)$ is a critical point. However, f does not attain a relative extrema there. If you move in the direction of $(0, 1)$, you decrease f, whereas if you move in the direction of $(1, 0)$, you increase f. ∎

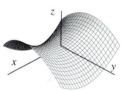

Figure 3. *The graph of* $f(x, y) = x^2 - y^2$.

Notice that in Example 2, if $\vec{r}(t) = (t, 0)$, then the function $f(\vec{r}(t))$ is concave up at $t = 0$. In the same way, if $\vec{r}(t) = (0, t)$, then the function $f(\vec{r}(t))$ is concave down at $t = 0$. We need a new word to describe this behavior.

Definition: Saddle Points

Let U be a subset of $\mathbb{R}^2$ and let $f : U \to \mathbb{R}$ be differentiable. If $\vec{r}_0$ is a critical point and if f attains neither a local maximum nor a local minimum at $\vec{r}_0$, then $\vec{r}_0$ is called a *saddle point* of f.

We still need something like the second derivative test to determine whether a critical point is a local maximum, a local minimum, or a saddle point. Before presenting the theorem we want, we need the following notation.

If U is a subset of $\mathbb{R}^2$ and $f : U \to \mathbb{R}$ has continuous second partial derivatives at $\vec{r}_0$, then we define

$$\mathcal{H}_1(f(\vec{r}_0)) = \frac{\partial^2 f}{\partial x^2}(\vec{r}_0)$$

and

$$\mathcal{H}_2(f(\vec{r}_0)) = \det \left. \begin{pmatrix} \frac{\partial^2 f}{\partial x^2} & \frac{\partial^2 f}{\partial x \partial y} \\ \frac{\partial^2 f}{\partial x \partial y} & \frac{\partial^2 f}{\partial y^2} \end{pmatrix} \right|_{\vec{r}_0}.$$

If U is a subset of $\mathbb{R}^3$ and $f : U \to \mathbb{R}$ has continuous second partial derivatives at $\vec{r}_0$, then we define $\mathcal{H}_1$ and $\mathcal{H}_2$ exactly as above, and we define

$$\mathcal{H}_3(f(\vec{r}_0)) = \det \left. \begin{pmatrix} \frac{\partial^2 f}{\partial x^2} & \frac{\partial^2 f}{\partial x \partial y} & \frac{\partial^2 f}{\partial x \partial z} \\ \frac{\partial^2 f}{\partial x \partial y} & \frac{\partial^2 f}{\partial y^2} & \frac{\partial^2 f}{\partial y \partial z} \\ \frac{\partial^2 f}{\partial x \partial z} & \frac{\partial^2 f}{\partial y \partial z} & \frac{\partial^2 f}{\partial z^2} \end{pmatrix} \right|_{\vec{r}_0}.$$

Theorem 2 (Second Derivative Test for Functions Defined on $\mathbb{R}^2$)

Suppose that U is an open subset of $\mathbb{R}^2$, the function $f : U \to \mathbb{R}$ has continuous second partial derivatives, and $\vec{r}_0$ is a critical point.

(a) *If $\mathcal{H}_1(f(\vec{r}_0)) > 0$ and $\mathcal{H}_2(f(\vec{r}_0)) > 0$, then f attains a local minimum at $\vec{r}_0$, and we say that f is concave up at $\vec{r}_0$.*

(b) *If $\mathcal{H}_1(f(\vec{r}_0)) < 0$ and $\mathcal{H}_2(f(\vec{r}_0)) > 0$, then f attains a local maximum at $\vec{r}_0$, and we say that f is concave down at $\vec{r}_0$.*

(c) *If $\mathcal{H}_2(f(\vec{r}_0)) < 0$, then $\vec{r}_0$ is a saddle point.*

If none of the conditions (a), (b), or (c) holds, and if $\mathcal{H}_1(f(\vec{r}_0)) = 0$ or $\mathcal{H}_2(f(\vec{r}_0)) = 0$, then we have no new information.

EXAMPLE 3: Returning to Example 1, we know that $f(x,y) = x^2 + xy + y^2 + x + y$ has its only critical point at $\left(-\frac{1}{3}, -\frac{1}{3}\right)$.

$$\mathcal{H}_1 \left(f \left(-\frac{1}{3}, -\frac{1}{3} \right) \right) = 2,$$

and

$$\mathcal{H}_2 \left(f \left(-\frac{1}{3}, -\frac{1}{3} \right) \right) = \det \begin{pmatrix} 2 & 1 \\ 1 & 2 \end{pmatrix} = 3.$$

Thus

$$f\left(-\frac{1}{3}, -\frac{1}{3}\right) = \frac{1}{9} + \frac{1}{9} + \frac{1}{9} - \frac{1}{3} - \frac{1}{3} = -\frac{1}{3}$$

is a local minimum.   ∎

Theorem 3 (Second Derivative Test for Functions Defined on $\mathbb{R}^3$)

Suppose that U is an open subset of $\mathbb{R}^3$, the function $f : U \to \mathbb{R}$ has continuous second partial derivatives, and $\vec{r}_0$ is a critical point.

  (a) *If $\mathcal{H}_1(f(\vec{r}_0)) > 0$, $\mathcal{H}_2(f(\vec{r}_0)) > 0$, and $\mathcal{H}_3(f(\vec{r}_0)) > 0$, then f attains a local minimum at $\vec{r}_0$, and we say that f is concave up at $\vec{r}_0$.*

  (b) *If $\mathcal{H}_1(f(\vec{r}_0)) < 0$, $\mathcal{H}_2(f(\vec{r}_0)) > 0$, and $\mathcal{H}_3(f(\vec{r}_0)) < 0$, then f attains a local maximum at $\vec{r}_0$, and we say that f is concave down at $\vec{r}_0$.*

  (c) *If $\mathcal{H}_1(f(\vec{r}_0))$ and $\mathcal{H}_3(f(\vec{r}_0))$ have opposite signs, or if $\mathcal{H}_2(f(\vec{r}_0)) < 0$, then $\vec{r}_0$ is a saddle point.*

If none of conditions (a), (b), or (c) holds, and if $\mathcal{H}_1(f(\vec{r}_0)) = 0$, $\mathcal{H}_2(f(\vec{r}_0)) = 0$, or $\mathcal{H}_3(f(\vec{r}_0)) = 0$, then we have no new information.

There are similar theorems for functions defined in higher dimensions, but we leave those to more advanced courses.

EXAMPLE 4: Determine the nature of the critical points of $f(x, y, z) = x^2 + y^2 + z^2 + xz$.

SOLUTION: $\vec{\nabla} f = (2x + z, 2y, 2z + x) = \vec{0}$ at $(0, 0, 0)$.

$$\frac{\partial^2 f}{\partial x^2}(0, 0, 0) = 2, \quad \frac{\partial^2 f}{\partial y^2}(0, 0, 0) = 2, \quad \frac{\partial^2 f}{\partial z^2}(0, 0, 0) = 2,$$

$$\frac{\partial^2 f}{\partial x \partial y}(0, 0, 0) = 0, \quad \frac{\partial^2 f}{\partial x \partial z}(0, 0, 0) = 1, \quad \frac{\partial^2 f}{\partial y \partial z}(0, 0, 0) = 0,$$

$$\mathcal{H}_1(f(0, 0, 0)) = 2,$$

$$\mathcal{H}_2(f(0, 0, 0)) = \det\begin{pmatrix} 2 & 0 \\ 0 & 2 \end{pmatrix} = 4,$$

and

$$\mathcal{H}_3(f(0,0,0)) = \det \begin{pmatrix} 2 & 0 & 1 \\ 0 & 2 & 0 \\ 1 & 0 & 2 \end{pmatrix} = 6.$$

Thus $f(0,0,0) = 0$ is a relative minimum.   ∎

EXAMPLE 5: Determine the nature of the critical points of $f(x,y,z) = x^2 + y^2 - z^2$.

SOLUTION: $\vec{\nabla}f = (2x, 2y, -2z) = \vec{0}$ at $(0,0,0)$.

$$\frac{\partial^2 f}{\partial x^2}(0,0,0) = 2, \quad \frac{\partial^2 f}{\partial y^2}(0,0,0) = 2, \quad \frac{\partial^2 f}{\partial z^2}(0,0,0) = -2,$$

$$\frac{\partial^2 f}{\partial x \partial y}(0,0,0) = 0, \quad \frac{\partial^2 f}{\partial x \partial z}(0,0,0) = 0, \quad \frac{\partial^2 f}{\partial y \partial z}(0,0,0) = 0.$$

We have

$$\mathcal{H}_1(f(0,0,0)) = 2,$$
$$\mathcal{H}_2(f(0,0,0)) = \det \begin{pmatrix} 2 & 0 \\ 0 & 2 \end{pmatrix} = 4,$$

and

$$\mathcal{H}_3(f(0,0,0)) = \det \begin{pmatrix} 2 & 0 & 0 \\ 0 & 2 & 0 \\ 0 & 0 & -2 \end{pmatrix} = -8.$$

Since $\mathcal{H}_1(f(0,0,0))$ and $\mathcal{H}_3(f(0,0,0))$ have opposite signs, $(0,0,0)$ is a saddle point.   ∎

Absolute Extrema

In general, finding absolute extrema of functions can be a very difficult problem. There is no general theory (yet) that works in every case. There are some situations that we can discuss here. The definition of absolute extrema is exactly what we would expect it to be from our studies of absolute extrema for functions defined on subsets of the real line.

Definition: Absolute Extrema

Suppose that U is a subset of $\mathbb{R}^n$, $f : U \to \mathbb{R}$ is a real-valued function defined on U, and $\vec{r}_0$ is a point in U. Then
$f(\vec{r}_0)$ is the *absolute maximum* for f provided that $f(\vec{r}_0) \geq f(\vec{r})$ for all $\vec{r}$ in U.
$f(\vec{r}_0)$ is the *absolute minimum* for f provided that $f(\vec{r}_0) \leq f(\vec{r})$ for all $\vec{r}$ in U.
$f(\vec{r}_0)$ is an *absolute extremum* for f provided that $f(\vec{r}_0)$ is either an *absolute maximum* or an *absolute minimum*.

The following theorems say that if the domain of f is open and connected, f attains a local extremum at $\vec{r}_0$, and f is concave up everywhere, then $f(\vec{r}_0)$ is an absolute miminum. Similarly, if f attains a local extremum at $\vec{r}_0$ and f is concave down everywhere, then $f(\vec{r}_0)$ is an absolute maximum.

Theorem 4 (Concavity Test for Functions Defined on $\mathbb{R}^2$)
Suppose that U is an open subset of $\mathbb{R}^2$, the function $f : U \to \mathbb{R}$ has continuous second partial derivatives, and $\vec{r}_0$ is a critical point.

(a) *If $\mathcal{H}_1(f(\vec{r})) > 0$ and $\mathcal{H}_2(f(\vec{r})) > 0$ for all $\vec{r}$ in U, then f attains an absolute minimum at $\vec{r}_0$.*

(b) *If $\mathcal{H}_1(f(\vec{r})) < 0$ and $\mathcal{H}_2(f(\vec{r})) > 0$ for all $\vec{r}$ in U, then f attains an absolute maximum at $\vec{r}_0$.*

EXAMPLE 6: From Example 3, we know that $f(x, y) = x^2 + xy + y^2 + x + y$ has its only relative minimum at $\left(-\frac{1}{3}, -\frac{1}{3}\right)$. Now for any $\vec{r}$,

$$\mathcal{H}_1(f(\vec{r})) = 2 \quad \text{and} \quad \mathcal{H}_2(f((\vec{r})) = \det \begin{pmatrix} 2 & 1 \\ 1 & 2 \end{pmatrix} = 3.$$

Thus $f\left(-\frac{1}{3}, -\frac{1}{3}\right) = \frac{1}{9} + \frac{1}{9} + \frac{1}{9} - \frac{1}{3} - \frac{1}{3} = -\frac{1}{3}$ is an absolute minimum. ∎

Theorem 5 (Concavity Test for Functions Defined on $\mathbb{R}^3$)
Suppose that U is an open and connected subset of $\mathbb{R}^3$, the function $f : U \to \mathbb{R}$ has continuous second partial derivatives, and $\vec{r}_0$ is a critical point.

(a) *If $\mathcal{H}_1(f(\vec{r})) > 0$, $\mathcal{H}_2(f(\vec{r})) > 0$, and $\mathcal{H}_3(f(\vec{r})) > 0$ for all $\vec{r}$ in U, then f attains an absolute minimum at $\vec{r}_0$.*

(b) *If $\mathcal{H}_1(f(\vec{r})) < 0$, $\mathcal{H}_2(f(\vec{r})) > 0$, and $\mathcal{H}_3(f(\vec{r})) < 0$ for all $\vec{r}$ in U, then f attains an absolute maximum at $\vec{r}_0$.*

EXAMPLE 7: From Example 4, we know that $f(x, y, z) = x^2 + y^2 + z^2 + xz$ attains a relative minimum at the origin. Also, we know that

$$\mathcal{H}_1(f(\vec{r})) = 2,$$

$$\mathcal{H}_2(f(\vec{r})) = \det \begin{pmatrix} 2 & 0 \\ 0 & 2 \end{pmatrix} = 4,$$

and

$$\mathcal{H}_3(f(\vec{r})) = \det \begin{pmatrix} 2 & 0 & 1 \\ 0 & 2 & 0 \\ 1 & 0 & 2 \end{pmatrix} = 6$$

for all $\vec{r}$ in $\mathbb{R}^3$. Thus $f(0,0,0) = 0$ is an absolute minimum.   ∎

EXERCISES 12.1

In Exercises 1–11, determine the nature of the critical points of the given function.

1. $f(x,y) = x^2 + y^2$.

2. $f(x,y) = x^2 + 3y^2 - xy + x + y$.

3. $f(x,y) = x^3 + y^2 - 3x - 2y$.

4. $f(x,y) = x^3 + y^3 + x + y$.

5. $f(x,y) = x^2 - 3xy + y^2$.

6. $f(x,y) = x^3 + y^2 + xy$.

7. $f(x,y,z) = x^2 - y^2 - z^2$

8. $f(x,y,z) = x^2 + y^2 + z^2 + 2x$.

9. $f(x,y,z) = x^2 - y^2 + z^2 + 2xy + x$.

10. $f(x,y,z) = x^2 + y^2 + z^2 + 2xy + x$.

11. $f(x,y,z) = y^3 + x^2 + z^2 - 3y^2$.

12. Use the methods of this section to find the distance from the point $(1,1,1)$ to the plane $x + y + z = 1$. (Hint: If $f(x,y) \geq 0$ for all (x,y) in the domain of f, then $\sqrt{f(x,y)}$ and $f(x,y)$ attain their minima at the same points.)

13. Find the point $\vec{r}_0 = (x_0, y_0, z_0)$ on the graph of $\sqrt{x^2 + y^2} = z$ of minimal distance to $\vec{v} = (1,1,0)$. Show that the vector $v - r_0$ is normal to the plane tangent to the graph at $\vec{r}_0$.

14. (Linear Regression) The following data were gathered in the field.

i	0	1	2	3	4	5
x_i	0.0	0.5	1.0	1.5	2.0	2.5
y_i	2.0	3.0	3.5	4.1	5.2	5.9

We know that the x–coordinates are accurate, but that the y coordinates may be in error. Find the straight line that "best approximates" the data, that is, if $y = mx + b$ is the equation of our straight line, then minimize the sum of the squares of the y–errors $\sum (y_i - (mx_i - b))^2$ to find m and b. Graph the data points and the line.

12.2   Extrema on Closed and Bounded Domains

While we do not have any general theorems for finding extrema of functions defined on subsets of $\mathbb{R}^n$, the situation changes if the domain of f is a closed[2] and bounded subset of $\mathbb{R}^n$. We appeal to a

---

[2] A subset U of $\mathbb{R}^n$ is *closed* if for every $\vec{r}$ *not* in U there is a ball centered at $\vec{r}$ containing no point of U.

fundamental theorem, which we present without proof.

Theorem 1 *If U is a closed and bounded subset of $\mathbb{R}^n$ and $f : U \to \mathbb{R}$ is continuous, then f has an absolute maximum and an absolute minimum.*

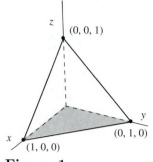

Figure 1.
$f(x, y) = 1 - x - y$, restricted to the triangle with vertices $(0, 0)$, $(1, 0)$, and $(0, 1)$

EXAMPLE 1:   The graph of $f(x, y) = 1 - x - y$ is the plane $x + y + z = 1$. It clearly does not have any relative maxima or minima, much less any absolute extrema. However, if we restrict the domain to the triangle with vertices $(0, 0)$, $(1, 0)$, and $(0, 1)$ as illustrated in Figure 1, then f attains its absolute maximum at the vertex $(0, 0)$, and it attains its absolute minimum at every point on the segment with endpoints $(1, 0)$ and $(0, 1)$.   ∎

To avoid complications, we restrict our attention to functions defined on subsets of the plane that are bounded by simple closed curves or finite sums of simple closed curves. For example, disks are bounded by circles and rectangles are bounded by four straight line intervals. Recall that when we were looking for absolute extrema for functions defined on intervals in $\mathbb{R}$, we found the extrema on the interior of the interval, then we evaluated the function on the endpoints of the interval (the boundary of the interval) and finally, we took the extrema of all of these values. The procedure for finding extrema of functions defined on closed and bounded subsets of $\mathbb{R}^n$ is exactly the same process.

Finding Absolute Extrema

Suppose that U is an open subset of $\mathbb{R}^2$ bounded by a closed curve C, and that $D = U \cup C$. Let $f : D \to \mathbb{R}$ have continuous second partial derivatives. To find the absolute extrema of f on D:

(a) Use Theorems 1 and 2 of Section 12.1 to find the local extrema for f on U;

(b) Use techniques from Chapter 5 to find the absolute extrema on the boundary C; and then

(c) Take the maximum or minimum of the values found in Steps (a) and (b).

EXAMPLE 2:   Find the absolute maximum and the absolute minimum of $f(x, y) = x + y$ on the unit disk $x^2 + y^2 \leq 1$.

SOLUTION: We take the open set U to be the interior of the disk

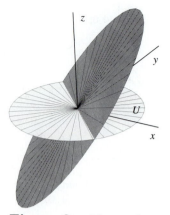

Figure 2.a *The graph of $f(x,y) = x + y$ restricted to $x^2 + y^2 \leq 1$.*

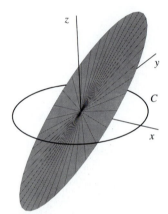

Figure 2.b *C is the boundary of the domain of f.*

$x^2 + y^2 \leq 1$. The boundary C is the circle $x^2 + y^2 = 1$. See Figures 2.a and 2.b. Since $\vec{\nabla} f(x,y) = (1,1)$, f has no critical points on U. Thus f has no extrema on U. Now we look at the boundary C.

The boundary C is parametrized by $\vec{r}(t) = (\cos t, \sin t)$, $0 \leq t \leq 2\pi$. To find any extrema of f on C, consider

$$g(t) = f(\vec{r}(t)) = \cos t + \sin t.$$

To find the extrema of f on C, we need only find the extrema of g, which is a function from $[0, 2\pi]$ into $\mathbb{R}$. This is a Chapter 5 type problem. We set $g'(t) = -\sin t + \cos t = 0$ to find the critical points and then use the second derivative to test concavity:

$$-\sin t + \cos t = 0 \implies \tan t = 1 \implies t = \frac{\pi}{4} \text{ and } t = \frac{5\pi}{4}.$$

$$g''(t) = -\cos t - \sin t \implies g''\left(\frac{\pi}{4}\right) = -\sqrt{2} \text{ and } g''\left(\frac{5\pi}{4}\right) = \sqrt{2}.$$

It follows that

- g is concave down at $\frac{\pi}{4}$, and $g\left(\frac{\pi}{4}\right) = \sqrt{2}$ is a local maximum; and

- g is concave up at $\frac{5\pi}{4}$, and $g\left(\frac{5\pi}{4}\right) = -\sqrt{2}$ is a local minimum.

Checking the endpoints 0 and 2π, $g(0) = g(2\pi) = 1$, we see that we have found the absolute maximum of g to be $g\left(\frac{\pi}{4}\right) = \sqrt{2}$ and the absolute minimum of g to be $g\left(\frac{5\pi}{4}\right) = -\sqrt{2}$.

It follows that $g\left(\frac{\pi}{4}\right) = f\left(\vec{r}\left(\frac{\pi}{4}\right)\right) = f\left(\frac{1}{\sqrt{2}}, \frac{1}{\sqrt{2}}\right) = \sqrt{2}$ is the absolute maximum of f, while $g\left(\frac{5\pi}{4}\right) = f\left(\vec{r}\left(\frac{5\pi}{4}\right)\right) = f\left(-\frac{1}{\sqrt{2}}, -\frac{1}{\sqrt{2}}\right) = -\sqrt{2}$ is the absolute minimum of f. ∎

EXAMPLE 3: Find the absolute extrema for $f(x,y) = x^2 + y^2$, $-1 \leq x \leq 1$, $-1 \leq y \leq 1$.

SOLUTION: We can determine the maximum and the minimum from the sketch of the graph of f in Figure 3.a. Clearly, $f(0,0) = 0$ is the absolute minimum on the domain of f, and f attains its maximum on the corners of the domain of f. Let us use the techniques of this section to verify our observations.

Let U be the interior of the domain, and let C denote the boundary. That is, (x,y) is in U if and only if $-1 < x < 1$, $-1 < y < 1$ as in Figure 3.c, and C is the union of the edges of the square as illustrated in Figure 3.d.

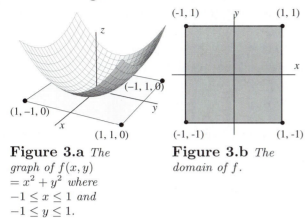

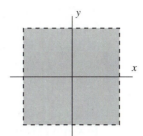

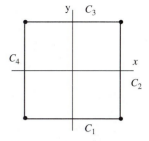

Figure 3.a *The graph of $f(x,y)$ $= x^2 + y^2$ where $-1 \le x \le 1$ and $-1 \le y \le 1$.*

Figure 3.b *The domain of f.*

Figure 3.c *The interior of the domain of f.*

Figure 3.d *The boundary of the domain of f.*

First we find the local extrema in U. The gradient $\vec{\nabla} f(x,y) = (2x, 2y)$. Setting this equal to $\vec{0}$, we have $(2x, 2y) = \vec{0}$, which implies that $(x,y) = (0,0)$, so the only critical point in U is the origin. The second derivative test shows that $f(0,0) = 0$ is a relative minimum, and in fact, f is concave up throughout U.

Now for the boundary. C is the union of the four edges $C_1 = [(-1,-1),(1,-1)]$, $C_2 = [(1,-1),(1,1)]$, $C_3 = [(1,1),(-1,1)]$, and $C_4 = [(-1,1),(-1,-1)]$ as illustrated in Figure 3.d. We parametrize C_i with h_i.

$$
\begin{aligned}
C_1\text{: } h_1(t) &= (t,-1) & -1 \le t \le 1, \\
C_2\text{: } h_2(t) &= (1,t) & -1 \le t \le 1, \\
C_3\text{: } h_3(t) &= (t,1) & -1 \le t \le 1\text{, and} \\
C_4\text{: } h_4(t) &= (-1,t) & -1 \le t \le 1.
\end{aligned}
$$

We need to apply the techniques of Chapter 5 to each part of the boundary. We first consider C_1, where

$$
f(h_1(t)) = t^2 + 1, \quad \frac{d}{dt} f(h_1(t)) = 2t \quad \text{and} \quad \frac{d^2}{dt^2} f(h_1(t)) = 2.
$$

Thus

$$
\frac{d}{dt} f(h_1(t)) = 0 \text{ at } t = 0 \quad \text{and} \quad \frac{d^2}{dt^2} f(h_1(t)) \text{ is always positive.}
$$

It follows that $f(h_1(t))$ attains its minimum at $t = 0$, and it attains its local maxima at its endpoints $t = -1$ and $t = 1$. So the absolute maximum for $f(h_1(t))$ is $f(h_1(-1)) = f(h_1(1)) = 2$. The maximum value for f on C_1 is 2, and this is attained at the corners $(-1,-1)$ and $(1,-1)$. The absolute minimum for f on C_1 is attained at $h_1(0) = (0,-1)$, where $f(0,-1) = 1$.

A similar analysis shows that the maximum value for f on each of C_2, C_3, and C_4 is 2, and the minimum value for f on C_2, C_3, and C_4 is 1.

Thus the absolute minimum value for f on its domain is attained at $(0,0)$, and $f(0,0) = 0$ is the absolute minimum. Similarly, the absolute maximum for f is 2, and that value is attained at each of the corners of the domain. ∎

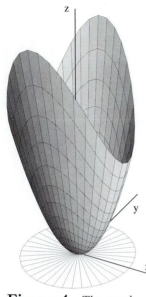

Figure 4. *The graph of $f(x,y) = 2x^2 + 4y^2$ restricted to the unit disk.*

EXAMPLE 4: Find the extrema of $f(x,y) = 2x^2 + 4y^2$ on the disk $x^2 + y^2 \leq 1$. See Figure 4.

SOLUTION: Let U be the interior of the disk $x^2 + y^2 \leq 1$, and let C be the circle $x^2 + y^2 = 1$. Just as in the previous example, the only critical point of f in the interior of the domain of f is the origin $(0,0)$, and f is always concave up on U. Thus $f(0,0) = 0$ is an absolute minimum of f.

The boundary of f is parametrized by $\vec{r}(t) = (\cos t, \sin t)$, $0 \leq t \leq 2\pi$. Thus $f(\vec{r}(t)) = 2\cos^2 t + 4\sin^2 t$, the derivative $\frac{d}{dt}(f(\vec{r}(t)) = 4\cos t \sin t$, and the second derivative $\frac{d^2}{dt^2}(f(\vec{r}(t)) = 4(\cos^2 t - \sin^2 t)$.

Checking for critical points, we see that $\frac{d}{dt}(f(\vec{r}(t)) = 4\cos t \sin t = 0$ at $t = \pi$, $\frac{\pi}{2}$ and $\frac{3\pi}{2}$. It is a temptation to include $t = 0$ and $t = 2\pi$, but these are endpoints of the domain of $\vec{r}$ and by our definitions, $\vec{r}'$ is not defined there! Admittedly this is just a technical difficulty, but we must check them as endpoints. Checking for concavity, we see that $\frac{d^2}{dt^2}(f(\vec{r}(t))) < 0$ at $t = \frac{\pi}{2}$ and $\frac{3\pi}{2}$. Thus $f \circ \vec{r}$ attains relative maxima at these points. Since $\frac{d^2}{dt^2}(f(\vec{r}(t))) > 0$ at $t = \pi$, we find that $f \circ \vec{r}$ attains a relative minimum there. Finally, checking $t = 0$, we see that $f(\vec{r}(0)) = f(\vec{r}(\pi)) = 2$. Comparing values, we see that $f\left(\vec{r}\left(\frac{\pi}{2}\right)\right) = f\left(\vec{r}\left(\frac{3\pi}{2}\right)\right) = 4$ is the absolute maximum of $f(x,y) = 2x^2 + 4y^2$ on the disk $x^2 + y^2 \leq 1$. ∎

EXAMPLE 5: A rectangular box must have a volume of 1 ft^3. The material for the lateral sides costs \$0.50/ft^2, the material for the bottom costs \$1.00/ft^2, and the material for the top costs \$0.25/ft^2. Find the dimensions of the box that minimizes the cost of the box.

SOLUTION: Let x and y denote the dimensions of the bottom, and let z denote the height of the box. Then

$$c(x,y,z) = 0.5(2xz + 2yz) + xy + 0.25xy = xz + yz + 1.25xy$$

is the cost function that we want to minimize. Since $xyz = 1$, we may express z in terms of x and y by $z = \frac{1}{xy}$. Substituting this expression for z into the cost function, we get

$$c(x,y) = \frac{1}{y} + \frac{1}{x} + 1.25xy.$$

This is the function we wish to minimize. The domain of c is where $x > 0$ and $y > 0$. However, if x or y is very small, then $\frac{1}{x} + \frac{1}{y}$ is large. Similarly, if x and y are large, then $c(x,y)$ is large. Let $a = \frac{1}{(\text{national debt})}$ and let $b = (\text{national debt})$. Let D be the rectangle $a \le x \le b$, $a \le y \le b$. Then we know that c restricted to D must attain its absolute minimum in the interior of D. This minimum must occur at a critical point! We set $\vec{\nabla} f(x,y) = \vec{0}$.

$$\vec{\nabla} f(x,y) = \left(-\frac{1}{x^2} + 1.25y, -\frac{1}{y^2} + 1.25x, \right) = 0.$$

This yields the two equations

$$-\frac{1}{x^2} + 1.25y = 0 \quad \text{and} \quad -\frac{1}{y^2} + 1.25x = 0.$$

Solving these equations simultaneously, we obtain $x = y = (1.25)^{-1/3}$. The only critical point for c is at $((1.25)^{-1/3}, (1.25)^{-1/3})$. The second derivative test shows that this is a local minimum. This is the absolute minimum on the domain D. Since $xyz = 1$, $z = (1.25)^{2/3}$ ft.   ∎

EXERCISES 12.2

In Exercises 1–3, find the absolute maximum and minimum of $f(x,y) = -x + \sqrt{3}y$ on the set S.

1. S is the disk $x^2 + y^2 \le 1$.

2. S is the disk $x^2 + 2x + y^2 - 2y \le 1$.

3. S is the rectangle $0 \le x \le 1$, $0 \le y \le 1$.

In Exercises 4–6, find the absolute maximum and minimum of $f(x,y) = x^2 + 2y^2 - x$ on the set S.

4. S is the disk $x^2 + y^2 \le 1$.

5. S is the set $x^2 + 2y^2 \le 1$. (Recall that an ellipse $\frac{x^2}{a^2} + \frac{y^2}{b^2} = 1$ can be parametrized by $\vec{r}(t) = (a\cos t, b\sin t)$.)

6. S is the rectangle $0 \le x \le 1$, $0 \le y \le 1$.

In Exercises 7–11, find the absolute maximum and minimum of $f(x,y) = x^2 - 2y^2$ on the set S.

7. S is the disk $x^2 + y^2 \le 1$.

8. S is the set $x^2 + 2y^2 \le 1$.

9. S is the rectangle $0 \le x \le 1$, $0 \le y \le 1$.

10. S is the rectangle $-1 \le x \le 1$, $-1 \le y \le 1$.

12.3   Lagrange Multipliers

Yet again, Luke Skywalker is in a hot spot. Fortunately, he knows the temperature as a function of position, and he wants to move to the coolest point possible. This time, he must conserve his energy. Thus he must stay on an equipotential surface so that he is doing no work against gravity. The temperature function is denoted by $T(x, y, z)$, and the equipotential surface S is defined by $\phi(x, y, z) = C$, where ϕ is a differentiable function from $\mathbb{R}^3$ to $\mathbb{R}$. He wants to find the position (x_0, y_0, z_0) on the surface S that yields the minimum value of T on the surface. That is, he wants to find the minimum of T restricted to S. Assuming that $\vec{\nabla}\phi(x_0, y_0, z_0) \neq \vec{0}$, the following theorem provides a relationship that ϕ and T must satisfy at (x_0, y_0, z_0) if T, restricted to S, attains a local extrema at (x_0, y_0, z_0).

Theorem 1 *Let ϕ and f be real valued functions defined on subsets of $\mathbb{R}^n$ with continuous partial derivatives. Let S be the equipotential or level surface defined by $\phi(\vec{r}) = c$, and assume that S is a subset of the domain of f. If the restriction of f to S has a local extremum at $\vec{r}_0$, and if $\vec{\nabla}\phi(\vec{r}_0) \neq \vec{0}$, then there is a number λ such that $\vec{\nabla}f(\vec{r}_0) = \lambda\vec{\nabla}\phi(\vec{r}_0)$. The position $\vec{r}_0$ is a* critical point *for f on the surface S.*

Before outlining the argument for Theorem 1, let us see how it is used.

EXAMPLE 1: Find the extrema of $f(x, y) = x^2 + xy + y^2$, where (x, y) is restricted to the line $x + 2y = 1$.

SOLUTION: Of course, we could use the equation of the line to obtain $x = 1 - 2y$, substitute this into f, and then use the techniques of Chapter 5. However, let us use Theorem 1. Let $\phi(x, y) = x + 2y$, and let S be the line defined by $\phi(x, y) = 1$.

$$\vec{\nabla}f(x, y) = (2x + y, 2y + x), \quad \text{and} \quad \vec{\nabla}\phi(x, y) = (1, 2).$$

Theorem 1 tells us that we want to find a number λ and a point $\vec{r}_0 = (x_0, y_0)$ so that

$$\vec{\nabla}f(x_0, y_0) = \lambda\vec{\nabla}\phi(x_0, y_0).$$

We now have three equations and three unknowns:

$$2x + y = \lambda, \tag{1}$$

$$2y + x = 2\lambda, \text{ and} \tag{2}$$

$$x + 2y = 1. \tag{3}$$

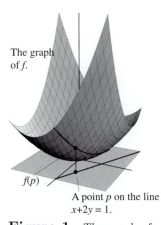

The graph of f.

$f(p)$

A point p on the line $x + 2y = 1$.

Figure 1. *The graph of f and the graph of f restricted to the line $x + 2y = 1$.*

Multiplying Equation (1) by 2 and subtracting from Equation (2), we have $3x = 0$, or $x = 0$. Substitution into Equation (1) yields $y = \lambda$. Using this information in Equation (3), we have that $\lambda = \frac{1}{2}$ and $y = \frac{1}{2}$. Thus we know that $\left(0, \frac{1}{2}\right)$ is a critical point for f restricted to S.

Since $\mathcal{H}_1\left(f\left(0, \frac{1}{2}\right)\right) > 0$ and $\mathcal{H}_2\left(f\left(0, \frac{1}{2}\right)\right) > 0$, we know that f is concave up at $\left(0, \frac{1}{2}\right)$ and hence, the restriction of f to S is also concave up. Thus $f\left(0, \frac{1}{2}\right) = \frac{1}{4}$ is an absolute minimum for f on S.

This is a good example to see why we would expect that $\vec{\nabla}(x^2 + xy + y^2)$ and $\vec{\nabla}(x + 2y)$ to have the same or opposite direction at points where extrema are achieved. The function $f(x, y)$, restricted to the line, attains its minimum of $\frac{1}{4}$. An inspection of Figure 2 indicates that the level curve $f(x, y) = \frac{1}{4}$ is tangent to the line $x + 2y = 1$ at the critical point $\left(0, \frac{1}{2}\right)$. Now, $\vec{\nabla}f\left(0, \frac{1}{2}\right)$ is normal to the level curve $f(x, y) = \frac{1}{4}$ at $\left(0, \frac{1}{2}\right)$, and therefore, it is normal to the tangent line $x + 2y = 1$. At the same time, the line $x + 2y = 1$ is a level curve of the function $g(x, y) = x + 2y$. Hence, $\vec{\nabla}g\left(0, \frac{1}{2}\right)$ is normal to the line. ■

Figure 2. *The level curve* $f(x, y) = \frac{1}{4}$ *is tangent to the line* $x + 2y = 1$ *at the critical point.*

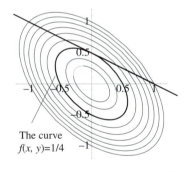

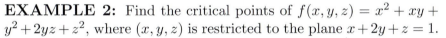

The curve
f(x, y)=1/4

EXAMPLE 2: Find the critical points of $f(x, y, z) = x^2 + xy + y^2 + 2yz + z^2$, where (x, y, z) is restricted to the plane $x + 2y + z = 1$.

SOLUTION: Because there are three unknowns and only two equations, there is no obvious path to reduce this problem to a single variable problem. We can, of course, use the equation of the plane to obtain $z = 1 - x - 2y$, and then substitute this into f and use the techniques of Section 12.1 to analyze the resulting function of two variables. However, let us use Theorem 1.

Let $\phi(x, y, z) = x + 2y + z$, and let S denote the plane defined by $\phi(x, y, z) = 1$.

$\vec{\nabla}f(x, y, z) = (2x + y, 2y + x + 2z, 2y + 2z)$, and $\vec{\nabla}\phi(x, y, z) = (1, 2, 1)$.

Theorem 1 tells us that we want to find a number λ and a point

$\vec{r}_0 = (x_0, y_0, z_0)$ so that

$$\vec{\nabla} f(x_0, y_0, z_0) = \lambda \vec{\nabla} \phi(x_0, y_0, z_0).$$

This gives us four equations and four unknowns:

$$
\begin{aligned}
2x + y &= \lambda, \\
2y + x + 2z &= 2\lambda, \\
2y + 2z &= \lambda, \text{ and} \\
x + 2y + z &= 1.
\end{aligned}
$$

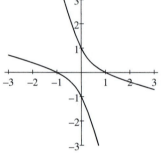

Figure 3. *The graph of* $x^2 + 4xy + y^2 = 1$.

Solving these equations simultaneously, we find that $(2, -2, 3)$ is the only critical point of T restricted to S. Since $\mathcal{H}_1(f(2, -2, 3)) > 0$, $\mathcal{H}_2(f(2, -2, 3)) > 0$, and $\mathcal{H}_3(f(2, -2, 3)) < 0$, $(2, -2, 3)$ is a saddle point for f in $\mathcal{R}^3$. This information does not tell us the nature of the critical point $(2, -2, 3)$. Why? ∎

EXAMPLE 3: Let S be the surface defined by $x^2 + 4xy + y^2 = 1$. Find the point(s) on S nearest the origin.

SOLUTION: The distance from a point to the origin is given by $d(x, y) = \sqrt{x^2 + y^2}$. We want to minimize the function d restricted to S. We can simplify things considerably by minimizing $f(x, y) = d^2(x, y) = x^2 + y^2$, subject to (x, y) satisfying $x^2 + 4xy + y^2 = 1$. Let $\phi(x, y) = x^2 + 4xy + y^2$, and set $\vec{\nabla} f(x, y) = \lambda \vec{\nabla} \phi(x, y)$. This gives us three equations:

$$
\begin{aligned}
x &= \lambda(x + 2y), && (4) \\
y &= \lambda(2x + y), \text{ and} && (5) \\
x^2 + 4xy + y^2 &= 1. && (6)
\end{aligned}
$$

Using Equations (4) and (5) to eliminate λ, we see that $x = \pm y$. However, $x = -y$ is not a solution for Equation (6). Substituting $x = y$ into Equation (6), we find the critical points $\left(\frac{1}{\sqrt{6}}, \frac{1}{\sqrt{6}}\right)$ and $\left(-\frac{1}{\sqrt{6}}, -\frac{1}{\sqrt{6}}\right)$. Both $\mathcal{H}_1(f)$ and $\mathcal{H}_2(f)$ are positive. Thus f is concave up everywhere, so ϕ restricted to S is concave up. It follows that

$$f\left(\frac{1}{\sqrt{6}}, \frac{1}{\sqrt{6}}\right) = f\left(-\frac{1}{\sqrt{6}}, -\frac{1}{\sqrt{6}}\right) = \frac{1}{3}$$

minimizes f restricted to S, and $\left(\frac{1}{\sqrt{6}}, \frac{1}{\sqrt{6}}\right)$ and $\left(-\frac{1}{\sqrt{6}}, -\frac{1}{\sqrt{6}}\right)$ are the points on S closest to the origin. This is consistent with the sketch of S in Figure 2. ∎

EXAMPLE 4: A rectangular box with sides parallel to the coordinate axes is inscribed inside the ellipsoid $x^2 + 4y^4 + 9z^2 = 1$. Find the dimensions of the box that produce the maximum volume.

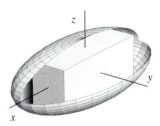

Figure 4. *A rectangular box (sides parallel to the coordinate axes) inscribed inside the ellipsoid.*

SOLUTION: Let (x, y, z) denote the vertex of the box lying in the first octant, as in Figure 4. The dimensions of the box are $2x$ by $2y$ by $2z$, so the volume of the box is given by $V(x, y, z) = 8xyz$. Furthermore, we know that (x, y, z) must be on the ellipsoid. We want to maximize V subject to $x^2 + 4y^4 + 9z^2 = 1$. The equations

$$x^2 + 4y^4 + 9z^2 = 1 \quad \text{and} \quad \vec{\nabla} 8xyz = \lambda \vec{\nabla}(x^2 + 4y^2 + 9z^2)$$

give us four equations with four unknowns.

$$
\begin{align}
8yz &= 2\lambda x, & (7) \\
8xz &= 8\lambda y, & (8) \\
8xy &= 18\lambda z, \text{ and} & (9) \\
x^2 + 4y^4 + 9z^2 &= 1. & (10)
\end{align}
$$

Eliminating λ from Equations (7) and (8) yields $4y^2 = x^2$. Eliminating λ from Equations (7) and (9) yields $9z^2 = x^2$. Substituting into Equation (10), we obtain $x = \pm\frac{1}{\sqrt{3}}$. Since we restricted (x, y, z) to the first octant, $x = \frac{1}{\sqrt{3}}$ is the only possible solution. Since V attains its maximum on the ellipsoid with (x, y, z) in the first octant, V must attain its maximum at $(x, y, z) = \left(\frac{1}{\sqrt{3}}, \frac{1}{2\sqrt{3}}, \frac{1}{3\sqrt{3}}\right)$. The dimensions of the box are $\frac{2}{\sqrt{3}}$ by $\frac{1}{\sqrt{3}}$ by $\frac{2}{3\sqrt{3}}$.

Argument for Theorem 1

Suppose that S is the surface defined by $\phi(x, y, z) = c$, and f is a continuously differentiable function from a subset of $\mathbb{R}^3$ into $\mathbb{R}$. Suppose further that f restricted to S attains a relative extremum at (x_0, y_0, z_0). Let $\vec{r}(t)$ parametrize any path lying in S that passes through (x_0, y_0, z_0). Let t_0 denote the time such that $\vec{r}(t_0) = (x_0, y_0, z_0)$. Then $f(\vec{r}(t))$ is a function from a subset of $\mathbb{R}$ into $\mathbb{R}$ that attains a relative extremum at $t = t_0$. It follows that

$$\frac{d}{dt} f(\vec{r}(t))\big|_{t=t_0} = 0.$$

Using the chain rule, we have

$$\frac{d}{dt}f(\vec{r}(t))|_{t=t_0} = \vec{\nabla}f|_{\vec{r}(t_0)} \cdot \vec{r}'(t)|_{t=t_0}.$$

Since

$$\vec{\nabla}f|_{\vec{r}(t_0)} \cdot \vec{r}'(t)|_{t=t_0} = 0,$$

we see that $\vec{\nabla}f|_{\vec{r}(t_0)}$ is orthogonal (perpendicular) to $\vec{r}'(t)|_{t=t_0}$. Thus we can conclude that if C is any curve lying in S and passing through (x_0, y_0, z_0), then $\vec{\nabla}f(x_0, y_0, z_0)$ is orthogonal to the curve C at that point. In Section 11.10, we argued that $\vec{\nabla}\phi(x_0, y_0, z_0)$ is orthogonal to S at (x_0, y_0, z_0). While we do not have the tools to provide a complete proof here, it turns out that this is enough to show that $\vec{\nabla}\phi(x_0, y_0, z_0)$ and $\vec{\nabla}f(x_0, y_0, z_0)$ point in the same or opposite directions. That is, we can find a number λ so that $\vec{\nabla}f(x_0, y_0, z_0) = \lambda\vec{\nabla}\phi(x_0, y_0, z_0)$.

EXERCISES 12.3

1. Find the extrema of $f(x,y) = xy$ on the unit circle $x^2 + y^2 = 1$.

2. Find the extrema of $f(x,y,z) = x - 2y + 3z$ on the sphere $x^2 + y^2 + z^2 = \frac{7}{2}$.

3. Find the points on the curve $xy = 1$ that are closest to the origin.

4. The temperature in the plane is given by $T(x,y) = x^2 - 4xy + 4y^2$. Find the positions on the unit circle that give the highest and lowest temperatures.

5. Find the extrema of $f(x,y,z) = 2x - 4z$ on the ellipsoid $\frac{x^2}{2} + \frac{y^2}{3} + z^2 = 1$.

6. Let S be the surface defined by $x^2 + 2xy - y^2 = 1$. Find the points on S nearest the origin.

7. A rectangular box with sides parallel to the coordinates axes is inscribed inside the region bounded by $z = x^2 + y^2 - 1$ and $z = 1 - x^2 - y^2$. Find the dimensions that maximize volume.

8. Find the point on the parabola $x = y^2$ that is closest to the line $y = \frac{x}{2} + 1$. (Hint: Use the square of the formula for the distance from a point to a line.)

Chapter 13

Change of Coordinate Systems

In Section 11.6 we viewed a function from $\mathbb{R}^n$ into $\mathbb{R}^n$ as a vector field. We can also use functions from $\mathbb{R}^n$ into $\mathbb{R}^n$ to transform one coordinate system into another. This is not unlike changing a lens on a camera in order to get a clearer picture. We can sometimes get a better description of a set in $\mathbb{R}^n$ by changing the way we describe the points in $\mathbb{R}^n$. We can also use functions from subsets of $\mathbb{R}^n$ into $\mathbb{R}^m$ to describe interesting geometric objects in $\mathbb{R}^m$, much in the same way as we used functions from intervals into $\mathbb{R}^m$ to describe simple curves. In this chapter we introduce some special functions from $\mathbb{R}^n$ into $\mathbb{R}^m$, and explore some of their properties.

13.1   Translations and Linear Transformations

A *translation* is a function from $\mathbb{R}^n$ onto $\mathbb{R}^n$ defined by adding a constant to every point in $\mathbb{R}^n$. Translations are used to shift objects in $\mathbb{R}^n$ in a rigid fashion without any rotations. The formal definition is

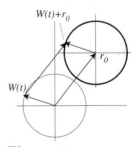

Figure 1. *The unit circle translated by $\vec{r}_0$.*

Definition: Translations

The statement that the function $\vec{T}$ is a translation by $\vec{r}_0$ means that $\vec{r}_0$ is a vector and $\vec{T}(\vec{r}) = \vec{r} + \vec{r}_0$.

EXAMPLE 1: The image of the wrapping function $\vec{W}(t)$ is the unit circle centered at the origin. The composition of $\vec{W}$ followed by a translation by $\vec{r}_0$ yields the function $\vec{f}(t) = \vec{W}(t) + r_0$, the image of which is the unit circle centered at $\vec{r}_0$. See Figure 1. ∎

571

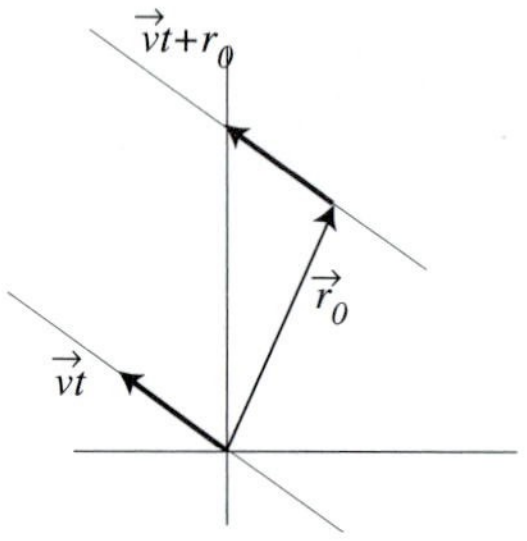

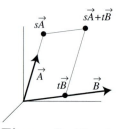

Figure 2. *The line $t\vec{v}$ translated by $\vec{r}_0$.*

Figure 3. *The plane determined by vectors $\vec{A}$ and $\vec{B}$.*

EXAMPLE 2: The image of the function $\vec{r}(t) = t\vec{v}$ is the line with direction $\vec{v}$ that contains the origin. If we follow $\vec{r}(t)$ with a translation by $\vec{r}_0$, we have $\vec{h}(t) = \vec{r}(t) + \vec{r}_0 = \vec{v}t + \vec{r}_0$, the image of which is the line with direction $\vec{v}$ that contains the point $\vec{r}_0$. See Figure 2. ∎

EXAMPLE 3: Consider two vectors $\vec{A}$ and $\vec{B}$ drawn emanating from the origin that do not have the same direction. That is, $\vec{A} \times \vec{B} \neq \vec{0}$. Then these vectors determine a plane containing the origin. In fact, if $\vec{r}$ is any point on that plane, then we can find a pair of numbers (s, t) so that $\vec{r} = s\vec{A} + t\vec{B}$, as illustrated in Figure 3. This defines a function $\vec{h}(s, t) = s\vec{A} + t\vec{B}$ from the st−plane onto the plane determined by the vectors $\vec{A}$ and $\vec{B}$. ∎

This is one example of a type of function called a linear function or linear transformation that is used extensively in mathematical modeling.

Definition: Linear Functions

A function $\vec{T}$ from $\mathbb{R}^n$ into $\mathbb{R}^m$ is said to be a *linear function* or a *linear transformation* if for each u in $\mathbb{R}$ and each $\vec{r}_1$ and $\vec{r}_2$ in $\mathbb{R}^n$,

- $\vec{T}(\vec{r}_1) + \vec{T}(\vec{r}_2) = \vec{T}(\vec{r}_1 + \vec{r}_2)$ (Superposition Property)

- $\vec{T}(u\vec{r}_1)) = u\vec{T}(\vec{r}_1)$.   (Scalability Property)

EXAMPLE 4: Show that the function $f(x) = 3x$ from $\mathbb{R}$ to $\mathbb{R}$ is linear.

SOLUTION: First, we check for the superposition property.

$$f(x_1 + x_2) = 3(x_1 + x_2) = 3x_1 + 3x_2 = f(x_1) + f(x_2).$$

Next, for scalability.

$$f(ux) = 3ux = u(3x) = uf(x).$$ ∎

EXAMPLE 5: Even though the graph of the function $f(x) = 3x + 1$ is a line, f is not linear. To show this, let $x_1 = 1$ and $x_2 = 1$. Then $f(1) + f(1) = (3 + 1) + (3 + 1) = 8$. However, $f(2) = 7$. Thus $f(1+1) \neq f(1) + f(1)$, and f fails to have the superposition property. ∎

We could easily dedicate the rest of this book to the study of linear transformations and still not exhaust the material that is important to the engineer or scientist. However, in this text we only introduce techniques for constructing linear transformations and concentrate on some special properties of these functions. The following theorem is our working characterization of linear transformations.

Theorem 1 *The function $\vec{T} : \mathbb{R}^n \to \mathbb{R}^m$ is a linear transformation if and only if there are vectors $\vec{A}_1, \vec{A}_2, \ldots, \vec{A}_n$ in $\mathbb{R}^m$ such that*
$$\vec{T}(x_1, x_2, \ldots, x_n) = x_1\vec{A}_1 + x_2\vec{A}_2 + \ldots + x_n\vec{A}_n.$$

Proof: We argue only the case that $\mathbb{R}^n = \mathbb{R}^m = \mathbb{R}^3$. The general case is similar. First, suppose that $\vec{T}$ is a linear transformation from $\mathbb{R}^3$ into $\mathbb{R}^3$. Let $\vec{A}_1 = \vec{T}(1, 0, 0)$, $\vec{A}_2 = \vec{T}(0, 1, 0)$, and $\vec{A}_3 = \vec{T}(0, 0, 1)$. Then from the definition,

$$
\begin{aligned}
\vec{T}(x_1, x_2, x_3) &= \vec{T}(x_1, 0, 0) + \vec{T}(0, x_2, x_3) \\
&= \vec{T}(x_1, 0, 0) + (\vec{T}(0, x_2, 0) + \vec{T}(0, 0, x_3)) \\
&= x_1\vec{T}(1, 0, 0) + x_2\vec{T}(0, 1, 0) + x_3\vec{T}(0, 0, 1) \\
&= x_1\vec{A}_1 + x_2\vec{A}_2 + x_3\vec{A}_3.
\end{aligned}
$$

Suppose that $\vec{A} = (a_x, a_y, a_z)$, $\vec{B} = (b_x, b_y, b_z)$, and $\vec{C} = (c_x, c_y, c_z)$ are vectors and $\vec{T}$ is defined by $\vec{T}(x, y, z) = x\vec{A} + y\vec{B} + z\vec{C}$. Let u and v be numbers, and let $\vec{r}_1 = (x_1, y_1, z_1)$ and $\vec{r}_2 = (x_2, y_2, z_2)$ be vectors.

$$
\begin{aligned}
\vec{T}(u\vec{r}_1 + v\vec{r}_2) &= \vec{T}(u(x_1, y_1, z_1) + v(x_2, y_2, z_2)) \\
&= \vec{T}(ux_1 + vx_2, uy_1 + vy_2, uz_1 + vz_2) \\
&= (ux_1 + vx_2)\vec{A} + (uy_1 + vy_2)\vec{B} + (uz_1 + vz_2)\vec{C} \\
&= (ux_1\vec{A} + uy_1\vec{B} + uz_1\vec{C}) + (vx_2\vec{A} + vy_2\vec{B} + vz_2\vec{C}) \\
&= u(x_1\vec{A} + y_1\vec{B} + z_1\vec{C}) + v(x_2\vec{A} + y_2\vec{B} + z_2\vec{C}) \\
&= u\vec{T}(\vec{r}_1) + v\vec{T}(\vec{r}_2).
\end{aligned}
$$

This shows that $\vec{T}$ is linear. ∎

EXAMPLE 6: Let $\vec{T}$ be defined by $\vec{T}(u, v) = u(1, 2) + v(0, 1) = (u, 2u + v)$. In this example, $\vec{A} = (1, 2)$ and $\vec{B} = (0, 1)$. It helps to visualize the function by realizing that $\vec{T}$ is that linear transformation that takes $(1, 0)$ onto $(1, 2)$ and $(0, 1)$ onto $(0, 1)$, and it takes the unit square with adjacent edges the vectors $\hat{i}$ and $\hat{j}$ onto the parallelogram with adjacent edges $\vec{A}$ and $\vec{B}$. See Figure 4. ∎

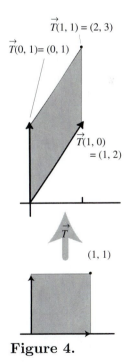

Figure 4.

$$
\begin{aligned}
\vec{T}(u, v) &= u(1, 2) + v(0, 1) \\
&= (u, 2u + v).
\end{aligned}
$$

EXAMPLE 7:

(a) Let $\vec{T}$ be defined by $\vec{T}(u,v) = u(1,0) + v(0,-1) = (u,-v)$. We could equivalently define $\vec{T}$ with the pair of equations $x = u$ and $y = -v$. The linear transformation $\vec{T}$ reflects the plane across the x–axis. See Figure 5.a.

(b) Let $\vec{T}$ be defined by $T(u,v) = u(0,1) + v(1,0) = (v,u)$. The linear transformation $\vec{T}$ reflects the plane across the line $y = x$. See Figure 5.b. ■

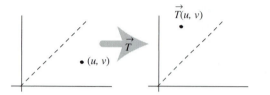

Figure 5.a *The transformation $\vec{T}(u,v) = (u,-v)$.*

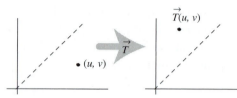

Figure 5.b *The transformation $\vec{T}(u,v) = (v,u)$.*

EXAMPLE 8:

(a) $\vec{T}(u,v,w) = u(1,1,1) + v(1,0,0) + w(1,0,1) = (u+v+w, u, u+w)$ defines a linear transformation from $\mathbb{R}^3$ into $\mathbb{R}^3$.

(b) $\vec{T}(u,v,w) = u(1,2) + v(0,1) + w(1,-1) = (u+w, 2u+v-w)$ defines a linear transformation from $\mathbb{R}^3$ into $\mathbb{R}^2$.

(c) $\vec{T}(u,v) = u(1,0,-1) + v(1,2,1) = (u+v, 2v, -u+v)$ defines a linear transformation from $\mathbb{R}^2$ into $\mathbb{R}^3$. ■

Important Facts about Linear Transformations

(a) If $\vec{T}$ is a linear transformation from $\mathbb{R}$ into $\mathbb{R}^n$ and $\vec{T}(1) = \vec{A}$, then $\vec{T}(u) = u\vec{A}$.

(b) If $\vec{T}$ is a linear transformation from $\mathbb{R}^2$ into $\mathbb{R}^n$ and $\vec{T}(1,0) = \vec{A}$ and $\vec{T}(0,1) = \vec{B}$, then $\vec{T}(u,v) = u\vec{A} + v\vec{B}$.

(c) If $\vec{T}$ is a linear transformation from $\mathbb{R}^3$ into $\mathbb{R}^n$ and $\vec{T}(1,0,0) = \vec{A}, \vec{T}(0,1,0) = \vec{B}$ and $\vec{T}(0,0,1) = \vec{C}$, then $\vec{T}(u,v,w) = u\vec{A} + v\vec{B} + w\vec{C}$.

EXAMPLE 9:

(a) Let $\vec{T}$ be defined by

$$\vec{T}(u,v,w) \;=\; \begin{pmatrix} 3u + 5v + 2w \\ u - 2v - w \\ u - v + w \end{pmatrix}$$

$$= u \begin{pmatrix} 3 \\ 1 \\ 1 \end{pmatrix} + v \begin{pmatrix} 5 \\ -2 \\ -1 \end{pmatrix} + w \begin{pmatrix} 2 \\ -1 \\ 1 \end{pmatrix}.$$

The domain of $\vec{T}$ is $\mathbb{R}^3$ and the range is $\mathbb{R}^3$. Let $\vec{A} = \vec{T}(1,0,0) = (3,1,1)$, $\vec{B} = \vec{T}(0,1,0) = (5,-2,-1)$, and $\vec{C} = \vec{T}(0,0,1) = (2,-1,1)$. Then

$$\vec{T}(u,v,w) = u\vec{A} + v\vec{B} + w\vec{C}.$$

(b) Let $\vec{T}$ be defined by

$$\vec{T}(u,v) = \begin{pmatrix} -2u + v \\ 3u \\ 5u - v \end{pmatrix} = u \begin{pmatrix} -2 \\ 3 \\ 5 \end{pmatrix} + v \begin{pmatrix} 1 \\ 0 \\ -1 \end{pmatrix}.$$

The domain of $\vec{T}$ is $\mathbb{R}^2$ and the range is $\mathbb{R}^3$. Let $\vec{A} = \vec{T}(1,0) = (-2,3,5)$ and $\vec{B} = \vec{T}(0,1) = (1,0,-1)$. Then

$$\vec{T}(u,v) = u\vec{A} + v\vec{B}.$$

(c) Let $\vec{T}$ be defined by

$$\vec{T}(u,v,w) = (3u + 2v - w).$$

The domain of $\vec{T}$ is $\mathbb{R}^3$ and the range is $\mathbb{R}$. Let $\vec{A} = \vec{T}(1,0,0) = (3)$, $\vec{B} = \vec{T}(0,1,0) = (2)$ and $\vec{C} = \vec{T}(0,0,1) = (-1)$ (notice that we are thinking of members of $\mathbb{R}$ as vectors). Then

$$\vec{T}(u,v,w) = u\vec{A} + v\vec{B} + w\vec{C}. \qquad \blacksquare$$

Example 9 points to a compact way to describe any given linear transformation. A *matrix* is a rectangular array of numbers. A $n \times m$–matrix is a matrix with n rows and m columns.

EXAMPLE 10:

(a) $A = \begin{pmatrix} 3 & 2 & -1 \\ 2 & 0 & -6 \end{pmatrix}$ is a 2×3–matrix.

(b) $A = \begin{pmatrix} 0 & -1 \\ 2 & 0 \end{pmatrix}$ is a 2×2–matrix.

(c) $A = \begin{pmatrix} 3 & 2 & -1 \end{pmatrix}$ is a 1×3–matrix.

(d) $A = \begin{pmatrix} 3 \\ 2 \\ -1 \end{pmatrix}$ is a 3×1–matrix. $\qquad \blacksquare$

A $1 \times n$–matrix as in Example 10.c is a *row matrix* while a $n \times 1$–matrix is a *column matrix*. If $\vec{A}$ is a vector, then we write $[A]$ to denote the column matrix corresponding to $\vec{A}$. Thus, if $\vec{A} = (-1, 2, 3)$, then $[A] = \begin{pmatrix} -1 \\ 2 \\ 3 \end{pmatrix}$.

If $\vec{T}$ is a linear transformation from $\mathbb{R}^2$ into $\mathbb{R}^n$, and $\vec{T}(1,0) = \vec{A}$ and $\vec{T}(0,1) = \vec{B}$, then $\vec{T}(u,v) = u\vec{A} + v\vec{B}$. We let $A_{\vec{T}}$ be the matrix with n rows and 2 columns, the first column of which is $[A]$ and the second column of which is $[B]$.

$$A_{\vec{T}} = \left(\begin{array}{cc} [A] & [B] \end{array} \right).$$

In Example 9(b), $\vec{T}$ was defined by $\vec{T}(u,v) = u\vec{A} + v\vec{B}$, where $\vec{A} = \vec{T}(1,0) = (-2, 3, 5)$ and $\vec{B} = \vec{T}(0,1) = (1, 0, -1)$. Then $A_{\vec{T}}$ is the matrix

$$A_{\vec{T}} = \left(\begin{array}{cc} [A] & [B] \end{array} \right) = \begin{pmatrix} -2 & 1 \\ 3 & 0 \\ 5 & -1 \end{pmatrix}.$$

Notice that the matrix $A_{\vec{T}}$ contains all of the information necessary to work with the linear transformation $\vec{T}$.

We define multiplication of $A_{\vec{T}}$ with the column vector $\begin{pmatrix} u \\ v \end{pmatrix}$ by

$$A_{\vec{T}} \begin{pmatrix} u \\ v \end{pmatrix} = \vec{T}(u,v) = u\vec{A} + v\vec{B}.$$

In Example 9(a), $\vec{T}$ was defined by $\vec{T}(u,v,w) = u\vec{A} + v\vec{B} + w\vec{C}$, where $\vec{A} = \vec{T}(1,0,0) = (3,1,1)$, $\vec{B} = \vec{T}(0,1,0) = (5,-2,-1)$, and $\vec{C} = \vec{T}(0,0,1) = (2,-1,1)$. The matrix $A_{\vec{T}}$ is

$$A_{\vec{T}} = \left(\begin{array}{ccc} [A] & [B] & [C] \end{array} \right) = \begin{pmatrix} 3 & 5 & 2 \\ 1 & -2 & -1 \\ 1 & -1 & 1 \end{pmatrix}.$$

In this case, we define multiplication of $A_{\vec{T}}$ with the column vector $\begin{pmatrix} u \\ v \\ w \end{pmatrix}$ by

$$A_{\vec{T}} \begin{pmatrix} u \\ v \\ w \end{pmatrix} = \vec{T}(u,v,w) = u\vec{A} + v\vec{B} + w\vec{C}.$$

EXAMPLE 11: Let $\vec{T}$ be the linear transformation with matrix

$$A_{\vec{T}} = \begin{pmatrix} 2 & 0 \\ 3 & -1 \\ 0 & 3 \end{pmatrix}.$$

Then $\vec{T}(u,v) = A_{\vec{T}} \begin{pmatrix} u \\ v \end{pmatrix} = u(2,3,0) + v(0,-1,3).$ ∎

Suppose that $\vec{T}(u,v,w) = u\vec{A} + v\vec{B} + w\vec{C}$ is a linear transformation from $\mathbb{R}^3$ into $\mathbb{R}^n$, where $\vec{T}(1,0,0) = \vec{A}$, $\vec{T}(0,1,0) = \vec{B}$ and $\vec{T}(0,0,1) = \vec{C}$. As in the case of functions from $\mathbb{R}^2$ into $\mathbb{R}^n$, we can completely describe a linear transformation from $\mathbb{R}^3$ into $\mathbb{R}^n$ with a matrix $A_{\vec{T}}$ with n rows and 3 columns, the columns of which are the vectors $\vec{A}, \vec{B}$, and $\vec{C}$.

$$A_{\vec{T}} = \left(\; [A] \quad [B] \quad [C] \; \right).$$

Exactly as before, we define multiplication of $A_{\vec{T}}$ with the column vector $\begin{pmatrix} u \\ v \\ w \end{pmatrix}$ by

$$A_{\vec{T}} \begin{pmatrix} u \\ v \\ w \end{pmatrix} = \vec{T}(u,v,w) = u\vec{A} + v\vec{B} + w\vec{C}.$$

EXAMPLE 12: The matrix for the transformation in Example 9(a) is

$$A_{\vec{T}} = \begin{pmatrix} 3 & 5 & 2 \\ 1 & -2 & -1 \\ 1 & -1 & 1 \end{pmatrix}.$$

Thus

$$\vec{T}(u,v,w) = A_{\vec{T}} \begin{pmatrix} u \\ v \\ w \end{pmatrix} = u \begin{pmatrix} 3 \\ 1 \\ 1 \end{pmatrix} + v \begin{pmatrix} 5 \\ -2 \\ -1 \end{pmatrix} + w \begin{pmatrix} 2 \\ -1 \\ 1 \end{pmatrix}.\;∎$$

EXAMPLE 13: Let $\vec{T}$ be the linear transformation with matrix

$$A_{\vec{T}} = \begin{pmatrix} 1 & -1 & 2 \\ 1 & 0 & -1 \\ 2 & -3 & 2 \end{pmatrix}.$$

Then

$$\vec{T}(u,v,w) = A_{\vec{T}} \begin{pmatrix} u \\ v \\ w \end{pmatrix} = u \begin{pmatrix} 1 \\ 1 \\ 2 \end{pmatrix} + v \begin{pmatrix} -1 \\ 0 \\ -3 \end{pmatrix} + w \begin{pmatrix} 2 \\ -1 \\ 2 \end{pmatrix}. \blacksquare$$

EXAMPLE 14: Let $\vec{T}$ be the linear transformation with matrix

$$A_{\vec{T}} = \begin{pmatrix} 1 & 0 & -1 \\ 2 & -3 & 2 \end{pmatrix}.$$

Then

$$\vec{T}(u,v,w) = A_{\vec{T}} \begin{pmatrix} u \\ v \\ w \end{pmatrix} = u \begin{pmatrix} 1 \\ 2 \end{pmatrix} + v \begin{pmatrix} 0 \\ -3 \end{pmatrix} + w \begin{pmatrix} -1 \\ 2 \end{pmatrix}. \blacksquare$$

Observation: Notice that the number of columns in $A_{\vec{T}}$ is the dimension of the domain of $\vec{T}$, and the number of rows in $A_{\vec{T}}$ is the dimension of the range. Thus, if $A_{\vec{T}}$ has 2 rows and 3 columns, then $\vec{T}$ is a function from $\mathbb{R}^3$ into $\mathbb{R}^2$. If $A_{\vec{T}}$ has 3 rows and 2 columns, then $\vec{T}$ is a function from $\mathbb{R}^2$ into $\mathbb{R}^3$. If $A_{\vec{T}}$ has 2 rows and 2 columns, then $\vec{T}$ is a function from $\mathbb{R}^2$ into $\mathbb{R}^2$.

In Example 3, we used a linear transformation to *parametrize* a plane determined by two vectors emanating from the origin. If $\vec{T}(u,v) = u\vec{A} + v\vec{B}$ is any linear transformation from $\mathbb{R}^2$ into $\mathbb{R}^3$ such that $\vec{A} \times \vec{B} \neq 0$, then $\vec{T}$ parametrizes the plane determined by $\vec{A}$ and $\vec{B}$, when drawn emanating from the origin. (The test $\vec{A} \times \vec{B} \neq \vec{0}$ is a check to be sure that the two vectors do not have the same direction.) It is also true that any plane containing the origin is the image of a linear transformation.

EXAMPLE 15: Find a linear transformation that takes the u,v-plane onto the plane with equation $2x + 3y - 4z = 0$. We proceed by solving for x of the variables in terms of y and z to obtain $x = 2z - \frac{3y}{2}$. We let $y(u,v) = u$ and $z(u,v) = v$, which gives $x(u,v) = 2v - \frac{3u}{2}$, and we have the coordinate functions for

$$\vec{T}(u,v) = \begin{pmatrix} 2v - 3/2u \\ u \\ v \end{pmatrix}. \qquad\qquad \blacksquare$$

> **Theorem 2** *If $\mathcal{P}$ is the plane with equation $ax + by + cz = 0$, then $\mathcal{P}$ is the image of a linear transformation with domain $\mathbb{R}^2$.*

Proof: At least one of the coordinates of (a,b,c) is not zero. Assume

$a \neq 0$. Then we solve $ax + by + cz = 0$ for x in terms of y and z to obtain $x = -\left(\frac{b}{a}\right) y - \left(\frac{c}{a}\right) z$. As in Example 13, we let $y(u, v) = u$ and $z(u, v) = v$ and define

$$\vec{T}(u, v) = \begin{pmatrix} -\left(\frac{b}{a}\right) u - \left(\frac{c}{a}\right) v \\ u \\ v \end{pmatrix}.$$

$\mathcal{P}$ is parametrized by $\vec{T}$.

If $a = 0$, then we can solve for either y in terms of x and z or z in terms of x and y.

We have described planes in $\mathbb{R}^3$ that contain the origin as the images of linear transformations with domain $\mathbb{R}^2$. Now, suppose that $\vec{A}$ and $\vec{B}$ are vectors in $\mathbb{R}^3$, drawn emanating from $\vec{r}_0$, such that $\vec{A} \times \vec{B} \neq 0$. The vector $\vec{A} \times \vec{B}$ is normal to the plane and $\vec{r}_0$ is a point in the plane. $\vec{T}(u, v) = u\vec{A} + v\vec{B}$ parametrizes a plane containing the origin, which is parallel to $\mathcal{P}$. We can translate the image of $\vec{T}(u, v)$ to $\mathcal{P}$ by composing $\vec{T}$ with the translation by $\vec{r}_0$ to obtain

$$\vec{h}(u, v) = \vec{T}(u, v) + \vec{r}_0.$$

EXAMPLE 16: The vectors $\vec{A} = (1, 2, -1)$ and $\vec{B} = (0, 1, 1)$ are drawn emanating from $\vec{C} = (1, 1, 2)$ to define a plane $\mathcal{P}$. Find a linear transformation $\vec{T}$ with domain $\mathbb{R}^2$ and a vector $\vec{r}_0$ such that $\vec{h}(u, v) = \vec{T}(u, v) + \vec{r}_0$ parametrizes $\mathcal{P}$.

SOLUTION: First, we parametrize the plane determined by the vectors $\vec{A}$ and $\vec{B}$ drawn emanating from the origin. Then we translate this plane by the vector $\vec{C}$. See Figure 6. Let

$$A_{\vec{T}} = \begin{pmatrix} 1 & 0 \\ 2 & 1 \\ -1 & 1 \end{pmatrix}.$$

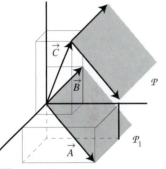

Figure 6. *The plane $\mathcal{P}$ is a translation of $\mathcal{P}_1$ by the vector $\vec{C} = (1, 1, 2)$.*

Then

$$\vec{T}(u, v) = A_{\vec{T}} \begin{pmatrix} u \\ v \end{pmatrix} = \begin{pmatrix} 1 & 0 \\ 2 & 1 \\ -1 & 1 \end{pmatrix} \begin{pmatrix} u \\ v \end{pmatrix} = \begin{pmatrix} u \\ 2u + v \\ -u + v \end{pmatrix}$$

is a parametrization of the plane $\mathcal{P}_1$ containing the origin that is parallel to $\mathcal{P}$. Now we compose $\vec{T}$ with a translation by $\vec{r}_0 = \vec{C}$ to obtain

$$\vec{h}(\vec{r}) \;=\; A_{\vec{T}}\vec{r}+\vec{C} = \begin{pmatrix} 1 & 0 \\ 2 & 1 \\ -1 & 1 \end{pmatrix} \begin{pmatrix} u \\ v \end{pmatrix} + \begin{pmatrix} 1 \\ 1 \\ 2 \end{pmatrix}$$

$$= \begin{pmatrix} u+1 \\ 2u+v+1 \\ -u+v+2 \end{pmatrix}. \qquad\blacksquare$$

EXAMPLE 17: Find a linear transformation composed with a translation that parametrizes the plane $x - y + z = 2$.

SOLUTION: First, we parametrize the plane $x - y + z = 0$ as in Example 13 with $\vec{T}(u,v) = (u-v, u, v)$ and then translate the result with any point in the plane, say $(0,0,2)$, to obtain $\vec{h}(u,v) = (u - v, u, v + 2)$. $\blacksquare$

Summary

$\vec{T}$ is a linear transformation from $\mathbb{R}^n$ into $\mathbb{R}^m$ if and only if $A_{\vec{T}}$ is a $m \times n$–matrix (n columns and m rows.)

If $\vec{T}(u_1, \ldots, u_n) = u_1 \vec{A}_1 + \cdots + u_n \vec{A}_n$, then $A_{\vec{T}} = \left(\; [A_1] \quad \ldots \quad [A_n] \; \right).$

Any plane in $\mathbb{R}^3$ can be parametrized with a function of the form $\vec{h}(\vec{r}) = A\vec{r} + \vec{r}_0$, where A is a matrix with two columns and three rows and $\vec{r}_0$ is an arbitrary point in the plane.

EXERCISES 13.1

In Exercises 1–6, determine the domain and range of the linear transformation associated with the matrix.

1. $\begin{pmatrix} 1 & 1 \\ 2 & -1 \end{pmatrix}.$

2. $\begin{pmatrix} 1 & -1 & 0 \\ 0 & 2 & -1 \\ 1 & 0 & 1 \end{pmatrix}.$

3. $\begin{pmatrix} 1 & -1 & 0 \end{pmatrix}.$

4. $\begin{pmatrix} 1 & 0 \\ 2 & -1 \\ 1 & 0 \end{pmatrix}.$

5. $\begin{pmatrix} 1 & -1 & 0 \\ 0 & 2 & -1 \end{pmatrix}.$ 6. $\begin{pmatrix} 1 & -1 \end{pmatrix}.$

In Exercises 7–14, find vectors $\vec{A}$ and $\vec{B}$, such that $\vec{T}(u,v) = u\vec{A} + v\vec{B}$, and find the associated $n \times 2$ matrix $A_{\vec{T}}$.

7. $\vec{T}(u,v) = (2u, 3v).$

8. $\vec{T}(u,v) = (u+v, u-v).$

9. $\vec{T}(u,v) = \left(3u - 6v, \frac{v}{2} - \frac{u}{3}\right).$

10. $\vec{T}(u,v) = \left(-v, u - \frac{v}{2}\right).$

11. $\vec{T}(u,v) = (-v, u-v).$

12. $\vec{T}(u,v) = (u-v, 2u+6v).$

13. $\vec{T}(u,v) = \left(v - 6u, -u - v, u + \frac{v}{3}\right).$

14. $\vec{T}(u,v) = (6u + 3v).$

In Exercises 15–22, find vectors $\vec{A}, \vec{B},$ and $\vec{C}$ such that $\vec{T}(u,v,w) = u\vec{A} + v\vec{B} + w\vec{C}$, and find the associated $n \times 3$ matrix $A_{\vec{T}}.$

15. $\vec{T}(u,v,w) = (u + v + w, u - v + w, v - w).$

16. $\vec{T}(u, v, w) = (u - 3v, 3u + v - \pi w)$.

17. $\vec{T}(u, v, w) = (6u + 3v - w, -u - v - w, u - w)$.

18. $\vec{T}(u, v, w) = (5w, v + w, v)$.

19. $\vec{T}(u, v, w) = (u + v + w, u - v)$.

20. $\vec{T}(u, v, w) = (u - v + w, u + v - 6w)$.

21. $\vec{T}(u, v, w) = (u + 6v - 10w)$.

22. $\vec{T}(u, v, w) = (15u - 6v + 2w)$.

In Exercises 23–30, find the matrix for the linear transformation $\vec{T}$.

23. $\vec{T}$ takes $(0, 1)$ onto $(1, 3)$, and $(1, 0)$ onto $(-1, 5)$.

24. $\vec{T}$ takes $(1, 0, 0)$ onto $(1, 3)$, $(0, 1, 0)$ onto $(-1, 5)$, and $(0, 0, 1)$ onto $(-1, 0)$.

25. $\vec{T}$ takes $(1, 0, 0)$ onto $(1, 3, 2)$, $(0, 1, 0)$ onto $(-1, 5, 1)$, and $(0, 0, 1)$ onto $(-1, 0, 3)$.

26. $\vec{T}$ takes $(1, 0, 0)$ onto -1, $(0, 1, 0)$ onto 1, and $(0, 0, 1)$ onto 3.

27. $\vec{T}$ takes $(1, 0, 0)$ onto 1, $(0, 1, 0)$ onto 2, and $(0, 0, 1)$ onto 3.

28. $\vec{T}$ takes $\hat{i}$ onto $\hat{j}$, $\hat{j}$ onto $-\hat{i}$ and $\hat{k}$ onto $-\hat{k}$.

29. $\vec{T}$ reflects the plane over the x–axis.

30. $\vec{T}$ reflects the plane over the line $x = -y$.

In Exercises 31–36, determine which of the planes is the image of a linear transformation. Where appropriate, find a linear transformation that parametrizes the plane.

31. $\mathcal{P}$ is the graph of $x + y + z = 0$.

32. $\mathcal{P}$ is the graph of $x - 2y + 3z = 0$.

33. $\mathcal{P}$ is the graph of $x + 2y + 3z = 3$.

34. The vector $(1, 2, 3)$ is normal to $\mathcal{P}$, and $\mathcal{P}$ contains the origin.

35. The vector $(1, 2, 3)$ is normal to $\mathcal{P}$, and $\mathcal{P}$ contains $(1, 1, 1)$.

36. The vector $(1, 2, 3)$ is normal to $\mathcal{P}$, and $\mathcal{P}$ contains $(-1, -1, 1)$.

In Exercises 37–42, find a function of the form
$$\vec{h}(s, t) = A \begin{pmatrix} s \\ t \end{pmatrix} + (x_0, y_0, z_0),\ \text{such that the plane } \mathcal{P}$$
is the image of $\vec{h}$.

37. $\mathcal{P}$ is the graph of $x + y + z = 3$.

38. $\mathcal{P}$ is the graph of $x + 2y + 3z = -1$.

39. $\mathcal{P}$ is the graph of $x - 2y + 3z = 3$.

40. The vector $(1, 2, 3)$ is normal to $\mathcal{P}$, and $\mathcal{P}$ contains $(1, 1, 1)$.

41. The vector $(1, 2, 3)$ is normal to $\mathcal{P}$, and $\mathcal{P}$ contains $(1, -1, 2)$.

42. The vector $(1, 2, 3)$ is normal to $\mathcal{P}$, and $\mathcal{P}$ contains $(-1, -1, 1)$.

In Exercises 43–47, prove that $\vec{T}$ is not linear by showing that $\vec{T}(\vec{r} + \vec{s}) \neq \vec{T}(\vec{r}) + \vec{T}(\vec{s})$ for some choice of vectors $\vec{r}$ and $\vec{s}$.

43. $\vec{T}(u, v) = (3u - v + 1, 2u + v - 6)$.

44. $\vec{T}(u, v, w) = (u - 2v + 3, u + w)$.

45. $\vec{T}(u, v) = (u^2 + 2u, u - v)$.

46. $\vec{T}(u, v, w) = (u^2 + v^2, u + v - 2, u + v)$.

47. $\vec{T}(u, v) = (\sin u, \cos v)$.

In Exercises 48–52, $\vec{T}_\theta$ is the transformation that rotates the plane θ radians, as in Figure 7.

48. Find the **unit** vectors $\hat{e}_1$ and $\hat{e}_2$ in terms of θ.

49. Find the matrix for $\vec{T}_\theta$.

50. Find $\vec{T}_{\pi/4}(1, 1)$.

51. Find $\vec{T}_{\pi/2}(1, 1)$.

52. Find $\vec{T}_{-\pi/4}(1, 1)$.

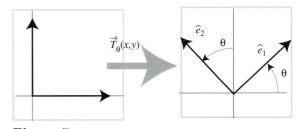

Figure 7.

53. Let $\vec{T}(u,v) = (u,-v)$ is the linear transformation that reflects the plane over the x–axis. Show that $\vec{T} \circ \vec{T}_{\pi/4} \neq \vec{T}_{\pi/4} \circ \vec{T}$ (that is, a rotation followed by a reflection gives a different result than does a reflection followed by a rotation.)

54. Let $\vec{T}$ be a function from $\mathbb{R}^2$ into $\mathbb{R}^2$. Show that if each of its coordinate functions is linear, $\vec{T}$ is linear.

55. Let $\vec{T}$ be a linear transformation from $\mathbb{R}^2$ into $\mathbb{R}^2$. Show that each of its coordinate functions is linear.

56. Show that if the function $\vec{T}$ from $\mathbb{R}^n$ into $\mathbb{R}^m$ is linear, then $\vec{T}(\vec{0}) = \vec{0}$. (Hint: $\vec{T}(\vec{0} + \vec{0}) = \vec{T}(\vec{0})$.)

13.2   Other Transformations

There are a number of functions or transformations that are not linear but still quite important. Linear transformations have the nice property that they take planes onto single points, lines, or other planes. However, it may be helpful to change our point of view more drastically.

Suppose $\vec{F}$ is a function from uvw–space into xyz–space, given by

$$\vec{F}(u,v,w) \;=\; \begin{pmatrix} x(u,v,w) \\ y(u,v,w) \\ z(u,v,w) \end{pmatrix}.$$

We define

$$\left.\frac{\partial \vec{F}}{\partial u}\right|_{(u,v,w)} = \left.\begin{pmatrix} \frac{\partial x}{\partial u} \\ \frac{\partial y}{\partial u} \\ \frac{\partial z}{\partial u} \end{pmatrix}\right|_{(u,v,w)}, \qquad \left.\frac{\partial \vec{F}}{\partial v}\right|_{(u,v,w)} = \left.\begin{pmatrix} \frac{\partial x}{\partial v} \\ \frac{\partial y}{\partial v} \\ \frac{\partial z}{\partial v} \end{pmatrix}\right|_{(u,v,w)},$$

and

$$\left.\frac{\partial \vec{F}}{\partial w}\right|_{(u,v,w)} = \left.\begin{pmatrix} \frac{\partial x}{\partial w} \\ \frac{\partial y}{\partial w} \\ \frac{\partial z}{\partial w} \end{pmatrix}\right|_{(u,v,w)}.$$

The Polar Transformation

EXAMPLE 1: Let $\vec{P}$ be the function from $r\theta$–space defined by

$$\vec{P}(r,\theta) = (r\cos\theta, r\sin\theta).$$

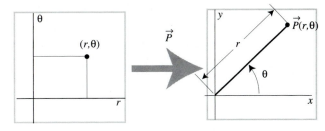

Figure 1. $\vec{P}(r,\theta)$ *is the point* r *units from the origin on the line with inclination* θ.

The function $\vec{P}$ is called the *polar transformation*. If $\vec{P}(r,\theta) = (x,y)$, then (r,θ) are called *polar coordinates* for (x,y), and (x,y) are the *rectangular coordinates* for (r,θ).

The transformation P takes a horizontal line of the form $\theta = \theta_0$ in $r\theta$–space onto a line passing through the origin in xy–space, as illustrated in Figure 2.a. Similarly, P takes a vertical line of the form $r = r_0$ in $r\theta$–space onto a circle of radius r_0 in xy–space, as in Figure 2.b.

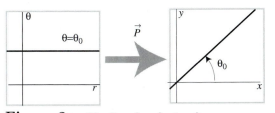 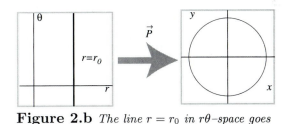

Figure 2.a *The line* $\theta = \theta_0$ *in* $r\theta$–*space goes onto the line containing the origin with inclination* θ_0 *in* xy–*space.*

Figure 2.b *The line* $r = r_0$ *in* $r\theta$–*space goes onto the circle centered at the origin with radius* r_0 *in* xy–*space.*

The partial derivatives of $\vec{P}$ are

$$\frac{\partial \vec{P}}{\partial r} = (\cos\theta, \sin\theta) \text{ and } \frac{\partial \vec{P}}{\partial \theta} = (-r\sin\theta, r\cos\theta).$$

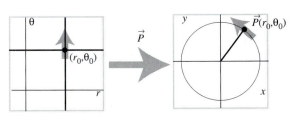 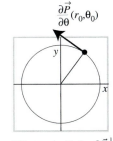

Figure 3.a *As the position in* $r\theta$–*space moves in the positive direction on the line* $r = r_0$, *its image in* xy–*space moves counterclockwise around the circle of radius* r_0.

Figure 3.b $\left.\frac{\partial \vec{P}}{\partial \theta}\right|_{(r_0,\theta_0)}$ *is a vector tangent to the circle* $x^2 + y^2 = r_0^2$ *at the point* $\vec{P}(r_0,\theta_0)$.

Consider the geometry associated with the partial derivatives of
P. Figure 3.a illustrates that as the position in $r\theta$–space moves in
the positive direction on the line $r = r_0$, its image in xy–space moves
counterclockwise around the circle of radius r_0. Thus, $\left.\dfrac{\partial \vec{P}}{\partial \theta}\right|_{(r_0,\theta_0)}$ is
a vector tangent to the circle $x^2 + y^2 = r_0^2$ at the point $\vec{P}(r_0,\theta_0) =$
$r_0\left(\cos(\theta_0), r_0 \sin(\theta_0)\right)$, as illustrated in Figure 3.b.

As illustrated in Figures 4.a and 4.b, we can discern the geometry
of $\frac{\partial P}{\partial r}$ by inspection.

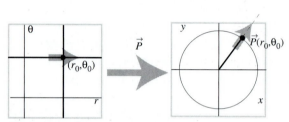
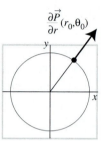

Figure 4.a *As the position in $r\theta$–space moves
in the positive direction on the line $\theta = \theta_0$, its
image in xy–space moves out the line with
inclination θ_0.*

Figure 4.b $\left.\dfrac{\partial \vec{P}}{\partial r}\right|_{(r_0,\theta_0)}$
*is a vector emanating from
the point (r_0,θ_0). The
vector points away from
the origin.*

The Cylindrical Transformation

EXAMPLE 2: Let $\vec{C}_z$ be the function from $r\theta z$–space into xyz–
space, defined by

$$\vec{C}_z(r,\theta,z) = (r\cos\theta, r\sin\theta, z).$$

In Figure 5.a we plot the image of the point (r,θ,z).

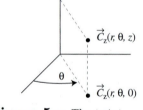

Figure 5.a *The point
$\vec{C}_z(r,\theta,z)$.*

The function $\vec{C}_z$ is called the *cylindrical transformation (about
the z–axis)* because it takes planes of the form $r = r_0$ onto the
lateral surface of the cylinder $x^2 + y^2 = r_0$. Figure 5.b illustrates how
planes parallel to the $r\theta$–plane in $r\theta z$–space are carried into xyz–
space by $\vec{C}_z$. If $\vec{C}_z(r,\theta,z) = (x,y,z)$, then (r,θ,z) are the *cylindrical
coordinates* for (x,y,z), and (x,y,z) are the *rectangular coordinates*
for (r,θ,z).

Thus

$$\frac{\partial \vec{C}_z(r,\theta,z)}{\partial r} = (\cos\theta, \sin\theta, 0),$$

$$\frac{\partial \vec{C}_z(r,\theta,z)}{\partial \theta} = (-r\sin\theta, r\cos\theta, 0), \text{ and}$$

$$\frac{\partial \vec{C}_z(r, \theta, z)}{\partial z} \;=\; (0, 0, 1).$$

The geometry associated with the cylindrical transformation and its partial derivatives is illustrated in Figures 6.a, 6.b, and 6.c.

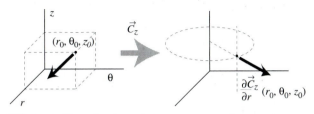

Figure 6.a *As the point moves away from* (r_0, θ_0, z_0) *in the r direction holding θ and z fixed,* $\vec{C}_z(r, \theta, z)$ *moves directly away from the z-axis.*

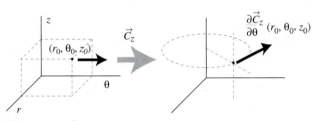

Figure 6.b *As the point moves away from* (r_0, θ_0, z_0) *in the θ direction holding r and z fixed,* $\vec{C}_z(r, \theta, z)$ *moves in direction tangent to the circle parametrized by*

$$\vec{h}(\theta) = \begin{pmatrix} r_0 \cos(\theta) \\ r_0 \sin(\theta) \\ z_0 \end{pmatrix} = \vec{C}_z(r_0, \theta, z_0)$$

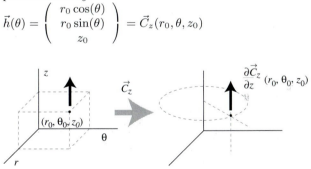

Figure 6.c *As the point moves away from* (r_0, θ_0, z_0) *in the z direction holding r and θ fixed,* $\vec{C}_z(r, \theta, z)$ *moves in the z direction in xyz-space.*

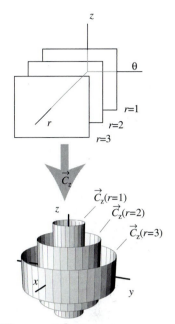

Figure 5.b *Images (in xyz–space) of planes that are parallel to the θz–coordinate plane in $r\theta z$–space.*

The Spherical Transformation

EXAMPLE 3: Let

$$\vec{S}(\rho, \phi, \theta) = \begin{pmatrix} \rho\cos(\theta)\sin(\phi) \\ \rho\sin(\theta)\sin(\phi) \\ \rho\cos(\phi) \end{pmatrix}.$$

$\vec{S}$ is called the *spherical transformation* since $\vec{S}$ takes the plane $\rho = \rho_0$ in $\rho\phi\theta$–space onto a sphere of radius ρ_0. See Figure 7.

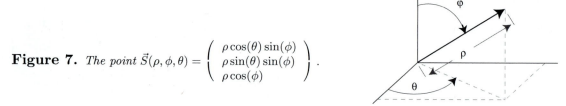

Figure 7. *The point* $\vec{S}(\rho, \phi, \theta) = \begin{pmatrix} \rho\cos(\theta)\sin(\phi) \\ \rho\sin(\theta)\sin(\phi) \\ \rho\cos(\phi) \end{pmatrix}.$

The spherical transformation appears formidable. However, as we illustrate in Figures 8.a and 8.b, the derivation of the function is a straightforward application of elementary trigonometry and geometry. ρ is the length of the hypotenuse of the right triangle $\vec{O}\vec{A}\vec{B}$. The z–coordinate of $\vec{S}(\rho, \phi, \theta)$ is $\rho\cos(\phi)$ since $\vec{O}\vec{A}$ is the side adjacent ϕ in the right triangle $\vec{O}\vec{A}\vec{B}$. The line segment $\vec{O}\vec{C}$ is the hypotenuse of the right triangle $\vec{O}\vec{C}\vec{D}$, and its length is $\rho\sin(\phi)$. $\vec{O}\vec{D}$ is the side adjacent the angle θ. Therefore, the x–coordinate of $\vec{S}(\rho, \phi, \theta)$ is $\cos(\theta)(\rho\sin(\phi))$. In the same way, $\vec{B}\vec{C}$ is the side opposite θ in the triangle $\vec{O}\vec{C}\vec{D}$, and the y–coordinate of $\vec{S}(\rho, \phi, \theta)$ is $\sin(\theta)(\rho\sin(\phi))$.

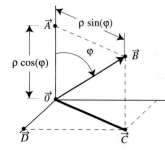

Figure 8.a *The z–coordinate of* $\vec{S}(\rho, \phi, \theta)$ *is* $\cos(\phi)$.

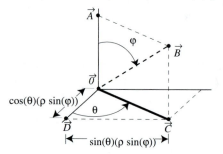

Figure 8.b *The x–coordinate of* $\vec{S}(\rho, \phi, \theta)$ *is* $\cos(\theta)(\rho\sin(\phi))$, *and the y–coordinate of* $\vec{S}(\rho, \phi, \theta)$ *is* $\sin(\theta)(\rho\sin(\phi))$.

Figures 9.a–c. illustrate how $\vec{S}$ takes planes parallel to the coordinate planes in uvw–space into xyz–space. If $\vec{S}(\rho, \phi, \theta) = (x, y, z)$, then (ρ, ϕ, θ) are called *spherical coordinates* for (x, y, z), and (x, y, z) are called the *rectangular coordinates* for (ρ, ϕ, θ).

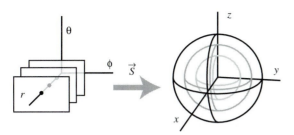

Figure 9.a $\vec{S}$ *takes planes parallel to the*
$\theta\phi$–coordinate plane onto spheres.

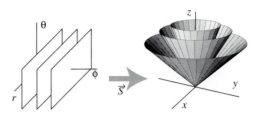

Figure 9.b $\vec{S}$ *takes planes parallel to the*
$r\theta$–coordinate plane onto cones.

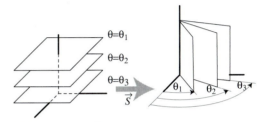

Figure 9.c $\vec{S}$ *takes planes parallel to the*
$r\phi$–coordinate plane onto planes containing the
z–axis, like pages of a book.

The geometry associated with the partial derivatives of the spherical transformation is illustrated in Figures 10.1, 10.b, and 10.c.

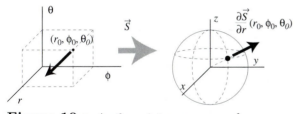

Figure 10.a *As the point moves away from*
(r_0, ϕ_0, θ_0) in the r direction holding ϕ and θ fixed,
$\vec{S}(r, \phi, \theta)$ moves radially away from the origin.

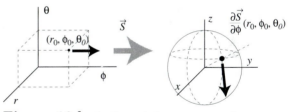

Figure 10.b *As the point moves away from*
(r_0, ϕ_0, θ_0) in the θ direction holding r and ϕ fixed,
$\vec{S}(r, \phi, \theta)$ moves in direction tangent to the latitude
parametrized by $\vec{h}(\theta) = (r_0, \phi_0, \theta)$.

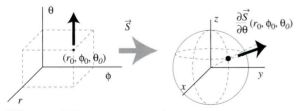

Figure 10.c *As the point moves away from*
(r_0, ϕ_0, θ_0) in the ϕ direction holding r and θ fixed,
$\vec{S}(r, \phi, \theta)$ moves around the longitude parametrized by
$\vec{h}(\phi) = (r_0, \phi, \theta_0)$.

EXERCISES 13.2

In Exercises 1–9, plot the points in xy–space that have the given polar coordinates, and compute the partial derivatives of the polar transformation $\vec{P}$ at the given point. (The angles are measured in radians.)

1. $(2, 0)$.   2. $(2, 2\pi)$.   3. $(2, -2\pi)$.

4. $(-2, \pi)$.   5. $\left(5, \frac{\pi}{4}\right)$.   6. $\left(-5, \frac{5\pi}{4}\right)$.

7. $\left(5, -\frac{\pi}{4}\right)$.   8. $\left(1, \frac{\pi}{6}\right)$.   9. $\left(-1, \frac{7\pi}{6}\right)$.

In Exercises 10–14, find all polar coordinates for the given point in xy–space. ($r^2 = x^2 + y^2$, and, if $x \neq 0$, $\tan(\theta) = \frac{y}{x}$.)

10. $\left(0, \frac{1}{2}\right)$.   11. $\left(\frac{1}{2}, 0\right)$.   12. $(1, 1)$.

13. $(-1, \sqrt{3})$.   14. $(0, 0)$.

In Exercises 15–23, find the point in xyz–space that has the given cylindrical coordinates, and compute partial derivatives of the cylindrical transformation $\vec{C}_z$ at the given point. (The angles are measured in radians.)

15. $(2, 0, 2)$.   16. $(2, 2\pi, 2)$.   17. $(2, -2\pi, 2)$.

18. $(-2, \pi, -1)$.   19. $\left(5, \frac{\pi}{4}, 2\right)$.   20. $\left(-5, \frac{5\pi}{4}, 2\right)$.

21. $\left(5, -\frac{\pi}{4}, 3\right)$.   22. $\left(1, \frac{\pi}{6}, -1\right)$ 23. $\left(-1, \frac{7\pi}{6}, -1\right)$

In Exercises 24–28, find the cylindrical coordinates for the given point in xyz–space. ($r^2 = x^2 + y^2$, and, if $x \neq 0$, $\tan(\theta) = \frac{y}{x}$.)

24. $\left(0, \frac{1}{2}, 2\right)$.   25. $\left(\frac{1}{2}, 0, 2\right)$.   26. $(1, 1, 1)$.

27. $(-1, \sqrt{3}, -1)$.   28. $(0, 0, 2)$.

In Exercises 29–36, the points are given in spherical coordinates. Find the corresponding rectangular coordinates, and the partial derivatives of the spherical transformation $\vec{S}$ at the given point.

29. $\left(2, 0, \frac{\pi}{4}\right)$.   30. $\left(-2, \pi, \frac{\pi}{4}\right)$.

31. $\left(-2, 0, \frac{7\pi}{4}\right)$.   32. $\left(2, \pi, \frac{7\pi}{4}\right)$.

33. $\left(2, \pi, -\frac{\pi}{4}\right)$.   34. $\left(1, \frac{\pi}{4}, \frac{\pi}{4}\right)$.

35. $\left(-1, \frac{5\pi}{4}, \frac{\pi}{4}\right)$.   36. $\left(1, \frac{5\pi}{4}, -\frac{\pi}{4}\right)$.

In Exercises 37–40, find values for (ρ, ϕ, θ) such that (i) $\rho \geq 0$, (ii) $0 \leq \theta < 2\pi$, and (iii) $0 \leq \phi \leq \pi$ and such that $\vec{S}(\rho, \phi, \theta)$ is the given point. ($\rho^2 = x^2 + y^2 + z^2$, $z = \rho \cos(\phi)$, and $\tan(\theta) = \frac{y}{x}$.)

37. $(1, 0, 1)$.   38. $(1, 1, 1)$.

39. $(1, \sqrt{2}, 1)$.   40. $(\sqrt{6}, 2, 2\sqrt{2})$

41. Show that if $\vec{p}$ is a point on the plane $\rho = \rho_0$ in $\rho\phi\theta$–space, then $\vec{S}$ takes $\vec{p}$ onto a point in the sphere of radius ρ_0 centered at the origin in xyz–space.

42. Show that if $\vec{q}$ is on the sphere of radius ρ_0, centered at the origin, then there is a point $\vec{p}$ in the plane $\rho = \rho_0$ in $\rho\phi\theta$–space such that $\vec{S}(\vec{p}) = \vec{q}$.

43. Show that the partial derivatives of the polar function are mutually perpendicular at (r, θ) provided that $r \neq 0$.

44. Show that the partial derivatives of the cyliderical transformation are mutually perpendicular at (r, θ, z) provided that $r \neq 0$.

45. Show that the partial derivatives of the spherical transformation are mutually perpendicular at (r, ϕ, θ) provided that $r \neq 0$.

In Exercises 46–57, sketch the set satisfying the given conditions.

46. $\vec{P}(A)$, where A is the line $r = 2$ in $r\theta$–space.

47. $\vec{P}(A)$, where A is the line $r = -2$ in $r\theta$–space.

48. $\vec{P}(A)$, where A is the line $\theta = \frac{\pi}{2}$ in $r\theta$–space.

49. $\vec{P}(A)$, where A is the line $\theta = \frac{\pi}{4}$ in $r\theta$–space.

50. $\vec{P}(A)$, where A is the line $\theta = \frac{3\pi}{4}$ in $r\theta$–space.

51. $\vec{P}(A) + (1, 1)$, where A is the line $r = 2$ in $r\theta$–space.

52. $\vec{P}(A) + (-1, 2)$, where A is the line $r = 4$ in $r\theta$–space.

53. $\vec{P}(A)$, where A is the rectangle in $r\theta$–space $0 \leq r \leq 1$, $0 \leq \theta, \leq 2\pi$.

54. $\vec{P}(A)$, where A is the rectangle in $r\theta$–space $0 \leq r \leq 1$, $0 \leq \theta \leq \frac{\pi}{4}$.

55. $\vec{P}(A)$, where A is the rectangle in $r\theta$–space $1 \leq r \leq 2$, $0 \leq \theta \leq 2\pi$.

56. $\vec{P}(A)$, where A is the rectangle in $r\theta$–space $1 \le r \le 2,\ 0 \le \theta \le \frac{\pi}{4}$.

57. $\vec{P}(A)$, where A is the rectangle in $r\theta$–space $1 \le r \le 2,\ -\frac{\pi}{4} \le \theta \le \frac{\pi}{4}$.

In Exercises 58–81, describe the set satisfying the given conditions.

58. $\vec{C}_z(A)$, where A is the plane $r = 3$ in $r\theta z$–space.

59. $\vec{C}_z(A)$, where A is the plane $r = -2$ in $r\theta z$–space.

60. $\vec{C}_z(A)$, where A is the plane $\theta = \frac{\pi}{2}$ in $r\theta z$–space.

61. $\vec{C}_z(A)$, where A is the plane $\theta = \frac{\pi}{4}$ in $r\theta z$–space.

62. $\vec{C}_z(A)$, where A is the plane $\theta = \frac{3\pi}{4}$ in $r\theta z$–space.

63. $\vec{C}_z(A) + (1, 2, 3)$, where A is the plane $r = 2$ in $r\theta z$–space.

64. $\vec{C}_z(A)$, where A is the rectangle in $r\theta z$–space $0 \le r \le 1,\ 0 \le \theta \le 2\pi,\ z = 0$.

65. $\vec{C}_z(A)$, where A is the rectangle in $r\theta z$–space $0 \le r \le 1,\ 0 \le \theta \le \frac{\pi}{4},\ z = 2$.

66. $\vec{C}_z(A)$, where A is the rectangle in $r\theta z$–space $1 \le r \le 2,\ 0 \le \theta, \le 2\pi,\ z = -3$.

67. $\vec{C}_z(A)$, where A is the rectangular box in $r\theta z$–space $0 \le r \le 2,\ 0 \le \theta, \le 2\pi,\ 0 \le z \le 1$.

68. $\vec{C}_z(A)$, where A is the rectangular box in $r\theta z$–space $1 \le r \le 2,\ 0 \le \theta, \le \pi,\ 1 \le z \le 2$.

69. $\vec{C}_z(A)$, where A is the rectangular box in $r\theta z$–space $1 \le r \le 2,\ -\pi \le \theta, \le \pi,\ 1 \le z \le 2$.

70. $\vec{S}(A)$, where A is the plane in $r\phi\theta$–space $r = 3$.

71. $\vec{S}(A)$, where A is the plane in $r\phi\theta$–space $r = -2$.

72. $\vec{S}(A)$, where A is the plane in $r\phi\theta$–space and $\theta = \frac{\pi}{2}$.

73. $\vec{S}(A)$, where A is the plane in $r\phi\theta$–space and $\theta = \frac{\pi}{4}$.

74. $\vec{S}(A)$, where A is the plane in $r\phi\theta$–space and $\phi = \frac{3\pi}{4}$.

75. $\vec{S}(A) + (1, 2, 3)$, where A is the plane in $r\phi\theta$–space and $r = 2$.

76. $\vec{S}(A)$, where A is the rectangle in $r\phi\theta$–space $0 \le r \le 1,\ 0 \le \theta \le 2\pi,\ \phi = 0$.

77. $\vec{S}(A)$, where A is the rectangle in $r\phi\theta$–space $0 \le r \le 1,\ 0 \le \theta \le \frac{\pi}{4},\ \phi = \frac{\pi}{3}$.

78. $\vec{S}(A)$, where A is the rectangle in $r\phi\theta$–space $1 \le r \le 2,\ 0 \le \theta, \le 2\pi,\ \phi = \frac{\pi}{3}$.

79. $\vec{S}(A)$, where A is the rectangular box in $r\phi\theta$–space $0 \le r \le 2,\ 0 \le \theta, \le 2\pi,\ 0 \le \phi \le \frac{\pi}{4}$.

80. $\vec{S}(A)$, where A is the rectangular box in $r\phi\theta$–space $1 \le r \le 2,\ 0 \le \theta, \le \pi,\ 0 \le \phi \le \pi$.

81. $\vec{S}(A)$, where A is the rectangular box in $r\phi\theta$–space $1 \le r \le 2,\ -\pi \le \theta, \le \pi,\ 0 \le \phi \le \frac{\pi}{2}$.

In Exercises 82–85, we define $\vec{C}_x$ similarly to $\vec{C}_z$ except that the x–axis is the axis of symmetry. That is $\vec{C}_x(x, r, \theta) = (x, r\cos\theta, r\sin\theta)$.

82. Show that the partial derivatives of $\vec{C}_x$ are mutually perpendicular, except at points where $r = 0$.

83. Describe $\vec{C}_x(A)$, where A is the plane $x = 2$.

84. Describe $\vec{C}_x(A)$, where A is the plane $r = 2$.

85. Describe $\vec{C}_x(A)$, where A is the plane $\theta = \frac{\pi}{2}$.

In Exercises 86–89, we define $\vec{C}_y(r, y, \theta) = (r\cos\theta, y, r\sin\theta)$.

86. Show that the partial derivatives of $\vec{C}_y$ are mutually perpendicular, except at points where $r = 0$.

87. Describe $\vec{C}_y(A)$, where A is the plane $y = 2$.

88. Describe $\vec{C}_y(A)$, where A is the plane $r = 2$.

89. Describe $\vec{C}_y(A)$, where A is the plane $\theta = \frac{\pi}{2}$.

90. Show that the functions $\vec{S}$, $\vec{C}_x$, $\vec{C}_y$, and $\vec{C}_z$ satisfy the right hand rule. That is, the partial derivative with respect to the first variable crossed with the partial derivative with respect to the second variable has the same direction as the partial derivative with respect to the third variable.

91. Calculate the partial derivatives of $\vec{f}(u,v) = (2\cos(u)\sin(v), 2\sin(u)\sin(v), 2\cos(v))$.

92. What is the image of $\vec{f}$ of Exercise 91?

13.3   The Derivative

Let $\vec{F}(u,v)$ be a function from $\mathbb{R}^2$ into $\mathbb{R}^2$. Let F_x denote the x–coordinate function, and let F_y denote the y–coordinate function, so that

$$\vec{F}(u,v) = (F_x(u,v), F_y(u,v)).$$

Let $\vec{r}(t)$ be a parametrization for a curve in $\mathbb{R}^2$, and let

$$\vec{g}(t) = (F_x(\vec{r}(t)), F_y(\vec{r}(t))).$$

Now,

$$\frac{d\vec{g}(t)}{dt} = \left(\frac{dF_x(\vec{r}(t))}{dt}, \frac{dF_y(\vec{r}(t))}{dt} \right).$$

By Theorem 1 of Section 11.8, we see that

$$\begin{aligned}
\frac{dF_x(\vec{r}(t))}{dt} &= \nabla F_x(\vec{r}(t)) \cdot \vec{r}\,'(t) \\
&= \left(\frac{\partial F_x}{\partial u}(\vec{r}(t)) \right) u'(t) + \left(\frac{\partial F_x}{\partial v}(\vec{r}(t)) \right) v'(t) \\
&= \frac{\partial F_x}{\partial u} \frac{du}{dt} + \frac{\partial F_x}{\partial v} \frac{dv}{dt},
\end{aligned}$$

and

$$\begin{aligned}
\frac{dF_y(\vec{r}(t))}{dt} &= \nabla F_y(\vec{r}(t)) \cdot \vec{r}\,'(t) \\
&= \left(\frac{\partial F_y}{\partial u}(\vec{r}(t)) \right) u'(t) + \left(\frac{\partial F_y}{\partial v}(\vec{r}(t)) \right) v'(t) \\
&= \frac{\partial F_y}{\partial u} \frac{du}{dt} + \frac{\partial F_y}{\partial v} \frac{dv}{dt}.
\end{aligned}$$

We can express the above two equations in matrix notation in the following way:

$$\begin{aligned}
\frac{d\vec{g}(t)}{dt} &= \begin{pmatrix} \frac{dF_x(\vec{r}(t))}{dt} \\ \frac{dF_y(\vec{r}(t))}{dt} \end{pmatrix} = \begin{pmatrix} \frac{\partial F_x}{\partial u} \frac{du}{dt} + \frac{\partial F_x}{\partial v} \frac{dv}{dt} \\ \frac{\partial F_y}{\partial u} \frac{du}{dt} + \frac{\partial F_y}{\partial v} \frac{dv}{dt} \end{pmatrix} \\
&= \begin{pmatrix} \frac{\partial F_x}{\partial u} & \frac{\partial F_x}{\partial v} \\ \frac{\partial F_y}{\partial u} & \frac{\partial F_y}{\partial v} \end{pmatrix} \begin{pmatrix} \frac{du}{dt} \\ \frac{dv}{dt} \end{pmatrix}.
\end{aligned}$$

Definition: The Derivative of $\vec{F}$ at (u, v)

Let $\vec{F} : \mathbb{R}^2 \to \mathbb{R}^2$ be a function

$$\vec{F}(u, v) = (F_x(u, v), F_y(u, v)).$$

If $\frac{\partial F_x}{\partial u}, \frac{\partial F_x}{\partial v}, \frac{\partial F_y}{\partial u}$, and $\frac{\partial F_y}{\partial v}$ exist, we define

$$D\vec{F}\big|_{(u,v)} = \begin{pmatrix} \frac{\partial F_x}{\partial u} & \frac{\partial F_x}{\partial v} \\ \frac{\partial F_y}{\partial u} & \frac{\partial F_y}{\partial v} \end{pmatrix}$$

to be the *derivative of $\vec{F}$ evaluated at* (u, v).

Notice that the first row of $D\vec{F}$ is simply ∇F_x, the second row is ∇F_y, and the derivative of $\vec{F}$ is defined so that we have the chain rule for functions from $\mathbb{R}^2$ into $\mathbb{R}^2$.

$$\frac{d}{dt}\vec{F}(\vec{r}(t)) = D\vec{F}\big|_{\vec{r}(t)}\,\vec{r}\,'(t).$$

EXAMPLE 1: Consider the polar transformation $\vec{P}(r, \theta) = (r\cos\theta, r\sin\theta)$. Then

$$D\vec{P}\big|_{(r,\theta)} = \begin{pmatrix} \cos\theta & -r\sin\theta \\ \sin\theta & r\cos\theta \end{pmatrix}.$$

If $\vec{r}(t) = (1, 2\pi t)$, then $\vec{P}(\vec{r}(t)) = (\cos(2\pi t), \sin(2\pi t))$ describes a particle moving around a circle in xy–space at a rate of one rotation/sec. We can either calculate the derivative directly as

$$\frac{d\vec{P}(\vec{r}(t))}{dt} = 2\pi(-\sin(2\pi t), \cos(2\pi t)),$$

or use the above to obtain

$$D\vec{P}\big|_{(r,\theta)} = \begin{pmatrix} \cos(2\pi t) & -\sin(2\pi t) \\ \sin(2\pi t) & \cos(2\pi t) \end{pmatrix},$$

and

$$\vec{r}\,'(t) = (0, 2\pi).$$

Thus

$$\frac{d\vec{P}(\vec{r}(t))}{dt} = D\vec{P}|_{\vec{r}(t)}\vec{r}'(t)$$

$$= \begin{pmatrix} \cos(2\pi t) & -\sin(2\pi t) \\ \sin(2\pi t) & \cos(2\pi t) \end{pmatrix} \begin{pmatrix} 0 \\ 2\pi \end{pmatrix}$$

$$= 2\pi(-\sin(2\pi t), \cos(2\pi t)). \qquad \blacksquare$$

In general, if $\vec{s}(t) = (r(t), \theta(t))$ is an expression for the polar coordinates of a particle at time t, then $\vec{P}(\vec{s}(t))$ will give the rectangular or Cartesian coordinates of the particle. The derivative $\vec{s}'(t)$ will then denote the polar coordinates of the velocity vector and

$$\frac{d}{dt}\vec{P}(\vec{s}(t)) = D\vec{P}|_{\vec{s}(t)}\vec{s}'(t)$$

$$= \begin{pmatrix} \cos(\theta(t)) & -r(t)\sin(\theta(t)) \\ \sin(\theta(t)) & r(t)\cos(\theta(t)) \end{pmatrix} \begin{pmatrix} r'(t) \\ \theta'(t) \end{pmatrix}$$

will give the Cartesian coordinates of the velocity vector.
Note: It is not uncommon to encounter the notation $d\vec{s}/dt$, which is meant to represent the rate of change of position in rectangular coordinates even though $\vec{s}$ represents the polar coordinates of the point.

EXAMPLE 2: Suppose that a particle is moving in the plane so that when the polar coordinates are $\left(2, \frac{\pi}{4}\right)$, its velocity, in polar coordinates, is $(1, \pi)$. What is the velocity in rectangular coordinates?

SOLUTION: If t_0 denotes the time at which the particle is at $\left(2, \frac{\pi}{4}\right)$, and if $\vec{s}$ is the parametrization giving the position of the particle at time t in polar coordinates, then $\vec{s}(t_0) = \left(2, \frac{\pi}{4}\right)$ and $\vec{s}'(t_0) = (1, \pi)$.

$$D\vec{P}|_{\vec{s}(t_0)}\vec{s}'(t_0) = \begin{pmatrix} \cos(\theta(t_0)) & -r(t_0)\sin(\theta(t_0)) \\ \sin(\theta(t_0)) & r(t_0)\cos(\theta(t_0)) \end{pmatrix} \begin{pmatrix} r'(t_0) \\ \theta'(t_0) \end{pmatrix}$$

$$= \begin{pmatrix} \cos\left(\frac{\pi}{4}\right) & -2\sin\left(\frac{\pi}{4}\right) \\ \sin\left(\frac{\pi}{4}\right) & 2\cos\left(\frac{\pi}{4}\right) \end{pmatrix} \begin{pmatrix} 1 \\ \pi \end{pmatrix}$$

$$= \begin{pmatrix} \frac{\sqrt{2}}{2} & -\sqrt{2} \\ \frac{\sqrt{2}}{2} & \sqrt{2} \end{pmatrix} \begin{pmatrix} 1 \\ \pi \end{pmatrix}$$

$$= \begin{pmatrix} \frac{\sqrt{2}}{2} - \pi\sqrt{2} \\ \frac{\sqrt{2}}{2} + \pi\sqrt{2} \end{pmatrix}.$$

so the particle's x–coordinate is decreasing at the rate of $\left(\frac{\sqrt{2}}{2}\right) - \pi\sqrt{2}$, while the y–coordinate is increasing at the rate of $\left(\frac{\sqrt{2}}{2}\right) + \pi\sqrt{2}$. ■

We have defined the derivative of a function $\vec{F}$ from uv–space into xy–space (evaluated at (u, v)) to be the 2×2 matrix whose first row is $\nabla F_x|_{(u,v)}$, and whose second row is $\nabla F_y|_{(u,v)}$. Notice also that the first column of $D\vec{F}|_{(u,v)}$ is $\frac{\partial \vec{F}}{\partial u}$, and the second column is $\frac{\partial \vec{F}}{\partial v}$.

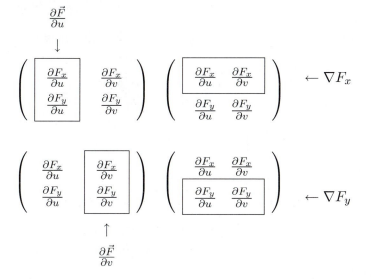

We define the derivative from $\mathbb{R}^n$ into $\mathbb{R}^m$ in a similar manner. If $\vec{F}$ is a function from uv–space into xyz–space, then we define the derivative of $\vec{F}$ to be the 3×2 matrix whose first row is the gradient of F_x, whose second row is the gradient of F_y, and whose third row is the gradient of F_z.

$$D\vec{F}|_{(u,v)} = \begin{pmatrix} \frac{\partial F_x}{\partial u} & \frac{\partial F_x}{\partial v} \\ \frac{\partial F_y}{\partial u} & \frac{\partial F_y}{\partial v} \\ \frac{\partial F_z}{\partial u} & \frac{\partial F_z}{\partial v} \end{pmatrix}_{\text{evaluated at } (u, v)}$$

If $\vec{F}$ is a function from uvw–space into xyz–space, then the derivative of $\vec{F}$ is a 3×3 matrix, the rows again being the gradients of the coordinate functions.

$$DF\vec{|}_{(u,v,w)} = \begin{pmatrix} \frac{\partial F_x}{\partial u} & \frac{\partial F_x}{\partial v} & \frac{\partial F_x}{\partial w} \\ \frac{\partial F_y}{\partial u} & \frac{\partial F_y}{\partial v} & \frac{\partial F_y}{\partial w} \\ \frac{\partial F_z}{\partial u} & \frac{\partial F_z}{\partial v} & \frac{\partial F_z}{\partial w} \end{pmatrix}_{\text{evaluated at } (u,v,w)}$$

Finally, if $\vec{F}$ is a function from uvw–space into xy–space, then the derivative of $\vec{F}$ is a 2×3 matrix, again with the rows being the gradients of the coordinate functions.

$$DF\vec{|}_{(u,v,w)} = \begin{pmatrix} \frac{\partial F_x}{\partial u} & \frac{\partial F_x}{\partial v} & \frac{\partial F_x}{\partial w} \\ \frac{\partial F_y}{\partial u} & \frac{\partial F_y}{\partial v} & \frac{\partial F_y}{\partial w} \end{pmatrix}_{\text{evaluated at } (u,v,w)}$$

In general, if $\vec{F}$ is a function from $\mathbb{R}^n$ into $\mathbb{R}^m$, then the derivative of $\vec{F}$ is an $m \times n$ matrix, with the rows being the gradients of the coordinate functions.

EXAMPLE 3: Let $\vec{T}$ be a linear transformation from $\mathbb{R}^3$ into $\mathbb{R}^3$ with the associated 3×3 matrix $A_{\vec{T}}$, so that $\vec{T}(\vec{p}) = A_{\vec{T}}\vec{p}$. Thus, if

$$\vec{p} = \begin{pmatrix} u \\ v \\ w \end{pmatrix} \text{ and } A_{\vec{T}} = \begin{pmatrix} a_1 & b_1 & c_1 \\ a_2 & b_2 & c_2 \\ a_3 & b_3 & c_3 \end{pmatrix},$$

then

$$A_{\vec{T}}\vec{p} = u \begin{pmatrix} a_1 \\ a_2 \\ a_3 \end{pmatrix} + v \begin{pmatrix} b_1 \\ b_2 \\ b_3 \end{pmatrix} + w \begin{pmatrix} c_1 \\ c_2 \\ c_3 \end{pmatrix}.$$

We see that

$$T_x(u,v,w) = (ua_1 + vb_1 + wc_1), \text{ and } \nabla T_x(u,v,w) = (a_1, b_1, c_1)$$

$$T_y(u,v,w) = (ua_2 + vb_2 + wc_2), \text{ and } \nabla T_x(u,v,w) = (a_2, b_2, c_2)$$

$$T_z(u,v,w) = (ua_3 + vb_3 + wc_3), \text{ and } \nabla T_x(u,v,w) = (a_3, b_3, c_3),$$

so $D\vec{T}|_{\vec{v}} = A_{\vec{T}}$.   ∎

Theorem 1 *If $\vec{T}$ is a linear transformation from $\mathbb{R}^n$ into $\mathbb{R}^m$ and $A_{\vec{T}}$ is the matrix associated with $\vec{T}$, then the derivative of $\vec{T}$ evaluated at any point in $\mathbb{R}^n$ is $A_{\vec{T}}$.*

EXAMPLE 4: Let $\vec{T}$ be defined by $\vec{T}(u, v) = (2u + v, v)$. Then

$$D\vec{T}|_{(u,v)} = A_{\vec{T}} = \begin{pmatrix} 2 & 1 \\ 0 & 1 \end{pmatrix}.$$

If a particle moves with velocity $(1, 2)$ in uv–space, then its velocity viewed in xy–space is

$$A_{\vec{T}}\begin{pmatrix} 1 \\ 2 \end{pmatrix} = \begin{pmatrix} 2 & 1 \\ 0 & 1 \end{pmatrix}\begin{pmatrix} 1 \\ 2 \end{pmatrix} = \begin{pmatrix} 4 \\ 2 \end{pmatrix}. \qquad \blacksquare$$

The derivative of a function from a subset of $\mathbb{R}^n$ into $\mathbb{R}^m$ satisfies the chain rule.

Theorem 2 (The Chain Rule: Functions from $\mathbb{R}^n$ into $\mathbb{R}^m$)
Suppose that S_1 is a subset of the reals, S_2 is a subset of $\mathbb{R}^n$, $\vec{r} : S_1 \to S_2$ and $\vec{F} : S_2 \to \mathbb{R}^m$ are differentiable. Let

$$\vec{g}(t) = \vec{F} \circ \vec{r}(t) = \vec{F}(\vec{r}(t)).$$

Then

$$\frac{d\vec{g}}{dt} = D\vec{F}|_{\vec{r}(t)}\vec{r}\,'(t).$$

EXAMPLE 5: If

$$\vec{r}(t) = (u(t), v(t), w(t))$$

is a differentiable function from a subset of the reals into uvw–space, and

$$\vec{F}(u, v, w) = (x(u, v, w), y(u, v, w), z(u, v, w))$$

is a differentiable function from a subset of uvw–space into xyz–space as in Figure 1, then

$$
\begin{aligned}
\frac{d\vec{g}}{dt} &= D\vec{F}|_{\vec{r}(t)}\vec{r}\,'(t) \\[2mm]
&= \begin{pmatrix} \frac{\partial x}{\partial u} & \frac{\partial x}{\partial v} & \frac{\partial x}{\partial w} \\[1mm] \frac{\partial y}{\partial u} & \frac{\partial y}{\partial v} & \frac{\partial y}{\partial w} \\[1mm] \frac{\partial z}{\partial u} & \frac{\partial z}{\partial v} & \frac{\partial z}{\partial w} \end{pmatrix} \begin{pmatrix} u'(t) \\[1mm] v'(t) \\[1mm] w'(t) \end{pmatrix} \\[2mm]
&= \begin{pmatrix} \frac{\partial x}{\partial u}\frac{du}{dt} + \frac{\partial x}{\partial v}\frac{dv}{dt} + \frac{\partial x}{\partial w}\frac{dw}{dt} \\[1mm] \frac{\partial y}{\partial u}\frac{du}{dt} + \frac{\partial y}{\partial v}\frac{dv}{dt} + \frac{\partial y}{\partial w}\frac{dw}{dt} \\[1mm] \frac{\partial z}{\partial u}\frac{du}{dt} + \frac{\partial z}{\partial v}\frac{dv}{dt} + \frac{\partial z}{\partial w}\frac{dw}{dt} \end{pmatrix}
\end{aligned}
$$

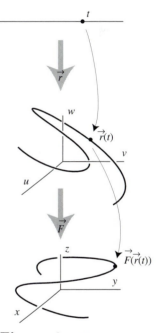

Figure 1. *The composite of $\vec{F}$ following $\vec{r}$.*

It is critical to remember that all of the entries in

$$
D\vec{F} = \begin{pmatrix} \frac{\partial x}{\partial u} & \frac{\partial x}{\partial v} & \frac{\partial x}{\partial w} \\ \frac{\partial y}{\partial u} & \frac{\partial y}{\partial v} & \frac{\partial y}{\partial w} \\ \frac{\partial z}{\partial u} & \frac{\partial z}{\partial v} & \frac{\partial z}{\partial w} \end{pmatrix}
$$

are evaluated at $\vec{r}(t)$. Often, you will see the coordinate functions written out separately as follows.

$$
\begin{aligned}
\frac{dx}{dt} &= \frac{\partial x}{\partial u}\frac{du}{dt} + \frac{\partial x}{\partial v}\frac{dv}{dt} + \frac{\partial x}{\partial w}\frac{dw}{dt} \\
\frac{dy}{dt} &= \frac{\partial y}{\partial u}\frac{du}{dt} + \frac{\partial y}{\partial v}\frac{dv}{dt} + \frac{\partial y}{\partial w}\frac{dw}{dt} \\
\frac{dz}{dt} &= \frac{\partial z}{\partial u}\frac{du}{dt} + \frac{\partial z}{\partial v}\frac{dv}{dt} + \frac{\partial z}{\partial w}\frac{dw}{dt},
\end{aligned}
$$

where all of the terms $\frac{\partial x}{\partial u}$, $\frac{\partial x}{\partial v}$, $\frac{\partial x}{\partial w}$, $\frac{\partial y}{\partial u}$, etc. are evaluated at $\vec{r}(t) = (u(t), v(t), w(t))$. ∎

EXAMPLE 6: Suppose that

$$
\vec{r}(t) = (u(t), v(t))
$$

is a differentiable function from a subset of the reals into uv–space, and

$$
\vec{F}(u, v) = (x(u, v), y(u, v), z(u, v))
$$

is a differentiable function from a subset of uv–space into xyz–space (as in Figure 1, except that the intermediate space is a plane rather than a three dimensional space.) Then

$$
\frac{d\vec{g}}{dt} = D\vec{F}\big|_{\vec{r}(t)}\,\vec{r}\,'(t) = \begin{pmatrix} \frac{\partial x}{\partial u} & \frac{\partial x}{\partial v} \\ \frac{\partial y}{\partial u} & \frac{\partial y}{\partial v} \\ \frac{\partial z}{\partial u} & \frac{\partial z}{\partial v} \end{pmatrix} \begin{pmatrix} \frac{du}{dt} \\ \frac{dv}{dt} \end{pmatrix} = \begin{pmatrix} \frac{\partial x}{\partial u}\frac{du}{dt} + \frac{\partial x}{\partial v}\frac{dv}{dt} \\ \frac{\partial y}{\partial u}\frac{du}{dt} + \frac{\partial y}{\partial v}\frac{dv}{dt} \\ \frac{\partial z}{\partial u}\frac{du}{dt} + \frac{\partial z}{\partial v}\frac{dv}{dt} \end{pmatrix}.
$$

The coordinate functions of $\frac{d\vec{g}}{dt}$ written separately give

$$
\frac{dx}{dt} = \frac{\partial x}{\partial u}\frac{du}{dt} + \frac{\partial x}{\partial v}\frac{dv}{dt}
$$

$$\frac{dy}{dt} = \frac{\partial y}{\partial u}\frac{du}{dt} + \frac{\partial y}{\partial v}\frac{dv}{dt}$$

$$\frac{dz}{dt} = \frac{\partial z}{\partial u}\frac{du}{dt} + \frac{\partial z}{\partial v}\frac{dv}{dt}. \qquad \blacksquare$$

The Chain Rule for Partial Derivatives

Suppose that $\vec{F}$ is a differentiable function from uv–space into xy–space, and that ϕ is a real-valued differentiable function defined on a subset of xy–space that contains the range of $\vec{F}$. Then $\phi \circ \vec{F} = \phi(\vec{F})$ is a real-valued function defined on uv–space. We want to find $\frac{\partial(\phi \circ \vec{F})}{\partial u} =$ the rate that ϕ changes as u changes.

To set this up, assume that

$$\vec{F}(u,v) = (x(u,v), y(u,v))$$

and let $\vec{r}(t) = (u_1 + t, v_1)$, for some point (u_1, v_1), and $\vec{g}(t) = \vec{F}(\vec{r}(t))$. By the definition of the partial derivative for real-valued differentiable functions, we see that

$$\left.\frac{\partial \phi \circ \vec{F}(u,v)}{\partial u}\right|_{(u_1,v_1)} = \left.\frac{d\phi(\vec{g}(t))}{dt}\right|_{t=0} = \nabla\phi|_{\vec{g}(0)} \cdot \left.\frac{d\vec{g}(t)}{dt}\right|_{t=0}.$$

We know that

$$\nabla\phi|_{\vec{g}(0)} = \left.\left(\frac{\partial\phi}{\partial x}, \frac{\partial\phi}{\partial y}\right)\right|_{\vec{g}(0)} = \left.\left(\frac{\partial\phi}{\partial x}, \frac{\partial\phi}{\partial y}\right)\right|_{\vec{F}(u_1,v_1)}.$$

We need to calculate $\frac{d\vec{g}(t)}{dt}|_{t=0}$.

$$\left.\frac{d\vec{g}(t)}{dt}\right|_{t=0} = \left. D\vec{F}\right|_{\vec{r}(0)} \vec{r}'(0)$$

$$= \left.\begin{pmatrix} \frac{\partial x}{\partial u} & \frac{\partial x}{\partial v} \\[2mm] \frac{\partial y}{\partial u} & \frac{\partial y}{\partial v} \end{pmatrix}\right|_{\vec{r}(0)} \begin{pmatrix} 1 \\[1mm] 0 \end{pmatrix}$$

$$= \left.\left(\frac{\partial x}{\partial u}, \frac{\partial y}{\partial u}\right)\right|_{\vec{r}(0)}$$

$$= \left.\left(\frac{\partial x}{\partial u}, \frac{\partial y}{\partial u}\right)\right|_{(u_1,v_1)}.$$

Putting it together we obtain

$$\frac{\partial \phi \circ \vec{F}(u,v)}{\partial u}\bigg|_{(u_1,v_1)} = \left(\frac{\partial \phi}{\partial x}, \frac{\partial \phi}{\partial y}\right)\bigg|_{\vec{F}(u_1,v_1)} \cdot \left(\frac{\partial x}{\partial u}, \frac{\partial y}{\partial u}\right)\bigg|_{(u_1,v_1)}.$$

More simply, we write

$$\frac{\partial \phi \circ \vec{F}}{\partial u} = \left(\frac{\partial \phi}{\partial x}, \frac{\partial \phi}{\partial y}\right) \cdot \left(\frac{\partial x}{\partial u}, \frac{\partial y}{\partial u}\right) = \frac{\partial \phi}{\partial x}\frac{\partial x}{\partial u} + \frac{\partial \phi}{\partial y}\frac{\partial y}{\partial u}.$$

Similarly we can use $\vec{r}(t) = (u_1, v_1 + t)$ and obtain

$$\frac{\partial \phi \circ \vec{F}}{\partial v} = \left(\frac{\partial \phi}{\partial x}, \frac{\partial \phi}{\partial y}\right) \cdot \left(\frac{\partial x}{\partial v}, \frac{\partial y}{\partial v}\right) = \frac{\partial \phi}{\partial x}\frac{\partial x}{\partial v} + \frac{\partial \phi}{\partial y}\frac{\partial y}{\partial v}.$$

It is understood that $\frac{\partial \phi}{\partial x}$ and $\frac{\partial \phi}{\partial y}$ are evaluated at $\vec{F}(u,v) = (x(u,v), y(u,v))$.

In much the same way, we obtain the three dimensional case, which we present as the following chain rule for partial derivatives without proof.

Theorem 3 (The Chain Rule for Partial Derivatives)
Suppose

$$\vec{F}(u,v,w) = (x(u,v,w), y(u,v,w), z(u,v,w))$$

is a differentiable function from uvw–space into xyz–space, and ϕ is a real valued differentiable function defined on a subset of xyz– space that contains the image of $\vec{F}$. Then

$$\frac{\partial(\phi \circ \vec{F})}{\partial u}(u,v,w) = \nabla \phi \cdot \frac{\partial \vec{F}}{\partial u}$$

$$= \frac{\partial \phi}{\partial x}\frac{\partial x}{\partial u} + \frac{\partial \phi}{\partial y}\frac{\partial y}{\partial u} + \frac{\partial \phi}{\partial z}\frac{\partial z}{\partial u},$$

$$\frac{\partial(\phi \circ \vec{F})}{\partial v}(u,v,w) = \nabla \phi \cdot \frac{\partial \vec{F}}{\partial v}$$

$$= \frac{\partial \phi}{\partial x}\frac{\partial x}{\partial v} + \frac{\partial \phi}{\partial y}\frac{\partial y}{\partial v} + \frac{\partial \phi}{\partial z}\frac{\partial z}{\partial v},$$

and

$$\frac{\partial(\phi \circ \vec{F})}{\partial w}(u, v, w) = \nabla\phi \cdot \frac{\partial \vec{F}}{\partial w}$$

$$= \frac{\partial\phi}{\partial x}\frac{\partial x}{\partial w} + \frac{\partial\phi}{\partial y}\frac{\partial y}{\partial w} + \frac{\partial\phi}{\partial z}\frac{\partial z}{\partial w}$$

$\nabla\phi$, $\frac{\partial\phi}{\partial x}$, $\frac{\partial\phi}{\partial x}$, and $\frac{\partial\phi}{\partial x}$ are evaluated at $\vec{F}(u, v, w)$. The partial derivatives of x, y, and z are evaluated at (u, v, w). It is common in the literature to write $\frac{\partial\phi}{\partial u}$ to mean $\frac{\partial(\phi \circ \vec{F})}{\partial u}$.

There are, of course, similar results if, for example,

$$\vec{F}(u, v, w) = \begin{pmatrix} x(u, v, w) \\ y(u, v, w) \end{pmatrix}$$

is a differentiable function from uvw–space into xy–space, and ϕ is a real-valued differentiable function defined on a subset of xy–space that contains the image of $\vec{F}$, or if

$$\vec{F}(u, v) = \begin{pmatrix} x(u, v) \\ y(u, v) \\ z(u, v) \end{pmatrix}$$

is a differentiable function from uv–space into xyz–space, and ϕ is a real-valued differentiable function defined on a subset of xyz–space that contains the domain of $\vec{F}$.[1]

---

[1]The general statement of the theorem is as follows.
If $\vec{F}(u_1, \ldots, u_n) = (x_1(u_1, \ldots, u_n), \cdots, x_m(u_1, \ldots, u_n))$ is a differentiable function from a subset of $\mathbb{R}^n$ into $\mathbb{R}^m$, and $\phi(x_1, \cdots, x_m)$ is a real-valued differentiable function defined on a subset of $\mathbb{R}^m$ that contains the image of $\vec{F}$, then

$$\frac{\partial(\phi \circ \vec{F})}{\partial u_i}(u_1, \ldots, u_n) = \nabla\phi\Big|_{\vec{F}(u_1,\ldots,u_n)} \cdot \frac{\partial \vec{F}}{\partial u_i}(u_1, \ldots, u_n)$$

$$= \frac{\partial\phi}{\partial x_1}\frac{\partial x_1}{\partial u_i} + \cdots + \frac{\partial\phi}{\partial x_m}\frac{\partial x_m}{\partial u_i}$$

EXAMPLE 7: Let $\vec{P}(r, \theta) = (r\cos\theta, r\sin\theta)$ and let $\phi(x, y) = x^2 + y^2$. Then

$$\nabla\phi(x, y) = (2x, 2y), \text{ so } \nabla\phi(P(r, \theta)) = (2r\cos\theta, 2r\sin\theta).$$

We also have

$$\frac{\partial P}{\partial r}(r, \theta) = (\cos\theta, \sin\theta) \text{ and } \frac{\partial P}{\partial\theta}(r, \theta) = (-r\sin\theta, r\cos\theta).$$

Thus

$$
\begin{aligned}
\frac{\partial\phi}{\partial r} &= \nabla\phi(P(r, \theta)) \cdot \frac{\partial P}{\partial r}(r, \theta) \\
&= (2r\cos\theta, 2r\sin\theta) \cdot (\cos\theta, \sin\theta) \\
&= 2r\cos^2\theta + 2r\sin^2\theta = 2r,
\end{aligned}
$$

and

$$
\begin{aligned}
\frac{\partial\phi}{\partial\theta} &= \nabla\phi(P(r, \theta)) \cdot \frac{\partial P}{\partial\theta}(r, \theta) \\
&= (2r\cos\theta, 2r\sin\theta) \cdot (-r\sin\theta, r\cos\theta) . \\
&= -2r^2\cos\theta\sin\theta + 2r^2\cos\theta\sin\theta = 0.
\end{aligned}
$$

As we would expect, a direct computation produces the same results. $\phi(P(r, \theta)) = r^2$. Thus

$$\frac{\partial}{\partial r}(r^2) = 2r \text{ and } \frac{\partial}{\partial\theta}(r^2) = 0. \qquad\blacksquare$$

Linear Approximations

Suppose that $\vec{h}(u, v) = (x(u, v), y(u, v), z(u, v))$ is a function from $\mathbb{R}^2$ into $\mathbb{R}^3$. Let $\vec{r}_0 = (u_0, v_0)$ be a fixed point. Let $\vec{r}(t)$ be defined by $\vec{r}_1(t) = \vec{h}(t, v_0)$ and $\vec{r}_2(t) = \vec{h}(u_0, t)$. Then

$$\frac{\partial h}{\partial u}(u_0, r_0) = \vec{r}'_1(u_0) \quad \text{and} \quad \frac{\partial h}{\partial v}(u_0, r_0) = \vec{r}'_2(v_0).$$

$\vec{r}'_1(u_0)$ is tangent to the curve parametrized by it, which in turn lines in the image of $\vec{h}$. It follows that $\vec{U} = \dfrac{\partial\vec{h}}{\partial u}(u_0, v_0)$ is tangent to the image of $\vec{h}$. Similarly, $\vec{V} = \dfrac{\partial\vec{h}}{\partial v}(u_0, v_0)$ is tangent to the image of $\vec{h}$.

Thus, as long as $\vec{U} \times \vec{V} \neq \vec{0}$, they determine the plane tangent to the surface at $\vec{h}(u_0, v_0)$. See Figure 2. This plane is the image of

$$
\begin{aligned}
\vec{T}(s,t) &= s\vec{U} + t\vec{V} + \vec{r}_0 \\[2mm]
&= D(\vec{h})\big|_{(u_0,v_0)} \begin{pmatrix} s \\ t \end{pmatrix} + \vec{h}(u_0, v_0) \\[2mm]
&= \begin{pmatrix} \frac{\partial x}{\partial u}(u_0, v_0) & \frac{\partial x}{\partial v}(u_0, v_0) \\[2mm] \frac{\partial y}{\partial u}(u_0, v_0) & \frac{\partial y}{\partial v}(u_0, v_0) \\[2mm] \frac{\partial z}{\partial u}(u_0, v_0) & \frac{\partial z}{\partial v}(u_0, v_0) \end{pmatrix} \begin{pmatrix} s \\ t \end{pmatrix} + \begin{pmatrix} x(u_0, v_0) \\[2mm] y(u_0, v_0) \\[2mm] z(u_0, v_0) \end{pmatrix}.
\end{aligned}
$$

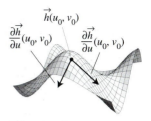

Figure 2. *The vectors* $\frac{\partial \vec{h}}{\partial u}(u_0, v_0)$ *and* $\frac{\partial \vec{h}}{\partial v}(u_0, u_0)$ *determine the plane tangent to the surface at* $\vec{h}(u_0, v_0)$

Note: $\vec{T}$ is a linear transformation translated to the point of tangency.

EXAMPLE 8: Let $\vec{S}(\phi, \theta) = (\sin(\phi)\cos(\theta), \sin(\phi)(\sin(\theta), \cos(\phi))$. Then the image of $\vec{S}$ is the unit sphere centered at the origin. As illustrated in Figures 3.a and 3.b, $\frac{\partial \vec{S}}{\partial \theta}(\phi_0, \theta_0)$ is tangent to a latitude circle and $\frac{\partial \vec{S}}{\partial \phi}(\phi_0, \theta_0)$ is tangent to a longitudinal circle. We leave it as an exercise to show that their cross product points away from the origin (its radial), so the plane determined by these vectors drawn emanating from the origin is tangent to the sphere. To find the plane tangent to the sphere at $\vec{S}\left(\frac{\pi}{4}, \frac{\pi}{4}\right)$, we find the derivative of $\vec{S}$ at that point.

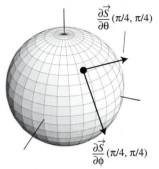

Figure 3.a *The vectors* $\frac{\partial \vec{S}}{\partial \theta}$ *and* $\frac{\partial \vec{S}}{\partial \phi}$ *determine the plane to the sphere at* $\vec{S}\left(\frac{\pi}{4}, \frac{\pi}{4}\right)$.

$$x(\phi, \theta) = \sin(\phi)\cos(\theta),$$

$$\frac{\partial x}{\partial \phi} = \cos(\phi)\cos(\theta), \quad \frac{\partial x}{\partial \phi}\left(\frac{\pi}{4}, \frac{\pi}{4}\right) = \frac{1}{2},$$

and

$$\frac{\partial x}{\partial \theta} = -\sin(\phi)\sin(\theta), \quad \frac{\partial x}{\partial \theta}\left(\frac{\pi}{4}, \frac{\pi}{4}\right) = -\frac{1}{2}.$$

Similarly, we get

$$\frac{\partial y}{\partial \phi}\left(\frac{\pi}{4}, \frac{\pi}{4}\right) = \frac{1}{2}, \quad \frac{\partial z}{\partial \phi}\left(\frac{\pi}{4}, \frac{\pi}{4}\right) = -\frac{1}{\sqrt{2}},$$

and

$$\frac{\partial y}{\partial \theta}\left(\frac{\pi}{4}, \frac{\pi}{4}\right) = \frac{1}{2}, \quad \frac{\partial z}{\partial \theta}\left(\frac{\pi}{4}, \frac{\pi}{4}\right) = 0.$$

Figure 3.b *The approximating tangent plane to the sphere.*

The tangent plane is the image of

$$\vec{T}(s,t) = \begin{pmatrix} \frac{1}{2} & -\frac{1}{2} \\ \frac{1}{2} & \frac{1}{2} \\ -\frac{1}{\sqrt{2}} & 0 \end{pmatrix} \begin{pmatrix} s \\ t \end{pmatrix} + \begin{pmatrix} \frac{1}{2} \\ \frac{1}{2} \\ \frac{1}{\sqrt{2}} \end{pmatrix} = \begin{pmatrix} \frac{s}{2} - \frac{t}{2} + \frac{1}{2} \\ \frac{s}{2} + \frac{t}{2} + \frac{1}{2} \\ -\frac{s}{\sqrt{2}} + \frac{1}{\sqrt{2}} \end{pmatrix}. \quad \blacksquare$$

We have been focusing on tangent planes because they are relatively easy to visualize and, inspecting the basic formula, they closely parallel the earlier work we have done with tangent lines. The parallel does not stop there. Recall that the first order Taylor approximation for a real valued function is given by

$$p_1(x) = f'(x_0)(x - x_0) + f(x_0).$$

In exactly the same way, we can approximate functions from $\mathbb{R}^n$ into $\mathbb{R}^m$. Let $\vec{h}$ be a function from $\mathbb{R}^n$ into $\mathbb{R}^m$. The function

$$\vec{p}(\vec{r}) = D(\vec{h})|_{\vec{r}_0}[\vec{r} - \vec{r}_0] + \vec{h}(\vec{r}_0)$$

is called the *first order Taylor approximation* for $\vec{h}$ at $\vec{r}_0$. Just as in the first order Taylor polynomials for real valued functions, if $\|\vec{r} - \vec{r}_0\|$ is small, then $\|\vec{p}(\vec{r}) - \vec{h}(\vec{r})\|$ is small. In fact, if all of the partial derivatives are continuous, then

$$\lim_{\|\vec{r} - \vec{r}_0\| \to 0} \frac{1}{\|\vec{r} - \vec{r}_0\|} \|\vec{p}(\vec{r}) - \vec{h}(\vec{r})\| = 0.$$

EXAMPLE 9: The first order Taylor polynomial for the function $\vec{S}$ from Example 8 at $\left(\frac{\pi}{4}, \frac{\pi}{4}\right)$ is

$$\vec{p}(\phi, \theta) = \begin{pmatrix} \frac{1}{2} & -\frac{1}{2} \\ \frac{1}{2} & \frac{1}{2} \\ -\frac{1}{\sqrt{2}} & 0 \end{pmatrix} \begin{pmatrix} \phi - \frac{\pi}{4} \\ \theta - \frac{\pi}{4} \end{pmatrix} + \begin{pmatrix} \frac{1}{2} \\ \frac{1}{2} \\ \frac{1}{\sqrt{2}} \end{pmatrix}.$$

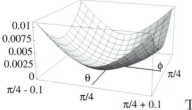

Figure 4. *The error function over the interval $\frac{\pi}{4} - 0.1 \le \phi \le \frac{\pi}{4} + 0.1$ and $\frac{\pi}{4} - 0.1 \le \theta \le \frac{\pi}{4} + 0.1$*

The approximating plane is displayed in Figure 3.b. In Figure 4, we sketch the error function $\|\vec{p}(\phi, \theta) - \vec{s}(\phi, \theta)\|$ over the rectangle $\frac{\pi}{4} - 0.1 \le \phi \le \frac{\pi}{4} + 0.1$ and $\frac{\pi}{4} - 0.5 \le \theta \le \frac{\pi}{4} + 0.5$. Inspection of the figure reveals that in this region, the error on the approximating function is small, near $\left(\frac{\pi}{4}, \frac{\pi}{4}\right)$. $\quad \blacksquare$

Summary

(a) If $\vec{F}$ is a function from $\mathbb{R}^n$ into $\mathbb{R}^m$, then the derivative of $\vec{F}$ evaluated at $\vec{r}$ is defined to be the $m \times n$ matrix such that for each i, with $1 \leq i \leq m$, the ith row of $D(\vec{F})|_{\vec{r}}$ is the gradient of the ith coordinate function of $\vec{F}$ evaluated at $\vec{r}$.

(b) If $\vec{r}$ is a function from $\mathbb{R}$ into $\mathbb{R}^n$ and $\vec{F}$ is a function from $\mathbb{R}^n$ into $\mathbb{R}^m$, then

$$\frac{d}{dt}\vec{F}(\vec{r}(t)) = D\vec{F}|_{\vec{r}(t)}\vec{r}\,'(t).$$

If

$$\vec{F}(u, v, w) = \begin{pmatrix} x(u, v, w) \\ y(u, v, w) \\ z(u, v, w) \end{pmatrix} \quad \text{and} \quad \vec{r}(t) = \begin{pmatrix} u(t) \\ v(t) \\ z(t) \end{pmatrix},$$

then

$$\frac{dx}{dt} = \nabla x(u, v, w)|_{\vec{r}(t)} \cdot \vec{r}\,'(t) = \frac{\partial x}{\partial u}\frac{du}{dt} + \frac{\partial x}{\partial v}\frac{dv}{dt} + \frac{\partial x}{\partial w}\frac{dw}{dt}$$

$$\frac{dy}{dt} = \nabla y(u, v, w)|_{\vec{r}(t)} \cdot \vec{r}\,'(t) = \frac{\partial y}{\partial u}\frac{du}{dt} + \frac{\partial y}{\partial v}\frac{dv}{dt} + \frac{\partial y}{\partial w}\frac{dw}{dt}$$

$$\frac{dz}{dt} = \nabla z(u, v, w)|_{\vec{r}(t)} \cdot \vec{r}\,'(t) = \frac{\partial z}{\partial u}\frac{du}{dt} + \frac{\partial z}{\partial v}\frac{dv}{dt} + \frac{\partial z}{\partial w}\frac{dw}{dt}.$$

(c) If $\vec{T}$ is a linear transformation from $\mathbb{R}^n$ into $\mathbb{R}^m$, and if $A_{\vec{T}}$ is the $m \times n$ matrix associated to $\vec{T}$, then the derivative of $\vec{T}$ is the constant matrix $A_{\vec{T}}$.

(d) If $\vec{F}$ is a differentiable function from a subset of uvw–space into xyz–space, and ϕ is a real-valued differentiable function from a subset of xyz–space containing the image of $\vec{F}$, then

$$\frac{\partial(\phi \circ \vec{F})}{\partial u}(u, v, w) = \nabla\phi|_{\vec{F}(u,v,w)} \cdot \frac{\partial \vec{F}}{\partial u}(u, v, w),$$

which is usually shortened to

$$\frac{\partial \phi}{\partial u} = \frac{\partial \phi}{\partial x}\frac{\partial x}{\partial u} + \frac{\partial \phi}{\partial y}\frac{\partial y}{\partial u} + \frac{\partial \phi}{\partial z}\frac{\partial z}{\partial u}.$$

(e)

$$\vec{p}(\vec{r}) = D(\vec{h})|_{\vec{r}_0} \cdot (\vec{r} - \vec{r}_0) + \vec{h}(\vec{r}_0)$$

is the *first order Taylor approximation* for $\vec{h}$ at $\vec{r}_0$.

EXERCISES 13.3

In Exercises 1–7, find the derivative of the linear transformations.

1. $\vec{T}(u, v) = (3u, -2v)$.

2. $\vec{T}(u, v) = (u + 3v, v - u)$.

3. $\vec{T}(u, v) = (u - v, 2v, u + v)$.

4. $\vec{T}(s, t) = (s - t, 3t + s, t - s)$.

5. $\vec{T}(u, v) = u + v$.

6. $\vec{T}(u, v, w) = (2u - v, u + w, u + v + w)$.

7. $\vec{T}(r, s, t) = (2r + s - t, r - s - 3t, r + s + t)$.

8. $\vec{T}(u, v, w) = (2u - w + v, u + v - 22w)$.

9. $\vec{T}(u, v, w) = (u + v + w, u + v + w)$.

10. $\vec{T}(u, v, w) = u + v + w$.

11. Find the derivative of the cylindrical transformation $\vec{C}_z$.

12. Find the derivative of the cylindrical transformation $\vec{C}_x$.

13. Find the derivative of the spherical transformation.

In Exercises 14–19, $\vec{F}$ is a function from $\mathbb{R}^n$ into $\mathbb{R}^m$. Given the derivative of $\vec{F}$ at $\vec{r}_0$, determine n and m.

14. $D\vec{F}|_{\vec{r}_0} = \begin{pmatrix} 1 & 2 \\ -1 & 0 \end{pmatrix}$.

15. $D\vec{F}|_{\vec{r}_0} = \begin{pmatrix} 0 & 2 \\ -1 & 3 \\ 1 & 0 \end{pmatrix}$.

16. $D\vec{F}|_{\vec{r}_0} = \begin{pmatrix} 1 & 2 & 1 \\ -1 & 0 & -1 \end{pmatrix}$.

17. $D\vec{F}|_{\vec{r}_0} = \begin{pmatrix} 1 \\ -1 \end{pmatrix}$.

18. $D\vec{F}|_{\vec{r}_0} = \begin{pmatrix} 1 & 2 & 3 \end{pmatrix}$.

19. $D\vec{F}|_{\vec{r}_0} = \begin{pmatrix} a_1 & a_2 & a_3 \\ b_1 & b_2 & b_3 \\ c_1 & c_2 & c_3 \end{pmatrix}$.

In Exercises 20–23, a particle's position $\vec{s}_0$ and velocity $\vec{v}_0$ are given in polar coordinates. Find the position and velocity in rectangular coordinates.

20. $\vec{s}_0 = (1, \pi), \quad \vec{v}_0 = (2, 3)$.

21. $\vec{s}_0 = \left(2, \frac{\pi}{2}\right), \quad \vec{v}_0 = (-1, 2)$.

22. $\vec{s}_0 = \left(1, \frac{\pi}{3}\right), \quad \vec{v}_0 = (1, -2)$.

23. $\vec{s}_0 = \left(1, \frac{7\pi}{4}\right), \quad \vec{v}_0 = (2, 3)$.

In Exercises 24–27, a particle's position $\vec{s}_0$ and velocity $\vec{v}_0$ are given in cylindrical coordinates. Find the position and velocity in rectangular coordinates.

24. $\vec{s}_0 = (1, \pi, 1), \quad \vec{v}_0 = (2, 3, 1)$.

25. $\vec{s}_0 = \left(-1, \frac{\pi}{4}, 2\right), \quad \vec{v}_0 = (-1, 2, 0)$.

26. $\vec{s}_0 = \left(1, \frac{\pi}{3}, 1\right), \quad \vec{v}_0 = (1, -2, -2)$.

27. $\vec{s}_0 = \left(1, \frac{7\pi}{4}, -2\right), \quad \vec{v}_0 = (2, 3, 1)$.

In Exercises 28–31, a particle's position $\vec{s}_0$ and velocity $\vec{v}_0$ are given in spherical coordinates. Find the position and velocity in rectangular coordinates.

28. $\vec{s}_0 = \left(1, \pi, \frac{\pi}{4}\right), \quad \vec{v}_0 = (2, 3, 1)$.

29. $\vec{s}_0 = \left(-1, \frac{\pi}{4}, \frac{\pi}{2}\right), \quad \vec{v}_0 = (-1, 2, 0)$.

30. $\vec{s}_0 = \left(1, \frac{\pi}{3}, \frac{\pi}{3}\right), \quad \vec{v}_0 = (1, -2, -2).$

31. $\vec{s}_0 = \left(1, \frac{7\pi}{4}, -\frac{\pi}{3}\right), \quad \vec{v}_0 = (2, 3, 1).$

32. Find the derivative of the function $\vec{h}(u, v) = (u\cos(v), u\sin(v), u).$

33. Find the derivative of the function $\vec{h}(u, v) = (u^2\cos(v), u^2\sin(v), u).$

In Exercises 34–36, $\phi(x, y, z) = e^{xy+z}$. Use the chain rule for partial derivatives to calculate $\frac{\partial\phi}{\partial u} = \frac{\partial(\phi \circ \vec{F})}{\partial u}$ and $\frac{\partial\phi}{\partial v} = \frac{\partial(\phi \circ \vec{F})}{\partial v}$ for the given function $\vec{F}$.

34. $\vec{F}(u, v) = (u + v, v, uv).$

35. $\vec{F}(u, v) = (u, u\cos v, u\sin v).$

36. $\vec{F}(u, v) = (\cos v \sin u, \sin v \sin u, \cos u).$

In Exercises 37–39, $\vec{P}(r, \theta) = (r\cos\theta, r\sin\theta)$ is the polar transformation. Use the chain rule to calculate $\frac{\partial(\phi \circ P)}{\partial\theta} = \frac{\partial(\phi)}{\partial\theta}$ for the given function ϕ.

37. $\phi(x, y) = x + y.$ 38. $\phi(x, y) = xy^2.$

39. $\phi(x, y) = \ln x + 3y.$

In Exercises 40–42, $\vec{S}(\rho, \phi, \theta) = (\rho\cos\theta\sin\phi, \rho\sin\theta \sin\phi, \rho\cos\phi)$ is the spherical transformation. Use the chain rule to calculate $\frac{\partial(\phi \circ P)}{\partial\theta} = \frac{\partial(\phi)}{\partial\theta}$ for the given function ϕ.

40. $\phi(x, y) = x + y + z.$ 41. $\phi(x, y) = x^2 + y^2 + z^2.$

42. $\phi(x, y) = xyz.$

In Exercises 43–45, you are given the derivative of $\vec{F}$, $\vec{r}_0$, $\vec{F}(\vec{r}_0)$, and $\Delta\vec{r}$. Use Taylor's first order approximation for $\vec{F}$ at $\vec{r}_0$ to approximate $\vec{F}(\vec{r}_0 + \Delta\vec{r})$.

43. $D\vec{F}\big|_{\vec{r}_0} = \begin{pmatrix} 1 & 2 \\ -1 & 0 \end{pmatrix}, \quad \vec{r}_0 = \begin{pmatrix} 1 \\ -1 \end{pmatrix},$

$\vec{F}(\vec{r}_0) = \begin{pmatrix} 2 \\ 3 \end{pmatrix}$ and $\Delta\vec{r} = \begin{pmatrix} 0.1 \\ -0.05 \end{pmatrix}.$

44. $D\vec{F}\big|_{\vec{r}_0} = \begin{pmatrix} 1 & 0 & 1 \\ -1 & 1 & 0 \end{pmatrix}, \quad \vec{r}_0 = \begin{pmatrix} 1 \\ -1 \\ 2 \end{pmatrix},$

$\vec{F}(\vec{r}_0) = \begin{pmatrix} -2 \\ 1 \end{pmatrix}$ and $\Delta\vec{r} = \begin{pmatrix} 0.1 \\ -0.1 \\ 0.02 \end{pmatrix}.$

45. $D\vec{F}\big|_{\vec{r}_0} = \begin{pmatrix} 1 & 2 \\ -1 & 0 \\ 0 & 1 \end{pmatrix}, \quad \vec{r}_0 = \begin{pmatrix} 0 \\ 1 \end{pmatrix},$

$\vec{F}(\vec{r}_0) = \begin{pmatrix} 2 \\ 1 \\ 3 \end{pmatrix}$ and $\Delta\vec{r} = \begin{pmatrix} 0.1 \\ -0.1 \end{pmatrix}.$

In Exercises 46–49, find the first order Taylor polynomial for the polar transformation $\vec{P}$ at $\vec{r}_0$.

46. $\vec{r}_0 = \left(2, \frac{\pi}{2}\right)$ 47. $\vec{r}_0 = \left(-2, \frac{\pi}{3}\right)$

48. $\vec{r}_0 = \left(1, \frac{5\pi}{3}\right)$ 49. $\vec{r}_0 = \left(-1, \frac{\pi}{6}\right)$

In Exercises 50–53, find the first order Taylor polynomial for the cylindrical transformation $\vec{C}_z$ at $\vec{r}_0$.

50. $\vec{r}_0 = \left(2, \frac{\pi}{2}, 3\right)$ 51. $\vec{r}_0 = \left(-2, \frac{3\pi}{2}, -1\right)$

52. $\vec{r}_0 = \left(1, \frac{5\pi}{3}, 0\right)$ 53. $\vec{r}_0 = \left(-1, \frac{\pi}{6}, 5\right)$

In Exercises 54–57, find the first order Taylor polynomial for the spherical transformation $\vec{S}$ at $\vec{r}_0$.

54. $\vec{r}_0 = \left(2, \frac{\pi}{2}, \frac{\pi}{4}\right)$ 55. $\vec{r}_0 = \left(-2, \frac{3\pi}{2}, \frac{\pi}{4}\right)$

56. $\vec{r}_0 = \left(1, \frac{5\pi}{3}, \frac{\pi}{3}\right)$ 57. $\vec{r}_0 = \left(-1, \frac{\pi}{6}, \frac{5\pi}{6}\right)$

58. Use the Taylor polynomial from Exercise 46 to approximate $\vec{P}(1.9, \frac{\pi}{2} + 0.2).$

59. Use the Taylor polynomial from Exercise 53 to approximate $\vec{C}_z(\vec{r}_0 + \Delta\vec{r})$, where $\Delta\vec{r} = (-0.1, 0.2, 0.1).$

60. Use the Taylor polynomial from Exercise 55 to approximate $\vec{S}(\vec{r}_0 + \Delta\vec{r})$, where $\Delta\vec{r} = (-0.1, 0.2, 0.1).$

13.4   Arc Length for Curves in Other Coordinate Systems

Arc Length for Polar Coordinates

Suppose that $\vec{u}(t) = (r(t), \theta(t))$, $a \leq t \leq b$, is a parametrization of a path in the plane given by polar coordinates. Then $(x(t), y(t)) = \vec{s}(t) = \vec{P}(\vec{u}(t))$ is a parametrization for the path in rectangular coordinates. The derivative is given by

$$\frac{d}{dt}(x(t), y(t)) = \frac{d\vec{s}(t)}{dt} = \begin{pmatrix} \cos(\theta(t)) & -r(t)\sin(\theta(t)) \\ \sin(\theta(t)) & r(t)\cos(\theta(t)) \end{pmatrix} \begin{pmatrix} r'(t) \\ \theta'(t) \end{pmatrix}.$$

Thus

$$\begin{pmatrix} x'(t) \\ y'(t) \end{pmatrix} = \vec{s}'(t) = \begin{pmatrix} r'(t)\cos(\theta(t)) - \theta'(t)r(t)\sin(\theta(t)) \\ r'(t)\sin(\theta(t)) + \theta'(t)r(t)\cos(\theta(t)) \end{pmatrix}.$$

We now have

$$\|\vec{s}'(t)\| = \sqrt{(x'(t))^2 + (y'(t))^2}$$

$$= \left(((r'(t)\cos(\theta(t)) - \theta'(t)r(t)\sin(\theta(t)))^2 + (r'(t)\sin(\theta(t)) + \theta'(t)r(t)\cos(\theta(t)))^2 \right)^{1/2}.$$

Suppressing t and expanding, we obtain

$$\|\vec{s}'\| = \left(r'^2\cos^2\theta + \theta'^2 r^2\sin^2\theta - 2r'\theta'r\cos\theta\sin\theta + r'^2\sin^2\theta + \theta'^2 r^2\cos^2\theta + 2r'\theta'r\sin\theta\cos\theta \right)^{1/2}$$

$$= \left(r'^2(\cos^2\theta + \sin^2\theta) + \theta'^2 r^2(\sin^2\theta + \cos^2\theta) \right)^{1/2}$$

$$= \sqrt{(r'(t))^2 + (\theta'(t)r(t))^2}.$$

Thus the length of the curve, where $\vec{u}(a) = \vec{A}$, $\vec{u}(b) = \vec{B}$, and the image of $\vec{u}$ is C, is given by

$$L = \int_{\vec{A}_C}^{\vec{B}} d\vec{s} = \int_a^b \sqrt{(r'(t))^2 + (\theta'(t)r(t))^2}\, dt. \qquad (1)$$

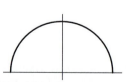

Figure 1. *The graph $\vec{P}(C)$ is the top half of the circle with radius k.*

EXAMPLE 1: Let C be the graph with polar equation $r = k$, for $0 \leq \theta \leq \pi$. Thus C is a vertical line in $r\theta$–space. The graph $\vec{P}(C)$ is the top half of the circle with radius k. (See Figure 1.)

A parametrization for C in polar coordinates is $\vec{u}(t) = (k, t), 0 \leq t \leq \pi$. The arc length is given by $\int_{A_C}^{B} d\vec{s}$, where $\vec{A} = \vec{P}(\vec{u}(0)) = (k, 0)$ and $\vec{B} = \vec{P}(\vec{u}(\pi)) = (-k, 0)$. $r'(t) = 0$ and $\theta'(t) = 1$. Using Equation (1), we obtain

$$L = \int_{(k,0)_C}^{(-k,0)} d\vec{s} = \int_0^{\pi} \sqrt{(r'(t))^2 + (\theta'(t)r(t))^2}\, dt$$

$$= \int_0^{\pi} \sqrt{k^2}\, dt = \pi k. \qquad \blacksquare$$

EXAMPLE 2: Let C be the graph with polar equation $r = e^{\theta}$, for $0 \leq \theta \leq \ln 2$. A parametrization for C in polar coordinates is $\vec{u}(t) = (e^t, t), 0 \leq t \leq \ln 2$. Thus $r'(t) = e^t$ and $\theta'(t) = 1$. The arc length is given by $\int_{A_C}^{B} d\vec{s}$, where $\vec{A} = \vec{P}(u(0)) = (1, 0)$ and $\vec{B} = \vec{P}(u(\ln 2)) = (2\cos(\ln 2), 2\sin(\ln 2))$. Using Equation (1), we obtain

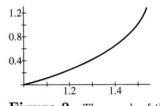

Figure 2. *The graph of the curve in Example 2.*

$$L = \int_{(1,0)}^{(2\cos(\ln 2), 2\sin(\ln 2))} d\vec{s} = \int_0^{\ln 2} \sqrt{(r'(t))^2 + (\theta'(t)r(t))^2}\, dt$$

$$= \int_0^{\ln 2} \sqrt{e^{2t} + e^{2t}}\, dt$$

$$= \sqrt{2} \int_0^{\ln 2} e^t\, dt$$

$$= 2\sqrt{2} - \sqrt{2} = \sqrt{2}. \qquad \blacksquare$$

Arc Length for Cylindrical Coordinates

In a similar fashion, we can compute arc length in cylindrical coordinates in $\mathbb{R}^3$. If $\vec{u}(t) = (r(t), \theta(t), z(t))$ is a parametrization of a path in $\mathbb{R}^3$, then $(x(t), y(t), z(t)) = \vec{s}(t) = \vec{C}(\vec{u}(t))$ is a parametrization of the path in rectangular coordinates. The derivative is given by

$$\frac{d}{dt}(x(t), y(t), z(t)) = \frac{d\vec{s}(t)}{dt}$$

$$= \begin{pmatrix} \cos(\theta(t)) & -r(t)\sin(\theta(t)) & 0 \\ \sin(\theta(t)) & r(t)\cos(\theta(t)) & 0 \\ 0 & 0 & 1 \end{pmatrix} \begin{pmatrix} r'(t) \\ \theta'(t) \\ z'(t) \end{pmatrix}.$$

Thus

$$\begin{pmatrix} x'(t) \\ y'(t) \\ z'(t) \end{pmatrix} = \vec{s}\,'(t) = \begin{pmatrix} r'(t)\cos(\theta(t)) - \theta'(t)r(t)\sin(\theta(t)) \\ r'(t)\sin(\theta(t)) + \theta'(t)r(t)\cos(\theta(t)) \\ z'(t) \end{pmatrix}.$$

Suppressing the variable t,

$$\begin{aligned} \|\vec{s}\,'\| &= \sqrt{x'^2 + y'^2 + z'^2} \\[2mm] &= \sqrt{(r'\cos\theta - \theta'r\sin\theta)^2 + (r'\sin\theta + \theta'r\cos\theta)^2 + z'^2} \\[2mm] &= \sqrt{r'^2{}^2\cos^2\theta + \theta'^2 r^2\sin^2\theta - 2r'\theta'r\cos\theta\sin\theta +} \\ &\qquad \overline{r'^2\sin^2\theta + \theta'^2 r^2\cos^2\theta + 2r'\theta'r\sin\theta\cos\theta + z'^2} \\[2mm] &= \sqrt{r'^2(\cos^2\theta + \sin^2\theta) + \theta'^2 r^2(\cos^2\theta + \sin^2\theta) + z'^2} \\[2mm] &= \sqrt{r'^2 + r^2\theta'^2 + z'^2}. \end{aligned}$$

Figure 3. *The helix from Example 3.*

Thus the length of the curve is given by

$$L = \int_{\vec{A}_C}^{\vec{B}} d\vec{r} = \int_a^b \sqrt{r'^2(t) + r^2(t)\theta'^2 + z'^2(t)}\ dt. \tag{2}$$

EXAMPLE 3: Let $\vec{u}(t) = (1, 2\pi t, t)$, for $0 \le t \le 3$, be a parametrization of a helix in cylindrical coordinates. Find the length of the helix. See Figure 3.

SOLUTION: $\vec{s}(t) = \vec{C}(\vec{u}(t))$. By Equation (2),

$$\begin{aligned} \|\vec{s}\,'(t)\| &= \sqrt{r'^2(t) + r^2(t)\theta'^2(t) + z'^2(t)} \\[2mm] &= \sqrt{(2\pi)^2 + 1} = \sqrt{4\pi^2 + 1} \end{aligned}$$

Thus the arc length is given by

$$L = \int_0^3 \sqrt{4\pi^2 + 1}\ dt = 3\sqrt{4\pi^2 + 1}. \qquad\blacksquare$$

Arc Length for Spherical Coordinates

For spherical coordinates we follow a similar procedure. Let $\vec{u}(t) = (\rho(t), \phi(t),\ \theta(t))$ be a parametrization of a path in $\mathbb{R}^3$ in spherical coordinates. Then $(x(t), y(t), z(t)) = \vec{s}(t) = \vec{S}(\vec{u}(t))$ will be a parametrization of the path in rectangular coordinates. The derivative is given by

$$\frac{d}{dt}(x(t), y(t), z(t)) = \frac{d\vec{s}(t)}{dt}$$

$$= \begin{pmatrix} \cos\theta(t)\sin\phi(t) & \rho(t)\cos\theta(t)\cos\phi(t) & -\rho(t)\sin\theta(t)\sin\phi(t) \\ \sin\theta(t)\sin\phi(t) & \rho(t)\sin\theta(t)\cos\phi(t) & \rho(t)\cos\theta(t)\sin\phi(t) \\ \cos\phi(t) & -\rho(t)\sin\phi(t) & 0 \end{pmatrix} \begin{pmatrix} \rho'(t) \\ \phi'(t) \\ \theta'(t) \end{pmatrix}.$$

Suppressing the variable t,

$$\begin{pmatrix} x' \\ y' \\ z' \end{pmatrix} = \vec{s}' = \begin{pmatrix} \rho'\cos\theta\sin\phi + \phi'\rho\cos\theta\cos\phi - \theta'\rho\sin\theta\sin\phi \\ \rho'\sin\theta\sin\phi + \phi'\rho\sin\theta\cos\phi + \theta'\rho\cos\theta\sin\phi \\ \rho'\cos\phi - \phi'\rho\sin\phi \end{pmatrix}.$$

We now have

$$\begin{aligned} \|\vec{s}'\| &= \sqrt{x'^2 + y'^2 + z'^2} \\ &= \big((\rho'\cos\theta\sin\phi + \phi'\rho\cos\theta\cos\phi - \theta'\rho\sin\theta\sin\phi)^2 \\ &\quad + (\rho'\sin\theta\sin\phi + \phi'\rho\sin\theta\cos\phi + \theta'\rho\cos\phi\sin\phi)^2 \\ &\quad + (\rho'\cos\phi - \theta'\rho\sin\phi)^2 \big) \\ &= \sqrt{\rho'^2 + \rho^2\phi'^2 + \rho^2\theta'^2\sin^2\phi}. \end{aligned}$$

By expanding and simplifying, we obtain

$$L = \int_{\vec{A}_C}^{\vec{B}} d\vec{r} = \int_a^b \sqrt{\rho'^2(t) + \rho^2(t)\phi'^2(t) + \rho^2(t)\theta'^2(t)\sin^2\phi(t)}\ dt.$$

EXERCISES 13.4

In Exercises 1–5, find the length, in xy–space, of the graphs with the given polar equation.

1. $r = e^{2\theta}, \quad 0 \le \theta \le \ln 3$.

2. $r = 2\cos(\theta), \quad 0 \le \theta \le 2\pi$.

3. $r = 3\sec(\theta), \quad 0 \le \theta \le \frac{\pi}{4}$.

4. $r = \cos^2(\theta/2), \quad 0 \le \theta \le 2\pi$.

5. $r = \sin^2(\theta/2), \quad 0 \le \theta \le 2\pi$.

6. Let $\vec{h}(t) = (\rho(t), \phi(t), \theta(t))$ describe, in spherical coordinates, the location of a particle at time t. Show that the magnitude of the particle's velocity in xyz–space is given by

$$\sqrt{\rho'^2(t) + \rho^2(t)\phi'^2(t) + \rho^2(t)\theta'^2(t)\sin^2\phi(t)}.$$

7. Find the arc length of the helix given (in xyz–coordinates) by

$$\vec{r}(t) = (\cos(2\pi t), \sin(2\pi t), t), 0 \le t \le 4,$$

using both the xyz–coordinate version of the arc length integral and the cylindrical coordinate version.

8. The path parametrized by

$$\vec{r}(t) = \begin{pmatrix} \cos(2\pi t)\sin\left(\frac{t\pi}{4}\right) \\ \sin(2\pi t)\sin\left(\frac{t\pi}{4}\right) \\ \cos\left(\frac{t\pi}{4}\right) \end{pmatrix}, \ 0 \le t \le 4.$$

in xyz–coordinates describes a "spiral" along the unit sphere centered at the origin, starting at the north pole and descending to the south pole. Compute the arc length of this path using both xyz–coordinates and spherical coordinates.

9. Let $\vec{r}(t) = (1, 2\pi t, t), 0 \le t \le 4$, represent the helix of Exercise 3 in cylindrical coordinates.

   a. Show that the helix can be represented as

   $$\vec{u}(t) = \left(\sqrt{t^2 + 1}, \text{Arcsin}\left(\frac{1}{\sqrt{t^2+1}} \right), 2\pi t \right),$$

   $0 \le t \le 4$, in spherical $\rho\phi\theta$–coordinates.

   b. Compute the arc length in spherical coordinates, and compare it to Exercise 7.

13.5   Change of Area with Linear Transformations

In this section, we introduce the idea of the rate that a linear transformation changes area or volume. First, it is helpful to recall some facts about areas of parallelograms and volumes of parallelepipeds.

- Suppose that the vectors $\vec{A}$ and $\vec{B}$ are drawn emanating from a common point in $\mathbb{R}^3$ forming adjacent edges of a parallelogram $\mathcal{P}$. Then $\|\vec{A} \times \vec{B}\|$ is the area of $\mathcal{P}$.

- If the vectors $\vec{A} = (a_1, b_1)$ and $\vec{B} = (b_1, b_2)$ are drawn emanating from the origin in $\mathbb{R}^2$ forming adjacent edges of a parallelogram $\mathcal{P}$, then $\|(a_1, a_2, 0) \times (b_1, ba_2, 0)\| = |a_1 b_2 - a_2 b_1|$ is the area of $\mathcal{P}$.

- $\left| \det \begin{pmatrix} a_1 & a_2 \\ b_1 & b_2 \end{pmatrix} \right| = |a_1 b_2 - a_2 b_1|$ (which is the area of $\mathcal{P}$).

- If the vectors $\vec{A} = (a_1, a_2, a_3)$, $\vec{B} = (b_1, b_2, b_3)$ and $\vec{C} = (c_1, c_2, c_3)$ are drawn emanating from the origin and they do not lie in a common plane (they are not co-planer), then they form the adjacent edges of a parallelepiped, $\mathcal{P}$. The magnitude of their triple product is the volume of $\mathcal{P}$. Computationally, the volume of $\mathcal{P}$ is

$$|\vec{A} \cdot (\vec{B} \times \vec{C})| = \left| \det \begin{pmatrix} a_1 & a_2 & a_3 \\ b_1 & b_2 & b_3 \\ c_1 & c_2 & c_3 \end{pmatrix} \right|.$$

The following theorem is be useful in this and subsequent sections.

Theorem 1

$$\det \begin{pmatrix} a_1 & a_2 \\ b_1 & b_2 \end{pmatrix} = \det \begin{pmatrix} a_1 & b_1 \\ a_2 & b_2 \end{pmatrix}$$

and

$$\det \begin{pmatrix} a_1 & a_2 & a_3 \\ b_1 & b_2 & b_3 \\ c_1 & c_2 & c_3 \end{pmatrix} = \det \begin{pmatrix} a_1 & b_1 & c_1 \\ a_2 & b_2 & c_2 \\ a_3 & b_3 & c_3 \end{pmatrix}.$$

While this theorem is easily proven with direct computation, it seems rather remarkable that the parallelepiped determined by the vectors (a_1, a_2, a_3), (b_1, b_2, b_3), and (c_1, c_2, c_3) has the same volume as the parallelepiped determined by (a_1, b_1, c_1), (a_2, b_2, c_2), and (a_3, b_3, c_3).

Let $\vec{T}$ be defined by

$$\vec{T}(u, v) = u\vec{A} + v\vec{B},$$

where $\vec{A}$ and $\vec{B}$ are vectors in $\mathbb{R}^3$. If $\vec{A}$ and $\vec{B}$ are drawn emanating from the origin, they are adjacent sides of a parallelogram P with area $\|\vec{A} \times \vec{B}\|$. Let R be the unit square in uv–space with adjacent sides the vectors $(1, 0)$ and $(0, 1)$. The area of R is one square unit, and $\vec{T}(R) = P$, which has an area of $\|\vec{A} \times \vec{B}\|$. (See Figure 1.) Thus $\vec{T}$ will take one square unit of area onto a parallelogram having area $\|\vec{A} \times \vec{B}\|$. It turns out that $\|\vec{A} \times \vec{B}\|$ can properly be thought of as *the rate that T changes area.* If C is a set in uv–space, then the area of $\vec{T}(C)$ is $\mathrm{Area}(C)\|\vec{A} \times \vec{B}\|$.

If $\vec{A} = (a_1, a_2)$ and $\vec{B} = (b_1, b_2)$ are in $\mathbb{R}^2$, then

$$\|(a_1, a_2, 0) \times (b_1, b_2, 0)\| = |a_1 b_2 - a_2 b_1| = \left| \det D\vec{T} \right|$$

gives the area of the parallelogram determined by $\vec{A}$ and $\vec{B}$. Thus, if $\vec{T}$ is a linear transformation from $\mathbb{R}^2$ into $\mathbb{R}^2$, then $|\det D(\vec{T})|$ is the rate that $\vec{T}$ changes area.

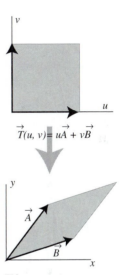

Figure 1. *The image of the unit square $0 \leq u \leq 1$, $0 \leq v \leq 1$ is the parallelogram with adjacent edges $\vec{A}$ and $\vec{B}$ drawn emanating from the origin.*

In the same fashion, if $\vec{T}$ is a linear transformation from $\mathbb{R}^3$ into $\mathbb{R}^3$, then $\left|\det(D\vec{T})\right|$ is the rate that $\vec{T}$ changes volume.

To recapitulate:

- If $\vec{T}(r) = r\vec{A}$ is a linear transformation from $\mathbb{R}$ into $\mathbb{R}^n$, then $\|\vec{A}\|$ is the rate that $\vec{T}$ changes length.

- If $\vec{T}(\vec{r}) = A_T\vec{r}$ is a linear transformation from $\mathbb{R}^2$ into $\mathbb{R}^2$, then $|\det A_T|$ is the rate that $\vec{T}$ changes area.

- If $\vec{T}(\vec{r}) = A_T\vec{r}$ is a linear transformation from $\mathbb{R}^2$ into $\mathbb{R}^3$, then $\left\|\vec{T}(1,0) \times \vec{T}(0,1)\right\|$ is the rate that $\vec{T}$ changes area.

- If $\vec{T}(\vec{r}) = A_T\vec{r}$ is a linear transformation from $\mathbb{R}^3$ into $\mathbb{R}^3$, then $|\det A_T|$ is the rate that $\vec{T}$ changes volume.

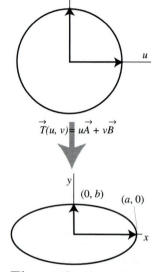

Figure 2. *The image of the unit disc $u^2 + v^2 \leq 1$ is the ellipse $\left(\frac{x}{a}\right)^2 + \left(\frac{y}{b}\right)^2 \leq 1$*

EXAMPLE 1: Let $\vec{A} = (a,0)$ and $\vec{B} = (0,b)$, where a and b are positive numbers. Let $\vec{T}$ be defined by

$$\vec{T}(u,v) = u\vec{A} + v\vec{B} = (au, bv).$$

Let C be the unit circle $u^2 + v^2 = 1$. Then $\vec{T}(C)$ is the ellipse with equation $\left(\frac{x}{a}\right)^2 + \left(\frac{y}{b}\right)^2 = 1$. See Figure 2. The area bounded by C is π and the area bounded by $\vec{T}(C)$ is $\|\vec{A} \times \vec{B}\|\pi = ab\pi$. Thus the area bounded by the ellipse $\left(\frac{x}{a}\right)^2 + \left(\frac{y}{b}\right)^2 = 1$ is $ab\pi$. This agrees with the value obtained by integration. ∎

EXAMPLE 2: Let $\mathcal{C}$ be the unit circle in uv–space.

(a) Let $\vec{T}(u,v) = \begin{pmatrix} 2u + 3v \\ u \\ u + v \end{pmatrix}$. Then $\vec{T}(1,0) = (2,1,1)$ and $\vec{T}(0,1) = (3,0,1)$. The rate that $\vec{T}$ changes area is $\|(2,1,1)\times(3,0,1)\| = \sqrt{11}$. The area of $\vec{T}(\mathcal{C})$ is $\pi\sqrt{11}$.

(b) Let $\vec{T}(u,v) = \begin{pmatrix} 2u + 3v \\ u \end{pmatrix}$. The rate that $\vec{T}$ changes area is

$$\left| \det\left(D\vec{T}\right) \right| = \left|\det\begin{pmatrix} 2 & 3 \\ 1 & 0 \end{pmatrix}\right| = 3.$$

Since the area of the unit circle is π, the area of $\vec{T}(\mathcal{C})$ is 3π. ∎

EXAMPLE 3: Let S denote the surface bounded between the graphs of $v = u^2$ and $v = u$. Let $\vec{T}(u, v) = (2u + v, v - u)$. Find the area of the image of S.

SOLUTION: The area of S is $\int_0^1 u - u^2\, du = \frac{1}{6}$. The rate that $\vec{T}$ changes area is $\left| \det \begin{pmatrix} 2 & 1 \\ 1 & -1 \end{pmatrix} \right| = 3$. Thus the area of $\vec{T}(S) = (3) * \left(\frac{1}{6} \right) = \frac{1}{2}$. ∎

EXAMPLE 4: Let $\vec{T}$ be the transformation defined by

$$\vec{T}(u, v, w) = (au, bv, cw).$$

Then $\vec{T}$ takes the unit sphere $u^2 + v^2 + w^2 \leq 1$ onto the ellipsoid E with equation $\left(\frac{x}{a} \right)^2 + \left(\frac{y}{b} \right)^2 + \left(\frac{z}{c} \right)^2 \leq 1$. See Figure 3 for the case that $a = 1$, $b = 2$, and $c = 3$.

$$DT = \begin{pmatrix} a & 0 & 0 \\ 0 & b & 0 \\ 0 & 0 & c \end{pmatrix}.$$

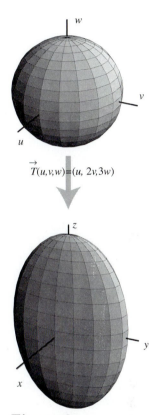

$\vec{T}(u,v,w) = (u, 2v, 3w)$

Thus $\vec{T}$ changes volume at a rate of abc, since the unit cube goes onto a box with sides of length a, b, and c. Since the volume bounded by a unit sphere is $\left(\frac{4}{3} \right)\pi$, the volume bounded by the ellipsoid E is $\frac{4}{3}abc\pi$. ∎

If $\vec{T}$ is a linear transformation from $\mathbb{R}^2$ into $\mathbb{R}$, then $\vec{T}$ takes an object with area onto an object with no area. Therefore, the rate that $\vec{T}$ changes area is zero. This is consistent with the fact that $\vec{T}(1, 0)$ and $\vec{T}(0, 1)$ must point in the same or opposite directions (they are either a positive number, a negative number, or zero.) Indeed, if the domain of $\vec{T}$ is $\mathbb{R}^2$, then:

Figure 3.
$\vec{T}(u, v, w) = (u, 2v, 3w)$ *takes the unit sphere onto the ellipse* $x^2 + \frac{y^2}{4} + \frac{z^2}{9} = 1$.

- If the range of $\vec{T}$ is $\mathbb{R}^1$, then the rate that $\vec{T}$ changes area is zero.

- If the range of $\vec{T}$ is $\mathbb{R}^2$ or $\mathbb{R}^3$, then $\vec{T}(0, 1) \times \vec{T}(1, 0) = \vec{0}$ if and only if the image of $\vec{T}$ is a line or the origin.

Similarly, if the domain of $\vec{T}$ is $\mathbb{R}^3$, then:

- If the range of $\vec{T}$ is either $\mathbb{R}$ or $\mathbb{R}^2$, then the rate that $\vec{T}$ changes volume is zero.

- If the range is $\mathbb{R}^3$ then $\det A_{\vec{T}} = 0$ if and only if the image of $\vec{T}$ is a plane, a line, or the origin.

Summary

If $\vec{T}$ is a linear transformation from $\mathbb{R}^2$ into $\mathbb{R}^2$, then the rate that $\vec{T}$ changes area is $|\det(A_{\vec{T}})| = |\det(DT)|$.

If $\vec{T}$ is a linear transformation from $\mathbb{R}^3$ into $\mathbb{R}^3$, then the rate that $\vec{T}$ changes volume is $|\det(A_{\vec{T}})| = |\det(DT)|$.

If $\vec{T}(u, v) = u\vec{A} + v\vec{B}$ is a linear transformation from $\mathbb{R}^2$ into $\mathbb{R}^3$, then the rate that $\vec{T}$ changes area is $\|\vec{A} \times \vec{B}\|$.

If $\vec{T}$ is a linear transformation from $\mathbb{R}^3$ into $\mathbb{R}^2$ or $\mathbb{R}$, then the rate that $\vec{T}$ changes volume is 0.

If $\vec{T}$ is a linear transformation from $\mathbb{R}^2$ into $\mathbb{R}$, then the rate that $\vec{T}$ changes volume is 0.

EXERCISES 13.5

In Exercises 1–5, determine the rate that each transformation changes area.

1. $\vec{T}(u, v) = (3u, -2v)$.

2. $\vec{T}(u, v) = (u + 3v, v - u)$.

3. $\vec{T}(u, v) = (u - v, 2v, u + v)$.

4. $\vec{T}(s, t) = (s - t, 3t + s, t - s)$.

5. $\vec{T}(u, v) = u + v$.

In Exercises 6–10, determine the rate that each transformation changes volume.

6. $\vec{T}(u, v, w) = (2u - v, u + w, u + v + w)$.

7. $\vec{T}(r, s, t) = (2r + s - t, r - s - 3t, r + s + t)$.

8. $\vec{T}(u, v, w) = (2u - w + v, u + v - 22w)$.

9. $\vec{T}(u, v, w) = (u + v + w, u + v + w)$.

10. $\vec{T}(u, v, w) = u + v + w$.

11. Find a linear transformation from the uv–plane into the xy–plane that takes the circle $u^2 + v^2 = 1$ onto the ellipse E with equation $4x^2 + 9y^2 = 36$. Find the area bounded by E using the techniques of this section.

12. Find the area bounded by the ellipse $2x^2 + 3y^2 = 5$.

13. Find a linear transformation from uvw–space onto xyz–space that takes the unit sphere centered at the origin onto the ellipsoid E with equation $3x^2 + 4y^2 + 2z^2 = 1$. Find the volume bounded by E.

14. Let $\mathcal{R}$ be the rectangle bounded between the lines $u = 2$, $u = -2$, $v = 0$, and $v = 4$ and let $\vec{T}(u, v) = (-u + v, u - 2v, 2u + 4v)$. Find the area of $\vec{T}(\mathcal{R})$.

15. Let $\mathcal{R}$ be the rectangle bounded between the lines $u = 2$, $u = -2$, $v = 2$, and $v = 6$ and let $\vec{T}(u, v) = (-u + v, 2u - 2v)$. Find the area of $\vec{T}(\mathcal{R})$.

16. Let $\mathcal{R}$ be the rectangle bounded between the lines $u = 2$, $u = -2$, $v = 2$, and $v = 6$ and let $\vec{T}(u, v) = (-u + v, u - 2v)$. Find the area of $\vec{T}(\mathcal{R})$.

17. Let $\mathcal{R}$ be the region bounded between the graphs of $v = u^2$ and $v = u^3$ and let $\vec{T}(u, v) = (u + v, u - 2v, 2u + v)$. Find the area of $\vec{T}(\mathcal{R})$.

18. Let $\mathcal{R}$ be the region bounded between the graphs of $u = v^2$ and $u = v^3$ and let $\vec{T}(u, v) = (3u + v, u - 2v)$. Find the area of $\vec{T}(\mathcal{R})$.

19. Let $\mathcal{C}$ be the box bounded between the planes $u = 2$, $u = -2$, $v = 3$, $v = 4$, $w = 0$, and $w = 10$. Let $\vec{T}(u, v, w) = (-u + v + 2w, u - 2v, 2u + 4v + w)$. Find the volume of $\vec{T}(\mathcal{C})$.

20. Let $\mathcal{E}$ be the ellipsoid $\frac{u^2}{4} + \frac{v^2}{9} + z^2 = 1$, and let $\vec{T}(u, v) = (u + v + 2w, u - 2v, 2u + v + w)$. Find the volume of $\vec{T}(\mathcal{E})$.

21. Let $\mathcal{V}$ be the solid obtained by rotating the region bounded between the u–axis and the graph of $v = u^2 + 1$, $-1 \le u \le 1$ about the u–axis. Let $\vec{T}(u, v, w) = (u + v + 2w, u - 2v, 2u + v + w)$. Find the volume of $\vec{T}(\mathcal{E})$.

13.6   The Jacobian

When a function from $\mathbb{R}^n$ into $\mathbb{R}^m$ is not linear, then the rate that the function changes area or volume becomes a local property. In this section, we learn how a function changes area or volume at a point. This idea is not really new. Let $\vec{f}$ be a function from $\mathbb{R}$ into $\mathbb{R}^n$. Recall the geometric interpretation of the derivative, in which $\|\vec{f}'(t)\|$ represents the rate that $\vec{f}$ changes arc length at t. To see that this is a reasonable interpretation, let $[t, t + h]$ be an interval in the domain of $\vec{f}$. (See Figure 1.) If h is small, then $\|\vec{f}(t + h) - \vec{f}(t)\|$ is an approximation of the length of $\vec{f}([t, t + h])$, and

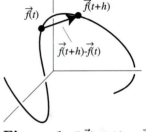

Figure 1. $\|\vec{f}(t + h) - \vec{f}(t)\|$ *approximates the arc length from $\vec{f}(t)$ to $\vec{f}(t + h)$.*

$$\|\vec{f}'(t)\| = \lim_{h \to 0} \frac{\text{length of } \vec{f}([t, t + h])}{\text{length of } [t, t + h]}$$

$$= \lim_{h \to 0} \frac{\|\vec{f}(t + h) - \vec{f}(t)\|}{h}.$$

The rate that $\vec{f}$ changes length at t is called the *Jacobian* of $\vec{f}$ at t. We denote the Jacobian of $\vec{f}$ at t by $J\vec{f}(t)$. Notice that if g is a function from a subset of $\mathbb{R}^n$ into $\mathbb{R}$ and if $\vec{r}$ is a parametrization for a curve C with endpoints $\vec{A}$ and $\vec{B}$ and domain $[a, b]$, then

$$\int_{\vec{A}_C}^{\vec{B}} g \, d\vec{r} = \int_a^b g(\vec{r}(t)) J\vec{r}(t) \, dt.$$

The Jacobian can also be defined for functions with domain in $\mathbb{R}^n$

for any positive integer n. However, we define it only for functions with domain in $\mathbb{R}^2$ or $\mathbb{R}^3$.

Let D be a subset of $\mathbb{R}^2$ (uv–space), and let $\vec{f}$ be a function from D into $\mathbb{R}^2$ or $\mathbb{R}^3$. We want the Jacobian of $\vec{f}$ at (u,v) to be the rate that $\vec{f}$ changes area at (u,v). Let S_h be the square having sides of length h with vertices $(u,v), (u+h,v), (u,v+h)$ and $(u+h,v+h)$. Assume that S_h is a subset of D. We approximate the surface $\vec{f}(S_h)$ with the parallelogram P_h with the adjacent sides the vectors $\vec{f}(u+h,v) - \vec{f}(u,v)$ and $\vec{f}(u,v+h) - \vec{f}(u,v)$, drawn emanating from $\vec{f}(u,v)$. See Figure 2.

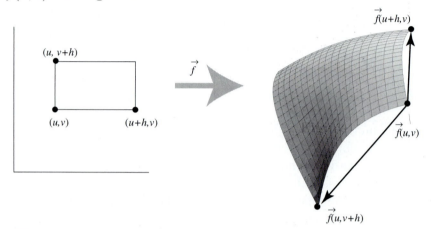

Figure 2. *The area of $\vec{f}(S_h)$ is approximated by the area of P_h.*

The area of P_h is $\|(\vec{f}(u+h,v) - \vec{f}(u,v)) \times (\vec{f}(u,v+h) - \vec{f}(u,v))\|$. The rate that $\vec{f}$ changes area at (u,v) is given by

$$
\begin{aligned}
\lim_{h \to 0} \frac{\text{area of } P_h}{\text{area of } S_h} &= \lim_{h \to 0} \frac{\|(\vec{f}(u+h,v) - \vec{f}(u,v)) \times (\vec{f}(u,v+h) - \vec{f}(u,v))\|}{h^2} \\
&= \lim_{h \to 0} \left\| \frac{\vec{f}(u+h,v) - \vec{f}(u,v)}{h} \times \frac{\vec{f}(u,v+h) - \vec{f}(u,v)}{h} \right\| \\
&= \left\| \frac{\partial \vec{f}(u,v)}{\partial u} \times \frac{\partial \vec{f}(u,v)}{\partial v} \right\|.
\end{aligned}
$$

Definition: The Jacobian for Functions from $\mathbb{R}^2$ to $\mathbb{R}^2$

If $\vec{f} \colon \mathbb{R}^2 \to \mathbb{R}^2$ is a differentiable function, we define the *Jacobian of $\vec{f}$*, denoted by $\vec{f}(u,v)$, to be

$$\left\| \frac{\partial \vec{f}(u,v)}{\partial u} \times \frac{\partial \vec{f}(u,v)}{\partial v} \right\|.$$

It represents the rate that $\vec{f}$ changes area at (u,v).

Of course, if $\vec{f}$ is a function from a subset of $\mathbb{R}^2$ into $\mathbb{R}$, then $J\vec{f}(u,v) = 0$.

EXAMPLE 1: Let $\vec{P}$ be the polar transformation from $r\theta$–space defined by

$$\vec{P}(r,\theta) = (r\cos\theta, r\sin\theta).$$

The partial derivatives of $\vec{P}$ are

$$\frac{\partial \vec{P}}{\partial r} = (\cos\theta, \sin\theta) \text{ and } \frac{\partial \vec{P}}{\partial \theta} = (-r\sin\theta, r\cos\theta).$$

Thus the Jacobian of $\vec{P}$ is given by

$$J\vec{P}(r,\theta) = \left\| \frac{\partial \vec{P}}{\partial r} \times \frac{\partial \vec{P}}{\partial \theta} \right\| = |r|,$$

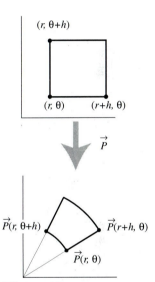

and is the rate that $\vec{P}$ changes area at (r,θ). This fits the geometry of the function very well. Recall that the area of a sector of a circle of radius R is $\frac{R^2(\Delta\theta)}{2}$, where $\Delta\theta$ is the angle of the sector. If we consider the square S in $r\theta$–space with sides of length h and one vertex (r,θ) as in Figure 3, then $\vec{P}(S)$ is the portion of a sector as in Figure 3.

The area of S is h^2, and the area of $\vec{P}(S)$ is $\frac{(r+h)^2 h}{2} - \frac{r^2 h}{2} = \frac{(2rh+h^2)h}{2}$. Thus, [area of $\vec{P}(S)$]/[area of S] $= \frac{(2rh+h^2)h/2}{h^2} = r + \frac{h}{2}$. It follows that as h gets close to 0, then [area of $\vec{P}(S)$]/[area of S] gets close to r. ∎

Figure 3. *The square S in $r\theta$–space. The sector $\vec{P}(S)$ in xy–space.*

EXAMPLE 2: Find the Jacobian of $\vec{h}(u,v) = (u, u^2\cos(v), u^2\sin(v))$.

SOLUTION:

$$\frac{\partial \vec{h}}{\partial u}(u,v) = \left(1, 2u\cos(v), 2u\sin(v)\right).$$

$$\frac{\partial \vec{h}}{\partial v}(u,v) = \left(0, -u^2\sin(v), u^2\cos(v)\right).$$

$$\frac{\partial \vec{h}}{\partial u} \times \frac{\partial \vec{h}}{\partial v} = \begin{pmatrix} -2u^3\cos^2(v) - 2u^3\sin^2(v) \\ -u^2\cos(v) \\ -u^2\sin^2(v) \end{pmatrix}.$$

$$J\vec{h}(u,v) \;=\; \left\|\frac{\partial \vec{h}}{\partial u} \times \frac{\partial \vec{h}}{\partial v}\right\| = \sqrt{4x^6 + x^4}. \qquad \blacksquare$$

Let D be a subset of $\mathbb{R}^3$, and let $\vec{f}$ be a function from D into $\mathbb{R}^3$. We are interested in how $\vec{f}$ changes volume at a point. We proceed exactly as we did with areas and functions from $\mathbb{R}^2$ into $\mathbb{R}^2$.

Let B_h be a square box with three adjacent edges the vectors $[(u+h,v,w)-(u,v,w)], [(u,v+h,w)-(u,v,w)]$, and $[(u,v,w+h)-(u,v,w)]$. Let P_h be the parallelepiped with three adjacent edges $[\vec{f}(u+h,v,w)-\vec{f}(u,v,w)], [\vec{f}(u,v+h,w)-\vec{f}(u,v,w)]$, and $[\vec{f}(u,v,w+h)-\vec{f}(u,v,w)]$. We now approximate the volume of $\vec{f}(B_h)$ with the volume of P_h (see Figure 4).

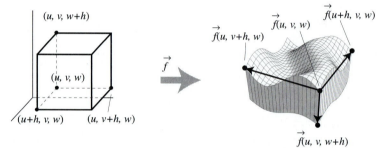

Figure 4. *The volume of $\vec{f}(B_h)$ is approximated by the volume of a parallelepiped.*

The rate that $\vec{f}$ changes volume is given by

$$\lim_{h\to 0} \frac{\text{volume of } P_h}{\text{volume of } B_h}$$

$$= \lim_{h\to 0} \left| \frac{\left(\vec{f}(u+h,v,w)-\vec{f}(u,v,w)\right)\left[\left(\vec{f}(u,v+h,w)-\vec{f}(u,v,w)\right) \times \left(\vec{f}(u,v,w+h)-\vec{f}(u,v,w)\right)\right]}{h^3} \right|$$

$$= \lim_{h\to 0} \left| \frac{\left(\vec{f}(u+h,v,w)-\vec{f}(u,v,w)\right)}{h} \cdot \left[\frac{\left(\vec{f}(u,v+h,w)-\vec{f}(u,v,w)\right)}{h} \times \frac{\left(\vec{f}(u,v,w+h)-\vec{f}(u,v,w)\right)}{h} \right] \right|$$

$$= \left| \frac{\partial \vec{f}}{\partial u} \cdot \left(\frac{\partial \vec{f}}{\partial v} \times \frac{\partial \vec{f}}{\partial w}\right) \right|.$$

> **Definition:** The Jacobian for Functions from $\mathbb{R}^3$ to $\mathbb{R}^3$
>
> ---
>
> If $\vec{f}\colon \mathbb{R}^3 \to \mathbb{R}^3$ is a differentiable function, we define the *Jacobian of* $\vec{f}$, denoted by $J\vec{f}(u,v,w)$, to be
>
> $$\left| \frac{\partial \vec{f}}{\partial u} \cdot \left(\frac{\partial \vec{f}}{\partial v} \times \frac{\partial \vec{f}}{\partial w} \right) \right|.$$
>
> It represents the rate that $\vec{f}$ changes volume at (u,v,w).

EXAMPLE 3: Let $\vec{C}_z$ be the cylindrical transformation from $r\theta z$–space into xyz–space defined by

$$\vec{C}_z(r,\theta,z) = (r\cos\theta, r\sin\theta, z).$$

Thus

$$\frac{\partial \vec{C}_z(r,\theta,z)}{\partial r} = (\cos\theta, \sin\theta, 0),$$

$$\frac{\partial \vec{C}_z(r,\theta,z)}{\partial \theta} = (-r\sin\theta, r\cos\theta, 0),$$

and

$$\frac{\partial \vec{C}_z(r,\theta,z)}{\partial z} = (0,0,1),$$

so

$$J\vec{C}_z(r,\theta,z) = \left| \frac{\partial \vec{C}_z}{\partial r} \cdot \left(\frac{\partial \vec{C}_z}{\partial \theta} \times \frac{\partial \vec{c}}{\partial z} \right) \right| = |r|.$$

EXAMPLE 4: Let S be the spherical transformation defined by

$$\vec{S}(\rho,\phi,\theta) = \begin{pmatrix} \rho\cos(\theta)\sin(\phi) \\ \rho\sin(\theta)\sin(\phi) \\ \rho\cos(\phi) \end{pmatrix}.$$

It is left as an exercise to show that $J\vec{S}(\rho,\phi,\theta) = \rho^2 \sin(\phi)$.

The following theorem follows from direct computation, and its proof is left as an exercise (Exercises 27 and 28).

Theorem 1 *If $\vec{F}$ is a differentiable function from a subset of $\mathbb{R}^2$ into $\mathbb{R}^2$ or from a subset of $\mathbb{R}^3$ into $\mathbb{R}^3$, and $\vec{r}$ is a point in $\mathbb{R}^2$ or $\mathbb{R}^3$, then*

$$J\vec{F}(\vec{r}) = \left| det\ D(\vec{F})|_{\vec{r}} \right|.$$

If $\vec{F}$ is a differentiable function and $\vec{G}(\vec{r}) = \vec{F}(\vec{r}) + \vec{r}_0$, then $J\vec{F} = J\vec{G}$.

The second part of Theorem 1 is not terribly surprising. A rigid motion such as a translation does not change area or volume. The next example leads to Theorem 2.

EXAMPLE 5: Let $\vec{T}_1(u,v) = \begin{pmatrix} 5u + v \\ -v \end{pmatrix}$, and let $\vec{T}_2(x,y) = \begin{pmatrix} 2x - 3y - 1 \\ x + y + 3 \end{pmatrix}$. Then

$$
\begin{aligned}
\vec{T}_2 \circ \vec{T}_1(u,v) &= \begin{pmatrix} 2(5u + v) - 3(-v) - 1 \\ (5u + v) + (-v) + 3 \end{pmatrix} \\
&= \begin{pmatrix} 10u + 5v - 1 \\ 5u + 3 \end{pmatrix}. \\
J(\vec{T}_2 \circ \vec{T}_1)(u,v) &= \left| det\, D\vec{T}_2 \circ \vec{T}_1(u,v) \right| \\
&= \left| det \begin{pmatrix} 10 & 5 \\ 5 & 0 \end{pmatrix} \right| = 25.
\end{aligned}
$$

Thus

$$J\vec{T}_1(u,v) = \left| det \begin{pmatrix} 5 & 1 \\ 0 & -1 \end{pmatrix} \right| = 5,$$

and

$$J\vec{T}_2(x,y) = \left| det \begin{pmatrix} 2 & -3 \\ 1 & 1 \end{pmatrix} \right| = 5.$$

Observe that

$$J(\vec{T}_2 \circ \vec{T}_1)(u,v) = JT_2(x,y)JT_2(u,v). \qquad \blacksquare$$

The observation in Example 5 is no coincidence. In general, the following theorem tells us that if $\vec{f} \circ \vec{g}$ is defined, then the rate that $\vec{f} \circ \vec{g}$ changes area (or volume) at $\vec{r}$ is the rate that $\vec{g}$ changes area (or volume) times the rate that $\vec{f}$ changes area (or volume) at $\vec{g}(\vec{r})$.

> **Theorem 2**
> *Suppose that $\vec{F}$ and $\vec{H}$ are functions from $\mathbb{R}^3$ into $\mathbb{R}^3$. Then $J(\vec{F} \circ \vec{H})(\vec{r}) = J\vec{F}(\vec{H}(\vec{r}))J\vec{H}(r).$*
>
> *Suppose that $\vec{F}$ is a function from $\mathbb{R}^2$ into $\mathbb{R}^2$ and $\vec{H}$ is a function from $\mathbb{R}^2$ into **either** $\mathbb{R}^2$ or $\mathbb{R}^3$. Then $J(\vec{F} \circ \vec{H})(\vec{r}) = J\vec{F}(\vec{H}(\vec{r}))J\vec{H}(r).$*

EXAMPLE 6: The polar function $\vec{P}$ changes area at a rate of r at the point (r, θ). The linear transformation $\vec{T}(x, y) = (2x + 3y, -x + y, x)$ changes area at the rate of $\sqrt{35}$. Then $J(\vec{T} \circ \vec{P})(r, \theta) = \sqrt{35}|r|$. If we compute $J(\vec{T} \circ \vec{P})(r, \theta)$ directly, we have

$$(\vec{T} \circ \vec{P})(r, \theta) = \begin{pmatrix} 2r\cos(\theta) + 3r\sin(\theta) \\ -r\cos(\theta) + r\sin(\theta) \\ r\cos(\theta) \end{pmatrix},$$

and

$$J(\vec{T} \circ \vec{P})(r, \theta)$$

$$= \left\| \frac{\partial}{\partial r}(\vec{T} \circ \vec{P})(r, \theta) \times \frac{\partial}{\partial \theta}(\vec{T} \circ \vec{P})(r, \theta) \right\|$$

$$= \left\| \begin{pmatrix} 2\cos(\theta) + 3\sin(\theta) \\ -\cos(\theta) + \sin(\theta) \\ \cos(\theta) \end{pmatrix} \times \begin{pmatrix} -2r\sin(\theta) + 3r\cos(\theta) \\ r\sin(\theta) + r\cos(\theta) \\ -r\sin(\theta) \end{pmatrix} \right\|$$

$$= \|(-r, 3r, -5r)\| = \sqrt{35r^2} = \sqrt{35}|r|. \qquad \blacksquare$$

EXERCISES 13.6

In Exercises 1–4, find the Jacobian of the polar transformation $\vec{P}$ at the given point. (The angles are measured in radians.)

1. $(2, 0)$.

2. $(2, 2\pi)$.

3. $(2, -2\pi)$.

4. $(-2, \pi)$.

In Exercises 5–8, find the Jacobian of the cylindrical transformation $\vec{C}_z$ at the given point. (The angles are measured in radians.)

5. $(2, 0, 2)$.

6. $(2, 2\pi, 2)$.

7. $(2, -2\pi, 2)$.

8. $(-2, \pi, -1)$.

In Exercises 9–12, find the Jacobian of the spherical transformation $S(\rho, \phi, \theta)$ at the given point. (The angles are measured in radians.)

9. $\left(2, 0, \frac{\pi}{4}\right)$.

10. $\left(-2, \pi, \frac{\pi}{4}\right)$.

11. $\left(-2, \frac{\pi}{4}, \frac{7\pi}{4}\right)$.

12. $\left(2, \frac{\pi}{3}, \frac{7\pi}{4}\right)$.

In Exercises 13–14, use Theorem 1 to find the Jacobian for the given function

13. $\vec{S}(r, \phi, \theta) + (2, 3, 4)$.

14. $\vec{P}(r, \theta) + (-1, 4)$.

In Exercises 15–17, you are given $J\vec{f}$ and $J\vec{g}$, find $J(\vec{f} \circ \vec{g})$ at $\vec{r}_0$.

15. $J\vec{g}(u, v) = |3u - 2v|$ and $J\vec{f}(x, y) = |x^2(y + 2) - x|$. $\vec{r}_0 = (1, 2)$ and $\vec{g}(\vec{r}_0) = (-1, 4)$.

16. $J\vec{g}(u, v) = \sqrt{u^2 + v^2}$ and $J\vec{f}(x, y) = |x^2 - xy|$. $\vec{r}_0 = (-1, 2)$ and $\vec{g}(\vec{r}_0) = (2, -1)$.

17. $J\vec{g}(u, v, w) = |w|\sqrt{u^2 + v^2}$ and $J\vec{f}(x, y, z) = |x^2 - xy + z|$. $\vec{r}_0 = (1, -1, 2)$ and $\vec{g}(\vec{r}_0) = (2, -1, 1)$.

Use Theorem 2 in Exercises 18–22 to find the Jacobian for $\vec{f} \circ \vec{g}$. $\vec{P}$ is the polar function and $\vec{S}$ is the spherical transformation.

18. $\vec{f} = \vec{P}$ and $\vec{g}(u, v) = (2u + 6v + 2, -u)$.

19. $\vec{f} = \vec{P}$ and $\vec{g}(u, v) = (-2u + 6v + 2, u + v - 1)$.

20. $\vec{f} = \vec{S}$ and $\vec{g}(u, v, w) = (-2u + 6v + 2w + 1, u + v + 2w - 1, u + v - 2w)$.

21. $\vec{f}(u, v) = (-2u + 6v + 2, u + v - 1)$ and $\vec{g} = \vec{P}$.

22. $\vec{f}(u, v, w) = (-2u + 6v + 2w + 1, u + v + 2w - 1, u + v - 2w)$ and $\vec{f} = \vec{S}$.

23. Show that $J(\vec{S}(\rho, \phi, \theta)) = \rho^2 \sin(\phi)$.

24. Calculate $J(\vec{f}(u, v))$ for $\vec{f}(u, v) = (2 \cos(u) \sin(v), 2 \sin(u) \sin(v), 2 \cos(v))$.

25. What is the image of $\vec{f}$ of Exercise 24?

26. Show that $J\vec{F} = J(\vec{F} + \vec{r}_0)$.

27. Let $\vec{F}$ be a differentiable function from a subset of $\mathbb{R}^2$ into $\mathbb{R}^2$. Show that

$$J(\vec{F}(u, v)) = |\det D(F)|\big|_{(u,v)}.$$

28. Let $\vec{F}$ be a differentiable function from a subset of $\mathbb{R}^3$ into $\mathbb{R}^3$. Show that

$$J(\vec{F}(u, v, w)) = |\det D(F)|\big|_{(u,v,w)}.$$

29. Suppose that $\vec{f}$ is a function from $\mathbb{R}^3$ into $\mathbb{R}^3$ and $\vec{g}$ is a function from $\mathbb{R}^2$ into $\mathbb{R}^3$. What can be said about the relationship between $J(\vec{f} \circ \vec{g})$ and the product $J(\vec{f})J(\vec{g})$?

Chapter 14

Multiple Integrals

14.1   The Integral over a Rectangle

Much of Chapter 11 was devoted to applying the integral to physical and geometric problems that are modeled on curves in space. For example, we learned to use line integrals to find mass centers and moments of inertia of wires in space. We now extend the notion of the integral and the associated notions of partition and selection to surfaces and solids in space. This allows us to calculate the mass of a surface or a solid. We start with the simple case of a rectangle in the plane.

EXAMPLE 1: Let R be the rectangle bounded by the lines $x = 0, x = 1, y = 0$, and $y = 2$. Suppose that the density of the surface R at the point (x, y) is given by $\rho(x, y) = x^2 + y^2$. Find the mass of the surface R.

SOLUTION: Let $\{x_0, x_1, \ldots, x_n\}$ be a partition of $[0, 1]$, and let $\{y_0, y_1, \ldots, y_m\}$ be a partition of $[0, 2]$, so that the lines $\{x = x_0, x = x_1, \ldots, x = x_n\} \cup \{y = y_0, y = y_1, \ldots, y = y_m\}$ partition R. For each $i \leq n$ and $j \leq m$, let $R_{i,j}$ be the rectangle bounded by the lines $x = x_{i-1}, x = x_i, y = y_{j-1}$, and $y = y_j$. This is illustrated in Figure 1. For each $i \leq n$ and $j \leq m$, let $M(R_{i,j})$ denote the mass of $R_{i,j}$.

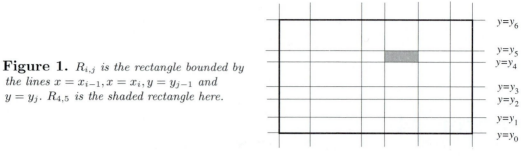

Figure 1. $R_{i,j}$ *is the rectangle bounded by the lines* $x = x_{i-1}, x = x_i, y = y_{j-1}$ *and* $y = y_j$. $R_{4,5}$ *is the shaded rectangle here.*

The mass of R is then given by

$$
\begin{aligned}
M(R) &= \quad M(R_{1,1}) + M(R_{1,2}) + \cdots + M(R_{1,m}) \\
&\quad + M(R_{2,1}) + M(R_{2,2}) + \cdots + M(R_{2,m}) \\
&\qquad \vdots \qquad\qquad \vdots \qquad\qquad\qquad \vdots \\
&\quad + M(R_{n,1}) + M(R_{n,2}) + \cdots + M(R_{n,m}) \\
&= \sum_{j=1}^{m} M(R_{1,j}) + \sum_{j=1}^{m} M(R_{2,j}) + \cdots + \sum_{j=1}^{m} M(R_{n,j}) \\
&= \sum_{i=1}^{n} \sum_{j=1}^{m} M(R_{i,j}).
\end{aligned}
$$

An approximation for $M(R_{i,j})$ is given by $\rho(x_i, y_j)(x_i - x_{i-1})(y_j - y_{j-1})$. So an approximation for $M(R)$ is

$$
\begin{aligned}
M(R) &= \sum_{i=1}^{n} \sum_{j=1}^{m} M(R_{i,j}) \approx \sum_{i=1}^{n} \sum_{j=1}^{m} \rho(x_i, y_j)(x_i - x_{i-1})(y_j - y_{j-1}) \\
&= \sum_{j=1}^{m} (x_1^2 + y_j^2)(x_1 - x_0)(y_j - y_{j-1}) \\
&\quad + \sum_{j=1}^{m} (x_2^2 + y_j^2)(x_2 - x_1)(y_j - y_{j-1}) \\
&\quad + \cdots + \sum_{j=1}^{m} (x_n^2 + y_j^2)(x_n - x_{n-1})(y_j - y_{j-1}) \\
&= \left(\sum_{j=1}^{m} (x_1^2 + y_j^2)(y_j - y_{j-1}) \right)(x_1 - x_0) \\
&\quad + \left(\sum_{j=1}^{m} (x_2^2 + y_j^2)(y_j - y_{j-1}) \right)(x_2 - x_1)
\end{aligned}
$$

$$+ \cdots + \left(\sum_{j=1}^{m} (x_n^2 + y_j^2)(y_j - y_{j-1}) \right) (x_n - x_{n-1}).$$

However,

$$\sum_{j=1}^{m} \left(x_i^2 + y_j^2 \right) (y_j - y_{j-1}) \approx \int_0^2 \left(x_i^2 + y^2 \right) \, dy$$

$$= \left. \left(x_i^2 y + \frac{y^3}{3} \right) \right|_{y=0}^{y=2} = 2x_i^2 + \frac{8}{3}.$$

Thus we have

$$M(R) \approx \left(\int_0^2 \left(x_1^2 + y^2 \right) \, dy \right) (x_1 - x_0) + \left(\int_0^2 \left(x_2^2 + y^2 \right) \, dy \right) (x_2 - x_1)$$

$$+ \cdots + \left(\int_0^2 \left(x_n^2 + y^2 \right) \, dy \right) (x_n - x_{n-1})$$

$$= \sum_{i=1}^{n} \left(\int_0^2 \left(x_i^2 + y^2 \right) \, dy \right) (x_i - x_{i-1})$$

$$= \sum_{i=1}^{n} \left(2x_i^2 + \frac{8}{3} \right) (x_i - x_{i-1})$$

$$\approx \int_0^1 \left(2x^2 + \frac{8}{3} \right) \, dx = \frac{2x^3}{3} + \frac{8x}{3} \bigg|_0^1 = \frac{10}{3}.$$

Or, to write this another way, we have

$$M(R) = \int_0^1 \left(\int_0^2 \left(x^2 + y^2 \right) \right) \, dy \right) \, dx$$

where, when we compute $\int_0^2 (x^2 + y^2) \, dy$, we treat x as though it were a constant. ∎

In general, if f is a function from a subset of $\mathbb{R}^2$ into $\mathbb{R}$, then $\int_a^b f(x, y) \, dy$ will be computed by treating x as a constant. Similarly, $\int_a^b f(x, y) \, dx$ will be computed by treating y as a constant.

EXAMPLE 2: Compute $\int_0^1 (x^5 y + e^{xy}) \, dy$ and $\int_0^1 (x^5 y + e^{xy}) \, dx$.

SOLUTION:

$$\int_0^1 \left(x^5 y + e^{xy} \right) \, dx = \frac{x^6}{6} y + \frac{e^{xy}}{y} \bigg|_{x=0}^{x=1} = \frac{y}{6} + \frac{e^y}{y} - \frac{1}{y},$$

and

$$\int_0^1 \left(x^5 y + e^{xy}\right) \, dy = \frac{x^5 y^2}{2} + \frac{e^{xy}}{x}\Big|_{y=0}^{y=1} = \frac{x^5}{2} + \frac{e^x}{x} - \frac{1}{x}. \qquad \blacksquare$$

Example 1 leads us to some new notation and terminology.

Definition: R–sums

Let R be the rectangle bounded by the lines $x = a, x = b, y = c,$
and $y = d$, where $a \le b$ and $c \le d$. Let f be a real valued function
defined on $\mathbb{R}$. If $\{x_0, \dots, x_n\}$ is a partition of $[a, b]$ and $\{y_0, \dots, y_m\}$
is a partition of $[c, d]$, then the collection of lines

$$\mathcal{P} = \{x = x_0, x = x_1, \dots, x = x_n\} \cup \{y = y_0, y = y_1, \dots, y = y_m\}$$

is called a *partition of R*.

We let $R_{i,j}$ denote the rectangle bounded by the lines $x = x_{i-1}, x = x_i, y = y_{j-1}$, and $y = y_j$. If, for each i and j, we choose a point $\vec{s}_{i,j}$
in $R_{i,j}$, then the collection $\mathcal{S} = \{\vec{s}_{i,j} \mid 1 \le i \le n \text{ and } 1 \le j \le m\}$ is
called a *selection for $\mathcal{P}$*. Let $A_{i,j}$ denote the area of $R_{i,j}$. Then

$$\sum_{i=1}^n \left(\sum_{j=1}^m f(\vec{s}_{i,j}) A_{i,j} \right) = \sum_{i=1}^n \left(\sum_{j=1}^m f(\vec{s}_{i,j})(x_i - x_{i-1})(y_j - y_{j-1}) \right)$$

is called the *R–sum for f determined by $\mathcal{P}$ and $\mathcal{S}$*.
The number L is called the *limit of the collection of R–sums for f*
if the collection of R–sums approximates L.[1]
If L is the limit of the R–sums for f, then L is denoted by

$$\iint_R f \, dA,$$

and it is called the *integral of f over R*.

---

[1]that is, if k is a positive number, then there is a partition $\mathcal{P}$ of R so that
if $\tilde{\mathcal{P}}$ refines $\mathcal{P}$ and if S is any R–sum for f determined by $\tilde{\mathcal{P}}$, then $|L - S| < k$.

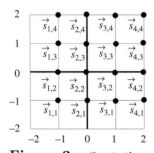

Figure 2.a $\vec{s}_{i,j}$ *is the
upper right corner of $R_{i,j}$.*

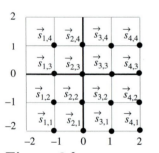

Figure 2.b $\vec{s}_{i,j}$ *is the lower
right corner of $R_{i,j}$.*

EXAMPLE 3: Let $f(x,y) = xy$. Figures 2.a and 2.b display
two selections for the same partition of the rectangle R bounded by
the lines $x = -2$, $x = 2$, $y = -2$, and $y = 2$. Find the R–sum
approximation for $\iint_R f(x,y) \, dA$.

SOLUTION: The partition is the set of lines $x = -2$, $x = -1$, $x =$
0, $x = 1$, $x = 2$, $y = -2$, $y = -1$, $y = 0$, $y = 1$, and $y = 2$. Each
rectangle $R_{i,j}$ has area 1(unit squared).

For the selection indicated in Figure 2.a:

$$
\begin{aligned}
f(\vec{s}_{1,1}) = f(-1,-1) = 1 \qquad & f(\vec{s}_{2,1}) = f(0,-1) = 0 \\
f(\vec{s}_{3,1}) = f(1,-1) = -1 \qquad & f(\vec{s}_{4,1}) = f(2,-1) = -2 \\
f(\vec{s}_{1,2}) = f(-1,0) = 0 \qquad & f(\vec{s}_{2,2}) = f(0,0) = 0 \\
f(\vec{s}_{3,2}) = f(1,0) = 0 \qquad & f(\vec{s}_{4,2}) = f(2,0) = 0 \\
f(\vec{s}_{1,3}) = f(-1,1) = -1 \qquad & f(\vec{s}_{2,3}) = f(0,1) = 0 \\
f(\vec{s}_{3,3}) = f(1,1) = 1 \qquad & f(\vec{s}_{4,3}) = f(2,1) = 2 \\
f(\vec{s}_{1,4}) = f(-1,2) = -2 \qquad & f(\vec{s}_{2,4}) = f(0,2) = 0 \\
f(\vec{s}_{3,4}) = f(1,2) = 2 \qquad & f(\vec{s}_{4,4}) = f(2,2) = 4
\end{aligned}
$$

$$
\begin{aligned}
\sum_{i=1}^{4}\left(\sum_{j=1}^{4} f(\vec{s}_{i,j})A_{i,j}\right) =\ & (1+0-1-2)+(0+0+0+0)+ \\
& (-1+0+1+2)+(-2+0+2+4) = 4.
\end{aligned}
$$

For the selection indicated in Figure 2.b:

$$
\begin{aligned}
f(\vec{s}_{1,1}) = f(-1,-2) = 2 \qquad & f(\vec{s}_{2,1}) = f(0,-2) = 0 \\
f(\vec{s}_{3,1}) = f(1,-2) = -2 \qquad & f(\vec{s}_{4,1}) = f(2,-2) = -4 \\
f(\vec{s}_{1,2}) = f(-1,-1) = 1 \qquad & f(\vec{s}_{2,2}) = f(0,-1) = 0 \\
f(\vec{s}_{3,2}) = f(1,-1) = -1 \qquad & f(\vec{s}_{4,2}) = f(2,-1) = -2 \\
f(\vec{s}_{1,3}) = f(-1,0) = 0 \qquad & f(\vec{s}_{2,3}) = f(0,0) = 0 \\
f(\vec{s}_{3,3}) = f(1,0) = 0 \qquad & f(\vec{s}_{4,3}) = f(2,0) = 0 \\
f(\vec{s}_{1,4}) = f(-1,1) = -1 \qquad & f(\vec{s}_{2,4}) = f(0,1) = 0 \\
f(\vec{s}_{3,4}) = f(1,1) = 1 \qquad & f(\vec{s}_{4,4}) = f(2,1) = 2
\end{aligned}
$$

$$
\begin{aligned}
\sum_{i=1}^{4}\left(\sum_{j=1}^{4} f(\vec{s}_{i,j})A_{i,j}\right) =\ & (2+0-2-4)+(1+0-1-2)+ \\
& (0+0+0+0)+(-1+0+1+2) \\
=\ & -4. \qquad\blacksquare
\end{aligned}
$$

It is illustrative to recall Example 1. Our approach there assumed that the mass of the rectangle R was the limit of the R–sums for the function ρ (which in the language we have just developed would be denoted by $\iint_{R}\rho\ dA$). Our next theorem tells us that this was a valid assumption as long as ρ is continuous. Furthermore, the next theorem tells us that our method for computing the integral $\iint_{R}\rho\ dA$ in Example 1 is one that works in general.

Theorem 1 *Let R be the rectangle bounded by the lines $x = a, x = b, y = c$, and $y = d$, where $a < b$ and $c < d$, and let f be a continuous function from R into $\mathbb{R}$. Then $\iint_R f\ dA$ exists and*

$$\iint_R f\ dA = \int_a^b \left(\int_c^d f(x,y)\ dy \right)\ dx = \int_c^d \left(\int_a^b f(x,y)\ dx \right)\ dy.$$

In general, if R is the rectangle bounded by the lines $x = a, x = b, y = c$, and $y = d$, if f is a function from R into $\mathbb{R}$, and if Q is a numerical quantity that can be approximated with R–sums for f, then $Q = \iint_R f\ dA$.

EXAMPLE 4: If R is a rectangle bounded by the lines $x = a, x = b, y = c$, and $y = d$, and if $\rho(x,y)$ denotes the mass density of R at (x,y), then the total mass of R is $\iint_R \rho\ dA$.

To see that this is true, let $\mathcal{P} = \{x = x_0, x = x_1, \ldots, x = x_n\} \cup \{y = y_0, y = y_1, \ldots, y = y_m\}$ be a partition of R, and let $\mathcal{S} = \{\vec{s}_{i,j} \mid 1 \le i \le n, 1 \le j \le m\}$ be a selection from $\mathcal{P}$ so that for each i and j, $\vec{s}_{i,j} = (x_i, y_j)$ is a point in the region $R_{i,j}$ bounded by the lines $x = x_{i-1}$, $x = x_i$ and by the lines $y = y_{j-1}$, $y = y_j$. We let $A_{i,j}$ denote the area of $R_{i,j}$ so that an approximation of the mass will be

$$\sum_{i=1}^{n} \sum_{j=1}^{m} \rho(\vec{s}_{i,j}) A_{i,j}, \text{ which also approximates } \iint_R \rho\ dA.$$

Thus $M = \iint_R \rho\ dA$. ∎

EXAMPLE 5: With the knowledge we have now, the problem from Example 1 is easy. As before, we let R be the rectangle bounded by the lines $x = 0$, $x = 1$, $y = 0$, and $y = 2$. Let $\rho(x,y) = x^2 + y^2$ be the mass density of R at (x,y). Then the mass of R is given by

$$M = \iint_R \rho\ dA = \int_0^2 \int_0^1 (x^2 + y^2)\ dx\,dy$$

$$= \int_0^2 \left(\frac{x^3}{3} + xy^2 \right) \Bigg|_{x=0}^{x=1}\ dy = \int_0^2 \frac{1}{3} + y^2\ dy = \frac{10}{3}. \quad ∎$$

EXAMPLE 6: Let R be the rectangle bounded by the lines $x = a$, $x = b$, $y = c$, and $y = d$. Suppose that $z = f(x,y)$ is a continuous function defined on R, and that $f(x,y) \ge 0$ on R. Then the graph of f, the planes $x = a$, $x = b$, $y = c$, $y = d$, and $z = 0$ bound a region B in xyz–space.

To approximate the volume of B, let

$$\mathcal{P} = \{x = x_0, x = x_1, \ldots, x = x_n\} \cup \{y = y_0, y = y_1, \ldots, y = y_m\}$$

be a partition of R and let $\{\vec{s}_{i,j}\}$ be a selection from $\mathcal{P}$. Let $V_{i,j}$ be the volume of the part of B bounded by the planes $x = x_{i-1}$, $x = x_i$, $y = y_{j-1}$, and $y = y_j$. (See Figure 2.) Then

$$V_{i,j} \text{ is approximated by } f(\vec{s}_{i,j})(x_i - x_{i-1})(y_j - y_{j-1}),$$

and the total volume is $\sum_{j=1}^{m} \sum_{i=1}^{n} V_{i,j}$, which is approximated by

$$\sum_{j=1}^{m} \sum_{i=1}^{n} f(\vec{s}_{i,j})(x_i - x_{i-1})(y_j - y_{j-1}) \approx \iint_R f \, dA.$$

Figure 3. *Let $V_{i,j}$ be the volume of the part of B bounded by the planes $x = x_{i-1}$, $x = x_i$, $y = y_{j-1}$, and $y = y_j$.*

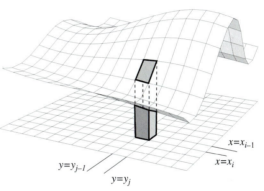

EXERCISES 14.1

Calculate the integrals in Exercises 1–7.

1. $\int_{-1}^{2} e^{xy} \, dx$.

2. $\int_{0}^{3} e^{xy} \, dy$.

3. $\int_{2}^{3} \cos(\pi x)\sin(\pi y) \, dx$.

4. $\int_{2}^{3} \cos(\pi x)\sin(\pi y) \, dy$.

5. $\int_{0}^{2} \int_{-1}^{1} xe^{xy} \, dy \, dx$.

6. $\int_{-1}^{1} \int_{0}^{2} xe^{xy} \, dx \, dy$.

7. $\int_{0}^{1} \int_{3}^{4} x^2 y \, dx \, dy$.

In Exercises 8–11, $\vec{f}(x,y) = x^2 y^2 + x$. Calculate the R–sum for the indicated partition and selection.

In Exercises 12–18, calculate $\iint_R f \, dA$, where R is the rectangle bounded by $x = a$, $x = b$, $y = c$, and $y = d$.

12. $f(x, y) = 2x^2 + 3y, a = 0, b = 1, c = -1, d = 2.$

13. $f(x, y) = 4x^2 y, a = -1, b = 1, c = 2, d = 3.$

14. $f(x, y) = \dfrac{x \ln y}{y}, a = -1, b = 1, c = 1, d = e^2.$

15. $f(x, y) = \dfrac{x}{y^2 + 2y - 3}, a = 0, b = 1, c = -1,$
    $d = 0.$

16. $f(x, y) = x\sqrt{x^2 + y}, a = 0, b = 1, c = 0, d = 1.$

17. $f(x, y) = x \cos(xy), a = 0, b = 1, c = -\pi, d = \pi.$

18. $f(x, y) = 2^{3x+y}, a = 0, b = \frac{1}{3}, c = 0, d = 1.$

In Exercises 19–21, B is the solid bounded by the planes $x = a$, $x = b$, $y = c$, $y = d$, $z = 0$, and the graph $z = f(x, y)$. Find the volume of B.

19. $f(x, y) = x^2 + y^2, a = 0, b = 1, c = 0, d = 1.$

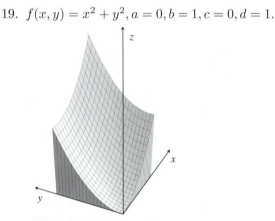

Figure 4. *The solid for Exercise 19.*

20. $f(x, y) = x^2 y, a = -1, b = 1, c = 0, d = 2.$

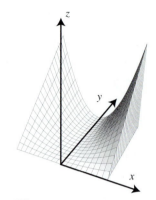

Figure 5. *The solid for Exercise 20.*

21. $f(x, y) = \cos(x) \cos(y), a = 0, b = \frac{\pi}{2},$
    $c = 0, d = \frac{\pi}{2}.$

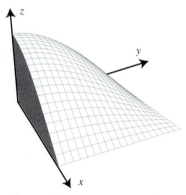

Figure 6. *The solid for Exercise 21.*

14.2   Simple Surfaces

We saw in the last section how to evaluate integrals over rectangular regions in $\mathbb{R}^2$. But if we can integrate over a rectangle, why not over a disk, or a triangle, or even a sphere?

It turns out that the methods of the last section can allow us to integrate over such sets. In the next several sections, we will see how to evaluate integrals over certain 2–dimensional sets, called simple surfaces. Since we already know how to integrate over rectangles, it is not surprising that our approach will involve relating these surfaces to rectangles through a parametrization. We begin with the definition of a Simple Surface, which is analogous to a Fundamental Curve.

Definition: Simple Surfaces

The statement that the set S in $\mathbb{R}^3$ (or $\mathbb{R}^2$) is a *simple surface* means that there is a rectangle R bounded by two horizontal lines and two vertical lines in the plane and a function $\vec{h}$ with domain R and image S such that:

(a) The partial derivatives of $\vec{h}$ are continuous;

(b) $\vec{h}$ is one–to–one in the interior of R; and

(c) $J(\vec{h}(x, y)) = \left\| \frac{\partial \vec{h}}{\partial x} \times \frac{\partial \vec{h}}{\partial y} \right\| \neq 0$ in the interior of R.

The function $\vec{h}$ is called a *parametrization for S*.

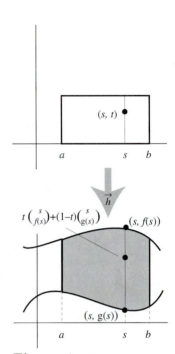

Figure 1. *The region bounded between the graphs of f and g over the interval $[a, b]$*

The idea of a simple surface is just the 2–dimensional version of the idea of a fundamental curve. Intuitively, a surface is simple if it can be obtained by stretching a rectangle out of shape and then perhaps pinching together some of the points of the boundary. Some specific examples should be helpful.

EXAMPLE 1: Let f and g be continuously differentiable functions from $\mathbb{R}$ into $\mathbb{R}$, and let $[a, b]$ be an interval so that if x is between a and b, then $f(x) > g(x)$. Show that the region S bounded between the graphs of f and g and the lines $x = a$ and $x = b$ is a simple surface. See Figure 1.

SOLUTION: Let R be the rectangle in st–space bounded by the lines $s = a, s = b, t = 0$, and $t = 1$. We construct a function $\vec{h}$ from R onto S satisfying conditions (a), (b), and (c) in our definition of a simple surface. The idea is to construct a parametrization $\vec{h}$ for S such that, for each s, the function $\vec{h}$ takes the straight line interval $[(s, 0), (s, 1)]$ in R onto the straight line interval $[(s, g(s)), (s, f(s))]$ in S as illustrated in Figure 1.

Assume that for a fixed s, $\vec{h}(s, t)$ is of the form

$$\vec{h}(s, t) = \vec{v}_s t + \vec{r}_s, \quad \text{where } \vec{h}(s, 0) = (s, g(s)) \text{ and } \vec{h}(s, 1) = (s, f(s)).$$

The condition

$$\vec{h}(s, 0) = (s, g(s)) \text{ implies that } \vec{r}_s = (s, g(s)).$$

So

$$\vec{h}(s, t) = \vec{v}_s t + (s, g(s)).$$

For $t = 1$, we have

$$\vec{h}(s, 1) = \vec{v}_s + (s, g(s)) = (s, f(s)).$$

This implies that

$$\vec{v}_s = (0, (f(s) - g(s)).$$

Thus

$$\vec{h}(s, t) = (s, g(s) + t(f(s) - g(s))).$$

Then $\vec{h}$ is a function that takes R onto S. We need to show that $\vec{h}$ is one-to-one on the interior of R, and that the partial derivatives of $\vec{h}$ are continuous. To see that $\vec{h}$ is one-to-one on the interior of R, suppose that $\vec{h}(s, t) = \vec{h}(s', t')$ where (s, t) and (s', t') are points in the interior of R. We can show that $s = s'$ and that $t = t'$.

$$\vec{h}(s, t) = (s, g(s) + t(f(s) - g(s)) = (s', g(s') + t'(f(s') - g(s')) = \vec{h}(s', t').$$

By considering the first coodinates we see that $s = s'$. By comparing the second coordinates we have that

$$g(s) + t(f(s) - g(s)) = g(s) + t'(f(s) - g(s)).$$

This tells us that

$$(t - t')(f(s) - g(s)) = 0.$$

Since s is between a and b, we know that $f(s) \neq g(s)$, so $t = t'$. This completes the proof that $\vec{h}$ is one-to-one in the interior of R. The parametrization $\vec{h}$ is clearly continuous. Since $\frac{\partial \vec{h}}{\partial}s = (1, g'(s) + t(f'(s) - g'(s)))$ and $\frac{\partial \vec{h}}{\partial t} = (0, f(s) - g(s))$, we see that both partial derivatives are also continuous. We leave it to the reader to show that $J(\vec{h}(s, t)) \neq 0$ on the interior of R.   ∎

EXAMPLE 2: Let $\vec{A} = (a_x, a_y, a_z), \vec{B} = (b_x, b_y, b_z)$, and $\vec{C} = (c_x, c_y, c_z)$ be vectors in $\mathbb{R}^3$, and let P be the parallelogram with adjacent sides $\vec{A}$ and $\vec{B}$ drawn emanating from the point $\vec{C}$. The linear transformation $\vec{T}$ with associated matrix

$$A = \begin{pmatrix} a_x & b_x \\ a_y & b_y \\ a_z & b_z \end{pmatrix}$$

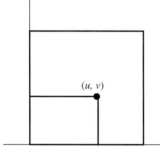

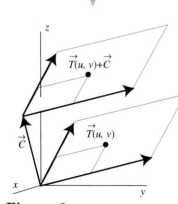

Figure 2.
$\vec{r}(u, v) = \vec{C} + u\vec{A} + v\vec{B}$ takes the unit square onto P.

takes the unit square in uv–space onto the parallelogram with adjacent sides $\vec{A}$ and $\vec{B}$ drawn emanating from the origin. Thus the function $\vec{r}$ defined by

$$\vec{r}(u, v) = \vec{C} + \vec{T}(u, v) = \vec{C} + u\vec{A} + v\vec{B}$$

takes the unit square onto the desired parallelogram D. See Figure 2. Recall that $J(\vec{r}(u,v)) = \|\vec{A} \times \vec{B}\|$. ∎

EXAMPLE 3: Show that the sphere with radius r is a simple surface.

SOLUTION: We employ the spherical transformation keeping r fixed. Let $\vec{P} = (x, y, z)$ be an arbitrary point on the sphere as in Figure 3.

Since we are keeping r fixed,

$$\vec{h}(\phi, \theta) = \vec{S}(r, \phi, \theta) = \begin{pmatrix} r \sin \phi \cos \theta \\ r \sin \phi \sin \theta \\ r \cos \phi \end{pmatrix}$$

is a function from $\phi\theta$–space to the sphere. Notice that for a fixed θ, the parametrization $\vec{h}(\phi, \theta)$ sweeps out a longitude as ϕ goes from 0 to π. Then as θ goes from 0 to 2π, the longitude sweeps out the entire sphere. Thus $\vec{h}$ takes the rectangle R defined by $0 \le \theta \le 2\pi$ and $0 \le \phi \le \pi$ onto the sphere of radius r.

In Figure 4, we see that $\vec{h}$ takes a vertical line of the form $\phi = a$ onto a latitude on the sphere, and $\vec{h}$ takes a horizontal line of the form $\theta = b$ onto a longitude of the sphere. Thus if (ϕ, θ) and (ϕ', θ') are points in the interior of R, and if $\phi \ne \phi'$, then $\vec{h}$ takes (ϕ, θ) and (ϕ', θ') onto different latitudes. If $\theta \ne \theta'$, then $\vec{h}$ takes these points onto different longitudes. Thus, $\vec{h}$ is one-to-one on the interior of R. It is easily checked that the partial derivatives of $\vec{h}$ are continuous, and that $J(\vec{h}(\phi, \theta)) = r^2 \sin \phi$. So the sphere is a simple surface.

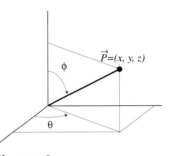

Figure 3.

$$\vec{P} = \begin{pmatrix} r \sin \phi \cos \theta \\ r \sin \phi \sin \theta \\ r \cos \phi \end{pmatrix}$$

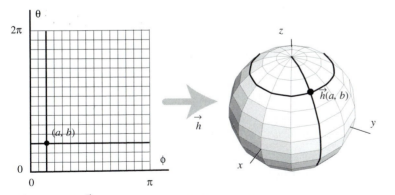

Figure 4. $\vec{h}$ *takes the rectangle* R *defined by* $0 \le \theta \le 2\pi$ *and* $0 \le \phi \le \pi$ *onto the sphere of radius* r.

Surfaces of Rotation

EXAMPLE 4: Suppose that $\vec{r}(t) = (x(t), y(t), 0)$ $a \leq t \leq b$ is a parametrization for a simple curve C lying in the xy–coordinate plane that does not intersect the x–axis. Let S be the set obtained by rotating C about the x–axis. Show that S is a simple surface.

SOLUTION: From Chapter 13, you are familiar with $\vec{C}_z$, the cylindrical transformation about the z–axis. The basic tool in our current example is the cylindrical transformation $\vec{C}_x$. $\vec{C}_x$ is a function from $xr\theta$–space into xyz–space that wraps a plane of the form $r = r_0$ into a cylinder of radius r_0 about the x–axis.

In particular, suppose that $\vec{C}_x$ acts on a vertical line segment connecting the points $(x_0, r_0, 0)$ and $(x_0, r_0, 2\pi)$ in $xr\theta$–space. The resulting image in xyz–space is a circle lying in the plane $x = x_0$, with radius r_0 and center $(x_0, 0, 0)$. (See Figure 5.)

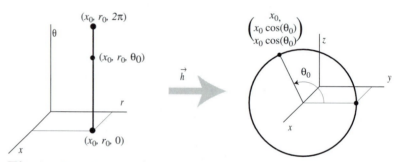

Figure 5. *The vertical line segment connecting the points $(x_0, r_0, 0)$ and $(x_0, r_0, 2\pi)$ in $xr\theta$–space is taken onto the circle in xyz–space lying in the plane $x = x_0$, with radius r_0 and center $(x_0, 0, 0)$.*

If we rotate the point $(x(t), y(t), 0)$ about the x–axis, the radius of rotation is $x(t)$, and

$$\vec{h}(\theta, t) = \begin{pmatrix} x(t) \\ x(t)\cos(\theta) \\ x(t)\sin(\theta) \end{pmatrix}$$

wraps the line segment connecting the points $(x(t), y(t), 0)$ and $(x(t), y(t), 2\pi)$ around a circle centered on the x–axis. Thus the function

$$\vec{h}(\theta, t) = \begin{pmatrix} x(t) \\ x(t)\cos(\theta) \\ x(t)\sin(\theta) \end{pmatrix}, \ a \leq x \leq b, \ 0 \leq \theta \leq 2\pi$$

rotates the image of $\vec{r}$ around the x–axis.   ■

EXAMPLE 5: Use the function $\vec{h}$ from Example 4 to parametrize the surface obtained by rotating the graph of $y = x^2 + 1$, $-1 \leq x \leq 1$ about the x–axis.

SOLUTION: A parametrization for the curve is $\vec{r}(t) = (t, t^2 + 1)$. The function

$$\vec{h}(t, \theta) = \left(t, (t^2 + 1)\cos(\theta), (t^2 + 1)\sin(\theta)\right), \ -1 \leq t \leq 1, \ 0 \leq \theta \leq 2\pi$$

is our desired parametrization. Figure 6 displays some animated frames that show how the surface sweeps out its rotation as θ increases. ∎

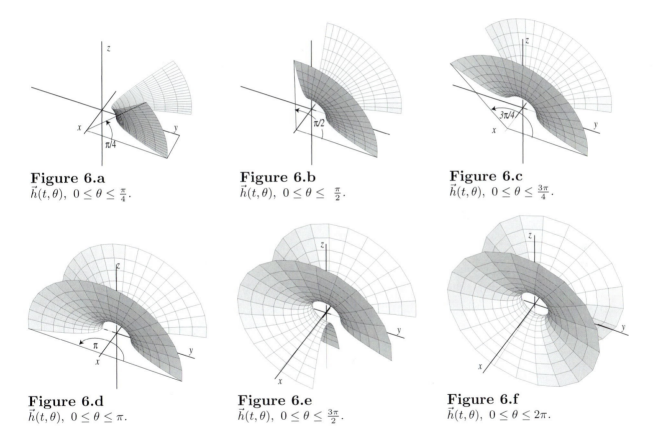

Figure 6.a
$\vec{h}(t, \theta), \ 0 \leq \theta \leq \frac{\pi}{4}$.

Figure 6.b
$\vec{h}(t, \theta), \ 0 \leq \theta \leq \frac{\pi}{2}$.

Figure 6.c
$\vec{h}(t, \theta), \ 0 \leq \theta \leq \frac{3\pi}{4}$.

Figure 6.d
$\vec{h}(t, \theta), \ 0 \leq \theta \leq \pi$.

Figure 6.e
$\vec{h}(t, \theta), \ 0 \leq \theta \leq \frac{3\pi}{2}$.

Figure 6.f
$\vec{h}(t, \theta), \ 0 \leq \theta \leq 2\pi$.

We can rotate a simple curve about the y–axis in exactly the same fashion. In this case, the x–coordinate of the point is the radius of rotation.

EXAMPLE 6: Suppose that $\vec{r}(t) = (x(t), y(t), 0)$, $a \leq t \leq b$ is a parametrization for a simple curve C lying in the xy–coordinate plane that does not intersect the y–axis. Let S be the set obtained by rotating C about the y–axis. Then

$$h(\theta, t) = \begin{pmatrix} x(t)\cos(\theta) \\ y(t) \\ x(t)\sin(\theta) \end{pmatrix}, \; a \leq t \leq b, \; 0 \leq \theta$$

parametrizes the surface.   ∎

EXAMPLE 7: The unit circle centered at $(2, 2)$ is rotated about the y–axis. Use the function from the previous example to parametrize the resulting surface.

SOLUTION: The circle is parametrized by $\vec{r}(t) = (\cos(t) + 2, \sin(t) + 2, 0)$, $0 \leq t \leq 2\pi$. According to Example 6, the desired parametrization is

$$\vec{h}(t, \theta) = \begin{pmatrix} (\cos(t) + 2)\cos(\theta) \\ \sin(t) + 2 \\ (\cos(t) + 2)\sin(\theta) \end{pmatrix}, \; 0 \leq t \leq 2\pi, \; 0 \leq \theta.$$

Figure 7 demonstrates how the surface grows as the circle is rotated about the y–axis.   ∎

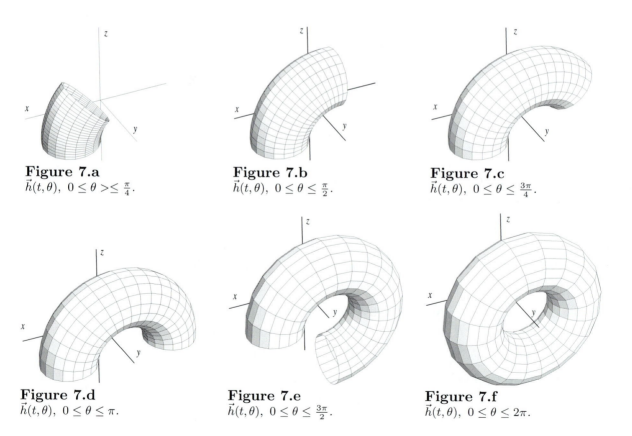

Figure 7.a
$\vec{h}(t, \theta), \; 0 \leq \theta >\leq \frac{\pi}{4}$.

Figure 7.b
$\vec{h}(t, \theta), \; 0 \leq \theta \leq \frac{\pi}{2}$.

Figure 7.c
$\vec{h}(t, \theta), \; 0 \leq \theta \leq \frac{3\pi}{4}$.

Figure 7.d
$\vec{h}(t, \theta), \; 0 \leq \theta \leq \pi$.

Figure 7.e
$\vec{h}(t, \theta), \; 0 \leq \theta \leq \frac{3\pi}{2}$.

Figure 7.f
$\vec{h}(t, \theta), \; 0 \leq \theta \leq 2\pi$.

We can easily extend our method to rotate curves about lines parallel to the x–axis or the y–axis, as the next example demonstrates.

EXAMPLE 8: Suppose that $\vec{r}(t) = (x(t), y(t), 0)$ $a \leq t \leq b$ is a parametrization for a simple curve C lying in the xy–coordinate plane that does not intersect the line $y = y_0$. Let S be the set obtained by rotating C about the line $y = y_0$. Show that S is a simple surface.

SOLUTION: The simplest approach is first to translate C by $(0, -y_0, 0)$ so that the resulting curve $\tilde{C} = C - (0, -y_0, 0)$ is geometrically related to the x–axis exactly as C is related to the line $y = y_0$. A parametrization for $\tilde{C}$ is $\vec{r}_1(t) = (x(t), y(t) - y_0, 0)$. In Figure 8 we illustrate the idea using the top half of a circle C for our curve to be rotated.

A parametrization for $\tilde{C}$ is $\vec{\rho}(t) = (x(t), y(t) - y_0, 0)$ $a \leq t \leq b$. Let $\tilde{S}$ be the surface obtained by rotating $\tilde{C}$ about the x–axis. Then $\tilde{S}$ has exactly the same shape as does S. A translation by $(0, y_0, 0)$ moves $\tilde{S}$ onto the desired surface S. See Figure 9.

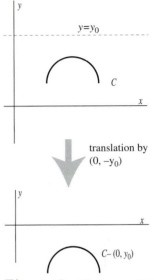

Figure 8. *The curve C is translated by $(0, -y_0)$.*

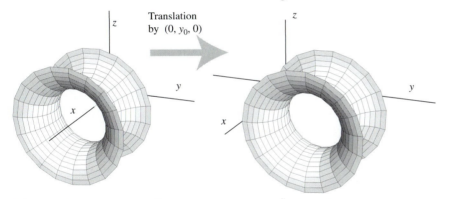

Figure 9. *The surface $\tilde{S}$ obtained by rotating $\tilde{C}$ about the x–axis and then the surface S obtained by translating $\tilde{S}$ by the vector $(0, y_0, 0)$.*

A parametrization for $\tilde{S}$ is

$$\vec{h}_1(t, \theta) = \begin{pmatrix} x(t) \\ (y(t) - y_0)\cos(\theta) \\ (y(t) - y_0)\sin(\theta) \end{pmatrix}, \ a \leq t \leq b, \ 0 \leq \theta \leq 2\pi.$$

Finally, translating by $(0, y_0, 0)$, we have

$$\vec{h}(t, \theta) = \begin{pmatrix} x(t) \\ (y(t) - y_0)\cos(\theta) \\ (y(t) - y_0)\sin(\theta) \end{pmatrix} + (0, y_0, 0)$$

$$= \begin{pmatrix} x(t) \\ (y(t) - y_0)\cos(\theta) + y_0 \\ (y(t) - y_0)\sin(\theta) \end{pmatrix}. \qquad \blacksquare$$

Graphs of Functions

Suppose that f is a real valued function defined on the rectangle $a \le x \le b$ and $c \le y \le d$. Then the graph of f is a surface and if f is differentiable, then the graph of f is a simple surface that can be parametrized by

$$\vec{h}(u, v) = (u, v, f(u, v)), \; a \le x \le b, c \le y \le d.$$

$$\frac{\partial \vec{h}}{\partial u}(u, v) = \left(1, 0, \frac{\partial f}{\partial u}(u, v)\right) \quad \text{and} \quad \frac{\partial \vec{h}}{\partial v}(u, v) = \left(0, 1, \frac{\partial f}{\partial v}(u, v)\right).$$

$$J\vec{h} = \left\| \left(1, 0, \frac{\partial f}{\partial u}\right) \times \left(0, 1, \frac{\partial f}{\partial v}\right) \right\|$$

$$= \sqrt{\left(\frac{\partial f}{\partial u}(u, v)\right)^2 + \left(\frac{\partial f}{\partial v}(u, v)\right)^2 + 1}.$$

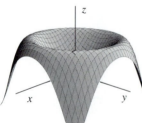

Figure 10. *The graph of* $f(x, y) = \sin(x^2 + y^2)$, $-1.5 \le x \le 1.5$, $-1.5 \le y \le 1.5$.

EXAMPLE 9: Parametrize the graph of $f(x, y) = \sin(x^2 + y^2)$, $-1.5 \le x \le 1.5$, $-1.5 \le y \le 1.5$, and find the rate that the parametrization changes area. The surface is sketched in Figure 10.

SOLUTION: We parametrize the graph by

$$\vec{h}(u, v) = \begin{pmatrix} u \\ v \\ \sin(u^2 + v^2) \end{pmatrix}, \; -1.5 \le u \le 1.5, \; -1.5 \le v \le 1.5.$$

We now have $\dfrac{\partial f}{\partial u}(u, v) = 2u\cos(u^2 + v^2)$ and $\dfrac{\partial f}{\partial v}(u, v) = 2v\cos(u^2 + v^2)$. Thus

$$J\vec{h}(u, v) = \left\|(1, 0, 2u\cos(u^2 + v^2) \times (0, 1, 2v\cos(u^2 + v^2))\right\|$$

$$= \left\|(-2u\cos(u^2 + v^2), -2v\cos(u^2 + v^2), 1)\right\|$$

$$= \sqrt{4(u^2 + v^2)\cos^2(u^2 + v^2) + 1}. \qquad \blacksquare$$

Tangent Planes

We can use a parametrization of a surface to find vectors normal to the surface and to find tangent planes.

EXAMPLE 10: Let $\vec{h}$ be the standard parametrization for the sphere of radius 1, centered at the origin in $\mathbb{R}^3$ and let $\vec{P} = \left(\frac{1}{2}, \frac{1}{2}, \frac{1}{\sqrt{2}} \right)$.

(a) Parametrize the plane tangent to the sphere at $\vec{P}$.

(b) Find a vector normal to the sphere at $\vec{P}$.

(c) Find an equation for the plane tangent to the sphere at $\vec{P}$.

(d) Find the rate that $\vec{h}$ changes area at $\vec{P}$.

SOLUTION: Recall that the parametrization is given by $\vec{h}(\phi, \theta) = (\sin\phi\cos\theta, \sin\phi\sin\theta, \cos\phi)$. The partial derivatives of $\vec{h}$ are given by

$$\frac{\partial \vec{h}}{\partial \phi} = (\cos\phi\cos\theta, \cos\phi\sin\theta, -\sin\phi),$$

and

$$\frac{\partial \vec{h}}{\partial \theta} = (-\sin\phi\sin\theta, \sin\phi\cos\theta, 0).$$

The point $\vec{P}$ in terms of $\phi\theta$–coordinates is $\left(\frac{\pi}{4}, \frac{\pi}{4} \right)$; that is, $\vec{P} = \vec{h}\left(\frac{\pi}{4}, \frac{\pi}{4} \right)$.

$$\frac{\partial \vec{h}}{\partial \phi}\left(\frac{\pi}{4}, \frac{\pi}{4} \right) = \left(\frac{1}{2}, \frac{1}{2}, \frac{\sqrt{2}}{2} \right),$$

$$\frac{\partial \vec{h}}{\partial \theta}\left(\frac{\pi}{4}, \frac{\pi}{4} \right) = \left(-\frac{1}{2}, \frac{1}{2}, 0 \right), \text{ and}$$

$$D\vec{h}\Big|_{\left(\frac{\pi}{4}, \frac{\pi}{4} \right)} = \begin{pmatrix} \frac{1}{2} & -\frac{1}{2} \\ \frac{1}{2} & \frac{1}{2} \\ \frac{\sqrt{2}}{2} & 0 \end{pmatrix}.$$

Taylor's first order approximation to $\vec{h}$ is

$$\vec{r}(\phi, \theta) = D\vec{h}\Big|_{\left(\frac{\pi}{4}, \frac{\pi}{4} \right)} \begin{pmatrix} \frac{\pi}{4} - \phi \\ \frac{\pi}{4} - \theta \end{pmatrix} + \vec{h}\left(\frac{\pi}{4}, \frac{\pi}{4} \right)$$

$$= \begin{pmatrix} \frac{1}{2} & -\frac{1}{2} \\ \frac{1}{2} & \frac{1}{2} \\ \frac{\sqrt{2}}{2} & 0 \end{pmatrix} \begin{pmatrix} \frac{\pi}{4} - \phi \\ \frac{\pi}{4} - \theta \end{pmatrix} + \begin{pmatrix} \frac{1}{2} \\ \frac{1}{2} \\ \frac{1}{\sqrt{2}} \end{pmatrix}.$$

$\vec{r}(\phi, \theta)$ parametrizes the tangent plane.

The vectors $\frac{\partial \vec{h}}{\partial \phi}\left(\frac{\pi}{4}, \frac{\pi}{4}\right)$ and $\frac{\partial \vec{h}}{\partial \theta}\left(\frac{\pi}{4}, \frac{\pi}{4}\right)$ lie in the plane tangent to the sphere at $\vec{P}$. The normal to the plane is given by

$$\frac{\partial \vec{h}}{\partial \phi}\Big|_{\left(\frac{\pi}{4}, \frac{\pi}{4}\right)} \times \frac{\partial \vec{h}}{\partial \theta}\Big|_{\left(\frac{\pi}{4}, \frac{\pi}{4}\right)} = \left(\frac{1}{2}, \frac{1}{2}, \frac{\sqrt{2}}{2}\right) \times \left(-\frac{1}{2}, \frac{1}{2}, 0\right)$$

$$= \left(\frac{-\sqrt{2}}{4}, -\frac{\sqrt{2}}{4}, \frac{1}{2}\right).$$

Thus the equation of the plane is given by $\vec{n} \cdot (\vec{r} - \vec{r}_0)$, or

$$-\frac{\sqrt{2}}{4}\left(x - \frac{1}{2}\right) - \frac{\sqrt{2}}{4}\left(y - \frac{1}{2}\right) + \frac{1}{2}\left(z - \frac{\sqrt{2}}{2}\right) = 0.$$

The rate at which $\vec{h}$ changes area is given by

$$\left\| \frac{\partial \vec{h}}{\partial \phi}\Big|_{\left(\frac{\pi}{4}, \frac{\pi}{4}\right)} \times \frac{\partial \vec{h}}{\partial \theta}\Big|_{\left(\frac{\pi}{4}, \frac{\pi}{4}\right)} \right\| = \left\| \left(-\frac{\sqrt{2}}{4}, -\frac{\sqrt{2}}{4}, \frac{1}{2}\right) \right\| = \frac{\sqrt{2}}{2}. \qquad \blacksquare$$

Summary

All of the examples we have been looking at in this section are examples of parametrizations of simple surfaces in the plane or in $\mathbb{R}^3$. These parametrizations are tools used to deform a rectangle in the plane to fit the surface we wish to study. The requirements on the parametrizations are to ensure that we can use the tools of calculus to explore the surfaces being modeled. Let

$$\vec{h}(u, v) = (x(u, v), y(u, v), z(u, v)), \ a \leq u \leq b, \ c \leq v \leq d$$

be a parametrization for a simple surface.

If $\frac{\partial \vec{h}}{\partial u}(u_0, v_0)$ and $\frac{\partial \vec{h}}{\partial v}(u_0, v_0)$ are drawn emanating from $\vec{h}(u_0, v_0)$, then

- Both $\frac{\partial \vec{h}}{\partial u}(u_0, v_0)$ and $\frac{\partial \vec{h}}{\partial v}(u_0, v_0)$ are vectors tangent to the surface at $\vec{h}(u_0, v_0)$.

- $\frac{\partial \vec{h}}{\partial u}(u_0, v_0) \times \frac{\partial \vec{h}}{\partial v}(u_0, v_0)$ is a vector normal or perpendicular to the surface at $\vec{h}(u_0, v_0)$.

- $J\vec{h}(u_0, v_0) = \left\| \frac{\partial \vec{h}}{\partial u}(u_0, v_0) \times \frac{\partial \vec{h}}{\partial v}(u_0, v_0) \right\|$ is the rate that $\vec{h}$ changes area at (u_0, v_0).

- The derivative of $\vec{h}$ is the matrix, the first column of which is $\frac{\partial \vec{h}}{\partial u}(u_0, v_0)$ and the second column of which is $\frac{\partial \vec{h}}{\partial v}(u_0, v_0)$. That is,

$$D\vec{f}(u_0, v_0) = \begin{pmatrix} \frac{\partial x}{\partial u} & \frac{\partial x}{\partial v} \\ \frac{\partial y}{\partial u} & \frac{\partial y}{\partial v} \\ \frac{\partial z}{\partial u} & \frac{\partial z}{\partial v} \end{pmatrix}.$$

- Let $\vec{r}_0 = (u_), v_0)$. Then

$$D\vec{h}|_{\vec{r}_0}(\vec{r} - \vec{r}_0) + \vec{h}(\vec{r}_0) = \begin{pmatrix} \frac{\partial x}{\partial u} & \frac{\partial x}{\partial v} \\ \frac{\partial y}{\partial u} & \frac{\partial y}{\partial v} \\ \frac{\partial z}{\partial u} & \frac{\partial z}{\partial v} \end{pmatrix} \begin{pmatrix} u - u_0 \\ v - v_0 \end{pmatrix} + \begin{pmatrix} x(u_0, v_0) \\ y(u_0, v_0) \\ z(u_0, v_0) \end{pmatrix}$$

is the first order Taylor approximation to the surface, which parametrizes a plane tangent to the surface at $\vec{h}(u_0, v_0)$.

EXERCISES 14.2

In Exercises 1–5, find a parametrization for the region in the xy–plane bounded by the given graphs, and calculate the rate that the parametrization changes area.

1. $x = 1, x = 2, f(x) = x$, and $g(x) = 0$.

2. $x = 0, x = 1, f(x) = -1$, and $g(x) = e^x$.

3. $x = 0, x = 1, f(x) = -x$, and $g(x) = e^x$.

4. $x = 1, x = 3, f(x) = x^2$, and $g(x) = -x^2$.

5. $x = 1, x = 2, f(x) = x$, and $g(x) = \ln x$.

In Exercises 6–8, find a parametrization for the parallelogram with adjacent edges $\vec{A}$ and $\vec{B}$ drawn emanating from the position $\vec{C}$. Calculate the derivative and the Jacobian of the parametrization.

6. $\vec{A} = (1, 2), \vec{B} = (-1, 0)$, and $\vec{C} = (1, 3)$.

7. $\vec{A} = (1, -1, 2), \vec{B} = (0, -1, 0)$, and $\vec{C} = (1, 0, 3)$.

8. $\vec{A} = (-3, -1, 0), \vec{B} = (0, -1, 1)$, and $\vec{C} = (1, 1, 3)$.

In Exercises 9–11, parametrize the parallelogram with adjacent sides the line segments $[\vec{A}, \vec{B}]$ and $[\vec{A}, \vec{C}]$.

9. $\vec{A} = (1, 2), \vec{B} = (-1, 0)$, and $\vec{C} = (1, 3)$.

10. $\vec{A} = (1, -1, 2), \vec{B} = (0, -1, 0)$, and $\vec{C} = (1, 0, 3)$.

11. $\vec{A} = (-3, -1, 0), \vec{B} = (0, -1, 1)$, and $\vec{C} = (1, 1, 3)$.

In Exercises 12–14, find a parametrization for the sphere of radius r and center $\vec{C}$, and calculate its derivative and Jacobian.

12. $r = 1, \vec{C} = (1, 1, 1)$.

13. $r = 6, \vec{C} = (-1, 3, 0)$.

14. $r = 2, \vec{C} = (1, -1, 1)$.

In Exercises 15–17, find a parametrization for the disc in the plane with radius r and center $\vec{C}$. Calculate its derivative and Jacobian.

15. $r = 1, \vec{C} = (1,1)$.

16. $r = 6, \vec{C} = (-1,0)$.

17. $r = 2, \vec{C} = (1,-3)$.

In Exercises 18–25, find a parametrization for the surface obtained by rotating $\mathcal{G}$ about the x–axis. Calculate its derivative and Jacobian.

18. $\mathcal{G}$ is the graph of $y = e^x, 0 \le x \le 2$.

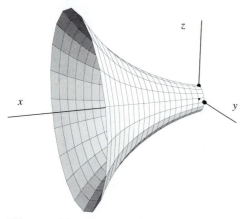

Figure 11. *The surface in Exercise 18.*

19. $\mathcal{G}$ is the graph of $y = \sin x, 0 \le x \le \pi$.

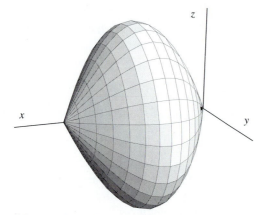

Figure 12. *The surface in Exercise 19.*

20. $\mathcal{G}$ is the graph of $y = \ln x, 1 \le x \le e^2$.

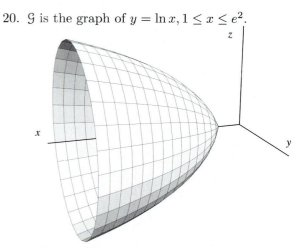

Figure 13. *The surface in Exercise 20.*

21. $\mathcal{G}$ is the image of $\vec{r}(t) = (\cos(t), 2\sin(t))$, $0 \le t \le \pi$.

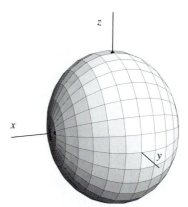

Figure 14. *The surface in Exercise 21.*

22. $\mathcal{G}$ is the image of $\vec{r}(t) = (4\cos(t), 2\sin(t) + 2)$, $0 \le t \le \pi$.

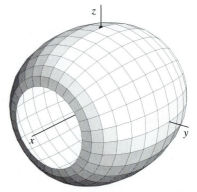

Figure 15. *The surface in Exercise 22.*

23. $\mathcal{G}$ is the image of $\vec{r}(t) = t(\cos(t), \sin(t))$, $0 \le t \le \pi$.

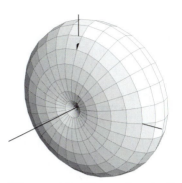

Figure 16. *The surface in Exercise 23.*

24. $\mathcal{G}$ is the graph of $x^2 + (y-2)^2 = 1$.

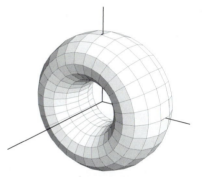

Figure 17. *The surface in Exercise 24.*

25. $\mathcal{G}$ is the graph of $(x+1)^2 + (y-2)^2 = 1$.

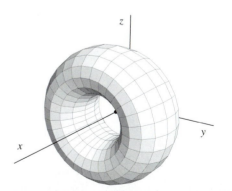

Figure 18. *The surface in Exercise 25.*

In Exercises 26–31, find a parametrization for the surface obtained by rotating $\mathcal{G}$ about the y–axis. Calculate its derivative and Jacobian.

26. $\mathcal{G}$ is the graph of $y = e^x, 0 \le x \le 2$.

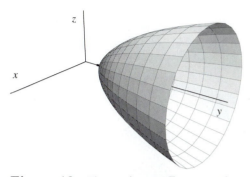

Figure 19. *The surface in Exercise 26.*

27. $\mathcal{G}$ is the graph of $y = \cos x, 0 \le x \le 2\pi$.

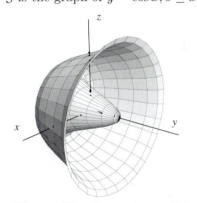

Figure 20. *The surface in Exercise 27.*

28. $\mathcal{G}$ is the graph of $y = \ln x, 1 \le x \le e^2$.

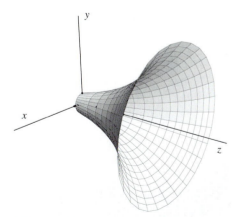

Figure 21. *The surface in Exercise 28.*

29. $\mathcal{G}$ is the image of $\vec{r}(t) = (\cos(t), 2\sin(t))$, $0 \le t \le \pi$.

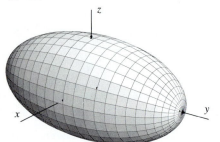

Figure 22. *The surface in Exercise 29.*

30. $\mathcal{G}$ is the image of $\vec{r}(t) = (4\cos(t)+4, 2\sin(t)+1)$, $0 \le t \le 2\pi$.

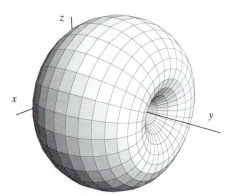

Figure 23. *The surface in Exercise 30.*

31. $\mathcal{G}$ is the image of $\vec{r}(t) = t(\cos(t), \sin(t))$, $\frac{\pi}{2} \le t \le \frac{3\pi}{2}$.

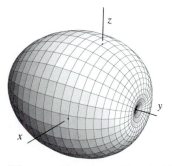

Figure 24. *The surface in Exercise 31.*

32. Show that if the curve parametrized by $\vec{r}(t) = (x(t), y(t))$ is rotated about the x–axis, then the vector

$$(y(t)y'(t), -y(t)x'(t)\cos(\theta), -y(t)x'(t)\sin(\theta))$$

is normal to the surface at $\vec{h}(t, \theta) = (x(t),\; y(t)\cos(\theta), y(t)\sin(\theta))$ and $J\vec{h}(t, \theta) = |y(t)|\sqrt{(x'(t))^2 + (y'(t))^2}$.

33. Show that if the graph of $f(x)$ is rotated about the x–axis, then the Jacobian is $J\vec{h}(x, \theta) = |f(x)|\sqrt{1 + (f'(x))^2}$.

34. Show that if the curve parametrized by $\vec{r}(t) = (x(t), y(t))$ is rotated about the y–axis, then the vector

$$(x(t)y'(t)\cos(\theta), -x(t)x'(t)\cos(\theta), x(t)y'(t)\sin(\theta))$$

is normal to the surface at $\vec{h}(t, \theta) = (x(t)\cos(\theta), y(t), x(t)\sin(\theta))$ and $Jh(t, \theta) = |x(t)|\sqrt{(x'(t))^2 + (y'(t))^2}$.

35. Show that if the graph of $f(x)$ is rotated about the y–axis, then the Jacobian is
$$J\vec{h}(x, \theta) = |x|\sqrt{1 + (f'(x))^2}.$$

In Exercises 36–41, $\mathcal{C}$ is the circle $(x - 2)^2 + (y + 3)^2 = 1$. Find a parametrization and the rate the parametrization changes area when $\mathcal{C}$ is rotated about the given line in the plane.

36. $y = 0$.   37. $x = 0$.   38. $x = -2$

39. $y = 5$.   40. $x = 5$.   41. $y = -5$

42. Parametrize the part of the surface of the sphere $x^2 + y^2 + z^2 = 1$ that lies in the first octant. (That is, $x \ge 0$, $y \ge 0$, and $z \ge 0$.)

43. Parametrize the top part of the surface of the sphere $x^2 + y^2 + z^2 = 1$.

44. Parametrize the part of the surface of the sphere $x^2 + y^2 + z^2 = 1$ such that $x \ge 0$ and $y \le 0$.

45. Let

$$\vec{h}(\phi, \theta) = \begin{pmatrix} a\sin(\phi)\cos(\theta) \\ b\sin(\phi)\sin(\theta) \\ c\cos(\phi) \end{pmatrix}, \quad \begin{matrix} 0 \le \theta \le 2\pi, \\ 0 \le \phi \le \pi. \end{matrix}$$

Show that $\vec{h}$ parametrizes the ellipsoid $\frac{x^2}{a^2} + \frac{y^2}{b^2} + \frac{z^2}{c^2} = 1$, and find the Jacobian of $\vec{h}$.

46. Use the results of the previous exercise to find a parametrization for the surface of the ellipsoid $\frac{x^2}{4} + \frac{y^2}{9} + \frac{z^2}{16} = 1$.

47. Use the parametrization $\vec{h}$ from Exercise 45 to find a vector normal to the ellipsoid $\frac{x^2}{4} + \frac{y^2}{9} + \frac{z^2}{16} = 1$ at $\vec{h}\left(\frac{\pi}{4}, \frac{\pi}{4}\right)$.

48. Find a vector normal to the graph of $f(x, y) = x^2 + y^2$ at $(2, 3, 13)$.

49. Find Taylor's first order approximation to the graph of $f(x, y) = \sin(xy)$ at $\left(\frac{\pi}{3}, \frac{1}{2}, \frac{1}{2}\right)$.

50. Find an equation for the plane tangent to the surface of Exercise 18 at the point $\left(1, \frac{e}{\sqrt{2}}, \frac{e}{\sqrt{2}}\right)$.

51. Find an equation for the plane tangent to the surface of Exercise 19 at the point $\left(\frac{\pi}{2}, \frac{1}{2}, \frac{\sqrt{3}}{2}\right)$.

52. The function

$$\vec{h}(t, \theta) = \begin{pmatrix} x \\ x^3\sin(\theta) \\ x^3\cos(\theta) \end{pmatrix}, \quad \begin{matrix} -1 \le t \le 1, \\ 0 \le \theta \le 2\pi \end{matrix}$$

parametrizes the surface obtained by rotating the graph of $y = x^3$, $-1 \le x \le 1$ about the x-axis. Show that $\vec{h}$ is not one-to-one on the interior of its domain. Is the Jacobian of $\vec{h} = 0$ at any point on the interior of its domain? Explain why $\vec{h}$ does not meet the requirements for its image to be a simple surface.

14.3   An Introduction to Surface Integrals

In the last section, we learned to parametrize several different types of surfaces. We use these techniques to integrate over surfaces that can be parametrized.

EXAMPLE 1: Find the surface area of the sphere with radius r.

SOLUTION: From the previous section we have a parametrization

$$\vec{h}(\phi, \theta) = (r\sin\phi\cos\theta, r\sin\phi\sin\theta, r\cos\phi), 0 \le \phi \le \pi, 0 \le \theta \le 2\pi$$

for the sphere of radius r. Let R be the rectangle bounded by the line $\phi = 0, \phi = \pi, \theta = 0$, and $\theta = 2\pi$, and let

$$\{\phi = \phi_0, \phi = \phi_1, \ldots, \phi = \phi_n\} \cup \{\theta = \theta_0, \theta = \theta_1, \ldots, \theta = \theta_m\}$$

be a partition of the rectangle R. Let $A_{i,j}$ denote the area of $\vec{h}(R_{i,j})$, where $R_{i,j}$ is the rectangle bounded by the lines $\phi = \phi_{i-1}, \phi = \phi_i, \theta = \theta_{j-1}$, and $\theta = \theta_j$. Then, the area $A_{i,j}$ is approximated by the area of the parallelogram with adjacent sides the straight line segments $\vec{h}(\phi_{i-1}, \theta_j) - \vec{h}(\phi_{i-1}, \theta_{j-1})$ and $\vec{h}(\phi_i, \theta_{j-1}) - \vec{h}(\phi_{i-1}, \theta_{j-1})$. See Figure 1. Thus

$$A_{i,j} \approx$$

$$\left\| \left(\vec{h}(\phi_{i-1}, \theta_j) - \vec{h}(\phi_{i-1}, \theta_{j-1}) \right) \times \left(\vec{h}(\phi_i, \theta_{j-1}) - \vec{h}(\phi_{i-1}, \theta_{j-1}) \right) \right\|$$

$$= \left\| \frac{\vec{h}(\phi_{i-1}, \theta_j) - \vec{h}(\phi_{i-1}, \theta_{j-1})}{\phi_i - \phi_{i-1}} \times \frac{\vec{h}(\phi_i, \theta_{j-1}) - \vec{h}(\phi_{i-1}, \theta_{j-1})}{\theta_j - \theta_{j-1}} \right\| \cdot$$

$$(\phi_i - \phi_{i-1})(\theta_j - \theta_{j-1})$$

$$\approx \left\| \frac{\partial \vec{h}(\phi_{i-1}, \theta_{j-1})}{\partial \phi} \times \frac{\partial \vec{h}(\phi_{i-1}, \theta_{j-1})}{\partial \theta} \right\| (\phi_i - \phi_{i-1})(\theta_j - \theta_{j-1})$$

$$= J(\vec{h}(\phi_{i-1}, \theta_{j-1}))(\phi_i - \phi_{i-1})(\theta_j - \theta_{j-1})$$

$$= \left(\begin{array}{c} \text{Rate } \vec{h} \\ \text{changes} \\ \text{Area} \end{array} \right) \times \left(\begin{array}{c} \text{Area} \\ \text{of} \\ R_{i,j} \end{array} \right).$$

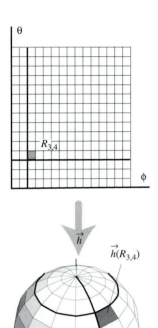

Figure 1. $\vec{h}$ takes the shaded rectangle $R_{i,j}$ in the domain of $\vec{h}$ onto the shaded "patch" on the sphere.

We now have

$$\frac{\partial \vec{h}}{\partial \phi} = (r \cos \phi \cos \theta, r \cos \phi \sin \theta, -r \sin \phi),$$

and

$$\frac{\partial \vec{h}}{\partial \theta} = (-r \sin \phi \sin \theta, r \sin \phi \cos \theta, 0).$$

Thus

$$J(\vec{h}(\phi, \theta)) = \left\| \frac{\partial \vec{h}(\phi, \theta)}{\partial \phi} \times \frac{\partial \vec{h}(\phi, \theta)}{\partial \theta} \right\| = r^2 \sin \phi.$$

The area of S is given by

$$\sum_{i=1}^{n} \sum_{j=1}^{m} A_{i,j} \approx \sum_{i=1}^{n} \sum_{j=1}^{m} \underbrace{r^2 \sin \phi_j}_{J(\vec{h}(\phi_{i-1}, \theta_{j-1}))} \underbrace{(\phi_i - \phi_{i-1})(\theta_j - \theta_{j-1})}_{\text{Area of } R_{i,j}}$$

$$\approx \iint_R r^2 \sin \phi \, dA$$

$$= \int_0^{2\pi} \int_0^{\pi} \underbrace{r^2 \sin \phi}_{\substack{\text{Rate } \vec{h} \\ \text{changes} \\ \text{area}}} \underbrace{d\phi \, d\theta}_{dA} = \int_0^{2\pi} 2r^2 \, d\theta = 4\pi r^2. \qquad \blacksquare$$

Notice the importance of the Jacobian in the above calculation. Because we knew how $\vec{h}$ changes area, we were able to transform knowledge about the areas of rectangles $R_{i,j}$ into knowledge about the corresponding areas on the sphere.

EXAMPLE 2: Let S be the portion of the sphere of radius r with center at the origin such that $x \geq 0, y \geq 0$, and $z \geq 0$. If the density of S at (x, y, z) is given by $\rho(x, y, z) = x + y + z$, find the mass of S.

SOLUTION: Suppose that we have the parametrization $\vec{h}$ as in the solution of Example 1, except that the domain is $0 \leq \phi \leq \frac{\pi}{2}$ and $0 \leq \theta \leq \frac{\pi}{2}$. Let $R_{i,j}$ be as in the solution of Example 1, and let $A_{i,j}$ be the area of $\vec{h}(R_{i,j})$. In Example 1, we took as an approximation for the area $A_{i,j}$ the expression

$$\left\| \frac{\partial \vec{h}(\phi_{i-1}, \theta_{j-1})}{\partial \phi} \times \frac{\partial \vec{h}(\phi_{i-1}, \theta_{j-1})}{\partial \theta} \right\| (\phi_i - \phi_{i-1})(\theta_j - \theta_{j-1}).$$

Thus we take as an approximation of the mass of S

$$\sum_{i=1}^{n} \sum_{j=1}^{m} \rho(\vec{h}(\phi_{i-1}, \theta_{j-1})) \left\| \frac{\partial \vec{h}(\phi_{i-1}, \theta_{j-1})}{\partial \phi} \times \frac{\partial \vec{h}(\phi_{i-1}, \theta_{j-1})}{\partial \theta} \right\| (\phi_i - \phi_{i-1})(\theta_j - \theta_{j-1})$$

$$= \sum_{i=1}^{n} \sum_{j=1}^{m} \rho(\vec{h}(\phi_{i-1}, \theta_{j-1})) J(\vec{h}(\phi_{i-1}, \theta_{j-1}))(\text{ Area of } R_{i,j}),$$

which is an approximation of

$$\iint_R \rho(\vec{h}(\phi, \theta)) \left\| \frac{\partial \vec{h}(\phi, \theta)}{\partial \phi} \times \frac{\partial \vec{h}(\phi, \theta)}{\partial \theta} \right\| dA$$

$$= \iint_R \rho(\vec{h}(\phi, \theta)) J(\vec{h}(\phi, \theta)) \, dA$$

$$= \int_0^{\pi/2} \int_0^{\pi/2} (r^3 \sin^2 \phi \cos \theta + r^3 \sin^2 \phi \sin \theta + r^3 \cos \phi \sin \phi) \, d\phi \, d\theta$$

$$= r^3 \int_0^{\pi/2} \left(\frac{\phi}{2} - \frac{1}{2} \sin \phi \cos \phi \right) \cos \theta + \left(\frac{\phi}{2} - \frac{1}{2} \sin \phi \cos \phi \right) \sin \theta +$$

$$\frac{1}{2} \cos^2 \phi \Big|_{\phi=0}^{\phi=\pi/2} \, d\theta$$

$$= r^3 \int_0^{\pi/2} \left(\frac{\pi}{4} \cos \theta + \frac{\pi}{4} \sin \theta - \frac{1}{2} \right) \, d\theta = \frac{\pi}{4} r^3. \qquad \blacksquare$$

The preceding examples lead us to the notion of a surface integral.

Definition: $\iint_S f \ dS$

Suppose that S is a simple surface and f is a continuous function from S into $\mathbb{R}$. Then the number L is called the *integral of f over S* provided that it is true that if $\vec{h}$ is a parametrization for S with domain the rectangle R, then

$$L \ = \ \iint_R f(\vec{h}(x,y)) \left\| \frac{\partial \vec{h}(x,y)}{\partial x} \times \frac{\partial \vec{h}(x,y)}{\partial y} \right\| dA$$

$$= \ \iint_R f(\vec{h}(x,y)) J(\vec{h}(x,y)) \ dA.$$

If L exists, then L will be denoted by $\displaystyle\iint_S f \ dS$.

This definition reveals the importance of requiring that the partial derivatives of a parametrization be continuous. In order to ensure that

$$\iint_S f dS = \iint_R f(\vec{h}(x,y)) \left\| \frac{\partial \vec{h}(x,y)}{\partial x} \times \frac{\partial \vec{h}(x,y)}{\partial y} \right\| dA$$

exists, we need to know that $(f \circ \vec{h}) \left\| \frac{\partial \vec{h}}{\partial x} \times \frac{\partial \vec{h}}{\partial y} \right\|$ is continuous.

Applications of surface integrals generally make the assumption that if $\vec{h}$ is a parametrization for the simple surface S with domain a rectangle R, and if $\{x = x_0, x = x_1, \ldots, x = x_n\} \cup \{y = y_0, y = y_1, \ldots, y = y_m\}$ is a sufficiently fine partition of R, and $\vec{s}_{i,j}$ is an arbitrary point of $R_{i,j}$, then

$$\sum_{i=1}^{n} \sum_{j=1}^{m} \left\| \frac{\partial \vec{h}(s_{i,j})}{\partial x} \times \frac{\partial \vec{h}(s_{i,j})}{\partial y} \right\| (x_i - x_{i-1})(y_j - y_{j-1})$$

is a good approximation of the area of the surface S. The proof that this is a valid assumption is beyond the scope of this book. However, with our definition of a simple surface, this assumption is valid.

In particular then, we have that $\iint_S dS$ is the area of the surface S.

EXAMPLE 3: Verify that the area of the sphere of radius ρ is $4\pi\rho^2$.

SOLUTION: Let

$$\vec{h}(\phi, \theta) = \begin{pmatrix} \rho \sin(\phi) \cos(\theta) \\ \rho \sin(\phi) \sin(\theta) \\ \rho \cos(\phi) \end{pmatrix}, \ 0 \leq \phi \leq \pi, \ 0 \leq \theta \leq 2\pi.$$

We know that $J\vec{h}(\phi, \theta) = \rho^2 \sin\phi$. The area is

$$\iint_S dS = \int_0^\pi \int_0^{2\pi} \rho^2 \sin(\phi) d\theta d\phi$$

$$= 2\pi\rho^2 \int_0^\pi \sin(\phi) d\phi = 4\pi\rho^2. \qquad \blacksquare$$

EXAMPLE 4: Let P be the parallelogram with adjacent edges the line segments $[\vec{A}, \vec{B}]$ and $[\vec{A}, \vec{C}]$. Calculate $\iint_P f \, dS$, where $f(x, y, z) = x + y + z$, $\vec{A} = (1, 0, 1)$, $\vec{B} = (1, 1, 1)$, and $\vec{C} = (0, 1, 1)$.

SOLUTION: A parametrization of our parallelogram is given by

$$\vec{h}(s, t) = \vec{A} + s(\vec{B} - \vec{A}) + t(\vec{C} - \vec{A})$$

$$= (1, 0, 1) + s(0, 1, 0) + t(-1, 1, 0) = (1 - t, s + t, 1),$$

$$0 \leq s \leq 1, 0 \leq t \leq 1.$$

So $\frac{\partial \vec{h}}{\partial s} = (0, 1, 0)$, and $\frac{\partial \vec{h}}{\partial t} = (-1, 1, 0)$. The Jacobian, $J(\vec{h}(s, t)) = \left\| \frac{\partial \vec{h}}{\partial s} \times \frac{\partial \vec{h}}{\partial t} \right\| = 1$, and $f(\vec{h}(s, t)) = (1 - t) + (s + t) + 1 = 2 + s$.

$$\iint_P f \, dS = \iint_R f(\vec{h}(s, t)) J(\vec{h}(s, t)) \, dA = \int_0^1 \int_0^1 (2 + s) \, dt \, ds$$

$$= \int_0^1 (2t + st) \Big|_{t=0}^{t=1} ds = \int_0^1 (2 + s) \, ds$$

$$= \left(2s + \frac{s^2}{2} \right) \Big|_{s=0}^{s=1} = \frac{5}{2}. \qquad \blacksquare$$

The following theorem is rather nice because it tells us that if S is the region in the plane bounded between the graphs of two functions (over some interval), then we can integrate on S without parametrizing it.

Theorem 1 *Suppose that f and g are continuous functions from the interval $[a, b]$ into $\mathbb{R}$ such that $f(x) > g(x)$ for all x between a and b. Let S be the region bounded by the graphs of f and g and the lines $x = a$ and $x = b$ as in Figure 2. If ρ is a continuous function defined on S, then*

$$\iint_S \rho \, dS = \int_a^b \int_{g(x)}^{f(x)} \rho(x, y) \, dy \, dx.$$

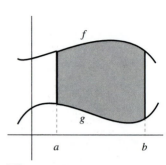

Figure 2. *The region bounded by the graphs of f and g and the lines $x = a$ and $x = b$.*

Proof: From Example 1 of the previous section, a parametrization for S is given by

$$\vec{h}(s, t) = (s, g(s) + t(f(s) - g(s))), a \leq s \leq b, 0 \leq t \leq 1.$$

The partial derivatives of $\vec{h}$ are

$$\frac{\partial \vec{h}(s, t)}{\partial s} = (1, g'(s) + t(f'(s) - g'(s)))$$

$$\frac{\partial \vec{h}(s, t)}{\partial t} = (0, f(s) - g(s))$$

Thus

$$J(\vec{h}(s, t)) = f(s) - g(s).$$

The integral has the form

$$\iint_S \rho \, dS = \iint_R \rho(\vec{h}(s, t)) J(\vec{h}(s, t)) \, dA$$

$$= \int_a^b \int_0^1 \rho(\vec{h}(s, t)) J(\vec{h}(s, t)) \, dt \, ds.$$

We now have

$$\int_0^1 \rho(\vec{h}(s, t)) J(\vec{h}(s, t)) \, dt = \int_0^1 \rho(s, g(s) + t(f(s) - g(s)))(f(s) - g(s)) \, dt.$$

Let $u = g(s) + t(f(s) - g(s))$. Then, since we are integrating with respect to t, $du = (f(s) - g(s)) \, dt$. When $t = 0, u = g(s)$ and when $t = 1, u = f(s)$. Thus

$$\int_0^1 \rho(s, g(s) + t(f(s) - g(s)))(f(s) - g(s)) \, dt = \int_{g(s)}^{f(s)} \rho(s, u) \, du.$$

This implies that

$$\iint_S \rho \, dS = \int_a^b \int_{g(s)}^{f(s)} \rho(s, u) \, du \, ds = \int_a^b \int_{g(x)}^{f(x)} \rho(x, y) \, dy \, dx. \quad \blacksquare$$

EXAMPLE 5: Calculate $\iint_S (x + y) \, dS$, where S is the region bounded between $y = x^2 + 1$ and $y = x$ and the lines $x = 0$ and $x = 2$. (See Figure 3.)

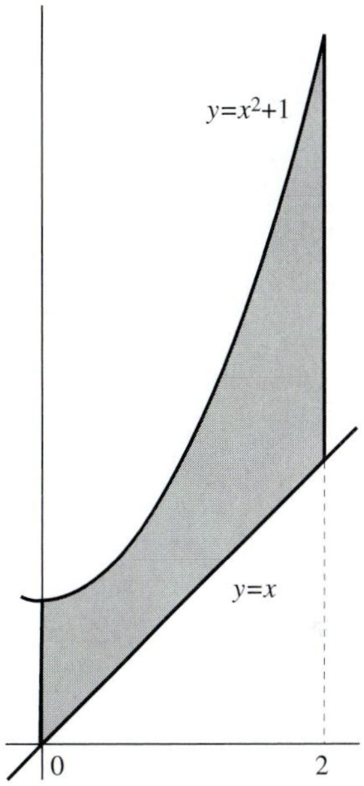

SOLUTION:

$$\iint_S (x + y) \, dS = \int_0^2 \int_x^{x^2+1} x + y \, dy \, dx = \int_0^2 xy + \frac{y^2}{2} \Big|_{y=x}^{y=x^2+1} dx$$

$$= \int_0^2 x(x^2 + 1) + \frac{(x^2 + 1)^2}{2} - \frac{3x^2}{2} \, dx$$

$$= \int_0^2 \frac{x^4}{2} + x^3 - \frac{x^2}{2} + x + \frac{1}{2} \, dx$$

$$= \frac{x^5}{10} + \frac{x^4}{4} - \frac{x^3}{6} + \frac{x^2}{2} + \frac{x}{2} \Big|_0^2$$

$$= \frac{32}{10} + \frac{16}{4} - \frac{8}{6} + \frac{4}{2} + \frac{2}{2} = \frac{133}{15}. \quad \blacksquare$$

Figure 3. *The region bounded by $y = x^2 + 1$, $y = x$, and the lines $x = 0$ and $x = 2$.*

EXERCISES 14.3

In Exercises 1–8, compute $\iint_S f \, dS$.

1. S is parametrized by $\vec{h}(s, t) = (s, t, s + 2t), 0 \leq s \leq 1$ and $-1 \leq t \leq 1$, and $f(x, y, z) = xyz$.

2. S is parametrized by $\vec{h}(s, t) = (\cos s, \sin s + \sin t, 1), 0 \leq s \leq \pi$ and $-\frac{\pi}{2} \leq t \leq \frac{\pi}{2}$, and $f(x, y, z) = 1$.

3. S is parametrized by $\vec{h}(s, t) = (s + 3t, t, 1), 0 \leq s \leq \pi$ and $0 \leq t \leq \frac{\pi}{2}$, and $f(x, y, z) = \cos x + \sin y + z$.

4. S is the unit disc centered at the origin, and $f(x, y) = x^2 + y^2$.

5. S is the disc of radius 4 centered at the origin, and $f(x, y) = (x^2 + y^2)^{3/2}$.

6. S is the upper half of the unit disc, and $f(x, y) = x^2(x^2 + y^2)^3$.

7. S is the region between the circles $x^2 + y^2 = 1$ and $x^2 + y^2 = 2$, and $f(x, y) = x^2(x^2 + y^2)$.

8. S is the upper half of the unit disc, and $f(x, y) = e^{-u}$, where $u = x^2 + y^2$.

In Exercises 9 and 10, compute $\iint_S f \, dS$, where S is the parallelogram with the line segments $[\vec{A}, \vec{B}]$ and $[\vec{A}, \vec{C}]$ as adjacent edges.

9. $\vec{A} = (0, 0), \vec{B} = (1, 1)$, and $\vec{C} = (1, 0)$ where $f(x, y) = xy$.

10. $\vec{A} = (0, 0, 0), \vec{B} = (1, 1, 1)$, and $\vec{C} = (1, 0, 1)$ where $f(x, y, z) = xyz$.

11. Compute the area $\iint_S dS$, where S is the lateral surface of the right circular cone with height 2 and base of radius 1.

Figure 4. *The right cylinder cone in Exercise 11.*

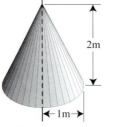

2m

←1m→

In Exercises 12–14, S is the region bounded by the graphs of f and g and the lines $x = a$ and $x = b$. Sketch S and compute $\iint_S \rho\, dS$.

12. $f(x) = x + 1$, $g(x) = x^3$, $a = -1$, $b = 1$, $\rho(x, y) = 3x - 2y$.

13. $f(x) = \sqrt{x}$, $g(x) = -2x$, $a = 1$, $b = 2$, $\rho(x, y) = x^2 y$.

14. $f(x) = x^3$, $g(x) = x$, $a = 1$, $b = 3$, $\rho(x, y) = e^{y/x}$.

15. Find $\iint_S x + y\, dS$, where S is the region bounded between $y = x^2$ and $y = x^3$.

16. For the given function f, let S be the graph of the function $f(x, y)$, $a \le x \le b$, $c \le y \le d$. Show that

$$\iint_S g(x, y, z)dS$$
$$= \iint_R g(x, y, f(x, y))\sqrt{\left(\frac{\partial f}{\partial x}\right)^2 + \left(\frac{\partial f}{\partial y}\right)^2 + 1}\, dA$$
$$= \int_c^d \int_a^b g(x, y, f(x, y))\sqrt{\left(\frac{\partial f}{\partial x}\right)^2 + \left(\frac{\partial f}{\partial y}\right)^2 + 1}\, dx\, dy.$$

In Exercises 17–22, S is the graph of the function $f(x, y)$, $a \le x \le b$, $c \le y \le d$. Express the surface integral as a double integral. If you have Mathematica©, Maple©, or a similar computational package available, you might try using it to complete your computations.

17. $\iint_S dS$, $f(x, y) = x^2 + y^2$, $a = -2$, $b = 2$, $c = -2$, and $d = 2$.

18. $\iint_S xyz\, dS$, $f(x, y) = x^2 + y^2$, $a = -2$, $b = 2$, $c = -2$, and $d = 2$.

19. $\iint_S dS$, $f(x, y) = \sin(x^2 + y^2)$, $a = -2$, $b = 2$, $c = -2$, and $d = 2$.

20. $\iint_S xyz\, dS$, $f(x, y) = \sin(x^2 + y^2)$, $a = -2$, $b = 2$, $c = -2$, and $d = 2$.

21. $\iint_S dS$, $f(x, y) = e^{xy}$, $a = -2$, $b = 1$, $c = 0$, and $d = 2$.

22. $\iint_S xyz\, dS$, $f(x, y) = e^{xy}$, $a = 1$, $b = 3$, $c = -1$, and $d = 2$.

23. The curve C in the plane is parametrized by $\vec{r}(t) = (x(t), y(t))$, $a \le t \le b$ does not intersect the x–axis except perhaps at the end points. Let S denote the surface obtained by rotating C about the x–axis. Show that

$$\iint_S g(x, y, z)dS$$
$$= \iint_R g\left(x(t), y(t)\cos(\theta), y(t)\sin(\theta)\right) \cdot$$
$$|y(t)|\|\vec{r}\,'(t)\|dA$$
$$= \int_0^{2\pi} \int_a^b g\left(x(t), y(t)\cos(\theta), y(t)\sin(\theta)\right) \cdot$$
$$|y(t)|\|\vec{r}\,'(t)\|dt\, d\theta.$$

24. Suppose that the graph C of the function $f(x)$, $a \le x \le b$ does not cross the x–axis between a and b. Let S be the surface obtained by rotating C about the x–axis. Show that

$$\iint_S g(x, y, z)\, dS$$
$$= \int_0^{2\pi} \int_a^b g\left(x, f(x)\cos(\theta), f(x)\sin(\theta)\right) \cdot$$
$$|f(x)|\sqrt{1 + (f'(x))^2}\, dx d\theta.$$

25. The graph of $y = x^2 + 1$, $-1 \le x \le 1$ is rotated about the x–axis. Express the area of the resulting surface as a double integral.

26. The part of the graph of $y = 1 - x^2$ lying above the x–axis is rotated about the x–axis forming a surface S. Express $\iint_S xyz\, dS$ as a double integral.

27. The circle $(x-2)^2 + (y-3)^2 = 1$ is rotated about the x–axis to obtain a surface S. Find the resulting area.

28. The surface S is obtained by rotating the ellipse $x^2 + \frac{(y+2)^2}{4} = 1$ about the x–axis. Express the area of S as a double integral. Hint: This is the ellipse $x^2 + \frac{y^2}{4} = 1$ translated by the vector $(0, -2)$.

29. The circle $x^2 + y^2 = 1$ is rotated about the x–axis to give a surface S. Through what angle must the circle be rotated to yield all of S?

30. Assume that the curve C of Exercise 23 does not intersect the y–axis. It is rotated about the y–axis to yield a surface S. Show that

$$\iint_S g(x, y, z)\,dS$$
$$= \iint_R g\left(x(t)\cos(\theta), y(t), x(t)\sin(\theta)\right) \cdot$$
$$|x(t)|\,\|\vec{r}'(t)\|\,dA$$
$$= \int_0^{2\pi} \int_a^b g\left(x(t)\cos(\theta), y(t), x(t)\sin(\theta)\right) \cdot$$
$$|x(t)|\,\|\vec{r}'(t)\|\,dt\,d\theta.$$

31. Suppose that f' is continuous on $[a, b]$, that f is positive on $[a, b]$, and G is the graph of f on $[a, b]$. Let S be the surface obtained by rotating G about the y–axis. Show that

$$\iint_S g(x, y, z)\,dS$$
$$= \int_0^{2\pi} \int_a^b G\left(x\cos(\theta), f(x), x\sin(\theta)\right)$$
$$|x|\sqrt{1 + (f'(x))^2}\,dt\,d\theta.$$

32. Let $f(x) = e^x$, $0 \le x \le 2$. Express the area of the surface obtained by rotating f about the y–axis as a double integral.

33. Let S be the surface obtained by rotating $f(x) = x^2$, $1 \le x \le 3$ about the y–axis. Express $\iint_S x + yz\,dS$ as a double integral.

34. The circle $(x-3)^2 + (y+2)^2 = 1$ is rotated about the y–axis to form the surface S. Express $\iint_S yze^x\,dS$ as a double integral.

14.4 Some Applications of Surface Integrals

In the development of the surface integral in Section 14.3, we parametrized the surface with a function $\vec{h}$ having as its domain a rectangle R. We then partitioned R into small rectangles $R_{i,j}$ for $i = 1, \ldots, n$, $j = 1, \ldots, m$ and then used the $R_{i,j}$'s, via $\vec{h}$, to divide the surface S into nonoverlapping pieces $S_{i,j}$, $i = 1, \ldots, n$, $j = 1, \ldots, m$, where $S_{i,j} = \vec{h}(R_{i,j})$. We then chose selection points $\vec{s}_{i,j}$ from $R_{i,j}$ and calculated

$$\underbrace{\sum \sum}_{\substack{\text{Sum of} \\ \text{a lot of} \\ \text{pieces of} \\ \text{the} \\ \text{surface}}} \underbrace{f(\vec{h}(\vec{s})_{i,j}))}_{\substack{f \text{ evaluated} \\ \text{at a point} \\ \text{in a piece of} \\ \text{the surface}}} \underbrace{[J(\vec{h}(\vec{s}_{i,j}))\text{Area}(R_{i,j})]}_{\substack{\text{Approximation of} \\ \text{the area of a small} \\ \text{piece of the surface}}} \text{ as our approximation for } \iint_S f\,dS.$$

While there is no rigor in the following, it is helpful in applications of surface integrals to associate $\iint_S f\,dS$ with the above mentioned sums; that is,

$$\underbrace{\sum \sum}_{\substack{\text{Sum of} \\ \text{many} \\ \text{pieces of} \\ \text{the} \\ \text{surface}}} \underbrace{f(\vec{h}(\vec{s}_{i,j}))}_{\substack{f \text{ evaluated} \\ \text{at a point} \\ \text{in a piece of} \\ \text{the surface}}} \underbrace{[J(\vec{h}(\vec{s}_{i,j}))\text{Area}(R_{i,j})]}_{\substack{\text{Approximation of} \\ \text{the area of a small} \\ \text{piece of the surface}}} \approx \underbrace{\iint_S}_{\substack{\text{Sum of} \\ \text{many} \\ \text{pieces of} \\ \text{the} \\ \text{surface}}} \underbrace{f}_{\substack{f \text{ evaluated} \\ \text{at a point} \\ \text{in a piece of} \\ \text{the surface}}} \underbrace{dS}_{\substack{\text{Approximation} \\ \text{of the area of} \\ \text{a small piece} \\ \text{of the surface}}}$$

A Rule of Thumb for Surface Integrals

Generally, surface integrals will be applied in the following situation:

(a) S is a simple surface and ρ is a real-valued function defined on S.

(b) Q is a physical quantity that can be approximated by:

   (i) Breaking S into "very small" non overlapping pieces $S_{i,j}, i = 1, \ldots, n, j = 1, \ldots, m$.

   (ii) Choosing an arbitrary "selection point" $\vec{s}_{i,j}$ from each $S_{i,j}$.

   (iii) Calculating the sum $\sum \sum \rho(\vec{s}_{i,j}) \, \text{Area}(S_{i,j})$ to approximate Q.

EXAMPLE 1: Suppose that S is a surface in $\mathbb{R}^3$ and $\rho(x, y, z)$ gives the mass density of S at (x, y, z). Assuming that ρ is continuous, if ΔS is a very small piece of S and $\vec{s}$ is a point of ΔS, then the product of $\rho(\vec{s})$ and Area(ΔS) will be an approximation of the mass of ΔS. Thus, using our rule of thumb above, $\iint_S \rho \, dS$ is the mass of S.

EXAMPLE 2: Let S be the surface in $\mathbb{R}^2$ bounded by the graphs of $y = x^2$, the x–axis, and the line $x = 1$. See Figure 1.

Let $\rho(x, y) = xy$ denote the mass density of S at (x, y). Then the mass of S is given by $\iint_S xy \, dS$. We can employ Theorem 1 of the previous section to calculate this integral without parametrizing S.

$$\iint_S xy \, dS = \int_0^1 \int_0^{x^2} xy \, dy \, dx = \int_0^1 \frac{xy^2}{2}\Big|_{y=0}^{y=x^2} \, dx$$

$$= \int_0^1 \frac{x^5}{2} \, dx = \frac{1}{12}.$$ ∎

In Section 11.5 we computed the center of mass for a thin wire that was modeled by a fundamental curve. We can apply these same techniques to find the center of mass for a surface in $\mathbb{R}^3$. Let S be a surface parametrized by $\vec{r}$. Partition S by $\mathcal{P} = \{S_{i,j} \mid 1 \leq i \leq n, 1 \leq j \leq m\}$, and let $\mathcal{S} = \{\vec{s}_{1,1}, \dots, \vec{s}_{n,m}\}$ be a selection from $\mathcal{P}$, where $\vec{s}_{i,j} = (x_{i,j}, y_{i,j}, z_{i,j})$. Assume that there is a density function $\rho(x, y, z)$ on S. If we let X be the number approximated by

$$\frac{\sum_{i=1}^n \sum_{j=1}^m x_{i,j} \rho(\vec{s}_{i,j})(\text{area of } S_{i,j})}{\text{total mass of } S},$$

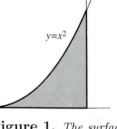

then

$$X = \frac{\iint_S x \rho(\vec{r}) \, dS}{\iint_S \rho(\vec{r}) \, dS}.$$

Figure 1. *The surface bounded by the graphs of $y = x^2$, the x-axis, and the line $x = 1$.*

Similarly, let

$$Y = \frac{\iint_S y \rho(\vec{r}) \, dS}{\iint_S \rho(\vec{r}) \, dS} \quad \text{and} \quad Z = \frac{\iint_S z \rho(\vec{r}) \, dS}{\iint_S \rho(\vec{r}) \, dS}.$$

The point (X, Y, Z) is called the *center of mass* of S. As in the case of curves, there is no reason that the center of mass need be on the surface.

EXAMPLE 3: Let S be the upper hemisphere of the sphere of radius R with center at the origin. Let $\rho(x, y, z) = 1$ be the mass density function. Find the center of mass of S.

SOLUTION: We parametrize S by $\vec{r}(s, t) = (R \cos(s) \sin(t), R \sin(s) \sin(t), R \cos(t))$ for $0 \leq s \leq 2\pi, 0 \leq t \leq \frac{\pi}{2}$. The Jacobian of $\vec{r}$ is $R^2 \sin(t)$. Thus

$$X = \frac{\iint_S x \rho(\vec{r}) \, dS}{\iint_S \rho(\vec{r}) \, dS} = \frac{\int_0^{\pi/2} \int_0^{2\pi} R \cos(s) \sin(t) R^2 \sin(t) \, ds \, dt}{\int_0^{\pi/2} \int_0^{2\pi} R^2 \sin(t) \, ds \, dt} = \frac{0}{2\pi R^2};$$

$$Y = \frac{\iint_S y\rho(\vec{r})\,dS}{\iint_S \rho(\vec{r})\,dS} = \frac{\int_0^{\pi/2}\int_0^{2\pi} R\sin(s)\sin(t)R^2\sin(t)\,ds\,dt}{\int_0^{\pi/2}\int_0^{2\pi} R^2\sin(t)\,ds\,dt} = \frac{0}{2\pi R^2};$$

$$Z = \frac{\iint_S z\rho(\vec{r})\,dS}{\iint_S \rho(\vec{r})\,dS} = \frac{\int_0^{\pi/2}\int_0^{2\pi} R\cos(t)R^2\sin(t)\,ds\,dt}{\int_0^{\pi/2}\int_0^{2\pi} R^2\sin(t)\,ds\,dt} = \frac{\pi R^3}{2\pi R^2} = \frac{R}{2}.$$

$$(X,Y,Z) = \left(0,0,\frac{R}{2}\right). \qquad \blacksquare$$

EXAMPLE 4: Suppose that S is a simple surface with a mass density function $\rho(\vec{r})$ and that S is rotating about the z–axis with an angular speed w rad/[unit of time]. If ΔS is a small piece of S and $\vec{s} = (x,y,z)$ is a point of ΔS, then $w(x^2+y^2)^{1/2}$ approximates the speed of ΔS and $\rho(x,y,z)\,\text{area}(\Delta S)$ approximates the mass of ΔS. Thus

$$\frac{1}{2}\underbrace{\left[w\sqrt{x^2+y^2}\right]^2}_{\substack{\text{Approximation}\\\text{of the speed}\\\text{squared of a}\\\text{piece of the}\\\text{surface.}}}\underbrace{\rho(x,y,z)\,\text{area}(\Delta S)}_{\substack{\text{Approximation}\\\text{of the mass}\\\text{of a piece of}\\\text{the surface.}}}$$

approximates the kinetic energy $\frac{mv^2}{2}$ of ΔS. If we break S up into small non overlapping pieces and add up the kinetic energy of each piece of S, then we have the total kinetic energy of the rotating mass. Therefore, by our rule of thumb,

$$\iint_S \frac{1}{2}[w^2(x^2+y^2)]\rho(x,y,z)\,dS \text{ is the kinetic energy of the sur-}$$
face rotating about the z–axis.

Of course we get similar formulas for the kinetic energy of a surface rotating about the x–axis or the y–axis:

$$\iint_S \frac{1}{2}[w^2(x^2+z^2)]\rho(x,y,z)\,dS \text{ is the kinetic energy of a surface}$$
rotating about the y–axis with angular speed w and:

$$\iint_S \frac{1}{2}[w^2(y^2+z^2)]\rho(x,y,z)\,dS \text{ is the kinetic energy of a surface}$$
rotating about the x–axis with angular speed w. $\qquad \blacksquare$

Note that the formulas for the kinetic energy of a rotating surface parallel those for fundamental curves as given in Section 11.5. We can define the moments of inertia similarly.

Definition: Moment of Inertia

$I_x = \iint_S (y^2 + z^2)\rho(x, y, z) \, dS = $ moment of inertia about the x–axis.

$I_y = \iint_S (x^2 + z^2)\rho(x, y, z) \, dS = $ moment of inertia about the y–axis.

$I_z = \iint_S (x^2 + y^2)\rho(x, y, z) \, dS = $ moment of inertia about the z–axis.

The kinetic energy of a surface rotating about an axis with angular speed w is given by $\frac{1}{2}w^2 I_a$, where $a = x, y,$ or z indicates the axis of rotation.

EXAMPLE 5: Let S and ρ be as in Example 2. If S is rotating about the x–axis at 2π rad/sec, what is the total kinetic energy (where distance is measured in meters)?

SOLUTION:

$$
\begin{aligned}
\text{Kinetic Energy} \;=\;& \frac{1}{2}w^2 I_x = \frac{1}{2}w^2 \iint_S (y^2 + z^2)\rho(x, y, z) \, dS \\[2mm]
=\;& \frac{1}{2}w^2 \iint_S (y^2 + z^2)xy \, dS \\[2mm]
=\;& \frac{1}{2}(2\pi)^2 \int_0^1 \int_0^{x^2} y^2 xy \, dy \, dx \\[2mm]
=\;& 2\pi^2 \int_0^1 \int_0^{x^2} y^2 xy \, dy \, dx \\[2mm]
=\;& 2\pi^2 \int_0^1 \left(\frac{y^4}{4}\right) x \Big|_{y=0}^{y=x^2} \, dx \\[2mm]
=\;& 2\pi^2 \int_0^1 \left(\frac{x^8}{4}\right) x \, dx = \frac{\pi^2}{20} \text{ Joules.} \qquad \blacksquare
\end{aligned}
$$

Volumes of Solids Bounded Between Graphs of Functions

Figure 2.a *V is the set of all points (x, y, z) such that (x, y) is in S and $f(x, y) \geq z \geq g(x, y)$.*

Suppose that S is a simple surface lying in xy–space, and that f and g are continuous functions from S into $\mathbb{R}$ such that $f(x, y) \geq g(x, y)$ for all (x, y) in S. Then the set of all points (x, y, z) such that (x, y) is in S and $f(x, y) \geq z \geq g(x, y)$ is a solid V as in Figures 2.a and 2.b.

Let ΔS be a small piece of S and let $\vec{s}$ be a point in ΔS. Then the part of V lying "above" ΔS can be approximated by $[f(\vec{s}) - g(\vec{s})]$ area(ΔS). By our rule of thumb, the total volume of V must be $\iint_S (f - g)\, dS$.

In order to calculate the volume of a solid using the above method, we must be able to identify the top and bottom boundaries of the solid. It is often useful to sketch the solid, but this is not aways easy, as illustrated by the next example.

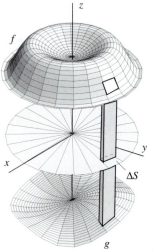

Figure 2.b *That part of V lying "above" ΔS.*

EXAMPLE 6: Find the volume of the region inside the cylinder $x^2 + y^2 = 1$ and between $z = 1 - x$ and $z = -(x^2 + y^2)$.

In Figure 3.a, we draw the top bounding function, and the bottom bounding graph is sketched in Figure 3.b. In Figures 4.a and 4.b, we give views of the solid from above and from below. It is the concave bottom bounding surface that makes the solid difficult to visualize.

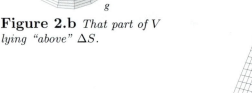

Figure 3.a *The top bounding function $z = 1 - x$.*

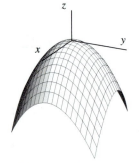

Figure 3.b *The bottom bounding function $z = -(x^2 + y^2)$.*

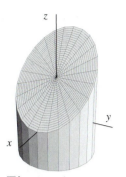

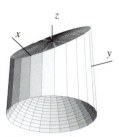

Figure 4.a *The solid viewed from above.*

Figure 4.b *The solid viewed from below.*

SOLUTION: Let S denote the disc $(x^2 + y^2) \leq 1$. Note that $-(x^2 + y^2) < 1 - x$, for all x and y in S. Thus the volume is given by $\iint_S (f - g)\, ds$, where $f(x,y) = 1 - x$, and $g(x,y) = -(x^2 + y^2)$. The polar transformation gives a convenient parametrization for S:

$$\vec{\rho}(r, \theta) = (r\cos(\theta), r\sin(\theta)), 0 \leq r \leq 1 \text{ and } 0 \leq \theta \leq 2\pi.$$

Since $J(\vec{\rho}(r, \theta)) = r$, we have

$$
\begin{aligned}
\iint_S (f - g)\, dS &= \iint_S (1 - x + x^2 + y^2)\, dS \\
&= \int_0^{2\pi} \int_0^1 (1 - r\cos(\theta) + r^2) r \, dr\, d\theta \\
&= \int_0^{2\pi} \left[\frac{r^2}{2} - \left(\frac{r^3}{3} \right) \cos(\theta) + \frac{r^4}{4} \right] \Big|_{r=0}^{r=1} d\theta \\
&= \int_0^{2\pi} \left[\frac{3}{4} - \left(\frac{1}{3} \right) \cos(\theta) \right] d\theta \\
&= \frac{3\pi}{2}. \qquad \blacksquare
\end{aligned}
$$

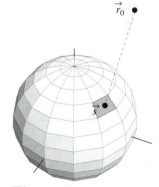

Figure 5. *The electric potential at $\vec{r}_0$ due to the charged piece of surface ΔS is approximated by*

$$\frac{\rho\, Area(\Delta S)}{\|\vec{s} - \vec{r}_0\|}.$$

EXAMPLE 7: Recall that if q is a point charge at $\vec{r}_0$, then the electric potential at a position $\vec{r}$ is $V(\vec{r}) = \frac{k}{\|\vec{r} - \vec{r}_0\|}$. Suppose that the unit sphere carries a charge that is uniformly distributed on the sphere. The charge density is a constant ρ measured in Coulombs/m^2. Let $\vec{r}_0$ be a fixed point in space and let $\vec{s}$ be a point in a small piece of the sphere ΔS, as illustrated in Figure 5. Then the electric potential at $\vec{r}_0$ due to the charged piece of surface ΔS is approximated by

$$\frac{\rho\,\text{Area}(\Delta S)}{\|\vec{s} - \vec{r}_0\|}.$$

Therefore, if we divide the sphere into non overlapping pieces $S_{i,j}$ and pick selection points $\vec{s}_{i,j}$, then the electric field at $\vec{r}_0$ due to the charged sphere will be approximated by the sum

$$\sum\sum\left(\frac{\rho}{\|\vec{s}_{i,j} - \vec{r}_0\|}\right)\,\text{Area}(S_{i,j}).$$

We conclude that the potential at $\vec{r}_0$ due to the charges sphere is

$$V(\vec{r}_0) = \iint_S \frac{\rho}{\|\vec{r}_0 - \vec{r}\|}\,dS = \iint_S \frac{\rho}{\sqrt{(x_0 - x)^2 + (y_0 - y)^2 + (z_0 - z)^2}}\,dS$$

$$= \rho\int_0^\pi \int_0^{2\pi} \frac{1}{\sqrt{(x_0 - \cos(\theta)\sin(\phi))^2 + (y_0 - \sin(\theta)\sin(\phi))^2 + (z_0 - \cos(\phi))^2}}\,\sin(\phi)\,d\theta\,d\phi.\,\blacksquare$$

EXERCISES 14.4

1. The mass of a surface is 10 kg. The mass is uniformly distributed over the surface. The surface area is 5 m². What is the mass density of the surface?

2. The mass of a surface is M. The mass is uniformly distributed over the surface. The surface area is A. What is the mass density of the surface?

In Exercises 3–11, S is the region in the xy–plane bounded by the graph of $y = x$, the x–axis, and the line $x = 2$. The mass density of S is given by $\rho(x, y) = x + y$.

3. Find the mass of S.

4. Find the center of mass of S.

5. Find the kinetic energy of S if it is rotating about the x–axis with an angular speed of π rad/sec.

6. Find the kinetic energy of S if it is rotating about the y–axis with an angular speed of $\frac{\pi}{2}$ rad/sec.

7. Find the kinetic energy of S if it is rotating about the origin in the xy–plane with an angular speed of 2π rad/sec.

8. Find the kinetic energy of S if it is rotating about the line $x = 2$ with an angular speed of 2π rad/sec.

9. Find the kinetic energy of S if it is rotating about the point $(-1, -1)$ in the xy–plane with an angular speed of 2π rad/sec.

10. Find the volume of the solid consisting of all points (x, y, z) where (x, y) is in S and $0 \le z \le e^{x+y}$.

Figure 6. *Solid in Exercise 10.*

11. Find the volume of the solid consisting of all points (x, y, z), where (x, y) is in S and $-x \leq z \leq x^2$.

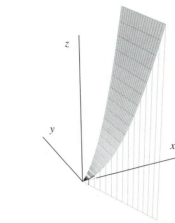

Figure 7. *Solid in Exercise 11.*

In Exercises 12–19, S is the region in the xy–plane bounded by the graphs of $y = x$ and $y = x^2$. The mass density of S is given by $\rho(x, y) = x^2 y$.

12. Find the mass of S.

13. Find the center of mass of S.

14. Find the kinetic energy of S if it is rotating about the x–axis with an angular speed of 2π rad/sec.

15. Find the kinetic energy of S if it is rotating about the y–axis with an angular speed of 2π rad/sec.

16. Find the kinetic energy of S if it is rotating about the z–axis with an angular speed of 2π rad/sec.

17. Find the kinetic energy of S if it is rotating about the line $x = 1$ in the xy–plane with an angular speed of 2π rad/sec.

18. Find the volume of the solid consisting of all points (x, y, z), where (x, y) is in S and $0 \leq z \leq x + y$.

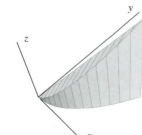

Figure 8. *The solid in Exercise 18.*

19. Find the volume of the solid consisting of all points (x, y, z), where (x, y) is in S and $x^2 \leq z \leq x$.

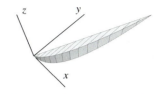

Figure 9. *The solid in Exercise 19.*

20. Let S be the region in the xy–plane bounded by $xy = 4$ and $x + y = 5$. Find the volume of the solid containing (x, y, z) if (x, y) is in S and $0 \leq z \leq x + y$.

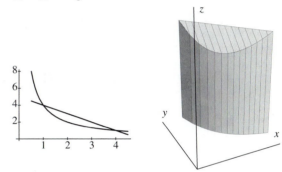

Figure 10. *The region S and the solid for Exercise 20.*

In Exercises 21–25, S is the parallelogram with two adjacent edges the line segments $[\vec{A}, \vec{B}]$ and $[\vec{A}, \vec{C}]$, where $\vec{A} = (1, 2, 1), \vec{B} = (0, 1, 0),$ and $\vec{C} = (1, 1, 1)$.

21. Find the mass of S if the mass density of S is given by $\rho(x, y, z) = x + z$.

22. Find the center of mass of S if the mass density of S is given by $\rho(x, y, z) = x + z$.

23. Find the kinetic energy of S if it is rotating about the z–axis with angular speed of 2π rad/sec and if it has a constant mass density k.

24. Find the kinetic energy of S if it is rotating about the x–axis with an angular speed of 2π rad/sec and if it has a constant mass density k.

25. Find the kinetic energy of S if it is rotating about the z–axis with an angular speed of 2π rad/sec and if its mass density is as in Exercise 21.

26. The surface of a sphere of radius 1 m with a constant mass density ρ is rotating about a line passing through the center of the sphere at a rate of f rotations/sec. Find its kinetic energy.

In Exercises 27 and 28, a disc of radius R with constant mass density ρ is rotating about a line passing through its center at a rate of f rotations/sec.

27. Find the disc's kinetic energy if the axis of rotation lies in the plane containing the disc.

28. Find the disc's kinetic energy if the axis of rotation is perpendicular to the disc.

29. Let P be the plane passing through the origin that is normal to the vector $(1, 0, 1)$. Let V be the solid containing the set of points (x, y, z) that are below P, above the xy–coordinate plane, and lying inside the cylinder $x^2 + y^2 \leq 1$. Find the volume of V.

Figure 11. *The solid V for Exercise 29.*

30. Let P be the plane passing through the point $(0, 0, 2)$ that is normal to the vector $(1, 1, 1)$. Let V be the solid containing the set of points (x, y, z) that are below P, above the xy–coordinate plane, and lying inside the cylinder $x^2 + y^2 \leq 1$. Find the volume of V.

Figure 12. *The solid V for Exercise 30.*

31. Express the volume of the intersection of the cylinders $x^2 + y^2 \leq 1$ and $x^2 + z^2 \leq 1$ as a double integral.

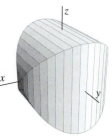

Figure 13. *The solid V for Exercise 31.*

32. An electric charge of 5 Coulombs is uniformly distributed on a surface with area 3 m^2. What is the charge density?

33. An electric charge is uniformly distributed on a sphere as in Example 7. Show that the electric potential is zero at the center of the sphere.

34. The vectors $(1, 2, 3)$ and $(1, 0, 0))$ are drawn emanating from the the vector $(1, 1, 1)$ to form adjacent edges of a parallelogram P. An electric charge is uniformly distributed on P. If $\vec{r}$ is **not** on the parallelogram, $V(\vec{r})$ denotes the electric potential at $\vec{r}$ due to the charge on P. If ρ denotes the charge density on P, express $V(\vec{r})$ as a double integral.

14.5   Change of Variables

In Section 6.2 we developed the chain rule in reverse for integration over a single variable. We now extend this result for integrals over a surface. The derivative of a parametrization for a fundamental curve, which is used in the chain rule in reverse, generalizes to the

Jacobian of a parametrization for a simple surface. Thus it should not be too surprising to see the Jacobian in the following Change of Variables Theorem.

Theorem 1 (Change of Variables) *Let P and Q be surfaces, and let $\vec{h}$ be a differentiable function from P to Q such that*

(a) *$\vec{h}$ is one-to-one on the interior of P;*

(b) *$J(\vec{h}(\vec{r}))$ is continuous; and*

(c) *$J(\vec{h}(\vec{r})) \neq 0$ on the interior of P.*

If f is a continuous function from Q into $\mathbb{R}$, then

$$\iint_Q f\ dS = \iint_P f(\vec{h}(\vec{r})) J(\vec{h}(\vec{r}))\ dS.$$

Notice that this theorem allows you to evaluate the surface integral by relating the surface to some other surface over which you can integrate. This is more flexible than using the definition of surface integral, which required that the surface in question be parametrized over a rectangle.

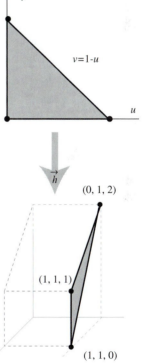

Figure 1. *The triangle in uv–space is taken onto the triangle in xy–space.*

EXAMPLE 1: Let Q be the triangle with vertices $\vec{A} = (1,1,1), \vec{B} = (0,1,2)$, and $\vec{C} = (1,1,0)$. Calculate $\iint_Q x + y + z\ dS$.

SOLUTION: We let P be the triangle in uv–space with vertices $(0,0), (1,0)$, and $(0,1)$, and let $\vec{T}$ be the linear transformation that takes $(1,0)$ onto $\vec{B} - \vec{A} = (-1,0,1)$ and $(0,1)$ onto $\vec{C} - \vec{A} = (0,0,-1)$. If $\vec{h}$ is defined by $\vec{h}(\vec{r}) = \vec{A} + \vec{T}(\vec{r})$, then $\vec{h}$ takes the unit square onto the parallelogram with adjacent edges the line segments $[\vec{A}, \vec{B}]$ and $[\vec{A}, \vec{C}]$, and $\vec{h}$ takes the triangle P in uv–space with vertices $(0,0), (1,0)$, and $(0,1)$ onto Q. See Figure 1.

$$\vec{h}(u,v) = \begin{pmatrix} 1 \\ 1 \\ 1 \end{pmatrix} + \begin{pmatrix} -1 & 0 \\ 0 & 0 \\ 1 & -1 \end{pmatrix} \begin{pmatrix} u \\ v \end{pmatrix} = \begin{pmatrix} 1 - u \\ 1 \\ 1 + u - v \end{pmatrix}$$

and

$$J(\vec{h}(u,v)) = 1.$$

Thus

$$\iint_Q f\ dS \;=\; \iint_P f(\vec{h}(\vec{r})) J(\vec{h}(\vec{r}))\ dS$$

$$= \int_0^1 \int_0^{1-u} (1-u) + (1) + (1+u-v) \, dv \, du$$

$$= \int_0^1 \int_0^{1-u} 3 - v \, dv \, du = \int_0^1 3(1-u) - \frac{(1-u)^2}{2} \, du$$

$$= \int_0^1 \frac{5}{2} - 2u - \frac{u^2}{2} \, du = \frac{5}{2} - 1 - \frac{1}{6} = \frac{4}{3}. \qquad \blacksquare$$

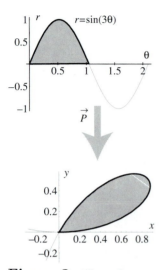

Figure 2. *The polar function $\vec{P}$ takes the region in $r\theta$–space onto the region in xy–space.*

EXAMPLE 2: Find the area of the region in xy–space that is bounded by the set with polar equation $r = \sin(3\theta), 0 \le \theta \le \frac{\pi}{3}$.

SOLUTION: We start by sketching the region in $r\theta$–space and in xy–space as shown in Figure 2. Let Q be the region in xy–space, and let P be the region in $r\theta$–space bounded by the graph of $r = \sin(3\theta), 0 \le \theta \le \frac{\pi}{3}$, and the r–axis. Then the polar transformation $\vec{P}$ takes P onto Q.

$$\text{Area} \quad = \quad \iint_Q dS = \iint_P J(\vec{P}(\vec{r})) \, dS = \int_0^{\pi/3} \int_0^{\sin(3\theta)} r \, dr \, d\theta$$

$$= \quad \int_0^{\pi/3} \frac{\sin^2 3\theta}{2} \, d\theta = \frac{\pi}{12}. \qquad \blacksquare$$

EXAMPLE 3: A region S lying in the xy–plane is bounded by the set with polar equation $r = 1 - \cos\theta$. Its mass density is the constant ρ, and it is rotating about the z–axis with an angular frequency of f rotations/sec. Find its kinetic energy.

SOLUTION: Let P be the region in $r\theta$–space bounded by the r–axis and the graph of $r = 1 - \cos\theta, 0 \le \theta \le 2\pi$, and let S be the associated region in xy–space as in Figure 3.

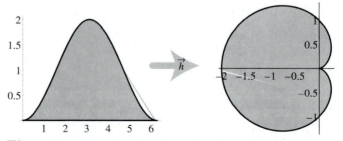

Figure 3. *The polar function takes the region P in $r\theta$–space onto the surface S in xy–space.*

The angular speed of S is $2\pi f$. Therefore, its kinetic energy is given by

$$
\begin{aligned}
\text{K. E.} &= \frac{1}{2}(2\pi f)^2 \iint_S \left(x^2 + y^2\right) \rho \, dS \\
&= \frac{1}{2}(2\pi f)^2 \rho \iint_P \left(r^2 \cos^2 \theta + r^2 \sin^2 \theta\right) J(\vec{P}(\vec{r})) \, dS \\
&= \frac{1}{2}(2\pi f)^2 \rho \int_0^{2\pi} \int_0^{1-\cos\theta} r^3 \, dr \, d\theta \\
&= \frac{1}{2}(2\pi f)^2 \frac{1}{4}\rho \int_0^{2\pi} (1-\cos\theta)^4 \, d\theta \\
&= \frac{1}{2}(2\pi f)^2 \frac{15\pi}{16}\rho = \frac{15\pi^3 f^2 \rho}{8}.
\end{aligned}
$$

$\blacksquare$

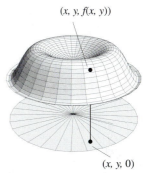

Figure 4. S is the graph of f and P is the domain of f.

EXAMPLE 4: Let P be a surface in xy–space, and let f be a function from P into $\mathbb{R}$ that is continuous and that has continuous partial derivatives. Let S be the set of all points of the form (x, y, z), where (x, y) is in P and $z = f(x, y)$. Let $\vec{h}$ be the function from P into S defined by $\vec{h}(x, y) = (x, y, f(x, y))$. (See Figure 4.)

It is easily seen that

$$
J(\vec{h}(x, y)) = \sqrt{\left(\frac{\partial f(x, y)}{\partial x}\right)^2 + \left(\frac{\partial f(x, y)}{\partial y}\right)^2 + 1}.
$$

Thus the area of such a surface S is given by

$$
\begin{aligned}
\text{Area} &= \iint_S dS = \iint_P J(\vec{h}(\vec{r})) \, dS \\
&= \iint_P \sqrt{\left(\frac{\partial f(x, y)}{\partial x}\right)^2 + \left(\frac{\partial f(x, y)}{\partial y}\right)^2 + 1} \, dS.
\end{aligned}
$$

$\blacksquare$

EXAMPLE 5: Let T be the triangle in the plane with vertices $(0,0), (1,0)$, and $(1,1)$. Find the surface area of the set of points (x, y, z) satisfying $z = y + x^2$ for (x, y) in T.

SOLUTION: We employ the formula developed in Example 4. Let $f(x, y) = y + x^2$. Then $\frac{\partial f}{\partial x} = 2x$, $\frac{\partial f}{\partial y} = 1$, and

$$
\sqrt{\left(\frac{\partial f(x, y)}{\partial x}\right)^2 + \left(\frac{\partial f(x, y)}{\partial y}\right)^2 + 1} = \sqrt{2 + 4x^2}.
$$

Thus the area is

$$
\iint_S dS = \iint_T \sqrt{2 + 4x^2}\, dS = \int_0^1 \int_0^x \sqrt{2 + 4x^2}\, dy\, dx
$$

$$
= \int_0^1 \left(y\sqrt{2 + 4x^2} \right)\Big|_{y=0}^{y=x} dx = \frac{1}{12}\left(6^{3/2} - 2^{3/2} \right). \quad \blacksquare
$$

EXERCISES 14.5

In Exercises 1–4, T is the triangle in $\mathbb{R}^n$ with vertices $\vec{A}, \vec{B},$ and $\vec{C}$. Compute $\iint_T f\, dS$.

1. $\vec{A} = (0,0), \vec{B} = (1,1), \vec{C} = (2,0),$ and $f(x,y) = xy^2$.

2. $\vec{A} = (-1,0), \vec{B} = (1,1), \vec{C} = (2,-1),$ and $f(x,y) = e^{(x+y)}$.

3. $\vec{A} = (1,0,0), \vec{B} = (0,1,0), \vec{C} = (0,0,1),$ and $f(x,y,z) = e^{(x+y+z)}$.

4. $\vec{A} = (1,0,0), \vec{B} = (0,1,0), \vec{C} = (0,0,1),$ and $f(x,y,z)$ is the square of the distance from (x,y,z) to the origin.

In Exercises 5–7, T is the triangle in $\mathbb{R}^n$ with vertices $\vec{A} = (0,1,0), \vec{B} = (1,1,0),$ and $\vec{C} = (1,2,1)$. Assume that the mass density of T is a constant $\rho\,\mathrm{kg/m}^2$.

5. Find the kinetic energy if T is rotating about the x–axis at a rate of 2 rotations/sec.

6. Find the kinetic energy if T is rotating about the y–axis at a rate of 2 rotations/sec.

7. Find the kinetic energy if T is rotating about the z–axis at a rate of 2 rotations/sec.

In Exercises 8–12, sketch the region in xy–space bounded by the curve with the given polar equation, and find its area.

8. $r = 3\sin\theta$.   9. $r = \sin(2\theta)$.

10. $r = 3 - \sin\theta$.   11. $r = 1 + \cos(2\theta)$.

12. $r = 4 + \sin\theta$.

In Exercises 13–17, R is the region in $r\theta$–space bounded by the given curves. Sketch R (as it appears in $r\theta$–space) and $\vec{P}(R)$, the image of R in xy–space. Find the area of the region in xy–space.

13. $r = \theta,\ \theta = \pi,$ and $r = 0$.

14. $r = e^\theta,\ \theta = 2\pi,$ and $r = 0$.

15. $r = \sin\theta,\ 0 \le \theta \le \pi,$ and $r = 0$.

16. $r = 1 + \sin\theta,\ 0 \le \theta \le 2\pi,$ and $r = 0$.

17. $r = \sin(3\theta),\ 0 \le \theta \le \frac{\pi}{3},$ and $r = 0$.

In Exercises 18–22, S is the region in xy–space bounded by the curve with the given polar equation. Find the volume of the solid that contains all points (x,y,z) that satisfy the conditions that (x,y) is in S and $g(x,y) \le z \le f(x,y)$.

18. $r = 1 + \sin\theta, g(x,y) = 0,\ f(x,y) = \sqrt{x^2 + y^2}$.

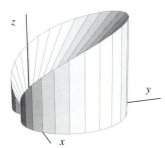

Figure 5. *The solid in Exercise 18.*

19. $r = \sin\theta, g(x,y) = 0, f(x,y) = y$.

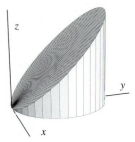

Figure 6. *The solid in Exercise 19.*

20. $r = \sin\theta, g(x,y) = -y, f(x,y) = x^2 + y^2$.

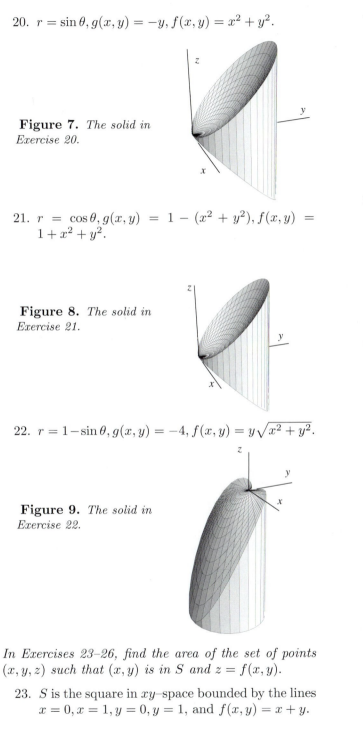

Figure 7. *The solid in Exercise 20.*

21. $r = \cos\theta, g(x,y) = 1 - (x^2 + y^2), f(x,y) = 1 + x^2 + y^2$.

Figure 8. *The solid in Exercise 21.*

22. $r = 1 - \sin\theta, g(x,y) = -4, f(x,y) = y\sqrt{x^2 + y^2}$.

Figure 9. *The solid in Exercise 22.*

In Exercises 23–26, find the area of the set of points (x, y, z) such that (x, y) is in S and $z = f(x, y)$.

23. S is the square in xy–space bounded by the lines $x = 0, x = 1, y = 0, y = 1$, and $f(x,y) = x + y$.

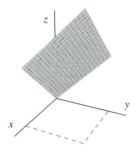

Figure 10. *The surface in Exercise 23.*

24. S is the square of Exercise 23, and $f(x,y) = y + \frac{x^2}{2}$.

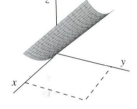

Figure 11. *The surface in Exercise 24.*

25. S is the triangle with vertices $(0,0)$, $(0,4)$, $(4,4)$, and $f(x,y) = y^2$.

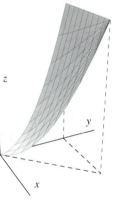

Figure 12. *The surface in Exercise 25.*

26. S is the disc of radius 4 centered at the origin, and $f(x,y) = xy$.

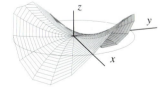

Figure 13. *The surface in Exercise 26.*

27. Let S be the part of the graph of $z = x^2 + y^2$ lying above the disk $x^2 + y^2 \leq 1$. Find the area of S.

28. Let S be the part of the graph of $z = x^2 + y^2$ lying above the disk $x^2 + y^2 \leq 1$. Express $\iint_S x + y^2 + z \, dS$ as a double integral.

29. Let S be the part of the graph of $z = x^2 + y^2$ lying above the disk $(x - 1)^2 + (y + 2)^2 \leq 3$. Express $\iint_S x + y^2 + z \, dS$ as a double integral.

30. Let S be the part of the graph of $z = x^2 + y^2$ lying above the triangle with verticies $(-1, 0)$,$(1, -2)$, and $(3, 3)$. Express the area of S as a double integral.

31. Let S be the part of the cylinder $x^2 + z^2 = 1$, $z \geq 0$ lying above the disc $x^2 + y^2 \leq 1$. Express the area of S as a double integral.

32. Let S be the total surface area of the intersection of $x^2 + y^2 \leq 1$ and $x^2 + z^2 \leq 1$. Express the area of S as the sum of four double integrals.

14.6   Simple Solids

The concept of parametrization has been a constant theme throughout this text. We have parametrized curves and surfaces in both 2–space and 3–space. For the purpose of integration, we put restrictions on the type of parametrization that we could use. These restrictions gave us the concepts of fundamental curve and simple surfaces. We now complete this discussion by defining simple solids.

Definition: Simple Solids

The set S in 3–space is a *simple solid* if there is a box B bounded by planes parallel to the coordinate planes and a differentiable function $\vec{h}$ from B onto S such that

(a) $\vec{h}$ is one-to-one in the interior of B;

(b) $\vec{h}$ and all of the first partial derivatives of $\vec{h}$ are continuous; and

(c) $J(\vec{h}(u, v, w))$ is not 0 in the interior of B.

The idea of a simple solid is the three dimensional version of the idea of a simple surface in 2– or 3–space, and the idea of a fundamental curve in 1–, 2–, or 3–space. One might think of a piece of malleable plastic or clay in the shape of the box B and use $\vec{h}$ to distort the box into S. Condition (a) means that no two points in the interior of the box B are pinched together in S. The condition that $J(\vec{h}(u, v, w)) \neq 0$ essentially means that no piece of volume is somehow shrunk to a point, a line, or a plane. That $\vec{h}$ is continuous means that the box is not broken when it is distorted by $\vec{h}$. We build solids from boxes just as we build surfaces from rectangles.

First, we extend the idea of parametrizing the surface of a sphere to that of parametrizing the solid ball.

EXAMPLE 1: Let $\vec{h}(r, \phi, \theta)$ be the restriction of the spherical transformation to the box B bounded by the planes $r = 0, r = R, \phi = 0, \phi = \pi, \theta = 0$, and $\theta = 2\pi$. Then $\vec{h}$ is one-to-one on the interior of B, and its first partial derivatives are continuous. $J(\vec{h}(r, \phi, \theta)) = r^2 \sin \phi \neq 0$ on the interior of B, since $r > 0$ and $0 < \phi < \pi$ there. Thus $\vec{h}$ is a parametrization of the ball of radius R. ■

The next example is an extension of the idea of a surface of rotation.

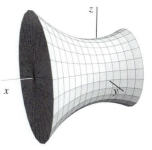

Figure 1.a *The solid obtained by rotating the region bounded between the graph of a function and the x–axis about the x–axis.*

EXAMPLE 2: Suppose that f is a function that is continuous and differentiable on the interval $[a, b]$ such that $f(x) \neq 0$ for every x in $[a, b]$. Let R be the region in the plane bounded by the graph of f, the x–axis, and the lines $x = a$ and $x = b$. Let S be the solid obtained by rotating R about the x–axis. (See Figure 1.) Find a parametrization for the solid S.

SOLUTION: For each point t in the interval $[a, b]$, let R_t be the surface obtained by intersecting the plane $x = t$ with the solid S. Then R_t is a disc of radius $f(t)$. A parametrization for R_t is given by $\vec{r}_t(s, \theta) = (t, sf(t) \cos \theta, sf(t) \sin \theta)$. (See Figure 1.b.)

Let $\vec{h}$ be defined by

$$\vec{h}(t, s, \theta) = (t, \vec{r}_t(s, \theta)) = (t, sf(t) \cos \theta, sf(t) \sin \theta)$$

for

$$a \leq t \leq b, 0 \leq s \leq 1, 0 \leq \theta \leq 2\pi.$$

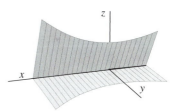

Figure 1.b *The cross section* R_t.

Figure 1.c displays the "slice" of the solid obtained by holding θ fixed at $\theta = \frac{\pi}{3}$ radians. In Figure 1.d, the wedge $0 \leq \theta \leq \frac{\pi}{3}$ is removed to reveal the cross sections of the solid.

The partial derivatives of $\vec{h}$ are:

$$\frac{\partial h(t, s, \theta)}{\partial t} = (1, sf'(t) \cos \theta, sf'(t) \sin \theta);$$

$$\frac{\partial h(t, s, \theta)}{\partial s} = (0, f(t) \cos \theta, f(t) \sin \theta);$$

and

$$\frac{\partial h(t, s, \theta)}{\partial \theta} = (0, -sf(t) \sin \theta, sf(t) \cos \theta).$$

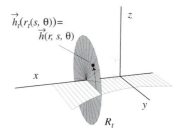

Figure 1.c *The "slice" obtained by holding θ constant.*

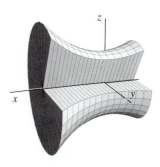

Figure 1.d *The image of* $\vec{h}(t, s, \theta), \frac{\pi}{3} \leq \theta \leq 2\pi$.

Thus all of the first partial derivatives of $\vec{h}$ are continuous.

$$J(\vec{h}(t,s,\theta)) = s[f(t)]^2, \text{ which is not 0 if } s > 0 \text{ and } a \le t \le b.$$

To see that $\vec{h}$ is one-to-one, suppose that $\vec{h}(t_1, s_1, \theta_1) = \vec{h}(t_2, s_2, \theta_2)$. This implies that

$$(t_1, s_1 f(t_1) \cos \theta_1, s_1 f(t_1) \sin \theta_1) = (t_2, s_2 f(t_2) \cos \theta_2, s_2 f(t_2) \sin \theta_2). \tag{1}$$

By comparing the first coordinates of each side of Equation (1), we see that $t_1 = t_2$. Since the sum of the squares of the second and third coordinates of each side of Equation (1) should be equal, we see that

$$(s_1 f(t_1))^2 = (s_2 f(t_2))^2.$$

However, $t_1 = t_2$ implies that $s_1^2 = s_2^2$. Since $0 < s_i < 1$ for $i = 1, 2$, we see that $s_1 = s_2$. By comparing the second and third coordinates of each side of Equation (1) we see that

$$\cos \theta_1 = \cos \theta_2 \quad \text{and} \quad \sin \theta_1 = \sin \theta_2.$$

This is only possible if $\theta_1 = \theta_2$. Thus $\vec{h}$ is one-to-one in the interior of the domain of $\vec{h}$. ∎

The solid of revolution in Example 2 is obtained by first parametrizing the cross sections of the solid in a "continuous" manner so that the parametrizing functions can be put together to parametrize the solid. This is an example of a general approach described in the following theorem.

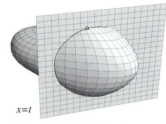

$x{=}t$

Figure 2.a *The plane $x = t$ intersecting the solid*

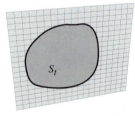

Figure 2.b *S_t is the common part of the solid and the plane $x = t$.*

Theorem 1 (Solids with Continuous Cross Sections)
Suppose that S is a solid and $[a, b]$ is an interval such that:

(a) *If $a < t < b$, then the cross section of S obtained by intersecting the plane $x = t$ with S is a surface S_t (as in Figure 2);*

(b) *S_t is parametrized by the function $\vec{h}_t(u, v) = (t, y_t(u, v), z_t(u, v))$ with domain the rectangle in $uv-$space bounded by the lines $u = c, u = d, v = e,$ and $v = f$, where $c < d$, and $e < f$;*

(c) *All of S lies between the planes $x = a$ and $x = b$;*

(d) *The function* $\vec{h}$, *defined by* $\vec{h}(t, u, v) = (t, \vec{h}_t(u, v)) = (t, y_t(u, v), z_t(u, v))$, *with domain the box* B *bounded by the planes* $t = a, t = b, u = c, u = d, v = e$, *and* $v = f$, *is continuous; and*

(e) $\dfrac{\partial \vec{h}(t, u, v)}{\partial t}$ *is continuous;*

then $\vec{h}$ *parametrizes* S, *and* $J(\vec{h}(t, u, v)) = J(\vec{h}_t(u, v))$.

Proof: We will show that $\vec{h}$ is one-to-one, that the partial derivatives of $\vec{h}$ are continuous, and that the Jacobian of $\vec{h}$ is not zero on the interior of B.

First we prove that $\vec{h}$ is one-to-one. Consider (t_1, u_1, v_1) and (t_2, u_2, v_2) in the interior of B such that $\vec{h}(t_1, u_1, v_1) = \vec{h}(t_2, u_2, v_2)$. If $t_1 \neq t_2$, then $\vec{h}(t_1, u_1, v_1)$ and $\vec{h}(t_2, u_2, v_2)$ differ in the first coordinate. Thus $t_1 = t_2$. This implies that $h_{t_1}(u_1, v_1) = h_{t_1}(u_2, v_2)$. However, h_{t_1} is a parametrization for S_{t_1} and is, by definition, one-to-one. Thus $(u_1, v_1) = (u_2, v_2)$, and $\vec{h}$ is one-to-one.

Second, we show that the partial derivatives of $\vec{h}$ are continuous. It is given in the hypothesis that $\frac{\partial \vec{h}}{\partial t}$ is continuous. The partial derivative of $\vec{h}$ with respect to u is given by

$$\frac{\partial \vec{h}(t, u, v)}{\partial u} = \left(0, \frac{\partial y_t(u, v)}{\partial u}, \frac{\partial z_t(u, v)}{\partial u} \right).$$

Since $\vec{h}_t$ is a parametrization for S_t (t is held constant for $\vec{h}_t$), we know that $\frac{\partial y_t(u,v)}{\partial u}$ and $\frac{\partial z_t(u,v)}{\partial u}$ are continuous. Thus $\frac{\partial \vec{h}(t,u,v)}{\partial u}$ is continuous. A similar argument shows that $\frac{\partial \vec{h}}{\partial v}$ is continuous.

Finally, we compute $J(\vec{h}(t, u, v))$. With the information given, it is not possible to compute $\frac{\partial \vec{h}(t,u,v)}{\partial t}$, but we do know that the first coordinate of $\frac{\partial \vec{h}(t,u,v)}{\partial t}$ is 1 so that $\frac{\partial \vec{h}(t,u,v)}{\partial t} = \left(1, \frac{\partial y(t,u,v)}{\partial t}, \frac{\partial z(t,u,v)}{\partial t} \right)$. Thus

$$D\vec{h}\big|_{(t,u,v)} = \begin{pmatrix} 1 & 0 & 0 \\ \frac{\partial y(t,u,v)}{\partial t} & \frac{\partial y_t(u,v)}{\partial u} & \frac{\partial y_t(u,v)}{\partial v} \\ \frac{\partial z(t,u,v)}{\partial t} & \frac{\partial z_t(u,v)}{\partial u} & \frac{\partial z_t(u,v)}{\partial v} \end{pmatrix}.$$

Expanding by cofactors we obtain:

$$J(\vec{h}(t, u, v)) = \left| \det \left(D\vec{h}\big|_{(t,u,v)} \right) \right|$$

$$= \left| \det \begin{pmatrix} \frac{\partial y_t(u,v)}{\partial u} & \frac{\partial y_t(u,v)}{\partial v} \\ \frac{\partial z_t(u,v)}{\partial u} & \frac{\partial z_t(u,v)}{\partial v} \end{pmatrix} \right| = J(\vec{h}_t(u,v)).$$

Since $J(\vec{h}_t(u,v)) \neq 0$, we see that $J(\vec{h}(t,u,v)) \neq 0$ in the interior of B. ■

In Example 2, the cross section of the solid in the plane $x = t$ is a circle of radius $f(t)$. We now look at an example where the cross sections are squares.

EXAMPLE 3: Let $f(t) = t^2, 0 \leq t \leq 1$. Let S be the solid in xyz–space such that the cross section of S lying in the plane $x = t$ is a square with vertices $(t,0,0), (t,f(t),0), (t,0,f(t))$, and $(t,f(t),f(t))$. Parametrize S. See Figure 3.

Let T_t denote the cross section obtained by intersecting the plane $x = t$ with S. Let R be the rectangle in uv–space bounded by the lines $u = 0, u = 1, v = 0$ and $v = 1$. A parametrization for T_t is given by

$$\vec{h}_t(u,v) = (uf(t), vf(t)) = (ut^2, vt^2), \; 0 \leq u \leq 1, \; 0 \leq v \leq 1.$$

Thus a parametrization for S is given by

$$\vec{h}(t,u,v) = (t, uf(t), vf(t)) = (t, ut^2, vt^2),$$

where

$$0 \leq t \leq 1, \; 0 \leq u \leq 1 \, 0 \leq v \leq 1.$$

The partial derivatives of $\vec{h}$ are given by

$$\frac{\partial h(t,u,v)}{\partial t} = (1, 2ut, 2vt);$$

$$\frac{\partial h(t,u,v)}{\partial u} = (0, t^2, 0);$$

and

$$\frac{\partial h(t,u,v)}{\partial v} = (0, 0, t^2).$$

Thus

$$J(\vec{h}(t,u,v)) = \left| \det \begin{pmatrix} 1 & 0 & 0 \\ 2ut & t^2 & 0 \\ 2vt & 0 & t^2 \end{pmatrix} \right| = t^4.$$ ■

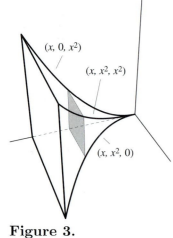

$(x, 0, x^2)$

(x, x^2, x^2)

$(x, x^2, 0)$

Figure 3.
The cross section T_t.

EXAMPLE 4: Let S be the solid in xyz–space bounded by the graphs of $y = x^2$ and $z = x^2$, the xy and xz–coordinate planes, and the plane $x = 1$. Find a parametrization for S.

SOLUTION: The problem here is to visualize what S looks like. A tactic that proves useful is to sketch the bounding surfaces. Figure 4.a illustrates the graph of $y = x^2$, and Figure 4.b is a sketch of the graph of $z = x^2$. Figure 4.c displays the intersecting bounding surfaces, and the solid S is shown in Figure 4.d. This is the same solid as the one in Example 3!   ■

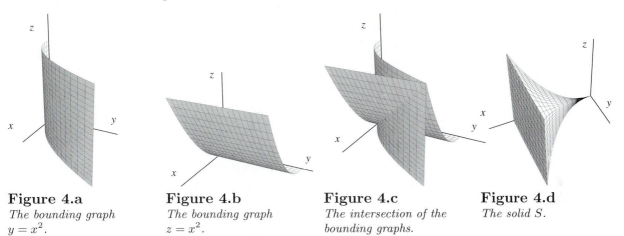

Figure 4.a
The bounding graph
$y = x^2$.

Figure 4.b
The bounding graph
$z = x^2$.

Figure 4.c
The intersection of the bounding graphs.

Figure 4.d
The solid S.

EXAMPLE 5: Let S be the solid in xyz–space bounded by the graph of $z = x^2$ (Figure 5.a), the plane $z - y = 0$ (Figure 5.b), the xy–coordinate plane, the xz–coordinate plane, and the plane $z = 1$ (Figure 5.c). Figure 5.d displays the intersecting planes, and Figure 5.e shows the solid S. Use continuous cross sections to parametrize S.

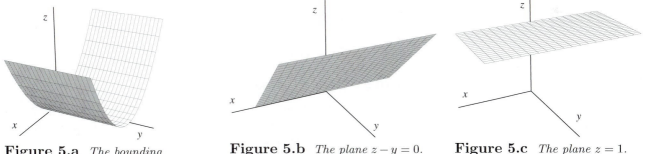

Figure 5.a   *The bounding graph $z = x^2$.*

Figure 5.b   *The plane $z - y = 0$.*

Figure 5.c   *The plane $z = 1$.*

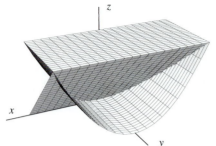

Figure 5.d *The intersecting bounding planes.*

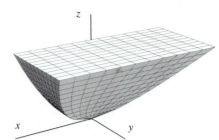

Figure 5.e *The solid S.*

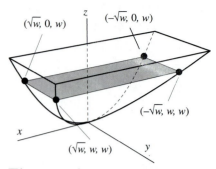

Figure 5.f *A cross section parallel to the xy–coordinate plane.*

SOLUTION: From the diagrams above, we see that the best cross sections are those parallel to the xy–coordinate plane, which are rectangles. (See Figure 5.f.)

Let w be a number between 0 and 1 and let R_w be the rectangular cross section obtained by intersecting the plane $z = w$ with the solid S. If (x, y, z) is a point in R_w, then $z = w, 0 \le y \le w$ and $-\sqrt{w} \le x \le \sqrt{w}$. The function

$$\vec{h}_w(u, v) = \begin{pmatrix} u\sqrt{w} \\ wv \\ w \end{pmatrix}, \ -1 \le u \le 1 \text{ and } 0 \le v \le 1,$$

is a parametrization for R_w, and

$$\vec{h}(u, v, w) = \begin{pmatrix} u\sqrt{w} \\ wv \\ w \end{pmatrix}, \ -1 \le u \le 1, \ 0 \le v \le 1, \text{ and } 0 \le w \le 1,$$

parametrizes S.   ■

Rotating Surfaces About an Axis

Suppose that $\vec{r}(s, t) = (x(s, t), y(s, t))$, $a \le s \le b$, $c \le t \le d$ parameterizes a surface R in the plane and that R does not intersect the x–axis. Then

$$\vec{h}(\theta, s, t) = \begin{pmatrix} x(s, t) \\ y(s, t) \cos(\theta) \\ y(s, t) \sin(\theta) \end{pmatrix}, \ a \le s \le b, \ c \le t \le d,$$

parametrizes the solid obtained by rotating the surface R about the x–axis. You are asked in the exercises to show that

$$J\vec{h}(\theta, s, t) = |y(s, t)| J\vec{r}(s, t) = |y \det Dh| = |y(s, t)| \left| \frac{\partial x}{\partial s} \frac{\partial y}{\partial t} - \frac{\partial x}{\partial t} \frac{\partial y}{\partial s} \right|.$$

Similarly, if the surface R does not intersect the y–axis, then the solid obtained by rotating R about the y–axis can be parametrized by

$$\vec{h}(\theta, s, t) = \begin{pmatrix} x(s,t)\cos(\theta) \\ y(s,t) \\ x(s,t)\sin(\theta) \end{pmatrix}, \quad a \le s \le b, \ c \le t \le d.$$

$$J\vec{h}(\theta, s, t) = |x(s,t)|J\vec{r}(s,t) = |x\det Dh| = |x(s,t)| \left| \frac{\partial x}{\partial s}\frac{\partial y}{\partial t} - \frac{\partial x}{\partial t}\frac{\partial y}{\partial s} \right|.$$

EXAMPLE 6: Let R be the unit disc centered at $(0, 2)$. Parametrize the solid obtained by rotating R about the x–axis, and find the Jacobian of the parametrization.

SOLUTION: $\vec{r}(\rho, \theta) = (\rho\cos(\theta), \rho\sin(\theta) + 2)$, $0 \le \theta \le 2\pi$, $0 \le \rho \le 1$, parametrizes the disc. It follows that

$$\vec{h}(\rho, \theta, \phi) = \begin{pmatrix} \rho\cos(\theta) \\ (\rho\sin(\theta) + 2)\cos(\phi) \\ (\rho\sin(\theta) + 2)\sin(\phi) \end{pmatrix}, \quad \begin{matrix} 0 \le \rho \le 1, \\ 0 \le \theta \le 2\pi, \\ 0 \le \phi \le 2\pi \end{matrix}$$

is a parametrization for the solid of rotation. The Jacobian is

$$J\vec{h}(\rho, \theta, \phi) = (\rho\sin(\theta) + 2)\rho = \rho^2 \sin(\theta) + 2\rho. \quad \blacksquare$$

EXERCISES 14.6

In Exercises 1–3, the vectors $\vec{A}, \vec{B}$, and $\vec{C}$ are drawn emanating from the position $\vec{D}$ to form adjacent edges of a solid parallelepiped P. Parametrize P and find the Jacobian of your parametrization.

1. $\vec{A} = (0, 1, 1), \vec{B} = (-1, 2, 1), \vec{C} = (1, 1, 1)$, and $\vec{D} = (0, 0, 0)$.

2. $\vec{A} = (0, 2, 1), \vec{B} = (-1, 2, 1), \vec{C} = (1, -1, 3)$, and $\vec{D} = (1, 1, 1)$.

3. $\vec{A} = (0, 1, 1), \vec{B} = (-1, -1, 1), \vec{C} = (1, 0, 4)$, and $\vec{D} = (-1, 2, 0)$.

In Exercises 4–8, parametrize the solid S obtained by rotating the region R (lying in the xy–plane) about the x–axis. Determine the rate that your parametrization changes volume (as a function of position.)

4. R is the region bounded by the graph of $y = x$, the x–axis, and the line $x = 1$.

5. R is the region bounded by the graphs of $y = x$ and $y = x^2$.

6. R is the rectangle bounded by the lines $x = 1$, $x = 3$, $y = 2$, and $y = 52$.

7. R is the disc of radius 3 centered at the origin.

8. R is the disc of radius 2 centered at the point $(3, 4)$.

In Exercises 9–13, the region R lies in the xy–coordinate plane. Parametrize the solid obtained by rotating R about the y–axis, and find the Jacobian of your parametrization.

9. R is the region bounded by the graph of $y = x^2$, the x–axis, and the line $x = 1$.

10. R is the region bounded by the graphs of $y = x^2$ and $y = x^3$.

11. R is the rectangle bounded by the lines $x = 1, x = 3, y = 2$, and $y = 52$.

12. R is the disc of radius 3 centered at the origin.

13. R is the disc of radius 2 centered at the point $(3, 4)$.

In Exercises 14–21, parametrize the solid S obtained by rotating the region R (lying in the xy–plane) about the line L.

14. R is the region bounded by the graph of $y = x$, the x–axis, and the line $x = 1$. L is the line $x = -1$.

15. R is the region bounded by the graph of $y = x$, the x–axis, and the line $x = 1$. L is the line $x = 3$.

16. R is the region bounded by the graph of $y = x$, the x–axis, and the line $x = 1$. L is the line $y = -1$.

17. R is the region bounded by the graph of $y = x$, the x–axis, and the line $x = 1$. L is the line $y = 6$.

18. R is the region bounded by the graphs of $y = x$ and $y = x^2$. L is the line $x = 4$.

19. R is the rectangle bounded by the lines $x = 1, x = 3, y = 2$, and $y = 52$. L is the line $y = -2$.

20. R is the disc of radius 3 centered at the origin. L is the line $y = -3$.

21. R is the disc of radius 2 centered at the point $(3, 4)$. L is the line $x = -1$.

In Exercises 22–27, S is a solid lying in the upper half of xyz–space, $z \geq 0$, and between the planes $x = 0$ and $x = 1$. The base of S lies in the xy–coordinate plane and is the region bounded by the x–axis, the graph of $y = x^2$, and the line $x = 1$. Let A_t denote the cross section of S lying in the plane $x = t$. Find a parametrization for S for the given shape of A_t.

22. A_t is a rectangle with height t.

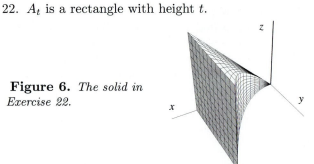

Figure 6. *The solid in Exercise 22.*

23. A_t is a rectangle with its base in the xy–plane and with height 2.

24. A_t is an isosceles right triangle with one leg in the xy–plane and the other in the xz–plane.

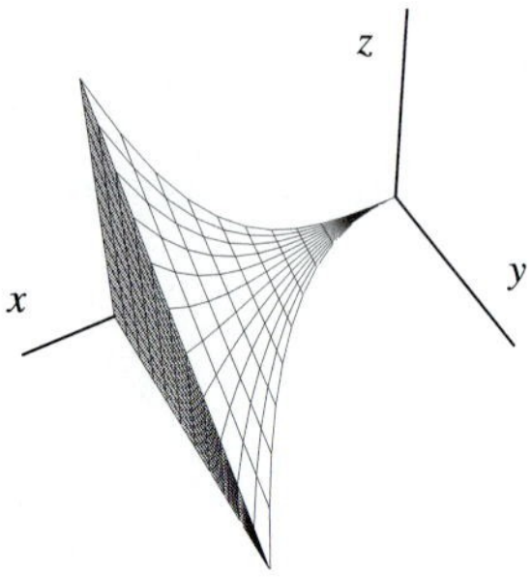

Figure 7. *The solid in Exercise 24.*

25. A_t is an equilateral triangle.

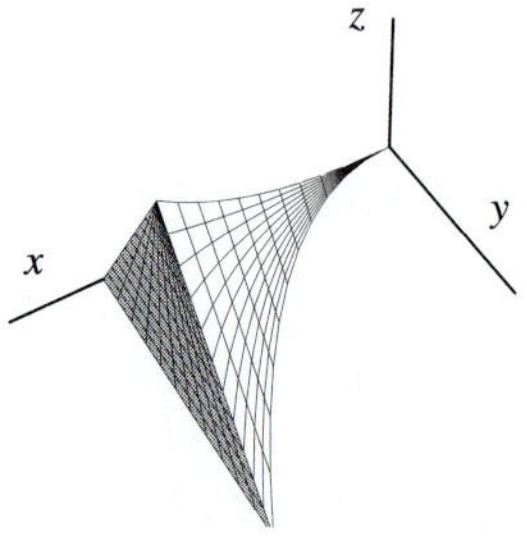

Figure 8. *The solid in Exercise 25.*

26. A_t is an isosceles triangle with height t^3.

27. A_t is the upper half of a disk.

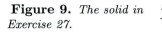

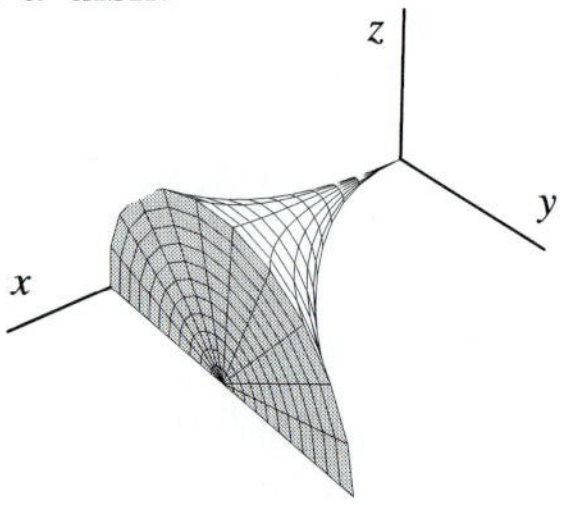

Figure 9. *The solid in Exercise 27.*

In Exercises 28–32, S is a solid lying in the upper half of xyz–space, $z \geq 0$, and between the planes $x = 0$ and $x = 1$. The base of S lies in the xy–coordinate plane and is the region bounded by the graphs of $y = x^2$ and $y = x^3$. Let A_t denote the cross section of S lying in the plane $x = t$. Find a parametrization for S for the given shape of A_t.

28. A_t is a square.

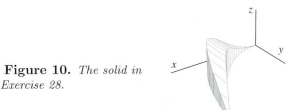

Figure 10. *The solid in Exercise 28.*

29. A_t is a rectangle with its base in the xy–plane and with height 2.

30. A_t is an isosceles right triangle with one leg in the xy–plane and the other in the xz–plane.

31. A_t is an equilateral triangle.

32. A_t is the upper half of a disk with diameter $t^2 - t^3$.

33. Parametrize the intersection of the cylinders $x^2 + y^2 \leq 1$ and $x^2 + z^2 \leq 1$.

In Exercises 34 and 35, $\vec{r}(s,t) = (x(s,t), y(s,t))$, $a \leq s \leq b$, $c \leq t \leq d$, parameterizes a surface R in the xy–plane .

34. Given that R does not intersect the x–axis, then

$$\vec{h}(\theta, s, t) = \begin{pmatrix} x(x,t) \\ y(s,t)\cos(\theta) \\ y(s,t)\sin(\theta) \end{pmatrix},$$

$$a \leq s \leq b,\ c \leq t \leq d$$

parametrizes the solid obtained by rotating the surface R about the x–axis. Show that $J\vec{h}(\theta, s, t) = |y(s,t)|J\vec{r}(s,t)$. Why are we concerned about the surface intersecting the x–axis?

35. If R does not intersect the y–axis, then the solid obtained by rotating R about the y–axis can be parametrized by

$$\vec{h}(\theta, s, t) = \begin{pmatrix} x(x,t)\cos(\theta) \\ y(s,t) \\ x(s,t)\sin(\theta) \end{pmatrix},$$

$$a \leq s \leq b,\ c \leq t \leq d.$$

Show that $J\vec{h}(\theta, s, t) = |x(s,t)|J\vec{r}(s,t)$. Why are we concerned about the surface intersecting the y–axis?

14.7   Triple Integrals

The development in this section is similar to the development of surface integrals. We start with an example which is a 3–dimensional version of Example 1 of Section 14.1.

EXAMPLE 1: Let B be the solid in xyz–space bounded by the planes $x = 0, x = 1, y = 0, y = 2, z = 0$, and $z = 1$. Suppose that the mass density of B at (x, y, z) is given by $\rho(x,y,z) = x^2 + y^2 + z^2$. Let $\{x_0, x_1, \ldots, x_n\}$ be a partition of the segment $[0, 1]$, let $\{y_0, y_1, \ldots, y_m\}$ be a partition of the segment $[0, 2]$, and let $\{z_0, z_1, \ldots, z_s\}$ be a partition of the segment $[0, 1]$. Then the planes

$$x = x_0, \ x = x_1, \ldots, \ x = x_n;$$
$$y = y_0, \ y = y_1, \ldots, \ y = y_m;$$
$$z = z_0, \ z = z_1, \ldots, \ z = z_s$$

partition B into small nonoverlapping boxes. For each $i \leq n, j \leq m$ and $k \leq s$, let $B_{i,j,k}$ be the box bounded by the planes $x = x_{i-1}, x = x_i, y = y_{j-1}, y = y_j, z = z_{k-1}$, and $z = z_k$. See Figures 1 and 2.

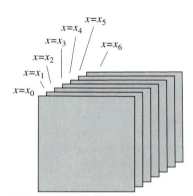

Figure 1.a *Partitioning planes perpendicular to the x-axis.*

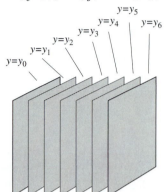

Figure 1.b *Partitioning planes perpendicular to the y-axis.*

Figure 1.c *Partitioning planes perpendicular to the z-axis.*

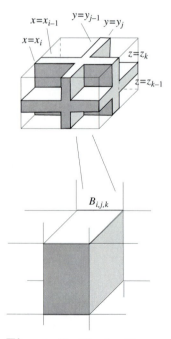

Figure 2. *The box $B_{i,j,k}$.*

$$\rho(x_i, y_j, z_k)\text{Volume}(B_{i,j,k})$$
$$= \ \rho(x_i, y_j, z_k)(x_i - x_{i-1})(y_j - y_{j-1})(z_k - z_{k-1})$$

is an approximation of the mass of $B_{i,j,k}$. Thus the sum

$$\sum_{k=1}^{s} \sum_{j=1}^{m} \sum_{i=1}^{n} \rho(x_i, y_j, z_k)(x_i - x_{i-1})(y_j - y_{j-1})(z_k - z_{k-1}) \qquad (2)$$

is an approximation of the mass of B. However,

$$\sum_{i=1}^{n} (x_i^2 + y_j^2 + z_k^2)(x_i - x_{i-1})$$

is an approximation of

$$\int_0^1 (x^2 + y_j^2 + z_k^2) \, dx,$$

and so Equation (2) is an approximation of

$$\left[\sum_{k=1}^{s} \left[\sum_{j=1}^{m} \left[\int_0^1 (x^2 + y_j^2 + z_k^2) \, dx \right] (y_j - y_{j-1}) \right] (z_k - z_{k-1}) \right].$$

This in turn is an approximation of

$$\left[\sum_{k=1}^{s}\left[\int_0^2\left[\int_0^1 (x^2 + y^2 + z_k^2)\ dx\right]\ dy\right](z_k - z_{k-1})\right],$$

which approximates

$$\left[\int_0^1\left[\int_0^2\left[\int_0^1 (x^2 + y^2 + z^2)\ dx\right]\ dy\right]\ dz\right].$$

Thus Equation (2) approximates

$$\left[\int_0^1\left[\int_0^2\left[\int_0^1 (x^2 + y^2 + z^2)\ dx\right]\ dy\right]\ dz\right],$$

or

$$
\begin{aligned}
\int_0^1\int_0^2\int_0^1 (x^2 + y^2 + z^2)\ dx\,dy\,dz &= \int_0^1\int_0^2\left(\frac{1}{3} + y^2 + z^2\right)\ dy\,dz \\
&= \int_0^1\left(\frac{2}{3} + \frac{8}{3} + 2z^2\right)\ dz \\
&= \frac{2}{3} + \frac{8}{3} + \frac{2}{3} = 4.
\end{aligned}
$$

So the mass of B is 4. ∎

As in Section 14.1, Example 1 leads us to terminology similar to that for integrals over rectangular boxes. We used points to partition an interval when we were defining an integral on a segment on the line in Chapter 6. We used lines to partiton a rectangle in the plane in Section 1 of this chapter. In the following definition, we use planes to partition a rectangular box in 3–space. As you read this definition, keep in mind that we are just extending to three dimensions what we have already done on the line and in the plane. It might be helpful to refer to Figures 1 and 2.

Definition: $\iiint_B \rho\ dV$

Let B be the solid box in xyz–space bounded by the planes $x = a, x = b, y = c, y = d, z = e$, and $z = f$. Suppose that $\rho(x, y, z)$ is a continuous function defined on B. Let $\{x_0, x_1, \ldots, x_n\}$ be a partition of the segment $[a, b]$, let $\{y_0, y_1, \ldots, y_m\}$ be a partition of the segment $[c, d]$, and let $\{z_0, z_1, \ldots, z_s\}$ be a partition of the segment $[e, f]$ so that the planes

$$x = x_0, \; x = x_1, \; \ldots, \; x = x_n;$$
$$y = y_0, \; y = y_1, \; \ldots, \; y = y_m;$$
$$z = z_0, \; z = z_1, \; \ldots, \; z = z_s$$

partition B into small nonoverlapping boxes. This set of planes is called a *partition of B*. For each $i \le n, j \le m$ and $k \le s$, let $B_{i,j,k}$ be the box bounded by the planes $x = x_{i-1}, x = x_i, y = y_{j-1}, y = y_j, z = z_{k-1}$, and $z = z_k$, and let $\vec{s}_{i,j,k}$ be a point in $B_{i,j,k}$. The set $\{\vec{s}_{i,j,k}\}$ is called a *selection for the partition of B*. If L is a number that can be approximated within any specified tolorance by sums of the type

$$\left[\sum_{k=1}^{s} \left[\sum_{j=1}^{m} \left[\sum_{i=1}^{n} \rho(\vec{s}_{i,j,k})(x_i - x_{i-1}) \right] (y_j - y_{j-1}) \right] (z_k - z_{k-1}) \right]$$

simply by insuring the partition divides the box B into "small enough" pieces, then the number L is called the integral of ρ over the solid box B. We write

$$L = \iiint_B \rho \, dV.$$

Computationally,

$$\iiint_B \rho \, dV = \int_0^1 \int_0^2 \int_0^1 \rho(x, y, z) \, dx \, dy \, dz$$
$$= \left[\int_0^1 \left[\int_0^2 \left[\int_0^1 \rho(x, y, z) \, dx \right] dy \right] dz \right].$$

EXAMPLE 2: Returning to Example 1, B is the solid in xyz–space bounded by the planes $x = 0, x = 1, y = 0, y = 2, z = 0$, and $z = 1$ and the mass density of B at (x, y, z) is given by $\rho(x, y, z) = x^2 + y^2 + z^2$. The mass of B is $\iiint_B (x^2 + y^2 + z^2) \, dV$. ∎

As with double integrals, the notation

$$\iiint_B \rho \, dV = \int_e^f \int_c^d \int_a^b \rho(x, y, z) \, dx \, dy \, dz$$

means that we first integrate with respect to x, then with respect to y, and finally with respect to z. It can be shown that if ρ is continuous, then the order of integration does not affect the value obtained. Thus

$$\int_e^f \int_c^d \int_a^b \rho(x,y,z)\,dx\,dy\,dz \;=\; \int_e^f \int_a^b \int_c^d \rho(x,y,z)\,dy\,dx\,dz$$

$$=\; \int_a^b \int_c^d \int_e^f \rho(x,y,z)\,dz\,dy\,dx, \text{ etc.}$$

Using triple integrals is quite similar to using double integrals, only now we are dividing a box into small nonoverlapping boxes, while with double integrals we divided a rectangle into small nonoverlapping rectangles. We have

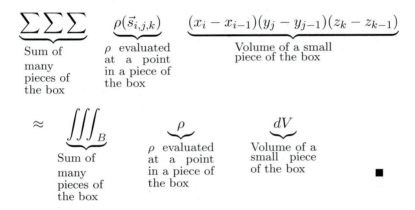

Now that we have the notion of the integral over a box, we can extend the notion to integrals over simple solids.

> **Definition:** $\iiint_S f\,dV$
>
> Suppose that f is a continuous function with domain the simple solid S, and L is a number. If, whenever $\vec{h}$ is a parametrization for the solid S with domain B,
>
> $$L = \iiint_B f\vec{h}((x,y,z))\,J(\vec{h}(x,y,z))\,dV,$$
>
> then the number L is called the *integral of f over S*, and it is denoted by $\iiint_S f\,dV$.

Notice that implicit in the above definition is the idea of breaking the solid S into nonoverlapping pieces. Let

$$\mathcal{P} = \begin{array}{l} x = x_0,\; x = x_1,\; \ldots,\; x = x_n \\ y = y_0,\; y = y_1, \ldots,\; y = y_m \\ z = z_0,\; z = z_1, \ldots,\; z = z_s \end{array}$$

be a partition of B. Then

$$\vec{h}(\mathcal{P}) = \begin{matrix} \vec{h}(x = x_0), \ \vec{h}(x = x_1), \ \ldots, \ \vec{h}(x = x_n) \\ \vec{h}(y = y_0), \ \vec{h}(y = y_1), \ldots, \ \vec{h}(y = y_m) \\ \vec{h}(z = z_0), \ \vec{h}(z = z_1), \ldots, \ \vec{h}(z = z_s) \end{matrix}$$

"partitions" the solid S as illustrated below.

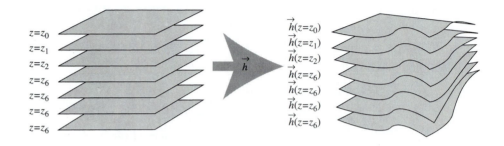

Figure 3. *The planes partitioning the box B are "distorted" by $\vec{h}$ to partition the solid S.*

The ijk^{th} box $B_{i,j,k}$ is distorted by $\vec{h}$ to the piece of the solid S, $\vec{h}(B_{i,j,k})$ as illustrated in Figure 4.

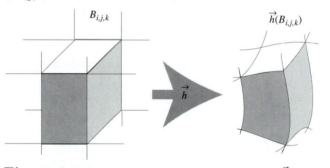

Figure 4. *The piece of B, $B_{i,j,k}$, is distorted by $\vec{h}$ to the piece of the solid S, $S_{i,j,k} = \vec{h}(B_{i,j,k})$.*

If $s_{i,j,k}$ is a point in $B_{i,j,k}$, then the volume of $S_{i,j,k} = \vec{h}(B_{i,j,k})$ is approximated by $\mathrm{vol}(B_{i,j,k})J(\vec{h}(\vec{s}_{i,j,k}))$, and $\vec{h}(s_{i,j,k})$ is a point in $\vec{h}(B_{i,j,k})$. The sum

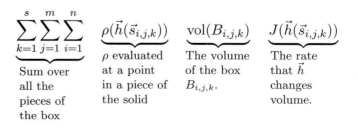

approximates both $\iiint_S \rho \, dV$ and $\iiint_B \rho(\vec{h}(\vec{s}))J(\vec{h}(\vec{s})) \, dV$. The following captures the geometry of the integral over a solid.

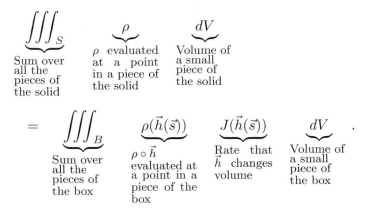

$$
\underbrace{\iiint_S}_{\substack{\text{Sum over} \\ \text{all the} \\ \text{pieces of} \\ \text{the solid}}} \quad \underbrace{\rho}_{\substack{\rho \text{ evaluated} \\ \text{at a point} \\ \text{in a piece of} \\ \text{the solid}}} \quad \underbrace{dV}_{\substack{\text{Volume of} \\ \text{a small} \\ \text{piece of} \\ \text{the solid}}}
$$

$$
= \underbrace{\iiint_B}_{\substack{\text{Sum over} \\ \text{all the} \\ \text{pieces of} \\ \text{the box}}} \quad \underbrace{\rho(\vec{h}(\vec{s}))}_{\substack{\rho \circ \vec{h} \\ \text{evaluated at} \\ \text{a point in a} \\ \text{piece of the} \\ \text{box}}} \quad \underbrace{J(\vec{h}(\vec{s}))}_{\substack{\text{Rate that} \\ \vec{h} \text{ changes} \\ \text{volume}}} \quad \underbrace{dV}_{\substack{\text{Volume of} \\ \text{a small} \\ \text{piece of} \\ \text{the box}}}.
$$

If $\rho(x,y,z)$ is a mass density function for a solid S, then the mass of S is

$$
M = \iiint_S \rho \, dS.
$$

We calculate the center of mass of a volume S with density function $\rho(x,y,z)$ by the now familiar formulas

$$
X = \frac{\iiint_S x\rho \, dV}{\iiint_S \rho \, dV}, \quad Y = \frac{\iiint_S y\rho \, dV}{\iiint_S \rho \, dV}, \quad \text{and} \quad Z = \frac{\iiint_S z\rho \, dV}{\iiint_S \rho \, dV}.
$$

Similarly, if a solid is rotating about an axis, then we can define the moments of inertia.

$I_x = \iiint_S (y^2 + z^2)\rho(x,y,z) \, dV = $ moment of intertia about the x–axis.

$I_y = \iiint_S (x^2 + z^2)\rho(x,y,z) \, dV = $ moment of intertia about the y–axis.

$I_z = \iiint_S (x^2 + y^2)\rho(x,y,z) \, dV = $ moment of intertia about the z–axis.

The kinetic energy of a solid rotating about an axis with angular speed ω is given by $\frac{1}{2}\omega^2 I_a$, where $a = x, y,$ or z indicates the axis of rotation.

EXAMPLE 3: A ball of radius 2 m is rotating about a line passing through its center at a rate of 2 rotations/sec. The mass density at a point of the ball is the square of the distance from the point to the center of the ball. Find the kinetic energy of the rotating ball.

SOLUTION: Assume that the ball is situated in xyz–space so that its center is at the origin and so that it is rotating about the z–axis.

Recall that the rotational kinetic energy is given by the formula $ke = \frac{1}{2}I_z\omega^2$, where ω is the angular speed, which is $(2\pi)2 = 4\pi$ radians/sec (2π times the frequency). Let S denote the ball. The kinetic energy of S is

$$\frac{1}{2}(4\pi)^2 \iiint_S (x^2 + y^2)\rho(x,y,z) \; dV$$
$$= \; 8\pi^2 \iiint_S \left(x^2 + y^2\right)\left(x^2 + y^2 + z^2\right) \; dV.$$

A convenient parametrization for the ball is obtained by using spherical coordinates. Let $\vec{S}(r, \phi, \theta) = (r \sin \phi \cos \theta, r \sin \phi \sin \theta, r \cos \phi), 0 \leq r \leq 2, 0 \leq \phi \leq \pi$, and $0 \leq \theta \leq 2\pi$. Let B be the box in $r\phi\theta$–space bounded by the planes $r = 0$, $r = 2$, $\theta = 0$, $\theta = 2\pi, \phi = 0$, and $\phi = \pi$.

$$8\pi^2 \iiint_S \left(x^2 + y^2\right)\left(x^2 + y^2 + z^2\right) \; dV$$

$$= \; 8\pi^2 \iiint_B \left((r\cos(\theta)\sin(\phi))^2 + (r\sin(\theta)\sin(\phi))^2\right) \cdot$$

$$\left((r\cos(\theta)\sin(\phi))^2 + (r\sin(\theta)\sin(\phi))^2 + r^2\cos^2(\phi)\right) \cdot$$

$$J(\vec{S}(r,\phi,\theta) \; dV$$

$$= \; 8\pi^2 \iiint_B r^6 \sin^3 \phi \; dV$$

$$= \; 8\pi^2 \int_0^\pi \int_0^{2\pi} \int_0^2 r^6 \sin^3 \phi \; dr \, d\theta \, d\phi$$

$$= \; 8\pi^2 \frac{2^7}{7} \int_0^\pi \int_0^{2\pi} \sin^3 \phi \; d\theta \, d\phi$$

$$= \; \frac{2^{11}\pi^3}{7} \int_0^\pi \sin^3 \phi \; d\phi = \frac{2^{13}\pi^3}{21}. \qquad\blacksquare$$

EXAMPLE 4: Let S be the solid of Example 3 of Section 14.6.

(a) Find the volume of S; and

(b) Calculate $\iiint_S (x + y + z)\, dV$.

SOLUTION: In the solution of Example 3 of Section 14.6, we obtained the parametrization of S and the Jacobian.

$$\vec{h}(t, u, v) \;=\; (t, ut^2, vt^2), 0 \le t \le 1, 0 \le u \le 1, \text{ and } 0 \le v \le 1$$

$$J(\vec{h}(t, u, v)) \;=\; t^4.$$

Letting B be the box in tuv–space bounded by the planes $t = 0, t = 1, u = 0, u = 1, v = 0,$ and $v = 1$, we have

(a) The volume of S is given by

$$\iiint_S dV = \iiint_B J(\vec{h}(t, u, v))\, dV = \int_0^1 \int_0^1 \int_0^1 t^4\, dt\, du\, dv = \frac{1}{5};$$

and

(b)

$$
\begin{aligned}
\iiint_S (x + y + z)\, dV &= \iiint_B \left(t + ut^2 + vt^2\right) J(\vec{h}(t, u, v))\, dV \\
&= \int_0^1 \int_0^1 \int_0^1 \left(t + ut^2 + vt^2\right) t^4\, dt\, du\, dv \\
&= \int_0^1 \int_0^1 \int_0^1 \left(t^5 + ut^6 + vt^6\right)\, dt\, du\, dv \\
&= \int_0^1 \int_0^1 \left(\frac{1}{6} + \frac{u}{7} + \frac{v}{7}\right)\, dt\, du\, dv \\
&= \frac{13}{42}. \quad\blacksquare
\end{aligned}
$$

EXAMPLE 5: Let S be the solid of Example 5 of the previous section. Find

(a) The volume of S; and

(b) $\iiint_S (xyz)\, dV$.

SOLUTION: From Example 5 of Section 14.6,

$$\vec{h}(u,v,w) = \begin{pmatrix} u\sqrt{w} \\ wv \\ w \end{pmatrix}; \ -1 \le u \le 1, 0 \le v \le 1, \text{ and } 0 \le w \le 1.$$

Let B be the box in uvw–space defined by $-1 \le u \le 1, 0 \le v \le 1$ and $0 \le w \le 1$. Direct computation shows that $J(\vec{h}(u,v,w)) = w^{3/2}$. Thus

(a) The volume of S is

$$\iiint_S dV = \iiint_B J(\vec{h}(u,v,w)) \, dV$$

$$= \int_0^1 \int_0^1 \int_{-1}^1 w^{3/2} \, du \, dv \, dw = \frac{4}{5};$$

and

(b)

$$\iiint_S (xyz) \, dV = \int_0^1 \int_0^1 \int_{-1}^1 uvw^4 \, du \, dv \, dw = 0. \qquad \blacksquare$$

EXERCISES 14.7

In Exercises 1–6, evaluate $\iiint_B f(x,y,z) \, dV$.

1. B is the box bounded by the planes $x = -1, x = 1, \ y = 0, \ y = 1, \ z = 2, \ z = 3,$ and $f(x,y,z) = xyz$.

2. B is the box bounded by the planes $x = 1, x = 2, \ y = 0, \ y = 1, \ z = -1, \ z = 3,$ and $f(x,y,z) = x + xy + xyz$.

3. B is the box bounded by the planes $x = -1, x = 2, y = -1, \ y = 1, \ z = -1, \ z = 3,$ and $f(x,y,z) = x + xz \cos(\pi xy)$.

4. B is the box bounded by the planes $x = 0, x = 2, \ y = -2, \ y = 1, \ z = -1, \ z = 0,$ and $f(x,y,z) = e^{x+y+z}$.

5. B is the box bounded by the planes $x = 0, x = 2, \ y = 1, \ y = 2, \ z = \pi, \ z = 2\pi,$ and $f(x,y,z) = \dfrac{e^x \ln y \sin(z)}{y}$.

6. B is the box bounded by the planes $x = 0, x = 2, \ y = 1, \ y = 2, z = 0, \ z = 3,$ and $f(x,y,z) = \dfrac{e^x \ln y \sin(z)}{y}$.

In Exercises 7 and 8, the vectors $\vec{A}$, $\vec{B}$, and $\vec{C}$ are drawn emanating from the position $\vec{D}$ to form adjacent edges of a solid parallelepiped P (See Exercise 1 of the previous section). Calculate $\iiint_P f(x,y,z) \, dV$.

7. $\vec{A} = (0,1,1), \vec{B} = (-1,2,1), \vec{C} = (1,1,1), \vec{D} = (0,0,0),$ and $f(x,y,z) = x + y + z$

8. $\vec{A} = (0,2,1), \ \vec{B} = (-1,2,1), \ \vec{C} = (1,-1,3), \vec{D} = (1,1,1),$ and $f(x,y,z) = xy$

9. The box B bounded by the planes $x = -1$ m, $x = 1$ m, $y = 0$ m, $y = 1$ m, $z = 2$ m, and $z = 3$ m, has constant mass density ρ kg/m^3 and it is rotating about the x–axis at a rate of f rotations/sec. Find its kinetic energy.

In Exercises 10–12, the mass density of a sphere of radius 1 m centered at the origin is given by $\rho(x,y,z) = \sqrt{x^2 + y^2}$ kg/m^3.

10. Find the mass of the sphere.

11. Find the center of mass of the sphere.

12. Find the kinetic energy of the sphere if it is rotating about the z–axis at a rate of f rotations/sec.

13. Find the volume of the solid of Exercise 4 of Section 14.6.

14. Find the volume of the solid of Exercise 5 of Section 14.6.

15. Find the volume of the solid of Exercise 6 of Section 14.6.

16. Find the volume of the solid of Exercise 7 of Section 14.6.

17. Find the volume of the solid of Exercise 8 of Section 14.6.

18. Find the volume of the solid of Exercise 22 of Section 14.6.

19. Find the volume of the solid of Exercise 23 of Section 14.6.

20. Calculate $\iiint_S (xyz)\, dV$, where S is the solid of Exercise 28 of Section 14.6.

21. Calculate $\iiint_S (x+y+z)\, dV$, where S is the solid of Exercise 23 of Section 14.6.

22. Calculate $\iiint_S x\, dV$, where S is the solid of Exercise 28 of Section 14.6.

23. Calculate $\iiint_S (x+y+z)\, dV$, where S is the solid of Exercise 29 of Section 14.6.

24. Let S be the solid consisting of the points that satisfy both of the inequalities $x^2 + y^2 \leq 1$ and $x^2 + z^2 \leq 1$. Find the volume of S.

14.8   More on Triple Integrals

As we would expect, there is a change of variable theorem for solid integrals that is quite similar to the one for surface integrals.

Theorem 1 (Change of Variables) *Let P and Q be simple solids, and let $\vec{h}$ be a continuous function from P to Q such that:*

(a) *$\vec{h}$ is one-to-one on the interior of P;*

(b) *$J(\vec{h}(\vec{r}))$ is continuous; and*

(c) *$J(\vec{h}(\vec{r})) \neq 0$ on the interior of P.*

If f is a continuous function from Q into $\mathbb{R}$, then

$$\iiint_Q f\, dV = \iiint_P f(\vec{h}(\vec{r})) J(\vec{h}(\vec{r}))\, dV.$$

We can use the change of variables theorem to derive some formulas for integrating over certain types of solids without having to resort to parametrizations for the solid.

> **Theorem 2** *Suppose that S is a surface in xy–space and that f and g are defined on S such that $f(x,y) > g(x,y)$ for all (x,y) in S, and f and g have continuous first partial derivatives on S. The set of points (x,y,z) satisfying the conditions that (x,y) is in S and $g(x,y) \leq z \leq f(x,y)$ defines a solid D (see Figure 1). If ρ is a continuous function on D, then*
>
> $$\iiint_D \rho\, dV = \iint_S \int_{g(x,y)}^{f(x,y)} \rho(x,y,z)\ dz\, dS.$$

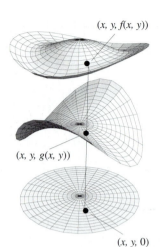

$(x, y, f(x, y))$

$(x, y, g(x, y))$

$(x, y, 0)$

Figure 1.a *Two bounding functions over a surface in the xy–plane.*

Proof: Let $\vec{h}$ be a parametrization for S with domain the rectangle bounded by $a \leq u \leq b$ and $c \leq v \leq d$.

First, we prove the theorem for the special case that $f(x,y) = 1$ and $g(x,y) = 0$. Let V be the resulting solid. Let $\vec{h}(u,v) = (x(u,v), y(u,v))$ be a parametrization for S with domain the rectangle bounded by $a \leq u \leq b$ and $c \leq v \leq d$. Let B be the box bounded by the planes $u = a, u = b, v = c, v = d, w = 0$, and $w = 1$. Let $\vec{s}$ be the function from B into V defined by

$$\vec{s}(u,v,w) = \begin{pmatrix} x(u,v) \\ y(u,v) \\ w \end{pmatrix}.$$

It is easily verified that $J(\vec{s}(u,v,w)) = J(\vec{h}(u,v))$. Thus

$$\begin{aligned}
\iiint_V \rho\, dV &= \int_c^d \int_a^b \int_0^1 \rho(u,v,w) J(\vec{s}(u,v,w))\ dw\, du\, dv \\
&= \int_c^d \int_a^b \int_0^1 \rho(u,v,w) J(\vec{h}(u,v))\ dw\, du\, dv \\
&= \iint_S \int_0^1 \rho(u,v,w)\ dw\, dS.
\end{aligned}$$

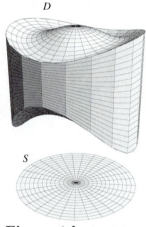

D

S

Figure 1.b *A solid bounded between two functions over a surface in the xy–plane.*

For the general case of the theorem, let g and f be arbitrary functions satisfying the conditions of the theorem, and let D be the resulting solid. Let $\vec{r}$ be the function defined by

$$\vec{r}(x,y,t) = \begin{pmatrix} x \\ y \\ g(x,y) + t[f(x,y) - g(x,y)] \end{pmatrix}.$$

It is a straightforward computation to show that $\vec{r}(x, y, t)$ takes the solid V (defined in the argument of the special case) onto the solid D, that $J(\vec{r}(x, y, t)) = [f(x, y) - g(x, y)]$, and that $\vec{r}$ satisfies the conditions of the Change of Variable Theorem. Therefore, we have from Theorem 1 that

$$\iiint_D \rho \, dV = \iiint_V \rho(\vec{r}(x, y, t)) J(\vec{r}(x, y, t)) \, dV$$

$$= \iint_S \int_0^1 \rho(x, y, g(x, y) + t[f(x, y) - g(x, y)])$$

$$[f(x, y) - g(x, y)] dt \, dS.$$

Letting $u(t) = g(x, y) + t[f(x, y) - g(x, y)]$, we see that $\frac{du}{dt} = f(x, y) - g(x, y)$, $u(0) = g(x, y)$, and $u(1) = f(x, y)$. Thus

$$\iiint_D \rho \, dV = \iint_S \int_{g(x,y)}^{f(x,y)} \rho(x, y, u) \, du \, dS. \qquad \blacksquare$$

EXAMPLE 1: Let D be the solid bounded between the graphs of $z = y$, $z = -y$, and $x^2 + y^2 = 1$ and such that $y \geq 0$. Calculate $\iiint_D z^2 \, dV$.

SOLUTION: If we let S be the half disc parametrized by

$$\vec{r}(\rho, \theta) = (\rho \cos(\theta), \rho \sin(\theta)), \ 0 \leq \rho \leq 1, \ 0 \leq \theta \leq \pi,$$

then D is the set of points (x, y, z) bounded by the graphs of $z = y$ and $z = -y$ such that (x, y) is in S. See Figures 2a and 2b. Applying Theorem 1, we have

$$\iiint_D z^2 \, dV = \iint_S \int_{-y}^y z^2 dz \, dS = \iint_S 2\frac{y^3}{3} \, dS$$

$$= \frac{2}{3} \int_0^\pi \int_0^1 r^3 \sin^3(\theta) r \, dr \, d\theta$$

$$= \frac{2}{15} \int_0^\pi \sin^3(\theta) \, d\theta$$

$$= \frac{2}{15} \int_0^\pi (1 - \cos^2(\theta)) \sin(\theta) \, d\theta \qquad \text{let } u = \cos(\theta)$$

$$= \frac{2}{15} \int_{u=1}^{u=-1} 1 - u^2 du = \frac{4}{45}. \qquad \blacksquare$$

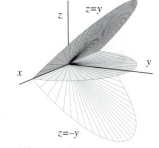

Figure 2.a *The bounding graphs for the solid in Example 1.*

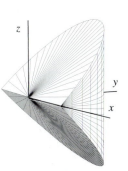

Figure 2.b *The solid in Example 1.*

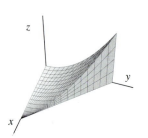

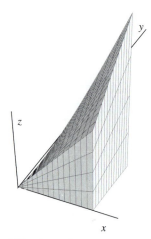

Figure 3. *The solid in Example 2.*

EXAMPLE 2: Let S be the triangle in xy–space with vertices $(0,1), (1,0),$ and $(0,0)$. Let $f(x,y) = y^2$ and $g(x,y) = 0$. Let D be the solid consisting of the points (x,y,z) satisfying the conditions that (x,y) is in S and $0 \le z \le y^2$ (See Figure 3). Calculate $\iiint_D xyz \, dV$.

SOLUTION: By Theorem 1,

$$\iiint_D xyz \, dV = \iint_S \int_0^{y^2} xyz \, dz \, dS$$

$$= \iint_S xy \frac{z^2}{2} \bigg|_{z=0}^{z=y^2} \, dS = \iint_S x \frac{y^5}{2} \, dS.$$

S is the surface in xy–space that is bounded by the lines $y = 1-x$, the x–axis and the y–axis. Thus

$$\iint_S x \frac{y^5}{2} \, dS = \int_0^1 \int_0^{1-x} x \frac{y^5}{2} \, dy \, dx = \int_0^1 x \frac{y^6}{12} \bigg|_{y=0}^{y=1-x} \, dx$$

$$= \frac{1}{12} \int_0^1 x \left(1 - 6x + 15x^2 - 20x^3 + 15x^4 - 6x^5 + x^6 \right) \, dx$$

$$= \frac{1}{672}. \qquad \blacksquare$$

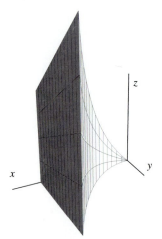

Figure 4.a *The solid in Example 3 viewed from the negative side of the xz–plane.*

EXAMPLE 3: Let S be the surface in xy–space bounded by the x–axis, the line $x = 1$, and the graph of $y = x^2$, and let V be the solid containing the points (x,y,z) that satisfy the conditions that (x,y) is in S and $0 \le z \le x + y$. Figures 4.a and 4.b display two views of S.

(a) Compute the volume of V; and

(b) Compute $\iiint_V (x + y + z) \, dV$.

SOLUTION:

(a) The volume of V is given by

$$\iiint_V dV = \iint_S \int_0^{x+y} dz \, dS$$

$$= \iint_S (x+y) \, dS = \int_0^1 \int_0^{x^2} (x+y) \, dy \, dx$$

$$= \int_0^1 \left(x^3 + \frac{x^4}{2} \right) \, dx = \frac{1}{4} + \frac{1}{10} = \frac{7}{20}.$$

Figure 4.b *The solid in Example 3 viewed from the positive side of the xz–plane.*

(b)

$$\iiint_V (x+y+z)\ dV \quad = \quad \iint_S \int_0^{x+y} (x+y+z)\ dz\ dS$$

$$= \quad \int_0^1 \int_0^{x^2} \int_0^{x+y} (x+y+z)\ dz\ dy\ dx$$

$$= \quad \int_0^1 \int_0^{x^2} \left(xz+yz+\frac{z^2}{2}\right)\Bigg|_{z=0}^{z=x+y}\ dy\ dx$$

$$= \quad \int_0^1 \int_0^{x^2} \left(2x^2+4xy+2y^2\right)\ dy\ dx$$

$$= \quad \int_0^1 \left(2x^2y+\frac{4xy^2}{2}+\frac{2y^3}{3}\right)\Bigg|_{y=0}^{y=x^2}\ dx$$

$$= \quad \int_0^1 \left(2x^4+2x^5+\frac{2x^6}{3}\right)\ dx = \frac{29}{45}. \qquad \blacksquare$$

Useful Observation

Examples 2 and 3 point toward a general situation where the problem of parametrizing certain types of solids can be avoided entirely. If

(a) a and b are numbers such that $a < b$.

(b) f and g are functions defined on a subset of the real numbers such that

    (i) for each x between a and b, we have $g(x) < f(x)$; and

    (ii) the set of points in xy–space bounded by the graphs of f and g and the lines $x = a$ and $x = b$ is a simple surface S.

(c) F and G are real valued functions defined on the surface S such that:

    (i) for each (x,y) in the interior of S, we have that $G(x,y) < F(x,y)$; and

    (ii) the set of points in xyz–space satisfying $a \le x \le y, g(x) \le y \le f(x)$, and $G(x,y) \le z \le F(x,y)$ is a simple solid V,

then

$$\iiint_V \rho(x,y,z)\ dV = \iint_S \int_{G(x,y)}^{F(x,y)} \rho(x,y,z)\ dz\ dS$$
$$= \int_a^b \int_{g(x)}^{f(x)} \int_{G(x,y)}^{F(x,y)} \rho(x,y,z)\ dz\ dy\ dx.$$

EXAMPLE 4: Describe the solid that is the domain of the integral

$$\int_0^1 \int_0^{x^2} \int_{\sin(xy)}^{1+x+y} f(x,y,z)dz\ dy\ dx.$$

SOLUTION: The integral

$$\int_0^1 \int_0^{x^2} \int_{\sin(xy)}^{1+x+y} f(x,y,z)dz\ dy\ dx$$

is of the form

$$\iint_S \int_{\sin(xy)}^{1+x+y} f(x,y,z)dz\ dy\ dx,$$

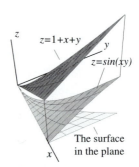

where S is the region in the plane bounded by the x–axis and the graphs of $x = 1$ and $y = x^2$. The solid is bounded above by the graph of $f(x,y) = 1 + x + y$ and below by $g(x,y) = \sin(xy)$. Thus the solid is the set of points (x,y,z) such that $\sin(xy) \le z \le (1 + x + y)$, $0 \le x \le 1$ and $0 \le y \le x^2$. See Figure 5. ∎

Figure 5. *The bounding surfaces for the solid in Example 4.*

EXAMPLE 5: The integral

$$\int_0^1 \int_{-z}^{z^2} \int_0^{y+z} f(x,y,z)dx\ dy\ dz$$

is of the form

$$\iint_S \int_0^{y+z} f(x,y,z)dx\ dS.$$

In this case, the surface S is the region in the yz–plane bounded between the graphs of $y = -z$, $y = z^2 + 1$ and the lines $z = 0$ and $z = 1$. The point (x,y,z) is in the domain of the integral provided $0 \le z \le 1$, $-z \le y \le 1 + y^2$, and $0 \le x \le y + z$. The bounding surfaces and the solid are illustrated in Figure 6.

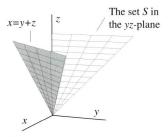

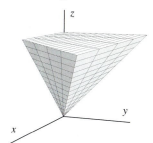

Figure 6.a *The bounding surfaces for the domain of the integral in Example 5.*

Figure 6.b *The domain of the integral in Example 5.*

In general:

An expression of the form $\int_S \int_\alpha^\beta f(x,y,z)dz\,dS$ only makes sense if S is a surface in the xy–coordinate plane. α and β must be functions of x and y.

An expression of the form $\int_S \int_\alpha^\beta f(x,y,z)\,dy\,dS$ only makes sense if S is a surface in the xz–coordinate plane. α and β must be functions of x and z.

An expression of the form $\int_S \int_\alpha^\beta f(x,y,z)\,dx\,dS$ only makes sense if S is a surface in the yz–coordinate plane. α and β must be functions of x and z.

EXAMPLE 6: The expression $\int_0^1 \int_{-x}^1 \int_0^z dz\,dy\,dx$ does not make sense. This would be expected to be an integral of the form $\int_S \int_\alpha^\beta f(x, y,z)dz\,dS$. However, β is not a function of x and y. ∎

EXERCISES 14.8

In Exercises 1–4, evaluate the triple integral.

1. $\displaystyle\int_0^1 \int_{-z}^0 \int_{-y}^z dx\,dy\,dz.$

2. $\displaystyle\int_{-1}^1 \int_0^z \int_{x+z}^{z^2} x\,dy\,dx\,dz.$

3. $\displaystyle\int_0^1 \int_{-x}^{x^2} \int_0^{y+x} x+y\;dz\,dy\,dx.$

4. $\displaystyle\int_0^1 \int_{y^3}^{y^2} \int_{-x}^y xy\,dz\,dx\,dy.$

In Exercises 5–9 the integral is of the form $\iint_S \int_{F(y,z)}^{G(y,z)} f(x,y,z)\,dx\,dS$, $\iint_S \int_{F(x,z)}^{G(x,z)} f(x,y,z)\,dy\,dS$, or $\int_S \int_{F(x,y)}^{G(x,y)} f(x,y,z)\,dz\,dS$. Describe the set S and determine which of the coordinate planes it is in.

5. $\displaystyle\int_0^1 \int_0^2 \int_{-1}^2 f(x,y,z)\;dx\,dy\,dz.$

6. $\displaystyle\int_0^1 \int_{-z}^0 \int_{-y}^z f(x,y,z)\;dx\,dy\,dz.$

7. $\displaystyle\int_0^2 \int_{-x}^{2+x^2} \int_{-y}^{x+y} f(x,y,z)\;dz\,dy\,dx.$

8. $\int_0^2 \int_{-x}^{2+x^2} \int_{-x}^{x+z} f(x,y,z)\, dy\, dz\, dx.$

9. $\int_0^2 \int_{-z}^{2+z^2} \int_{-z}^{y+z} f(x,y,z)\, dx\, dy\, dz.$

In Exercises 10–15, $f(x,y,z)$ is a continuous function. Determine whether the expression makes sense. If the expression is integrable, describe the domain of the integral.

10. $\int_0^1 \int_{-z}^0 \int_{-y}^z f(x,y,z)\, dx\, dy\, dz.$

11. $\int_0^1 \int_{-z}^0 \int_{-y}^z f(x,y,z)\, dz\, dx\, dy.$

12. $\int_0^1 \int_{-z}^0 \int_{-y}^z f(x,y,z)\, dy\, dx\, dz.$

13. $\int_0^1 \int_{-z}^0 \int_0^{x+y+z} f(x,y,z)\, dx\, dy\, dz.$

14. $\int_0^1 \int_{-z}^0 \int_{-y^2}^{y+z} f(x,y,z)\, dz\, dy\, dx.$

15. $\int_0^1 \int_{-z}^0 \int_{-y^2}^{y+x} f(x,y,z)\, dx\, dy\, dz.$

In Exercises 16–22, find the volume of the solid V consisting of all points (x,y,z) such that (x,y) is in the surface S in xy–space and $G(x,y) \le z \le F(x,y)$.

16. S is the region bounded by the x–axis, the graph of $y = x^2$, and the line $x = 2$. $G(x,y) = -x$ and $F(x,y) = x + y$.

17. S is the region bounded by the graphs of $y = x^2$ and $y = -x^2$, and the line $x = 2$. $G(x,y) = -x$ and $F(x,y) = x + y$.

18. S is the region bounded by the graphs of $y = x$ and $y = x^2$. $G(x,y) = -xy$ and $F(x,y) = x^2$.

19. S is the region bounded by the circle $x^2 + y^2 = 1$. $G(x,y) = 0$ and $F(x,y) = 2 - x$.

20. S is the parallelogram with two adjacent edges the line segments $[\vec{A}, \vec{B}]$ and $[\vec{A}, \vec{C}]$, where $\vec{A} = (0,0), \vec{B} = (1,1)$, and $\vec{C} = (1,2)$. $G(x,y) = -2$ and $F(x,y) = x + y$.

21. S is the parallelogram with two adjacent edges the line segments $[\vec{A}, \vec{B}]$ and $[\vec{A}, \vec{C}]$, where $\vec{A} = (0,0), \vec{B} = (1,1)$, and $\vec{C} = (1,3)$. $G(x,y) = x - y$ and $F(x,y) = x^2 + y + 1$.

22. S is the triangle with vertices $\vec{A} = (0,0), \vec{B} = (1,1)$, and $\vec{C} = (1,2)$. $G(x,y) = -2$ and $F(x,y) = x + y$.

Find the volume of the solid V in Exercises 23–28.

23. S is the surface in xz–space bounded by the x–axis, the graph of $z = x^2$, and the line $x = 2$. V is the set of points (x,y,z) such that (x,z) is in S and $-x \le y \le x + z$.

24. S is the surface in yz–space bounded by the y–axis, the graph of $z = y^2$, and the line $y = 2$. V is the set of points of (x,y,z) such that (y,z) is in S and $-y \le x \le y + z$.

25. S is the region in xz–space bounded by the graphs of $x = z^2$, $x = -z^2$, and the line $z = 2$. V is the set of points (x,y,z) such that (x,z) is in S and $-z \le y \le z$.

26. S is the region in yz–space bounded by the graphs of $y = z^2$ and $y = 8 - z^2$. V is the set of points (x,y,z) such that (y,z) is in S and $-y \le x \le y$.

27. S is the triangle in yz–space with vertices $(1,0), (2,2)$, and $(3,1)$. V is the set of points (x,y,z) such that (y,z) is in S and $-y \le x \le z$.

28. S is the set of points in xz–space satisfying $\frac{x^2}{4} + z^2 \le 1$, and V is the set of points (x,y,z) such that (x,z) is in S and $0 \le y \le x + 5$.

In Exercises 29–33, find the volume of the solid bounded by the graphs (in xyz–space) of the given equations.

29. $y = x^2, y + z = 1$, and $z = 0$.

30. $x + y + z = 1$ and the three coordinate planes.

31. $z = x^2, z = 1 - x^2, y = 0$, and $y = z$.

32. $x^2 + y^2 = 1$ and $y^2 + z^2 = 1$.

33. $x^2 + y^2 - 1 = z$ and $1 - x^2 - y^2 = z$.

In Exercises 34–40, calculate $\iiint_V f(x, y, z)\, dV$.

34. V is defined in Exercise 16, $f(x, y, z) = x$.

35. V is defined in Exercise 16, $f(x, y, z) = xy$.

36. V is defined in Exercise 17, $f(x, y, z) = x+y+z$.

37. V is defined in Exercise 20, $f(x, y, z) = x^2 + y^2$.

38. V is defined in Exercise 22, $f(x, y, z) = xy + z$.

39. V is defined in Exercise 23, $f(x, y, z) = x+y+z$.

40. V is defined in Exercise 25, $f(x, y, z) = x+y+z$.

Chapter 15

Divergence and Stokes' Theorem

The area of a disc of radius r is given by the familiar formula $A = \pi r^2$. Notice that the derivative of A is $A' = 2\pi r$, which is the length of the boundary of the disc. Likewise the volume of the sphere of radius r is given by $V = \frac{4}{3}\pi r^3$. The derivative of V is $V' = 4\pi r^2$, which is the surface area of the bounding sphere. These are not coincidences. In the present chapter we present theorems that relate the integral over a volume with an integral over the volume's boundary surface, and the integral over a surface with an integral over the surface's boundary curve. The last section of this chapter provides us with a tool to distinguish between conservative and nonconservative vector fields.

15.1 Oriented Surfaces and Surface Integrals

Our task in this section is to model a fluid flowing through a screen. Although the idea of the rate of flow of a fluid has an intuitive meaning to us, we need a formal definition here. Before defining the rate of flow of a fluid, we review some basic facts about cross and triple products and we introduce the idea of the oriented area of a parallelogram.

Suppose that vectors $\vec{A}$ and $\vec{B}$ are drawn emanating from a common point. Recall that $\|\vec{A} \times \vec{B}\|$ is the area of the parallelogram with $\vec{A}$ and $\vec{B}$ as its adjacent sides. The vectors $\vec{A} \times \vec{B}$ and $\vec{B} \times \vec{A}$ are normal to the surface of the parallelogram. See Figure 1.

The surface of the parallelogram has two sides. In some problems, it is important to decide which side we think of as the top of the parallelogram and which side we think of as the bottom, that is,

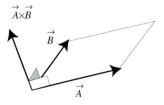

Figure 1.a *The vector $\vec{A} \times \vec{B}$.*

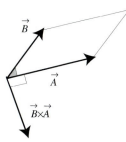

Figure 1.b *The vector $\vec{B} \times \vec{A}$.*

which way the parallelogram is to be oriented in our thinking. The vectors $\vec{A} \times \vec{B}$ and $\vec{B} \times \vec{A}$ are often called the *oriented areas* of the parallelogram.

Suppose that the angle between $\vec{C}$ and $\vec{B} \times \vec{A}$ is acute as in Figure 2. Then, as we showed in Section 1.5, $\vec{C} \cdot (\vec{B} \times \vec{A})$ is the volume of the parallelepiped, and $\vec{C} \cdot (\vec{A} \times \vec{B})$ is the negative of the volume. In any case, the volume of a parallelepiped with adjacent edges $\vec{A}, \vec{B}$, and $\vec{C}$ drawn emanating from a common point is given by

$$|\vec{C} \cdot (\vec{A} \times \vec{B})| = |\vec{A} \cdot (\vec{B} \times \vec{C})| = |\vec{B} \cdot (\vec{A} \times \vec{C})|, \text{ etc.}$$

If the angle between $\vec{C}$ and $\vec{A} \times \vec{B}$ is acute, then $\vec{C} \cdot (\vec{A} \times \vec{B})$ is positive. If the angle is obtuse, then $\vec{C} \cdot (\vec{A} \times \vec{B})$ is negative.

This interpretation of a triple product as the volume of the parallelepiped determined by the three vectors makes the triple product $\vec{C} \cdot (\vec{A} \times \vec{B})$ extremely useful.

Now to the problem at hand. Let $\vec{A}$ and $\vec{B}$ be mutually perpendicular vectors emanating from a common point. Assume that $\|\vec{A}\| = \|\vec{B}\| = \Delta h$. Then $\vec{A}$ and $\vec{B}$ are adjacent edges of a square $\mathcal{S}$ with area Δh^2. Assume that $\mathcal{S}$ models a screen through which water can freely flow and that $\mathcal{S}$ is placed in a river. As illustrated in Figure 3, the amount of water flowing through the screen depends on the orientation of the screen relative to the flow of the water.

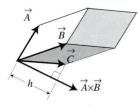

Figure 2. *$\vec{A}, \vec{B}$ and $\vec{C}$ drawn emanating from a common point form adjacent edges of a parallelepiped with volume $|\vec{C} \cdot (\vec{A} \times \vec{B})|$.*

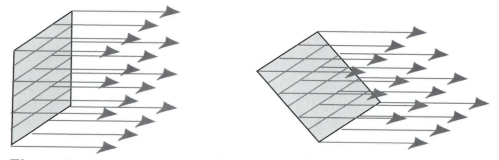

Figure 3. *The rate of flow of a fluid through a screen depends on the orientation of the screen in the fluid.*

Let the screen be placed in a flowing fluid at position $\vec{r}_0$ so that the vector $\vec{A} \times \vec{B}$ points in the direction of the flow at $\vec{r}_0$. Let $\hat{n}$ be a unit vector pointing in the direction of the flow at $\vec{r}_0$, and let flow($\mathcal{S}$) denote the volume of fluid passing through $\mathcal{S}$ every unit of time. Then

$$\vec{F}(\vec{r}_0) = \lim_{h \to 0} \frac{\text{flow}(\mathcal{S})}{h^2} \hat{n}$$

is the rate of flow of the fluid (per unit area) at $\vec{r}_0$.

Assuming that the rate of flow of a fluid is constant, we then

extend the idea of the rate of flow of a fluid at a point to the idea of
the rate of flow through a parallelogram.

EXAMPLE 1: Suppose that a river is flowing with constant velocity $\vec{v}$ and a screen cut in the shape of a parallelogram is placed in
the current. Two adjacent sides of the screen can be represented by
the vectors $\vec{A}$ and $\vec{B}$. The oriented area of the screen is $\vec{A} \times \vec{B}$. The
rate at which the water passes through the screen is $\vec{v} \cdot (\vec{A} \times \vec{B})$. See
Figure 4.

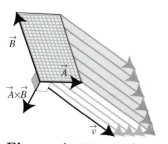

The volume of water passing through the screen per unit time is
$|\vec{v} \cdot (\vec{A} \times \vec{B})|$. The rate of flow of the water will be positive if the angle
between $\vec{v}$ and $(\vec{A} \times \vec{B})$ is acute, and negative if the angle between $\vec{v}$
and $(\vec{A} \times \vec{B})$ is obtuse.   ∎

Figure 4. *Volume of water flowing through the screen per unit time.*

We now extend these ideas to flows through surfaces in general.

EXAMPLE 2: Suppose that we represent a river by D, and let
$\vec{F}$ be a function from D into $\mathbb{R}^3$ so that $\vec{F}(\vec{r})$ denotes the rate of flow
of the water at position $\vec{r}$ in the river. Suppose also that a screen in
the river can be represented by a simple surface S. Find the rate of
the flow of water passing through the screen.

SOLUTION: Let R be the rectangle in the uv–plane that is bounded
by the lines $u = a, u = b, v = c$, and $v = d$, where $a < b$ and $c < d$.
Let $\vec{h}$ be a parametrization of S with domain R.

Let $\{u = u_0, u = u_1, \ldots, u = u_n\} \cup \{v = v_0, v = v_1, \ldots, v = v_m\}$
be a partition of R, and for each $0 \leq i \leq n$ and $0 \leq j \leq m$, let $\vec{s}_{i,j}$
be a selection point from $R_{i,j}$, where $R_{i,j}$ is the region in R bounded
by the lines $u = u_{i-1}, u = u_i, v = v_{j-1}$, and $v = v_j$. Let $\sigma_{i,j}$ be the
portion of the screen that is the image of $R_{i,j}$ under $\vec{h}$. As in the
previous section, we approximate the area of $\sigma_{i,j}$ by the area of a
parallelogram; however, this time we preserve the orientation of the
parallelogram. See Figure 5.

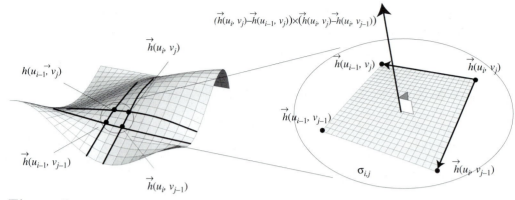

Figure 5. *The directed area of $\sigma_{i,j}$ is approximated by the directed area of the parallelogram with adjacent edges $\vec{h}(u_i, v_j) - \vec{h}(u_{i-1}, v_j)$ and $\vec{h}(u_i, v_j) - \vec{h}(u_i, v_{j-1})$.*

Let $\vec{A}_{i,j}$ be the oriented area of the parallelogram approximating $\sigma_{i,j}$; that is,

$$\vec{A}_{i,j} = [\vec{h}(u_i, v_j) - \vec{h}(u_{i-1}, v_j)] \times [\vec{h}(u_i, v_j) - \vec{h}(u_i, v_{j-1})].$$

The magnitude of $\vec{A}_{i,j}$ is an approximation of the area of $\sigma_{i,j}$, and $\vec{F}(\vec{s}_{i,j}) \cdot \vec{A}_{i,j}$ approximates the volume per unit time of water passing through the $\sigma_{i,j}$ portion of the screen.

Now

$$\frac{\partial \vec{h}(\vec{s}_{i,j})}{\partial u}(u_i - u_{i-1}) \quad \approx \quad \frac{\vec{h}(u_i, v_j) - \vec{h}(u_{i-1}, v_j)}{u_i - u_{i-1}}(u_i - u_{i-1})$$

$$\frac{\partial \vec{h}(\vec{s}_{i,j})}{\partial v}(v_i - v_{i-1}) \quad \approx \quad \frac{\vec{h}(u_i, v_j) - \vec{h}(u_i, v_{j-1})}{v_j - v_{j-1}}(v_j - v_{j-1}).$$

So an approximation for the volume per unit time of water passing through the $\sigma_{i,j}$ portion of the screen is given by

$$\vec{F}(\vec{s}_{i,j}) \cdot \left(\frac{\partial \vec{h}(\vec{s}_{i,j})}{\partial u} \times \frac{\partial \vec{h}(\vec{s}_{i,j})}{\partial v} \right) (u_i - u_{i-1})(v_j - v_{j-1}).$$

Thus the rate of flow of water through the screen is approximated by

$$\sum_{j=1}^{m} \sum_{i=1}^{n} \vec{F}(\vec{s}_{i,j}) \cdot \left(\frac{\partial \vec{h}(\vec{s}_{i,j})}{\partial u} \times \frac{\partial \vec{h}(\vec{s}_{i,j})}{\partial v} \right) (u_i - u_{i-1})(v_j - v_{j-1}),$$

which approximates

$$\iint_R \vec{F} \cdot \left(\frac{\partial \vec{h}}{\partial u} \times \frac{\partial \vec{h}}{\partial v} \right) dA.$$

Obviously, the rate of flow through the screen will be positive or negative according to the direction of the flow through the screen. This observation leads us to the notion of the orientation of the surface.

Essentially, an oriented surface is a surface with two sides, one of which is chosen to be the positive side. This statement probably seems a little ridiculous at first. Our intuition *falsely* tells us that any surface would surely have two sides, but then recalls the Möbius band. A model of this surface can be obtained by the construction illustrated in Figure 6.

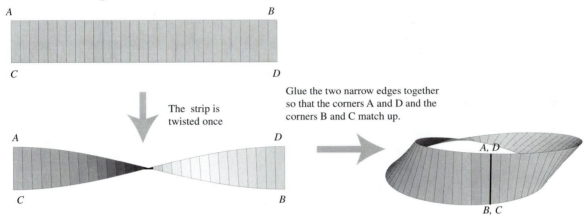

Figure 6. *A Möbius band can be constructed by twisting a strip of paper once and then gluing two ends together as illustrated.*

Start at any point of the Möbius band and trace a line around the surface. By going around twice, you will wind up at the starting point. This is a perfectly valid surface that has only one side. A rigorous definition of the idea of a two-sided surface is a technical matter. We assume that all our surfaces are two-sided, and tackle the easier problem of deciding which side is which.

Let $\vec{h}$ be a parametrization for the surface S with domain in uv–space. Then $\frac{\partial \vec{h}}{\partial u} \times \frac{\partial \vec{h}}{\partial v}$ is a vector that is normal to S at the point $\vec{h}(u, v)$. Refer to Figure 5.

We use this vector to assign an orientation to the surface. We say that $\frac{\partial \vec{h}}{\partial u} \times \frac{\partial \vec{h}}{\partial v}$ points in the direction of the *positive side* of the surface. Note that $\frac{\partial \vec{h}}{\partial v} \times \frac{\partial \vec{h}}{\partial u}$ points in the opposite direction, which we call the *negative side* of the surface. ∎

EXAMPLE 3: If $\vec{A}$ and $\vec{B}$ are vectors drawn emanating from $\vec{r}_0$, and if $\vec{h}$ is the parametrization defined by

$$\vec{h}(u, v) = u\vec{A} + v\vec{B} + \vec{r}_0, 0 \leq u \leq 1, 0 \leq v \leq 1,$$

Figure 7.
$\vec{h}(u, v) = u\vec{A} + v\vec{B} + \vec{r}_0.$

then $\vec{h}$ is a parametrization for the parallelogram with adjacent sides $\vec{A}$ and $\vec{B}$. See Figure 7.

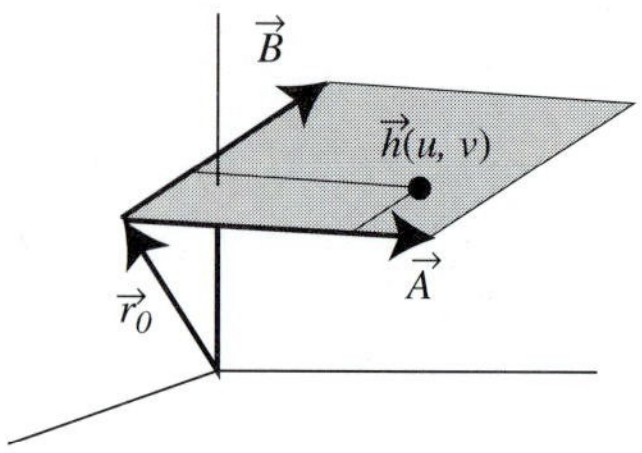

The partial derivatives of $\vec{h}$ are given by

$$\frac{\partial \vec{h}(u, v)}{\partial u} = \vec{A} \quad \text{and} \quad \frac{\partial \vec{h}(u, v)}{\partial v} = \vec{B},$$

so

$$\left(\frac{\partial \vec{h}}{\partial u} \times \frac{\partial \vec{h}}{\partial v} \right) \Bigg|_{(u,v)} = \vec{A} \times \vec{B}.$$

Thus the orientation of the parallelogram given by the parametrization $\vec{h}$ points in the direction of $\vec{A} \times \vec{B}$. ∎

We have the following general definition.

Definition: $\iint_S \vec{F} \cdot d\vec{S}$

If $\vec{F}$ is a function from $\mathbb{R}^3$ into $\mathbb{R}^3$, and if $\vec{h}$ is a parametrization of the surface S with domain R (R is a rectangle in $\mathbb{R}^2$) and S is a subset of the domain of $\vec{F}$, then we define

$$\iint_S \vec{F} \cdot d\vec{S} = \iint_R \vec{F}(\vec{h}(u, v)) \cdot \left(\frac{\partial \vec{h}}{\partial u} \times \frac{\partial \vec{h}}{\partial v} \right) \, dA.$$

$\iint_S \vec{F} \cdot d\vec{S}$ is called the *flux of $\vec{F}$ through S.*

The integral $\iint_S \vec{F} \cdot d\vec{S}$ can be thought of as the rate of flow of $\vec{F}$ through S. Intuitively, $\iint_S \vec{F} \cdot d\vec{S}$ will be positive if $\vec{F}$ is flowing predominantly in the positive direction through S.

Another way to express $\iint_S \vec{F} \cdot d\vec{S}$ is by observing that

$$\vec{n} = \frac{\frac{\partial \vec{h}}{\partial u} \times \frac{\partial \vec{h}}{\partial v}}{\left\| \frac{\partial \vec{h}}{\partial u} \times \frac{\partial \vec{h}}{\partial v} \right\|}$$

is the unit vector normal to S, so that

$$\iint_R \vec{F}(\vec{h}(u, v)) \cdot \left(\frac{\partial \vec{h}}{\partial u} \times \frac{\partial \vec{h}}{\partial v} \right) \, dA$$

$$= \iint_R \left(\vec{F}(\vec{h}(u, v)) \cdot \vec{n}(u, v) \right) \left\| \frac{\partial \vec{h}}{\partial u} \times \frac{\partial \vec{h}}{\partial v} \right\| \, dA$$

$$= \iint_R \underbrace{\vec{F}(\vec{h}(u,v)) \cdot \vec{n}(u,v)}_{\substack{\text{The component} \\ \text{of } \vec{F} \text{ normal to } S}} \underbrace{J(\vec{h}(u,v))}_{\substack{\text{The rate} \\ \text{that } \vec{h} \\ \text{changes} \\ \text{area}}} \, dA$$

$$= \iint_S \left(\vec{F} \cdot \vec{n} \right) \, dS.$$

This interpretation of $\iint_S \vec{F} \cdot d\vec{S}$ as $\iint_S (\vec{F} \cdot \vec{n}) \, dS$ fits well with our rule of thumb for applications of the surface integral, since it is true that if ΔS is a small piece of surface and $\vec{n}$ is a unit vector normal to ΔS at a point of ΔS, and if $\vec{F}(\vec{r})$ is the velocity of the fluid at that position in ΔS, then $\vec{F} \cdot \vec{n}$ times the area of ΔS is an approximation of the rate of flow of the fluid through the piece of screen ΔS. See Figure 8.

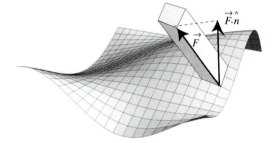

Figure 8. *$\vec{F} \cdot \vec{n}$ times the area of ΔS is an approximation of the rate of flow of the fluid through the piece of screen ΔS.*

EXAMPLE 4: Let P be the parallelogram with adjacent sides the vectors $\vec{A} = (1, 1, 1)$ and $\vec{B} = (0, 1, 0)$ drawn emanating from the point $\vec{r}_0 = (-1, 0, 1)$. Let the orientation of P point in the direction of $\vec{A} \times \vec{B}$. Compute the flux of $\vec{F}$ through P, where $\vec{F}(x, y, z) = (x, y, z)$.

SOLUTION: In Example 3, a parametrization for P is given by

$$\vec{h}(u, v) = u\vec{A} + v\vec{B} + \vec{r}_0 = \begin{pmatrix} u - 1 \\ u + v \\ u + 1 \end{pmatrix}.$$

The cross product of the partial derivatives of $\vec{h}$ is given by

$$\frac{\partial \vec{h}}{\partial u} \times \frac{\partial \vec{h}}{\partial v} = \begin{pmatrix} 1 \\ 1 \\ 1 \end{pmatrix} \times \begin{pmatrix} 0 \\ 1 \\ 0 \end{pmatrix} = \begin{pmatrix} -1 \\ 0 \\ 1 \end{pmatrix}.$$

$\vec{F}$ evaluated at $\vec{h}$ is given by

$$\vec{F}(\vec{h}(u, v)) = \begin{pmatrix} u - 1 \\ u + v \\ u + 1 \end{pmatrix}.$$

Thus

$$\vec{F}(\vec{h}(u,v)) \cdot \left(\frac{\partial \vec{h}}{\partial u} \times \frac{\partial \vec{h}}{\partial v} \right) \Bigg|_{(u,v)} = -(u-1) + 0 + (u+1) = 2.$$

So the flux is

$$\begin{aligned}
\iint_S \vec{F} \cdot d\vec{S} &= \int_0^1 \int_0^1 2 \, du \, dv \\
&= \int_0^1 2u \Big|_{u=0}^{u=1} dv \\
&= \int_0^1 2 \, dv = 2v \Big|_0^1 = 2. \qquad \blacksquare
\end{aligned}$$

If the idea of integration over an oriented surface is well understood by the reader, it is often the case that seemingly difficult integration problems can be made almost trivial by using the geometry of the problem. In fact, this will almost always be the case in applications of integration over oriented surfaces to solutions of problems in physics and engineering. The main thing to look for is the following situation:

$$\vec{F} \cdot \vec{n} = \vec{F} \cdot \frac{\frac{\partial \vec{h}}{\partial u} \times \frac{\partial \vec{h}}{\partial v}}{\left\| \frac{\partial \vec{h}}{\partial u} \times \frac{\partial \vec{h}}{\partial v} \right\|} = C \quad \text{(a constant)}.$$

In this situation, we have that

$$\iint_S \vec{F} \cdot d\vec{S} = \iint_S \vec{F} \cdot \frac{\frac{\partial \vec{h}}{\partial u} \times \frac{\partial \vec{h}}{\partial v}}{\left\| \frac{\partial \vec{h}}{\partial u} \times \frac{\partial \vec{h}}{\partial v} \right\|} \, dS = C \iint_S dS = C(\text{Area of } S).$$

EXAMPLE 5: Suppose that $\vec{F}(\vec{r}) = \frac{1}{\|\vec{r}\|^2} \vec{r}$ and S is the unit sphere centered at the origin with orientation pointing away from the region bounded by S. Note that the magnitude of $\vec{F}$ is a constant on the sphere, and $\vec{F}$ always points away from the origin. Thus we have that $\vec{F} \cdot \vec{n} = 1$. So

$$\iint_S \vec{F} \cdot \left(\frac{\frac{\partial \vec{h}}{\partial u} \times \frac{\partial \vec{h}}{\partial v}}{\left\| \frac{\partial \vec{h}}{\partial u} \times \frac{\partial \vec{h}}{\partial v} \right\|} \right) \, dS = \iint_S dS = 4\pi. \qquad \blacksquare$$

Managing the Orientation of the Parametrization

The method of determining the orientation of a surface is a matter of bookkeeping. We have been careful to determine the orientation by taking the cross product of the partial derivative of the parametrization with respect to the first variable with the partial derivative with respect to the second variable. As illustrated in Figure 9, we can often determine the orientation without computing the partial derivatives.

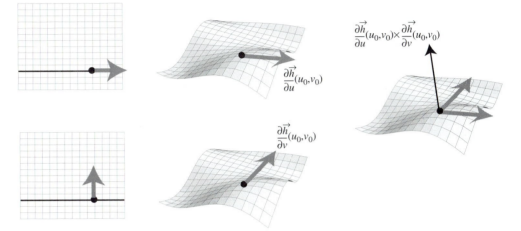

Figure 9. *The direction of a point as it moves through $\vec{h}(u_0, v_0)$ along the curve parametrized by $\vec{h}(u, v_0)$ is the same as $\frac{\partial \vec{h}}{\partial u}$ at (u_0, v_0). Similarly, the direction of a point as it moves through $\vec{h}(u_0, v_0)$ along the curve parametrized by $\vec{h}(u_0, v)$ is the same as $\frac{\partial \vec{h}}{\partial v}$ at (u_0, v_0).*

EXAMPLE 6: Our standard parametrization for the unit sphere is

$$\vec{h}(\phi, \theta) = \begin{pmatrix} \sin(\phi)\cos(\theta) \\ \sin(\phi)\sin(\theta) \\ \cos(\phi) \end{pmatrix}, \ 0 \le \phi \le \pi, \ 0 \le \theta \le 2\pi.$$

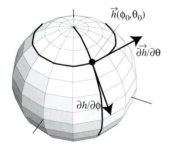

Figure 10. *The partial derivatives for the standard parametrization of the sphere.*

If a point moves along the surface of the sphere through $\vec{h}(\phi_0, \theta_0)$ holding $\theta = \theta_0$, then the point is moving down a longitude. If you watch a point moving through $\vec{h}(\phi_0, \theta_0)$ holding ϕ_0 from above (from the "north pole"), then the point will be moving in a counterclockwise direction. See Figure 10. The right hand rule shows that the orientation is away from the interior of the sphere.

If we change the order in which we list the variables, we change the orientation. Thus, if we write

$$\vec{h}(\theta, \phi) = \begin{pmatrix} \sin(\phi)\cos(\theta) \\ \sin(\phi)\sin(\theta) \\ \cos(\phi) \end{pmatrix}, \ 0 \le \phi \le \pi, \ 0 \le \theta \le 2\pi,$$

then the orientation will point into the interior of the sphere.   ■

Pressure

If $\phi(\vec{r})$ is a real valued function defined on an oriented surface S, then if we compute $\iint_S \phi \, d\vec{S}$, the result is a vector.

EXAMPLE 7: Let $\phi(\vec{r}) = x+y+z$ and let $\vec{S}$ be the parallelogram parametrized by

$$\vec{h}(s,t) = \begin{pmatrix} 2s + 3t + 1 \\ s - t \\ s + t - 3 \end{pmatrix}, \ 0 \le s \le 1, \ 0 \le t \le 1.$$

Find $\iint_S \phi \, d\vec{S}$.

SOLUTION:

$$\iint_S (x + y + z) \, d\vec{S}$$

$$= \int_0^1 \int_0^1 ((2s + 3t + 1) + (s - t) + (s + t - 3)) \frac{\partial \vec{h}}{\partial s} \times \frac{\partial \vec{h}}{\partial t} \, ds \, dt$$

$$= \int_0^1 \int_0^1 (4s + 3t - 2)(2, 1, -5) \, ds \, dt$$

$$= \frac{3}{2}(2, 1, -5).$$

   ■

EXAMPLE 8: Let $\phi(\vec{r}) = kz$ and let $\vec{S}$ be the sphere parametrized by

$$\vec{h}(\phi, \theta) = \begin{pmatrix} \cos(\theta) \sin(\phi) \\ \sin(\theta) \sin(\phi) \\ \cos(\phi) \end{pmatrix}, \ 0 \le \phi \le \pi, \ 0 \le \theta \le 2\pi.$$

Find $\iint_S \phi \, d\vec{S}$.

SOLUTION:

$$\iint_S z \, d\vec{S} = \int_0^{2\pi} \int_0^\pi \cos(\phi) \frac{\partial \vec{h}}{\partial s} \times \frac{\partial \vec{h}}{\partial t} \, ds \, dt$$

$$= \int_0^{2\pi} \int_0^\pi \cos(\phi) \begin{pmatrix} \cos(\theta) \sin^2(\phi) \\ \sin(\theta) \sin^2(\phi) \\ \cos(\phi) \sin(\phi) \end{pmatrix}$$

$$= \begin{pmatrix} \int_0^{2\pi} \int_0^{\pi} \cos(\phi)\cos(\theta)\sin^2(\phi)\,d\phi\,d\theta \\ \int_0^{2\pi} \int_0^{\pi} \cos(\phi)\sin(\theta)\sin^2(\phi)\,d\phi\,d\theta \\ \int_0^{2\pi} \int_0^{\pi} \cos^2(\phi)\sin(\phi)\,d\phi\,d\theta \end{pmatrix}$$

$$= \begin{pmatrix} 0 \\ 0 \\ -\frac{4\pi}{3} \end{pmatrix}. \qquad\qquad \blacksquare$$

Suppose that

(a) S is an oriented surface,

(b) $\hat{n}(\vec{r})$ denotes the unit vector that is normal to S at $\vec{r}$ and points in the positive direction of S, and

(c) $\rho(\vec{r})$ is a nonnegative real valued function defined on S.

The function $\rho(\vec{r})$ is called a *pressure function*. $\vec{p}(\vec{r}) = -\rho(\vec{r})\hat{n}(\vec{r})$ can be thought of as the force per unit area exerted on the surface by a fluid. In the mks system, the units of the pressure function are Pascals, which are abbreviated by Pa. One Pascal is one Newton/meter^2, and the force due to pressure is exerted against the surface on its positive side. The equations we establish assume that the surface containing the fluid is oriented so that its positive side points into the fluid. The vector $\vec{P} = \iint_S \vec{p}(\vec{r})dS$ is the total force exerted on S due to the pressure ρ. A little algebra will give this force as a more familiar expression. Recall that if $\vec{h}(u,v)$ is a parametrization of S, then the unit vector normal to the surface is

$$\hat{n}(\vec{h}(u,v)) = \frac{\frac{\partial \vec{h}(u,v)}{\partial u} \times \frac{\partial \vec{h}(u,v)}{\partial v}}{\left\| \frac{\partial \vec{h}(u,v)}{\partial u} \times \frac{\partial \vec{h}(u,v)}{\partial v} \right\|}.$$

$$\vec{P} = \iint_S -\rho(\vec{r})\hat{n}(\vec{r})\,dS$$

$$= -\iint_R \rho(\vec{h}(u,v)) \left(\frac{\frac{\partial \vec{h}(u,v)}{\partial u} \times \frac{\partial \vec{h}(u,v)}{\partial v}}{\left\| \frac{\partial \vec{h}(u,v)}{\partial u} \times \frac{\partial \vec{h}(u,v)}{\partial v} \right\|} \right) \left\| \frac{\partial \vec{h}(u,v)}{\partial u} \times \frac{\partial \vec{h}(u,v)}{\partial v} \right\| dA$$

$$= -\iint_R \rho(\vec{h}(u,v)) \left(\frac{\partial \vec{h}(u,v)}{\partial u} \times \frac{\partial \vec{h}(u,v)}{\partial v} \right) dA$$

$$= -\iint_S \rho(\vec{r})d\vec{S}$$

In physics, you learn that the pressure function in a fluid (either a gas or a liquid) can be quite complicated. However, if the fluid is static and the temperature is constant, then the pressure and density are functions of depth. Thus the pressure function can be written $\rho(x, y, z) = \rho(z)$ and the density function is $\lambda(z)$. In this ideal situation, the density and pressure functions satisfy the relation:

$$\frac{d\rho}{dz} = -g\lambda(z), \tag{1}$$

where g is the acceleration due to gravity. Let (x_0, y_0, z_0) be a reference point in the fluid and let $\rho_0 = \rho(x_0, y_0, z_0)$. If the density function is constant ($\lambda(z) = \lambda$), we say that the fluid is *incompressible* and Equation (1) becomes

$$\rho'(z) = -g\lambda$$

or

$$\rho(z) = -g\lambda z + K. \tag{2}$$

EXAMPLE 9: A dam is shaped so that the graph of $z = y^2$, $y \geq 0$, in xyz–space is the face of the dam, as in Figure 11. Given that the dam is 5m wide and and the water is 3m deep, find the net force on the dam due to water pressure.

SOLUTION: We assume that water is an incompressible fluid. Recall that the density of water is 1×10^3 kg/m^3. The atmospheric pressure at sea level is $\lambda = 1 \times 10^5$ N/m^2.

$$\rho(z) = -g\lambda z + K = -9.8 \times 10^3 z + K.$$

We have placed our origin at the base of the dam. Therefore, the pressure function at the top of the water is generated by the atmosphere, and $\rho(3) = -9.8 \times 10^3 \times 3 + K = -29.4 \times 10^3 + K = 1 \times 10^5$. This gives us that $K = 1.3 \times 10^5$ and

$$\rho(z) = -9.8 \times 10^3 z + 1.3 \times 10^5.$$

We parametrize the surface of the dam by $h(x, y) = (x, y, y^2)$, $0 \leq x \leq 5$, $0 \leq y \leq \sqrt{3}$. Note that as x increases, $h(x, y)$ moves in the positive x–direction, and as y increases, $h(x, y)$ moves in the positive y–direction and up. Thus the pressure due to the water is against the positive side of the surface of the dam.

We have

$$\rho(\vec{h}(x, y)) = -9.8 \times 10^3 y^2 + 1.3 \times 10^5.$$

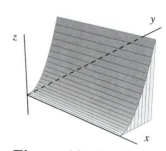

Figure 11. *The dam for Example 9.*

For ease of computation, we work with

$$\rho(\vec{h}(x,y)) = Cy^2 + K$$

and substitute the values for C and K after the calculations are done. The force due to water pressure acting on the dam is

$$
\begin{aligned}
\vec{P} &= \iint_S -\rho(\vec{r})\,d\vec{S} = -\int_0^{\sqrt{3}}\int_0^5 (Cy^2 + K)\left(\frac{\partial\vec{h}}{\partial x}\times\frac{\partial\vec{h}}{\partial x}\right)dx\,dy \\[2mm]
&= -\int_0^{\sqrt{3}}\int_0^5 (Cy^2 + K)\,(0, -2y, 1)\,dx\,dy \\[2mm]
&= -\begin{pmatrix} 0 \\ -2\int_0^{\sqrt{3}}\int_0^5 (Cy^3 + Ky)\,dx\,dy \\ \int_0^{\sqrt{3}}\int_0^5 (Cy^2 + K)\,dx\,dy \end{pmatrix} \\[2mm]
&= -5\begin{pmatrix} 0 \\ -2\int_0^{\sqrt{3}} (Cy^3 + Ky)\,dy \\ \int_0^{\sqrt{3}} (Cy^2 + K)\,dy \end{pmatrix} \\[2mm]
&= -5\begin{pmatrix} 0 \\ -9C/2 - 3K \\ \sqrt{3}(C + K) \end{pmatrix} \\[2mm]
&= -5\begin{pmatrix} 0 \\ (-9)(-9.8)(10^3/2) - (3)(1.3)(10^5) \\ \sqrt{3}(-9.8)(10^3) + (1.3)(10^5)) \end{pmatrix} \\[2mm]
&= \begin{pmatrix} 0 \\ 1.7\times 10^6 \\ 10^6 \end{pmatrix} \text{ N} \qquad\blacksquare
\end{aligned}
$$

EXERCISES 15.1

In Exercises 1–3, P is the parallelogram with adjacent sides the vectors $\vec{A}$ and $\vec{B}$ drawn emanating from $\vec{C}$. Find parametrizations $\vec{h}_1$ and $\vec{h}_2$ for P such that the orientation for P determined by $\vec{h}_1$ points in the direction $\vec{A} \times \vec{B}$, and the orientation determined by $\vec{h}_2$ points in the direction of $\vec{B} \times \vec{A}$.

1. $\vec{A} = (1, 2, 3), \vec{B} = (1, 1, 1)$, and $\vec{C} = (1, 0, 0)$.

2. $\vec{A} = (1, -1, 3), \vec{B} = (-1, 2, 2)$, and $\vec{C} = (1, -1, 0)$.

3. $\vec{A} = (-1, 4, 0), \vec{B} = (1, 0, 1)$, and $\vec{C} = (1, 0, 9)$.

4. Calculate $\iint_P \vec{F} \cdot d\vec{S}$, where P is the parallelogram of Exercise 1 with the orientation determined by $\vec{h}_1$ and $\vec{F}(x, y, z) = (x, y, z)$.

5. Calculate $\iint_P \vec{F} \cdot d\vec{S}$, where P is the parallelogram of Exercise 1 with the orientation determined by $\vec{h}_2$ and $\vec{F}(x, y, z) = (x, y, z)$.

6. Calculate $\iint_P \vec{F} \cdot d\vec{S}$, where P is the parallelogram of Exercise 2 with the orientation determined by $\vec{h}_2$ and $\vec{F}(x, y, z) = (xy, y + x, x + z)$.

7. Calculate $\iint_P \vec{F} \cdot d\vec{S}$, where P is the parallelogram of Exercise 2 with the orientation determined by $\vec{h}_1$ and $\vec{F}(x, y, z) = (xy, y + x, x + z)$.

8. Find a parametrization of the upper half of the unit sphere, centered at the origin, with orientation pointed toward the origin. Calculate $\iint_S \vec{F} \cdot d\vec{S}$, where $\vec{F}(x, y, z) = (2, 3, -1)$.

9. Find a parametrization of the upper half of the unit sphere, centered at the origin, with orientation pointed away from the origin. Calculate $\iint_S \vec{F} \cdot d\vec{S}$, where $\vec{F}(x, y, z) = (x, y, z)$.

10. Let L be the line segment from $(0, 0, 0)$ to $(1, 1, 0)$, and let S be the surface obtained by rotating L about the y–axis, with orientation pointing away from the y–axis. Calculate $\iint_S \vec{F} \cdot d\vec{S}$, where $\vec{F}(x, y, z) = (2, 3, -1)$.

11. Parametrize that portion of the paraboloid $z = x^2 + y^2$ with $z \leq 4$ so that the surface is oriented away from the z–axis, and calculate $\iint_P \vec{F} \cdot d\vec{S}$, where $\vec{F}(x, y, z) = (0, x^2 z, x^2 + y^2 + z)$.

In Exercises 12–14, use the geometry of the problem to compute $\iint_P \vec{F} \cdot d\vec{S}$.

12. P is the sphere of radius R centered at the origin, with orientation pointing away from the center, and $\vec{F}(\vec{r}) = \frac{k\vec{r}}{\|\vec{r}\|^3}$.

13. P is the unit cube centered at the origin, with orientation pointing away from the region bounded by P, and $\vec{F}(\vec{r}) = \vec{c}$.

14. P is the cylindrical surface $x^2 + z^2 = 1, -1 \leq y \leq 1$, with orientation pointing away from the y–axis, and $\vec{F}(x, y, z) = \frac{(x, 0, z)}{x^2 + z^2}$.

15. Suppose that $\vec{F}$ is a constant vector field and that P is the surface of a parallelepiped. What must be true about the flux of $\vec{F}$ through P?

16. The graph of a differentiable function $f(x, y)$ is parametrized by $\vec{h}(x, y) = (x, y, f(x, y))$. Show that the orientation always tends to point up. How would you obtain a parametrization that orients the surface in a direction that tends to point down?

17. The function $\vec{h}(x, \theta) = (x, f(x) \cos(\theta), f(x) \sin(\theta))$ rotates the graph of f about the x–axis. Does the orientation point away from the x–axis?

18. The function $\vec{h}(x, \theta) = (x \cos(\theta), f(x), x \sin(\theta))$ rotates the graph of f about the y–axis. Does the orientation point away from the y–axis?

19. The function $\vec{h}(x, \theta) = (x \sin(\theta), f(x), x \cos(\theta))$ rotates the graph of f about the y–axis. Does the orientation point away from the y–axis?

In Exercises 20–23, calculate $\iint_S \phi d\vec{S}$.

20. S is the parallelogram with adjacent edges $\vec{A} = (1, 1, 1)$ and $\vec{B} = (0, -1, 2)$ drawn emanating from $\vec{C} = (-1, 1, 2)$, oriented in the direction of $\vec{A} \times \vec{B}$ and $\phi(x, y, z) = x$.

21. S is the parallelogram with adjacent edges $\vec{A} = (1, 0, 1)$ and $\vec{B} = (0, -1, 2)$ drawn emanating from $\vec{C} = (0, 1, 0)$, oriented in the direction of $\vec{B} \times \vec{A}$ and $\phi(x, y, z) = x + y + z$.

22. S is cylinder $x^2 + y^2 = 1$, $-1 \leq z \leq 1$, oriented away from the z–axis and $\phi(x, y, z) = z$.

23. S is cylinder $(x - 1)^2 + y^2 = 1$, $0 \leq z \leq 1$, oriented away from the z–axis and $\phi(x, y, z) = z$.

24. A dam is shaped so that the graph of $z = 3y$, $y \geq 0$, in xyz–space is the face of the dam. The magnitude of the pressure of water on the surface of the dam is $p(\vec{r}) = kd + c$, where d is the depth of the water at $\vec{r}$. Given that the dam is 25m wide and and the water is 10m deep, find the force on the dam due to the pressure.

25. A dam is shaped so that the graph of $y = 100z - zx^2$, $0 \leq z \leq 100$, $-10 \leq x \leq 10$, is the face of the dam, as illustrated here. The magnitude of the pressure of water on the surface of the dam is $p(\vec{r}) = kd + c$, where d is the depth of the water at $\vec{r}$. Find the force on the dam due to the pressure.

Figure 12.
Illustration for Exercise 25.

15.2   Gaussian Surfaces

Some surfaces are easily described in terms of a single parametrization. The parametrization $\vec{h}(s, t) = (s \cos t, s \sin t)$ for $0 \leq s \leq 1$, $0 \leq t \leq \pi$ describes the upper half of the unit disc. We can easily orient this disc using the techniques of the previous section. However, a cube is impossible to parametrize by a single parametrization, due to its corners and edges. We remedy this problem by constructing the cube out of six squares. We can parametrize each square, and can then "glue" these parametrizations together. We must be careful that the orientation of each square is compatible with its neighbors. To this end we present the following definition.

Definition: Gaussian Surfaces

Suppose that $\{S_1, S_2, \ldots, S_n\}$ is a collection of smooth oriented surfaces in $\mathbb{R}^3$ such that:

(a) $S_1 \cup S_2 \cup \cdots \cup S_n$ forms the boundary for a bounded and connected volume in $\mathbb{R}^3$.

(b) If S_i intersects S_j ($i \neq j$), then $S_i \cap S_j$ is a subset of the boundary of S_i and the boundary of S_j.

(c) The orientation of each S_i points away from the volume bounded by S.

Then $S = S_1 \cup S_2 \cup \cdots \cup S_n$ is called a *Gaussian surface*.

The sphere, which is parametrized by

$$\vec{h}(\phi, \theta) = \begin{pmatrix} r \sin \phi \cos \theta \\ r \sin \phi \sin \theta \\ r \cos \phi \end{pmatrix} \quad 0 \le \phi \le \pi, 0 \le \theta \le 2\pi,$$

is an example of a Gaussian surface, which we already have at hand. ∎

EXAMPLE 1: The surface of the cube $\{(x, y, z) \mid 0 \le x \le 1, 0 \le y \le 1, 0 \le z \le 1\}$ can be expressed as the union of six oriented surfaces (see Figure 1). ∎

Figure 1. *The six faces of the cube:*

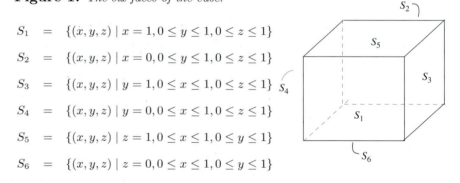

$$S_1 = \{(x, y, z) \mid x = 1, 0 \le y \le 1, 0 \le z \le 1\}$$

$$S_2 = \{(x, y, z) \mid x = 0, 0 \le y \le 1, 0 \le z \le 1\}$$

$$S_3 = \{(x, y, z) \mid y = 1, 0 \le x \le 1, 0 \le z \le 1\}$$

$$S_4 = \{(x, y, z) \mid y = 0, 0 \le x \le 1, 0 \le z \le 1\}$$

$$S_5 = \{(x, y, z) \mid z = 1, 0 \le x \le 1, 0 \le y \le 1\}$$

$$S_6 = \{(x, y, z) \mid z = 0, 0 \le x \le 1, 0 \le y \le 1\}$$

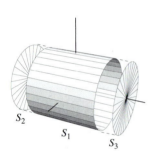

Figure 2. *S_1 is the lateral surface and S_2 and S_3 are the ends of the cylinder.*

EXAMPLE 2: The surface of the cylinder, $\{(x, y, z) \mid x^2 + z^2 \le 1, -1 \le y \le 1\}$, can be expressed as the union of the three oriented surfaces S_1, S_2, and S_3, where S_1 is the lateral surface and S_2 and S_3 are the ends of the cylinder. (See Figure 2). ∎

EXAMPLE 3: Let P be the parallelepiped with adjacent edges $\vec{A}$, $\vec{B}$, and $\vec{C}$ drawn emanating from $\vec{D}$, as illustrated in Figure 3.a below.

For this to be a Gaussian surface, each face must be oriented to point away from the enclosed volume.

Let f_B be the "bottom" face of P, the face that has the vectors $\vec{A}$ and $\vec{B}$ drawn emanating from $\vec{D}$, as illustrated in Figure 3.b. f_B can be parametrized by

$$\vec{h}(s, t) = s\vec{A} + t\vec{B} + \vec{D}, \quad \left\{ \begin{array}{c} 0 \le s \le 1 \\ 0 \le t \le 1 \end{array} \right\}.$$

Since $\frac{\partial \vec{h}}{\partial s} \times \frac{\partial \vec{h}}{\partial t} = \vec{A} \times \vec{B}$, $\vec{h}$ orients f_B in the direction of $\vec{A} \times \vec{B}$. If $(\vec{A} \times \vec{B}) \cdot \vec{C}$ is negative, then h orients f_B away from the interior of the parallelepiped. If $(\vec{A} \times \vec{B}) \cdot \vec{C}$ is positive, then we simply reverse the rolls of s and t. That is, we parametrize f_B by

$$\vec{h}(s,t) = t\vec{A} + s\vec{B} + \vec{D}, \quad \left\{ \begin{array}{c} 0 \leq s \leq 1 \\ 0 \leq t \leq 1 \end{array} \right\}.$$

Now $\frac{\partial \vec{h}}{\partial s} \times \frac{\partial \vec{h}}{\partial t} = \vec{B} \times \vec{A}$, so $\vec{h}$ orients f_B in the direction of $\vec{B} \times \vec{A}$.

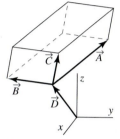

Figure 3.a *The parallelepiped P with vectors $\vec{A}$, $\vec{B}$ and $\vec{C}$ drawn emanating from $\vec{D}$.*

Figure 3.b *The face of P that has the vectors $\vec{A}$ and $\vec{B}$ drawn emanating from $\vec{D}$.*

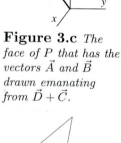

Figure 3.c *The face of P that has the vectors $\vec{A}$ and $\vec{B}$ drawn emanating from $\vec{D} + \vec{C}$.*

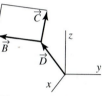

Figure 3.d *The face of P that has the vectors $\vec{C}$ and $\vec{B}$ drawn emanating from $\vec{D}$.*

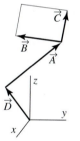

Figure 3.e *The face of P that has the vectors $\vec{C}$ and $\vec{B}$ drawn emanating from $\vec{D} + \vec{D}$.*

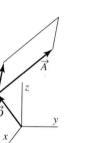

Figure 3.f *The face of P that has the vectors $\vec{A}$ and $\vec{C}$ drawn emanating from $\vec{D}$.*

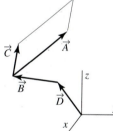

Figure 3.g *The face of P that has the vectors $\vec{A}$ and $\vec{C}$ drawn emanating from $\vec{D} + \vec{B}$.*

In the same way, let f_T be the "top" face of P, the face that has the vectors $\vec{A}$ and $\vec{B}$ drawn emanating from $\vec{C} + \vec{D}$, as illustrated in Figure 3.c. We want to orient f_T in the direction of $\vec{A} \times \vec{B}$ if $(\vec{A} \times \vec{B}) \cdot \vec{C}$ is positive. In this case, we may parametrize f_T by

$$\vec{h}(s,t) = s\vec{A} + t\vec{B} + \vec{C} + \vec{D}, \quad \left\{ \begin{array}{c} 0 \leq s \leq 1 \\ 0 \leq t \leq 1 \end{array} \right\}.$$

Otherwise, we want the orientation in the direction of $\vec{B} \times \vec{A}$. In either case, f_T is oriented in the opposite direction from the orientation of f_B.

Let f_L be the "left" face of P, as illustrated in Figure 3.g. We want f_L oriented in the direction of $\vec{C} \times \vec{B}$ if $(\vec{C} \times \vec{B}) \cdot \vec{A}$ is negative. Otherwise, we orient f_L in the direction of $\vec{B} \times \vec{C}$. Similarly, the orientation for f_R (illustrated in Figure 3.fb) is opposite the orientation of f_L. We continue this process for the "front" face and "back" face.

■

If $S = S_1 \cup S_2 \cup \cdots \cup S_n$ is a Gaussian surface and $\vec{f}$ is a continuous vector field, then

$$\iint_S \vec{f} \cdot d\vec{S} = \iint_{S_1} \vec{f} \cdot d\vec{S} + \iint_{S_2} \vec{f} \cdot d\vec{S} + \cdots + \iint_{S_n} \vec{f} \cdot d\vec{S}.$$

In Example 1, $\iint_S \vec{f} \cdot d\vec{S}$ would be computed by integrating $\vec{f}$ over all six faces and adding the results together. In Example 2, $\iint_S \vec{f} \cdot d\vec{S}$ would be computed by integrating $\vec{f}$ over the lateral surface and each of its ends and adding the three integrals together. In Example 3, $\iint_S \vec{f} \cdot d\vec{S}$ would be computed by integrating $\vec{f}$ over all six faces of the parallelepiped and adding the results together.

EXERCISES 15.2

1. According to Example 7 of Section 14.2,

$$\vec{h}(t, \theta) = \begin{pmatrix} (\cos(t) + 2) \cos(\theta) \\ \sin(t) + 2 \\ (\cos(t) + 2) \sin(\theta) \end{pmatrix},$$

$$0 \le t \le 2\pi,\ 0 \le \theta \le \pi,$$

parametrizes the surface obtained by rotating the unit circle centered at $(2, 2)$ about the y–axis. In which direction (in or out of the bounded solid) does the orientation of the surface point?

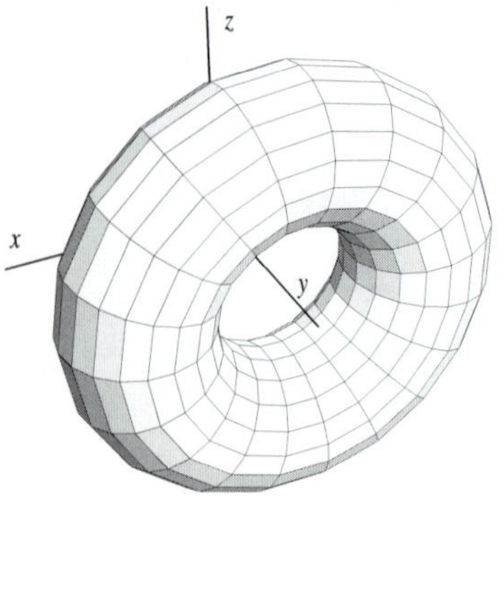

Figure 4.
Illustration for Exercise 1.

2. The function

$$\vec{h}(t, \theta) = \begin{pmatrix} (\cos(-t) + 2) \cos(\theta) \\ \sin(-t) + 2 \\ (\cos(-t) + 2) \sin(\theta) \end{pmatrix},$$

$$0 \le t \le 2\pi,\ 0 \le \theta \le \pi,$$

also parametrizes the surface obtained by rotating the unit circle centered at $(2, 2)$ about the y–axis. Explain the difference between this parametrization and the one of Exercise 1. In which direction (in or out of the bounded solid) does the orientation of the surface point?

3. Another parametrization for the surface obtained by rotating the unit circle centered at $(2, 2)$ about the y–axis is given by

$$\vec{h}(t, \theta) = \begin{pmatrix} (\cos(t) + 2) \cos(-\theta) \\ \sin(t) + 2 \\ (\cos(t) + 2) \sin(-\theta) \end{pmatrix},\quad \begin{array}{l} 0 \le t \le 2\pi, \\ 0 \le \theta \le \pi. \end{array}$$

Explain the difference between this parametrization and the one of Exercise 1. In which direction (in or out of the bounded solid) does the orientation of the surface point?

4. A simple closed curve C is parametrized by $\vec{r}(t) = (x(t), y(t))$ so that when viewed from the positive z–direction, the point $\vec{r}(t)$ moves in a counterclockwise direction, as illustrated below. $\vec{h}(t, \theta) = (x(t)\cos(\theta), y(t), x(t)\sin(\theta))$ rotates C about the y–axis, as illustrated. Is this a Gaussian surface?

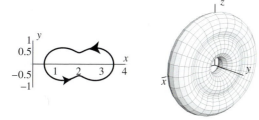

Figure 5. *Illustration for Exercise 4.*

The curve C of Exercise 4 is reparametrized by $\vec{r}_1(t) = (x_1(t), y_1(t))$ so that the direction of motion around C is reversed. See Figure 6. In Exercises 5 and 6, determine the orientation of the surface determined by the given parametrization.

5. $\vec{h}(t, \theta) = (x(t)\cos(\theta), y(t), x(t)\sin(\theta))$.

6. $\vec{h}(t, \theta) = (x(t)\cos(-\theta), y(t), x(t)\sin(-\theta))$.

Figure 6.
Illustration for Exercises 5 and 6.

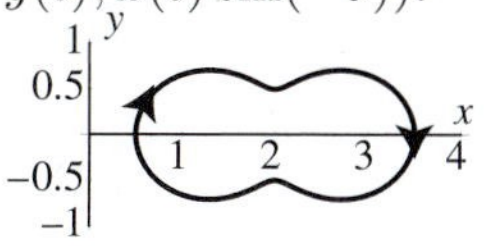

7. Let C be a simple closed curve parametrized by $\vec{r}(t) = (x(t), y(t))$, $a \le t \le b$. If C does not intersect the x–axis, then a surface S for a simple solid is obtained by rotating C about the x–axis. A parametrization for S is

$$\vec{h}(t, \theta) = \begin{pmatrix} x(t) \\ y(t)\cos(\theta) \\ y(t)\sin(\theta) \end{pmatrix}, \quad \begin{array}{l} a \le t \le b, \\ 0 \le \theta \le 2\pi. \end{array}$$

Given that the curve C is oriented as below, determine whether the surface is a Gaussian surface.

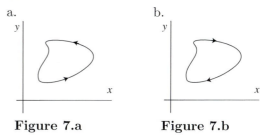

a.   b.

Figure 7.a   **Figure 7.b**

8. Find parametrizations for each of the sides of the cube $\{(x, y, z) \mid -1 \le x \le 1, -1 \le y \le 1, -1 \le z \le 1\}$ so that the orientation of each side points away from the interior of the cube.

9. Find parametrizations for each of the sides of the cube $\{(x, y, z) \mid -3 \le x \le 1, 0 \le y \le 4, -10 \le z \le 10\}$ so that the orientation of each side points away from the interior of the cube.

10. Find parametrizations for the smooth surfaces S_1, S_2, and S_3 of Example 2 so that the orientation of each of the surfaces points away from the interior of the cylinder.

11. Find parametrizations for the smooth surfaces S_1, S_2, and S_3 of Example 2 so that the orientation of each of the surfaces points into the interior of the cylinder.

In Exercises 12 and 13, P is the surface of the parallelepiped with adjacent edges the vectors $\vec{A}, \vec{B}$, and $\vec{C}$ all drawn emanating from the point $\vec{D}$. Find parametrizations for each of the faces of P that orient the faces away from the region bounded by P.

12. $\vec{A} = (0, 1, 1), \vec{B} = (1, 1, 1), \vec{C} = (1, 1, 0)$, and $\vec{D} = (1, 1, 1)$.

13. $\vec{A} = (1, 2, 0), \vec{B} = (1, -1, 1), \vec{C} = (0, 1, 2)$, and $\vec{D} = (-1, 2, 1)$.

14. Let $\vec{f}(x, y, z) = (1, 1, 1)$. Calculate $\iint_P \vec{f} \cdot d\vec{S}$, where P is the surface of the parallelepiped of Exercise 12.

15. Let $\vec{f}(x, y, z) = (-1, 2, 3)$. Calculate $\iint_P \vec{f} \cdot d\vec{S}$, where P is the surface of the parallelepiped of Exercise 13.

16. Give a geometric argument that if $\vec{f}$ is a constant vector field and P is the surface of a parallelepiped, then $\iint_P \vec{f} \cdot d\vec{S} = 0$.

17. Let $\vec{f}(x, y, z) = (1, 1, 1)$ and let S be the surface of the unit sphere, oriented so that it is a Gaussian surface. Compute $\iint_S \vec{f} \cdot d\vec{S}$. Give a geometric reason for your answer.

In Exercises 18-27, assume that the water pressure at depth z is $\rho(z) = kz + c$.

18. A rectangular box with height h, width w and depth d is submerged in water so that its top is r m deep. Its orientation in the water is as illustrated in Figure 8. What is the force on the box due to the pressure?

Figure 8. *Figure for Exercise 18*

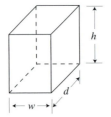

19. A cylindrical can (with a top and a bottom) with height h, and radius r is submerged in water so that its top is d m deep and its axis is vertical. Find the force on the can due to the pressure.

20. A cylindrical can (with a top and a bottom) with height h, and radius r is submerged in water so that its axis is h m deep and its axis is horizontal. Find the resulting force on the can.

21. A spherical tank of radius r meters is submerged in water so that its center is h m deep. What is the resulting force on the tank?

In Exercises 22–30, $C(h)$ is a cylindrical container of radius R and height h immersed in a static fluid with its axis vertical (parallel to the z–axis) and its base lying in the plane $z = z_0$. Its surface $S(h)$ is oriented as a Gaussian surface. Let $S_T(h)$ be the top surface, $S_B(h)$ the bottom surface, and $S_L(h)$ be the lateral side as in Figure 8, so that $S(h) = S_T(h) \cup S_B(h) \cup S_L(h)$. $\rho(d)$ denotes the pressure at depth d.

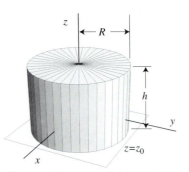

Figure 9. *The cylindrical container.*

22. Show that the force due to the pressure on the lateral side is $\vec{0}$.

23. Show that the force due to the pressure on the top is $-\pi R^2 \rho(h + z_0)\hat{k}$.

24. Show that the force due to the pressure on the bottom is $\pi R^2 \rho(z_0)\hat{k}$.

You have shown that the force due to the pressure on the container is

$$\vec{P}(h) = \pi R^2 (\rho(z_0) - \rho(h + z_0))\hat{k}.$$

Assume that the container is porous so that it is full of the same fluid that surrounds it and the pressures and densities within the container are the same as those outside the container at the same depth. Let $\lambda(z)$ denote the mass density function in Exercises 25–30.

25. Show that the weight of the fluid within the container is

$$\vec{w}(h) = -g\pi R^2 \int_{z_0}^{z_0+h} \lambda(z)\, dz\ \hat{k}.$$

26. Explain why Exercise 25 shows that
$$\frac{d\vec{w}}{dh} = -g\pi R^2 \lambda(z_0 + h)\hat{k}.$$

Assume that the container is not moving relative to the fluid. Explain why this assumption together with the results of Exercises 25 and 26 gives us the results in Exercises 27–30.

27. $\vec{P}(h) + \vec{w}(h) = \vec{0}$.

28. $\rho(z_0 + h) = \rho(z_0) - g \int_{z_0}^{z_0+h} \lambda(z)\, dz.$

29. $\rho'(z_0 + h) = -g\lambda(z_0 + h).$

30. $\rho'(z) = g\lambda(z).$ (Compare this final equation with Equation 1 of the previous section.)

15.3 Divergence

Let B be the box in xyz–space bounded by planes $x = a$, $x = b$, $y = c$, $y = d$, $z = e$, and $z = f$ (where $a < b, c < d$, and $e < f$), and let $\vec{F}$ be a vector field on a subset of $\mathbb{R}^3$ that has continuous partial derivatives on B. Suppose that the faces of B are oriented away from the interior of B so that the surface of B is a Gaussian surface. We will let S_a be the face of B lying in the plane $x = a$, S_b the face of B in the plane $x = b$, S_c the face of B in the plane $y = c$, etc. (See Figure 1.)

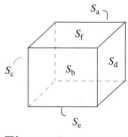

Figure 1. *The six faces of the box bounded by the planes* $x = a, x = b, y = c, y = d, z = e,$ *and* $z = f$

We can evaluate the integral $\iint_{\text{Surface of } B} \vec{F} \cdot d\vec{S}$ by evaluating the associated integrals over the six faces.

$$\iint_{\text{Surface of } B} \vec{F} \cdot d\vec{S}$$

$$= \iint_{S_a} \vec{F} \cdot d\vec{S} + \iint_{S_b} \vec{F} \cdot d\vec{S} + \iint_{S_c} \vec{F} \cdot d\vec{S} + \iint_{S_d} \vec{F} \cdot d\vec{S}$$

$$+ \iint_{S_e} \vec{F} \cdot d\vec{S} + \iint_{S_f} \vec{F} \cdot d\vec{S}.$$

We first evaluate $\iint_{S_a} \vec{F} \cdot d\vec{S} + \iint_{S_b} \vec{F} \cdot d\vec{S}$.

Let $\vec{h}_a(v, u) = (a, u, v)$ and $\vec{h}_b(u, v) = (b, u, v), c \le u \le d$, and $e \le v \le f$. Then $\vec{h}_a$ is a parametrization for S_a and $\vec{h}_b$ is a parametrization for S_b. Notice that we are calling v the first variable for h_a and u the first variable for h_b. We see below that this will orient both of the surfaces away from the interior the box. The partial derivative of $\vec{h}_a$ with respect to its first variable is

$$\frac{\partial \vec{h}_a(v, u)}{\partial v} = (0, 0, 1),$$

and the partial of h_a with respect to its second variable is

$$\frac{\partial \vec{h}_a(u, v)}{\partial u} = (0, 1, 0),$$

so

$$\frac{\partial \vec{h}_a}{\partial v} \times \frac{\partial \vec{h}_a}{\partial u} = -(1, 0, 0).$$

Similarly, the partials of $\vec{h}_b$ with respect to its first variable and its seond variable are

$$\frac{\partial \vec{h}_b(u, v)}{\partial u} = (0, 1, 0), \quad \text{and} \quad \frac{\partial \vec{h}_b(u, v)}{\partial v} = (0, 0, 1).$$

Thus

$$\frac{\partial \vec{h}_b}{\partial u} \times \frac{\partial \vec{h}_b}{\partial v} = (1, 0, 0).$$

The orientations of S_a and S_b due to $\vec{h}_a$ and $\vec{h}_b$ point away from the interior of B, as promised.

$$\iint_{S_a} \vec{F} \cdot d\vec{S} + \iint_{S_b} \vec{F} \cdot d\vec{S}$$

$$= \int_e^f \int_c^d \left[F_x(\vec{h}_a(v, u)), F_y(\vec{h}_a(v, u)), F_z(\vec{h}_a(v, u)) \right] \cdot (-1, 0, 0) \; du \, dv$$

$$+ \int_e^f \int_c^d \left[F_x(\vec{h}_b(u, v)), F_y(\vec{h}_b(u, v)), F_z(\vec{h}_b(u, v)) \right] \cdot (1, 0, 0) \; du \, dv$$

$$= \int_e^f \int_c^d -F_x(a, u, v) \; du\, dv + \int_e^f \int_c^d F_x(b, u, v) \; du \, dv$$

$$= \int_e^f \int_c^d [F_x(b, u, v) - F_x(a, u, v)] \; du \, dv$$

$$= \int_e^f \int_c^d \int_a^b \frac{\partial F_x(w, u, v)}{\partial w} \; dw \, du \, dv$$

$$= \iiint_V \frac{\partial F_x(w, u, v)}{\partial w} \; dV = \iiint_V \frac{\partial F_x(x, y, z)}{\partial x} \; dV$$

Similarly, we can show that

$$\iint_{S_c} \vec{F} \cdot d\vec{S} + \iint_{S_d} \vec{F} \cdot d\vec{S} = \iiint_V \frac{\partial F_y(x, y, z)}{\partial y} \; dV,$$

and

$$\iint_{S_e} \vec{F} \cdot d\vec{S} + \iint_{S_f} \vec{F} \cdot d\vec{S} = \iiint_V \frac{\partial F_z(x, y, z)}{\partial z} \; dV.$$

Thus we have that

$$\iint_{\text{Surface of } B} \vec{F} \cdot d\vec{S}$$

$$= \iiint_V \frac{\partial \vec{F}_x(x,y,z)}{\partial x} \, dV + \iiint_V \frac{\partial \vec{F}_y(x,y,z)}{\partial y} \, dV + \iiint_V \frac{\partial \vec{F}_z(x,y,z)}{\partial z} \, dV$$

$$= \iiint_V \left[\frac{\partial \vec{F}_x(x,y,z)}{\partial x} + \frac{\partial \vec{F}_y(x,y,z)}{\partial y} + \frac{\partial \vec{F}_z(x,y,z)}{\partial z} \right] \, dV.$$

We have established a special case of Gauss' Theorem.

Theorem 1 (Gauss' Theorem) *Suppose that $\vec{F}$ is a function from a subset of $\mathbb{R}^3$ into $\mathbb{R}^3$ such that:*

(a) the domain contains the solid V;

(b) the partial derivatives of $\vec{F}$ are continuous on V; and

(c) the surface of V is a Gaussian surface (i.e., the surface of V is oriented away from the solid V).

Then

$$\iint_{\text{Surface of } V} \vec{F} \cdot d\vec{S}$$

$$= \iiint_V \left[\frac{\partial \vec{F}_x(x,y,z)}{\partial x} + \frac{\partial \vec{F}_y(x,y,z)}{\partial y} + \frac{\partial \vec{F}_z(x,y,z)}{\partial z} \right] \, dV.$$

Notice that the integral on the left in Gauss' Theorem is the flux of $\vec{F}$ through S. Gauss' Theorem states that the flux can be calculated by integrating specific derivatives over the entire volume V bounded by S.

Theorem 1 leads us to the definition of the *divergence* of a force field.

Definition: $\text{Div}(\vec{F})$

If $\vec{F}(x,y,z) = (F_x(x,y,z), F_y(x,y,z), F_z(x,y,z))$, then the *divergence of $\vec{F}$* is defined to be

$$\text{Div}(\vec{F}(x,y,z)) = \frac{\partial F_x(x,y,z)}{\partial x} + \frac{\partial F_y(x,y,z)}{\partial y} + \frac{\partial F_z(x,y,z)}{\partial z}.$$

Notation: If we let $\vec{\nabla} = \left(\frac{\partial}{\partial x}, \frac{\partial}{\partial y}, \frac{\partial}{\partial z} \right)$, then we can write

$$\text{Div}(\vec{F}) = \vec{\nabla} \cdot \vec{F} = \left(\frac{\partial}{\partial x}, \frac{\partial}{\partial y}, \frac{\partial}{\partial z} \right) \cdot (F_x, F_y, F_z) = \left(\frac{\partial F_x}{\partial x} + \frac{\partial F_y}{\partial y} + \frac{\partial F_z}{\partial z} \right).$$

Writing $\text{Div}(\vec{F})$ as $\vec{\nabla} \cdot \vec{F}$ is commonly done, easy to remember and consistent with our using $\vec{\nabla}\phi$ to denote the gradiant of ϕ. With this notation, Theorem 1 becomes

$$\iint_{\text{Surface of } V} \vec{F} \cdot d\vec{S} = \iiint_V \vec{\nabla} \cdot \vec{F} \, dV.$$

EXAMPLE 1: Let V be the solid bounded by $x^2 + y^2 = 9$ and the planes $z = 0$, and $z = 1$ and let S be the boundary of V. Let $\vec{F}$ be the vector field defined by

$$\vec{F}(x, y, z) = \begin{pmatrix} x^3 - z^{|y|} \\ y^3 + x \ln(2 + z) \\ z \end{pmatrix}.$$

Calculate $\iint_S \vec{F} \cdot d\vec{S}$.

SOLUTION: $\text{Div}(\vec{F}(x, y, z)) = (3x^2 + 3y^2 + 1)$, and thus

$$\iint_S \vec{F} \cdot d\vec{S} = \iiint_V \vec{\nabla} \cdot \vec{F} \, dV = \iiint_V (3x^2 + 3y^2 + 1) \, dV.$$

If we let D be the disk in xy–space centered at the origin with radius 3, then

$$\iiint_V (3x^2 + 3y^2 + 1) \, dV = \iint_D \int_0^1 (3x^2 + 3y^2 + 1) \, dz \, dS$$

$$= \iint_D (3x^2 + 3y^2 + 1) dS.$$

Using polar coordinates to parametrize D, we get

$$\iiint_V (3x^2 + 3y^2 + 1) \, dV = \int_0^3 \int_0^{2\pi} (3r^2 + 1)r \, d\theta \, dr$$

$$= 2\pi \int_0^3 (3r^3 + r) dr$$

$$= \frac{261\pi}{2}. \qquad \blacksquare$$

EXAMPLE 2: Let $\vec{F}(x, y, z) = (y^3, \sin(xz), xye^{xy})$ and let S be an arbitrary Gaussian surface. Calculate the flux of $\vec{F}$ through S.

SOLUTION: $\text{Div}(\vec{F}(x, y, z)) = 0$, and so

$$\iint_S \vec{F} \cdot d\vec{S} = \iiint_{\substack{\text{Volume} \\ \text{bounded} \\ \text{by } S}} \text{Div}(\vec{F}(x, y, z))\, dV = 0.$$

∎

Some Useful Intuition: A Physical Interpretation of the Divergence

Let V_h be the box given by $a \leq x \leq a + h$, $b \leq y \leq b + h$, and $c \leq z \leq c + h$. It can be shown that

$$\text{Div}(\vec{F})\Big|_{(a,b,c)} = \lim_{h \to 0} \frac{\int_{\text{Surface of } V_h} \vec{F} \cdot d\vec{S}}{\text{Volume of } V_h}$$

$$= \lim_{h \to 0} \frac{\text{Flux of } \vec{F} \text{ through the boundary of } V_h}{\text{Volume of } V_h}.$$

Thus the divergence can be thought of as a quantification of the rate that $\vec{F}$ is expanding or contracting per unit volume at the point (a, b, c). If $\text{Div}(\vec{F})$ is positive at (a, b, c), then $\vec{F}$ is thought of as expanding at that point. If $\text{Div}(\vec{F})$ is negative at (a, b, c), then $\vec{F}$ is thought of as contracting at that point. If $\text{Div}(\vec{F}) = 0$ at (a, b, c), then $\vec{F}$ is said to be *incompressible* at (a, b, c).

A slightly different perspective is obtained if we think of the vector field $\vec{F}$ as the rate of flow of a fluid. If S is a Gaussian surface and if $\iint_S \vec{F} \cdot d\vec{S} > 0$, then the net flow is out of the surface. Thus, if $\nabla \cdot \vec{F}(\vec{r}) > 0$, then we can think of $\vec{r}$ as a source of the flow. If $\iint_S \vec{F} \cdot d\vec{S} < 0$, then the net flow is into the bounded region. And so, if $\nabla \cdot \vec{F}(\vec{r}) < 0$, then we think of $\vec{r}$ as a sink. If the vector field seems to be radiating away from $\vec{r}$, then we would expect $\nabla \cdot \vec{F}(\vec{r})$ to be positive. If the vector field seems to be radiating toward $\vec{r}$, then we would expect $\nabla \cdot \vec{F}(\vec{r})$ to be negative. If the vector field seems to be just "passing through" $\vec{r}$, then we would expect $\nabla \cdot \vec{F}(\vec{r})$ to be zero. See Figures 2.a–2.c.

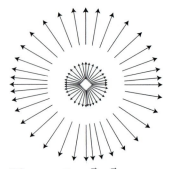

Figure 2.a $\vec{\nabla} \cdot \vec{F}$ *is positive at the center of the figure.*

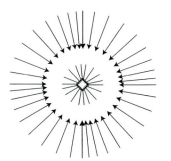

Figure 2.b $\vec{\nabla} \cdot \vec{F}$ *is negative at the center of the figure.*

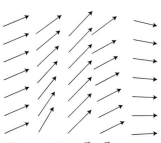

Figure 2.c $\vec{\nabla} \cdot \vec{F} = 0$.

EXERCISES 15.3

In Exercises 1–3, compute the divergence of the vector fields.

1. $\vec{F}(x, y, z) = (xyz, x^2y, z\sin(x))$.

2. $\vec{F}(x, y, z) = (x + 2^{xy}, y + 2^{xy}, z + 2^{xy})$.

3. $\vec{F}(x, y, z) = \dfrac{k(x, y, z)}{(x^2 + y^2 + z^2)^{3/2}}$.

In Exercises 4–8, find the flux of $\vec{F}$ through S using Gauss' Theorem (assume that S is oriented away from the interior of S).

4. $\vec{F}(x, y, z) = (x^2, xy, z)$, and S is the surface of the cube bounded by the planes $x = 1, x = 2, y = 0, y = 2, z = -1$, and $z = 2$.

5. $\vec{F}(x, y, z) = (xy - \sin z, xy + xz, xz)$, and S is the surface of the solid bounded by the coordinate planes, and the plane $x + y + z = 1$.

6. $\vec{F}(x, y, z) = \left(xy^2, \dfrac{yz^2}{2}, \dfrac{z^3}{6}\right)$, and S is the sphere of radius 1 centered at the origin.

7. $\vec{F}(x, y, z) = \left(xy^2, \dfrac{yz^2}{2}, \dfrac{z^3}{6}\right)$, and S is the surface of the solid bounded by the graph of $y^2 + z^2 = 1$, and the planes $x = 0$ and $x + y = 4$.

8. $\vec{F}(x, y, z) = (x - yz, y - xz, z - xy)$, and S is the unit sphere centered at the origin.

9. Show that if $\vec{F}(x, y, z) = (x + h(y, z), y + k(x, z), z + s(x, y))$, then the flux of $\vec{F}$ through an arbitrary Gaussian surface is three times the volume of the bounded solid.

10. Let $\vec{F}(r) = \dfrac{k\vec{r}}{\|\vec{r}\|^3}$ so that the magnitude of $\vec{F}$ is inversely proportional to the distance from the origin, and $\vec{F}$ is directed away from the origin. (Readers with a background in physics will recognize $\vec{F}$ as an electric field due to a point charge at the origin.) Let S be a Gaussian surface that does not contain the origin in its interior.

    a. Show that $\text{Div}(\vec{F}) = 0$.

    b. Let S be the unit sphere centered at the origin. Compute directly $\iint_S \vec{F} \cdot d\vec{S}$.

    c. Do your answers to a. and b. contradict Gauss' Theorem? Explain.

11. Suppose that S is the Gaussian surface bounding the solid D and that ϕ is a differentiable real valued function defined on D. Show that

$$\iint_S \phi \, d\vec{S} = \iiint_D \vec{\nabla}\phi \, dV.$$

12. Suppose that all of the second partial derivatives of the real valued function ϕ are defined. We define $\vec{\nabla}^2\phi = \vec{\nabla} \cdot \vec{\nabla}\phi$, which is the divergence of the gradiant of ϕ. Show that

$$\vec{\nabla}^2\phi = \dfrac{\partial^2\phi}{\partial x^2} + \dfrac{\partial^2\phi}{\partial y^2} + \dfrac{\partial^2\phi}{\partial z^2}.$$

13. Explain why the expression $\vec{\nabla}(\vec{\nabla} \cdot \phi)$ does not make sense.

14. Suppose that S is a Gaussian surface bounding the solid D and that all of the second partial derivatives of the real valued functions ϕ and ρ are defined on D. Show that

$$\iint_S \phi \vec{\nabla} \rho \, d\vec{S} = \iiint_D \phi \vec{\nabla}^2 \rho + \vec{\nabla} \phi \cdot \vec{\nabla} \rho \, dV.$$

This is called *Green's first identity*.

15. Suppose that S is a Gaussian surface bounding the solid D and that all of the second partial derivatives of the real valued functions ϕ and ρ are defined on D. Show that

$$\iint_S \left(\phi \vec{\nabla} \rho - \rho \vec{\nabla} \phi \right) d\vec{S} = \iiint_D \left(\phi \vec{\nabla}^2 \rho - \rho \vec{\nabla}^2 \phi \right) dV.$$

This is *Green's second identity*.

16. Suppose that S_1 and S_2 are Gaussian surfaces and S_2 lies in the solid bounded by S_1 as illustrated in Figures 3.a and b. Let D_1 be the solid bounded by S_1 and D_2 the solid bounded by S_2. Let D_3 be the solid lying between S_1 and S_2. Show that

$$\iiint_{D_3} \vec{\nabla} \vec{F} \, dV = \iint_{S_1} \vec{F} \cdot d\vec{S} - \iint_{S_2} \vec{F} \cdot d\vec{S}.$$

17. **Gauss's Law** Let S_2 of Exercise 16 be a sphere or radius r centered at the origin as pictured in Figure 3.b. Let $\vec{V}(\vec{r}) = \dfrac{k\vec{r}}{\|\vec{r}\|^3}$. This might be the electric potential field for a point charge at the origin as illustrated in Figure 3.c.

   a. Show that $\iint_{S_2} \vec{V} \cdot d\vec{S} = 4\pi k$.

   b. Show that $\iint_{D_3} \vec{\nabla} \cdot \vec{V} \, dV = 0$.

   c. Show that $\iint_{S_1} \vec{V} \cdot d\vec{S} = 4\pi k$.

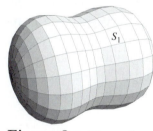

Figure 3.a *The surface S_1 for Exercise 16.*

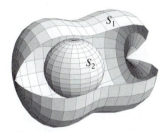

Figure 3.b *The surface S_2 embedded in the region bounded by S_1.*

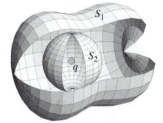

Figure 3.c *A charged particle at the center of S_2.*

15.4   Surfaces and Their Boundaries

Earlier we called a fundamental curve from $\vec{A}$ to $\vec{B}$ a simple closed curve if $\vec{A} = \vec{B}$. We want to extend this idea so that it will include boundaries of simple surfaces. Suppose that $\vec{a}_0, \vec{a}_1, \vec{a}_2, \ldots, \vec{a}_k$ are distinct points in $\mathbb{R}^n$ and that $C_1, C_2, \ldots, C_k, C_{k+1}$ are fundamental curves such that, for $1 \le i \le k$, C_i is a path from $\vec{a}_{i-1}$ to $\vec{a}_i$ and C_{k+1} is a path from $\vec{a}_k$ to $\vec{a}_0$. Assume that C_i and C_j do not intersect, except that C_i intersects C_{i-1} at $\vec{a}_i$ and C_{k+1} intersects C_1 at $\vec{a}_0$. (See Figure 1.)

The C_is form a path that a particle might follow, starting out at $\vec{a}_0$, then passing through $\vec{a}_1, \vec{a}_2, \vec{a}_3, \ldots$, until it finally returns to

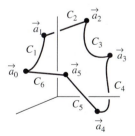

Figure 1. *The curve $C_1 \cup C_2 \cup \cdots \cup C_6$ forms a simple closed curve.*

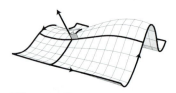

Figure 2. *If a particle is moving around βS in the counterclockwise direction (observed from the positive side of S), then the particle is moving in the positive direction.*

$\vec{a}_0$. The union of such a collection of paths is called a *simple closed curve*. We are interested in simple closed curves that are boundaries of simple surfaces.

EXAMPLE 1: Let C be the boundary of the triangle with vertices $\vec{A}, \vec{B}$, and $\vec{C}$. Then C is the union of the three line segments $[\vec{A}, \vec{B}], [\vec{B}, \vec{C}]$, and $[\vec{C}, \vec{A}]$. ∎

Notice that if C is a simple closed curve, then there are two directions that a particle can travel on C. We say that C is *oriented* if there is a "preferred (or positive) direction around C." Throughout this section, we assume that if S is an oriented surface and if βS is the boundary of S, then the orientation on S induces an orientation on βS. That is, if a particle is observed moving around βS (the boundary of S) in the counterclockwise direction (from the positive side of S), then the particle is moving in the positive direction. A particle moving in the positive direction around βS will "see" the positive surface on its left. See Figure 2.

EXAMPLE 2: Suppose that a parallelogram is determined by the vectors $\vec{A}$ and $\vec{B}$ drawn emanating from a common point. Suppose further that an ant is moving around the perimeter of the parallelogram, beginning at the point of emanation and walking along the vector $\vec{A}$. Thus the ant will move in a counterclockwise direction when viewed from the $\vec{A} \times \vec{B}$ side of the parallelogram. (See Figure 3.) However, when viewed from the $\vec{B} \times \vec{A}$ side, the ant will move in the clockwise direction. ∎

Figure 3. *The ant moves in the counterclockwise direction when viewed from the $\vec{A} \times \vec{B}$ side of the parallelogram.*

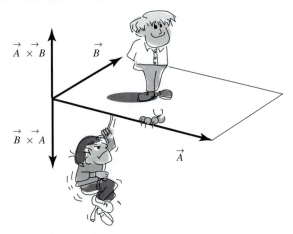

This is consistent with the "right hand rule." If you place your right hand on the surface with your "little finger side" on the surface and your fingers pointing in the positive direction of the boundary then your thumb will point in the positive direction of the surface.

Just as we use the parametrization of a surface to determine the orientation of the surface, we use the parametrization of the boundary to determine its orientation. If $\vec{r}(t)$ is the parametrization of the boundary of S, then we want $\vec{r}(t)$ to move in the positive direction around the curve as t increases.

First, we parametrize the edges of a rectangle in the plane that models a point moving counterclockwise around its boundary.

EXAMPLE 3: Let R be the rectangle $a \leq x \leq b$, $c \leq y \leq d$ as in Figure 4. Let C_1 be the line segment from $\vec{v}_1 = (a, c)$ to $\vec{v}_2 = (b, c)$, C_2 the line segment from $\vec{v}_2 = (b, c)$ to $\vec{v}_3 = (b, d)$, C_3 the line segment from $\vec{v}_3 = (b, d)$ to $\vec{v}_4 = (b, c)$, and C_4 the line segment from $\vec{v}_4 = (b, c)$ to $\vec{v}_1 = (a, c)$.

Then

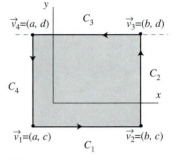

Figure 4. *The rectangle in Example 3.*

$$\vec{r}_1(t) = \vec{v}_1 + t(\vec{v}_2 - \vec{v}_1) = (a + t(b - a), c), \quad 0 \leq t \leq 1,$$
parametrizes C_1,

$$\vec{r}_2(t) = \vec{v}_2 + t(\vec{v}_3 - \vec{v}_2) = (b, c + t(d - c)), \quad 0 \leq t \leq 1,$$
parametrizes C_2,

$$\vec{r}_3(t) = \vec{v}_3 + t(\vec{v}_4 - \vec{v}_3) = (b + t(a - b), d), \quad 0 \leq t \leq 1,$$
parametrizes C_3, and

$$\vec{r}_4(t) = \vec{v}_4 + t(\vec{v}_1 - \vec{v}_4) = (a, d + t(c - a)), \quad 0 \leq t \leq 1,$$
parametrizes C_4.

EXAMPLE 4: Suppose that $f(x, y)$ is a continuously differentiable function from $\mathbb{R}^2$ into $\mathbb{R}$ and R is the rectangle bounded by the lines $x = a$, $x = b$, $y = c$, and $z = c$. Then $\vec{h}(x, y) = (x, y, f(x, y))$ parametrizes S, the graph of f over the rectangle R as in Figure 5.b. Notice that the boundary of R goes onto the boundary of the graph of f. Let C_1, C_2, C_3, and C_4 denote the sides of R as in Figure 5.a and Example 3. Then $\vec{h}(C_1) \cup \vec{h}(C_2) \cup \vec{h}(C_3) \cup \vec{h}(C_4)$ is the boundary of the S as illustrated in Figure 5.c.

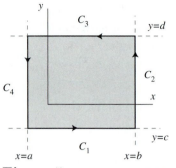

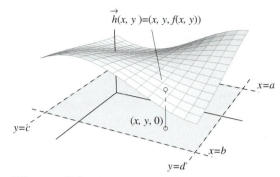

Figure 5.a *The rectangle R.*

Figure 5.b *The graph of f over the rectangle R.*

Let $\vec{r}_1$, $\vec{r}_2$, $\vec{r}_3$, and $\vec{r}_4$ be the parametrizations of C_1, C_2, C_3, and C_4, respectively as in Example 3. Then, as illustrated in Figure 5.c,

$$\vec{\rho}_1(t) = \vec{h}(\vec{r}_1(t)),\ 0 \le t \le 1 \text{ parametrizes } \vec{h}(C_1),$$

$$\vec{\rho}_2(t) = \vec{h}(\vec{r}_2(t)),\ 0 \le t \le 1 \text{ parametrizes } \vec{h}(C_2),$$

$$\vec{\rho}_3(t) = \vec{h}(\vec{r}_3(t)),\ 0 \le t \le 1 \text{ parametrizes } \vec{h}(C_3), \text{ and}$$

$$\vec{\rho}_4(t) = \vec{h}(\vec{r}_4(t)),\ 0 \le t \le 1 \text{ parametrizes } \vec{h}(C_4). \qquad \blacksquare$$

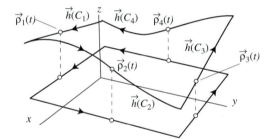

Figure 5.c *The boundary of S.*

EXAMPLE 5: Let S be the graph of $f(x,y) = x^2 + y^2$, $-1 \le x \le 1$, $-1 \le y \le 1$ as illustrated in Figure 6.b. Then $\vec{h}(x,y) = (x,y,f(x,y)$ parametrizes S. From Examples 3 and 4,

$$a + t(b - a) = 1 + t(1 - (-1)) = 1 + 2t, \text{ so}$$

$$\vec{r}_1(t) = \begin{pmatrix} -1 + 2t \\ -1 \end{pmatrix},\ 0 \le t \le 1,$$

parametrizes C_1 and

$$\vec{\rho}_1(t) = \vec{h}(\vec{r}_1(t)) = \begin{pmatrix} -1 + 2t \\ -1 \\ (-1 + 2t)^2 + 1 \end{pmatrix},\ 0 \le t \le 1,$$

parametrizes $\vec{h}(C_1)$.

$$c + t(d - c) = 1 + t(1 - (-1)) = 1 + 2t, \text{ so}$$

$$\vec{r}_2(t) = \begin{pmatrix} 1 \\ -1 + 2t \end{pmatrix}, \ 0 \le t \le 1,$$

parametrizes C_1 and

$$\vec{\rho}_2(t) = \vec{h}(\vec{r}_2(t)) = \begin{pmatrix} 1 \\ -1 + 2t \\ (-1 + 2t)^2 + 1 \end{pmatrix}, \ 0 \le t \le 1,$$

parametrizes $\vec{h}(C_2)$. C_3 and C_4 can be similarly parametrized.

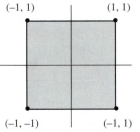

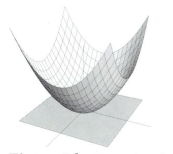

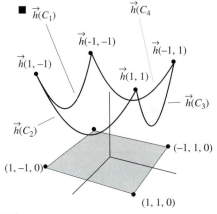

Figure 6.a *The domain of the parametrization.*

Figure 6.b *The surface S.*

Figure 6.c *The boundary of S.*

While it is always the case that if $\vec{h}$ parametrizes a surface S over the rectangle R, then the boundary of S will be a subset of the image of the boundary of R, and parts of the image of the boundary of R can be in the interior of S. You can think of these points going into a seam of S. A familiar example is the parametrization of a sphere using spherical coordinates.

EXAMPLE 6: The function

$$\vec{h}(\phi, \theta) = \begin{pmatrix} \sin(\phi)\cos(\theta) \\ \sin(\phi)\sin(\theta) \\ \cos(\phi) \end{pmatrix}, \ \begin{pmatrix} 0 \le \phi \le \pi \\ 0 \le \theta \le 2\pi \end{pmatrix}$$

parametrizes the unit sphere. The parametrization $\vec{h}$ takes all of the points on the left edge of R onto the north pole and all of the points on the right edge of R onto the south pole. The bottom and top edges are taken onto the same longitudinal arc, but in opposite directions. It is helpful to move back from the boundary and look at the images of the line segments

$\tilde{C}_1 =$ the segment from (δ, δ) to $(\pi - \delta, \delta)$,

$\tilde{C}_2 =$ the segment from $(\pi - \delta, \delta)$ to $(\pi - \delta, 2\pi - \delta)$,

$\tilde{C}_3 =$ the segment from $(\pi - \delta, 2\pi - \delta)$ to $(\delta, 2\pi - \delta)$,

and

$\tilde{C}_4 =$ the segment from $(\delta, 2\pi - \delta)$ to (δ, δ),

which form a closed curve approximating the boundary of the rectangle as illustrated in Figure 7.d. The image of this curve may be easier to trace. See Figure 7.e. ■

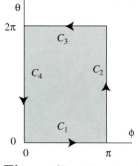

Figure 7.a *The domain of the parametrization of the sphere.*

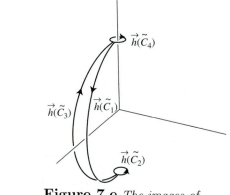

Figure 7.b *The sphere.*

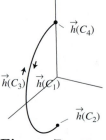

Figure 7.c *The image of the boundary of the domain of $\vec{h}$ is a seam in the sphere.*

Figure 7.d *The line segments $\tilde{C}_1$, $\tilde{C}_2$, $\tilde{C}_3$, and $\tilde{C}_4$*

Figure 7.e *The images of the line segments $\tilde{C}_1$, $\tilde{C}_2$, $\tilde{C}_3$, and $\tilde{C}_4$*

Examples 5 and 6 lead us to the following observation. Suppose that R, the rectangle bounded by the lines $x = a$, $x = b$, $y = c$, and $y = d$, is the domain of a parametrization $\vec{h}$ of the surface S as in Figure 5. Let C_1, C_2, C_3, and C_4 be the edges of R so that $C_1 \cup C_2 \cup C_3 \cup C_4$ is the boundary of R. Then $\Psi = \vec{h}(C_1 \cup C_2 \cup C_3 \cup C_4)$

contains the boundary of S. Those parts of $\vec{h}(C_1 \cup C_2 \cup C_3 \cup C_4)$ that are not in the boundary of S are in the seam of S. If $\vec{h}$ takes a part of $C_1 \cup C_2 \cup C_3 \cup C_4$ onto a curve C from points $\vec{A}$ to $\vec{B}$ and C is part of the seam of S, then it will take another part of $C_1 \cup C_2 \cup C_3 \cup C_4$ onto the same curve C, but from $\vec{B}$ to $\vec{A}$ (the reverse direction). In this case, if $\vec{F}$ is any vector function that is integrable on Ψ, then

$$\int_{\Psi} \vec{F} \cdot d\vec{r} = \int_{\text{the boundary of } S} \vec{F} \cdot d\vec{r}.$$

Let $\vec{r}_1$, $\vec{r}_2$, $\vec{r}_3$, and $\vec{r}_4$ be the parametrizations for C_1, C_2, C_3, and C_4, respectively, that are consistent with the orientation of the surface. Even though the parametrizations may not be one-to-one, we have that

$$\int_{\beta S} \vec{F} \cdot d\vec{r} = \int_0^1 \vec{F}(\vec{r}_1(t)) \cdot \vec{r}_1{}'(t)dt + \int_0^1 \vec{F}(\vec{r}_2(t)) \cdot \vec{r}_2{}'(t)dt$$

$$+ \int_0^1 \vec{F}(\vec{r}_3(t)) \cdot \vec{r}_3{}'(t)dt + \int_0^1 \vec{F}(\vec{r}_4(t)) \cdot \vec{r}_4{}'(t)dt.$$

EXAMPLE 7: Let $\mathcal{C}$ be the unit circle with center $(2,0)$ and let S be the surface obtained by rotating $\mathcal{C}$ $90°$ about the y–axis oriented as in Figure 8.b. The function

$$\vec{\gamma}(\theta) = \begin{pmatrix} x(\theta) \\ y(\theta) \end{pmatrix} = \begin{pmatrix} \cos(\theta) + 2 \\ \sin(\theta) \end{pmatrix}, \quad 0 \le \theta \le 2\pi$$

parametrizes the circle. To rotate the circle, we use the function

$$\vec{h}(\theta, \phi) = \begin{pmatrix} x(\theta)\cos(\phi) \\ y(\theta) \\ x(\theta)\sin(\phi) \end{pmatrix} = \begin{pmatrix} (\cos(\theta) + 2)\cos(\phi) \\ \sin(\theta) \\ (\cos(\theta) + 2)\sin(\phi) \end{pmatrix}, \quad \begin{matrix} 0 \le \theta \le 2\pi, \\ 0 \le \phi \le \frac{\pi}{4} \end{matrix} \cdot$$

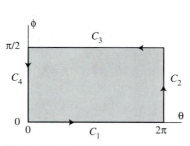

Figure 8.a *The domain of the parametrization.*

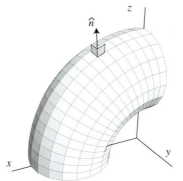

Figure 8.b *The oriented surface S.*

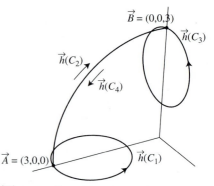

Figure 8.c *The oriented boundary of S.*

Figures 9.a and 9.b display the surface and its boundary pulled apart at the seam.

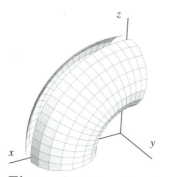

Figure 9.a *The surface in Example 7 pulled apart at its seam.*

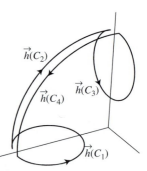

Figure 9.b *The boundary of the surface in Example 7 pulled apart at its seam.*

We see that the edges C_2 and C_4 of the domain are taken by $\vec{h}$ into a seam of the surface. $\vec{h}(C_1)$ is the circle in the xy–plane being rotated and C_3 is taken onto a circle in the yz–plane. The boundary is oriented so that if you move around the boundary in the positive direction, then the surface is to your left. If $\vec{F}$ is any integrable vector field, then

$$\int_{\vec{h}(C_2)} \vec{F} \cdot d\vec{r} = -\int_{\vec{h}(C_4)} \vec{F} \cdot d\vec{r},$$

so

$$\int_{\vec{h}(C_2)} \vec{F} \cdot d\vec{r} + \int_{\vec{h}(C_4)} \vec{F} \cdot d\vec{r} = 0.$$

It follows that

$$\int_{\text{boundary of } S} \vec{F} \cdot d\vec{r} = \int_{\vec{h}(C_1)} \vec{F} \cdot d\vec{r} + \int_{\vec{h}(C_3)} \vec{F} \cdot d\vec{r}.$$

To parametrize $\vec{h}(C_1)$ and $\vec{h}(C_3)$, we refer to Examples 3. The values for the bounds on R are $a = 0$, $b = 2\pi$, $c = 0$, and $d = \frac{\pi}{4}$.

$$\vec{r}_1(t) = (2\pi t, 0), \ 0 \leq t \leq 1$$

parametrizes C_1, and

$$\vec{r}_3(t) = \left(2\pi - 2\pi t, \frac{\pi}{4}\right), \ 0 \leq t \leq 1$$

parametrizes C_3.

$$\rho_1(t) = \vec{h}(\vec{r}_1(t)) = \begin{pmatrix} (\cos(2\pi t) + 2 \\ \sin(2\pi t) \\ 0 \end{pmatrix}, \ 0 \leq t \leq 2\pi$$

parametrizes $\vec{h}(C_1)$, while

$$\rho_3(t) = \vec{h}(\vec{r}_3(t)) = \begin{pmatrix} 0 \\ \sin(2\pi - 2\pi t)\cos(2\pi - 2\pi t) + 2 \end{pmatrix}, \ 0 \leq t \leq 2\pi$$

parametrizes $\vec{h}(C_3)$.   ∎

EXERCISES 15.4

In these exercises, C_1, C_2, C_3, and C_4 are the edges of a rectangle as in Example 3. In Exercises 1–3, $\mathcal{R}$ is the rectangle in the plane bounded by the lines $x = a$, $x = b$, $y = c$, and $y = d$. Parametrize the edges C_1, C_2, C_3, and C_4 to model a point moving counterclockwise around $\mathcal{R}$.

1. $a = 0$, $b = 2$, $c = 0$, and $d = 3$.

2. $a = -1$, $b = 2$, $c = 1$, and $d = 3$.

3. $a = -\pi$, $b = \pi$, $c = 0$, and $d = 4\pi$.

In Exercises 4–7, vectors A and B emanate from C to form adjacent edges for a parallelogram $\mathcal{P}$. Parametrize each of the edges of $\mathcal{P}$ consistently with the orientation of P.

4. $\vec{A} = (1,1,1)$, $\vec{B} = (0,1,1)$, and $\vec{C} = (0,0,0)$. $\mathcal{P}$ is oriented in the direction of $\vec{A} \times \vec{B}$.

5. $\mathcal{P}$ is the parallelogram of Exercise 4, but oriented in the direction of $\vec{B} \times \vec{A}$.

6. $\vec{A} = (1,0,1)$, $\vec{B} = (2,1,1)$, and $\vec{C} = (1,1,1)$. $\mathcal{P}$ is oriented in the direction of $\vec{A} \times \vec{B}$.

7. $\mathcal{P}$ is the parallelogram of Exercise 6, but oriented in the direction of $\vec{B} \times \vec{A}$.

In Exercises 8–11, $\vec{\rho}(r, \theta) = (r\cos(\theta), r\sin(\theta))$.

8. $\vec{\rho}(r, \theta)$, $0 \leq r \leq 2$, $0 \leq \theta \leq 2\pi$, parametrizes a disk of radius 2 centered at the origin. Sketch the disk and the images of C_1, C_2, C_3, and C_4. Which of the edges C_1, C_2, C_3, and C_4 are taken into the boundary of the disk by $\vec{\rho}$? Which of the edges C_1, C_2, C_3, and C_4 are taken into a seam of the disk by $\vec{\rho}$?

9. $\vec{\rho}(r,\theta)$, $1 \leq r \leq 2$, $0 \leq \theta \leq 2\pi$, parametrizes an annulus centered at the origin as illustrated below. Sketch the images of C_1, C_2, C_3, and C_4. Which of the edges C_1, C_2, C_3, and C_4 are taken into the boundary of the disk by $\vec{\rho}$? Which of the edges C_1, C_2, C_3, and C_4 are taken into a seam of the annulus by $\vec{\rho}$? The boundary is oriented so that if you move in the positive direction around the boundary, then the surface is to your left. Find functions that parametrize the pieces of the boundary consistently with the orientation of the boundary.

Figure 10. *The annulus of Exercise 9.*

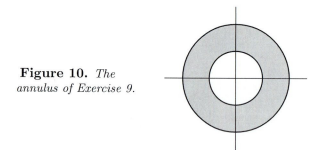

10. $\vec{\rho}(r,\theta)$, $1 \leq r \leq 2$, $0 \leq \theta \leq \frac{\pi}{4}$ parametrizes part of an annulus centered at the origin. Sketch the images of C_1, C_2, C_3, and C_4. The boundary is oriented so that if you move in the positive direction around the boundary, then the surface is to your left. Find functions that parametrize the boundary consistently with the orientation of the boundary.

11. Let $\mathcal{S}$ be the surface of Exercise 10. Find functions that parametrize the boundary of $\mathcal{S}$ so that as you move around the boundary in the positive direction, $\mathcal{S}$ is to your right.

12. The function

$$\vec{S}(\theta,\phi) = \begin{pmatrix} \cos\theta\sin(\phi) \\ \sin(\theta)\sin(\phi) \\ \cos(\psi) \end{pmatrix}, \quad \begin{array}{l} 0 \leq \theta \leq 2\pi, \\ 0 \leq \phi \leq \frac{\pi}{2} \end{array}$$

parametrizes the top half of the unit sphere centered at the origin. Sketch the images of C_1, C_2, C_3, and C_4. Find a function that parametrizes the boundary consistently with the orientation of the surface.

In Exercises 13 and 14, $\mathcal{S}$ is the graph of the function $f(x,y)$, $a \leq x \leq b$ and $c \leq y \leq d$. The orientation of $\mathcal{S}$ established by

$$\vec{h}(x,y) = \begin{pmatrix} x \\ y \\ f(x,y) \end{pmatrix},$$

which points in the positive z direction. Find functions that parametrize the boundary of $\mathcal{S}$ consistently with the orientation of $\mathcal{S}$.

13. $a = -1$, $b = 1$, $c = 0$, and $d = 2$. $f(x,y) = x^2 y$.

14. $a = 0$, $b = 2\pi$, $c = 0$, and $d = 2\pi$. $f(x,y) = \sin(x) + \cos(y)$.

In Exercises 15 and 16, $\mathcal{S}$ is the graph of the function $f(x,y)$, $x^2 + y^2 \leq 1$. The orientation of $\mathcal{S}$ established by

$$\vec{h}(x,y) = \begin{pmatrix} x \\ y \\ f(x,y) \end{pmatrix}$$

points in the positive z direction. Find a function that parametrizes the boundary of $\mathcal{S}$ consistently with the orientation of $\mathcal{S}$. Hint: First parametrize the boundary of the disk in the plane.

15. $f(x,y) = x^2 + y^2$.

16. $f(x,y) = e^{x+y}$.

17. Let S be the graph of the equation $z = 4 - x^2 - y^2$ where $z \geq 0$, oriented away from the z–axis. Parametrize the boundary of S (consistently with the orientation of S).

15.5 Stokes' Theorem

Recall that in Chapter 11, we introduced the curl of a vector field in connection with our study of conservative forces, work, and energy.

In particular, recall that if

$$\vec{F}(x, y, z) = (F_x(x, y, z), F_y(x, y, z), F_z(x, y, z))$$

is a force field, then

$$
\mathrm{Curl}(\vec{F}) \;=\; \vec{\nabla} \times F = \det \begin{pmatrix} \hat{\imath} & \hat{\jmath} & k \\[4pt] \frac{\partial}{\partial x} & \frac{\partial}{\partial y} & \frac{\partial}{\partial z} \\[4pt] F_x & F_y & F_z \end{pmatrix}
$$

$$
= \begin{pmatrix} \frac{\partial F_z}{\partial y} - \frac{\partial F_y}{\partial z} \\[6pt] \frac{\partial F_x}{\partial z} - \frac{\partial F_z}{\partial x} \\[6pt] \frac{\partial F_y}{\partial x} - \frac{\partial F_x}{\partial y} \end{pmatrix}.
$$

In Theorem 2 of Section 11.11, we learned that if the force field $\vec{F}$ is conservative, then $\vec{\nabla} \vec{F}(\vec{r}) = \vec{0}$. Under certain conditions, the converse of this theorem is true. In this section, we set the stage for learning what these conditions are and apply them in Section 15.6. There are also some nice geometric and physicial interpretations of the curl that we explore here.

We begin by stating Stokes' Theorem.

Theorem 1 (Stokes' Theorem) *Suppose that*

- *S is a simple (oriented) surface,*

- *the boundary βS is a simple closed curve or the union of a finite number of simple closed curves oriented consistently with the orientation on S, and*

- *$\vec{F}$ is a vector field such that all of the partial derivatives of the coordinate functions of $\vec{F}$ are continuous;*

then

$$\int_S \mathrm{Curl}(\vec{F}) \cdot d\vec{S} = \int_{\beta S} \vec{F} \cdot d\vec{r}.$$

EXAMPLE 1: Let S be the graph of the equation $z = 4 - x^2 - y^2$ where $z \geq 0$, oriented away from the z–axis. Then the boundary of S, denoted by βS, is the circle of radius 2 centered at the origin. Verify Stokes' Theorem for the vector field $\vec{F}(x, y, z) = (2y, -z, 1)$.

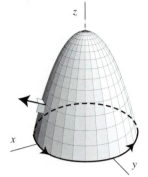

Figure 1. *The boundary of S is the circle of radius 2 in the xy–plane. The orientation of S points away from the x–axis.*

SOLUTION: Figure 1 illustrates βS and S. We parametrize S by rotating the graph of $z = 4 - x^2$ about the z–axis. Let

$$\vec{h}(u, \theta) = \begin{pmatrix} u\cos\theta \\ u\sin\theta \\ 4 - u^2 \end{pmatrix} \text{ for } 0 \leq u \leq 2, \ 0 \leq \theta \leq 2\pi.$$

An inspection of S indicates that if $\vec{n}$ is a vector normal to S pointing away from the z–axis, then the z–component of $\vec{n}$ should be positive.

$$\frac{\partial \vec{h}}{\partial u} \times \frac{\partial \vec{h}}{\partial \theta} = \det \begin{pmatrix} \vec{\imath} & \vec{\jmath} & \vec{k} \\ \cos\theta & \sin\theta & -2u \\ -u\sin\theta & u\cos\theta & 0 \end{pmatrix} = \begin{pmatrix} 2u^2\cos\theta \\ 2u^2\sin\theta \\ u \end{pmatrix},$$

and

$$\text{Curl}(\vec{F}) = \begin{pmatrix} \frac{\partial F_z}{\partial y} - \frac{\partial F_y}{\partial z} \\ \frac{\partial F_x}{\partial z} - \frac{\partial F_z}{\partial x} \\ \frac{\partial F_y}{\partial x} - \frac{\partial F_x}{\partial y} \end{pmatrix} = \begin{pmatrix} 1 \\ 0 \\ -2 \end{pmatrix}.$$

Thus

$$\text{Curl}(\vec{F}) \cdot \left(\frac{\partial \vec{h}}{\partial u} \times \frac{\partial \vec{h}}{\partial \theta} \right) = 2u^2\cos\theta - 2u.$$

$$\begin{aligned} \int_S \text{Curl}(\vec{F}) \cdot d\vec{S} &= \int_0^{2\pi} \int_0^2 2u^2\cos\theta - 2u \ du \ d\theta \\ &= \int_0^{2\pi} \frac{2}{3}u^3\cos\theta - u^2 \Big|_{u=0}^{u=2} \ d\theta \\ &= \int_0^{2\pi} \frac{16}{3}\cos\theta - 4 \ d\theta \\ &= \frac{16}{3}\sin\theta - 4\theta \Big|_{\theta=0}^{\theta=2\pi} = -8\pi. \end{aligned}$$

A parametrization that agrees with the orientation of βS is

$$\vec{h}(t) = (2\cos\theta, 2\sin\theta, 0), \ 0 \leq \theta \leq 2\pi.$$

$$\begin{aligned} \int_{\beta S} \vec{F} \cdot d\vec{r} &= \int_0^{2\pi} [4\sin\theta, 0, 1] \cdot [-2\sin\theta, 2\cos\theta, 0] \ d\theta \\ &= -\int_0^{2\pi} 8\sin^2\theta \ d\theta = -8\pi. \quad \blacksquare \end{aligned}$$

EXAMPLE 2: Suppose that S is the cylindrical surface described by $x^2 + y^2 = 1, 0 \le z \le 1$. The surface S is oriented away from the z–axis. Verify Stokes' Theorem for $\vec{F}(x, y, z) = (-y, x, z)$.

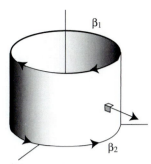

SOLUTION: The boundary, βS of S, is the union of two circles, β_1 at the top and β_2 at the bottom. Now β_1 and β_2 must be oriented so that as a particle moves around the boundary in the positive direction, the positive side of the surface S will be on the particle's left. See Figure 2.

Let $\vec{h}_1(t) = (\cos(t), -\sin(t), 1)$ and $\vec{h}_2(t) = (\cos(t), \sin(t), 0)$, for $0 \le t \le 2\pi$. Then $\vec{h}_1'(t) = (-\sin(t), -\cos(t), 0)$ and $\vec{h}_2'(t) = (-\sin(t), \cos(t), 0)$. We see that

Figure 2. *The boundary of S is the union of two circles, β_1 at the top and β_2 at the bottom. S is oriented away from the z–axis.*

$$
\begin{aligned}
\int_{\beta_1} \vec{F} \cdot d\vec{r} &= \int_0^{2\pi} (\sin(t), \cos(t), 1) \cdot (-\sin(t), -\cos(t), 0)\ dt \\
&= \int_0^{2\pi} -\sin^2(t) - \cos^2(t)\ dt = -\int_0^{2\pi} dt = -2\pi,
\end{aligned}
$$

and

$$
\begin{aligned}
\int_{\beta_2} \vec{F} \cdot d\vec{r} &= \int_0^{2\pi} (-\sin(t), \cos(t), 0) \cdot (-\sin(t), \cos(t), 0)\ dt \\
&= \int_0^{2\pi} \sin^2(t) + \cos^2(t)\ dt = \int_0^{2\pi} dt = 2\pi.
\end{aligned}
$$

Thus

$$
\int_{\beta S} \vec{F} \cdot d\vec{r} = \int_{\beta_1} \vec{F} \cdot d\vec{r} + \int_{\beta_2} \vec{F} \cdot d\vec{r} = -2\pi + 2\pi = 0.
$$

To compute $\int_S \mathrm{Curl}(\vec{F}) \cdot d\vec{S}$:

$$
\mathrm{Curl}(\vec{F}) = \begin{pmatrix} 0 \\ 0 \\ 2 \end{pmatrix}.
$$

The function $\vec{h}(\theta, t) = (\cos(\theta), \sin(\theta), t)$, for $0 \le \theta \le 2\pi$ and $0 \le t \le 1$, is a parametrization for S. The partial derivatives are $\frac{\partial \vec{h}}{\partial \theta} = (-\sin(\theta), \cos(\theta), 0)$ and $\frac{\partial \vec{h}}{\partial t} = (0, 0, 1)$, so $\frac{\partial \vec{h}}{\partial \theta} \times \frac{\partial \vec{h}}{\partial t} = (\cos(\theta), \sin(\theta), 0)$, and $\text{Curl}(\vec{F}) \cdot \left(\frac{\partial \vec{h}}{\partial \theta} \times \frac{\partial \vec{h}}{\partial t} \right) = 0$. Thus $\int_S \text{Curl}(\vec{F}) \cdot d\vec{S} = 0$. ∎

Some Useful Intuition:
A Physical Interpretation of the Curl

Suppose that $\vec{F} : \mathbb{R}^3 \to \mathbb{R}^3$ is a vector field. Let D_h be an oriented disk of radius h centered at $\vec{r}_0$ with the unit normal vector $\hat{\eta}$ pointing in the positive direction of D_h. See Figure 3.a.

It can be shown that

$$(\vec{\nabla} \times \vec{F}) \cdot \hat{\eta} = \text{Curl}(\vec{F}) \cdot \hat{\eta} = \lim_{h \to 0} \frac{\int_{\text{Boundary of } D_h} \vec{F} \cdot d\vec{r}}{\text{Area of } D_h}.$$

Thus $\vec{\nabla} \times \vec{F}|_{\vec{r} = \vec{r}_0}$ is a quantification of the tendency for $\vec{F}$ to "rotate at $\vec{r}_0$." Suppose that $\vec{F}$ is viewed as a velocity field (i.e., $\vec{F}(\vec{r}) =$ velocity of a particle at position $\vec{r}$). Suppose further that we place a paddle wheel centered at $\vec{r}_0$ with its axis pointed in the direction $\hat{\eta}$. See Figure 3.b.

If $\hat{\eta}$ points in the direction of $\vec{\nabla} \times \vec{F}$, then the maximum rotation of the wheel will be obtained. Indeed, $\|\vec{\nabla} \times \vec{F}\|$ is 2ω, where ω is the rate of rotation in radians per unit time.

It is for this reason that the curl of $\vec{F}$ is often referred to as the rotation of $\vec{F}$. If $\text{Curl}(\vec{F}) = 0$, then $\vec{F}$ is said to be *irrotational*.

To see that this interpretation of the curl is reasonable, suppose that we have a rigid body rotating with constant angular velocity $\vec{\omega}$ where the axis of rotation contais the origin. We learned in Section 5.6 that the velocity of a point on the mass at postion $\vec{r} = (x, y, z)$ is given by $\vec{v}(\vec{r}) = \vec{\omega} \times \vec{r}$. Let $\vec{\omega} = (\omega_x, \omega_y, \omega_z)$. Then

$$\vec{\omega} \times \vec{r} = \det \begin{pmatrix} \hat{\imath} & \hat{\jmath} & \hat{k} \\ \omega_x & \omega_y & \omega_z \\ x & y & z \end{pmatrix} = \begin{pmatrix} z\omega_y - y\omega_z \\ x\omega_z - z\omega_x \\ y\omega_x - x\omega_y \end{pmatrix}.$$

Direct computation shows that

$$\vec{\nabla} \times \vec{v}(\vec{r}) = (2\omega_x, 2\omega_y, 2\omega_z) = 2\vec{\omega}.$$

EXAMPLE 3: The vector field $\vec{F}(x, y, z) = (0, x, 0)$ is illustrated in Figure 4.a. Let D_h be the disk of radius h in the xy–plane centered at $\vec{r}_0 = (x_0, y_0, 0)$. The area of D_h is πh^2. The unit normal vector for D_h that points in the positive direction is $\hat{\eta} = (0, 0, 1)$. If we parametrize the boundary of D_h by $\vec{r}(t) = (x_0 + h \cos t, y_0 + h \sin t, 0)$ for $0 \le t \le 2\pi$, then

$$\lim_{h \to 0} \frac{\int_{\text{Boundary of } D_h} \vec{F} \cdot d\vec{r}}{\text{Area of } D_h}$$

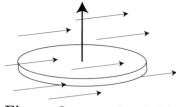

Figure 3.a *An oriented disk D_h of radius h centered at $\vec{r}_0$ with the normal unit vector $\hat{\eta}$.*

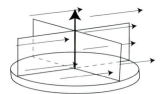

Figure 3.b *A paddle wheel centered at $\vec{r}_0$ with its axis pointed in the direction $\hat{\eta}$.*

$$= \lim_{h \to 0} \frac{\int_0^{2\pi} (0, y_0 + h \cos t, 0) \cdot (-h \sin t, h \cos t, 0) \ dt}{\pi h^2}$$

$$= \lim_{h \to 0} \frac{\int_0^{2\pi} y_0 h \cos t + h^2 \cos^2 t \ dt}{\pi h^2}$$

$$= \lim_{h \to 0} \frac{y_0 h \sin t + h^2 \frac{2 + \sin 2t}{4} \big|_0^{2\pi}}{\pi h^2}$$

$$= \lim_{h \to 0} \frac{h^2 \pi}{\pi h^2} = 1.$$

On the other hand, $\mathrm{Curl}(\vec{F}) = (0, 0, 1)$ and $\hat{\eta} = (0, 0, 1)$. Thus $(\vec{\nabla} \times \vec{F}) \cdot \hat{\eta} = 1$.

Viewing Figure 4.a, we are not surprised that there is a tendency to rotate at points along the y–axis. It may take a little thought to see that it is reasonable to expect a positive rotation at points away from the y–axis. However, as illustrated in Figure 4.b, it is apparent that more "work" is done by $\vec{F}$ on a point mass moving up the right side of the circle than is done "against" $\vec{F}$ as the mass moves down the left side. ■

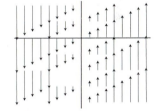

Figure 4.a *The vector field in Example 3.*

Green's Theorem

If we restrict our attention to vector fields in the plane, then Stokes' Theorem takes on a particularly simple but useful form. Consider the following example.

EXAMPLE 4: Let S be the disk in the plane $x^2 + y^2 \leq 1$. Verify Stokes' Theorem for the vector field $\vec{F}(x, y) = (xy, y^2)$.

SOLUTION: In order to use Stokes' theorem, we must translate the 2–dimensional problem into a 3–dimensional one: we write $\vec{F}(x, y, z) = (xy, y^2, 0)$, so that $\mathrm{Curl}(\vec{F}) = (0, 0, -x)$. A parametrization for S is $\vec{h}(r, \theta) = (r \cos(\theta), r \sin(\theta), 0)$, and $\frac{\partial \vec{h}}{\partial r} \times \frac{\partial \vec{h}}{\partial \theta} = (0, 0, r)$. The orientation of S determined by $\vec{h}$ points in the positive z–direction.

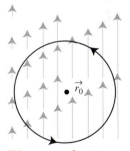

Figure 4.b *The circular path in Example 3.*

$$\int_S \mathrm{Curl}(\vec{F}) \cdot d\vec{S} = \int_0^{2\pi} \int_0^1 (0, 0, -r \cos \theta) \cdot (0, 0, r) \ dr \ d\theta$$

$$= \int_0^{2\pi} \int_0^1 -r^2 \cos \theta \ dr \ d\theta = 0.$$

A parametrization for βS that is compatible with the orientation of S is given by $\vec{h}(\theta) = (\cos(\theta), \sin(\theta), 0)$.

$$\int_{\beta S} \vec{F} \cdot d\vec{r} = \int_0^{2\pi} (\cos(\theta) \sin(\theta), \sin^2(\theta), 0) \cdot (-\sin(\theta), \cos(\theta), 0) \ d\theta$$

$$= \int_0^{2\pi} [-\cos(\theta)\sin^2(\theta) + \cos(\theta)\sin^2(\theta) + 0] \, d\theta = 0. \quad \blacksquare$$

Notice that in Example 4 the computations simplify greatly. This is because the problem is planar. In general, suppose that

$$\vec{F}(x,y) = (F_x(x,y), F_y(x,y))$$

and that the surface S lies in the $xy-$plane. We assume that the boundary of S is oriented in the counterclockwise direction. Let

$$\vec{h}(u,v) = (x(u,v), y(u,v)), \ a \leq u \leq b, \ \text{and} \ c \leq v \leq d$$

be a parametrization of S such that $\det Dh(u,v) > 0$. As in Example 4, to apply Stokes' Theorem we translate the problem into a 3–dimensional one by writing

$$\vec{F}(x,y,z) = (F_x(x,y), F_y(x,y), 0) \ \text{and} \ \vec{h}(u,v) = (x(u,v), y(u,v), 0).$$

The requirement that $\det Dh(u,v) > 0$ gives us that $d\vec{S} = \hat{k}Jh(u,v) = \hat{k}dS$. Assuming that $\vec{F}$ is continuously differentiable, we may apply Stokes' Theorem to obtain

$$\iint_S \text{Curl}(\vec{F}) \cdot d\vec{S} = \iint_S \left(\frac{\partial F_y}{\partial x} - \frac{\partial F_x}{\partial y} \right) \, dS = \int_{\beta S} (F_x, F_y, 0) \cdot d\vec{r}.$$

$$(3)$$

However,

$$\int_{\beta S} (F_x, F_y, 0) \cdot d\vec{r} = \int_{\beta S} F_x \, dx + F_y \, dy. \qquad (4)$$

Combining Equations (3) and (4) we obtain

$$\iint_S \left(\frac{\partial F_y}{\partial x} - \frac{\partial F_x}{\partial y} \right) \, dS = \int_{\beta S} F_x \, dx + F_y \, dy. \qquad (5)$$

Equation (5) is known as the *Curl form of Green's Theorem*.

If we let $\vec{G}(x,y) = (-F_y(x,y), F_x(x,y))$ and apply the above analysis, we obtain

$$\iint_S \left(\frac{\partial F_x}{\partial x} + \frac{\partial F_y}{\partial y} \right) \, dS = \int_{\beta S} -F_y \, dx + F_x \, dy. \qquad (6)$$

However,

$$\frac{\partial F_x}{\partial x} + \frac{\partial F_y}{\partial y} = \text{Div}(\vec{F}).$$

Thus Equation (6) can be expressed as

$$\iint_S \text{Div}(\vec{F})\, dS = \int_{\beta S} -F_y\, dx + F_x\, dy. \qquad (7)$$

Equation (7) is referred to as the *Divergence form of Green's Theorem*. Recall that if we integrate the divergence of a vector field over a solid, then we are calculating the flux of the field through the boundary of the solid. The two dimensional version is completely analogous. The right side of Equation (7) is the flux of $\vec{F}$ through the curve βS. To see that this is a reasonable interpretation, consider the following example.

EXAMPLE 5: Suppose that $\vec{F}(x,y) = (F_x(x,y), F_y(x,y))$ gives the rate of flow of a fluid on the xy–plane. Let S be a simple surface lying in the plane and let βS denote the oriented boundary of the plane. See Figure 5.a. As usual, the orientation of βS is such that if you move in the positive direction on βS, then S is to the left.

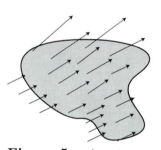

Figure 5.a *A vector field representing a flow over S.*

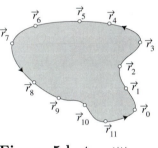

Figure 5.b *A partition of the boundary of S.*

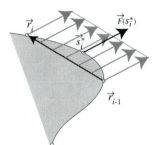

Figure 5.c *The area of the rectangle shaded with arrows is an approximation of the rate of flow through the part of the curve from $\vec{r}_{i-1}$ to $\vec{r}_i$.*

We partition βS with $\mathcal{P} = \{\vec{r}_0, \vec{r}_1, \ldots, \vec{r}_n\}$ as illustrated in Figure 5.b. Consider a section of the boundary from $\vec{r}_{i-1}$ to $\vec{r}_i$ as in Figure 5.c. Let $\vec{s}_i^*$ be a selection point on the curve between $\vec{r}_{i-1}$ and $\vec{r}_i$. Then the area A_i of the parallelogram with adjacent edges $\vec{r}_i - \vec{r}_{i-1}$ and $\vec{F}(\vec{s}_i^*)$ approximates the magnitude of the rate of flow through the segment of the curve. See Figure 5.c. If we move the problem into 3–space by writing $\vec{r}_i = (x_i, y_i, 0)$, $\vec{s}_i^* = (x_i^*, y_i^*, 0)$ and $\vec{F}(x,y) = (F_x(x,y), F_y(x,y), 0)$, then

$$A_i = \left\| \vec{F}(\vec{s}_i^*) \times (\vec{r}_i - \vec{r}_{i-1}) \right\|.$$

Now,

$$\vec{F}(\vec{s}_i^*) \times (\vec{r}_i - \vec{r}_{i-1}) = \left((y_i - y_{i-1})F_x(\vec{s}_i^*) - (x_i - x_{i-1})F_y(\vec{s}_i^*)\right)\hat{k}.$$

Notice that the right hand rule implies that $(y_i - y_{i-1})F_x(\vec{s}_i^*) - (x_i - x_{i-1})F_y(\vec{s}_i^*)$ is positive if $\vec{F}(\vec{s}_i^*)$ points out of the bounded region (the flow is out of S at $\vec{s}_i^*$) and $(y_i - y_{i-1})F_x(\vec{s}_i^*) - (x_i - x_{i-1})F_y(\vec{s}_i^*)$ is negative if $\vec{F}(\vec{s}_i^*)$ points into the bounded region (the flow is into S at $\vec{s}_i^*$.) An approximation of the flow through βS is

$$\sum_{i=1}^{n}(y_i - y_{i-1})F_x(\vec{s}_i^*) - (x_i - x_{i-1})F_y(\vec{s}_i^*)$$

$$= \sum_{i=1}^{n}(-F_y(\vec{s}_i^*), F_x(\vec{s}_i^*)) \cdot (\vec{r}_i - \vec{r}_{i-1})$$

$$\approx \int_{\beta S}(-F_y, F_x) \cdot d\vec{r}$$

$$= \int_{\beta S}-F_y\,dx + F_x\,dy. \qquad\blacksquare$$

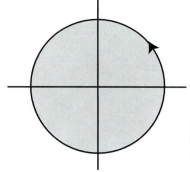

Figure 6.a *An oriented disk in the plane.*

When computing either $\int_{\beta S}F_x\,dx + F_y\,dy$ or $\int_{\beta S}-F_y\,dx + F_x\,dy$, you must be careful in your choice of orientation. The procedure we follow is that as we integrate along one of the simple closed curves, the surface S must always be towards our left. So, for example, if the surface is a disk D of radius one centered at the origin, you must integrate along the boundary circle in a counterclockwise direction. See Figure 6.a. However, if the surface were an annulus, for example, the region $A = \{(x,y) \mid 1 \le \sqrt{x^2 + y^2} \le 2\}$, then the outer circle would be parametrized in a counterclockwise direction, whereas the inner circle would be parametrized in a clockwise direction. See Figure 6.b.

Note: Some texts use the symbol $\oint$ to denote the oriented line integral around a simple closed curve.

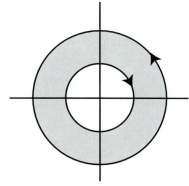

Figure 6.b *An oriented annulus in the plane.*

EXAMPLE 6: Verify both forms of Green's Theorem for $\vec{F}(x,y) = (x^2, xy)$ on $S = \{(x,y) \mid x^2 + y^2 \le 1, y \ge 0\}$.

SOLUTION: The boundary of S consists of two pieces, the semicircle C_1, which can be parametrized by $\vec{h}_1(t) = (\cos t, \sin t)$, for $0 \le t \le 2\pi$, and the straight line segment C_2 from $x = -1$ to $x = 1$ on the x–axis, parametrized by $h_2(t) = (t, 0)$ for $-1 \le t \le 1$. We parametrize the surface S by $\vec{h}(s,t) = (s\cos t, s\sin t)$, for $1 \le s \le 2$ and $0 \le t \le 2\pi$. We then calculate the Jacobian of $\vec{h}$ obtaining $J(\vec{h}) = s$. We also have that $\mathrm{Div}(\vec{F}) = 3x$ and $\mathrm{Curl}(\vec{F}) = y$.

Thus from the Divergence form of Green's Theorem we have

$$\int_S \text{Div}(\vec{F}(x,y))\, dS \;=\; \int_0^\pi \int_0^1 3s^2 \cos t\, ds dt$$

$$=\; \int_0^\pi s^3 \Big|_0^1 \cos t\, dt = \sin t \Big|_0^\pi = 0,$$

and

$$\int_{\beta S} -xy\, dx + x^2\, dy \;=\; \int_0^\pi \cos t \sin^2 t + \cos^3 t\, dt + \int_{-1}^1 0\, dt$$

$$=\; \int_0^\pi \cos t \sin^2 t + \cos t(1 - \sin^2 t)\, dt$$

$$=\; \int_0^\pi \cos t\, dt = \sin t \Big|_0^\pi = 0.$$

From the Curl form of Green's theorem, we have

$$\int_S \text{Curl}(\vec{F}(x,y))\, dS \;=\; \int_0^\pi \int_0^1 s^2 \sin t\, ds dt$$

$$=\; \int_0^\pi \frac{s^3}{3} \Big|_0^1 \sin t\, dt$$

$$=\; -\frac{1}{3} \cos(t) \Big|_0^\pi = \frac{2}{3},$$

and

$$\int_{\beta S} x^2\, dx + xy\, dy \;=\; \int_0^\pi -\cos^2 t \sin t\, dt + \cos^2 t \sin t\, dt + \int_{-1}^1 t^2\, dt$$

$$=\; \int_{-1}^1 t^2\, dt = \frac{2}{3}. \qquad\qquad \blacksquare$$

Let $\vec{F}(x,y) = (x,y)$. Then we have the following direct consequence of the divergence form of Green's Theorem.

Theorem 2 *Suppose that C is a simple closed curve that is the boundary of a surface S lying in the xy–plane. Then*

$$\frac{1}{2}\int_C (-y, x) \cdot d\vec{r} = \int_S dS = \textit{Area of } S,$$

where the orientation of C is taken to be counterclockwise.

EXAMPLE 7: Verify Theorem 2 for the disk of radius r.

SOLUTION: We show that $\frac{1}{2}\int_C (-y, x) \cdot d\vec{r} = \pi r^2$, where C is the circle with radius r centered at the origin and oriented in a counterclockwise fashion. Let $\vec{h}(t) = (r\cos(t), r\sin(t))$ be the usual parametrization for C.

$$\int_C (-y, x) \cdot d\vec{r} = \int_0^{2\pi} (-r\sin(t), r\cos(t)) \cdot (-r\sin(t), r\cos(t)) \, dt$$

$$= \int_0^{2\pi} r^2 \, dt = 2\pi r^2. \qquad \blacksquare$$

EXERCISES 15.5

In Exercises 1–4, verify Stokes' Theorem for the vector field and surface.

1. $\vec{F}(x, y, z) = (yz, xz, yz)$, and S is the parallelepiped with adjacent edges $\vec{a} = (1, 1, 1)$ and $\vec{b} = (1, 0, 0)$, drawn emanating from the origin, with orientation in the direction of $\vec{a} \times \vec{b}$.

2. $\vec{F}(x, y, z) = (yz, xz, yz)$, and S is the hemisphere $x^2 + y^2 + z^2 = 1, z \geq 0$, with orientation pointing away from the origin.

3. $\vec{F}(x, y, z) = (y^2, 3x, 2y)$, and S is the hemisphere $x^2 + y^2 + z^2 = 9, z \geq 0$, with orientation pointing away from the origin.

4. $\vec{F}(x, y, z) = (2x - 2y, y - x, z)$, and S is that part of the plane $x + y + z = 1$ that is bounded by the coordinate planes. The orientation of S is pointing away from the origin.

In Exercises 5–8, use Theorem 2 to find the area of the surface S in the plane.

5. S is the region bounded between the graphs of $y = x^2$ and $y = x$.

6. S is the region bounded by the ellipse $\frac{x^2}{a^2} + \frac{y^2}{b^2} = 1$.

7. S is the region bounded between the graphs of $y = x^2$ and $x^2 + y^2 = 2$.

8. S is the region bounded by the set of points with polar equation $r = \sin(2\theta), 0 \leq \theta \leq \frac{\pi}{2}$.

9. Let C be the straight line interval from $\vec{P_1} = (x_1, y_1)$ to $\vec{P_2} = (x_2, y_2)$. Show that

$$\int_{C_{\vec{P_1}}}^{P_2} (-y, x) \cdot d\vec{r} = x_1 y_2 - x_2 y_1.$$

10. Use Exercise 9 to find the area of the triangle with vertices $(0, 0), (1, -1)$, and $(5, 1)$.

11. Use Exercise 9 to find the area of the polygon bounded by the line segments $[\vec{a_1}, \vec{a_2}]$, $[\vec{a_2}, \vec{a_3}]$, $[\vec{a_3}, \vec{a_4}]$, and $[\vec{a_4}, \vec{a_1}]$, where $\vec{a_1} = (0, 0)$, $\vec{a_2} = (1, -1)$, $\vec{a_3} = (5, 2)$, and $\vec{a_4} = (-1, 0)$.

12. Use Exercise 9 to find the area of the polygon illustrated below.

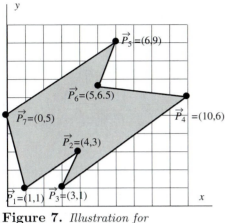

Figure 7. *Illustration for Exercise 12.*

13. Use Green's theorem to evaluate $\int_C xy^2\ dx + x^2y\ dy$ over the curve C, the unit circle.

In Exercises 14–16, T is the triangle with vertices $(0,0)$, $(1,0)$, and $(0,1)$. Verify Green's Theorem (using both the Curl form and the Divergence form) for the vector field $\vec{F}$.

14. $\vec{F}(x,y) = (x^3, y^3)$.

15. $\vec{F}(x,y) = (y^3, x^3)$.

16. $\vec{F}(x,y) = (x^2 + y^2, 2xy)$.

17. Let $\vec{F}$ be a vector field in $\mathbb{R}^2$ whose coordinate functions have continuous partial derivatives. Assume that $\text{Div}(\vec{F}) = 0$. Show that if C_1 and C_2 are any two circles centered at the origin, then the flux through C_1 equals the flux through C_2.

15.6   Stokes' Theorem and Conservative Fields

One of the most important applications of Stokes' Theorem is in the theory of conservative vector fields. Before starting this section, we review some of the basic facts about conservative fields:

(a) A vector field $\vec{F}$ is conservative if and only if there is a real valued function f such that $\vec{\nabla}f = \vec{F}$ ($-f$ is called a *potential function for $\vec{F}$*).

(b) If C is a simple curve from $\vec{a}$ to $\vec{b}$ lying in the domain of $\vec{F}$, and $\vec{\nabla}f = \vec{F}$, then $\int_C \vec{F} \cdot d\vec{r} = f(\vec{b}) - f(\vec{a})$.

(c) If C_1 and C_2 are simple curves from $\vec{a}$ to $\vec{b}$ lying in the domain of $\vec{F}$, then $\int_{C_1} \vec{F} \cdot d\vec{r} = \int_{C_2} \vec{F} \cdot d\vec{r}$.

(d) If $\vec{F}$ is conservative, then $\text{Curl}(\vec{F}) = \vec{0}$. (Theorem 2 of Section 11.10).

In fact, we know that conditions (a)–(c) are equivalent to each other.

We would like to have a simple way of determining whether a given vector field is conservative. Verifying any of conditions (a)–(c) requires some effort. We might wonder whether the converse of (d) is true. Would knowing that $\text{Curl}(\vec{F}) = \vec{0}$ allow us to conclude that $\vec{F}$ is conservative?

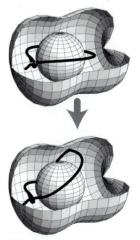

Figure 1. *The loop can be pulled through the knot without leaving the set S.*

It turns out that if the domain of the vector field $\vec{F}$ is "well-behaved," and if the Curl of $\vec{F}$ is $\vec{0}$, then $\vec{F}$ is conservative. This is a particularly useful theorem since it gives us a relatively simple test to determine whether or not a vector field is conservative. First, we must decide what we mean by "well-behaved." What must we require of the domain of $\vec{F}$?

A property of the domain of $\vec{F}$ that is sufficient to ensure that $\vec{F}$ is conservative is that of being simply connected. We say that a subset in $\mathbb{R}^n$ is *simply connected* if it has the "slip knot property."

The idea of the *slip knot property* can perhaps be best explained in the following intuitive fashion. Imagine that you have a rope lying entirely in the set S and suppose that the rope has a loop passing through a slip knot.

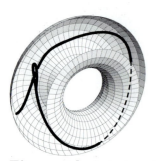

Figure 2.a *The torus is not simply connected.*

If, regardless of how the rope is positioned in S, the loop can be pulled through the slip knot without leaving S, then we say that S has the slip knot property. S will **not** satisfy this property if it has certain "bad kinds" of holes. An example of a set with the slip knot property would be 3–space minus a finite number of points or "bubbles." As we see in Figure 1, in this set, the loop can be pulled through the knot, even though it may have to move around the missing bubbles.

However, if a hole extends all the way through the set S, we cannot pull the loop through the knot. For example, if S is a torus (as in Figure 2), then if the loop goes around the hole, it cannot be pulled through the knot without leaving S.

Figure 2.b *A section of the surface of the torus is cut away to reviel a loop that cannot be pulled through its knot without leaving the torus.*

Apparently, 3–space minus a line will not have the slip knot property. Also, the plane minus a point does not have the slip knot property, but the upper-half plane $\{(x, y) \mid y \geq 0\}$ does.

Now we are ready to define a simply connected open set.

Definition: Simply Connected Open Set

A set S is said to be an *simply connected open set* if

(a) S has the slip knot property.

(b) If $\vec{A}$ and $\vec{B}$ are points of S, then there is a simple curve or path from $\vec{A}$ to $\vec{B}$ lying in S.

(c) If $\vec{A}$ is a point of S, then there is an open ball centered at $\vec{A}$ and lying entirely in S.

We can now state the desired theorem.

Theorem 1 *If $\vec{F}$ is a vector field with continuous partial derivatives and if the domain of $\vec{F}$ is open and simply connected, then $\vec{F}$ is conservative if and only if $\mathrm{Curl}(\vec{F}) = \vec{0}$.*

Theorem 1 is a basic tool in applications of vector fields, because it gives a fairly simple test to determine whether a given vector field is conservative. The only complication in its use is determining whether the domain of the vector field is open and simply connected.

EXAMPLE 1: Let $\vec{F}(x, y, z) = (3x^2 + y^2, 2xy, -3z^2)$.

$$\mathrm{Curl}(\vec{F}) = \det \begin{pmatrix} \vec{\imath} & \vec{\jmath} & \vec{k} \\ \frac{\partial}{\partial x} & \frac{\partial}{\partial y} & \frac{\partial}{\partial z} \\ 3x^2 + y^2 & 2xy & -3z^2 \end{pmatrix} = \begin{pmatrix} 0 \\ 0 \\ 0 \end{pmatrix}.$$

Since the coordinate functions of $\vec{F}$ are polynomials, their partial derivatives are continuous, and the domain of $\vec{F}$ is all of xyz–space. Thus the domain of $\vec{F}$ is open and simply connected. By Theorem 1, $\vec{F}$ is conservative. ∎

EXAMPLE 2: Let D denote xy–space minus the origin. Let $\vec{F}$ be the function from D into $\mathbb{R}^2$, which is defined by $\vec{F}(x, y) = \left(-\dfrac{y}{x^2 + y^2}, \dfrac{x}{x^2 + y^2} \right)$. In Exercise 5, the reader is asked to show that $\mathrm{Curl}(\vec{F}) = \vec{0}$, but $\vec{F}$ is not conservative. The problem, of course, is that D is not simply connected.

Summary: Suppose that $\vec{F}$ is a vector field such that all of the partial derivatives of $\vec{F}$ are continuous and that the domain of $\vec{F}$ is open and simply connected. Then the following conditions are equivalent:

(a) $\vec{F}$ is conservative.

(b) There is a real-valued function f such that $\vec{\nabla} f = \vec{F}$. (Note: $-f$ is called a *potential function for $\vec{F}$*.)

(c) If C_1 and C_2 are simple curves from $\vec{a}$ to $\vec{b}$ lying in the domain of $\vec{F}$, then

$$\int_{C_1} \vec{F} \cdot d\vec{r} = \int_{C_2} \vec{F} \cdot d\vec{r}.$$

(d) $\mathrm{Curl}(\vec{F}) = \vec{0}$.

In fact, if f is a function such that $\vec{\nabla} f = \vec{F}$, and if C is a simple curve from $\vec{a}$ to $\vec{b}$ lying in the domain of $\vec{F}$, then $\int_C \vec{F} \cdot d\vec{r} = f(\vec{b}) - f(\vec{a})$.

EXERCISES 15.6

In Exercises 1–4, determine whether the function is conservative, and if it is, find a function f such that $\vec{\nabla} f = \vec{F}$. (Refer to Example 5 and Exercises 4–8 of Secton 11.9.)

1. $\vec{F}(x, y, z) = (yz, xz, xy)$.

2. $\vec{F}(x, y, z) = (ye^{xy+z}, xe^{xy+z}, e^{xy+z})$.

3. $\vec{F}(x, y, z) = (xe^{xy+z}, ye^{xy+z}, ze^{xy+z})$.

4. $\vec{F}(\vec{r}) = \dfrac{k\vec{r}}{\|\vec{r}\|^3}$.

5. Let $\vec{F}$ be the function defined by $\vec{F}(x, y) = \left(-\dfrac{y}{x^2 + y^2}, \dfrac{x}{x^2 + y^2} \right)$. (See Example 2.)

   a. Let C_1 be the top half of the unit circle from $(1, 0)$ to $(-1, 0)$, and let C_2 be the bottom half of the same circle. Show that $\int_{C_1} \vec{F} \cdot d\vec{r} \neq \int_{C_2} \vec{F} \cdot d\vec{r}$.

   b. Show that $\mathrm{Curl}(\vec{F}) = \vec{0}$.

Appendix A

Mathematical Induction

Suppose that you want to teach a child how to climb a ladder. How would you instruct him or her? One possible scenario is the following. You show the child how to get onto the first rung. Once the child is on the first rung, each step after that is the same. First you move one foot to the next rung, then the other foot to the next rung, then you move your hands up one rung. This part of the instruction is independent of which rung he or she is on. Whether on the second rung or the tenth rung, the method is the same. So if the child knows how to get to the 9th rung, he or she will know how to get to the 10th rung. Even if the ladder were $1,000$ rungs tall, it is possible to get to that $1,000^{\text{th}}$ rung. Mathematical induction is quite similar to this. If you are given a formula you are trying to prove for all positive integers n, you first prove that it is true for $n = 1$. This may or may not require a lot of work. Then you show that if you know that the formula is true for $n = i$, you can prove that it is true for $n = i + 1$. You are proving that the process of moving up the ladder at the $i + 1$ rung is similar to the process of moving up the ladder at the ith rung. It is time for an example.

EXAMPLE 1: Using mathematical induction, prove that the sum $1 + 2 + \cdots + n = \frac{n(n+1)}{2}$.

SOLUTION: It is easiest to think of this problem in terms of a succession of statements or propositions. Each proposition represents a "rung" of the ladder. Let $P(1)$ be the proposition of the example for $n = 1$, that is, $1 = \frac{1(1+1)}{2}$. Let $P(2)$ be the proposition that $1 + 2 = \frac{2(2+1)}{2}$. In general, let $P(n)$ be the proposition that $1 + 2 + \cdots + n = \frac{n(n+1)}{2}$. We first prove that $P(1)$ is true. This is easy; $1 = \frac{1(1+1)}{2}$. If we could prove that $P(i)$ being true implied that

747

$P(i+1)$ were true, then we could let $i = 1$, which would imply that $P(2)$ is true. Turning our attention to $i = 2$, we would also see that $P(3)$ is true, etc. We now start with the left side of $P(i+1)$.

$$1 + 2 + \cdots + i + (i+1) = (1 + 2 + \cdots + i) + (i+1)$$
$$= \frac{i(i+1)}{2} + (i+1).$$

This is the *induction step*. It uses the fact that $P(i)$ is true, so we substitute $\frac{i(i+1)}{2}$ for $1 + 2 + \cdots + i$.

Now we perform some algebraic operations.

$$1 + 2 + \cdots + i + (i+1) = \frac{i(i+1)}{2} + (i+1)$$
$$= \frac{i(i+1)}{2} + \frac{2(i+1)}{2}$$
$$= \frac{i(i+1) + 2(i+1)}{2}$$
$$= \frac{(i+1)(i+2)}{2}$$
$$= \frac{(i+1)((i+1)+1)}{2},$$

which is precisely the formula for $P(i+1)$. Thus $P(n)$ is true for all positive integers n.   ■

EXAMPLE 2: Using mathematical induction, show that $n^3 - n$, for positive integers n, is divisible by 3.

SOLUTION: As in Example 1, $P(n)$ is the proposition that $n^3 - n$ is divisible by 3. Note that $P(1)$ is true since $1^3 - 1 = 0$, which is divisible by 3. Assume that $P(i)$ is true, that is, $i^3 - i$ is divisible by three. Let us look at $P(i+1)$. The expression

$$(i+1)^3 - (i+1) = (i^3 + 3i^2 + 3i + 1) - (i+1) = i^3 + 3i^2 + 2i$$

can be rewritten as

$$(i+1)^3 - (i+1) = (i^3 - i) + 3i^2 + 3i.$$

Note that $3i^2$ and $3i$ are each divisible by 3, and $(i^3 - i)$ is divisible by 3 since $P(i)$ is true. Thus $P(i+1)$ is true.   ■

In proving theorems by mathematical induction you need to:

(a) Prove the first step. In some cases, this can be a lengthy process.

(b) Find the $P(i)$ terms embedded in the $P(i+1)$ terms. In Example 1 we used the terms $1 + 2 + \cdots + i$ and replaced them with $\frac{i(i+1)}{2}$. In Example 2 we used the fact that $i^3 - i$ was divisible by 3. In some examples we alter this step and generate $P(i+1)$ directly from $P(i)$.

Mathematical induction need not always begin with $n = 1$. It can start with any integer as the next example illustrates.

EXAMPLE 3: Prove Bernoulli's inequality, that $(1+x)^n > 1+nx$ for n an integer with $n \geq 2$ if $x > -1, x \neq 0$.

SOLUTION: We start our induction with $n = 2$ obtaining $1 + 2x + x^2 > 1 + 2x$, since $x \neq 0$. We assume that Bernoulli's inequality is true for some $i \geq 2$ and will prove that it is true for $i + 1$. Thus $(1+x)^i > 1 + ix$. Now, since $x > -1$, we can multiply both sides by $1 + x$ to obtain

$$(1 + x)^{i+1} > (1 + ix)(1 + x) = 1 + (i + 1)x + ix^2 > 1 + (i + 1)x,$$

which is $P(i + 1)$. ∎

You should be aware that mathematical induction is not a panacea. Some problems that can be solved using mathematical induction can be solved more easily by traditional means. In Example 2, if we had noticed that $n^3 - n$ factors as $(n - 1)n(n + 1)$, then it is obvious that the product of three consecutive integers must have 3 as a divisor.

EXERCISES

1. Prove that
$$1^2 + 2^2 + \cdots + n^2 = \frac{n(n + 1)(2n + 1)}{6}.$$

2. Prove that
$$1^3 + 2^3 + \cdots + n^3 = \left(\frac{n(n + 1)}{2}\right)^2.$$

3. Prove that
$$\frac{1}{1 \cdot 2} + \frac{1}{2 \cdot 3} + \cdots + \frac{1}{n(n + 1)} = \frac{n}{n + 1}.$$

4. Prove that $x^n - y^n$ has $x - y$ as a factor for all positive integers n. (Hint: Subtract and add $x^n y$.)

5. Prove that $1 + 3 + 5 + \cdots + (2n - 1) = n^2$.

6. Prove that $(\cos \theta + i \sin \theta)^n = \cos n\theta + i \sin n\theta$, where $i = \sqrt{-1}$.

7. Show that $n! > 2^n$ for $n \geq 4$.

8. Prove that 2 divides into $n^2 + 5n$ for every positive integer n.

9. Prove that $n < 2^n$.

Appendix B

Continuity for Functions from $\mathbb{R}^n$ into $\mathbb{R}^m$

In Section 11.4 we considered functions $f(\vec{s})$ that appeared to be functions from $\mathbb{R}^n$ into $\mathbb{R}$, but in practice $\vec{s}$ was a fundamental curve that was a function from $\mathbb{R}$ into $\mathbb{R}^n$. Thus, in all of our computations, we were using functions from $\mathbb{R}$ into $\mathbb{R}$. We begin our discussion of continuity for functions from $\mathbb{R}^n$ into $\mathbb{R}^m$ by considering the slightly simpler case of functions from $\mathbb{R}^n$ into $\mathbb{R}$.

Our definition of limits for functions from $\mathbb{R}^n$ to $\mathbb{R}$ is a generalization of the definition of a limit from Section 3.5.

Definition: Limits for Functions $f\colon \mathbb{R}^n \to \mathbb{R}$

Suppose that f is a function from a subset of $\mathbb{R}^n$ into $\mathbb{R}$. The statement that the number L is the *limit of $f(\vec{x})$ as $\vec{x}$ goes to $\vec{a}$* means that:

(a) $\vec{a}$ is a point (not necessarily in the domain of f).

(b) If $\epsilon > 0$, then there is a number $\delta > 0$ such that if $0 < \|\vec{a} - \vec{x}\| < \delta$, then

   (i) $\vec{x}$ is in the domain of f, and
   (ii) $|L - f(\vec{x})| < \epsilon$.

We denote this limit by

$$\lim_{\vec{x} \to \vec{a}} f(\vec{x}) = L.$$

We use this definition of limit to define the notion of continuity.

Definition: Continuity for Functions $f\colon \mathbb{R}^n \to \mathbb{R}$

A function f defined on a subset $\mathcal{D}$ of $\mathbb{R}^n$ with range in $\mathbb{R}$ is said to be *continuous at a point* $\vec{a}$ *in* $\mathcal{D}$ if $\lim_{\vec{x}\to\vec{a}} f(\vec{x}) = f(\vec{a})$. The function f is said to be *continuous* if and only if it is continuous at each point $\vec{a}$ in $\mathcal{D}$.

The following theorem shows that a translation of a continuous function is continuous.

Theorem 1 *If* $f\colon \mathbb{R}^n \to \mathbb{R}$ *is a continuous function, and a is a real number, then* $f + a\colon \mathbb{R}^n \to \mathbb{R}$, *given by* $(f + a)(\vec{x}) = f(\vec{x}) + a$, *is continuous.*

The proof is left to the reader.

We can also show that linear maps are continuous.

Theorem 2 *Let* $\vec{A} = (a_1, a_2, \ldots, a_n)$, *and let* $\vec{x} = (x_1, x_2, \ldots, x_n)$. *Then the function* $f\colon \mathbb{R}^n \to \mathbb{R}$ *given by* $f(\vec{x}) = \vec{A}\cdot\vec{x} = a_1 x_1 + a_2 x_2 + \cdots + a_n x_n$ *is continuous.*

Proof:

Case 1: If $\vec{A} = \vec{0}$, then for any $\vec{y}$ in $\mathbb{R}^n$, we have that $|f(\vec{x}) - f(\vec{y})| = 0 < \epsilon$. Thus δ can be any real number.

Case 2: If $\vec{A} \neq \vec{0}$, then for any $\vec{y}$ in $\mathbb{R}^n$, let $\delta = \frac{\epsilon}{\|\vec{A}\|}$.

$$
\begin{aligned}
|f(\vec{x}) - f(\vec{y})| &= |\vec{A}\cdot\vec{x} - \vec{A}\cdot\vec{y}| = |\vec{A}\cdot(\vec{x} - \vec{y})| \leq \|\vec{A}\|\|\vec{x} - \vec{y}\| \\
&< \|\vec{A}\|\delta = \|\vec{A}\|\frac{\epsilon}{\|\vec{A}\|} = \epsilon.
\end{aligned}
$$
∎

Though technically more tedious, it can also be shown that polynomials in several variables, such as $f(x_1, x_2) = 3x_1^2 - 4x_1 x_2^3$, are continuous as well as more complicated functions involving fractional powers, trigonometric functions, and exponential functions, as long as we are careful about defining the domain.

We can also show that the standard operations with continuous functions produce continuous functions.

Theorem 3 *Let* $f\colon \mathbb{R}^n \to \mathbb{R}$ *and* $g\colon \mathbb{R}^n \to \mathbb{R}$ *be two functions that are continuous on a domain* $\mathcal{D}$ *in* $\mathbb{R}^n$. *Let c be a constant. Then*

(a) $cf\colon \mathcal{D} \to \mathbb{R}$ *is continuous.*

(b) $f \pm g\colon \mathcal{D} \to \mathbb{R}$, *defined by* $(f+g)(\vec{x}) = f(\vec{x}) + g(\vec{x})$, *is continuous.*

(c) $fg\colon \mathcal{D} \to \mathbb{R}$, *defined by* $(fg)(\vec{x}) = f(\vec{x})g(\vec{x})$, *is continuous.*

(d) $f/g\colon \mathcal{C} \to \mathbb{R}$, *defined by* $\left(\dfrac{f}{g}\right)(\vec{x}) = \dfrac{f(\vec{x})}{g(\vec{x})}$, *is continuous where* $\mathcal{C}$ *is the subset of* $\mathcal{D}$ *with* $g(x) \neq 0$.

A function from $\mathbb{R}^n$ to $\mathbb{R}^m$ can be thought of as a collection of functions from $\mathbb{R}^n$ to $\mathbb{R}$.

EXAMPLE 1: Let $\vec{f}\colon \mathbb{R}^3 \to \mathbb{R}^2$ be given by $\vec{f}(x_1, x_2, x_3) = (x_1^2 + x_2, x_1 x_3)$. We can express $\vec{f}$ by the ordered pair $(f_1(x_1, x_2, x_3), f_2(x_1, x_2, x_3))$ where each of the functions f_1 and f_2 are functions from $\mathbb{R}^3$ to $\mathbb{R}$ given by $f_1(x_1, x_2, x_3) = x_1^2 + x_2$ and $f_2(x_1, x_2, x_3) = x_1 x_3$. ∎

In general, any function $\vec{f}\colon \mathbb{R}^n \to \mathbb{R}^m$ can be expressed as an m–tuple of functions $(f_1, f_2, \ldots, f_m)$, each from $\mathbb{R}^n$ to $\mathbb{R}$.

We define continuity of a function $\vec{f}$ by requiring that each of the functions f_i be continuous.

Definition: Continuity for Functions $\vec{f}\colon \mathbb{R}^n \to \mathbb{R}^m$

A function $\vec{f} = (f_1, f_2, \ldots, f_m)$ defined on a subset $\mathcal{D}$ of $\mathbb{R}^n$ with range in $\mathbb{R}^m$ is said to be *continuous at a point* $\vec{a}$ in $\mathcal{D}$ if $\lim_{\vec{x} \to \vec{a}} f_i(\vec{x}) = f_i(\vec{a})$ for each i with $1 \leq i \leq m$.
The function $\vec{f}$ is said to be *continuous* if and only if it is continuous at each point $\vec{a}$ in $\mathcal{D}$.

In particular, if each f_i is a linear function, then by Theorem 2, the function $\vec{f} = (f_1, f_2, \ldots, f_m)$ is continuous.

EXAMPLE 2: Let $\vec{A}_1 = (2, 3, 1)$ and $\vec{A}_2 = (1, -1, 3)$. Also, let $\vec{x} = (x_1, x_2, x_3)$. The function $\vec{f}\colon \mathbb{R}^3 \to \mathbb{R}^2$ given by $\vec{f}(\vec{x}) = (\vec{A}_1 \cdot \vec{x}, \vec{A}_2 \cdot \vec{x}) = (2x_1 + 3x_2 + x_3, x_1 - x_2 + 3x_3)$ is continuous. Note that $\vec{f}$ is the function given in matrix notation by

$$\begin{pmatrix} 2 & 3 & 1 \\ 1 & -1 & 3 \end{pmatrix} \begin{pmatrix} x \\ y \\ z \end{pmatrix}.$$ ∎

As in Theorem 3, it can be shown that operations involving continuous functions produce continuous functions.

Theorem 4 *Let $\vec{f} \colon \mathbb{R}^n \to \mathbb{R}^m$ and $\vec{g} \colon \mathbb{R}^n \to \mathbb{R}^m$ be functions that are continuous on a subset $\mathcal{D}$ of $\mathbb{R}^n$. Let c be a constant. Then*

(a) $c\vec{f} \colon \mathcal{D} \to \mathbb{R}^m$ *is continuous.*

(b) $\vec{f} \pm \vec{g} \colon \mathcal{D} \to \mathbb{R}^m$ *is continuous.*

(c) $\vec{f} \cdot \vec{g} \colon \mathcal{D} \to \mathbb{R}$ *is continuous.*

(d) $\vec{f} \times \vec{g} \colon \mathcal{D} \to \mathbb{R}^3$ *is continuous for $n = m = 3$.*

Appendix C

Conic Sections in $\mathbb{R}^n$

C.1 Parabolas, Ellipses, and Hyperbolas

You have encountered parabolas, ellipses, and hyperbolas in your algebra and geometry courses. In particular, you know that graphs of equations of the form $y = ax^2 + bx + c$ or $x = ay^2 + by + c$ are parabolas. Recall that graphs of equations of the form $y = ax^2$ or $x = ay^2$ are parabolas *in standard position*. The *axis* of the parabola given by $y = ax^2$ is the y–axis. If $a > 0$, then y is always non-negative and the parabola will open in the positive y direction. See Figure 1.a. If $a < 0$, then y is always non-positive and the parabola will open in the negative y direction, as illustrated in Figure 1.b. The *axis* of the parabola given by $x = ay^2$ is the x–axis. The graphs of the cases where $a > 0$ and $a < 0$ are illustrated in Figures 1.c and 1.d, respectively.

Figure 1.a
A parabola with equation of the form $y = ax^2$, $a > 0$.

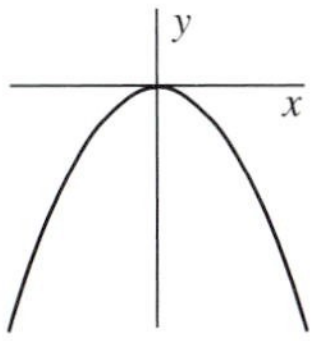

Figure 1.b
A parabola with equation of the form $y = ax^2$, $a < 0$.

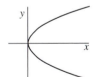

Figure 1.c
A parabola with equation of the form $x = ay^2$, $a > 0$.

Figure 1.d
A parabola with equation of the form $x = ay^2$, $a < 0$.

We use the following geometric construction to obtain a general definition of a parabola. Let D be a line and $\vec{F}$ be a point not on D, as in Figure 2. Let $\mathcal{P}$ be the set of all points (x, y) in the plane that have the property that the distance from (x, y) to $\vec{F}$ is equal to the distance from (x, y) to the line D.

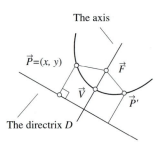

The axis

$\vec{P}=(x,y)$

$\vec{F}$

$\vec{V}$

$\vec{P}'$

The directrix D

Figure 2. *The parabola* $\mathcal{P}$.

- $\mathcal{P}$ is called a *parabola*.

- The point $\vec{F}$ will be called the *focus* of $\mathcal{P}$.

- The line D will be called the *directrix* for $\mathcal{P}$.

- We call the line that contains the focus and is perpendicular to the directrix the *axis*.

- The point of intersection of the axis with $\mathcal{P}$ is called the *vertex*.

To construct a parabola in standard position, let $c > 0$, $\vec{F} = (0, c)$ and let D be the line $y = -c$. (See Figure 3.a). Then the point $\vec{P} = (x, y)$ will be in $\mathcal{P}$ if and only if

The distance from $\vec{P}$ to the line D = The distance from $\vec{P}$ to $\vec{F}$.
$$|y + c| = \sqrt{x^2 + (y - c)^2}.$$

Squaring both sides and simplifying, we obtain

$$x^2 = 4cy,$$

which is of the form $y = ax^2$, where $a = \frac{1}{4c}$. We get the parabola in standard position that opens in the negative y direction if $c < 0$, $\vec{F} = (0, c)$ and D is the line $y = -c$. See Figure 3.b. The other cases are illustrated in Figures 3.c and 3.d.

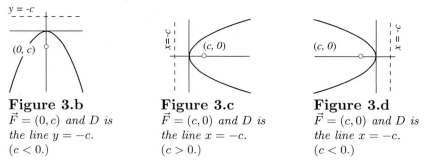

Figure 3.a
$\vec{F} = (0, c)$ *and* D *is the line* $y = -c$.
$(c > 0.)$

Figure 3.b
$\vec{F} = (0, c)$ *and* D *is the line* $y = -c$.
$(c < 0.)$

Figure 3.c
$\vec{F} = (c, 0)$ *and* D *is the line* $x = -c$.
$(c > 0.)$

Figure 3.d
$\vec{F} = (c, 0)$ *and* D *is the line* $x = -c$.
$(c < 0.)$

EXAMPLE 1: Find the focus and directrix of the parabola $y = 6x^2$.

SOLUTION: The equation is of the form $y = 4cx^2$, where $c = \frac{3}{2}$. Thus the focus is $\left(0, \frac{3}{2}\right)$ and its directrix is the line $y = -\frac{3}{2}$. ∎

EXAMPLE 2: Find an equation for the parabola in standard position with focus $(0, -2)$.

SOLUTION: The axis is the y–axis and $c = -2$. Thus the parabola has an equation of the form $x^2 = 4cy$. An equation for the parabola is $x^2 = -8y$. ∎

EXAMPLE 3: Find equations of all parabolas in standard position that contain the point $(1, -1)$.

SOLUTION: There are two possibilities: (a) a parabola with equation of the form $x = ay^2$ and (b) a parabola with equation of the form $y = ax^2$.

(a) a is chosen so that $(1, -1)$ satisfies $x = ay^2$, or $(1) = a(-1)^2$. In this case, $a = 1$ and the desired equation is $y^2 = x$.

(b) a is chosen so that $(1, -1)$ satisfies $y = ax^2$, or $(-1) = a(-1)^2$. Then $a = -1$ and the desired equation is $y = -x^2$.   ∎

Parabolas are of interest throughout science and engineering. If air resistance is neglected, then the trajectory of a baseball will be a parabola (see Theorem 3 in Section 6.3). Parabolas are used in the design of mirrors and other reflectors. In Exercise 19 of Section 4.1 you are asked to show that a light beam emanating from the focus of a parabola will be reflected in a direction parallel to the parabola's axis. This property is used to form your car's headlight's beams. The process is reversed in satellite dishes. A cross section of the dish's reflector is a parabola and the receiver is placed at the focus.

If the dish is aimed directly at the satellite, then rays of the satellite's signal arriving at the dish will be nearly parallel to the axis of the parabolic cross section. Thus the signal hitting the dish will be reflected back to the receiver.

The Ellipse

We start our discussion of ellipses with a geometric definition.

Definition: The Ellipse

The set $\mathcal{E}$ in the plane is an ellipse if and only if there are points $\vec{F_1}$ and $\vec{F_2}$ a distance of $2c$ apart and a number $a > c$ such that $\vec{P}$ is in $\mathcal{E}$ if and only if

$$\text{dist}(\vec{F_1}, \vec{P}) + \text{dist}(\vec{F_2}, \vec{P}) = 2a.$$

- The points $\vec{F_1}$ and $\vec{F_2}$ are the *foci* of the ellipse of $\mathcal{E}$.

- We will call the line containing the foci the *axis*.

- The points of intersection of the axis with $\mathcal{E}$ are called the *vertices* of the ellipse.

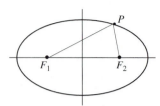

Figure 4.a *The locations of the tacks, $\vec{F_1}$ and $\vec{F_2}$, and the end of the pencil, $\vec{P}$, represented in the plane.*

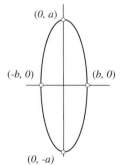

Figure 4.b *The ellipse can be constructed with a string and a pencil.*

Figure 5.a *The graph of $\frac{x^2}{a^2} + \frac{y^2}{b^2} = 1$. The axis is the x-axis.*

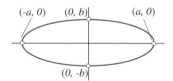

Figure 5.b *The graph of $\frac{y^2}{a^2} + \frac{x^2}{b^2} = 1$. The axis is the y-axis.*

Ellipses occur naturally throughout nature just as parabolas occur. Perhaps the best known occurrence is that of Kepler's observation that the earth's trajectory is an ellipse with the sun at one of its foci.

We say that the ellipse is in *standard position* provided that its foci are on one of the axes equidistant from the origin. That is, the foci are of the form $\vec{F_1} = (-c, 0)$ and $\vec{F_2} = (c, 0)$, or $\vec{F_1} = (0, -c)$ and $\vec{F_2} = (0, c)$.

To obtain an equation for an ellipse in standard position, think of tying the ends of a string of length $2a$ to two tacks placed at positions $\vec{F_1} = (-c, 0)$ and $\vec{F_2} = (c, 0)$ on a piece of paper. We then use the point of the pencil to pull the string tight on the paper. The sum of the distances from the point of the pencil to the tacks is the length of the string. See Figures 4.a and 4.b. Keeping the string tight, we trace out an ellipse.

Let $\mathcal{E}$ denote the set of points such that $\vec{P} = (x, y)$ is in $\mathcal{E}$ if and only if

$$\text{dist}(\vec{F_1}, \vec{P}) + \text{dist}(\vec{F_2}, \vec{P}) = 2a$$

or

$$\sqrt{(x+c)^2 + y^2} + \sqrt{(x-c)^2 + y^2} = 2a$$

or

$$\sqrt{(x+c)^2 + y^2} = 2a - \sqrt{(x-c)^2 + y^2}.$$

Squaring both sides, we obtain

$$cx + a^2 = a\sqrt{(x+c)^2 + y^2}.$$

Again squaring both sides and simplifying yields:

$$x^2(c^2 - a^2) - a^2 y^2 = a^2(c^2 - a^2).$$

Now, if we let $b^2 = a^2 - c^2$, and divide both sides by $a^2 b^2$, we get the equation

$$\frac{x^2}{a^2} + \frac{y^2}{b^2} = 1, \text{ where } b^2 = a^2 - c^2.$$

See Figure 5.a. If we go through the same process with $\vec{F_1} = (0, -c)$ and $\vec{F_2} = (0, c)$, we obtain the equation

$$\frac{y^2}{a^2} + \frac{x^2}{b^2} = 1, \text{ where } b^2 = a^2 - c^2.$$

See Figure 5.b.

The preceding construction yields following theorem.

Theorem 1 *Suppose that $a^2 > b^2$ and that $c^2 = a^2 - b^2$. Then*

$$\frac{x^2}{a^2} + \frac{y^2}{b^2} = 1$$

is an equation for the ellipse with foci $(-c, 0)$ and $(c, 0)$, and

$$\frac{y^2}{a^2} + \frac{x^2}{b^2} = 1$$

is an equation for the ellipse with foci $(0, -c)$ and $(0, c)$.

EXAMPLE 4: $4x^2 + 9y^2 = 25$ and $4y^2 + 9x^2 = 25$ are equations for ellipses in standard position. In both cases $a^2 = \frac{25}{4}$ and $b^2 = \frac{25}{9}$. Rewriting the given equations in the form suggested by Theorem 2, we have

$$4x^2 + 9y^2 = 25 \quad \text{or} \quad \frac{x^2}{\frac{25}{4}} + \frac{y^2}{\frac{25}{9}} = 1 \quad \text{(Figure 6.a)}$$

and

$$4y^2 + 9x^2 = 25 \quad \text{or} \quad \frac{y^2}{\frac{25}{4}} + \frac{x^2}{\frac{25}{9}} = 1 \quad \text{(Figure 6.b)}.$$

The foci for the graph of $4x^2 + 9y^2 = 25$ are $\left(\pm\frac{5\sqrt{5}}{6}, 0\right)$. The foci for the graph of $4y^2 + 9x^2 = 25$ are $\left(0, \pm\frac{5\sqrt{5}}{6}\right)$. ∎

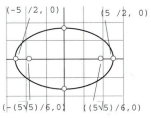

Figure 6.a *The graph of $4x^2 + 9y^2 = 25$.*

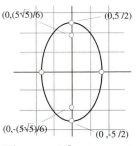

Figure 6.b *The graph of $4y^2 + 9x^2 = 25$.*

EXAMPLE 5: Find an equation for the ellipse with foci $(-1, 1)$ and $(1, 2)$, and $a = 2$.

SOLUTION: The point $\vec{P} = (x, y)$ is in $\mathcal{E}$ if and only if

$$\text{dist}(\vec{P}, \vec{F}_1) + \text{dist}(\vec{P}, \vec{F}_2) = 4$$

or

$$\sqrt{(x+1)^2 + (y-1)^2} + \sqrt{(x-1)^2 + (y-2)^2} = 4$$

or

$$\sqrt{(x+1)^2 + (y-1)^2} = 4 - \sqrt{(x-1)^2 + (y-2)^2}.$$

Squaring both sides and simplifying, we obtain

$$4x + 2y - 19 = -8\sqrt{(x-1)^2 + (y-2)^2}.$$

Again, squaring both sides:

$$48x^2 - 16xy + 60y^2 + 24x - 180y - 41 = 0. \qquad \blacksquare$$

The above example shows that equations for ellipses that are not in standard position can be unwieldy and complicated.

The Hyperbola

If a comet falls under the sun's gravitational field and if it is traveling fast enough, then its trajectory will be a hyperbola. The definition we give for the hyperbola is similar to the one given for the ellipse.

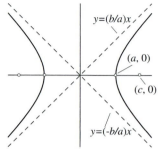

Figure 7.a *The graph of $\frac{x^2}{a^2} - \frac{y^2}{b^2} = 1$.*

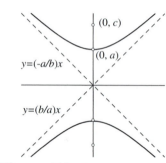

Figure 7.b *The graph of $\frac{y^2}{a^2} - \frac{x^2}{b^2} = 1$.*

Definition: The Hyperbola

The set $\mathcal{H}$ in the plane is a hyperbola if and only if there are points $\vec{F}_1$ and $\vec{F}_2$ a distance of $2c$ apart, and a number $a < c$ such that $\vec{P}$ is in $\mathcal{H}$ if and only if

$$|\text{dist}(\vec{F}_1, \vec{P}) - \text{dist}(\vec{F}_2, \vec{P})| = 2a.$$

- The points $\vec{F}_1$ and $\vec{F}_2$ are the *foci* of the hyperbola $\mathcal{H}$.

- The line containing the foci is the *axis*.

- The points of intersection of the axis with $\mathcal{H}$ are the *vertices* of the hyperbola.

- If the foci are of the form $\vec{F}_1 = (-c, 0)$ and $\vec{F}_2 = (c, 0)$, or of the form $\vec{F}_1 = (0, -c)$ and $\vec{F}_2 = (0, c)$, then the hyperbola is in *standard position*. Figures 7.a and 7.b illustrate the two possible cases for an ellipse in standard position.

We have a theorem for hyperbolas that corresponds to Theorem 1 for ellipses.

Theorem 2 *Suppose that $c^2 = a^2 + b^2$. Then*

$$\frac{x^2}{a^2} - \frac{y^2}{b^2} = 1 \tag{1}$$

is an equation for the hyperbola with foci $(-c, 0)$ and $(c, 0)$, as illustrated in Figure 7.a. Similarly,

$$\frac{y^2}{a^2} - \frac{x^2}{b^2} = 1 \tag{2}$$

is an equation for the hyperbola with foci $(0, -c)$ and $(0, c)$, as illustrated in Figure 7.b.

The argument for this theorem is similar to the one for Theorem 1 and is left as an exercise.

EXAMPLE 6: $4x^2 - 9y^2 = 25$ and $4y^2 - 9x^2 = 25$ are equations for hyperbolas in standard position. $a^2 = \frac{25}{4}$ and $b^2 = \frac{25}{9}$. The axis for the graph of $4x^2 - 9y^2 = 25$ is the x–axis and the axis for the graph of $4y^2 - 9x^2 = 25$ is the y–axis. The foci for the graph of $4y^2 - 9x^2 = 25$ are $\left(0, \pm\frac{5\sqrt{13}}{6}\right)$.   ∎

We can gain some insight into the shape of the graph of a hyperbola by setting the right-hand side of Equation (1) equal to 0. This yields an equation for a pair of lines

$$y = \pm\frac{bx}{a}.$$

These lines are called the *asymptotes* for the hyperbola for Equation (1). Similarly, if we set the right-hand side of Equation (2) equal to 0, we obtain an equation for its asymptotes:

$$y = \pm\frac{ax}{b}.$$

The relationship between a hyperbola and its asymptotes is illustrated in Figures 7.a and 7.b.

If we solve for y in both of the equations

$$\frac{y^2}{a^2} - \frac{x^2}{b^2} = 1 \quad \text{and} \quad \frac{y^2}{a^2} - \frac{x^2}{b^2} = 0,$$

we obtain

$$y = \pm a\sqrt{\frac{x^2}{b^2} - 1}, \tag{3}$$

and

$$y = \pm a\sqrt{\frac{x^2}{b^2}} = \pm a\frac{|x|}{b}. \tag{4}$$

If the values for $|x|$ in Equations (3) and (4) are very large, then there is very little difference in the corresponding values for y. For example, what is the difference between

$$\sqrt{(\text{the national debt})^2 - 1} \quad \text{and} \quad (\text{the national debt})?$$

Therefore, we see that the hyperbola is well approximated by its asymptotes for large values of $|x|$. ∎

EXERCISES

In Exercises 1–6, find the equations of all parabolas in standard position satisfying the given conditions.

1. Focus $(\pi, 0)$.

2. Focus $(0, \pi)$.

3. Focus $(-\pi, 0)$.

4. Focus $(0, -\pi)$.

5. Contains the point $\left(2, \frac{5}{2}\right)$.

6. Contains the point $(-1, 2)$.

In Exercises 7–10, find the equation of the ellipse in standard position satisfying the given conditions.

7. Contains the points $(8, 0)$ and $(0, 3)$.

8. Foci $(0, \pm 2)$ and contains the point $(0, 3)$.

9. Foci $(\pm 2, 0)$ and contains the point $(1, -1)$.

10. Contains the points $(0, -2)$ and $(1, 1)$.

Find the foci and vertices of the ellipses in Exercises 11–13.

11. $x^2 + 4y^2 = 6$.

12. $\frac{x^2}{2} + 3y^2 = 5$.

13. $4x^2 + 9y^2 = 144$.

In Exercises 14–16, find the vertices, foci, and asymptotes of the following hyperbolas, and sketch their graphs.

14. $5y^2 - 4x^2 = 40$.

15. $4x^2 - 25y^2 = 100$.

16. $16y^2 - 9x^2 - 324 = 0$.

In Exercises 17–21, find equations for all the hyperbolas satisfying the given conditions. (a and b are as in Equations (1) and (2), and $c^2 = a^2 + b^2$.)

17. $a = 2$, $b = 5$.

18. $b = 3$, $c = 1$.

19. Foci $(0, 2)$ and $(0, -2)$ and $a = 1$.

20. $a = 1$, and one asymptote is $y = 3x$.

21. $b = 1$, and one asymptote is $y = 3x$.

In Exercises 22–25, $\mathcal{C}$ is the graph of $Ax^2 + By^2 = 1$. Given that $\mathcal{C}$ contains $\vec{P}_1$ and $\vec{P}_2$, determine whether $\mathcal{C}$ is a hyperbola or an ellipse.

22. $\vec{P}_1 = (1, 1)$, $\vec{P}_2 = (2, 5)$.

23. $\vec{P}_1 = (1, 1)$, $\vec{P}_2 = (5, 2)$.

24. $\vec{P}_1 = (2, 5)$, $\vec{P}_2 = (5, 2)$.

25. $\vec{P}_1 = (1, 1)$, $\vec{P}_2 = (-1, 1)$.

26. Show that an equation of the parabola with vertex (h, k) and focus $(h + c, k)$ is $(y - k)^2 = 4c(x - h)$. Similarly, if the focus is $(h, k+c)$, show that an equation of the parabola is $(x - h)^2 = 4c(y - k)$.

27. Find an equation for the set of points xyz–space that are equidistant from the point $(1, 0, 0)$ and the plane $x = -1$.

28. Find an equation for the set of points in xyz–space that are equidistant from the point $(0, 1, 0)$ and the plane $y = -1$.

29. Find an equation for the set of points $\vec{P}$ in xyz–space that are twice as far from $(0, 1, 0)$ as from the plane $y = -1$.

30. Show that an equation for the set of points in xyz–space that are equidistant from the point $(c, 0, 0)$ and the plane $x = -c$ is $y^2 + z^2 = 4cx$.

31. Show that an equation for the set of points in xyz–space that are equidistant from the point $(0, c, 0)$ and the plane $y = -c$ is $x^2 + z^2 = 4cy$.

32. Show that an equation for the set of points in xyz–space that are equidistant from the point $(0, 0, c)$ and the plane $z = -c$ is $x^2 + y^2 = 4cz$.

33. Find an equation for the set of all points in xyz–space such that the sum of the distances from (x, y, z) to the points $(1, 0, 0)$ and $(-1, 0, 0)$ is 4.

34. Find an equation for the set of all points in xyz–space such that the sum of the distances from (x, y, z) to the points $(0, 1, 0)$ and $(0, -1, 0)$ is 4.

35. Find an equation for the set of all points in xyz–space such that the sum of the distances from (x, y, z) to the points $(0, 0, 1)$ and $(0, 0, -1)$ is 4.

36. Show that $\frac{x^2}{a^2} + \frac{y^2 + z^2}{a^2 - c^2} = 1$ is an equation for the set of all points in xyz–space such that the sum of the distances from (x, y, z) to the points $(c, 0, 0)$ and $(-c, 0, 0)$ is $2a$.

37. Show that $\frac{y^2}{a^2} + \frac{x^2 + z^2}{a^2 - c^2} = 1$ is an equation for the set of all points in xyz–space such that the sum of the distances from (x, y, z) to the points $(0, c, 0)$ and $(0, -c, 0)$ is $2a$.

38. Let $c > 0$, $\vec{F}_1 = (0, c)$, and $\vec{F}_2 = (0, -c)$, and let $a > 0$ be a number less than c. Find an equation for the set of points (x, y) for which the absolute difference between the distances from (x, y) to $\vec{F}_1$ and from (x, y) to $\vec{F}_2$ is equal to $2a$. Argue that the set so described is a hyperbola in standard position.

39. Let $c > 0$, $\vec{F}_1 = (0, c)$, and $\vec{F}_2 = (0, -c)$, and let $a > 0$ be a number greater than c. Find an equation for the set of points (x, y) for which the absolute difference between the distances from (x, y) to $\vec{F}_1$ and from (x, y) to $\vec{F}_2$ is equal to $2a$. Argue that the set so described is an ellipse.

40. Let $\vec{F}_1 = (2, 0, 0)$ and $\vec{F}_2 = (-2, 0, 0)$. Find an equation for the set of points $\vec{r} = (x, y, z)$ in xyz–space such that $|\text{dist}(\vec{F}_1, \vec{r}) - \text{dist}(\vec{F}_2, \vec{r})| = 1$.

41. Let $\vec{F}_1 = (0, 2, 0)$ and $\vec{F}_2 = (0, -2, 0)$. Find an equation for the set of points $\vec{r} = (x, y, z)$ in xyz–space such that $|\text{dist}(\vec{F}_1, \vec{r}) - \text{dist}(\vec{F}_2, \vec{r})| = 1$.

42. Let $\vec{F}_1 = (0, 0, 2)$ and $\vec{F}_2 = (0, 0, -2)$. Find an equation for the set of points $\vec{r} = (x, y, z)$ in xyz-space such that $|\text{dist}(\vec{F}_1, \vec{r}) - \text{dist}(\vec{F}_2, \vec{r})| = 1$.

Kepler's Laws 2: *The remainder of the exercises present conic sections in terms of their eccentricity. This is a viewpoint that sets the stage for deriving Kepler's Third Law from Newton's Law of Gravitation.*

43. Let $\vec{F} = (0.5, 0)$, and let $\mathcal{D}$ be the line $x = 2$. Let $\mathcal{C}$ be the set of all points $\vec{q}$ such that the distance from $\vec{q}$ to $\vec{F}$ is half the distance from $\vec{q}$ to the line $\mathcal{D}$. Find an equation for $\mathcal{C}$ and characterize its shape.

44. Let $\vec{F} = (2, 0)$, and let $\mathcal{D}$ be the line $x = 0.5$. Let $\mathcal{C}$ be the set of all points $\vec{q}$ such that the distance from $\vec{q}$ to $\vec{F}$ is twice the distance from $\vec{q}$ to the line $\mathcal{D}$. Find an equation for $\mathcal{C}$ and characterize its shape.

In Exercises 45–48, $e \neq 1$ and a are positive numbers. Let $\vec{F} = (ae, 0)$, and $\mathcal{D}$ be the line $x = \frac{a}{e}$. Notice that if $e < 1$, then $\vec{F}$ is between the origin and $\mathcal{D}$. If $e > 1$, then $\mathcal{D}$ is between the origin and $\vec{F}$. Let $\mathcal{C}$ be the set of all points $\vec{q}$ such that the distance from $\vec{q}$ to $\vec{F}$ is e times the distance from $\vec{q}$ to the line $\mathcal{D}$.

45. Show that the equation for $\mathcal{C}$ can be written as

$$\frac{x^2}{a^2} + \frac{y^2}{a^2(1 - e^2)} = 1.$$

46. Show that if we place the focus at the point $(0, ae)$ and let the directrix be the line $y = \frac{a}{e}$, then we get equation

$$\frac{y^2}{a^2} + \frac{x^2}{a^2(1 - e^2)} = 1.$$

47. Argue that if $0 < e < 1$, then $\mathcal{C}$ is an ellipse, and if $e > 1$, then $\mathcal{C}$ is a hyperbola.

Place the focus at the origin, and let the directrix be the line $x = d$. Assume that $d > 0$. Exercises 48–50 will use the eccentricity to obtain vector equations for the conics.

48. Argue that an equation for the set of all vectors $\vec{r}$ in the plane of length e times the distance from the position $\vec{r}$ to the line $x = d$ is

$$\|\vec{r}\| = |ed - (e, 0) \cdot \vec{r}| = e|d - \|\vec{r}\| \cos \theta|. \quad (5)$$

49. Argue that if $e \leq 1$, then Equation (5) can be written as

$$\|\vec{r}\| = \frac{ed}{1 + e \cos \theta}.$$

50. Argue that if $e > 1$, then $\|\vec{r}\| = \frac{ed}{1 + e \cos \theta}$ gives only one branch of the hyperbola. In this case, what are the restrictions on θ? Show that the other branch can be obtained with the equation $\|\vec{r}\| = \frac{ed}{e \cos \theta - 1}$. What are the restrictions on θ in this equation?

C.2   Some 3–Dimensional Graphs

In this section, we learn how to visualize some sets in 3-dimensional space by intersecting these sets with planes that are parallel to the coordinate planes.

EXAMPLE 1: The graph of $x^2 + y^2 = 4z$ is illustrated in Figure 1.a. It will intersect the plane $z = z_0$ only if $z_0 \geq 0$. Let $z_0 > 0$. To see how this graph will intersect the plane $z = z_0$, we set $z = z_0$ in the equation $x^2 + y^2 = 4z$. This yields an equation of a circle: $x^2 + y^2 = 4z_0$. See Figure 1.b. The graph of $x^2 + y^2 = 4z$ will intersect the yz-plane in the parabola $y^2 = 4z$. This is illustrated in Figure 1.c. ∎

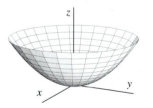

Figure 1.a *The graph of $x^2 + y^2 = 4z$.*

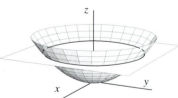

Figure 1.a *The intersection of the graph of $x^2 + y^2 = 4z$ with the plane $z = z_0$.*

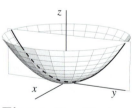

Figure 1.c *The intersection of the graph of $x^2 + y^2 = 4z$ with the yz–plane.*

EXAMPLE 2: Consider the graph of

$$\frac{x^2}{a^2} + \frac{y^2}{b^2} = z.$$

If $a^2 = b^2$, then, as in Example 1, this is a "paraboloid" that intersects every horizontal plane $z = z_0$, $(z_0 > 0)$ in the circle $x^2 + y^2 = a^2 z_0$. If $a^2 \neq b^2$, then the graph will intersect the horizontal plane $z = z_0$, $(z_0 > 0)$ in the ellipse $\frac{x^2}{a^2} + \frac{y^2}{b^2} = z_0$. The next example considers the case where $a = 2$ and $b = 1$. ∎

EXAMPLE 3: The graph of $\frac{x^2}{4} + y^2 = z$ intersects the xz–coordinate plane in the parabola $x^2 = 4z$. We may think of this as the "side view" of the graph. To get a "front view" of the graph, we look at its intersection with the yz–coordinate plane to obtain the parabola $y^2 = z$. However, to obtain a "top view" of the graph, we will look at its intersection with a plane of the form $z = z_0$, $(z_0 > 0)$. This yields an equation for an ellipse $\frac{x^2}{4} + y^2 = z_0$. See Figure 2. ∎

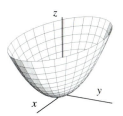

Figure 2.a
The graph of
$\frac{x^2}{4} + y^2 = z$.

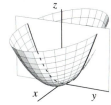

Figure 2.b
The graph of
$\frac{x^2}{4} + y^2 = z$
intersecting the plane
$z = z_0$.

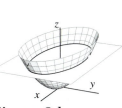

Figure 2.c
The graph of
$\frac{x^2}{4} + y^2 = z$
intersecting the plane
$x = 0$.

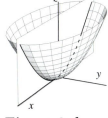

Figure 2.d
The graph of
$\frac{x^2}{4} + y^2 = z$
intersecting the plane
$y = 0$.

Figure 2.e *"Top view."*

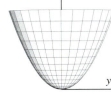

Figure 2.f *"Front view."*

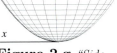

Figure 2.g *"Side view."*

EXAMPLE 4: As we illustrate in Figure 3, the graph of $\frac{z^2}{4} + y^2 = x$ intersects the plane $x = x_0$, $(x_0 > 0)$ in the ellipse $\frac{z^2}{4} + y^2 = x_0$ while it intersects each of the xz and xy–coordinate planes in parabolas. ∎

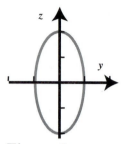

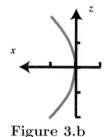

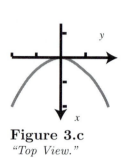

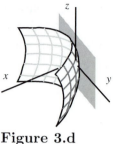

Figure 3.a
"Front View."

Figure 3.b
"Side View."

Figure 3.c
"Top View."

Figure 3.d
The graph of
$\frac{z^2}{4} + y^2 = x.$

Figure 3.a provides the "front view," which is the intersection of the graph of $\frac{z^2}{4} + y^2 = x$ with the plane $x = 1$. Figure 3.b provides the "side view," which is the intersection of the graph with the plane $y = 0$. In Figure 3.c, we have the intersection of the graph with the plane $z = 0$.

EXAMPLE 5: The graph of $\frac{z^2}{4} + \frac{x^2}{9} + y^2 = 1$ intersects each of the coordinate planes in a different ellipse as in Figure 4.

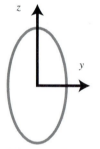

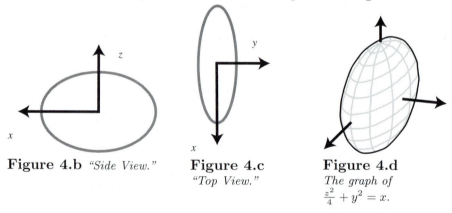

Figure 4.a
"Front View."

Figure 4.b *"Side View."*

Figure 4.c
"Top View."

Figure 4.d
The graph of
$\frac{z^2}{4} + y^2 = x.$

Figure 4.a provides the "front view," which is the intersection of the graph of $\frac{z^2}{4} + \frac{x^2}{9} + y^2 = 1$ with the plane $x = 1$. Figure 4.b provides the "side view," which is the intersection of the graph with the plane $y = 0$. In Figure 4.c, we have the intersection of the graph with the plane $z = 0$.

EXAMPLE 6: The graph of $\frac{x^2}{4} - y^2 = z$ intersects the plane $y = 0$ in the parabola $x^2 = 4z$, and it intersects the plane $x = 0$ in the parabola $y^2 = -z$. Thus the front and side views of the graph are fairly simple. However, the shape of the "top view" of the graph depends on the level of the horizontal plane used to "slice" the graph. If we look at the intersection of the graph with the plane $z = 0$, then we obtain the pair of lines $\frac{x^2}{4} - y^2 = 0$. These are the asymptotes for the other "horizontal slices," which are hyperbolas. This is illustrated in Figure 5. The computer drawing of this graph shows it to have a "saddle-like" appearance. ∎

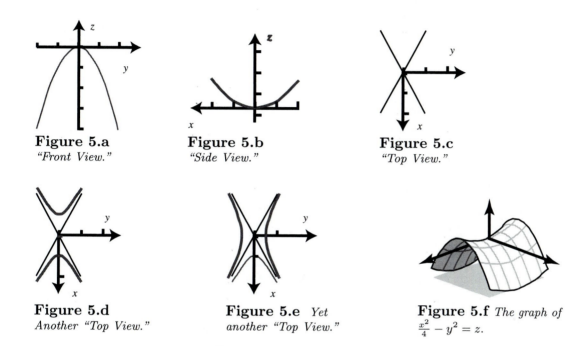

Figure 5.a
"Front View."

Figure 5.b
"Side View."

Figure 5.c
"Top View."

Figure 5.d
Another "Top View."

Figure 5.e *Yet another "Top View."*

Figure 5.f *The graph of* $\frac{x^2}{4} - y^2 = z.$

Figure 5.a provides the "front view," which is the intersection of the graph of $\frac{x^2}{4} - y^2 = z$ with the plane $x = 0$. Figure 5.b provides the "side view," which is the intersection of the graph with the plane $y = 0$. In Figure 5.c, we have the intersection of the graph with the plane $z = 0$. Notice that this is the graph of the asymptotes of the hyperbolas obtained by intersecting the graph with the plane $z = 1$ (Figure 5.d) and the plane $z = -1$ as in Figure 5.e. The graph is sketched in Figure 5.f.

EXERCISES

In Exercises 1–7, determine how the graph of $x^2 + y^2 = z^2$ intersects the given plane.

1. $x = 0$.   2. $x = 1$.   3. $x = -1$.   4. $y = 0$.

5. $y = 3$.   6. $z = 0$.   7. $z = -1$.

In Exercises 8–13, determine how the graph of $\frac{x^2}{4} + \frac{y^2}{9} = z$ intersects the given planes.

8. The xy–plane.   9. The xz–plane.

10. The yz–plane.   11. $x = 1$.

12. $z = 1$.   13. $z = -1$.

In Exercises 14–18, determine how the graph of $x^2/4 + z^2/9 = y$ intersects the given plane.

14. The xy–plane.   15. The xz–plane.

16. The yz–plane.   17. $x = 1$.

18. $y = 1$.

19. Determine how the graph of $\dfrac{x^2}{4} + \dfrac{y^2}{9} + z^2 = 1$ intersects each of the coordinate planes.

20. Determine how the graph of $\dfrac{x^2}{4} - \dfrac{y^2}{9} - z^2 = 1$ intersects each of the coordinate planes.

21. Match the equation with its graph.

    a. $x^2 - 2z^2 + y = 2$.

    b. $x^2 - 2y^2 + z = 0$.

    c. $x^2 - y^2 + 4z^2 - 4 = 0$.

    d. $x^2 - 4y^2 + 2z^2 - 1 = 0$.

    e. $x^2 - 4y^2 - 2z^2 - 1 = 0$.

    f. $2x^2 - 4y + z = 0$.

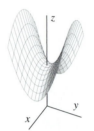

Figure 6.a   Figure 6.b   

Figure 6.c

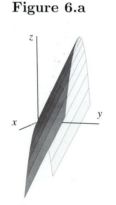

Figure 6.d   Figure 6.e   Figure 6.f

C.3   Translation in $\mathbb{R}^n$

Translation of a set $\mathcal{A}$ by a vector $\vec{v}$ is a movement of $\mathcal{A}$ rigidly in space without rotating the set. This is done by adding $\vec{v}$ to each point in $\mathcal{A}$.

Definition: Translation of a Set by a Vector

Suppose that $\mathcal{A}$ is a set in $\mathbb{R}^n$. Then we define a new set $\mathcal{A} + \vec{v}$ to be that set obtained by adding $\vec{v}$ to each member of $\mathcal{A}$. The set $\mathcal{A} + \vec{v}$ is called the *translation of $\mathcal{A}$ by $\vec{v}$.*

EXAMPLE 1: If $\mathcal{A} = \{(1,3,2),\ (5,\pi,-3),\ (4,-1,6)\}$, and $\vec{v} = (1,2,3)$, then $\mathcal{A} + \vec{v} = \{(2,5,5), (6,\pi+2,0),\ (5,1,9)\}$. ∎

EXAMPLE 2: Let $\mathcal{C}$ be the unit circle with center at the origin, and let $\vec{v} = (6,1)$. Then $\mathcal{C} + \vec{v}$ is the unit circle with center $(6,1)$. (See Figure 1.)

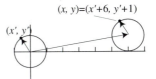

Figure 1. *The unit circle centered at the origin, translated by $(6,1)$.*

We have that (x,y) is in $\mathcal{C} + \vec{v}$ if and only if there is an (x',y') in $\mathcal{C}$ such that

$$(x,y) = (x',y') + \vec{v} = \begin{pmatrix} x' + 6 \\ y' + 1 \end{pmatrix}$$

But (x',y') is in $\mathcal{C}$ if and only if $x'^2 + y'^2 = 1$. Thus (x,y) is in $\mathcal{C} + \vec{v}$ if and only if $(x-6)^2 + (y-1)^2 = 1$. This, of course, agrees with the equation for a circle with radius 1 and center $(6,1)$. ∎

There is nothing unique about the equation of a circle. Indeed, we have the following theorem.

Theorem 1 *If $\mathcal{E}(x_1, x_2, \ldots, x_n)$ is an expression for the set $\mathcal{A}$ in $\mathbb{R}^n$ and $\vec{h} = (h_1, \ldots, h_n)$ is a vector, then $\mathcal{E}(x_1 - h_1, x_2 - h_2, \ldots, x_n - h_n)$ is an expression for the set $\mathcal{A} + \vec{h}$*

EXAMPLE 3: Let $\mathcal{S}$ be the sphere given by the equation $x^2 + y^2 + z^2 = 2$. Then $(x-1)^2 + (y-2)^2 + (z-3)^2 = 2$ is an equation for $\mathcal{S} + (1,2,3)$. ∎

EXAMPLE 4: Let $\mathcal{L}$ be the line $y = 2x$. Then $(y-3) = 2(x-1)$ is an equation for the line $\mathcal{L} + (1,3)$. ∎

Translation of Conic Sections

Consider $\mathcal{C}$, the graph of the equation $Ax^2 + By^2 + C = 0$. We know that $\mathcal{C}$ is a conic section in standard position, a pair of lines passing through the origin, the origin, or the empty set. An equation for $\mathcal{C}$ translated by $\vec{v} = (h,k)$ is

$$A(x-h)^2 + B(y-k)^2 + C = 0.$$

If we multiply everything out, we get something of the form

$$Ax^2 + By^2 + Dx + Ey + F = 0. \tag{6}$$

Now, suppose that we start off with Equation (6).

Case 1: $A = B = 0$. Then Equation (6) yields a straight line.

Case 2: $A \neq 0$ and $B \neq 0$. We start by completing the square in x and y to obtain

$$A\left(x^2 + \frac{Dx}{A} + \left(\frac{D}{2A}\right)^2\right) + B\left(y^2 + \frac{Ey}{B} + \left(\frac{E}{2B}\right)^2\right)$$

$$= -F + \frac{D^2}{4A} + \frac{E^2}{4B}.$$

Dividing by AB and factoring:

$$\frac{\left(x + \frac{D}{2A}\right)^2}{B} + \frac{\left(y + \frac{E}{2B}\right)^2}{A} = -\frac{F}{AB} + \frac{D^2}{4A^2B} + \frac{E^2}{4B^2A} \tag{7}$$

If $-\frac{F}{AB} + \frac{D^2}{4A^2B} + \frac{E^2}{4B^2A} = 0$ and A and B are of opposite signs, then Equation (7) describes a pair of lines. In the event that A and B are of the same sign, Equation (7) describes a single point $\left(-\frac{D}{2A}, -\frac{E}{2B}\right)$.

If $-\frac{F}{AB} + \frac{D^2}{4A^2B} + \frac{E^2}{4B^2A} \neq 0$, then we can divide both sides of Equation (7) by the right hand side to obtain:

$$\frac{(x-h)^2}{U} + \frac{(y-k)^2}{V} = 1, \tag{8}$$

where

$$h = -\left(\frac{D}{2A}\right)^2$$

$$k = -\left(\frac{E}{2B}\right)^2$$

$$U = B\left(-\frac{F}{AB} + \frac{D^2}{4A^2B} + \frac{E^2}{4B^2A}\right)$$

$$V = A\left(-\frac{F}{AB} + \frac{D^2}{4A^2B} + \frac{E^2}{4B^2A}\right)$$

Of course, Equation (8) is a translation of $\frac{x^2}{U} + \frac{y^2}{V} = 1$, which we know from earlier sections to be either an ellipse, a hyperbola, or the empty set.

Case 3: If either A or B are 0, but not both of them, then we can use a similar argument to show that we have a translation of a parabola in standard position, the empty set, one of the coordinate axes, or a pair of parallel lines. (See Exercise 7.)

We have proved:

Theorem 2 *The graph of $Ax^2 + By^2 + Dx + Ey + F = 0$ is a translation of a conic section in standard position, a line, a pair of lines, a point, or the empty set.*

EXAMPLE 5: Find an equation for a conic section $\mathcal{C}$ in standard position and a vector $\vec{r}$ such that the graph of $2x^2 + 3y^2 + 8x - 18y + 29 = 0$ is a translation of $\mathcal{C}$ by $\vec{r}$.

SOLUTION: The approach to this type of problem is to complete the square in x and in y.

$$
\begin{aligned}
2x^2 + 3y^2 + 8x - 18y + 29 &= 2(x^2 + 4x) + 3(y^2 - 6y) + 29 \\
&= 2(x^2 + 4x + 4) + 3(y^2 - 6y + 9) - 6 \\
&= 2(x + 2)^2 + 3(y - 3)^2 - 6 \\
&= 0.
\end{aligned}
$$

Reorganizing the last equation gives $\frac{(x+2)^2}{3} + \frac{(y-3)^2}{2} = 1$. Let $\mathcal{C}$ be the graph of $\frac{(x)^2}{3} + \frac{(y)^2}{2} = 1$ and let $\mathcal{C}'$ be the graph of $\frac{(x+2)^2}{3} + \frac{(y-3)^2}{2} = 1$. $\mathcal{C}'$ is simply $\mathcal{C}$ translated by $(-2, 3)$. The graphs of $\mathcal{C}'$ and $\mathcal{C}$ are sketched in Figure 2.

The center of $\mathcal{C}$ is $(0, 0)$, and the center of $\mathcal{C}'$ is $(0, 0) + (-2, 3) = (-2, 3)$. The vertices of $\mathcal{C}$ are $(\sqrt{3}, 0)$ and $(-\sqrt{3}, 0)$, and the vertices of $\mathcal{C}'$ are $(\sqrt{3}, 0) + (-2, 3) = (\sqrt{3} - 2, 3)$ and $(-\sqrt{3}, 0) + (-2, 3) = (-\sqrt{3} - 2, 3)$. The foci of $\mathcal{C}$ are $(1, 0)$ and $(-1, 0)$, and the foci of $\mathcal{C}'$ are $(1, 0) + (-2, 3) = (-1, 3)$ and $(-1, 0) + (-2, 3) = (-3, 3)$. $\blacksquare$

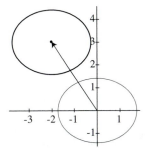

Figure 2. *The graphs of $2x^2 + 3y^2 + 8x - 18y + 29 = 0$ and $2x^2 + 3y^2 = 6$.*

EXAMPLE 6: Find an equation of the parabola $\mathcal{P}'$ with vertex $(5, 2)$ and focus $(3, 2)$. See Figure 3.

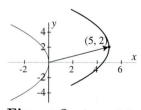

Figure 3. *A parabola translated by* $(5,2)$.

SOLUTION: The axis of $\mathcal{P}'$ is the line $y = 2$. Because the focus is to the left of the vertex, $\mathcal{P}'$ must open to the left. Therefore, $\mathcal{P}'$ is a translation by $(5,2)$ of a parabola $\mathcal{P}$ with equation of the form $y^2 = -4cx$, where c is the distance from the focus to the vertex of $\mathcal{P}$. Thus $c = 2$, or $p = 4$. An equation for $\mathcal{P}$ is $y^2 = -8x$, and the corresponding equation for $\mathcal{P}'$ is $(y - 2)^2 = -8(x - 5)$. ∎

EXAMPLE 7: Find the equation of the ellipse $\mathcal{E}$ with foci $(3, 2)$ and $(5, 2)$, which contains the point $(1, 1)$.

SOLUTION: The center of $\mathcal{E}$ will be $(4, 2)$. The axis of $\mathcal{E}$ is parallel to the x–axis, and so, $\mathcal{E}$ is the translation of an ellipse $\mathcal{E}'$ in standard position by the vector $(4, 2)$. The axis for $\mathcal{E}'$ is the x–axis, and $\mathcal{E}'$ has an equation of the form

$$\frac{x^2}{a^2} + \frac{y^2}{b^2} = 1.$$

Now, $2a$ is the sum of the distances between $(1,1)$ and the foci of $\mathcal{E}$. Thus

$$2a = \sqrt{(3 - 1)^2 + (2 - 1)^2} + \sqrt{(5 - 1)^2 + (2 - 1)^2} = \sqrt{5} + \sqrt{17},$$

or

$$a = \frac{\sqrt{5} + \sqrt{17}}{2} \quad \Rightarrow \quad a^2 = \frac{11 + \sqrt{85}}{2}.$$

Since the distance between the foci is $2c$, we have that c=1. Thus

$$b^2 = a^2 - c^2 = a^2 = \frac{11 + \sqrt{85}}{2} - 1 = \frac{9 + \sqrt{85}}{2}.$$

An equation for $\mathcal{E}'$ is given by

$$\frac{2x^2}{11 + \sqrt{85}} + \frac{2y^2}{9 + \sqrt{85}} = 1.$$

Thus an equation for $\mathcal{E} = \mathcal{E}' + (4, 2)$ is given by

$$\frac{2(x - 4)^2}{11 + \sqrt{85}} + \frac{2(y - 2)^2}{9 + \sqrt{85}} = 1.$$ ∎

EXERCISES

In Exercises 1 and 2, determine $\mathcal{A} + \vec{r}$.

1. $\mathcal{A} = \{(1, 3),\ (5, 20),\ (-3, 1)\}$, $\vec{r} = (1, -2)$.

2. $\mathcal{A} = \{(-1, \frac{5}{6}, 3),\ (-5, -1, 2),\ (3, 6, -1),\ (1, -3, 1)\}$, $\vec{r} = (-\frac{1}{3}, \frac{3}{4}, 2)$.

In Exercises 3–11, $\mathcal{G}$ is the graph of the given equation. Sketch $\mathcal{G}$ and $\mathcal{G} + \vec{r}$, and find an equation for $\mathcal{G} + \vec{r}$.

3. $x + 3y = 0$, $\vec{r} = (1, 2)$.

4. $x + 3y + 2 = 0$, $\vec{r} = (1, 2)$.

5. $x^2 + y^2 = 1$, $\vec{r} = (3, -1)$.

6. $4x^2 + 9y^2 = 36$, $\vec{r} = \left(\frac{1}{3}, \frac{1}{2}\right)$.

7. $2y^2 + 3x = 0$, $\vec{r} = (2, 1)$.

8. $4x^2 - 9y^2 = 0$, $\vec{r} = (3, 2)$.

9. $4x^2 + 9y^2 = 0$, $\vec{r} = (3, 2)$.

10. $4x^2 + 9y^2 = 1$, $\vec{r} = (3, 2)$.

11. $4x^2 - 9y^2 = 1$, $\vec{r} = (3, 2)$.

For the given line in Exercises 12–14, find an equation of the line L passing through the origin and a vector $\vec{r}$ so that the given line is a translation of L by $\vec{r}$.

12. $x + y = 3$.

13. $2x + 3y = 5$.

14. $3x + \frac{y}{2} = 2$.

In Exercises 15–23, you are given an equation of a conic section $\mathcal{C}'$. Find an equation of a conic section $\mathcal{C}$ in standard position and a vector $\vec{r}$ so that the given conic section is a translation of $\mathcal{C}$ by $\vec{r}$. Find the vertices and the foci of both $\mathcal{C}$ and $\mathcal{C}'$. Sketch $\mathcal{C}'$.

15. $x^2 - 2x - 6y + 8 = 0$.

16. $4x^2 + 9y^2 - 16x + 36y + 16 = 0$.

17. $x^2 - 6x - 4y + 5 = 0$.

18. $x^2 - 10x - 6y + 7 = 0$.

19. $y^2 - 4y + 8x + 12 = 0$.

20. $4x^2 + 32x + 15y^2 + 60y + 68 = 0$.

21. $3x^2 - 6x + 4y^2 + 24y + 38 = 0$.

22. $3x^2 - 4y^2 - 6x - 24y - 34 = 0$.

23. $16x^2 - 64x + 81y^2 + 162y + 109 = 0$.

Find an equation for each of the parabolas given in Exercises 24–27.

24. Focus $(0, 0)$; Vertex $(1, 0)$.

25. Vertex $(-2, -4)$; Directrix $y = 2$.

26. Vertex $(-2, -4)$; Directrix $y = -6$.

27. Vertex $(-2, 1)$, contains the point $(1, 4)$, and its axis is parallel to the x–axis.

Find an equation for each ellipse satisfying the conditions given in Exercises 28 and 29.

28. Vertices $(3, 5)$, $(3, -5)$; Focus $(3, 3)$.

29. Foci $(3, 4)$, $(3, -4)$ and contains the point $(0, 0)$.

In Exercises 30–33, $\mathcal{C}$ is the graph of the equation $Ax^2 + By^2 + Cx + Dy + F = 0$. What must be true about A, B, C, D, and F in order that

30. $\mathcal{C}$ is a parabola?

31. $\mathcal{C}$ is one of the coordinate axes?

32. $\mathcal{C}$ is a pair of parallel lines?

33. $\mathcal{C}$ is a pair of intersecting lines containing $(0, 0)$?

Kepler's Laws 3: *In the remaining exercises, we derive Kepler's First Law from Newton's Law of Gravitation: The Planets describe ellipses with the sun at one focus. Our only assumption will be that if $\vec{r} \times \vec{r}'$ is constant, then the motion is in a plane. These exercises assume that you have worked through Chapter 6.*

34. Show that $\vec{r}(t) \cdot \vec{r}'(t) = \|\vec{r}\| \frac{d}{dt} \|\vec{r}(t)\|$.   Hint: $\|\vec{r}\| = \sqrt{\vec{r} \cdot \vec{r}}$.

35. Show that $\vec{w} \times (\vec{u} \times \vec{v}) = (\vec{w} \cdot \vec{v})\vec{u} - (\vec{w} \cdot \vec{u})\vec{v}$.

36. Let $\mu = GM_1$, where M_1 is the mass of the sun. Assume that the origin is at the sun's center of mass and $\vec{r}(t)$ is the position of a planet and the only force acting on the planet is the sun's gravitational attraction. Show that Newton's Law of Gravitation implies that

$$\vec{r}''(t) = -\mu \frac{\vec{r}(t)}{\|\vec{r}(t)\|^3}. \qquad (9)$$

37. Let

$$\vec{h} = \vec{r} \times \vec{r}'. \qquad (10)$$

Which part of the previous exercise implies that $\vec{h}$ is constant?

38. Argue that $\frac{d}{dt}(\vec{r}'(t) \times \vec{h}) = \vec{r}''(t) \times \vec{h}$.

39. Cross multiply the left side of Equation (9) by $\vec{h}$ and the right side by $\vec{r}(t) \times \vec{r}'(t)$ to obtain $\vec{r}''(t) \times \vec{h} = -\dfrac{\mu}{\|\vec{r}\|^3}\vec{r} \times (\vec{r} \times \vec{r}')$. Use Exercises 34 and 35 to show that

$$\vec{r}''(t) \times \vec{h} = \mu \frac{d}{dt} \frac{\vec{r}}{\|\vec{r}\|}. \qquad (11)$$

40. Argue that integrating both sides of Equation (11) yields

$$\vec{r}' \times \vec{h} = \mu \left(\frac{\vec{r}}{\|\vec{r}\|} + \vec{e} \right), \qquad (12)$$

where $\mu\vec{e}$ is the constant of integration.

41. Dot-multiply both sides of Equation (10) by $\vec{h}$ and Equation (12) by $\vec{r}'(t)$, and use the fact that $\vec{r} \times \vec{r}' \cdot \vec{h} = \vec{r} \cdot \vec{r}' \times \vec{h}$ to obtain

$$\|\vec{h}\|^2 = \mu \left(\|\vec{r}\| + \|\vec{e}\|\|\vec{r}\| \cos\theta \right). \qquad (13)$$

42. Show that if $e < 1$, then the graph of Equation (13) is an ellipse. In this case, are there any constraints on θ?

43. Show that if $e = 1$, then the graph of Equation (13) is a parabola. In this case, are there any constraints on θ?

44. Show that if $e > 1$, then the graph of Equation (13) is a branch of a hyperbola. In this case, are there any constraints on θ?

Appendix D

Geometric Formulas

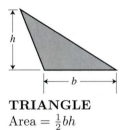

TRIANGLE
Area $= \frac{1}{2}bh$

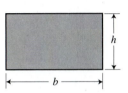

RECTANGLE
Perimeter $= 2b + 2w$
Area $= bw$

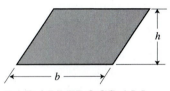

PARALLELOGRAM
Area $= bh$

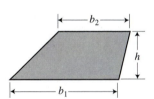

TRAPEZOID
Area $= \frac{1}{2}(b_1 + b_2)h$

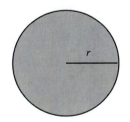

CIRCLE
Perimeter $= 2\pi r$
Area $= \pi r^2$

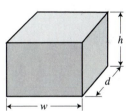

RECTANGULAR BOX
Lateral Area $=$
$\quad 2hw + 2wd + 2hd$
Volume $= hwd$

SPHERE
Surface Area $= 4\pi r^2$
Volume $= \frac{4}{3}\pi r^3$

CYLINDER
Volume $= \pi r^2 h$

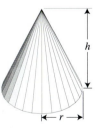

RIGHT   CIRCULAR CONE
Lateral Area $= \pi r\sqrt{r^2 + h^2}$
Volume $= \frac{1}{3}\pi r^2 h$

Appendix E

Answers to Odd Exercises

SECTION 1.1

1. $\vec{p}_2 = (1,0,3)$, $\vec{p}_3 = (1,0,0)$, $\vec{p}_4 = (1,2,0)$,
   $\vec{p}_5 = (0,2,0)$, $\vec{p}_6 = (0,2,3)$, and $\vec{p}_7 = (0,0,3)$.

3. $\vec{p}_1 = (1,3,3)$, $\vec{p}_2 = (1,0,3)$, $\vec{p}_3 = (1,0,0)$,
   $\vec{p}_5 = (0,3,0)$, and $\vec{p}_6 = (0,3,3)$.

5. $\sqrt{85}$

7. $\dfrac{\sqrt{11}}{2}$ 9. 1   11. $(0, \pm\sqrt{3})$

13. $a_1 = \dfrac{-1 \pm \sqrt{21}}{2}$, $a_2 = 1 \pm \sqrt{6}$

15. $\left(\dfrac{1}{4}, -\dfrac{\sqrt{15}}{4} \right)$ 17. $\left(-\dfrac{3}{2}, -\dfrac{3\sqrt{3}}{2} \right)$

19. 3.06456 in

21. $\left(\dfrac{1}{2}, \dfrac{11}{8}, -\dfrac{\sqrt{119}}{8} \right)$

23. a. $(2\sqrt{3}, 3, 2)$, b. $(3, 4, 3\sqrt{3})$

25. a. $(3\cos(1.45), -3\sin(1.45), 7)$
    b. $\left(2\cos\left(\dfrac{\pi}{8}\right), -2\sin\left(\dfrac{\pi}{8}\right), 3 \right)$

27. $(528, -5292, 3560)$

SECTION 1.2

1. No.

3. Infinitely many cantain $(1,0,0)$ and $(0,1,0)$.
   Only one plane contains all three points.

5. $(x+1)^2 + (y-2)^2 + (z-3)^2 = 36$

7. $(x-1)^2 + (y-1)^2 > 4$ 9. $w = 3$

11. $(x-h)^2 + (y-k)^2 + (z-l)^2 = r^2$

13. $(-2, -1)$, $r = \sqrt{7}$

15. $(0,0,0)$, $r = 2$

17. $\left(\dfrac{5}{2}, -\dfrac{3}{2}, -\dfrac{3}{2} \right)$, $r = \dfrac{\sqrt{19}}{2}$

19. $y = 3x - 11$ 21. $y = x - 2$

23. $y = -4x + 2$ 25. $x = 1$

27. $y = -x + 1$ 29. $y = 2z$

31. $(0,0,0)$

33. Cylinder centered about the line $x = -2$, $y = -1$
    with radius $\sqrt{11}$

35. Cylinder centered about y–axis with radius $\sqrt{5}$

37. Cylinder centered about the line $y = -1$, $z = -2$
    with radius $\sqrt{5}$

39. $2x + y - z - 2 = 0$

SECTION 1.3

1. $\vec{A} = \dfrac{\vec{C} - 2\vec{B}}{3}$, $\vec{B} = \dfrac{\vec{C} - 3\vec{A}}{2}$

3. $\|\vec{r}\| = 2$, $c\vec{r} = (-\sqrt{2}, \sqrt{2}, 2)$,
   $\vec{r} + \vec{v} = (0, 1 + \sqrt{2}, \sqrt{2} + 3)$,
   $\vec{r} - \vec{v} = (-2, 1 - \sqrt{2}, \sqrt{2} - 3)$

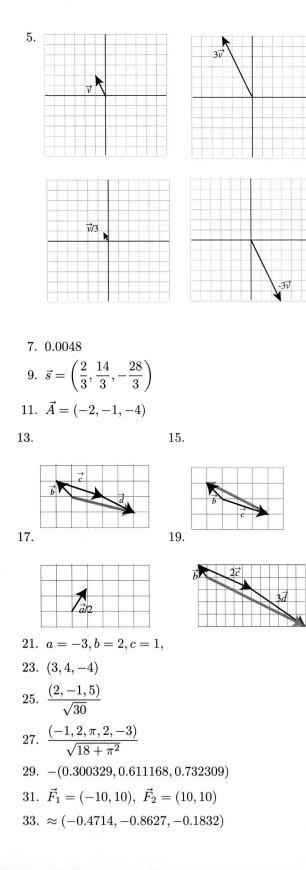

7. 0.0048

9. $\vec{s} = \left(\dfrac{2}{3}, \dfrac{14}{3}, -\dfrac{28}{3} \right)$

11. $\vec{A} = (-2, -1, -4)$

13.

15.

17.

19.

21. $a = -3, b = 2, c = 1,$

23. $(3, 4, -4)$

25. $\dfrac{(2, -1, 5)}{\sqrt{30}}$

27. $\dfrac{(-1, 2, \pi, 2, -3)}{\sqrt{18 + \pi^2}}$

29. $-(0.300329, 0.611168, 0.732309)$

31. $\vec{F}_1 = (-10, 10), \ \vec{F}_2 = (10, 10)$

33. $\approx (-0.4714, -0.8627, -0.1832)$

35. $\left(\dfrac{1}{2}, \dfrac{\sqrt{3}}{2} \right)$

37. $(1, \sqrt{3})$

SECTION 1.4

1. Not defined.

3. a. Positive   b. Negative   c. Postive
   d. Negative   e. Negative   f. Positive

5. $1 + 2\pi$ 7. 9   9. -1 11. 0

13. $2\sqrt{6}$ 15. 0   17. Obtuse

19. 1.04404×10^6 21. 0.559107

23. x–axis: $-\dfrac{1}{\sqrt{14}}$, y–axis: $\dfrac{3}{\sqrt{14}}$, z–axis: $\dfrac{2}{\sqrt{14}}$

25. x–axis: 0, y–axis: $-\dfrac{2}{\sqrt{13}}$, z–axis: $\dfrac{3}{\sqrt{13}}$

27.

a.   b.   c.

d.   e.   f.

29. $\operatorname{proj}_{\vec{q}}(\vec{p}) = (0, 0, 0),$
$\operatorname{orth}_{\vec{q}}(\vec{p}) = (5, 2, -1)$

31. $\operatorname{proj}_{\vec{q}}(\vec{p}) = \left(\dfrac{1}{15}, -\dfrac{2}{15}, -\dfrac{1}{3} \right)$
$\operatorname{orth}_{\vec{q}}(\vec{p}) = \left(\dfrac{14}{15}, -\dfrac{43}{15}, \dfrac{4}{3} \right)$

33. $\operatorname{proj}_{\vec{q}}(\vec{p}) = (0.1352, 0.0376, 0.085),$
$\operatorname{orth}_{\vec{q}}(\vec{p}) = (0.09686, -0.9476, 0.2650)$

35. $\pm \dfrac{(3, -1, 0)}{\sqrt{10}}$

SECTION 1.5

1. 0   3. -10 5. 7

7. $(2, -1, 5), \ \sqrt{30}$

9. $(2, 2, -1), \ 3$

11. $(8973, 15234, 39846)$

13. $\sqrt{30}$ 15. 3

17. $\pm(1, 0, -1)$

19. $\pm\left(\dfrac{1}{\sqrt{2}}, 0, -\dfrac{1}{\sqrt{2}}\right)$

21. $\dfrac{1}{3}$ 23. 3   25. 0

27. -47.7191.

29. a. $\vec{u} \cdot \vec{v}$ is not a vector

   b. 1

   c. $(-1, 0, 0)$

   d. $(0, 0, 0)$

31. $\vec{A}$ or $\vec{B}$ is $\vec{0}$, or $\vec{A}$ and $\vec{B}$ are colinear; that is, $\vec{A}$ and $\vec{B}$ have the same or opposite direction.

33. $(0, 0, -1)$Nm   35. $(2, 2, 0)$Nm

37. $\left(\dfrac{1}{\sqrt{2}}, \dfrac{1}{\sqrt{2}}, 0\right)$

39. 2m   41. $\dfrac{\sqrt{7}}{3}$ 43. $\sqrt{5}$

45. $\vec{A}$ or $\vec{B}$ is $\vec{0}$, or $\vec{A}$ and $\vec{B}$ are orthogonal.

47. $\vec{A}$, $\vec{B}$ or $\vec{C}$ is $\vec{0}$, or $\vec{A}$, $\vec{B}$ and $\vec{C}$ lie in a plane.

SECTION 1.6

1.

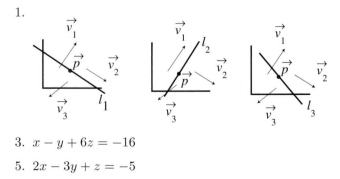

3. $x - y + 6z = -16$

5. $2x - 3y + z = -5$

7. $-4x + 21y + 11z - 4 = 0$

9. $x - y + z = 0$

11. $10x - 3y - 4z = 1$

13. To the xy–plane, $|c|$ units.
    To the xz–plane, $|b|$ units.
    To the yz–plane, $|a|$ units.

15. $\sqrt{2}$ 17. $\dfrac{4}{\sqrt{6}}$

19.

The projections

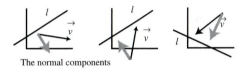

The normal components

21. $\text{proj}_{\mathcal{P}}\vec{F} \approx (-2.73, 0.944)$,
    $\text{proj}_{\vec{n}}\vec{F} \approx (-4.73, 2.006)$

23. $\vec{F}_{plane} = (-7.695, 43.643)$,
    $\vec{F}_{rope} = (7.695, 1.357)$

SECTION 2.1

1. Function of y. Domain $= \{2, 3, 4\}$, Image $= \{1, 2\}$.

3. Domain $= \{1, 2, 3\}$, Image $= \{6\}$.

5. y is a function of x. Domain $|x| \geq 1$, Image $0 \leq y$.

7. y is a function of x. Domain is $\mathbb{R}$, Image $y \geq 0$.

9. y is a function of x. Domain: all real numbers except $x = -2$ and $x = -1$.

11. y is a function of x. Domain $(-\infty, -2] \cup [-1, \infty)$.

13. $4, 1, \pi^2 + 2\pi + 1$ 15. $1, 4, 9$

17. Domain $= \mathbb{R}$, Image $=$ all nonnegative real numbers.

19. Domain $= \{x \mid x \geq 0\}$, Image $=$ all nonnegative real numbers

21. Domain $= \mathbb{R}$, Image $=$ all nonnegative real numbers

23. Domain $= \mathbb{R}$, Image $=$ all numbers which lie in the intervals $[2i, 2i+1)$ where $i = 0, \pm 1, \pm 2, \ldots$

25. $f(x) = x\left(50 - \dfrac{x}{2}\right)$

27. $A(h) = 6V^{2/3}$

29. $l = \dfrac{2d}{3}$

31. $-1, \dfrac{1}{2}, -\dfrac{1}{3}, \dfrac{1}{4}, -\dfrac{1}{5}$

33. $1, 4, 10, 20, 35$

35. $2, 1, 0.5, 0.25, 0.125$

37. $1, 1, 2, 3, 5$

43. $-1, 0, -1, 0, -1$

SECTION 2.2

1. 3.

5. 7.

9.

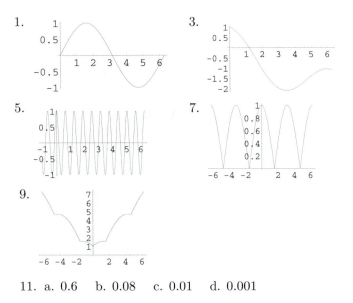

11. a. 0.6 b. 0.08 c. 0.01 d. 0.001

13. -0.811 15. -0.598 17. 0.134

19. a. $2.663, 2.938$ b. $0.34, 2.80$

21. $n = 2$

SECTION 2.3

1. $\vec{r}(t) = (1 + t, 2 - 2t, -1 + 3t)$

3. $\vec{r}(t) = (t, -2t, 3t)$

5. $\vec{r}(t) = (t, t, t)$

7. $\vec{r}(t) = (1, 2 + t)$

9. $\vec{r}(t) = (1 + t, 5 + 2t)$

11. $(1 + 2t, -1 + t, 3 - 3t)$

13. $(-5 - 4t, -1 + t, 2 + 2t)$

15. $(2, 3)$ 17. $(2, 3, -1)$ 19. $(\pi, -1, 1)$

21. $\vec{r}(z) = \left(\dfrac{1}{2} - z, -2z - \dfrac{1}{2}, z\right)$

23. $\vec{r}(x) = (x, 1, 1 - x)$

25. L intersects L_1 at $(0, 1, 3)$

27. L intersects L_1 at $(0, 1, 3)$

29. Skew   31. Not Perpendicular   33. Skew

SECTION 2.4

1. a. $(1, 0)$ b. $\left(\dfrac{1}{2}, \dfrac{\sqrt{3}}{2}\right)$ c. $\left(\dfrac{\sqrt{3}}{2}, \dfrac{1}{2}\right)$

   d. $\left(-\dfrac{\sqrt{2}}{2}, -\dfrac{\sqrt{2}}{2}\right)$ e. $(0, 1)$ f. $\left(-\dfrac{1}{2}, -\dfrac{\sqrt{3}}{2}\right)$

3. a. $(0, 0, 2)$ b. $(3, 2, -2)$ c. $(27, 4, -2)$
   d. $(2700, 0, 2)$

5.

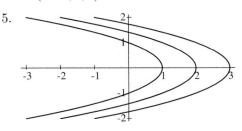

7. $\left(\cos t, \dfrac{1}{3}\sin t\right),$

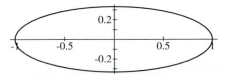

9. $(3\cos t, 2\sin t)$

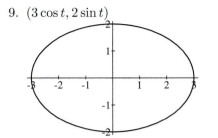

11. $(2\tan t, 3\sec t)$

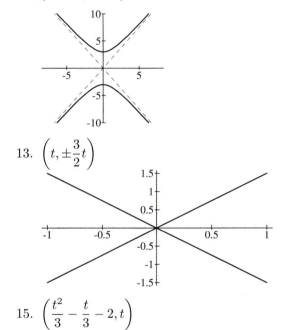

13. $\left(t, \pm\dfrac{3}{2}t\right)$

15. $\left(\dfrac{t^2}{3} - \dfrac{t}{3} - 2, t\right)$

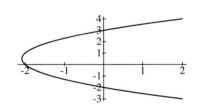

17. a. Curve iv   b. Curve ii   c. Curve iii
    d. Curve vi   e. Curve v   f. Curve i

19.

21.

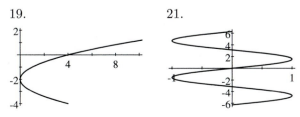

23.   a. $\left(\cos t, \dfrac{1}{2}\sin t\right),\ 0 \le t \le 2\pi$

    b. $(2\cos t, \sin t),\ 0 \le t \le 2\pi$

    c. $(4\cos t, 2\sin t),\ 0 \le t \le 2\pi$

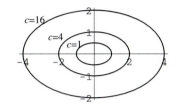

25.

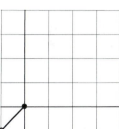

SECTION 2.5

1.   3.

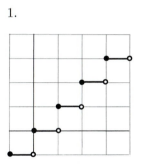 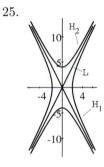

5. $t_1 : x$ neither, y increasing
    $t_2 : x$ increasing, y increasing
    $t_3 : x$ decreasing, y decreasing
    $t_4 : x$ decreasing, y neither
    $t_5 : x$ decreasing, y increasing

7.   a. $(-2.5, -1.5), (-1, 1), (0, 2.5), (1, 1), (2.5, -1)$

    b. $[-5, 5]$ to the right

    c. $[-5, 0)$ up, $(0, 5]$ down

d.

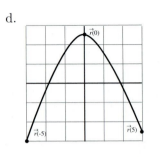

9.   a. $(1.5, 1), (0, 2), (-2, 1), (0, 0), (2, -1)$

    b. $[-5, -3) \cup (0, 4)$ to the right, $(-3, 0) \cup (4, 5]$ to the left

    c. $[-5, -2)$ up, $(-2, 5]$ down

    d.

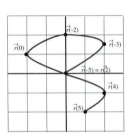

11.   a. $(0, 2)$ to the right, $(-2, 0)$ to the left

    b. $[-2, 2]$ up

    c.

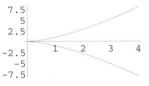

13.   a. $(0, 2]$ to the right, $[-2, 0)$ to the left

    b. $(0, 2]$ up, $[-2, 0)$ down

    c.

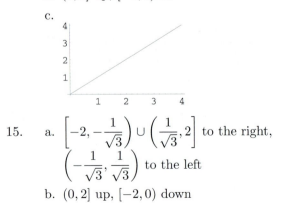

15.   a. $\left[-2, -\dfrac{1}{\sqrt{3}}\right) \cup \left(\dfrac{1}{\sqrt{3}}, 2\right]$ to the right, $\left(-\dfrac{1}{\sqrt{3}}, \dfrac{1}{\sqrt{3}}\right)$ to the left

    b. $(0, 2]$ up, $[-2, 0)$ down

c.

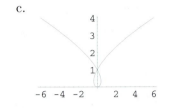

SECTION 2.6

1. $\dfrac{\sqrt{3}}{2}$ 3. 0   5. 4   7. $4(x+1)^2$

9. $(x^4 - 4x^3 + 12x^2 - 16x + 16, x^4 - 4x^3 + 10x^2 - 12x + 7)$

11. All numbers except $x = -1$ and $x = 1$.

13. $(-\infty, -1] \cup [1, \infty)$.

15. $\ldots [-2\pi, -\pi] \cup [0, \pi] \cup [2\pi, 3\pi] \cup [4\pi, 5\pi] \ldots$

17. Moves counterclockwise 3 revolutions per unit time.

19. Moves $\frac{\omega}{2\pi}$ revolutions per unit time, clockwise if ω is negative, counterclockwise if ω is positive.

21. Oscillates from $\theta = \frac{\pi}{2} + 1^R$ to $\theta = \frac{\pi}{2} - 1^R$.

23. Oscillates from angle A^R to $-A^R$ and back to A^R each $\frac{2\pi}{\omega}$ units of time.

25. Image is all of $\mathcal{C}$. Domain is $-2.25 \le t \le 1.5$.

27. Domain is $-\infty < t < \infty)$. Image is sketched below.

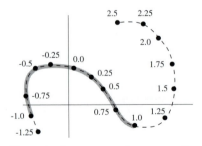

29. Domain is $-\infty < t < \infty$. Image is sketched below.

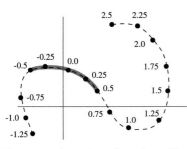

31. Starts at $(-1, 0, -1)$ and oscillates along the line from $(-\pi, 2\pi, -\pi-1)$ to $(-\pi-1, -2\pi, \pi-1)$ every $\frac{2\pi}{\omega}$ units of time. If $\omega < 0$, then the direction is reversed.

33. Oscillates along the parabola $y = x^2$ from $(6, 36)$ to $(-6, 36)$ every 2π units of time.

35. Moves along the parabola $y = x^2$ from (A, A^2) to $(-A, A^2)$ every $\frac{2\pi}{\omega}$ units of time. If $\omega < 0$ then the direction is reversed.

37. $-1 \le x < \infty$ for F and $0 \le x < \infty$ for F^{-1}

39. $-\infty < x < \infty$ for both

41. The domain of f^{-1} is $[-1, 3]$ and the range is $[2, 5]$.

43. $\left(-\frac{\pi}{2}, \frac{\pi}{2}\right)$ 45. $\left(-\frac{\pi}{2}, \frac{\pi}{2}\right)$

47. 36   49. $\frac{\sqrt{24}}{5}$

51. $\dfrac{x}{\sqrt{1-x^2}}$

53. $\left(x, \dfrac{x}{\sqrt{1-x^2}}, \dfrac{1}{x}\right)$

55.

$\sin(x)$ $\qquad$ $\cos(x)$

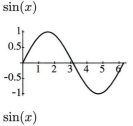

$\sin(x)$ $\qquad$ $\cos(x)$

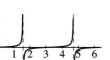

57.

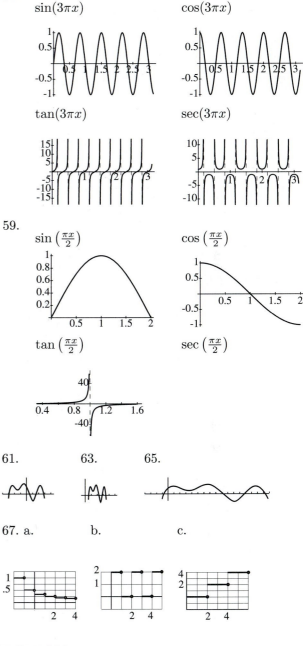

59.

61.   63.   65.

67. a.   b.   c.

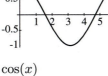

SECTION 2.7

1. $f + g:$ $\quad (-2, 0), (0, -2), (-\sqrt{2}, \sqrt{2}), (1, -\sqrt{3})$
   $f \cdot g; \quad 1, 1, 1, 1$

3. $(6\pi^2 + 2\pi, -3\pi - 1), \quad \left(\dfrac{3\pi^2}{2} + \pi, 0\right),$
   $\left(\dfrac{50\pi^2}{3} + \dfrac{10\pi}{3}, \dfrac{5\pi}{2} + \dfrac{1}{2}\right)$

5. $\vec{h}(x) = \dfrac{x^2 + 1, 3x, x^3 + 2}{\sqrt{x^6 + x^4 + 4x^3 + 11x^2 + 5}}$

7. $\vec{h}(x) = \dfrac{\cos x, \sin x, x}{\sqrt{x^2 + 1}}$

9. $\vec{h}(x) = \dfrac{(a\cos x)\imath + (b\sin x)\jmath}{\sqrt{a^2 \cos^2 x + b^2 \sin^2 x}}$

11. $r(t) = (-2t + 2, 1 + 2t)$. Moving along the line through $(2, 1)$ with direction $(-2, 2)$.

13. $r(t) = (8\cos t, 8\sin t)$. Travelling around a circle centered at the origin with radius 8.

15. i. Rotating about a circle as its center moves along the x axis.
ii. Rotating about a circle as its center moves along the y axis.

17. i. Spiralling along the line $y = x$ with one rotation every 2π units of time.
ii. Spiralling along the line $y = x$ with one rotation every 1 unit of time.

19. $r(t) = (8t - 2, 1 - 4t)$

21. $r(t) = (-2\cos t, -2\sin t)$

23. $\vec{r}(t) = (1 + 2t + 2\cos(6\pi t), t + 2\sin(6\pi t))$

25.   27.

29. When n is odd, there are n leaves. When n is even, there are $2n$ leaves.

31. $t_1 = 0, \quad t_2 = \frac{6\pi}{25}, \quad t_3 = \frac{18\pi}{25}, \quad t_4 = \frac{7\pi}{25}$.

33.   i. $\vec{r}_{3i}(t) = (-40\pi t, 0) + (1 - t)(\sin(40\pi t), -\cos(40\pi t))$

     ii. $\vec{r}_{3ii}(t) = (-40\pi t, 1) + (1 - t)(\sin(40\pi t), -\cos(40\pi t))$

     iii. $\vec{r}_{3iii}(t) = (-40\pi t, -10) + (1 - t)(\sin(40\pi t), -\cos(40\pi t))$

35. $\vec{r}(t) = \begin{pmatrix} (R - r)\cos(\theta(t)) + r\cos\left(\frac{R-r}{r}\theta(t)\right) \\ (R - r)\sin(\theta(t)) - r\sin\left(\frac{R-r}{r}\theta(t)\right) \end{pmatrix}$

SECTION 3.1

1. a. $(3, 2)$ m, $\quad (1, -1)$ m
b. $(3, 2)$ m/s, $\quad (2, -2)$ m/s

3. 150 mi   5. $(25, 0)$ mi/hr

7. $\sqrt{11} + \sqrt{3}$ ft   9. $\left(0, -\frac{2}{5}, \frac{2}{5}\right)$ ft/s

11. $(4, 0)$ m/s   13. $\left(\frac{8}{3}, 0\right)$ m/s

15. $(1, -5)$ m/s   17. $\frac{5\pi}{8}$ m/s   19. $\frac{5\pi}{8}$ m/s

21. 1.25 in   23. 1.025 in   25. 5 m/s

27. 58.45 ft^2 29. 250 mm^2/hr

31. 211.588 mm^2/hr   33. 29.5572 mm^2/hr

35.

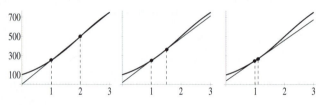

SECTION 3.2

1. 6   3. -84 5. -3

7. -3 9. $\frac{1}{4}$ 11. 0

13. 1   15. 0   17. 0

SECTION 3.3

1. $1, 2, 4$ 3. $(1, 2) \cup (4, 5]$

5. $-1.5, -1, 4, 4.5$

7. $[-2, -1.5) \cup (-1, -0.5) \cup (1, 3) \cup (4.4.5)$

9. $(1, -3)$ m/s

11. $\frac{5\pi}{8}$ m/s.   13. $\left(0, \frac{5\pi}{8}\right)$ m/s.

17. $6x$

19. $2(2x^2 + x)(4x + 1)$

21. $-\dfrac{4x + 3}{(2x^2 + 3x + 1)^2}$

23. $-3x^{-4}$ 25. $\dfrac{1}{2\sqrt{x + 2}}$

27. $\dfrac{x}{\sqrt{x^2+1}}$

29. $-\dfrac{x}{(x^2+1)^{3/2}}$

31. $2x$ 33. $y=0$

35. $y=5x-3$

37. $y=-\dfrac{x}{16}+\dfrac{11}{16}$

39. $-\dfrac{\sqrt{3}x}{2}+\dfrac{\sqrt{3}\pi}{6}+\dfrac{1}{2}$

41. $(0,3,0)$ 43. $\left(\dfrac{1}{2\sqrt{t}},1\right)$

45. $\left(6t^2,1,\dfrac{1}{2\sqrt{t+2}}\right)$

47. $(3t+3,t+21,\cos(t),-\sin(t))$

49. $(2t+1,5t+5,1-t)$

51. $\left(\dfrac{1}{2}-\dfrac{t}{4},4+7t\right)$

53. $(1+t,\sin(1)+t\cos(1),\cos(1)-t\sin(1))$

55. $y=2x-1$

57. $y=0$ and $y=-2x-2$

SECTION 3.4

1.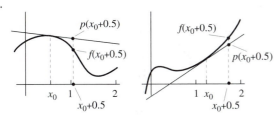

3. $x_0=9$, $p(9.2)=3.0333$

5. $x_0=0.0064$, $p(9.2)=0.083125$

7. $1.1\ \text{ft}^2$ 9. 4.5 m   11. $16.2\pi\ \text{ft}^3$.

13. a. Method 1: 3.31; Method 2: 3.01
    b. Method 1: 3.0301; Method 2: 3.0001
    c. Method 1: 3.003001; Method 2: 3.00001

15. a. Method 1: 4.641; Method 2: 4.04
    b. Method 1: 4.0604; Method 2: 4.0004
    c. Method 1: 4.006004; Method 2: 4.000004

17. 2.23   19. 9.94   21. $p(x)=2-x$

23. $0.642<x<1.558$ to three decimal places.

SECTION 3.5

1. $1,3.5$ 3. -1 5. -1 7. 1

9. 3 11. 3 13. 2 15. 1

17. Does not exist since $\lim_{x\to4^-}f(x)$ does not exist.

19. 2.5 21. None   23. Integers   25. $\pm\sqrt{\text{integer}}$

27. Integers   29. $(-\infty,-1]\cup[1,4]$

31. $(-10,0]\cup(2,\infty)$ 33. $\frac{1}{3}$ 35. 1

37. $\dfrac{1-t^2}{(t^2+1)^2}$ 39. 0 41 $(1,2)$

43. $\frac{4}{3}$ 45. Integers

47. $f\circ g$: integers or $\pm\sqrt{\text{positive integers}}$
    $g\circ f$: integers

49. Root ≈0.25. Error at most $\frac{1}{8}$

51. Root ≈1.875. Error at most $\frac{1}{8}$

SECTION 3.6

1. 2π m.   3. 0.0003 5. 0.00019

7. 0.01 9. 0.01.   11. 0.1

13. 0.01 15. 0.005 17. 0.0002

SECTION 4.1

1. 1 3. $4x+16$

5. $160x^{15}+20+\cos(x)-\sin(x)$

7. $(100x^{99}+10x^4+32\pi x^{31})(x^6-6x^2+x+22)+(x^{100}+2x^5+\pi x^{32})(6x^5-12x+1)$

9. $(2x^5+x^{31}+5x^2+2)(12x+6x^2+4x^3+2)+(6x^2+2x^3+x^4+2x+1)(10x^4+31x^{30}+10x)$

11. $-\dfrac{x^7+\pi x^4}{2(x+1)^{3/2}}+(x+1)^{-1/2}(7x^6+4\pi x^3)$

13. 0

15. $2(2x^{31}+5x^{20}+2x+3)(62x^{30}+100x^{19}+2)$

17. $3(5t^2+2t^3+t+1)^2(10t+6t^2+1)$

19. $(6t^2+6t+1,100t^{99}-10t)$

21. $(0, 6x^2 + 1, 0)$

23. $g''(x) = 6ax + 2b, g'''(x) = 6a$

25. $q''(t) = -4\sin t \cos t, q'''(t) = 4\sin^2 t - 4\cos^2 t$

27. $\lambda''(t) = 2\cos^2 t - 2\sin^2 t, \lambda'''(t) = -8\sin t \cos t$

29. $y - 6 = 31(x - 1)$

31. $y - \dfrac{3\sqrt{3}}{8} = \dfrac{9}{8}(x - \pi/3)$

33. $2t\cos t + 2t - t^2\sin t, \quad (0, \cos t - t\sin t - 3t^2, 0)$

35. $(-5, 3, -4)$ 37. $(3, -1, -3)$ 39. -10

41. $(6 + 9t, 12 + 11t, 7 + 205t)$

43. $((\pi + 2)^{25}, 61 + 3t)$

45. $x = \pm\sqrt{\dfrac{1}{3}}, \quad y \pm \dfrac{2}{3\sqrt{3}} = 0$

47. No real solution

49. $y + \sqrt{\dfrac{2}{27}} = x - \dfrac{2}{\sqrt{3}}$, and $y - \sqrt{\dfrac{2}{27}} = x + \dfrac{2}{\sqrt{3}}$

51. $y = x$

53. $(6t, 2)\,\mathrm{m/s^2}$

55. $\dfrac{1}{\sqrt{1 + 8t^2}}(2t, 1, -2t)$ and $\left(\dfrac{1}{\sqrt{2}}, 0, -\dfrac{1}{\sqrt{2}}\right)$

57. $\dfrac{1}{\sqrt{10}}(-3\sin(t), 1, -3\cos(t)), \quad (-\cos t, 0, \sin t)$

59. -1.6180 61. 1.16805 63. $f'''(x) = 720x^7$

65. $f'''(\theta) = -\cos(\theta)$ 67. $\dfrac{(-2 - 4\pi)}{3\pi}$

69. $(3t^2 + 2t + 1, 2t, 50t^9)$ ft/s

71. a. $\dfrac{-50t^9 + 3t^2 + 6t + 1}{6}(1, 2, -1)$ ft/s
    b. $(3t^2 + 2t + 1, 2t, 50t^9)$
    $-\dfrac{-50t^9 + 3t^2 + 6t + 1}{6}(1, 2, -1)$ ft/s

73. $\vec{p}(t) = (6t^5 + 5t^4, 42t^6 + 36t^5, 4t^3 + 3t^2)$ m kg/s,
    $\vec{p}'(t) = (30t^4 + 20t^3, 252t^5 + 180t^4, 12t^2 + 6t)$ m kg/s²

75. $(7t^6 + 4t^3 + 6t^2 + 2t, -12t^5 + 8t^3 + 1, 5t^4 - 6t^2 - 2t)$ m kg/s

77. If $y = xf(x)$, then $y^{(n)} = xf^{(n)}(x) + nf^{(n-1)}(x)$ for $n \geq 1$.

79. 4.0067

SECTION 4.2

1. $-101x^{-102} + 990x^{-100} + 2x$

3. $5\pi x^4 - 2x^{-3} - \dfrac{3x^2}{(x^3 + 1)^2}$ 5. 0

7. $(1 + \sqrt{2})(100t^{99} - 5t^{-6} - 31t^{-32})$

9. $\dfrac{-3t^8 - 15t^6 + 45t^{-4} + 105t^{-6}}{(5t^{-6} + t^6)^2}$

11. $\dfrac{(x^3 + 2x)\sec(x)\tan(x) - \sec(x)(3x^2 + 2)}{(x^3 + 2x)^2}$

13. $\Big[(x^2 - \pi x)\big((2x + 2)\tan(x)$
    $+(x^2 + 2x + 3 - \sqrt{2})\sec(x)\tan(x)\big)$
    $-(2x - \pi)(x^2 + 2x + 3 - \sqrt{2})\tan(x)\Big] / (x^2 - \pi x)^2$

15. $\dfrac{(ax^{-2} + bx + c)(2kx + d) - (kx^2 + dx)(-2ax^{-3} + b)}{(ax^{-2} + bx + c)^2}$

17. $\Bigg(\dfrac{-35t^{-36}\tan(t) - (t^{-35} + 1)\sec^2(t)}{\tan^2(t)},$
    $\dfrac{(t^5 - 1)\sec(t)\tan(t) - 5t^4\sec(t)}{(t^5 - 1)^2}\Bigg)$

23. $\sec x \tan^2 x + \sec^3 x$

25. $2(x + 1)^{-3}$ 27. $-4(x + 1)^{-3}$

29. $y - 1 = 96(x - 1)$

31. $y - 1 - \sqrt{2} = (2 + \sqrt{2})\left(\theta - \dfrac{\pi}{4}\right)$

33. $y = 0$ 35. $\left(0, \dfrac{2 - t}{12}, 0\right)$

37. $\left(\dfrac{1}{2}, 1, -\dfrac{5\pi\sqrt{2}}{4}\right) + u\left(0, -2, -\sqrt{2} + \dfrac{5\pi\sqrt{2}}{4}\right)$

39. $(0, 1, 0)$ 41. $\dfrac{4}{9}$ 43. $\dfrac{8}{3}$

45. Always increasing on domain.

47. Increasing when $x > 0$ and decreasing when $x < 0$.

49. $\left[0, \frac{\pi}{2}\right) \cup \left(\frac{\pi}{2}, \pi\right] \cup \left[2\pi, \frac{5\pi}{2}\right) \cup \left(\frac{5\pi}{2}, 3\pi\right] \cup$
$\left[4\pi, \frac{9\pi}{2}\right) \cup \left(\frac{9\pi}{2}, 5\pi\right] \ldots$

51. $\left(-\frac{\pi}{4}, \frac{\pi}{4}\right) \cup \left(\frac{3\pi}{4}, \frac{5\pi}{4}\right) \cup \left(\frac{7\pi}{4}, \frac{9\pi}{4}\right) \cup \ldots$

53. $x = 0$

57.

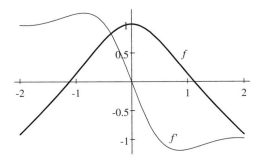

a. Increasing on $-2 \le x < 0$, decreasing on $0 < x \le 2$

b. f' is positive on $-2 \le x < 0$, f' is negative on $0 < x \le 2$

59.

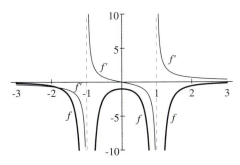

a. Increasing on $(-1, 0) \cup (1, 3]$, decreasing on $[-3, -1) \cup (0, 1)$

b. $f' > 0$ on $(-1, 0) \cup (1, 3]$, $f' < 0$ on $[-3, -1) \cup (0, 1)$

SECTION 4.3

1. $\frac{12}{5}x^{-1/5} - \frac{22}{7}x^{-29/7} - \frac{27}{2}x^{-5/2}$

3. $-\frac{3}{5}(x^{-5/2} + x^{-3/8})^{-8/5}\left(-\frac{5}{2}x^{-7/2} - \frac{3}{8}x^{-11/8}\right)$

5. $(x^{-1/7} + 2x^{5/2} - 4x)^{-2/5}\left(-\frac{2}{9}x^{-11/9} - 3x^2\right) +$
$(x^{-2/9} - x^3)\left(-\frac{2}{5}\right)(x^{-1/7} + 2x^{5/2} - 4x)^{-7/5}$
$\left(-\frac{1}{7}x^{-8/7} + 5x^{3/2} - 4\right)$

7. $3.2(x^{0.76} + x^{1.6})^{2.2}(0.76x^{-.24} + 1.6x^{0.6})$

9. $\frac{2}{3}[((x^{2/3} + 1)^{2/3} + 1)^{2/3} + 1]^{-1/3}\left(\frac{2}{3}\right)((x^{2/3} + 1)^{2/3} + 1)^{-1/3}\left(\frac{2}{3}\right)(x^{2/3} + 1)^{-1/3}\left(\frac{2}{3}x^{-1/3}\right)$

11. $\frac{1}{6}(x^{-1/2} + \pi x^{3/2})^3(x^{5/2} - 6x^{91/3})^{-5/6}$
$\left(\frac{5}{2}x^{3/2} - 182x^{88/3}\right) + (x^{5/2} - 6x^{91/3})^{1/6}$
$3(x^{-1/2} + \pi x^{3/2})^2\left(-\frac{1}{2}x^{-3/2} + \frac{3\pi}{2}x^{1/2}\right)$

13. $\frac{3}{4}[(x^{-1/2} + \pi x^{3/2})^3(x^{5/2} - 6x^{91/3})^{1/6}]^{-1/4}$
$[(x^{-1/2} + \pi x^{3/2})^3\left(\frac{1}{6}\right)(x^{5/2} - 6x^{91/3})^{-5/6}$
$((5/2)x^{3/2} - 182x^{88/3}) + (x^{5/2} - 6x^{91/3})^{1/6}$
$(3)(x^{-1/2} + \pi x^{3/2})^2\left(-\frac{1}{2}x^{-3/2} + \frac{3\pi}{2}x^{1/2}\right)]$

15. $2x\cos(x^2 + 1)$

17. $-(x^{-1/2} + 2x)(\sin(2x^{1/2} + x^2 + 1))$

19. $x(x^2 + 1)^{-1/2}\sec^2[(x^2 + 1)^{1/2}]$

21. $-0.4(2x + 3)^{-0.8}\csc^2(2x + 3)^{0.2}$

23. $-12(x + 1)^3\cot^2((x^2 + 2x + 1)^2)\csc^2[(x^2 + 2x + 1)^2]$

25. $-x^{-1/2}(x^{1/2} + \pi^2)\csc[(x^{1/2} + \pi^2)^2]\cot[(x^{1/2} + \pi^2)^2]$

27. $[-\sin((x^{1/2} - x^3 + 1)(3x^3 + 2x^{-5/2} + \pi))][(x^{1/2} - x^3 + 1)(9x^2 - 5x^{-7/2}) + \left(\frac{1}{2}x^{-1/2} - 3x^2\right)(3x^3 + 2x^{-5/2} + \pi)]$

29. $\frac{1}{2}[\cot^{-1/2}(5x^{1.2} + 6x - 3)][-\csc^2(5x^{1.2} + 6x - 3)](6x^{.2} + 6)$

31. $4(x + 1)\cos(x^2 + 2x + 3)\sin(x^2 + 2x + 3)$

33. $\frac{3}{4}\left(\cos^2 x + \tan\left(x^2 + 2\right) - \sec^3(3x) + 6\cot^2(\pi x)\right)^{-1/4}(-2\cos x\sin x + 2x\sec^2\left(x^2 + 2\right) - 9\sec^3(3x)\tan(3x) - 12\pi\cot(\pi x)\csc^2(\pi x))$

35. $(2\sqrt{3}, -2)$ 37. $y = 0$

39. $y - \frac{3}{4} = \left(\frac{\sqrt{3}\pi}{2}\right)(x - 1)$

41. No horizontal tangents.

43. $x = 1$ 45. $x = \pm 2$

47. No, it is getting further away! $x_{n+1} = -2x_n$

49.   a. $\vec{v} = (-a\omega \sin(\omega t), b\omega \cos(\omega t))$

   b. $v = |\omega| \sqrt{a^2 \sin^2(\omega t) + b^2 \cos^2(\omega t)}$

   c. $\dfrac{\vec{v}}{v}$

53. 0.6349   55. $\left(-\frac{5}{2}, \frac{5}{2}, \frac{5}{\sqrt{2}}\right)$ m/s

57. $h' = 2\,\text{m/s}$ $P' = \frac{32}{\sqrt{73}}$ m/s

59. $a' = -\dfrac{3h}{\sqrt{h^2 - 4}}$ m/s

61. ≈ 4.75 ft/s   63. $\frac{1}{20\pi}$ ft/s   65. $(-2, -4)$ ft/s

67. $v = (\cos t, 2\sin t \cos t)$ ft/s
   $\frac{1}{2}(\sin^4 t - \sin^2 t + 1)^{-1/2}(4\sin^3 t \cos t - 2\sin t \cos t)$

69. $2\sqrt{5}$ ft/s

71.   a. $\vec{v}(t) = (-40 + 40\pi \cos(40\pi t),$
      $40\pi \sin(40\pi t)),$
      $\vec{a}(t) = (-1600\pi^2 \sin(40\pi t), 1600\pi^2 \cos(40\pi t))$

   b. $(40\pi^2 \sin(40\pi t))/(1600 + 1600\pi^2)$
      $(-40 + 40\pi \cos(40\pi t), 40\pi \sin(40\pi t))$

   c. $\dfrac{\vec{v}(t)}{\|\vec{v}(t)\|} = \dfrac{-1 + \pi \cos(40\pi t), \pi \sin(40\pi t))}{\sqrt{1 + \pi^2}}$

73. $-\omega L \sin(\omega t) \begin{pmatrix} \cos(2\pi kt) \\ \sin(2\pi kt) \end{pmatrix}$
   $+ 2\pi k(a + L\cos(\omega t)) \begin{pmatrix} -\sin(2\pi kt) \\ \cos(2\pi kt) \end{pmatrix}$

75.   a.   i. $V'(t) = -\pi\rho^2 \omega R \cos(\omega t)$
         $\left[1 + \dfrac{R\sin(\omega t)}{\sqrt{L^2 - R^2 \cos^2(\omega t)}}\right]$

      ii. $V'(t) = -\pi\rho^2 t R \cos(t^2)$
         $\left[1 + \dfrac{R\sin(t^2)}{\sqrt{L^2 - R^2 \cos^2(t^2)}}\right]$

      iii. $V'(t) = -\pi\rho^2 \cos(t) R \cos(\sin(t))$
         $\left[1 + \dfrac{R\sin(\sin(t)}{\sqrt{L^2 - R^2 \cos^2(\sin(t))}}\right]$

   b. $V'(\theta) = -\pi\rho^2 R \cos(\theta)$
      $\left[1 + \dfrac{R\sin(\theta)}{\sqrt{L^2 - R^2 \cos^2(\theta)}}\right].$
      In the limiting case $(L \to R),$
      $V'(\theta) = -\pi\rho^2 R \cos(\theta) \left[1 + \dfrac{\sin(\theta)}{|\sin(\theta)|}\right]$

c. $V'' = \pi\rho^2 R \left[\sin(\theta) - \right.$
   $R\left(\dfrac{L^2 \cos(2\theta) - r^2 \cos^4(\theta)}{(L^2 - R^2 \cos^2 \theta)^{3/2}}\right)\left.\right]$
   In the limiting case $(L \to R),$
   $V'' = \pi\rho^2 R \left[1 - \dfrac{\cos(2\theta)}{|\sin^3(\theta)|}\right].$

77. 1   79. 42   81. $18t^5$

SECTION 4.4

1. $y' = \dfrac{3x^2 + y - y^3}{3xy^2 - x^3 - x}$

3. $y' = \dfrac{y}{x} + \dfrac{(x - y)^3}{4x(x + y)}$

5. $y' = -\dfrac{b^2 x}{a^2 y}$

7. $y'' = \dfrac{4y^3 - 6xy^4}{(1 - 2xy)^3}$

9. $y'' = \dfrac{a^2 b^2 y - b^2 x^2}{a^4 y^3}$

11. $y = -\frac{13x}{5} + \frac{31}{13}, \quad y = \frac{13x}{5} + \frac{31}{13}$

13. $y - 3 = -\frac{4}{9}(x - 9)$

15. a. $(-3, 1), (3, 1)$
    b. $(-2, 3), (0, 1), (2, 3), (0, -3)$

17. a. $(-4, 0), (0, 0), (4, 0)$
    b. $(-2, 2), (-2, -2), (2, -1), (2, 1)$

19. a. $(-3, 1), (-1, -1), (3, -2)$
    b. $(-3, 1), (-2, -2), (-1, -1), (2, -4), (1, 3)$

21. Horizontal line test near $(0, 1)$ and $(0, -1)$

SECTION 4.5

1. $p_1(x) = 6 + 31(x - 1), \ p_2(x) = 6 + 31(x - 1) +$
   $306(x - 1)^2$
   $f(1.01) = 6.3403, \ p_1(1.01) = 6.31, \ p_2(1.01) =$
   6.3406

3. $p_1(x) = 0, \ p_2(x) = 0$

5.   a. $p_1(x) = -\frac{\sqrt{3}}{2} - \left(\frac{\sqrt{3}}{2} + \frac{\pi}{6}\right)(x - 1)$
     b. $p_1(1.1) = -1.00499$

c.

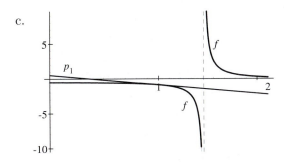

d. $\text{Error}_1(0.9) \approx 0.283 \quad \text{Error}_1(1.1) \approx 0.463$

e.

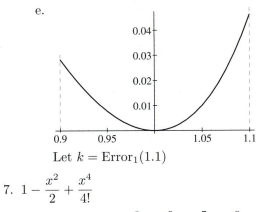

Let $k = \text{Error}_1(1.1)$

7. $1 - \dfrac{x^2}{2} + \dfrac{x^4}{4!}$

9. a. $p_9(x) = x - \dfrac{x^3}{3!} + \dfrac{x^5}{5!} - \dfrac{x^7}{7!} + \dfrac{x^9}{9!}$

b.

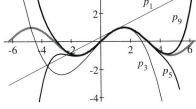

c. They are the same.

SECTION 5.1

1. ∞ 3. ∞ 5. $-\infty$ 7. ∞

9. $1/4$ 11. 0   13. 0   15. 0

17. $-\dfrac{\pi}{2}$

19. Vertical asymptotes $x = \dfrac{(2n+1)\pi}{2}$ 21. $\dfrac{1}{4}$

23. 0   25. 1   27. $\dfrac{\pi}{2}$ 29. $-\infty$

31. ∞ 33. Unknown   35. ∞ 37. 0

39. 0   41. 3.5   43. 0

SECTION 5.2

1. $x = 1.5$ b. Increasing on $(-1, 1.5)$ and decreasing on $(1.5, 4)$. c. Relative maximum at $x = 1.5$

3. a. $x = 0, 3$ b. Increasing on $(-1, 0)$ and $(3, 4)$. Decreasing on $(0, 3)$. c. Relative maximum at $x = 0$, relative minimum at $x = 3$.

The folloing short hand notation is used in solutions to Exercises 5–23: Horizontal Asymptotes–HA, Vertical Asymptotes–VA, Increasing–INC, Decreasing–DEC, Relative Maximima–MAX, Relative Minima–MIN.

5. INC $(0, \infty)$, DEC $(-\infty, 0)$, MIN $(0, 1)$

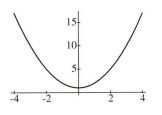

7. INC $(0, \infty)$, DEC $(-\infty, 0)$, MIN $(0, -1)$

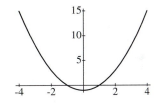

9. HA $y = 1$, VA $x = 1$, DEC $(-\infty, 1) \cup (1, \infty)$

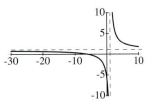

11. HA $y = 0$, VA $x = \pm 1$,
INC $x < -1, -1 < x < 1, x > 1$

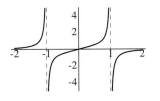

13. VA $x = 1$, DEC $x < 1, x > 1$

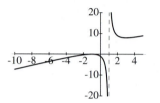

15. VA $x = 1$, INC $x < -1, x > 3$,
    DEC $-1 < x < 1$, $1 < x < 3$,
    MAX $(-1, 0)$, MIN $(3, 8)$

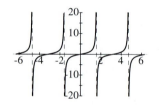

17. VA $x = \frac{\pi}{2} + n\pi$, INC $\left(-\frac{\pi}{2} + n\pi, \frac{\pi}{2} + n\pi\right)$

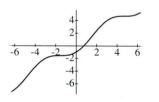

19. INC $(-\infty, \infty)$

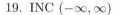

21. INC $\left(-\frac{3\pi}{4} + 2n\pi, \frac{\pi}{4} + 2n\pi\right)$,
    DEC $\left(\frac{\pi}{4} + 2n\pi, \frac{5\pi}{4} + 2n\pi\right)$,
    MAX $\left(\frac{\pi}{4} + 2n\pi, \sqrt{2}\right)$, MIN $\left(\frac{5\pi}{4} + 2n\pi, -\sqrt{2}\right)$

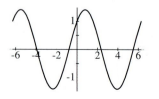

23. INC $\left(-\frac{\pi}{4} + n\pi, \frac{\pi}{4} + n\pi\right)$,
    DEC $\left(\frac{\pi}{4} + n\pi, \frac{3\pi}{4} + n\pi\right)$,
    MAX $\left(\frac{\pi}{4} + n\pi, 1\right)$, MIN $\left(-\frac{\pi}{4} + n\pi, -1\right)$

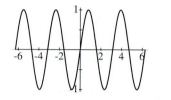

25. a. Minimum at $(0, 1)$,
       maximum at $(-2, 5), (2, 5)$

    b. Minimum at $(0, 1)$, maximum at $(4, 17)$.

27. a. No minimum nor maximum.

    b. Minimum at $\left(4, -\frac{255}{4}\right)$, maximum at $(1, 0)$.

29. a. Minimum at $(0, -1)$,
       maximum at $(\pi, 1 + \pi)$

    b. Minimum at $(-\pi, 1 - \pi)$,
       maximum at $(2\pi, 2\pi - 1)$.

31. $c = \frac{2}{\sqrt{3}}$ 33. $\frac{(-2+\sqrt{7})}{3}$

35. a. $x'(t) = 0$ only at points of intersection of
       $y = t$ and $y = \cot(t)$ which occures exactly
       once in each segment.

    b. 3.4256, 6.4373, 9.5292

    c.

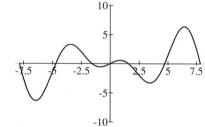

SECTION 5.3

1. a. $(4, 2), (0, 1)$

   b. $(1, 0), (1, 3), (2, 4), (3, 3), (3, 1)$

3. Right, down, decreasing

5. Left, up, decreasing

7. $x(t)$ is increasing on $0 < t < 2$, $3 < t < 5$ and decreasing on the other open intervals. $y(t)$ is increasing on $0 < t < 1$, $4 < t < 6$, $7 < t < 8$ and decreasing on the other open intervals.

9. a. $x'(t_1) < 0$, $x'(t_2) < 0$, $x'(t_3) < 0$, $x'(t_4) > 0$, $x'(t_5) > 0$

   b. $y'(t_1) = 0$, $y'(t_2) < 0$, $y'(t_3) > 0$, $y'(t_4) = 0$, $y'(t_5) < 0$

11. a. $(-2, -0.5)$, $(-0.5, 1)$, $(1, 2)$, $(2, 4)$

    b. Horizontal tangents at $t = -1, 1, 2.5$. Vertical tangents at $t = -0.5, 1, 2$

    c. Increasing on $(-1, -0.5)$, $(2, 2.5)$. Decreasing on other open intervals.

13. $\sqrt{3}y + x - 2 = 0$

15. $y - \frac{3}{4} = \frac{\sqrt{3}}{5}\left(x - \frac{\sqrt{3}}{4}\right)$

17. a. $t \geq 0$, $t \leq 0$

    b. No horizontal, vertical $t = 0$

    c. Increasing $t > 0$, decreasing $t < 0$

    d.
    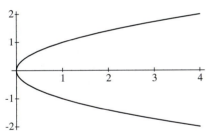

19. a. $t \geq -3$ $t \leq -3$

    b. No horizontal, vertical $t = -3$

    c. Increasing $t > -3$, decreasing $t < -3$

    d.
    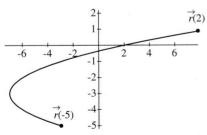

21. a. $t \geq -3$ $t \leq -3$

    b. Horizontal $t = 0$, vertical $t = -3$

    c. Increasing $(-3, 0)$, $t > 0$, decreasing $t < -3$

23. a. $\mathbb{R}$

    b. Horizontal $t = \frac{3\pi}{2} + 2n\pi$, vertical $t = \frac{\pi}{2} + 2n\pi$

    c. Increasing $\left(2n\pi + \frac{\pi}{2}, 2n\pi + \frac{3\pi}{2}\right)$, decreasing $\left(2n\pi - \frac{\pi}{2}, 2n\pi + \frac{\pi}{2}\right)$

    d.
    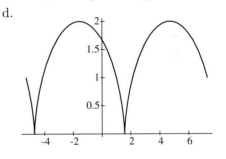

25. a. $t = n\pi$ b. $m = 3$

    c. $\left[\frac{(2n+1)\pi}{2}, \frac{(2n+3)\pi}{2}\right]$, $n = 0, \pm 1, \pm 2, \ldots$

    d. Increasing on $\left(\frac{(2n-1)\pi}{6}, \frac{(2n+1)\pi}{6}\right)$, $n = 0, \pm 1, \pm 2, \ldots$

    e. Figure 14.d.

27. a. $t = n\pi$ b. $m = \pm 4$

    c. $\left[\frac{(2n+1)\pi}{2}, \frac{(2n+3)\pi}{2}\right]$, $n = 0, \pm 1, \pm 2, \ldots$

    d. Decreasing on $\left(\frac{\pi}{8}, \frac{3\pi}{8}\right)$, $\left(\frac{\pi}{2}, \frac{5\pi}{8}\right)$, $\left(\frac{7\pi}{8}, \frac{9\pi}{8}\right)$, $\left(\frac{11\pi}{8}, \frac{3\pi}{4}\right)$, $\left(\frac{13\pi}{8}, \frac{15\pi}{8}\right)$

    e. Figure 14.a.

29. a. $t = \left(\frac{n\pi}{2}, \frac{(n+1)\pi}{2}\right)$, $n = 0, \pm 1, \pm 2, \ldots$

    b. Increasing when $\csc(t) > 0$. $\left(n\pi, \frac{3n\pi}{2}\right)$, $n - 0, \pm 1, \pm 2, \ldots$

31. Fig. 15.i   33. Fig. 15.f   35. Fig. 15.d

37. Fig. 15.j   39. Fig. 15.c   41. Fig. 15.h

43. a. $[0, b_1]$, $[b_1, b_2]$, $[b_2, b_3]$, $[b_3, b_4]$, $[b_4, b_5]$, $[b_5, b_6]$ and $[b_6, b_7]$,

d.
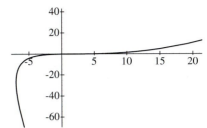

b. Increasing on $[0, b_1]$, $[d_1, b_2]$, $[d_2, b_3]$, $[d_3, b_4]$, $[d_4, b_5]$, $[d_5, b_6]$ and $[d_6, 1]$. Decreasing on $[b_1, d_1]$, $[b_2, d_2]$, $[b_3, d_3]$, $[b_4, d_4]$, $[b_5, d_5]$ and $[b_6, d_6]$.

d. If $f > 0$ then ρ is always decreasing. If $f < 0$ then ρ is always increasing.

45. a. $t = \dfrac{\theta}{2\pi k}$

b. $\rho = \left(a + L\cos\left(\frac{\omega\theta}{2\pi k}\right)\right)$

c.

d.

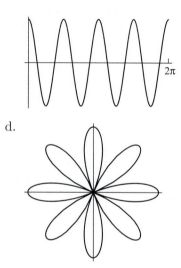

SECTION 5.4

1. a. No inflection points

b. Neither concave up on down.

3. a. Inflection point at x=-0,5.

b. Concave up on $(-2, -0.5)$ concave down on $(-0.5, 4)$.

5. a. Decreasing on $(-2, 2)$ and increasing on $(2, 4)$.

b. Relative min at $x = 2$

c. No inflection points.

d. Always concave up.

7. a. Increasing on $(-1, 3)$. Decreasing on $(-2, -1)$ and $(3, 4)$.

b. Local min at $x = -1$, local max at $x = 3$.

c. Inflection point at $x = 1$.

d. Concave up on $(-2, 1)$ and concave down on $(1, 4)$.

9. a. Inflection point at $x = 2$.

b. Concave down on $(-2, 2)$ and concave up on $(2, 4)$.

11. a. Inflection points at $x = -1$ and $x = 3$.

b. Concave up on $(-1, 3)$ and concave down on $(-2, -1)$ and (3.4).

13. There is a c_3 so that $f''(c_3) = \frac{f'(c_2) - f'(c_1)}{c_2 - c_1} > 0$. Thus f is concave up at c_3.

15. Maximum at $x = 1$, minimum at $x = 3$

17. 2nd Derivative Test fails.

19. Maximum at $x = 1$, minimum at $x = -1$

21. Maximum at $t = 0$

23. Always increasing, inflection point $(-1, 4)$.

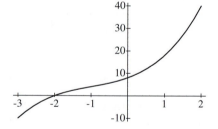

25. Maximum $(-6, 0)$, minimum $\left(-\frac{14}{5}, -\frac{(16)^2(24)^3}{5^5}\right)$ inflection points at $\left(\frac{-14-4\sqrt{6}}{5}, -\frac{(16-4\sqrt{6})^2(24+4\sqrt{6})^3}{5^5}\right)$, $\left(\frac{-14+4\sqrt{6}}{5}, -\frac{(16+4\sqrt{6})^2(24-4\sqrt{6})^3}{5^5}\right)$, and $(2, 0)$.

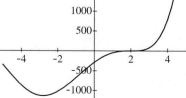

27. No relative extrema. Inflection points at $\left(-\sqrt{2\sqrt{2}}, \frac{16}{\sqrt{2\sqrt{2}}}\right)$ and $\left(\sqrt{2\sqrt{2}}, -\frac{16}{\sqrt{2\sqrt{2}}}\right)$.

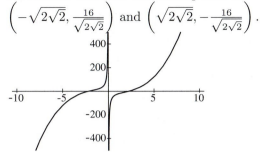

29. No relative extrema. No inflection points.

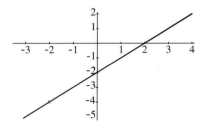

31. Maximum at $x = -4 - 2\sqrt{3}$, minimum at $x = -4 + 2\sqrt{3}$. No inflection points.

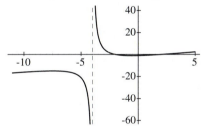

33. Relative maximum at $(0, -1)$. No inflection points.

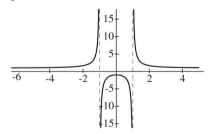

35. Maximum $\left(\frac{(2n+1)\pi}{2}, 1\right)$, minimum $(n\pi, 0)$ inflection points $\left(\frac{\pi}{4} + \frac{n\pi}{2}, \frac{1}{2}\right)$.

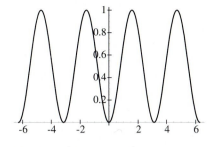

37. Maxima $\left(\frac{\pi}{6} + 2n\pi, 2\right)$, minima $\left(\frac{\pi}{6} + (2n+1)\pi, -2\right)$, inflection points $\left(\frac{5\pi}{6} + n\pi, (-1)^{n+1}\right)$.

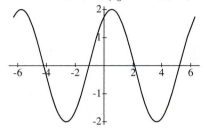

39. Maximum $(0, 1)$, minima $(-1, 0)$ and $(1, 0)$, inflection points $(-1, 0)$ and $(1, 0)$.

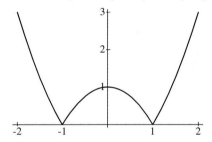

41. Minimum $(0, 0)$, no inflection points.

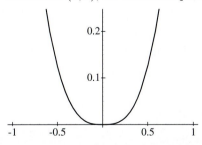

43. Absolute maximum $\left(1, \frac{1}{2}\right)$, absolute minimum $\left(-1, -\frac{1}{2}\right)$

45. No extrema

47. Absolute maximum $\left(-1, \frac{1}{3}\right)$, absolute minimum $\left(1, \frac{1}{3}\right)$

51. Function a. – Derivative e. – 2nd Derivative d.
Function b. – Derivative a. – 2nd Derivative c.
Function c. – Derivative b. – 2nd Derivative a.
Function d. – Derivative c. – 2nd Derivative e.
Function e. – Derivative d. – 2nd Derivative b.

53. $-\dfrac{8\sqrt{2}(32+\pi^2)}{(\pi-4)^3}$
    55. $-\dfrac{36}{125}$

57. Concave up $((2n-1)\pi, 2n\pi)$,
concave down $(2n\pi, (2n+1)\pi)$

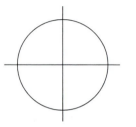

59. Nowhere concave up nor down.

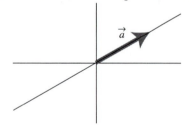

61. $t = \dfrac{2}{\pi(2n+1)}$ b. ≈ 0.152, $\approx 0.104 \approx 0.079$

SECTION 5.5

1. a. Speeding up, right
    b. Slowing down, left
    c. Slowing down
    d. Slowing down, right
    e. Right

3.

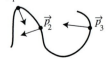

5. $\vec{a}_T = (0,-1,0)$, $\vec{a}_N = (2,0,1)$, $\hat{t} = (0,1,0)$,
$\hat{n} = \dfrac{(2,0,1)}{\sqrt{5}}$, $\hat{b} = \dfrac{(1,0,-2)}{\sqrt{5}}$

7. $\vec{a}_T = \dfrac{(-2,1,3)}{2}$, $\vec{a}_N = \dfrac{(6,-3,5)}{2}$, $\hat{t} = \dfrac{(-2,1,3)}{\sqrt{14}}$,
$\hat{n} = \dfrac{(6,-3,5)}{\sqrt{70}}$, $\hat{b} = \dfrac{(1,2,0)}{\sqrt{5}}$

9. $\vec{a}_T = \frac{144}{145}(1,12)$, $\vec{a}_N = \left(-\frac{144}{145}, \frac{12}{145}\right)$

11. $\vec{a}_T = (0,0)$, $\vec{a}_N = (0,0)$

13. $a_T = (0,0)$, $a_N = \left(-\frac{1}{2}, -\frac{\sqrt{3}}{2}\right)$

15. $a_T = (0,0)$, $a_N = (0,2)$

17. $\hat{t} = \dfrac{(1,1,1)}{\sqrt{3}}$, $\hat{n} = \dfrac{(0,1,-1)}{\sqrt{2}}$, $\hat{b} = \dfrac{(-2,1,1)}{\sqrt{6}}$

19. $\hat{t} = (0,0,1)$, $\hat{n} = \dfrac{(1,1,0)}{\sqrt{2}}$, $\hat{b} = \dfrac{(-1,1,0)}{\sqrt{2}}$

21.   a. $\hat{b}_1 = -\hat{k}$, $\hat{b}_2 = \hat{k}$, $\hat{b}_3 = -\hat{k}$

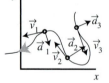

b. $\hat{b}_1 = \hat{k}$, $\hat{b}_2 = -\hat{k}$, $\hat{b}_3 = \hat{k}$

SECTION 5.6

1. The angular speed is a scalar quantity that is the absolute value of the rate of change of the angle, whereas the angular acceleration is a vector quality that gives the rate of change of the angular velocity.

3. $v = 30$, $\|\vec{a}_T\| = 6$, $\|\vec{a}_N\| = 300$

5. $\|\vec{a}_N\| = \cos^2 t$

7. $\vec{v}(t_0) = (-1, -5, -2)$, $\vec{a}_N(t_0) = (12, 0, -6)$, $\vec{a}_T(t_0) = (2, 10, 4)$, $\vec{a}(t_0) = (14, 10, -2)$

9. 2   11. 108   13. $-3\pi\hat{k}$ 15. $-3\pi\hat{k}$

17. $-3\pi\hat{\imath}$ 19. $\sqrt{3}\pi(1, 1, 1)$

21. $\vec{v}$ must be in the plane whose normal vector is $(1, 1, 1)$ or $\vec{v}$ and $\vec{r}$ must be perpendicular.

23. $\vec{\omega}$ and $\vec{r}$ must be perpendicular.

25. $\|\vec{a}_N\| = \frac{14}{5}$

27. $\rho = \frac{14}{\sqrt{3}}$, $c = \begin{pmatrix} 1 \\ 2 \\ 2 \end{pmatrix} + \frac{14}{\sqrt{3}} \begin{pmatrix} 1 \\ -2 \\ -1 \end{pmatrix}$.

29. $\frac{14}{5}$

31. $\rho = \dfrac{196}{\sqrt{966}}$, $c = \dfrac{(-595, 679, 3325)}{483}$

33. $\kappa = \frac{\sqrt{20880}}{21025}$, $c = (2, 8) + \frac{4205}{348}(-12, 1)$

35. $\kappa = 0$

39. $\vec{a}_T = (0, 0)$, $\vec{a}_N = (-2, 0)$, $\kappa = 2$, center $= \left(\frac{3}{2}, 0\right)$

41. $\vec{a}_T = (0, 0, 0)$, $\vec{a}_N = (0, -1, 0)$, $\kappa = \frac{1}{2}$, center $= (0, -1, 0)$

43. $\frac{2}{125}$ mi

45.

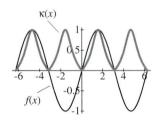

49.   a.

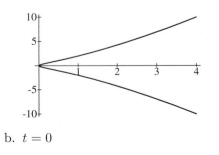

   b. $t = 0$

d.

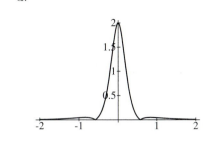

SECTION 5.7

1. 230 and 0   3. 115 and 115   5. 115 and 115

7. $-\frac{3}{14}$ 9. Cube with side of length $\frac{\sqrt{2}}{2}$

11. Radius is $\sqrt[3]{\frac{25}{12\pi}}$, height $5\sqrt[3]{\frac{144}{625\pi}}$

13. $\frac{4}{\sqrt{3}}$ by $\frac{8}{3}$

15. Top length is R, height is $\frac{\sqrt{3}R}{2}$

17. Radius is $R\sqrt{\frac{3}{2}}$, height is $R^2\sqrt{3}$

19. \$117   21. $\frac{5}{\sqrt{3}}$ miles from point P

SECTION 5.8

1. 0   3. 1.9×10^{-5} 5. 0.8   7. 3.5×10^{-4}

9.

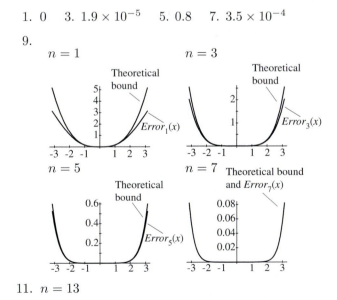

11. $n = 13$

13. For $h = 0.01$, 0.86601, -0.49999. For $h = 0.00000001$, 0.86602, 18189.894.

15. $(\pm 0.9, 2.297)$, $(\pm 0.8, 1.667)$, $(\pm 0.7, 1.4)$, $(\pm 0.6, 1.25)$, $(\pm 0.5, 1.155)$, $(\pm 0.4, 1.091)$, $(\pm 0.3, 1.048)$, $(\pm 0.2, 1.021)$, $(\pm 0.1, 1.005)$, $(0, 1)$

17. $(0.1, 10.034)$, $(0.2, 5.004)$, $(0.3, 3.35)$, $(0.4, 2.501)$, $(0.5, 2)$, $(0.6, 1.667)$, $(0.7, 1.429)$, $(0.8, 1.25)$, $(0.9, 1.111)$, $(1, 1)$

19.

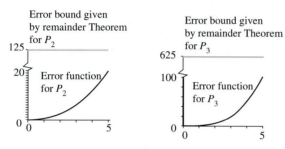

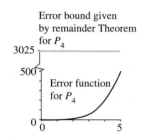

SECTION 6.1

1. $\dfrac{x^3}{3} + C$

3. $\dfrac{x^3}{3} + \dfrac{3}{4}x^{4/3} + C$

5. $\dfrac{5}{101}x^{101} + \dfrac{2}{5}x^{-5} + 10x + C$

7. $\dfrac{10}{7}t^{7/2} - 3t^2 - 2t + \dfrac{\pi t}{3} + C$

9. $\dfrac{1}{2}\sin(2x) + C$

11. $\dfrac{x^4}{4} - \cos(0.312)x + C$

13. $\dfrac{1}{5}\sin(5x - \pi) + C$

15. $\dfrac{t^4}{2} + t^3 - t^2 - 3t + C$

17. $\dfrac{2}{3}x^{3/2} + 4x^{1/2} + C$

19. $\begin{pmatrix} \frac{x^4}{4} + \frac{3}{2}x^2 + 2x + C_x \\ \frac{5}{3}x^3 - \frac{x^2}{2} - 2x + C_y \\ 3x^2 - 3x + C_z \end{pmatrix}$

21. $\begin{pmatrix} \frac{5}{2}x^2 - \frac{2}{5}x^{5/2} + C_x \\ -\frac{x^3}{3} - \frac{x^2}{2} - x + C_y \\ \frac{2}{3}(x+1)^{3/2} + C_z \end{pmatrix}$

23. $\dfrac{x^4}{8} - \dfrac{5}{2}x^2 + x^3 - 3x + C$

25. $\dfrac{a^2 x^4}{4} - 2bx + C$

27. $\dfrac{3tx^2}{2} + 5x + C$

29. $(2 - t^2)\left(\dfrac{5x^2}{2} + 3x\right) + C$

31. $(2 - t^2)(5x + 3)s + C$

33. $\dfrac{x^3}{3} - x^2 + x + C$

35. $\dfrac{(x-1)^{11}}{11} + C$

37. $\begin{pmatrix} \frac{t^4}{4} - \frac{t^3}{3} + \frac{t^2}{2} + 3t + C_x \\ -\cos(t-1) + C_y \\ \sin(t+a) + C_z \end{pmatrix}$

43. $f(x) = x^3 + \dfrac{5}{2}x^2 - \dfrac{3}{2}$

45. $f(t) = -2t^{-1/2} - 3t^{-1} + 5t - 1$

47. $f(s) = \dfrac{20}{99}s^{11/2} - \dfrac{s^3}{2} - 3s^2 + \dfrac{133}{18})s - \dfrac{23}{11}$

49. $f(x) = \dfrac{4}{15}(x-1)^{5/2} + \dfrac{x}{3} - \dfrac{14}{15}$

51. $\vec{f}(x) = \begin{pmatrix} -\frac{1}{2}\cos(2x) \\ \frac{1}{4}\sin(4x) + 3 \\ 25x - 1 - \frac{25\pi}{4} \end{pmatrix}$

53.

55. $F(x) = -x^{-1}$ and $G(x) = -x^{-1} - 1$ for $x > 0$ and $G(x) = -x^{-1} - 26$ for $x < 0$ is one example of such a pair.

59. ≈ 0.69182

61. and 63.

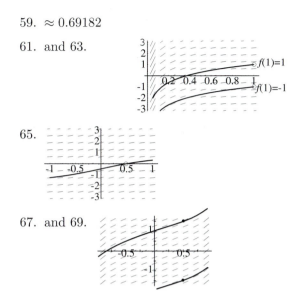

65.

67. and 69.

SECTION 6.2

1. $\sqrt{(x^2+1)^2 + (x^2+1)} + C$

3. $\sqrt{\sin^2(x) + \sin(x)} + C$

5. $\dfrac{\sqrt{\sec^2(3x) + \sec(3x)}}{3} + C$

7. $\dfrac{(x^2+x+6)^{26}}{26} + C$

9. $\dfrac{3}{4}\left(x^3 + x^{1/2} - \dfrac{1}{2}x\right)^{4/3} + C$

11. $-\dfrac{1}{2}(x^2+1)^{-1} + C$

13. $-\dfrac{1}{8}(x^2+1)^{-4} + C$

15. $-\dfrac{1}{2}\cos(x^2) + C$

17. $\dfrac{1}{2}\sin((x+1)^2) + C$

19. $\dfrac{1}{3}\sin^3(x) + C$

21. $\dfrac{1}{2}\sin^2(x) + \sin(x) + C$

23. $\sec(x+1) + C$

25. $\tan(x-3) + C$

27. $-\dfrac{1}{3}\cos(t^3+6) + C$

29. $\sec\left(\dfrac{t^6}{2} + t\right) + C$

31. $f(x) = -\dfrac{1}{2}\cos(x^2) + 32.5$

33. $f(t) = \dfrac{1}{2}\tan(t^2) + 1$

35. $f(w) = \dfrac{1}{4}\sin^2(w^2 - 4) + \pi$

37. $f(\beta) = \sin(\beta) - \dfrac{1}{3}\sin^3(\beta)$

39. $y = -1 \pm \sqrt{2x^2 + C}$

41. $\cos(y^2 + 2) = -\dfrac{4}{3}(x+2)^{3/2} + C$

43. $\sin y = \left(\dfrac{x^2}{4} + C\right)^2$

45. $\sin(y^2)\sec^2 x + C$

47. $(y+2)^{3/4} = -\dfrac{3}{8}(x+1)^{-2} + 3^{3/4} + \dfrac{3}{8}$

49.

SECTION 6.3

1. $\begin{pmatrix} t^3 + t^2 + t - 2 \\ \frac{5t^2}{2} + 2t - \frac{5}{2} \\ \frac{2t^{3/2}}{3} - \frac{3t^2}{2} + 5t - \frac{31}{6} \end{pmatrix}$

3. $\begin{pmatrix} -\frac{1}{2}\cos(t^2 + 2\pi t) + \frac{3}{2} \\ \frac{1}{3}\sin(t^3) - 1 \\ \frac{1}{2}\sin^2(t) \end{pmatrix}$

5. $(2t+2, -t, 2t)$

7. $(t, 2t, 3t)$

9. $(225t - 428, -16t^2 + 96t - 78, 100t - 90)$

11. $\begin{pmatrix} \frac{t^4}{12} + \frac{2t}{3} + \frac{19}{12} \\ \frac{t^3}{3} + \frac{3t^2}{2} - 4t - \frac{25}{6} \\ \frac{4(t+1)^{5/2}}{15} - t - \frac{4\sqrt{2}t}{3} - \frac{4\sqrt{2}}{3} \end{pmatrix}$

13. $\vec{r}(t) = (1 - 2t, 2 + 3t, 3 - t)$

15. $\vec{r}(t) = \left(1 - \frac{2t}{3}, 2 + t, 3 - \frac{t}{3}\right)$

17. $\vec{v}(t) = (3t + 5, 0)$, $\vec{r}(t) = \left(\frac{3t^2}{2} + 5t + 2, 0\right)$, $\vec{v}(2) = (11, 0)$, $\vec{r}(2) = (18, 0)$

19. $\vec{v}(t) = (5, 2t + 2)$, $\vec{r}(t) = (5t + 2, t^2 + 2t)$, $\vec{v}(2) = (5, 6)$, $\vec{r}(2) = (12, 8)$

21. $\vec{v}(t) = (2t + 2)\hat{\imath} - 6t\hat{\jmath} + t\hat{k}$, $\vec{r}(t) = (t^2 + 2t)\hat{\imath} - 3t^2\hat{\jmath} + \left(\frac{t^2}{2} - 2\right)\hat{k}$, $\vec{v}(2) = 6\hat{\imath} - 12\hat{\jmath} + 2\hat{k}$, $\vec{r}(2) = 8\hat{\imath} - 12\hat{\jmath}$

23. -5 ft/sec^2

25. $\vec{v}(2) = (320, 6)$, $\vec{a}(2) = (640, 2)$

27. 122.5 m

29. ≈ 2 sec $\quad \approx 61$ m

33. a. $\frac{20}{7}$ sec $\quad$ b. $\sqrt{3284}$ m/sec

35. a. ≈ 1.13 sec $\quad$ b. ≈ 57.3 m/sec

SECTION 6.4

1. 2 $\quad$ 3. 2 $\quad$ 5. 2

7. $\frac{3}{10}$ $\quad$ 11. $\frac{8}{\pi}$ $\quad$ 13. $\frac{16\pi^5}{5}$

15. π^2 $\quad$ 17. $\frac{\sqrt{2}t^2}{2}$

SECTION 6.5

1. -2 $\quad$ 3. 6 $\quad$ 5. $\frac{61}{16}$ $\quad$ 7. $\frac{592}{125}$

9. 95 $\quad$ 11. 49.875 $\quad$ 13. 35

15. $f(x) = 2x^2 + 3x + 1$, $[1, 2]$

17. $-\frac{3}{2}$ $\quad$ 19. $\frac{3\pi}{2}$

1. a. i. $\mathcal{P}_1 = \{0, 1, 2\}$ $\quad$ $\mathcal{P}_2 = \{0, \frac{1}{2}, 1, \frac{3}{2}, 2\}$
   $\mathcal{P}_3 = \{0, \frac{1}{4}, \frac{1}{2}, \frac{3}{4}, 1, \frac{5}{4}, \frac{3}{2}, \frac{7}{4}, 2\}$

   ii. $\mathcal{P}_1 = \{-2, 0, 2\}$
   $\mathcal{P}_2 = \{-2, -1, 0, 1, 2\}$
   $\mathcal{P}_3 = \{-2, -\frac{3}{2}, -1, -\frac{1}{2}, 0, \frac{1}{2}, 1, \frac{3}{2}, 2\}$

   iii. $\mathcal{P}_n = \{a, a + \frac{b-a}{2^n}, a + 2(b-a)^n, \ldots\}$

   b. i. By Theorem 4 of Section 5.2

   ii. $\mathcal{S}_{*_1} = \{0, 1\}$
   $\mathcal{S}_{*_2} = \{0, \frac{1}{2}, 1, \frac{3}{2}\}$
   $\mathcal{S}_{*_3} = \{0, \frac{1}{4}, \frac{1}{2}, \frac{3}{4}, 1, \frac{5}{4}, \frac{3}{2}, \frac{7}{4}\}$

   iii. $\mathcal{S}_{*_1} = \{0, 0\}$
   $\mathcal{S}_{*_2} = \{-1, 0, 0, 1, \}$
   $\mathcal{S}_{*_3} = \{-\frac{3}{2}, -1, -\frac{1}{2}, 0, 0, \frac{1}{2}, 1, \frac{3}{2}\}$

   c. i. $L_1 = 1, L_2 = 1.75, L_3 = 2.1875$
   ii. $L_1 = 0, L_2 = 2, L_3 = 3.5$

   d. i. $\mathcal{S}_1^* = \{1, 2\}$
   $\mathcal{S}_2^* = \{\frac{1}{2}, 1, \frac{3}{2}, 2\}$
   $\mathcal{S}_3^* = \{\frac{1}{4}, \frac{1}{2}, \frac{3}{4}, 1, \frac{5}{4}, \frac{3}{2}, \frac{7}{4}, 2\}$

   ii. $\mathcal{S}_1^* = \{-2, 2\}$ $\quad$ $\mathcal{S}_2^* = \{-2, -1, 1, 2\}$
   $\mathcal{S}_3^* = \{-2, -\frac{3}{2}, -1, -\frac{1}{2}, \frac{1}{2}, 1, \frac{3}{2}, 2\}$

   e. i. $L_1 = 5, L_2 = 3.75, L_3 = 3.1875$
   ii. $L_1 = 16, L_2 = 10, L_3 = 7.5$

SECTION 6.6

1. $-\frac{3}{2}$ $\quad$ 3. $\frac{3\pi}{2}$ $\quad$ 5. 0

7. $\frac{21}{2}$ $\quad$ 9. $\frac{39}{2}$ $\quad$ 11. 0

13. 0 $\quad$ 15. $\frac{\pi^2}{2}$ $\quad$ 17. 1 $\quad$ 19. 1

21. $\int_{-1}^{3} x^3 \cos(x) \, dx$

23. -2 $\quad$ 25. 8 $\quad$ 27. 1, 1 and 2

31. -8

SECTION 6.7

1. $\int_0^2 \pi x^4 \, dx$

3. $\int_{-1}^{1} \pi(1 - x^2)^2 \, dx$

5. $\int_{-1}^{1} \pi x^6 \, dx$

7. $\int_{-1}^{2} \pi(x^3 - 2x^2 - x + 2)^2 \, dx$

9. $2\int_0^1 \pi(x^2 - x^6) \, dx$

11. $\int_0^2 2\pi x^3 \; dx$ 13. $\int_0^1 2\pi(x - x^3) \; dx$

15. $2\int_0^1 2\pi x^4 \; dx$

17. $\int_{-1}^0 2\pi(-x^4 + 2x^3 + x^2 - 2x) \; dx$
$\quad - \int_1^2 2\pi(x^4 - 2x^3 - x^2 + 2x) \; dx$

19. $2\int_0^1 2\pi(x^2 - x^4) \; dx$

21. $\int_0^2 \pi\left((x^2 + 2)^2 - 4\right) \; dx$

23. $\int_0^1 \pi\left((1 - x^3)^2 - (1 - x^2)^2\right) \; dx$

25. $\int_0^1 \frac{\pi}{2}(1 - z)^2 \; dz$

27. $\int_0^1 4x\sqrt{1 - x^2} \; dx$

29. $\int_0^1 4\sqrt{1 - x^2} - x^2 \; dx$

SECTION 6.8

1. $\frac{8}{3}$ 3. $\frac{2}{3}$ 5. 414 7. $\sqrt{2} - 1$

9. $\frac{1}{2}$ 11. $\sqrt{2} - 1$ 13. $\dfrac{2(\sqrt{37} - 1)}{3}$

15. $\sqrt{2} - 1$ 17. $\dfrac{1 - \sqrt{2}}{2}$ 19. $\frac{\pi}{2}$

21. $\frac{2}{3}$ 23. $\dfrac{1}{1 + x^2}$ 25. $-\frac{1}{2}$

27. $-\sin(x)\sqrt{1 - \cos^3 x} - \cos(x)\sqrt{1 - \sin^3 x}$

29. $\frac{1}{2}$ 31. $\frac{\pi^2}{2}$ 33. $\frac{4}{3}$ 35. $\frac{1}{6}$

37. $\frac{3}{2}$ 39. $\frac{\pi}{5}$ 41. $\frac{\pi}{6}$

43. $\frac{\pi}{2}$ 45. $\frac{\pi}{10}$ 47. $\frac{7324\pi}{105}$

49. a. 0.71877 b. 0.66877 c. 0.69377

SECTION 7.1

1. $x > -1$

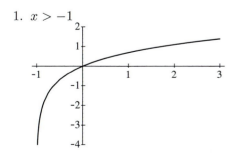

3. $x > -1$

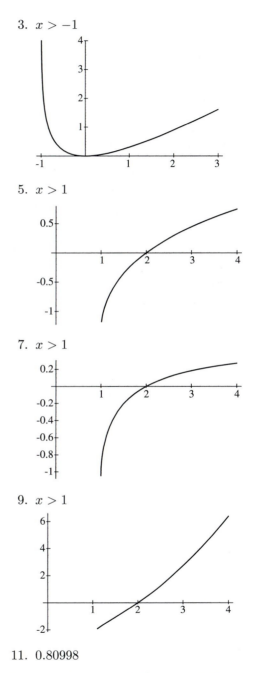

5. $x > 1$

7. $x > 1$

9. $x > 1$

11. 0.80998

13. a. 0.05 b. 0.51989 c. 0.00371
d. 2 is not in the domain of F

SECTION 7.2

1. Trapezoidal 4.2, 4.0625   Simpson 4.0, 4.0

3. Trapezoidal 1.01269, 1.0032
   Simpson 1.00004, 1.00

5. a. 0.49183, 1.22554, 2.32012, 3.95302, 6.3890, 10.0232, 15.4447

   b. No, there must be at least 3 points.

   c. 1.22950,   1.50175,   1.83425,   2.24045, 2.73650,   3.34205,   4.08225,   4.98635, 6.09000,   7.43840,   9.08525,   11.09675, 13.55375

   d.

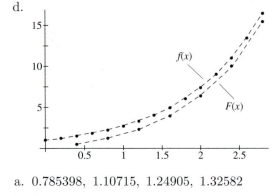

7. a. 0.785398, 1.10715, 1.24905, 1.32582

   b. 1, 2, 3, 4

   c. $\text{Arctan}(x)$

SECTION 7.3

1. $x > 3$ 3. $x > 0$ 5. $|x| < 1$

7. $\frac{11}{x}$ 9. 0   11. $y \approx 2.177(x - 2)$

13. $\ln|x + 1| + C$

15. $\dfrac{\ln^6 x}{6} + C$

17. $\dfrac{\ln^2|\sin t|}{2} + C$

19. $-\frac{1}{2}\csc^2 x + C$

21. $\ln|\sin x| + C$

23. $\ln|\sec x + \tan x| + C$

25. $p_4(x) = ax - \dfrac{(ax)^2}{2} + \dfrac{(ax)^3}{3} - \dfrac{(ax)^4}{4}$

27. $p_n(x) = ax - \dfrac{(ax)^2}{2} + \dfrac{(ax)^3}{3} - \cdots + (-1)^{n+1}\dfrac{(ax)^n}{n}$

29. a. Same

    b.

    c. $0, 1, 2, 3, 4, 5$

    d. They all equal $\frac{1}{\ln(2)}$

31. a.

    b. $n = 2$

    c. 0.367879

37. $\dfrac{1}{7}\ln\left|\dfrac{x - 2}{2x + 3}\right| + C$

39. $\dfrac{1}{5}\ln\left|\dfrac{x - 1}{x + 4}\right| + C$

41. $\ln\left|\dfrac{2x + 1}{3x + 2}\right| + C$

43. a. $x > 0$

    c. Increasing when $x > A$, decreasing when $x < A$

    d. $f''(2) \approx 1.962$, $f''(8) \approx 0.0011$

    e. $n = 7$

    f. $C \approx 7.39$

    g. Concave up when $x < C$, concave down when $x > C$

    h.

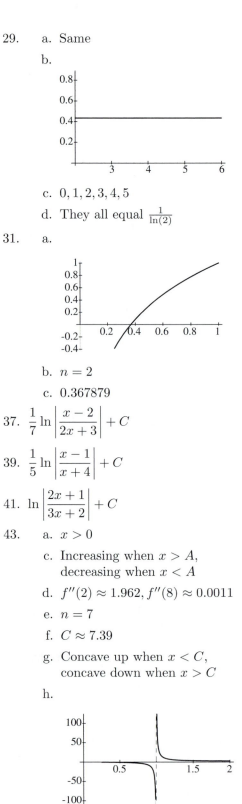

45. 2.48751

SECTION 7.4

1.

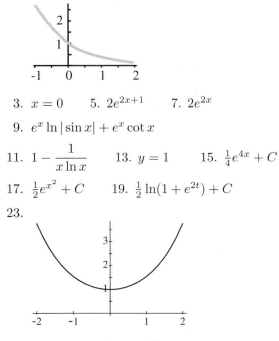

3. $x = 0$ 5. $2e^{2x+1}$ 7. $2e^{2x}$

9. $e^x \ln|\sin x| + e^x \cot x$

11. $1 - \dfrac{1}{x \ln x}$ 13. $y = 1$ 15. $\frac{1}{4}e^{4x} + C$

17. $\frac{1}{2}e^{x^2} + C$ 19. $\frac{1}{2}\ln(1 + e^{2t}) + C$

23.

25. Maximum at $(2, 4e^{-2})$, minimum at $(0,0)$ and inflection points at $(2 + \sqrt{2}, (6 + 4\sqrt{2})2e^{-2-\sqrt{2}})$ and $(2 - \sqrt{2}, (6 - 4\sqrt{2})e^{-2+\sqrt{2}})$.

27. $e - e^{-1}$ 29. 2, 2.5937, 2.7048, 2.7168

31. $y(x) = -\sqrt{5e^{2x^3/3 + 2x - 8/3} - 1}$

33. $y(x) = \text{Arcsin}(Ce^{x^2/2}) - 2$

35. $p_4(x) = 1 + ax + \dfrac{(ax)^2}{2} + \dfrac{(ax)^3}{6} + \dfrac{(ax)^4}{24}$

37. $p_n(x) = 1 + ax + \dfrac{(ax)^2}{2} + \cdots + \dfrac{(ax)^n}{n!}$

39.   a.

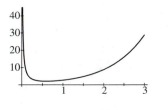

b. $f''(1) = e$, $f''(0.1) \approx -90.9$

c. ≈ 0.59186, Yes

41.   a. $0, 1, 2, 3, 4, 5$

b. $\dfrac{1}{\ln(2)}$

45. $x^2 e^x - xe^x + 2e^x + C$

47. $x \ln x - x + C$

49. ≈ 2.9262 51. ≈ 5.0048

SECTION 7.5

1. $2^x \ln 2$

3. $2^{2x+1} \ln 2$

5. $\dfrac{3x^2 + 2}{(x^3 + 2x) \ln 3}$

7. $x^{x+\sin x}\left(1 + \dfrac{\sin x}{x} + (\ln x)(1 + \cos x)\right)$

9. $\dfrac{2^x}{\ln 2} + C$

11. $\dfrac{\ln 2(\log_2 x)^2}{2} + C$

13. 3   15. $\frac{1}{2}$

17.

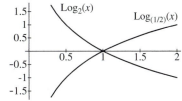

19. Increasing when $x > 1$ and decreasing when $0 < x < 1$

21. The line $y = 1$.

SECTION 7.6

1. a. i b. -1

3. i. $8 - 3i$ ii. $13 - 13i$ iii. $\sqrt{5}$ iv. 26

5. i. $e^{-2-i\pi/6} + e^{1-i3\pi/4}$ ii. $e^{-1-i\pi 11/12}$

iii. $\sqrt{e^{-4} + \dfrac{e^2}{2} + (\sqrt{6} - \sqrt{2})\dfrac{e^{-1}}{2} + \dfrac{e^2}{2}}$

iv. e^{-4}

7. $e^{\ln 2 + i\pi/4}$

9. $e^{\ln(\sqrt{3}) + i\,\mathrm{Arctan}(2/\sqrt{3})}$

11. $\dfrac{e^2\sqrt{2}(-1+i)}{2}$

13. a. $p_{2m}(x) = 1 + x + \dfrac{x^2}{2} + \cdots + \dfrac{x^{2m}}{(2m)!}$

   b. $p_{2m}(3x) = 1 + 3x + \dfrac{(3x)^2}{2} + \cdots + \dfrac{(3x)^{2m}}{(2m)!}$

   c. $p_{2m}(cx) = 1 + cx + \dfrac{(cx)^2}{2} + \cdots + \dfrac{(cx)^{2m}}{(2m)!}$

   d. $p_{2m}(cx) = 1 + ix - \dfrac{x^2}{2} - i\dfrac{x^3}{6} + \dfrac{x^4}{24} + \cdots$
   $+ \dfrac{(ix)^{2m}}{(2m)!}$

   e. $p_{2m}(x) = x - \dfrac{x^3}{3!} + \dfrac{x^5}{5!} - \dfrac{x^7}{7!} + \cdots$
   $+ (-1)^m \dfrac{x^{2m+1}}{(2m+1)!}$

   f. $p_{2m}(x) = 1 - \dfrac{x^2}{2} + \dfrac{x^4}{4!} - \dfrac{x^6}{6!} + \cdots$
   $+ (-1)^m \dfrac{x^{2m}}{(2m)!}$

SECTION 7.7

1.

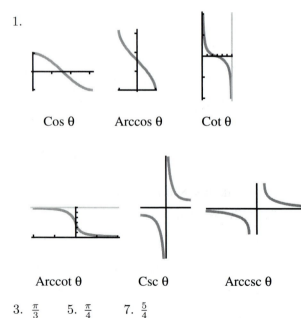

Cos θ   Arccos θ   Cot θ

Arccot θ   Csc θ   Arccsc θ

3. $\frac{\pi}{3}$ 5. $\frac{\pi}{4}$ 7. $\frac{5}{4}$

SECTION 7.8

1. $\frac{1}{2}\,\mathrm{Arctan}(2x) + C$

3. $\frac{1}{8}\ln|4x^2 + 1| + C$

5. $\frac{1}{2}\,\mathrm{Arcsin}\left(\dfrac{2x}{3}\right) + C$

7. $\dfrac{1}{\sqrt{6}}\,\mathrm{Arctan}\left(\sqrt{\dfrac{2x}{3}}\right) + C$

9. $\dfrac{1}{ab}\,\mathrm{Arctan}\left(\dfrac{bx}{a}\right) + C$

11. $\dfrac{1}{b}\,\mathrm{Arcsec}\left(\left|\dfrac{ax}{b}\right|\right) + C$

13. $-\dfrac{\sqrt{a^2 - b^2 x^2}}{b^2} + C$

15. $\ln|1 + e^x| + C$

17. $\frac{1}{2}\ln|1 + e^{2x}| + C$

19. $\mathrm{Arcsec}(|\ln x|) + C$

21. $\mathrm{Arctan}(x + 1) + C$

23. $\dfrac{2x + 2}{\sqrt{1 - (x^2 + 2x + 3)^2}}$

25. $-\dfrac{1}{x\sqrt{1 - \ln^2 x}}$

27. $-\dfrac{1}{(\ln 3)(\mathrm{Arccot}\,x)(1 + x^2)}$

29. $\dfrac{3x^2 + \pi x^{\pi-1} + \pi^x \ln \pi}{\sqrt{1 - (x^3 + x^\pi + \pi^x)^2}}$

31. $y = \tan(\ln|\sec x + \tan x| + C)$

33. $y = 4\sec\left(\frac{1}{2}\arctan(2x) + C\right)$

SECTION 7.9

1. $\dfrac{x^2 e^{2x}}{2} - \dfrac{(2x-1)e^{2x}}{2} + C$

3. $\dfrac{x^2 \ln(5x)}{2} - \dfrac{x^2}{4} + C$

5. $\dfrac{e^x - 1}{2}\sqrt{2e^x - e^{2x}} + \dfrac{1}{2}\,\mathrm{Arcsin}\left(\dfrac{e^x - 1}{2}\right) + C$

7. $e^x\,\mathrm{Arccsc}(3e^x) + \dfrac{1}{3}\ln\left(3e^x + \sqrt{9e^{2x} - 1}\right) + C$

9. $\sqrt{4 - e^{2x}} - 2\ln\left|\dfrac{2 + \sqrt{4 - e^{2x}}}{e^x}\right| + C$

11. $\dfrac{\left(\sec^5\theta\right)\left(\ln 2\sec\theta\right)}{5} - \dfrac{\sec^5\theta}{25} + C$

13. $x^2 e^x + e^x + C$

15. $\ln\left|e^x + \sqrt{1 + e^{2x}}\right| + C$

17. $\dfrac{(x+1)\sqrt{x^2 + 2x + 3}}{\sqrt{2}}$
$+ \dfrac{3}{2}\ln\left|(x+1) + \sqrt{x^2 + 2x + 3}\right| + C$

19. $2\sqrt{17} + \frac{1}{4}\ln(4 + \sqrt{17}) - \sqrt{5} - \frac{1}{4}\ln(2 + \sqrt{5})$

SECTION 8.1

1. $y(t) = 10e^{2t}$, $\mathbb{R}$

3. $y(t) = \frac{23}{2}e^{2t} - \frac{3}{2}$, $\mathbb{R}$

5. $y(x) = -3e^{(4/3)(x^3 - 1)}$, $\mathbb{R}$

7. $y(x) = 2x - 2$, $\mathbb{R}$

9. Because $\frac{y'}{y-3}$ is undefined at $y = 3$

11. $k = -\frac{1}{2}\ln 2$ $y = 4e^{-\left(\frac{1}{2}\ln 2\right)t}$

13.
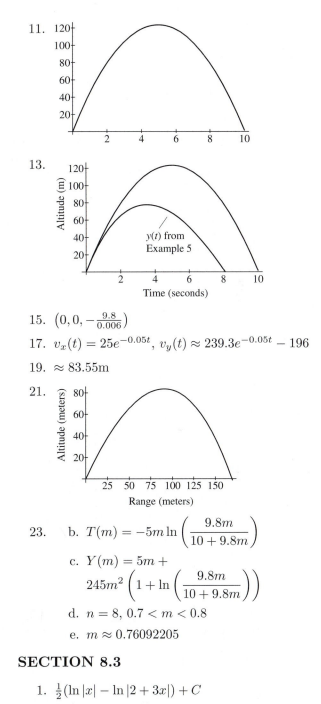

15. $\approx 16{,}275{,}473$ 17. 1600 19. $\approx 99.65\%$

21. $\approx \$1{,}161.8$ 23. $\approx 6.9\%$ 25. ≈ 0.0693

27. 0.05 29. $\approx 18{,}196$ 31. 0.3

33. ≈ 134 35. -0.3 37. ≈ 1.5

SECTION 8.2

1. $i(t) = \frac{6}{5}\left(1 - e^{-1000t}\right)$

3. $q(t) = (2 - VC)e^{-t/(RC)} + VC$

5. ≈ 4.03 hours

7. $v(t) = 100(10.3e^{-0.01t} - 9.8)$

9. ≈ 123.4m

15. $\left(0, 0, -\frac{9.8}{0.006}\right)$

17. $v_x(t) = 25e^{-0.05t}$, $v_y(t) \approx 239.3e^{-0.05t} - 196$

19. ≈ 83.55m

23.   b. $T(m) = -5m\ln\left(\dfrac{9.8m}{10 + 9.8m}\right)$

c. $Y(m) = 5m +$
$245m^2\left(1 + \ln\left(\dfrac{9.8m}{10 + 9.8m}\right)\right)$

d. $n = 8$, $0.7 < m < 0.8$

e. $m \approx 0.76092205$

SECTION 8.3

1. $\frac{1}{2}\left(\ln|x| - \ln|2 + 3x|\right) + C$

3. $\frac{1}{2}\left(x - \ln|2 + 3e^x|\right) + C$

5. $\ln|\tan x| - \ln|1 + 3\tan x| + C$

7. $x = 0$ is not in the domain of $\frac{1}{3x - 2x^2}$

9. $y(x) = \dfrac{6}{9 - 11e^{-2x}}$

11. $y(t) = \dfrac{100}{5 - e^{-0.2t}}$

13. $y(t) = -\dfrac{100}{9e^{-0.2t} + 5}$

15. $c_2 = \frac{1}{6000}$

17. i. 2000   ii. ≈ 1464 iii. Increasing
iv. $(-\infty, \infty)$

19.

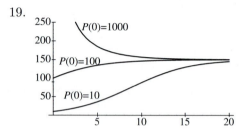

21. a. 106   b. ≈ 26.34 years   c ≈ 20.09 years.

23. $c_1 \approx 0.9885$, $c_2 \approx 0.0059$

$$P(t) = \dfrac{197.7}{1.18 - 0.1915e^{-0.9885t}}.$$

The equilibrium population is 168

SECTION 9.1

1.	4	3.	$+\infty$	5.	$+\infty$
7.	$+\infty$	9.	0	11.	0
13.	1	15.	0	17.	$-\frac{1}{2}$
19.	$+\infty$	21.	0	23.	1
25.	∞	27.	$-\infty$	29.	Nothing
31.	0	33.	0	35.	1
37.	1	39.	1	41.	1
43.	0				

SECTION 9.2

1. 3   3. $-\frac{1}{4}$ 5. Diverges

7. π 9. 3   11. Diverges

13. $\frac{\pi}{4}$ 15. $\frac{2}{\ln 2}$ 17. Diverges

19. Diverges

21. Diverges, $e^x x^{4/3} \geq e^x$ on $[1, \infty)$

23. Converges, $0 \leq \cos^2(x)x^{-3} \leq x^{-3}$ on $[1, \infty)$

25. Converges, $0 \leq [[x^{-3}]] \leq x^{-3}$ on $[1, \infty)$

27. $\frac{3}{2}$ 29. Diverges   31. 3π

SECTION 9.3

1.   3   3.   5   5.   Diverg

7.   Diverges   13.   Diverges   15.   Diverg

17. Converges by Comparison to $\left(\frac{1}{n}\right)^2$

19. Diverges by the Integral Test

21. Converges, Geometric Series

23. Converges by the Comparison Test. $\ln n < n$ for $n \geq 1$ and for $n > 1$, $\frac{\ln n}{n!} < \frac{1}{(n-1)!}$. For $n > 2$, $\frac{1}{(n-1)!} < \frac{1}{2^{(n-2)}}$.

25. Converges by the p–Test

27. Converges. Converges by the Integral Test.

29. Converges by the Limit Comparison Test using $\frac{1}{i^2}$.

33. $\frac{1}{1}$ 35. $\frac{2}{3}$ 37. $\frac{992}{165}$

SECTION 9.4

1. Converges Absolutely

3. Converges conditionally

5. Converges conditionally

7. Converges Absolutely

9. Converges Absolutely

11. Diverges   13. Diverges

15. Diverges   17. Diverges.

19. $\sum_{k=1}^{\infty}(-1)^k\left(\frac{1}{k}\right) = -1 + \left(\frac{1}{2} - \frac{1}{3}\right) + \left(\frac{1}{4} - \frac{1}{5}\right) + \cdots$ and each term in parentheses is positive.

SECTION 9.5

1. Converges   3. Diverges   5. Diverges

7. Converges   9. Converges

11.　1　$-1 < x \leq 1$　　　13.　1　$0 < x < 2$

15.　∞　All x　　　17.　1　$4 \leq x < 6$

19.　$\frac{1}{2}$　$2 \leq x < 3$　　　21.　∞　All x

23.　2　$-2 < x < 2$　　　25.　1　$-1 < x < 1$

27.　Yes, it can converge at an end point.

29.　No, $R \leq 1$.　　　31.　$|x| > 1$

SECTION 9.6

1.　$1 - x + x^2 + x^3 - x^4 + \cdots + (-1)^n x^n + \cdots = \sum_{n=0}^{\infty} (-1)^n x^n$

3.　$1 + x^2 + x^4 + x^6 + \cdots + x^{2n} + \cdots = \sum_{n=0}^{\infty} x^{2n}$

5.　$-x - \dfrac{x^2}{2} - \dfrac{x^3}{3} - \dfrac{x^4}{4} - \cdots - \dfrac{x^n}{n} - \cdots = -\sum_{n=1}^{\infty} \dfrac{x^n}{n}$

7.　$1 - \dfrac{\left(\frac{x}{2}\right)^2}{2!} + \dfrac{\left(\frac{x}{2}\right)^4}{4!} - \cdots + (-1)^n \dfrac{\left(\frac{x}{2}\right)^{2n}}{(2n)!} + \cdots$
$= \sum_{n=0}^{\infty} (-1)^n \dfrac{(x)^{2n}}{(2n)! 4^n}$

9.　$\sum_{n=0}^{\infty} \dfrac{(-1)^n x^{6n+4}}{(2n+1)!}$

11.　$\sum_{n=1}^{\infty} \dfrac{(-1)^n 2^{2n-1} x^{2n-1}}{(2n)!}$

15.　$4 \leq x < 6$

17.　Yes. If $R = 3$, $x = -3$ is an end point.

19.　No, $R \leq 1$.　　21.　$A(n) = \dfrac{n^2 4^n}{n!}$, $C = \dfrac{3}{2}$

23.　$A(n) = \dfrac{n^2 \alpha^n}{n!}$, $C = \dfrac{4 - \beta}{\alpha}$

25.　$C = -\frac{1}{3}$

27.　$-\frac{4}{3} < x \leq \frac{2}{3}$

SECTION 9.7

1.　$\sqrt{13}$　　　3.　$\sqrt{5}$　　　　5.　$\sqrt{5}$

7.　$\sqrt{3}$　　　9.　1

SECTION 10.1

1.　$e - 2$

3.　$\frac{1}{2} x^2 e^{2x} - \frac{1}{2} x e^{2x} + \frac{1}{4} e^{2x} + C$

5.　$\frac{1}{2} e^{x^2} + C$

7.　$x \ln |x| - x + C$

9.　$\dfrac{x^2 \ln |x|}{2} - \dfrac{x^2}{4} + C$

11.　$\dfrac{2x^{3/2}}{3 \ln |x|} - \dfrac{4}{9} x^{3/2} + C$

13.　$\dfrac{(\ln |x|)^2}{2} + C$

15.　$\cos x - \cos x \ln |\cos x| + C$

17.　$-x^3 \cos x + 3x^2 \sin x + 6x \cos x - 6 \sin x + C$

19.　$1 - \dfrac{\sqrt{3}}{2}$

21.　$\frac{\pi}{4}$

23.　$\frac{2}{3} - \dfrac{3\sqrt{3}}{8}$

25.　$-\dfrac{\sqrt{3}}{8} + \dfrac{1}{3}$

27.　$\dfrac{2}{15} - \dfrac{11\sqrt{3}}{160}$

29.　$\frac{1}{2} e^x (\sin x + \cos x) + C$

31.　$\frac{\pi}{16}$

33.　$\frac{\pi}{4n}$

SECTION 10.2

1.　$-\dfrac{\sqrt{4 + x^2}}{4x} + C$

3.　$\frac{25}{6} \operatorname{Arcsin} \left(\frac{3x}{5}\right) + x\sqrt{25 - 9x^2} + C$

5.　$\dfrac{x}{\sqrt{9 - x^2}} - \operatorname{Arcsin} \left(\dfrac{x}{3}\right) + C$

7.　$\frac{1}{2} x \sqrt{b^2 x^2 + a^2} + \frac{a^2}{2} \operatorname{Arcsinh} \left(\frac{xb}{a}\right) + C$

9.　$\frac{1}{4} \ln |4x + \sqrt{16x^2 - 4}| + C$

11.　$\sqrt{x^2 - 4x + 3} + 2 \ln \left|x - 2 + \sqrt{x^2 - 4x + 3}\right| + C$

13. $\frac{1}{2}\left(x\sqrt{x^2-9}+9\ln\left(x+\sqrt{x^2-9}\right)\right)+C$

15. $-\frac{1}{27}(25-9x^2)^{3/2}+C$

17. $\frac{x^2}{2}+25\ln|x|+C$

19. $-\frac{1}{a}\ln\left|\dfrac{a+\sqrt{a^2-b^2x^2}}{bx}\right|+C$

21. $\frac{1}{2}\ln\left|(x-3)+\sqrt{(x-3)^2-\frac{9}{4}}\right|+C$

23. $\frac{1}{2}\ln\left|\dfrac{1-e^x}{1+e^x}\right|+C$

29. $\frac{4\sqrt{2}\pi^3}{3}$

SECTION 10.3

1. $\ln|x-1|+2\ln|x+1|+C$

3. $\ln|x|+3\ln|x-1|+2\ln|x+1|+C$

5. $\frac{11}{10}\ln|x-3|+\frac{2}{5}\ln|x+2|-\frac{1}{2}\ln|x-1|+C$

7. $\frac{x^2}{2}+\frac{1}{2}\ln|x^2-1|+C$

9. $2\ln|x|-(x+1)^{-1}+C$

11. $\text{Arctan }x+\ln|x-1|+C$

13. $\frac{1}{2}\left(\ln|2+x|-\frac{1}{2}\ln|x^2+4|+\frac{1}{2}\text{Arctan}\left(\frac{x}{2}\right)\right)+C$

15. $\frac{1}{2}\ln|x+1|-\frac{1}{4}\ln|x^2+1|+\frac{1}{2}\text{Arctan}(x)+C$

17. $-\dfrac{1}{x^2+x+1}+\ln|x+1|+C$

SECTION 10.4

1. $x'(t)=\sqrt{\frac{2}{x}}$

3. $x'(t)=\sqrt{2-\frac{2}{x}}$

5. $x'(t)=\sqrt{3x^2+2x-81}$

7. $x(t)=2(1-8t)^{-1/2}$

SECTION 11.1

1. -4J   3. -8J   5. $-\frac{\pi}{4}$J

7. $1-e$ J   9. 10J   11. $e^{0.1}-0.975$J

13. $6.43125\pi\,10^7$ J

15. $\pi\left(\dfrac{-245+245\,e^{15\,k}-3675\,k+27562.5\,k^2\,\mu}{k^2}\right)$
J.

17. $3.675\pi\,10^6$ 19. 816.667πJ   21. 4.9J

23. $\dfrac{9.8\,e^k}{k^2}+\dfrac{4.9\,(-2.73205+k)\,(0.732051+k)}{k^2}$J

25. $-\dfrac{kq_1q_2}{2}$J   27. $kq_1q_2(1-\sqrt{2})$J

SECTION 11.2

1. 4 m   3. $10\sqrt{19.6}$m/s   5. 20 cm

7. $\ln(v_0^2+1)$ m

9. $x+\ln\sqrt{\dfrac{x+1}{x-1}}=v_0^2$

11. $\sqrt{6.36}$ m/s   13. $\sqrt{7.96}$ m/s

15. $\sqrt{4.08}$ m/s

17. $h(v_0)=\dfrac{v_0^2R^2}{2kM_E-v_0^2R}$

SECTION 11.3

1. $\vec{s}(t)=(1-2t,2t,2-2t)$
   $\vec{r}(t)=(-1+2t,2-2t,2t)$

3. $\vec{s}(t)=(1-t,3-4t)$ $\vec{r}(t)=(t,-1+4t)$

5. $\vec{r}(t)=(-1,2,6-3t)$

7. $\vec{r}(t)=(-23+6t,15-4t,-4+2t)$

9. $\vec{s}(t)=(-3+6t,-\pi,et)$

SECTION 11.4

1. 26.7 3. 67.4

5. $2\sqrt{2}\pi^2$ 7. $\sqrt{2}$

9. $\dfrac{\sqrt{3}}{2}+\ln\left(\dfrac{1+\sqrt{3}}{\sqrt{2}}\right)$

11. $\dfrac{13\sqrt{13}-8}{27}$

13. $4\sqrt{2}$ 15. $\frac{14\sqrt{2}}{3}$ 17. $-\frac{2\sqrt{2}}{3}$

19. You are dividing by zero at $t=0.5$.

21. 8   23. 4

SECTION 11.5

1. Mass $= \frac{3}{2}\sqrt{10}$, $(\overline{X}, \overline{Y}, \overline{Z}) = \left(\frac{14}{9}, \frac{1}{3}, 3\right)$,
   $I_x = \frac{59\sqrt{10}}{4}$, $I_y = \frac{69\sqrt{10}}{4}$, $I_z = 5\sqrt{10}$

3. Mass $= 5\sqrt{10}$, $(\overline{X}, \overline{Y}, \overline{Z}) = \left(\frac{22}{15}, \frac{3}{5}, 3\right)$,
   $I_x = \frac{101\sqrt{10}}{2}$, $I_y = \frac{337\sqrt{10}}{6}$, $I_z = 20\sqrt{10}$

5. $621\pi^2\sqrt{5}$ 7. $\left(0, \frac{2}{\pi}\right)$

9. Mass $= c \int_0^1 \sqrt{1 + 4x^2}\, dx$,
   $\overline{X} = c \int_0^1 x\sqrt{1 + 4x^2}\, dx / \text{Mass}$,
   $\overline{Y} = c \int_0^1 (x^2 + 1)\sqrt{1 + 4x^2}\, dx / \text{Mass}$,
   $I_y = c \int_0^1 x^2\sqrt{1 + 4x^2}\, dx$

11. $(\overline{X}, \overline{Y}, \overline{Z}) = (\pi, 0, 0)$

SECTION 11.6

1.

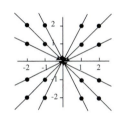
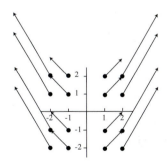

3.

5.

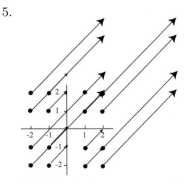

7. $\vec{E} = \dfrac{2k}{3\sqrt{3}}(1, 1, 1)$

9. $\vec{E} = \dfrac{k}{17\sqrt{34}}(5, 0, 3)$

11. $\vec{E} = \dfrac{2k}{27\sqrt{3}}(9 - \sqrt{3}, 9 - 2\sqrt{3}, 9 + 2\sqrt{3})$

15. Every point on the charged wire, except the end point, is to the right of $(-1, -1, -1)$.

17. Every point on the charged wire, except the end-point, is to the left of $(1, 1, 1)$.

19. No. If $(c, 0, 0)$ is any point on the charged wire, then the field at $(2, 3, 4)$ due to a small segment containing $(c, 0, 0)$ will have the same direction as the vector $(2 - c, 3, 4)$. So we would expect that $\dfrac{E_y(2, 3, 4)}{E_z(2, 3, 4)} = \dfrac{3}{4}$.

21. For any point $(a, 0, 0)$ on the wire, the x–component of $\vec{E}$ generated by a small region about $(a, 0, 0)$ is minus one times the x–component of $\vec{E}$ generated by a small region about $(-a, 0, 0)$ as illustrated below.

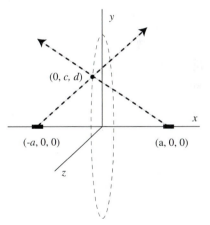

23. For any point $\vec{r}$ on the wire, the x and y components of $\vec{E}$ generated by a small region about $\vec{r}$ is minus one times the x and y components of $\vec{E}$ generated by a small region about the point on the opposite side of the wire as illustrated below.

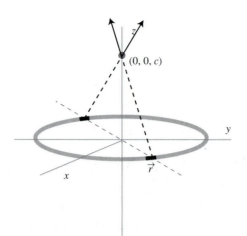

SECTION 11.7

1. 4   3. 5   5. $1 - \frac{1}{\sqrt{2}}$

7. $\dfrac{1}{\sqrt{x_0^2 + y_0^2 + z_0^2}} - \dfrac{1}{\sqrt{x_1^2 + y_1^2 + z_1^2}}$

9. 6.75   11. 2   13. 3

15. 0   17. 0

19. $\vec{r}'(t)$ is tangent to the cylinder; and hence, it is perpendicular to the force.

SECTION 11.8

1. $\frac{\partial f}{\partial x} = yzx^{(yz-1)}$, $\frac{\partial f}{\partial y} = x^{yz} z \ln x$, $\frac{\partial f}{\partial z} = x^{yz} y \ln x$

3. $\dfrac{\partial f}{\partial x} = \dfrac{x \cos(x^2 + y^2 + z^2)}{(x^2 + y^2 + z^2)^{1/2}}$,
$\dfrac{\partial f}{\partial y} = \dfrac{y \cos(x^2 + y^2 + z^2)}{(x^2 + y^2 + z^2)^{1/2}}$,
$\dfrac{\partial f}{\partial z} = \dfrac{z \cos(x^2 + y^2 + z^2)}{(x^2 + y^2 + z^2)^{1/2}}$

5. $-\dfrac{K\vec{r}}{\|\vec{r}\|^3}$ 7. $\left(\frac{1}{x}, \frac{1}{y}, \frac{1}{z}\right)$

9. $\left(\dfrac{1}{\sqrt{1 - x^2 y^2 z^2}}\right) \begin{pmatrix} yz \\ xz \\ xy \end{pmatrix}$ 11. 2

13. $\vec{V}_1 = \frac{K}{\|\vec{r}\|} + 10 - \frac{K}{\sqrt{3}}$ $\vec{V}_2 = \frac{K}{\|\vec{r}\|} + 10 - \frac{K}{\sqrt{2}}$.

15. Positive, $f(a, y)$ is a function of y that is increasing at $y = b$.

17. Negative, $f(c, y)$ is a function of y that is decreasing at $y = d$.

19. Positive, $f(a, y)$ is a function of y that is increasing at $y = b$.

21. Negative, $f(c, y)$ is a function of y that is decreasing at $y = d$.

23. $(0,0)$. $f(0, y)$ is a minimum at $y = 0$ and $f(x, 0)$ is a minimum at $x = 0$.

SECTION 11.9

1. 3   3. $\frac{1}{\sqrt{3}} - 1$ 5. $\dfrac{x^3 + y^3 + z^3}{3}$

7. $xy^2 + yx^2$ 9. 10   11. -10

13. $\frac{9}{\sqrt{6}}$ 15. $\frac{6}{\sqrt{3}}$ 17. $\frac{13}{\sqrt{14}}$

19. $\left(\frac{2}{\sqrt{14}}, \frac{3}{\sqrt{14}}, \frac{1}{\sqrt{14}}\right)$

SECTION 11.10

1. $-\frac{2}{\sqrt{6}}$ 3. 0   5. $\frac{1}{\sqrt{2}}$

7. $\frac{1}{\sqrt{11}}(-3, 1, 1)$ 9. $\left(-\frac{1}{\sqrt{2}}, -\frac{1}{\sqrt{2}}\right)$

11. $x + y + z = 2$ 13. $x + y + z = 2$.

15. $x = 0$

17. Hyperbolas of the form $\frac{x^2}{4C} - \frac{y^2}{C} = 1$ for $C \neq 0$.

19.

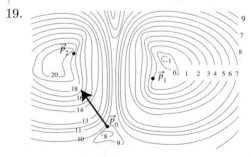

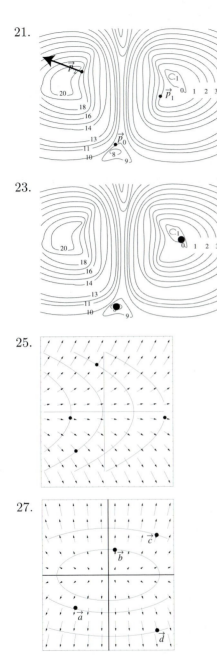

21.

23.

25.

27.

29. Graph 1–Contour 2–Gradient 3,
Graph 2–Contour 1–Gradient 4,
Graph 3–Contour 4–Gradient 2,
Graph 4–Contour 3–Gradient 1

SECTION 11.11

1. $\dfrac{\partial^2 f}{\partial x^2} = 2y\cos(xy) - xy^2\sin(xy)$

   $\dfrac{\partial^2 f}{\partial x \partial y} = 2x\cos(xy) - x^2 y\sin(xy)$

3. $\dfrac{\partial^2 f}{\partial x^2} = y^x \ln^2 y$

   $\dfrac{\partial^2 f}{\partial x \partial y} = (1 + x\ln y)y^{x-1}$

5. $(0, 0, y\cos(x)\cos(y) - xe^{xy})$

7. $(xz - xe^{xy}, -yz, xyze^{xy} - x\sin(y)$
   $-xy\cos(y) + ze^{xy})$

9. $\vec{0}$ 11. $\vec{0}$ 13. Not conservative

15. Not conservative   17. No information

19. $\dfrac{\partial h(0,0)}{\partial y} = 0, \quad \dfrac{\partial^2 h(0,0)}{\partial x \partial y} = -1$

21. No. The second partial derivatives are not continuous at $(0,0)$. $\displaystyle\lim_{x\to 0}\dfrac{\partial^2 h(x,0)}{\partial x \partial y} \neq \lim_{y\to 0}\dfrac{\partial^2 h(0,y)}{\partial x \partial y}.$

SECTION 12.1

1. $(0,0)$ saddle point

3. $(1,1)$ local minimum, $(-1,1)$ saddle point

5. $(0,0)$ saddle point

7. $(0,0,0)$ saddle point

9. $\left(-\frac{1}{4}, -\frac{1}{4}, 0\right)$ saddle point

11. $(0,0,0)$ saddle point, $(0,2,0)$ local minimum

13. $\vec{r}_0 = \left(\frac{1}{2}, \frac{1}{2}, \frac{1}{\sqrt{2}}\right)$

SECTION 12.2

1. Absolute minimum of $-\sqrt{3}$ at $(0,-1)$
   Absolute maximum of $\sqrt{3}$ at $(0,1)$

3. Absolute minimum of -1 at $(1,0)$
   Absolute maximum of $\sqrt{3}$ at $(0,1)$

5. Absolute minimum of $-\frac{1}{4}$ at $\left(\frac{1}{2},0\right)$
   Absolute maximum of 2 at $(-1,0)$

7. Absolute minimum of -2 at both $(0,1)$ and $(0,-1)$
Absolute maximum of 1 at both $(1,0)$ and $(-1,0)$

9. Absolute minimum of -2 at $(0,1)$
Absolute maximum of 1 at $(1,0)$

SECTION 12.3

1. Maxima at $\left(\frac{1}{\sqrt{2}}, \frac{1}{\sqrt{2}}\right)$ and $\left(-\frac{1}{\sqrt{2}}, -\frac{1}{\sqrt{2}}\right)$
Minima at $\left(-\frac{1}{\sqrt{2}}, \frac{1}{\sqrt{2}}\right)$ and $\left(\frac{1}{\sqrt{2}}, -\frac{1}{\sqrt{2}}\right)$.

3. $(1,1,0)$ and $(-1,-1,0)$

5. Maximum at $\left(\frac{\sqrt{6}}{3}, 0, -\frac{\sqrt{6}}{3}\right)$
Minimum at $\left(-\frac{\sqrt{6}}{3}, 0, \frac{\sqrt{6}}{3}\right)$.

7. $x = \frac{2}{3}$, $y = \frac{2}{3}$, $z = \frac{1}{9}$ for dimensions $\frac{4}{3} \times \frac{4}{3} \times \frac{2}{9}$

SECTION 13.1

1. Domain$= \mathbb{R}^2$, Range$=\mathbb{R}^2$

3. Domain$= \mathbb{R}^3$, Range$=\mathbb{R}$

5. Domain$= \mathbb{R}^3$, Range$=\mathbb{R}^2$

7. $\vec{A} = (2,0)$, $\vec{B} = (0,3)$, $A_{\vec{T}} = \begin{pmatrix} 2 & 0 \\ 0 & 3 \end{pmatrix}$

9. $\vec{A} = \left(3, -\frac{1}{3}\right)$, $\vec{B} = \left(-6, \frac{1}{2}\right)$,
$A_{\vec{T}} = \begin{pmatrix} 3 & -6 \\ -\frac{1}{3} & \frac{1}{2} \end{pmatrix}$

11. $\vec{A} = (0,1)$, $\vec{B} = (-1,-1)$, $A_{\vec{T}} = \begin{pmatrix} 0 & -1 \\ 1 & -1 \end{pmatrix}$

13. $\vec{A} = (-6,-1,1)$, $\vec{B} = \left(1,-1,\frac{1}{3}\right)$,
$A_{\vec{T}} = \begin{pmatrix} -6 & 1 \\ -1 & -1 \\ 1 & \frac{1}{3} \end{pmatrix}$

15. $\vec{A} = (1,1,0)$, $\vec{B} = (1,-1,1)$, $\vec{C} = (1,1,-1)$,
$A_T = \begin{pmatrix} 1 & 1 & 1 \\ 1 & -1 & 1 \\ 0 & 1 & -1 \end{pmatrix}$

17. $\vec{A} = (6,-1,1)$, $\vec{B} = (3,-1,0)$,
$\vec{C} = (-1,-1,-1)$,
$A_T = \begin{pmatrix} 6 & 3 & -1 \\ -1 & -1 & -1 \\ 1 & 0 & -1 \end{pmatrix}$

19. $\vec{A} = (1,1)$, $\vec{B} = (1,-1)$, $\vec{C} = (1,0)$,
$A_T = \begin{pmatrix} 1 & 1 & 1 \\ 1 & -1 & 0 \end{pmatrix}$

21. $\vec{A} = (1)$, $\vec{B} = (6)$, $\vec{C} = (-10)$,
$A_T = \begin{pmatrix} 1 & 6 & -10 \end{pmatrix}$

23. $\begin{pmatrix} -1 & 1 \\ 5 & 3 \end{pmatrix}$ 25. $\begin{pmatrix} 1 & -1 & -1 \\ 3 & 5 & 0 \\ 2 & 1 & 3 \end{pmatrix}$

27. $\begin{pmatrix} 1 & 2 & 3 \end{pmatrix}$.   29. $\begin{pmatrix} 1 & 0 \\ 0 & -1 \end{pmatrix}$

31. $\vec{T}(u,v) = (u,v,-u-v)$

33. Not the image of a linear transformation. Plane does not contain the origin.

35. Not the image of a linear transformation.

37. $\begin{pmatrix} 1 & 0 \\ 0 & 1 \\ -1 & -1 \end{pmatrix} \begin{pmatrix} s \\ t \end{pmatrix} + \begin{pmatrix} 0 \\ 0 \\ 3 \end{pmatrix}$

39. $\begin{pmatrix} 1 & 0 \\ 0 & 1 \\ -\frac{1}{3} & \frac{2}{3} \end{pmatrix} \begin{pmatrix} s \\ t \end{pmatrix} + \begin{pmatrix} 0 \\ 0 \\ 1 \end{pmatrix}$

41. $\begin{pmatrix} 1 & 0 \\ 0 & 1 \\ -\frac{1}{3} & \frac{2}{3} \end{pmatrix} \begin{pmatrix} s \\ t \end{pmatrix} + \begin{pmatrix} 0 \\ 0 \\ \frac{5}{3} \end{pmatrix}$

49. $\begin{pmatrix} \cos(\theta) & -\sin(\theta) \\ \sin(\theta) & \cos(\theta) \end{pmatrix}$ 51. $\begin{pmatrix} -1 \\ 1 \end{pmatrix}$

SECTION 13.2

1. $(x,y) = (2,0)$, $\frac{\partial P}{\partial r} = (1,0)$, $\frac{\partial P}{\partial \theta} = (0,2)$

3. $(x,y) = (2,0)$, $\frac{\partial P}{\partial r} = (1,0)$, $\frac{\partial P}{\partial \theta} = (0,2)$

5. $(x,y) = \left(\frac{5}{\sqrt{2}}, \frac{5}{\sqrt{2}}\right)$, $\frac{\partial P}{\partial r} = \left(\frac{\sqrt{2}}{2}, \frac{\sqrt{2}}{2}\right)$,
$\frac{\partial P}{\partial \theta} = \left(-\frac{5\sqrt{2}}{2}, \frac{5\sqrt{2}}{2}\right)$

7. $(x,y) = \left(\frac{5}{\sqrt{2}}, -\frac{5}{\sqrt{2}}\right)$, $\frac{\partial P}{\partial r} = \left(\frac{\sqrt{2}}{2}, -\frac{\sqrt{2}}{2}\right)$,
$\frac{\partial P}{\partial \theta} = \left(\frac{5\sqrt{2}}{2}, \frac{5\sqrt{2}}{2}\right)$

9. $(x,y) = \left(\frac{\sqrt{3}}{2}, \frac{1}{2}\right)$, $\frac{\partial P}{\partial r} = \left(-\frac{\sqrt{3}}{2}, -\frac{1}{2}\right)$,
$\frac{\partial P}{\partial \theta} = \left(-\frac{1}{2}, \frac{\sqrt{3}}{2}\right)$

11. $\left(\frac{1}{2}, 0\right)$ 13. $\left(-2, \frac{2\pi}{3}\right)$

15. $(2,0,2)$, $\frac{\partial C_z}{\partial r} = (1,0,0)$, $\frac{\partial C_z}{\partial \theta} = (0,2,0)$, $\frac{\partial C_z}{\partial z} = (0,0,1)$

17. $(2,0,2)$, $\frac{\partial C_z}{\partial r} = (1,0,0)$, $\frac{\partial C_z}{\partial \theta} = (0,2,0)$, $\frac{\partial C_z}{\partial z} = (0,0,1)$

19. $\left(\frac{5}{\sqrt{2}}, \frac{5}{\sqrt{2}}, 2\right)$, $\frac{\partial C_z}{\partial r} = \left(\frac{1}{\sqrt{2}}, \frac{1}{\sqrt{2}}, 0\right)$, $\frac{\partial C_z}{\partial \theta} = \left(-\frac{5}{\sqrt{2}}, \frac{5}{\sqrt{2}}, 0\right)$, $\frac{\partial C_z}{\partial z} = (0,0,1)$

21. $\left(\frac{5}{\sqrt{2}}, -\frac{5}{\sqrt{2}}, 3\right)$, $\frac{\partial C_z}{\partial r} = \left(\frac{1}{\sqrt{2}}, -\frac{1}{\sqrt{2}}, 0\right)$, $\frac{\partial C_z}{\partial \theta} = \left(\frac{5}{\sqrt{2}}, \frac{5}{\sqrt{2}}, 0\right)$, $\frac{\partial C_z}{\partial z} = (0,0,1)$

23. $\left(\frac{\sqrt{3}}{2}, \frac{1}{2}, -1\right)$, $\frac{\partial C_z}{\partial r} = \left(-\frac{\sqrt{3}}{2}, -\frac{1}{2}, 0\right)$, $\frac{\partial C_z}{\partial \theta} = \left(-\frac{1}{2}, \frac{\sqrt{3}}{2}, 0\right)$, $\frac{\partial C_z}{\partial z} = (0,0,1)$

25. $(r,\theta,z) = \left(\frac{1}{2}, 0, 2\right)$ 27. $(r,\theta,z) = \left(-2, \frac{2\pi}{3}, -1\right)$

29. $(x,y,z) = (0,0,2)$, $\frac{\partial \vec{S}}{\partial r} = (0,0,1)$, $\frac{\partial \vec{S}}{\partial \theta} = (0,0,0)$, $\frac{\partial \vec{S}}{\partial \phi} = (\sqrt{2}, \sqrt{2}, 0)$

31. $(x,y,z) = (0,0,-2)$ $\frac{\partial \vec{S}}{\partial r} = (0,0,1)$, $\frac{\partial \vec{S}}{\partial \theta} = (0,0,0)$, $\frac{\partial \vec{S}}{\partial \phi} = (-\sqrt{2}, \sqrt{2}, 0)$

33. $(x,y,z) = (0,0,-2)$ $\frac{\partial \vec{S}}{\partial r} = (0,0,-1)$, $\frac{\partial \vec{S}}{\partial \theta} = (0,0,0)$, $\frac{\partial \vec{S}}{\partial \phi} = (-\sqrt{2}, \sqrt{2}, 0)$

35. $(x,y,z) = \left(\frac{1}{2}, \frac{1}{2}, \frac{1}{\sqrt{2}}\right)$, $\frac{\partial \vec{S}}{\partial r} = \left(-\frac{1}{2}, -\frac{1}{2}, -\frac{\sqrt{2}}{2}\right)$, $\frac{\partial \vec{S}}{\partial \theta} = \left(-\frac{1}{2}, \frac{1}{2}, 0\right)$, $\frac{\partial \vec{S}}{\partial \phi} = \left(\frac{1}{2}, \frac{1}{2}, -\frac{\sqrt{2}}{2}\right)$

37. $(r,\phi,\theta) = \left(\sqrt{2}, \frac{\pi}{4}, 0\right)$

39. $(r,\phi,\theta) = \left(2, \frac{\pi}{3}, \operatorname{Arctan}(\sqrt{2})\right)$

47. 49.

51. 53.

55. 57.

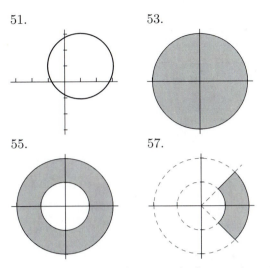

59. Cylinder or radius 2 with the z–axis as its axis.

61. The plane $x = y$.

63. A cylinder with axis the line obtained by intersecting the planes $x = 1$ and $y = 2$.

65. One eighth of a unit disc lying in the plane $z = 2$ and centered on the z–axis.

67. A solid cylinder with radius 2 and height 1. Centered on the z–axis and lying between the planes $z = 0$ and $z = 1$.

69. A cylinderical annulus centered on the z–axis with height 1, outer radius 2, and inner radius 1. It lies between the planes $z = 1$ and $z = 2$.

71. Sphere of radius 2 centered at the origin.

73. The plane $x = y$.

75. Sphere of radius 2 centered at $(1, 2, 3)$.

77. One eighth of a cone of radius 1.

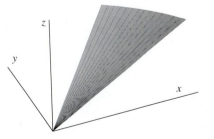

79. A conical sector of a sphere of radius 1.

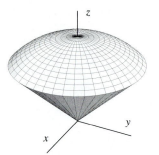

81. Upper half of a ball with a hollow core.

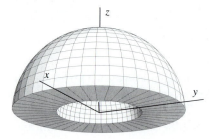

83. A plane in xyz–space parallel to the xz–plane

85. The plane $y = 0$

87. A plane parallel to the xz–coordinate plane

89. The plane $x = 0$

91. $\frac{\partial \vec{f}}{\partial u}(u,v) = 2(-\sin(u)\sin(v), \cos(u)\sin(v), 0)$,
$\frac{\partial \vec{f}}{\partial v}(u,v) = 2(\cos(u)\cos(v), \sin(u)\cos(v), -\sin(v))$

SECTION 13.3

1. $\begin{pmatrix} 3 & 0 \\ 0 & -2 \end{pmatrix}$ 3. $\begin{pmatrix} 1 & -1 \\ 0 & 2 \\ 1 & 1 \end{pmatrix}$ 5. $\begin{pmatrix} 1 & 1 \end{pmatrix}$

7. $\begin{pmatrix} 2 & 1 & -1 \\ 1 & -1 & -3 \\ 1 & 1 & 1 \end{pmatrix}$ 9. $\begin{pmatrix} 1 & 1 & 1 \\ 1 & 1 & 1 \end{pmatrix}$

11. $\begin{pmatrix} \cos\theta & -r\sin\theta & 0 \\ \sin\theta & r\cos\theta & 0 \\ 0 & 0 & 1 \end{pmatrix}$

13. $\begin{pmatrix} \cos\theta\sin\phi & r\cos\phi\cos\theta & -r\sin\phi\sin\theta \\ \sin\phi\sin\theta & r\cos\phi\sin\theta & r\sin\phi\cos\theta \\ \cos\phi & -r\sin\phi & 0 \end{pmatrix}$

15. $n = 2, \ m = 3$ 17. $n = 1, \ m = 2$

19. $n = 3, \ m = 3$

21. $\vec{s}_0 = (0, 2)$ $\vec{v}_0 = (-4, -1)$

23. $\vec{s}_0 = \left(\frac{\sqrt{2}}{2}, \frac{\sqrt{2}}{2} \right)$
$\vec{v}_0 = \left(\sqrt{2} + \frac{3\sqrt{2}}{2}, -\sqrt{2} + \frac{3\sqrt{2}}{2} \right)$

25. $\vec{s}_0 = \left(-\frac{\sqrt{2}}{2}, -\frac{\sqrt{2}}{2}, 2 \right)$
$\vec{v}_0 = \left(\frac{\sqrt{2}}{2}, -\frac{3\sqrt{2}}{2}, 0 \right)$

27. $\vec{s}_0 = \left(\frac{\sqrt{2}}{2}, -\frac{\sqrt{2}}{2}, -2 \right)$
$\vec{v}_0 = \left(\frac{5}{\sqrt{2}}, \frac{1}{\sqrt{2}}, 1 \right)$

29. $\vec{s}_0 = \left(0, -\frac{\sqrt{2}}{2}, -\frac{\sqrt{2}}{2} \right)$
$\vec{v}_0 = \left(\frac{\sqrt{2}}{2}, -\frac{3\sqrt{2}}{2}, 0 \right)$

31. $\vec{s}_0 = \left(-\frac{\sqrt{2}}{4}, \frac{\sqrt{6}}{4}, -\frac{\sqrt{2}}{2} \right)$
$\vec{v}_0 = \left(\frac{\sqrt{2}-5\sqrt{6}}{4}, -\frac{1+\sqrt{3}}{2\sqrt{2}}, \frac{2+\sqrt{3}}{2} \right)$

33. $\begin{pmatrix} 2u\cos v & -u^2\sin v \\ 2u\sin v & u^2\cos v \\ 1 & 0 \end{pmatrix}$.

35. $\frac{\partial \phi}{\partial u} = (2u\cos v + \sin v)e^{u^2\cos v + u\sin v}$
$\frac{\partial \phi}{\partial v} = (-u^2\sin v + u\cos v)e^{u^2\cos v + u\sin v}$

37. $\frac{\partial \phi}{\partial \theta} = r(\cos\theta - \sin\theta)$

39. $\frac{\partial \phi}{\partial \theta} = -\frac{1}{r}\tan\theta + 3r\cos\theta$

41. $\frac{\partial \phi}{\partial \theta} = 0$

43. $(2, 2.9)$ 45. $(1.9, 0.9, 2.9)$

47. $\begin{pmatrix} \frac{1}{2} & \sqrt{3} \\ \frac{\sqrt{3}}{2} & -1 \end{pmatrix} \begin{pmatrix} r + 2 \\ \theta - \frac{\pi}{3} \end{pmatrix} + \begin{pmatrix} -1 \\ -\sqrt{3} \end{pmatrix}$

49. $\begin{pmatrix} \frac{\sqrt{3}}{2} & \frac{1}{2} \\ \frac{1}{2} & -\frac{\sqrt{3}}{2} \end{pmatrix} \begin{pmatrix} r + 1 \\ \theta - \frac{\pi}{6} \end{pmatrix} - \begin{pmatrix} \frac{\sqrt{3}}{2} \\ \frac{1}{2} \end{pmatrix}$

51. $\begin{pmatrix} 0 & -2 & 0 \\ -1 & 0 & 0 \\ 0 & 0 & 1 \end{pmatrix} \begin{pmatrix} r + 2 \\ \theta - \frac{3\pi}{2} \\ z + 1 \end{pmatrix} + \begin{pmatrix} 0 \\ 2 \\ -1 \end{pmatrix}$

53. $\begin{pmatrix} \frac{\sqrt{3}}{2} & \frac{1}{2} & 0 \\ \frac{1}{2} & -\frac{\sqrt{3}}{2} & 0 \\ 0 & 0 & 1 \end{pmatrix} \begin{pmatrix} r + 2 \\ \theta - \frac{\pi}{6} \\ z - 5 \end{pmatrix} + \begin{pmatrix} -\sqrt{3} \\ -\frac{1}{2} \\ 5 \end{pmatrix}$

55. $\begin{pmatrix} -\frac{1}{\sqrt{2}} & 0 & -\sqrt{2} \\ -\frac{1}{\sqrt{2}} & 0 & \sqrt{2} \\ 0 & -2 & 0 \end{pmatrix} \begin{pmatrix} r+2 \\ \phi - \frac{3\pi}{2} \\ \theta - \frac{\pi}{4} \end{pmatrix}$
$+ \begin{pmatrix} \sqrt{2} \\ \sqrt{2} \\ 0 \end{pmatrix}$

57. $\begin{pmatrix} -\frac{\sqrt{3}}{4} & \frac{3}{4} & \frac{1}{4} \\ \frac{1}{4} & -\frac{\sqrt{3}}{4} & \frac{\sqrt{3}}{4} \\ \frac{\sqrt{3}}{2} & \frac{1}{2} & 0 \end{pmatrix} \begin{pmatrix} r+1 \\ \phi - \frac{\pi}{6} \\ \theta - \frac{5\pi}{6} \end{pmatrix}$
$+ \begin{pmatrix} \frac{\sqrt{3}}{4} \\ -\frac{1}{4} \\ -\frac{\sqrt{3}}{2} \end{pmatrix}$

59. $(-0.853, -0.723, 5.1)$

SECTION 13.4

1. $4\sqrt{5}$ 3. 3 5. 4

7. $4\sqrt{4\pi^2 + 1}$ 9. $4\sqrt{4\pi^2 + 1}$

SECTION 13.5

1. 6 3. $2\sqrt{3}$ 5. 0

7. 2 9. 0

11. $T(u, v) = (3u, 2v)$, Area: 6π

13. $T(u, v, w) = \left(\frac{u}{\sqrt{3}}, \frac{v}{2}, \frac{w}{\sqrt{2}} \right)$, Area: $\frac{2\pi}{3\sqrt{6}}$.

15. 0 17. $\frac{\sqrt{35}}{12}$ 19. 680 21. $\frac{322\pi}{15}$

SECTION 13.6

1. 2 3. 2 5. 2 7. 2

9. 0 11. $2\sqrt{2}$ 13. $r^2 \sin(\phi)$.

15. 7 17. $14\sqrt{2}$ 19. $|-16u + 48v + 16|$

21. $32(-2u + 6v + 2w + 1)^2 \sin(u + v + 2w)$

25. Sphere of radius 2 centered at the origin.

29. Nothing. Example: let $g(u, v) = (u, 0, v)$ and consider $C_z \circ g$.

SECTION 14.1

1. $\dfrac{e^{2y} - e^{-y}}{y}$ 3. 0 5. $e^2 + e^{-2} - 2$

7. $\frac{37}{6}$ 9. 272 11. 256

13. $\frac{20}{3}$ 15. $-\frac{1}{8}\ln(3)$ 17. $\frac{4}{\pi}$

19. $\frac{2}{3}$ 21. 1

SECTION 14.2

1. $\vec{h}(s, t) = \begin{pmatrix} s \\ st \end{pmatrix}$, $\begin{cases} 0 \le t \le 1 \\ 1 \le s \le 2 \end{cases}$
$J(\vec{h}(s, t)) = s$

3. $\vec{h}(s, t) = \begin{pmatrix} s \\ t(e^s + s) - s \end{pmatrix}$, $\begin{cases} -1 \le t \le e \\ 0 \le s \le 1 \end{cases}$
$J(\vec{h}(s, t)) = e^s + s$

5. $\vec{h}(s, t) = \begin{pmatrix} s \\ t(s - \ln s) + \ln s \end{pmatrix}$, $\begin{cases} 0 \le t \le 2 \\ 1 \le s \le 2 \end{cases}$
$J(\vec{h}(s, t)) = s - \ln s$

7. $\vec{h}(s, t) = \begin{pmatrix} 1 + s \\ -s - t \\ 2s + 3 \end{pmatrix}$, $\begin{cases} 0 \le t \le 1 \\ 0 \le s \le 1 \end{cases}$
$D(\vec{h}(s, t)) = \begin{pmatrix} 1 & 0 \\ -1 & -1 \\ 2 & 0 \end{pmatrix}$
$J(\vec{h}(s, t)) = \sqrt{5}$

9. $\vec{h}(s, t) = \begin{pmatrix} 1 - 2s \\ 2 - 2s + t \end{pmatrix}$, $\begin{cases} 0 \le t \le 1 \\ 0 \le s \le 1 \end{cases}$
$D(\vec{h}(s, t)) = \begin{pmatrix} -2 & 0 \\ -2 & 1 \end{pmatrix}$
$J(\vec{h}(s, t)) = 2$

11. $\vec{h}(s, t) = \begin{pmatrix} 3s + 4t - 3 \\ -1 + 2t \\ s + 3t \end{pmatrix}$, $\begin{cases} 0 \le t \le 1 \\ 0 \le s \le 1 \end{cases}$
$D(\vec{h}(s, t)) = \begin{pmatrix} 3 & 4 \\ 0 & 2 \\ 1 & 3 \end{pmatrix}$
$J(\vec{h}(s, t)) = \sqrt{65}$

13. $\vec{h}(\phi, \theta) = \begin{pmatrix} 6\sin\phi\cos\theta - 1 \\ 6\sin\phi\sin\theta + 3 \\ 6\cos\phi \end{pmatrix}$, $\begin{cases} 0 \le \theta \le 2\pi \\ 0 \le \phi \le \pi \end{cases}$
$D(\vec{h}(\phi, \theta)) = \begin{pmatrix} 6\cos\phi\cos\theta & -6\sin\phi\sin\theta \\ 6\cos\phi\sin\theta & 6\sin\phi\cos\theta \\ -6\sin\phi & 0 \end{pmatrix}$,
$J(\vec{h}(\phi, \theta)) = 36|\sin\phi|$

15. $\vec{h}(r,\theta) = \begin{pmatrix} r\sin\theta + 1 \\ r\cos\theta + 1 \end{pmatrix}$, $\begin{cases} 0 \le r \le 1 \\ 0 \le \theta \le 2\pi \end{cases}$

$D(\vec{h}(s,t)) = \begin{pmatrix} \sin\theta & r\cos\theta \\ \cos\theta & -r\sin\theta \end{pmatrix}$

$J(\vec{h}(s,t)) = r$

17. $\vec{h}(r,\theta) = \begin{pmatrix} r\sin\theta + 1 \\ r\cos\theta - 3 \end{pmatrix}$, $\begin{cases} 0 \le r \le 2 \\ 0 \le \theta \le 2\pi \end{cases}$

$D(\vec{h}(r,\theta)) = \begin{pmatrix} \sin\theta & r\cos\theta \\ \cos\theta & -r\sin\theta \end{pmatrix}$

$J(\vec{h}(r,\theta)) = r$

19. $\vec{h}(t,\theta) = \begin{pmatrix} t \\ \sin t\cos\theta \\ \sin t\sin\theta \end{pmatrix}$, $\begin{cases} 0 \le t \le \pi \\ 0 \le \theta \le 2\pi \end{cases}$

$D(\vec{h}(t,\theta)) = \begin{pmatrix} 1 & 0 \\ \cos t\cos\theta & -\sin t\sin\theta \\ \cos t\sin\theta & \sin t\cos\theta \end{pmatrix}$,

$J(\vec{h}(t,\theta)) = \sin t\sqrt{\cos^2(t) + 1}$

21. $\vec{h}(t,\theta) = \begin{pmatrix} \cos t \\ 2\sin t\cos\theta \\ 2\sin t\sin\theta \end{pmatrix}$, $\begin{cases} 0 \le t \le \pi \\ 0 \le \theta \le 2\pi \end{cases}$

$D(\vec{h}(t,\theta)) = \begin{pmatrix} 0 & -\sin t \\ -2\sin t\sin\theta & 2\cos t\cos\theta \\ 2\sin t\cos\theta & 2\cos t\sin\theta \end{pmatrix}$,

$J(\vec{h}(t,\theta)) = 2\sin t\sqrt{4\cos^2 t + \sin^2 t}$

23. $\vec{h}(t,\theta) = \begin{pmatrix} t\cos t \\ t\sin t\cos\theta \\ t\sin t\sin\theta \end{pmatrix}$, $\begin{cases} 0 \le t \le \pi \\ 0 \le \theta \le 2\pi \end{cases}$

$D(\vec{h}(t,\theta)) =$
$\begin{pmatrix} 0 & \cos t - t\sin t \\ -t\sin t\sin\theta & (\sin t + t\cos t)\cos\theta \\ t\sin t\cos\theta & (\sin t + t\cos t)\sin\theta \end{pmatrix}$,

$J(\vec{h}(t,\theta)) = t\sin t\sqrt{1 + t^2}$

25. $\vec{h}(t,\theta) = \begin{pmatrix} \cos(t) - 1 \\ (\sin(t) + 2)\cos\theta \\ (\sin(t) + 2)\sin\theta \end{pmatrix}$, $\begin{cases} 0 \le t \le 2\pi \\ 0 \le \theta \le 2\pi \end{cases}$

$D(\vec{h}(t,\theta)) =$
$\begin{pmatrix} 0 & -\sin t \\ -(\sin(t) + 2)\sin\theta & \cos t\cos\theta \\ (\sin(t) + 2)\cos\theta & \cos t\sin\theta \end{pmatrix}$,

$J(\vec{h}(t,\theta)) = 2 + \sin t$

27. $\vec{h}(t,\theta) = \begin{pmatrix} t\cos\theta \\ \cos t \\ t\sin\theta \end{pmatrix}$, $\begin{cases} 0 \le t \le 2\pi \\ 0 \le \theta \le 2\pi \end{cases}$

$D(\vec{h}(t,\theta)) = \begin{pmatrix} -t\sin\theta & \cos\theta \\ 0 & -\sin t \\ t\cos\theta & \sin\theta \end{pmatrix}$,

$J(\vec{h}(t,\theta)) = t\sqrt{1 + \sin^2 t}$

29. $\vec{h}(t,\theta) = \begin{pmatrix} \cos t\cos\theta \\ 2\sin t \\ \cos t\sin\theta \end{pmatrix}$, $\begin{cases} 0 \le t \le \pi \\ 0 \le \theta \le 2\pi \end{cases}$

$D(\vec{h}(t,\theta)) = \begin{pmatrix} -\cos t\sin\theta & -\sin t\cos\theta \\ 0 & 2\cos t \\ \cos t\cos\theta & -\sin t\sin\theta \end{pmatrix}$,

$J(\vec{h}(t,\theta)) = |\cos(t)|\sqrt{\sin^2 t + 4\cos^2 t}$

31. $\vec{h}(t,\theta) = \begin{pmatrix} t\cos t\cos\theta \\ t\sin t \\ t\cos t\sin\theta \end{pmatrix}$, $\begin{cases} \frac{\pi}{2} \le t \le \frac{3\pi}{2} \\ 0 \le \theta \le 2\pi \end{cases}$

$D(\vec{h}(t,\theta)) =$
$\begin{pmatrix} -t\cos t\sin\theta & (\cos t - t\sin t)\cos\theta \\ 0 & \sin t + t\cos t \\ t\cos t\cos\theta & (\cos t - t\sin t)\sin\theta \end{pmatrix}$,

$J(\vec{h}(t,\theta)) = t|\cos t|\sqrt{1 + t^2}$

37. $\vec{h}(s,\theta) = \begin{pmatrix} (\cos(s) + 2)\cos\theta \\ \sin(s) - 3 \\ (\cos(s) + 2)\sin\theta \end{pmatrix}$,

$0 \le s \le 2\pi, \quad 0 \le \theta \le 2\pi$,

$J(\vec{h}(s,\theta)) = 2 + \cos s$

39. $\vec{h}(s,\theta) = \begin{pmatrix} \cos(s) + 2 \\ (\sin(s) - 8)\cos(\theta) + 5 \\ (\sin(s) - 8)\sin\theta \end{pmatrix}$,

$0 \le s \le 2\pi, \quad 0 \le \theta \le 2\pi$,

$J(\vec{h}(s,\theta)) = 8 - \sin s$

41. $\vec{h}(s,\theta) = \begin{pmatrix} \cos(s) + 2 \\ (\sin(s) + 2)\cos(\theta) - 5 \\ (\sin(s) + 2)\sin\theta \end{pmatrix}$,

$0 \le s \le 2\pi, \quad 0 \le \theta \le 2\pi$,

$J(\vec{h}(s,\theta)) = 2 + \sin s$

43. $\vec{h}(s,\theta) = \begin{pmatrix} \cos\theta\sin\phi \\ \sin\theta\sin\phi \\ \cos\phi \end{pmatrix}$,

$0 \le \theta \le \frac{\pi}{2}, \quad 0 \le \theta \le \frac{\pi}{2}$

45. $\dfrac{\sin(\phi)\sqrt{b^2c^2\sin^2\phi\cos^2\theta + a^2c^2\sin^2\phi\sin^2\theta}}{+a^2b^2\cos^2\phi}$

47. $(3\sqrt{2}, 2\sqrt{2}, 6)$

49. $\begin{pmatrix} 1 & 0 \\ 0 & 1 \\ \frac{\sqrt{3}}{4} & \frac{\pi\sqrt{3}}{6} \end{pmatrix}\begin{pmatrix} x - \frac{\pi}{3} \\ y - \frac{1}{2} \end{pmatrix} + \begin{pmatrix} \frac{\pi}{3} \\ \frac{1}{2} \\ \frac{1}{2} \end{pmatrix}$

51. $y + \sqrt{3}z - 2 = 0$

SECTION 14.3

1. $\frac{2\sqrt{6}}{3}$ 3. $\frac{\pi^2}{2} + \pi - \frac{2}{3}$ 5. $\frac{2048\pi}{5}$

7. $\frac{21\pi}{2}$ 9. $\frac{7}{12}$ 11. π

13. $-\frac{421}{40}$ 15. $\frac{11}{40}$

17. $\int_{-2}^{2}\int_{-2}^{2}\sqrt{1 + 4x^2 + 4y^2}\,dx\,dy$

19. $\int_{-2}^{2}\int_{-2}^{2}\sqrt{1 + (4x^2 + 4y^2)\cos(x^2 + y^2)}\,dx\,dy.$

21. $\int_{0}^{2}\int_{-2}^{1}\sqrt{1 + (x^2 + y^2)e^{2xy}}\,dx\,dy.$

25. $\int_{0}^{2\pi}\int_{-1}^{1}(x^2 + 1)\sqrt{1 + 4x^2}\,dx\,d\theta$

27. $\int_{0}^{2\pi}\int_{0}^{2\pi}(\sin(t) + 3)\,dt\,d\theta = 12\pi^2.$

33. $\int_{0}^{2\pi}\int_{1}^{3}(x\cos(\theta) + x^3\sin(\theta))x\sqrt{1 + 4x^2}\,dx\,d\theta.$

SECTION 14.4

1. $2\mathrm{kg/m}^2$

3. $\iint_S (x + y)\,dS = \int_{0}^{2}\int_{0}^{x}(x + y)\,dy\,dx = 4$

5. $\frac{28\pi^2}{15}$

7. $2\pi^2\int_{0}^{2}\int_{0}^{x}(x^2 + y^2)(x + y)\,dx\,dy = \frac{80\pi^2}{3}$

9. $\frac{1}{2}(2\pi)^2\int_{0}^{2}\int_{0}^{x}((x + 1)^2 + (y + 1)^2)$
     $(x + y)\,dy\,dx = 80\pi^2$

11. $\int_{0}^{2}\int_{0}^{x}(x^2 + x)\,dy\,dx = \frac{20}{3}$

13. $\left(\frac{35}{48}, \frac{35}{54}\right)$ 15. $\frac{2\pi^2}{63}$ 17. $\frac{\pi^2}{180}$

19. $\frac{1}{30}$ 21. $\int_{0}^{1}\int_{0}^{1}((1-t) + (1-t)\sqrt{2})\,ds\,dt = \sqrt{2}$

23. $3k\pi^2\sqrt{2}$ 25. $4\pi^2\sqrt{2}$

27. $\frac{1}{2}(2\pi f)^2\int_{0}^{R}\int_{0}^{2\pi}(s\cos t)^2\rho s\,dt\,ds = \frac{1}{2}\pi^3 R^4\rho f^2$

29. $\int_{0}^{1}\int_{\pi/2}^{3\pi/2} -r^2\cos(\theta)\,d\theta\,dr = \frac{2}{3}$

31. $\int_{0}^{2\pi}\int_{0}^{1} 2s\sqrt{1 - s^2\cos^2 t}\,ds\,dt$

SECTION 14.5

1. $\frac{1}{6}$ 3. $\frac{\sqrt{3}e}{2}$

5. $8\sqrt{2}\pi^2\rho$ 7. $\frac{56\pi^2\rho\sqrt{2}}{3}$

9. $\frac{\pi}{2}$ 11. $\frac{3\pi}{2}$

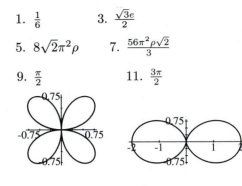

13. $\frac{\pi^3}{6}$

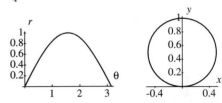

15. $\frac{\pi}{4}$

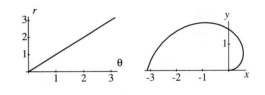

17. $\frac{\pi}{12}$

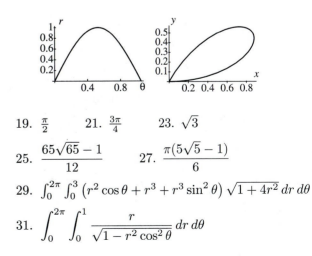

19. $\frac{\pi}{2}$ 21. $\frac{3\pi}{4}$ 23. $\sqrt{3}$

25. $\frac{65\sqrt{65} - 1}{12}$ 27. $\frac{\pi(5\sqrt{5} - 1)}{6}$

29. $\int_{0}^{2\pi}\int_{0}^{3}\left(r^2\cos\theta + r^3 + r^3\sin^2\theta\right)\sqrt{1 + 4r^2}\,dr\,d\theta$

31. $\int_{0}^{2\pi}\int_{0}^{1}\dfrac{r}{\sqrt{1 - r^2\cos^2\theta}}\,dr\,d\theta$

SECTION 14.6

1. $\vec{h}(u, v, w) =$
$$\begin{pmatrix} -v + w \\ u + 2v + w \\ u + v + w \end{pmatrix}, \quad \begin{cases} 0 \le u \le 1 \\ 0 \le v \le 1 \\ 0 \le w \le 1 \end{cases}$$
$J(u, v, w) = 1$

3. $\vec{h}(u, v, w) =$
$$\begin{pmatrix} -1 + u + 2w \\ 2 - u - 3v - 2w \\ u + v + 4w \end{pmatrix}, \quad \begin{cases} 0 \le u \le 1 \\ 0 \le v \le 1 \\ 0 \le w \le 1 \end{cases}$$
$J(u, v, w) = 6$

5. $\vec{h}(t, s, \theta) =$
$$\begin{pmatrix} s \\ (s^2 - ts^2 + st)\cos\theta \\ (s^2 - ts^2 + st)\sin\theta \end{pmatrix}, \quad \begin{cases} 0 \le t \le 1 \\ 0 \le s \le 1 \\ 0 \le \theta \le 2\pi \end{cases}$$
$J(s, t, \theta) = (s^3 - s^2)(st - s - t)$

7. $\vec{h}(r\phi, \theta) =$
$$\begin{pmatrix} r\cos\phi \\ r\cos\theta\cos\phi \\ r\sin\theta\cos\phi \end{pmatrix}, \quad \begin{cases} 0 \le r \le 3 \\ 0 \le \phi \le \pi \\ 0 \le \theta \le 2\pi \end{cases}$$
$J(r, \phi, \theta) = r^2 \sin\phi$

9. $\vec{h}(s, t, \theta) = \begin{pmatrix} s\cos\theta \\ s^2 t \\ s\sin\theta \end{pmatrix} \quad \begin{cases} 0 \le t \le 1 \\ 0 \le s \le 1 \\ 0 \le \theta \le 2\pi \end{cases}$
$J(s, t, \theta) = s^3$

11. $\vec{h}(s, t, \theta) = \begin{pmatrix} s\cos\theta \\ t \\ s\sin\theta \end{pmatrix} \quad \begin{cases} 2 \le t \le 52 \\ 1 \le s \le 3 \\ 0 \le \theta \le 2\pi \end{cases}$
$J(s, t, \theta) = s$

13. $\vec{h}(r, \phi, \theta) =$
$$\begin{pmatrix} \cos(\theta)(3 + r\cos\phi) \\ 4 + r\sin\phi \\ \sin(\theta)(3 + r\cos\phi) \end{pmatrix} \quad \begin{cases} 0 \le r \le 2 \\ 0 \le \phi \le 2\pi \\ 0 \le \theta \le 2\pi \end{cases}$$
$J(r, \phi, \theta) = r(3 + r\cos\phi)$

15. $\vec{h}(r, t, \theta) =$
$$\begin{pmatrix} 3 + (r - 3)\cos\theta \\ rt \\ (r - 3)\sin\theta \end{pmatrix} \quad \begin{cases} 0 \le r \le 1 \\ 0 \le t \le 1 \\ 0 \le \theta \le 2\pi \end{cases}$$
$J(r, t, \theta) = |r(r - 3)|$

17. $h(r, t, \theta) =$
$$\begin{pmatrix} r \\ 6 + (-6 + rt)\cos\theta \\ (-6 + rt)\sin\theta \end{pmatrix} \quad \begin{cases} 0 \le r \le 1 \\ 0 \le t \le 1 \\ 0 \le \theta \le 2\pi \end{cases}$$
$J(r, t, \theta) = |r(-6 + rt)|$

19. $h(r, t, \theta) =$
$$\begin{pmatrix} r \\ -2 + (2 + t)\cos\theta \\ (2 + t)\sin\theta \end{pmatrix} \quad \begin{cases} 1 \le r \le 3 \\ 2 \le t \le 52 \\ 0 \le \theta \le 2\pi \end{cases}$$
$J(r, t, \theta) = |t + 2|$

21. $\vec{h}(t, s, \theta) =$
$$\begin{pmatrix} -1 + (4 + r\cos\phi)\cos\theta \\ 4 + r\cos\phi \\ (4 + r\sin\phi)\sin\theta \end{pmatrix} \quad \begin{cases} 0 \le r \le 3 \\ 0 \le \phi \le w\pi \\ 0 \le \theta \le 2\pi \end{cases}$$
$J(r, \phi, \theta) = r(r + r\cos\phi)$

23. $\vec{h}(t, s, u) = \begin{pmatrix} t \\ st^2 \\ 2u \end{pmatrix} \quad \begin{cases} 0 \le t \le 1 \\ 0 \le s \le 1 \\ 0 \le u \le 1 \end{cases}$

25. $\vec{h}(t, s, u) =$
$$\begin{pmatrix} t \\ t^2\left(\frac{u}{2} + (1 - u)v\right) \\ \frac{1}{2}t^2 u\sqrt{3} \end{pmatrix} \quad \begin{cases} 0 \le t \le 1 \\ 0 \le s \le 1 \\ 0 \le u \le 1 \end{cases}$$

27. $\vec{h}(t, s, \theta) =$
$$\begin{pmatrix} t \\ \frac{t^2}{2} + \frac{t^2 s}{2\cos\theta} \\ \frac{t^2 s}{2\sin\theta} \end{pmatrix} \quad \begin{cases} 0 \le t \le 1 \\ 0 \le s \le 1 \\ 0 \le \theta \le \pi \end{cases}$$

SECTION 14.7

1. 0   3. $12 - \frac{16}{\pi^2}$ 5. $(1 - e^2)\ln^2(2)$

7. $\frac{7}{2}$ 9. $\frac{8\rho\pi^2 f^2}{3}$ 11. $(0, 0, 0)\text{m}$

13. $\frac{\pi}{3}$ 15. 5400π 19. $\frac{2}{3}$

21. $\frac{41}{30}$ 23. $\frac{31}{105}$

SECTION 14.8

1. $\frac{1}{6}$ 3. $\frac{209}{420}$

5. The rectangle $0 \le y \le 2$, $0 \le z \le 1$.

7. The region in the yz-plane bounded between the graphs of $y = -z$, $y = 0$, $z = 0$, and $z = 1$.

9. The region in the yz-plane bounded between the graphs of $y = -z$, $y = 2 + z^2$, $z = 0$, and $z = 2$.

11. The integration limits do not define a solid.

13. The integration limits do not define a solid.

15. The integration limits do not define a solid.

17. 16   19. 2π 21. $\frac{31}{6}$

23. $\frac{48}{5}$ 25. 16   27. $\frac{9}{2}$

29. $\frac{8}{15}$ 31. $\frac{\sqrt{2}}{3}$ 33. π

35. $\frac{608}{21}$ 37. $\frac{87}{6}$ 39. $\frac{38496}{945}$

SECTION 15.1

1. $h_1(s,t) = \begin{pmatrix} s+t+1 \\ 2s+t \\ 3s+t \end{pmatrix}$,

   $h_2(s,t) = \begin{pmatrix} s+t+1 \\ s+2t \\ s+3t \end{pmatrix}$

3. $h_1(s,t) = \begin{pmatrix} 1-s+t \\ 4s \\ 9+t \end{pmatrix}$,

   $h_2(s,t) = \begin{pmatrix} 1-t+s \\ 4t \\ 9+s \end{pmatrix}$

5. 1   7. 7

9. $\vec{h}(r, \phi, \theta) = \begin{pmatrix} \cos\theta \sin\phi \\ \sin\theta \sin\phi \\ \cos\phi \end{pmatrix}$ $\begin{array}{l} 0 \le \theta \le 2\pi \\ 0 \le \phi \le \frac{\pi}{2} \end{array}$

   $\iint_S \vec{F} \cdot d\vec{S} = 2\pi$

11. $\vec{h}(\theta, r) = \begin{pmatrix} r\cos\theta \\ r\sin\theta \\ r^2 \end{pmatrix}$ $\begin{array}{l} 0 \le r \le 2 \\ 0 \le \theta \le 2\pi \end{array}$,

    $\iint_P \vec{F} \cdot d\vec{r} = -16\pi$

13. 0   15. 0   17. No

19. Not necessarily. For example, if $f(x) = c$, then the orientation is in the $-\hat{k}$ direction.

21. $\left(-\frac{5}{2}, 5, \frac{5}{2}\right)$ 23. $(0,0,0)$

25. $\left(0, (2c+100k) \times 10^3, \frac{4c}{3} \times 10^5\right)$

SECTION 15.2

1. Out of the bounded region.

3. Oriented into the bounded region.

5. Oriented into the bounded region.

7. a. Yes   b. No

9. Faces labeled as in Figure 1:
   $S_1 : \vec{h}_1(s,t) = (1, 4s, -10+20t)$
   $S_2 : \vec{h}_2(s,t) = (-3, 4t, -10+20s)$
   $S_3 : \vec{h}_3(s,t) = (-3+4t, 4, -10+20s)$
   $S_4 : \vec{h}_4(s,t) = (-3+4s, 0, -10+20t)$
   $S_5 : \vec{h}_5(s,t) = (-3+4s, 4t, 10)$
   $S_6 : \vec{h}_6(s,t) = (-3+4t, 4s, -10)$
   $0 \le s \le 1, \ 0 \le t \le 1$

11. $\vec{h}_1(t, \theta) = (\cos(\theta), \sin(\theta), t), \ \begin{cases} 0 \le \theta \le 2\pi \\ -1 \le t \le 1 \end{cases}$

    $\vec{h}_2(t, \theta) = (t\cos(\theta), t\sin(\theta), -1), \ \begin{cases} 0 \le \theta \le 2\pi \\ 0 \le t \le 1 \end{cases}$

    $\vec{h}_3(\theta, t) = (t\cos(\theta), t\sin(\theta), 1), \ \begin{cases} 0 \le \theta \le 2\pi \\ 0 \le t \le 1 \end{cases}$

13. Faces labeled as in Figure 1:
    $S_1 : \vec{h}_1(s,t) = (s, 1+2s+t, 2+2t)$
    $S_2 : \vec{h}_2(s,t) = (-1+s+t, 2-s+2t, 1+s)$
    $S_3 : \vec{h}_3(s,t) = (t, 4+s-t, 1+2s+t)$
    $S_4 : \vec{h}_4(s,t) = (-1+s, 2-s+t, 1+s+2t)$
    $S_5 : \vec{h}_5(s,t) = (-1+s+t, 3-s+2t, 3+s)$
    $S_6 : \vec{h}_6(s,t) = (-1+s+t, 2+2s-t, 1+t)$
    $0 \le s \le 1, \ 0 \le t \le 1.$

15. 0

17. 0. Let $\hat{n}(p_0)$ be the unit normal vector at p_0 pointing out of the sphere and $\hat{n}(p_1)$ be the unit normal vector at p_1, the antipodal point of p_0. Then $(1,1,1) \cdot \hat{n}(p_0) = -(1,1,1) \cdot \hat{n}(p_1)$. Thus the flux of $(1,1,1)$ through a piece of the surface surrounding p_0 is negated by the flux through a piece of the surface surrounding p_1.

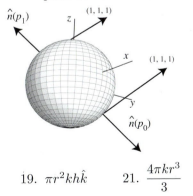

19. $\pi r^2 kh\hat{k}$ 21. $\dfrac{4\pi k r^3}{3}$

SECTION 15.3

1. $yz + x^2 + \sin(x)$ 3. 0 5. $\frac{1}{8}$

7. 2π

9. $\text{Div}(\vec{F}(x,y,z)) = 3$ so $\int_S \vec{F} \cdot d\vec{S} =$
   $\int_V \text{Div}(\vec{F}(x,y,z))\, dV = 3 \int_V\, dV = 3V.$

11. Hint: Let $\vec{F}(\vec{r}) = (\phi(\vec{r}), 0, 0)$ and apply Gauss' Theorem.

13. $\vec{\nabla} \cdot \phi$ cannot be computed since ϕ is a real valued function.

15. Use the results from Exercise 14.

17. c. Use the results from parts a and b and Exercise 16.

SECTION 15.4

1. $\vec{r}_1(t) = (2t, 0),$ $\vec{r}_2(t) = (2, 3t)$
   $\vec{r}_3(t) = (2 - 2t, 3),$ $\vec{r}_4(t) = (0, 3 - 3t)$
   $0 \le t \le 1$

3. $\vec{r}_1(t) = (-\pi + 2\pi t, 0), \vec{r}_2(t) = (\pi, 4\pi t)$
   $\vec{r}_3(t) = (\pi - 2\pi t, 4\pi), \vec{r}_4(t) = (-\pi, 4\pi(1 - t)),$
   $0 \le t \le 1$

5. $\vec{r}_1(t) = (0, t, t), \vec{r}_2(t) = (t, 1 + t, 1 + t),$
   $\vec{r}_3(t) = (1, 2 - t, 2 - t),$
   $\vec{r}_4(t) = (1 - t, 1 - t, 1 - t),$
   $0 \le t \le 1$

7. $\vec{r}_1(t) = (1 + 2t, 1 + t, 1 + t),$
   $\vec{r}_2(t) = (3 + t, 2, 2 + t),$
   $\vec{r}_3(t) = (4 - 2t, 2 - t, 3 - t),$
   $\vec{r}_4(t) = (2 - t, 1, 2 - t),$
   $0 \le t \le 1$

9.

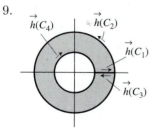

C_2 and C_4 are taken into the boundary while C_1 and C_3 are taken into the seam.
$\vec{h}_1(t) = (1 + t, 0),\ \vec{h}_2(t) = 2(\cos(2\pi t), \sin(2\pi t)),$
$\vec{h}_3(t) = (2 - t, 0),$
$\vec{h}_4(t) = (\cos(-2\pi t), \sin(-2\pi t)),\ 0 \le t \le 1$

11. $\vec{r}_1(t) = (2 - t, 0)$
    $\vec{r}_2(t) = \left(\cos\left(\frac{\pi t}{4}\right), \sin\left(\frac{\pi t}{4}\right)\right)$
    $\vec{r}_3(t) = \frac{1}{\sqrt{2}}(1 + t, 1 + t)$
    $\vec{r}_4(t) = 2\left(\cos\left(\frac{\pi}{4}(1 - t)\right), \sin\left(\frac{\pi}{4}(1 - t)\right)\right)$
    $0 \le t \le 1$

13. $\vec{r}_1(t) = (2t - 1, 0, 0),$
    $\vec{r}_2(t) = (1, 2t, 2t),$
    $\vec{r}_3(t) = \left(1 - 2t, 2, 2(1 - 2t)^2\right),$
    $\vec{r}_4(t) = (-1, 2 - 2t, 2 - 2t),$
    $0 \le t \le 1$

15. $(\cos(t), \sin(t), 0),\ 0 \le t \le 2\pi$

17. $\vec{h}(t) = (2\cos t, 2\sin t, 0),\ 0 \le t \le 2\pi$

SECTION 15.5

1. $\iint_S \vec{\nabla} \times \vec{F} \cdot d\vec{S} = \int_{\beta S} \vec{F} \cdot d\vec{r} = \frac{1}{2}$

3. $\iint_S \vec{\nabla} \times \vec{F} \cdot d\vec{S} = \int_{\beta S} \vec{F} \cdot d\vec{r} = -27\pi$

5. $\frac{1}{6}$ 7. $\frac{1}{3} + \frac{\pi}{2}$ 11. 9 13. 0

SECTION 15.6

1. $f(x, y, z) = xyz$ 3. $\text{Curl}\vec{F} \ne \vec{0}$

Appendix C.1

1. $y^2 = 4\pi x$ 3. $y^2 = -4\pi x$

5. $8y = 5x^2$ and $25x = 8y^2$

7. $\frac{x^2}{64} + \frac{y^2}{9} = 1$

9. $\dfrac{x^2}{5 + \sqrt{5}} + \dfrac{y^2}{\sqrt{5} - 1} = 1$

11. Foci: $\left(\pm\frac{3}{\sqrt{2}}, 0\right)$, Vertices: $(\pm\sqrt{6}, 0)$

13. Foci: $(0, \pm 2\sqrt{5})$, Vertices: $(0, \pm 6)$

15. Foci: $(\pm\sqrt{29}, 0)$, Vertices: $(\pm 5, 0)$,
    Asymptotes: $y = \pm\frac{2x}{5}$

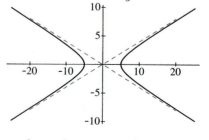

17. $25x^2 - 4y^2 = 100, 25y^2 - 4x^2 = 100$

19. $3y^2 - x^2 = 3$

21. $9x^2 - y^2 = 1, y^2 - 9x^2 = 9$

23. Hyperbola

25. Either

27. $y^2 + z^2 = 4x$

29. $x^2 - 3y^2 + z^2 - 10y - 3 = 0$

33. $3x^2 + 4y^2 + 4z^2 - 12 = 0$

35. $4x^2 + 4y^2 + 3z^2 - 12 = 0$

39. $\dfrac{x^2}{a^2 - c^2} + \dfrac{y^2}{a^2} = 1$

41. $4x^2 - 60y^2 + 4z^2 + 15 = 0$

43. $3x^2 + 4y^2 = 3$, Ellipse

Appendix C.2

1. Two lines, $y = z$ and $y = -z$

3. Hyperbola, $z^2 - y^2 = 1$

5. Hyperbola, $z^2 - x^2 = 9$

7. Circle with center $(0, 0, -1)$ and radius 1

9. Parabola, $z = \dfrac{x^2}{4}$

11. Parabola, $z = \dfrac{y^2}{9} + \dfrac{1}{4}$

13. No intersection

15. Point, $(0, 0, 0)$

17. Parabola, $y = \dfrac{z^2}{9} + \dfrac{1}{4}$

19. xy–plane: Ellipse $\dfrac{x^2}{4} + \dfrac{y^2}{9} = 1$;
    xz–plane: Ellipse $\dfrac{x^2}{4} + z^2 = 1$;
    yz–plane: Ellipse $\dfrac{y^2}{9} + z^2 = 1$

21. a–6.a, b–6.c, c–6.b, d–6.f, e–6.e, f–6.d

Appendix C.3

1. $(2, 1), (6, 18), (-2, -1)$

3. $x + 3y = 7$

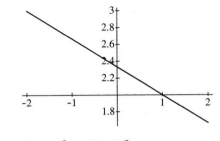

5. $(x - 3)^2 + (y + 1)^2 = 1$

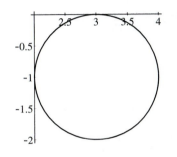

7. $2(y - 1)^2 + 3(x - 2) = 0$

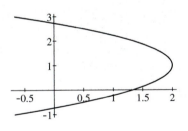

9. $4(x - 3)^2 - 9(y - 2)^2 = 0$

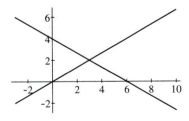

11. $4(x-3)^2 - 9(y-2)^2 = 1$

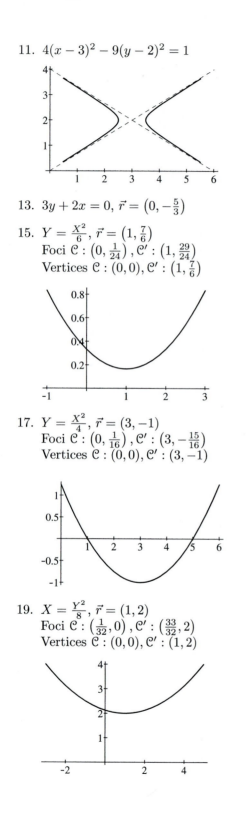

13. $3y + 2x = 0$, $\vec{r} = \left(0, -\frac{5}{3}\right)$

15. $Y = \frac{X^2}{6}$, $\vec{r} = \left(1, \frac{7}{6}\right)$
    Foci $\mathcal{C} : \left(0, \frac{1}{24}\right)$, $\mathcal{C}' : \left(1, \frac{29}{24}\right)$
    Vertices $\mathcal{C} : (0,0)$, $\mathcal{C}' : \left(1, \frac{7}{6}\right)$

17. $Y = \frac{X^2}{4}$, $\vec{r} = (3, -1)$
    Foci $\mathcal{C} : \left(0, \frac{1}{16}\right)$, $\mathcal{C}' : \left(3, -\frac{15}{16}\right)$
    Vertices $\mathcal{C} : (0,0)$, $\mathcal{C}' : (3, -1)$

19. $X = \frac{Y^2}{8}$, $\vec{r} = (1, 2)$
    Foci $\mathcal{C} : \left(\frac{1}{32}, 0\right)$, $\mathcal{C}' : \left(\frac{33}{32}, 2\right)$
    Vertices $\mathcal{C} : (0,0)$, $\mathcal{C}' : (1, 2)$

21. $3X^2 + 4Y^2 = 1$, $\vec{r} = (1, -3)$
    Foci $\mathcal{C} : \left(\pm\frac{1}{2\sqrt{3}}, 0\right)$, $\mathcal{C}' : \left(1 \pm \frac{1}{2\sqrt{3}}, -3\right)$
    Vertices $\mathcal{C} : \left(\pm\frac{1}{\sqrt{3}}, 0\right)$, $\mathcal{C}' : \left(1 \pm \frac{1}{\sqrt{3}}, -3\right)$

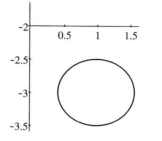

23. $\frac{4X^2}{9} + \frac{9Y^2}{4} = 1$, $\vec{r} = (2, -1)$
    Foci $\mathcal{C} : \left(\pm\frac{\sqrt{65}}{6}, 0\right)$, $\mathcal{C}' : \left(2 \pm \frac{\sqrt{65}}{6}, -1\right)$
    Vertices $\mathcal{C} : \left(\pm\frac{2}{3}, 0\right)$, $\mathcal{C}' : \left(2 \pm \frac{2}{3}, -1\right)$

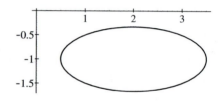

25. $y = -24(x + 2)^2 - 4$

27. $x = \frac{(y-1)^2}{3} - 2$

29. $\frac{(x-3)^2}{9} + \frac{y^2}{25} = 1$

31. Either $A = B = D = F = 0, C \neq 0$ or
    $A = B = C = F = 0, D \neq 0$

33. $C = D = F = 0$ and $AB < 0$

Index

TABLE OF INTEGRALS

Elementary Functions

1. $\int cf(x)\ dx = c\int f(x)\ dx$

2. $\int f(x) \pm g(x)\ dx = \int f(x)\ dx \pm \int g(x)\ dx$

3. $\int x^n\ dx = \dfrac{x^{n+1}}{n+1}, \quad n \neq -1$

4. $\int x^{-1}\ dx = \ln|x|$

5. $\int \dfrac{f'(x)}{f(x)}\ dx = \ln|f(x)|$

Trigonometric Functions

6. $\int \sin ax\ dx = -\dfrac{\cos ax}{a}$

7. $\int \cos ax\ dx = \dfrac{\sin ax}{a}$

8. $\int \tan ax\ dx = \dfrac{\ln|\sec ax|}{a}$

9. $\int \cot ax\ dx = \dfrac{\ln|\sin ax|}{a}$

10. $\int \sec ax\ dx = \dfrac{\ln|\sec ax + \tan ax|}{a}$

11. $\int \csc ax\ dx = -\dfrac{\ln|\csc ax + \cot ax|}{a}$

12. $\int \sin^2 ax\ dx = \dfrac{x}{2} - \dfrac{\sin 2ax}{4a}$

13. $\int \cos^2 ax\ dx = \dfrac{x}{2} + \dfrac{\sin 2ax}{4a}$

14. $\int \tan^2 ax\ dx = \dfrac{\tan ax}{a} - x$

15. $\int \cot^2 ax\ dx = -\dfrac{\cot ax}{a} - x$

16. $\int \sec^2 ax\ dx = \dfrac{\tan ax}{a}$

17. $\int \csc^2 ax\ dx = -\dfrac{\cot ax}{a}$

18. $\int \sin^n ax\ dx = -\dfrac{\cos ax \sin^{n-1}}{na} + \dfrac{n-1}{n}\int \sin^{n-2} ax\ dx$

19. $\int \cos^n ax\ dx = \dfrac{\sin ax \cos^{n-1}}{na} + \dfrac{n-1}{n}\int \cos^{n-2} ax\ dx$

20. $\int \sec^n ax\ dx = \dfrac{\tan ax \sec^{n-2} ax}{a(n-1)} + \dfrac{n-2}{n-1}\int \sec^{n-2} ax\ dx$

21. $\int \csc^n ax\ dx = -\dfrac{\cot ax \csc^{n-2} ax}{a(n-1)} + \dfrac{n-2}{n-1}\int \csc^{n-2} ax\ dx$

22. $\int \tan^n ax\ dx = \dfrac{\tan^{n-1} ax}{a(n-1)} - \int \tan^{n-2} ax\ dx$

23. $\int \cot^n ax\ dx = -\dfrac{\cot^{n-1} ax}{a(n-1)} - \int \cot^{n-2} ax\ dx$

24. $\int x^n \sin ax\ dx = -\dfrac{x^n \cos ax}{a} + \dfrac{n}{a}\int x^{n-1}\cos ax\ dx$

25. $\int x^n \cos ax\ dx = \dfrac{x^n \sin ax}{a} - \dfrac{n}{a}\int x^{n-1}\sin ax\ dx$

26. $\int \sin ax \cos bx\ dx = -\dfrac{\cos(a+b)x}{2(a+b)} - \dfrac{\cos(a-b)x}{2(a-b)}, \quad a^2 \neq b^2$

27. $\int \sin ax \sin bx\ dx = \dfrac{\sin(a-b)x}{2(a-b)} + \dfrac{\sin(a+b)x}{2(a+b)}, \quad a^2 \neq b^2$

28. $\int \cos ax \cos bx\ dx = \dfrac{\sin(a-b)x}{2(a-b)} + \dfrac{\sin(a+b)x}{2(a+b)}, \quad a^2 \neq b^2$

29. $\int \sin^n ax \cos^m ax\ dx = \dfrac{\sin^{n+1} ax \cos^{m-1} ax}{a(m+n)}$
$+ \dfrac{m-1}{m+n}\int \sin^n ax \cos^{m-2} ax\ dx, \quad m \neq -n$

30. $\int \text{Arcsin } ax\ dx = x\text{Arcsin } ax + \dfrac{\sqrt{1-a^2x^2}}{a}$

31. $\int \text{Arccos } ax\ dx = x\text{Arccos}(ax) - \dfrac{\sqrt{1-a^2x^2}}{a}$

32. $\int \text{Arctan}(ax)\ dx = x\text{Arctan}(ax) - \dfrac{\ln(a+a^2x^2)}{2a}$

33. $\int \text{Arccot } ax\ dx = x\text{Arccot } ax + \dfrac{\ln(a+a^2x^2)}{2a}$

34. $\int \text{Arcsec } ax\ dx = x\text{Arcsec } ax - \dfrac{\ln(ax+\sqrt{a^2x^2-1})}{a}$

35. $\int \text{Arccsc } ax\ dx = x\text{Arccsc } ax + \dfrac{\ln(ax+\sqrt{a^2x^2-1})}{a}$

Exponential and Logarithmic Functions

36. $\int e^{ax}\ dx = \dfrac{e^{ax}}{a}$

37. $\int xe^{ax}\ dx = (ax-1)\dfrac{e^{ax}}{a}$

38. $\int x^n e^{ax}\ dx = \dfrac{x^n e^{ax}}{a} - \dfrac{n}{a}\int x^{n-1} e^{ax}\ dx$

39. $\int b^{ax}\ dx = \dfrac{b^{ax}}{a \ln b}$

40. $\int \ln ax\ dx = x\ln ax - x$

41. $\int x^n b^{ax}\ dx = \dfrac{x^n b^{ax}}{a \ln b} - \dfrac{n}{a \ln b}\int x^{n-1} b^{ax}\ dx, \quad b > 0, b \neq 1$

42. $\int e^{ax}\sin bx\ dx = \dfrac{e^{ax}}{a^2+b^2}(a \sin bx - b \cos bx)$

43. $\int e^{ax}\cos bx\ dx = \dfrac{e^{ax}}{a^2+b^2}(a \cos bx + b \sin bx)$

44. $\int \dfrac{\ln ax}{x}\ dx = \dfrac{1}{2}(\ln ax)^2$

45. $\int \dfrac{dx}{x \ln ax}\ dx = \ln|\ln ax|$

46. $\int x^n \ln ax\ dx = \dfrac{x^{n+1}}{n+1}\ln ax - \dfrac{x^{n+1}}{(n+1)^2}, \quad n \neq -1$

Hyperbolic Trigonometric Functions

47. $\int \sinh ax\ dx = \dfrac{\cosh ax}{a}$

48. $\int \cosh ax\ dx = \dfrac{\sinh ax}{a}$

49. $\int \tanh ax\ dx = \dfrac{\ln|\cosh ax|}{a}$

50. $\int \coth ax\ dx = \dfrac{\ln|\sinh ax|}{a}$